# ENVIRONMENTAL SCIENCE

# ENVIRONMENTAL SCIENCE

## toward a sustainable future

*ninth edition*

**Richard T. Wright** Gordon College

Upper Saddle River, New Jersey 07458

PEARSON
Prentice Hall

Library of Congress Cataloging-in-Publication Data

Wright, Richard T., 1933-
Environmental science: toward a sustainable future.-- 9th ed./Richard T. Wright.
p. cm.
Includes bibliographical references and index.
ISBN 0-13-144200-7
1. Environmental sciences. I. Title.
GE105.W75 2004
363.7--dc22
2004044564

Executive Editor: *Dan Kaveney*
Editor in Chief, Science: *John Challice*
Production Editor: *Donna King*
Assistant Managing Editor: *Beth Sweeten*
Executive Managing Editor: *Kathleen Schiaparelli*
Marketing Manager: *Shari Meffert*
Associate Editor: *Amanda Griffith*
Editorial Assistant: *Margaret Ziegler*
Editor in Chief, Development: *Carol Trueheart*
Development Editor: *John Murzdek*
Managing Editor, AV Production & Management: *Patty Burns*
Director of Creative Services: *Paul Belfanti*
Creative Director: *Carole Anson*
Art Director: *Kenny Beck*
Interior Design: *Kenny Beck/Cica 86, Inc.*
AV Editor: *Jessica Einsig*
Art Studio: *Artworks: Royce Copenheaver, Ryan Currier, Daniel Knopsnyder, Nathan Storck*
Director, Image Resource Center: *Melinda Reo*
Manager, Rights and Permissions: *Zina Arabia*
Manager, Visual Research: *Beth Brenzel*
Manager, Cover Visual Research & Permissions: *Karen Sanatar*
Image Permission Coordinator: *Nancy Seise*
Photo Researcher: *Stephen Forsling*
Manufacturing Manager: *Trudy Pisciotti*
Manufacturing Buyer: *Alan Fischer*
Media Editor: *Chris Rapp*
Assistant Managing Editor, Science Media: *Nicole Bush*
Assistant Managing Editor, Science Supplements: *Becca Richter*
Media Production Editor: *Rich Barnes*
Vice President of Production and Manufacturing: *David W. Riccardi*
Cover Design: *Geoffrey Cassar*
Cover Photo: *Charles Krebs/Corbis*


Pearson Prentice Hall
Pearson Education, Inc.
Upper Saddle River, NJ 07458



Printed in the United States of America
10 9 8 7 6 5 4 3

ISBN 0-13-144200-7 (Student Edition)
ISBN 0-13-192021-9 (School Edition)

Pearson Education Ltd., *London*
Pearson Education Australia Pty. Ltd., *Sydney*
Pearson Education Singapore, Pte. Ltd.
Pearson Education North Asia Ltd., *Hong Kong*
Pearson Education Canada, Inc., *Toronto*
Pearson Educatión de Mexico, S.A. de C.V.
Pearson Education—Japan, *Tokyo*
Pearson Education Malaysia, Pte. Ltd.

**Printed on Recycled Paper**

# about the author

**Richard T. Wright** is Professor Emeritus of Biology at Gordon College in Massachusetts, where he taught environmental science for 28 years. He earned a B.A. from Rutgers University and a M.A. and Ph.D. in biology from Harvard University. For many years Wright received grant support from the National Science Foundation for his work in marine microbiology, and, in 1981, he was a founding faculty member of Au Sable Institute of Environmental Studies in Michigan, where he also served as Academic Chairman for 11 years. He is a Fellow of the American Association for the Advancement of Science, Au Sable Institute, and the American Scientific Affiliation. In 1996, Wright was appointed a Fulbright Scholar to Daystar University in Kenya, where he taught for 2 months. He is a member of many environmental organizations, including The Nature Conservancy, Habitat for Humanity, the Union of Concerned Scientists, the Audubon Society and is a supporting member of the Trustees of Reservations. Wright continues to be actively involved in writing and speaking about the environment, and spends his spare time hiking, fishing, birding, golfing, and enjoying his three children and seven grandchildren.

# brief contents

# essays

## earth watch



## ethics



## global perspective



## career link



## guest essay

# contents

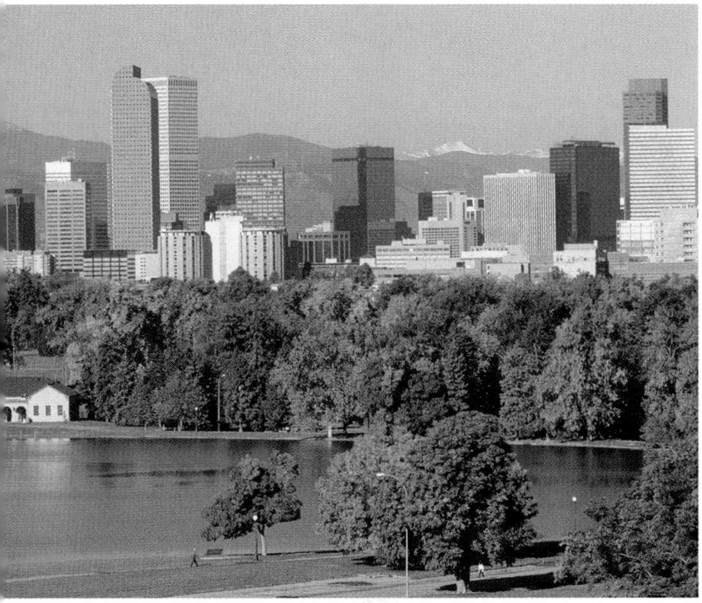

# preface

As we plunge into a new century and a new millenium, the environment is being called on to supply the growing needs of an expanding human population in the developing countries and increasing affluence in the developed countries. In many areas, we are already taking more from Earth's systems than they can provide in a sustainable fashion. Our ecological footprint weighs heavily on Earth's natural resources—the "ecosystem capital" that provides the goods and services that sustain human life and economic well-being. Also, there are still billions of people who are not adequately housed, fed, or provided with health care or a paying job. Yet we must, as soon as possible, make a transition to a sustainable civilization, one in which a stable human population recognizes the finite limits of Earth's systems to produce resources and absorb wastes, and acts accordingly. This is hard to picture at present, but it is the only future that makes any sense. If we fail to achieve it by our deliberate actions, the natural world will impose it on us in highly undesirable ways.

Environmental science stands at the interface between humans and Earth and explores the interactions and relations between them. This relationship will need to be considered in virtually all future decision making. This text considers a full spectrum of views and information in an effort to establish a solid base of understanding and a sustainable formula for the future. What you have in your hands is a readable guide and up-to-date source of information that will help you to explore the issues in more depth. It will also help you to connect them to a framework of ideas and values that will equip you to become part of the solution to many of the environmental problems confronting us.

As the field of environmental science evolves and continues to change, so has this text. In this new edition, I hope to continue to reflect accurately the field of environmental science; in so doing, I have constantly attempted to accomplish each of the following objectives:

- To write in a style that makes learning about environmental science both interesting to read and easy to understand, without overwhelming the student with details.
- To present well-established scientific principles and concepts that form the knowledge base for an understanding of our interactions with the natural environment.
- To organize the text in a way that promotes sequential learning, yet allows individual chapters to stand on their own.
- To address all of the major environmental issues that confront our society and help to define the subject matter of environmental science.
- To present the latest information available by making full use of the resources of the Internet, books, and journals.
- To give an assessment of options or progress in solving environmental problems.
- To support the text with excellent supplements for the instructor and the student that strongly enhance the teaching and learning processes.

Because I believe that learning how to live in the environment is one of the most important subjects in any student's educational experience, I have made every effort to put in your hands a book that will help the study of environmental science come alive.

## A Guide to the Ninth Edition of Environmental Science

### Overview

The **ninth edition** is more than just an update. The main new feature of this edition is the extensive use of six **unifying themes** that help the reader to focus on the significance of the many issues that are presented. The themes of **sustainability**, **sound science**, and **stewardship** are retained from the eighth edition and now identified as **strategic themes**. To these I have added three more themes, which I call **integrative themes: ecosystem capital, policy and politics**, and **globalization.** These six themes provide important threads linking the different subjects and chapters of the text. To make the connections clear, I have added at the end of each chapter a section called **Revisiting the Themes**, where each theme is discussed and connected to the chapter matter. In this edition, I continue to provide a balance between pure science and the political, social, and historical perspectives of environmental affairs. I am also careful to reflect differences in interpretation of environmental concerns where they exist, while maintaining the standard of sound science for judging those concerns.

Most important, the ninth edition reflects the changing environmental scene in the United States, as well as in the rest of the world. Information from new books, journal articles, and Internet-based reports from governmental and nongovernmental organizations has been incorporated into every chapter. New illustrations have been introduced—69 new photos and 49 new diagrams.

Each chapter opens with a case study or an illustrative story to catch the reader's interest and lead into the chapter's subject. A new feature in this edition is the **Guest Essays**; these provide challenging insights from the perspective of professionals in their fields. Another new feature is the companion CD, **Global City**. This is a set of 9 exercises built around the concept of a large city struggling with many of the environmental problems presented in the book: soil erosion, water pollution, energy use, urban sprawl, and municipal waste, among others. Those familiar with the eighth edition will see a new, crisp layout, with the use of many more **subheads**. These enable students to read the text in clearly identified bite-sized segments.

## Introduction

After taking a look at the plight of Easter Island, **Chapter 1 (Introduction: Toward a Sustainable Future)** presents a global environmental picture, starting with the concept of the ecological footprint. In discussing ecosystem decline, the Millennium Ecosystem Assessment is presented as a new effort to understand the links between human well-being and the goods and services provided by ecosystems. The chapter then introduces the three **strategic themes**: **sustainability**—the practical goal that our interactions with the natural world should be working toward; **stewardship**—the ethical and moral framework that should inform our public and private actions; and **sound science**—the basis for our understanding of how the world works and how human systems interact with it. Each theme is thoroughly defined and explored. Following this, the three **integrative themes** are introduced in depth: **ecosystem capital**—the natural and managed ecosystems that provide essential goods and services to human enterprises; **policy and politics**—the human decisions that determine what happens to the natural world and the political processes that lead to those decisions; and **globalization**—the accelerating interconnectedness of human activities, ideas, and cultures. A new Earth Watch essay critiques Bjørn Lomborg's controversial book, *The Skeptical Environmentalist*.

## Part One. Ecosystems: Basic Units of the Natural World

**Part One (Chapters 2–4)** explores natural ecosystems—what they are, how they function, and how they change. I no longer employ the five Sustainability Principles (an approach that has outlived its usefulness), although the actual processes they drew attention to are still very much in place. Chapter 5 (Ecosystems and Evolutionary Change) from the older editions has been removed, but material on natural selection and speciation has been kept in **Chapter 4 (Ecosystems: How They Change)**. Some of the new material that appears in these chapters includes the work of the Heinz Center on the State of the Nation's Ecosystems, the concepts of top-down and bottom-up regulation of populations, and a guest essay by biologist David Lahti on the village weaverbird. The end of Chapter 4 returns to the idea of the ecological footprint to address the issue of human carrying capacity.

## Part Two. The Human Population

**Chapter 5 (The Human Population: Dimensions)** first looks at the dynamics of the human population. The pressures on natural systems as a result of the growth of that population are examined, with a focus on the differences between the developed and developing countries. The chapter ends with the **demographic transition**—the shift from high birth and death rates to low birth and death rates that has brought stable populations to the industrialized world. **Chapter 6 (Population and Development)** describes the developing countries' difficulties in moving through this transition. A major new focus in the material on development is the **Millennium Development Goals**, presented in the context of the Human Development Report of 2002. Development aid and the World Bank are explored in detail, and the political controversy over the so-called "gag rule" is detailed. A guest essay by economist Chris Barrett discusses his concept of the "poverty trap" in the context of Kenyan subsistence farmers and natural resource management.

## Part Three. Renewable Resources

**Part Three (Chapters 7–11)** addresses the science and policies surrounding our use of the natural resources of water, soil, agriculture, and wildlife. Issues concerning the use of such resources in food production, forest growth, and fisheries management are examined in light of increasing population growth and increasing pressure on those resources: again, we all-the-while keep our eyes on sustainability. Some examples of issues receiving a **new emphasis** are: the Aral Sea as a major environmental disaster, the "water wars" of the Western United States, a challenge to the conventional approach to measuring erosion (GLASOD), a detailed look at genetically modified food and the controversies it has generated, the Klamath River controversy, restoration of the Everglades ecosystem, the Bush administration's "Healthy Forests initiative," and a new Global Perspective box on the work of Gordon Sato, the "mangrove man" who has helped Eritreans to reclaim some of their coastal mangrove ecosystems.

## Part Four. Energy

**Part Four (Chapters 12–14)** presents the energy resources currently available and the consequences each can have on the environment. **Chapter 12 (Energy From Fossil Fuels)** opens with the story of the Arctic National Wildlife Refuge and the controversy over drilling there for oil. New perspectives on energy security are added in light of recent terrorism and the U.S.-Iraqi war of 2003. The Cheney report on National Energy Policy is critiqued; supply-side and demand-side responses to energy futures are detailed. **Chapter 13 (Energy From Nuclear Power)** examines the possible future of nuclear power, especially as the Bush administration has moved to encourage this option. The various generations of nuclear power plants are explained, with new information on generations III and IV plants. **Chapter 14** (**Renewable Energy**) examines (among many other options) the topic of renewable energy for transportation, especially the potential for fuel cell vehicles. Overall energy policy is summarized at the chapter's end, where the Union of Concerned Scientists' Clean Energy Blueprint is discussed. A guest essay features the work of Ken Touryan and the National Renewable Energy Laboratory in Denver, Colorado, where he works.

## Part Five. Pollution and Prevention

**Part Five** (**Chapters 15–21**) begins with a chapter on environmental health (**Chapter 15: Environmental Hazards and Human Health**). Risk-based public policy is presented here, and extended to international public health risks, using the World Health Organizations precedent-setting World Health Report 2002. The text goes on to investigate the pollution of water, land, and air that results from human activities and our interactions with the environment that were discussed in earlier chapters. The coverage ranges from the use of pesticides to protect our crops, through sewage treatment and contamination of water, to municipal and hazardous wastes, and on to major atmospheric changes and more local and regional air pollution. Examples of some new issues introduced in this edition are (a) the controversy over arsenic in the context of water quality standards versus drinking water standards, (b) a new Earth Watch essay describing a remarkable approach to municipal solid waste management as practiced on Nantucket Island, (c) the Brownfields Act of 2002 and problems with the Superfund Trust Fund are discussed, (d) a thoroughly rewritten presentation of global climate change featuring the work of the Intergovernmental Panel on Climate Change's third assessment, and the responses of mitigation and adaptation, and (e) a thoroughly reorganized chapter on air pollution (**Chapter 21: Atmospheric Pollution**) adds the perspectives of costs and benefits of air pollution regulation and the controversy over the New Source Review.

## Part Six. Toward a Sustainable Future

**Part Six** (**Chapters 22 and 23**) directly addresses the relationship that exists among economics, public policy, and the environment, focusing especially on our present environmental concerns. The concept of Net National Product is put forward in **Chapter 22** as a solution to the shortcomings of the Gross National Product, and the 2003 OMB report (*Informing Regulatory Decisions*) is cited as a prime example of the importance of environmental regulation in promoting health. A new guest essay by former EPA Ombudsman Robert Martin presents his philosophy of a "transformational environmental policy." In **Chapter 23** (**Sustainable Communities and Lifestyles**) several new reports on urban sprawl are examined; the negative as well as positive impacts of urban sprawl are discussed. The text then goes on to examine urban blight as an outcome of urban sprawl in the developed countries, and as an outcome of population pressure and a lack of employment opportunities in the developing countries. Part VI closes with a look at personal involvement, lifestyles, and values as vital components of a vision for a sustainable future.

## Individual Text Elements

**Essays:** *Environmental Science* features five kinds of essays: Earth Watch, Ethics, Global Perspective, Career Link, and Guest Essays. Lists of essays are found at the end of the outline for each chapter.

> **Earth Watch** essays provide further information that enhances the student's understanding of particular aspects of the topic being covered.
>
> **Ethics** essays focus on the fact that many environmental issues do not involve clear-cut rights or wrongs, but present ethical dilemmas.
>
> **Global Perspective** essays help the student appreciate the global nature and extent of the topic in question.
>
> **Career Link** essays present an individual who has chosen to work in some area of environmental concern. The essays discuss how the person got into his or her career.
>
> **Guest Essays** provide challenging insights from the perspective of professionals in their fields.

**Making a Difference:** I believe that no amount of text-based learning about the environment truly becomes useful until students challenge themselves and those around them to begin making a difference. With this in mind, each of the six parts of the text concludes with a section that suggests courses of action that each student can take to bring about the needed changes to foster sustainability.

**Chapter Opening:** Each chapter begins with a set of "**Key Issues and Questions**"—succinct statements regarding key aspects of the issue being covered and questions inviting the student to explore those issues.

**Chapter Outline:** Chapter outlines may be found in the **Table of Contents**. Importantly, the text of each chapter is organized according to a logical outline of first-, second-, and third-order headings to assist student outlining, note taking, and learning.

**Review Questions:** Each chapter concludes with a set of "**Review Questions**" addressing each aspect of the topic covered. Of course, these questions may serve as learning objectives, as test items, or for review.

**Thinking Environmentally:** A set of questions, "**Thinking Environmentally**," is included at the end of each chapter.

These questions invite the student to make connections between knowledge gleaned from the chapter and other areas of the environmental arena and to apply knowledge gained to specific environmental problems. The questions may be used also for testing or to focus class discussion.

**Vocabulary:** Each new term will be found in boldface type where it is first introduced and defined. All such items are found in the glossary at the end of the book.

**Appendices:** At many points in the text, reference is made to the work being done by various environmental organizations. A listing of major national environmental organizations is given in **Appendix A**. Most of these organizations and agencies have a home page on the Internet and can be located via the Web site that supports this text.

A conversion chart for various English and metric units is found in **Appendix B**.

For students who need some grounding in chemistry, a discussion of atoms, molecules, atomic bonding, and chemical reactions is provided in **Appendix C**.

**Glossary and Index:** A comprehensive glossary provides definitions of virtually all of the special terms, treaties, legislation, and programs identified in the text in boldface type. The index gives page references for all of these terms and for thousands of other topics and issues dealt with in the text.

## Text Elements on the Web

### Video Case Studies

Selected from the archives of several major video producers, these timely and relevant video segments offer students an overview of a particular environmental issue or controversy. A listing of the new video segments and a brief synopsis of each video are found on the ninth edition Web site at **http://www.prenhall.com/wright.** Since videos from earlier volumes are also made available to instructors who adopt the ninth edition of *Environmental Science,* a list of these case studies is also provided on the Web site.

### Bibliography and Additional Reading

An updated listing of articles and books used in preparing this new edition of the text, organized according to chapter, can be found on the ninth edition Web site at **http://www.prenhall.com/wright.** Included are many general references dealing with environmental topics.

## For the Instructor

### Instructor's Resource Center CD-ROM (0-13-144210-4)

Everything instructors need where they want it. The Prentice Hall Instructor Resource Center makes helps make instructors more effective by saving them time and effort. All digital resources can be found in one, well-organized, easy-to-access place. The IRC on CD includes:

Figures—JPEGs of all illustrations and select photos from the text.

PowerPoint™—Pre-authored slides outline the concepts of each chapter with embedded art and can be used as is for lecture, or customized to fit instructors' lecture presentation needs.

TestGen—The TestGen-EQ software, questions, and answers.

Electronic files of the Instructor's Manual and Test Item File.

Links to Companion Website, Research Navigator™, WebCT, BlackBoard, and Course Compass, and their test question cartridges.

### Annotated Instructor's Edition (0-13-147541-X)

This new supplement will help instructors integrate the media and other available resources more easily, saving time and effort in classroom preparation.

### Instructor's Resource Box (0-13-144967-2)

*Environmental Science's* resources are now organized for the way you use them. For every chapter in *Environmental Science 9e,* you'll find a file in the Instructor's Resource Box containing:

- Transparencies
- Print-outs of the PowerPoint presentations
- Print-outs of graphics and animations available on the Instructor Resource Center CD.

Included in the front of the box are text-wide resources:

- Instructor's Resource Center on CD-ROM
- Instructor's Manual with Tests
- TestGen EQ test generation software.

### Instructor's Resource Manual (0-13-144206-6)

By Nancy Ostiguy (Pennsylvania State University)

This thorough resource manual features a chapter outline, instructional goals, concepts and connections, a suggested lecture format, and answers to the chapter-opening "Key Issues and Questions," as well as creative discussion questions, activities, and labs.

### Test Item File (0-13-144208-2)

By M. Stephen Ailstock (Anne Arundel Community College)

Contains over 1,800 test questions, including multiple-choice, short-answer, and essay questions.

### Test Gen EQ (0-13-144207-4)

TestGen-EQ is a computerized test generator that lets instructors view and edit testbank questions, transfer questions to tests, and print the test in a variety of customized formats. Included in each package is the QuizMaster-EQ program that lets instructors administer tests on a computer network, record student scores, and print diagnostic reports.

**Transparency Pack (0-13-144211-2)**
Includes over 200 illustrations and photographs from the text, all enlarged for excellent classroom visibility, 65 transparency masters. All images within the transparency pack are also available electronically on the Instructor's Resource Center on CD (0-13-144210-4).

**Prentice Hall Video Library, Volume VI (0-13-144209-0)**
This unique video series contains is selected from the archives of major video producers, each video includes a written summary that ties the segment to particular sections of the text, making it easier to enhance your classroom presentation with timely and relevant video programs.

**Course Management Systems**
Prentice Hall offers three platforms for managing your course online: CourseCompass, WebCT, or BlackBoard. Each platform allows you to easily post your syllabus, communicate with students, administer homework and quizzes, and record results in a gradebook. The Test Item File for *Environmental Science: Toward a Sustainable Future 9e* is available for download into each of the platforms. Please contact your local Prentice Hall representative for details.

## For the Student

**Global City CD-ROM**
Included free with every copy of the text, this CD provides a series of problem-solving activities utilizing real data, helping students apply and critically evaluate information related to current environmental issues.

Global City features the following exercises:

**Stop the Pest:** Understand a pest species population growth.

**Save the Species:** Help conserve a population of endangered sea turtles.

**Water Conservation:** Assess water use in Global City.

**Soils:** Determine which soils are appropriate for activities.

**Power Up Global City:** Help decide how to provide and conserve energy.

**Risky Business:** Compare risks as Global City faces a disease outbreak.

**Fisheries:** Conserve the Rosy Salmon.

**Find the Sample:** Match contaminated water samples with their sources.

**Breathe Free, Global City:** Identify the sources of air pollution.

**Study Guide (0-13-144202-3)**
By Clark E. Adams (Texas A & M University)
This study guide helps students identify the important points from the text and then provides them with review exercises, study questions, self-check exercises, and vocabulary review.

**Online Study Guide**
**http://www.prenhall.com/wright.** This unique tool is designed to launch student exploration of environmental science resources on the World Wide Web. The home page is regularly updated and linked specifically to chapters in the text. In addition to providing a juried guide to many interesting Web-based resources, the site features review exercises (from which the students receive immediate feedback), updates of environmental issues by region, a guide to environmental careers, and a guide to help students learn how to start making a positive difference for Earth's environment.

**Student Lecture Notebook (0-13-144963-X)**
The ideal lecture companion, this three-hole punched, soft cover notebook includes the same color illustrations as seen in the text, with ample room for note taking.

**Research Navigator™**
**Research Navigator™** is the easiest way for students to start a research assignment or research paper. Complete with extensive help on the research process and four exclusive databases of credible and reliable source material including the EBSCO Academic Journal and Abstract Database, *New York Times* Search by Subject Archive, "Best of the Web" Link Library, and *Financial Times* Article Archive and Company Financials, Research Navigator helps students quickly and efficiently make the most of their research time. Access to Research Navigator is free when purchased with a copy of this textbook. Please contact your local Prentice Hall representative for more details.

## Reviewers

I offer my sincere thanks to those who reviewed the seventh edition and previous editions of this text. Their comments, suggestions, and constructive criticisms have all been carefully considered and in many instances have led to significant improvements in the text. I thank the following people:

Hans Beck Northern
*Illinois University*

Patricia J. Beyer
*Bloomsburg University*

John Brock
*Arizona State University East*

Robin Gibson-Brown
*Carven Community College*

Joe Haverly
*Rock Valley College*

Alan Holyoak
*Manchester College*

Birgit Koehler
*Williams College*

Daniel Meer
*Cardinal Stritch University*

Stephen R. Overmann
*Southwest Missouri State University*

Scott Robinson
*University of Illinois*

Janice Simpkin
*College of Southern Idaho*

Steve Trombulak
*Middlebury College*

Townsend E. Weeks
*Brookdale Community College*

Lorne Wolfe
*Georgia Southern University*

Ian A. Worley
*University of Vermont*

## Reviewers of Previous Editions

M. Stephen Ailstock
*Anne Arundel Community College*

John Blachley
*College of the Desert*

Robert H. Blodgett
*Austin Community College*

Darren Divine
*University of Nevada, Las Vegas*

Norm Dronen
*Texas A & M University*

David Gardner
*Owens Community College*

Ray Grizzle
*University of New Hampshire*

William P. Hayes
*Catholic University*

Kathleen Keating
*Cook-Rutgers University*

Robert Kistler
*Bethel College*

David Liscio
*Endicott College*

Alberto L. Mancinelli
*Columbia University*

Kenneth E. Mantai
*State University of New York, Fredonia*

Nancy Ostiguy
*Pennsylvania State University*

Stephen R. Overmann
*Southeast Missouri State University*

Julia D. Schroeder
*John A. Logan College*

Morris L. Sotonoff
*Chicago State University*

Christy N. Stather
*Illinois State University*

Max R. Terman
*Tabor College*

Phillip L. Watson
*Ferris State University*

Narayanaswamy Bharathan
*Northern State University, Aberdeen*

Roger G. Bland
*Central Michigan University*

Jack L. Butler
*University of South Dakota*

Ann S. Causey
*Auburn University*

Robert W. Christopherson
*American River College*

Lynnette Danzl-Tauer
*Rock Valley College*

Phil Evans
*East Carolina University, Pitt Community College*

Gian Gupta
*University of Maryland, Eastern Shore*

John P. Harley
*Eastern Kentucky University*

Vern Harnapp
*University of Akron*

Stanley Hedeen
*Xavier University*

Clyde W. Hibbs
*Ball State University*

John C. Jahoda
*Bridgewater State College*

Karolyn Johnston
*California State University, Chico*

Guy R. Lanza
*East Tennessee State University*

John Mathwig
*College of Lake County*

Richard J. McCloskey
*Boise State University*

SuEarl McReynolds
*San Jacinto College, Central*

Eric Pallant
*Allegheny College*

David J. Parrish
*Virginia Polytechnic Institute*

Carol Skinner
*Edinboro University of Pennsylvania*

Richard E. Terry
*Brigham Young University*

## Acknowledgments

Although the content and accuracy of this text are the responsibility of the author, it would never have seen the light of day without the dedicated work of many other people. I want to express my heartfelt thanks to all those at Prentice Hall who have contributed to the book in so many ways.

A special thanks goes to my executive editor, Daniel Kaveney, for his encouragement and attention to the oversight of this entire project. Marketing manager Shari Meffert has provided valuable marketing support for this and the previous edition of this book. I thank her for her efforts and for her travels on behalf of this project. John and Jane Murzdek performed yeoman duty as developmental editors in going over each new chapter draft and the illustration program. Donna King was my production editor, keeping me focused on the details of transcribing manuscripts into a published product, and doing it so cheerfully. Dorothy Boorse of Gordon College worked long and hard to produce the excellent Global City CD that accompanies this book. In this effort, she was ably assisted by Prentice Hall media editor, Chris Rapp. I thank them both for this valuable addition to the book. Thanks also go to Stephen Forsling, photo researcher, who scoured the Web and other media sources for the pictures I was looking for. In addition, I thank Clark Adams for writing the study guide, Nancy Ostiguy for doing a fine job with the instructor's guide, M. Stephen Ailstock for the good work on the test bank, and Isobel Heathcote (of the University of Guelph) for her fine work on the "Environment on the Web" essays, and for crafting and keeping up the book's home page on the World Wide Web.

Thirteen years ago, Prentice Hall editor David Brake asked me if I would be interested in helping Bernard Nebel write the fourth edition of his environmental science text. Because of my longtime concern about environmental issues and my interest in writing, I accepted the offer. As the years passed, my commitment to environmental stewardship and deep concerns about our society's interactions with the environment have led me to direct more and more of my energy and ability to writing and speaking about environmental issues.

As I have accepted more of the responsibility for writing this text, I have realized what an amazing job Bernie did in producing the first three editions alone while also teaching full time. He did it because he was frustrated with existing environmental science texts and was convinced he could produce a more readable and effective book—and he did! Bernie Nebel and I share very similar philosophical and educational values and enjoyed collaborating over the years. Although I alone have been responsible for the text since the seventh edition, I am deeply indebted to Bernie for his wonderful sense of organization and beautiful and clear prose that still form part of the book. Both of us have offered this book in its successive editions as our contribution to the students who are now encountering this new century, in the hope that they will join us in helping to bring about the environmental revolution that must come—hopefully sooner than later.

I wish to offer some very personal thanks to my wife, Ann, who has been with me since the beginning of my work in biology and has provided the emotional base and companionship without which I would be far less of a person and a biologist. Her love and patience have sustained me in immeasurable ways. Finally, I offer my gratefulness to God, who is the author of the amazing Creation I love so much. I count it a privilege to be involved in promoting the care of His Creation.

*Richard T. Wright*

# Introduction: Toward a Sustainable Future

### Key Topics

1. The Global Environmental Picture
2. Three Strategic Themes: Sustainability, Stewardship, and Sound Science
3. Three Integrative Themes: Ecosystem Capital, Policy/Politics, and Globalization
4. The Environment in the 21st Century

On Easter Sunday, 1722, three Dutch ships commanded by admiral Jacob Roggeveen sighted a previously uncharted island in the South Pacific. They named it Easter Island. The island, some 166 $km^2$ (65 $mi^2$), is perhaps the most remote spot on the planet—1,200 miles from the nearest island and 2,300 miles from the coast of South America. The Dutch found the island to be inhabited by Polynesians who were living primitively. Roggeveen wrote, "We originally, from a further distance, have considered the said Easter Island as sandy; the reason for that is this, that we counted as sand the withered grass, hay, or other scorched and burnt vegetation, because its wasted appearance could give no other impression than of a singular poverty and barrenness." The sailors were amazed to find many large stone statues scattered about the island (Fig. 1–1), evidence of a once sophisticated civilization. When asked about the statues, the islanders could reply only, "Our legends say that our ancestors made them, but we don't know how or why." The past culture and civilization of the island had vanished.

*Aerial View of Easter Island*

***The Past.*** Working from the legends that islanders told, and conducting excavations for evidence, archaeologists have pieced together the following possible chronology of events: The original inhabitants of Easter Island were Polynesians who probably arrived on the island as part of a deliberate colonization mission sometime between 400 and 800 A.D. The evidence from pollen grains and plant remains found in coring from the bottom of volcanic lakes shows that these early arrivals encountered an island abundantly forested with a wide variety of trees, including palms, conifers, and sandalwood. As their population grew and flourished, they cut trees for agriculture, for structural materials, and to move the huge stone heads from the quarries where they were shaped to the sites at which they would be erected. By 1600, all the trees were gone. Without plant roots, the cleared land failed to hold water, and the soil washed into the sea. The eroded soil baked hard and dry after rains, offering little support for agriculture.

As the forest was depleted and soil and water resources were degraded, the work necessary for existence became harder and the rewards fewer. The gap between the ruling religious elites and secular

**Figure 1–1 Easter Island.** The great stone heads and other artifacts found on Easter Island show that a prosperous culture once existed there. The present barren, eroded landscape indicates that the civilization collapsed from the overexploitation of forest and soil resources. Is the story of Easter Island a parable for modern civilization?

warrior–workers widened, apparently becoming intolerable. Some time during the 17th century, there was a revolt of the workers. In the great war that ensued, virtually the entire ruling religious class was killed. Still, the situation worsened. Anarchy broke out among the workers, who splintered into groups and continued to fight among themselves. Starvation and disease became epidemic. Without any trees, no one could escape the island by boat. A population that had once numbered ten to twenty thousand was down to a few thousand at the time of its "discovery" by the Dutch. Many have puzzled over the reasons why the Easter Islanders, who could walk around their island in a day, were unable to foresee the consequences of their practices.

The Easter islanders, who call themselves the Rapa Nui, suffered terribly from their subsequent contacts with the "civilized" world. In the 19th century they were visited repeatedly by whalers, in search of food and water, who infected the islanders with venereal diseases. Then, in midcentury, Peruvian slavers raided the islands and took over 2,000 islanders captive for the South American slave trade. Eventually, smallpox spread to the island, and by 1877 only 111 Rapa Nui remained. Easter Island was then annexed by Chile, which fostered sheep farming while enclosing all of the islanders in one village. Visits by archeologists in the mid-20th century brought the island to the world's attention, and things began to improve for the islanders. Today many of the stone statues have been restored, the islanders have regained a measure of control over their own destiny, and a significant tourist trade has brought hotels, inns, restaurants, and curio shops to Hanga Roa, the only town on the island (Fig. 1–2). The islanders now depend on imported food, however, and unemployment and alcoholism have grown serious.

***Lessons.*** This is a sad story. It is not over, however, and some of the island's people are working to preserve Easter Island's cultural heritage and restore the soil to allow the cultivation of some agricultural specialties that could diversify the island's economy. The story also carries some powerful lessons, as we look back at what happened. For example, environmentally unsustainable practices probably led to the socially unstable society that evolved. When a society fails to care for the environment that sustains it, when its population increases beyond the capacity of the land and water to provide adequate food for all, and when the disparity between *haves* and *have-nots* widens

**Figure 1–2 Hanga Roa.** The tiny harbor in Hanga Roa, where nearly all of Easter Island's population of 3,000 live. It is the only area with running water and electricity.

into a gulf of social injustice, the result is disaster. The civilization collapses. History is replete with the ruins of other civilizations, such as the Mayans, Greeks, Incas, and Romans, that failed to recognize the constraints of their environment.

The more developed world had a devastating impact on Easter Island, too. Foreign contact brought disease, slavery, and subjugation that lasted for two centuries. More recently, outsiders have brought economic and social help to the islanders, but the island's language and culture are in danger of being lost. The future is uncertain for Easter Island. Much depends on the efforts of the Rapa Nui themselves, for they have the most to gain and the most to lose as they undergo continued development.

We will revisit this story at the end of the chapter and draw some additional lessons from it.

***Moving On.*** In the meantime, this chapter briefly explores the current condition of our planet and then examines a number of themes that address the goals of the text:

- To understand how the natural world works
- To understand how human systems are interacting with natural systems
- To accurately assess the status and trends of crucial natural systems
- To promote a long-term, sustainable relationship with the natural world

## 1.1 The Global Environmental Picture

The arrival of the new millennium became the occasion for taking stock in many areas of human concern, including, in particular, the global environment. The picture of the state of our planet that emerged from a number of surveys was troubling. Four global trends were of particular concern: (a) population growth and economic development, (b) a decline of vital life-support ecosystems, (c) global atmospheric changes, and (d) a loss of biodiversity. Each of these issues is explored in greater depth in later chapters.

### Population Growth and Economic Development

*Ecological Footprint.* The world's human population, over 6.4 billion persons in 2004, has grown by 2 billion in just the last 25 years. It is continuing to grow, adding nearly 77 million persons per year. Even though the growth rate is gradually slowing, the world population in 2050 could be 8.9 billion, according to the most recent projections from the U.N. Population Division (Fig. 1–3). Each person creates a certain demand on Earth's resources, and that demand tends to increase with greater affluence. A simple, but useful,

**Figure 1–3 World population explosion.** World population started a rapid growth phase in the early 1800s and has increased sixfold in the last 200 years. It is growing by nearly 77 million people per year. (See Chapter 5.) Future projections are based on assumptions that birthrates will continue to decline. (Data from U.N. Population Division, 2002 revision.)

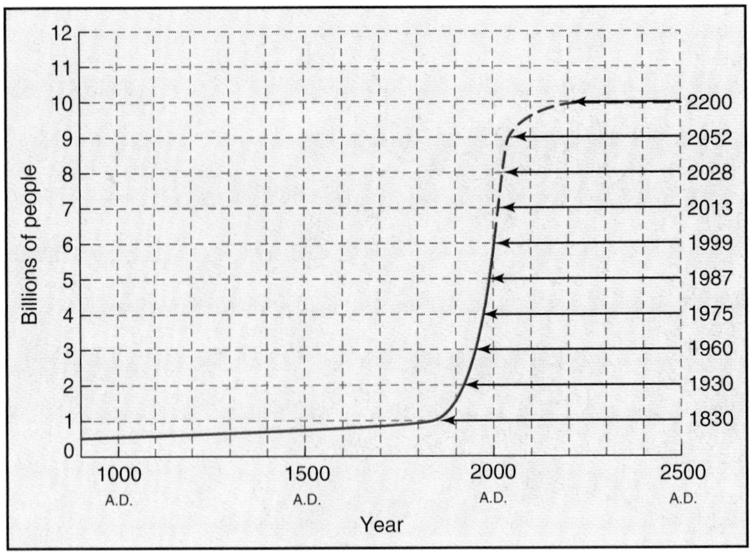


tool for calculating this demand is the **ecological footprint**, a concept developed by a team of scholars at the University of British Columbia (Fig. 1–4). A "footprint analysis" calculates the natural areas required to satisfy human needs and demands in food, housing, transportation, consumer goods, and various services such as absorbing wastes. Thus, the ecological footprint of a typical American is 5.1 hectares (12.6 acres) per person, while that of a typical Indian is 0.4 hectare (1 acre) per person. For the U.S. population, this works out to $1\frac{1}{2}$ times our national area—an indication that we are drawing on resources and services that extend outside the country.

The 2.5 billion persons added to the human population by 2050 will all have to be fed, clothed, housed, and, hopefully, supported by gainful employment. Virtually all of the increase will be in the developing countries; in the poorest of these, birthrates are still much higher than death rates. In the same countries, 1.2 billion experience severe poverty, lacking sufficient income to meet their basic needs for food, clothing, and shelter. Over 800 million—one out of every five—remain undernourished. At the same time, global economic production continues to rise, having tripled since 1980. In spite of that growth, the gap between the average incomes of the wealthiest 20% and the poorest 20% in the world is growing wider. In 1990 it was 60 to 1, and in 1999 it widened to 74 to 1. (It was 30 to 1 in 1960.) It is no surprise that the poorer countries urgently desire to improve their status and undergo economic development as rapidly as possible. As they do, their ecological footprint will grow, and the net environmental impact of those added to the population in the next half century is almost unimaginable. Stabilizing population growth in the developing countries is vitally important for the natural world. It is also the most important prerequisite for closing the economic gap between those nations and the industrialized countries.

## The Decline of Ecosystems

Natural and managed ecosystems support human life and economies with a range of goods and services. These vital resources are stressed by the dual demands of increasing population and affluence-driven consumption per person. Around the world we see groundwater supplies depleted, agricultural soils degraded, oceans overfished, and forests cut faster than they can regrow. A recent report from the United Nations, entitled *Pilot Analysis of Global Ecosystems*, or PAGE, addressed these trends as it examined the status of the five major ecosystems which deliver the goods and services that support human life and the economy: coastal/marine systems, freshwater systems, agricultural lands, grasslands, and forests. PAGE represents an international collaboration of scientists, the first step in an ongoing project designed to measure the long-term impact of human actions on these vital ecosystems. To quote the report's summary, "nearly every measure we use to assess the health of ecosystems tells us we are drawing on them more than ever and degrading them at an accelerating pace."

One of the most disturbing findings of PAGE is that human activities are now beginning to significantly affect the natural chemical cycles—water, carbon, nitrogen, and phosphorus—on which all ecosystems depend. In spite of the critical importance of these cycles to human affairs, we still lack much of the knowledge needed to assess present conditions and to move toward deliberate and wise management of natural ecosystems. Few enterprises could be more important than the pursuit of that knowledge.

**Millennium Ecosystem Assessment.** Toward that end, a four-year U.N. effort, called the *Millennium Ecosystem Assessment* (MEA), was launched on World Environment Day, 2001. U.N. Secretary-General Kofi

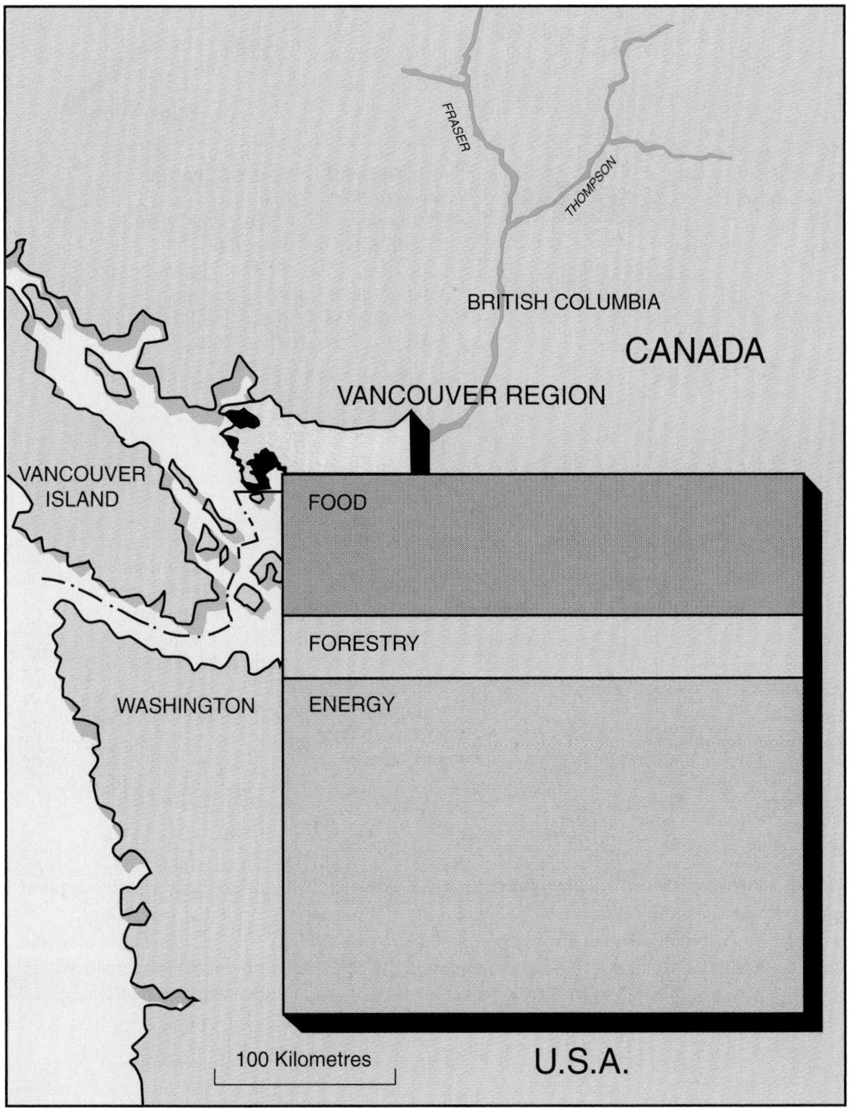


**Figure 1–4 Ecological footprint.** Canadians living in Western Canada's Lower Fraser Valley depend on the productivity of a natural area 19 times their home region in order to satisfy their needs for food, clothing, energy, and shelter. (Adapted from Mathis Wackernagel and William Rees: *Our Ecological Footprint: Reducing Human Impact on the Earth.* Gabriola Island, B.C.: New Society Publishers, 1996. Used with permission)

Annan stated that this assessment would "map the health of our planet, and so fill important gaps in the knowledge that we need to preserve it." Like PAGE, this study embraces the ecosystem approach, recognizing that the ecosystem concept is the best framework for assessing the health of our planet. The project is a major effort to provide an integrated understanding of ecosystems throughout the world—their status, the connections between them, and the impacts we are having on them. This effort focuses especially on the linkages between ecosystem goods and services and human well-being, working at global, regional, and local scales. The MEA conceptual framework, shown in Figure 1–5, illustrates the various dimensions of this project. The goal is to halt the decline of ecosystem health and integrity around the world and promote the restoration and wise management of ecosystems. MEA intends to provide the scientific framework for building the knowledge base required for sound policy decisions and management interventions; it remains for policy makers and managers to act on that knowledge.

## Global Atmospheric Changes

Historically, pollution has been a relatively local problem, affecting a given river, lake, or bay, or the air in a city. Today, scientists are analyzing pollution on a global scale. For example, concern about depletion of the stratospheric ozone layer has already led to international action—the Montreal Protocol in 1987—aimed at curbing pollution from the release of chlorofluorocarbon refrigerants into the atmosphere. A more serious problem today is the danger of global climate change due to carbon dioxide ($CO_2$), an unavoidable by-product of burning fossil fuels—crude oil, coal, and natural gas. Because of the large amount of fossil fuels currently being burned, $CO_2$ levels in the atmosphere have grown from about 280 parts per million (ppm) in 1900 to over 375 ppm in 2004. Indeed, the level of atmospheric $CO_2$ is increasing by 0.4% per year, and given our dependency on fossil fuels, there is no end in sight.

Carbon dioxide is a natural component of the lower atmosphere, along with nitrogen and oxygen. It is required

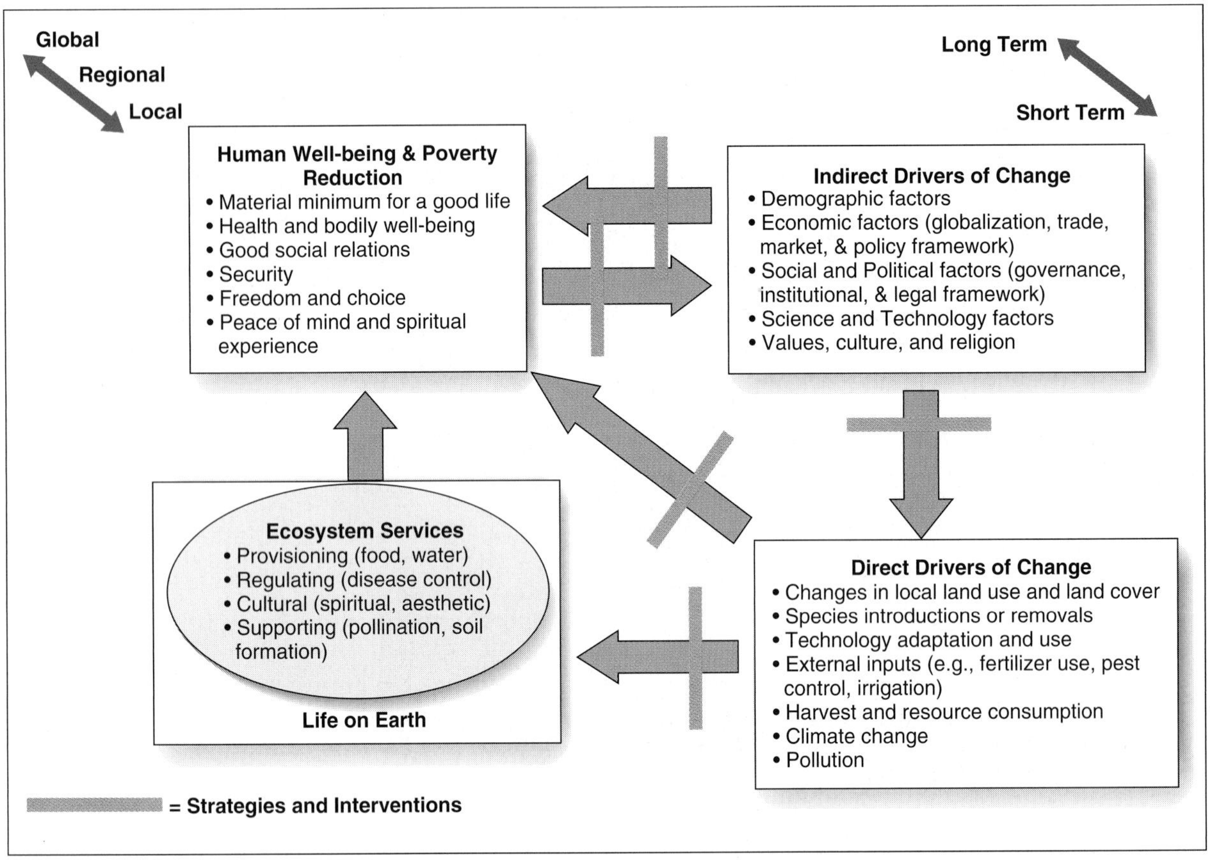


**Figure 1–5 Conceptual framework of the Millennium Ecosystem Assessment.** The left-hand side of this scheme shows the basic relationship between ecosystem services and human well-being. On the right are the factors that can lead to changes in ecosystems, some acting indirectly, others directly. The changes can be positive or negative. The red crossbars indicate where actions can be taken to enhance positive changes or respond to negative changes. All of these interactions can take place at different spatial (global, regional, local) and temporal (long-term, short-term) scales. (*From Ecosystems and Human Well-being: A Framework for Assessment.* Washington: Island Press, 2003. Used With Permission.)

by plants for photosynthesis and is important to the Earth–atmosphere energy system: Carbon dioxide gas is transparent to incoming light from the Sun, but absorbs infrared (heat) energy radiated from Earth's surface, thus slowing the loss of this energy to space. The absorption of infrared energy by carbon dioxide warms the lower atmosphere in a phenomenon known as the *greenhouse effect.* Although the concentration of $CO_2$ is a small percentage of the atmosphere, even slight increases in the volume of the gas affects temperatures. Figure 1–6 graphs air temperatures from 1880 to the present and illustrates the clear warming trend. Referring to this trend, the latest report of the Intergovernmental Panel on Climate Change (IPCC), released in 2000, stated that anthropogenic greenhouse gases (those due to human activities) have "contributed substantially to the observed warming over the last 50 years."

**The Kyoto Protocol.** Concern about global climate change led representatives of 166 nations to meet in Kyoto, Japan, in December of 1997 to negotiate a treaty to reduce emissions of carbon dioxide and other greenhouse gases. At that meeting, most of the industrialized nations (including the United States) agreed to reduce emissions to below 1990 levels, a goal that is to be achieved by the year 2010. The treaty had to be ratified before it came into effect, but by 2003, not enough industrialized nations had ratified it to bring it into force. Significantly, the United States, led by the Bush administration, has reneged on its pledge and has withdrawn from the Kyoto process. Kyoto, however, is only a first step; even if the treaty is adhered to by all parties, the levels of greenhouse gases will continue to rise indefinitely. At issue for many countries are the conflicting concerns between the short-term economic impacts of reducing the use of fossil fuels and the long-term consequences of climate change for the planet and all its inhabitants. Stabilizing the atmospheric concentration of $CO_2$, however, is essential to stabilizing the global climate itself, and this cannot be accomplished without serious reductions in the current rate of use of fossil fuels. This is one of the defining environmental issues of the 21st century.

## Loss of Biodiversity

The rapidly growing human population, together with increasing consumption, is accelerating the conversion of forests, grasslands, and wetlands to agriculture and urban

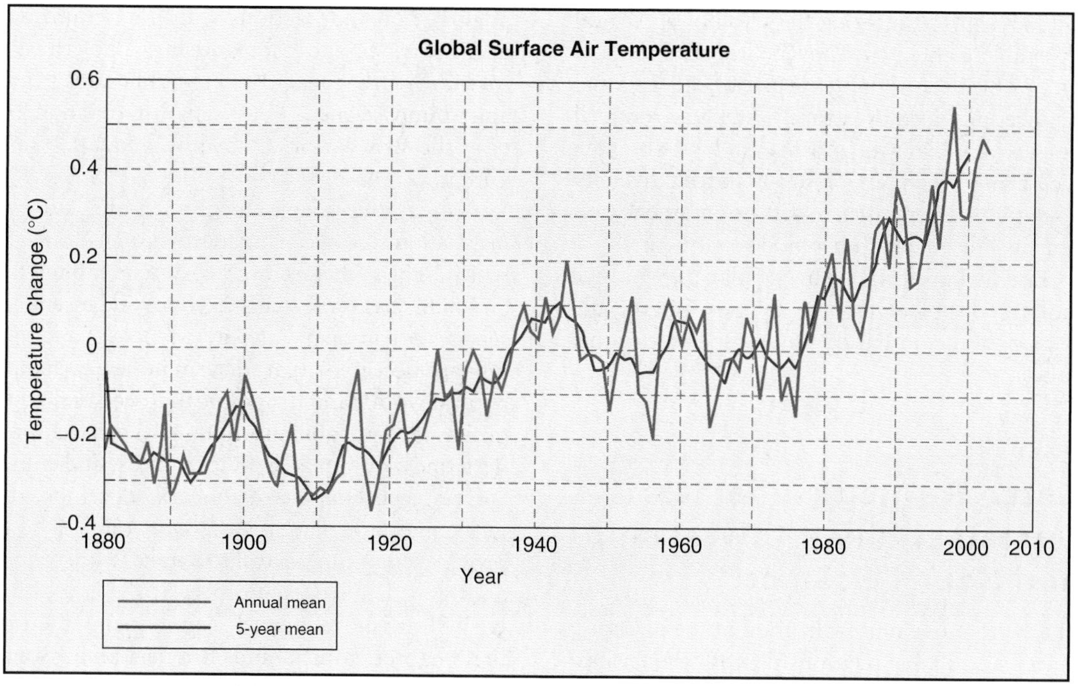


**Figure 1-6 Annual mean global surface atmospheric temperatures.** The baseline, or zero point, is the 1951–1980 average temperature. The warming trend since 1975 is conspicuous. (*Sources:* Goddard Institute for Space Studies and the World Meteorological Organization.)

development (Fig. 1–7). The inevitable result is the loss of most of the wild plants and animals that occupy those natural habitats. If the species involved have no populations at other locations, they are doomed to extinction by this process of **habitat alteration**. Pollution also degrades habitats—particularly aquatic and marine habitats—destroying the species they support. Further, hundreds of species of mammals, reptiles, amphibians, fish, birds, and butterflies, as well as innumerable plants, are exploited for their commercial value. Even when species are protected by law, they continue to be hunted, killed, and marketed illegally.

**Figure 1-7 Natural ecosystems giving way to development.** Continuing growth requires a massive reorganization and exploitation of natural resources, bringing record levels of degradation of natural ecosystems. Here we see a mature forest in the northeastern United States being stripped away.

As a result, Earth is rapidly losing many of its species, although no one knows exactly how many. **Biodiversity** refers to the total diversity of living things—plants, animals, and microbes—that inhabit the planet. About 1.75 million species have been described and classified, but scientists estimate that at least 14 million species may exist on Earth. Because so many species remain unidentified, the exact number of species becoming extinct can only be estimated.

**Risks of Losing Biodiversity.** Why is losing biodiversity so critical? For one thing, all domestic plants and animals used in agriculture are derived from wild species, and we still rely on introducing genes from wild species into our domestic species to keep them vigorous and capable of adapting to different conditions. For another, some 80% of the human population depends on traditional medicines, which in turn are highly dependent on biodiversity. Many modern prescription drugs were originally derived from plants, even though only a small percentage of plants has been thoroughly studied for their medicinal properties. Biodiversity is the mainstay of agricultural crops and of medicines. The loss of biodiversity can only curtail development in these areas. Biodiversity is also a critical factor in maintaining the stability of natural systems and enabling them to recover after disturbances such as fires or volcanic eruptions. Most of the essential goods and services provided by natural systems are derived directly from various living organisms, and we threaten our own well-being when we diminish the biodiversity within those natural systems. There are also aesthetic and moral arguments for maintaining biodiversity. Shall we continue to erase living species from the planet, or do we have a moral responsibility to protect and preserve the amazing diversity of life on Earth? Once a species is gone, it is gone forever.

## 1.2 Three Strategic Themes: Sustainability, Stewardship, and Sound Science

What will it take to move our civilization in the direction of a "long-term sustainable relationship with the natural world" (one of the four goals set forth in the introduction to this chapter)? The answer to this question is complex, but Figure 1–8 outlines two sets of unifying themes—strategic and integrative—that apply to changing or giving direction to the interactions between human and natural systems. *Strategic* themes deal with how we should *conceptualize* our task of forging a sustainable future. These themes are **sustainability**—the practical goal that our interactions with the natural world should be working toward; **stewardship**—the ethical and moral framework that informs our public and private actions; and **sound science**—the basis for our understanding of how the world works and how human systems interact with it.

*Integrative* themes deal with the *current status of interactions* between human systems and the natural world. These themes are **ecosystem capital**—the natural and managed ecosystems that provide essential goods and services to human enterprises; **policy and politics**—the human decisions that determine what happens to the natural world, and the political processes that lead to those decisions; and **globalization**—the accelerating interconnectedness of human activities, ideas, and cultures.

As each chapter of this text develops, these themes will come into play at different points. The themes will provide common threads that hold the topics together and bind them to each other. At the end of each chapter we will revisit the themes, summarizing their relevance to the chapter topics and often adding some editorial thoughts. We begin with sustainability, stewardship, and sound science, the three *strategic themes*.

### Sustainability

**Sustainable Systems.** A system or process is *sustainable* if it can be continued indefinitely, without depleting any of the material or energy resources required to keep it

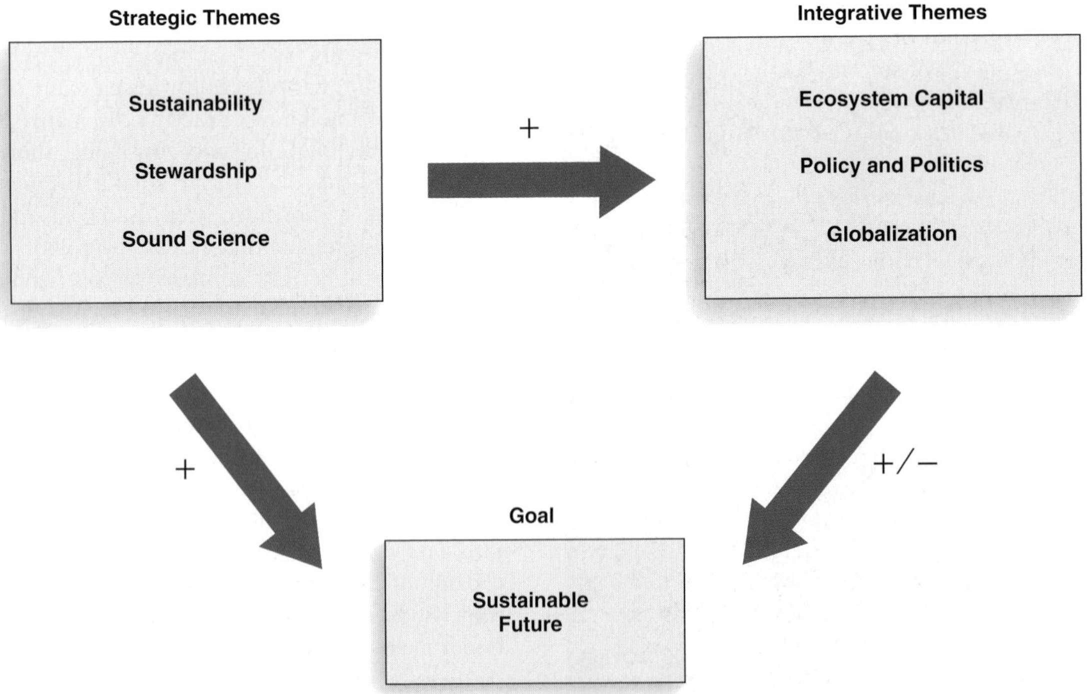


**Figure 1–8 Unifying themes.** Sustainability, stewardship, and sound science are three *strategic themes* that must be embraced by our society in order to move toward a sustainable future. Ecosystem capital, policy and politics, and globalization are three *integrative themes* that deal with the current status of interactions between human societies and natural systems. Each integrative theme can have positive or negative affects on achieving a sustainable future, and each can be positively influenced by the application of the strategic concepts.

running. The term was first applied to the idea of **sustainable yields** in human endeavors such as forestry and fisheries. Trees, fish, and other biological species normally grow and reproduce at rates faster than that required just to keep their populations stable. This built-in capacity allows every species to increase or replace a population following some natural disaster.

Thus, it is possible to harvest a certain percentage of trees or fish every year without depleting the forest or reducing the fish population below a certain base number. As long as the number harvested stays within the capacity of the population to grow and replace itself, the practice can be continued indefinitely. The harvest then represents a sustainable yield. It becomes unsustainable only when trees are cut or fish are caught at a rate that exceeds the capacity of their present population to reproduce and grow. The concept of a sustainable yield can also be applied to freshwater supplies, soils, and the ability of natural systems to absorb pollutants without being damaged.

The notion of sustainability can be extended to include ecosystems. *Sustainable ecosystems* are entire natural systems that persist over time by recycling nutrients and maintaining a diversity of species in balance and by using the Sun as a source of sustainable energy.

**Sustainable Societies.** Applying the concept of sustainability to human systems, we say that a *sustainable society* is a society in balance with the natural world, continuing generation after generation, neither depleting its resource base by exceeding sustainable yields nor producing pollutants in excess of nature's capacity to absorb them. Many primitive societies were sustainable in this sense for thousands of years.

When the concept of sustainability is applied to modern societies, we generally picture certain desirable, healthy characteristics of the people, their communities, and the ecosystems on which they depend. Many of our interactions with the environment are *not sustainable,* however, as is demonstrated by such global trends as the declining health of essential ecosystems and the increased emissions of greenhouse gases. Although population growth in the industrialized countries has almost halted, these countries are using energy and other resources at unsustainable rates, producing pollutants that are accumulating in the atmosphere, water, and land. In contrast, developing countries are experiencing a continued rapid population growth, yet are often unable to meet the needs of many of their people in spite of heavy exploitation of their natural resources. On the basis of expectations of continued economic growth and progress, the crux of the problem is seen to be modern society's inexperience with sustainability. No modern civilization on Earth has ever done it. How do we resolve this dilemma? One answer is the concept of sustainable development.

**Sustainable Development.** **Sustainable development** is a term that was first brought into common use by the World Commission on Environment and Development, a group appointed by the United Nations. The commission made sustainable development the theme of its final report, *Our Common Future,* published in 1987. The report defined the term as a form of development or progress that "meets the needs of the present without compromising the ability of future generations to meet their own needs." The concept arose in the context of a debate between the *environmental* and *developmental* concerns of different groups of countries. **Development** refers to the continued improvement of living standards by economic growth, usually in the developing countries. Both groups of countries (developed and developing) have embraced the concept of sustainable development, although the industrialized countries are usually more concerned about environmental sustainability, while the developing countries are more concerned about economic development. The basic idea, however, is to maintain and improve the well-being of both humans and ecosystems.

The concept of sustainable development is now so well entrenched in international circles that it has become almost an article of faith. It sounds comforting, so people want to believe that it is possible, and it appears to incorporate some ideals that are sorely needed, such as **equity**—whereby the needs of the present are actually met and where future generations are seen as equally deserving as those living now. Sustainable development means different things to different people, however, as illustrated by the viewpoints of three important disciplines traditionally concerned with the processes involved. **Economists** are concerned mainly with growth, efficiency, and the maximum use of resources. **Sociologists** focus on human needs and on concepts like equity, empowerment, social cohesion, and cultural identity. **Ecologists** show their greatest concern for preserving the integrity of natural systems, for living within the carrying capacity of the environment, and for dealing effectively with pollution. It can be argued, however, that *sustainable solutions* will be found only where the concerns of these three groups intersect, as illustrated in Figure 1–9.

***An Ideal.*** There are many dimensions to sustainable development—environmental, social, economic, political—and no societies today have achieved anything resembling it. Nevertheless, as with justice, equality, and freedom, it is important to uphold sustainable development as an ideal—a goal toward which all human societies need to be moving, even if we have not achieved it completely anywhere. For example, policies and actions in a society that reduce infant mortality, increase the availability of family planning, improve the air quality, provide more abundant and pure water, preserve and protect natural ecosystems, reduce soil erosion, reduce the release of toxic chemicals to the environment, restore healthy coastal fisheries, and so on, are all moving that society in the right direction—toward a sustainable future. At the same time, these actions are all quite measurable, so progress in achieving sustainable development can be assessed. Communities and nations are developing "sustainability indicators" to track their progress, such as

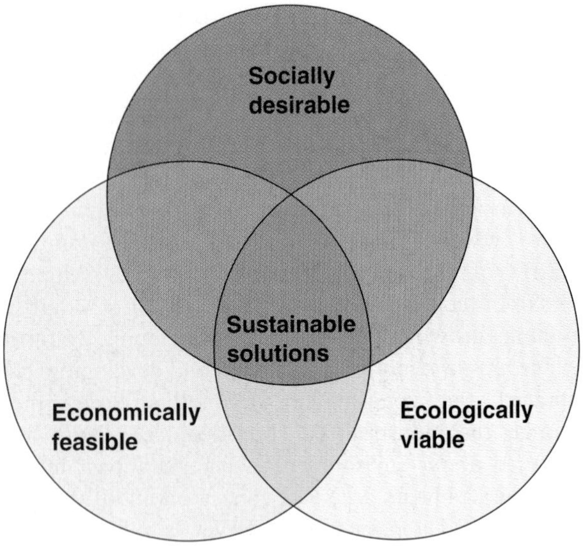


**Figure 1–9 Sustainable solutions.** The concerns of sociologists, economists, and ecologists must intersect in order to achieve sustainable solutions in a society.

the "ecological footprint." Progress in this case means reducing the ecological footprint in an industrial country and perhaps increasing it in a poorer, developing country.

**An Essential Transition.** The transition to a truly sustainable civilization is hard to picture at present. It requires achieving a stable human population that recognizes the finite limits of Earth's systems to produce resources and absorb wastes and that acts accordingly. However, if we fail to achieve sustainability by our deliberate actions, *the natural world will impose it on us in highly undesirable ways,* such as through famine, disease, and deprivation. To achieve sustainability will require a special level of dedication and commitment to care for the natural world and to act with justice and equity toward one another. What will it take to make the transition to a sustainable future? There is broad agreement on the following major points:

- A demographic transition from a continually increasing human population to one that is stable.
- A resource transition to an economy that relies on nature's income and protects ecosystem capital from depletion.
- A technology transition from pollution-intensive economic production to environmentally benign processes.
- A political/sociological transition to societies that embrace a stewardly and just approach to people's needs and in which large-scale poverty is eliminated.
- A community transition from the present car-dominated urban sprawl of developed countries to the "smart growth" concepts of smaller, functional settlements and more livable cities.

This is not an exhaustive list, but it does begin to show what a sustainable future should look like.

## Stewardship

The second of our strategic themes is stewardship—the ethical and moral framework that should inform our public and private actions. Stewardship is a concept that emerged from the institution of slavery. A steward was a slave put in charge of the master's household, responsible for maintaining the welfare of the people and the property of the owner. Because a steward did not own the property himself, the steward's ethic involved a faithful caring for something on behalf of someone else.

Applying this concept to the world today, stewards are those who care for something—from the natural world or from human culture—that is not theirs and that they will pass on to the next generation. Modern-day stewardship, therefore, is an ethic that guides actions taken to benefit the natural world and other people. Stewardly care is compatible with the goal of sustainability, but it is different from it, too, because stewardship deals more directly with *how* sustainability is to be achieved—what values and ethical considerations must be foremost as different choices are weighed. (See "Ethics," p. 11.)

**Some Stewards.** How is stewardship achieved? Sometimes stewardship leads people to try to stop the destruction of the environment or to stop the pollution that is degrading human neighborhoods and health. Examples of people performing this kind of stewardly action are Lois Gibbs and other homeowners at Love Canal (a neighborhood in Niagara Falls, New York), who drew attention in the 1970s to the chemical wastes buried in their neighborhood; Rachel Carson, who, in her 1962 book *Silent Spring,* alerted the public about the dangers of pesticides; Rodolfo Montiel, who organized fellow peasants to block loggers from cutting virgin forests near his Mexican village and, as a result, was tortured and imprisoned in 1999; and Ken Saro-Wiwa, who was executed in 1995 by the Nigerian dictatorship for defending the Ogoni people of Nigeria when oil spills and toxic wastes were devastating their farming and fishing (Fig. 1–10).

More often, stewardship is a matter of everyday people caring enough for each other and for the natural world that they do the things that are compatible with that care. These include participating fully in recycling efforts, purchasing cars that pollute less and use less energy, turning off the lights in an empty room, refusing to engage in the conspicuous consumption constantly being urged on them by commercial advertising, supporting organizations that promote sustainable practices, staying informed on environmentally sensitive issues, and expressing their citizenship by voting for candidates who are sympathetic to environmental concerns and the need for sustainable development.

**Justice and Equity.** The stewardship ethic is concerned not only with the care of the natural world, but also with establishing just relationships among humans. This concern

**ethics**

## What is the Stewardship Ethic?

Ethics is about the *good*—those values and virtues we should encourage—and about the *right*—our moral duties as we face practical problems. Ethics, therefore, is a "normative" discipline—it tells us what we ought to do. Some things are right, and some things are wrong. How do we know what we ought to do? In other words, how does an ethic work? A fully developed ethic has four ingredients:

1. **Cases.** These refer to specific acts and ask whether a particular act is morally justified. The answer must be based on moral rules.
2. **Moral rules.** These are general guidelines that can be applied to various areas of concern, such as the rules that govern how we should treat endangered species.
3. **Moral principles.** Moral rules are based on more general principles, which are the broadest ethical concepts. They are considered to be valid in all cases. An example is the principle of distributive justice, which states that all persons should be treated equitably.
4. **Bases.** Ethical principles are justified by reference to some philosophical or theological basis. This is the foundation for an ethical system.

As we have defined it, a stewardship ethic is concerned with right and wrong as they apply to taking care of the natural world and the people in it. Because a steward is someone who cares for the natural world on behalf of others, we might ask, "To whom is the steward responsible?" Many would answer, "The steward is responsible to present and future generations of people who depend on the natural world as their life-support system." For people with religious convictions, stewardship stems from a belief that the world and everything in it belongs to a higher being; thus, they are stewards on behalf of God. For others, stewardship becomes a matter of concern that stems from a deep understanding and love of the natural world and the necessary limitations on our use of that world.

Is there a well-established stewardship ethic? Insofar as human interests are concerned, ethical principles and rules are fairly well established, even if they are often violated by public and private acts. However, there is no firmly established ethic that deals with care for natural lands and creatures for their own sake. Most of our ethic concerning natural things really deals with how those things serve human purposes; that is, our current ethic is highly anthropocentric. For example, a basic ethical principle from the U.N. Declaration on Human Rights and the Environment is "All persons have the right to a secure, healthy, and ecologically sound environment."

Many organizations have developed stewardship principles for their own guidance, and these principles can sometimes serve as broader guidelines for society, although they usually address specific areas of concern and are generally anthropocentric. For example, the U.S. Forest Service has published a "Land and Service Ethic," which is supposed to guide Forest Service personnel as they carry out their activities. The Forest Service's Land Ethic principle is to "promote the sustainability of ecosystems by ensuring their health, diversity, and productivity." The agency goes on to explain, "Through ecosystem sustainability, present and future generations will reap the benefits that healthy, diverse, and productive ecosystems provide." (*The Forest Service: Ethics and Course to the Future.* USDA Forest Service, Washington, DC, October 1994.)

A stewardship ethic is concerned with the following issues, among others:

1. How to define the common good in cases where conflicting needs emerge, such as the economic need to extract resources from natural systems versus the need to maintain those systems in a healthy state.
2. How to balance the needs of present generations versus those of future generations.
3. How to preserve species when doing so clearly means limiting some of the property rights commonly enjoyed by people and organizations.
4. How to encourage people and governments to exercise compassion and care for others who suffer profoundly from a lack of access to the basic human needs of food, shelter, health, and gainful work.
5. How to promote virtues that are conducive to stewardly care of the natural world, such as benevolence, accountability, frugality, and responsibility. (There are many others.)
6. How to limit consumption while at the same time allowing people the freedom to choose their lifestyles. In other words, how do you balance individual environmental rights and responsibilities with those of the community?

These are some of the issues that a stewardship ethic must grapple with in order to be truly helpful in accomplishing our stated goal of providing the "ethical and moral framework that informs our public and private actions." The work of establishing a well-recognized, broadly applied stewardship ethic remains to be done.

for justice has been applied to the United States in what is called the *environmental justice movement.* The major problem addressed by the movement is **environmental racism**—the placement of waste sites and other hazardous facilities in towns and neighborhoods in which most of the residents are nonwhite. The flip side of this problem is seen when wealthier, more politically active, and often predominantly white communities receive a disproportionately greater share of facilities, such as new roads, public buildings, and water and sewer projects.

People of color are seizing the initiative to correct these wrongs, creating citizen groups and watchdog agencies to

**Figure 1–10 Stewardship at work.** Ken Saro-Wiwa was a Nigerian writer who lost his life defending the Ogoni people against government-backed interests that were degrading their environment. He is an example of unusual environmental stewardship at work.

bring effective action and to monitor progress. An example can be found in Piney Woods and Alton Park, suburbs of Chattanooga, Tennessee, populated largely by African-Americans. In those neighborhoods, there are 42 known hazardous waste sites, 12 of which have been listed as Superfund sites (polluted sites deemed serious enough to require governmental intervention and remediation). Cancer and asthma rates there were unusually high, so people organized the group Stop TOxic Pollution (STOP) to draw attention to the toxic waste and health problems and, eventually, to acquire a technical assistance grant from the Environmental Protection Agency (EPA) to help fence off the worst sites.

**Justice for the Developing World.** Justice is especially crucial for the developing world, where unjust relationships often leave people without land, with inadequate food, and in poor health. Abject poverty is the condition of at least 1.2 billion people, whose poverty is often brought on by injustices within societies where wealthy elites maintain political power and, through corruption and nepotism, steal money and create corporations that receive preferential treatment. For example, Ferdinand Marcos (of the Philippines), Mobutu Sese Seko (Zaire), and Suharto (Indonesia) diverted billions of dollars from their respective country's resources for their private gain, while ignoring the needs of the desperately poor and powerless majorities in their countries (Fig. 1–11). These leaders plundered their countries and enriched their families until they were finally driven out of office.

Some of the poverty of the developing countries can be attributed to unjust economic practices of the wealthy industrialized countries. The history of European colonialism demonstrates how a dominant culture can enrich itself by seizing power in an undeveloped region and then exploiting the people and their natural resources. Although colonialism is now rare, its legacy is still seen in the dependency fostered in so many of the former colonies and the poorly developed technological, educational, and social conditions there. Another source of the continued poverty of many of the developing countries is the current pattern of international trade. By imposing restrictive tariffs and import quotas, industrialized countries have maintained inequities that discriminate against the developing countries. The United Nations reports that the rate of protection against exports from developing countries is significantly higher than the rate against exports from industrial countries. Although these barriers are falling, they are still in place for many manufactured goods coming from the developing world. Such barriers deprive people in developing countries of jobs and money that would go far to improve their living conditions. Although international justice is a difficult issue, it is part of the mission of stewardly action to address it.

(a) (b) (c)

**Figure 1–11 Leaders who stole from their people.** (a) Mobutu Sese Seko, former president of Zaire; (b) Suharto, former president of Indonesia; and (c) Ferdinand Marcos, former president of the Philippines. These national leaders diverted billions of dollars from the people of their countries to enrich themselves and their families.

## Sound Science

This book describes **environmental science,** which employs the methods of sound science to provide the information needed by human societies to improve human welfare and to promote the health of the natural systems that sustain those societies. Why *sound* science? The term *sound science* is used to distinguish legitimate science from *junk science*—information that is presented as valid science, but that does not conform to the rigors of the methods and practice of legitimate science. Unfortunately, science can be done poorly or, worse, can be deliberately misused. This can take the form of biased presentations of scientific information, in which undue prominence is given to some data while unwelcome data are ignored. Or information may be presented as science that does not conform to the accepted methods of scientific research. When some special-interest group has an agenda to push, when some politically motivated decision needs to be justified, when the goal is to convince the media and the public of some desired perspective (and often to counter an opposing scientific argument), sound science is generally abandoned, and the result can appropriately be called *junk science.* The "Earth Watch" essay on p. 14 presents a recent prominent example of this phenomenon. The first step in learning how to distinguish sound science from junk science is to understand the scientific method.

**Science and the Scientific Method.** In its essence, science is simply a way of gaining knowledge; that way is called the **scientific method.** The term *science* further refers to all the knowledge gained through that method. What is "the scientific method?" In previous schooling, you may have learned that the scientific method consists of the following sequence: observation, hypothesis, test (experiment), and theory (Fig. 1–12). This sequence, although basically correct, is an oversimplification, because it leaves out the unique thought processes often involved in the steps it comprises. In addition, it omits what is really the most fundamental aspect of science, namely, that *science is a thoroughly human enterprise.*

A more contemporary view of how science works can be seen through the writings of Del Ratzsch, a philosopher of science. According to Ratzsch, there are three major components to the structure of a science: *data, theories,* and *shaping principles.*

***Data.*** The *data* consist of information gathered from observations and measurements drawn from the natural world or from human interactions with it and from testing ideas through experimentation. Data must be acquired through the senses, directly or through the use of instruments, and data and observations must be recorded with the highest possible degree of accuracy. It is important to remember that these are inevitably the result of a conscious choice by the scientist—that is, which kinds of data, which experiments, and which instruments shall be used?

How can we be sure that data are accurate? As a matter of fact, not every reported observation is accurate, for reasons ranging from honest misperceptions to calculated mischief. Therefore, an important aspect of science, and a trait of scientists, is to be skeptical of any new report until it is confirmed or verified. Such confirmation usually means that other investigators must repeat and check out the data of the first investigator and validate (or invalidate) their accuracy. As observations are confirmed by more and more investigators, they gain the status of *factual* data. In other words, **facts** are things or events that have been confirmed by more than one observer and remain open to be reconfirmed by additional people. Things or events that do not allow this kind of confirmation—UFOs, for example—remain in the realm of speculation from a scientific standpoint.

***Theories.*** Theories are the major objective of scientific reasoning—the explanations of how things work in the natural world, on the basis of the data collected. Theories are models that, hopefully, accurately represent how a system works. For example, we observe that water evaporates and leads to moist air and that water from moist air condenses on a cool surface. We also observe clouds and precipitation. Putting these observations together logically, we derive the concept of the hydrologic cycle. Water evaporates and then condenses as air is cooled, condensation forms clouds, and precipitation follows. Water thus makes a cycle from the surface of Earth into the atmosphere and back to Earth (Fig. 9–3). Note how the example broadens our everyday experiences of water evaporating and falling as rain into an understanding of a cycle involving both. When scientists construct theories, they must be objective and rational. *Objectivity* is achieved when *all* data and observations are considered, not just those which conform to the current model. *Rationality* refers to the need to make clear, logical connections between data and the theory-forming process. Often, those connections will take the form of mathematical relationships. As a result, theories are very much a human endeavor: They do not simply emerge from the data, but are often a reflection of how creative or imaginative the scientist is.

***Shaping Principles.*** Shaping principles are those conscious and unconscious values and assumptions scientists bring to their work. They can profoundly influence the course of scientific investigations, both in gathering data and, especially, in forming theories. One of the most important shaping principles is the *worldview* a scientist brings to his or her field of study. A person's **worldview** is a set of assumptions and values that the person believes to be true about how the world works and about his or her place in it. Worldviews involve both crucial and trivial issues and often determine the direction of a person's work and the choices the person makes every day. Worldviews are often strongly influenced by the culture in which an individual lives.

Other shaping principles are assumptions about the *uniformity of nature*—that the natural world obeys certain fundamental laws and does so without exception

**earth watch**

### A Danish Missile

In 2001, Bjorn Lomborg, associate professor of statistics at the University of Aarhus, Denmark, attacked the environmental "establishment" in his book *The Skeptical Environmentalist: Measuring the Real State of the World.* The book's basic message is that things are far better than we have been told and that we should not worry so much about the state of the world. Lomborg employed masses of data to support these claims.

Lomborg begins his book with what he calls "the litany," the doomsday statements coming from environmentalists and scientists and often repeated by the media: The environment is in poor shape. Our resources are running out, population is continuing to grow, and there is less and less to eat. The air and water are getting more polluted, species are being extinguished in vast numbers, forests are disappearing, fish stocks are collapsing, and the coral reefs are dying. Topsoil is disappearing, we are destroying the wilderness, decimating the biosphere, and breaking down the world's ecosystem. We are fast approaching the limits of growth and will end up killing ourselves.

According to Lomborg, this is all wrong. He analyzes statistical trends in human welfare and the resources supporting human life and prosperity and reports that, in practically all cases, things are getting better, not worse. Lomborg believes that there are problems, such as continued hunger and poverty in many developing countries, but things are better than they used to be. We are not overexploiting renewable natural resources like forests and fisheries, and we have no serious energy problems. Similarly, Lomborg believes that, although there are still some problems with pollution, such as air pollution in the developing world, acid rain, and nutrient overload, many things have improved. Rivers have been cleaned up, pesticides in human tissues have been lowered, and air pollution in developed countries has diminished. In short, we should be optimistic about the future.

#### Why Bad News?

Lomborg believes that there are three reasons why we hear so much bad news about the environment: (1) Scientists depend on public funding for their research, and it is possible that some of them exaggerate the problems so that they can keep getting funded. (2) We assume that environmentalist organizations like Greenpeace, the World Wildlife Fund, and the Worldwatch Institute are acting in the public interest when they alert their constituents about environmental problems. But maybe they're not, says Lomborg: These groups depend heavily on attracting members who provide their financial support, and they can get more members if they frighten people a little. (3) The media need to sell their product, and bad news sells better than good news.

*The Skeptical Environmentalist* is an interesting, but flawed, book. Lomborg correctly states that some environmentalists have made claims about impending environmental disaster that have not held up, and he does admit that there is still much that needs to be done, particularly to alleviate hunger and poverty. He is even right that many things are getting better. Nevertheless, he makes claims that are difficult to support. For example, he says that people are paralyzed by fear about the environment. In addition, he is highly critical about the amount we spend on the environment today; he says we may be committing "statistical murder" by diverting funds from programs that could save people's lives to those which lower pollutants like pesticides and nutrients, which are actually doing little damage to people. Lomborg's analysis of the improvements in the environment is particularly flawed because these improvements are due largely to the work of the very scientists, environmental organizations, and media institutions that he blames for the bad news. Environmental public policy does not just happen; it results from a combination of sound science, advocacy by citizen-led groups, and media attention that convinces state legislatures or Congress to address a problem.

A branch of the Danish Research Agency has accused Lomborg of "scientific dishonesty." They argue that his book displays "systematic one-sidedness" as he presents his analysis of the environmental scene. The agency did not believe that Lomborg deliberately tried to mislead readers, but rather that he used poor judgement in selecting the data he presented and the conclusions he drew. Lomborg's book violates several of the basic principles of *sound science.* For example, it represents a biased selection of scientific results; it was not subjected to the review and correction by scientists in the field before publication; it was written by a statistician, not a natural scientist, yet it addresses a host of issues for which Lomborg cannot claim competence; and it does not represent the consensus in most of the fields concerned. (For a detailed critique of Lomborg's book, see the January 2002 issue of *Scientific American;* For Lomborg's rebuttal, see the May 2002 issue.)

(for example, gravity, the thermodynamic laws, and so on). Many scientists believe that *quantifiability* is a requirement for data: If you can't measure it, it isn't scientifically valid. This is a shaping principle. Prior commitment to a particular theory or paradigm can also be a powerful shaping principle that influences one's scientific work. These three components of science—data, theories, and shaping principles—interact strongly, forming a framework that is hard to break and that often involves deep commitment from the scientists who have built the framework. The bottom line, according to Del Ratzsch, is "that science is a decidedly human pursuit. Science is seen as no

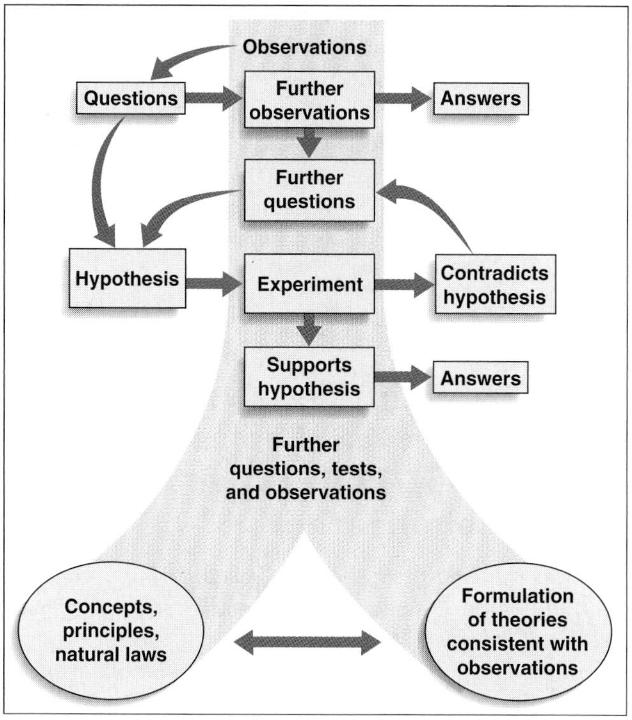


**Figure 1–12 Steps of the scientific method.** The process leading to the formation of theories and the postulation of natural laws and concepts is a continual interplay between observations, hypotheses, tests (experiments), theories, and further refinement.

more ruggedly and rigidly objective and logical than the humans who do it."

Science has successfully brought us to a high degree of understanding of the natural world and has also brought a host of technologies based on that understanding. Doesn't this success mean that our theories about how the world works must be true? A philosopher of science would answer, "*No, not necessarily.*" The best we can do is to establish a theory beyond a reasonable doubt. Theories are always less than absolutely certain, because we can never know whether some other theory exists that will do a better job of explaining the data. Also, scientists always deal with models, not the real thing. Generally speaking, *our confidence in scientific knowledge should be proportional to the evidence supporting it.*

**The Scientific Community.** The scientific method is a way of gaining understanding. It starts with basic data that are then developed into a logically coherent picture of broader phenomena through the formation of theories. The value of this approach is seen in the scientific and technological progress it has made possible. Keep in mind, though, that science and its outcomes take place in the context of a **scientific community** and a larger society. There is no single authoritative source that makes judgments on the soundness of scientific theories. Instead, it is the collective body of scientists working in a given field who, because of their competence and experience, establish what is sound science and what is not. They do so by communicating their findings to each other and to the public as they publish their work in peer-reviewed journals. The process of *peer review* is crucial; in it, experts in a given field review the analyses and results of their colleagues' work. Careful scrutiny is given to grant proposals and published research, with the objective of rooting out poor or sloppy science and affirming work that is clearly meritorious.

**Controversies in Science.** Many environmental issues are embroiled in controversies that are so polarized that no middle ground seems possible. On the one hand are persons who argue from apparently sound facts and proven theories. On the other are persons who disagree and present opposing theories to interpret the facts. Both groups may have motives for arguing their case that are not at all apparent to the public. In the face of such controversy, many people are understandably left confused. Why does so much controversy still prevail? There are four main reasons:

1. **New information.** We are continually confronted by new observations—the hole in the ozone layer, for instance, or the dieback of certain forests. It takes some time before all the hypotheses regarding the cause of what we have observed can be adequately tested. During this time, there may be honest disagreement as to which hypothesis is most likely. Such controversies are gradually settled by further observations and testing, but the process leads into the second reason for continuing controversy.
2. **Complex phenomena.** Certain phenomena, such as the hole in the ozone layer or the loss of forests, do not lend themselves to simple tests or experiments. Therefore, it is difficult and time consuming to prove the causative role of one factor or to rule out the involvement of another. Gradually, different lines of evidence come to support one hypothesis and exclude another, enabling the issue to be resolved. When is there enough evidence to say unequivocally that one hypothesis is right and another wrong? Deciding that there is enough evidence to be convincing involves subjective judgment.
3. **Bias.** The biases or vested interests of a person may affect the amount of information the person requires to be convinced. For example, the Tobacco Institute, a lobbying association for the tobacco industry, argued for years that the connection between smoking and illness had not been proved and that more studies were necessary. By harping on the absence of absolute proof (a scientific impossibility anyway) and simply ignoring the overwhelming body of evidence supporting the connection between smoking and illness, the tobacco lobby

succeeded in keeping the issue controversial and thereby delayed regulatory restrictions on smoking. Thus, the third reason for controversy is that there are many vested interests which wish to maintain and promote disagreement, because they stand to profit by doing so. The need to watch out for this kind of behavior when one is evaluating the two sides of a controversy is crucial.

4. **Subjective values.** The fourth reason for controversy is that subjective value judgments, as well as subjective judgments of fact, may be involved. This is particularly true in environmental science, because the discipline deals with the human response to environmental issues. For example, there is virtually no controversy regarding nuclear power, as long as it is considered at the purely scientific level of physics. However, when it comes to the environmental level of deciding whether to promote the further use of nuclear energy to generate electrical power, controversy arises because different people have different subjective feelings about the relative risks and benefits involved.

Some controversy is the inevitable outcome of the scientific process itself, but much of it is attributable to less noble causes, such as junk science. Unfortunately, the media and the public may be unaware of the true nature of the information—whether it is sound or junk science—and will often give equal credibility to opposing views on an issue. The result is that the junk science is given equal respectability with the sound science in the public arena.

**Evaluating Science.** Whether scientist or layperson, we can use facets of the scientific method to judge the relative validity of alternative viewpoints and to develop our own capacity for logical reasoning. The basic questions to ask are the following:

- What are the observations (*data*) underlying the conclusion (*theory*)? Can they be satisfactorily verified?
- Do the explanations and theories follow logically from the data (*rationality*)?
- Does the explanation account for all of the observations (*objectivity*)? (If the conclusion is logically inconsistent with any observations, it must be judged as questionable at best.)
- Are there reasons that a particular explanation is favored? Who profits from having that explanation accepted broadly?
- Is the conclusion supported by the community of scientists with the greatest competence to judge the work? If not, it is highly suspect.

In sum, *sound science* is absolutely essential to forging a sustainable relationship with the natural world. Our planet is dominated by human beings—our activities have reached such an intensity and such a scale that we are now one of the major forces affecting nature. Our influence on the natural ecosystems that support most of the world economy and process our wastes is strong and widespread, and we need to know how to manage the planet so as to maintain a sustainable relationship with it. As a result, the information gathered by scientists needs to be accurate, credible, and communicated clearly to decision makers and the public.

## 1.3 Three Integrative Themes: Ecosystem Capital, Policy/Politics, and Globalization

Whereas our first three themes—sustainability, stewardship, and sound science—are primarily *strategic*, shaping our thinking about the natural world and how to proceed in order to establish a sustainable future, the second three themes—ecosystem capital, policy and politics, and globalization—are *integrative*. They describe some of the dimensions of what we are dealing with as our human systems draw their sustenance from the natural world and inevitably affect it. We begin by discussing **ecosystem capital.**

### Ecosystem Capital

**Goods and Services.** Natural and managed ecosystems provide human enterprises with essential goods and services. The world economy depends heavily on many renewable resources as we exploit these systems for **goods**, such as fresh water, all of our food, much of our fuel, wood for lumber and paper, leather, furs and raw materials for fabrics, oils and alcohols, and much more. Just three sectors—agriculture, forestry, and fishing—are responsible for 50% of all jobs worldwide and 70% of all jobs in sub-Saharan Africa, eastern Asia, and the Pacific islands.

These same ecosystems also provide a flow of **services** that support human life and economic well-being, such as the breakdown of waste, regulation of the climate, erosion control, pest management, the maintenance of crucial nutrient cycles, and so forth. In a very real sense, these goods and services can be thought of as *capital—ecosystem capital.* The products of this capital—its income, so to speak—generate wealth. As a result, the stock of ecosystem capital in a nation and its income-generating capacity represents a major form of the wealth of the nation. These goods and services are provided free of charge, and they can be sustained year after year. Following the MEA paradigm, Figure 1–13 shows the most prominent ecosystem goods and services that represent what we refer to as ecosystem capital.

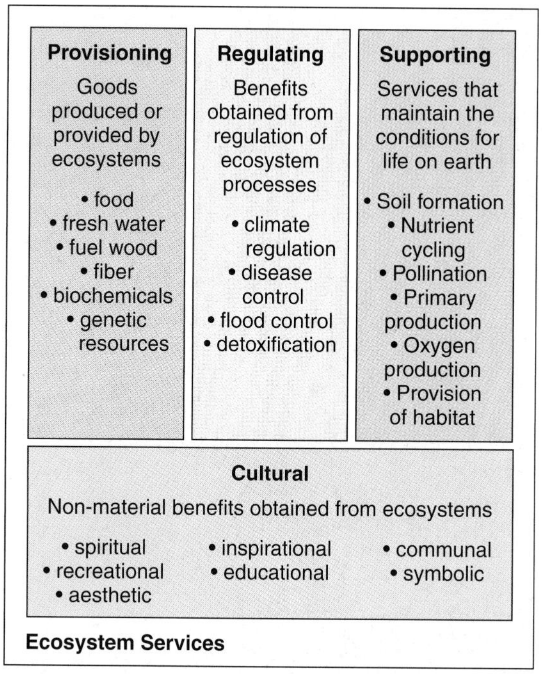


**Figure 1–13** **Ecosystem services.** Provisioning goods, regulating benefits, cultural (nonmaterial) benefits, and supporting services that originate with natural ecosystems and that support life and human enterprises. (From the Millennium Ecosystem Assessment, Millennium Assessment Methods document.)

**Exploitation.** Much of a nation's ecosystem capital is renewable and therefore may be exploited. For example, fisheries and forests are harvested for food and timber, agricultural soils are used for producing food and fiber, groundwater is removed for irrigation and domestic use, and many natural systems are employed for recreation, hunting, and tourism. Just because ecosystem capital is renewable, however, does not necessarily mean that it will be exploited sustainably. In addition, there are many threats to this ecosystem capital that come from other human activities. (See Fig. 1–4.) For example, pollutants are discharged to air and water, exotic species are introduced, land is transformed to other uses (such as highways, mining, and housing tracts), wetlands are drained, and so forth. Unsustainable exploitation and its damaging impacts represent a loss of the goods and services provided by the these ecosystems. Far too often, the people most affected by this loss are the poor. Their water supply is polluted, and they suffer from infectious waterborne diseases; timber is removed by a multinational company, and there is a sudden loss of traditional food, fuel, and other forest products.

Considering the vital importance of ecosystem capital for human well-being, why isn't it being better protected and preserved? Many of the factors that drive changes in ecosystem capital are related to exploitation. There may be market forces that drive intense exploitation, such as a building boom that encourages timber harvesting. There may be social needs, like whole communities that depend on a local fishery that is in decline, forcing the fishers to continue to exploit the fishery in order to pay their bills. There may simply be a lack of information on how human use is linked to the degradation of an ecosystem, or a lack of understanding of the vital goods and services provided by, for example, coastal wetlands or some other ecosystem. There may be unintended consequences of local decisions on broader concerns, such as the impact of driving large vehicles on long-term global climate. Frequently there is no effective market mechanism for valuing the goods and services provided by natural systems, so these assets are often seriously undervalued. As a result, efforts like the MEA are now being carried out to correct these failures.

**Protecting Ecosystem Capital.** There is good news, however. In *The New Economy of Nature,* Gretchen Daily and Katherine Ellison describe examples in which ecosystem capital is being protected by using market-based mechanisms. For instance, the government in Costa Rica pays private landowners to maintain functioning ecosystems such as forests. The city of New York, faced with pressure from the EPA to build a $6 billion water filtration plant, decided instead to take measures to protect the Catskill–Delaware watershed that delivers most of the city's water. By paying farmers and foresters to provide protective vegetation near water supplies, purchasing land around reservoirs, enforcing limits on growth, and taking other measures, the city was able to cleanse its water for less than $2 billion. The overall message of this book is that combinations of private enterprise and government policies have the potential to protect ecosystem capital and make it a profitable enterprise.

## Policy and Politics

The second integrating theme is *policy and politics*—the human decisions that determine what happens to the natural world, and the political processes that lead to those decisions. Few would question the value of gathering knowledge about what ecosystems contribute to human well-being, the current condition of those ecosystems, the impacts of human use on the ecosystems, and the possible future changes as human populations grow and demand more from the natural world. However, this knowledge is of little use unless it becomes the basis of *public policy.*

The purpose of environmental public policy is *to promote the common good.* Just what the common good consists of may be a matter of debate, but at least two goals stand out: *the improvement of human welfare* and *the protection of the natural world.* Environmental public policy addresses two sets of environmental issues: (1) the prevention or reduction of air, water, and land pollution and (2) the use of natural resources like forests, fisheries, oil, land, and so forth. All public policy is developed in a sociopolitical context that we will simply call *politics.*

**Local Level.** Some policies are developed at local levels to solve local problems. Cities and towns, for example, establish zoning regulations in order to protect citizens from haphazard and incompatible land uses. Some municipalities and regions go further and establish broader policies like "smart growth" to address the evils of urban sprawl. Others have embraced the wisdom of sustainability and have moved to create sustainable cities and communities. None of this happens in a political vacuum, however, and behind every progressive zoning law or smart-growth initiative is often a political battle between entities with strong interests in maintaining the status quo and those promoting environmentally sustainable changes. The outcome in any local or regional battle can go either way.

**Broader Levels.** Many problems, however, are broader in scope and must be addressed at higher levels of government. For example, the problems of air pollution transcend local, state, and even national boundaries. The processes that contribute to air pollution are complex, including a myriad of activities fundamental to an industrial society—from people heating their homes and driving cars, to local dry-cleaning establishments and car-repair shops providing their services, to major power plants generating electricity for entire regions. It becomes the responsibility of national governments to address these broader problems. In the United States, many environmental policies have emerged as the outcome of grassroots politics in the late 1900s, whereby private citizens and nongovernmental organizations (such as the Audubon Society and the Nature Conservancy) demanded action on crucial problems and Congress responded. Today, the EPA exists as a monument to the environmental movement. This agency enforces regulations established by Congress, such as the Clean Air Act, which regulates air pollution.

As with local environmental policies, however, there are political battles at the national level surrounding almost every environmental issue. Bitter conflicts have emerged over issues involving access to publicly owned resources such as water, grazing land, and timber. The Endangered Species Act, which protects wild plants and animals whose existence is threatened, has been a lightning rod for controversy. When the EPA develops regulations to address air and water pollution, for example, business special interests oppose the regulations and environmental special interests favor them. The result is often some compromise that leaves everyone unhappy.

**Current Politics.** The two dominant political parties in the United States—the Democrats and the Republicans—have very different views on environmental issues. When Republican George W. Bush took office in 2001 (Fig. 1–14), his administration and their allies in Congress began a process of policy redirection that the *New York Times* has called the "perfect storm" for wrecking environmental programs. With the support of a host of antienvironmental entities, such as the energy, timber, and mining industries, the construction industry, road builders, and "free-market" think tanks, many of the Clinton (Democratic) era environmental policies have already been weakened or cancelled. Most prominently, President Bush has reneged on the Kyoto Protocol, withdrawn support for international family planning, softened air and water quality standards, and crafted an energy policy that favors heavy exploitation of fossil fuels, particularly on the sensitive Arctic National Wildlife Refuge. This activity points out the fact that *politics* always accompanies *policy*, and as we proceed in this text to deal with the different environmental issues, we will employ that theme to gain insight into how governments are responding to the issues.

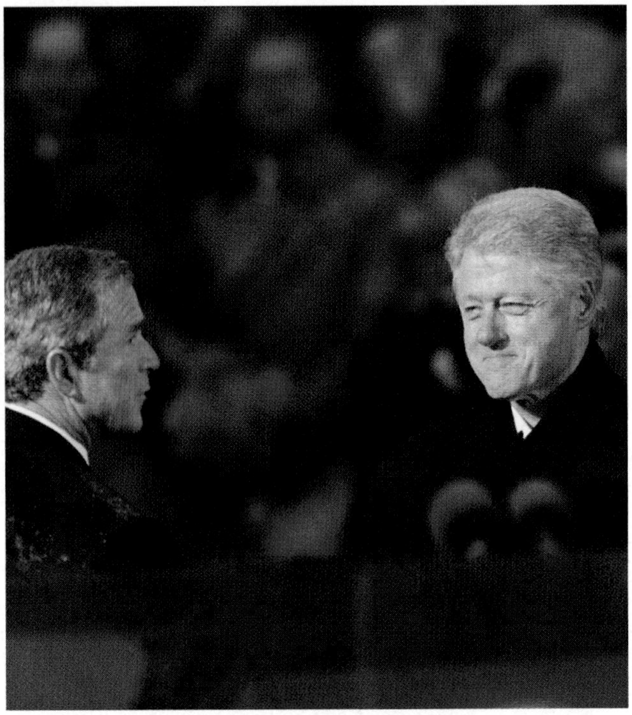

**Figure 1–14 Bush inaugurated President in 2001.** Republican George W. Bush (on the left) is seen here accepting congratulations from ex-president Bill Clinton after his inauguration as the 43rd president of the United States.

## Globalization

*Globalization*—the accelerating interconnectedness of human activities, ideas, and cultures—is the third integrative theme. For centuries, global connections have brought people together through migration, trade, and the exchange of ideas. What is new about the current scene is the increasing intensity, speed, extent, and impact of connectedness, to the point where its consequences are profoundly changing many human enterprises. These changes are most evident in the globalization of economies, cultural patterns, political arrangements, environmental resources, and pollution.

For many in the world, globalization has brought undeniable improvements in well-being and health, as economic exchanges have connected people to global markets, information exchanges have improved public-health practices, and agricultural research has improved crop yields. For some, however, globalization has brought the dilution and even destruction of cultural and religious ideals and norms and has done little to improve economic well-being. Whether helpful or harmful, globalization is taking place, and it is important not only to recognize its impacts, but also to correct them when they are harmful or undesirable.

**Economic Changes.** The major element of globalization may be the economic reorganization of the world. This transformation has been facilitated by the globalization of communication, whereby people are instantly linked through the Internet, satellites, and cables. Other components of the reorganization are the relative ease of transportation and financial transactions; the dominance of transnational corporations with unprecedented wealth and power; trade arrangements between major economic players, such as the North American Free Trade Agreement (NAFTA) and the General Agreement on Tariffs and Trade (GATT); and the power of the World Trade Organization (WTO). Moreover, economic changes bring cultural, environmental, and technological changes (to name a few) in their wake. Western diets, styles, and culture are marketed throughout the world, to the detriment of local customs and diversity. The Western lifestyle, with its devotion to suburban living, automobile ownership, and rich diets, has become the ideal to which many millions in the developing world aspire.

**Environmental Changes.** On the one hand, the increased dissemination of information has made it possible for environmental organizations and government agencies to connect with the public and enable people to become politically involved with current issues. It has also enabled consumers to find more environmentally friendly consumer goods and services, such as shade-grown coffee and sustainably harvested timber. On the other hand, globalization has contributed to some notoriously harmful outcomes, such as the worldwide spread of emerging diseases like the SARS virus and AIDS, the global dispersion of exotic species, the trade in hazardous wastes, the spread of persistent organic pollutants, the radioactive fallout from nuclear accidents, the exploitation of oceanic fisheries, the destruction of the ozone layer, and global climate change.

**Protests.** The foregoing and other perceived negative impacts of globalization have led to a significant protest movement. For instance, whenever the WTO holds its meetings, protesters appear in the thousands because of their concerns about the impact of the agency's policies on environmental resources, human rights, and labor organizations (Fig. 1–15). Meetings involving developments in biotechnology and the marketing of genetically modified organisms for food have invariably drawn protests over concerns about the unknown potential of these developments for environmental harm and threats to food safety. Many who criticize globalization cite its ideological commitment to "trickle-down" economics and "laissez-faire" capitalism, whereby the rich apparently get richer and the poor get the leavings. They are concerned that globalization is often promoted as an unquestionable good, yet does not seem to be at all capable of promoting other goods, such as a safe environment, financial stability, sustainable development, or social justice. For these reasons and others already mentioned, we will use the theme of globalization to illustrate and explore the interconnectedness that so characterizes the 21st century.

**Figure 1–15 World Trade Organization protest.** An estimated 30,000 people from around the world gathered in downtown Seattle to protest the WTO millennium meeting.

## 1.4 The Environment in the 21st Century

### World Summits

**UNCED: Rio, 1992.** Sustainable development was the primary focus of the 1992 world summit meeting of leaders and representatives from 180 nations—the United Nations Conference on Environment and Development (UNCED)—held in Rio de Janeiro, Brazil. The major outcome of this conference, a "blueprint" intended to guide development in sustainable directions into and through the 21st century, was published in book form as *Agenda 21*. Other important accomplishments included the establishment of conventions on climate change and biodiversity, the declaration of principles for sustainable forest development, and a commitment to double the funding for help in the developing countries. Following the conference, major regions, nations, and communities consulted the concepts promoted in the document and began to develop their own versions of how sustainable development would be achieved in their various jurisdictions. For example, **Baltic 21** is the program for 11 countries surrounding the Baltic Sea. It addresses economic, social, and environmental objectives—the three sectors shown in Figure 1–9—as they apply to establishing sustainability in different sectors, such as agriculture, energy, fisheries, forests, industry, tourism, and transportation.

Five years later, in June 1997, the U.N. General Assembly held a special session to review the progress made since UNCED. Sadly, many observers concluded that there were more instances of failure than progress, citing a continued deterioration in the global environment, a decline in aid to the developing countries, and a failure to set targets for reducing $CO_2$ emissions. The conference did, however, recognize some achievements and served to revitalize the UNCED commitments to sustainable development.

**Figure 1–16 World Summit on Sustainable Development.** UN General Secretary Kofi Annan addresses the WSSD in Johannesburg, South Africa in September, 2002. Unfortunately, delegates failed to agree on many crucial issues, and the meeting was judged to be a failure.

**WSSD: Johannesburg, 2002.** In September 2002, in Johannesburg, South Africa, sustainable development once again became the focus of a U.N.-sponsored world summit: the World Summit on Sustainable Development (WSSD) (Fig. 1–16). The meeting attempted to address the undeniable fact that most of the agreements made at UNCED—in particular, development assistance—were implemented weakly at best. Preparations for the summit focused on issues vital to the process of sustainable development, such as generating clean water and sanitation, providing energy services (especially renewable energy), reversing the deterioration of agricultural lands, protecting marine fisheries, addressing toxic chemicals and human health, and protecting biodiversity. The conveners hoped that the meetings would result in greater implementation of what had been started in UNCED.

Unfortunately, WSSD did very little of what was hoped for. The resulting Plan of Implementation bore little resemblance to that of UNCED, with only weak statements of support for the crucial issues. The United States fought against many of the proposed targets and timetables, such as a goal of having renewable energy provide 15% of countries' energy by 2015. Some agreements were reached: By 2015, the proportion of people who lack access to basic sanitation would be cut in half; by 2015, ocean fisheries would be restored; and by 2010, the rate of loss of biodiversity would be reduced. Observers did agree that, for the first time, all three dimensions of sustainable development (Figure 1–9) were represented. James Gustave Speth, former administrator of the U.N. Development Program and currently dean of the Yale School of Forestry and Environmental Studies, was intimately involved with the planning of the summit and later recorded his reflections. According to Speth, the meeting revealed that "our world is badly divided on key issues: corporate accountability, globalization and WTO, trade and subsidies, climate and energy, development priorities and aid, and many others.... In the end, delegates could only agree on platitudes." Thankfully, environmental progress does not depend on global summit meetings.

## A New Commitment

People in all walks of life—scientists, sociologists, workers and executives, economists, government leaders, and clergy, as well as traditional environmentalists—are recognizing that "business as usual" is not sustainable. Many global trends are on a collision course with the fundamental systems that maintain our planet as a tolerable place to live. A finite planet cannot continue to grow by 77 million persons annually without significant detrimental effects. The current degradation of ecosystems, atmospheric changes, losses of species, and depletion of water resources inevitably lead to a point where resources are no longer adequate to support the human population and where, consequently, civil order will break down. As one observer put it, "If we don't change direction, we will end up where we are heading."

**Good News.** The news is not all bad, however. Food production has improved the nutrition of millions in the developing world, and the percentage of individuals who are undernourished has declined from 35% to 20% over the past 30 years. Population growth rates continue to decline in many of the developing countries. A rising tide of environmental awareness in the industrialized countries has led to the establishment of policies, laws, and treaties that have improved the protection of natural resources and significantly reduced the pollution load. Thus, environmental degradation *can* be slowed down and reversed, people *can* be freed from hunger and poverty, and peoples' behavior toward the environment *can* be transformed from exploitative to conserving. Numerous caring people are beginning to play an important role in changing society's treatment of Earth. For example, many are now adding their voices to the literally hundreds of traditional environmental and professional organizations devoted to controlling pollution and protecting wildlife. People in business have formed the Business Council for Sustainable Development, economists have formed the International Society for Ecological Economics, religious leaders have formed the National Religious Partnership for the Environment, and philosophers are speaking out for a new ethic of "caring for creation." Further, the attention of the world and its leaders has been drawn to the environment and sustainable development through a series of world summit meetings.

We end this chapter by revisiting the strategic and integrative themes, a practice we will maintain throughout the text.

## revisiting the themes

### Sustainability

The Easter Island story graphically illustrated the consequences for a society that was unable to adapt sustainably to its limited resources. The result was not only the loss of natural goods and services that the island's ecosystems could provide, but also the loss of an entire culture. As we examine the global environment, ask yourself whether there are any parallels between what happened on Easter Island and what is happening now across the entire planet. The human population continues to expand. With expansion comes a larger ecological footprint, so natural and managed ecosystems are being pressed to provide increasing goods and services. We may have reached the limit in some resources, and most of the ways we use the other resources are currently unsustainable. In addition, we extract and burn fossil fuels much faster than they could ever be replenished, bringing on global climate change, and we tolerate the continued loss of biodiversity. Finally, the world summits that have featured sustainable development demonstrate how difficult it is to achieve. This is still a world of *haves* and *have-nots,* and the haves still seem unwilling to take the necessary steps to help those suffering from hunger and the other effects of poverty achieve the economic development that would improve their well-being and the sustainable well-being of the entire planet.

### Stewardship

The religion and culture of the Easter Islanders did not include an effective stewardship ethic. If they had, their story would have had a happier ending. Although current cultures and religions place some value on stewardship, it is not a fundamental part of the way we think about caring for the natural world and for our fellow humans. If it were, the world summits would have had a more positive outcome.

### Sound Science

Our brief look at the global environment was based largely on the information uncovered by sound science. The ongoing work of the Millenium Ecosystem Assessment demonstrates the need for more scientific study of the status and trends in global ecosystems and the ways in which human systems affect them. As we continue to draw on ecosystem capital, sound science must guide us in making the appropriate management interventions and policy decisions.

### Ecosystem Capital

The goods and services provided by Easter Island could likely have sustained a population of many thousands of people if those people had only known how to draw on the interest and leave the capital "invested." There were forests, soils, and coastal fisheries that could have been exploited wisely. It didn't happen. And now, we in the 21st century appear to be making the same mistakes, drawing down our ecosystem capital in case after case, in the mistaken belief that there is always more around the corner or over the hill.

### Policy and Politics

We will never know what kind of policies (if any) the ancient Easter Islanders devised for using natural resources. If there were such policies, they were countered by a political climate that led to class warfare and eventual decline. Even though most modern countries have policies that govern the use of resources and the control of pollution, they are not working very well. If they were, the global environment would be in better shape. The various earth summits demonstrate, for example, that politics trumps policy. Nations are tempted to act entirely on the basis of their self-interest, as the United States did in the WSSD. Where governments backed by antienvironmental special interests are in power, the environment and sustainability can take a beating. The hopeful news is that the vast majority of people really do care about the environment and will resist the way the Bush administration is acting to roll back several decades of environmental progress.

### Globalization

Contact with the outside world brought nothing but misery for the Easter Islanders for centuries. Recently, however, some positive economic and social help has come to the islanders, although there is the accompanying danger that their culture will be swamped by the impact of globalization. The Internet is one positive aspect of globalization. It has brought unprecedented access to information, making it both easier to gain useful knowledge about the environment and harder to cover up antienvironmental political action. The world economy has the potential to contribute to sustainable development, as developing countries are able to enter the market with their goods and services. Globalization carries many negative elements, however, such as allowing the concentration of economic power in big corporations, encouraging developing countries to overexploit their natural resources, and elevating economics over environmental concerns. Thus, those who care about the environment are right to view globalization with skepticism.

## review questions

1. What factors brought about the collapse of the Easter Island civilization? How did later contact with the rest of the world affect the islanders?
2. Cite four global trends which indicate that we are "still losing the environmental war."
3. Define *sustainability* and *sustainable society.*
4. Define *sustainable development* and state some of its features.
5. List five transitions that should lead to a future sustainable civilization.
6. Describe the origins of the stewardship ethic and its modern usage.
7. What are the concerns of the environmental justice movement?
8. How does justice become an issue between the industrialized countries and the developing countries?
9. What is the "scientific method?" How realistic is it?
10. What are Del Ratzsch's three major components of the structure of science?
11. Give several reasons for the existence of scientific controversies.
12. Define *ecosystem capital.* Why is it not better protected?
13. Describe and give examples of the interaction between policy and politics.
14. What is globalization? What are its most significant elements?
15. Compare and contrast UNCED and WSSD.
16. What are the new directions, new focus, and new adherents of modern environmental concern?

## thinking environmentally

1. Have a class debate between people representing the developing countries and people representing the developed countries. Characterize the two sides in terms of the issues surrounding population growth, energy use, resource use, and sustainable development. Find common interests between the two.
2. Some people say that the concept of sustainable development either is an oxymoron or represents going back to some kind of primitive living. Argue instead that neither is the case, but that sustainable development is the only course that will allow the continued advancement of civilization.

3. List all the prerequisites for a sustainable society that really works. Why is it necessary that people from all walks of life be involved? State the roles (in general) that each needs to play in achieving a sustainable society.

4. Study Figure 1–5; consider any significant human activity and fit it into the conceptual framework of the figure, following through with all of its impacts on human well-being and ecosystem services.

5. Set up a debate between proponents and opponents of smoking. Use the concepts of sound science and junk science to conduct the debate, and show how the two kinds of science will lead to different conclusions. Other interesting debate options include the "ozone hoax" and global warming; the Internet can provide ample material for both sides of the debate. (See also Chapter 20.)

# part one

# ecosystems: basic units of the natural world

Tropical rain forests seem humid, warm, dense, and full of unusual plants and animals when you first step into them. You may be unaware, however, that they are home to a greater diversity of living things than anywhere else on Earth, that they are the result of millions of years of adaptive evolution, and that they store more carbon than is in the entire atmosphere. Nor can you sense how energy flows through these forests or how nitrogen and phosphorus are cycled and recycled. This information comes from the work of many scientists who have been asking basic questions about how such natural systems function, how they came to be, and how they relate to the rest of the world.

How important is the information that comes from these scientists? In this first part of the text, we hope to convince you that information about how the natural world works is absolutely crucial to the human enterprise of living on Earth. *Ecosystems*—the term used to describe natural units like the tropical rain forest—are not just the backdrop for human activities; they are the basic context of life on Earth, including human life. Ecosystems are self-sustaining systems, and if you believe that the human enterprise should also be self-sustaining (and currently is not), then studying ecosystems will help you learn how to construct a sustainable society. Understanding environmental science begins with understanding what ecosystems are, how they work, and how they change over time.

◄ **Tropical rain forest.**

chapter 2

# Ecosystems: What They Are

**Key Topics**

1. Ecosystems: A Description
2. The Structure of Ecosystems
3. From Ecosystems to Global Biomes
4. The Human Factor
5. Revisiting the Themes

Barrier islands are found along the east and Gulf coasts of the United States. Plum Island, in Massachusetts, is a nine-mile-long barrier island in the northern part of the state. (See opposite page.) The southern two-thirds of the island is part of the Parker River National Wildlife Refuge, while the northern third is occupied by a host of summer cottages and year-round houses. On the refuge part of the island, sand dunes emerge at the upper edge of the ocean beach and extend back several hundred meters. Occupying the sand dunes is a mosaic of vegetation. The primary dunes, closest to the ocean, are colonized by beach grass (*Ammophila brevigulata*). Behind these dunes is a heathlike low growth, dominated by false heather (*Hudsonia tomentosa*). Farther back, a shrub community is found, where poison ivy and bayberry dominate. Even farther back is the maritime forest, with pitch pine, poplar, wild cherry, and a few other deciduous trees. Here and there is a depression that reaches the water table and is occupied by cranberry plants.

**A barrier Island** **Plum Island in northern Massachusetts is a nine-mile-long sand island separating the ocean from a bay, a salt marsh, and the upland. The inset shows the northern, developed end of the island.**

Animals on the barrier island—mice, rabbits, deer, skunks, red foxes, and coyotes—are seldom seen by humans, but leave their tracks on the sand. A diverse bird community can be found here, drawing birders who often find rarities like the peregrine falcon and snowy owl. Life for the plants—especially those close to the ocean—is harsh. The sand holds little water or nutrients, salt spray from the ocean stresses leaf surfaces, and wind either blows loose sand away from plants' roots or piles it up around the plants and buries them. Nevertheless, the barrier island plant community is a functioning system that has endured for centuries.

***Cottage Colony.*** On the northern third of the island, scores of streets were laid out in the early 1900s, and hundreds of small cottages were built. The dunes were either removed or built upon. Now, the only vegetation is a few beach grass plants that lie between the beach and the first houses. There are no dunes to speak of, no shrubs or maritime forest, and no deer or foxes. The entire island would probably end up like this if the wildlife refuge didn't prevent further development. The contrast between these "communities" is stark and unforgettable.

From time to time, the environment tests this barrier island, as it tests all barrier islands, with storms. On the southern section, the dunes absorb the powerful waves that pound the beach and wash upward toward the land. After a storm, the beach grass and primary dunes are intact, the back dunes untouched. The dune communities continue to thrive, forming a true barrier to the ocean that protects the fragile salt marshes and land behind the island. On the northern section, however, storms often wash beach sand by the houses and onto the streets, and occasionally a house is swept into the ocean. No one visits the northern Plum Island dunes to view birds or animals, because there are no dunes, birds, or animals to visit or view. In comparison, thousands of visitors enjoy the Parker River National Wildlife Refuge each year for its wildlife and beauty. Moreover, the ecosystems on the refuge provide valuable protection, as well as opportunities for recreational hunting.

***The Ecosystem Approach.*** Humans live in environments that, like Plum Island, have a mix of natural and degraded ecosystems. Are there enough ecosystems in good health to continue to make their vital contributions to human well-being and support current human populations? What are the trends in ecosystem manipulation and conversion throughout the world, such as defor-

## earth watch

### Taking Stock

Considering how important ecosystems are to human affairs, you might assume that we have a good inventory of them. It turns out, however, that we do not know *what* we have, or much about the *goods and services* specific ecosystems can produce, or how they *respond to changes* brought on by human activities. Nevertheless, we are going to have to manage natural ecosystems more intensely and deliberately if we want to enjoy their benefits in the future. The question is, How can you manage something you do not even know you have, let alone understand how you are affecting it?

In 1995, recognizing this problem, the White House Office of Science and Technology Policy (OSTP) asked the Heinz Center to "create a nonpartisan, scientifically grounded report on the state of the nation's environment." The OSTP asked the center to focus on ecosystems as the best approach to its work. The Heinz Center published its first full report in 2002, called *The State of the Nation's Ecosystems: Measuring the Lands, Waters, and Living Resources of the United States*. [The report is available from Cambridge University Press and on the Internet (www.heinzctr.org/ecosystems).] The report, a collaborative effort involving almost 150 experts from academia, government, environmental organizations, and business, was funded by federal, corporate, and foundation sources.

The report "is written for decision makers and opinion leaders concerned about the 'big picture' of the nation's ecosystems," with the goal of providing unbiased information to be used to form public policy. It divides the lands and waters of the United States into six major types of ecosystem: **coasts and oceans, farmlands, forests, fresh waters, grasslands and shrub lands**, and **urban and suburban areas.** For each type of ecosystem, unique "indicators" are specified—key characteristics that may be used to describe the current status and trends within the ecosystem, to be tracked over time. The indicators, about 100 or so, are organized into 10 categories.

The forest ecosystem illustrates how the 10 categories are made specific for a given type (not all indicator categories are used for each type of ecosystem):

#### System Dimensions

- *Ecosystem Extent*–forest area and ownership, forest types, forest management categories
- *Fragmentation and Landscape Pattern*–forest pattern and fragmentation

#### Chemical and Physical Conditions

- *Nutrients, Carbon, and Oxygen*–nitrate in forest streams, carbon storage
- *Contaminants*–not applicable
- *Physical*–not applicable

#### Biological Components

- *Plants and Animals*–at-risk native species, area covered by nonnative plants
- *Communities*–forest age, forest disturbance (fire, insects, disease), fire frequency, forest community types with significantly reduced area
- *Ecological Productivity*–not applicable

#### Human Uses

- *Food, Fiber, and Water*–timber harvest, timber growth, and harvest
- *Recreation and Other Services*–recreation in forests

The main body of the Heinz Center report presents the indicator data for each type of ecosystem and, in many cases, explains why there are no adequate data to report. Nearly half the indicators show missing data. This is only the first report; future reports will bring the indicator data up to date and address gaps in the data. For now, however, the report represents a landmark for taking the pulse of our nation's ecosystems.

estation and overgrazing? What impacts will the next 50 years of population growth have on ecosystem goods and services? The MEA is attempting to answer these important questions, and it is doing so by adopting the "ecosystem approach," as mentioned in Chapter 1.

Closer to home, in 2002 the Heinz Center of Washington, DC, published a report on ecosystems in the United States, called *The State of The Nation's Ecosystems.* (See "Earth Watch," p. 28.) This report came in response to a government request to create a scientifically based assessment of the state of the nation's environment, and it was carried out by hundreds of scientists from academia, the government, environmental organizations, and the private sector. Like the MEA, the report was organized according to the ecosystem approach. These studies demonstrate that the ecosystem concept has gained universal recognition in the scientific and policy-making communities. To better understand why, we begin this chapter by examining the structure of natural ecosystems.

## 2.1 Ecosystems: A Description

**Biotic Communities.** The grouping or assemblage of *plants, animals,* and *microbes* we observe when we study a natural forest, a grassland, a pond, a coral reef, or some other undisturbed area is referred to as the area's **biota** (*bio,* living) or **biotic community**. The plant portion of the biotic community includes all vegetation, from large trees down through microscopic algae. Likewise, the animal portion includes everything from large mammals, birds, reptiles, and amphibians through earthworms, tiny insects, and mites. Microbes encompass a large array of microscopic bacteria, fungi, and protozoans. Thus, the *biotic* community comprises a *plant* community, an *animal* community, and a *microbial* community.

The particular kind of biotic community found in a given area is, in large part, determined by **abiotic** (nonliving, chemical, and physical) factors, such as the amount of water or moisture present, the temperature, the salinity, or the type of soil in the area. These abiotic factors both support and limit the particular community. For example, a relative lack of available moisture prevents the growth of most species of plants, but supports certain species, such as cacti; these kinds of areas are deserts. Land with plenty of available moisture and a suitable temperature supports forests. The presence of water is the major factor that sustains aquatic communities.

**Species.** The first step in investigating a biotic community may be simply to catalogue all the *species* present. **Species** are the different kinds of plants, animals, and microbes in the community. A given species includes all those individuals which have a strong similarity in appearance to one another and which are distinct in appearance from other such groups (robins vs. redwing blackbirds, for example). Similarity in appearance suggests a close genetic relationship. Indeed, the *biological definition* of a species is the entirety of a population that can interbreed and produce fertile offspring, whereas members of different species generally do not interbreed. Breeding is often impractical or impossible to observe, however, so for purposes of identification, appearance usually suffices.

**Populations.** Each species in a biotic community is represented by a certain **population**—that is, by a certain number of individuals that make up the interbreeding, reproducing group. The distinction between *population* and *species* is that *population* refers only to those individuals of a certain species that live within a given area, whereas *species* is all inclusive, referring to all the individuals of a certain kind, even though they may exist in different populations in widely separated areas.

**Associations.** One reason to identify the biotic community is to understand how it fits into the landscape or how it differs from other biotic communities. To identify a biotic community may require an assessment of the corresponding plant community. Vegetation is readily measured and is a strong indicator of the environmental conditions of a site. The most basic kind of plant community is the **association**, defined as a plant community with a definite composition, uniform habitat characteristics, and uniform plant growth. On Plum Island, for example, the **pitch pine/false heather** (*Pinus rigida/Hudsonia tomentosa*) **woodland** association represents a subgroup of the maritime forest (Fig. 2–1).

**Figure 2–1 A plant association.** The **pitch pine/false heather** (*Pinus rigida/Hudsonia tomentosa*) **woodland** association on Plum Island.

The species within a community depend on and support one another. In particular, certain animals will not be present unless certain plants that provide their necessary food and shelter are present. Thus, the plant community supports (or limits by its absence) the animal community. In addition, every plant and animal species is adapted to cope with the abiotic factors of the region. For example, every species that lives in temperate regions is adapted in one way or another to survive the winter season, which includes a period of freezing temperatures (Fig. 2–2). These interactions among organisms and their environments are discussed in Section 2.2. For now, keep in mind that the populations of different species within a biotic community are constantly interacting with each other and with the abiotic environment.

**Figure 2–2 Winter in the forest.** Many trees and other plants of temperate forests are so adapted to the winter season that they actually *require* a period of freezing temperature in order to grow again in the spring.

**Ecosystems.** This brings us to the concept of an *ecosystem*, which joins together the biotic community *and* the abiotic conditions that it lives in. The ecosystem concept considers the ways populations interact with each other and the abiotic environment to reproduce and perpetuate the entire grouping. As a result, an **ecosystem** is a grouping of plants, animals, and microbes occupying an explicit unit of space and interacting with each other and their environment. For study purposes, an ecosystem is any more or less distinctive biotic community living in a certain environment. Thus, a forest, a grassland, a wetland, a marsh, a pond, a sand dune, and a coral reef, each with its respective species in a particular environment, can be studied as distinct ecosystems. Often, an ecosystem contains a group of associations, like those on the dunes of Plum Island.

Because no organism can live apart from its environment or from interacting with other species, ecosystems are the functional units of sustainable life on Earth. The study of ecosystems and the interactions that occur among organisms and between organisms and their environment belongs to the science of **ecology**, and the investigators who conduct such studies are called **ecologists**.

***Ecotone.*** While it is convenient to divide the living world into different ecosystems, you will find that there are

**Figure 2–3 Ecotones on land.** Ecosystems are not isolated from one another. One ecosystem blends into the next through a transitional region—an ecotone—that contains many species common to both systems.

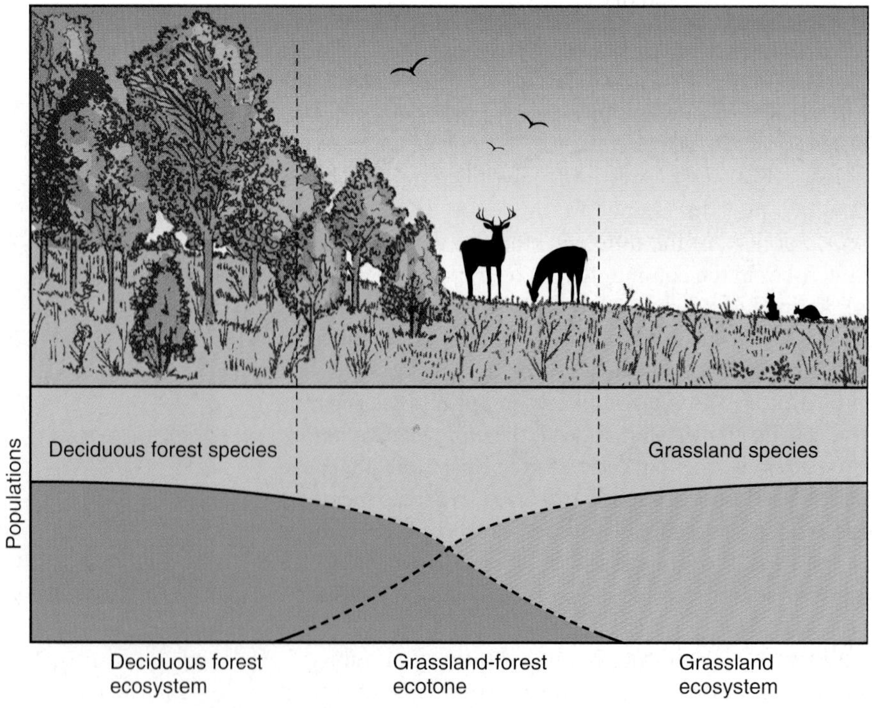

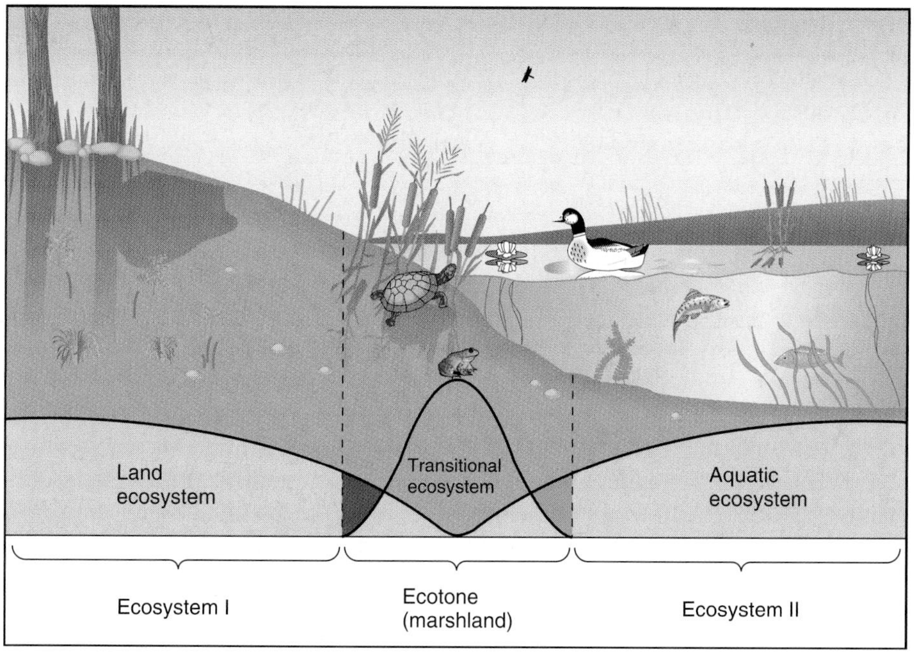


**Figure 2–4 Terrestrial-to-aquatic-system ecotone.** An ecotone may create a unique habitat that harbors specialized species not found in either of the ecosystems bordering it. Typically, cattails, reeds, and lily pads grow in the ecotone shown, along with several species of frogs and turtles, as well as egrets and herons.

seldom distinct boundaries between ecosystems, and they are never totally isolated from one another. Many species will occupy (and thus be a part of) two or more ecosystems at the same time, or they may move from one ecosystem to another at different times, as in the case of migrating birds. In passing from one ecosystem to another, the one may grade into the other through a transitional region, known as an **ecotone**, that shares many of the species and characteristics of both ecosystems (Fig. 2–3). The ecotone between adjacent systems may also include *unique* conditions that support distinctive plant and animal species; consider, for example, the marshy area that often occurs between the open water of a lake and dry land (Fig. 2–4). Ecotones may be studied as distinct ecosystems in their own right.

**Landscapes.** What happens in one ecosystem affects other ecosystems. For this reason, ecologists have begun using the concept of **landscapes**—a group of interacting ecosystems. Thus, a barrier island, a saltwater bay, and the salt marsh behind it constitute a landscape. Landscape ecology is the science that studies the interactions among ecosystems.

*Biomes.* Similar or related ecosystems or landscapes are often grouped together to form major kinds of ecosystems called **biomes**. Tropical rain forests, grasslands, and deserts are biomes. While more extensive than an ecosystem in its breadth and complexity, a biome is still basically a distinct type of biotic community supported and limited by certain abiotic environmental factors. As with ecosystems, there are generally no sharp boundaries between biomes. Instead, one grades into the next through transitional regions.

Likewise, there are major categories of aquatic and wetland ecosystems that are determined primarily by the depth, salinity, and permanence of water in them. Among these ecosystems are lakes, marshes, streams, rivers, estuaries, bays, and ocean systems. As units of study, these aquatic systems may be viewed as ecosystems, as parts of landscapes, or as major biomelike features such as seas or oceans. (The biome category is reserved exclusively for terrestrial systems.) Table 2–1 lists the six major aquatic systems and their primary characteristics.

**Biosphere.** Regardless of how we choose to divide (or group) and name different ecosystems, they all remain interconnected and interdependent. Terrestrial biomes are connected by the flow of rivers between them and by migrating animals. Sediments and nutrients washing from the land may nourish or pollute the ocean. Seabirds and mammals connect the oceans with the land, and all biomes share a common atmosphere and water cycle.

Therefore, all the species on Earth, along with all their environments, make up one vast ecosystem, often called the **biosphere**. Although the separate local ecosystems are the individual units of sustainability, they are all interconnected to form the biosphere. The concept is analogous to the idea that the cells of our bodies are the units of living systems, but are all interconnected to form the whole body. Carrying the analogy further, to what degree can individual ecosystems be degraded or destroyed before an entire biome, or even the biosphere, is affected? Conversely, to what degree can basic global parameters, such as the atmosphere and the temperature, be altered before major ecosystems on Earth are affected? To begin to understand ecosystems in more depth, in Section 2.2 we discuss how they are structured.

The key terms introduced in Section 2.1 are summarized in Table 2–2.

**table 2-1 Major Aquatic Systems**

| Aquatic Systems | Major Environmental Parameters | Dominant Vegetation | Dominant Animal Life | Distribution |
|---|---|---|---|---|
| **Lakes and Ponds (freshwater)** | Bodies of standing water; low concentration of dissolved solids; seasonal vertical stratification of water | Rooted and floating plants, phytoplankton | Zooplankton, fish, insect larvae, ducks, geese, herons | Physical depressions in the landscape where precipitation and groundwater accumulate |
| **Streams and Rivers (freshwater)** | Flowing water; low level of dissolved solids; high level of dissolved oxygen, often turbid | Attached algae, rooted plants | Insect larvae, fish, amphibians, otters, raccoons, wading birds, ducks, geese, swans | Landscapes where precipitation and groundwater flow by gravity toward oceans or lakes |
| **Inland Wetlands (freshwater)** | Standing water, at times seasonally dry; thick organic sediments; high nutrients | Marshes: grasses, reeds, cattails. Swamps: water-tolerant trees. Bogs: sphagnum moss, low shrubs | Amphibians, snakes, numerous invertebrates wading birds, ducks, geese, alligators, turtles | Shallow depressions, poorly drained, often occupy sites of lakes and ponds that have filled in |
| **Estuaries** | Variable salinity; tides create two-way currents, often rich in nutrients, turbid | Phytoplankton in water column, rooted grasses like salt-marsh grass, mangrove swamps in tropics with salt-tolerant trees and shrubs | Zooplankton, rich shellfish, worms, crustaceans, fish, wading birds, sandpipers, ducks, geese | Coastal regions where rivers meet the ocean; may form bays behind sandy barrier islands |
| **Coastal Ocean (saltwater)** | Tidal currents promote mixing; nutrients high | Phytoplankton, large benthic algae, turtle grass, symbiotic algae in corals | Zooplankton, rich bottom fauna of worms, shellfish, crustaceans, echinoderms; coral colonies, jellyfish, fish, turtles, gulls, terns, ducks, sea lions, seals, dolphins, penguins, whales | From coastline outward over continental shelf; coral reefs abundant in tropics |
| **Open Ocean** | Great depths (to 11,000 meters); all but upper 200 m dark and cold; poor in nutrients except in upwelling regions | Exclusively phytoplankton | Diverse zooplankton and fish adapted to different depths; seabirds, whales, tuna, sharks, squid, flying fish | Covering 70% of Earth, from edge of continental shelf outward |

## 2.2 The Structure of Ecosystems

*Structure* refers to parts and the way they fit together to make a whole system. There are two key aspects to every ecosystem, namely, the biota, or biotic community, and the abiotic environmental factors. The way different categories of organisms fit together is referred to as the **biotic structure**, and the major feeding relationships between organisms constitute the **trophic structure** (*trophic*, feeding). All ecosystems have the same three basic categories of organisms that interact in the same ways.

### Trophic Categories

The major categories of organisms are (1) *producers*, (2) *consumers*, and (3) *detritus feeders* and *decomposers*. Together, these groups produce food, pass it along food chains, and return the starting materials to the abiotic parts of the environment, respectively.

**Producers.** Producers are organisms that capture energy from the Sun or from chemical reactions to convert carbon dioxide ($CO_2$) to organic matter. Most producers are green plants, which use light energy to convert $CO_2$ and water to organic compounds such as the sugar glucose and then release oxygen as a by-product. This chemical conversion, which is driven by light energy, is called **photosynthesis**. Plants are able to manufacture all the complex organic molecules that make up their bodies via photosynthesis, along with additional *mineral nutrients* such as nitrogen, phosphorus, potassium, and sulfur, which they absorb from the soil or from water (Fig. 2–5).

Plants use a variety of molecules to capture light energy in photosynthesis, but the most predominant of

**table 2-2 Important Terms**

| | |
|---|---|
| Species | All the members of a specific kind of plant, animal, or microbe; a kind given by similarity of appearance or capacity for interbreeding and producing fertile offspring. |
| Population | All the members of a particular species occupying a given area. |
| Association | A plant community with a definite composition, uniform habitat characteristics, and uniform plant growth. |
| Biotic community | All the populations of different plants, animals, and microbes occupying a given area. |
| Abiotic factors | All the factors of the physical environment: moisture, temperature, light, wind, pH, type of soil, salinity, etc. |
| Ecosystem | The biotic community together with the abiotic factors; includes all the interactions among the members of the biotic community, and between the biotic community and the abiotic factors, within an explicit unit of space. |
| Landscape | A group of interacting ecosystems in a particular area. |
| Biome | A grouping of all the ecosystems of a similar type (e.g., tropical forests or grasslands). |
| Biosphere | All species and physical factors on Earth functioning as one unified ecosystem. |

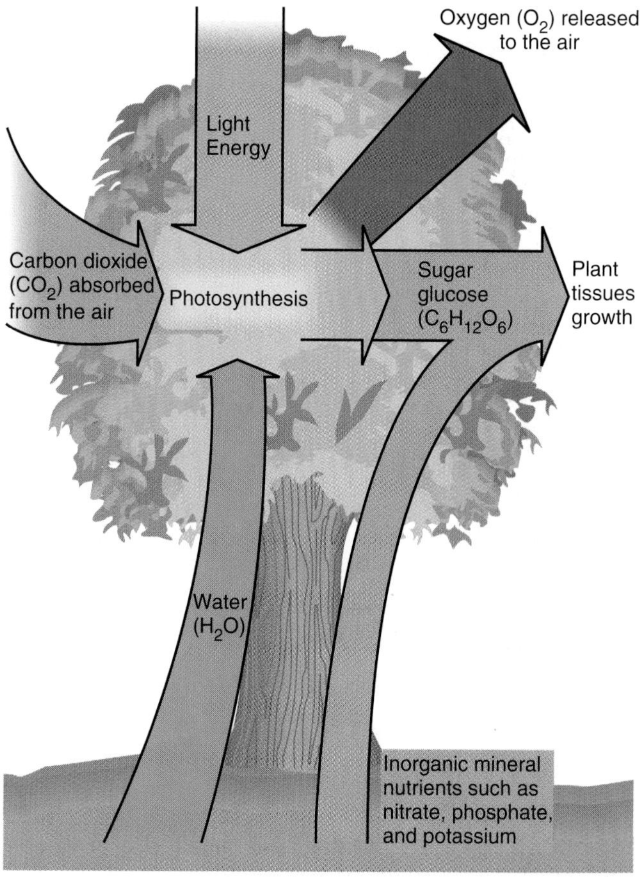


**Figure 2–5 Green-plant photosynthesis.** The producers in all major ecosystems are green plants.

these is **chlorophyll**, a green pigment. Hence, plants that photosynthesize are easily identified by their green color, such as the beach grass, shrubs, and maritime forest plants on Plum Island. In some plants, additional red or brown photosynthetic pigments (in red and brown algae, for example) may overshadow the green. Producers range in diversity from microscopic photosynthetic bacteria and single-celled algae through medium-sized plants such as grass, daisies, and cacti, to gigantic trees. Every major ecosystem, both aquatic and terrestrial, has its particular producers, which are actively engaged in photosynthesis.

***Organic vs. Inorganic.*** The term **organic** refers to all those materials that make up the bodies of living *organ*isms—molecules such as proteins, fats or lipids, and carbohydrates. Likewise, materials that are specific products of living organisms, such as dead leaves, leather, sugar, wood, coal, and oil, are considered *organic*. By contrast, materials and chemicals in air, water, rocks, and minerals, which exist apart from the activity of living organisms, are considered **inorganic** (Fig. 2–6). Interestingly, there are bacteria that are able to use the energy in some inorganic chemicals to form organic matter from $CO_2$ and water. This process is called **chemosynthesis**, and these organisms are producers, too.

The key feature of *organic* materials and molecules is that they are constructed in large part from bonded carbon and hydrogen atoms, a structure that is not found among *inorganic* materials. This carbon–hydrogen structure has its origins in photosynthesis, in which hydrogen atoms taken from water molecules and carbon atoms taken from carbon dioxide are joined together to form organic compounds (Fig. 2–5). Green plants use light as the energy source to produce all the complex organic molecules their bodies need from the simple inorganic chemicals that are present in the environment. As this conversion from inorganic to organic occurs, some of the energy from light is stored in the organic compounds.

***Autotrophs vs. Heterotrophs.*** All organisms in the ecosystem *other than the producers* feed on organic matter as their source of energy. These organisms include not only all animals, but also **fungi** (mushrooms, molds, and similar organisms), most bacteria, and even a few higher plants that do not have chlorophyll and thus cannot photosynthesize.

As a result, green plants, which carry on photosynthesis, are absolutely essential to every ecosystem (with the exception of a few, such as deep-sea hydrothermal

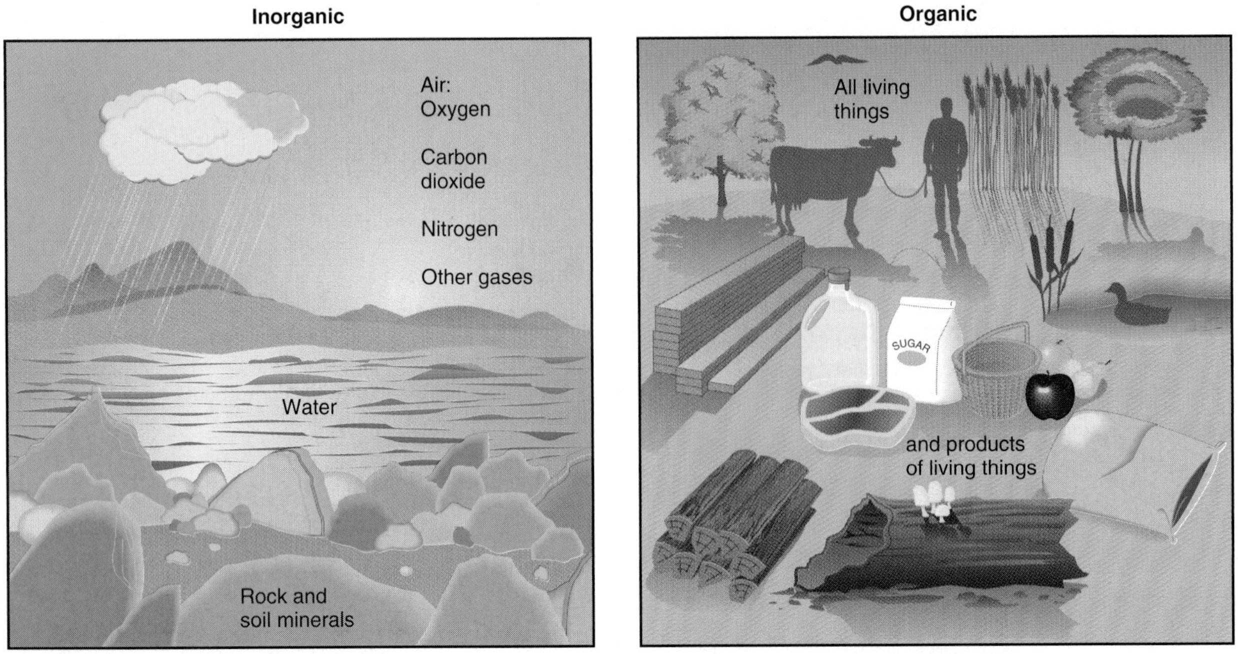


**Figure 2–6 Organic and inorganic.** Water and the simple molecules found in air, rocks, and soils are *inorganic.* The complex molecules that make up plant and animal tissues are *organic.*

vents, which depend on chemosynthesis). The photosynthesis and growth of green plants constitute the *production* of organic matter, which sustains all other organisms in the ecosystem.

Indeed, all organisms in the biosphere can be categorized as either *autotrophs* or *heterotrophs*, depending on whether they do or do not produce the organic compounds they need to survive and grow. Chemosynthetic bacteria are **autotrophs** (*auto*, self; *troph*, feeding), because they produce their own organic material from inorganic constituents in their environment through the use of an external energy source. The most important and common autotrophs by far, however, are green plants, which use chlorophyll to capture light energy for photosynthesis. All other organisms, which must *consume* organic material to obtain energy, are **heterotrophs** (*hetero*, other). Heterotrophs may be divided into numerous subcategories, the two major ones being **consumers** (which eat living prey) and **detritus feeders** and **decomposers**, both of which feed on dead organisms or their products.

**Consumers.** Consumers encompass a wide variety of organisms ranging in size from microscopic bacteria to blue whales. Among consumers are such diverse groups as protozoans, worms, fish and shellfish, insects, reptiles, amphibians, birds, and mammals (including humans).

For the purpose of understanding ecosystem structure, consumers are divided into various subgroups according to their food source. Animals—as large as elephants or as small as mites—that feed directly on producers are called **primary consumers** or **herbivores** (*herb*, grass).

Animals that feed on primary consumers are called **secondary consumers**. Thus, elk, which feed on vegetation, are primary consumers, whereas wolves are secondary consumers because they feed on elk (Fig. 2–7). There may also be third (tertiary), fourth (quaternary), or even higher levels of consumers, and certain animals may occupy more than one position on the consumer scale. For instance, humans are primary consumers when they eat vegetables, secondary consumers when they eat beef, and tertiary consumers when they eat fish that feed on smaller fish that feed on algae. Secondary and higher order consumers are also called **carnivores** (*carni*, meat). Consumers that feed on both plants and animals are called **omnivores** (*omni*, all).

***Predators, Parasites, Pathogens.*** In any relationship in which one organism feeds on another, the organism that does the feeding is called the **predator**, and the organism that is fed on is called the **prey**. Predation thus ranges from the classic predator–prey interactions between carnivores and herbivores, to herbivores feeding on plants, and parasites feeding on their hosts. In fact, **parasites** are another important category of consumers. Parasites are organisms—either plants or animals—that become intimately associated with their "prey" and feed on it over an extended period of time, typically without killing it, but sometimes weakening it so that it becomes more prone to being killed by predators or adverse conditions. The plant or animal that is fed upon is called the **host**.

A tremendous variety of organisms may be parasitic. Various worms are well-known examples, but certain protozoans, insects, and even mammals (vampire bats) and plants (dodder) (Fig. 2–8a) are also parasites. Many serious plant diseases and some animal diseases (such as athlete's foot) are caused by parasitic fungi. Indeed, virtually every major

**Figure 2–7 Secondary consumers.** Gray wolves have brought down an elk.

group of organisms has at least some members that are parasitic. Parasites may live inside or outside their hosts, as the examples shown in Fig. 2–8 illustrate.

In medicine, a distinction is generally made between bacteria and viruses that cause disease (known as **pathogens**), on the one hand, and parasites, which are usually larger organisms, on the other. Ecologically, however, there is no real distinction. Bacteria are foreign organisms, and viruses are organismlike entities feeding on, and multiplying in, their hosts over a period of time and doing the same damage as do other parasites. Therefore, disease-causing bacteria and viruses can be considered highly specialized parasites. Representative examples of producers and consumers, and the feeding relationships among them, are shown in Fig. 2–9.

**Detritus Feeders and Decomposers.** Dead plant material, such as fallen leaves, branches and trunks of dead trees, dead grass, the fecal wastes of animals, and dead animal bodies, are called **detritus**. Many organisms are specialized to feed on detritus, and these consumers are called detritus feeders or *detritivores*. Earthworms, millipedes, fiddler crabs, termites, ants, and wood beetles are all detritus feeders. As with regular consumers, there are *primary* **detritus feeders** (those which feed directly on detritus), *secondary* **detritus feeders** (those which feed on primary detritus feeders), and so on.

An extremely important group of primary detritus feeders is the *decomposers*, namely, fungi and bacteria. Much of the detritus in an ecosystem—particularly dead

(a) (b) (c)

**Figure 2–8 Diversity of parasites.** Nearly every major biological group of organisms has at least some members that are parasitic on others. Shown here is (a) dodder, a plant parasite that has no leaves or chlorophyll. The orange "strings" are dodder stems, which suck sap from the host plant. (b) Nematode worms (*Ascaris lumbricoides*), the largest of the human parasites, reach a length of 14 inches (35 cm). (c) Lampreys attached to a whitefish. Lampreys parasitize many fish species.

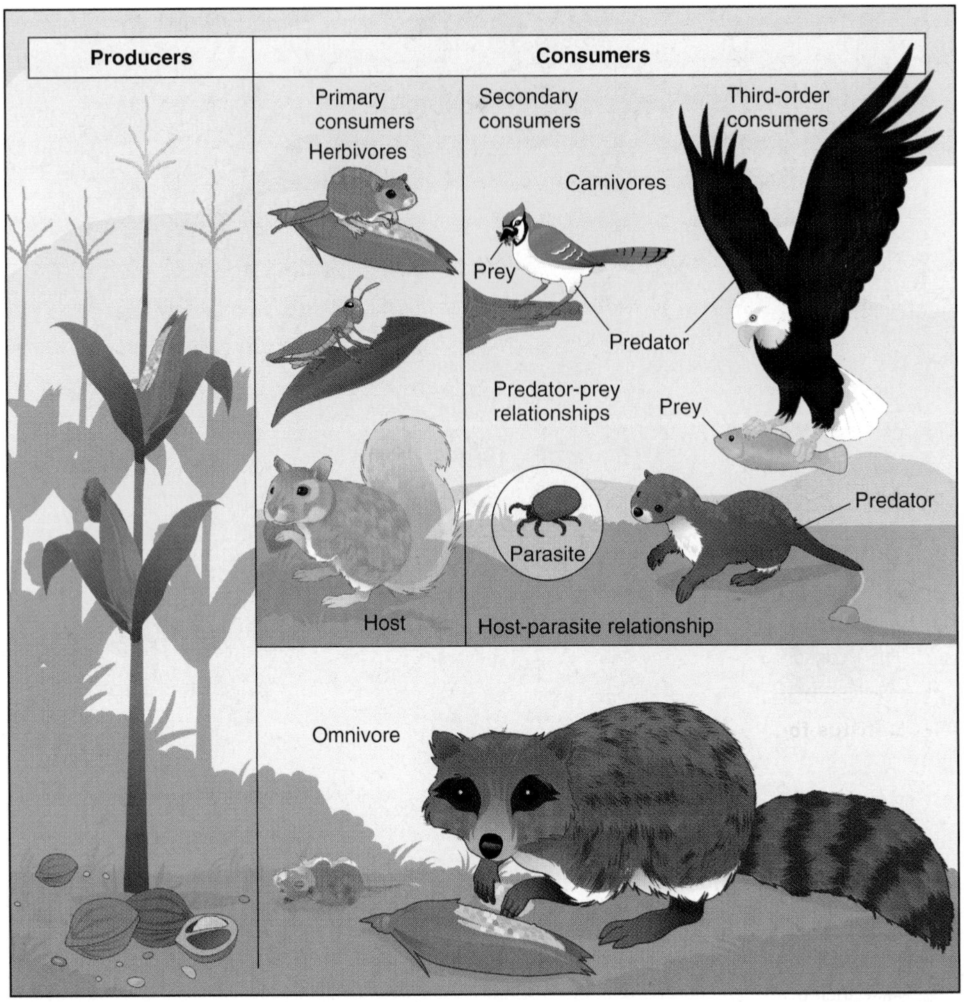


**Figure 2–9 Trophic relationships among producers and consumers.**

leaves and the wood of dead trees or branches—does not appear to be eaten as such, but rots away. Rotting is the result of the metabolic activity of fungi and bacteria. These organisms secrete digestive enzymes that break down wood, for example, into simple sugars that the fungi or bacteria then absorb for their nourishment. Thus, the rotting you observe is really the result of material being consumed by fungi and bacteria. Even though these organisms are called decomposers because of their unique behavior, they are grouped with detritus feeders because their function in the ecosystem is the same. Secondary detritus feeders, such as protozoans, mites, insects, and worms (Fig. 2–10), feed, in turn, upon decomposers. When a fungus or other decomposer dies, its body becomes part of the detritus and the source of energy and nutrients for still more detritus feeders and decomposers.

In sum, despite the diversity of ecosystems, they all have a similar *biotic structure*. All consist of (1) autotrophs, or producers, which produce organic matter that becomes the source of energy and nutrients for (2) heterotrophs, which are various categories of consumers, detritus feeders, and decomposers (Fig. 2–11).

## Trophic Relationships: Food Chains, Food Webs, and Trophic Levels

A caterpillar eats an oak leaf, a warbler eats the caterpillar, and a hawk eats the warbler. This a **food chain**. While it is interesting to trace these pathways, it is important to recognize that food chains seldom exist as isolated entities. Caterpillars feed on several kinds of plants, are preyed upon by several kinds of birds, and so on. Consequently, virtually all food chains are interconnected and form a complex *web* of feeding relationships—the **food web**.

Despite the number of theoretical food chains and the complexity of food webs, they all basically lead through a series of steps or levels, namely, from producers to primary consumers (or primary detritus feeders) to secondary consumers (or secondary detritus feeders), and so on. These *feeding levels* are called **trophic levels**. All producers belong to the first trophic level, all primary consumers (in other words, all herbivores) belong to the second trophic level, organisms feeding on these herbivores belong to the third level, and so forth.

Whether you visualize the biotic structure of an ecosystem in terms of food chains, food webs, or trophic

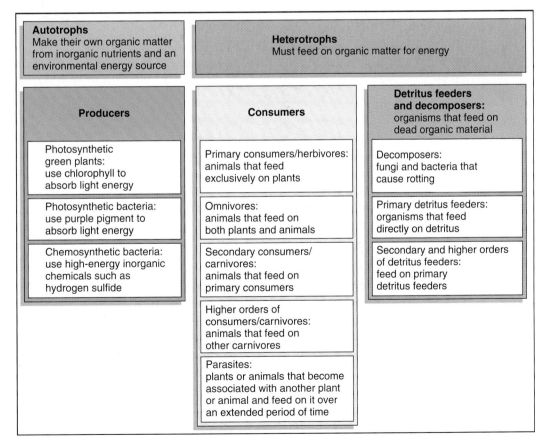

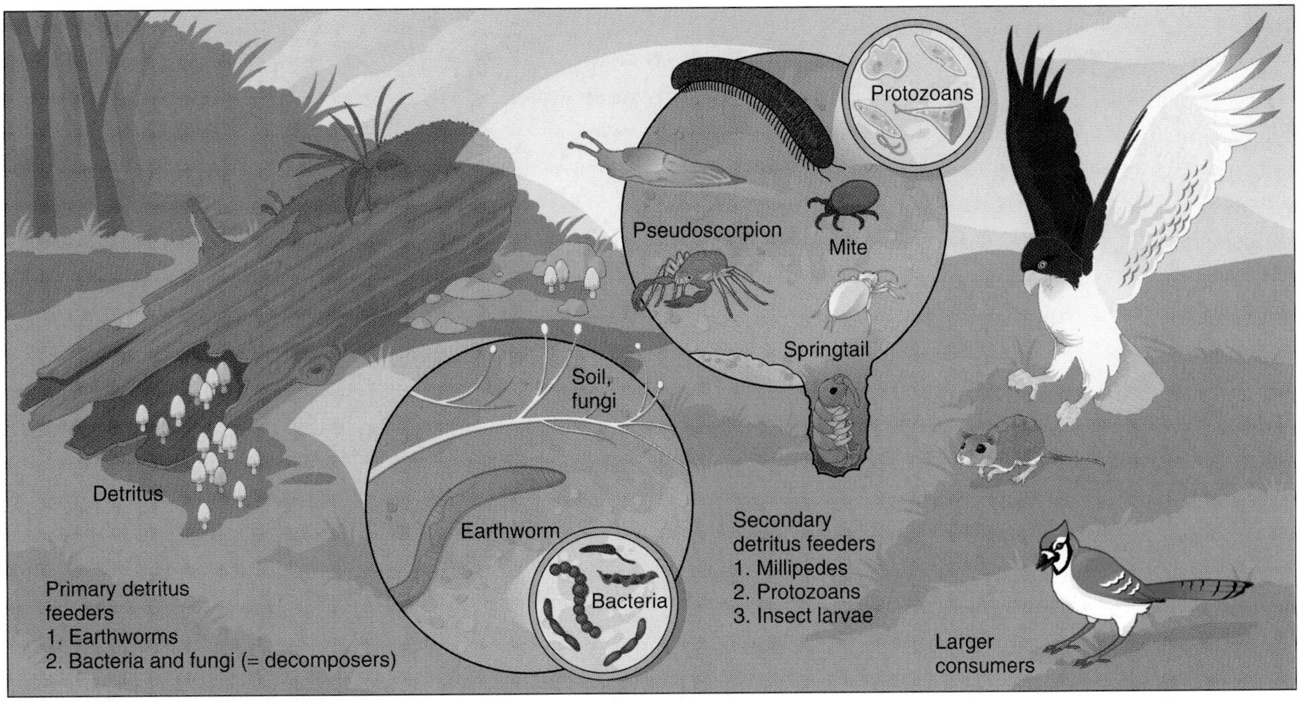


**Figure 2–10 Detritus food web.** The feeding (trophic) relationships among primary detritus feeders, secondary detritus feeders, and consumers.

| **Autotrophs** Make their own organic matter from inorganic nutrients and an environmental energy source | **Heterotrophs** Must feed on organic matter for energy | |
|---|---|---|
| **Producers** | **Consumers** | **Detritus feeders and decomposers:** organisms that feed on dead organic material |
| Photosynthetic green plants: use chlorophyll to absorb light energy | Primary consumers/herbivores: animals that feed exclusively on plants | Decomposers: fungi and bacteria that cause rotting |
| Photosynthetic bacteria: use purple pigment to absorb light energy | Omnivores: animals that feed on both plants and animals | Primary detritus feeders: organisms that feed directly on detritus |
| Chemosynthetic bacteria: use high-energy inorganic chemicals such as hydrogen sulfide | Secondary consumers/ carnivores: animals that feed on primary consumers | Secondary and higher orders of detritus feeders: feed on primary detritus feeders |
| | Higher orders of consumers/carnivores: animals that feed on other carnivores | |
| | Parasites: plants or animals that become associated with another plant or animal and feed on it over an extended period of time | |

**Figure 2–11 Trophic categories.** A summary of how living organisms are ecologically categorized according to feeding attributes.

(a)

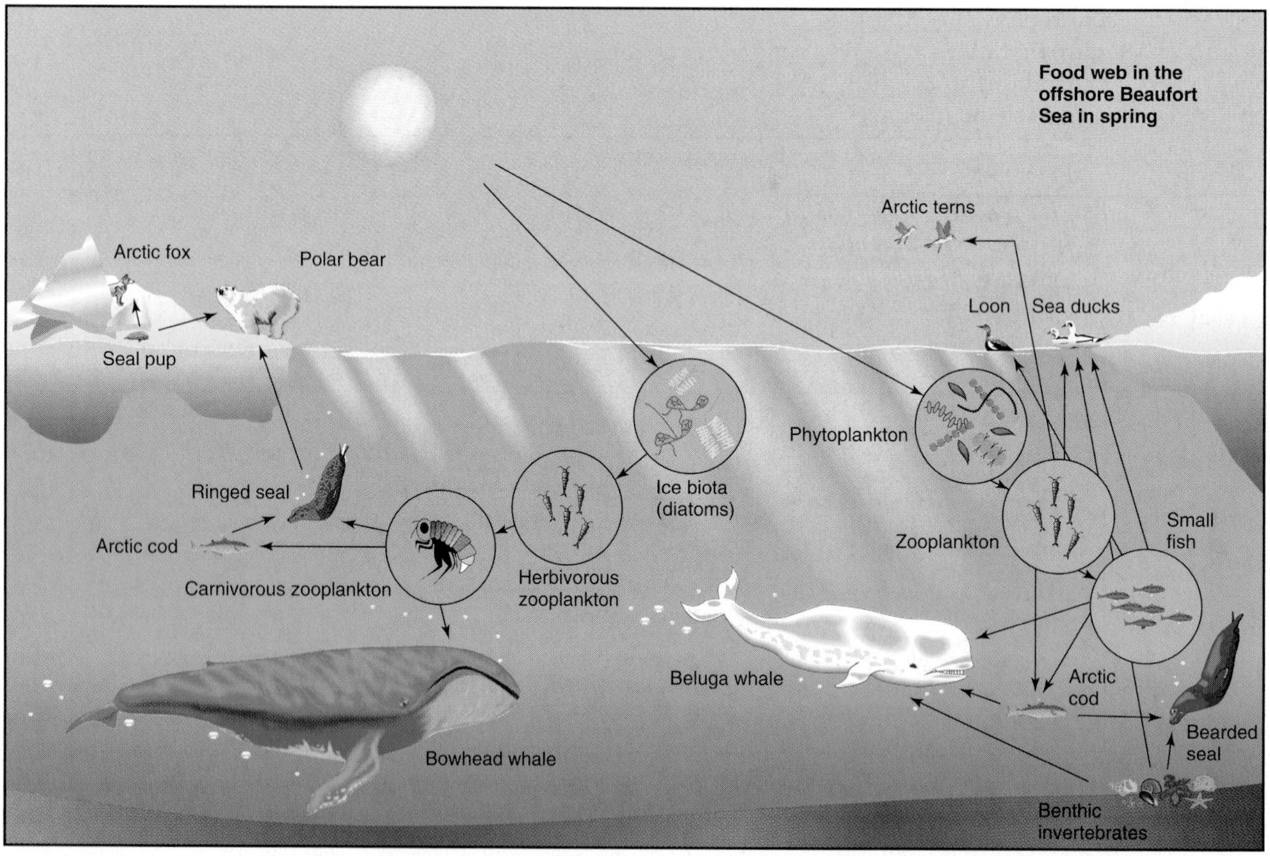


(b)

**Figure 2–12 Food webs.** (a) Specific pathways, such as that from nuts to squirrels to foxes (shown by green arrows), are referred to as *food chains*. A *food web* is the collection of all food chains, which are invariably interconnected (all arrows). Trophic levels, indicated by shading at the left, show that food always flows from producers to herbivores to carnivores. (b) A marine food web.

levels, there is a fundamental movement of the chemical nutrients and stored energy they contain from one organism or level to the next. These movements of energy and nutrients are described in more detail in Chapter 3. A visual comparison of food chains, food webs, and trophic levels is shown in Fig. 2–12.

***Limits on Trophic Levels.*** How many trophic levels are there? Usually, there are no more than three or four in terrestrial ecosystems and sometimes five in marine systems. (See Fig. 2-12b.) This answer comes from straightforward observations. The **biomass**, or total combined (net dry) weight (often, per unit area or volume), of all the organisms at each trophic level can be estimated by collecting (or trapping) and weighing suitable samples. In terrestrial ecosystems, the biomass is roughly 90% less at each higher trophic level. For example, if the biomass of producers in a grassland is 1 ton (2,000 lb) per acre, the biomass of herbivores will be about 200 pounds per acre, and that of primary carnivores will be about 20 pounds per acre. At this rate, you can't go through very many trophic levels before the biomass approaches zero. Depicting these relationships graphically gives rise to what is commonly called a **biomass pyramid** (Fig. 2–13).

The biomass decreases so much at each trophic level for three reasons. First, much of the food that is consumed by a heterotroph is not converted to the body tissues of the heterotroph; rather, it is broken down, and the stored energy it contains is released and used by the heterotroph. Second, much of the biomass—especially at the producer level—is never eaten by herbivores and goes directly to the decomposers. Third, carnivores that eat carnivores as prey must be larger than their prey, and there are limits to the size and distribution of ever-larger carnivores roaming over an ever-larger area.

As organic matter is broken down, its chemical elements are released back to the environment, where, in the inorganic state, they may be reabsorbed by autotrophs (producers). Thus, a continuous *cycle of nutrients* is sustained, from the environment through organisms and back to the environment. As organisms eat other organisms, they expend energy to grow and reproduce, and their numbers as species are sustained. The spent energy, on the other hand, is lost as heat given off from bodies (Fig. 2–14). In sum, all food chains, food webs, and trophic levels *must start with producers*, and producers must have suitable environmental conditions to support their growth. Populations of all heterotrophs, including humans, are ultimately limited by what plants produce, in accordance with the concept of the biomass pyramid. Should any factor cause the productive capacity of green plants to be diminished, all other organisms at higher trophic levels will be diminished accordingly.

## Nonfeeding Relationships

**Mutually Supportive Relationships.** The overall structure of ecosystems is characterized by their feeding relationships. You may think that one species benefits and

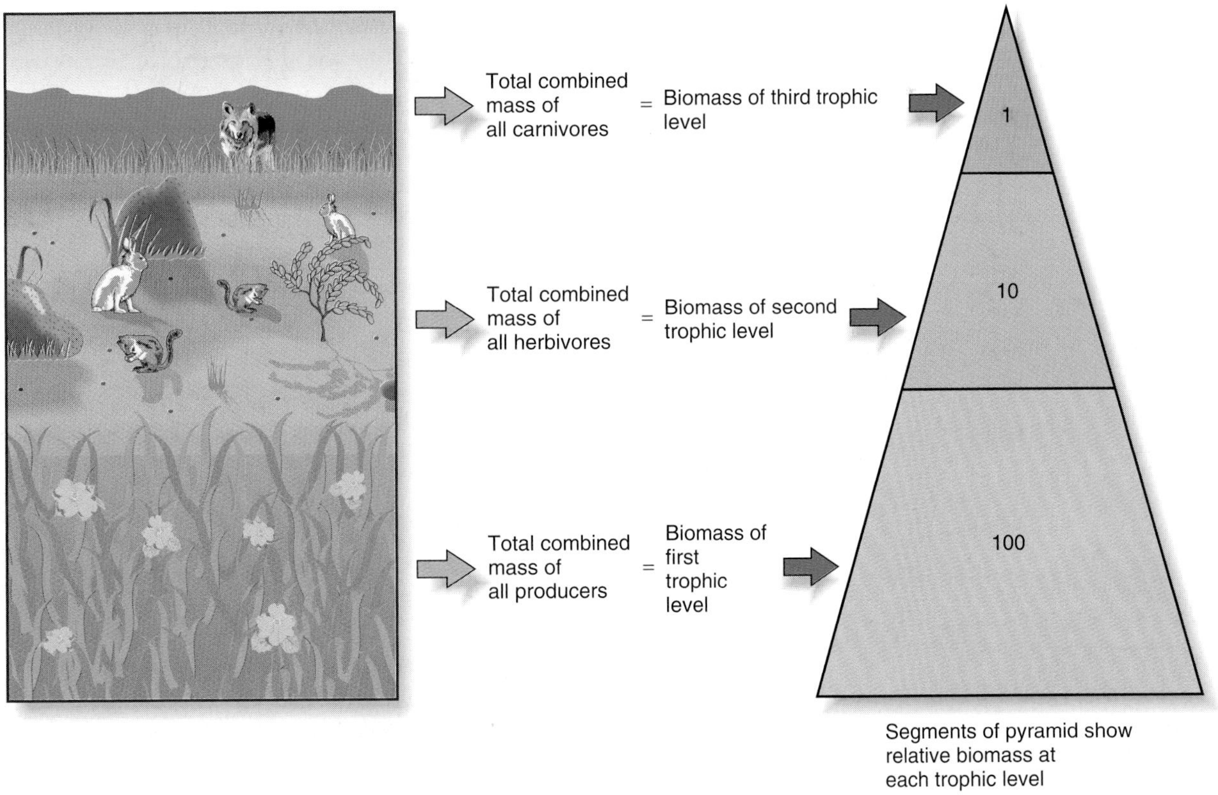


**Figure 2–13 Biomass pyramid.** A graphic representation of the biomass (the total combined mass of organisms) at successive trophic levels has the form of a pyramid.

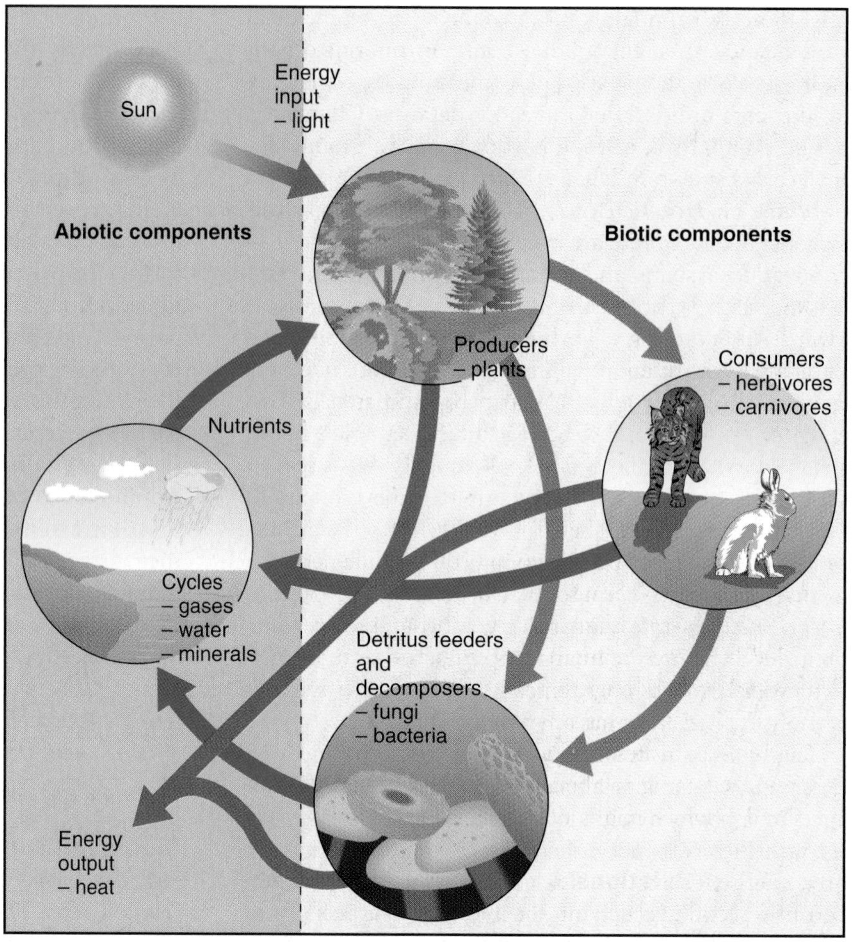


**Figure 2–14 Nutrient cycles and energy flow.** The movement of nutrients (blue arrows), energy (red arrows), and both (brown arrows) through the ecosystem. Nutrients follow a cycle, being used over and over. Light energy absorbed by producers is released and lost as heat energy as it is "spent."

the other is harmed to some extent in any feeding relationship. However, many relationships provide a mutual benefit to both species. This phenomenon is called **mutualism**, and it is exemplified by the relationship between flowers and pollinating insects. The insects benefit by obtaining nectar from the flowers, and the plants benefit by being pollinated in the process (Fig. 2–15).

In some cases, the mutualistic relationship has become so close that the species involved are no longer capable of living alone. The group of plants known as *lichens* (Fig. 2–16), for example, are actually composed of two organisms—a fungus and an alga. The fungus provides protection for the alga, enabling it to survive in dry habitats where it could not live by itself, and the alga,

**Figure 2–15 A mutualistic relationship.** Insects like this honeybee are attracted to flowers by their nectar, a good food source for the bees. As they move from plant to plant, the bees pollinate the flowers, enabling them to set fertile seed for the next generation.

**Figure 2–16 Lichens.** The crusty-appearing "plants" commonly seen growing on rocks or the bark of trees are actually composed of a fungus and an alga growing in a symbiotic relationship.

which is a producer, provides food for the fungus, which is a heterotroph. These two species living together in close union have a *symbiotic* relationship. However, **symbiosis** by itself simply means that the two organisms "live together" in close union (*sym*, together; *bio*, living); it does not specify a mutual benefit or harm. Therefore, symbiotic relationships may include parasitic relationships as well as mutualistic relationships.

While not categorized as mutualistic, many relationships in an ecosystem may aid its overall sustainability. For example, plant detritus provides most of the food for decomposers and soil-dwelling detritus feeders such as earthworms. Thus, these organisms benefit from plants, but the plants also benefit because the activity of the organisms releases nutrients from the detritus and returns them to the soil, where they can be reused by the plants. Similarly, insect-eating birds benefit from vegetation by finding nesting materials and places among trees, while the plant community benefits because the birds reduce the populations of many herbivorous insects. Even in predator–prey relationships, some mutual advantage may exist. The killing of individual prey that are weak or diseased may benefit the population as a whole by keeping it healthy. Predators and parasites may also prevent herbivore populations from becoming so abundant that they overgraze their environment (which might jeopardize the entire ecosystem).

**Competitive Relationships.** Considering the complexity of food webs, you might think that species of animals would be in a great "free-for-all" competition with each other. In fact, fierce competition rarely occurs, because each species tends to be specialized and adapted to its own *habitat* or *niche*.

**Habitat** refers to the kind of place—defined by the plant community and the physical environment—where a species is biologically adapted to live. For example, a deciduous forest, a swamp, and a grassy field are types of habitats. Different types of forests (for instance, coniferous vs. deciduous) provide markedly different habitats and support different species of wildlife.

Even when different species occupy the same habitat, competition may be slight or nonexistent because each species has its own *niche*. An animal's **ecological niche** refers to what the animal feeds on, where it feeds, when it feeds, where it finds shelter, how it responds to abiotic factors, and where it nests. Seeming competitors can coexist in the same habitat, but have separate niches. Competition is minimized because potential competitors are using different resources. (It's like a shopping mall where the stores avoid competition by offering different goods.) For example, woodpeckers, which feed on insects in deadwood, do not compete with birds that feed on seeds. Bats and swallows both feed on flying insects, but they do not compete, because bats feed on night-flying insects and swallows feed during the day. Sometimes the "resource" can be the space used by different species as they forage for food, as in the case of five species of warblers that coexist in the spruce forests of Maine (Fig. 2–17). The birds, which feed at different levels of the forest and on different parts of the trees, exemplify what is called *resource partitioning*. By adapting to each other's presence over time, these species avoid competition, and all of them benefit.

Depending on how a set of resources is "divided up" among species, there may be unavoidable overlap between the niches of the species. All green plants require water, nutrients, and light, and where they are growing in the same location, one species may eliminate others through competition. (That's why maintaining flowers and vegetables against the advance of weeds is a constant struggle.) However, different plant species are also adapted and specialized to particular conditions. Thus, each species is able to hold its own against competition where conditions are well suited to it. The same concepts hold true for species in aquatic and marine ecosystems.

If two species compete directly in many respects, as sometimes occurs when a species is introduced from another continent, one of the two generally perishes in the competition. This is the *competitive exclusion principle*. For example, the introduction of the European rabbit to Australia has led to the decline and disappearance of several small marsupial animal species, due to direct competition for food and burrows.

## Abiotic Factors

We now turn to the *abiotic* side of the ecosystem. As noted before, the environment involves the interplay of many physical and chemical factors—**abiotic factors**—that different species respond to. Abiotic factors can be categorized as **conditions** or **resources.** *Conditions* are abiotic factors that vary in space and time, but are not used up or made unavailable to other species. Conditions include temperature (extremes of heat and cold, as well as average temperature), wind, pH (acidity), salinity (saltiness), and fire. Within aquatic systems, for example, the key conditions are salinity (freshwater vs. saltwater), temperature, the texture of the bottom (rocky vs. silty), the depth and turbidity (cloudiness) of the water (determining how much, if any, light reaches the bottom), and currents.

*Resources* are any factors—biotic or abiotic—that are consumed by organisms. Abiotic resources include water, chemical nutrients (like nitrogen and phosphorus), light (for plants), and oxygen. Abiotic resources also include spatial needs, such as a place on the intertidal rocks or a hole in a tree. Resources, unlike conditions, can be the objects of competition between individuals or species.

The degree to which each abiotic factor is present (or absent) profoundly affects the ability of organisms to survive. However, each species may be affected differently by each factor. This difference in response to environmental factors determines which species may or may not occupy a given region or a particular area within a region. In turn, the organisms that do or do not survive determine the nature of a given ecosystem.

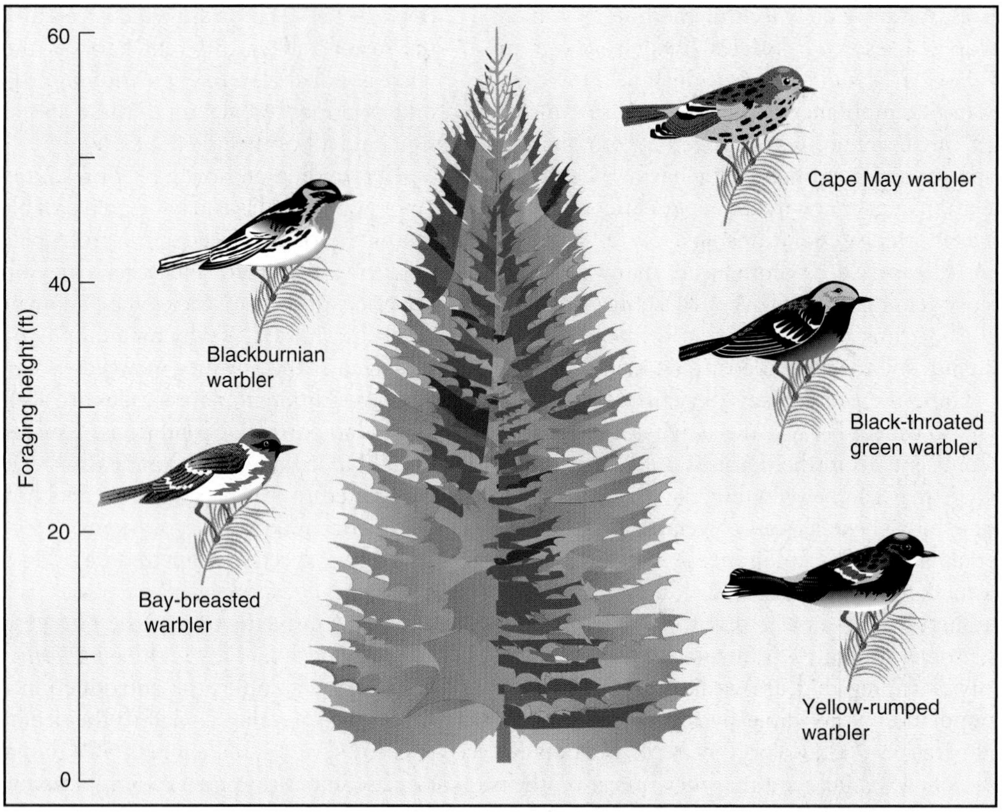


**Figure 2–17 Resource partitioning.** Five species of North American warblers reduce the competition among themselves by feeding at different levels and on different parts of trees.

**Optimum, Zones of Stress, and Limits of Tolerance.** *Different species thrive under different environmental regimes.* This principle applies to all living things, both plants and animals. Some survive where it is very wet, others where it is relatively dry. Some thrive in warmth, others in cooler situations. Some tolerate freezing, while others do not. Some require bright sun; others do best in shade. Aquatic systems are divided into freshwater and saltwater regimes, each with its respective fish and other organisms.

Laboratory experiments demonstrate that different species are best adapted to different factors. Organisms can be grown under controlled conditions in which one factor is varied while other factors are held constant. Such experiments demonstrate that, for every factor, there is an **optimum**, a certain level at which the organisms do best. At higher or lower levels the organisms do less well, and at further extremes they may not be able to survive at all. This concept is shown graphically in Figure 2–18. Temperature is shown as the variable in the figure, but the idea pertains to any abiotic factor that might be tested.

The point at which the best response occurs is called the optimum, but this may be a range of several degrees (or other units), so it is common to speak of an *optimal range*. The entire span that allows any growth at all is called the **range of tolerance**. The points at the high and low ends of the range of tolerance are called the **limits of tolerance**. Between the optimal range and the high or low limit of tolerance are **zones of stress**. That is, as the factor is raised or lowered from the optimal range, the organisms experience increasing stress, until, at either limit of tolerance, they cannot survive.

***A Fundamental Principle.*** Not every species has been tested for every possible factor. Based on the consistency of such observations, however, the following is considered to be a fundamental biological principle: *Every species (both plant and animal) has an optimum range, zones of stress, and limits of tolerance with respect to every abiotic factor.*

This line of experimentation also demonstrates that different species vary in characteristics with respect to the values at which the optimum and the limits of tolerance occur. For instance, what may be an optimal amount of water for one species may stress a second and kill a third. Some plants cannot tolerate any freezing temperatures, others can tolerate slight, but not intense, freezing, and some actually require several weeks of freezing temperatures in order to complete their life cycles. Also, some species have a very broad range of tolerance, whereas others have a much narrower range. While optimums and limits of tolerance may differ from one species to another, there may be great overlap in their ranges of tolerance.

The concept of a range of tolerance affects more than just the growth of individuals: Because the health

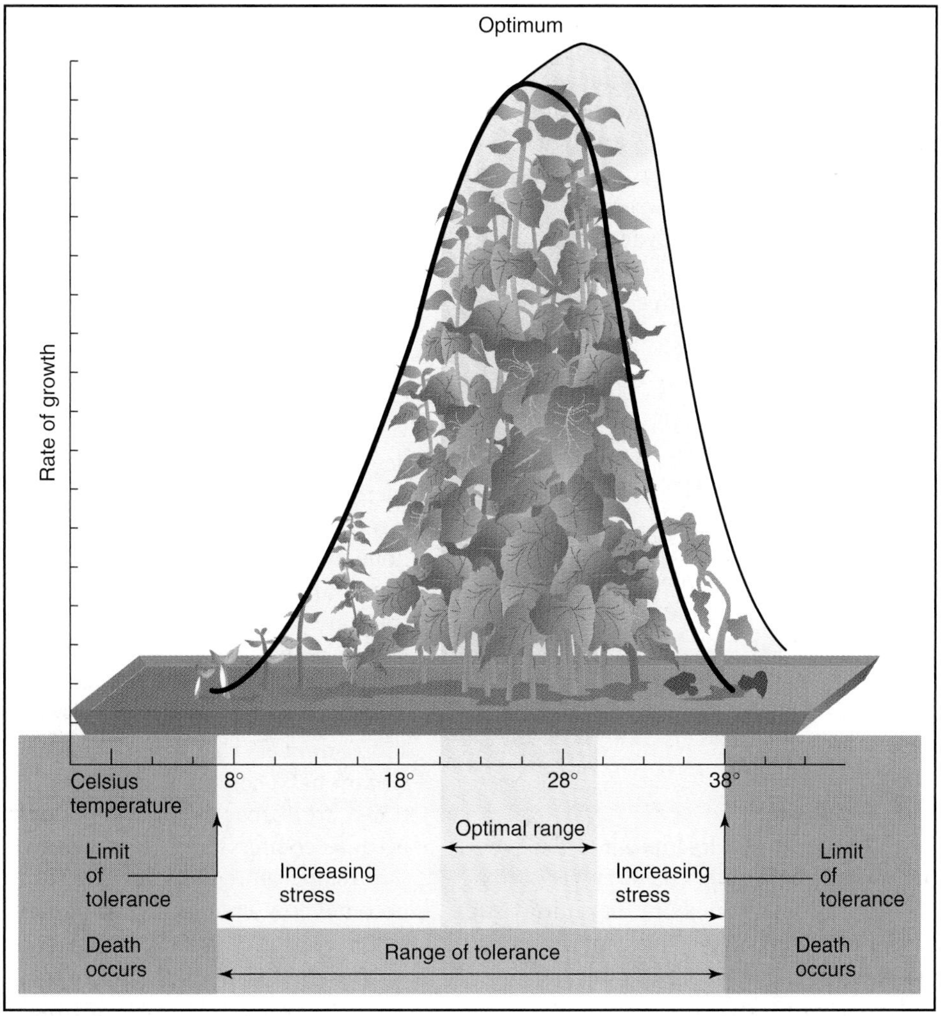


**Figure 2–18** **Survival curve.** For every factor influencing growth, reproduction, and survival, there is an optimum level. Above and below the optimum, stress increases, until survival becomes impossible at the limits of tolerance. The total range between the high and low limits is the range of tolerance.

and vigor of individuals affect reproduction and the survival of the next generation, the population is also influenced. Consequently, the population density (individuals per unit area) of a species is greatest where all conditions are optimal, and it decreases as any one or more conditions depart from the optimum. Different ranges of tolerance for different factors contribute significantly to the identity of an ecological niche for a given species.

**Law of Limiting Factors.** In 1840, Justus von Liebig studied the effects of chemical nutrients on plant growth. He observed that restricting any one of the many different nutrients at any given time had the same effect: *It limited growth.* A factor that limits growth is called, naturally, a **limiting factor**. *Any one factor* being outside the optimal range will cause stress and limit the growth, reproduction, or even survival of a population. This observation is referred to as the **law of limiting factors**, or Liebig's law of minimums.

The limiting factor may be a problem of *too much*, as well as a problem of *too little*. For example, plants may be stressed or killed not only by underwatering or underfertilizing, but also by overwatering or overfertilizing, which are common pitfalls for beginning gardeners. Note also that the limiting factor may change from one time to another. For instance, in a single growing season, temperature may be limiting in the early spring, nutrients may be limiting later, and then water may be limiting if a drought occurs. Also, if one limiting factor is corrected, growth will increase only until another factor comes into play. The organism's genetic potential is an ultimate limiting factor. Thus, a mouse will never grow to the bulk of an elephant, no matter how much you feed it!

***Biotic Factors.*** Observations made since Liebig's time show that his law has a much broader application. That is, growth may be limited not only by abiotic factors, but also by biotic factors. Thus, the limiting factor for a population may be competition or predation from another species. With agricultural crops, for example, there is a constant struggle to keep them from being limited or even eliminated by weeds and insects.

Finally, while one factor may be determined to be limiting at a given time, several factors outside the optimum may combine to cause additional stress or even death. In particular, pollutants may act in a way that causes organisms to become more vulnerable to disease or drought. Such cases are examples of **synergistic effects**, or **synergisms**, which are defined as two or more factors interacting in a way that causes an effect much greater than one would anticipate from the effects of each of the two acting separately.

## 2.3 From Ecosystems to Global Biomes

We can now use the concepts of optimums and limiting factors to gain a better understanding of why different regions or even localized areas may have distinct biotic communities, creating an amazing variety of ecosystems, landscapes, and biomes.

### The Role of Climate

The **climate** of a given region is a description of the average temperature and precipitation—the weather—that may be expected on each day throughout the entire year. (See Chapter 20.) Climates in different parts of the world vary widely. Equatorial regions are continuously warm, with high rainfall and no discernible seasons. Above and below the equator, temperatures become increasingly seasonal (characterized by warm or hot summers and cool or cold winters); the farther we go toward the poles, the longer and colder the winters become, until at the poles it is perpetually winterlike. Likewise, colder temperatures are found at higher elevations, so that there are even snowcapped mountains on or near the equator.

Annual precipitation in any area also may vary greatly, from virtually zero to well over 100 inches (250 cm) per year. Precipitation may be evenly distributed throughout the year or concentrated in certain months, dividing the year into wet and dry seasons.

Different temperature and rainfall conditions may occur in almost any combination, yielding a wide variety of climates. In turn, a given climate will support only those species that find the temperature and precipitation levels optimal or at least within their ranges of tolerance. As indicated in Figure 2–18, population densities will be greatest where conditions are optimal and will decrease as any condition departs from the optimum. A species will be excluded from a region (or local areas) where any condition is beyond its limit of tolerance. How will this variation affect the biotic community?

***Biome Examples.*** To illustrate, let us consider six major types of biomes and their global distribution. Table 2–3 describes these terrestrial biomes and their major characteristics, and Figure 2–19 shows the distribution of, and variations in, the biomes as they occur globally. Within the temperate zone (between 30° and 50° of latitude), the amount of rainfall is the key limiting factor. The **temperate deciduous forest biome** is found where annual precipitation is 30–80 in. (75–200 cm). Where rainfall tapers off or is highly seasonal (10–60 in., or 25–150 cm, per year), **grassland and prairie biomes** are found, and regions receiving an average of less than 10 inches (25 cm) per year are occupied by a **desert biome.**

The effect of temperature, the other dominant parameter of climate, is largely superimposed on that of rainfall. That is, 30 inches (75 cm) or more of rainfall per year will usually support a forest, but temperature will determine the *kind* of forest. For example, broad-leafed evergreen species, which are extremely vigorous and fast growing, but cannot tolerate freezing temperatures, predominate in the **tropical rain forest.** By dropping their leaves and becoming dormant each autumn, deciduous trees are well adapted to freezing temperatures. Therefore, wherever rainfall is sufficient, deciduous forests predominate in temperate latitudes. Most deciduous trees, however, cannot tolerate the extremely harsh winters and short summers that occur at higher latitudes and higher elevations. Therefore, northern regions and high elevations are occupied by the **coniferous forest biome,** because conifers are better adapted to those conditions.

Temperature by itself limits forests only when it becomes low enough to cause **permafrost** (permanently frozen subsoil). Permafrost prevents the growth of trees, because roots cannot penetrate deeply enough to provide adequate support. However, a number of grasses, clovers, and other small flowering plants can grow in the topsoil above permafrost. Consequently, where permafrost sets in, the coniferous forest biome gives way to the **tundra biome** (Table 2–3). At still colder temperatures, the tundra gives way to permanent snow and ice cover.

The same relationship of rainfall effects being primary and temperature effects secondary applies in deserts. Any region receiving less than about 10 inches (25 cm) of rain per year will be a desert, but the unique plant and animal species found in hot deserts are different from those found in cold deserts.

A summary of the relationship between biomes, and temperature and rainfall conditions, is given in Figure 2–20. The average temperature for a region varies with both latitude and altitude, as shown in Figure 2–21.

#### Microclimate and Other Abiotic Factors

A specific site may have temperature and moisture conditions that are significantly different from the overall, or average, climate of the region in which it is located. For example, a south-facing slope, which receives more direct sunlight in the northern hemisphere, will be relatively warmer and hence also drier than a north-facing slope (Fig. 2–22). Similarly, the temperature range in a sheltered ravine will be narrower than that in a more

**table 2-3 Major Terrestrial Biomes**

| Biome | Climate and Soils | Dominant Vegetation | Dominant Animal Life | Geographic Distribution |
|---|---|---|---|---|
| Deserts | Very dry; hot days and cold nights; rainfall less than 10 in./yr; soils thin and porous | Widely scattered thorny bushes and shrubs, cacti | Rodents, lizards, snakes, numerous insects, owls, hawks, small birds | N. and S.W. Africa, parts of Middle East and Asia, S.W. United States, northern Mexico |
| Grasslands and Prairies | Seasonal rainfall, 10 to 60 in./yr; fires frequent; soils rich and often deep | Grass species, from tall grasses in areas with higher rainfall to short grasses where drier; bushes and woodlands in some areas | Large grazing mammals: bison, goats; wild horses; kangaroos; antelopes, rhinos, warthogs, prairie dogs, coyotes, jackals, lions, hyenas; termites important | Central North America, central Asia, subequatorial Africa and South America, much of southern India, northern Australia |
| Tropical Rain Forests | Nonseasonal; annual average temperature 28°C; rainfall frequent and heavy, average over 95 in./yr; soils thin and poor in nutrients | High diversity of broad-leafed evergreen trees, dense canopy, abundant epiphytes and vines; little understory | Enormous biodiversity; exotic, colorful insects, amphibians, birds, snakes; monkeys, small mammals, tigers, jaguars | Northern South America, Central America, western central Africa, islands in Indian and Pacific Oceans, S.E. Asia |
| Temperate Forests | Seasonal; temperature below freezing in winter; summers warm, humid; rainfall from 30–80 in./yr; soils well developed | Broad leafed deciduous trees, some conifers; shrubby undergrowth, ferns, lichens, mosses | Squirrels, raccoons, opossums, skunks, deer, foxes, black bears, snakes, amphibians, rich soil microbiota, birds | Western and central Europe, eastern Asia, eastern North America |
| Coniferous Forests | Seasonal; winters long and cold; precipitation light in winter, heavier in summer; soils acidic, much humus and litter | Coniferous trees (spruce, fir, pine, hemlock), some deciduous trees (birch, maple); poor understory | Large herbivores such as mule deer, moose, elk; mice, hares, squirrels; lynx, bears, foxes, fishers, marten; important nesting area for neotropical birds | Northern portions of North America, Europe, Asia, extending southward at high elevations |
| Tundra | Bitter cold, except for an 8- to 10-week growing season with long days and moderate temperatures; precipitation low, soils thin and underlain with permafrost | Low-growing sedges, dwarf shrubs, lichens, mosses, and grasses | Year round: lemmings, arctic hares, arctic foxes, lynx, caribou, musk ox; summers: abundant insects, many migrant shorebirds, geese, and ducks | North of the coniferous forest in northern hemisphere, extending southward at elevations above the coniferous forest |

exposed location, and so on. The conditions found in a specific localized area are referred to as the **microclimate** of that location. In the same way that different climates determine the major biome of a region, different microclimates result in variations of ecosystems within a biome.

Soil type and topography may also contribute to the diversity found in a biome, because these two factors affect the availability of moisture. In the eastern United States, for example, oaks and hickories generally predominate on rocky, sandy soils and on hilltops, which retain little moisture, whereas beeches and maples are found on richer soils, which hold more moisture, and red maples and cedars inhabit low, swampy areas. In the transitional region between desert and grassland [10–20 inches (25–50 cm) of rainfall per year], a soil capable of holding water will support grass, but a sandy soil with little ability to hold water will support only desert species.

In certain cases, an abiotic factor other than rainfall or temperature may be the primary limiting factor. For example, the strip of land adjacent to a coast frequently receives a salty spray from the ocean, a factor that relatively few plants can tolerate. Consequently, an association of salt-tolerant plants frequently occupies this strip, as on a barrier island. Relative acidity or alkalinity (pH) may also have an overriding effect on a plant or animal community.

**Figure 2–19 World distribution of the major terrestrial biomes.** (*Source:* Figure 20–4 from *Geosystems: An Introduction to Physical Geography*, 4th ed., by Robert W. Christopherson. Copyright © 2000 by Prentice Hall, Inc. Reprinted by permission of Pearson Education, Inc. Upper Saddle River, NJ 07458.)

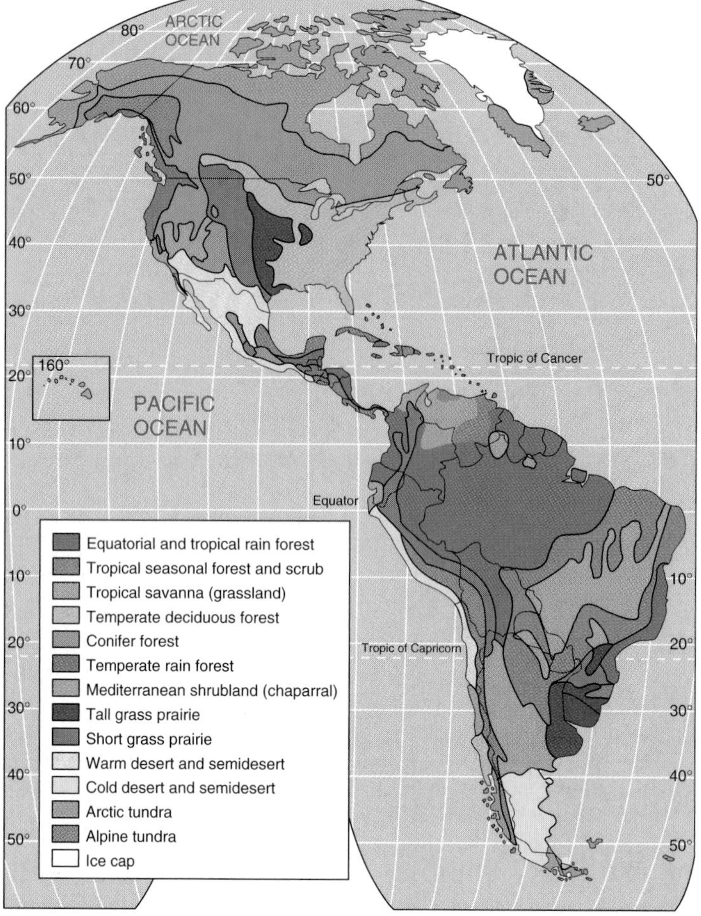


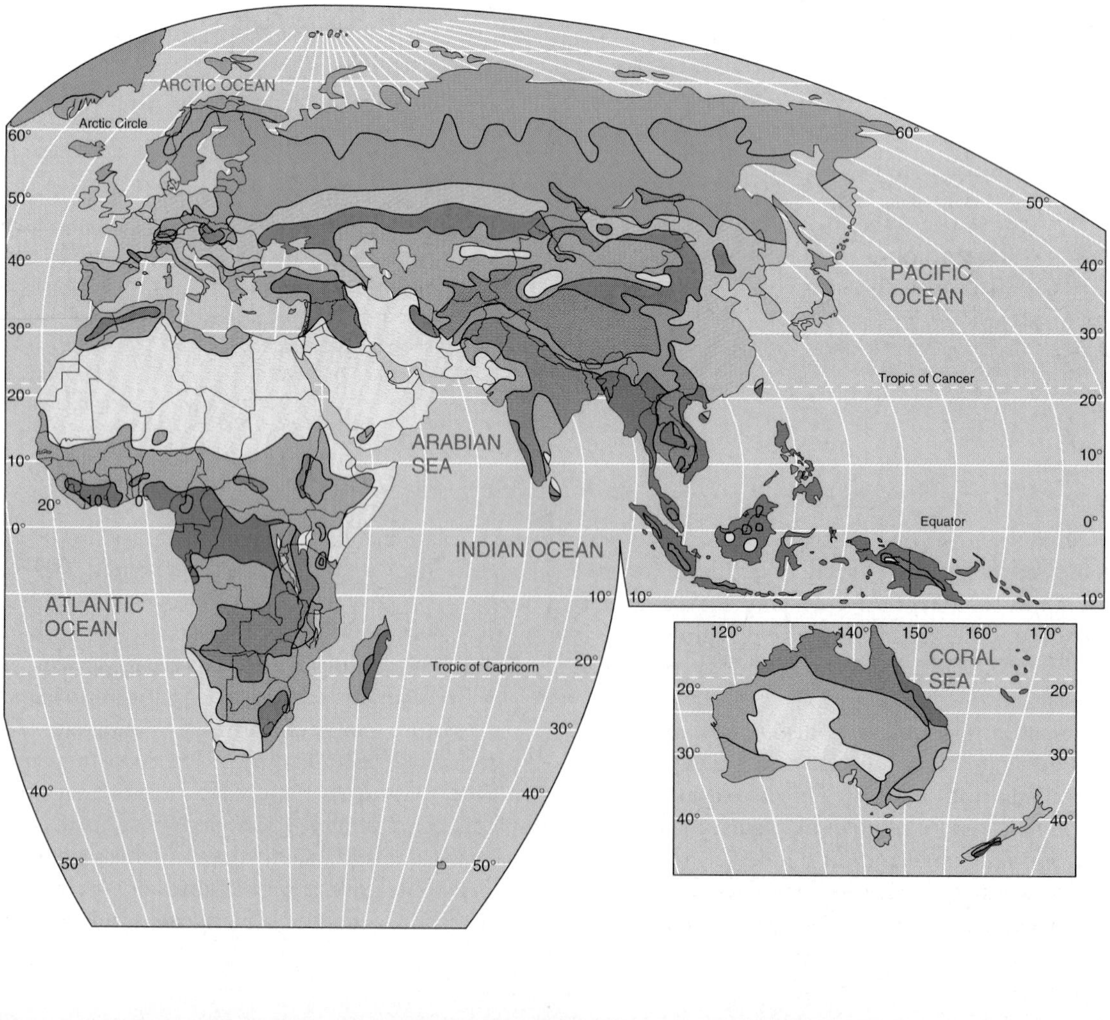

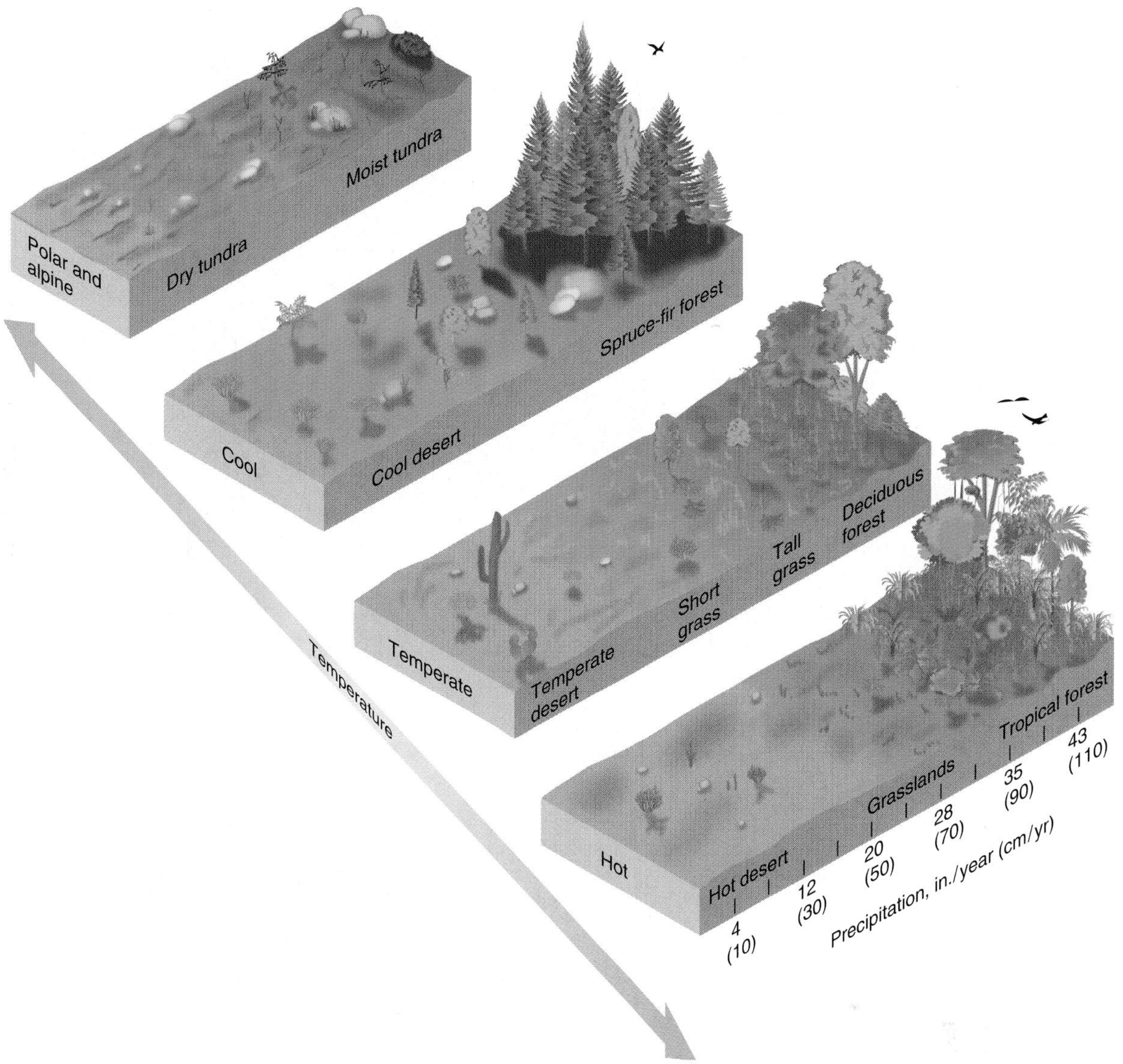


**Figure 2–20 Climate and major biomes.** Moisture is generally the overriding factor determining the type of biome that may be supported in a region. Given adequate moisture, an area will likely support a forest. Temperature, however, determines the *kind* of forest. The situation is similar for grasslands and deserts. At cooler temperatures, there is a shift toward less precipitation because lower temperatures reduce evaporative water loss. Temperature becomes the overriding factor only when it is low enough to sustain permafrost. (*Source:* Redrawn from *Geosystems*, 5th ed., by Robert W. Christopherson. Copyright 2003 by Prentice Hall, Inc. Upper Saddle River, NJ 07458.)

## Biotic Factors

Some biotic factors—that is, factors caused by other species—may be limiting. Grasses thrive when rainfall is more than 30 inches (75 cm). When the rainfall is great enough to support trees, however, increased shade may limit grasses. Thus, the factor that limits grasses from taking over high-rainfall regions is biotic—namely, the overwhelming competition from taller species. The distribution of plants may also be limited by the presence of certain herbivores; elephants, for example, are notorious for destroying woodlands, so their presence leads predominantly to grasslands. (See Fig. 5–14.)

Limiting factors also apply to animals. As with plants, the limiting factor may be abiotic—cold temperatures or lack of open water, for instance—but it is more frequently biotic, such as the absence of a plant community that would otherwise provide suitable food or habitat.

## Physical Barriers

A final factor that may limit species to a particular region is the existence of a physical barrier, such as an ocean, a desert, or a mountain range, that species are unable to cross. Thus, species making up the communities on separate continents or remote islands are usually quite different, despite living in similar climates.

When physical barriers are overcome—as when humans knowingly or unknowingly transport a species from one continent to another—the introduced species

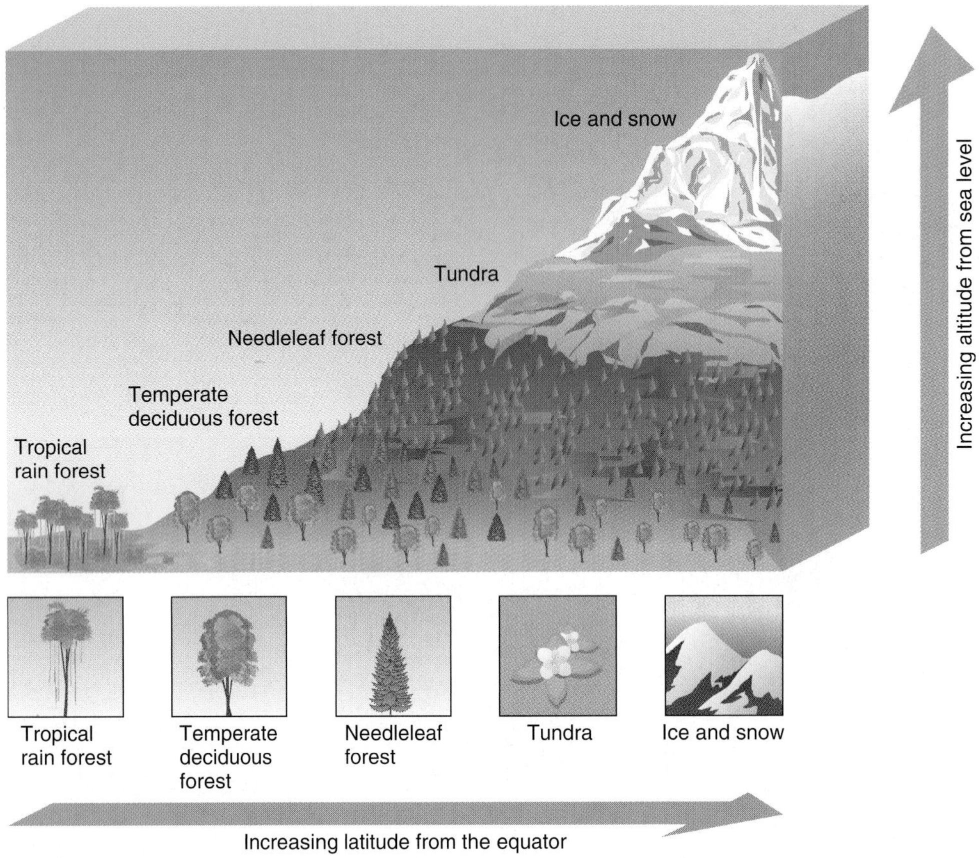


**Figure 2–21 Effects of latitude and altitude.** Decreasing temperatures, which result in the biome shifts noted in Fig. 2–20, occur with both increasing latitude (distance from the equator) and increasing altitude. (*Source:* Redrawn from *Geosystems*, 5th ed., by Robert W. Christopherson. Copyright 2003 by Prentice Hall, Inc. Upper Saddle River, NJ 07458.)

may make a successful *invasion*. The starling, for example, originally confined to Eurasia, has successfully invaded North America. Additional examples of imported species and their impacts are discussed further in Chapter 4. Note also that humans erect barriers—dams, roadways, aboveground pipelines, cities, and farms—that may block the normal movement of populations and cause their demise. For example, the Atlantic salmon is now an endangered species in some New England states because of the many dams that prevent the migration they require for spawning.

## Summary

The biosphere consists of a great variety of environments, both aquatic and terrestrial. In each environment, plant, animal, and microbial species are adapted to all the abiotic factors. In addition, they are adapted to each other, in various feeding and nonfeeding relationships. Each environment supports a more or less unique grouping of organisms interacting with each other and with the environment in a way that perpetuates or sustains the entire group. That is, each environment, together with the species it supports, is an *ecosystem*. Every ecosystem is tied to others through species that migrate from one system to another and through exchanges of air, water, and minerals common to the whole planet. At the same time, each species—and, as a result, each ecosystem—is kept within certain bounds by limiting factors. The spread of each species is limited at some point because that species is unable to tolerate particular conditions, to compete with other species, or to cross some physical barrier. The distribution of species is always due to one or more limiting factors.

## 2.4 The Human Factor

So far, we have been looking at ecosystems as natural-functioning systems, unaffected by humans. In reality, the impact of humans must be taken into consideration because we have become such a dominant presence on Earth. We have replaced many natural systems with agriculture and urban and suburban developments; we make heavy use of most of the remaining "natural" systems, for wood, food, and other commercial products; and the by-products of our economic activities have polluted and degraded ecosystems everywhere. Our involvement with natural ecosystems is so

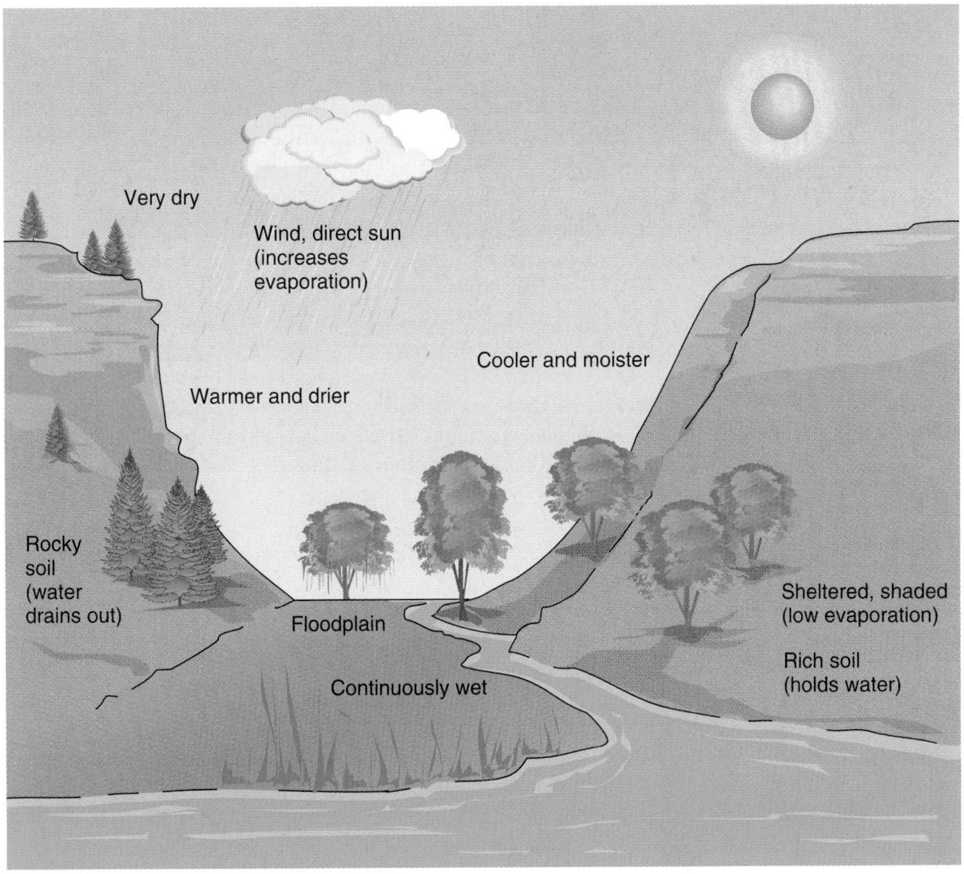


**Figure 2–22 Microclimates.** Abiotic factors such as terrain, wind, and type of soil create different microclimates by influencing temperature and moisture in localized areas.

pervasive that if we want to continue to enjoy their goods and services, we must learn how to *manage* and sometimes *restore* those ecosystems to keep them healthy and productive (see Ethics, p. 50). A brief look into the past may help you understand the changes that have occurred to make humans such a dominant part of the landscape and may better equip you to deal with the future.

## Three Revolutions

**Neolithic Revolution.** Natural ecosystems have existed and perpetuated themselves on Earth for hundreds of millions of years, while humans are relative newcomers to the scene. Archaeological and anthropological evidence indicates that hominid ancestry goes back at least four million years. The fossil record reveals as many as 14 hominid species. The earliest fully modern humans appeared about 40,000 years ago, when they invaded Europe, and their culture is known as the **Paleolithic** (40,000 to 10,000 years ago).

Paleolithic humans survived in small tribes as hunter–gatherers, catching wildlife and collecting seeds, nuts, roots, berries, and other plant foods (Fig. 2–23). Settlements were never large and were of relatively short duration because, as one area was picked over, the tribe was forced to move on. As hunter–gatherers, these people were much like other omnivorous consumers in natural ecosystems. Populations could not expand beyond the sizes that natural food sources supported, and deaths from predators, disease, and famine were common. In time, however, the hunter–gatherer culture was successful enough to foster population increases, and the pressures of rising populations led to the next stage.

About 12,000 years ago, a highly significant change in human culture occurred when humans in the Middle East began to develop animal husbandry and agriculture. Evidence suggests that this change—the domestication of wild species—can be traced to a shift in the climate—the Younger Dryas episode—which brought first unusually warm and wet, and then cooler and drier, conditions. The cooler and drier conditions made hunting and gathering increasingly difficult.

The development of agriculture provided a more abundant and reliable food supply, but it was a turning point in human history for other reasons as well. Because of its profound effect, it is referred to as a *revolution*—specifically, the **Neolithic Revolution**. Conducting agriculture does not just allow, but *requires*, permanent (or at least long-term) settlements and the specialization of labor. Some members of the settlement specialize in tending crops and producing food, freeing others to specialize in other endeavors. With this specialization of labor in permanent settlements, there

**ethics**

## Can Ecosystems Be Restored?

The human capacity for destroying ecosystems is well established. To some degree, however, we also have the capacity to restore them. In many cases, restoration simply involves stopping the abuse. For example, it has been found that after pollution is curtailed, water quality improves and fish and shellfish gradually return to previously polluted lakes, rivers, and bays. Similarly, forests may gradually return to areas that have been cleared. Humans can speed up the process by seeding, planting seedling trees, and reintroducing populations of fish and animals that have been eliminated.

In some cases, however, specific ecosystems have been destroyed or disturbed to such an extent that they require the efforts of a new breed of scientist: the *restoration ecologist.* Two types of ecosystem that have suffered most in this regard are the *prairie* and *wetland ecosystems.* The potential for restoration of any ecosystem rests on the following three assumptions: (a) Abiotic factors must have remained unaltered or, if not, can at least be returned to their original state. (b) Viable populations of the species formerly inhabiting the ecosystem must still exist. (c) The ecosystem must not have been upset by the introduction of one or more foreign species that cannot be eliminated and that may preclude the survival of reintroduced native species. If these conditions are met, revival efforts have the potential to restore the ecosystem to some semblance of its former state.

Suppose, for example, that the Natural Lands Trust has acquired land in the Great Plains and wishes to restore the prairie that once flourished there. The problems are many. The lack of grazing and regular fires have led to much woody vegetation, exotic species abound in the region and can continuously disperse seeds on the experimental prairie, and there may be no remnants of the original prairie grasses and herbs on the site. How does one proceed? It may help to draw on a primer on ecological restoration published by the Society for Ecological Restoration.[1]

First, an *inventory* must be taken of what is on the site and, from historical records, what used to be present. Then, a *model* must be developed of the desired ecosystem in structural and functional terms, usually based on the use of an existing *reference ecosystem. Goals* are set for the restoration efforts by defining the desired future condition of the site—in particular, how the restored ecosystem will be integrated into its landscape setting. Next, an *implementation plan* is designed that can convert the goals into specific actions. For example, it may be necessary to remove all herbaceous vegetation with herbicides and then plow and plant the land with native grasses. To maintain the prairie, it may be necessary to burn it regularly or, in the case of larger landholdings, to introduce bison. Finally, the results of the implementation plan should be *monitored* and strategies developed for the long-term *protection and maintenance* of the restored ecosystem.

Why should we restore ecosystems? Restoration ecologists Steven Apfelbaum and Kim Chapman cite several compelling reasons.[2] First, we should do so for *aesthetic reasons.* Natural ecosystems are often beautiful, and the restoration of something beautiful and pleasing to the eye is a worthy project that can be uplifting to many people. Second, we should do so for *the benefit of human use.* The ecosystem services of a restored wetland, for example, can be enjoyed by present and future generations. Finally, we should do so for *the benefit of the species and ecosystems themselves.* Nature has value and a right to a continued existence, so people should act to preserve and restore ecosystems and species in order to preserve that right. Do you find these reasons compelling?

[1] "The SER Primer on Ecological Restoration," www.ser.org/, Society for Ecological Restoration Science & Policy Working Group. 2002.

[2] "Ecological Restoration: A Practical Approach," Ch. 15 in *Ecosystem Management: Applications for Sustainable Forest and Wildlife Resources*, ed. by Mark S. Boyce and Alan Haney (New Haven, CT: Yale University Press, 1997).

is more incentive and potential for technological development, such as better tools, better dwellings, and better means of transporting water and other materials. Trade with other settlements begins, and commerce is thus born. Also, living in settlements permits better care and protection for everyone; therefore, the number of early deaths is reduced. This reduced mortality rate, coupled with more reliable food production, supports population growth, which in turn supports (and is supported by) expanding agriculture. In short, modern civilization had its origins in the invention of agriculture about 12,000 years ago.

**Industrial Revolution.** For another 11,000-plus years, the human population increased and spread throughout the Earth. Agriculture and natural ecosystems supported the growth of a civilization and culture that increased in knowledge and mastery over the natural world. With the birth of modern science and technology in the 17th and 18th centuries, the human population—by 1800 almost a billion strong—was on the threshold of another revolution: the **Industrial Revolution** (Fig. 2–24). The Industrial Revolution created the modern world, with its global commerce, factories, large cities, and pollution. The Industrial Revolution and its technological marvels were energized by fossil fuels—first coal and then oil and gas. Pollution and exploitation took on new dimensions as the industrial world turned to the extraction of raw materials from all over the world (hence the desire for colonies). In time, every part of Earth was affected by this revolution and continues to be affected even today. As a result, we

**Figure 2–23 Hunter–gatherer culture.** Before the advent of agriculture, all human societies had to forage for their food, as these bushmen from Namibia are doing.

now live in a time of uninterrupted population growth and economic expansion, with all of the environmental problems outlined in Chapter 1.

In this historical progression, natural systems are displaced by the **human system**, a term that refers to our total system, including animal husbandry, agriculture, and all human social and cultural developments. Keep in mind, however, that the human system still depends heavily on goods and services provided by natural ecosystems. If the human system functioned as a true ecosystem, it would recognize this dependence and would also establish sustainable practices in its own right.

Indeed, the human system does have some features in common with natural ecosystems, such as the series of trophic levels from crop producers to human consumers. In other respects, however, it is far off the mark—failing

**Figure 2–24 Industrial Revolution.** The Industrial Revolution began in England in the 1700s. Coal was the energy source, and economic growth and pollution were the consequences.

to break down and recycle its "detritus" (for example, trash and chemical wastes) and other by-products. Suffering the consequences of pollution is one example of this gap between the human system and natural ecosystems. Remaining highly dependent on fossil fuels and thereby suffering the buildup of carbon dioxide in the atmosphere is another. A third revolution—an environmental one—is needed.

**Environmental Revolution.** In Chapter 1, we suggested that a "business as usual" approach in human affairs will not work and that a number of transitions will be necessary to bring about a sustainable society. Some observers have referred to this shift as the **Environmental Revolution**, because the transitions necessary to move the human system from its present state to one that is sustainable are indeed revolutionary. The necessary transitions are discussed in succeeding chapters.

Revolutions suggest overthrowing something, and indeed, what is involved is an overthrow of prevalent attitudes toward our economy and the environment. This does not have to be a violent revolution; it could take place so peacefully that it would take a future generation to look back and realize that a major revolution had occurred. Yet it appears that the options are limited as we look to the future. We can choose to undergo the changes necessary to achieve sustainability by planning properly and learning as we go, or we can ignore the signs of unsustainability and increase our impact on the environment by driving bigger cars (and more of them), living in bigger houses, flying off to more vacations, and, in general, expecting to enjoy more of everything. And the developing world, as it tries desperately to catch up to our living standards, could make the same mistakes we are making, with devastating consequences because there are so many more people there than in the developed world. If we choose to ignore the signs that our current practices are unsustainable, a different kind of environmental revolution will be thrust upon us by the inability of the environment to support an irresponsible human population.

Lester Brown, former president of WorldWatch Institute, said, "There is no middle path. The challenge is either to build an economy that is sustainable or to stay with our unsustainable economy until it declines. It is not a goal that can be compromised. One way or another, the choice will be made by our generation, but it will affect life on Earth for all generations to come." (Worldwatch Institute, State of the World 2000, p. 21)

## revisiting the themes

As we revisit the themes at each chapter's end, some of them may only tangentially touch on the subject matter. In this chapter, for example, policy and politics and globalization, two of the three integrative themes, are not particularly relevant to the study of what ecosystems are, so they have been omitted from our review.

### Sustainability

Ecosystems are the functional units of sustainable life on earth. They are, in effect, models of sustainability. The continued production of organic matter by primary producers sustains all other life in an ecosystem. That is, nutrient cycling is sustained, and so are the populations of species. Much of this sustainability is accomplished by trophic relationships, with organisms feeding and being fed on. Many other relationships between species aid in overall sustainability. The characteristics of ecosystems that contribute to sustainability are discussed in greater detail in the next two chapters.

### Stewardship

If the environmental revolution is accomplished by our deliberate actions to achieve a sustainable future, it will require a broad commitment to the ethic of stewardship, whereby people seek the common good and exercise stewardly care for the environment and for the needs of fellow humans.

### Sound Science

The work of ecologists in building up our knowledge of ecosystems and their current status is, in the best sense, sound science. The Heinz Center report, *The State of the Nation's Ecosystems,* lays out a blueprint for identifying the major indicators of ecosystem health and exposes those areas needing more study, as well as evaluating ecosystem statuses and trends.

### Ecosystem Capital

The structure of ecosystems and biomes in general represents the fundamental characteristics of natural systems which enable them to provide the goods and services—the ecosystem capital—that humans depend so much upon. In a real sense, humans are trophic consumers, using the organic matter provided by the producers and other consumers. We are part of the food web of many ecosystems, so we can influence (both positively and negatively) the way food webs function. Thus, we must learn to manage these systems in order to keep them healthy and productive.

## review questions

1. What is the difference between the biotic community and the abiotic environmental factors of an ecosystem?
2. Define and compare the terms *species*, *population*, *association*, and *ecosystem*.
3. Compared with an ecosystem, what are an ecotone, landscape, biome, and biosphere?
4. Identify and describe the biotic and the abiotic components of the biome of the region in which you live.
5. Name and describe the roles of the three main trophic categories that make up the biotic structure of every ecosystem. Give examples of organisms from each category.
6. How do the terms *organic* and *inorganic* relate to the biotic and abiotic components of an ecosystem?
7. Name and describe the attributes of the two categories into which all organisms can be divided.
8. Give four categories of consumers in an ecosystem and the role that each plays.
9. State the similarities and differences between detritus feeders and decomposers, based on what they do, how they do it, and the kinds of organisms that occupy each category.
10. Differentiate between the concepts of *food chain, food web*, and *trophic levels*.
11. Relate the concept of the biomass pyramid to the fact that all heterotrophs depend upon autotrophic production.
12. Describe three nonfeeding relationships that exist between organisms.
13. How is competition among different species of an ecosystem reduced?
14. Differentiate between the two types of abiotic factors. What is the effect on a population when any abiotic factor shifts from the optimum to the limit of tolerance and beyond? What things in addition to abiotic factors may act as limiting factors?
15. Describe how differences in climate cause the Earth to be partitioned into six major biomes.
16. What are three situations that might cause microclimates to develop within an ecosystem?
17. Use what you have learned about ecosystem structure to describe the barrier island ecosystem found on Plum Island.
18. What is significant about each of the following revolutions: Neolithic, Industrial, and Environmental?

## thinking environmentally

1. From local, national, and international news, compile a list of the many ways humans are altering abiotic and biotic factors on a local, regional, and global scale. Analyze ways that local changes may affect ecosystems on larger scales and ways that global changes may affect ecosystems locally.
2. Write a scenario of what would happen to an ecosystem or to the human system in the event of one of the following: (a) All producers are killed through a loss of fertility of the soil or through toxic contamination. (b) All parasites are eliminated. (c) Decomposers and detritus feeders are eliminated. Support all of your statements with reasons drawn from your understanding of the way ecosystems function.
3. Consider the various kinds of relationships humans have with other species, both natural and domestic. Give examples of relationships that (a) benefit humans, but harm other species, (b) benefit both humans and other species, and (c) benefit other species, but harm humans. Give examples in which the relationship may be changing—for instance, from exploitation to protection. Discuss the ethical issues involved in changing relationships.
4. Explain how the human system can be modified into a sustainable ecosystem in balance with (i.e., preserving) other natural ecosystems without losing the benefits of modern civilization.

Lake Victoria
KENYA
OCTOBER
Maasai Mara area
SEPTEMBER
AUGUST
TANZANIA
NOVEMBER
JULY
Serengeti Ecosystem
JUNE
DECEMBER
JANUARY
FEBRUARY
MARCH
APRIL
MAY
Ngorongoro area
Serengeti Ecosystem
International Boundary
kilometers 0 10
miles 0 10
N

# Ecosystems: How They Work

**Key Topics**

1. Matter, Energy, and Life
2. Energy Flow in Ecosystems
3. The Cycling of Matter in Ecosystems
4. Implications for Human Societies

Pictured in the opening photo, the Serengeti is a vast tropical savanna ecosystem of 25,000 $km^2$ in northern Tanzania and southern Kenya. The rain falls bimodally: Short rains normally occur in November and December, long rains during March through May. A rainfall gradient passes from the drier southeastern plains (50 cm/yr) to the wet northwest in Kenya (120 cm/yr). The southeastern plains are a treeless grassland, and as the volcanic soils thicken to the north and west, the grasses gradually shift to woodlands—mixed *Acacia* trees and grassy patches. Large herbivores (wildebeest, zebra, and Thomson's gazelle) dominate the ecosystem. Herds of over 1.5 million animals are common on the plains during the rainy seasons. Just below the surface of the land, hordes of termites in extensive galleries process organic detritus as fast as it appears and restore nutrients to the soil.

***Migration.*** As the rains fail, the large herds move to the woodlands in the north, where forage is available year-round. There, the animals face heavier predation from lions and hyenas, which need the woodlands for cover for raising young and for successful stalking. This migration, in which the animals travel 200 or more kilometers and back again each year, is energetically costly. Why do they do it if there is sufficient forage in the woodlands? The answer is still being investigated, but evidence points to two possible factors: (1) The vegetation in the plains is high in phosphorus, which the herbivores need for successful growth and lactation. Their annual presence there maintains the high phosphorus content as their wastes are broken down and nutrients are returned to the soils. (2) The presence of high numbers of predators in the woodlands may force the herds to migrate to the grasslands, where they are less vulnerable, especially when giving birth to their young.

In this ecosystem known as the Serengeti, producers, herbivores, carnivores, and scavengers or detritus feeders interact in a sustainable set of relationships. However, the Serengeti and its spectacular wildlife will continue to exist only as long as the land is protected. This World Heritage site is pressed on all sides by poaching, the population growth of pastoralists (cattle and goat herders), and

***The Serengeti*** **A large herd of wildebeest and zebra graze on the lush grasses, with *Acacia* woodlands in the distance. Inset shows a map of the Serengeti ecosystem and the migration route taken by the large herds.**

mechanized agriculture, but the Kenyan and Tanzanian governments appear to be committed to maintaining the core national parks (which consist of 14,700 $km^2$).

You will learn in this chapter how ecosystems like the Serengeti work, starting at the fundamental level of chemicals and energy. You will also learn how natural ecosystems sustain human life, and you will look at an estimate of what ecosystem goods and services are worth to humankind. Studying how ecosystems work will give you some insight into why natural systems are sustainable and may suggest ways to make our human system more sustainable.

## 3.1 Matter, Energy, and Life

### Matter in Living and Nonliving Systems

*Atoms.* The basic building blocks of all **matter** (all gases, liquids, and solids in both living and nonliving systems) are **atoms.** Only 94 different kinds of atoms occur in nature, and these are known as the naturally occurring **elements.** In addition, chemists and physicists have created 21 more in the laboratory, but they are so unstable that they break down into simpler elements. (See Table C–1, p. 670.)

How can these relatively few *building blocks* make up the countless materials of our world, including the tissues of living things? Picture each kind of atom as a different-sized Lego® block. Like Legos, atoms can build a great variety of things. Also like Legos, natural materials can be taken apart into their separate constituent atoms, and the atoms can then be reassembled into different materials. All chemical reactions, whether they occur in a test tube, in the environment, or inside living things, and whether they occur very slowly or very fast, involve rearrangements of atoms to form different kinds of matter.

Atoms do not change during the disassembly and reassembly of different materials. A carbon atom, for instance, will always remain a carbon atom. Furthermore, atoms are neither created nor destroyed during chemical reactions. The same number and kind of different atoms exist before and after any reaction. This constancy of atoms is regarded as a fundamental natural law, the *law of conservation of matter.*

A more detailed discussion of atoms—how they differ from one another, how they bond to form various gases, liquids, and solids, and how chemical formulas can be used to describe different chemicals—is given in Appendix C (p. 669). Studying Appendix C now may help you better understand the material we are about to cover.

How are atoms put together? Which atoms make up living organisms? Where are those atoms found in the environment? How do they become part of living organisms? Answers to these questions are presented next.

*Molecules and Compounds.* A **molecule** consists of two or more atoms bonded together in a specific way. The properties of a material depend on the specific way in which atoms are bonded to form molecules, as well as on the atoms themselves. Similarly, a **compound** consists of two or more *different kinds* of atoms bonded together. A *molecule*, therefore, may consist of two or more of the *same kind* of atoms, or two or more different kinds of atoms, bonded together, whereas at least two *different kinds* of atoms are always involved in a *compound.* For example, the fundamental units of oxygen gas, which consists of two oxygen atoms bonded together, are molecules, but not a compound. Water, by contrast, is both a molecule and a compound, because the fundamental units are two hydrogen atoms bonded to an oxygen atom.

On the chemical level, then, the cycle of growth, reproduction, death, and decay of organisms is a continuous process of taking various atoms from the environment (food), assembling them into living organisms (growth), disassembling them (decay), and repeating the process. Driving the cycle is the genetically programmed urge living things have to grow and reproduce.

**Four Spheres.** During growth and decay, atoms move from the environment into living things and then return to the environment. To picture this process, think of the environment as three open systems, or "spheres," occupied by living things—the **biosphere** (Fig. 3–1). The **lithosphere** is Earth's crust, made up of rocks and minerals. The **hydrosphere** is water in all of its liquid and solid compartments: oceans, rivers, ice, and groundwater. The **atmosphere** is the thin layer of gases (including water vapor) separating Earth from outer space. Matter is constantly being exchanged within and between these four spheres.

*Key Elements.* Living things are characterized by six key elements: **carbon (C), hydrogen (H), oxygen (O), nitrogen (N), phosphorus (P),** and **sulfur (S).** These six elements are the essential ones in the organic molecules that make up the tissues of plants, animals, and microbes. By looking at the chemical nature of the spheres, you can see where the six key elements and others occur in the environment (Table 3–1).

*Atmosphere.* The lower atmosphere is a mixture of molecules of three important gases—oxygen ($O_2$), nitrogen ($N_2$), and carbon dioxide ($CO_2$)—along with water vapor and trace amounts of several other gases that have no immediate biological importance (Fig. 3–2). The gases in the atmosphere are normally stable, but under some circumstances they react chemically to form new compounds (for example, ozone is produced from oxygen in the upper atmosphere, as described in Chapter 20).

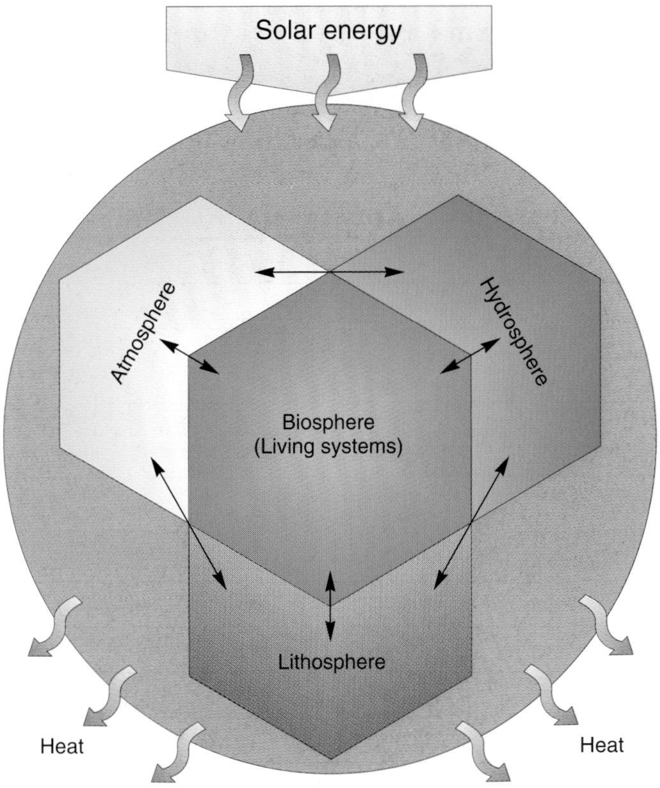


**Figure 3–1 The four spheres of Earth's environment.** The biosphere is all of life on Earth. It depends on, and interacts with, the atmosphere (air), the hydrosphere (water), and the lithosphere (soil and rocks). Adapted from *Geosystems,* 5e by Robert W. Christopherson. Copyright 2003 by Pearson Education, Inc.

***Hydrosphere.*** While the atmosphere is a major source of *carbon* and *oxygen* for all organisms (and a source of *nitrogen* for a few of them), the hydrosphere is the source of *hydrogen*. Each molecule of water consists of two hydrogen atoms bonded to an oxygen atom, so the chemical formula for water is $H_2O$. A weak attraction known as *hydrogen bonding* exists between water molecules. At temperatures below freezing, hydrogen bonding holds the molecules in position with respect to one another, and the result is a solid (ice or snow). At temperatures above freezing, but below vaporization, hydrogen bonding still holds the molecules close, but allows them to move past one another, producing the liquid state. Vaporization occurs as hydrogen bonds break and water molecules move into the air independently. As temperatures are lowered again, all of these changes of state go in the reverse direction (Fig. 3–3). Despite the changes of state, the water molecules themselves retain their basic chemical structure of two hydrogen atoms bonded to an oxygen atom. Only the *relationship* between the molecules changes.

***Lithosphere.*** All the other elements required by living organisms, as well as the 72 or so elements that are not required by them, are found in the lithosphere, in the form of rock and soil minerals. A **mineral** is any hard, crystalline, inorganic material of a given chemical composition. Most rocks are made up of relatively small crystals of two or more minerals, and soil generally consists of particles of many different minerals. Each mineral is made up of dense clusters of two or more kinds of atoms bonded together by an attraction between positive and negative charges on the atoms, as explained in Appendix C and shown in Fig. 3–4.

***Interactions.*** Air, water, and minerals interact with each other in a simple, but significant, manner. Gases from the air and ions (charged atoms) from minerals may dissolve in water. Therefore, natural water is inevitably a *solution* containing variable amounts of dissolved gases and minerals. This solution is constantly subject to change, because any dissolved substances in it may be removed by various processes, or additional materials may dissolve in it. Molecules of water enter the air by evaporation and leave it again via condensation and precipitation. (See the hydrologic cycle, Chapter 7.) Thus, the amount of moisture in the air fluctuates constantly. Wind may carry dust or mineral particles, but the amount changes constantly, because the particles gradually settle out from the air. The various interactions are summarized in Fig. 3–5.

**Organic Compounds.** The chemical compounds making up the tissues of living organisms are referred to as **organic.** Unlike the relatively simple molecules that occur in the environment (such as $CO_2$, $H_2O$, and $N_2$), the key chemical elements in living organisms (C, H, O, N, P, S) bond to form very large, complex organic molecules, such as proteins, carbohydrates (sugars and starches), lipids (fatty substances), and nucleic acids (DNA and RNA). Some of these molecules may contain millions of atoms, and their potential diversity is infinite. Indeed, the diversity of living things reflects the diversity of these molecules.

The molecules that make up the tissues of living things are constructed mainly from carbon atoms bonded together into chains, with hydrogen and oxygen atoms attached. Nitrogen, phosphorus, and sulfur may be present

**table 3-1 Elements Found in Living Organisms and the Locations of Those Elements in the Environment**

| Element (Kind of Atom) | Biologically Important Molecule or Ion in Which the Element Occurs[a] | | | Location in the Environment[b] | | |
|---|---|---|---|---|---|---|
| | Symbol | Name | Formula | Atmosphere | Hydrosphere | Lithosphere |
| Carbon | C | Carbon dioxide | $CO_2$ | X | X | X($CO_3^-$) |
| Hydrogen | H | Water | $H_2O$ | X | (Water itself) | |
| Atomic oxygen (required in respiration) | O | Oxygen gas | $O_2$ | X | X | |
| Molecular oxygen (released in photosynthesis) | $O_2$ | Water | $H_2O$ | | (Water itself) | |
| Nitrogen | N | Nitrogen gas<br>Ammonium ion<br>Nitrate ion | $N_2$<br>$NH_4^+$<br>$NO_3^-$ | X | X<br>X<br>X | Via fixation<br>X<br>X |
| Sulfur | S | Sulfate ion | $SO_4^{2-}$ | | X | X |
| Phosphorus | P | Phosphate ion | $PO_4^{3-}$ | | X | X |
| Potassium | K | Potassium ion | $K^+$ | | X | X |
| Calcium | Ca | Calcium ion | $Ca^{2+}$ | | X | X |
| Magnesium | Mg | Magnesium ion | $Mg^{2+}$ | | X | X |
| **Trace Elements**[c] | | | | | | |
| Iron | Fe | Iron ion | $Fe^{2+}$, $Fe^{3+}$ | | X | X |
| Manganese | Mn | Manganese ion | $Mn^{2+}$ | | X | X |
| Boron | B | Boron ion | $B^{3+}$ | | X | X |
| Zinc | Zn | Zinc ion | $Zn^{2+}$ | | X | X |
| Copper | Cu | Copper ion | $Cu^{2+}$ | | X | X |
| Molybdenum | Mo | Molybdenum ion | $Mo^{2+}$ | | X | X |
| Chlorine | Cl | Chloride ion | $Cl^-$ | | X | X |

Note: These elements are found in *all* living organisms—plants, animals, and microbes. Some organisms require certain elements in addition to the ones listed. For example, humans require sodium and iodine.

[a]A molecule is a chemical unit of two or more atoms bonded together. An ion is a single atom or group of bonded atoms that has acquired a positive or negative charge as indicated.

[b]"X" means that element exists in indicated "sphere."

[c]Only small or trace amounts of these elements are required.

also, but the key common denominator is carbon–carbon and carbon–hydrogen or carbon–oxygen bonds (Fig. 3–6). Hence, the carbon-based molecules that make up the tissues of living organisms are called *organic molecules.* (The similarity between the words *organic* and *organism* is deliberate, not coincidental.) **Inorganic,** then, refers to all other molecules or compounds—that is, those with neither carbon–carbon nor carbon–hydrogen bonds.

All plastics and countless other human-made compounds are based on carbon–carbon bonding and are, chemically speaking, organic compounds. To resolve any confusion this may cause, the compounds making up living organisms are referred to as **natural organic compounds** and the human-made ones as **synthetic organic compounds.**

In conclusion, the elements essential to life (C, H, O, and so on) are present in the atmosphere, hydrosphere, or lithosphere in relatively simple molecules. In living *organisms* of the biosphere, on the other hand, they are *organized* into highly complex *organ*ic compounds. These organic compounds in turn make up the various parts of cells, which in their turn make up the tissues and organs of the organism, which in its turn is part of a population (Fig. 3–7). During growth and reproduction, then, the atoms from simple molecules in the environment are used to construct the complex organic molecules of an organism. Decomposition and decay is the reverse process. Each of these processes is discussed in more detail later in the chapter; first, however, we must consider another factor: *energy.*

**Clean, dry air is a mixture of molecules of three important gases.**

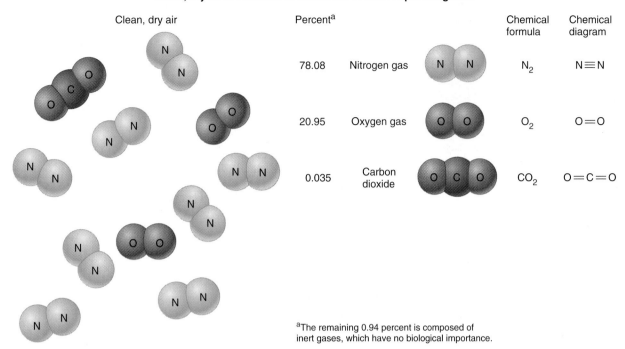

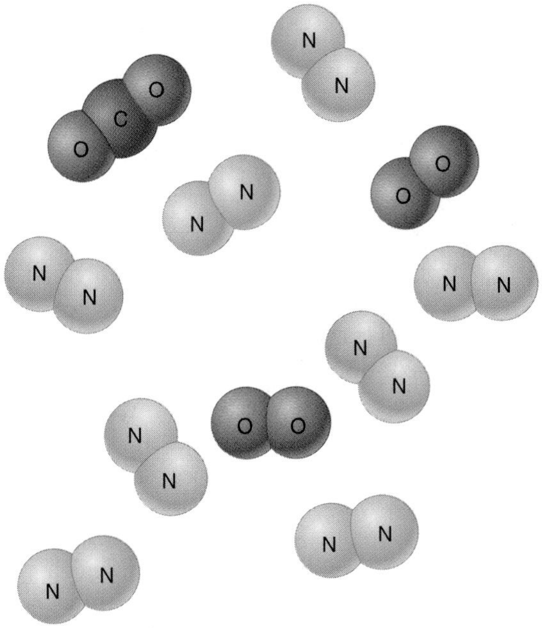


| Percent[a] | | Chemical formula | Chemical diagram |
|---|---|---|---|
| 78.08 | Nitrogen gas | $N_2$ | N≡N |
| 20.95 | Oxygen gas | $O_2$ | O=O |
| 0.035 | Carbon dioxide | $CO_2$ | O=C=O |

[a]The remaining 0.94 percent is composed of inert gases, which have no biological importance.

**Figure 3–2 The major gases of clean, dry air.** From a biological point of view, the three most important gases of the lower atmosphere are nitrogen, oxygen, and carbon dioxide. (Note that the proportion of carbon dioxide is deliberately overrepresented in the diagram.)

## Energy Basics

In addition to rearranging atoms, chemical reactions absorb or release energy. To grasp this concept, you must be able to distinguish between matter and energy.

**Matter and Energy.** The universe is made up of *matter* and *energy*. A more technical definition of **matter** than the one given earlier in this chapter is *anything that occupies space and has mass*—that is, anything that can be weighed when gravity is present. This definition

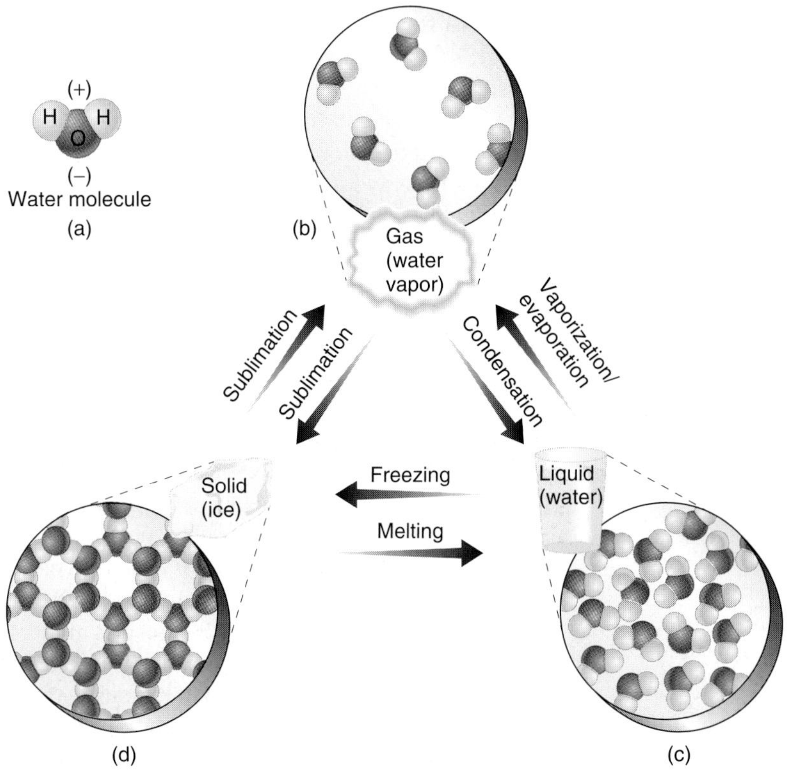


**Figure 3–3 Water and its three states.** (a) Water consists of molecules, each of which is formed when two hydrogen atoms bond to an oxygen atom ($H_2O$). (b) In water vapor, the molecules are separate and independent. (c) In liquid water, the weak attraction between water molecules known as hydrogen bonding gives the water its liquid property. (d) At freezing temperatures, hydrogen bonding holds the molecules firmly, giving the solid state—ice.

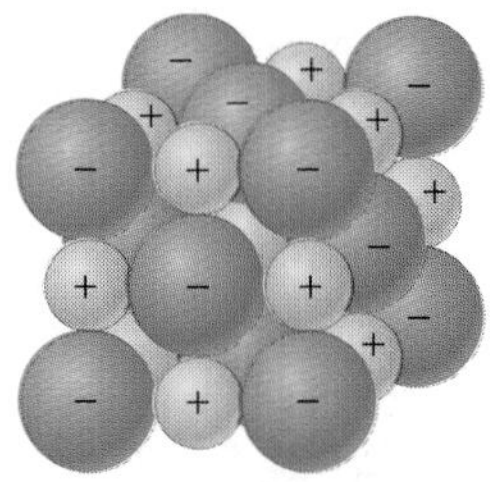

**Figure 3–4 Minerals.** Minerals (hard crystalline compounds) are composed of dense clusters of atoms of two or more elements. The atoms of most elements gain or lose one or more electrons, becoming negative (−) or positive (+) ions. Salt (sodium chloride, NaCl) is held together by the attraction between the positive sodium (Na+) and negative chloride (Cl−) charges.

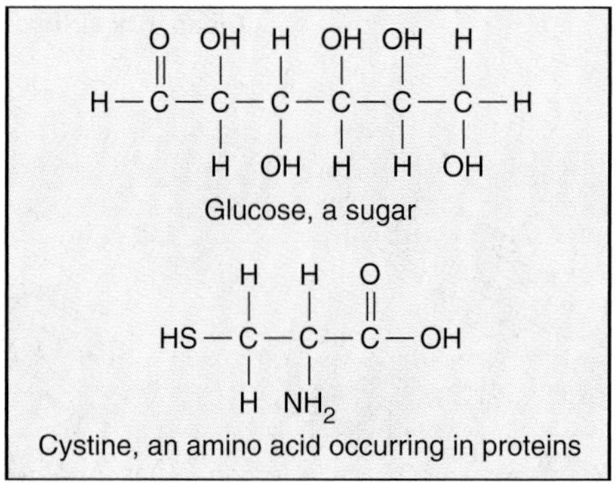

Glucose, a sugar
Cystine, an amino acid occurring in proteins


**Figure 3–6 Organic molecules.** The organic molecules that make up living organisms are larger and more complex than the inorganic molecules found in the environment. Glucose, a sugar, and cysteine, an amino acid, show this relative complexity.

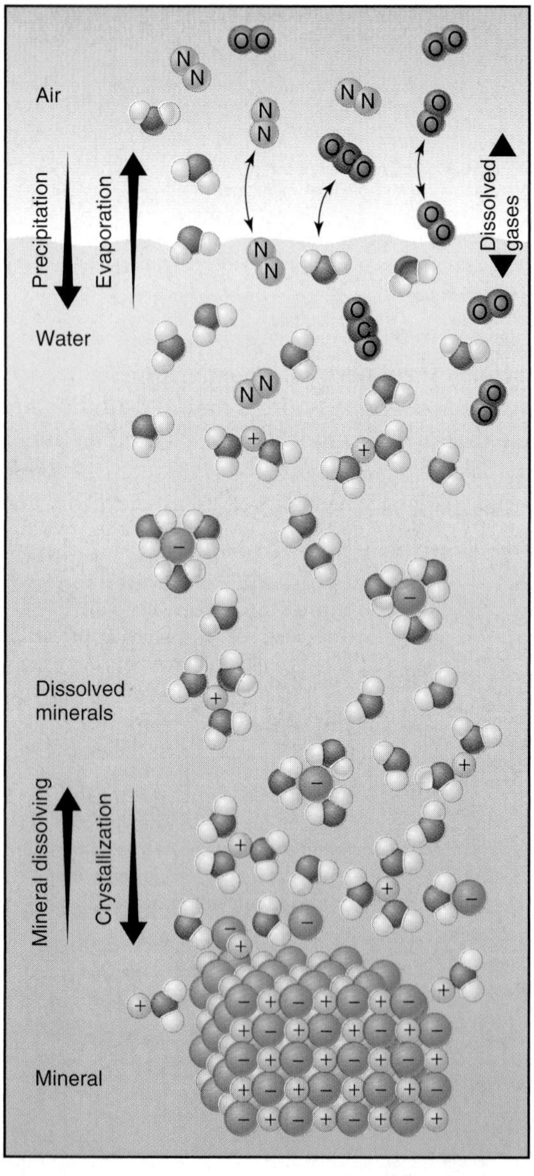


**Figure 3–5 Interrelationship among air, water, and minerals.** Minerals and gases dissolve in water, forming solutions. Water evaporates into air, causing humidity. These processes are all reversible: Minerals in solution recrystallize, and water vapor in the air condenses to form liquid water.

covers all solids, liquids, and gases, and living as well as nonliving things.

Atoms are made up of protons, neutrons, and electrons, which in turn are made of still smaller particles. Because atoms are the basic units of all elements and remain unchanged during chemical reactions, they can be treated as the basic units of matter.

In contrast to matter, *light, heat, movement*, and *electricity* do not have mass, nor do they occupy space. (Note that *heat*, as used here, refers not to a hot object, but to the heat energy you can feel radiating from the hot object.) These are the common forms of energy with which you are probably familiar. What do the various forms of energy have in common? They *affect* matter, causing changes in its *position* or its *state*. For example, the release of energy in an explosion causes things to go flying—a change in position. Heating water causes it to boil and change to steam, a change of state. On a molecular level, changes of state are actually movements of atoms or molecules. For instance, the degree of heat energy contained in a substance is a measure of the relative vibrational motion of the atoms and molecules of the substance. Therefore, we can define **energy** as *the ability to move matter*.

***Kinetic and Potential Energy.*** Energy can be categorized as either *kinetic* or *potential* (Fig. 3–8). **Kinetic energy** is *energy in action or motion*. Light, heat energy, physical motion, and electrical current are all forms of kinetic energy. **Potential energy** is *energy in storage*. A substance or system with potential energy has the capacity, or *potential*, to release one or more forms of kinetic energy. A stretched rubber band has potential energy; it can send a paper clip flying. Numerous chemicals, such as gasoline and other fuels, release kinetic energy—heat energy, light, and movement—when ignited. The potential energy contained in such chemicals and fuels is called **chemical energy.**

Energy may be changed from one form to another in innumerable ways (Fig. 3–9). Besides understanding that potential energy can be converted to kinetic energy, it is especially important to recognize that kinetic energy can be converted to potential energy. (Consider, for example, charging a battery or pumping water into

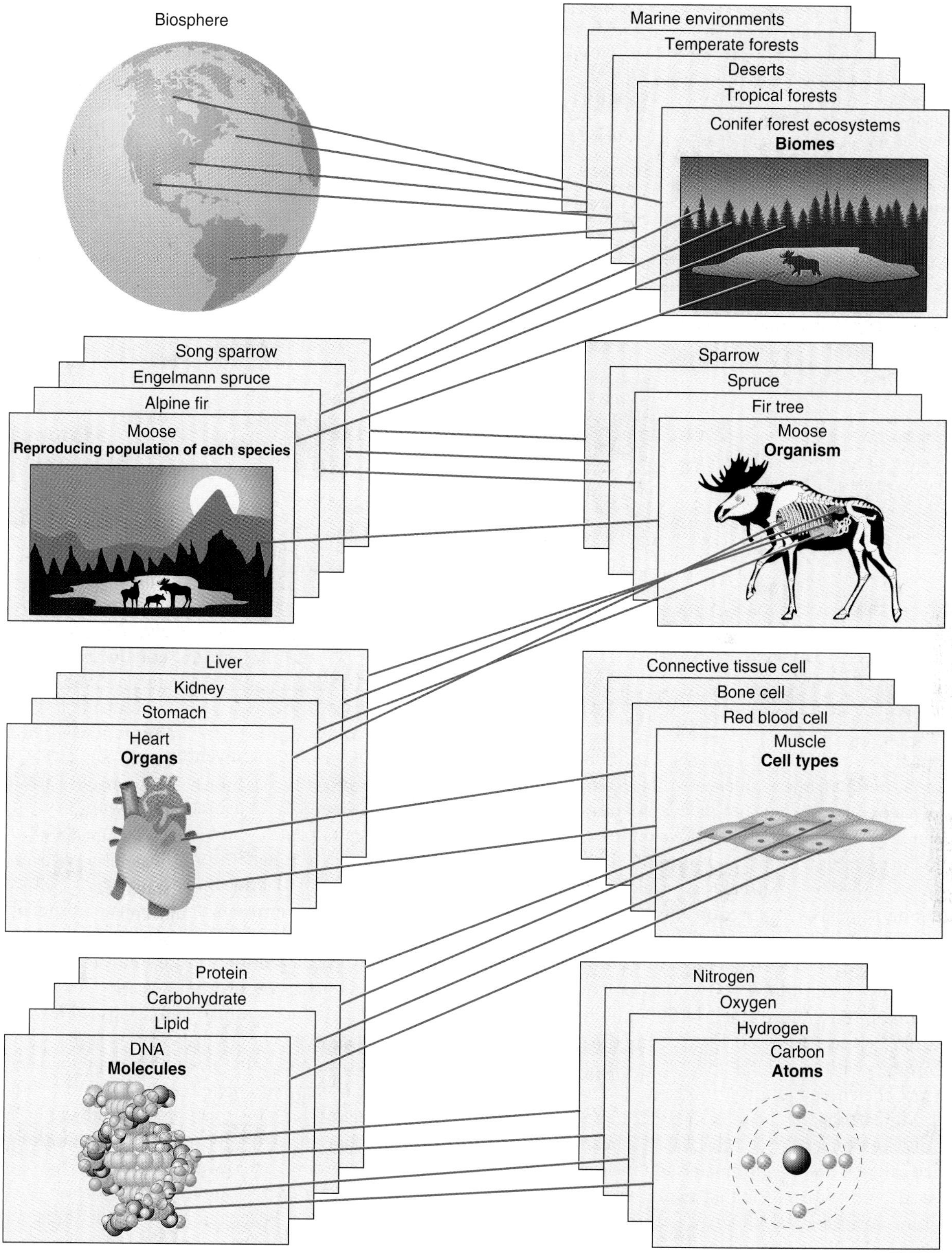


**Figure 3–7 Life as a hierarchy of organization of matter.** In the inorganic sphere, elements are arranged simply in molecules of the air, water, and minerals. In living organisms, they are arranged in complex organic molecules, which in turn make up cells that constitute tissues, organs, and, thus, the whole organism. Levels of organization continue up through populations, species, ecosystems, and, finally, the whole biosphere.

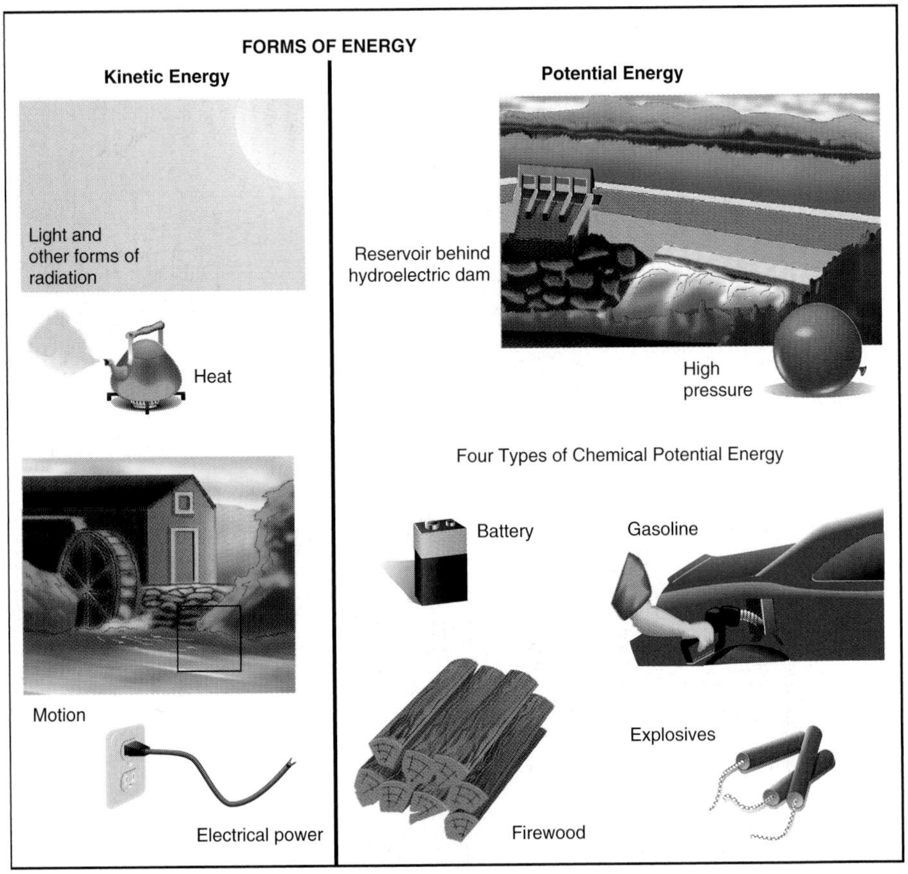


**Figure 3–8 Forms of energy.** Energy is distinct from matter in that it neither has mass nor occupies space. It has the ability to act on matter, though, changing the position or the state of the matter. Kinetic energy is energy in one of its active forms. Potential energy is the potential that systems or materials have to release kinetic energy.

a high-elevation reservoir.) We shall see later in this section that photosynthesis does just that.

Because energy does not have mass or occupy space, it cannot be measured in units of weight or volume, but it can be measured in other kinds of units. One of the most common units is the **calorie,** which is defined as the amount of heat required to raise the temperature of 1 gram (1 milliliter) of water 1 degree Celsius. This is a very small unit, so it is frequently more convenient to use kilocalories (1 kilocalorie = 1,000 calories), the amount of heat required to raise 1 liter (1,000 milliliters) of water 1 degree Celsius. (Kilocalories are sometimes denoted as "Calories" with a capital "C." Food Calories, which measure the energy in given foods, are actually kilocalories.) Many forms of chemical energy can be measured in calories by converting a substance to heat energy in a device called a bomb calorimeter and measuring the heat released by the corresponding rise in temperature. Temperature measures the molecular motion in a substance caused by the kinetic energy present in it.

If energy is defined as the ability to move matter, then no matter can be moved *without* the absorption or release of energy. Indeed, no *change* in matter—from a few atoms coming together or apart in a chemical reaction to a major volcanic eruption—can be separated from its respective change in energy.

**Energy Laws: Laws of Thermodynamics.** Because energy can be converted from one form to another, numerous would-be inventors over the years have tried to build machines or devices that would produce more energy than they consumed. A common idea is to use the output from a generator to drive a motor that, in turn, drives the generator to keep the cycle going and yields additional power in the bargain. Unfortunately, all such devices have one feature in common: They don't work. When all the inputs and outputs of energy are carefully measured, they are found to be *equal*. There is no net gain or loss in total energy. This observation is now accepted as a fundamental natural law, **the law of conservation of energy.** It is also called the **first law of thermodynamics,** and it can be expressed as follows: *Energy is neither created nor destroyed, but may be converted from one form to another.* This law really means that you can't get something for nothing.

Imaginative "energy generators" fail for two reasons: First, in every energy conversion, a portion of the energy is converted to heat energy (thermal infrared). Second, there is no way of trapping and recycling heat energy without expending even more energy in doing so. Consequently, in the absence of energy inputs, any and every system will sooner or later come to a stop as its energy is converted to heat and lost. This is now accepted as another natural law, the **second law of thermodynamics,** and it can be expressed as follows: *In any energy conversion, some of the usable energy is always lost.* Thus, you can't get something for nothing (the first law) and, in fact, you can't even break even (the second law)!

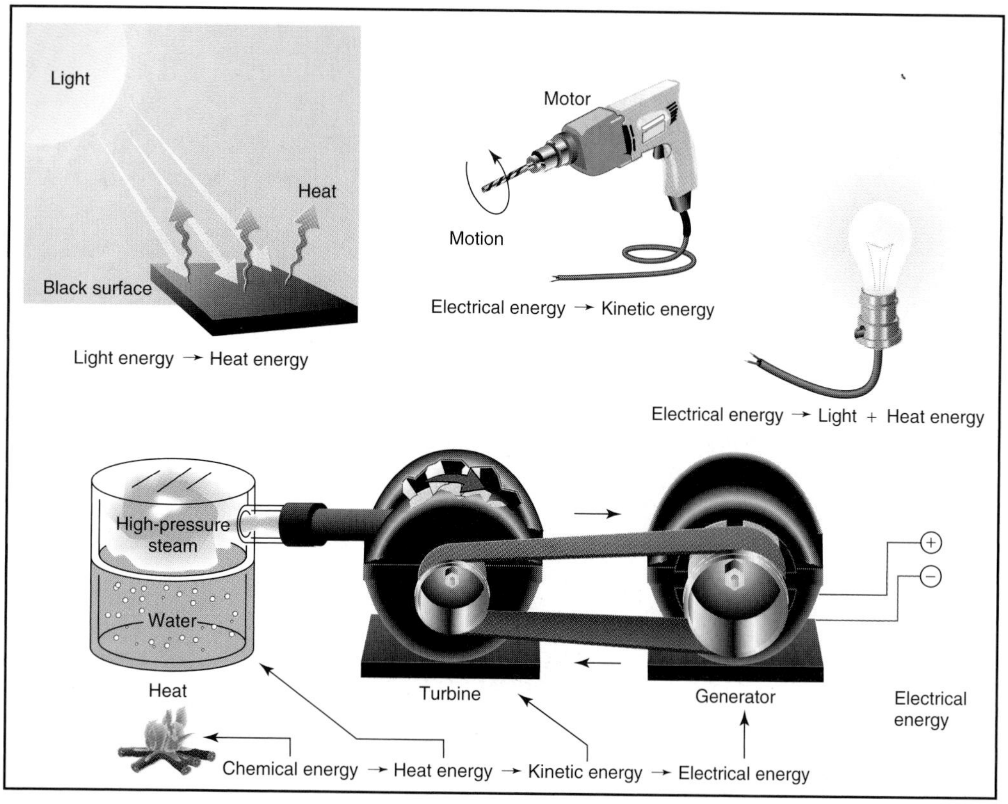


**Figure 3–9 Energy conversions.** Any form of energy except heat energy can spontaneously transform into any other form. Heat is a form of energy that flows from one system or object to another because the two are at different temperatures; therefore, heat can spontaneously transform only to something cooler.

*Entropy.* Underlying the loss of usable energy to heat is the principle of increasing *entropy*. **Entropy** is a measure of the degree of disorder in a system, so increasing entropy means increasing disorder. Without energy inputs, everything goes in one direction only—toward increasing entropy. This principle of ever-increasing entropy is the reason that all human-made things tend to deteriorate. The increasing disorder of your dormitory room as the semester wears on is one example of entropy.

The conversion of energy and the loss of usable energy to heat are both aspects of increasing entropy. Heat energy is the result of the random vibrational motion of atoms and molecules. Thus, it is the lowest (most disordered) form of energy, and its spontaneous flow to cooler surroundings is a way for that disorder to spread. Therefore, the second law of thermodynamics may be more generally stated as follows: *Systems will go spontaneously in one direction only—toward increasing entropy*. The second law also says that systems will go spontaneously only toward *lower* potential energy, a direction that releases heat from the systems (Fig. 3–10).

The word *spontaneously* is very important in this statement of the second law. It is possible to pump water uphill, charge a battery, stretch a rubber band, compress air, or otherwise increase the potential energy of a system. The verbs *pump*, *charge*, *stretch*, and *compress* tell us, however, that energy is being put into the system. In contrast, flow in the opposite direction, which releases energy, occurs spontaneously (Fig. 3–11a).

Whenever you see something gaining potential energy, therefore, keep in mind that the energy is being obtained from somewhere else (the first law). Moreover, the amount of energy lost from that "somewhere else" is greater than the amount gained (the second law). Let us now relate these concepts of matter and energy to organic molecules, organisms, ecosystems, and the biosphere.

## Energy Changes in Organisms

All organic molecules, which make up the tissues of living organisms, contain *high potential energy*. When these molecules are burned, the heat and light of the flame are the potential energy being released as kinetic energy. By contrast, try as you might, you will not be able to get energy by burning inorganic compounds like carbon dioxide, water, or rock-based minerals. (Some minerals, such as magnesium and sulfur, can be burned.) Indeed, many of these materials are used as fire extinguishers. They are nonflammable because they have very *low potential energy*. Thus, the production of organic material from inorganic material represents a *gain* in potential energy. Conversely, the breakdown of organic matter *releases* energy.

This relationship between the formation and breakdown of organic matter on the one hand, and the gain

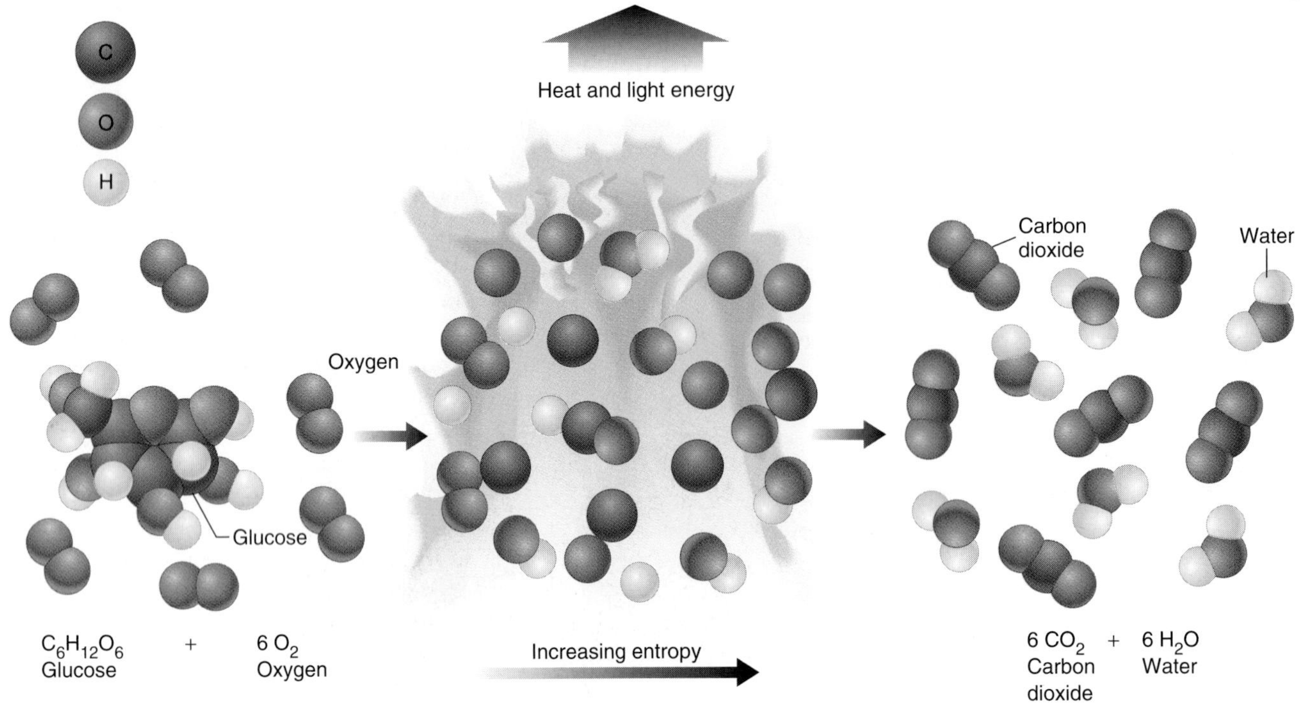


**Figure 3–10 Entropy.** Systems go spontaneously only in the direction of increasing entropy. When glucose, a major constituent of wood, is burned, heat is released, and the atoms become more disordered. Both of these phenomena are aspects of increasing entropy.

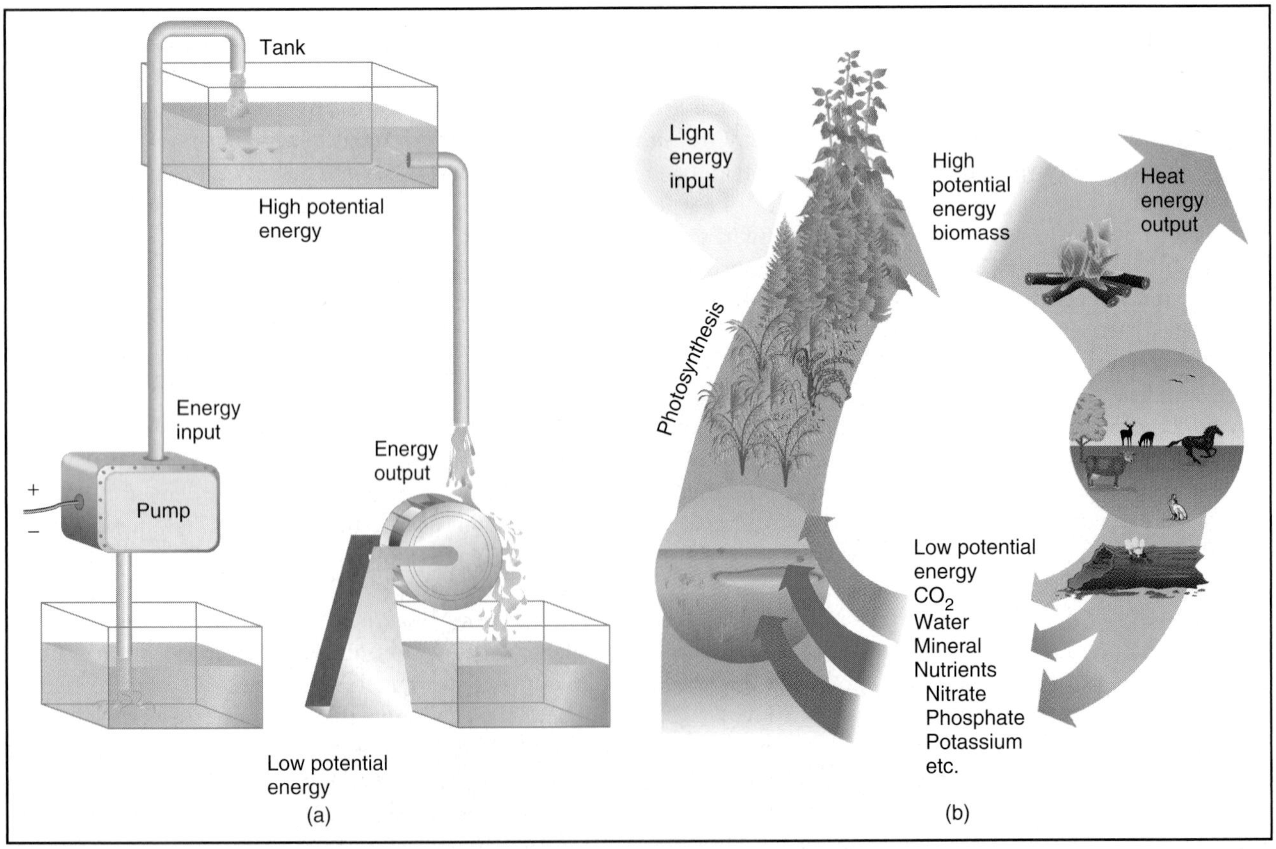


**Figure 3–11 Storage and release of potential energy.** (a) A simple physical example of the storage and release of potential energy. (b) The same principle applied to ecosystems.

and release of energy on the other, forms the basis of the *energy dynamics of ecosystems*. *Producers* (green plants) make high-potential-energy organic molecules for their needs from low-potential-energy raw materials in the environment—namely, carbon dioxide, water, and a few dissolved compounds of nitrogen, phosphorus, and other elements. This "uphill" conversion is possible because producers use chlorophyll to absorb light energy. On the other hand, all *consumers*, *detritus feeders*, and *decomposers* obtain energy for movement and growth from feeding on and breaking down organic matter made by producers (Fig. 3–11b).

**Producers and Photosynthesis.** Recall from Chapter 2 that producers (green plants) use *photosynthesis* to make sugar (glucose, stored chemical energy) from carbon dioxide, water, and light energy. This process, which also releases oxygen gas as a by-product, is described by the following chemical equation:

**Photosynthesis**
(An energy-demanding process)

$$\underset{\substack{\text{carbon dioxide}\\\text{(gas)}}}{6\,CO_2} + \underset{\text{water}}{12\,H_2O} \xrightarrow[\substack{\uparrow\\\text{light energy}\\\text{input}}]{} \underset{\text{glucose}}{C_6H_{12}O_6} + \underset{\substack{\text{oxygen}\\\text{(gas)}}}{6\,O_2} + \underset{\text{water}}{6\,H_2O}$$

(low potential energy) (high potential energy)

Chlorophyll in the cells of the plant absorbs the kinetic energy of light and uses it to remove the hydrogen atoms from water ($H_2O$) molecules. The hydrogen atoms combine with carbon atoms from carbon dioxide to form a growing chain of carbons that eventually turns into a glucose molecule. After the hydrogen atoms are removed from water, the oxygen atoms that remain combine with each other to form oxygen gas, which is released into the air. Water appears on both sides of the equation because 12 molecules are consumed and 6 molecules are newly formed during photosynthesis.

The key energy steps in photosynthesis remove the hydrogen from water molecules and join carbon atoms together to form the carbon–carbon and carbon–hydrogen bonds of glucose. These steps convert the low-potential-energy bonds in water and carbon dioxide molecules to the high-potential-energy bonds of glucose. The laws of thermodynamics are not violated in this process because much more energy goes into the reaction than comes out. Careful measurements show that the rate of photosynthesis (which determines the amount of glucose formed) is related to the intensity of light, but seems terribly inefficient: At best, 2 calories of sugar are formed for each 100 calories of light energy falling on the plant (2% efficiency).

***Within the Plant.*** The glucose produced in photosynthesis serves three purposes in the plant: (1) Either by itself or combined with nitrogen, phosphorus, sulfur, and other mineral nutrients absorbed by the plant's roots, glucose is the raw material used for making all the other organic molecules (proteins, carbohydrates, and so on) that make up the stem, roots, leaves, flowers, and fruit of the plant. (2) The synthesis of all these organic molecules requires additional energy, as do the plant's absorption of nutrients from the soil and certain other functions. This energy is obtained when the plant breaks down a portion of the glucose to release its stored energy in a process called *cell respiration*, which is discussed later. (3) A portion of the glucose produced may be stored for future use. For storage, the glucose is generally converted to starch, as in potatoes, or to oils, as in seeds. These conversions are summarized in Fig. 3–12.

None of these reactions, from the initial capture of light by chlorophyll to the synthesis of plant structures,

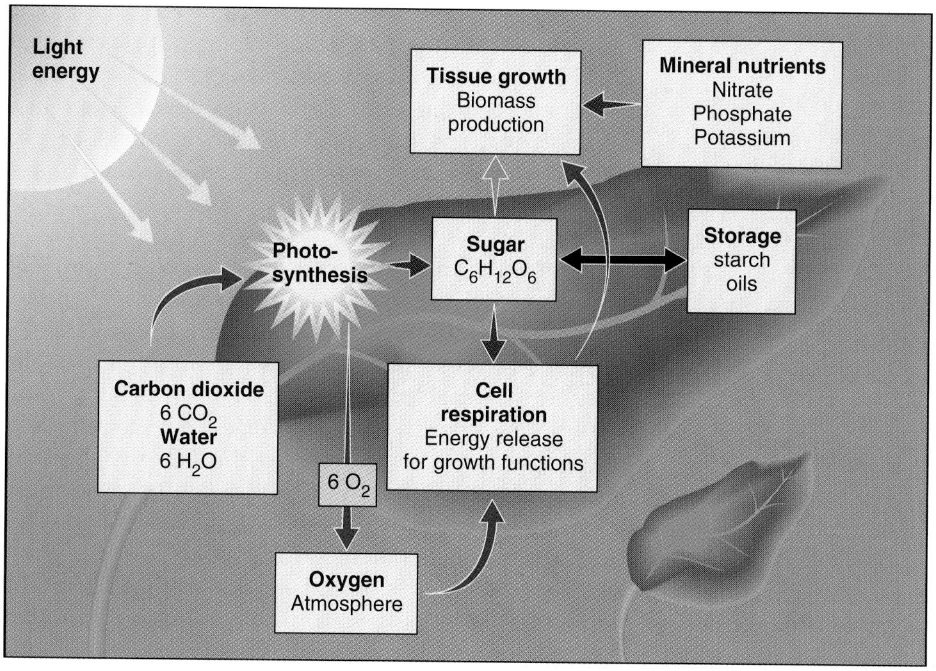


**Figure 3–12 Producers as chemical factories.** Using light energy, producers make glucose from carbon dioxide and water, releasing oxygen as a by-product. Breaking down some of the glucose to provide additional chemical energy, they combine the remaining glucose with certain nutrients from the soil to form other complex organic molecules that the plant then uses for growth.

take place automatically. Each step is catalyzed by specific **enzymes,** *proteins that promote the synthesis or breaking of chemical bonds.* The same is true of cell respiration.

*Setting the Table.* As the plants in an ecosystem convert sunlight into new organic matter, they are "setting the table" for the rest of the ecosystem—the herbivores, carnivores, and decomposers described in the food webs in Chapter 2. Because the plants are creating *new* organic matter for the ecosystem, they are the *primary* producers. Given suitable conditions and resources, the producers of an ecosystem will maintain their photosynthetic activity over time in the process called *primary production.* The total amount of photosynthetic activity in producers is called *gross primary production*; subtracting the energy consumed by the plants themselves yields the *net primary production.* Thus, net primary production is the *rate* at which new organic matter is made available to consumers in an ecosystem. In the Serengeti, the grasses in the plains have to produce at the rate of 560 kg dry weight per $km^2$ per day in order to keep up with the rate at which they are being grazed.

**Consumers.** Consumers need energy to move about and to perform such internal functions as pumping blood. In addition, consumers need energy to synthesize all the molecules required for growth, maintenance, and repair of their bodies. This energy comes from the breakdown of organic molecules in food (or from the body's own tissues if food is unavailable). Between 60 and 90% of the food that we and other consumers eat and digest acts as "fuel" to provide energy.

*Digestion.* First, the starches, fats, and proteins that you eat are digested—broken down into simpler molecules—in the stomach or intestine. Starches are broken down into sugar (glucose), for example. These simpler molecules are then absorbed from the intestine into the bloodstream and transported to the body's individual cells.

*Respiration.* Inside each cell, organic molecules may be broken down through a process called **cell respiration** to release the energy required for the work done by that cell. Most commonly, cell respiration involves the breakdown of glucose, and the overall chemical equation is basically the reverse of that for photosynthesis:

Cell Respiration
(An energy-releasing process)

$$\underset{\substack{\text{glucose}\\\text{(high potential energy)}}}{C_6H_{12}O_6} + \underset{\text{oxygen}}{6\,O_2} \rightarrow \underset{\text{carbon dioxide}}{6\,CO_2} + \underset{\substack{\text{water}\\\text{(low potential energy)}}}{6\,H_2O} + \text{energy}$$

The purpose of cell respiration is to release the potential energy contained in organic molecules to perform the activities of the organism. Note that *oxygen* is *released* in photosynthesis, but *consumed* in cell respiration to break down glucose to carbon dioxide and water. Oxygen is absorbed through the lungs with every inhalation (or through the gills, in the case of fish) and is transported to all the body's cells via the circulatory system. Carbon dioxide, which is formed as a waste product, moves from the cells into the circulatory system and is eliminated through the lungs (or gills) with every exhalation.

In keeping with the second law of thermodynamics, converting the potential energy of glucose to the energy required to do the body's work is not 100% efficient. Considerable waste heat is produced, and this is the source of *body heat.* This heat output can be measured in animals (cold-blooded or warm-blooded) and in plants. It is more noticeable in warm-blooded animals only because much of their body heat is used to maintain their body temperature.

*Gaining Weight.* The basis of weight gain or loss becomes apparent here. Organic matter is broken down in cell respiration only as it is needed to meet the energy demands of the body. This is why your breathing rate, the outer reflection of cell respiration, varies with changes in your level of exercise and activity. If you consume more calories from food than your body needs, the excess may be converted to fat and stored, and the result is a gain in weight. In contrast, the principle of dieting is to eat less and exercise more, to create an energy demand that exceeds the amount of energy contained in your food. This imbalance forces the body to break down its own tissues to make up the difference, and the result is a weight loss. Carried to an extreme, such an imbalance leads to *starvation* and even death when the body runs out of anything expendable to break down for its energy needs.

*Oxidation.* The overall reaction for cell respiration is the same as that for simply burning glucose. Thus, it is not uncommon to speak of "burning" our food for energy. Such a breakdown of molecules is also called **oxidation.** The distinction between burning and cell respiration is that in cell respiration the oxidation takes place in about 20 small steps, each catalyzed by a specific enzyme. The energy is released in small "packets" that can be captured to drive the functions of each cell. If all the energy from glucose molecules were released in a single "bang," as occurs in burning, it would be like heating and lighting a room with large firecrackers—energy indeed, but hardly useful.

**The Fate of Food.** Whereas 60–90% of the food that consumers eat, digest, and absorb is oxidized for energy, the remaining 10–40%, which is converted to the body tissues of the consumer, is no less important. This is the fraction that enables the body to grow, maintain, and repair itself. A portion of what is ingested by consumers is not digested, but simply passes through the digestive system and out as fecal wastes. For consumers that eat plants, this waste is largely **cellulose,** the material of plant cell walls. It is often referred to as *fiber, bulk,* or *roughage,* and some of it is a necessary part of the diet. The intestines need to push some fiber through them so that they can keep clean and open. Waste products can also

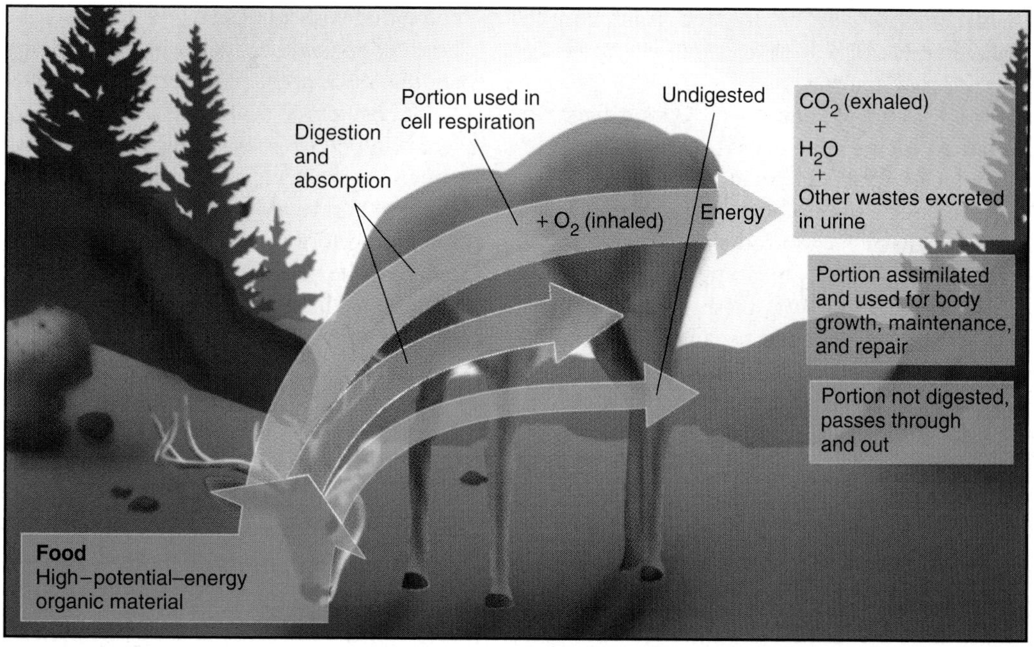


**Figure 3–13** **Consumers.** Only a small portion of the food ingested by a consumer is assimilated into body growth, maintenance, and repair. A larger amount is used in cell respiration to provide energy; waste products are carbon dioxide, water, and various mineral nutrients. A third portion is not digested and becomes fecal waste.

include compounds of nitrogen, phosphorus, and any other elements present, in addition to the usual carbon dioxide and water. These by-products are excreted in the urine (or as similar waste in other kinds of animals) and returned to the environment.

In sum, organic material (food) eaten by any consumer follows one of three pathways: (1) More than 60% of what is digested and absorbed is oxidized to provide energy, and waste products are released back to the environment; (2) the remainder of what is digested and absorbed goes into body growth, maintenance and repair, or storage (fat); and (3) the portion that is not digested or absorbed passes out as fecal waste (Fig. 3–13). In an ecosystem, therefore, only that portion of the food which becomes the body tissue of the consumer can become food for the next organism in the food chain. This process is often referred to as *secondary* production, and like primary production, it also can be expressed as a rate (amount of growth of the consumer, or consumer trophic level) over time.

**Detritus Feeders and Decomposers: The Detritivores.** Detritus is largely cellulose because it consists mostly of dead leaves, the woody parts of plants, and animal fecal wastes. Nevertheless, it is still organic and high in potential energy for those organisms which can digest it—namely, the decomposers described in Chapter 2. Beyond having the ability to digest cellulose, decomposers (various species of fungi and bacteria, as well as a few other microbes) act as any other consumer, using the cellulose as a source of both energy and nutrients. Termites and some other detritus feeders can digest woody material because they maintain decomposer microorganisms in their guts in a mutualistic symbiotic relationship. The termite (a detritus feeder) provides a cozy home for the microbes (decomposers) and takes in the cellulose, which the microbes digest for both their own and the termites' benefit (Fig. 3–14).

Most decomposers use oxygen for cell respiration, which breaks the detritus down into carbon dioxide, water, and mineral nutrients. Likewise, there is a release of waste heat, which you may observe as the "steaming" of a manure or compost pile on a cold day. The release of

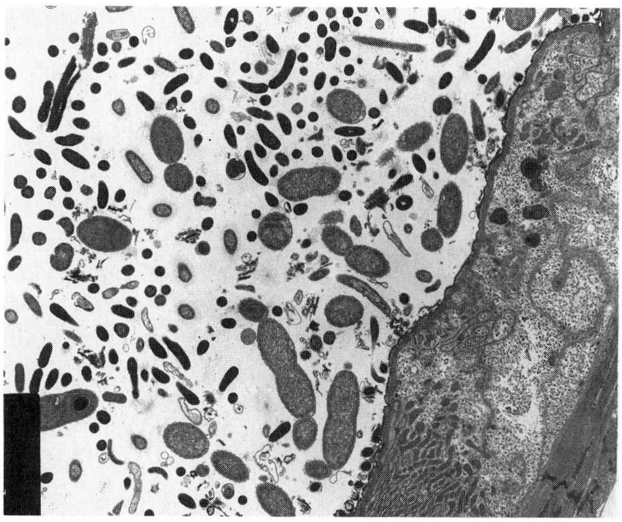

**Figure 3–14** **Termite gut.** Termites can live on cellulose-based woody matter because their guts contain a consortium of symbiotic microbes able to digest cellulose. In this electron micrograph the symbionts are on the left, and the termite intestinal wall on the right.

nutrients by decomposers is vitally important to the primary producers, because it is the major source of nutrients in most ecosystems.

*Fermentation.* Some decomposers (certain bacteria and yeasts) can meet their energy needs through the partial breakdown of glucose that can occur in the absence of oxygen. This modified form of cell respiration, called **fermentation,** results in such end products as ethyl alcohol ($C_2H_6O$), methane gas ($CH_4$), and acetic acid ($C_2H_4O_2$). The commercial production of these compounds is achieved by growing the particular organism on suitable organic matter in a vessel without oxygen. In nature, **anaerobic,** or *oxygen-free*, environments commonly exist in the sediments of lakes, marshes, or swamps and in the guts of animals, where oxygen does not penetrate readily. Methane gas is commonly produced in these locations. A number of large grazing animals, including cattle, maintain fermenting bacteria in their digestive systems in a mutualistic, symbiotic relationship similar to that just described for termites. As a result, both cattle and termites produce methane.

For simplicity, the focus of this chapter is on terrestrial ecosystems. Keep in mind, though, that exactly the same processes occur in aquatic ecosystems. As aquatic plants and algae absorb dissolved carbon dioxide and mineral nutrients from the water, they use photosynthesis to produce food and dissolved oxygen that sustain consumers and other heterotrophs. Likewise, aquatic heterotrophs return carbon dioxide and mineral nutrients to the aquatic environment. Aquatic and terrestrial systems are never entirely isolated from one another, and they exchange materials all the time.

To summarize, the different biotic components of ecosystems function on the basis of two common processes: (a) the *flow of energy*, using sunlight as the basic energy source, and (b) the *cycling of nutrients.* These processes are examined in more depth at the ecosystem level in Sections 3.2 and 3.3.

## 3.2 Energy Flow in Ecosystems

### Primary Production

In most ecosystems, sunlight, or solar energy, is the initial source of energy absorbed by producers through the process of photosynthesis. (The only exceptions are ecosystems near the ocean floor, where the producers are chemosynthetic bacteria.) As mentioned in Section 3.1, primary production captures only about 2%, at most, of incoming solar energy. Even though this seems like a small fraction, the resulting net production—estimated at some 115 billion tons of organic matter per year—is enough to fuel all of life on Earth. In a given ecosystem, the actual biomass of primary producers at any given time is referred to as the *standing-crop biomass.* Both biomass and primary production vary greatly in different ecosystems. For example, a forested ecosystem maintains a very large biomass compared with a tropical grassland, yet the rate of primary production could be higher in the grassland, where animals continually graze newly produced organic matter.

*Ecosystems Compared.* The productivity of different types of ecosystems (e.g., terrestrial biomes and aquatic ecosystems) has been examined to evaluate their contribution to global productivity and to investigate why some are more productive than others. Figure 3–15 presents (a) the average net primary productivity, (b) the percentage of different ecosystems over Earth's surface, and, subsequently, (c) the percentage of global net primary productivity attributed to 19 of the most important ecosystems. Some key relationships between abiotic factors and specific ecosystems can be seen in the data. Tropical rain forests are both highly productive and contribute considerably to global productivity; they cover a large area of the land and are characterized by ideal climatic conditions for photosynthesis—warm temperatures and abundant rainfall. The open oceans cover 65% of Earth's surface, so they account for a large portion of global productivity, yet their actual rate of production is low enough that they are veritable biological deserts. Although light, temperature, and water are abundant, primary production in the oceans is limited by the scarcity of nutrients—a good lesson in the significance of limiting factors (Chapter 2). (See Global Perspective, p. 70). The seasonal effects of differences in latitude can also be seen by comparing productivity in tropical, temperate, and boreal (coniferous) forests.

### Energy Flow and Efficiency

As primary producers are consumed by herbivores, energy is transferred from producer to consumer. Recall from Chapter 2, Section 2.2, that each of these components is a *trophic level.* Thus, energy flow in an ecosystem can be characterized by how the energy moves from one trophic level to another. Figure 3–16 shows how energy flows through three trophic levels of a grazing food web. At each trophic level, some energy goes into growth (production), some is converted to heat (respiration), and some is given off as waste or is not consumed. As energy flows from one trophic level to the next, only a small fraction is actually passed on. This is due to three things: (1) Much of the preceding trophic level is standing biomass and is not consumed; (2) much of what is consumed is used for energy; and (3) some of what is consumed is undigested and passes through the organism.

Figure 3–16 shows that a very large proportion of the primary-producer trophic level is not consumed in the grazing food web. As this material dies (leaves drop, grasses wither and die, and so on), it is joined by the fecal wastes and dead bodies from higher trophic levels and represents the starting point for a separate food web, the detritus food web, pictured earlier in Fig. 2–10. In many cases, the majority of the energy in an ecosystem flows through the detritus food web.

*Efficiency.* Because energy is lost when it is transferred to the next higher trophic level, each successive

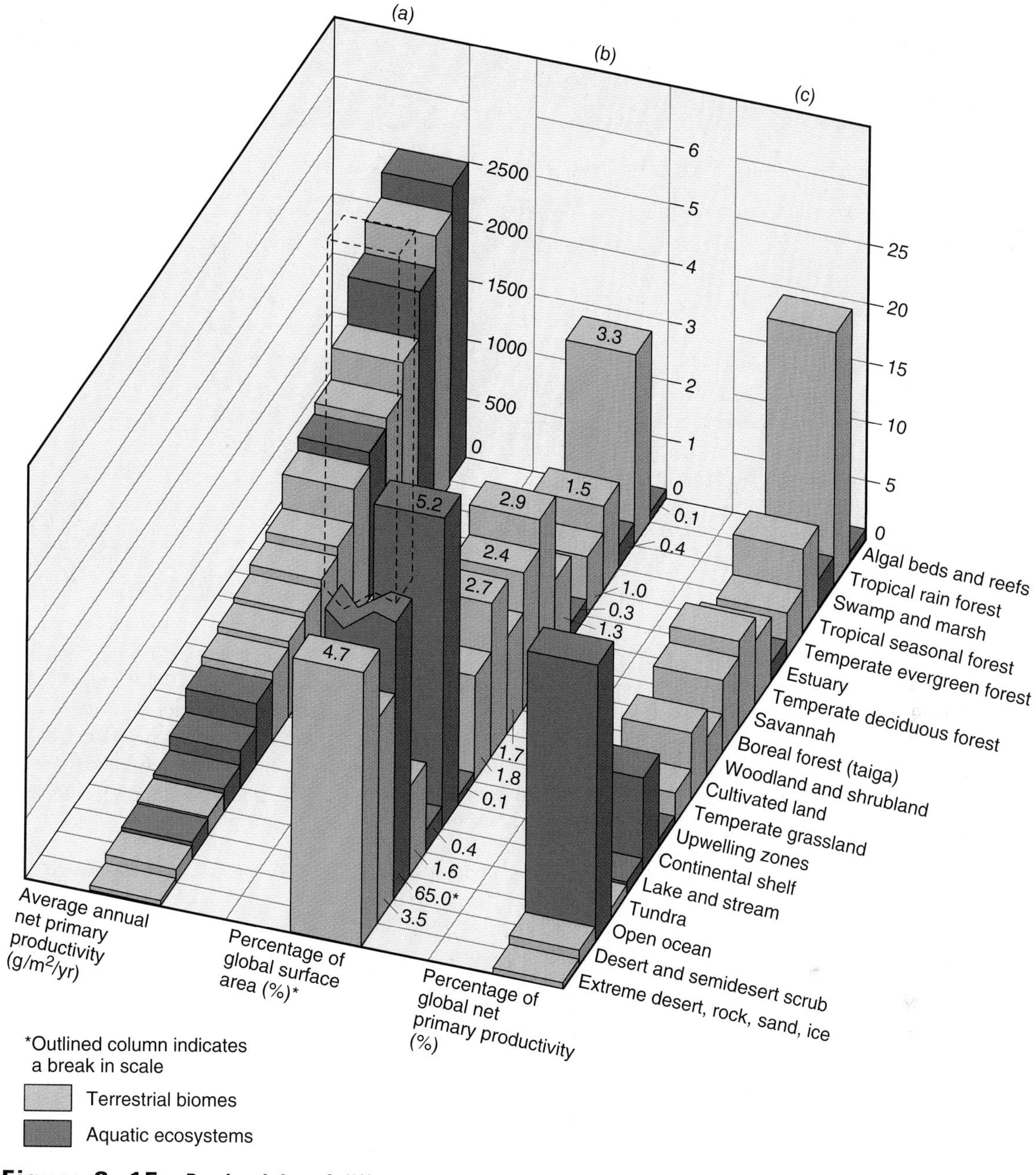


**Figure 3–15 Productivity of different ecosystems.** (a) The annual net primary productivity of different ecosystems; (b) the percentage of different ecosystems over Earth's surface area; and (c) the percentage of global net primary productivity.

trophic level captures only a fraction of the energy that entered the previous trophic level and is usually represented by a much smaller biomass. Calculations show that the efficiency of transfer in a number of ecosystems ranges from 5 to 20%, with 10% being the average. Thus, there is an approximate 90% *loss* of energy as it moves from one trophic level to the next. This loss gets quite critical at increasingly higher trophic levels and is the reason carnivores are much less abundant than herbivores, carnivores that eat other carnivores are even less abundant, and so forth. In any given ecosystem, therefore, there are usually only three to five trophic levels; there simply isn't enough energy left to pass along to "supercarnivores."

What happens to all the solar energy entering ecosystems? Most of it is absorbed by the atmosphere, oceans, and land, thus heating them in the process. The small fraction (2–5%) captured by living plants is either passed along to the next trophic level or degraded into the lowest and most disordered form of energy—heat—as the plant decomposes. Eventually, all of the energy entering ecosystems escapes as heat. According to the laws of thermodynamics, no energy will actually be lost. So many energy conversions are taking place in ecosystem trophic activities, however, that entropy is increased and all the energy is degraded to a form unavailable to do further work. The ultimate result is that *energy flows in*

## global perspective

### Light and Nutrients: The Controlling Factors in Marine Ecosystems

Ecosystems run on solar energy and they recycle nutrients. The major limiting factors in marine ecosystems are the availabilities of light and nutrients. Light diminishes as water depth increases, because even clear water absorbs light. The layer of water from the surface down to the greatest depth at which there is adequate light for photosynthesis is known as the **euphotic zone**. Below the euphotic zone, photosynthesis does not occur. In clear water, the euphotic zone may be as deep as 600 feet (180 m), but in turbid (cloudy) water, it may be only a few centimeters. In coastal waters, where the euphotic zone extends to the bottom, the bottom may support abundant plant life in the form of aquatic vegetation attached to or rooted in sediments. If the euphotic zone does not extend to the bottom, the bottom will be barren of plant life.

Whether shallow or deep, the euphotic zone supports a diverse ecosystem. Phytoplankton—algae and photosynthetic bacteria that grow as single cells or in small groups of cells—can maintain themselves close to the surface in the euphotic zone. Phytoplankton support a diverse food web, from the zooplankton (small crustaceans and protozoans) that feed on them to many species of fish and sea mammals (whales and porpoises) at the higher trophic levels.

An entire ecosystem can also operate in the cold, dark depths below the euphotic layer. This ecosystem is nourished by detritus raining down from above and, closer to the ocean floor, by vents and fissures that produce mineral-rich water and warmth.

In a phytoplankton-based system, nutrients dissolved in the water become critically important. If the water contains too few dissolved nutrients such as phosphorus or nitrogen compounds, the growth of phytoplankton and, hence, the rest of the ecosystem will be limited. If the bottom receives light, it may support vegetation despite nutrient-poor water, because this kind of vegetation draws nutrients from the bottom sediments. Indeed, in some estuaries, nutrient-rich water inhibits the growth of bottom vegetation, because the dissolved nutrients support the growth of phytoplankton instead, which makes the water turbid and shades out the bottom vegetation.

The most productive areas of the ocean—the areas supporting the most abundant marine life of all sorts—are generally found within 200 miles (320 km) of shore. Either the bottom is within the euphotic zone and thus supports abundant vegetation, or nutrients washing in from the land support an abundant primary production of phytoplankton.

In the open ocean, there is less marine life the further you move from shore. Indeed, marine biologists consider most of the open ocean to be a "biological desert." Life is scarce here because the bottom is well below the euphotic zone and the water is nutrient poor.

When plants and animals die in the ocean, they sink to the bottom. Nutrients carried to the bottom with this settling detritus are released into solution by decomposers, primarily bacteria. This nutrient-rich bottom water is carried along by deep-running ocean currents. Where the currents hit underwater mountains or continental rims, the nutrient-rich water is forced to the surface. Phytoplankton flourish in these areas of **upwelling** (rising) and thus support a rich diversity of fish and marine mammals. As a result, the world's oceans are far from being uniformly stocked with fish. By far the richest marine fishing areas are continental shelves and regions of upwelling, as shown on the accompanying map of the North Atlantic. The yellow and green areas overlie Georges Bank and the Gulf of Maine and show the effects of nutrients from upwelling and local estuaries.

*a one-way direction through ecosystems*; it is not recycled, so it must be continually resupplied by sunlight.

## Running on Solar Energy

No system can run without an input of energy, and living systems are no exception. For all major ecosystems, both terrestrial and aquatic, the initial source of energy is *sunlight*. As a basic energy source, sunlight is highly sustainable because it is both *nonpolluting* and *nondepletable*.

***Nonpolluting.*** Light from the Sun is a form of pure energy; it contains no substance that can pollute the environment. All the matter and pollution involved in the production of light energy are conveniently left behind on the Sun some 93 million miles (150 million kilometers) away in space.

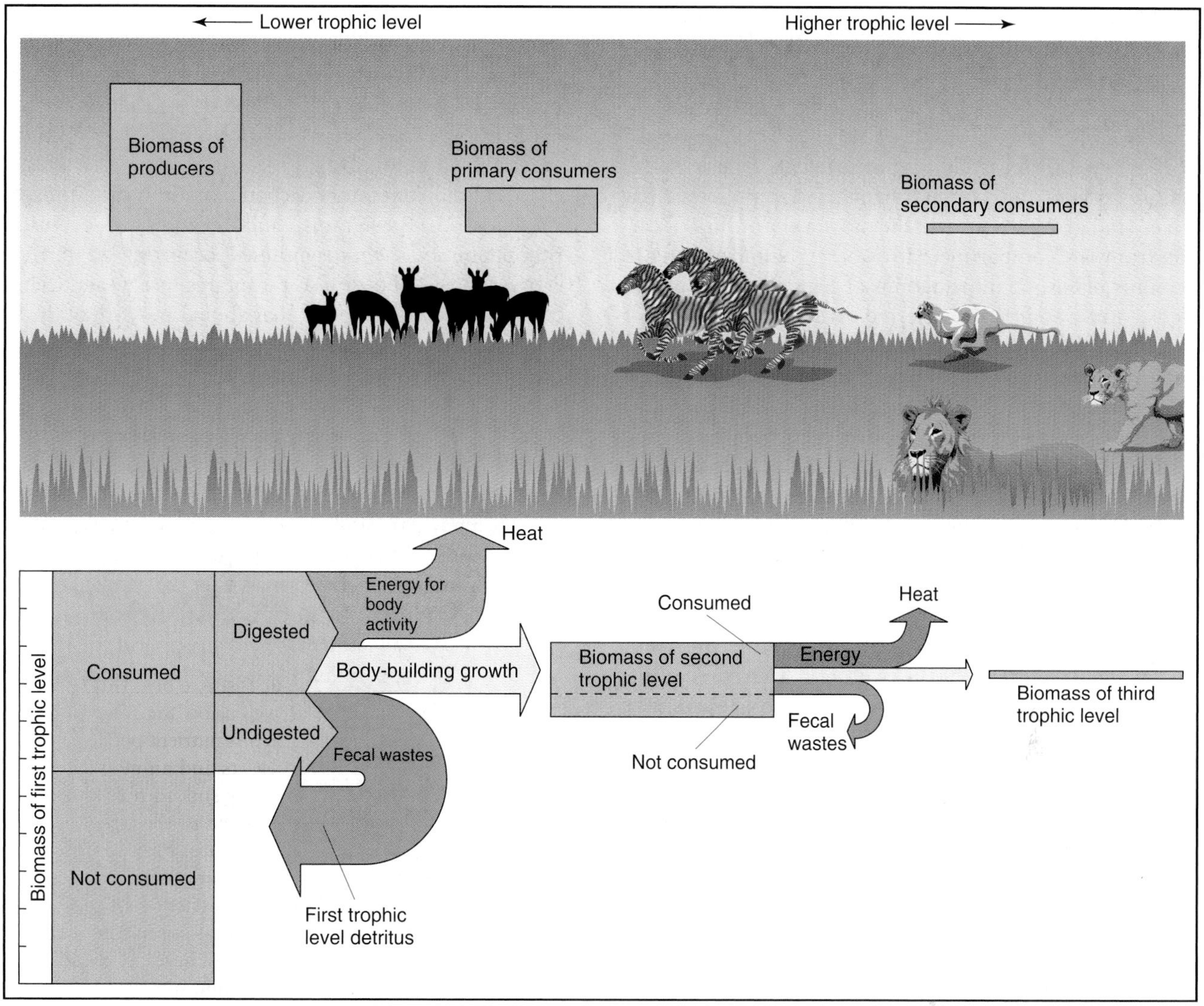


**Figure 3–16** **Energy flow through trophic levels in a grazing food web.** Each trophic level is represented as biomass boxes, and the pathways taken by the energy flow are indicated with arrows.

*Nondepletable.* The Sun's energy output is remarkably constant; the radiant energy from the sun (called the solar constant) strikes Earth's atmosphere at about 2 calories per $cm^2$ per minute. How much or how little of this energy is used on Earth will not influence, much less deplete, the Sun's output. Even though cosmologists predict that the Sun will expire some day (a few billion years from now!), for all practical purposes the Sun is an everlasting source of energy.

Energy flow is one of the two fundamental processes that make ecosystems work. The second process is the cycling of nutrients and other elements.

## 3.3 The Cycling of Matter in Ecosystems

The various inputs and outputs of producers, consumers, detritus feeders, and decomposers fit together remarkably well. The products and by-products of each group are the food or essential nutrients for the other. Specifically, the organic material and oxygen produced by green plants are the food and oxygen required by consumers and other heterotrophs. In turn, the carbon dioxide and other wastes generated when heterotrophs break down their food are exactly the nutrients needed by green plants. This kind of recycling is fundamental to sustainability, for two reasons: (a) It prevents the accumulation of wastes that would cause problems (for example, ammonia is excreted by many animals and is toxic at relatively low concentrations), and (b) it guarantees that the ecosystem will not run out of essential elements.

According to the law of conservation of matter, atoms cannot be created, destroyed, or changed, so recycling is the only possible way to maintain a dynamic system. To see how well the biosphere has mastered recycling, we now focus on the pathways of three key elements heavily affected by human activities: carbon, phosphorus, and nitrogen. Because these pathways all lead in circles and involve biological, geological, and chemical processes, they are known as **biogeochemical cycles.** (Recall that energy is not recycled.)

## The Carbon Cycle

The global carbon cycle is illustrated in Figure 3–17; major "pools" or compartments are represented by boxes, and arrows represent the movement or flux of carbon from one compartment to another. Numbers refer to amounts of carbon in petagrams (1 petagram = 1 billion metric tons, sometimes referred to as 1 gigaton). For descriptive purposes, it is convenient to start the carbon cycle with the "reservoir" of carbon dioxide ($CO_2$) molecules present in the air and bicarbonate ($HCO_3^-$) molecules present in water. Through photosynthesis and further metabolism, carbon atoms from $CO_2$ become the carbon atoms of the organic molecules making up a plant's body. The carbon atoms then move into food webs and become part of the tissues of all the other organisms in the ecosystem. At any point, a given atom may be respired and returned to the atmosphere; in aquatic systems, it will be returned to the inorganic carbonate in solution. Processes other than trophic transfer are significant. The figure indicates two in particular: (1) geological sedimentation and burial on the land and under the ocean and limestone formation of carbon in the ocean, and (2) weathering of calcium carbonate and combustion of fossil-fuel carbon laid down millions of years ago by biological systems. Notice how all four "spheres" are involved in this cycle.

The total amount of carbon dioxide in the atmosphere is about 750 petagrams, and the amount of primary production (via photosynthesis) occurring in terrestrial ecosystems is about 121 petagrams per year. Scientists have concluded, therefore, that about one-sixth of the total atmospheric carbon dioxide is taken up in photosynthesis in a year, but an equal amount is returned to the atmosphere, through cell respiration. This means that, on the average, a carbon atom cycles from the atmosphere, through one or more living things, and back to the atmosphere every six years.

Human intrusion into the carbon cycle is significant: We are diverting or canceling out 40% of terrestrial primary production in order to support human enterprises. Indeed, by burning fossil fuels and destroying forests, we have increased atmospheric carbon dioxide by 35% over preindustrial levels, a topic discussed in Chapter 20.

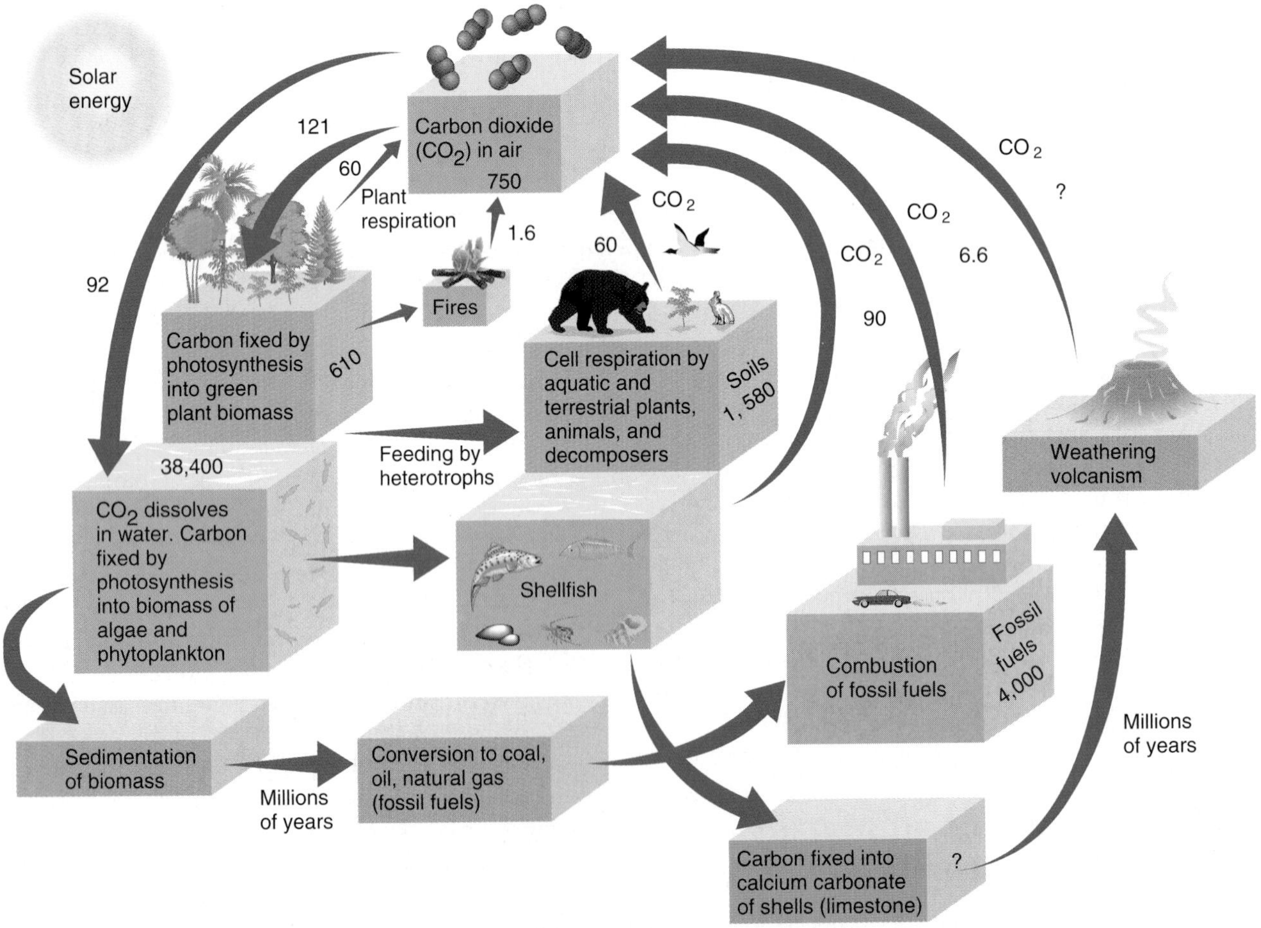


**Figure 3–17 The global carbon cycle.** Boxes in the figure refer to pools of carbon, and arrows refer to the movement, or flux, of carbon from one pool to another. Numbers are recorded in petagrams of carbon (1 petagram = 1 billion metric tons).

## The Phosphorus Cycle

The phosphorus cycle is representative of the cycles of all the biologically important mineral nutrients—those elements which have their origin in the rock and soil minerals of the lithosphere. (See Table 3–1.) We focus on phosphorus because its shortage tends to be a limiting factor in a number of ecosystems and its excess can seriously stimulate unwanted algal growth in freshwater systems.

The phosphorus cycle is illustrated in Fig. 3–18. Like the carbon cycle, it is depicted as a set of pools and fluxes to indicate key processes. Phosphorus exists in various rock and soil minerals as the inorganic ion *phosphate* ($PO_4^{3-}$). As rock gradually breaks down, phosphate and other ions are released. This slow process is the normal means of replenishing phosphorus that is lost to runoff. Plants absorb $PO_4^{3-}$ from the soil or from a water solution, and once the phosphate is incorporated into organic compounds by the plant, it is referred to as **organic phosphate.** Moving through food chains, organic phosphate is transferred from producers to the rest of the ecosystem. As with carbon, at each step it is highly likely that the organic compounds containing phosphate will be broken down in cell respiration or by decomposers, releasing $PO_4^{3-}$ in urine or other waste material. The phosphate may then be reabsorbed by plants to start the cycle again.

Phosphorus enters into complex chemical reactions with other substances that are not shown in this simplified version of the cycle. For example, $PO_4^{3-}$ forms insoluble chemical precipitates with a number of cations (positively charged ions), such as iron ($Fe^{3+}$), aluminum ($Al^{3+}$), and calcium ($Ca^{3+}$). If these cations are in sufficiently high concentration in soil or aquatic systems, the phosphorus can be bound up in chemical precipitates and rendered largely unavailable to plants. The precipitated phosphorus can slowly release $PO_4^{3-}$ as plants withdraw naturally occurring $PO_4^{3-}$ from soil, water, or sediments.

There is an important difference between the carbon cycle and the phosphorus cycle. No matter where $CO_2$ is released, it will mix into and maintain the concentration of $CO_2$ in the atmosphere. Phosphorus, however, does not have a gas phase, so it is recycled only if the wastes containing it are deposited in the *ecosystem from which it came*. The same holds true for other mineral nutrients. In natural ecosystems, wastes (urine, detritus) are deposited in the same area, so recycling occurs efficiently. As mentioned earlier, the rich growth of grasses on the plains of the Serengeti is traced to

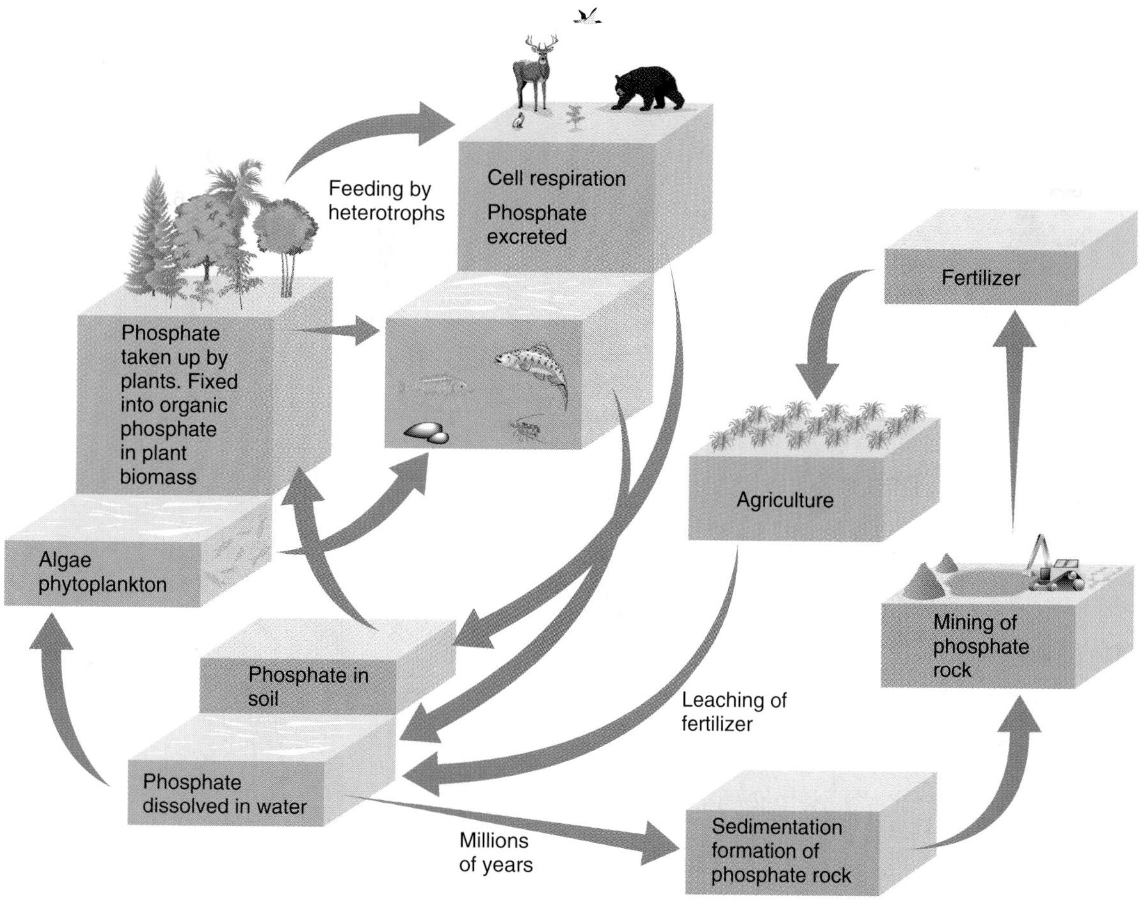


**Figure 3–18** **The global phosphorus cycle.**

phosphorus brought and maintained there by the dense herbivore herds. Humans have been extremely prone to interrupt this cycle, however.

***Human Impacts.*** The most serious intrusion into the phosphorus cycle comes from the use of phosphorus-containing fertilizers. Phosphorus is mined in several locations around the world (in the United States, Florida is a prominent source) and is then made into fertilizers, animal feeds, detergents, and other products. Much of the phosphorus applied to agricultural croplands and lawns makes its way into waterways—either directly, in runoff from the land, or indirectly, in sewage effluents. There is essentially no way to return this waterborne phosphorus to the soil, so the bodies of water end up overfertilized. This leads, in turn, to a severe water pollution problem known as eutrophication. (See Chapter 17.)

When we use manure, compost (rotted plant wastes), or sewage sludge (instead of chemical fertilizer) on crops, lawns, or gardens, the natural cycle is imitated. In too many cases, however, it is not. Human applications have almost quadrupled the amount of phosphorus annually entering soils and surface waters (normally around 3.5 teragrams per year [1 teragram = 1 million metric tons], to the present 13 teragrams per year). Therefore, we are accelerating the natural phosphorus cycle as we mine it from the earth and as it subsequently moves from the soil into aquatic ecosystems.

## The Nitrogen Cycle

The nitrogen cycle (Fig. 3–19) has aspects of both the carbon cycle and the phosphorus cycle. Like carbon, nitrogen possesses a gas phase; like phosphorus, it acts as a limiting factor. The nitrogen cycle is otherwise unique. Most notably, bacteria in soils, water, and sediments perform many of the steps of the cycle. Like phosphorus, nitrogen is in high demand by both aquatic and terrestrial plants.

The main reservoir of nitrogen is the air, which is about 78% nitrogen gas ($N_2$). Plants cannot use nitrogen gas directly from the air. Instead, the nitrogen must be in mineral form, such as ammonium ions ($NH_4^+$) or nitrate ions ($NO_3^-$). Beginning with the uptake of nitrates by green plants, nitrogen is incorporated into essential organic compounds such as proteins and nucleic acids.

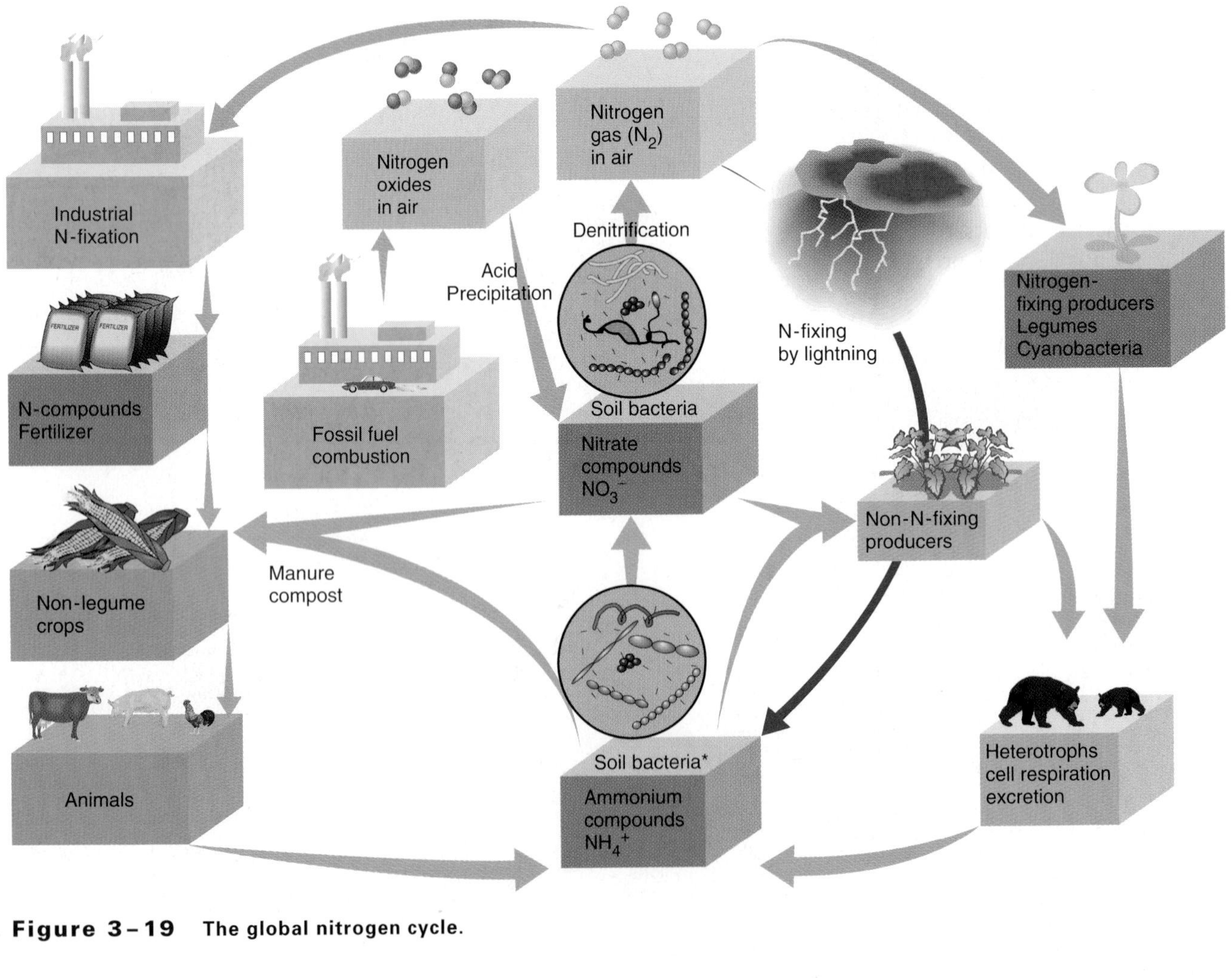


**Figure 3–19** **The global nitrogen cycle.**

The nitrogen then follows the classic energy-flow pattern from producers to herbivores to carnivores and, finally, to decomposers (referred to as heterotrophs in Fig. 3–19). At various points, nitrogen wastes are released, primarily as ammonium compounds. A group of soil bacteria, the nitrifying bacteria, oxidizes the ammonium to nitrate in a chemosynthetic process that yields energy for the bacteria. At this point, the nitrate is once again available for uptake by green plants—a local ecosystem cycle within the global cycle. In most ecosystems, the supply of nitrate or ammonium nitrogen is quite limited, yet there is an abundance of nitrogen gas—if it can be accessed.

***Nitrogen Fixation.*** A number of bacteria and cyanobacteria (chlorophyll-containing bacteria, formerly referred to as blue-green algae) can convert nitrogen gas to the ammonium form, a process called biological **nitrogen fixation.** In terrestrial ecosystems, the most important among these nitrogen-fixing organisms is a bacterium in the genus *Rhizobium*, which lives in nodules on the roots of legumes, the plant family that includes peas and beans (Fig. 3–20). [This is another example of mutualistic symbiosis. The legume provides the bacterium with a place to live and with food (sugar) and gains a source of nitrogen in return.] From the legumes, nitrogen enters the food web. Many ecosystems are "fertilized" by nitrogen-fixing organisms; legumes, with their symbiotic bacteria, are by far the most important. The legume family includes a huge diversity of plants, ranging from clovers (common in grasslands) to desert shrubs and many trees. Every major terrestrial ecosystem, from tropical rain forest to desert and tundra, has its representative legume species, and legumes are generally the first plants to recolonize a burned-over area. Without them, all production would be sharply impaired due to a lack of available nitrogen. The nitrogen cycle in aquatic ecosystems is similar. There, cyanobacteria are the most significant nitrogen fixers.

Three other important processes also "fix" nitrogen. One is the conversion of nitrogen gas to the ammonium form by discharges of lightning in a process known as *atmospheric nitrogen fixation*; the ammonium then comes down with rainfall. The second is the *industrial fixation* of nitrogen in the manufacture of fertilizer. (The Haber–Bosch process converts nitrogen gas and hydrogen to ammonia.) The third is a consequence of the *combustion of fossil fuels*, during which nitrogen from coal and oil is oxidized; some nitrogen gas is also oxidized during high-temperature combustion. Both of these processes lead to nitrogen oxides ($NO_x$) in the atmosphere, which are converted to nitric acid and then brought down to Earth as acid precipitation.

***Denitrification.*** **Denitrification** is a microbial process that occurs in soils and sediments, where oxygen is unavailable for normal bacterial decomposition. A number of microbes can take nitrate (which is highly oxidized) and use it as a substitute for oxygen. In so doing, the nitrogen is reduced (it gains electrons) to nitrogen gas and released back into the atmosphere. Large amounts of organic matter are decomposed in this manner. Farmers seek to avoid denitrification because it reduces soil fertility. Accordingly, they plow as early as possible in the spring in order to restore oxygen to the soil. In sewage treatment systems, denitrification is a desirable process and is promoted to remove nitrogen from the wastewater before it is released in soluble form to the environment (Chapter 17).

***Human Impacts.*** Human involvement in the nitrogen cycle is significant and a major cause for concern. Many agricultural crops are legumes (peas,

**Figure 3–20 Nitrogen fixation.** Bacteria in root nodules of legumes convert nitrogen gas in the atmosphere to forms that can be used by plants.

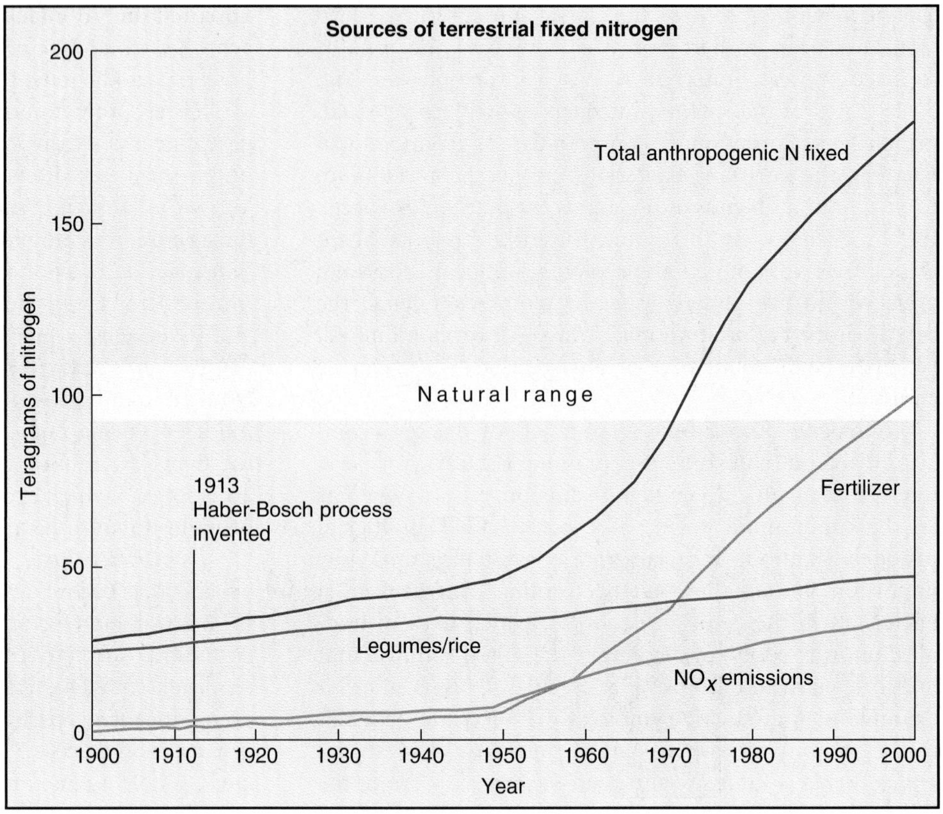


**Figure 3–21 Terrestrial fixed nitrogen.** Nitrogen fixed by human-promoted (anthropogenic) processes have surpassed the natural levels of nitrogen fixation, essentially fertilizing the global ecosystem.

beans, soybeans, alfalfa), so they draw nitrogen from the air, thus increasing the normal rate of nitrogen fixation on land. Crops that are nonleguminous (corn, wheat, potatoes, cotton, and so on) are heavily fertilized with nitrogen derived from industrial fixation. Also, fossil-fuel combustion fixes nitrogen from the air. All told, these processes add some 150 teragrams of nitrogen to terrestrial ecosystems annually (Fig. 3–21). This is approximately 1.5 times the natural rate of nitrogen fixation. In effect, we are more than doubling the rate at which nitrogen is moved from the atmosphere to the land.

The consequences of this global nitrogen fertilization are serious. Acid deposition has destroyed thousands of lakes and ponds and caused extensive damage to forests (Chapter 21). The surplus nitrogen has led to "nitrogen saturation" of many natural areas, whereby the nitrogen can no longer be incorporated into living matter and is released into the soil. There, it leaches cations (positively charged mineral ions) such as calcium and magnesium from the soil, which leads to mineral deficiencies in trees and other vegetation. Washed into surface waters, the nitrogen makes its way to estuaries and coastal oceans, where it promotes rich "blooms" of algae, some of which are toxic to fish and shellfish. When the algal blooms die, they sink to deeper water or sediments, where they reduce the oxygen supply and kill bottom-dwelling organisms like crabs, oysters, and clams, creating "dead zones." (See Chapter 17.) Of course, these are just the observable effects of nitrogen enrichment; there may be other effects that have not yet been well documented, such as a loss of biodiversity by encouraging luxuriant growth of a few dominant plant species.

Although we have focused on the cycles of carbon, phosphorus, and nitrogen, cycles exist for oxygen, hydrogen, and all the other elements that play a role in living things. Also, while the routes taken by distinct elements may differ, all of the cycles are going on simultaneously, and all come together in the tissues of living things. As the elements cycle through ecosystems, energy flows in from the Sun and through the living members of the ecosystems. The links between these two fundamental processes of ecosystem function are shown in Fig. 3–22.

## 3.4 Implications for Human Societies

### Ecosystem Sustainability

Ecosystems have existed for thousands of years or more, maintaining natural populations of the biota and the processes that they carry out, processes that in turn sustain the ecosystems. In this chapter, we have focused on energy flow and nutrient cycling—how natural ecosystems work, in *theory*. In *reality*, however, it is the Earth's specific ecosystems that we depend on for goods and services (ecosystem capital). Is our use of natural and managed ecosystems a serious threat to their long-term sustainability? We will look briefly at how we are affecting energy flow and nutrient cycling, but a more detailed

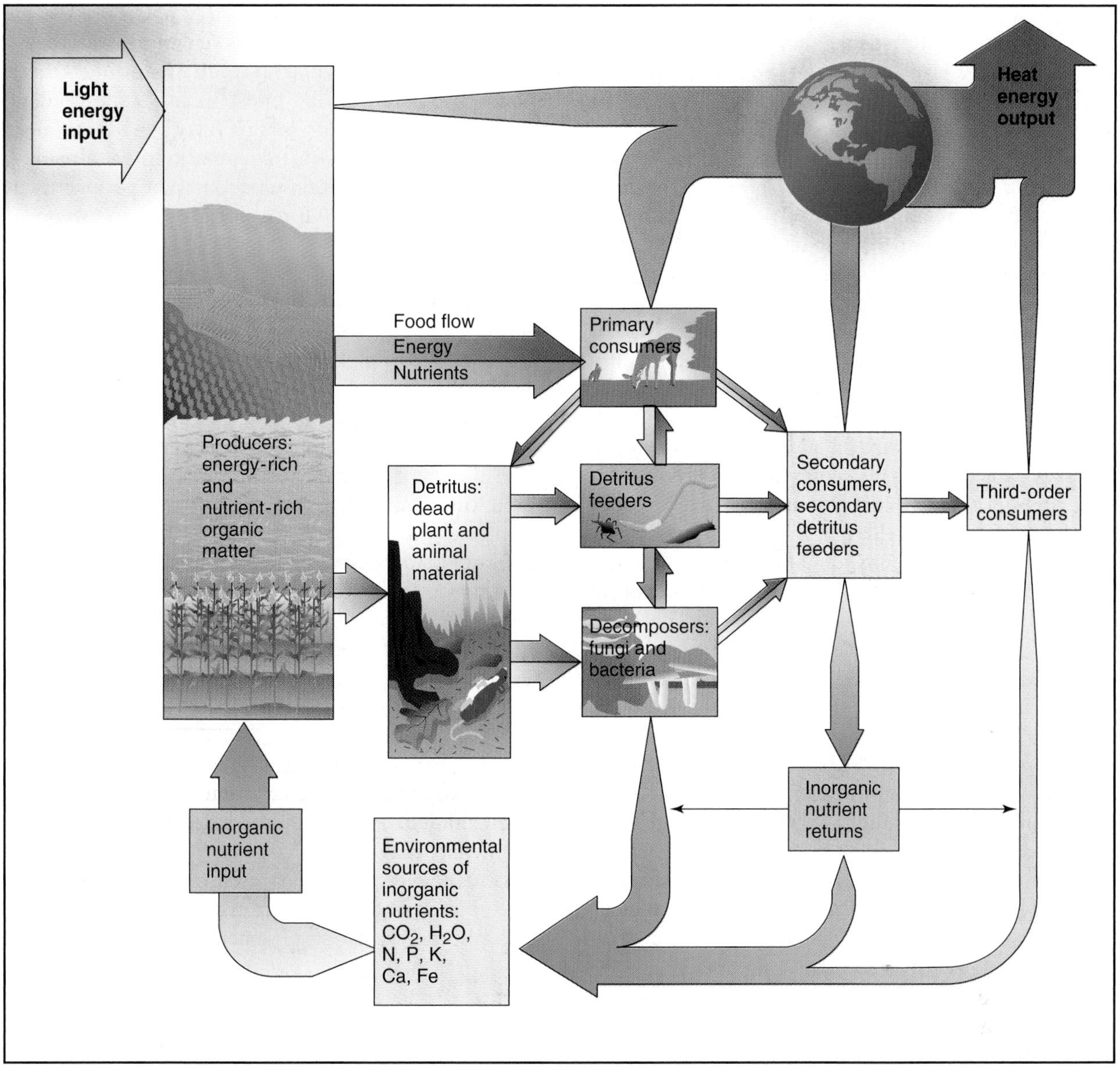


**Figure 3–22 Nutrient recycling and energy flow through an ecosystem.** Arranging organisms by feeding relationships and depicting the energy and nutrient inputs and outputs of each relationship shows a continuous recycling of nutrients (blue) in the ecosystem, a continuous flow of energy through it (red), and a decrease in biomass in it (thickness of arrows).

investigation of our use of specific types of ecosystems must wait until Chapter 11.

One of the reasons for studying natural ecosystems is that they are models of sustainability. As a result, we might benefit from understanding what it is that makes them sustainable and, where possible, how to emulate them. As Figure 3–22 shows, it is the Sun that energizes the processes of energy flow and nutrient cycling, and then the biological, geological, and chemical interactions within and between ecosystems are the drivers of change. One key to their sustainability, therefore, is that *ecosystems use sunlight as their source of energy*.

**Significance of Energy Flow.** Humans make heavy use of the energy that starts with sunlight and flows through natural and agricultural ecosystems. Agriculture, for example, provides most of our food. To accomplish this, we have converted almost 11% of Earth's land area from forest and grassland biomes to agricultural ecosystems. Grasslands provide animals for labor, meat, wool, leather, and milk. Forest biomes provide us with 3.3 billion cubic meters of wood annually for fuel, building material, and paper. Finally, some 15% of the world's energy consumption is derived directly from plant material.

Calculations of the total annual global net primary production of land ecosystems average out at 120 petagrams of dry matter, including agricultural as well as the more natural ecosystems. Two independent groups of researchers have calculated that humans currently appropriate 32% of this total production for agriculture, grazing, forestry, and human-occupied lands. Although these kinds of calculations require making a lot of estimates based on limited data, they do indicate that humans are using a large fraction of the whole, and that it is likely to grow. Further, because humans convert many natural and agricultural lands to urban and suburban housing, highways, dumps, factories, and the like, we cancel out an additional 8% of potential primary production. Thus, we appropriate 40% of the land's primary production to support human needs. In so doing, we have become the dominant biological force on Earth. As ecologist Stuart Pimm puts it, "Man eats Planet! Two-fifths already gone." Is this level of use sustainable? If so, and if our use increases, when do we reach the limits of sustainable use?

**Another Energy Source.** In addition to running on solar energy, which creates the ecosystem productivity that humans appropriate, the current human system depends heavily on fossil fuels—coal, natural gas, and crude oil. Crude oil is refined to produce liquid fuels such as gasoline, diesel fuel, fuel oil, and so on. Even in the production of food, which depends fundamentally on sunlight and photosynthesis, it is estimated that we use about 10 calories of fossil fuel for every calorie of food consumed. This additional energy is depleted in the course of preparing fields, fertilizing the plants, controlling pests, harvesting the food, and processing, preserving, transporting and, finally, cooking it.

The annual combustion of fossil fuels releases 6.6 petagrams of $CO_2^-$ carbon, as well as some 35 teragrams of nitrogen and 50 teragrams of sulfur. The biosphere has a limited capacity to absorb these by-products, however, so air pollution problems result, including urban smog, acid rain, and the potential for global climate change. Also, problems stemming from the depletion of fuels—particularly, crude oil—are on the horizon. For these reasons, most people concerned about sustainability are solar-energy advocates. Solar energy is extremely abundant. Just as important, we already have the technology to obtain much more of our energy needs from sunlight and the forces it causes, such as wind. (See Chapter 14.)

**Sustainability and Nutrient Cycling.** *Ecosystems dispose of wastes and replenish nutrients by recycling the elements.* This maintains their sustainability, indefinitely. Human systems, by contrast, are based in large part on a *one-directional flow* of elements (Fig. 3–23). For example, the fertilizer–nutrient phosphate, which is mined from deposits, ends up going into waterways via land runoff and effluents from sewage treatment. The same one-way flow occurs with such metals as aluminum, mercury, lead, and cadmium, which are the "nutrients" of our industry. At one end, these resources are mined from the Earth; at the other, they end up in dumps and landfills as items containing them are discarded. As a result, there are depletion problems at the resource end and pollution problems at the other. The Earth has substantial (but not unlimited) deposits of most minerals; however, the capacity of ecosystems (even the whole biosphere) to absorb wastes without being disturbed is comparatively limited. This limitation is aggravated, furthermore, by the fact that many of the products we use are nonbiodegradable.

Human intrusion into the carbon, nitrogen, and phosphorus cycles is substantial and will undoubtedly

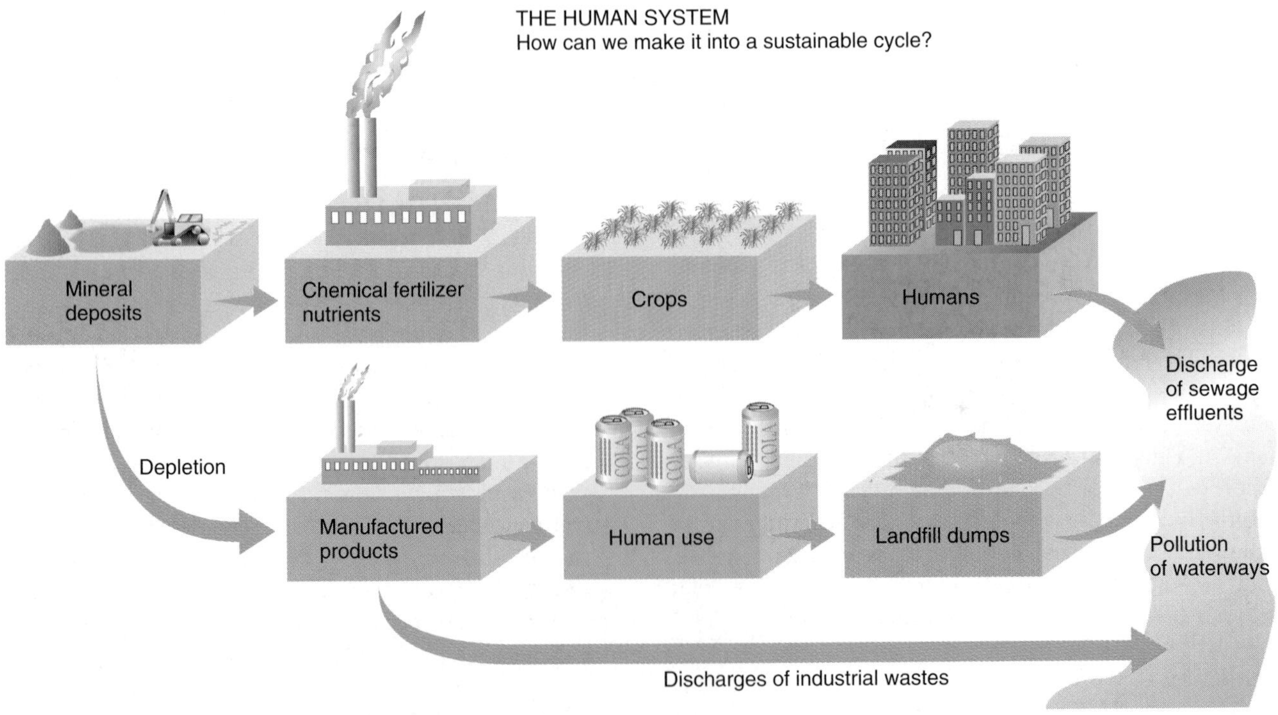


**Figure 3–23** **One-directional nutrient flow in human society.**

intensify in the coming years. Is this level of impact sustainable? If so, and if it increases, when do we reach the limits of sustainable impact?

## Value of Ecosystem Capital

How much are natural ecosystems worth to us? Recall from Chapter 1 that we have defined the goods and services we derive from natural systems as *ecosystem capital*. In a first-ever attempt of its kind, a team of 13 natural scientists and economists[1] collaborated to produce a report entitled "The Value of the World's Ecosystem Services and Natural Capital." Their reason for making such an effort was that the goods and services provided by natural ecosystems are not easily seen in the market (meaning the market economy that normally allows us to place value on things) or may not be in the market at all. Thus, things such as clean air to breathe, the formation of soil, the breakdown of pollutants, and the like never pass through the market economy. People are often not even aware of their importance. Because of this, these things are undervalued or not valued at all.

The team identified 17 major ecosystem goods and services that provide vital functions we depend on. The team also identified the ecosystem functions that actually carry out the vital human support and gave examples of each (Table 3–2), making the point that it is useless to consider human welfare without these ecosystem services, so in one sense, their value as a whole is infinite. However, the **incremental value** of each type of service can be calculated. That is, changes in the quantity or quality of various types of services may influence human welfare, and an economic value can be placed on that relationship. For example, removing a given forest will affect the ability of the forest to provide lumber in the future, as well as perform other services, such as soil formation and promotion of the hydrologic cycle. The economic value of this effect can be calculated. So, by calculating incremental values, making many approximations, and collecting data from other researchers who have worked on individual processes, the research team tabulated the annual global value of ecosystem services performed. According to their calculations, the total value to human welfare of a year's services amounts to $38 trillion (in 2000 dollars), and that is considered a conservative estimate! This is significantly more than the $25 trillion calculated for the gross world product of the world economy.

The real power of the team's analysis lies in its use for making local decisions. Thus, the value of a wetlands cannot be represented solely by the amount of soybeans that could be grown on the land if it were drained. Instead, wetlands provide other vital ecosystem services, and these should be balanced against the value of the soybeans in calculating the costs and benefits of a proposed change in land use. The bottom line of their analysis is, in their words, "that ecosystem services provide an important portion of the total contribution to human welfare on this planet." For this reason, the ecosystem capital stock (the ecosystems and the populations in them, including the lakes and wetlands) must be given adequate weight in public-policy decisions involving changes to them. Because these services are outside the market and uncertain, they are too often ignored or undervalued, and the net result is human changes to natural systems whose social costs far outweigh their benefits. For example, coastal mangrove forests in Thailand are often converted to shrimp farms; an analysis showed that the economic value of the forests (for timber, charcoal, storm protection, and fisheries support) exceeded the value of the shrimp farms by 70%.

***A New Look.*** In 2002, a new team looked at the 1997 team's calculation of value and examined the benefit–cost consequences of converting ecosystems to more direct human uses (e.g., a wetland to a soybean field).[2] In every case, the net balance of value was a loss. That is, services lost outweighed services gained. The team examined five different biomes and found consistent losses in ecosystem capital, running at −1.2% per year. Based on the calculated total ecosystem value of $38 trillion, this percentage represents an annual loss of $250 billion through habitat conversion alone. If so, the authors asked, why is conversion still happening? They suggested that the benefits of conversion were often exaggerated through various government subsidies, seriously distorting the market analysis. Thus, the market does not adequately measure many of the benefits from natural ecosystems, nor does it deal well with the fact that many major benefits are on regional or global scales, while the benefits of conversion are narrowly local. The authors argued strongly for a much greater global effort to conserve natural ecosystems. Beyond conservation, our tremendous dependence on natural systems should lead, logically, to proper management of those systems, a topic discussed in Chapter 4.

## The Future

On a global scale, the future growth of the human population and rising consumption levels will severely challenge ecosystem sustainability. Over the next 50 years, for example, the world's population is expected to increase by at least 2.5 billion people. The needs and demands of this expanded human population will create unprecedented pressures on the ability of Earth's systems to provide goods and services. Estimates based on current trends indicate that impacts on the nitrogen and phosphorus cycles alone will raise the amounts of these nutrients being added to the land and water two to three times above current levels. Demands on agriculture will likely require at least 15% more agricultural land; irrigated land area is expected to double, greatly stressing an already strained hydrologic cycle. Such growing demands cannot be met without

[1]Robert Costanza et al., "The Value of the World's Ecosystem Services and Natural Capital," *Nature* 387 (1997): 253–260.

[2]Andrew Balmford et al., "Economic Reasons for Conserving Wild Nature," *Science* 297 (August 9, 2002): 950–953.

## table 3-2 Ecosystem Services and Functions

| Ecosystem Service | Ecosystem Functions | Examples |
|---|---|---|
| Gas regulation | Regulation of atmospheric chemical composition | $CO_2$–$O_2$ balance, $O_3$ for UVB protection, and $SO_x$ levels |
| Climate regulation | Regulation of global temperature, precipitation, and other biologically mediated climatic processes at global or local levels | Greenhouse gas regulation, dimethylsulfoxide production affecting cloud formation |
| Disturbance regulation | Capacitance, damping, and integrity of ecosystem response to environmental fluctuations | Storm protection, flood control, drought recovery, and other aspects of habitat response to environmental variability controlled mainly by vegetation structure |
| Water regulation | Regulation of hydrological flows | Provisioning of water for agricultural (such as irrigation) or industrial (such as milling) processes or transportation |
| Water supply | Storage and retention of water | Provisioning of water by watersheds, reservoirs, and aquifers |
| Erosion control and sediment retention | Retention of soil within an ecosystem | Prevention of loss of soil by wind, runoff, or other removal processes; storage of silt in lakes and wetlands |
| Soil formation | Soil-formation processes | Weathering of rock and the accumulation of organic material |
| Nutrient cycling | Storage, internal cycling, processing, and acquisition of nutrients | Nitrogen fixation, N, P, and other elemental or nutrient cycles |
| Waste treatment | Recovery of mobile nutrients and removal or breakdown of excess nutrients and compounds | Waste treatment, pollution control, detoxification |
| Pollination | Movement of floral gametes | Provisioning of pollinators for the reproduction of plant populations |
| Biological control | Trophic-dynamic regulations of populations | Keystone predator to control prey species, reduction of herbivory by top predators |
| Refugia | Habitat for resident and transient populations | Nurseries, habitat for migratory species, regional habitats for locally harvested species, or overwintering grounds |
| Food production | That portion of primary production extractable as food | Production of fish, game, crops, nuts, and fruits by hunting, gathering, subsistence farming, or fishing |
| Raw materials | That portion of primary production extractable as raw materials | The production of lumber, fuel, or fodder |
| Genetic resources | Sources of unique biological materials and products | Medicine, products for materials science, genes for resistance to plant pathogens and crop pests, ornamental species (pets and horticultural varieties of plants) |
| Recreation | Provision of opportunities for recreational activities | Ecotourism, sport fishing, and other outdoor recreational activities |
| Cultural | Provision of opportunities for noncommercial uses | Aesthetic, artistic, educational, spiritual, or scientific values of ecosystems |

*Source:* Reprinted, by permission, from Robert Costanza, Ralph d'Arge, Rudolf de Groot, Stephen Farber, Monica Grasso, Bruce Hannon, Karin Limburg, Shahid Naeem, Robert V. O'Neill, Jose Paruelo, Robert G. Raskin, Paul Sutton, and Marjan van den Belt, "The Value of the World's Ecosystem Services and Natural Capital," *Nature* 387 (1997): 253–260.

## earth watch

### Biosphere 2

The proof of a theory lies in testing it. If the biosphere functions as we have described—running on solar energy and recycling all the elements from the environment through living organisms, and back to the environment—then we might be able to create an artificial biosphere that functions similarly.

The largest such experiment to date is Biosphere 2, constructed in Arizona, 30 miles north of Tucson. Biosphere 2 was developed entirely with private venture capital ($200 million), with a view toward gaining information and experience that might be used to create permanent space stations on the Moon or other planets or in long-distance space travel. In addition, it was hoped that Biosphere 2 would yield information that would further our understanding of our own biosphere—Biosphere 1.

As originally constructed, Biosphere 2 was a supersealed "greenhouse" enclosing an area of 2.5 acres (1 ha). Entry and exit were through a double air lock. Different environmental conditions within the containment supported several ecosystems. Accordingly, there were areas of tropical rain forest, savanna, desert, fresh- and saltwater marshes, and a miniocean complete with a coral reef, each stocked with representative species—over 4,000 in all. An agricultural area and living quarters for a crew of up to 10 "Biospherians" completed the arrangements.

All water, air, and nutrient recycling took place within the structure; wastes, including human and animal excrement, were treated and recycled to support the growth of plants. A crew of four men and four women began a two-year mission in September 1991. In addition to monitoring and collecting data on the natural systems, their main occupation was engaging in intensive organic

agriculture to produce plant foods both for themselves and for feeding a few goats and chickens, which produced eggs, milk, and a little meat. The living quarters included the comforts and conveniences of modern living, but all communication with the outside world was via electronics. The crew's environment, including its plants and animals, was totally sealed from that of the outside world.

Not everything went perfectly; in fact, quite a few things went wrong. At one point, additional oxygen had to be introduced because oxygen was being absorbed by the huge concrete structure that supported the greenhouse. The amount of nitrous oxide rose to dangerous levels and had to be controlled in order to preserve the health of the inhabitants. Food production failed, so food had to be imported. Nearly all the birds and animals thought to be able to tolerate the enclosure died off, with the exception of cockroaches and ants.

At the end of their two-year experimental sojourn, the Biospherians emerged somewhat thinner, but all in good health. Grave doubts were cast on the scientific value of the experiment, however, because of the necessary interventions and the loss of so many species. Perhaps the greatest value of the experiment was to demonstrate that even $200 million could not reproduce a functioning, self-contained ecosystem capable of supporting just eight people.

Biosphere 2 received new life when Columbia University's Lamont–Doherty Earth Observatory agreed to take it over in 1996. The facility became an educational center for undergraduates and classroom teachers. It also became an important tourist site, entertaining 180,000 visitors annually. Its ecosystems, with the capability of controlling many environmental parameters, were used for major research projects. Next to Earth, it is the world's largest greenhouse—a suitable place to explore the greenhouse effect. In particular, researchers were investigating the effects of increased levels of atmospheric $CO_2$ on the coral reef ecosystem and on the tropical rain forest. Early results indicated that such increases will greatly affect these systems—negatively in the case of the coral reefs and only slightly positively in that of the rain forests. Unfortunately, Columbia University decided in September; 2003, to withdraw its support from Biosphere 2, leaving this huge facility to an uncertain future.

facing serious trade-offs between different goods and services. More agricultural land means more food, but less forest, and therefore less of the important services forests perform. More water for irrigation means more rivers will be diverted, so less water will be available for domestic use and for sustaining the riverine ecosystems. These dire projections are not necessarily predictions, and they don't have to be inevitable outcomes. By looking ahead, we may choose other alternatives if we clearly understand the consequences of simply continuing present practices.

The pressures on ecosystems illustrate why it is crucially important to understand how ecosystems work and how human societies interact with them (see Earth Watch, p. 81). Although these are broad-scale, even global, pressures, the decisions that will most directly determine ecosystem sustainability are local or regional. Figure 1–5, which shows the conceptual framework of the Millennium Ecosystem Assessment project, identifies a number of "drivers of change," both direct and indirect. People living within and adjacent to an ecosystem usually are those who benefit most directly from the ecosystem. For example, they may remove vegetation, harvest wood, add fertilizer, withdraw water, harvest game animals, and so on. Their local decisions will have both intended and unintended consequences. If they harvest firewood from a forest (intended), soil erosion may increase (unintended). If many local decision makers make similar decisions, the impacts may have cumulative regional and, eventually, global unintended and undesirable consequences. The private gain achieved by the local decision to exploit an ecosystem may have to be modified or overruled by governmental decision makers who are more concerned with maintaining the ecosystem as a public good. Thus, responsible public policy is needed to address how ecosystems will be exploited and managed.

In Chapter 4, we discuss how ecosystems are balanced and how they change over time. As we do so, we will examine the consequences of such ecosystem behavior for management purposes and will also encounter additional examples of ecosystem properties that maintain sustainability.

# revisiting the themes

## Sustainability

Ecosystems have existed for millennia because they are sustainable. Although ecosystems are highly diverse, they share two fundamental characteristics that are crucial to their sustainability. The first is their energy source: They depend on a nondepletable, nonpolluting source—the Sun. The other is the efficient recycling of nutrients and other chemicals through the activities of the organisms and numerous geological and chemical processes. As a result, wastes do not accumulate in ecosystems, and essential elements for plant primary production are continuously resupplied.

The human system makes heavy use of ecosystems, and to that extent, we also depend on solar energy and nutrient recycling. Because of this dependency, our use of ecosystem productivity and our intrusion into the nutrient cycles must both come under the scrutiny of sustainability. When do we reach the limits of sustainable use? When does our impact on nutrient cycles become unsustainable? These are hard questions to answer, particularly because the most direct impacts on ecosystems occur at the local level, yet it is the cumulative impacts of many local decisions that lead to global changes.

## Sound Science

Virtually all of the information presented in this chapter was acquired through the application of sound scientific principles by countless scientists over many years. The basics of matter and energy and of ecosystem functioning have been well established for many years. Sound science is the basis of recent attempts to evaluate the human impact on ecosystem processes, too. A good example is the work of the teams trying to calculate the overall economic worth of ecosystem capital.

## Stewardship

The Serengeti ecosystem is vulnerable because it exists in a setting of an expanding human population and increasing conversion to mechanized agriculture. So far, the countries involved are protecting the core of the ecosystem, although pressures are threatening the land on the margins. Sustaining this irreplaceable ecosystem will be a strong challenge to the stewardship of the governments and the people of Kenya and Tanzania.

## Ecosystem Capital

This chapter, which describes the basics of how ecosystems work, is all about ecosystem capital. The goods and services ecosystems provide are essentially priceless, but the human system is so economically driven that it is helpful to estimate their economic value. When this is done, the calculated value exceeds the value of the entire gross world product, on an annual basis. Moreover, when natural ecosystems are converted to more managed systems like agriculture, studies indicate that this inevitably ends up as a net financial loss.

## Policy and Politics

If public-policy decisions were made on the basis of the true value of the goods and services provided by ecosystems, then efforts to conserve natural ecosystems would be intensified. This is seldom the case, however; local decisions heavily favor exploitation and the conversion of ecosystems, often because unwise subsidies encourage these processes, as well as the desire for short-term profit. To defend the public good by maintaining healthy ecosystems, governmental intervention is frequently the only recourse.

### Globalization

The development of fossil-fuel energy and its worldwide applications pose many threats to the long-term sustainability of human civilization. It will take global accord to make a successful transition to energy sustainability, and this will not happen until the powers controlling the global economy are convinced that it is needed. Other elements of globalization can be seen in the cycles of carbon, phosphorus, and nitrogen, which are influenced by global economic processes.

## review questions

1. What are the six key elements in living organisms, and where does each occur—in the atmosphere, hydrosphere, or lithosphere?
2. What is the "common denominator" that distinguishes between organic and inorganic molecules?
3. In one sentence, define matter and energy, and demonstrate how they are related.
4. Give four examples of potential energy. In each case, how can the potential energy be converted into kinetic energy?
5. State the two energy laws. How do they relate to entropy?
6. What is the chemical equation for photosynthesis? Examine the origin and destination of each molecule referred to in the equation. Do the same for cell respiration.
7. Food ingested by a consumer follows three different pathways. Describe what happens to the food in each pathway and what products and by-products are produced in each case.
8. Compare and contrast the decomposers with other consumers in terms of the matter and energy changes that they perform.
9. What factors affect the rate and amount of primary production?
10. What three factors account for decreasing biomass at higher trophic levels? That is, how do those factors account for the food pyramid?
11. Describe the biogeochemical cycle of carbon as it moves into and through organisms and back to the environment. Do the same for phosphorus and nitrogen.
12. What are the major human intrusions into each of the carbon, phosphorus, and nitrogen cycles?
13. What are two essential keys to ecosystem sustainability?
14. Compare human use of solar energy with our use of fossil-fuel energy. What problems are prominent?
15. What value has been assigned to the goods and services provided annually by ecosystem capital? How significant is this estimated value?

## thinking environmentally

1. Use the laws of conservation of matter and energy to describe the consumption of fuel by a car. That is, what are the inputs and outputs of matter and energy? (*Note*: Gasoline is a mixture of organic compounds containing carbon–hydrogen bonds.)
2. Relate your level of exercise—breathing hard and "working up an appetite"—to cell respiration in your body. What materials are consumed, and what products and by-products are produced?
3. Using your knowledge of photosynthesis and cell respiration, draw a picture of the hydrogen cycle and the oxygen cycle. (*Hint*: Consult the three cycles in the book for guidance.)
4. Give two reasons that tundra and desert ecosystems support a much smaller biomass of animals than do tropical rain forests.
5. Evaluate the sustainability of parts of the human social, cultural, and economic system, such as transportation, manufacturing, agriculture, and waste disposal, by relating them to the sustainability of natural ecosystems as set forth in this chapter. How can such things be modified to make them more sustainable?

chapter 4

# Ecosystems: How They Change

**Key Topics**

1. Dynamics of Natural Populations
2. Mechanisms of Population Equilibrium
3. Mechanisms of Species Adaptation
4. Ecosystem Responses to Disturbance
5. Lessons to Learn

In May and June of 1988, following an extended drought, lightning started a large number of fires in Yellowstone National Park, a 2.2-million-acre gem in the northwest corner of Wyoming. The fires burned slowly for a while, and then, fanned by high winds behind a series of cold fronts with no rain, they broke out and swept across hundreds of thousands of acres. Prior to the 1988 fires, the Park Service policy on fires was to allow them to burn their course unless they were near human habitations. This itself was a reversal of the Smokey the Bear policy of earlier years, during which all forest fires were to be extinguished, preferably before they even got started. The Yellowstone fires touched off a great political controversy over Park Service policy and led to fruitless efforts to put the fires out. In spite of the largest fire-fighting effort in U.S. history, the fires were finally put out only by a snowfall, leading to a standing joke at the time: "How do you put out the Yellowstone fires? Pour a hundred million dollars on it and wait for it to snow."

**Yellowstone National Park This scene illustrates the patchy landscape of the park as it appeared shortly after the devastating fires of 1988.**

***Recovery.*** The fires burned nearly 10% of the park area, leaving behind a patchy landscape, with heavily or lightly burned areas interspersed among untouched areas. Within two weeks, however, grasses and herbaceous vegetation began sprouting from the ashes, and a year later, abundant vegetation covered the burned areas (Fig. 4–1a). Herbivores such as bison and elk fed on the lush new growth. Thirteen years later, lodgepole pines several feet tall carpeted much of the burned-over area (Fig. 4–1b). It is expected that, some 25 years after the fires, the diversity of plants and animals in the burned areas will have completely recovered; in the meantime, the total diversity of the park has been unaffected.

It is hard to imagine anything more devastating to a forest ecosystem than a roaring crown fire. In the long run, however, as events at Yellowstone have shown, the total landscape generally recovers, populations of herbivores and predators reacted positively to the disturbance, and biodiversity is even enhanced. In fact, in Yellowstone, many populations of plants have turned out to be remarkably adaptable to fire, flourishing far more than in the unburned areas.

In Chapters 2 and 3, the focus was on the structure and function of ecosystems—viewing them as

(a)

(b)

**Figure 4–1 Recovery from fire.** In 1988, fires swept through Yellowstone National Park. (a) One year after the fires, ground vegetation was well established. (b) Thirteen years after the fires, lodgepole pines were replacing the herbaceous vegetation.

sustainable units in the natural landscape.
We considered the environmental conditions and resources that act to limit the broad distribution of ecosystems on Earth. We saw how important ecosystems are in human affairs, and we looked at efforts to assign value to the goods and services we derive from them. In addition, we found that ecosystems serve as models of sustainability, providing insight into how we might make our human system more sustainable.

***Ecosystem Balance?*** In this chapter, we take a closer look at the populations of different species that make up the living community of ecosystems. How, for example, are populations sustained over time? What kinds of interactions occur between populations in ecosystems? and How important are these interactions for ecosystems? In order to understand the answers to these questions, we encounter questions of balance, or equilibrium, in ecosystems: What does it mean for an ecosystem to be "in balance"? Are populations always in a state of equilibrium? What happens when a major disturbance like fire interrupts the "balance" in an ecosystem? Can ecosystems and the natural populations in them change over time, yet maintain the important processes that make them sustainable units?

You may believe that nature is stable and will be in balance if left alone. You will learn in this chapter, however, that an ecosystem is a dynamic system in which changes are constantly occurring. The notion of a balanced ecosystem, then, must be carefully defined. The objective in this chapter is to examine more of the basic mechanisms that underlie the sustainability of all ecosystems, including human systems and the rest of the biosphere. The chapter concludes with some perspectives on managing ecosystems, in light of the fact that ecosystems are essential to human life and welfare, and human impacts on ecosystems are extensive and potentially destructive. We first consider the perspective of populations and how they are controlled; after all, populations are the living components of ecosystems, responsible for all the important processes that take place there. Then we examine ways in which changes occur over time in ecosystems and the reasons for those changes.

## 4.1 Dynamics of Natural Populations

Each species in an ecosystem exists as a population; that is, each exists as a reproducing group. Studies indicate that, over time, the populations of most species in an ecosystem tend to remain more or less constant in size and geographic distribution. In short, deaths equal births, on average; otherwise the population would shrink or grow accordingly. The balance between births and deaths is called the **population equilibrium.** Nevertheless, populations are capable of growth under the right conditions. How, then, is growth related to equilibrium?

## Population Growth Curves

***Exponential Increase.*** Every species has the capacity to increase its population when conditions are favorable. Furthermore, the growth of a population under absolutely ideal conditions will be *exponential.* For example, a pair of rabbits producing 20 offspring, 10 of which are female, may grow by a factor of 10 each generation: 10, 100, 1,000, 10,000 ($10^1$, $10^2$, $10^3$, $10^4$), and so on. Such a series is called an **exponential increase.** This sort of growth results in a **population explosion.** A basic feature of an exponential increase is that the numbers increase faster and faster as the population doubles and redoubles, with each doubling occurring in the same amount of time. If we plot numbers over time during an exponential increase, the pattern produced is called a *J*-curve (Fig. 4–2).

There are situations that lead to population explosions in a species. Suppose, for example, that some abnormally severe years have reduced a population to a low level. If conditions then return to normal, the population may increase exponentially for a time, but then one of two things may occur: (1) Natural mechanisms may cause the population to level off and continue in a dynamic equilibrium. This pattern is known as an *S*-curve (Fig. 4–2). (2) In the absence of natural enemies, the population keeps growing until it exhausts essential resources—usually food—and then dies off precipitously due to starvation and, perhaps, diseases related to malnutrition, producing a reverse of the *J*-curve.

What follows the *J*-curve crash? For a herbivore species, any one of three scenarios may unfold. First, if the ecosystem has not been too seriously damaged, the plant food resource may recover, allowing the herbivore population to recover, and the *J*-curve may be repeated. This scenario is seen in periodic outbreaks of certain pest insects, even in natural ecosystems. Second, after the initial *J*, natural mechanisms may come into play as the ecosystem recovers, thus bringing the population into an *S*-balance. Just such a balance seems to have been established in the eastern United States for the introduced gypsy moth (Fig. 4–3). Stands of oak trees that were devastated by the initial invasion of gypsy moths a few years ago have recovered, and the insect remains at low levels. In the third scenario, damage to the ecosystem is so severe that recovery is limited, and small surviving populations eke out an existence in a badly degraded environment.

***Equilibrium Populations.*** The outstanding feature of natural ecosystems—ecosystems that are more or less undisturbed by human activities—is that they are made up of populations that are usually in the dynamic equilibrium represented by *S*-curves. *J*-curves come about when there are unusual disturbances, such as the introduction of a foreign species, the elimination of a predator, or the sudden alteration of a habitat. The increases represented by *J*-curves are only temporary in animal populations, because the animals inevitably die off as resources are exhausted. Nevertheless, under the right conditions, population increase is always possible for species.

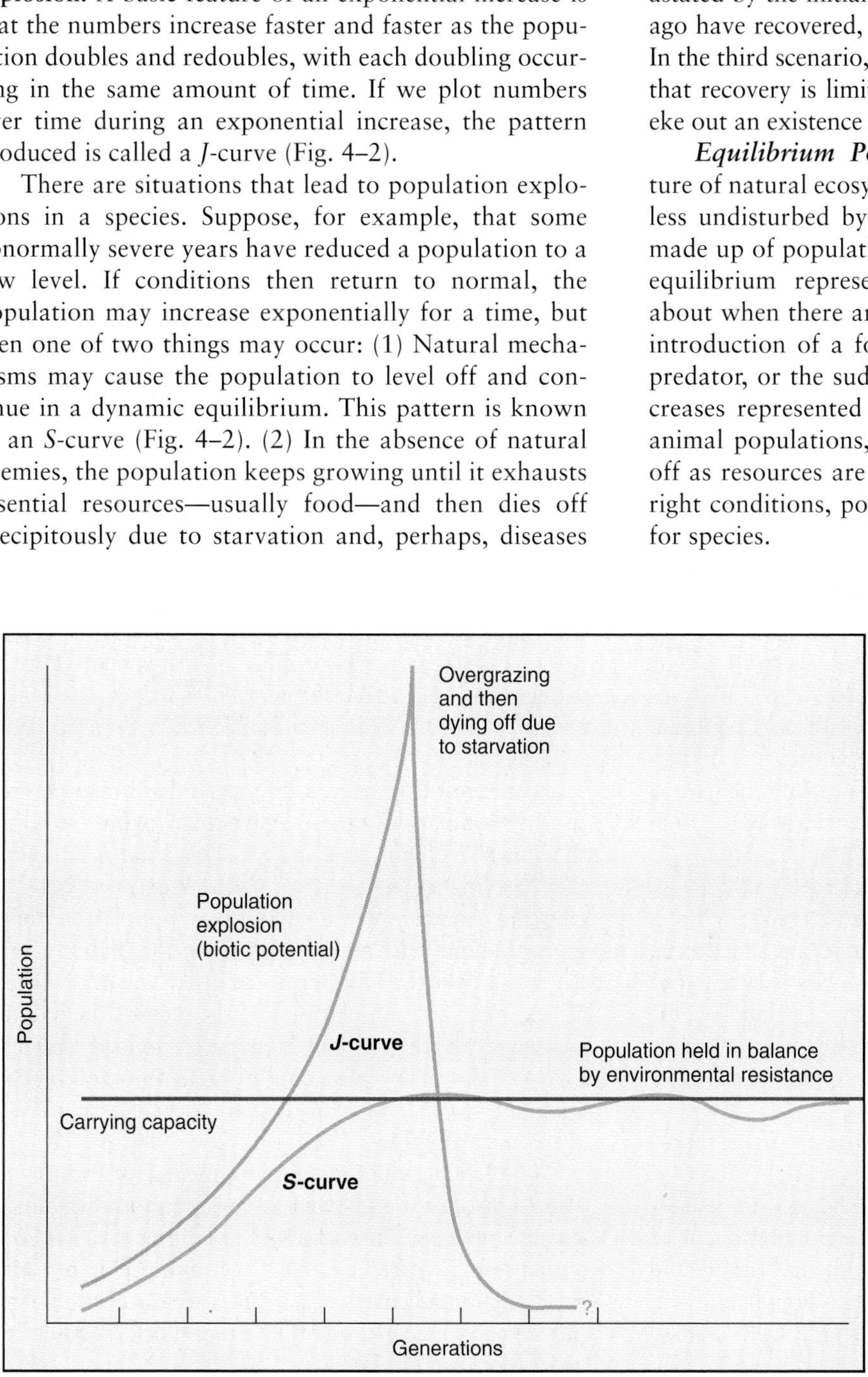


**Figure 4–2 Two types of growth curves.** The *J*-curve (blue) demonstrates population growth under optimal conditions, with no restraints. The *S*-curve (green) shows a population at equilibrium. The horizontal line (red) shows the carrying capacity of the environment for that population. Notice how the *J*-curve spikes well above, and then crashes well below, the carrying capacity, whereas the *S*-curve rises up to the carrying capacity and then oscillates between slightly above and slightly below it.

**Figure 4–3 Gypsy moth caterpillar.** The gypsy moth, an introduced species that has often caused massive defoliation of trees, now seems to have been brought under natural control in forests.

## Biotic Potential versus Environmental Resistance

The ability of populations to increase is known as **biotic potential.** Thus, biotic potential is the number of offspring (live births, eggs laid, or seeds or spores set in plants) that a species may produce under ideal conditions. The biotic potential of different species varies tremendously, averaging from less than one birth per year in certain mammals and birds to many millions per year for many plants, invertebrates, and fish. To have any effect on the size of subsequent generations, however, the young must survive and reproduce in turn. Survival through the early growth stages to become part of the breeding population is called **recruitment.**

*Reproductive Strategies.* There are two common **reproductive strategies** in the natural world. The first is to produce massive numbers of young, but then leave survival to the whims of nature. This strategy often results in very low recruitment. Thus, despite a high biotic potential, a population may not increase at all because of low recruitment. (Note that "low recruitment" is a euphemism for high mortality of the young.) However, this strategy is highly successful if a species is adapted to an environment that can suddenly change and become very favorable, like a rain-fed temporary pond. Organisms with this strategy are usually small, with rapid reproductive rates and short life spans.

The second strategy is to have a much lower reproductive rate (that is, a lower biotic potential), but then care for and protect the young until they can compete for resources with adult members of the population. This strategy works best where the environment is stable and already well populated by the species. Organisms with such a strategy are larger, longer lived, and well adapted to normal environmental fluctuations.

Additional factors that influence population growth and geographic distribution are the ability of animals to migrate, or of seeds to disperse, to similar habitats in other regions; the ability to adapt to and invade new habitats; defense mechanisms; and resistance to adverse conditions and disease. All of these factors are components of a species' **life history,** and represent a particular strategy for reproduction and survival that enables the species to be successful in a unique ecological niche in an ecosystem.

*Environmental Resistance.* Population explosions are seldom seen in natural ecosystems, because biotic and abiotic factors tend to cause mortality in populations. Among the biotic factors are predators, parasites, competitors, and lack of food. Among the abiotic factors are unusual temperatures, moisture, light, salinity, pH, lack of nutrients, and fire. The combination of all the biotic and abiotic factors that may limit a population's increase is referred to as **environmental resistance.**

In the natural world, conditions are always changing. When they are favorable, populations increase. When they are unfavorable, populations decrease. If these population numbers are plotted over time, the resulting pattern is an *S*-curve (Fig. 4–2).

In general, the reproductive ability of a species (its biotic potential) remains fairly constant, because that ability is part of the genetic endowment of the species. What varies substantially is recruitment. It is in the early stages of growth that individuals (plants or animals) are most vulnerable to predation, disease, lack of food (or nutrients) or water, and other adverse conditions. Consequently, environmental resistance effectively reduces recruitment. Some adults also perish—particularly the old or weak. If recruitment is at the **replacement level**—that is, just enough to replace these adults—then the population will remain constant (at equilibrium). If recruitment is insufficient to replace losses in the breeding population, then the population will decline.

*Carrying Capacity.* There is a definite upper limit to the population of any particular plant or animal that an ecosystem can support. This limit is known as the carrying capacity (Fig. 4–2). More precisely, the **carrying capacity** is the maximum population that a given habitat can support without the habitat being degraded over the long term—in other words, a sustainable system. If a population greatly exceeds the habitat's carrying capacity, it will undergo a *J*-curve crash, as Fig. 4–2 shows. Because conditions within habitats also change from year to year, the habitat's carrying capacity varies accordingly.

In certain situations, environmental resistance may affect reproduction, as well as causing mortality directly. For example, the loss of suitable habitat often prevents animals from breeding. Also, certain pollutants affect reproduction adversely. These situations are still environmental resistance, because they either block a population's growth or cause its decline.

In sum, whether a population grows, remains stable, or decreases is the result of an interplay between its biotic potential and environmental resistance (Fig. 4–4). In general, a population's biotic potential remains constant, so it is changes in environmental resistance that allow populations to increase or cause them to decrease. Population balance is a **dynamic balance**, which means that additions (births) and subtractions (deaths) are occurring continually and the population may fluctuate around a median (Fig. 4–2). Some populations fluctuate very little, whereas others fluctuate widely. As long as decreased populations restore their numbers and the ecosystem's carrying capacity is not exceeded, however, the population is considered to be at equilibrium. Still, questions remain. For example, what maintains the equilibrium within a certain range? What prevents a population from "exploding" or, conversely, becoming extinct?

## Density Dependence and Critical Number

The size of a population generally remains within a certain range when environmental resistance factors are **density dependent.** That is, as **population density** (the number of individuals per unit area) increases, environmental resistance becomes more intense and causes such an increase in mortality that population growth ceases or declines. Conversely, as population density decreases, environmental resistance lessens, allowing the population to recover. This balancing act will become clearer when specific mechanisms of maintaining a population at equilibrium are discussed in Section 4.2.

Factors in the environment that cause mortality can also be **density independent.** That is, their effect is independent of the density of the population. This is frequently true of abiotic factors. A sudden deep freeze in spring, for example, can kill many early germinating plants, regardless of their density. Similarly, a fire that sweeps through a forest may kill all small mammals in its wake. Although density-independent factors can be important sources of mortality, they are not involved in maintaining population equilibria.

***Critical Number.*** There are no guarantees that a population will recover from low numbers. Extinctions can and do occur in nature. The survival and recovery of a population depends on a certain minimum population base, which is referred to as the population's **critical number.** You can see the idea of critical number at work in a herd of deer, a pack of wolves, a flock of birds, or a school of fish. Often, the group is necessary to provide protection and support for its members. In some cases, the critical number is larger than a single pack or flock, because interactions between groups may be necessary as well. In any case, if a population is depleted below the critical number needed to provide such supporting interactions, the surviving members actually become more vulnerable, breeding fails, and extinction is almost inevitable.

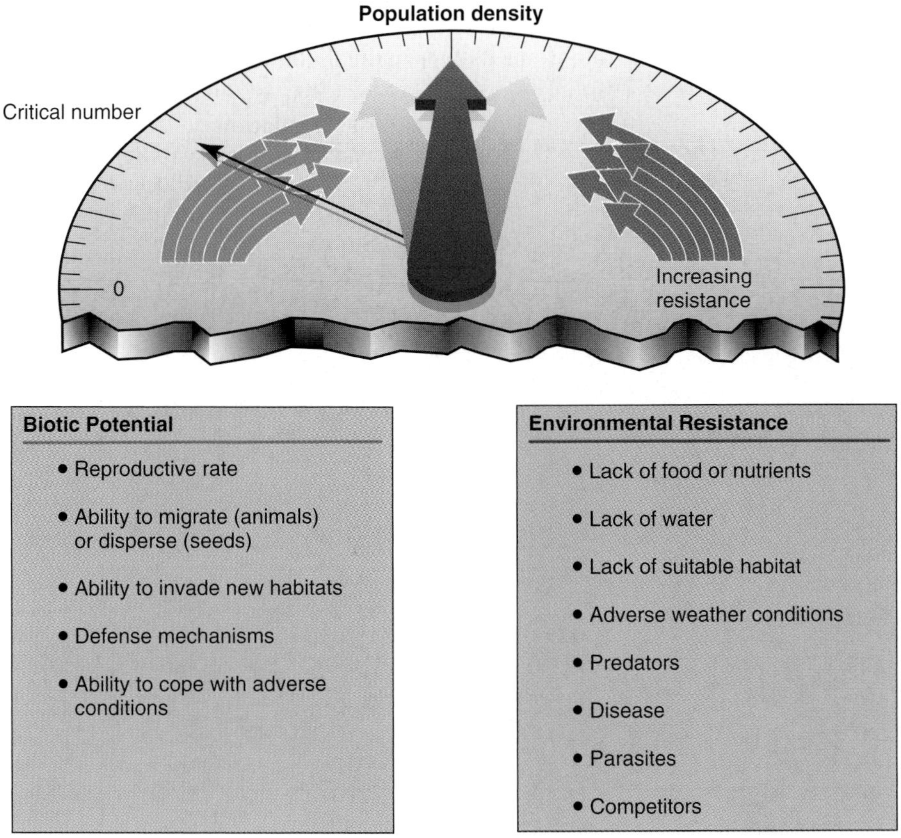

Population density
Critical number
0
Increasing resistance
Biotic Potential
• Reproductive rate
• Ability to migrate (animals) or disperse (seeds)
• Ability to invade new habitats
• Defense mechanisms
• Ability to cope with adverse conditions
Environmental Resistance
• Lack of food or nutrients
• Lack of water
• Lack of suitable habitat
• Adverse weather conditions
• Predators
• Disease
• Parasites
• Competitors


**Figure 4–4 Biotic potential and environmental resistance.** A stable population in nature is the result of the interaction between factors tending to increase population (biotic potential) and factors tending to decrease population (environmental resistance).

In Chapter 1, the loss of biodiversity was cited as one of the most disturbing global environmental trends. Human activities are clearly responsible for the decline, and even the extinction, of many plants and animals. This is happening because human impacts, such as altering habitats, introducing alien species, pollution, hunting, and other forms of exploitation, are not density dependent; they can even intensify as populations decline. Concern for these declines eventually led to the Endangered Species Act, which calls for the recovery of two categories of species. Species whose populations are declining rapidly are classified as **threatened.** If the population is near what scientists believe to be its critical number, the species may be classified as **endangered.** These definitions, when officially assigned by the U.S. Fish and Wildlife Service, set into motion a number of actions aimed at the recovery of the species in question. (See Chapter 10.)

## 4.2 Mechanisms of Population Equilibrium

With the general understanding of population equilibrium as a dynamic interplay between biotic potential and environmental resistance, we now focus on some specific kinds of population interactions. You need to understand the kinds of forces that affect natural populations in order to fully appreciate the concept of environmental resistance. In the natural world, a population is subjected to the total array of all the biotic and abiotic environmental factors around it. Many of these factors may cause *mortality* in a population, but only those that are density dependent are capable of actually *regulating* the population, keeping it around an equilibrium.

***Top-down or Bottom-up.*** Environmental scientists distinguish between top-down and bottom-up regulation. *Top-down regulation* is the control of a population (or species) by predation. In *bottom-up regulation,* the most important control of a population occurs as a result of the scarcity of some resource. This section begins with a discussion of predation (top-down regulation).

### Predator–Prey Dynamics

**Herbivores and Predators.** Predation is a conspicuous process in all ecosystems. Except for detritus feeders, all animals eat (prey on) other organisms, whether other animals or plants. Many studies have shown that herbivores are often regulated by their predators. A well-documented example is the interaction between wolves and moose on Isle Royale, a 45-mile-long island in Lake Superior that is now a national park.

***Wolves and Moose.*** During a hard winter early in this century, a small group of moose crossed the ice to the island and stayed. Their population grew considerably in the absence of predators. Then, in 1949, a pack of wolves also managed to reach the island. Nine years later, in 1958, wildlife biologists began carefully tracking the populations of the two species (Fig. 4–5). As seen in the figure, a rise in the moose population was followed by a rise in the wolf population, followed by a decline in the moose population and then a decline in the wolf population. The data can be interpreted as follows: A paucity of wolves represents low environmental resistance for the moose, so the moose population increases. Then, the abundance of moose represents optimal conditions (low environmental

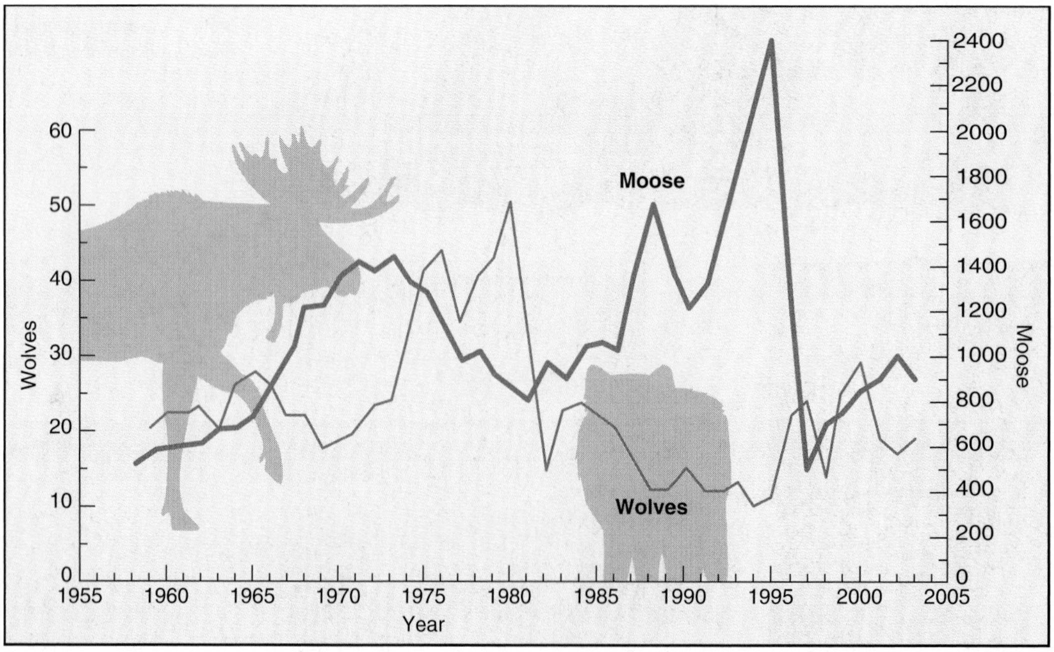


**Figure 4–5 Predator–prey relationship.** Wolf and moose populations on Isle Royale from 1955 to 2002.

resistance) for the wolves, so the wolf population increases. The growing wolf population means higher predation on the moose (high environmental resistance), so the moose population falls. The decline in the moose population is followed by a decline in the wolf population, because now there are fewer prey (high environmental resistance for the wolves). This cycle can be repeated indefinitely, providing a dynamic balance between the populations of moose and wolves.

The most recent data from Isle Royale show another cycle in progress, but the wolf population has not increased as rapidly as expected. Apparently, the wolves suffered a decline in the 1980s due to a canine virus introduced by human visitors and their dogs. Also, the most recent dramatic fall in the moose population cannot be attributed entirely to predation by the small number of wolves. Deep snow and an infestation of ticks in 1996 caused substantial mortality. The sharp decline in moose is thought to be responsible for keeping the wolf population low, as there were few calves for them to catch. (Wolves are incapable of bringing down an adult moose in good physical condition.) The animals they kill are the young and those weakened by another factor, such as sickness or old age. The most recent three years have seen a gradual rise in the moose population, while the wolves have suffered a mild decline. This is attributed to a series of mild winters, during which moose have been able to defend themselves and their calves more easily.

This long-term study shows that, in both predator and prey species, other factors may influence the observed fluctuations in population densities. For example, the shortage of vegetation that occurs as a moose population increases may stress the animals—especially the old, sick, and young—and make them more vulnerable to predators, parasites, and disease. Weather clearly plays a role in these interactions. The observation that wolves are often incapable of killing moose that are mature and in good physical condition is extremely significant. This is what often prevents predators from eliminating their prey. As the prey population is culled down to those healthy individuals which can escape attack, the predator population will necessarily decline, unless it can switch to other prey. The predators are limited by the availability of their crucial food resource. Meanwhile, the survivors of the prey population are healthy and can readily reproduce the next generation. Thus, a predator–prey relationship involves both top-down (on the prey) and bottom-up (on the predator) population regulation. Density-independent factors (weather) can also play a role in causing mortality.

**Parasites.** Much more abundant and also important as predators in population control is a huge diversity of parasitic organisms. Recall from Chapter 2 that parasites range from tapeworms, which may be a foot or more in length, to microscopic disease-causing bacteria, viruses, protozoans, and fungi. All species of plants and animals (including the predators), and even microbes themselves, may be infected with parasites.

Parasitic organisms affect the populations of their host organisms in much the same way that predators do their prey—in a density dependent manner. As the population density of the host increases, parasites and their *vectors* (agents that carry the parasites from one host to another), such as disease-carrying insects, have little trouble finding new hosts, and infection rates increase, causing higher mortality. Conversely, when the population density of the host is low, the transfer of infection is less efficient, so the levels of infection are greatly reduced, thus allowing the host population to recover.

A parasite can work in conjunction with a predator to control a given herbivore population: Parasitic infection breaks out in a dense population of herbivores; individuals weakened by infection are more easily removed by predators, leaving a smaller, but healthier, population.

The wide swings observed in the populations of moose and wolves on Isle Royale are typical of very simple ecosystems involving relatively few species. Most food webs are more complicated than that, however, because a population of any given organism is affected by a number of predators and parasites simultaneously. As a result, the population density of a species can be thought of more broadly as a consequence of the relationships the species has with its food sources and all of its natural enemies. Relationships between a prey population and its natural enemies are generally much more stable and less prone to wide fluctuations than when only a single predator or parasite is involved, because different predators or parasites come into play at different population densities. Also, when the preferred prey is at a low density, the population of the predator may be supported by switching to something else. Thus, the lag time between an increase or decrease in the prey population and that of the predator is diminished. These factors have a great damping effect on the rise and fall of the prey population.

**Plant–Herbivore Dynamics.** Predation was defined in Chapter 2 as any situation wherein one kind of organism feeds on another. Herbivores, therefore, are predators on plants. Just as too many wolves can bring the moose population dangerously low, too many herbivores can do the same to their plant food.

***Overgrazing.*** If herbivores eat plants faster than the plants can grow, the plants will eventually be depleted, and the animals will suffer. This is known as **overgrazing.** The best way to appreciate the potential for herbivore overgrazing is to observe what occurs when the herbivore's natural enemies are not present. A classic example with good documentation is the case of reindeer on St. Matthew Island, a 128-square-mile island in the Bering Sea midway between Alaska and Russia. In 1944, a herd of 29 reindeer (5 males and 24 females) was introduced onto the island, where they had no predators. From these 29 animals, the herd multiplied some two-hundredfold over the next 19 years. Early in the cycle, the animals were observed to be healthy and well nourished, as supporting vegetation was abundant. By 1963, however, when the

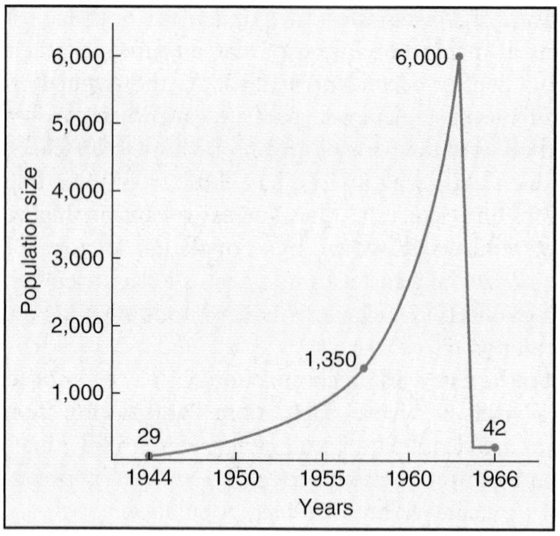

**Figure 4–6 Plant–herbivore interaction.** In 1944, a population of 29 reindeer (5 males and 24 females) was introduced onto St. Matthew Island, where they increased exponentially to about 6,000 and then died off due to overgrazing.

size of the herd had reached an estimated 6,000, the animals were malnourished. Lichens, an important winter food source, had been virtually eliminated and replaced by unpalatable sedges and grasses. During the winter of 1963–64, the absence of sufficient lichens, combined with harsh weather, resulted in death by starvation of nearly the entire herd; there were only 42 surviving animals in 1966 (Fig. 4–6).

The reindeer on St. Matthew Island demonstrate that no population can escape ultimate limitation by environmental resistance, although the form of environmental resistance and the consequences may differ. If a population is not held in check, it may explode, overgraze, and then crash (exhibiting *J*-curves all the way) as a result of starvation. Keep in mind, too, that the consequences of overgrazing are not just to the herbivore in question. One or more types of vegetation may be eliminated and replaced by other forms or not replaced at all, leaving behind a seriously degraded ecosystem. Other herbivores that depended on the original vegetation, and secondary and higher levels of consumers dependent on them, also are eliminated as food chains are severed. For example, innumerable extinctions have occurred among the unique flora and fauna of islands because sailors introduced goats to create a convenient food supply for return trips.

***Predator Removal.*** Eliminating predators or other natural enemies upsets basic plant–herbivore relationships in the same way as introducing an animal without natural enemies does. Examples of this type of folly abound as well. In much of the United States, for example, deer populations were originally controlled by wolves, mountain lions, and bear, most of which were killed because they were believed to be a threat to livestock and even humans. At present, deer populations in most areas would increase to the point of overgrazing if humans didn't hunt them in place of these natural predators. Indeed, drastic population increases do occur where hunting is prevented. As another example, sea urchins can harm the coastal marine ecosystems of eastern Canada when lobsters, which prey on the urchins, are heavily exploited for human consumption. Sea urchin populations increase and graze down the seaweeds of the subtidal rocky coasts, creating large areas of bare rock incapable of supporting the complex subtidal community ordinarily found there.

A second factor influencing plant–herbivore balance is that large herbivores—bison in the American west or elephants in Africa, for example—were originally able to roam vast regions. As forage was reduced in one area, a herd would simply migrate before overgrazing could occur, as with the large Serengeti herbivore populations (Chapter 3). Because humans have fenced in such regions for agriculture and cattle ranching, however, wild herbivores are increasingly confined to areas such as parks and reserves. Overgrazing in these presumably "protected" areas has become an increasing danger.

**Keystone Species.** In the west coast rocky intertidal zone, a starfish species, *Pisaster ochraceus,* feeds on mussels (herbivores that feed on plankton), thus keeping the mussels from blanketing the rocks (Fig. 4–7). As a result, barnacles, limpets, anemones, whelks, and other invertebrates are able to colonize the rich inter-tidal habitat. When ecologist Robert Paine experimentally removed *Pisaster* from a rocky shoreline, the mussels crowded everything else out and general species diversity was greatly reduced. Paine referred to the starfish as a "keystone species," in recognition of its crucial role in maintaining ecosystem biotic structure. (In architecture, the keystone is a fundamental part of the support and structure of a building). The lobster is a keystone species for the coastal seaweed ecosystems of eastern Canada.

In undisturbed ecosystems, herbivore populations are held in check by a number of important factors, including es-

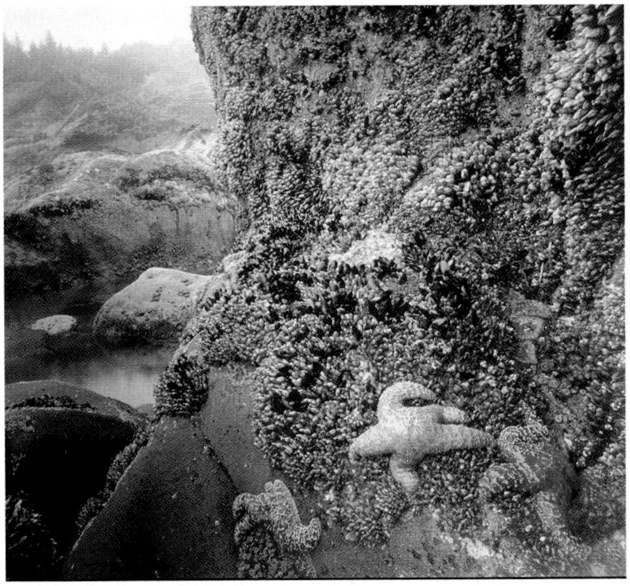

**Figure 4–7 Keystone species.** The starfish *Pisaster ochraceus* is crucial for maintaining a diverse functioning community in the rocky intertidal zone of the Pacific northwest.

pecially predators and parasites. As a consequence, populations of herbivores rarely increase enough to overgraze their food. The result is that primary producers (plants) are able to maintain a substantial standing biomass and sustain their production of the organic matter that is so important to the entire ecosystem. As general proof of the role of top-down control, note that the world is green, probably because predators generally keep herbivores from overeating their food supply.

## Competition

No *species* lives in isolation. Recall from Chapter 2 that species may compete for some scarce resource. When they do, their ecological niches are said to overlap. Can this kind of **interspecific competition** play a role in maintaining population equilibria?

No *individual* within a species lives in isolation, either. Instead, it is part of a population of individuals having identical requirements for success. (Their niches overlap completely.) If any requirement—any resource—is in short supply, the most intense competition for the resource will come from members of the same species. Some will get what they need, and some won't. This is called **intraspecific competition.**

Competition is a form of bottom-up regulation, because it occurs only when a resource is in limited supply. Ecologists have found, moreover, that interspecific and intraspecific competition affect a species differently. Let us examine this distinction.

### Intraspecific Competition

*Territoriality.* In lean times, a carnivore population (such as the wolf) has to switch to other prey or starve. Similarly, an herbivore population (such as the elk) has to migrate to find a more abundant source of the same or similar food. It turns out, though, that another factor—territoriality—often controls the populations of carnivores and some herbivores, including a great number of species that range from fish to birds and mammals. **Territoriality** refers to individuals or groups (such as a pack of wolves) defending a territory against the encroachment of others of the same species. Territoriality, therefore, is intraspecific competition. On Isle Royale, for example, there are three packs of wolves, and researchers recently observed a confrontation between two of the packs that led to the death of one of the wolves.

The males of many species of songbirds claim a territory—some limited space that they will defend vigorously—at the time of nesting. Their song warns other males to keep away (Fig. 4–8). The males of many carnivorous mammals, including dogs, "stake out" a territory by marking it with urine, the smell of which warns others to stay away. If others encroach, there may be a fight, but in most species a large part of the battle is intimidation—an actual fight rarely results in death.

*The Spoils.* In territoriality, what is really being protected by the defender or sought after by the "invader" is the claim to an area suitable for nesting, for establishing a harem, or for adequate food resources. Hence, the territory is defended only against others that would cause the most direct competition for those resources—members of the same species. As a consequence of territoriality, some members of the population are able to nest, mate, or gain access to sufficient food resources to rear the next generation. Thus, a healthy population of the species survives. If, by contrast, there were an even rationing of inadequate resources to all the members, with all of them trying to raise broods, the entire population would become malnourished and might perish. Through territoriality, breeding is restricted to only those individuals capable of claiming and defending territory; thus, population growth is curtailed in a density-dependent manner.

Individuals unable to claim a territory are usually the young of the previous generation(s). Some may hang out

**Figure 4–8 Territoriality.** A red-winged blackbird is announcing its claim on a territory.

on the fringes and seize their opportunity as they mature and older members with territories weaken or perish. Some, chased out of one territory after another, fall prey to various environmental resistance factors. Finally, some may be driven to disperse, either to find another region where they can successfully breed or to perish in conditions beyond their limit of tolerance along the way. In any case, territoriality is a powerful force behind the dispersal, as well as the stabilization, of populations.

***Self-thinning.*** Garden plants often need to be "thinned out" (that is, some need to be removed); otherwise, the flowers or vegetables will show the effects of crowding: thinner stalks, slower growth, and poorer fruit or flowers. These effects are due to resource limitation, be it of light, water, or nutrients. Studies have shown that when plants produce a large number of seedlings in a limited area, *self-thinning* is the result. As the plants grow, some thrive and some die. The number that survive and their growth rate are directly influenced by the density of the seedlings. (Hence, self-thinning is density dependent.)

Animals may also exhibit self-thinning. Crowded conditions always lead to competition for resources, whether it is between barnacles growing on rocks in the intertidal zone of marine coastal ecosystems or between the large numbers of young fish that resulted from a particularly good year for spawning. Some of the competitors thrive, and some don't make it.

***Impact on the Species.*** It was this sort of competition for scarce resources that led Charles Darwin to identify the "survival of the fittest" as one of the forces in nature leading to evolutionary changes in species. Those individuals in a competing group of young plants, animals, or microbes which are able to survive and reproduce while others do not, demonstrate superior *fitness* to the environment. Indeed, every factor of environmental resistance is a selective pressure resulting in the survival and reproduction of those individuals with a genetic endowment that enables them to better cope with their surroundings. This is the essence of **natural selection**, which is discussed in Section 4.3.

Intraspecific competition, therefore, has two distinct kinds of impact on the population of a species. In the short term, it can lead to the density-dependent regulation of a species population, through such factors as territoriality and self-thinning. It can also lead, however, to long-term improvements, in that the species adapts to its environment because those better able to compete are the ones who survive and reproduce, and their superior traits are passed on to successive generations.

### Interspecific Competition

***In Plants.*** How can interspecific competition (competition between members of different species) maintain population equilibria? Consider a natural ecosystem that contains hundreds or even thousands of species of green plants, all competing for nutrients, water, and light. What prevents one plant species from driving out others?

Recall from Chapter 2 that differences in topography, type of soil, and so on mean that the landscape is far from uniform, even within a single ecosystem. Instead, it is composed of numerous microclimates or microhabitats. That is, the specific conditions of moisture, temperature, light, and so on differ from location to location. Thus, the adaptation of a species to specific conditions enables it to thrive and overcome its competitors in one location, but not in another. Consider, for example, the distribution of trees along streams and rivers. In the Great Plains states, trees grow only along waterways, because elsewhere the environment is too dry. This pattern of growth creates what are called **riparian** woodlands (Fig. 4–9). In the eastern United States, sycamore and red maple, which can thrive in water-saturated soil, grow along riverbanks. Oaks and pines, which require well-drained soil, occupy higher elevations. In the West, white alder, willow, and cottonwoods can survive in water-saturated soils.

A second factor affecting the competition between plant species is the fact that a single species generally cannot utilize all of the resources in a given area. Therefore, any resources that remain may be claimed by other species having different adaptations. For example, grasslands contain both grasses that have a fibrous root system and plants that have taproots (Fig. 4–10). These different root systems enable the plants to coexist because they obtain their water and nutrients from different layers of the soil. Trees in a forest, moreover, compete with each other for light in the canopy (the "layer" of treetops), but leave lots of space near the ground, which may be occupied by plants (ferns and mosses, for example) that can tolerate the reduced light intensity. Another adaptation is the plethora of spring wildflowers that inhabit temperate deciduous forests. Sprouting

**Figure 4–9 Riparian woodlands.** Competition between plant communities is often maintained by differing amounts of available moisture. Riparian woodlands are shown growing only along a river in this prairie region because areas beyond the banks of the river are generally too dry to support the growth of trees.

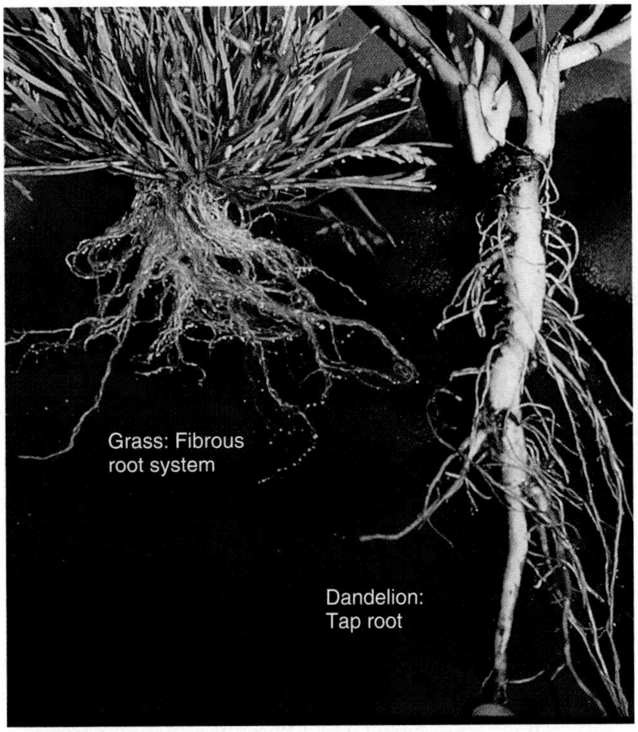


**Figure 4–10 Coexistence in plants.** Plants with fibrous roots may coexist with plants having taproots because each draws water and nutrients from a different part of the soil.

from perennial roots or bulbs in the early part of that season, these plants take advantage of the light that can reach the forest floor before the trees grow leaves.

Another option is mutualism, an arrangement between two species whereby both benefit. In warm, humid climates, for example, the branches of trees are often covered with **epiphytes**, or air plants. Epiphytes are not parasitic; indeed, there is some evidence that they help to gather the minute amounts of nutrients that come with rainfall and make them accessible to the tree on which the epiphytes are located.

*Plants: Bottom Line.* Plants usually exist in communities with a rich mixture of species, so that a given individual is likely to be in proximity with any of a number of species, as well as with individuals of its own species. It is thus more likely that the *size* of neighboring plants will play a larger role than the actual species in determining the intensity of competition. In trees, for example, the species in the canopy suppress individuals of all tree species trying to grow beneath them if the canopy is closed and light is effectively intercepted. Other plants may grow near the ground, but they must be able to cope with low light intensities. Resources vary in intensity and over space and time, so there is a heterogeneous environment for plants, allowing different species to coexist because they use common resources in different ways. Ordinarily, no one species is able to outcompete all the others, because the different species have become specialized in different ways in their use of resources. Individual plants may be suppressed by a neighboring plant, but the impact might be identical whether the neighbor is of the same species or a different one.

*In Animals.* Many experimental studies have examined the competition between animal species and its importance in the natural world. For example, the Russian ecologist G. F. Gause studied several species of the protozoan *Paramecium* grown in laboratory flasks. He showed that one species would outcompete and eliminate the other in media that would ordinarily support each species separately. This work led to the *competitive exclusion principle,* discussed in Chapter 2. Another classic study, by J. H. Connell on two species of barnacle living on the rocky coast of Scotland, showed that one species smothered another when they occurred on rocks in the middle of the intertidal zone. Both of these studies showed, however, that two species could coexist under conditions that favored the poorer competitor. In barnacles, for example, the "loser" did better than the "winner" high up in the intertidal zone, where heat and drying were most intense.

Recall from Chapter 2 that five species of warbler coexist in the spruce forests of Maine, due to *resource partitioning.* (See Fig. 2–17.) These five species are very similar in size, in the food they consume, and even in the trees they inhabit. They are able to coexist, however, because each makes use of a different part of the tree in foraging and in establishing nest sites. How did they manage to divide up the resource so efficiently? It is assumed that, because competition between species is a harmful process, natural selection operating over many years caused each species to specialize in its foraging and nest site preferences. Many other studies have pointed to similar examples of resource partitioning, so the phenomenon is well established in ecological theory. Still, it is also possible that all of the species in a group like the warblers developed their preferences independently.

The importance of interspecific competition has been shown as well both in field experiments in which competitors were removed and in natural situations where a competitor was present in one place (say, on one island), but absent from another. In these cases, *competitive release* occurs. That is, once the strong competitor is absent, the weaker one thrives and exploits resources it is normally unable to acquire because of the presence of its competitor. Many studies support this conclusion.

*Animals: Bottom Line.* Interspecific competition between animals may be a strong factor influencing the distribution and abundance of species. In the short run, interspecific competition can reduce the success of species and even eliminate them from an environment; thus, it can help regulate a population. In the long run, interspecific competition helps drive natural selection, bringing about greater specialization of ecological niches and allowing resources to be divided up among species.

## Introduced Species

Another way of seeing how important predation and competition are in maintaining population equilibria is to observe what happens when species from foreign ecosystems are introduced. Over the past 500 years, and

## The Village Weaverbird: Marvel or Menace?

David Lahti
*University of Massachusetts*

*After attending Gordon College, David Lahti earned a Ph.D. in philosophy at the Whitefield Institute at Oxford in 1998, for his work on the relationship between natural science and morality. In 2003 he received a Ph.D. from the University of Michigan in ecology and evolutionary biology, and moved to the University of Massachusetts where he is a Darwin Postdoctoral Research Fellow. He has studied weaverbirds for the last five years in Africa, the Indian Ocean, and the Caribbean, with his wife April.*

A lone African tree flickers yellow and white as if it were on fire. Someone approaching it from a distance eventually makes out the forms of dozens of birds, hanging upside-down by their feet from globular nests, swaying frantically as they babble a complex song and flutter their outstretched wings. They are male village weaverbirds, advertising their nestbuilding accomplishments in synchrony to a nearby flock of females. The males do have something to boast about, as weavers are considered to be the most proficient nest-builders in the bird world. They weave their nests as tightly as baskets, and use particular hitches and knots to tie off the ends of certain strands and to attach the nests to branches. Once they've chosen a nest, each individual female lays eggs of similar appearance throughout her lifetime, but the eggs of the population as a whole have among the widest variety of colors and spotting patterns of any bird. Laying a distinctive egg helps a female ensure that the eggs in her nest are her own, and not those of some African cuckoos, which mimic the eggs of other birds and lay eggs in their nests.

For most people who live with the village weaver, however, appreciating its nestbuilding and egg artistry can be difficult. This species is a serious agricultural pest throughout most of its range. A margin of trees near a field of rice or sorghum can be filled with village weaver colonies, as many as three hundred birds nesting in a single tree. This bird, together with another weaver species (the quelea) poses the most significant threat to agriculture in West Africa. Rafael Jawo, a farmer in The Gambia, loses up to a third of his annual yield from dense flocks of village weavers descending on his ricefields.

The village weaver has been introduced to islands in the Caribbean and the Indian Ocean, where it has become invasive, overpopulated, and a threat to native bird species. There too the weaver ravages agricultural lands, and is the most damaging agricultural pest in Mauritius and Haiti. The village weaver has many traits that make it an excellent invader of new areas, such as the ability to raise several broods of young per season, a generalized diet, and a preference for human-altered habitats. Moreover, it is regularly transported to new areas in the cage bird trade, and often escapes from captivity. In the last two decades the village weaver has been sighted in the wild for the first time in North and South America, and Europe. Many of these birds attempted to breed, and some were successful.

Throughout the world, humans have unwittingly made a menace out of the village weaver. In its natural range and with only wild food sources, it would have less volatile population dynamics and little adverse impact on human economies. This was probably the case throughout its range before monocultures became established in Africa following European colonization. Even today in some areas, such as central Uganda and eastern South Africa, the village weaver is a functioning element in the ecosystem without undergoing population explosions or creating serious problems for agriculture. In both of these areas, small grains suitable for the weaver's diet are not in widespread cultivation.

What can be done to remedy the situation elsewhere in Africa and on the islands? Methods of physical control such as fire, scarecrows, rattles, shooting, nest-robbing, and felling trees have been attempted but have met with little success. Poison kills large numbers of individuals, but this does not affect long-term population sizes. When populations are dense, competition for food is more intense, fewer young are raised, and population growth slows. When managers poison the population, the large quantities of suitable grains mean that food is abundant for the survivors, and the population again increases exponentially. In Africa, then, destruction of birds may become a permanent part of agricultural practice.

Long-term alternatives may exist, however. One possibility is economic and agricultural diversification. Large stretches of land in many African nations are dedicated to the production of a single kind of crop. If a network of ricefields in central Gambia was converted into a patchwork of rice, maize, groundnuts, mangos, and palm oil, only a portion of the area would supply food for village weavers, and the more diverse ecosystem might support their competitors and predators.

In the introduced populations, eradication of the village weaver is another option, although the practicality of this varies. A common pattern in an introduced species is a period of slow or no population growth, followed by an unpredicted population explosion and spread. Although established populations like those in Mauritius and Haiti are difficult to control, both of these populations were once small and more manageable. During the period when the village weaver was not invasive, however, few observers would have predicted its future spread or impact. Today there are small, recently established populations of the village weaver in Venezuela and Martinique. We cannot know whether the village weaver will eventually become invasive in those areas or others. Nevertheless, it is possible that what is done (or not done) today may determine the economic and ecological future of those areas.

For areas not yet invaded, prevention is the best strategy. Transport and housing of caged birds are still largely unregulated in some European countries, such as France, which is the largest importer of weavers. If fewer birds were allowed to escape, the likelihood of a population becoming established would decrease accordingly.

The village weaver is a species with some remarkable traits. More importantly, it has been an integral part of the biological communities of sub-Saharan Africa for millions of years. To view this bird solely as an agricultural pest and an invasive species is short-sighted and anthropocentric, although these problems are very real and troubling. The village weaver is named for its preference for living near humans. Our challenge is to find a way to make this a peaceful coexistence.

especially now that there is a vast global commerce, thousands of species of plants, animals, and microbes have been accidentally or deliberately introduced onto new continents and islands (see Guest Essay, p. 96). A recent attempt to estimate the economic losses due to introduced species in the United States calculated the total cost to be in excess of $138 billion per year. A couple of examples suffice to make the point.

***Rabbits.*** In 1859, rabbits were introduced into Australia from England, to be used for sport shooting. The Australian environment proved favorable to the rabbits and contained no carnivore or other natural enemies capable of controlling them. As a result, the rabbit population exploded and devastated vast areas of rangeland by overgrazing (Fig. 4–11). The devastation was extremely damaging to both native kangaroos and ranchers' sheep. It was finally brought under control by introducing a disease-causing virus in rabbits. Over time, however, the rabbits adapted to the virus and began to repeat their explosive growth. After the recent release of a different virus, rabbits have undergone a 95% decline in some regions, and kangaroos and rare plants are thriving once again.

***American Chestnut.*** Prior to 1900, the dominant tree in the eastern deciduous forests of the United States was the American chestnut, which was highly valued for both its high-quality wood and its prolific production of chestnuts, eaten by wildlife and people alike. In 1904, however, a fungal disease called the chestnut blight was accidentally introduced when some Chinese chestnut trees carrying the disease were planted in New York. The fungus spread through the forests, killing nearly every American chestnut tree by 1950. Although oaks filled in where the chestnuts died, the ecological and commercial loss was incalculable. There is hope, however, because researchers have recently crossbred the American and Chinese chestnut, creating a hybrid that is both 94% native and resistant to the blight. It will be several decades, though, before enough of the hybrids are available for sale to nurseries and wider distribution.

***Pests.*** Most of the important insect pests in croplands and forests—Japanese beetles, fire ants, and gypsy moths, for example—are species introduced from other ecosystems. Domestic cats introduced into island ecosystems have often proved to be effective predators and have exterminated many species of wildlife unique to the islands. They are also responsible for greatly diminished songbird populations in urban and suburban areas, including parks (Chapter 10). Because of their voracious appetites, goats introduced onto islands have been devastating to both native plants and the animals that depend on that vegetation.

Despite all that is known about plants and animals, the problem of introduced species is increasing, not decreasing, due to expanding world trade and travel. In the 1980s, for example, the zebra mussel and the closely related quagga mussel were introduced into the Great Lakes with the discharge of ballast water from European ships (Fig. 4–12a). The mussels are now spreading through the Mississippi River basin and may go much farther, causing untold ecological and commercial damage as they displace native mussel species and clog water-intake pipes. These problems may be exported as well as imported. In 1982, several species of jellyfishlike animals known as ctenophores were similarly transported from the east coast of the United States to the Black Sea and the Sea of Azov in Eastern Europe. The ctenophores have cost Black Sea fisheries an estimated $250 million and have totally shut down fisheries in the Sea of Azov, because they kill larval fish directly and deprive larger fish of food (Fig. 4–12b).

***Plants.*** Plant species have also been moved all over the world, often by horticulturists. In some cases, they have wreaked havoc. In 1884, for example, the water hyacinth,

**Figure 4–11 Rabbits overgrazing in Australia.** On one side of a rabbitproof fence, there is lush pasture; on the other side, the land is barren.

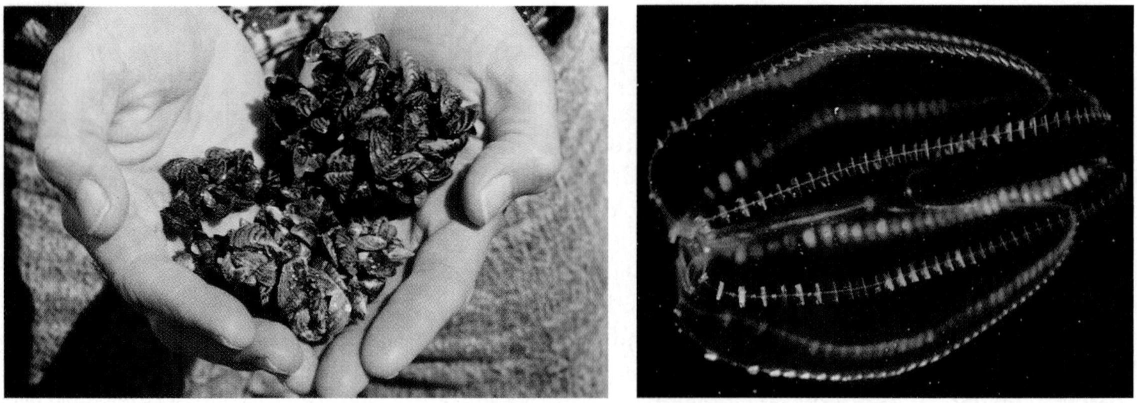

(a) (b)

**Figure 4–12 Introduced species.** (a) Zebra mussels introduced from Europe are now proliferating throughout the Great Lakes and the Mississippi valley. (b) Ctenophores originating on the South Atlantic Coast of the United States and introduced into Europe have destroyed fishing in the Black Sea and the Sea of Azov.

a plant originally from South and Central America, was introduced into Florida as an ornamental flower. It soon escaped into waterways, where it had little competition and few natural enemies. It eventually proliferated to the extent that navigation was difficult or impossible on some waters. Lakes and ponds were covered with up to 200 tons of hyacinths per acre (Fig. 4–13a). With a combination of herbicides, harvesting, and biological control by several insect species, Florida's water hyacinth problem is now under "maintenance control"; the hyacinth is still present, but constant surveillance keeps it from proliferating. Sadly, this same species has now spread throughout the major rivers and lakes of Africa, where it is presently causing enormous economic and biological damage.

Kudzu, a vigorous vine introduced from Japan in 1876, was widely planted on farms throughout the southeastern United States, so it could be used for cattle fodder and erosion control. From wherever it is planted, however, kudzu invades and climbs over adjacent forests. The species now occupies 7 million acres of the deep South (Fig. 4–13b). Because it is a threat to forest growth, considerable efforts are being exerted by the Forest Service to contain it.

Spotted knapweed (Fig. 4–13c), which was probably unwittingly introduced into this country with alfalfa seed imported from Europe in the 1920s, has spread over millions of acres of rangelands in the northwestern United States and southwestern Canada. Bitter tasting and virtually inedible, knapweed is endangering wildlife such as elk and rendering lands worthless for grazing by domestic cattle as it displaces native plants and grasses. Recent studies have indicated that the knapweed thrives because it has no herbivore enemies and because it secretes chemicals from its roots that inhibit the growth of neighboring plants. Knapweed has also been found to promote greater soil erosion where it is abundant. To date, control efforts have been unable to check the spread of this invader.

Wetlands throughout temperate regions of the United States and Canada, already reduced by over 50% by development pressures, are now being further degraded by the invasion of purple loosestrife, which was introduced from Europe in the 1800s as an ornamental and medicinal plant (Fig. 4–13d). By outcompeting and eliminating native wetland vegetation and being inedible itself, purple loosestrife is threatening many wildlife species, including waterfowl. Control of these kinds of weeds by introducing plant-eating insects is being investigated; however, it is hard to find an insect (or any other organism, for that matter) that will control the target weed, but not attack desired species. (See Chapter 16.)

***Lessons.*** The ecological lessons to be learned from the introduction of undesirable species are two. First, the regulation of populations is a matter of complex interactions among the members of the biotic community. Second, and just as important, the relationships are specific to the organisms in each particular ecosystem. Therefore, when a species is transported over a physical barrier from one ecosystem to another, it is unlikely to fit into the framework of relationships in the new biotic community. In most cases, it finds the environmental resistance of the new system too severe and dies out. No harm is then done. In some instances, the introduced species simply joins the native flora or fauna and does no measurable harm (as in the case of the ring-necked pheasant, a valuable game bird). The new species has then become *naturalized.*

In the worst cases, the transported species becomes *invasive.* It finds physical conditions and a food supply that are hospitable, together with an insufficient number of natural enemies to stop its population growth. Then its population explodes, and it drives out native species by outcompeting them for space, food, or other resources (if not by predation). Such disruptions may be caused by any category of organism—plant, herbivore, carnivore, or parasite—large or small. Fortunately, only a small percentage of successfully transplanted species has become invasive—some 10% of the 7,000 alien species in the United States.

(a) (b) (c) (d)

**Figure 4–13 Introduced plant species.** (a) Water hyacinth overgrowing waterways. (b) Kudzu overgrowing forests. (c) Spotted knapweed taking over rangeland in the northwestern United States and southwestern Canada. (d) Purple loosestrife, a native of Eurasia, is now abundant in wetlands of the northeastern United States and Canada.

*Remedies?* One solution to the takeover by an invasive species may be to introduce a natural enemy. Indeed, this approach has been used in a number of cases, including rabbit control in Australia. Other approaches are discussed in Chapter 16 in connection with the biological control of pests. Unfortunately, control with a natural enemy is more easily said than done. Recall that control is achieved through the interplay among all the factors of environmental resistance, which often include several natural enemies as well as *all* the abiotic factors. Thus, a single natural enemy that will control a pest simply may not exist. In fact, there is not even any guarantee that the natural enemy, when introduced into the new ecosystem, will focus its attention on the target pest. Control of the rabbit population in Australia was initially attempted by introducing foxes. The foxes soon learned, however, that they could catch other Australian wildlife more easily than rabbits and thus went their own way. In short, to prevent doing more harm than good, a great deal of research needs to be done before a natural enemy is introduced.

Oddly, most invasive species are not problems in their native lands. Their impact is so different in a new setting, though, because ecosystems on different continents or remote islands have been isolated by physical barriers for millions of years. Consequently, the species within each ecosystem have developed adaptations to other species *within their own ecosystem, and these are independent of adaptations that have developed in other ecosystems.* How these kinds of adaptations (and many others) have come about is discussed in the next section.

## 4.3 Mechanisms of Species Adaptation

### Change through Natural Selection

Predation and competition are two important mechanisms that keep natural populations under control. Although it was not discussed in Section 4.2, predators and prey become well adapted to each other's presence. Predators rarely are able to eliminate their prey species, largely because the prey have various defenses against their predators. Intraspecific competition also represents a powerful force that can lead to improved adaptations of a species to its environment. Interspecific competition, by contrast, promotes adaptations in the competitors that allow them to specialize in exploiting a resource. This specialization can lead to resource partitioning that allows many potential competitors to share a basic resource (like light in a forest or space in spruce trees). In this section, we discuss how these adaptations have come about.

***Selective Pressure.*** Most young plants and animals in nature do not survive; instead, they fall victim to various environmental resistance factors. These factors—predators, parasites, drought, and other—are known as **selective pressures.** That is, each factor can affect which individuals survive and reproduce and which are eliminated. If a predator is present, for example, prey animals having traits that protect them or that allow them to escape from their enemies (such as coloration that blends in with the background) tend to survive and reproduce (Fig. 4–14), and those without such traits tend to become the predator's dinner. Any individual with a gene that slows it down or makes it conspicuous will tend to be eaten. Thus, predators may be seen as a selective force favoring the survival of genes that enhance the prey's ability to escape or protect itself and causing the elimination of any genes handicapping those functions. The need

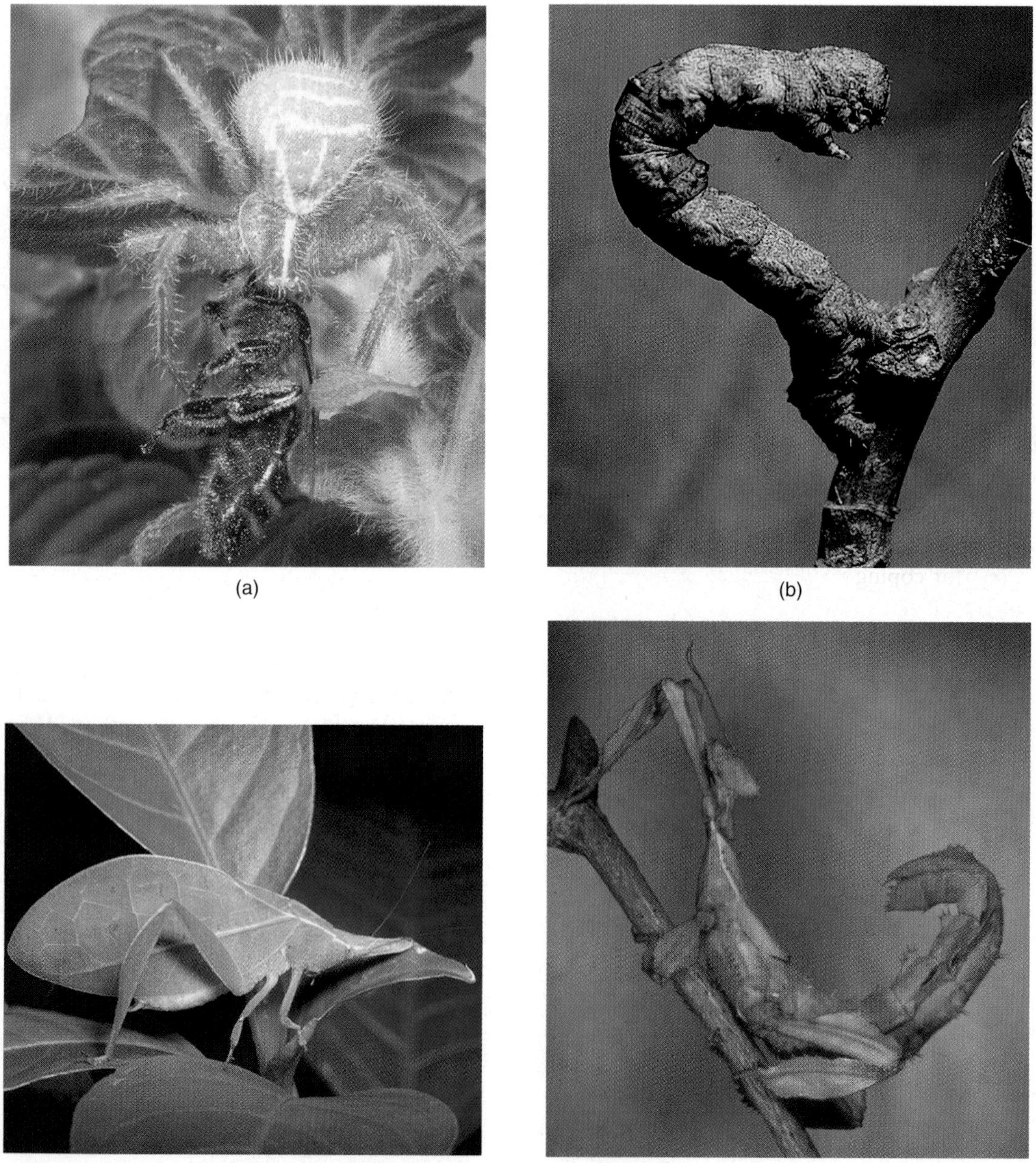


**Figure 4–14 Adaptation of certain species.** Modifications of body shape and color that allow species to blend into the background and thus protect their populations from predation are among the most amazing adaptations. Shown are (a) the Thomisus spider, (b) the spanworm, (c) the leaf katydid, and (d) the giant Australian stick insect.

for food can also be seen as a selective pressure acting on the predator, enhancing those characteristics benefiting predation—such as keen eyesight and swift speed.

*Every* factor of environmental resistance is a selective pressure resulting in the survival and reproduction of those individuals with a genetic endowment that enables them to cope with their surroundings. In nature, there is a constant selection and, consequently, a modification of a species' gene pool toward features that enhance survival and reproduction *within the existing biotic community and environment.* Because the process occurs naturally, it is known as **natural selection.**

These concepts were first presented by Charles Darwin in his book *The Origin of Species by Natural Selection* (1859). The modification of the gene pool that occurs through natural selection over the course of many generations is the sum and substance of *biological evolution.* Darwin deserves tremendous credit for constructing his theory purely from his own observations, without any knowledge of genes or genetics—information that wasn't discovered until several decades later. Today, our modern understanding of DNA, mutations, and genetics fully supports Darwin's theory of evolution by natural selection.

**Adaptations to the Environment.** Under the selective pressures exerted by the factors of environmental resistance, the gene pool of each population is continually tested. Indeed, virtually all traits of any organism can be seen as features that adapt the organism for survival and reproduction, or, in Darwinian terms, **fitness.** Essentially all characteristics of organisms can be grouped as follows:

- Adaptations for coping with climatic and other abiotic factors.
- Adaptations for obtaining food and water (for animals) or for obtaining nutrients, energy, and water (for plants).
- Adaptations for escaping from or protecting against predation and for resistance to disease-causing or parasitic organisms.
- Adaptations for finding or attracting mates (in animal populations) or for pollinating and setting seed (in plant populations).
- Adaptations for migrating (animals) or for dispersing seeds (plants).

The fundamental question about any trait is, Does it facilitate survival and reproduction of the organism? If the answer is yes, the trait will be maintained through natural selection. Consequently, various organisms have evolved different traits to accomplish the same function. For example, the ability to run fast, to fly, or to burrow, and protective features such as quills, thorns, and an obnoxious smell or taste, all help reduce predation and can be found in various organisms (Fig. 4–15).

**The Limits of Change.** When facing a new, powerful selective pressure—such as a different climatic condition or a new species invading the ecosystem—species have three, and only three, alternatives:

1. **Adaptation.** The population of survivors may gradually adapt to the new condition through natural selection.
2. **Migration.** Surviving populations may migrate and find an area where conditions are suitable to them.
3. **Extinction.** Failing the first two possibilities, extinction is inevitable.

Migration and extinction need no further explanation. The critical question is, What factors determine whether a species will be able to adapt to new conditions, instead of failing and becoming extinct?

Recall that adaptation occurs by selective pressures eliminating all those individuals that cannot tolerate the new condition. *For adaptation to occur, there must be some individuals with traits (alleles) that enable them to survive and reproduce under the new conditions.* There must also be enough survivors to maintain a viable breeding population. If a breeding population is maintained, the process of natural selection should lead to increased adaptation over successive generations. If a viable population is *not* maintained at any stage, extinction is assured. For example, the California condor population had declined to about 20 birds by the early 1980s. To save the species from extinction, wildlife biologists captured the remaining wild condors and initiated a breeding program. The population is now over 200 individuals, and about 80 birds are now living in the wild. This is still not a viable breeding population, but the wild condors are now forming pairs and nesting, and biologists eagerly anticipate the survival of the first wild-born condors since they were taken into captivity.

***Keys to Survival.*** There are four key variables among species that will affect whether or not a viable population of individuals is likely to survive new conditions: (1) geographical distribution, (2) specialization to a given habitat or food supply, (3) genetic variation within the gene pool of the species, and (4) the reproductive rate relative to the rate of environmental change.

Consider how these factors work. It is unlikely that any change will affect all locations uniformly or equally. Therefore, a species such as the housefly—which is present over most of Earth, does not have requirements for a specialized habitat or food supply, and has a high degree of genetic variation—is likely to survive almost any conceivable alteration in Earth's environment. The California condor, on the other hand, requires an extensive habitat of mountains and ravines and the carcasses of large animals to eat and has little genetic variation in its population.

Assuming the survival of a viable population, how fast can that population evolve further adaptations to

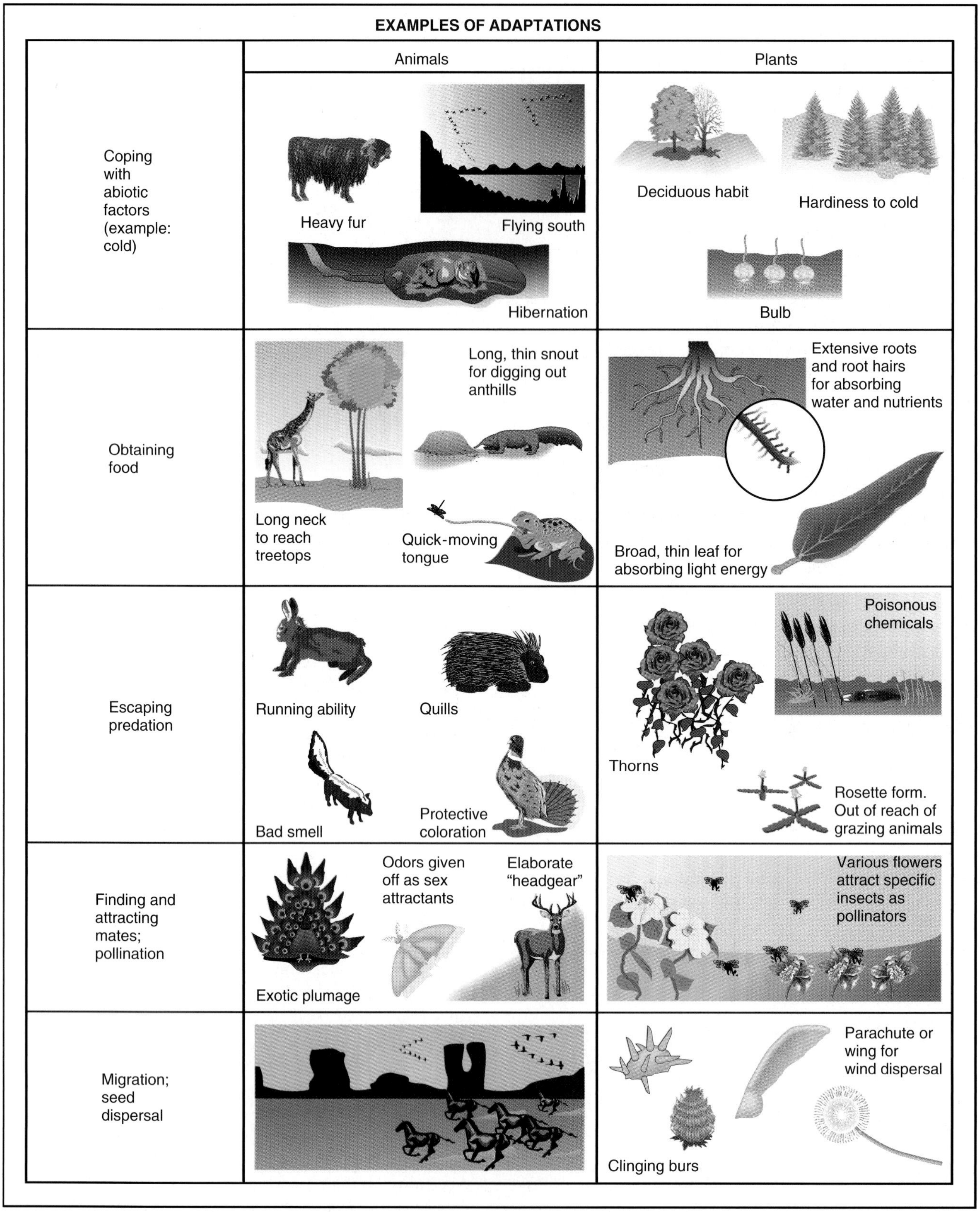

EXAMPLES OF ADAPTATIONS
Animals
Plants
Coping with abiotic factors (example: cold)
Heavy fur
Flying south
Hibernation
Deciduous habit
Hardiness to cold
Bulb
Obtaining food
Long neck to reach treetops
Long, thin snout for digging out anthills
Quick-moving tongue
Extensive roots and root hairs for absorbing water and nutrients
Broad, thin leaf for absorbing light energy
Escaping predation
Running ability
Quills
Bad smell
Protective coloration
Thorns
Poisonous chemicals
Rosette form. Out of reach of grazing animals
Finding and attracting mates; pollination
Exotic plumage
Odors given off as sex attractants
Elaborate "headgear"
Various flowers attract specific insects as pollinators
Migration; seed dispersal
Clinging burs
Parachute or wing for wind dispersal


**Figure 4–15 Adaptation for survival and reproduction.** The five general features listed at the left are essential for the continuation of every species. Each feature is found in each species as particular adaptations that enable the species to survive or reproduce. Across the various species, a multitude of adaptations will accomplish the same function. Thus, a tremendous diversity of species exists, with each species adapted in its own special way.

enable it to better cope with new or changing conditions? There is no genetic change, and hence no genetic adaptation, over the lifetime of the *individual.* Genetic change occurs only as the genes are "shuffled" during the process of sexual reproduction and as new mutations are introduced in the process. There must be enough offspring to afford the loss of those offspring with unfavorable traits. Again, species such as houseflies, which produce a new generation every few weeks and which produce several hundred offspring in each generation, can evolve much more rapidly than condors, which produce, at best, one or two chicks a year.

The rate at which changes in the environment occur is an important consideration. If environmental changes are very slow, populations of slowly reproducing species may be able to adapt. When changes in environmental conditions occur more rapidly, however, more and more species will be eliminated in the contest to adapt accordingly. These factors are summarized in Fig. 4–16. Currently, one of the most devastating trends is the loss of biodiversity due to human impacts and intrusions on the environment. These environmental changes are happening more rapidly than ever before, and their impact is being seen in the toll of wild plants and animals becoming scarce or extinct.

**Speciation.** There may be as many as 14 million species of plants, animals, and microbes currently in existence, all living and functioning in ecosystems and all contributing to an amazing biodiversity. That biodiversity is diminishing, however, because present species are becoming extinct faster than new species are appearing. In order to understand why, you need to understand how new species appear.

The infusion of new variations from mutations and the pressures of natural selection serve to adapt a species to the biotic community and the environment *in which it exists.* In this process of adaptation, the final "product"—giraffe, anteater, redwood tree—may be so different from the population that started the process that it is considered a different species. This is one aspect of the process of speciation.

***Prerequisites.*** The same process also may result in two or more species developing from one. There are only two prerequisites. The first is that the original population must separate into smaller populations that do not interbreed with one another. This **reproductive isolation** is crucial, because if the subpopulations continue to interbreed, all the genes will continue to mix through the entire population, keeping it as one species. The second prerequisite is that separated subpopulations must be exposed to different selective pressures. As the separated populations adapt to these pressures, they may gradually become so different as to be considered different species and thus be unable to interbreed with one another, even if they come together again later.

Consider, for example, the arctic and gray foxes of today. It is believed that thousands of years ago a northern subpopulation became separated from the main population, perhaps due to the effects of glaciation in North America. In the Arctic, selective pressures favor individuals that have heavier fur; a shorter tail, legs, ears, and nose (the shortness helps conserve body heat); and white fur (which helps the animals hide in the snow). In the southern regions, selective pressures favor individuals with the opposite traits. Mutations creating genes adaptive for the Arctic would actually be harmful in southern animals, because in warmer climates animals need a thinner coat to dissipate excessive body heat and white fur would make the animals more conspicuous. The geographic isolation of the two subpopulations over many generations, however, made it possible for one ancestral fox population to develop into two separate species (Fig. 4–17).

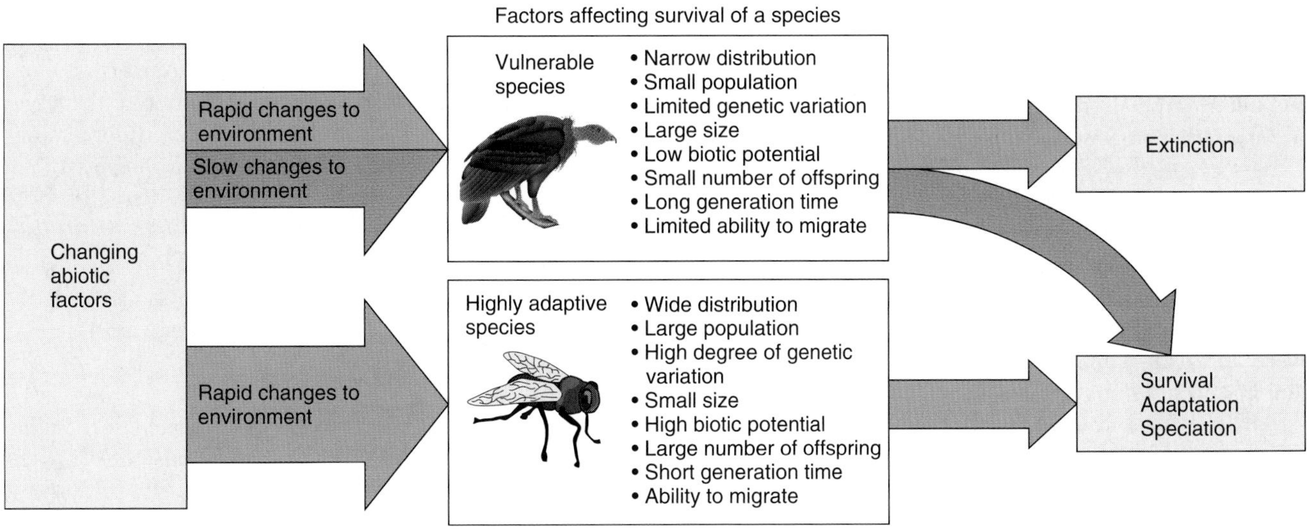


**Figure 4–16 Vulnerability of different organisms to environmental changes.** A summary of factors supporting the survival and adaptation of species, as opposed to their extinction.

**Figure 4–17 Evolution of new species.** A population spread over a broad area may face various selective pressures. If the population splits so that interbreeding among the subpopulations does not occur, the different pressures may result in the subpopulations evolving into new species, as shown here for the arctic fox and the gray fox.

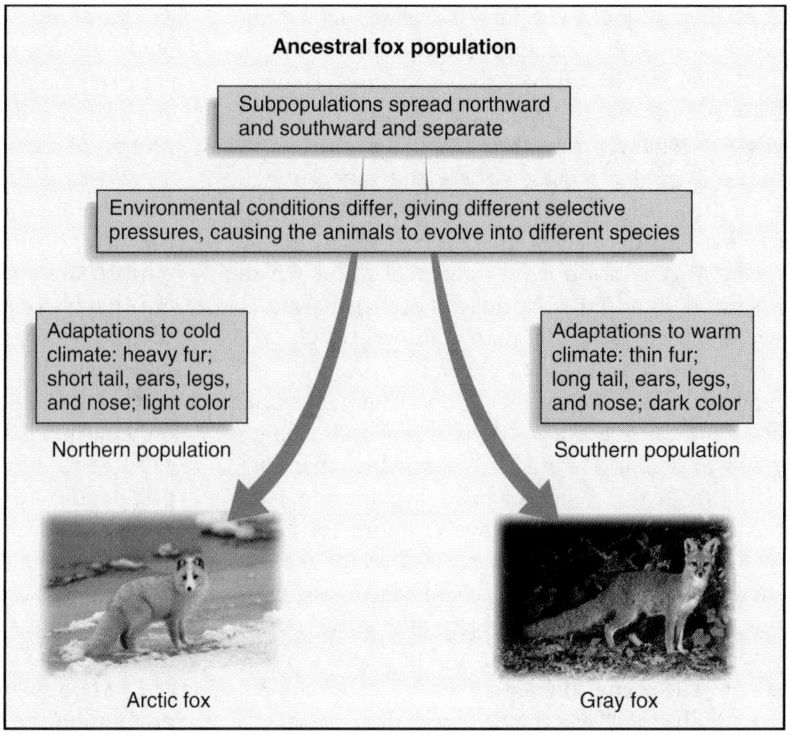


*Darwin's Finches.* A renowned example of speciation is one first observed by Darwin himself: the 14 species of finches living on the Galápagos Islands. Darwin was struck by the differences between finches on different islands. Their overall body structures were similar, however, so Darwin could see that all were members of the finch family. He wondered, though, whether the differences in size and beak structure could have appeared after the subpopulations were isolated from one another on separate islands (Fig. 4–18).

Although Darwin himself did not work out the details of how the different Galápagos finches evolved, it is likely that at some time in the past a few finches from the South American mainland were blown westward by a strong storm and became the first terrestrial birds to inhabit the relatively new (10,000-year-old) volcanic islands. As the initial population grew and members faced increased intraspecific competition, some birds dispersed to nearby islands, where they were separated from the main population. These subpopulations encountered different selective pressures and became specialized for feeding on different things (cactus fruit, insects). In time, when the changed populations dispersed back to their original islands, they were different enough from the parent species that they were distinguishable as new species, so interbreeding among them did not occur.

In sum, the key feature of speciation is that new species are not formed from scratch; they are formed only by the gradual modification of existing species. In addition, the gene pool of a species may be molded in many different directions when different populations within the species are isolated and subjected to different selective pressures, as illustrated by Darwin's finches. This concept is entirely consistent with the observation that groups of closely related species are generally found in nature, not distinct species with no close relatives. (Recall the five species of warblers in Fig. 2–17.) It is also consistent with, and explains why, all living things can be classified into a few relatively major groups. All the members of these major groups are variations on the same theme and, hence, derived from common ancestors.

Although it is difficult to imagine, the present array of plants, animals, and microbes, in all of its diversity, is likely to have originated through speciation over long periods of time and in every geographic area on Earth. This, then, is the source of our current biodiversity.

In the next section, we turn our attention to the larger picture of what happens in ecosystems when populations interact with each other and with the environment and, in the process, determine the structure and function of those ecosystems.

## 4.4 Ecosystem Responses to Disturbance

According to *equilibrium theory,* ecosystems are stable environments in which species interact constantly in well-balanced predator–prey and competitive relationships. Thus, *biotic interactions* determine the structure of living communities within ecosystems. This approach has led to

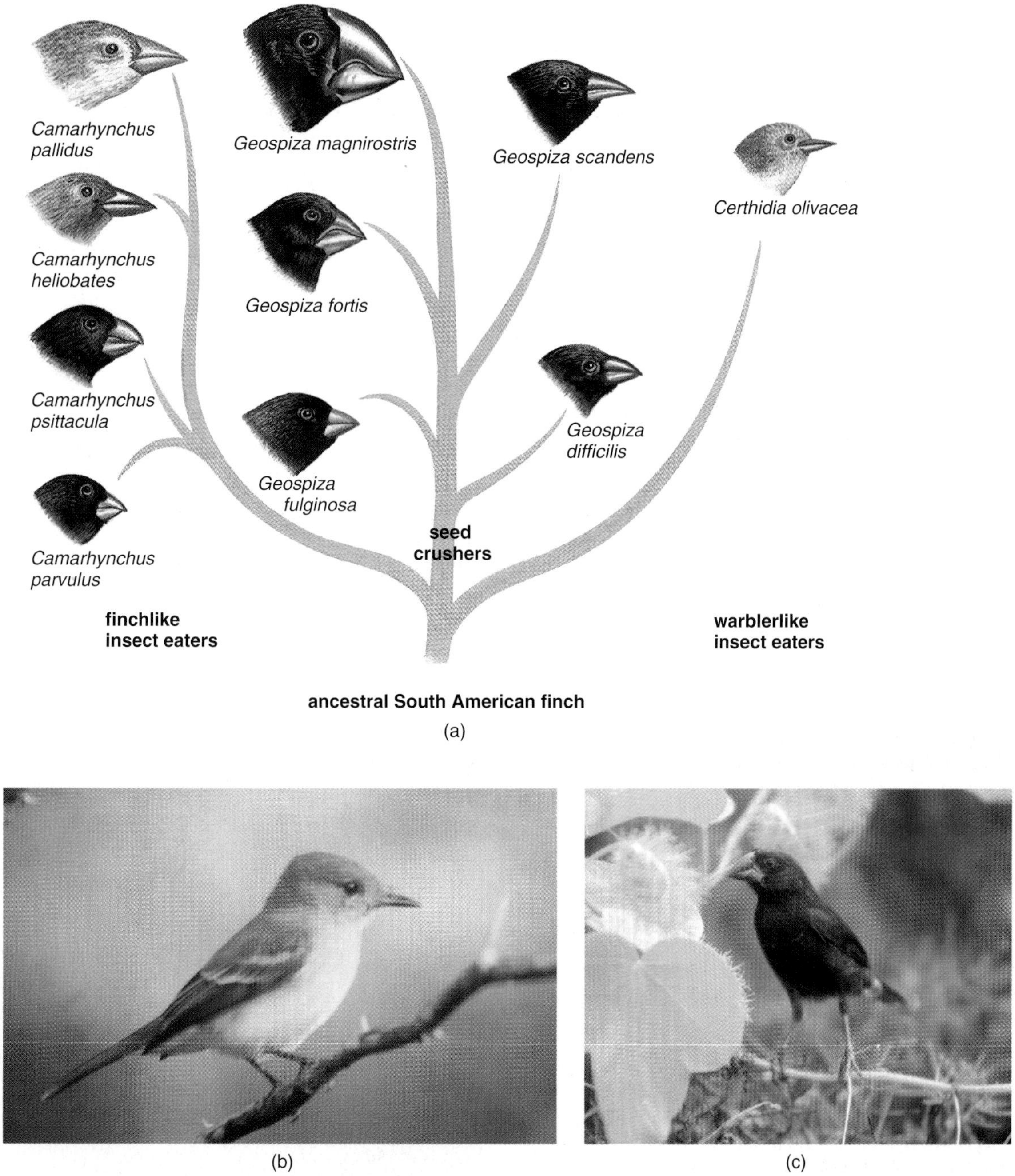


**Figure 4–18 Some of Darwin's finches.** (a) The similarities among these birds attest to their common ancestor. Selective pressures to feed on different foods have caused modification and speciation in adapting subpopulations. (b) The insect-eating *Certhidia olivacea.* (c) The large seed-crushing beak of *Geospiza magnirostris.* Notice the difference in beak between (b) and (c). (From p. 310 of *Biology: Life on Earth*, 4th ed., by Teresa Audesirk & Gerald Audesirk. Copyright © 1994 by Prentice Hall, Inc. Reprinted by permission of Pearson Education, Inc., Upper Saddle River, NJ 07458.)

the popular idea of "the balance of nature": Natural systems maintain a delicate balance over time that lends them great stability. Up to now, though, the dimension of *time* has been largely ignored. Are species and ecosystems at equilibrium all the time? If the environment were entirely stable and uniform, the answer would probably be yes. This is not the case, however, so how do ecosystems respond to disturbances?

*Nonequilibrium Systems.* To answer the preceding question, we must examine the possibility that time brings disturbances that give us a different picture of the sustainability of ecosystems. Indeed, many studies have shown that ecosystems are, in fact, very patchy environments. That is, conditions and resources within them vary with temperature, moisture, exposure to sunlight, soil conditions, and the like. The distribution of species is similarly patchy, reflecting the patchiness in conditions. In a forest, for example, the species of trees commonly vary independently in space, such that it is difficult to predict what species will tend to be found associated together in a given stand at a given time. But if neither conditions nor distributions of species are uniform even within ecosystems, then perhaps the concept of a stable equilibrium should be questioned. Indeed, the patchiness of ecosystems has caused many ecologists to think of them as nonequilibrium systems that seldom exhibit the characteristics of a true equilibrium. This controversy is well illustrated in the phenomenon known as ecological succession.

## Ecological Succession

Over the course of years, one biotic community may gradually give way to a second, the second perhaps to a third, and even the third to a fourth. In these cases, the ecosystem is changing. This phenomenon of transition from one biotic community to another is called **ecological**, or **natural, succession.** Succession occurs because the physical environment may be gradually modified by the growth of the biotic community itself, such that the area becomes more favorable to another group of species and less favorable to the current occupants. Pioneer species start the process, but as these grow, they create conditions that are favorable to more longer lived colonizers, and the process in general is driven by the changing conditions that pave the way for other species. This process is known as **facilitation.**

The succession of species does not go on indefinitely. A stage of development is eventually reached in which there appears to be a dynamic balance between all of the species and the physical environment. This final state is called a **climax ecosystem**, because the assemblage of species continues on in space and time. The major biomes discussed up to this point in the text have been climax ecosystems. Keep in mind, though, that all balances are relative to the current biotic community and the *existing* climatic conditions. Therefore, even climax ecosystems are subject to change if climatic conditions change or if new species are introduced or old ones are removed. Nevertheless, natural succession may be seen as a progression toward a relatively more stable climax—one that no longer changes over time. Sometimes, there may be several "final" stages, or a polyclimax condition, with adjoining ecosystems in the same environment at different stages. Three classic examples are presented next.

**Primary Succession.** If the area has not been occupied previously, the process of initial invasion and then progression from one biotic community to the next is called *primary succession.* An example is the gradual invasion of a bare rock surface by what eventually becomes a climax forest ecosystem. Bare rock is an inhospitable environment. It has few places for seeds to lodge and germinate, and if they do, the seedlings are killed by lack of water or by exposure to wind and sun on the rock surface. However, certain species of moss are uniquely adapted to this environment. Their tiny spores, specialized cells that function reproductively, can lodge and germinate in minute cracks, and moss can withstand severe drying simply by becoming dormant. With each bit of moisture, moss grows and gradually forms a mat that acts as a sieve, catching and holding soil particles as they are broken from the rock or as they blow or wash by. Thus, a layer of soil, held in place by the moss, gradually accumulates (Fig. 4–19).

The mat of moss and soil provides a suitable place for seeds of larger plants to lodge, and the greater amount of water held by the mat supports their germination and growth. The larger plants in turn collect and build additional soil, and eventually there is enough soil to support shrubs and trees. In the process, the fallen leaves and other litter from the larger plants smother and eliminate the moss and most of the smaller plants that initiated the process. Thus, there is a gradual succession from moss through small plants and, finally, to trees that form a climax forest ecosystem. The nature of the climax ecosystem differs according to the prevailing abiotic factors of the region, yielding the biomes typical of different climatic regions, as described in Chapter 2.

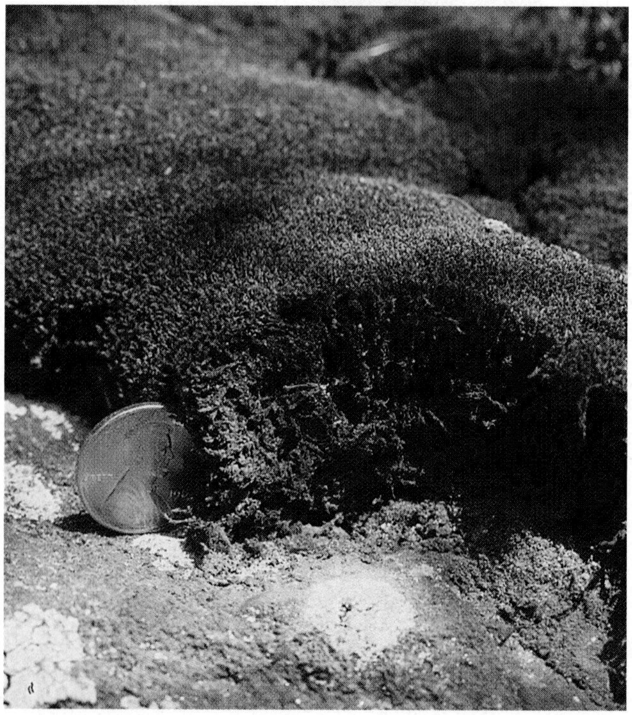

**Figure 4–19 Primary succession on bare rock.** Moss invades bare rock and acts as a collector, accumulating a layer of soil sufficient for additional plants to become established.

Because bare rock substrate can be exposed by retreating glaciers, earthquakes, landslides, and volcanic eruptions, there are always places for primary succession to start anew.

**Secondary Succession.** When an area has been cleared by fire or by humans and then left alone, plants and animals from the surrounding ecosystem may gradually reinvade the area—not at once, but through a series of distinct stages called *secondary succession.* The major difference between primary and secondary succession is that secondary succession starts with preexisting soil. Thus, the early, prolonged stages of soil building are bypassed. Still, a clear area has a microclimate quite the opposite from the cool, moist, shaded conditions beneath a forest canopy. Those plant species which propagate themselves in the microclimate of the forest floor cannot tolerate the harsh conditions of the clearing. Hence, the process of reinvasion begins with different species. The steps leading from abandoned agricultural fields in the eastern United States back to deciduous forests provide a classic example of secondary succession (Fig. 4–20a).

On an abandoned agricultural field, crabgrass predominates among the initial invaders. Crabgrass is particularly well adapted to invading bare soil. Its seeds germinate in the spring, and it grows and spreads rapidly by means of runners; moreover, it is exceptionally resistant to drought. Despite its vigor on bare soil, crabgrass is easily shaded out by taller plants. Consequently, taller weeds and grasses, which take a year or more to develop, eventually take over from the crabgrass. Next, young pine trees, which are well adapted to thrive in the direct sunlight and heat of open fields, gradually move in and shade out the smaller, sun-loving weeds and grasses, eventually forming a pine forest. But pine trees also shade out their own seedlings, which need bright sun to grow. Thus, the seedlings of deciduous trees, not pines, develop in the cool shade beneath the pine trees (Fig. 4–20b). Moreover, the pine trees have a limited life span, so they are replaced by oaks, hickories, beeches, maples, and other species of hardwoods. The seedlings of these deciduous trees continue to flourish beneath the cover of their parents, providing a stable balance—the climax eastern deciduous forest ecosystem.

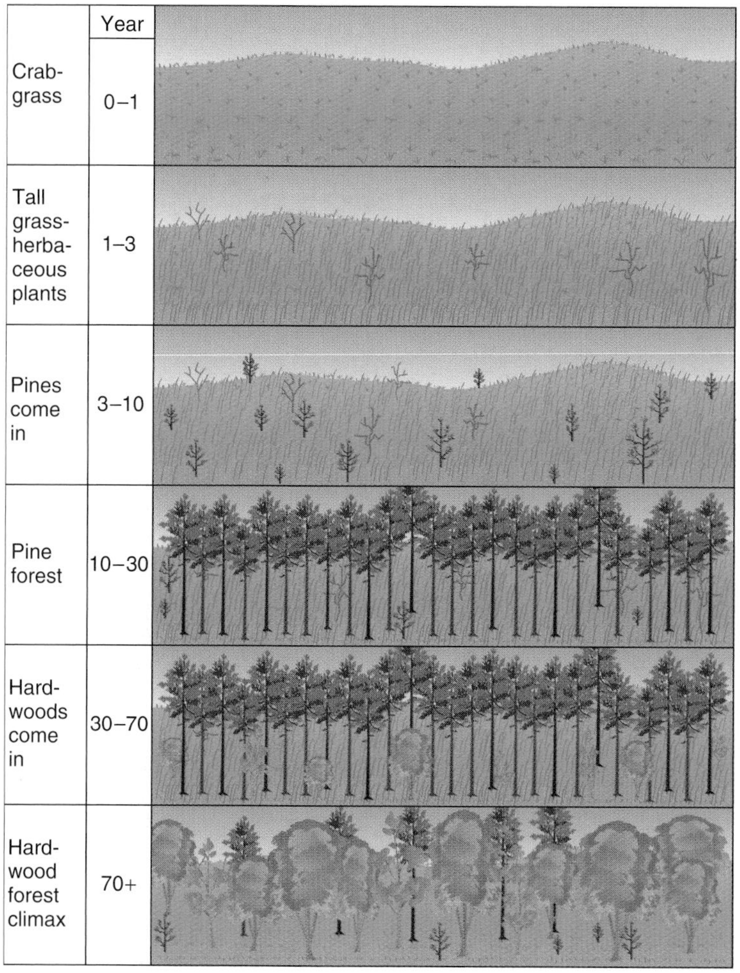


(a)

(b)

**Figure 4–20** **Secondary succession.** (a) Reinvasion of an agricultural field by a forest ecosystem occurs in the stages shown. (b) Hardwoods (species of oak) growing up underneath and displacing pines in eastern Maryland.

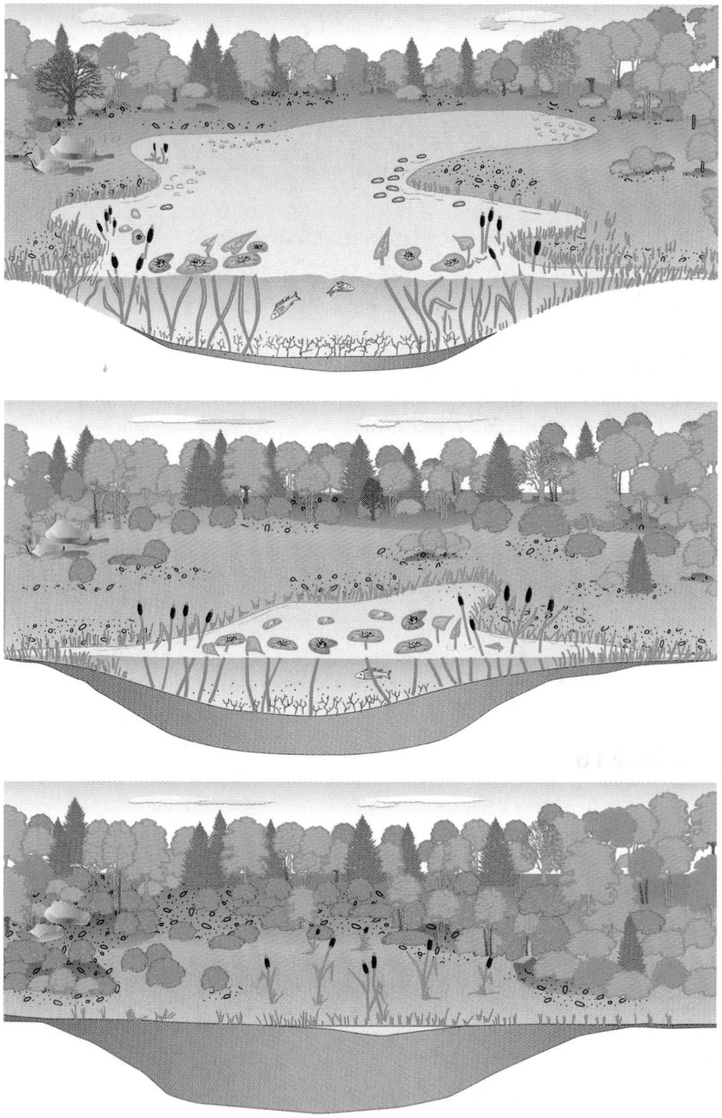

(a)

**Figure 4–21 Aquatic succession.** (a) Ponds and lakes are gradually filled and invaded by the surrounding land ecosystem. (b) In this photograph, taken in Banff National Park in the Canadian Rockies, you can visualize the lake that used to exist in the low-level area. It is now filled with sediment and covered by scrub willow. Spruce and fir forest is gradually encroaching.

**Aquatic Succession.** Natural succession also takes place in lakes or ponds. Succession occurs because soil particles inevitably erode from the land and settle out in ponds or lakes, gradually filling them. Aquatic vegetation produces detritus that also contributes to the filling process. As the buildup occurs, terrestrial species from the surrounding ecosystem can advance, and aquatic species must move farther out into the lake. In time, the shoreline gradually advances toward the center of the lake until, finally, the lake disappears altogether (Fig. 4–21). Here, the climax community may be a bog or a forest.

**Climate Change.** Pollen records indicate that the climate itself has changed over spans of hundreds and thousands of years, as glaciation and changes in the Earth's orbit have occurred (Chapter 20). In response to past climate changes, the forests of the temperate zone have shifted from coniferous to deciduous, with many changes in the dominant species of trees. Thus, a so-called climax ecosystem may only be a figment of our limited perspective. Note, however, that these changes have taken place over many hundreds or thousands of years. What happens if the climate changes more rapidly? Can forests adapt to a warming that takes place over decades?

Currently, one of the most serious environmental issues is global climate change. The burning of fossil fuels has increased the level of carbon dioxide in the atmosphere to the point where, by the mid-21st century, it will have doubled since the beginning of the Industrial Revolution. Because carbon dioxide is a powerful greenhouse gas (Chapter 20), a general warming of the atmosphere is expected, leading to many other effects on the Earth's climate. Numerous ecosystems will be affected by the changes, and it is questionable whether they can maintain their functional sustainability in the face of such rapid changes. The threat of disturbance on such an

(b)

**Figure 4–21b**

enormous scale is a major reason many nations of the world have begun to take steps to reduce emissions of greenhouse gases.

## Disturbance and Resilience

In order for natural succession to occur, the spores and seeds of the various invading plants and the breeding populations of the various invading animals must already be present in the vicinity. Ecological succession is not a matter of new species developing or even old species adapting to new conditions; it is a matter of populations of existing species taking advantage of a new area as conditions become favorable. Where do these early-stage species come from if their usual fate is to be replaced by late-stage or climax species? The answer is a key to the nonequilibrium theory: They come from other surrounding ecosystems in early stages of succession. Similarly, the late-stage species are recruited from ecosystems in later stages of succession. In any given landscape, therefore, all stages of succession are likely to be represented in the ecosystems. This is so because disturbances constantly create gaps or patches in the landscape. When a variety of successional stages is present in a landscape, as opposed to one single climax stage, a greater diversity of species can be expected; in other words, biodiversity is enhanced by disturbance! Consequently, natural succession is affected.

If certain species have been eliminated, natural succession will be blocked or modified. For example, beginning with the early colonization of Iceland by Norsemen in the 11th and 12th centuries, the native forests were cut for fuel. This process only accelerated with further European colonization in the 18th and 19th centuries. By 1850 not a tree was left standing, and Iceland remained a barren, tundralike habitat. Natural regeneration was prevented due to sheep grazing and the lack of a remaining source of seeds (Fig. 4–22). Tree seedlings are now being imported and planted in Iceland (4.5 million are planted annually) in the hope that a natural succession may be reestablished.

**Fire and Succession.** Fire is an abiotic factor that has particular relevance to succession. It is a major form of disturbance common to terrestrial ecosystems. About 80 years ago, forest managers interpreted the potential destructiveness of fire to mean that all fire was bad, whereupon they embarked on fire-prevention programs that eliminated fires from many areas. Unexpectedly, fire prevention did not preserve all ecosystems in their existing state. In pine forests of the southeastern United States, for instance, economically worthless scrub oaks and other broad-leafed species began to displace the more valuable pines. Grasslands were gradually taken over by scrubby, woody species that hindered grazing. Pine forests of the western United States that were once clear and open became cluttered with the trunks and branches

**Figure 4–22 Iceland.** Forests that originally covered much of this island nation were totally stripped for fuel in the 18th and 19th centuries. With natural succession impossible, Iceland has remained barren and tundralike, as seen here.

of trees that had died in the normal aging process. This deadwood became the breeding ground for wood-boring insects that proceeded to attack live trees. In California, the regeneration of redwood seedlings began to be blocked by the proliferation of broad-leafed species.

Scientists now recognize that fire, which is often started by lightning, is a natural and important abiotic factor. As with all abiotic factors, different species have different degrees of tolerance to fire. In particular, the growing buds of grasses and pines are located deep among the leaves or needles, where they are protected from most fires. By contrast, the buds of broad-leafed species, such as oaks, are exposed, so they are sensitive to damage from fire. Consequently, in regions where these species coexist and compete, periodic fires tip the balance in favor of pines, grasses, or redwood trees. In relatively dry ecosystems, where natural decomposition is slow, fire may also help release nutrients from dead organic matter. Some plant species even depend on fire. The cones of lodgepole pine, for example, will not release their seeds until they have been scorched by fire (Fig. 4–1b).

*Fire Climax Ecosystems.* Ecosystems that depend on the recurrence of fire to maintain their existence are now referred to as **fire climax ecosystems.** The category includes various grasslands and pine forests. Fire is being increasingly used as a tool in the management of such ecosystems. In pine forests, if ground fires occur every few years, relatively little deadwood accumulates. With only small amounts of fuel, fires usually just burn along the ground, harming neither pines nor wildlife significantly (Fig. 4–23). In forests where fire has not occurred for many decades, however, so much deadwood has accumulated that if a fire does break out, it will almost certainly become a crown fire. That is, so much heat is generated that entire living trees are ignited and destroyed. This long-term absence of fire was a major factor in the fires in Yellowstone National Park in the summer of 1988.

Crown fires do occur in nature, because not every area is burned on a regular basis and exceedingly dry conditions can make a forest vulnerable to crown fires even when no large amounts of deadwood are present. Thus, humans have only increased the potential for crown fires by fire-prevention programs. Even crown fires, however, serve to clear the deadwood and sickly trees that provide a breeding ground for insects, to release nutrients, and to provide for a fresh ecological start. Burned areas, such as those in Yellowstone National Park, soon become productive meadows as secondary succession starts anew. Thus, periodic crown

**Figure 4–23 Ground fire.** Periodic ground fires are necessary to preserve the balance of pine forests. Such fires remove excessive fuel and kill competing species.

fires create a patchwork of meadows and forests at different stages of succession that lead to a more varied, healthier habitat which supports a greater diversity of wildlife than does a uniform, aging conifer forest.

**Nonequilibrium Systems.** In sum, the concept most important to recognize is that disturbances such as fires, floods, windstorms, and droughts are important in structuring ecosystems. The disturbances remove organisms, reduce populations, and create opportunities for other species to colonize the ecosystem. Much of the patchiness observed in natural landscapes is evidence of periodic disturbances. Some ecosystems are more stable than others and experience disturbances only infrequently. Biotic relationships then become important in maintaining the stability of the ecosystem. Thus, the sustainability of ecosystems depends upon equilibria among the populations of the species in the biotic community, and it also depends upon existing relationships between the biotic community and abiotic factors of the environment, such as disturbances. We have seen that the natural biotic community itself may induce changes in abiotic factors that in turn result in changes in the biotic community (succession). Given this dynamic succession, a shift in any one or more physical factors in the environment may again push portions of the biotic community into a state of flux in which certain species that are stressed by the new conditions die out and other species that are better suited to those conditions thrive and become more abundant. This is the essence of the nonequilibrium theory of ecosystem structure.

***Resilience.*** Although the focus in this chapter has been primarily on populations and their interactions, do not lose sight of the importance of such ecosystem functions as the trophic interactions of energy flow and the efficient cycling of nutrients, discussed in Chapter 3. How does the nonequilibrium theory of ecosystem structure relate to these functions? Actually, disturbances and shifting biotic relationships not only may have little detrimental effect on an ecosystem, but may actually contribute to its ongoing functioning. Ecologists refer to this condition as **resilience.** Thus, a resilient ecosystem is an ecosystem that maintains its normal functioning—its integrity—even through a disturbance. Fire, for example, can appear to be a highly destructive disturbance to a forested landscape. Nevertheless, fire releases nutrients that nourish a new crop of plants, and in a short time the burned area is repopulated with trees and is indistinguishable from the surrounding area. The processes of replenishment of nutrients, dispersion by surrounding plants and animals, rapid regrowth of plant cover, and succession to a forest can all be thought of as **resilience mechanisms** (Fig. 4–24). Resilience, therefore, helps maintain the sustainability of ecosystems.

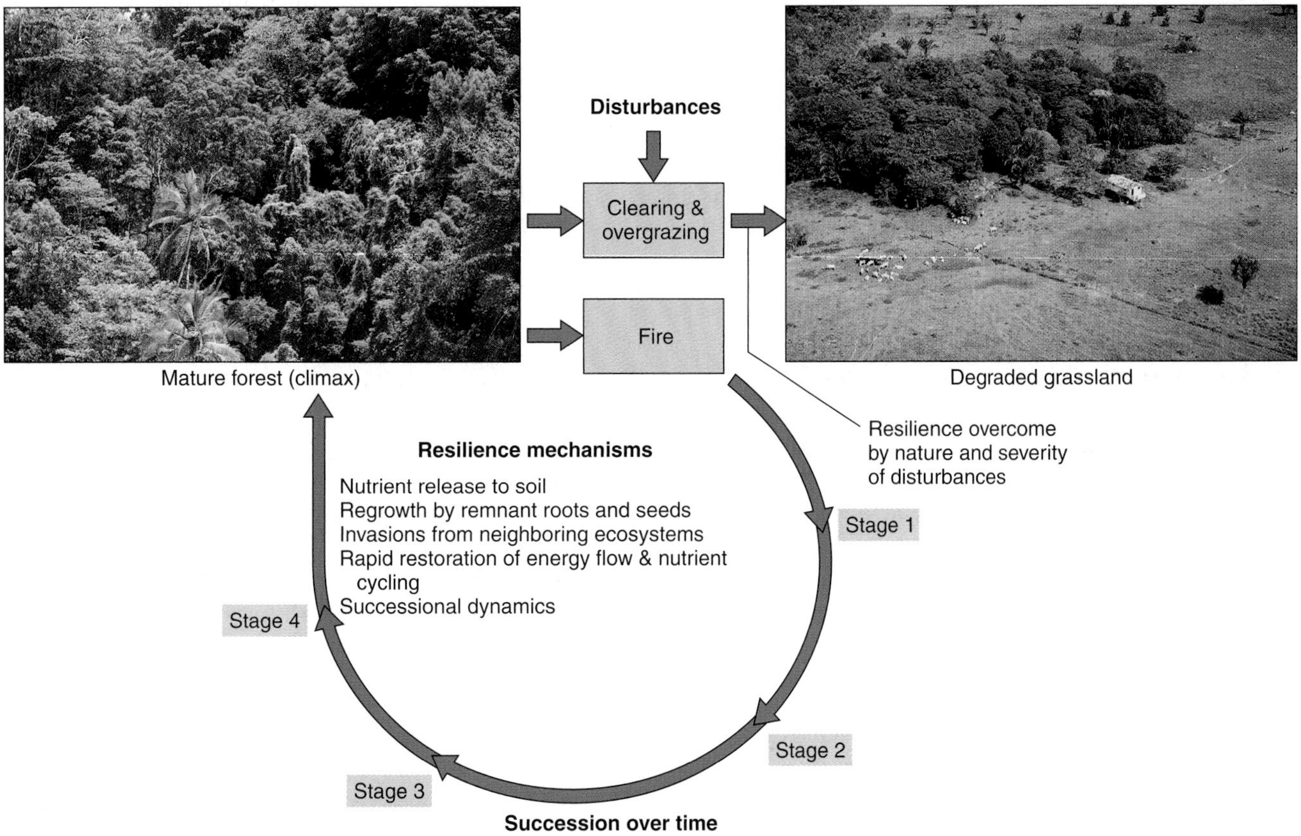


**Figure 4–24 Resilience in ecosystems.** Disturbances in ecosystems are usually ameliorated by a number of resilience mechanisms that restore normal ecosystem function in a short time.

earth watch

## An Endangered Ecosystems Act?

The Endangered Species Act formally recognizes the importance of saving species that are in serious trouble. The act has enjoyed some successes (Chapter 10), but is considered seriously flawed by many environmentalists because it does not directly address the major reason species become endangered: the loss of crucial ecosystems that are their habitats. Thus, *ecosystem* diversity, a level of diversity higher than biodiversity, is also important to maintain.

Ecosystem diversity is based on the idea that there should be a variety of ecosystems in a given area, be it local, national, or global. By protecting many different types of ecosystems, many endangered species will be protected, and the bulk of other species will be prevented from becoming endangered. However, maintaining the goods and services provided to us by ecosystems requires going beyond preserving basic types of ecosystems to protecting landscapes and regions of ecosystems large enough to sustain the functions we value. Thus, protecting ecosystem diversity requires both the qualitative and the quantitative preservation of ecosystems.

Is there a need for this kind of preservation in the United States? Wildlife ecologists Reed Noss and Michael Scott performed an extensive review of ecosystem declines and presented their results in terms of critically endangered, endangered, and threatened ecosystems of the United States.[1] They considered critically endangered ecosystems to be those which have suffered greater than a 98% decline in the area they formerly occupied. Twenty-eight ecosystems fit this description, including native grasslands in California, spruce—fir forest in the southern Appalachians, and tallgrass prairie east of the Missouri River. In addition, over 40 ecosystems were considered endangered (exhibiting an 85–98% decline), including all tallgrass prairies, coastal redwood forests, and large streams and rivers in all major regions of the United States. Many more made the category of threatened ecosystems (showing a 70–84% decline).

What has happened to these ecosystems? In large measure, they have been altered considerably by urban development, agriculture, exploitation of their resources, and pollution. Even though these transformations have served many legitimate human needs, Noss and Scott maintain that much could have been done to reduce the damage. Further, there are pressures—largely economic—to continue the pattern of transformation for many of the endangered ecosystems. In response to such real and continuing losses, many scientists and environmentalists have called for a national **Endangered Ecosystems Act.**

Implementing an Endangered Ecosystems Act would involve steps similar to those taken to protect endangered species—that is, taking an inventory of the status of ecosystems to identify those in trouble, protecting endangered ecosystems from further activities that would damage them, establishing recovery plans for many of the most critical ecosystems, and promoting research and monitoring to maintain surveillance and to gain knowledge for better management and protection. What do you think about this suggestion? Moreover, what could be done to make the proposed new act less controversial than the Endangered Species Act?

[1] Reed F. Noss and J. Michael Scott, "Ecosystem Protection and Restoration: The Core of Ecosystem Management," Ch. 12 in Mark S. Boyce and Alan Haney, eds. *Ecosystem Management: Applications for Sustainable Forest and Wildlife Resources* (New Haven, CT: Yale University Press, 1997.) See also Robert L. Peters and Reed F. Noss, "America's Endangered Ecosystems," *Defenders Magazine,* Fall 1995.

Ecosystem resilience has its limits, however. If forests are removed from a landscape by human intervention and the area is prevented from reforestation by overgrazing, the soil may erode away, leaving a degraded state that has little of the original ecosystem's functioning (Fig. 4–24). As in Iceland (Fig. 4–22), some disturbances may be so profound that they can overcome normal resilience mechanisms and create an entirely new and far less useful (in terms of ecosystem goods and services) ecosystem. (See Earth Watch, this page.) This new ecosystem can also have its own "pathological" resilience mechanisms, which may resist any restoration to the original state. Lakes, for example, show a resilience that maintains all of their desirable qualities. If they are subjected to heavy and persistent pollution, however, the degraded lakes have a new set of factors that act to maintain their degraded state.[1]

## Evolving Ecosystems?

To understand the concept of change through natural selection, we focused on how selective pressures may modify species. Each species always lives and reproduces within the context of an entire ecosystem, however, so it must simultaneously adapt to a host of biotic and abiotic factors present in the ecosystem. By the same token, all

[1] Stephen R. Carpenter and Kathryn L. Cottingham, "Resilience and Restoration of Lakes," *Conservation Ecology* [on-line] 1(1):2 (1997). URL: http://www.consecol.org/vol1/iss1/art2.

the species in the ecosystem are *simultaneously adapting to each other*! Does this mean, then, that there is a process of natural selection acting at the ecosystem level as well as at the species level?

This question has been debated for many years. Most scientists believe that natural selection operates only on individuals, eliminating or favoring them and the genes that they bear. The species is affected by this process when a significant number of individuals are subjected to similar selection; thus, the gene pool of the species is changed. Although it is unlikely that natural selection acts at the ecosystem level, ecosystems do change over time in response to the impacts of biotic and abiotic factors (succession). Hence, an ecosystem is affected by the species found within it, and the basic characteristics of any ecosystem are determined by the interactions of the species and the abiotic factors within it.

***Return to Serengeti.*** The Serengeti region of East Africa is occupied by large herbivores, some of which feed on grasses and herbs (wildebeest, zebra, Cape buffalo) and some of which feed on bushes and trees (giraffe, elephant). Elephants are notorious for pulling down trees to feed on the bark and leaves (Fig. 4–25), and their presence in the Serengeti normally maintains a healthy mixture of forested areas and grasslands. If elephant populations are removed—as they have been recently by ivory poachers—whole regions in the Serengeti will change from grasslands to woodlands. Conversely, if elephants become too numerous, as they have in some of Kenya's national parks, they will cause the forests to decline, leaving pure grasslands within a few years.

Thus, if relationships within an ecosystem are affected by the removal or introduction of a dominant organism (a keystone species) or by a change in climatic conditions, the entire ecosystem may develop into one that bears little resemblance to its precursor. When this occurs, environmental scientists do not say that the ecosystem has evolved; rather, it has developed into a different *kind* of ecosystem because of changing physical conditions or biota. It still has all of the functional components of an ecosystem, namely, producers, herbivores, carnivores, decomposers, and so on, but it now has a different cast of characters—a different biotic community—adapted to the new set of conditions. They may have come by various means of dispersal, possibly from similar ecosystems not too distant. Over time, in these changed ecosystems or biomes (forests, grasslands, tundra), species continue to evolve in ways that adapt them to particular abiotic factors (such as temperature, moisture, and light) and to particular roles in the entire system as a result of selective pressures brought on by the presence of other organisms. These adaptations to abiotic factors or to other organisms contribute to the remarkable resilience of ecosystems. Indeed, such adaptations are the primary source of ecosystem resilience.

In conclusion, ecosystems do *not* evolve in the Darwinian sense, but they do undergo changes as a result of

**Figure 4–25 Dominant organism in an ecosystem.** In some parts of the Serengeti in East Africa, elephants are numerous enough to be major ecosystem engineers. In time, their activities can change a woodland ecosystem to a grassland ecosystem. Here, one elephant is seen feeding on a tree.

the evolution of the species found in them. Why, then, have species evolved so differently on different continents and remote islands? Different landmasses reflect different histories and have been isolated from each other for millions of years. They will therefore possess unique groups of organisms and reflect unique ecosystems because of their long isolation and the unique dispersal of certain organisms to them.

It is clear, then, why species introduced from one region to another by humans may bring about major disruptions: because the species already present in the region have not evolved adaptations to the intruder.

## 4.5 Lessons to Learn

How can you use the information presented in Chapters 2 and 3, as well as that in the current chapter, to create a sustainable future? The answer to this question has two aspects. One is to protect or manage the natural environment in a way that maintains the goods and services vital to the human economy and to life support and that also maintains the beauty, interest, biodiversity, and other intrinsic values of the natural world. The second aspect is to establish a balance between our own species and the rest of the biosphere. The two aspects are interrelated.

### Managing Ecosystems

Virtually no ecosystems can escape human impact. Much of our impact stems from our dependence on ecosystems for vital goods and services. By using ecosystems, we are in fact managing them. Forests are a good example. Management can be a matter of fire suppression or, more recently, the use of controlled burns. It can involve clear-cutting or selective harvesting. The objectives of management can be to maximize profit from logging or to maintain the forest as a sustainable and diverse ecosystem. Management can be wise, devious, or even stupid. (See Ethics, p. 115)

**Ecosystem Management.** Since 1992, the U.S. Forest Service, the U.S. Bureau of Land Management, the U.S. Fish and Wildlife Service, the U.S. National Park Service, and the U.S. Environmental Protection Agency have all officially adopted ecosystem management as their management paradigm. This concerted adoption, in the words of Jerry Franklin, a leading forester, is "a paradigm shift of massive proportions."[2] What, then, is ecosystem management?

According to Franklin, **ecosystem management** comprises the following principles:

- It takes an integrated view of terrestrial and aquatic ecosystems. In particular, watersheds and riparian stands of trees are nurtured so as to minimize the impact of logging on streams and rivers.
- It integrates ecological concepts at a variety of spatial scales. Leaving dead logs and soil litter in place (called the ecosystem's "biological legacy") enhances the biodiversity of small organisms. At larger scales, corridors for animal migration are preserved between stands of forests, and fragmentation is minimized.
- It incorporates the perspectives of landscape ecology so that the range of possible landscapes in an ecosystem is recognized and preserved.
- It is an evolving paradigm. Managers are given the opportunity to learn from experiments and to employ new knowledge from forestry and landscape science in their practice.
- It incorporates the human element. The goods and services of forested ecosystems are recognized, and, through active involvement in monitoring and management, local people are included as important elements in stewardship of the resources.

Above all, the paradigm incorporates the objective of ecological sustainability. It should be evident that there is no way to accomplish this kind of ecosystem management without the involvement of sound science.

The U.S. Forest Service is responsible for managing some 192 million acres of national forests and is a key player in overseeing public natural resources. Recently, the Forest Service convened an interdisciplinary committee of scientists to advise the service on what sound science would recommend for managing national lands. According to the committee's report, "the first priority for management is to retain and restore the ecological sustainability of... watersheds, forests, and rangelands for present and future generations. The Committee believes that *the policy of sustainability should be the guiding star for stewardship of the national forests and grasslands* to assure the continuation of this array of benefits" (referring to the goods and services provided to the American people by the national lands).

Because there are uncertainties wherever the natural world is concerned, ecosystem management must be **adaptive.** That is, practices and policies should be seen as tentative and experimental, with a readiness to change when necessary. This principle is especially important, given that changes will occur wherever humans intervene; in addition, as time goes on, surprises will be inevitable, and new uncertainties will appear.

### The Pressures of Population

In the long run, efforts to protect natural ecosystems will be overwhelmed and thwarted by increasing pressures from humans if current trends in population growth and the exploitation of natural ecosystems

---

[2]Franklin, Jerry. "Ecosystem Management: An Overview," p. 21, Chapter 2 in: *Ecosystem Management: Applications for Sustainable Forest and Wildlife Resources*, Mark S. Boyce and Alan Haney, eds. (New Haven: Yale University Press, 1997).

**ethics**

### The Dilemma of Advocacy

What should a scientist do who has uncovered knowledge that points very clearly to the need for a change in some policy or practice involving the environment? If she published her results in a respected journal, presented them at scientific meetings, and taught them to her students, she would have met all of the traditional expectations of her university and her scientific field. The trouble is, doing so would likely have absolutely no impact on the problem she has uncovered.

An alternative course of action might be to do all of these things *and* to become an advocate for the policy implications of her research. This could involve contacting Congress, writing letters, working with the news media, and, in general, being dissatisfied until her message is heard by policy makers and corrective actions are taken. Increasingly, some scientists are becoming advocates as they work on serious problems like the loss of biodiversity, declining ecosystems, and changing biogeochemical cycles and as they see how things might be different. These scientists are taking a number of risks: (1) They may be criticized by those of their colleagues who feel that it is appropriate to do the research and point to the problems, but not to cross the line that separates science from social activism. (2) Their actions may blur the distinction between environmental scientists and environmental activists; thus, they may be perceived by policy makers and the public as just another group with an agenda to push. (3) They may be tempted to step out beyond the immediate implications of their work and make projections or predictions of disaster that never come true, thus harming the credibility of sound science.

More and more leading scientists, however, are calling for a closer engagement of science with societal problems. The issues, they argue, are so important that to ignore them is to renege on the traditional social contract between science and the public. That is, the public supports scientific research in exchange for the knowledge science uncovers and the expectation that scientists will produce something useful from their work. Thus, the argument goes, in view of the serious nature of our impacts on the environment, more scientists should be working on issues such as the depletion of resources, endangered species, climate change, and the like. Moreover, when they do work on these issues, they have a profound responsibility to see that their findings are communicated clearly to the public and to policy makers and resource managers. They should do so carefully, however, making sure that their listeners know when they are speaking as a scientist and when they are speaking as a private citizen taking a stand based on personal values. Some scientists are getting used to wearing two hats—one as a scientist and one as an activist. Others, however, warn of a "creeping advocacy syndrome" that threatens to ruin the objectivity that is so important to sound science. What do you think?

continue. A graph of the human population over the last 200 years (see Fig. 1–3) strongly resembles the upsweeping portion of the *J*-curve for the reindeer population on St. Matthew Island (Fig. 4–6). In addition, humans have overcome, for the most part, those factors which generally keep natural populations in balance, namely, disease, parasites, and predators. It appears, then, that the human population is behaving much like any population in the absence of natural enemies.

***Carrying Capacity?*** Recall from Section 4.1 that the carrying capacity is the maximum population that a given habitat can support without the habitat being degraded over the long term. What is the carrying capacity of our planet for human beings? Put another way, how many humans can the Earth support sustainably? The answer to this question must take the following factors into consideration: (1) the range of tolerable living standards, (2) the technologies that are developed and used, (3) the amount and efficiency of global trade, and (4) the corresponding ecosystem goods and services available. Because these factors both vary greatly for different countries and change over time, Earth's carrying capacity is difficult to ascertain.

A challenging way to address this issue is to return to the concept of an *ecological footprint* (Section 1.1). Using this concept, you can assess how much land area is required per person, assuming that all energy consumption, economic production, and waste discharge requires the goods and services of a finite area of land and water. Thus, ecological footprint analysis takes the four factors listed in the previous paragraph for carrying capacity and turns them into a land area required for each regional population. The calculated footprint may be related to a region's actual area. If the region depends greatly on trade to satisfy its needs, the ecological footprint may easily surpass the actual area of a region. In a sense, this is an *ecological deficit,* because the region or country depends on other regions to meet its needs. In effect, the region has exceeded its local carrying capacity.

Returning now to the question of a planetary carrying capacity, Wackernagel and Rees (in *Our Ecological Footprint*[3]) have calculated that the current human consumption of fossil fuels, forest products, fisheries, and agricultural products exceeds the available productive land by almost 30%. In other words, the global

[3]Mathis Wackernagel and William Rees. *Our Ecological Footprint: Reducing Human Impact on the Earth.* (Philadelphia: New Society Publishers, 1996).

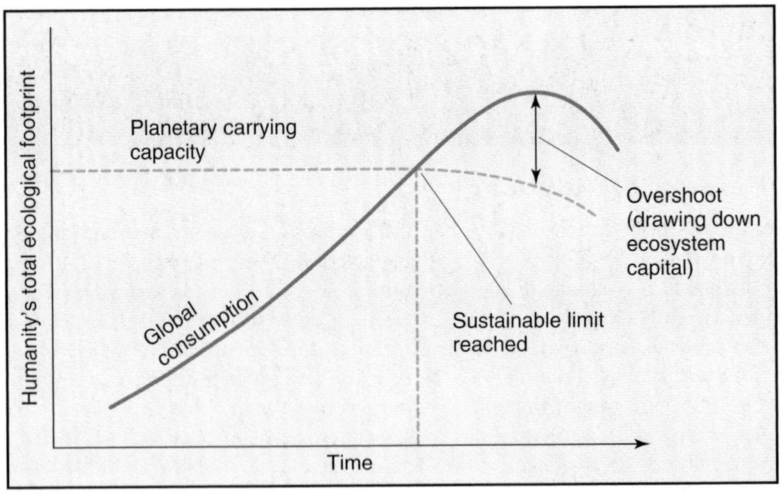


**Figure 4–26 Carrying capacity and overshoot.** Continued consumption of ecosystem goods and services occurs only by drawing down the capital stock itself. Ecological footprint analysis indicates that humans have already exceeded Earth's carrying capacity. (After Wackernagel and Rees, 1996)

population's ecological footprint is larger than the global carrying capacity. However, as the authors point out, no bells rang or alarms sounded when we crossed the line (Fig. 4–26), because we enjoy a large stock of ecosystem capital that can be depleted for some time to come. We are living off the future, in other words. Human well-being apparently now depends on drawing down the ecosystem capital that provides the goods and services as agricultural soils erode, fish stocks decline, forests shrink, and pollution of land, water, and air increases. This situation is unsustainable. Achieving a sustainable future will require all of the responses discussed in Chapter 1: *a demographic transition, a resource transition, a technology transition, a political–sociological transition, and a community transition.* All of these transitions are addressed in the chapters that follow. Chapters 5 and 6, for example, focus on the problems associated with stabilizing the human population.

# revisiting the themes

## Sustainability

Exponential growth is unsustainable, but so is linear growth. Natural populations are kept from continued growth by environmental resistance. This means hard times for youngsters and the elderly in a population, as they are preyed upon or thinned out by processes that relentlessly restrain populations from doing what comes naturally (biotic potential). The result, however, is that populations of species in ecosystems keep going, ultimately yielding highly sustainable ecosystems and the sustainable maintenance of biodiversity. Even in the face of disastrous disturbances like crown fires or volcanic eruptions, pioneer species appear, and, in time, the landscape is once again a functioning ecosystem. The natural world is amazingly resilient. It sounds like an oxymoron, but nonequilibrium systems are sustainable. Change is a fundamental property of ecosystems.

For the human system, an ecological footprint analysis indicates that we crossed the line into unsustainable consumption some time ago and are now depleting ecosystem capital stock instead of living off its sustainable goods and services.

## Stewardship

Following up on the latter idea, we do not have a license to go out and do whatever we wish to ecosystems because they are so resilient. Indeed, stewardship means, first of all, understanding how ecosystems do just what they do and then stepping back and restraining our destructive capabilities to let ecosystems continue to provide us with essential goods and services. Human agency has become the dominant force on earth affecting living systems. We can completely wipe out ecosystems, or we can convert them from productive units to sad, pathological systems with their own degraded resilience. In the process, we can drastically reduce biodiversity, a complete reversal of what we should be doing as stewards of life. Note that ecosystem management is, above all, a stewardship approach to the natural world. As the Forest Service

Committee of Scientists put it, "the policy of sustainability should be the guiding star for stewardship of the national forests and grasslands," fittingly linking sustainability and stewardship.

### Sound Science

As in the discussions of ecosystems in Chapters 2 and 3, the information presented here in Chapter 4 is the cumulative knowledge gathered over many years by thousands of scientists who have used the principles of sound science to obtain an accurate picture of how the natural world works. As scientists work on serious problems like declining ecosystems and the loss of biodiversity, they often become advocates for public policies that will turn things around. Scientists care, and they fear that, unless they speak out, others will never take a stand on these crucial environmental issues.

### Ecosystem Capital

This chapter, like Chapter 3, is all about ecosystem capital. There has been a paradigm change, though, from viewing ecosystems as static in time to viewing them as dynamic and changing with time. If we decide that we want forests to produce pines and not hardwoods, then we must manage the forests so as to prevent them from reaching a successional climax. As with any capital, management is essential. Adaptive ecosystem management incorporates experience and almost mandates midcourse corrections as we learn better how to both maintain the sustainability of ecosystems and also reap the benefits of their goods and services.

### Policy and Politics

The discussion of the Yellowstone fires brought public policy into focus from the very beginning of the chapter. Smokey the Bear said, "Only you can prevent forest fires." He didn't imagine forest personnel actually going out and lighting fires, but that is what is happening today as a result of our recent history of devastating fires. It is not wise policy to prevent all fires; otherwise fuel will continue to build up in our forests in the form of underbrush and dead trees. Policy here teeters on a knife edge, because a prescribed burn can turn into a property-damaging fire in a matter of minutes. To address this policy scene, in late 2002 President George W. Bush proposed the "Healthy Forests Initiative," in an attempt to reduce the risks of wildfires. The proposal, however, is highly controversial, and many environmental scientists have condemned it because it has all the appearances of giving a green light to the timber industry to increase commercial logging greatly as a reward for its heavy support of Republican candidates. This issue is addressed again in Chapter 11.

### Globalization

Worldwide trade has increased the spread of alien species considerably. This is a form of globalization we could do without, because it is enormously costly to human enterprises and damaging to natural ecosystems and biodiversity. The best medicine and first line of defense is simply prevention. For alien species already on board, mitigation might include attempts at eradicating, or at least suppressing the invasive organisms and restoring native biota in ecosystems affected by the invaders. To address this problem, then-President Clinton established the National Invasive Species Council in 1999. The council, which includes many members of the president's cabinet, in recognition of the complexities of dealing with such an international problem, lives on in the Bush administration.

## review questions

1. What are the two fundamental kinds of population growth curves? What are the causes and consequences of each?
2. Define biotic potential and environmental resistance, and give factors of each. Which generally remains constant, and which controls a population's size?
3. Differentiate between the terms *critical number* and *carrying capacity*. What is density dependence?
4. Distinguish between mortality and regulation in populations.
5. Describe the predator–prey relationship between the moose and wolves of Isle Royale. What other factors influence these two populations?
6. How are herbivores normally kept from overgrazing their food?
7. Distinguish between intraspecific and interspecific competition. How do they affect species as a form of environmental resistance?
8. What is meant by territoriality, and how does it control certain populations in nature?

9. What problems arise when a species is introduced from a foreign ecosystem? Why do these problems occur?
10. What is selective pressure, and how does it relate to natural selection?
11. Describe several of the basic traits that adapt an organism for survival and reproduction.
12. What factors determine whether a species will adapt to a change or whether the change will render it extinct?
13. How may natural selection lead to the development of new species (speciation)?
14. Define the terms *ecological succession* and *climax ecosystem*. How do disturbances allow for ecological succession?
15. What role may fire play in ecological succession, and how may fire be used in the management of certain ecosystems?
16. What is the nonequilibrium theory of ecosystem stability? What evidence is there for the theory?
17. What is meant by *ecosystem resilience*? How is this accomplished, and what can cause it to fail?
18. Succinctly describe ecosystem management? What is meant by *adaptive management*?

## thinking environmentally

1. Describe, in terms of biotic potential and environmental resistance, how the human population is affecting natural ecosystems.
2. Give two density-dependent factors of environmental resistance that would act on a field mouse population. Give two density-independent factors.
3. Select a species and describe how its various traits support its survival and reproduction. How could these traits have been affected by natural selection?
4. Consider the plants, animals, and other organisms present in a natural area near you, and then do the following: (a) Imagine how the area may have undergone ecological succession. (b) Analyze the population-balancing mechanisms that are operating among the various organisms. (c) Choose one species, and predict what will happen to it if two or three other species native to the area are removed from it. Then predict what will happen to your chosen species if two or three foreign species are introduced into the area.
5. Evaluate, in terms of supporting sustainable ecosystems, such practices as legal hunting, controlling pests with chemical sprays, using or preventing fires, and poaching endangered species.

## making a difference part one: chapters 2, 3, 4

1. Find out how your school, college, community, or city gets its water and power. Find out how it disposes of sewage and refuse, and what is the trend of land development and preservation around the area.
2. Find out whether there is an environmental organization or club on your campus or in your community. Join it if there is, or create one if there isn't.
3. Learn the common names of trees and other plants, birds, and mammals found in your area. This knowledge is a prerequisite to understanding and describing an actual ecosystem. Then, pass your knowledge on to children, who are often curious about the natural world.
4. Take courses offered by colleges or local environmental organizations that give you a greater appreciation for the natural world; amaze and influence your friends by becoming a local expert in some taxonomic group, such as birds, wildflowers, or butterflies.
5. Consider small animals, birds, butterflies, and other wildlife that might be present in your yard, school, or campus grounds were it not for human-created limiting factors. Investigate and begin a project to create natural habitats that will attract and support additional wildlife.
6. Create a list of the things you do and the things that you use and throw away. Briefly evaluate each in the context of sustainability. Consider how you might change your habits to move toward a more sustainable option. Begin by making one such change.
7. Rather than using chemical pesticides, fertilizers, and enormous amounts of water to maintain your lawn, which is an "unnatural" monoculture, introduce clover and other low-growing flowering plants that will create a sustainable balance with only mowing. Alternatively, replace your lawn with plants native to your area, which should be able to thrive on the amount of water available from rainfall and on the nutrients currently available from the soil.
8. Read local papers, make contact with environmental organizations, and become aware of efforts to protect natural areas locally. Support these efforts to protect such areas locally, regionally, and globally.
9. Select a current environmental issue related to biodiversity or endangered species, and write your representatives to express your concern and ask for their support for legislation that addresses the issue effectively.

10. Support and join efforts and organizations, such as the World Wildlife Fund and Conservation International, that are devoted to protecting endangered and threatened species.

11. Use the Internet to investigate an environmental issue, and create an annotated bibliography of Web sites that address the issue. Post your bibliography on an appropriate list-server that others can access.

12. Continue your education toward a profession that is important for environmental concerns, and make your lifework such that it helps, rather than hinders, the environmental revolution.

13. Use your citizenship to support environmentally sound practices and policies by voting in your local, state, and national elections for candidates who are clearly environmentally aware.

# part two

CHAPTER 5 The Human Population: Dimensions

CHAPTER 6 Population and Development

# the human population

**◄Downtown Lagos, Nigeria. The city, with a population of 15 million, is a reflection of the 134 million people in Nigeria, Africa's most populous country.**

From a population perspective, the 20th century was remarkable: We started the century with fewer than 2 billion people and ended it with 6 billion and counting. The United Nations Population Division suggests that our numbers may reach 8.9 billion by the middle of the century—a 50% increase in 50 years. Virtually all of the increase will be in the developing countries, which are already densely populated and straining to meet the needs of their people for food, water, health care, shelter, and employment.

In the industrialized countries, populations have almost stopped growing. Yet these countries, with only 20% of the world's people, account for two-thirds of the world's energy use and create pressures on water, forests, and fisheries that are contributing to the depletion of these renewable resources. When we try to imagine a world of 9 billion, with a greater proportion of people straining to improve their economic standards and catch up to the developed world, we may be staring into the face of disaster.

In Chapter 5, you learn about the dynamics of the human population, focusing especially on the demographic transition—the shift from high birth and death rates to low birth and death rates that has brought stable populations to the industrialized world. In Chapter 6, you learn what needs to be done to bring the developing countries through this transition—by way of sustainable development. It is important to take up population concerns before dealing with other significant issues, such as resource use and pollution, because population affects every environmental issue and it is impossible to achieve sustainability in these other areas until population stability is achieved.



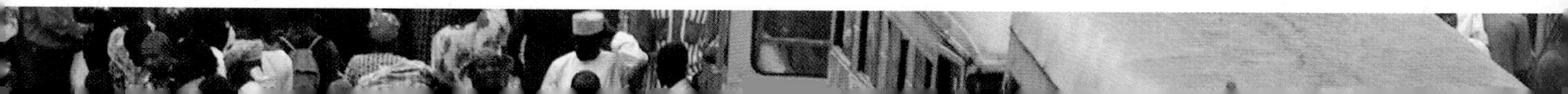

# The Human Population: Dimensions

**Key Topics**

1. Human Population Expansion and Its Cause
2. Different Worlds
3. Consequences of Population Growth and Affluence
4. Dynamics of Population Growth

The Mombasa highway leads out from the center of Nairobi, Kenya, and is lined with factories for several miles. Outside every factory gate every morning is a crowd of people—mostly men—waiting for the gates to open. Dressed in tattered clothes and worn-out shoes, they are the "casual workers," hired for periods of up to three months, but never under any employment contract. Many are there just hoping that they will be hired for the day. One such worker—let's call him Charles—has worked at one of these factories for almost a year. He is paid 100 Kenyan shillings—about $1.70—a day. He walks 10 km to work from his home in one of the shantytowns on the edge of the city. He can't afford transportation and can only occasionally buy lunch, which is a bowl of *githeri*—a 10-shilling dish of corn and beans sold in the kiosks that cluster around every factory. Charles has no work contract, nor do any of his coworkers. Virtually all of the factories pay workers 100 shillings a day; working six days a week for 10 hours a day, they earn, at most, $50 a month.

***Casual Workers in Kenya*** **High population growth and slow economic development combine to produce a large unemployed and underemployed workforce. Competition for scarce, low-paying factory jobs leaves many men like these without employment.**

Any worker who becomes a union representative is fired. Any worker who complains about an injury on the job is fired. For every job, there are scores of applicants, so no one dares attract attention. They work in dusty, dangerous conditions. Cement-factory workers have no masks to prevent them from breathing the dust. Metalworkers lack protective gear and often lose limbs or their lives due to the nature of the jobs, which are often monotonous, making the workers lose their concentration. There are 14,000 of these factories in cities like Nairobi, Mombasa, Kisumu, Nakuru, and Eldoret, in Kenya.

Scenes like those along the Mombasa highway can be found around most cities in the developing world. The staggering growth of cities is one of the stark realities of recent population growth in that world. The cities grow because the rural countryside provides only subsistence farming and livestock herding for most of its people—and with population growth rates as high as 3% per year, there are continually more people than there is land to till and cattle to herd. This is the tragic story described in *Cry the Beloved Country* (a novel by Alan Paton, set in

South Africa), multiplied by hundreds of thousands. Thus, the young people and the men migrate to cities, such as Nairobi, to find work. Kenya has an enormous unemployed and underemployed workforce—the legacy of decades of extremely high population growth rates.

***In the World.*** In the last century, the global human population has undergone an unprecedented expansion, more than tripling its numbers. Now the rate of growth is slowing down, but the increase in absolute numbers continues to be great. Remarkable changes in technology and substantial improvements in human well-being have accompanied this growth. In just the last 30 years, average income per capita has increased by 37%, infant mortality has been cut in half, and adult illiteracy has dropped by 22%.

There is a downside to this remarkable growth, however. As the Mombasa highway story indicates, extreme poverty is still widespread. An estimated 1.2 billion people live on less than $1 per day, and the income gap between the richest countries and the poorest ones is enormous and growing. Population growth and increasing affluence have placed great demands on the natural environment and will continue to do so: Ecosystems have been transformed for human uses, resulting in tremendous loss of biodiversity. Resources are being overexploited; forests and fisheries continue to decline, and soils are being degraded. Finally, the waste products of our economy are polluting the air, land, and water, with grim consequences for the natural world as well as for human well-being.

What are the implications of these trends for sustainability, whereby some things are improving and some are getting worse? There is broad consensus that the only way to begin to eliminate poverty and undergo the transition to a sustainable society is to halt population growth. The focus is especially on the developing world, where 98% of the net world population growth is occurring. This is not an easy task, because the basic decisions on population are made by millions of couples, not by public policy.

However, public policy, health care, and reproductive technology can influence population growth. Indeed, efforts to provide family planning to couples have been mounted in almost every country. In turn, these efforts and other socioeconomic factors have brought about a remarkable decline in reproduction rates during the second half of the 20th century. There is every expectation that population growth rates will continue to decline during the present century.

Your objective in this chapter is to learn about the dynamics of population growth and its social and environmental consequences. A continually growing population is unsustainable, so the focus in this and the next chapter is population stability and what it takes to get there.

## 5.1 Human Population Expansion and Its Cause

Considering all the thousands of years of human history, the recent rapid expansion of the global human population is a unique event—a phenomenon of just the past 100 years (Fig. 5–1). Let us look more closely at this event and why it occurred.

From the dawn of human history until the beginning of the 1800s, population increased slowly and variably, with periodic setbacks. It was roughly 1830 before world population reached the 1 billion mark. By 1930, however, just 100 years later, the population had doubled to 2 billion. Barely 30 years later, in 1960, it reached 3 billion, and in only 15 more years, by 1975, it had climbed to 4 billion. Thus, the population doubled in just 45 years, from 1930 to 1975. Then, 12 years later, in 1987, it crossed the 5 billion mark! In 1999, world population passed 6 billion, and it is currently growing at the rate of nearly 77 million people per year. This rate is equivalent to adding to the world every year the combined populations of New York, Los Angeles, Chicago, Philadelphia, Detroit, Dallas, Boston, and 10 other U.S. metropolitan areas.

On the basis of current trends (which assume a continued decline in fertility rates), the U.N. Population Division projects that world population will pass the 7 billion mark in 2013, the 8 billion mark in 2028, and the 9 billion mark in 2052, before population finally levels off at around 10 billion by the end of the 22nd century (Fig. 5–1).

### Reasons for the Patterns of Growth

The main reason for the slow and fluctuating population growth prior to the early 1800s was the prevalence of diseases that were often fatal, such as smallpox, diphtheria, measles, and scarlet fever. These diseases hit infants and children particularly hard. It was not uncommon for a woman who had seven or eight live births to have only one or two children reach adulthood. In addition, epidemics of diseases such as the black plague of the 14th century, typhus, and cholera eliminated large numbers of adults. Famines also took their toll periodically.

Prior to the 1800s, therefore, the human population was essentially in a dynamic balance with natural enemies—mainly diseases—and other aspects of environmental resistance. High reproductive rates were largely balanced by high mortality, especially among infants and children. With high birth and death rates, the population growth rate was low in these preindustrial societies.

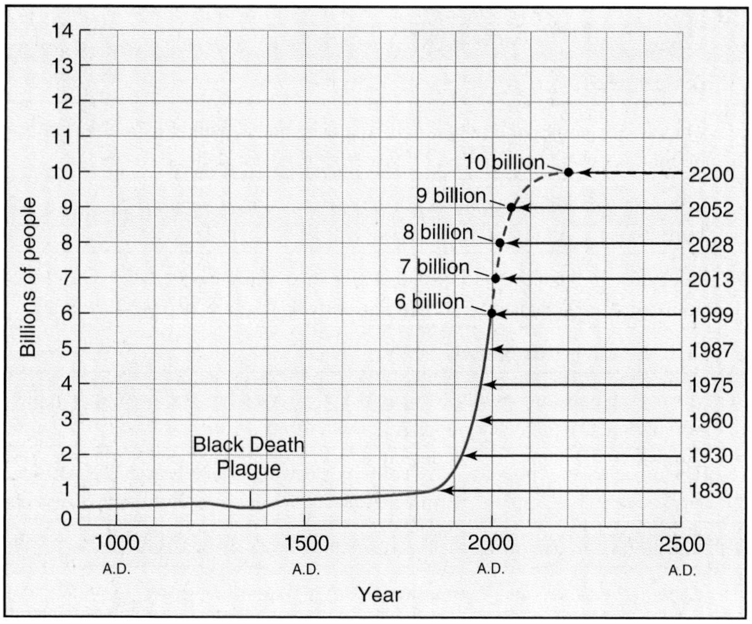


**Figure 5-1 World population over the centuries.** Population grew slowly for most of human history, but in modern times it has expanded greatly. (*Sources:* Basic plot from Joseph A. McFalls, Jr., "Population: A Lively Introduction," *Population Bulletin* 46, no. 2 [1991]: 4; updated from UN Population Division, median projection.)

***Breakthroughs.*** In the late 1800s, Louis Pasteur and others discovered that diseases were caused by infectious agents (now identified as various bacteria, viruses, and parasites) and that these organisms were transmitted via water, food, insects, and rodents. Soon vaccinations were developed for the different diseases, and whole populations were immunized against such scourges as smallpox, diphtheria, and typhoid fever. At the same time, cities and towns began treating their sewage and drinking water. Later, in the 1930s, the discovery of penicillin, the first in a long line of antibiotics, resulted in cures for otherwise often-fatal diseases such as pneumonia and blood poisoning. Improvements in nutrition began to be significant as well. In short, better sanitation, medicine, and nutrition brought about spectacular reductions in mortality, especially among infants and children, while birthrates remained high. From a biological point of view, the human population began growing almost exponentially, as does any natural population once it is freed from natural enemies and other environmental restraints.

***Declines.*** During the 1960s, the world *population growth rate* peaked at 2.1 percent per year, after having risen steadily for decades. Then it began a steady decline (Fig. 5–2). Twenty years later, the number added per year peaked, at 87 million. The declines are primarily a consequence of the decline in the world *total fertility rate*—that is, the average number of babies born to a woman over her lifetime. (Table 5–1 gives a list of definitions of technical terms used in this chapter.) In the 1960s the total fertility rate was an average of 5.0 children per

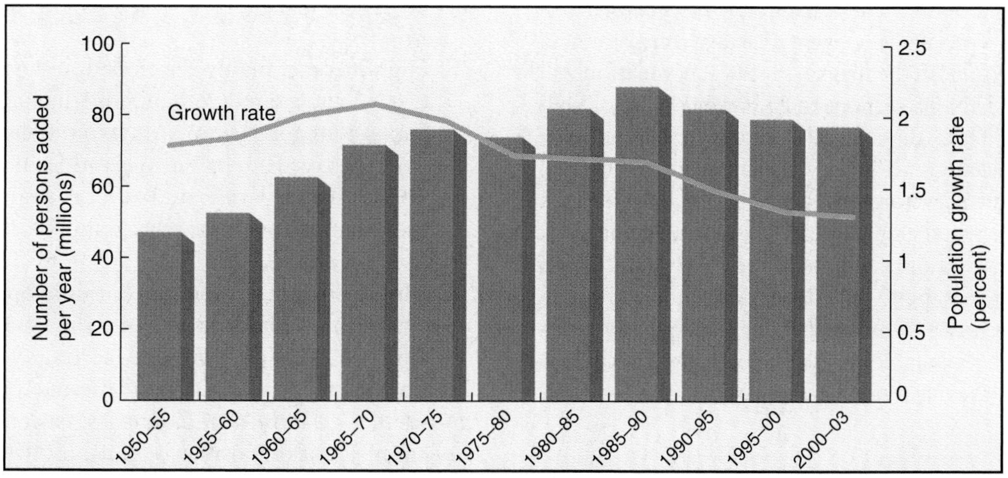


**Figure 5-2 World population growth rate and absolute growth.** Declining fertility rates in the last three decades have resulted in a decreasing rate of population growth. Absolute numbers, however, are still adding 77 million per year. [*Source:* Data from Shiro Horiuchi, "World Population Growth Rate," *Population Today* (June 1993): 7 and June/July 1996): 1; updated from U.S. Census Bureau, International Data Base.]

**table 5-1 Demographic Terms Used in This Chapter**

| Term | Definition |
|---|---|
| **Growth Rate (annual rate of increase)** | The rate of growth of a population, as a percentage. Multiplied by the existing population, this rate gives the net yearly increase for the population. |
| **Total Fertility Rate** | The average number of children each woman has over her lifetime, expressed as a yearly rate based on fertility occurring during a particular year. |
| **Replacement-level Fertility** | A fertility rate that will just replace a woman and her partner, theoretically 2.0, but adjusted slightly higher because of mortality and failure to reproduce. |
| **Infant Mortality** | Infant deaths per thousand live births. |
| **Population Profile (age structure)** | A bar graph plotting numbers of males and females for successive ages in the population, starting with youngest at the bottom. |
| **Population Momentum** | The tendency of a population to continue growing even after replacement-level fertility has been reached, due to continued reproduction by already existing age groups. |
| **Crude Birthrate** | The number of live births per thousand in a population in a given year. |
| **Crude Death Rate** | The number of deaths per thousand in a population in a given year. |
| **Doubling Time** | The time it takes for a population increasing at a given growth rate to double in size. |
| **Epidemiologic Transition** | The shift from high death rates to low death rates in a population as a result of modern medical and sanitary developments. |
| **Fertility Transition** | The decline of birthrates from high levels to low levels in a population. |
| **Demographic Transition** | The tendency of a population to shift from high birth and death rates to low birth and death rates as a result of the epidemiologic and fertility transitions. The result is a population that grows very slowly, if at all. |

woman: it has since declined to its present value of 2.8 children per woman.

Extrapolating the trend of lower fertility rates leads to the U.N. Population Division's projection that the global human population will reach 8.9 billion by 2050, but will continue to increase, leveling off at around 10 billion well into the 22nd century. This projection is the U.N.'s *medium* scenario; other scenarios are based on different fertility assumptions (Fig. 5–3). The assumption of declining fertility rates is crucial; if current fertility rates remain unchanged (*constant* in the figure), the 2050 population will be 12.8 billion. U.N. projections to 2150 range from 3.2 billion to 24.8 billion, demonstrating that very long-range population projections are largely exercises in mathematics.

The projected leveling off at around 10 billion raises the question of whether Earth can sustain such numbers. Where are the additional billions of people going to live, and how are they going to be fed, clothed, housed, educated, and otherwise cared for? Will enough energy and material resources be available for them to enjoy a satisfying life? What will the natural environment look like by then?

## 5.2 Different Worlds

To begin to answer these questions, you must first understand the tremendous disparities among nations. In fact, people in wealthy and poor countries live almost in separate worlds, isolated by radically different economic and demographic conditions.

### Rich Nations, Poor Nations

The World Bank, an arm of the United Nations, divides the countries of the world into three main economic categories, according to average per capita gross national income (Fig. 5–4):

1. **High-income, highly developed, industrialized countries.** This group (964 million) includes the United States, Canada, Japan, Australia, New Zealand, the countries of western Europe and Scandinavia, Singapore, Taiwan, Israel, and several Arab states. (2001 gross national income per capita, $9,206 and above; average of $26,710.)
2. **Middle-income, moderately developed countries.** These (2.7 billion) are mainly the countries of Latin America (Mexico, Central America, and South America), northern and southern Africa, China and some smaller eastern Asian countries, eastern Europe, and countries of the former U.S.S.R. (2001 gross national income per capita ranges from $745 to $9,205; average of $1,850.)
3. **Low-income, developing countries.** This group (2.65 billion) comprises the countries of eastern, western, and central Africa, India and other countries of central

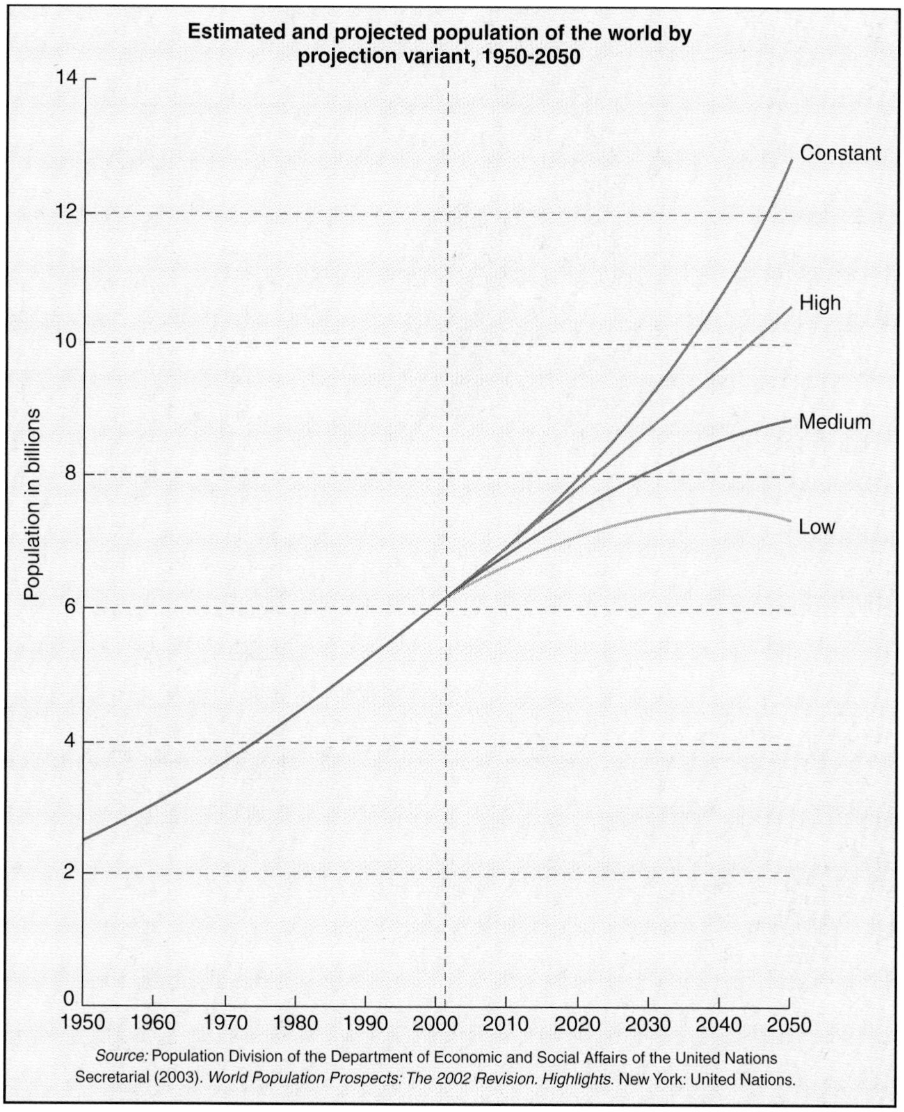


**Figure 5–3 United Nations population projections.** The most recent population projections demonstrate the vital role played by different fertility assumptions. The *Constant* projection assumes that fertility rates remain at their present level in all countries. The *Medium* projection assumes a gradual decline in fertility in the developing countries, where 3/4 drop below replacement level fertility by 2050 and the world total fertility rate is 2.02 children per woman. This projection is given the highest probability by the U.N. Population Division. the *High* projection assumes fertility rates 1/2 child greater than the *Medium* projection, and the *Low* projection assumes fertility rates 1/2 child lower than the *Medium* projection. All of these projections include the impact of the AIDS epidemic on increasing mortality. [*Source:* United Nations Population Division, 2002 revision.]

Asia, and a few former Soviet republics. (2001 gross national income per capita, less than \$745; average of \$430.)

The high-income nations are commonly referred to as **developed countries,** whereas the middle- and low-income countries are often grouped together and referred to as **developing countries.** The terms *more developed countries* (MDCs), *less developed countries* (LDCs), and *Third World countries* are being phased out, although you may still hear them used. (The First World was the high-income countries, whereas the Second World was the former Communist bloc, which no longer exists. Therefore, referring to the developing countries as the Third World is obsolete.)

***Disparities.*** The disparity in distribution of wealth among the countries of the world is mind boggling. The highly developed countries make up just 16% of the world's population, yet they control about 81% of the world's wealth, calculated on the basis of gross national income. The low-income developing countries, with 41% of the world's population, control only 3.4% of the world's gross national income. This amounts to a difference in per capita income of 62 to 1! The distribution of wealth *within* each country is also disproportionate, so the poorest people living in these poor nations are really badly off.

The disparity of wealth is difficult to understand just by looking at general income figures. Therefore, the United Nations Development Program (UNDP) has devised the Human Poverty Index (HPI), based on information about life expectancy, literacy, and living standards (Fig. 5–5). Table 5–2 compares the HPI with the percentage of people living on \$1 per day or less, for a number of low-income and middle-income countries. Different criteria are used to calculate the HPI for the high-income countries, and the data show that there are poor people in every country in the world. In fact, between 10% and 15% of the people in developed countries are poor (unable to afford adequate food, shelter, or clothing), compared with about 45% of those in developing countries. These data have been used to focus attention on the most deprived people in a

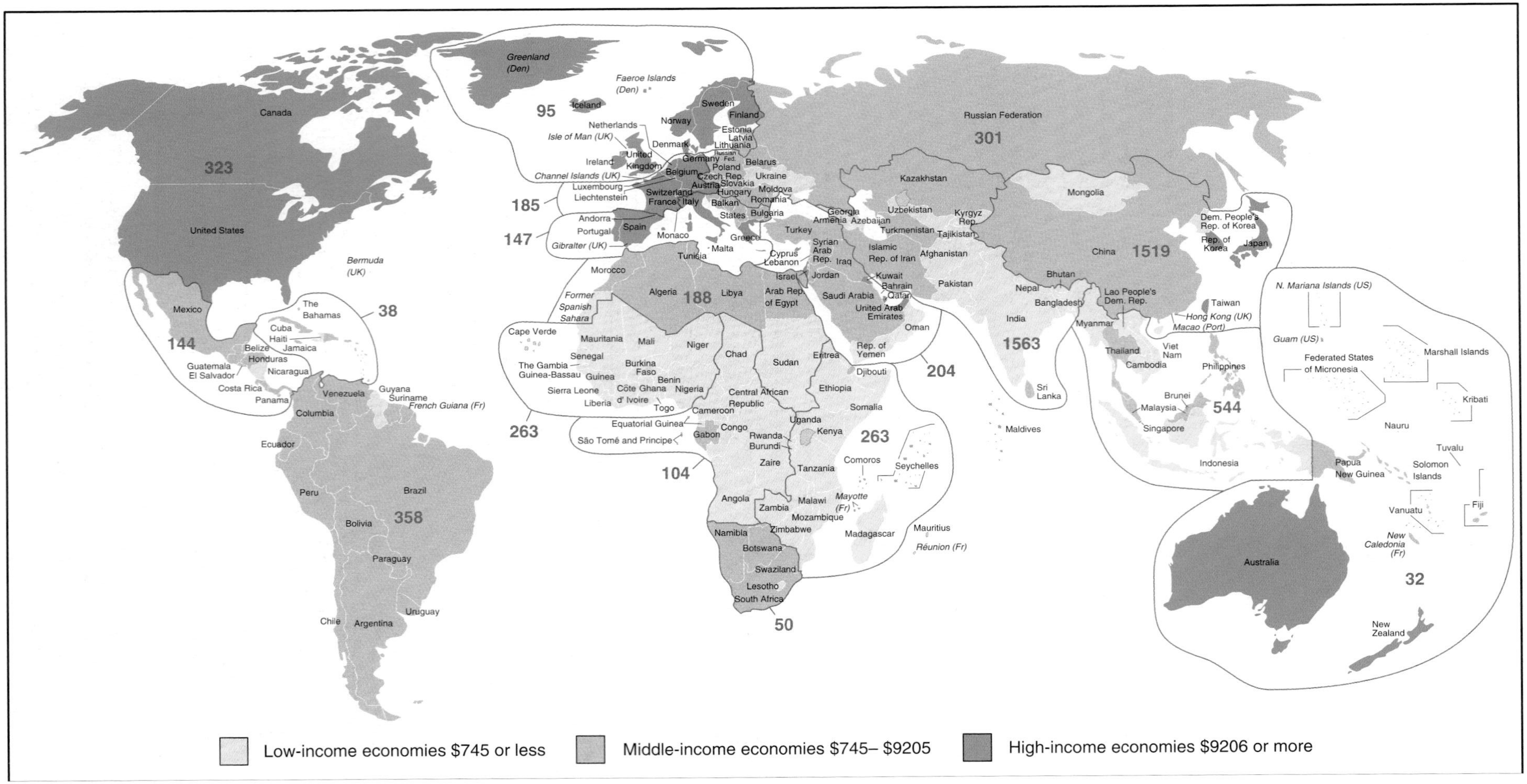


**Figure 5–4 Major economic divisions of the world.** Nations of the world are grouped according to gross national income per capita. The population of various regions (in millions) is also shown by magenta lines and numbers. [*Sources: World Development Report, 2003* (New York: Oxford University Press, Inc.), Copyright © 2003 by the International Bank for Reconstruction and Development/The World Bank. Populations from the Population Reference Bureau, *World Population Data Sheet, 2003.*]

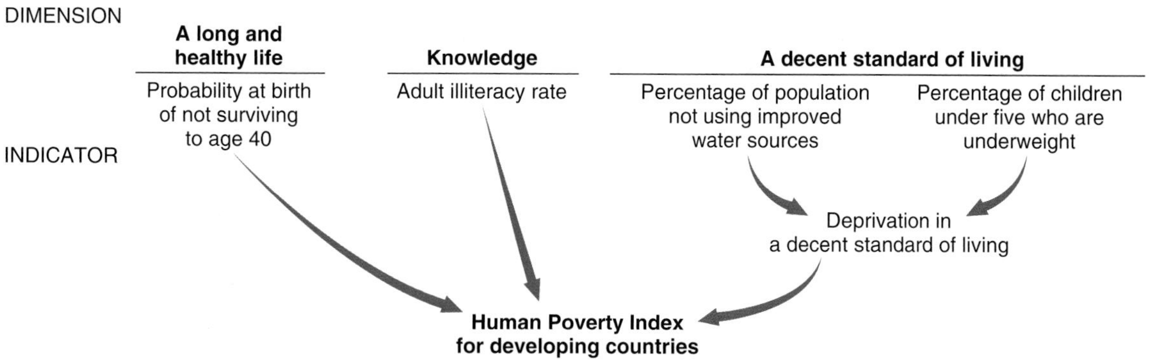


**Figure 5–5 Human Poverty Index for developing countries.** The index is based on information about life expectancy, literacy, and living standards. (*Source: Human Development Report 2002,* U.N. Development Program.)

country, to help countries develop appropriate policies, and to mark progress toward sustainable development. Clearly, there is work to do in all countries in order to address the general problem of poverty.

## Population Growth in Rich and Poor Nations

The population growth shown in Figs. 5–1 and 5–2 is for the world as a whole. If you look at population growth in developed versus developing countries, you find a discrepancy that parallels the great difference in wealth between these two groups of countries. The developed world, with a population of 965 million in mid-2003, is growing at a rate of 0.1% per year. These countries will add less than 1 million to the world's population in a year. The remaining countries, whose mid-2003 population was 5.35 billion, are increasing at a rate of almost 1.6% per year, adding over 76 million in a year. Consequently, *over 98% of world population growth is occurring in the developing countries.* What's lies behind this discrepancy?

***Fertility.*** Population growth occurs when births outnumber deaths. In the absence of high *mortality,* the major determining factor for population growth is births, conventionally measured using the *total fertility rate*—the average number of children each woman in a population has over her lifetime. A total fertility rate of 2.0 will give a stable population, because two children per woman will just replace a couple when they eventually die. Fertility rates greater than 2.0 will give a growing population, because each generation is replaced by a larger one, and, barring immigration, a total fertility rate less than 2.0 will lead to a declining population, because each generation will eventually be replaced by a smaller one. Given that infant and childhood mortality are not in fact zero, and that some women do not reproduce, **replacement-level fertility**—the fertility rate that will just replace the population of parents—is 2.1 for developed countries and higher for developing countries, which have higher infant and childhood mortality.

Total fertility rates in developed countries have declined over the past several decades to the point where they now average 1.5. The one major exception is the United States, with a total fertility rate of 2.0 in 2003. In developing countries, fertility rates have come down considerably in recent years, but they still average 3.1. Some rates are as high as 5 or more, however, which will cause the populations of those countries to double in just 20 to 40 years (Table 5–3). Thus, the populations of developing countries, half of which are the poor (low-income) countries, will continue growing, while the populations of developed countries will stabilize or even decline. As a consequence, the percentage of the world's population living in developing

**table 5-2 Poverty Indexes for Selected Countries**

| Country | HPI | Percentage at $1 per Day or Less |
|---|---|---|
| **Low Income** | | |
| Pakistan | 41 | 31 |
| Bangladesh | 42 | 29 |
| Nigeria | 35 | 70 |
| Mali | 47 | 73 |
| Ethiopia | 57 | 31 |
| Gambia | 49 | 59 |
| Yemen | 42 | 16 |
| India | 33 | 44 |
| **Middle Income** | | |
| Algeria | 23 | <2 |
| Guatemala | 24 | 10 |
| Mexico | 9.4 | 16 |
| Brazil | 12 | 12 |
| Turkey | 13 | 2.4 |
| China | 15 | 19 |
| Costa Rica | 4.0 | 13 |

**table 5-3 Population Data for Selected Countries**

| Country | Total Fertility Rate | Doubling Time of Population (Years) |
|---|---|---|
| World | 2.8 | 54 |
| **Developing Countries** | | |
| Average | 3.1 | 44 |
| (excluding China) | 3.5 | 37 |
| Egypt | 3.5 | 33 |
| Kenya | 4.4 | 35 |
| Madagascar | 5.8 | 23 |
| India | 3.1 | 41 |
| Iraq | 5.4 | 28 |
| Vietnam | 2.3 | 54 |
| Haiti | 4.7 | 39 |
| Brazil | 2.2 | 54 |
| Mexico | 2.8 | 29 |
| **Developed Countries** | | |
| Average | 1.5 | 700 |
| United States | 2.0 | 117 |
| Canada | 1.5 | 233 |
| Japan | 1.3 | 700 |
| Denmark | 1.7 | 700 |
| Germany | 1.3 | — |
| Italy | 1.2 | — |
| Spain | 1.2 | 700 |

*Note:* Dash indicates doubling time cannot be calculated, because growth is negative.
*Source:* Data from *2003 World Population Data Sheet* (Washington, DC: Population Reference Bureau, 2003).

countries—already 84%—is expected to climb steadily to over 90% by 2075 (Fig. 5–6). Nevertheless, it is not just the developing countries that have problems.

## Different Populations, Different Problems

Some time ago, ecologists Paul Ehrlich and John Holdren proposed a formula to account for the human factors that contribute to environmental pollution and the depletion of resources. They reasoned that human pressure on the environment was the outcome of three factors: *population, affluence,* and *technology.* They offered the following formula:

$$I = P \times A \times T.$$

According to this equation, called the *IPAT formula,* environmental impact ($I$) is proportional to population ($P$), multiplied by the affluence of the average lifestyle ($A$), and multiplied by the level of technology of the society ($T$). Although the equation is a simplification, there is broad agreement that these three factors play the major role in putting pressure on the environment. Given the high level of technology in the industrialized countries, and the affluent lifestyle that accompanies it, a fairly small population can have a very large impact on the environment. As a result, it is hypocritical to criticize developing countries for continuing to grow their populations. Those who live in wealthy, but population-stable, developed countries are equally guilty of environmental misuse.

***Big Footprints.*** For example, it is estimated that, because of differences in consumption, the average American places at least 20 times the demand on Earth's resources, including its ability to absorb pollutants, as does the average person in Bangladesh, a poor Asian country. Major world pollution problems, including the depletion of the ozone layer, the impacts of global climate change, and the accumulation of toxic wastes in the environment, are largely the consequence of the high consumption associated with affluent lifestyles in the developed countries. For instance, the United States, with only 5% of the worlds population, is currently responsible for over 24% of the total global emissions of carbon dioxide, the major greenhouse gas. Likewise, much of the global deforestation and loss of biodiversity is due to consumer demands in developed countries. When it comes to the ecological footprint, ours in the developed world is extremely large and heavy (Fig. 5–7).

The developing countries do have a population problem, and it is making their progress toward sustainable development that much more difficult. Their people are plagued by all of the consequences of poverty. They are often hungry, poorly educated, out of work, sickened by common diseases, and vulnerable to natural hazards such as droughts, floods, and forest fires. They may look with hope to the future, but often that hope is crushed by decades of grinding poverty. Their needs are great: economic growth, more employment, wise leaders, effective public policies, fair treatment by other nations, and, especially, technological and financial help from the wealthy nations. Much of this is a matter of justice, one of the key components of stewardly care.

***Enter Stewardship.*** Despite a fairly stable population, the developed countries have an equally daunting problem, but one having to do with consumption, affluence, damaging technologies, and burgeoning wastes. These issues must be addressed to achieve sustainability, and this, too, requires wise leaders and effective public policies. Fortunately, the environmental impacts of affluent lifestyles may be moderated to a large extent by practicing environmental stewardship. For example, suitable attention to wildlife conservation, pollution control, energy conservation and efficiency, and recycling

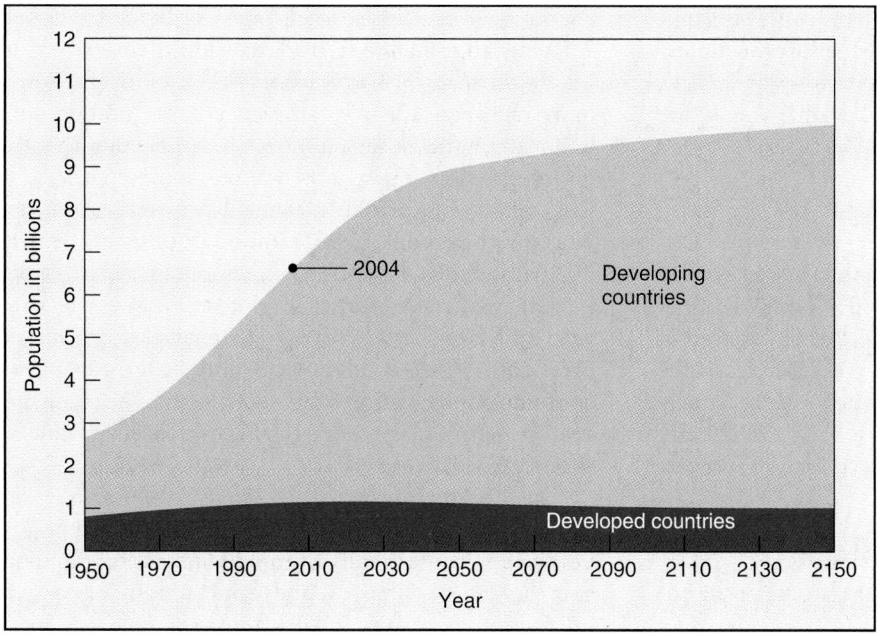


**Figure 5–6 Population increase in developed and developing countries.** Because of higher populations and higher birthrates, developing countries represent a larger and larger share of the world's population.

may offset, to some extent, the negative impact of a consumer lifestyle. In fact, a life devoted to conservation or other aspects of environmental stewardship might entirely offset the negatives and have a highly positive effect overall. As a result, the IPAT formula might be modified to

$$I = \frac{P \times A \times T}{S},$$

where $S$ stands for stewardly concern and practice.

Some people argue that population growth is the main problem, others claim that our highly consumption-oriented lifestyle is chiefly to blame, and still others maintain that it is our inattention to stewardship that is the prime shortcoming. In order to reach sustainability, however, all *three* areas must be addressed. That is, the population must stabilize, consumption must decrease, and stewardly action must increase. The next section focuses on the factors and consequences of population growth, principally in developing countries.

## 5.3 Consequences of Population Growth and Affluence

Expanding populations and increasing affluence—it sounds like trouble for the environment, and it is. It also means trouble for people, particularly in the developing countries.

### The Developing Countries

Prior to the Industrial Revolution, most of the human population survived through subsistence agriculture. That is, families lived on the land and produced enough food for their own consumption and perhaps some extra to barter for other essentials. Natural forests provided firewood, structural materials for housing, and wild game for meat. With a small, stable population, this system was basically sustainable. As the older generation

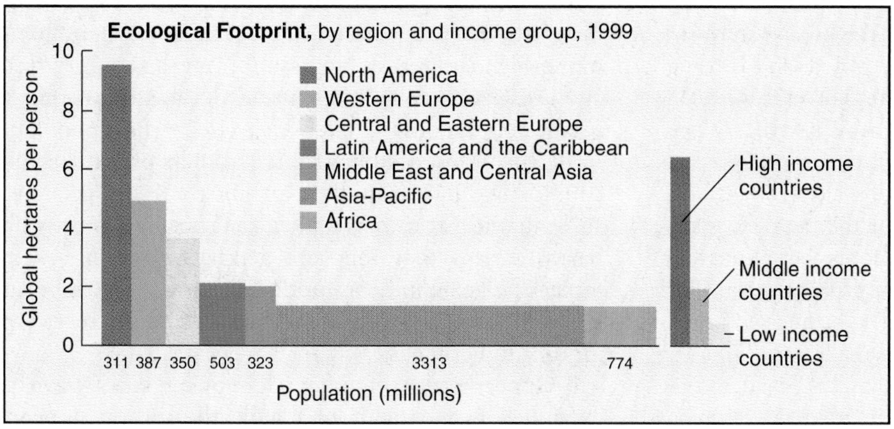


**Figure 5–7 Ecological footprints of world regions.** The footprint of developed countries is compared with that of developing regions. Populations are shown on the horizontal axis. (*Source: Living Planet Report 2002,* World Wide Fund for Nature.)

passed away, the land and natural systems could still support the next generation. Indeed, many cultures sustained themselves in this way over thousands of years. Now, however, except for a few primitive tribes, that way of life is history.

**Rural Populations.** After World War II, modern medicines—chiefly vaccines and antibiotics—were introduced into developing nations, whereupon death rates plummeted and populations grew rapidly. Today, 60% of the developing world—over 2.5 billion people—is directly engaged in agriculture, most at the subsistence level. What are the impacts of rapid growth on a population that is largely engaged in subsistence agriculture? Six basic options are possible, all of which are being played out to various degrees by people in these societies:

1. Reform the system of land ownership.
2. Intensify cultivation of existing land to increase production per unit area.
3. Open up new land to farm.
4. Move to cities and seek employment.
5. Engage in illicit activities for income.
6. Emigrate to other countries, either legally or illegally.

In addition, rapid population growth especially affects women and children. Let us look at each of the options and their consequences in a little more detail.

*Land Reform.* Rising population growth in rural developing countries has put increasing pressure on the need to reform the system of land ownership. Collectivization and ownership by the wealthy few are two patterns of agricultural land ownership that have historically kept rural peoples in poverty. Collectivization emerged from 20th century communism, and ownership by the wealthy few was the result of colonialism in the 19th and 20th centuries. South Africa, recently emerging from its apartheid past, demonstrates the kind of inequities perpetuated by colonialism. Whites still own 87% of the land, and the average landholding of whites is 3,800 acres, as opposed to $2\frac{1}{2}$ acres for blacks. Resolving these kinds of inequities can disrupt the social order, as events in Zimbabwe have shown. In that country, a chaotic land reform program has virtually destroyed the country's agricultural economy.

Collective agriculture was one of the great failures of the former Soviet Union. Today, state-owned land is being privatized, with the result that agricultural production is on the rise (although it still has a long way to go!). When China abandoned collective agriculture in 1978 and assigned most agricultural land to small-scale farmers, farm output grew more than 6% a year for the next 15 years, paving the way for China's recent economic boom. Private ownership, however, can have its own problems, too. For example, plots are often subdivided to heirs, and with each succeeding generation, the plots get smaller and smaller until they are too small to feed a family. This practice continues to be an enormous problem in many developing countries. (Three-fifths of all farms in India are less than 2.5 acres, too small to feed a large family.)

*Intensifying Cultivation.* The introduction of more highly productive varieties of basic food grains has had a dramatic beneficial effect in supporting the growing population, but is not without some concerns. (See Chapter 9.) For example, intensifying cultivation means working the land harder. Traditional subsistence farming in Africa involved rotating cultivation among three plots. In that way, the soil in each plot was cultivated for one year and then had two years to regenerate. With pressures to increase productivity, plots have been put into continuous production with no time off. The results have been a deterioration of the soil, decreased productivity (ironically), and erosion.

In addition, the increasing intensity of grazing is damaging the land, causing desertification. (See Chapter 8.) Given the countertrends of rapidly increasing population and the deterioration of land from overcultivation, food production per capita in Africa, for example, is currently decreasing.

*Opening Up New Lands for Agriculture.* Opening up new lands for agriculture may sound like a good idea, but there is really no such thing as "new land," and most good agricultural land is already in production. Opening up new land always means converting natural ecosystems to agricultural production, which means losing the goods and services those ecosystems were contributing. Even then, converted land is often not well suited for agriculture, unless it is irrigated. Irrigation, in turn, is costly and has its own environmental problems. Most of the tropical deforestation in South and Central America has occurred in order to increase agricultural production (Fig. 5–8). Much of this deforestation is done by poor, young people who are seeking an opportunity to get ahead, but are unskilled and untrained in the unique requirements of maintaining tropical soils. In a short time, the cleared land becomes unproductive for agriculture, leaving the people again in poverty. It is then taken over by large livestock companies.

*Migration to Cities.* Faced with the poverty and hardship of the countryside, many hundreds of millions of people in developing nations continue to migrate to cities in search of employment and a better life. By 2000, there were 292 "million-plus" cities in the developing world, many of which have become "megacities" of 10 million or more (Fig. 5–9a). The migration from the country to the city is expected to continue in the developing world. The urban population will likely surpass the rural population by 2020, almost doubling in size from its present 2 billion people. The most rapidly expanding cities, especially in sub-Saharan Africa, have fallen so far behind in providing basic services that they are getting worse, not better. Streets, for example, are potholed, sanitation and drinking water are poor, electricity and telephone service are erratic, and crime is rampant.

Opportunities in many developing cities have not expanded fast enough to handle the influx of people.

**Figure 5-8 Deforestation in the tropics.** Millions of acres of rain forest in Central and South America are being cut down each year to make room for agriculture, as shown in this photograph from Peru.

Many are forced to live in sprawling, wretched shantytowns and slums that do not even provide adequate water and sewers, much less other services (Fig. 5–9b). Diseases like malaria and malnutrition are endemic, and the incidence of HIV/AIDS is much higher in the cities than in the countryside, a consequence of much higher numbers of single men seeking work and also seeking the "services" of prostitutes or multiple sex partners—both of which are high risk factors for the spread of HIV/AIDS.

Worse, these cities often do not even provide the jobs people are looking for. Indeed, the high numbers of rural immigrants in the cities dilute the value of the one thing they have to sell: their labor. As we saw earlier, a common wage for a day's unskilled work is often equivalent to no more than a dollar or two—not enough for food, much less housing, clothing, and other amenities. Thousands, including many children, make their living by scavenging in dumps to find items they can salvage, repair, and sell. Many survive by begging—or worse.

***Illicit Activities.*** Anyone who doesn't have a way to grow sufficient food must gain enough income to buy it—and sometimes, desperate people break the law to do this. Although it is difficult to draw the line between the need and the greed that also draws people into illicit activities, it is undeniable that the shortage of adequate employment exacerbates the problem. Besides the rampant petty thievery and corruption that pervade many developing countries, income is also obtained from illegal activities such as raising drug-related crops and poaching wildlife.

***Emigration and Immigration.*** The gap between high- and low-income countries is reflected in the perception of many in the poorer countries who believe they can improve their well-being by migrating to a wealthier country. The fact that populations in the wealthy countries are aging suggests a strategy that, on the surface, looks appealing. That is, the wealthy countries need more and younger workers, so they should welcome the migrants from the low-income countries who are looking desperately for work. To some extent, and in some countries, this is exactly what is happening: Each year, many millions migrate to the United States and Europe in search of a better life, a shift that has been called "replacement migration." More millions migrate to neighboring countries to escape civil wars and ethnic persecution. Some 150 million people now live outside of their countries of birth.

Immigration, however, has its problems, too. Prejudice against foreigners is common, especially in countries with strong ethnic and cultural homogeneity, like Japan

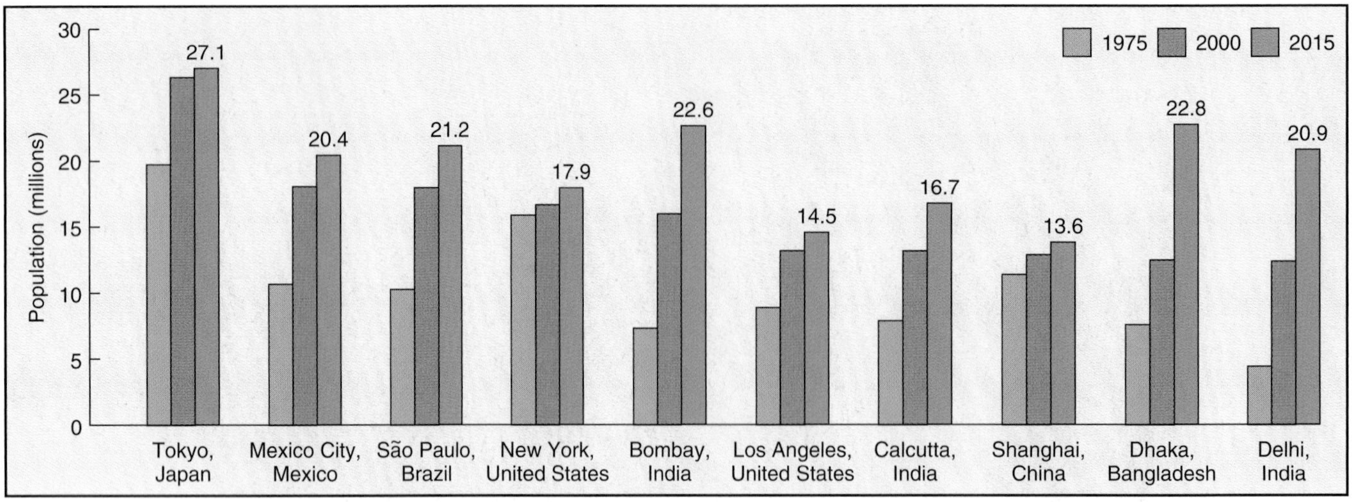


(a)

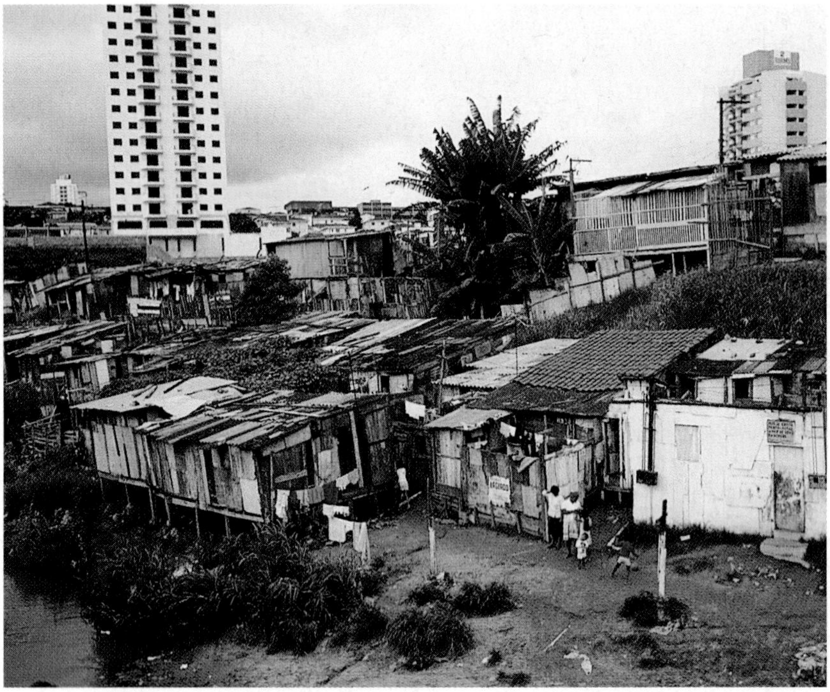

(b)

**Figure 5-9A Growing cities.** (a) The top 10 world metropolitan areas. Since 1975, cities in the developing world have grown phenomenally, and a number of them are now among the world's largest. (b) Slums on the outskirts of São Paulo, Brazil. Thirty-two percent of the city's population lives in these blighted areas. [*Source:* Data for part (a) from *World Urbanization Prospects: The 2001 Revision.* U.N. Population Division.]

and Germany. Even though the United States is a nation largely built on immigration, opposition to foreigners has intensified since the September 11, 2001, terrorist attacks. Nor have European countries welcomed immigrants: Although Western Europe's population is 42% larger than that of the United States, its immigration is only about half that of ours. (See the Ethics essay, "The Dilemma of Immigration," p. 135).

Refugee immigration leads to temporary refugee camps, where diseases and hunger often take a terrible toll on human life. Some "migrants" are little more than slaves. Recruiters from plantations in the Ivory Coast pay parents in neighboring Burkina Faso and Mali to send their children to work on the plantations, where they are kept under guard, sometimes for years.

**Impoverished Women and Children.** The hardships and deprivation of poverty fall most heavily on women and children. Men are freer to roam and pick up whatever work is available, and they may keep their wages for themselves. Some men take no responsibility at all for the women they impregnate, much less the children they sire. Even many married men, under the stress of poverty, abandon wives and children. Few developing countries

**ethics**

## The Dilemma of Immigration

For people trapped by poverty or lack of opportunity in their homeland, emigration to another country has always seemed a way to achieve a better life. As the New World opened up, many millions of people recognized this dream by emigrating to the United States and other countries. The United States and Canada are countries composed largely of immigrants and their descendants. Until 1875, all immigration into the United States was legal, so all who could manage to arrive could stay and become citizens. This openness was inscribed on the Statue of Liberty. The inscription reads, in part, "Give me your tired, your poor, your huddled masses yearning to breathe free, the wretched refuse of your teeming shore.... Send these, the homeless, tempest-tossed to me."

### History

Emigration from the Old World created a flood of migrants immigrating to the New World. This emigration relieved population pressures in European countries and aided in the development of the New World. A totally open policy toward immigration today, however, would be untenable. The United States, with its current population of 292 million, is no longer a vast, open land awaiting development. Still, hundreds of millions of people would immigrate if they could. How much immigration should be permitted, and should some groups be favored over others? In 1882, for example, the U.S. Congress passed the Chinese Exclusion Act, which barred the immigration of Chinese laborers, but not Chinese teachers, diplomats, students, merchants, or tourists. This act remained in effect until 1943, when China and the United States became allies in World War II. However, the current immigration policy still makes it easier for trained people to gain citizenship and relatively difficult for untrained people to do so. This policy has created what is commonly referred to as a "brain drain." Brainpower, many point out, is the "export" that developing nations can least afford.

### Now

Current immigration laws officially admit almost 900,000 new immigrants per year, a number larger than we have received at any time since the 1920s and larger than is accepted by all other countries combined. At present, immigration accounts for about 40% of U.S. population growth, which translates into 2.7 million people per year. The remainder of the population growth, called the natural increase, is the excess of the number of births over the number of deaths. If the fertility rate remains low, immigration will account for a growing proportion of our population growth. If the children born to new immigrants are added up (immigrants often have large families), immigration accounted for two-thirds of the growth of the American population in the 1990s.

### Illegal Immigration

The preceding discussion addresses only legal immigration. Illegal immigration is another matter. Hundreds of thousands, unable to gain access through legal channels, seek ways to enter the country illegally. The United States maintains an active border patrol, especially along the border with Mexico, which several thousand people try to cross each night. Most are caught and returned, but an undetermined number—estimated at 200,000 to 300,000 per year—slip through. The Illegal Immigration Reform Act of 1996 addressed this problem by strengthening the border patrol and stepping up efforts to locate and deport illegal aliens—actions that are supported by most observers. However, there are currently far more illegal agricultural workers than legal "guest workers"; the number of illegal farm workers is estimated at 650,000. Employers often prefer the illegal workers, because they work for lower wages and place fewer demands on employer services.

### September 11, 2001

The terrible events of September 11, 2001, have had their impact on immigration into the United States and on border issues, because some of the terrorists had entered from Canada and their visas revealed many irregularities. Prospective tourists or students now seeking U.S. visas are far more thoroughly screened, especially if they originate from Middle Eastern countries. Also, border security has been tightened at the many entry points from Canada and Mexico, and it has become more difficult to enter the country illegally.

A recent report by the National Research Council examined the economic impacts of our immigration policies and concluded that legal immigration basically benefits the U.S. economy and has little negative impact on native-born Americans. A few areas of the country where immigrants are especially numerous (for example, California, Texas, and Florida) experience some challenges in assimilating new immigrants, but the overall impact is positive, according to the Council.

### Immigration Reform

In 1990, Congress established the U.S. Commission on Immigration Reform and directed it to submit a report at the end of 1997. The Commission report recommended cutting back immigration to a core level of 550,000 per year and allowing 150,000 additional visas annually for spouses and minor children of legal permanent residents. (Currently, spouses and minor children of legal residents make up the largest component of immigration, 580,000 per year). This policy is to be phased out when the existing backlog is eliminated—judged to take up to eight years. Interestingly, the Commission recommended discontinuing the immigration of unskilled laborers, while continuing to welcome more highly skilled immigrants (albeit under quotas). The Commission has now disbanded, and it is up to Congress to take up the report and act on it—or not to do so. (To date, it has *not*.) The commission's recommendations, if adopted, will result in only a small reduction in the current rate of legal immigration. Neither political party in the United States seems willing to take on this issue; in fact, both parties constantly seek the votes of immigrant communities. It appears that immigration will continue to be a large and increasing part of growth in the United States well into the future.

As population pressures in developing countries continue to mount, the questions of how many immigrants to accept, from what countries, and where to draw the line regarding asylum seem certain to become more and more pressing. In addition to compassion, the social, economic, and environmental consequences—both national and global—of the alternatives must be weighed in making the final decision. Where do you stand?

have a welfare system that will provide care in such situations. Too often, the women cannot cope, so their children are abandoned, and the women turn to begging, stealing, or prostitution.

***Street Children.*** What happens to the children of these women? If they survive at all, it is by begging, scrounging through garbage, stealing, and finding shelter in any hole or crevice they can find (Fig. 5–10). The problem is so great that nearly every sizable city in the developing world has thousands of these "street children"—on the order of 25 million in all by some estimates—and their numbers are growing. Forced child labor, child prostitution, and the sale of children for adoption are all too common in developing countries. One can only speculate as to the kinds of adults such children become—if they survive. At the very least, all of these factors tend to lock impoverished women and children into the vicious cycle of illiteracy and squalid conditions that defines and perpetuates absolute poverty.

A summary of the consequences of rapid population growth in developing countries is given in Fig. 5–11. Population growth, poverty, and environmental degradation are not separate issues, but are, to the contrary, very much interrelated.

## Affluence

The United States has the dubious distinction of leading the world in the consumption of many resources. We are a large country and we are affluent. We consume the largest share of 11 of 20 major commodities: aluminum, coffee, copper, corn, lead, oil, oilseeds, natural gas, rubber, tin, and zinc. We lead in per capita consumption of many other items, such as meat. The average American eats more than three times the global average of meat. We lead the world in paper consumption, too, at 725 lb per person per year. All of these factors (and many more like them) contribute to the enormous ecological footprint each of us makes on the world.

Despite the adverse effects of affluence, increasing the average wealth of a population can affect the environment positively. An affluent country like ours provides such amenities as safe drinking water, sanitary sewage systems and sewage treatment, and the collection and disposal of refuse. Thus, many forms of pollution are held in check, and the environment improves with increasing affluence. In addition, if we can afford gas and electricity, we are not destroying our parks and woodlands for firewood. In short, we can afford conservation and management, better agricultural practices, and pollution control, thereby improving our environment.

***The Dark Side.*** Still, because the United States consumes so many resources, we also lead the world in the production of many pollutants. For example, by using such large quantities of fossil fuel (coal, oil, and natural gas) to drive our cars, heat and cool our homes, and generate electricity, the United States is responsible for a large share of the carbon dioxide produced. As mentioned, with about 5% of the world's population, the United States generates 24% of the emissions of carbon dioxide that may be changing global climate. Similarly, emissions of chlorofluorocarbons (CFCs) that have degraded the ozone layer,

**Figure 5–10 Foraging in trash.** In cities of the developing world, many poor people, including mothers and children, subsist only by scrounging through refuse for bits of food and items they can resell.

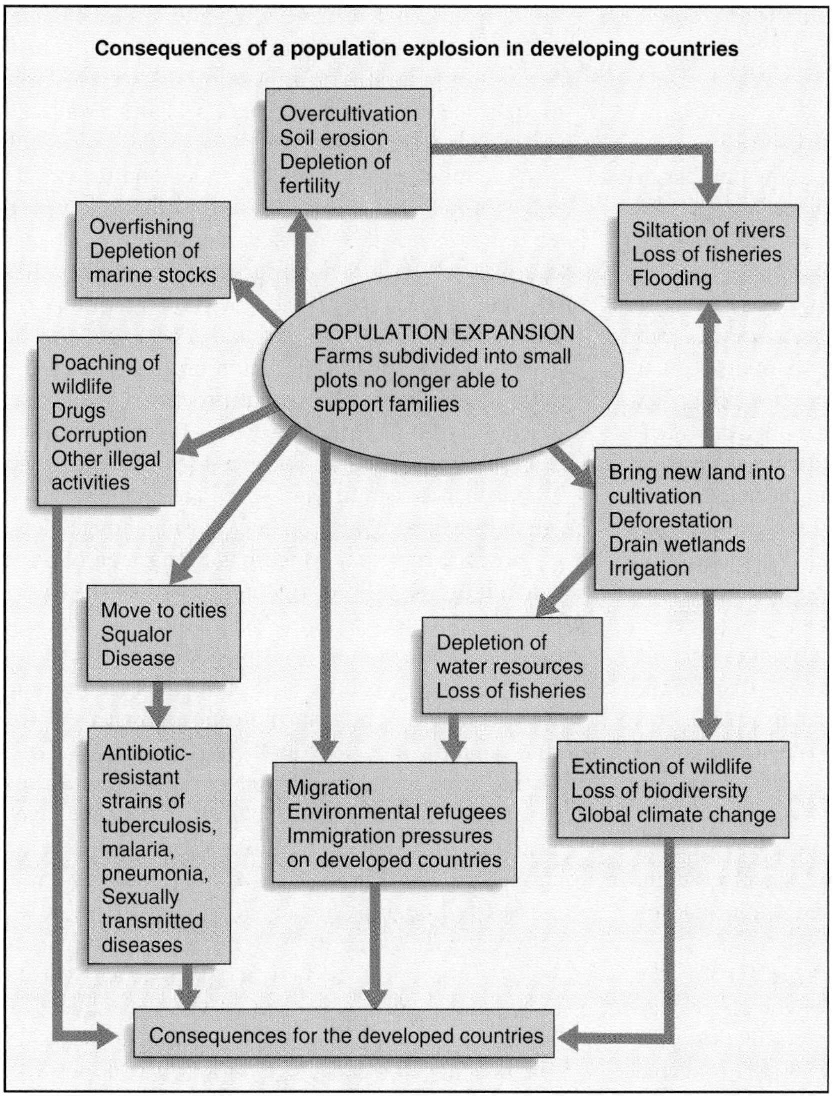


**Figure 5–11 Consequences of expanding populations in the developing world.** The diagram shows the numerous connections between unchecked population growth and social and environmental problems.

emissions of chemicals that cause acid rain, emissions of hazardous chemicals, and the production of nuclear wastes are all largely the by-products of affluent societies.

Economic factors place further demands on the environment and the developing world. The world's wealthy 20% is responsible for 86% of all private consumption and 80% of world trade. As a consequence, 11 of the 15 major world fisheries are either fully exploited or overexploited, and old-growth forests in southern South America are being clear-cut and turned into chips to make fax paper. Oil spills are a "by-product" of our appetite for energy. Tropical forests are harvested to satisfy the desires of the affluent for exotic wood furnishings. Metals are mined, timber harvested, commodities grown, and oil extracted, all far from the industrialized countries where these goods are used. Every one of these activities has a significant environmental impact. As increasing numbers of people strive for and achieve greater affluence, it seems more than likely that such pressures, and other ones like them, will mount.

*Out of Touch.* One way of generalizing the effect of affluence is to say that it enables the wealthy to clean up their immediate environment by transferring their wastes to more distant locations. It also allows them to obtain resources from more distant locations, so they neither see nor feel the impacts of getting those resources. In many respects, therefore, the affluent isolate themselves and may become totally unaware of the environmental stresses they cause with their consumption-oriented lifestyles. Still, affluence also provides people with opportunities to exercise lifestyle choices that are consistent with the concerns for stewardship and sustainability.

With this picture of population growth and its impacts in view, the next section discusses some additional dimensions of the problem, in order to provide a more thorough understanding of the issues.

## 5.4 Dynamics of Population Growth

In studying population growth, you must consider more than just the increase in numbers, which is simply births minus deaths. You must also consider how the number of births ultimately affects the entire population over the **longevity**, or lifetimes, of the individuals.

## Population Profiles

A **population profile** is a bar graph showing the number or proportion of people (males and females separately) at each age for a given population. The data are collected through a census of the entire population, a process in which each household is asked to fill out a questionnaire concerning the status of each of its members. Various estimates are made for those who do not maintain regular households. In the United States and most other countries, a detailed census is taken every 10 years. Between censuses, the population profile may be adjusted by using data regarding births, deaths, immigration, and the aging of the population. The field of collecting, compiling, and presenting information about populations is called **demography**; the people engaged in this work are **demographers.**

**Profile for the United States.** A population profile shows the **age structure** of the population—that is, the proportion of people in each age group at a given date. It is a snapshot of the population at a given time. Population profiles of the United States for 1990 and 2000 are shown in Figs. 5–12a and b. Leaving out the complication of emigration and immigration for the moment, each bar in the profile started out as a *cohort* of babies at a given point in the past, and that cohort has only been diminished by deaths as it has aged.

***Boom or Bust?*** In developed countries such as the United States, the proportion of people who die before age 60 is relatively small. Therefore, the population profile below age 60 is an "echo" of past events that affected birthrates. Figure 5–12a shows, for example, that smaller numbers of people were born between 1931 and 1935 (ages 55–59 in 1990). This is a reflection of lower birthrates during the Great Depression. The dramatic increase in people born in 1946 and for 14 years thereafter (ages 30–44 in 1990) is a reflection of returning veterans and others starting families and choosing to have relatively large numbers of children following World War II—the "baby boom." The general drop in numbers of people born from 1961 to 1976 (ages 10–29 in 1990) is a reflection of sharply declining fertility rates, with people choosing to have significantly fewer children—the "baby bust." The rise in numbers of people born in more recent years (ages 0–9 in 1990) is termed the "baby boom echo" and is due to the large baby-boom generation producing a similarly large number of children, even though the actual total fertility rate remained near 2.0. These changes in fertility are shown in Figure 5–13.

***Planning Tool.*** More than a view of the past, a population profile provides governments and businesses with a means of realistic planning for future demand for various goods and services, ranging from elementary schools to retirement homes. Consumer demands are largely age specific; that is, what children need and want differs from what teenagers, young adults, older adults, or, finally, people entering retirement want and need. Using a population profile, you can see the projected populations of particular age cohorts and plan to expand or retrench accordingly.

A number of industries expanded and then contracted as the baby-boom generation moved through a particular age range, and this phenomenon is not yet past. In sequence, schools, then colleges and universities, and then the job market were affected by the large influx of baby boomers. As the large baby-boom generation, now in middle age, moves up the population profile (see Fig. 5–12b), any business or profession that provides goods or services to seniors is looking forward to a period of growth. For example, look for a dramatic increase in the construction of retirement homes and long-term-care facilities in the near future. (See Earth Watch, "Are We Living Longer?" p. 140.)

***Social Security.*** Much has been made of the future demand for Social Security outlays to retirees, especially as the baby boomers retire. This is sometimes wrongly pictured as an impending disaster. The current condition of Social Security is robust, with annual cash flow surpluses of $150 billion. These surpluses are invested in Treasury bonds, and the income from them is used to pay benefits to existing retirees (and borrowed by the government to fund all sorts of other expenditures). These benefits have nothing to do with our rising national debt. However, as the number of retirees rises, the combined income from reserves and ongoing Social Security payments of workers will eventually (in about 40 years) be insufficient to pay 100% of the promised benefits. This deficit will develop slowly, and there is every reason to believe that Congress will address it long before it is expected to occur. Even if Congress does nothing, revenues will continue to cover approximately 73% of benefits.

## Future Populations

Current population growth in a country is calculated from three vital statistics: births, deaths, and migration. For instance, the population of the United States grew by nearly 2.7 million people during the year following July 1, 2000. This growth was the result of 4.05 million births, 2.43 million deaths, and a net migration of 1.1 million. Thus, about 40% of the population increase was due to immigration and 60% to natural increase.

***Predictions.*** Demographers used to make population *forecasts*. Most often, they took current birth and death rates, factored in expected immigration, and then extrapolated the data into the future. They were virtually always wrong. For example, in 1964, the total fertility rate (TFR) in the United States was around 3.0. The U.S. Census Bureau forecast that the TFR would range between 2.5 and 3.5 up to 2000 and projected a 2000 population of 362 million under the higher fertility assumption—81 million more than the actual number. The agency totally missed the baby bust of the late sixties and seventies, although the trend in 1964 was definitely downward.

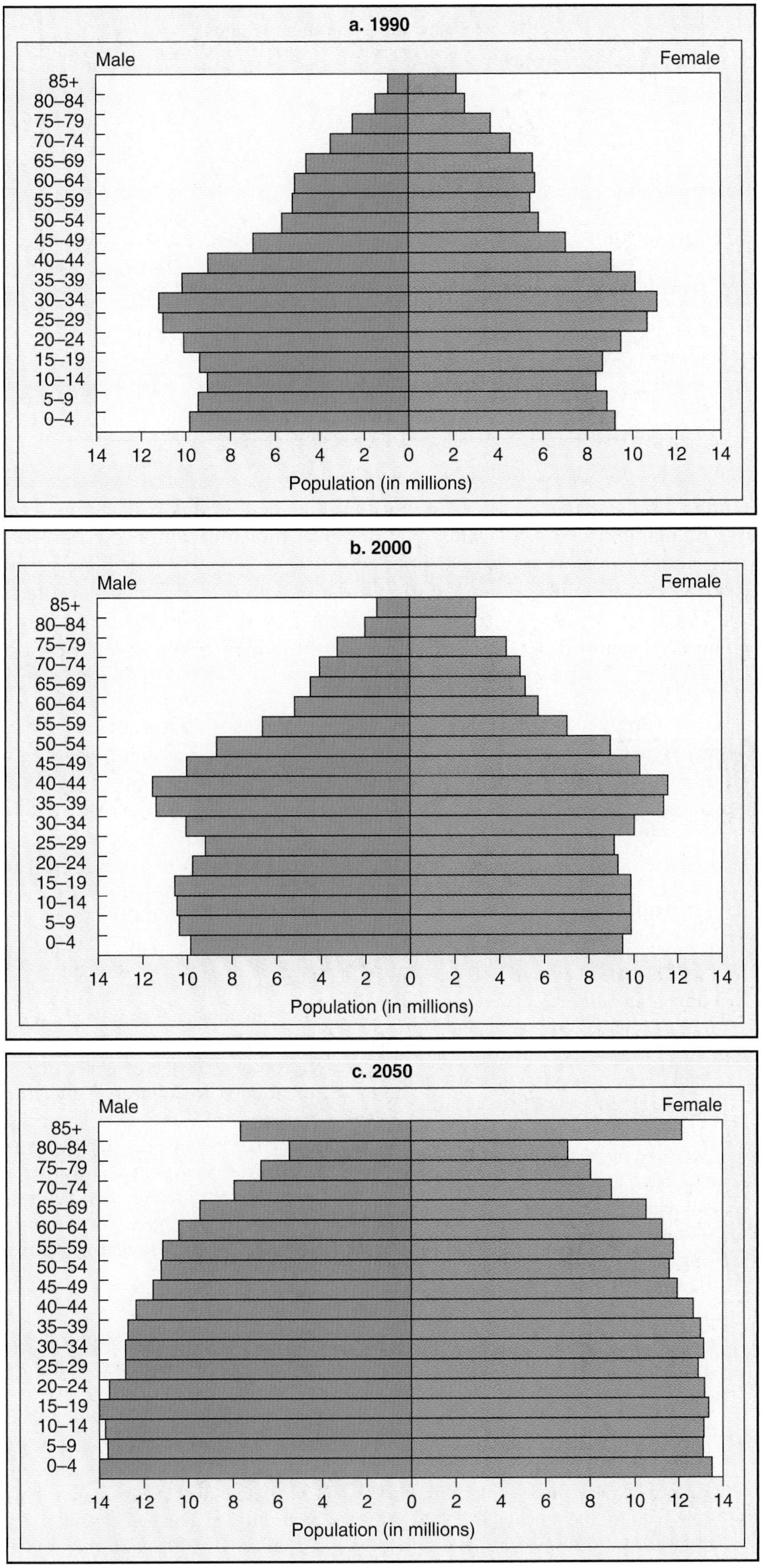


**Figure 5–12 Population profiles of the United States.** The age structure of the U.S. population (a) in 1990, (b) in 2000, and (c) projected to 2050. (*Source:* U.S. Census Bureau, International Data Base.)

**Figure 5–13 Total fertility rate, United States, 1920–2002.** The changes in fertility led to the baby boom and baby bust in the United States. The rate now hovers about the replacement level. [*Source:* Population Reference Bureau, Population Bulletin 57 (4): *What Drives U.S. Population Growth?* December 2002.]

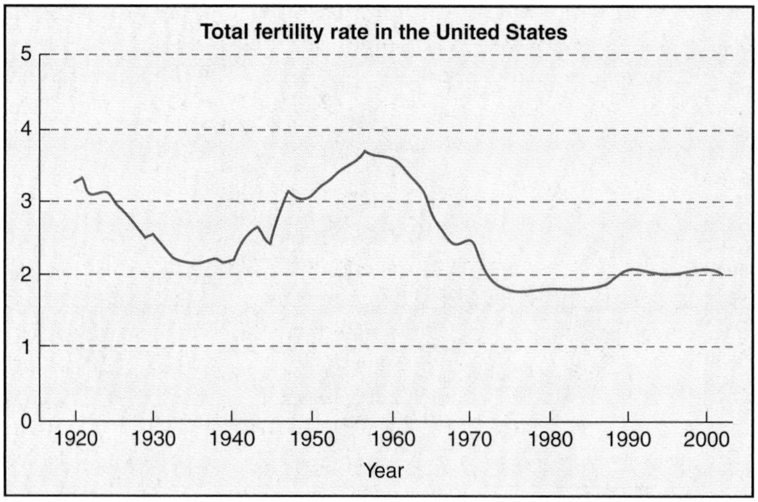


***Projections.*** Nowadays demographers only make *projections,* and they cover their bases by making their assumptions about fertility, mortality, and migration very clear. As noted at the beginning of the chapter, the projection that the world population will level off at around 10 billion is based on the assumption that fertility rates will continue their gradually declining trend. The United Nations gives three different projections of future world population (Fig. 5–3). The medium-fertility scenario assumes that a below-replacement-level fertility of 1.85 will be reached by three-fourths of the developing countries by 2050, the high-fertility scenario assumes a

## Are We Living Longer?

Most of us would like to live a long and prosperous life, hopefully just slipping away in our sleep at some advanced age. With the advent of modern medicine and disease control, this has become possible for more and more people. Average **life expectancy** (*the number of years a newborn can expect to live under current mortality rates*) has been increasing dramatically. Over the past two centuries, world life expectancy increased from about 25 years to 65 for men and 69 for women. Life expectancy and mortality rates are linked, and as infant mortality in particular has declined, life expectancy in the developing countries has climbed sharply. Life expectancies in those countries are now within 10 years of life expectancies in the developed countries. (The latter are now 72 for men and 79 for women.) What is not well known is that life expectancy has also continued to climb about $2^{1}{}_{2}$ years every decade for the low-mortality developed countries. Mortality rates have declined for every age.

### Longevity

Can we hope that this trend will continue, until perhaps a century from now human life expectancies will be approaching or going beyond 100? There is intense debate over this question. First, however, you must understand the difference between life expectancy and longevity. **Longevity** refers to the *maximum life span for a species.* The verifiable record for humans is held by Madame Jeanne Louise Calment, who died in France on August 4, 1997, at 122 years of age. The record for the United States is that of a man who died at 115 years of age in 1882! These figures suggest that human longevity has not really increased over the years. Indeed, many workers in the field of gerontology (the study of aging) are convinced that, in humans (as in all species), there is an aging process that cannot be prevented—that built into our cells are the biochemical seeds of destruction. Cells, tissues, and organs wear out at variable rates in different people. Thus, we will not likely find life expectancies to rise beyond about 85 years anywhere or anytime in the future.

Other workers argue that even though absolute longevity seems to be a limiting process, we can expect life expectancies to continue the increase they have shown in recent years. Why should life expectancies stop at 85, they ask? The trend of a $2\frac{1}{2}$-year increase per decade is based on reducing many of the factors that cause mortality at different ages. This trend should continue, given the medical prowess of our 21st-century science.

### Take a Test

So, how long do you want to live? How long do you expect to live? It is encouraging that there are many steps you can take to increase your chances of living longer. Check the Web site http://www.livingto100.com, and calculate your life expectancy. Following your answer, you can consult a key that explains why the questions you have answered are related to life expectancy. How did you do?

total fertility rate of 2.5 by 2050 and the continuation of that rate, and the low-fertility scenario assumes that the total fertility rate will reach 1.54 by 2050 and will be maintained. Note how each fertility assumption generates profoundly different world populations.

**Population Projections for Developed Countries.** The 2000 population profile of Italy, a developed country in southern Europe, shown in Fig. 5–14a, reflects the fact that Italian women have had a low fertility rate for some time. If the 2000 total fertility rate of 1.2 remains constant for the next 25 years, the profile presented in Fig. 5–14b is obtained. This profile shows a dramatic increase in the number of older people, a great reduction in the number of children and young people, and a net decline of 1.5 million.

***Graying of the Population.*** For the next 25 years, Italy's population will be **graying**, a term used to indicate that the proportion of elderly people is increasing. Overall, a net population decrease of 6% is expected to occur. What opportunities and risks does the changing population profile imply for Italy? If you were an adviser to the Italian government, what would your advice be for the short term? For the longer term? Unless a smaller population is the goal, it might be wise to encourage, and to provide incentives for, Italian couples to bear more children. It might also be a good idea to allow more immigration (*replacement migration*). But allowing the declining numbers of Italian people to be replaced by a non-Italian immigrant population has implications for Italian culture, religion, etc.

The very low fertility rate and the expected declining population seen in Italy are typical of an increasing number of highly developed nations. Who will produce the goods and services needed by their aging populations? Europe as a whole is on a trajectory of population decline

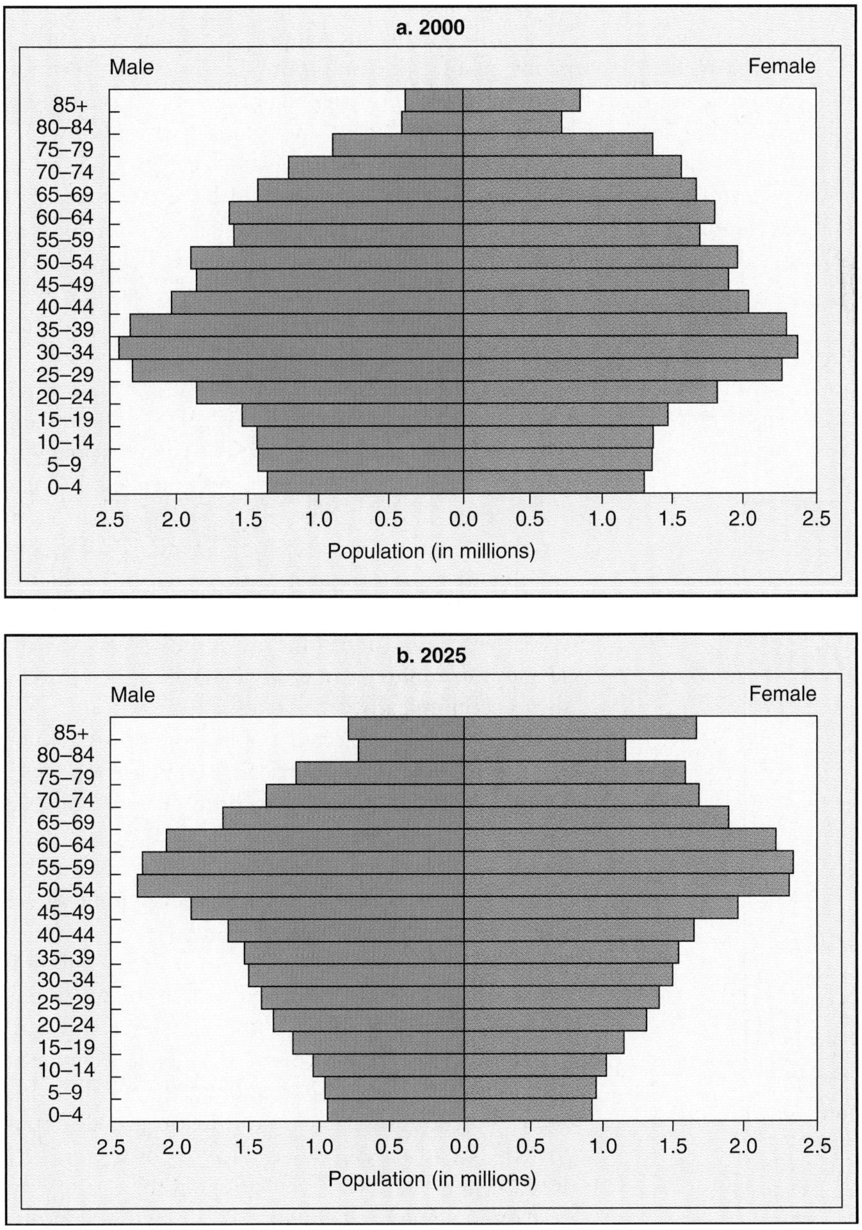


**Figure 5–14 Projecting future populations: developed country.** A population profile of Italy, representative of a highly developed country, (a) in 2000 and (b) projected to 2025. Note how larger numbers of persons are moving into older age groups and the number of children is diminishing. (*Source:* U.S. Census Bureau, International Data Base.)

if only natural increase is considered. In fact, Europe is in the midst of a migration crisis. Some 20 million foreigners now live and work in the countries of Western Europe, most of whom do not have resident status. In many of these countries, fear and mistrust of these "guest workers" has led to violent attacks on them. Many Europeans do not want more immigration, even to prevent the loss of the labor force and population decline. Yet just to maintain their current populations, most of these countries will have to triple their current immigration levels in the near future.

*No Graying Here.* In contrast to other developed countries, the fertility rate in the United States reversed directions in the late 1980s and started back up. On the basis of the lower fertility rate, the U.S. population had been projected to stabilize at between 290 and 300 million toward the middle of the next century. With a higher fertility rate of 2.0, the U.S. population is projected to be 420 million by 2050 (see Fig. 5–12c for a profile) and to continue growing indefinitely (Fig. 5–15). For this projection, immigration is assumed to remain constant at current levels—880,000 per year.

These projections for the United States show how differences in the total fertility rate and high immigration profoundly affect estimates of population when they are extrapolated 50 or more years forward. In light of the concerns for sustainable development, what do you think the population policy of the United States should be? Can you picture some scenarios that could lead to lower future growth?

**Population Projections for Developing Countries.** Developing countries are in a situation vastly different from that of developed countries. Fertility rates in developing countries are generally declining, but they are still well above replacement level. The average TFR (excluding China, where it is 1.7) is currently 3.5, which is comparable to the United States at the peak of the baby boom. Because of even higher past fertility rates, the population profiles of developing countries have a pyramidal shape.

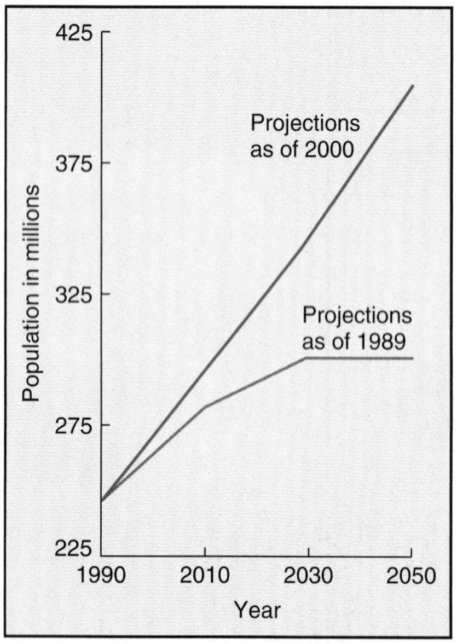


**Figure 5–15 Population projections for the United States.** Projections shift drastically with changes in fertility. Contrast the 1988 projection, based on a fertility rate of 1.8, with the 2000 projection, based on an increased fertility rate of 2.1. (*Source:* U.S. Census Bureau, International Data Base.)

*Iraq.* For example, the 2000 population profile for the western Asian country of Iraq, which has a fertility rate of 5.4, is shown in Fig. 5–16a. Even assuming that this fertility rate will gradually decline to 2.7 by 2025, the population will increase from 23 to 40 million, yielding the profile shown in Fig. 5–16b. Even the high infant mortality rate (currently 103 per thousand) comes nowhere close to offsetting the high fertility rate. Thus, the pyramidal form of the profile remains the same, because, for many years, the rising generation of young adults produces an even larger generation of children. The pyramid gets wider and wider, until the projected declining fertility rate begins to take effect.

While highly developed countries are facing the problems of a graying population, the high fertility rates in developing countries maintain an exceedingly young population. An "ideal" population structure, with equal numbers of persons in each age group and a life expectancy of 75 years, would have one-fifth (20%) of the population in each 15-year age group. By comparison, 40%–50% of the population is below 15 years of age in many developing countries, whereas less than 20% of the population is below the age of 15 in most developed countries (Table 5–4).

*Growth Impacts.* What do these differing population structures mean in terms of the need for new schools, housing units, hospitals, roads, sewage collection and treatment facilities, telephones, etc.? One of the things they mean is that if a country such as Iraq is simply to maintain its current standard of living, the amount of housing and all other facilities (not to mention food production) must be almost doubled in as little as 25 years. As a result, the population growth of a developing country can easily cancel out its efforts to get ahead economically.

Present and projected population profiles for developed and developing countries are compared in Fig. 5–17. The figure shows that, while little growth will occur in developed countries over the next 50 years, enormous growth is in store for the developing world, and this is assuming that fertility rates in the developing world continue their current downward trend!

Keep in mind that these or any other population projections should *not* be confused with predicting the future. They are intended only to show where we will end up if we continue the present course. As the old saying goes, "If you don't like where you are going, change direction." In other words, if we feel that the projected population growth is undesirable, we can try to bring fertility rates down faster. However, even bringing the fertility rates of developing countries down to 2.0 will not stop their

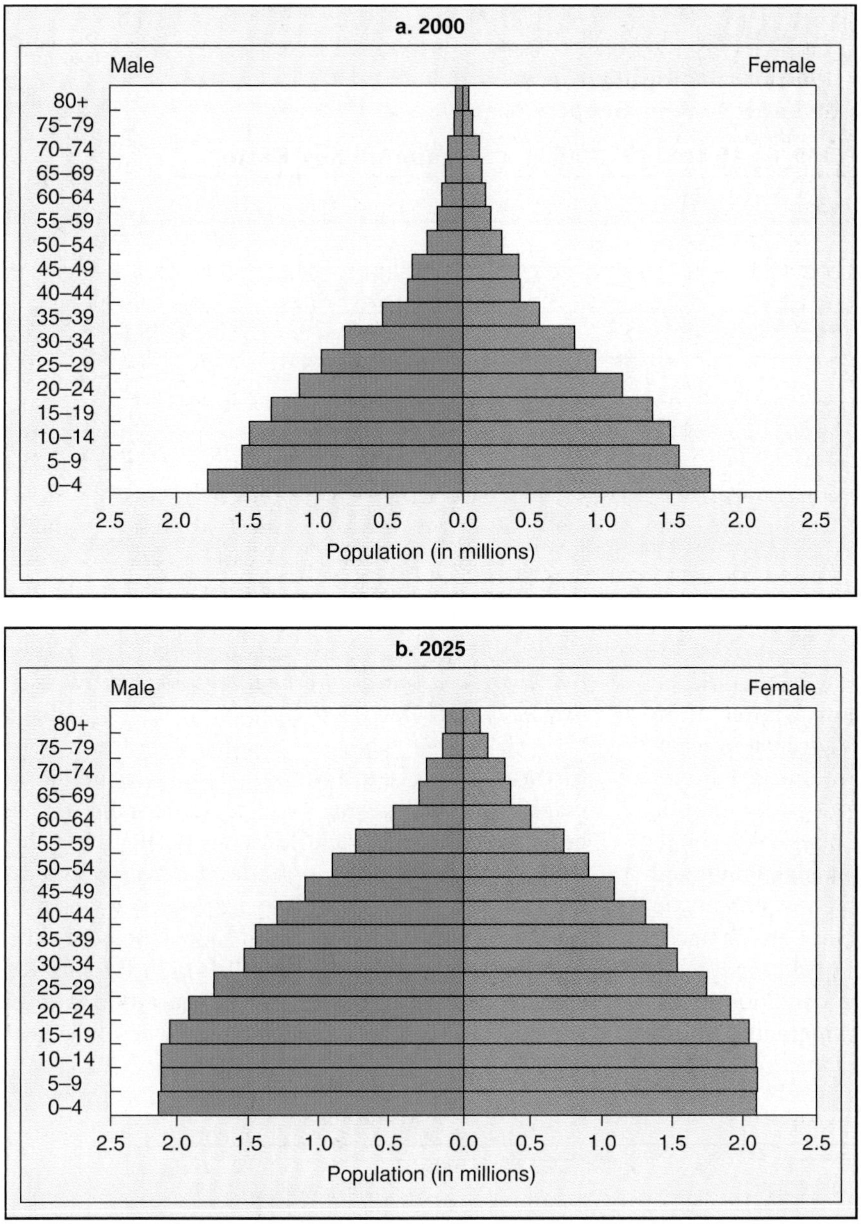


**Figure 5–16 Projecting future populations: developing country.** (a) The 2000 population profile of Iraq, a developing country. (b) Projection for the year 2025, based on the assumption that the total fertility rate of 5.7 will decline to 2.7 by 2025. (*Source:* U.S. Census Bureau, International Data Base.)

growth immediately. This is because of a phenomenon known as *population momentum.*

## Population Momentum

Countries with a pyramid-shaped population profile, such as Iraq, will continue to grow for 50–60 years, *even after the total fertility rate is reduced to the replacement level.* This phenomenon, called **population momentum,** occurs because such a small portion of the population is in the upper age groups (where most deaths occur) and many children are entering their reproductive years. Even if these rising generations have only two children per woman, the number of births will far exceed the number of deaths. The imbalance will continue until the current children reach the Iraqi limits of life expectancy—50 to 60 years. In other words, only a population at or below replacement-level fertility for many decades will achieve a stable population.

Despite population momentum, efforts to stabilize population are not fruitless. It just means that population growth cannot be halted quickly. Like a speeding train, there is a long time between applying the brakes and stopping completely. The earlier the fertility rates are reduced, the greater is the likelihood of achieving a steady-state population and a sustainable society in the near future.

## The Demographic Transition

The concept of a stable, nongrowing global human population based on people freely choosing to have smaller families is possible because it is already happening in developed countries. If we can understand the factors that have brought this about in those countries, then perhaps we can figure out how to make it occur in developing countries, too.

**table 5-4 Populations by Age Group**

| Region or Country | Percent of Population in Specific Age Groups <15 | 15 to 65 | >65 | Dependency Ratio* |
|---|---|---|---|---|
| sub-Saharan Africa | 44 | 53 | 3 | 89 |
| Latin America | 32 | 62 | 6 | 61 |
| Asia | 30 | 64 | 6 | 56 |
| Iraq | 47 | 50 | 3 | 100 |
| Europe | 17 | 68 | 15 | 47 |
| Germany | 15 | 68 | 17 | 47 |
| China | 22 | 71 | 7 | 41 |
| United States | 21 | 66 | 13 | 52 |

*Number of individuals below 15 and above 65, divided by the number between 15 and 65 and expressed as a percentage.
*Source:* Data from *2003 World Population Data Sheet* (Washington, DC: Population Reference Bureau, 2003).

Early demographers observed that the modernization of a nation brings about more than just a lower death rate resulting from better health care: A decline in fertility rate also occurs as people choose to limit the size of their families. Thus, as economic development occurs, human societies move from a primitive population stability, in which high birthrates are offset by high infant and childhood mortality, to a modern population stability, in which low infant and childhood mortality are balanced by low birthrates. This gradual shift in birth and death rates from the primitive to the modern condition in the industrialized societies is called the **demographic transition.** *The basic premise of the demographic transition is that there is a causal link between modernization and a decline in birth and death rates.*

**Birthrates and Death Rates.** To understand the demographic transition, you need to understand the **crude birthrate (CBR)** and **crude death rate (CDR).** The CBR and the CDR are *the number of births and deaths, respectively, per thousand of the population per year.* By giving the data per thousand of the population, populations of different countries can be compared regardless of their total size. The term *crude* is used because no consideration is given to what proportion of the population is old or young, male or female. Subtracting the CDR from the CBR gives the

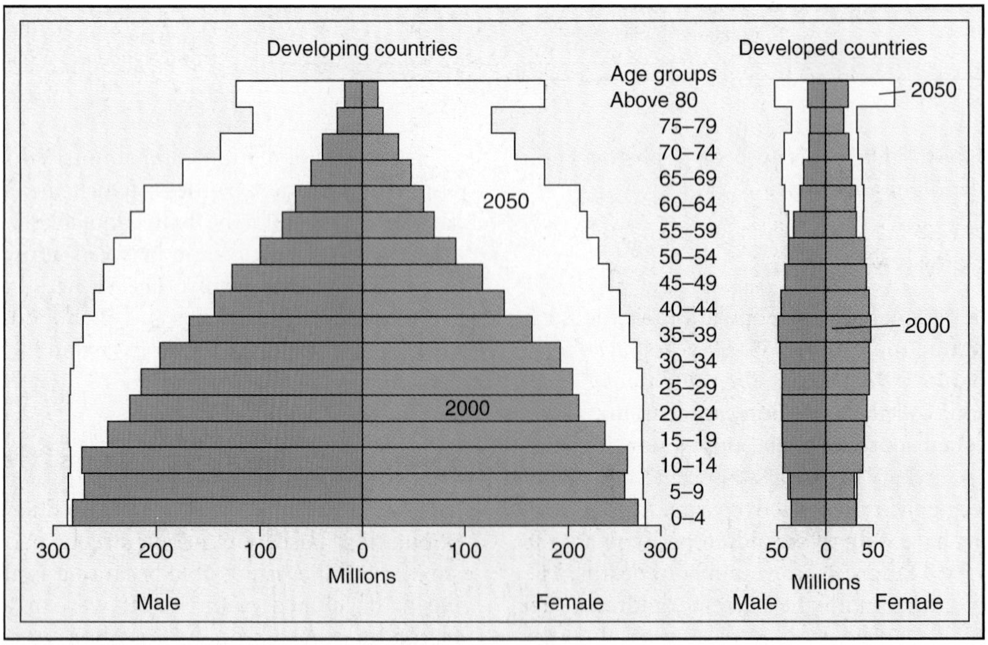


**Figure 5–17 Comparing projected populations.** 2000 population profiles for developed and developing countries, projected to the year 2050. (*Source:* U.S. Census Bureau, International Data Base.)

increase (or decrease) per thousand per year. Dividing this result by 10 then yields the *percent* increase (or decrease) of the population. Mathematically,

$$\begin{matrix}\text{Number of} & & \text{Number of} & & \text{Natural increase} & & \text{Percent increase}\\ \textit{births}\text{ per} & & \textit{deaths}\text{ per} & & \text{(or decrease)} & & \text{(or decrease)}\\ 1{,}000 & - & 1{,}000 & = & \text{in population} & \div 10 = & \text{in population}\\ \text{per year} & & \text{per year} & & \text{per }1{,}000 & & \text{per year}\\ \text{(CBR)} & & \text{(CDR)} & & \text{per year} & & \end{matrix}$$

A stable population is achieved if, and only if, the CBR and CDR are equal.

The **doubling time** is the number of years it will take a population growing at a constant percentage per year to double. It is calculated by dividing the percentage rate of growth into 70. (The 70 has nothing to do with population per se; it is derived from an equation for population *growth*.) The CBR, the CDR, and the doubling time of various countries are shown in Table 5–5.

**Epidemiologic Transition.** Throughout most of human history, crude death rates were high—40 or more per thousand for most societies. By the middle of the 19th century, however, the epidemics and other social conditions responsible for high death rates began to recede, and death rates in Europe and North America declined. The decline was gradual in the now-developed countries, lasting for many decades and finally stabilizing at a CDR of about 11 per thousand. At present, cancer and cardiovascular disease and other degenerative diseases account for most mortality, and many people survive to old age. This pattern of change in mortality factors has been called the *epidemiologic* transition (Fig. 5–18) and represents one element of the demographic transition. (Epidemiology is the study of diseases in human societies.)

**Fertility Transition.** Another pattern of change over time can be seen in crude birthrates. In the now-developed countries, birthrates have declined from a high of 40 to 50 per thousand to 8 to 12 per thousand—a *fertility* transition. As Fig. 5–18 shows, this did not happen at the same time as the epidemiologic transition; instead, it was

**table 5-5 Crude Birth and Death Rates for Selected Countries**

| Country or Region | Crude Birthrate | Crude Death Rate | Annual Rate of Increase (%) | Doubling Time (Years) |
|---|---|---|---|---|
| World | 22 | 9 | 1.3 | 54 |
| **Developing Nations** | | | | |
| Average (excluding China) | 28 | 9 | 1.9 | 37 |
| Egypt | 27 | 6 | 2.1 | 33 |
| Kenya | 35 | 15 | 2.0 | 35 |
| Madagascar | 43 | 13 | 3.0 | 23 |
| India | 25 | 8 | 1.7 | 41 |
| Iraq | 35 | 10 | 2.5 | 28 |
| Vietnam | 19 | 6 | 1.3 | 54 |
| Haiti | 32 | 14 | 1.8 | 39 |
| Brazil | 20 | 7 | 1.3 | 54 |
| Mexico | 29 | 5 | 2.4 | 29 |
| **Developed Nations** | | | | |
| Average | 11 | 10 | 0.1 | 700 |
| United States | 14 | 9 | 0.6 | 117 |
| Canada | 11 | 7 | 0.3 | 233 |
| Japan | 9 | 8 | 0.1 | 700 |
| Denmark | 12 | 11 | 0.1 | 700 |
| Germany | 9 | 10 | −0.1 | - |
| Italy | 9 | 10 | −0.1 | - |
| Spain | 10 | 9 | 0.1 | 700 |

*Note:* Dash indicates doubling time cannot be calculated, because growth is negative.
*Source:* Data from *2003 World Population Data Sheet* (Washington, DC: Population Reference Bureau, 2003).

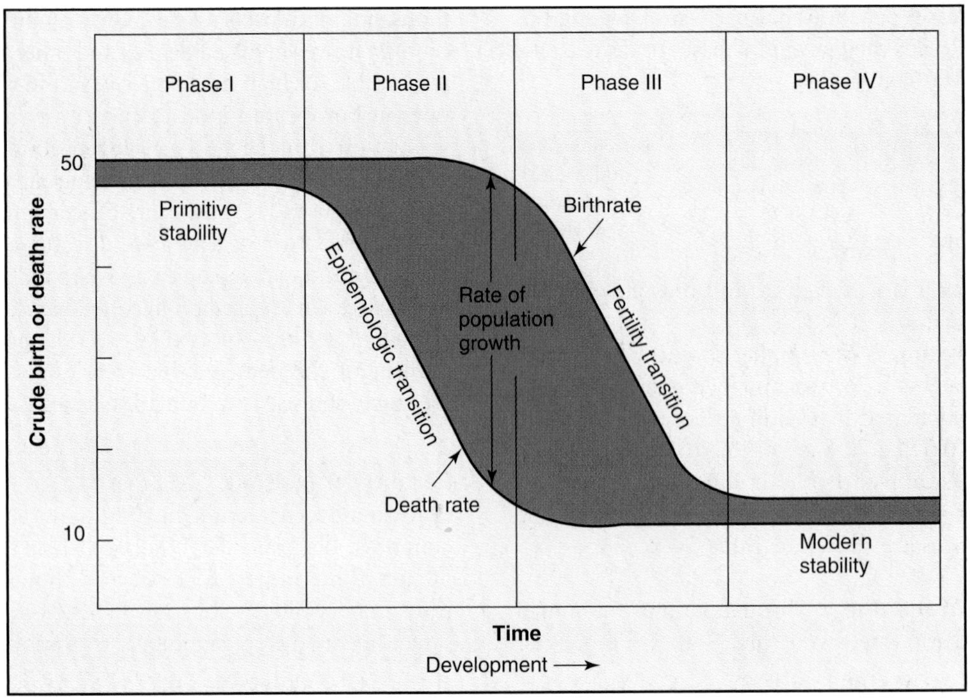


**Figure 5–18 The demographic transition.** The epidemiologic transition and the fertility transition combined to produce the demographic transition in the developed countries over many decades. (Adapted from Joseph A. McFalls, Jr., "Population: A Lively Introduction," *Population Bulletin* 53, no. 3[1998]: 39. Courtesy of Population Reference Bureau, 2000.)

delayed by decades or more. Because net growth is the difference between the CBR and the CDR, the time during which these two patterns are out of phase is a time of rapid population growth. The developed countries underwent such growth during the 19th and early 20th centuries, partly a result of massive emigration from the Old World to the less populated New World.

**Phases of the Demographic Transition.** The demographic transition is typically presented as occurring in the four phases shown in Fig. 5–18. **Phase I** is the primitive stability resulting from a high CBR being offset by an equally high CDR. **Phase II** is marked by a declining CDR—the epidemiologic transition. Because fertility and, hence, the CBR remain high, population growth accelerates during Phase II. The CBR declines during **Phase III** due to a declining fertility rate, but population growth is still significant. Finally, **Phase IV** is reached, in which modern stability is achieved by a continuing low CDR, but an equally low CBR.

Developed countries have generally completed the demographic transition, so they are in Phase IV. Developing countries, by contrast, are still in Phases II and III. Death rates have declined markedly, and fertility and birthrates are declining, but remain considerably above replacement levels. Therefore, populations in developing countries are still growing rapidly.

*Where They Are.* Using the concept of the demographic transition, you can plot the current birth and death rates of the major regions of the world to visualize where they are in the transition (Fig. 5–19). The vertical dividing line drawn through this plot separates the nations (to the right of the line) that are on a fast track to completing the demographic transition or are already there from the nations (to the left of the line) that are in the middle of the demographic transition. Nations to the left of the line (about half the world) are mostly in the fourth decade of rapid population increase. It is as if they are trapped there, with serious consequences.

*Why Worry?* On the basis of the demographic transition, some argue that we do not need to worry about population. Instead, it will stabilize by itself as developing countries reach Phase IV. Therefore, the argument goes, we need only encourage free enterprise, democratization, globalization, and other factors that will speed economic growth in developing countries. The major flaws in this argument are as follows:

1. The demographic transition occurred in the developed countries over a period of many decades; modernization did not happen overnight.
2. Many of the most populous developing countries are still very far behind the developed countries economically (Fig. 5–4) and are making very slow progress toward modernization. If they must modernize before population growth comes under control, their population growth and its demands for resources and services will undercut the very economic growth that is so necessary—a catch-22 with profound consequences.

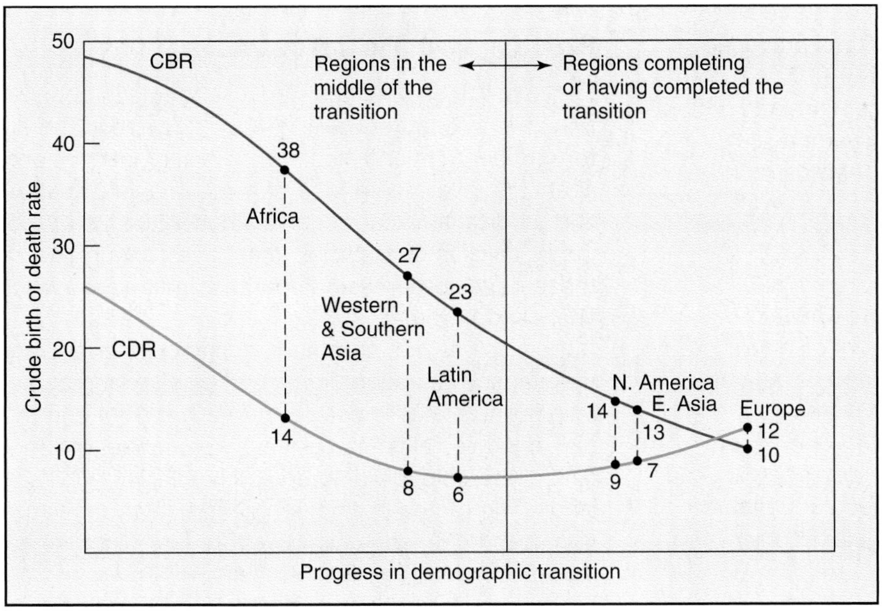


**Figure 5-19 World regions in the process of demographic transition.** Crude birthrates (CBRs) and crude death rates (CDRs) are shown for major regions of the world. A dividing line separates countries at or well along in the demographic transition from those still in the middle of the transition. (*Source:* Data from *2003 World Population Data Sheet* [Washington, DC: Population Reference Bureau, 2003].)

3. Present stresses on the biosphere are largely a consequence of the consumption-oriented lifestyles of the current 965 million people living in the high-income nations. Thus, the severe stresses being caused by the lifestyles of a billion people make any notion of a world with 10 billion people living with the same lifestyle utterly absurd.
4. Finally, and most important, the demographic transition really shows only a *correlation* between development and changing birth and death rates; it does not *prove* that development is necessary for the demographic transition to occur. Other factors may be much more significant.

In Chapter 6, you will investigate the factors that influence birth and death rates, and you will explore the relative contributions of economic development and family planning for bringing about the fertility transition. In the process, you may begin to picture what sustainable development might look like for the developing world, in particular, and what it will take to get there.

# revisiting the themes

## Sustainability

A sustainable world is a world whose population is generally stable—it cannot increase steadily, as the world's population currently does. World population may level off at 10 billion or even 9 billion, but will we be exceeding earth's carrying capacity? Three or four billion more people will need to be fed, clothed, housed, and employed, all in an environment that continues to provide essential goods and services. Many agree that this will be possible only if the additional people maintain a modest ecological footprint, much less than that of the developed countries. Because the additional people will be mostly in the developing countries, it will apparently be up to these countries to hold down their demands on resources and their pollution levels. Given their existing poverty and need for economic growth, this sounds like imperialism, and it is.

In a sense, the United States is playing a small role in the process by accepting so many immigrants from developing countries—about 1% of the annual population growth in these countries (hardly a significant percentage). More significant is the impact of immigration on the United States itself, which is currently on a trajectory of indefinite growth because of immigration. We could add 100 million to our population in the next 50 years.

## Stewardship

Calling on the developing countries to keep down their future demands on the environment is blatantly unjust. Still, there is no way for these countries to meet their future needs by mimicking what we in the developed world are currently doing to resources and the environment. Justice demands (and sustainability requires) that the rich countries reduce their environmental impact. Recall from Chapter 1 that several transitions must take place within the developed world, all of which would reduce our ecological footprint: a resource transition, a technology transition, and a community transition.

If enough people in the rich countries practice environmental stewardship, the pathway to such a future would be easier. Stewardly concern and action can do much to alleviate the negative impacts of affluence and technology, prominent items in the IPAT formula.

### Sound Science

Demography, the science of human population phenomena, reflects the diligent work of many scientists whose concern is to document data and interpret it. Their work forms much of the data and information presented in this chapter. Their past work has clarified the nature of the demographic transition, and their population projections enable us to see what the future will look like if current trends continue.

### Ecosystem Capital

Some workers have suggested that our ecological footprint already exceeds the carrying capacity of global ecosystems, and the only reason our system still works is that we are drawing down the existing abundant stock of ecosystem capital. For example, continued population growth in the subsistence farming countries puts increased pressure on agricultural resources, so expect to see more soil degradation and erosion as already intensely farmed lands are pressed to produce more. The loss of tropical forests will likely continue, as the soil under those forests will support agriculture (at least for a limited time). The goods and services once contributed by those forests, however, will be lost.

The developed countries are not off the hook. The high levels of consumption that our affluence allows lead to the loss of forests, the exploitation of fisheries, and the depletion of oil and commodities, all largely from ecosystems belonging to the developing world. Again, there will be a loss of goods and services, so the ecosystem capital stock will keep getting lower.

### Policy and Politics

Public policy and politics play a huge role in population affairs. Chapter 6 will explore this role in the different countries and also at the level of international agencies. The emergence of modern public health and medical advances is also a consequence of the public policies of countries that have encouraged these developments and countries that have in recent years adopted them.

Politics gave us communism and colonialism, and their legacy has been unjust and unwise land-ownership policies. Public policies in many former communist and colonial countries have introduced reform in land ownership, dividing up and privatizing public lands. The results have been quite uneven. Privatization does not always work well, as farms get increasingly subdivided and less efficient.

Immigration policies are heavily influenced by politics. A nation of immigrants, the United States is finding it virtually impossible to reduce the rate of immigration. As a result, we are looking into the face of heavy, unsustainable growth in population. Other countries are finding it difficult to accomplish what seems like a win–win exchange: A graying and eventually declining population admits immigrants from low-income countries looking to better their lives by migrating to where the jobs are.

### Globalization

The big footprints of the industrial (developed) countries have led to an unwanted globalization, as we have polluted the atmosphere with climate-changing gases. Widespread deforestation and loss of biodiversity are also a consequence of high consumption by the affluent developed world. Our affluence promotes a global economy in which goods (especially commodities and raw materials) move from developing countries to developed countries. This is an exchange that can help both groups of countries, as long as fair prices are paid for the goods and the producer countries are not degrading their ecosystems to produce them.

## review questions

1. How has the global human population changed from early times to 1800? From 1800 until the present? What is projected over the next 50 years?
2. How is the world divided in terms of relative per capita incomes? Fertility rates? Population growth rates?
3. What three factors are multiplied to give total environmental impact? Are developed nations exempt from environmental impact? Why not? What fourth factor modifies environmental impact, and how does it do so?
4. What are the environmental consequences of rapid population growth in rural developing countries? How are developed countries involved?
5. What information is given by a population profile? How is the information presented?

6. How do the population profiles and fertility rates of developed countries differ from those of developing countries?

7. How do future population projections for developed and developing countries contrast?

8. Discuss the immigration issues pertaining to developed and developing countries.

9. What is meant by population momentum, and what is its cause?

10. Define the crude birthrate (CBR) and crude death rate (CDR). Describe how these rates are used to calculate the percent rate of growth and the doubling time of a population.

11. What is meant by the demographic transition? Relate the epidemiologic transition and the fertility transition, two elements of the demographic transition, to its four phases.

12. How do the current positions of the developed and developing nations differ in the demographic transition?

## thinking environmentally

1. Make a cause-and-effect "concept map" showing the many social and environmental consequences linked to unabated population growth. Include crossovers between the developed and developing worlds.

2. It has been proposed that excess human populations be accommodated by building orbiting space stations that would house about 10,000 persons each. Each station would be able to produce its own food. How many space stations would be required to accommodate the world's projected population growth over the next 10 years? What kind of population policy would have to be enforced on the space stations? Are space stations a logical solution to the population problem?

3. Starting with a hypothetical population of 14,000 people and an even age distribution (1,000 in each five-year age group from 1–5 to 66–70), assume that this population initially has a total fertility rate of 2.0 and an average life expectancy of 70 years. Project how the population will change over the next 60 years under each of the following conditions:
   a. Total fertility rate and life expectancy remain constant.
   b. Total fertility rate changes to 4.0; life expectancy remains constant.
   c. Total fertility rate changes to 1.0; life expectancy remains constant.
   d. Total fertility rate remains at 2.0; life expectancy increases to 100.
   e. Total fertility rate remains at 2.0; life expectancy decreases to 50.

4. From the 2003 crude birth and crude death rates given in the following table, calculate the rate of population growth and the population doubling time for each of the countries shown:

| | CBR | CDR |
|---|---|---|
| Algeria | 23 | 5 |
| Ethiopia | 40 | 15 |
| Argentina | 19 | 8 |
| Iran | 18 | 6 |
| Russia | 9 | 16 |
| France | 13 | 9 |
| Germany | 9 | 10 |

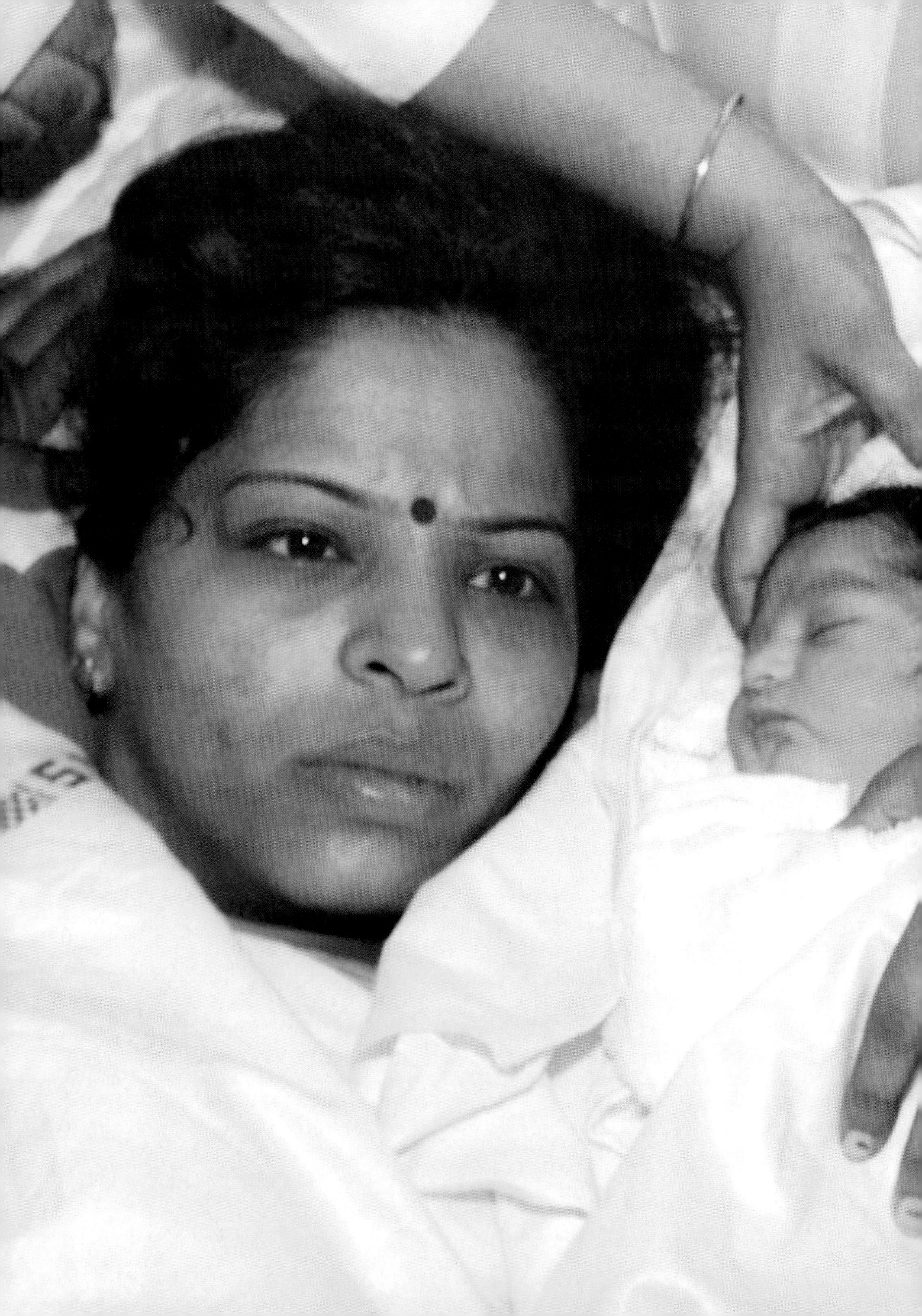

chapter 6

# Population and Development

**Key Topics**

1. Reassessing the Demographic Transition
2. Promoting Development
3. A New Direction: Social Modernization
4. The Cairo Conference

On May 11, 2000, Astha Arora, a baby girl born in Safdarjang Hospital in New Delhi, became India's one billionth person. With Astha, whose name means "faith" in Hindi, India joined China in the exclusive club of population billionaires. In the 57 years since India gained its independence, its population has tripled. It is now growing at a rate of almost 18 million per year and is expected to surpass China's population by midcentury. In spite of many decades of family-planning programs, India's population has continued its relentless growth, canceling out most of the country's gains in food production, health care, and literacy. More than half of India's children are undernourished, a third of the population lives below the poverty line, and half of its adults remain illiterate. With a per capita gross national income of $460 per year, India has a very long way to go to achieve the kind of development that characterizes the industrialized countries. It might be tempting to consider this a hopeless situation were it not for Kerala, one region of India where things are quite different.

**Astha Arora India's billionth person, with her mother, Anjana. Forty-two thousand babies are born every day in India.**

***The Kerala Story.*** Kerala is the southernmost state of India, with a population of 33 million occupying an area of 39.9 thousand square miles (about the size of Kentucky). This makes Kerala the second most densely populated state in India and close to the most densely populated region in the world (Fig. 6–1). Situated only 10° north of the equator, Kerala is tropical, with lush plantations of coffee, tea, rubber, and spices in the highlands and rice, coconut, sugarcane, tapioca, ginger, and bananas typically grown in the lowlands. With the Arabian Sea lapping at its shore, fishing is an important industry in Kerala. Kerala is very much like the rest of India in some ways. That is, it is crowded, per capita income is low, and food intake is around 2,200 calories per day (considered adequate, but on the low end). Here, the similarities end, however.

The people of Kerála have a life expectancy of 71 years, compared with 63 for all of India. Infant mortality in the state is 12 per thousand, versus 69 per thousand for India. The fertility rate is 1.8 (below replacement level), as against 3.2 for India. Literacy is over 95%, almost all villages in the state have access to school and modern health services, and women have achieved high offices in the land and are as well educated as the men. Even though

**Figure 6–1** Kerala state, in southwestern India, has achieved remarkable progress toward a stable population.

Kerala must be considered a poor region by every economic measurement, its people are well on the way to achieving a stable population.

What is different about Kerala? For one thing, there is a strong public policy commitment to health care and education. Also, land distribution is relatively equitable, food distribution is efficient, and India's old caste system has all but disappeared. These investments in social policy have paid off. Kerala's total fertility rate dropped from 3.7 to 1.8 in just two decades; in the last few years, Kerala has been making significant progress in per capita income growth. In short, Kerala shows that it is possible to bring a developing region from the midphase of the demographic transition to the threshold of its completion without the thorough economic development that characterized the industrial countries as they underwent their demographic transition.

This chapter follows up on the demographic information presented in Chapter 5. We begin with a reassessment of the demographic transition.

## 6.1 Reassessing the Demographic Transition

Recall from Section 5.4 that many regions of the developing world seem to be trapped in the middle phases of the demographic transition (Fig. 5–19), resulting in continuing rapid population growth. The countries in these regions would all like to experience the economic development that has brought South Korea, Indonesia, Malaysia, Brazil, and others into the middle- and even high-income nation groups. If they did so, it is likely that they would then move through the demographic transition. There are great disparities in economic growth among developing countries, however, and some 100 countries have been experiencing economic stagnation or decline. These same countries have the most rapid growth in population.

How did the now-developed nations come through the demographic transition without getting caught in the poverty–population trap? The improvements in disease control that lowered death rates (the epidemiological transition) occurred gradually through the 1800s and early 1900s. Industrialization, which introduced factors that lowered fertility, occurred over the same period. Therefore, there was never a huge discrepancy between birth and death rates (Fig. 6–2a). In contrast, modern medicine was introduced into the developing world relatively suddenly, bringing about a precipitous decline in death rates, while the fertility-lowering effects of development have been slow in coming (Fig. 6–2b).

*Key Question. What do the developing countries need to do to undergo the demographic transition?* This **key question** has been debated for some time. In 1798, Thomas Malthus, a British economist, pointed out that populations tend to grow exponentially, but there are definite limits to the expansion of agriculture. Thus, Malthus foresaw a world headed toward calamity if something was not done to control population. But Malthus could foresee neither the tremendous expansion of agriculture that would come with the Industrial Revolution nor the demographic transition—that is, that fertility rates would decline with the Industrial Revolution.

From early on, there were two basic conflicting schools of thought regarding the demographic transition:

1. We need to concentrate on population policies and family-planning technologies to bring down birthrates.
2. If we concentrate on development, population growth will slow down "automatically," as it did in the developed countries.

**Population Conferences.** These two schools of thought were reflected at several U.N. population conferences, one in Bucharest, Romania, in 1974, and a second in Mexico City in 1984. At the Bucharest conference, the United States was a strong advocate of population control through family planning, while the developing nations argued that "development was the best contraceptive." Their resistance to family planning was also bolstered by feelings that the developed world's promotion of population control was another form of economic imperialism or even genocide.

At the conference in Mexico City, the sides were somewhat reversed. Developing countries facing real problems of excessive population growth were asking for more assistance with family planning, whereas the United States, under pressure from right-to-life advocates, took the position that development was the answer and terminated all contributions to international family-planning

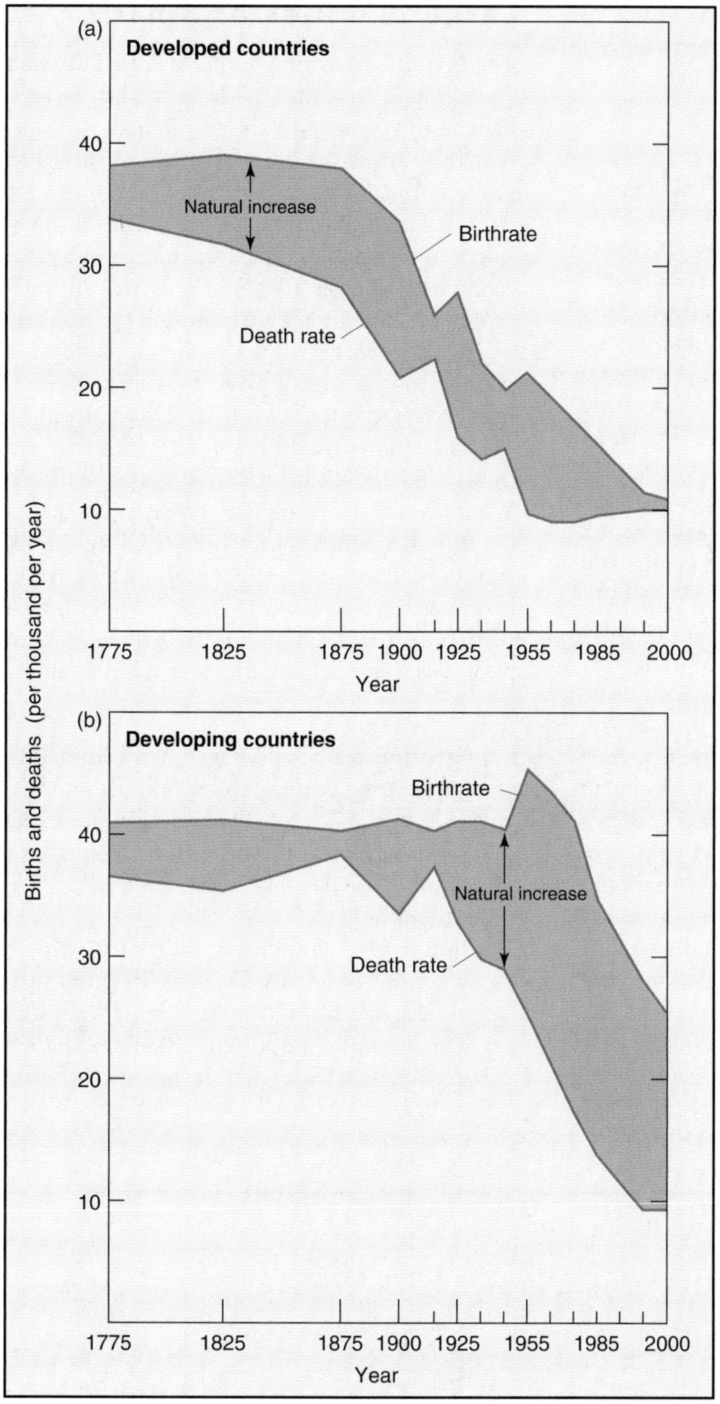


**Figure 6–2 Demographic transition in developed and developing countries.** (a) In developed countries, the decrease in birthrates proceeded soon after, and along with, the decrease in death rates, so very rapid population growth never occurred. (b) In developing countries, both birth and death rates remained high until the mid-1900s. Then the sudden introduction of modern medicine caused a precipitous decline in death rates. Birthrates remained high, however, resulting in very rapid population growth. (Redrawn with permission of Population Reference Bureau.)

efforts, a policy that remained in effect until 1993. The other developed countries, however, remained convinced that family planning was essential and continued to support international efforts to aid the developing countries in their attempts to implement policies designed to bring fertility rates down.

***End of Debate.*** A more recent population conference, the *International Conference on Population and Development (ICPD)*, was held in Cairo in 1994. Poverty, population growth, and development were clearly linked at this conference, and an important focus was also placed on resources and environmental degradation. The ideological split between developed and developing countries was a thing of the past. *The debate is over;* all agreed that population growth must be dealt with in order to make progress in reducing poverty and promoting economic development. The responsibility for bringing fertility rates down was placed firmly on the developing countries themselves, although the developed countries pledged to help with technology and other forms of aid. There was also broad agreement that (1) development must be linked to a reduction in poverty; (2) the existing poverty in the developing countries was an affront to human dignity that should not be tolerated; and (3) both poverty and development were a threat to the health of the environment, and only sustainable development would prevent a future of unprecedented biological and human impoverishment.

**Window of Opportunity.** An interesting thing happens as a country goes through the demographic transition. As birthrates decline, the working-age population increases relative to the younger and older members of the population. This relationship is known as the **dependency ratio** and is defined as the ratio of the non-working population (under 15 and over 65) to the working-age population (see Table 5-4). For a time, therefore, the society can spend less on new schools and old-age medical expenses and more on factors that will alleviate poverty and generate economic growth. This demographic window is open only once, however, as the numbers of younger children decrease, and stays open for only a generation or so, until the numbers of older people increase.

Figure 6–3 shows how dependency ratios have changed over time, and will likely change in the future, for several world regions. A number of East Asian and Latin American countries, such as South Korea, Brazil, and Mexico, have taken advantage of this window. As a result, they experienced rapid economic development once they put in place population and economic policies that were consistent with poverty reduction and economic expansion. The countries of South Asia (for example, Pakistan, India, and Bangladesh) are currently approaching this window of opportunity. Will they make the necessary investments in health, education, and economic opportunities to take full advantage of the demographic window? To do so, they must directly help the poorest people in their societies.

## Large Families or Small?

The fertility transition is the most vital element in the demographic transition. Many studies have shown that high fertility and poverty are linked, and one indication that they are is the correlation between fertility rate and gross national income per capita (Fig. 6–4). Countries on the far left side of this plot are all in the early or middle stages of the demographic transition; most are low-income economies. Those to the right are all middle-income economies and are moving through the transition.

To the observer in a developed country, it may seem strange that so many poor women in developing countries have large families. It is obvious from our perspective that more children spread a family's income more thinly and handicap efforts to get ahead economically. What we fail to recognize, however, is that the poor in developing countries live in a very different sociocultural situation. They are familiar only with poverty, and they make their choices accordingly. Numerous studies and surveys reveal the following as primary reasons that the poor in developing countries have large families:

1. **Security in one's old age.** A traditional custom and expectation in developing countries is that old people will be cared for by their children. Therefore, a primary reason given by poor women in developing nations for desiring many children is "to assure my care in old age."
2. **Infant and childhood mortality.** Closely coupled with the desire for security in one's old age is the experience of high infant and childhood mortality. The common and often personal experience of children dying leads people to try to make sure that

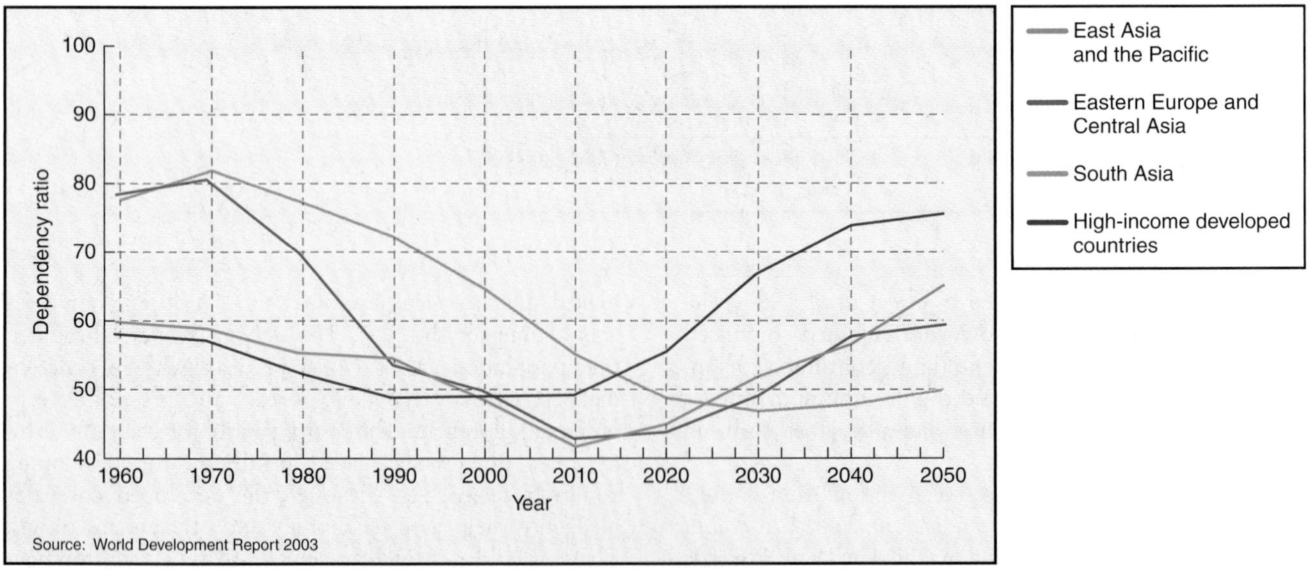


**Figure 6–3** **The demographic window.** As countries experience gradually decreasing fertility, the dependency ratio declines and opportunities for development increase. The plot compares several developing regions with some developed countries. (*World Development Report 2003.*)

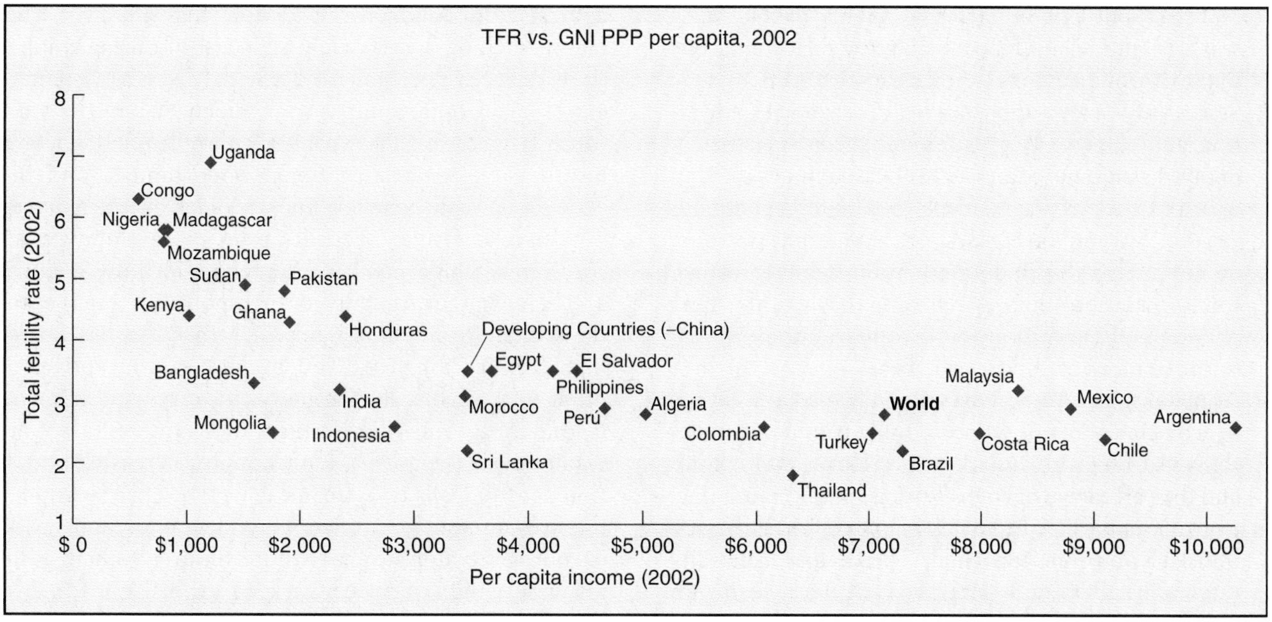


**Figure 6–4 Fertility rate and income in selected developing countries.** There is a correlation between income and lower total fertility. Factors that affect fertility more directly are health care, education for women, and the availability of information and services related to contraception. (*Source:* Data from *2002 World Population Data Sheet* [Washington, DC: Population Reference Bureau, 2002].)

some of their children will survive as an old-age "insurance policy."

3. **Helping hands.** Women of developing nations desire many children "to help me with my work." In subsistence-agriculture societies, women do most of the work relating to the direct care and support of the family. A child as young as five can begin to help with many of the chores, and 12-year-olds can do an adult's work (Fig. 6–5). In short, children are seen as an economic asset.
4. **Importance of education.** In traditional, subsistence-agriculture societies, education often seems unnecessary, and this remains the case for many children in the developing world, especially girls. Children who are

**Figure 6–5 Children as an economic asset.** Children working with adults in the fields in Bali, Indonesia. In most developing countries, children perform adult work and thus contribute significantly to the income of the family.

sent to school soon become an economic liability (they still eat, but they no longer help grow their food), one that many in the poor countries cannot afford.

5. **Status of women: opportunities for women's education and careers.** The traditional social structure in many developing countries still discourages and, in many cases, bars women from obtaining higher education, owning businesses or land, and pursuing many careers. Such discrimination against women forces them into doing what only they can do: bear children. Often, respect for a woman is proportional to the number of children she bears.
6. **Availability of contraceptives.** Providing contraceptives to women is a major facet of family planning. Studies show a strong correlation between lower fertility rates and the percentage of couples using contraception (Fig. 6–6). In fact, each 12% increase in contraceptive use translates into one less child. The decline in fertility rates in the 1980s and 1990s is directly correlated with a rise in the percentage of couples using contraception (38% in 1980 versus 60% in the late 1990s).

Perhaps the most profound finding in surveys of women in the developing world is that large numbers state that they want to delay having their next child or that they do not want any more children. Many of these women are not using contraceptives, however, because contraceptives are frequently unavailable or too expensive.

**Conclusions.** The six factors supporting large families are common to preindustrialized, agrarian societies. With industrialization and development, however, generally come factors conducive to having small families. These factors include the relatively high cost of raising children, the existence of pensions and a Social Security system, the existence of opportunities for women to join the workforce, free access to inexpensive contraceptives, adequate health care, wide educational opportunities and high educational achievement, and an older age at marriage.

***Vicious Cycle.*** Fertility rates in developing countries remain high not because people in those countries are behaving irrationally or irresponsibly, but because the sociocultural climate in which they live has favored high fertility for years and because contraceptives are often unavailable. Furthermore, poverty, environmental degradation, and high fertility drive one another in a vicious cycle (Fig. 6–7). For example, increasing population density leads to a greater depletion of rural community resources like firewood, water, and land, which encourages couples to have more children to help gather resources, and so on (see Guest Essay, p. 157).

Thus, it is not economic development by itself that leads to declining fertility rates. Rather, fertility rates decline insofar as development provides (1) security in one's old age apart from the ministrations of children, (2) lower infant and childhood mortality, (3) universal education for children, (4) opportunities for higher education and careers for women, and (5) unrestricted access to contraceptives. These are all outcomes of public policies, and countries and regions (like Kerala) that have made them a high priority have moved into and through the demographic transition.

The next section discusses efforts that are being made to bring about development in the countries that most need it.

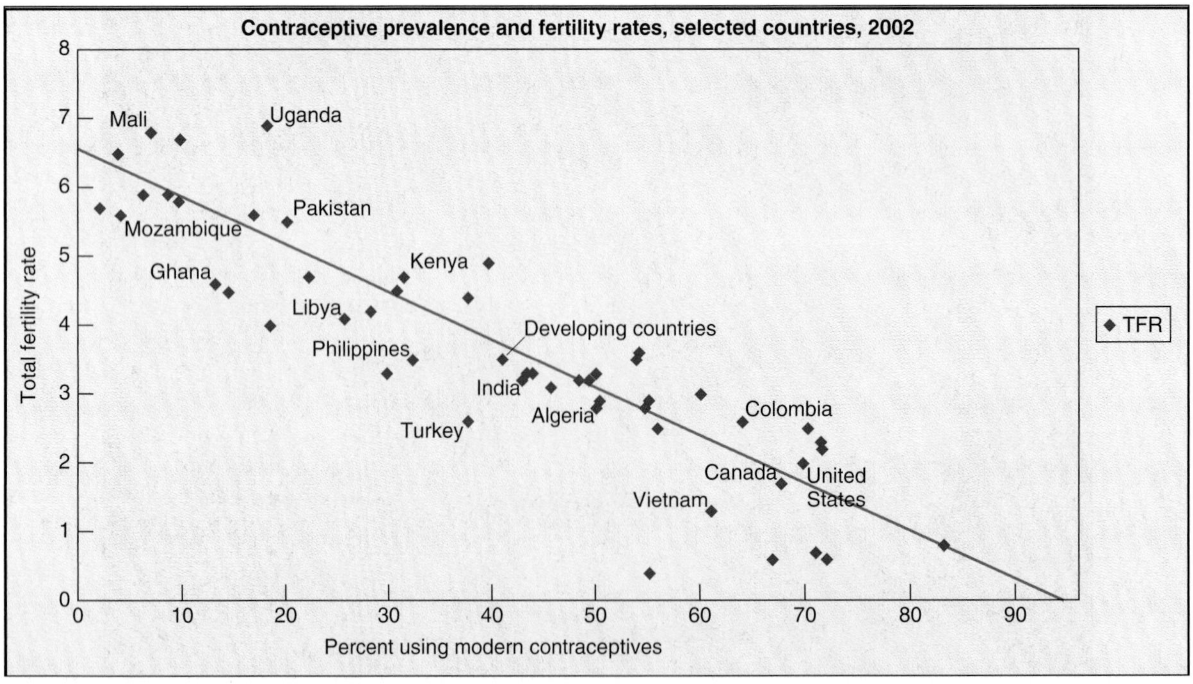


**Figure 6–6 Prevalence of contraception and fertility rates.** More than any other single factor, lower fertility rates are correlated with the percentage of the population using contraceptives. (*Source:* Data from *2002 Family Planning Worldwide* [Washington, DC: Population Reference Bureau, 2002].)

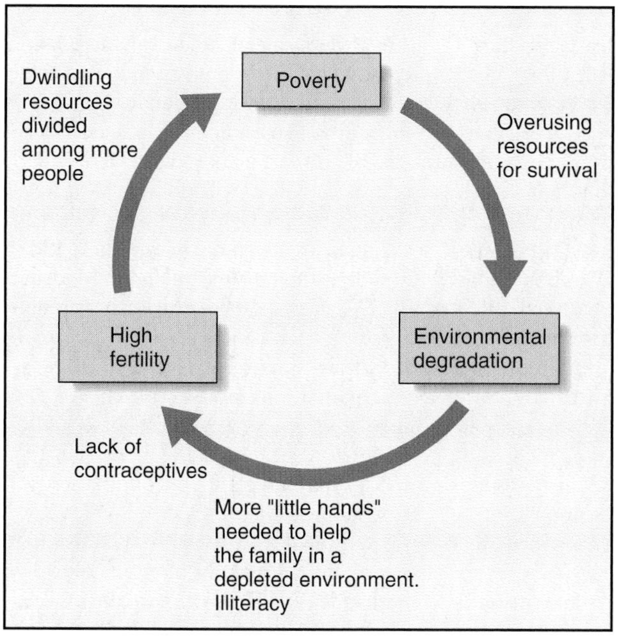


**Figure 6–7** **The poverty cycle.** Poverty, environmental degradation, and high fertility rates become locked in a self-perpetuating vicious cycle.

## 6.2 Promoting Development

### Good and Bad News

*Good News.* Many developing countries have made remarkable *economic* progress. The gross national products of some countries have increased as much as fivefold, bringing them from low- to medium-income status, and some medium-income nations have achieved high-income status. Although the world economy is still strongly dominated by the developed countries, the developing countries have become more and more involved in what is now an integrated global economy. Foreign investment is playing a large role in this development, increasing fourfold in developing countries in the last decade.

Tremendous *social* progress has been made in many developing countries, too. Efforts from many branches of the United Nations, such as the World Bank, World Health Organization (WHO), Food and Agricultural Organization (FAO), U.N. Educational, Scientific, and Cultural Organization (UNESCO), U.N. Population Fund (UNFPA), UN Development Agency, and United Nations Children's Fund (UNICEF), have augmented the work of various government programs. Private charitable organizations and nongovernmental organizations (NGOs) have also played a large role. Literacy rates, the percentage of the population with access to clean drinking water and sanitary sewers, and other social indicators of development have generally improved. Furthermore, the fertility rates of most developing countries have declined, although they are still far from the replacement level (Table 6–1).

*Bad News.* The successes just described, however, are dulled by some sobering facts. For example, a fifth of the world's population (1.2 billion people) lives on less than $1 a day, over 1 billion lack access to clean water, 2.4 billion lack access to sanitary facilities, and over 800 million are malnourished. These are the world's poorest people. As a consequence of the deficiencies they face, the poor suffer from higher rates of disease, lower life expectancy, higher infant mortality, higher illiteracy, poor living conditions, social exclusion, and powerlessness. In short, the vicious cycle of high fertility, poverty, and environmental degradation persists, presenting the international community with an enormous challenge.

Is the solution to these problems simply to facilitate economic growth, and the alleviation of poverty will

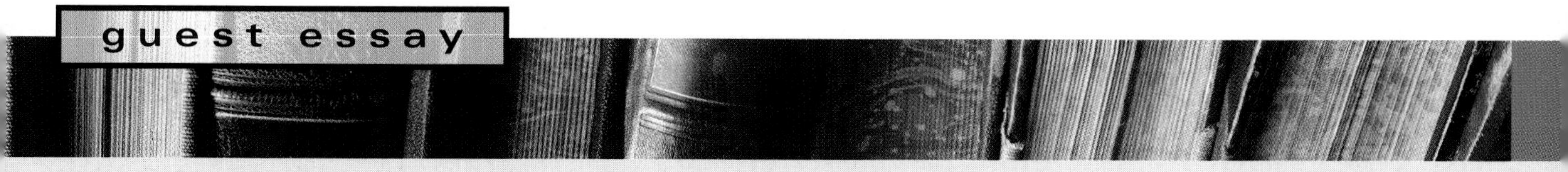


### Poverty Traps and Natural Resources Management

CHRIS BARRETT
*Cornell University*

The poorest people in the world rely disproportionately on the natural resource base, earning their living from agriculture, fishing, forestry, and hunting. Thus, questions of human development and the alleviation of poverty cannot be separated from issues of environmental management. The two are inextricable.

Some poverty, like some environmental problems, is short-lived. People lose a job unexpectedly, suffer from a drought, or endure some other shock that sets them back temporarily. In most of these cases, though, people can and usually do recover. There is remarkable inherent resilience in people, just as there is in many ecosystems.

Many of the rural poor are not so lucky. Their poverty is chronic, not transitory, reflecting low productivity due to their lack of productive assets such as land, livestock, and capital, and the meager returns they earn on the few assets they do own. Lacking access to credit, their only path out of long-term poverty would require unrealistic levels of personal sacrifice and savings in order to make productivity-enhancing investments in livestock, new technologies, or education. Indeed, in many cases, they cannot even afford to maintain current productivity by investing in replenishment of the soils on which poor farmers heavily depend. Instead, they must rely ever more heavily on the natural resource base, mining it for current subsistence at the cost of future resource productivity and continued poverty. Partly as a consequence, nearly two-fifths of the world's agricultural land is seriously degraded, with this figure higher and growing in many of the poorest areas of the world. The chronic poor thus appear trapped in a downward spiral of poverty and resource degradation.

Poverty traps ultimately depend on the existence of threshold effects. Threshold effects exist when people with very similar initial positions follow different trajectories. While farmers above a threshold can accumulate capital and grow their way out of poverty over time, those caught below the threshold are unable to accumulate productive capital or adopt improved technologies or natural resources management practices. They are then more likely to exit farming or to be forced to farm ever more marginal or fragile land. A vicious cycle of resource degradation and productivity decline ensues.

We see such patterns among farmers in Madzuu, a poor community in Vihiga District, western Kenya, where high population density—more than 1,100 people/km$^2$—and a lack of employment options outside of farming has decreased average farm size to only 0.3 hectares (about three-quarters of an acre). Seventy-one percent of Madzuu households survived on the equivalent of U.S.50¢/day per person or less in 2002. Madzuu maize farmers harvest only about one-seventh as much corn per acre, on average, as U.S. farmers. This is largely due to soils that have become heavily degraded through excessive continuous cultivation and few nutrient amendments through inorganic fertilizers and the application of manure.

The poor cannot afford to make such investments. Income distribution data shown in the accompanying graphic indicate that the 1989 incomes of households suffering soil degradation between 1989 and 2002 were strictly worse than the incomes of those who enjoyed improving soil quality over the same period.

Part of this is due to the strong correlation between educational attainment and both household income and farm productivity. Those with more education are more likely to find good paying jobs off-farm that provide a steady source of cash with which to buy fertilizers and invest in higher-value products such as exotic dairy cattle and tea bushes. As compared to other Madzuu households, those who had completed secondary school were 5 times more likely to own improved dairy stock, 18 times more likely to grow tea—the main cash crop in the region—and applied 4 times as much nitrogen fertilizer per hectare (although still only 19 kilograms/hectare, about one-sixth the rate of American farmers). With scant access to credit with which to invest in key assets such as livestock or productivity-enhancing and soil-sustaining inputs such as fertilizer, the off-farm labor market from educated, skilled workers becomes a key source of financing to keep people from falling below the poverty trap threshold and degrading their farmland in Madzuu.

The situation becomes further complicated by a parasitic weed, *Striga hermonthica,* that infests nutrient-depleted unirrigated soils in western Kenya. *Striga* is responsible for yield losses of 20–90% on infected fields, costing more than U.S.$1 billion/year in lost output across sub-Saharan Africa. Once established, "witchweed", as it is called, is resistant to conventional methods of weed control via herbicides and hand or mechanical weeding because a single plant produces thousands of seeds that remain dormant but viable in the soil for many years. A single surviving plant can recolonize a large area in a single season.

The best current method of *Striga* eradication involves planting a non-host crop, such as soybean or the fodder legume *Aeschynomene histrix,* which replenishes soil nutrients through

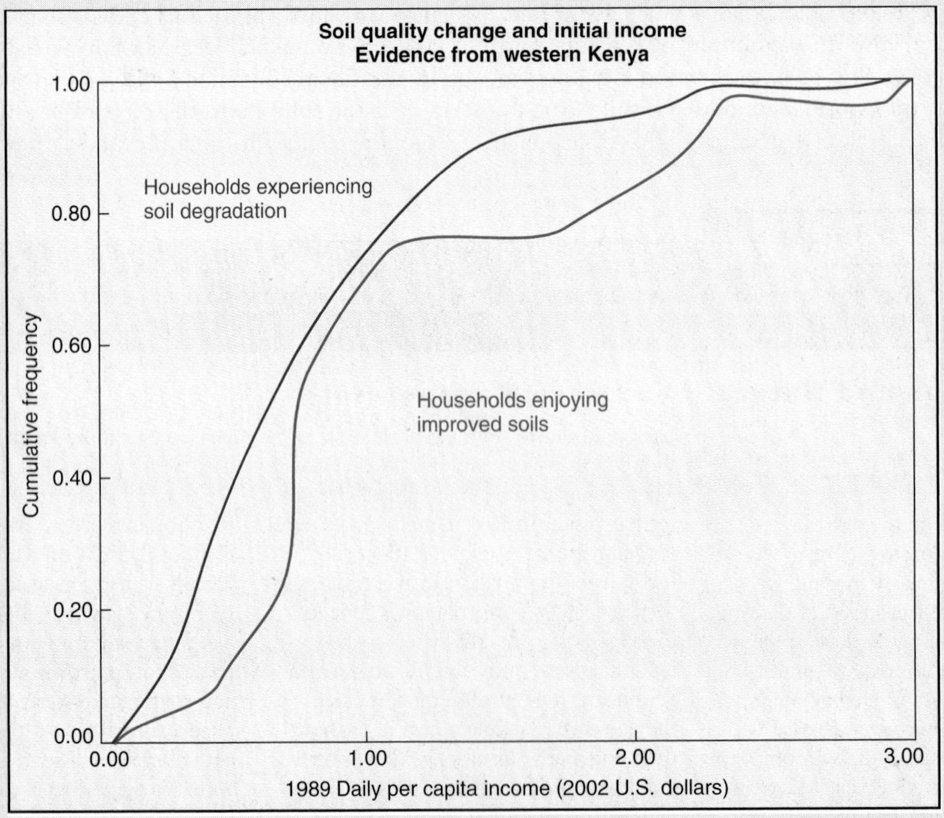


**Soil Quality Change and Initial Income Evidence from western Kenya**

atmospheric nitrogen fixation and provides nutritious feed for grazing livestock. These "trap crops" are completely resistant to *Striga,* inducing suicidal germination by witchweed seeds that cannot parasitize it and thus die. Fields planted wholly in a non-host crop for a fallow period of one or two years become cleansed of the weed. But most Madzuu farmers cannot afford to take land out of maize production for a year or two. Furthermore, *Striga* eradication requires coordination with adjoining farms, otherwise treated farms quickly become reinfected. The inability of large numbers of poor farmers to invest in *Striga* control effectively precludes effective treatment by even better-off farmers, creating a community-level poverty trap.

The problems of natural resources management in poor rural communities are closely linked to the challenge posed by poverty traps. Sustaining the valuable natural capital on which the poorest people in the world especially depend requires ongoing investment in maintaining forests, soils, water, and wildlife. The poor's inability to undertake such investments—and the failure of compensatory action by governments, private businesses, or nonprofit agencies to fill the breach—creates a vicious cycle of chronic poverty and land degradation in many parts of the tropics.

*Dr. Chris Barrett is Professor of Applied Economics and Management and co-Director of the African Food Security and Natural Resources Management program at Cornell University. He holds degrees from Princeton, Oxford, and Wisconsin-Madison and previously served on the faculty at Utah State University and as a staff economist with the Institute for International Finance. He is co-editor of the* American Journal of Agricultural Economics *and serves as an associate editor or editorial board member of* Environment and Development Economics, *the* Journal of African Economies, *and* World Development.

**Witchweed (Striga hermonthica), a parasitic weed that can devastate maize crops in Africa.** *Source:* L. J. Musselman from the Striga Photo Gallery of Species (www.science.siu.edu/parasitic-plants/Scrophulariaceae/Striga.Gallery.html).

follow? The UNFPA puts it the following way in *State of the World Population 2002:* "Overall economic growth is not enough: it requires directing development efforts to the poor.... Economic growth will not by itself end poverty. . . . Ending extreme poverty calls for commitment to the task, and specific action directed to it."

## Millennium Development Goals

In 1997, representatives from the United Nations, the World Bank, and the Organization for Economic Cooperation and Development (OECD) met to formulate a set of goals for international development that would, among other things, address poverty and its various aspects. The goals were sharpened and then adopted by all U.N. members at the U.N. Millennium Summit in 2000 as a framework for measuring progress toward development. Table 6–2 lists these *Millennium Development Goals,* or MDGs, as well as targets and indicators for monitoring progress. The indicators are measurable and provide feedback to agencies and donor countries concerned about whether progress is being made toward the different goals.

The goals reinforce each other and should all be considered as priorities and worked on together. For example, the indicators of progress for Goal 4 (*reduce child mortality*) are the infant mortality rate, the under-five mortality rate, and the proportion of one-year-old children immunized against measles. Goals 5 and 6 (*improve maternal health* and *combat HIV/AIDS, malaria, and other diseases*) should go far toward accomplishing Goal 4.

Most importantly, the MDGs are a clear set of targets for the developing countries themselves. Each country is expected to work on its own needs, but it is expected that all of the countries will do so in partnership with developed countries and development agencies, consistent with Goal 8 (*Develop a global partnership for development*). The low-income countries simply lack the resources to go it alone.

**table 6-1 Decline in Total Fertility Rates**

| | 1985 | 1990 | 1995 | 2000 | 2003 |
|---|---|---|---|---|---|
| Africa | 6.3 | 6.2 | 5.8 | 5.3 | 5.2 |
| Latin America and Caribbean | 4.2 | 3.5 | 3.1 | 2.8 | 2.7 |
| Asia (excluding China) | 4.6 | 4.1 | 3.5 | 3.3 | 3.1 |
| China | 2.1 | 2.3 | 1.9 | 1.8 | 1.7 |
| Developed countries | 2.0 | 2.0 | 1.6 | 1.5 | 1.5 |

[*Source:* Data from various *World Population Data Sheets* (Washington, DC: Population Reference Bureau).]

**table 6-2 Millennium Development Goals: Conditions in 1990 are Compared with the Year 2015 as the Date for Reaching the Targets**

**Goal 1. Eradicate extreme poverty and hunger.**
- Reduce by half the proportion of people whose income is less than $1 a day.
- Reduce by half the proportion of people who suffer from hunger.

**Goal 2. Achieve universal primary education.**
- Enroll all children in primary school.

**Goal 3. Promote gender equality and empower women.**
- Eliminate gender disparities in primary and secondary education by 2005.
- Eliminate gender disparity in all levels of education.

**Goal 4. Reduce child mortality.**
- Reduce infant and child mortality rates by two-thirds.

**Goal 5. Improve maternal health.**
- Reduce maternal mortality rates by three-quarters.

**Goal 6. Combat HIV/AIDS, malaria, and other diseases.**
- Have halted and begun to reverse the spread of HIV/AIDS.
- Have halted and begun to reverse the incidence of malaria and other major diseases.

**Goal 7. Ensure environmental sustainability.**
- Integrate the principles of sustainability into country policies and programs, and reverse the loss of environmental resources.
- Reduce by half the proportion of people without sustainable access to safe drinking water.
- Reduce by half the proportion of people without adequate sanitation.
- Achieve a significant improvement in the lives of at least 100 million slum dwellers by 2020.

**Goal 8. Forge a global partnership for development.**
- Develop a trading system that is open, rule based, and nondiscriminatory—one that is committed to sustainable development, good governance, and a reduction in poverty.
- With Official Development Assistance (ODA), address the special needs of the least-developed countries (including the elimination of tariffs and quotas for their exports and an increase in programs for debt relief).
- Deal with the debt problems of all developing countries by taking measures that will make debt sustainable in the long run.
- In cooperation with the private sector, make the benefits of new technologies available, especially information and communications technologies.

*Sources:* United Nations, World Bank, and Organization for Economic Cooperation and Development.

**Human Development Reports.** Since 1990, the U.N. Development Program has issued an annual Human Development Report focusing on a specific topic and bringing world attention to it. The 2002 report addresses the MDGs and documents country-by-country progress on the goals two years after the Millenium Summit. According to this report, only 55 countries, representing 1.45 billion people, are currently on track to reach as many as three-quarters of the MDGs. Thirty-three countries, with 1.6 billion people, are behind on more than half of the targets.

*Extreme Poverty.* The first goal (eradicate extreme poverty and hunger) aims to reduce by half the proportion of people whose income is less than $1 a day. This means reducing the current 29% of extremely poor people in the low- and middle-income economies to 14.5% by 2015. Figure 6–8 indicates the changes in poverty rates (percentage below $1 per day income) in different world regions, for the 1990 decade and for the rate of progress needed to achieve the goal in 2015. There is clear progress in many regions, but a regression in the transition economies of Europe and Central Asia and in sub-Saharan Africa. The World Bank projects that this goal can be reached by 2015 if growth in per capita income accelerates to an average of 3.6% a year—nearly twice the rate of the past decade, but a rate that has been exceeded by some regions. (China averaged almost 9% growth per capita during the 1980s and 1990s). Sadly, even if the MDG is achieved, there will still be at least 600 million people living in extreme poverty.

*Child Mortality.* The goal that seems most achievable, because of progress already made, is Goal 2, *achieve universal primary education.* The goal that seems least likely to be achieved is Goal 4, *reduce child mortality,* with the target of reducing under-five mortality rates by two-thirds. Some 85 countries with over 60% of the world's people are falling behind this goal, while 15 countries (primarily in sub-Saharan Africa) are actually slipping backwards. Progress toward the goal means improving nutrition, sanitation, and especially immunization against leading diseases. It also means that only 3.6 million will die of preventable causes every year, instead of the present 11 million.

*Country-level Reports.* Global and regional-level reports are important in providing a general picture of progress being made toward the MDGs, but the real action

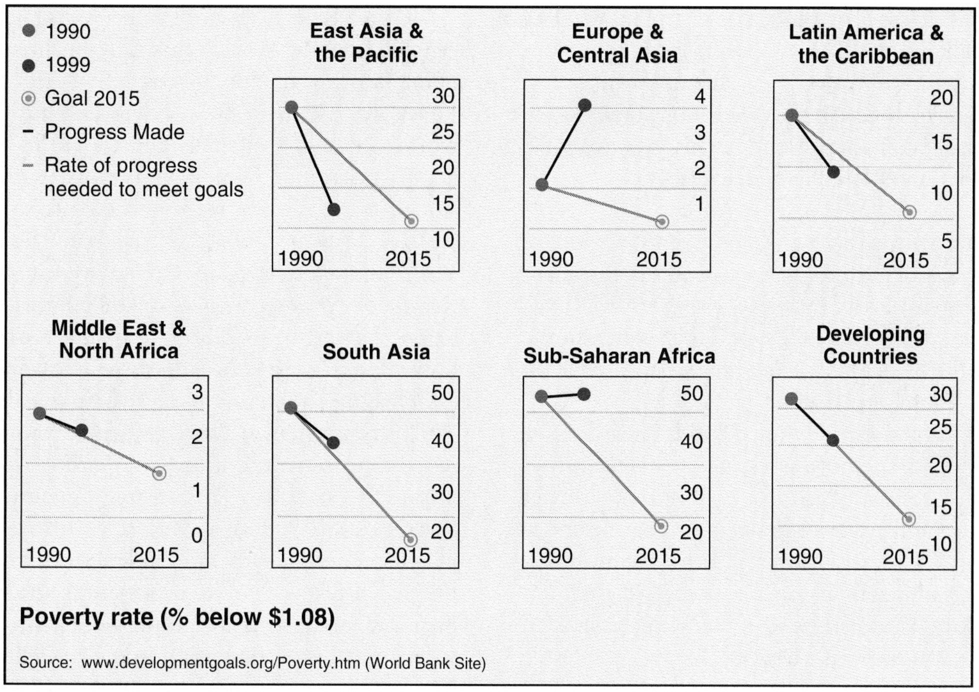


**Figure 6–8 Millennium Development Goal.** Poverty rates have declined in many regions, except for the transition economies of eastern Europe and central Asia. (World Bank Millennium Development Goals Web site, 2003.)

is in the individual countries. Toward that end, country-level MDG reports are now appearing in many countries, a sign that the MDG process is taking root and bearing fruit. Bolivia, Cambodia, Cameroon, Chad, Madagascar, Tanzania, and Viet Nam have already published reports, and more are coming. Albania is carrying the MDG process to the local level, in the *Albania Human Development Report 2002*. This report highlighted significant disparities among the country's different regions and is being used to target the neediest regions for development activities.

How much will it cost to achieve the MDGs by 2015, and what means will best deliver the help needed, especially by the low-income countries? Currently, the developed countries are providing around $56 billion a year in official development Aid (ODA). According to the U.N. Development Program, this amount will need to be doubled! Even more important is foreign investment and participation in global markets, which directly stimulate and help grow the economies of these countries. More important yet to the poorest people, however, is the work of various U.N. and private organizations that directly provide assistance and funnel aid to the developing countries. These agencies are discussed next, followed by the issues of development aid and the debt crisis.

## World Agencies at Work

In 1944, during World War II, delegates from around the world met in Bretton Woods, New Hampshire, and conceived a vision of development for the countries ravaged by World War II. They established the International Bank for Reconstruction and Development (now more commonly known as the **World Bank**). The World Bank functions as a special agency under the U.N. umbrella, owned by the countries that provide its funds. As the developed countries of Europe and Asia got back on their feet, the World Bank directed its attention more toward the developing countries. With deposits from governments and commercial banks in the developed world, the World Bank now lends money to governments (and only governments) of developing nations for a variety of projects at interest rates somewhat below the going market rates. In effect, the World Bank helps governments borrow large sums of money for projects they otherwise could not afford.

Annual loans from the World Bank have climbed steadily, from $1.3 billion in 1949 to $19.6 billion in 2002. With the power to approve or disapprove loans, and through the amount of money it lends, the World Bank has been the major single agency providing aid to developing countries for the past 50 years. How does the agency carry out its mandate, and how does its record look?

**What is the World Bank?** The World Bank is actually five closely associated agencies. The most prominent ones are the International Bank for Reconstruction and Development (IBRD) and the International Development Association (IDA). The IBRD aims at reducing poverty in middle-income and some of the stronger low-income countries through loans and advisory services. The IDA enables the World Bank to provide interest-free loans to the poorest countries. These countries are so poor that they would be unable to borrow on market terms. The International Finance Corporation (IFC) works through the private sector to promote economic development in countries that are investment risks. For example, the IFC loaned $100 million in 2002 to Suramericana, a conglomerate in Colombia involved in cement, textiles, and financial services. The IFC also provided substantial advice and guidance to the company. None of this was available from the traditional international financial sector.

**Bad Bank, Good Bank.** It is inaccurate either to credit the World Bank for all the progress made in development or to blame it for all areas in which progress has been lacking. However, critics cite many examples wherein the bank's projects have actually exacerbated the cycle of poverty and environmental decline.

***Bad Bank.*** For example, the World Bank loaned India nearly a billion dollars to create a huge electric-generating facility consisting of five coal-burning power plants at Singrauli and to develop open-pit coal mines to support the plants. The new power goes to distant cities, however, and has done little for the poor, who cannot afford electrical hookups. Worse, the project displaced over 200,000 rural poor people who had farmed the fertile soil of the region for generations, and it moved them to a much less fertile area without allowing them any say in the matter and giving them little, if any, compensation. In addition, the project has caused extensive air and water pollution. Hydroelectric dams in a number of countries have similarly displaced people and exacerbated their poverty.

This and many other past projects funded by the World Bank have been destructive to the environment and often to the poorer segments of society. For example, the World Bank funneled $1.5 billion into Latin America from 1963 to 1985 to clear millions of acres of tropical forests. Most of the cleared land was given over to large cattle operations that produced beef for export (Fig. 6–9), to many fastfood restaurants in the United States, among other establishments and locales. In other countries, projects have emphasized growing cash crops for export and fostering huge mechanized plantations, while leaving the poor marginalized. At the time, World Bank procedures considered environmental concerns as an add-on, rather than as an integral component of its policies.

***Good Bank.*** There has been a sea change in recent years in how the World Bank does business with the developing world. In the words of World Bank President James D. Wolfensohn, "Poverty amidst plenty is the world's greatest challenge, and we at the Bank have made it our mission to fight poverty with passion and professionalism. This objective is at the center of all of the work we do." Toward that end, the Bank helped initiate the Millenium Development Goals. It now requires *Poverty Reduction Strategy Papers* for all countries receiving IDA loans. The Bank's *Country Assistance Strategy* identifies how Bank assistance addresses reducing poverty and establishing and maintaining sustainable development.

***Environmental Strategy.*** The Bank also adopted a new environmental strategy, *Making Sustainable Commitments,* in 2002. This strategy includes a new *Forest Strategy* and *Operational Policy* on forests, designed to provide guidance to all Bank-supported activities that involve forests. The environmental strategy goal "is to promote environmental improvements as a fundamental element of development and poverty reduction strategies and actions," as stated in *Making Sustainable Commitments.* This goal includes the following three interrelated objectives and means of their accomplishment:

1. **Improving the Quality of Life**
   - Enhancing livelihoods, by protecting the long-term productivity and resilience of natural resources and ecosystems on which so many of the poor depend.
   - Reducing health risks, by reducing people's exposure to air pollution, waterborne diseases, and toxic chemicals.
   - Reducing vulnerability to natural hazards, by helping to prevent natural disasters and dealing with their impacts if they occur, and by supporting better weather forecasting and management of coastal-zone areas.
2. **Improving the Quality of Growth**
   - Supporting sustainable environmental management, by promoting good governance and people's participation in decision making and by strengthening environmental policy frameworks.

**Figure 6–9 Cattle ranch in Eastern Brazil.** Large World Bank loans went to clear rain forest and convert the land into rangeland, as seen here.

- Supporting sustainable private-sector development, by aiding markets in achieving environmental objectives and creating outlets for environmental goods and services.

3. **Improving the Quality of the Regional and Global Commons**
   - Building global environmental initiatives, by focusing on positive linkages between reducing poverty and protecting the environment, by building on overlaps between local benefits and global benefits, and by helping developing countries meet the costs of global environmental benefits through market and aid mechanisms.

Thus, the World Bank's new direction takes seriously the three intersecting concerns in sustainable development: development that balances economic, social, and environmental factors (Fig. 1–9).

### Other Agencies

*U.N. Development Program.* To quote its mission statement, the "UNDP is the UN's global development network, advocating for change and connecting countries to knowledge, experience and resources to help people build a better life." The agency has offices in 131 countries and works in 166! It helps developing countries attract and use development aid effectively. Its network also coordinates global and national efforts to reach the MDGs.

The FAO and the WHO are other important agencies; their essential work will be discussed in future chapters. We turn now to the debt crisis, to some extent an outcome of World Bank policies, but also a consequence of development aid in general.

## The Debt Crisis

The World Bank, many private lenders, and wealthy nations lend money to developing countries. Theoretically, development projects are intended to generate additional revenues that would be sufficient for the recipients to pay back their development loans with interest. Unfortunately, however, a number of things have gone wrong with this theory, such as corruption, mismanagement, and honest miscalculations.

Over time, developing countries as a group have become increasingly indebted. Their total debt reached $2.44 trillion in 2001. Interest obligations climb accordingly, and any failure to pay interest gets added to the debt, increasing the interest owed—the typical credit-debt trap. In 2001, the developing countries received $253 billion in new loans and paid back $260 billion in principal and $122 billion in interest on existing loans, a total of $382 billion and a *net flow to the creditor countries* of $138 billion! Who is really benefiting from this "development aid?"

*A Disaster.* The debt situation continues to be an economic, social, and ecological disaster for many developing countries. In order to keep up even partial interest payments, poor countries have done one or more of the following:

1. **Focused agriculture on growing cash crops for export.** This has occurred at the expense of peasant farmers growing food. Thus, hunger and malnutrition have increased, and so has poverty, as peasants have been pushed from the land.
2. **Adopted austerity measures.** Government expenditures have been drastically reduced so that income can go to pay interest. But what is cut? Usually, it is funds for schools, health clinics, police protection in poor areas, building and maintenance of roads in rural areas, and other goods and services that benefit not only the poor, but the country as a whole. In 2000, for example, Zambia (a small East African country) was indebted to the tune of $4 billion and was scheduled to spend over $385 million annually servicing its debt, an amount almost 70% of net government revenues and 2 ½ times what the country spends on social services.
3. **Invited the rapid exploitation of natural resources (for example, logging of forests and extraction of minerals) for quick cash.** With the emphasis on quick cash, few, if any, environmental restrictions are imposed. Thus, the debt crisis has meant disaster for the environment. Worse, this forced exploitation of material resources has placed such oversupplies of commodities on the market that prices have been severely depressed, so that actual earnings are low.

*Ecosystem Capital.* In essence, these measures are all examples of liquidating ecosystem capital to raise cash for short-term needs. They do not represent sustainability. As is typical with the credit trap, many countries have paid back in interest many times what they originally borrowed, yet the debt remains. Is there any point—humanitarian, ecological, or economic—to keeping such debts in place, especially when it is clear that they can never be paid off?

**Debt Relief.** The World Bank is addressing the problems of debt and poverty directly through two new initiatives: the *Consultative Group to Assist the Poorest* (CGAP) and the *Heavily Indebted Poor Country* (HIPC) initiative. The CGAP is designed to increase access to financial services for very poor households through what is called "microfinancing" (to be discussed shortly). The HIPC initiative addresses the debt problem of 42 of the poorest developing countries—amounting to $190 billion in 2000—by providing direct debt relief to a level deemed sustainable. To qualify, the countries have to demonstrate a track record of carrying out economic and social reforms that lead to greater stability. By 2003, 26 countries had qualified for debt relief totaling $40 billion. Through the HIPC initiative, debt service for these countries has fallen by almost $1 billion a year. In many countries that qualified for debt relief, spending on education and health has risen substantially.

*Zambia.* To demonstrate how HIPC works, consider the case of Zambia. That country's total debt has been reduced to $2.2 billion, thus reducing the amount needed to service the debt annually from $385 million to $150 million. Although this sounds like a tremendous break for Zambia, the country is currently experiencing food shortages, has a per capita gross national income of $320, and will have no chance of reaching the MDGs if $150 million of its income has to go to its rich creditors. Because of this, and other similar situations, many throughout the world are calling for total debt cancellation.

***Jubilee.*** A renewed emphasis on debt relief can be traced to *Jubilee 2000,* a worldwide coalition of people and organizations concerned about poverty. The name is derived from the biblical principle of debt cancellation every 50 years. Through demonstrations, petitions, and lobbying, Jubilee 2000 focused world attention on the problem of debt relief and extracted promises from the G7 group of industrialized nations (the United States, Japan, Germany, France, Britain, Canada, and Italy) to cancel their bilateral loans to the HIPCs. Some of the debt of the HIPCs is bilateral, consisting of low-interest loans or export credit debts (to promote exports of the lending country).

***Promises, Promises.*** In 1999, Germany promised to write off about $5 billion, Britain $3.3 billion, France $7 billion, and the United States about $6 billion owed them by the HIPCs. The total debt relief for the HIPCs, from the G7 countries and the World Bank, was supposed to amount to $110 billion. In most cases, it is contingent on the debtor countries applying the benefits they will receive to relieving poverty in their lands. Jubilee Research is taking up the work of Jubilee 2000 and is following the promised debt relief closely. According to Jubilee Research, though, less than $36 billion was written off by 2003, only half of what was promised by that date. Jubilee Research maintains that the World Bank and the other creditors should step up their activities, face reality, and cancel debts that they can never really expect to collect.

## Development Aid

College students and their parents know all about the difference between scholarships, grants, and loans, as they apply for "financial aid" from admissions offices. Paying back college loans is no picnic for the thousands of students who accept them in order to afford their education. It is all called aid, but the best kind of aid is the grant or scholarship, which does not have to be repaid. The developing countries are well aware of the differences, too, and have been receiving aid from private and public donors that is essential to their continued development. How much does this amount to?

***Official Development Assistance.*** The best records are for Official Development Assistance (ODA), coming from donor countries (Fig. 6–10). The total amount in 2000 was $53 billion, only about half of which actually went to low-income countries. The flow has declined over the last decade, probably as a result of the end of the Cold War. Studies have indicated that the allocation of aid often follows the interests of the donor countries. Among these interests are helping former colonies, solidifying regional ties (for example, Japan's aid to Asian countries), and pursuing strategic aims (for instance, receiving votes in the UN). As Figure 6–10 shows, aid as a percentage of the GNP (gross national product) of donor countries has declined to 0.22%. At the Rio Earth Summit of 1992, the aid target for donor countries was 0.7% of GNP.

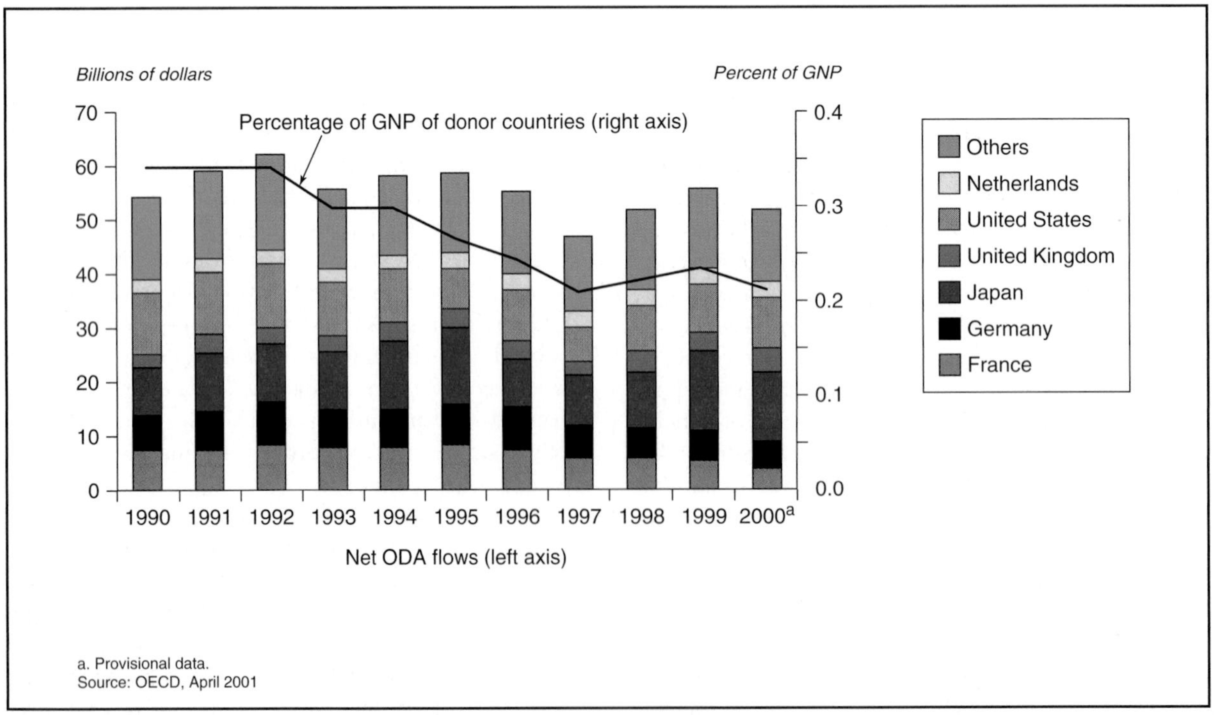


**Figure 6–10 Official development assistance.** Development assistance from donor countries in relation to their gross national product, 1990–2000.

*Donor Fatigue.* Some of the decline in aid is explained as "donor fatigue." That is, rich countries have tired of handing out multimillion-dollar rescue packages, only to have a large percentage go into the pockets of corrupt leaders or cover administrative costs. During the Cold War days, aid was used to erect palaces, build torture chambers, and line the Swiss bank accounts of many corrupt leaders. However, steps have recently been taken to minimize this misdirection of funds. Allocation of aid is increasingly being tied to policy reform in the recipient countries. The *Poverty Reduction Strategy Paper* (*PRSP*) approach of the World Bank is one example. In order to continue receiving aid, governments must prepare reports addressing policy reform and other procedures for simplifying and streamlining processes.

How much aid does the United States give? Our ODA donations have amounted to some $11 billion a year, which seems like a lot of money, but is only 0.1% of our GNP. Norway, by contrast, gives the equivalent of 1.2% of its GNP in development aid. President Bush recently pledged an additional $5 billion, to be added incrementally until 2006, when the total should be $16 billion. This increase has been called the *Millennium Challenge Account.* Its goal is to reward sound policy decisions that support economic growth and reduce poverty in recipient countries. To put it in its proper perspective, though, contributions in 2000 from religious organizations, foundations, corporations, non-governmental organizations, and individuals provided some $33.6 billion in aid, far outstripping our ODA.

*Need.* The World Bank has estimated that an additional $40–60 billion a year in aid is needed to finance the programs necessary to achieve the MDGs. Although critics have dismissed the effectiveness of its aid, the Bank has responded that the donor countries were responsible for much of the ineffectiveness of the aid because of the strings they continually tied to it. In addition, the Bank addressed subsidies that protect exports in the rich countries and tariffs on exports from developing countries that work directly against the objectives of development aid.

The United Nations and its various agencies frequently come under attack in the United States, where critics charge that the organization is useless to us. These same critics would like to reduce or eliminate foreign aid. The fact is, without the work of the World Bank and U.N. agencies like the UNDP and the FAO, the world would be in much worse shape than it is. We are fortunate that we do not need development aid, but billions of others around the world who are much worse off depend on such aid for continued economic development, especially as it is reflected in the MDGs. This is not a luxury.

The developing countries themselves must also address the root causes of poverty, and this usually means public policy changes at more local levels. To illustrate, the discussion in the next section assesses the relative roles played by development and family planning.

## 6.3 A New Direction: Social Modernization

As the case of Kerala demonstrates, demographic experts are recognizing that the shift from high to low fertility rates in the poorer developing countries does not require the economic trappings of a developed country. Instead, what is needed are efforts within the country made on behalf of the poor, with particular emphasis on the following points:

1. Improving education—especially literacy and the education of girls and women.
2. Improving health—especially lowering infant mortality.
3. Making family planning accessible (that is, both available and affordable).
4. Enhancing income through employment opportunities.
5. Improving resource management (reversing environmental degradation).

In all of these areas, the focus should be on women, because they not only bear the children, but also are the primary providers of nutrition, child care, hygiene, and early education. In short, it is women who are most relevant in determining both the numbers and the welfare of subsequent generations.

Fortunately, the world is by no means starting from scratch in any of these areas. Numerous programs, both private and government funded, have been in existence for many years, and a wealth of experience and knowledge has been gained. The result of success in these efforts is called **social modernization.** We shall look at each area in somewhat more detail.

### Improving Education

The education in question is not college or graduate school, nor even advanced high school. It is basic literacy—learning to read, write, and do simple calculations (Fig. 6–11). Illiteracy rates among poor women in developing countries are commonly between 50 and 70%, in part because the education of women is not considered important and in part because expanding populations have overwhelmed school systems and transportation systems. Providing basic literacy will empower people to glean information from pamphlets on everything from treating diarrhea, to conditioning soils with compost, to baking bread. An educated populace is an important component of the wealth of a nation.

Investing in the education of children represents a key element of the public-policy options of a developing country and is one that returns great dividends. In 1960, for example, Pakistan and South Korea both had similar incomes and population growth rates (2.6%), but very different school enrollments—94% in Korea versus 30% in Pakistan. Within 25 years, Korea's economic growth was three times that of Pakistan's, and now the rate of population growth in Korea has declined to 0.8% per year, while Pakistan's is still

**Figure 6–11 Education in developing countries.** Providing education in developing countries can be quite cost effective, because any open space can suffice as a classroom and few materials are required. This is a class in Bombay, India.

2.1% per year. Kerala, where literacy is greater than 95%, is another case in point. The lowest net enrollment of children in primary schools (about 40%) is in sub-Saharan Africa, where 45 million children are out of school. This region has made no progress in enrollment ratios since 1980, a clear consequence of rapid population growth.

## Improving Health

Like education, the health care needed most by poor communities in the developing world is not high-tech bypass surgery or chemotherapy. Instead, it is the basics of good nutrition and hygiene—steps such as boiling water to avoid the spread of disease and properly treating infections and common ailments such as diarrhea. (In developing countries, diarrhea is a major killer of young children, but it is easily treated by giving suitable liquids, a technique called *oral rehydration therapy.*) The discrepancy in infant mortality rates (infant deaths per 1,000 live births) between the developed countries (7/1,000) and developing countries (64/1,000) speaks for itself.

Health care in the developing world must emphasize pre- and postnatal care of the mother, as well as that of the children. Many governmental, charitable, and religious organizations are involved in providing basic health care, and when this is extended to rural countrysides in the form of clinics, it is one of the most effective ways of delivering family-planning information and contraceptives to women. Maternal mortality is over 1,000 per 100,000 live births in a number of developing countries in which maternal health care is almost nonexistent.

*Reproductive Health.* A fairly new concept in population matters, reproductive health was strongly emphasized in the International Conference on Population Development held in Cairo, Egypt, in 1994. Reproductive health focuses on women and infants and includes the following elements (as presented by the Population Reference Bureau):

- Prenatal care
- Safe childbirth and postnatal care
- Information and services pertaining to contraception
- Prevention and treatment of sexually transmitted diseases (STDs)
- Abortion services (where legal) and care afterwards
- Prevention and treatment of infertility
- Elimination of violence against women (coercive sex, rape), sexual trafficking, and female circumcision and infibulation (traditional practices in some African societies)
- Additional women's health services (for example, diagnosis and treatment for breast cancer and cancers of the reproductive tract)

Reproductive health care needs to be addressed particularly to the needs of young people. This is often a controversial issue, but sex education for adolescents is a very real need, and not just in the developing world. Premarital sex is on the rise in all regions of the world, resulting in unplanned pregnancies, out-of-wedlock births, unsafe abortions, and STDs. An increasing number of young people, though, are opting for sexual abstinence until marriage, often taking a pledge to do so. These young people have learned that there is no stigma connected to waiting, in spite of peer pressure and the constant portrayal of sexual situations in the popular media. The chilling impact of AIDS also has put a new urgency on health care and sex education.

## AIDS

One of the greatest challenges to health care in the developing countries is the human immunodeficiency virus (HIV), which leads to the sexually transmitted disease known as acquired immune deficiency syndrome (AIDS). Unfortunately, the global epidemic of AIDS is most severe in many of the poorest developing countries, which are least able to cope with the consequences (Fig. 6–12). More than 90% of all HIV-infected people (50 million by 2004) live in the developing countries, and most of these people are unaware that they are infected—thus guaranteeing that the epidemic will continue. In sub-Saharan Africa, the virus is spread primarily by heterosexual contact, resulting in high incidences of female infection, which often gets transmitted to their children. The incidence of infected adults is over 20% in countries such as Zimbabwe, Botswana, and South Africa. AIDS is now the leading cause of death in sub-Saharan Africa.

***Impact.*** The impact of this epidemic is horrendous for the developing world. Mortality rates are climbing in many countries. Life expectancy in Botswana has declined from 61 years in the late 1980s to 39 years at present. The long-term impact of AIDS can be seen in the projected population age structure of Botswana for 2020 (Fig. 6–13). In many countries, AIDS is tearing apart the very structure of the society. More than 1 million elementary school students in sub-Saharan Africa have lost teachers to the epidemic; farmers weakened by the virus are unable to plant or harvest crops. There may be as many as 25 million AIDS orphans in the developing world by 2010. Already inadequate health care systems are being swamped by the victims. There still is no cure for AIDS; the disease is invariably fatal. Efforts to develop a vaccine have so far failed, but researchers are frantically working on the problem, and there is hope that a vaccine will be found.

A policy of treatment and prevention is desperately needed for the developing world. Although anti-HIV drug therapy has helped lengthen the lives of infected people in the developed countries, the drugs are expensive and require a health care infrastructure that is largely lacking in the poorer countries. In 2002, the *Global Fund to Fight AIDS, Tuberculosis and Malaria* was launched. This is a multinational organization targeting the developing countries' needs for treatment and prevention. The United States has pledged $500 million to the fund, a good start, but much more is needed: The estimated funding required to prevent, treat, and care for people with HIV/AIDS ranges from $7 billion to 10 billion annually.

The best immediate hope is to convince people to change their sexual behavior—to practice "safe sex," limit their partners, avoid prostitution, and delay sexual activity until marriage. That these things can be accomplished is proven by the case of Uganda, where significant declines in HIV infection have been registered in recent years, especially among pregnant women. Again, education and literacy are important components of combating this social problem. The Ugandan government has

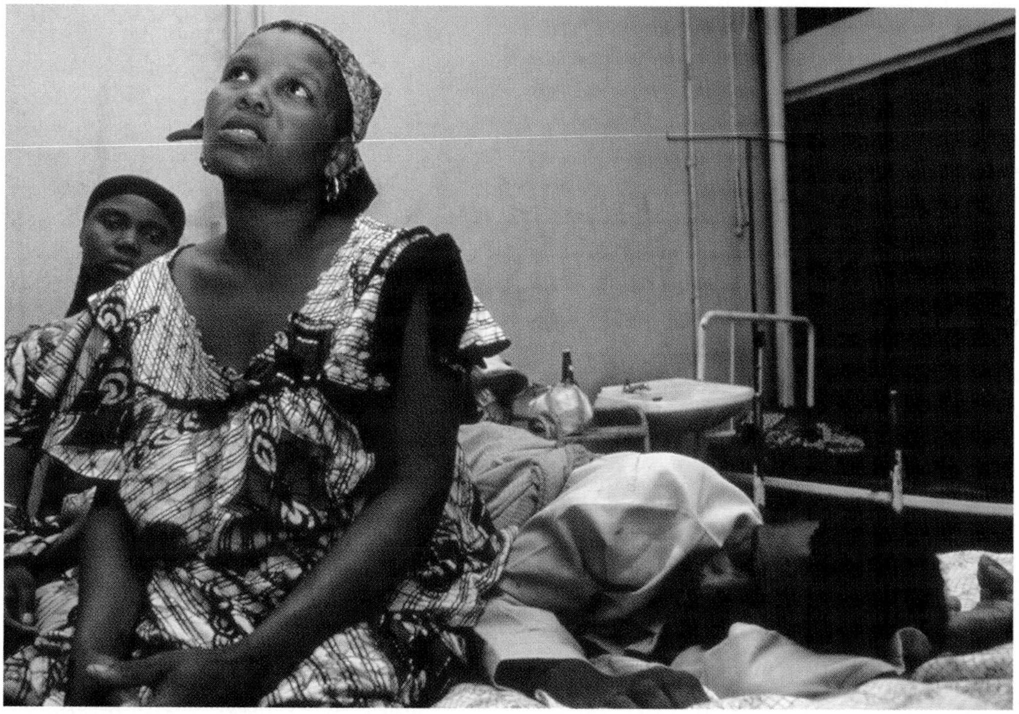

**Figure 6–12** **AIDS.** Four members of a Central African family are victims of AIDS. The disease is often acquired from other family members who are unaware that they are infected.

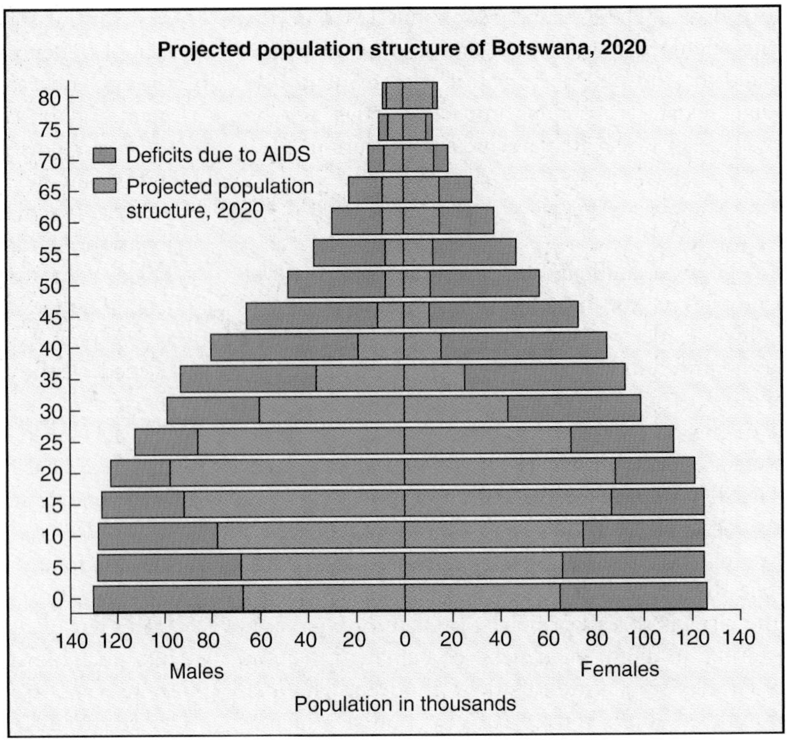


**Figure 6–13 Effect of AIDS on the future population structure of Botswana.** The AIDS epidemic will sharply reduce the total population—especially the number of children born and surviving. Similar effects may be expected in many of the sub-Saharan African countries. [*Source:* Data from "US Census Bureau World Population Profile 2000," reprinted as "AIDS in a New Millennium," by Bernard Schwartlander, et al., *Science,* 289 (July 7, 2000). Copyright © 2000 by American Assn. for the Advancement of Science. Reprinted by permission.]

mounted a national AIDS control program that includes a major education campaign.

## Family Planning

For those who can pay, information on contraceptives, related materials, and treatments are readily available from private doctors and health-care institutions. The poor, however, must depend on family-planning agencies, which are supported by a combination of private donations, government funding, and small amounts the clients may be able to afford. The stated policy of family-planning agencies (or similar private services) is, as the name implies, to *enable people to plan their own family size—that is, to have children only if and when they want them.* In addition to helping people avoid unwanted pregnancies, family planning often involves determining and overcoming fertility problems for those couples who are having reproductive difficulties. Family planning is a critical component of reproductive health care.

More specifically, family-planning services include the following:

- Counseling and education for singles, couples, and groups regarding human reproduction, the hazards of sexually transmitted diseases (AIDS, in particular), and the benefits and risks of various contraceptive techniques.
- Counseling and education on achieving the best possible pre- and postnatal health for mother and child. The emphasis is on good nutrition, sanitation, and hygiene.
- Counseling and education to avoid high-risk pregnancies. Pregnancies that occur when a woman is too young or too old and pregnancies that follow too closely on a previous pregnancy are considered high risk; they seriously jeopardize the health, and even the life, of the mother. Any existing children are also at risk if the mother suffers injury or death.
- Providing contraceptive materials or treatments after people have been properly instructed about all alternatives.

The vigorous promotion and provision of contraceptives has proven all by itself to lower fertility rates, as seen in Fig. 6–6. Those countries which have implemented effective family-planning programs have experienced the most rapid decline in fertility. For example, Thailand initiated a vigorous family-planning program as part of a national population policy in 1971. Population growth subsequently declined from 3.1% per year to the present 0.8% per year, and the Thai economy has posted one of the most rapid rates of increase over the same years. Encouraging and implementing family planning is the first and most important step a country can take to improve its chances of developing economically.

***Unmet Need.*** A society's fertility is determined not only by the choices couples make to have children, but also by births that are unplanned. Women who are not currently using contraception, but who want to postpone or prevent childbearing, are said to have an *unmet need.* Thus, both the spacing of children and the limiting of family size are involved. One major goal of family-planning policies and agencies is to ensure that all those who want or need reproductive health services actually have access to them. This goal is far from achieved in many parts of the developing world. The unmet need for family planning ranges from 7% to almost 50% of women in the different developing countries and affects an estimated 150 million women. Where women's unmet need is minimal, fertility rates are invariably lower. The governments of countries that want to lower their fertility know that meeting the unmet need for family-planning services is perhaps their most crucial objective. Indeed, much international aid is targeted at helping these countries provide reproductive health services to their families.

**Abortion.** Unfortunately, to many people, family planning conjures up images of the abortion clinic. Abortions, by definition, are terminations of unwanted pregnancies. Nearly everyone agrees that an abortion is the least desirable way to avoid having an unwanted child—especially in view of the other alternatives that are available. The document resulting from the Cairo population conference explicitly states that abortions should *never* be used as a means of family planning. Therefore, it is particularly important to understand that the primary functions of family planning are education and providing services directed at avoiding unwanted or high-risk pregnancies. If family-planning services were universally available and people availed themselves of them, there would be far fewer unwanted pregnancies and, hence, fewer abortions.

*How Many?* According to the WHO, 210 million women become pregnant each year. Of these, 46 million resort to abortion (22%). Almost half (20 million) of these abortions are unsafe, performed by people lacking the necessary skills or in medically deficient settings. As a result, 80,000 women die annually from such unsafe abortions, more than 95% in the developing countries. This disturbing recourse to abortions can be seen as a consequence of the lack of family-planning education and services. All studies show that cutbacks in family-planning services result in *more* unwanted pregnancies and *more* demand for abortions, not *less*. Indeed, after the Reagan administration cut off family-planning monies in 1984, abortions in the United States increased an estimated 70,000 a year.

In the United States, the abortion rate declined in the late 1990s and is currently at 21 per 1,000 women of childbearing age (1.31 million abortions performed in 2000). Much of this decline is traced to the increased availability of *emergency contraception*—methods of preventing pregnancy shortly after unprotected sexual intercourse. The most common technique is a pill based on the same hormones used in ordinary birth control; two such pills are levonorgestrel, sold under the brand name Plan B, and a combination of levonorgestrel and ethinyl estradiol, sold under the brand name Preven®. These pills are far less controversial than the drug mifepristone (formerly known as RU-486), a nonsurgical alternative to early abortions. This drug was approved by the U.S. Food and Drug Administration in 2000 and has become a lightning rod for anti-abortion activists. Congressional action to restrict the use of the drug is constantly being threatened.

*Family-Planning Agencies.* Planned Parenthood, which operates clinics throughout the world, is probably the best known family-planning agency. Another significant player is UNFPA, which provides financial and technical assistance to developing countries at their request. The emphasis of UNFPA is on combining family-planning services with maternal and child health care and expanding the delivery of such services in rural and marginal urban areas. Support for this U.N. agency and other family-planning agencies has become a political football in the United States, where the Clinton administration—through the Agency for International Development—tried to continue their funding and many Republican members of the House opposed it. The argument, used since the Reagan years, is that these agencies promote abortions or, in the case of China, condone government efforts that have forced couples to have no more than one child. (See the "Ethics" box, p. 170.) In the latest chapter of this battle, the United States cut off its 2002 contribution of $34 million to UNFPA because of China's one-child policy.

*Gag Rule, etc.* The first act of President George W. Bush in 2001 was to reinstate the Mexico City Policy, known less formally as the "Global Gag Rule." This policy was first imposed by President Reagan in 1984. The gag rule prohibits any U.S. Government aid from being given to foreign family-planning agencies if they provide abortions, counsel women about abortion if they are dealing with an unwanted pregnancy, or advocate for abortion law reforms in their own country. The rule is directed at NGOs like the International Planned Parenthood Federation. This rule is viewed by many as a serious blow to attempts to increase the availability of contraceptive services in the developing countries, because the United States has been the largest international donor to family-planning programs. More recently, the Bush administration has withdrawn its support for the entire 1994 ICPD Program of Action, because it uses terms such as "reproductive services" and "reproductive health care." These are deemed to imply a right to abortion. Critics of this move pointed out (unsuccessfully) that the conference documents made clear that "in no case should abortion be promoted as a method of family planning" and that the Cairo agreement was critical to the fight against HIV/AIDS and prevented the very unwanted pregnancies that could lead to abortions.

## Employment and Income

The bottom line of any economic system is the exchange of goods and services. At its simplest level, this entails a barter economy in which people agree on direct exchanges of certain goods or services. Barter economies are still widespread in the developing world.

The introduction of a cash economy facilitates the exchange of a wider variety of goods and services, and everyone may prosper, as they have a wider market for what they can provide and a wider choice of what they can get in return. In a poor community, everyone may have the potential to provide certain goods or services

ethics

## The Ethical Dilemma of China's Population Policies

What are the options associated with the limits of Earth's carrying capacity for the human species? What are the ethical and moral implications of each option?

1. You can argue, as some optimists do, that the problems do not really exist—that technology will always have a solution to make life better and better for more and more people.
2. Perhaps a disease (AIDS?), a famine, or a natural disaster—or even war—will come along and take care of the situation for us in "nature's way."
3. Some consider forced sterilization or abortion as a viable solution to the population problem. These three options lack credibility or compassion and are unacceptable to most people. Do you agree?
4. The most acceptable option—the one that has been the theme of this chapter—is to create an economic and social climate in which people, of their own volition, will desire to have fewer children. Along with such a climate, society will provide the means (family planning) to enable people to exercise that choice. This option has been realized "unconsciously" in developed countries, but it raises certain ethical dilemmas when you attempt to apply it to developing countries. For example, how far can you go in manipulating the social or economic environment before choice becomes coercion?

With its current population of 1.3 billion (a fifth of the world's people), China provides the most comprehensive example of a country that offers extensive economic incentives and disincentives aimed at reducing population growth. Some years ago, China's leaders recognized that, unless population growth was stemmed, the country would be unable to live within the limits of its resources. Because of inevitable population momentum, the leaders felt that the country could not even afford a total fertility rate of 2.0. Thus, they set a goal of one child per family, and to achieve that goal, they instituted an elaborate array of incentives and deterrents. The prime *incentives* were as follows:

- Paid leave to women who have fertility-related operations—namely, sterilization or abortion procedures.
- A monthly subsidy to one-child families.
- Job priority for only children.
- Additional food rations for only children.
- Housing preferences for single-child families.
- Preferential medical care to parents whose only child is a girl. (There is a strong preference for sons in China, and parents generally wish to have children until at least one son is born.)

*Penalties* for an excessive number of children in China included the following:

- Bonuses received for the first child must be repaid to the government if a second is born.
- A tax (now called a "social compensation fee") must be paid for having a second child.
- Higher prices must be paid for food for a second child.
- Maternity leave and paid medical expenses apply only to the first child.

Besides improving economic opportunities, these incentives and deterrents have helped China achieve a precipitous drop in its fertility rate, from about 4.5 in the mid-1970s to an official government rate of 1.7 in 2003 (although recent work by Chinese demographers have pegged the actual fertility rate somewhat higher). The policy is unpopular: In practice, the one-child policy has been pursued only in urban areas, some 29% of the population. People in rural areas and minorities (who are exempted from the policy) have a higher fertility rate. In the last decade, rapid economic growth and more personal freedoms have made the Chinese people harder to control. Many are willing to pay the penalties. It was the continued enforcement of the policy (including some enforced abortions and sterilizations) that prompted the Bush administration to withdraw funding from the UNFPA.

Despite these efforts, the population of China is still growing due to momentum. (A large percentage of the population is still at or below reproductive age.) Recently, China has begun to phase out its one-child policy, concerned about a future in which there will be insufficient adult children to help care for aging parents. One disturbing consequence of China's policy is the skewed ratio of males to females. Males are preferred in Chinese families, so some parents try to make sure that their one child is a male, through such extreme measure as abortion and even infanticide of female children.

Is China's population policy morally just or unjust? Is it possible that some other countries will have to resort to a China-like policy in the near future as they face severe limits on resources? What about India? At what point are such policies necessary? What do you think?

and may want other things in return, but there may be no money to get the system off and running. In a going economy, people who wish to start a new business venture generally begin by obtaining a bank loan to "set up shop." The poor, however, are considered high credit risks. Furthermore, they may want a smaller loan than what a commercial bank wants to deal with, and many of the poor may be women, who are denied credit because of gender discrimination alone. For these three reasons, poor communities have trouble getting start-up capital. Fortunately, however, the situation is changing.

***Grameen Bank.*** In 1976, Muhammad Yunus (Fig. 6–14), an economics professor in Bangladesh, conceived and created a new kind of bank (now known as the

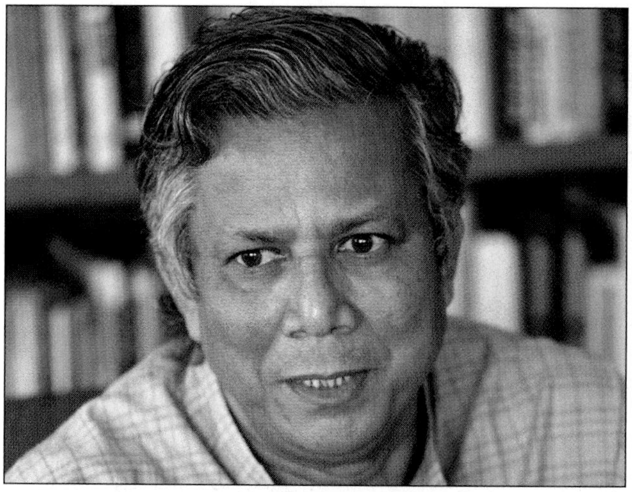

**Figure 6–14** **Muhammad Yunus.** An economics professor in Bangladesh, Yunus created the Grameen bank, which initiated the microlending movement.

Grameen Bank) that would engage in **microlending** to the poor. As the name implies, microloans are small—they average just $67—and are short term, usually just four to six months. Nevertheless, they provide such basic things as seed and fertilizer for a peasant farmer to start growing tomatoes, some pans for a baker to start baking bread, a supply of yarn for a weaver, some tools for an auto mechanic, and so on.

Yunus secured his loans by having the recipients form **credit associations**—groups of several people who agreed to be responsible for each other's loans. With this arrangement, the Grameen Bank experienced an exceptional rate of payback—greater than 97%. Small-scale agriculture loans from the Grameen Bank have had outstanding results. In a rural area of Bangladesh, small loans, along with horticultural advice, are now enabling peasant farmers to raise tomatoes and other vegetables for sale to the cities. These people have doubled their incomes in three years.

Microlending has been found to have the greatest social benefits when it is focused on women, because, as Yunus observed, "When women borrow, the beneficiaries are the children and the household. In the case of a man, too often the beneficiaries are himself and his friends." The credit associations also create another level of cooperation and mutual support within the community, particularly when the loans are directed toward women.

*More Microlending.* The unqualified success of microlending in stimulating economic activity and enhancing the incomes of people within poor communities has been so remarkable that the concept has been adopted with various modifications by a considerable number of private organizations dedicated to alleviating hunger and poverty. Among these organizations are Oxfam and Freedom from Hunger, the latter of which has projects in nine countries around the world and combines its lending with "problem-solving education" in a program called Credit with Education.

Recently, the U.S. Agency for International Development (AID) has begun providing grants to other organizations that wish to do microlending. Even the World Bank is realizing the need to be more sensitive to local communities in its lending practices and is making $200 million available for microlending. A major development in this arena has been the work of the Microcredit Summit Campaign, an outgrowth of the 1995 U.N. Fourth Conference on Women in Beijing. An update from this campaign indicates that, by the end of 2001, some 2,186 institutions had granted loans to 54.9 million "poorest clients." The ultimate goal of the campaign is to reach 100 million families with credit for self-employment and other business services by 2005! At the current growth rate of 37% per year, the figure may indeed be reached.

## Resource Management

The world's poor depend on local ecosystem capital resources—particularly water, soil for growing food, and forests for firewood. They often live in marginal lands of high ecological sensitivity: steep slopes, drylands, and so forth. Many lack access to enough land to provide an income and often depend on foraging in woodlands, forests, grasslands, and coastal ecosystems. For example, over 2 billion people depend for their heating and cooking needs on biomass fuels (wood, grasses, etc.) that they gather from natural areas. Such pressures have stripped the forests from the Garo Hills in northeast India and are destroying the mangrove swamps of West Africa.

A considerable part of the problem lies in utilizing resources poorly, including failing to replant trees and preventing soil erosion. The technologies for managing natural resources often can be found in traditional knowledge, which can be shared broadly with the help of appropriate organizations (Fig. 6–15). "Farmer-to-farmer" advisory and training services are spreading with the help of NGOs and U.N. agency work. Empowering the poor to manage community- or state-owned lands is one approach that has worked in Nepal. There, forestry user groups are given the right to own trees, but not the land, and there are now some 6,000 user groups managing 450,000 hectares (1.1 million acres). The groups develop forestry management plans, set timber sale prices, and manage the surplus income from the operations. Protecting forest resources and tackling land degradation in drylands are two sustainable steps that can be addressed by country policies, one of the targets of MDG Goal 7: *Ensure environmental sustainability.*

## Putting It All Together

Each of the five components of social modernization—education, improving health, family planning, employment and income, and resource management—both depends on and supports the other components. For example, better health and nutrition support better economic productivity, better economic productivity supports obtaining a better education, and a better education leads to a delay in

**Figure 6–15 Improving resource management** A major step in enhancing incomes and protecting the environment is to encourage better resource management. Here, local people are learning the skills required to raise tree seedlings that will later be transplanted in a reforestation project. This is part of the Greenbelt Movement in Kenya.

marriage and the desire to have fewer children. The availability of family-planning services is essential to realizing the desire of parents to have fewer children. In short, all the components work together to alleviate the conditions of poverty, reverse environmental degradation, and reduce population growth (Fig. 6–16). Conversely, the lack of any component—especially family-planning services—will undercut the ability to achieve all the other components.

The elements of social modernization are entirely compatible with the MDGs (Table 6–2); in fact, there is a good deal of overlap between them. Although the MDGs do not directly address slowing population growth, or the demographic transition, every one of the goals will be more reachable with continued progress in social modernization. The MDGs particularly address the needs of the poorer developing countries—the very ones that still have high fertility and stagnant economies. Indeed, the Cairo population conference, more than any other UN conference, stimulated the formation of the MDGs. The next section takes a brief look at that conference and its sequel, ICPD + 5.

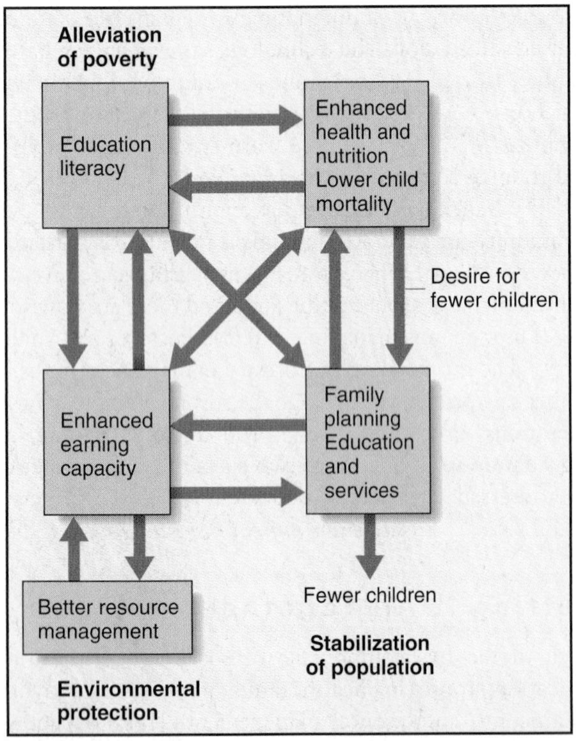


**Figure 6–16 Social development.** This diagram shows how the five main aspects of enhancing the well-being of the poor are mutually supportive of, and dependent on, one another.

## 6.4 The Cairo Conference

In September 1994, some 15,000 leaders and representatives from 179 nations and nearly 1,000 NGOs met in Cairo, Egypt, for the *ICPD*. Before the delegates was a draft document of a "Program of Action" to address the world's persistent problems of poverty and population.

The Vatican and Muslim countries raised some objections to the wording of certain sections of the draft document alluding to birth control methods and abortions. Differences were resolved, however, and in the end, all 179 nations—large and small, rich and poor—signed the final document, committing themselves to achieve the basic goals set forth therein by the year 2015. For the first time in history, therefore, the political, religious, and scientific communities of the world reached a consensus on the population issue.

***Consensus.*** Essentially, *all nations agreed that population is an issue of crisis proportions that must be confronted forthrightly.* This sentiment was summed up in the words of Lewis Preston, then president of the World Bank: "Putting it bluntly, if we do not deal with rapid population growth, we will not reduce poverty—and development will not be sustainable." This view reflects a fundamental change in attitude on the part of numerous leaders and organizations, because the World

Bank has traditionally argued that economic development alone will take care of the population problem.

The goals of the 20-year Program of Action are not cast simply in terms of reducing fertility or population growth per se. Instead, the 1994 ICPD document asks that "interrelationships between population, resources, the environment and development should be fully recognized, properly managed and brought into a harmonious, dynamic balance." Thus, the goals are set in terms of creating an economic, social, and cultural environment in which *all people,* regardless of race, gender, or age, can share equitably in a state of well-being. By providing opportunities for people to improve their quality of life, it is assumed that they will choose to have fewer children and that population growth will level off to the medium projection shown in Fig. 5–3. This means that no more than 7.5 billion persons will have to share the planet by 2020 and 9.3 billion by the year 2050.

***Program of Action.*** By signing the Program of Action, the governments of 179 nations have indicated their commitment to achieving a host of objectives over the next 20 years. In particular, this commitment means empowering women, meeting people's needs for education and health (including reproductive health), advancing gender equality, eliminating violence against women, enabling women to control their own fertility, and attaining other goals. To accomplish these, an estimated $17 billion per year was required by 2000, with additional amounts increasing to $21.7 billion by 2015. Two-thirds would come from the developing countries and the rest from the developed countries. The developed countries agreed to set aside 0.7% of their gross national product to achieve the objectives of the Program of Action, and they also agreed to assist the developing countries with the technologies needed to act locally.

***ICPD + 5.*** The ICPD was reviewed and appraised during 1999, five years after the original Cairo conference. The process involved several preliminary forums and roundtables held in different parts of the world, culminating in a special session of the U.N. General Assembly. The review demonstrated that the Program of Action was indeed being implemented in many developing countries. Changes in laws and programs have occurred. For example, national population policies in countries like India have switched their focus from strict family planning and targeted numbers to providing a wide range of reproductive health services, delivered to local levels. Further, many African countries have outlawed female genital mutilation, the right to contraceptives has been established in many countries, and in Bangladesh government policy has encouraged the fuller participation of girls in school through the secondary level. Finally, population and development have been more fully integrated in many nations, the use of family-planning methods is on the rise, and mortality in most countries has continued to fall.

Unfortunately, there are still some serious exceptions. Women and girls continue to face discrimination in many societies, the HIV/AIDS epidemic has led to a rise in mortality in numerous countries, infectious diseases continue to take an unacceptable toll in sickness and death, and millions of couples still lack access to reproductive health information and services. At the review, a new set of benchmark indicators was adopted, giving some quantitative measures for future progress. These were for the most part included in the MDGs listed in Table 6–2.

***Funding.*** Funding remains a problem. By 2000, the developing countries had provided $8.6 billion, fulfilling 75% of their commitment. The industrialized donor countries, by contrast, have not kept their promises to fund the rest—providing less than half of their one-third share ($5.7 billion) of the costs. If the 0.7%-of-gross-national-product recommendation made at the ICPD conference were followed, the United States, with by far the world's largest gross national product (almost $10 trillion in 2002), would be a major donor (0.7% of $10 trillion is $70 million). Instead, the Bush administration and Congress have acted to reduce U.S. support of UNFPA and the entire ICPD Program of Action! It is significant—and perhaps shameful for the nation—that two U.S. billionaires—Ted Turner and Bill Gates—have stepped into the gap with multimillion-dollar donations to support U.N. population activities.

Regardless of what happens in the United States, the international community is taking the ICPD Program of Action seriously and is, on the whole, engaged in implementing its recommendations. This is good news for everyone—even us—because it means, in the end, greater world security and a much better chance of achieving an environmentally sustainable lifestyle.

# revisiting the themes

## Sustainability

The demographic transition is essential for a sustainable future. This chapter is all about what has to be done economically and socially to bring about development that is sustainable. Economic progress is being made in many developing countries, and it is accompanied by most of the elements of social modernization. However, economic progress is still lacking in many others. For these, social modernization is the most feasible approach, because it will bring fertility rates down and enable countries to open the demographic window that may make economic progress possible.

The World Bank's new environmental policy, *Making Sustainable Commitments,* places sustainability at the center of the agency's work with developing countries, especially the poorer ones. This means protecting the environment locally, because it directly supports the livelihoods of so many. Globally, the Bank seeks those policies that mutually benefit developing countries and the entire world.

One of the key MDGs is #7: *Ensure environmental sustainability.* This goal is facilitated by integrating principles of sustainability into country policies and programs and by reversing the loss of environmental resources. Improving the management of these resources is an important component of social modernization.

## Stewardship

Justice for the developing world was one of the key ethical issues identified in Chapter 1 in the discussion of stewardship. *Distributive justice* is an ethical ideal that considers the gross inequality of the human condition as unjust—the kind of inequality represented by the billion-plus people who live on less than $1 a day and typical people in the wealthy countries. Although equal distribution of the world's goods and services to everyone is neither attainable nor necessary, the existing poverty in the poor countries is an affront to human dignity. The alleviation of poverty must be made a priority of public policy to address this injustice. Aid from the rich countries to the poor countries should not be seen as an opportunity for advancing political goals, but as part of the obligation we have as human beings to look after and care for each other. Reread each of the MDGs and then ask yourself if it isn't in your heart to affirm these goals as you think of those whose well-being will be improved as the goals are being met.

## Ecosystem Capital

The debt crisis is also a crisis for ecosystems, because poor countries are forced to liquidate their ecosystem capital in order to service the debts they have incurred. Because these countries are often areas of high biodiversity, liquidating the capital often means a loss of those habitats that preserve countless species that may be found nowhere else on earth. It also means a loss of those global services that the ecosystems provide, such as storing carbon. Even without the pressures of international debt, the rising populations of the poorer countries depend heavily on natural ecosystems for their food and work and often exploit them beyond their sustainable limits, to those countries' own detriment. Preserving ecosystem capital while drawing down only its interest is difficult to accomplish, but it is a necessary element of social modernization and sustainable development.

## Policy and Politics

So much of what has been discussed in this chapter depends on putting into place and implementing public policies that will promote economic growth and alleviate poverty. These policies involve investing in educational opportunities to reduce illiteracy, extending basic health services to the poor, and providing reproductive health services to women. Many countries are taking the MDGs seriously and are making progress toward those goals, recognizing that the most important steps are the ones taken to change public policies within the countries themselves. The developed countries have a vital role in this process, too, though: They need to extend Official Development Aid, invest in economic enterprises in the poor countries, and help resolve the debt crisis in those countries. Achieving these aims also requires reforms in public policy—reforms that, unfortunately, can be overturned by politics, as shown by the United States in the area of population aid.

## Globalization

What role can globalization play in bringing about social modernization, the demographic transition, and economic growth in the developing countries? A small number of developing countries (China, Malaysia, and Indonesia, for example) have successfully entered the global market, but more need to. Their exports have paved the way for economic progress, and these countries are becoming more connected to the economic and information-rich high-tech world. The poorer countries are mostly populated by self-employed peasants, however, who must make the transition to becoming wage earners in a market economy. This is not easy and they can't do it by themselves. The jobs have to be there, and that often requires investment from outside agencies—businesses and NGOs with the knowledge and capital needed. Also, there has to be some hope of marketing the products, which means entering the global market on a level playing field. At present, however, the playing field is far from level, because developed countries protect their exports with subsidies, quotas, and tariffs, and all of these trade barriers work against the developing countries. Toward that end, Goal 8 of the MDGs, *develop a global partnership for development,* calls for eliminating these barriers and encouraging the poor countries themselves to develop a market-based trading system committed to sustainable development and attractive to foreign investment.

Alternatively, people can become small-scale entrepreneurs with the aid of microlending agencies and never worry about the global market. This is an attractive alternative and one that is growing exponentially.

Finally, globalization can be harmful to the poor, because it opens markets for agricultural products and can make subsistence farming uneconomical. It can widen the gap between the rich and poor within a society, as the wealthier members capitalize on the information flow and quick transfer of goods and services, leaving the poor farther behind.

## review questions

1. What have been the two basic schools of thought regarding the demographic transition? How were these reflected in the three most recent global population conferences?
2. Discuss the six specific factors that influence the number of children a poor couple desires.
3. Discuss the proposition that social and economic progress has been made in recent years in the developing countries.
4. What are the MDGs? Cite an example of one goal and the target used to measure progress in attaining it.
5. What has been the major agency and mechanism for promoting development in poor nations in recent years? Which agency is most important to poorer countries?
6. What is meant by the debt crisis of the developing world? What is being done to help resolve this crisis?
7. What is development aid, and how does it measure up against the need for such aid?
8. What are the five interdependent components that must be addressed to bring about social modernization?
9. What is family planning, and why is it critically important to all other aspects of development?
10. What is meant by an unmet need? The gag rule?
11. What is microlending? How does it work?
12. What was the significance of the 1994 Cairo Conference? Has there been progress in the ensuing years? What is the biggest obstacle to progress?

## thinking environmentally

1. Is the world population below, at, or above the optimum? Defend your answer by pointing out things that may improve and things that may worsen as the population increases.
2. Suppose you are the head of an island nation with a poor, growing population and the natural resources of the island are being degraded. What kinds of policies would you initiate, and what help would you ask for to try to provide a better, sustainable future for your nation's people?
3. List and discuss the benefits and harms of writing off debts owed by developing nations.
4. Describe what you think will be the long-term results (for the people, the society, and the environment) of limiting family-planning services in developing nations.
5. Explore the consequences of donating computers to every family in a developing country.

## making a difference part two: chapters 5 and 6

1. Think carefully about your own reproduction. What concerns will you and your mate weigh as you plan your family?
2. Become involved in the abortion debate, pro or con. Whether the United States supports international family planning, whether abortions remain legal in your state, and other issues will be determined by votes cast by you or your representatives.
3. Become involved in, and support, programs promoting effective sex education and responsible sexual behavior. Consider the advantages of abstinence and monogamy as ways to avoid needing an abortion or contracting a sexually transmitted disease.
4. Share your knowledge and energy by joining the Peace Corps or another organization engaged in providing appropriate technology to a developing country.
5. Encourage sustainability by buying products that originate from the developing world.

# part three

# renewable resources

A mountainside on the island of Bali is terraced for rice cultivation; tallgrass prairies in Illinois and Iowa are plowed under to raise corn; tropical rain forests in Brazil are converted to pasture for cattle; desert in Israel is irrigated to raise vegetables. Forests are harvested for wood and paper pulp; coastal oceans are fished; grasslands are grazed by sheep and cattle. We get food and fiber from a host of natural ecosystems, some thoroughly managed and some not. All of these activities depend on water, nutrients, and sunlight—the basis for productivity in natural systems. All are renewable resources; that is, they are replenished as energy flows and water and nutrients cycle in ways that have long sustained life on Earth. These same systems provide us with priceless services, such as soil formation, waste degradation, pest control, climate modification, flood control, and many others. This is our ecosystem capital.

At the same time, we embrace the beauty of wild ecosystems and turn to them for enjoyment and renewal. Can we have it both ways? Can we hope to preserve nature while we also make use of its goods and services, especially in light of continuing population growth and, therefore, continuing pressure on all resources? In this part of the text, you will dig more deeply into the science of water, soil, food production, forest growth, and fisheries. You will examine all of these renewable resources and ways of managing them as stewards, while again keeping your eyes on a sustainable future.

◀ **Delaware River valley, New Jersey and Pennsylvania**

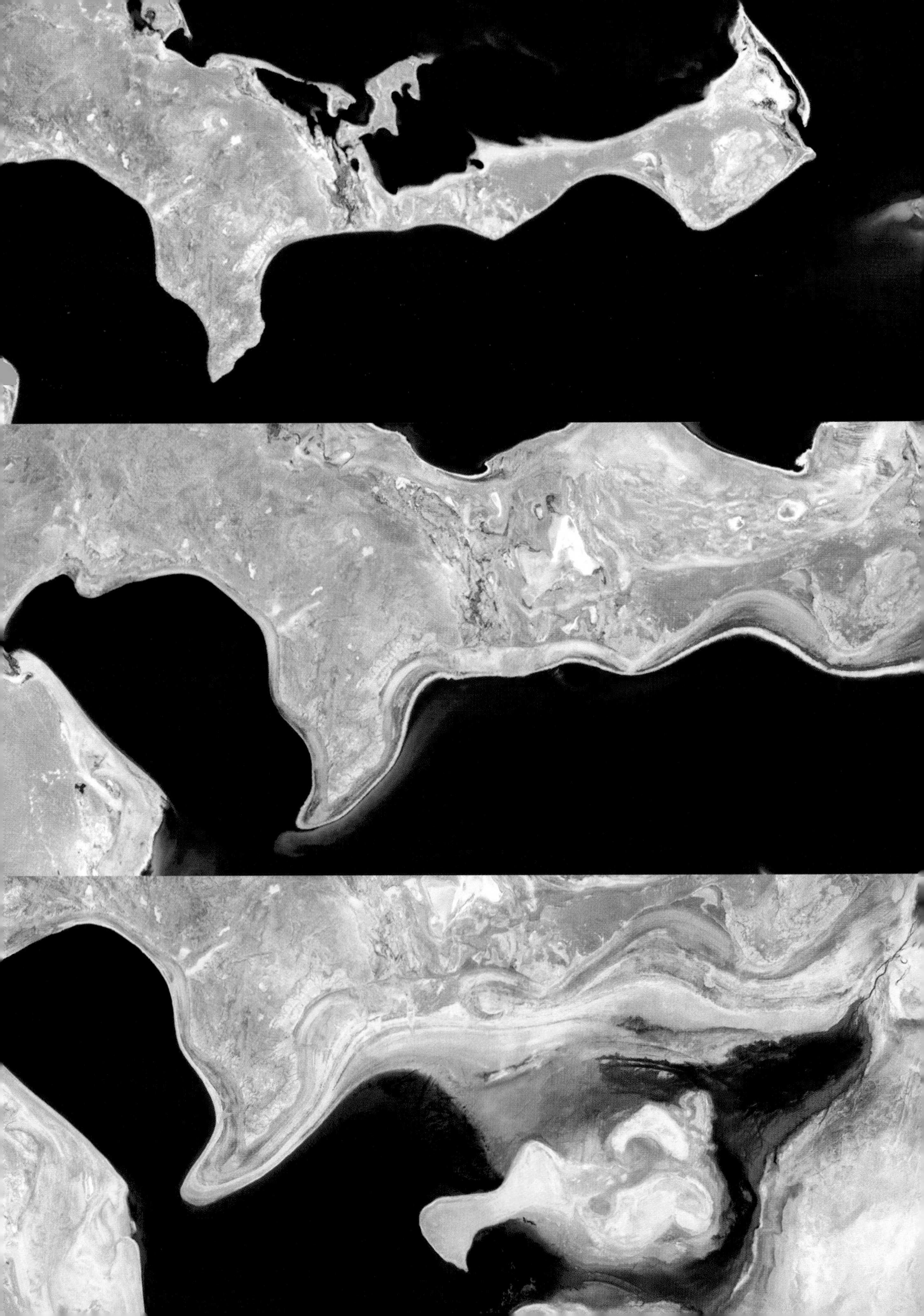

# Water: Hydrologic Cycle and Human Use

**Key Topics**

1. Water: A Vital Resource
2. Hydrologic Cycle: Natural Cycle, Human Impacts
3. Water: A Resource to Manage, a Threat to Control
4. Water Stewardship: Public-Policy Challenges

The former Soviet Union was responsible for one of the world's worst environmental disasters: the death of the Aral Sea (a freshwater lake located in present-day Kazakhstan and Uzbekistan). Central planning in the 1930s decided that the area surrounding the sea could grow cotton if it were irrigated. In turn, the cotton would earn valuable hard currency as an export crop. In time, the Amu Darya and the Syr Darya, the two large rivers flowing into the Aral, were tapped for irrigation water. By 1960, millions of hectares of land came under irrigation, the Soviet Union became the world's second-largest cotton exporter, and the project was judged a huge success. However, from the 50 cubic kilometers of water the sea was receiving in 1965 (an already reduced amount from what it had received in the 1930s), inflowing water was reduced to zero by the early 1980s. Predictably, the sea, once 68,000 km$^2$ (26,000 mi$^2$) in area, began to shrink. (See the photos on the opposite page.)

***Impacts.*** The most immediate impact was felt in the large fishery, an industry that had employed 60,000 people. Because essentially no new water was entering the lake, the salinity began to increase as water evaporated, devastating the lake's ecology. The lake is now three times as salty as the ocean. The commercial fishery collapsed by the early 1980s. In time, the lake lost 80% of its original volume, the water level dropped more than 53 feet, and, as the photos show, the surface area of the lake shrank greatly. There were additional impacts as well. The exposed lake bed [more than 28,000 km$^2$ (10,800 mi$^2$)] now contains accumulated salts, and the dry winds of the area pick up the salt and dust, dropping them on the irrigated land, choking people for miles around, and carrying the Aral pollutants as far as the Arctic and Pakistan. Many health impacts have been reported from the area, including increased cancer rates and higher infant mortality. The local climate changed; the growing season shortened, forcing many farmers to switch from cotton to rice. Many animal species unique to the Aral Sea have become extinct.

**Three satellite views of the Aral Sea; top, 1973; middle, 1987; bottom, 2000. The sea is getting smaller every year.**

***Regime Change.*** With the collapse of the Soviet Union in 1991, the future of the Aral Sea is now in the hands of the republics of Uzbekistan and Kazakhstan.

Turkmenistan, Tajikistan, and Kyrgyzstan take water from the Amu Darya and the Syr Darya for irrigation, so they are also involved. These five republics are all struggling economically, and the diverted waters support billions of dollars worth of agriculture annually, employing millions of people. It is hard to imagine, therefore, that those countries will agree on restoring the Aral. In any event, there is little hope for restoring the Aral Sea to its former status, because it would take stopping all irrigation from the two rivers for half a century just to increase the sea's area to twice its present size. In the process, the surrounding countries' economies would be destroyed.

The Aral Sea is in the process of breaking up into several smaller lakes. One or two of these might regain a fishery if the rivers are allowed to fill them directly. The key is to reduce the diversion of water and prevent the increasing salinization of the sea. Because the irrigation systems involved are highly inefficient, this is a possibility. The republics have little to invest, however, so they must depend on international agencies for help. The World Bank, through its International Bank for Reconstruction and Development program, has provided a $64 million loan to Kazakhstan to improve conditions in the northern Aral Sea to the point where it could support a small fishery.

***Other Dry Runs.*** The Aral Sea's fate is not unique. The Rio Grande, the second-longest river in the United States, disappears about 300 feet from the Gulf of Mexico, because its waters are drawn down all along its course for domestic use and irrigation. The mighty Colorado River is almost completely drained by the United States before it can reach Mexico, leaving a delta in the Gulf of California that has become a moonscape of baked mud. The Dead Sea is "dying" because its inflow of fresh water is diverted by both Israel and Jordan. The Dead Sea may disappear by 2050, unless seawater is pumped into it from the Red Sea, a prohibitively expensive project. In these and many other parts of the world, domestic use, industrial use, and, especially, agricultural use for irrigation are competing for an increasingly scarce supply of the life-giving water. Chapter 7, therefore, the first of five chapters on renewable resources, focuses on water.

## 7.1 Water: A Vital Resource

Water is absolutely fundamental to life as we know it. It is difficult even to imagine a form of life that might exist without water. Happily, Earth is virtually flooded with water. A total volume of some 325 million cubic miles (1.4 billion cubic kilometers) covers 71% of Earth's surface. About 97.5% of this volume is the salt water of the oceans and seas. The remaining 2.5% is **fresh water**—water with a salt content of less than 0.1% (1,000 ppm). This is the water upon which most terrestrial biota, ecosystems, and humans depend. Of the 2.5%, though, two-thirds is bound up in the polar ice caps and glaciers. Thus, only 0.77% of all water is found in lakes, wetlands, rivers, groundwater, biota, soil, and the atmosphere (Fig. 7–1). Nevertheless, evaporation from the oceans combines with precipitation to resupply that small percentage continually through the solar-powered hydrologic cycle, described in detail in Section 7.2. Thus, *fresh water is a continually renewable resource.*

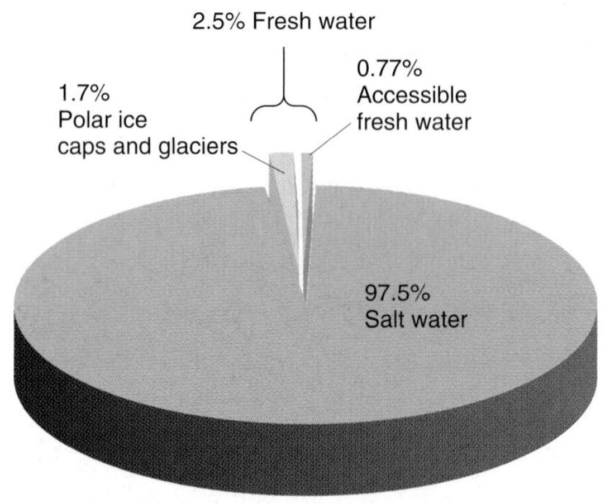


**Figure 7–1 Earth's water.** The Earth has an abundance of water, but terrestrial ecosystems, humans, and agriculture depend on accessible fresh water, which constitutes only 0.77% of the total.

Streams, rivers, ponds, lakes, swamps, estuaries, groundwater, bays, oceans, and the atmosphere all contain water, and they all represent ecosystem capital—goods and services vital to human interests. They provide drinking water, water for industries, and water to irrigate crops. Bodies of water furnish energy through hydroelectric power and control flooding by absorbing excess water. They provide transportation, recreation, waste processing, and habitats for aquatic plants and animals. Fresh water is a vital resource for all land ecosystems, modulating the climate through evaporation and essential global warming (when the fresh water is in the atmosphere as water vapor). During the last two centuries, many of these uses (and some threats to them), have led us to construct a huge infrastructure designed to bring water under control. We have built dams, canals, reservoirs, aqueducts, sewer systems, treatment plants, water towers, elaborate pipelines, irrigation systems, and desalination plants. As a result, waterborne diseases have been brought under control, vast cities thrive in deserts, irrigation makes it possible to grow 40% of the world's food, and one-fifth of all electricity is generated through hydropower. These great benefits are especially available to people in the developed countries (Fig. 7–2a).

In the developing world (Fig. 7–2b), by contrast, over 1 billion people still lack access to clean drinking water, 2 ½ billion do not have access to adequate sanitation services, and over 3 million deaths each year are traced to waterborne diseases (mostly in children under 5). In addition, because of the infrastructure that is used to control water, whole seas are being lost, rivers are running dry, tens of million people have been displaced to make room for reservoirs, groundwater aquifers are being pumped down, and disputes over water have raised tensions from local to international levels. Fresh water is a limiting resource in many parts of the world and is certain to become even more so as the 21st century unfolds.

There are two ways to consider water issues. The focus in this chapter is on *quantity*—that is, on the global water cycle and how it works, on the technologies we use to control water and manage its use, and on public policies we have put in place to govern our different uses of water. Chapter 17 focuses on water *quality*—that is, on water pollution and its consequences, on sewage treatment technologies, and on public policies for dealing with water pollution issues. The next section discusses the hydrologic cycle.

## 7.2 Hydrologic Cycle: Natural Cycle, Human Impacts

Earth's **water cycle**, also called the **hydrologic cycle**, is represented in Fig. 7–3. The basic cycle consists of water rising to the atmosphere through *evaporation* and *transpiration* (the loss of water vapor as it moves from the soil through green plants and exits through leaf pores) and returning to the land and oceans through *condensation* and *precipitation*. Modern references to the hydrologic cycle distinguish between water vapor and liquid water as *green water* and *blue water*, respectively. Green water and blue water are vitally linked in the hydrologic cycle. Terms commonly used to describe water are listed and defined in Table 7–1.

### Evaporation, Condensation, and Purification

Recall from Chapter 3 that a weak attraction known as hydrogen bonding tends to hold water molecules ($H_2O$) together. Below 32°F (0°C), the kinetic energy of the molecules is so low that the hydrogen bonding is strong enough to hold the molecules in place with respect to one another, and the result is *ice*. At temperatures above freezing, but below boiling (212°F or 100°C), the kinetic energy of the molecules is such that hydrogen bonds keep

(a)

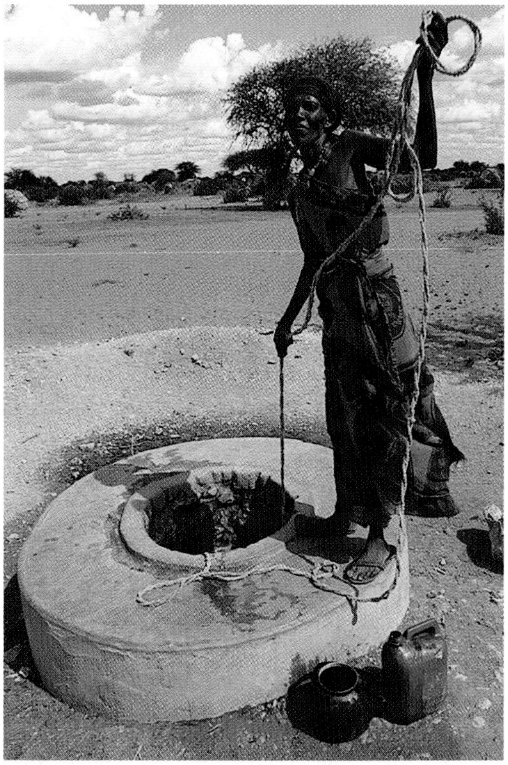

(b)

**Figure 7–2 Differences in availability.** Human conditions range from (a) being able to luxuriate in water to (b) having to walk long distances to obtain enough simply to survive. In many cases, however, the abundance of water is illusory, because water supplies are being overdrawn.

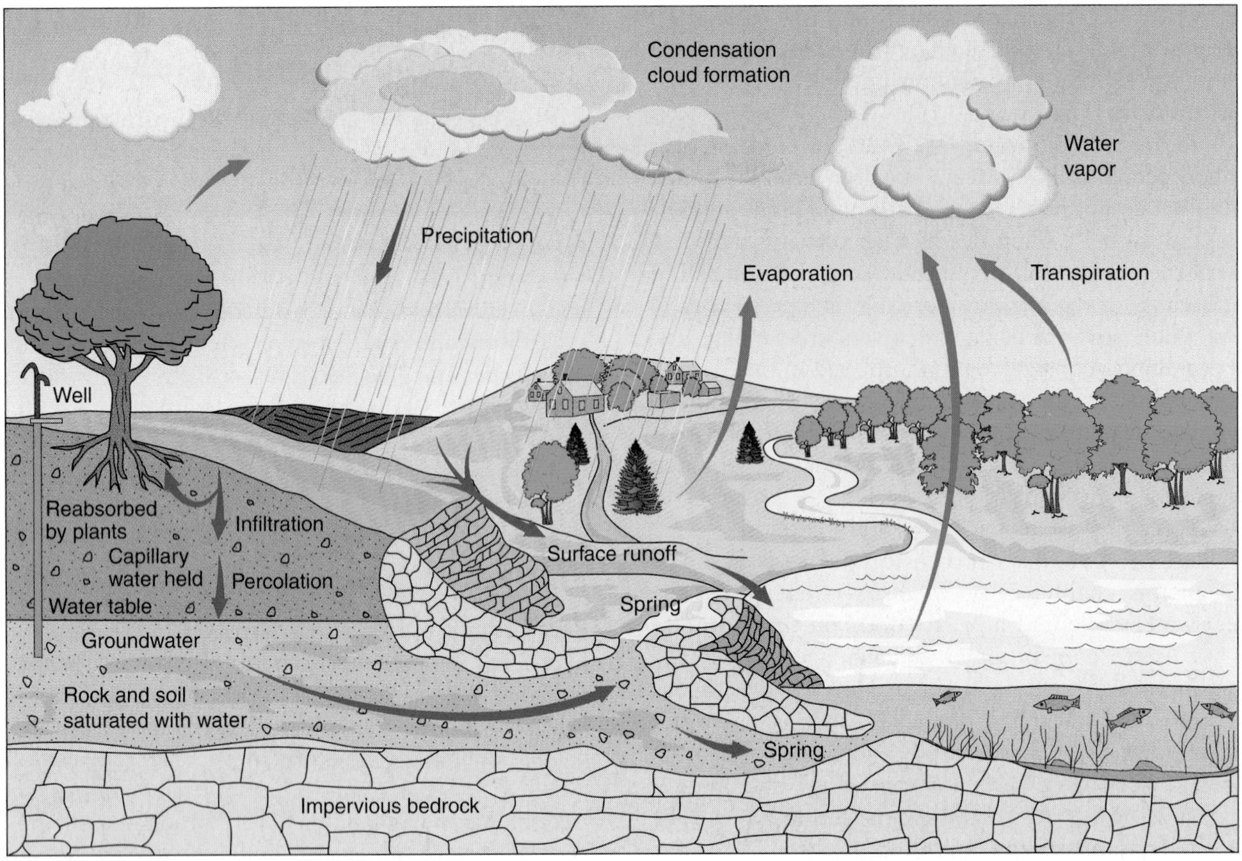


**Figure 7–3 The hydrologic cycle.** The Earth's fresh waters are replenished as water vapor enters the atmosphere by evaporation or transpiration from vegetation, leaving salts and other impurities behind. As precipitation hits the ground, three additional pathways are possible. Colored arrows distinguish between blue and green water flows. (See text.)

**table 7-1 Terms Commonly Used to Describe Water**

| Term | Definition |
|---|---|
| Water quantity | The amount of water available to meet demands. |
| Water quality | The degree to which water is pure enough to fulfill the requirements of various uses. |
| Fresh water | Water having a salt concentration below 0.1%. As a result of purification by evaporation, all forms of precipitation are fresh water, as are lakes, rivers, groundwater, and other bodies of water that have a throughflow of water from precipitation. |
| Salt water | Water, typical of oceans and seas, that contains at least 3% salt (30 parts salt per 1,000 parts water). |
| Brackish water | A mixture of fresh and salt water, typically found where rivers enter the ocean. |
| Hard water | Water that contains minerals, especially calcium or magnesium, that cause soap to precipitate, producing a scum, curd, or scale in boilers. |
| Soft water | Water that is relatively free of minerals. |
| Polluted water | Water that contains one or more impurities, making the water unsuitable for a desired use. |
| Purified water | Water that has had pollutants removed or is rendered harmless. |
| Storm water | Water from precipitation that runs off of land surfaces in surges. |
| Green water | Water in the soil and in organisms that eventually ends up as water vapor—the main source of water for natural ecosystems and rain-fed agriculture. |
| Blue water | Renewable surface water runoff and groundwater recharge—the focus of management and the main source of water for human withdrawals. |

breaking and re-forming with different molecules. The result is *liquid water*. As the water molecules absorb energy from sunlight or an artificial source, the kinetic energy they gain may be enough to allow them to break away from other water molecules entirely and enter the atmosphere. This process is known as **evaporation**, and the result is **water vapor**—water molecules in the gaseous state.

You will learn in Chapter 20 that water vapor is a powerful greenhouse gas. It provides about two-thirds of the total warming from all greenhouse gases. The amount of water vapor in the air is the **humidity**. Humidity is generally measured as **relative humidity**, the amount of water vapor as a percentage of what the air can hold *at a particular temperature*. For example, a relative humidity of 60% means that the air contains 60% of the maximum amount of water vapor it could hold at that particular temperature. The amount of water vapor air can hold increases and decreases with the temperature. When warm, moist air is cooled, and its relative humidity rises until it reaches 100%, further cooling causes the excess water vapor to *condense* back to liquid water, because the air can no longer hold as much water vapor (Fig. 7–4).

**Condensation** is the opposite of evaporation. It occurs when water molecules rejoin by hydrogen bonding to form liquid water. If the droplets form in the atmosphere, the result is fog and clouds. (Fog is just a very low cloud.) If the droplets form on the cool surfaces of vegetation, the result is dew. Condensation is greatly facilitated by the presence of *aerosols* in the atmosphere. **Aerosols** are microscopic liquid or solid particles originating from land and water surfaces. They provide sites that attract water vapor and promote the formation of droplets of moisture. Aerosols originate naturally, from sources such as volcanoes, wind-stirred dust and soil, and sea salts. Anthropogenic sources—sulfates, carbon, and dust—contribute almost as much as the natural sources (more in some locations) and can have a significant impact on regional and global climates.

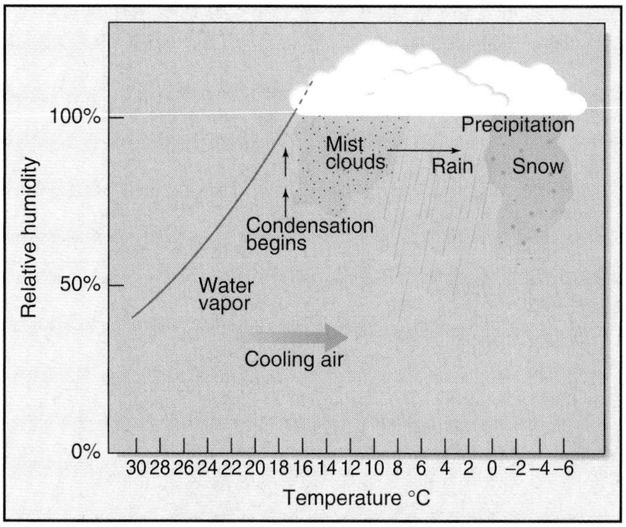


**Figure 7–4 Condensation.** The amount of water vapor that air can hold increases and decreases with the temperature. The red line follows an air mass that starts with 40% relative humidity (RH) at 30°C and is cooled. When the mass of air cools to 17°C, it has reached 100% RH, and condensation begins, forming clouds. If we started with a higher RH, clouds would form sooner when the air is cooled. Further cooling and condensation results in precipitation.

***Purification.*** The processes of evaporation and condensation purify water naturally. When water in an ocean or a lake evaporates, only the water molecules leave the surface; the dissolved salts and other solids remain behind in solution. When the water vapor condenses again, it is thus purified water—except for the pollutants and other aerosols it may pick up from the air. The water in the atmosphere turns over every 10 days, so water is constantly being purified. (The most chemically pure water for use in laboratories is obtained by distillation, a process of boiling water and recondensing the vapor.)

Thus, evaporation and condensation are the source of all natural fresh water on Earth. Fresh water from precipitation falling on the land gradually makes its way through aquifers, streams, rivers, and lakes to the oceans. In the process, it carries along salts from the land, which eventually accumulate in the oceans. Salts also accumulate in inland seas or lakes, such as the Great Salt Lake in Utah. The salinization of irrigated croplands (see Chapter 8) is a noteworthy human-made example of this process.

## Precipitation

Warm air rises from the Earth's surface because it is less dense than the cooler air above. As it encounters the lower atmospheric pressure at increasing altitudes, the warm air gradually cools as it expands—a process known as *adiabatic cooling*. When the relative humidity reaches 100% and cooling continues, condensation occurs and clouds form. As condensation intensifies, water droplets become large enough to fall as precipitation. *Adiabatic warming* occurs as the air descends and is compressed by the higher air pressure in the lower atmosphere.

Precipitation on Earth ranges from near zero in some areas to more than 100 inches (2.5 m) per year in others. (Figure 7–5 shows the pattern over land.) The distribution depends basically on patterns of rising or falling air currents. As air rises, it cools, condensation occurs, and precipitation results. As air descends, it tends to become warmer, causing evaporation to increase and dryness to result. A rain-causing event that you see in almost every television weather report is the movement of a cold front. As the cold front moves into an area, the warm, moist air already there is forced upward because the cold air of the advancing front is denser. The rising warm air cools, causing condensation and precipitation along the leading edge of the cold front.

**Figure 7–5** **Global precipitation.** Note the high rainfall in equatorial regions and the regions of low rainfall to the north and south.

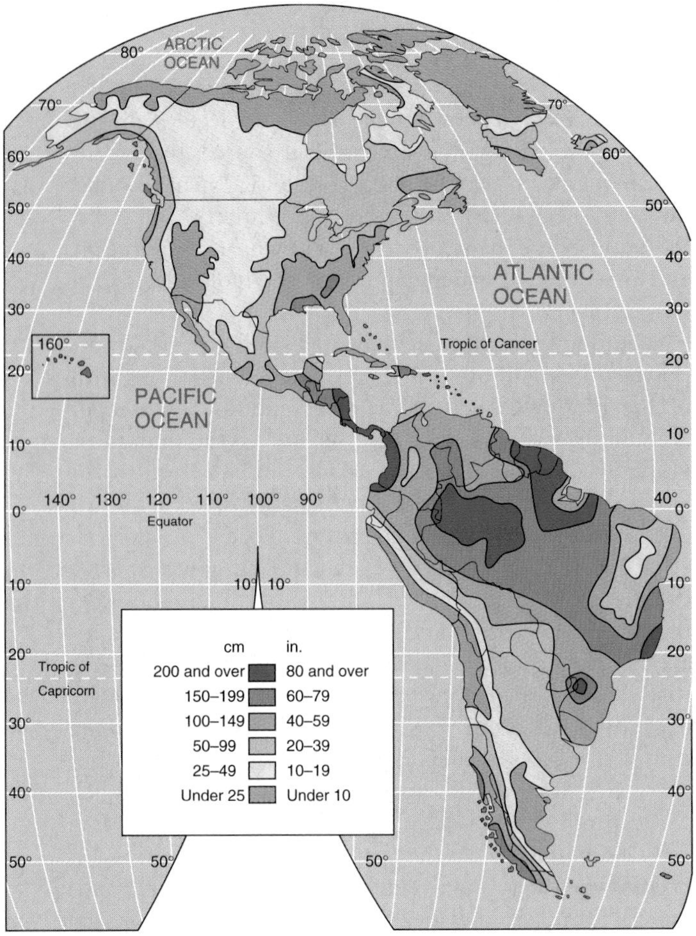


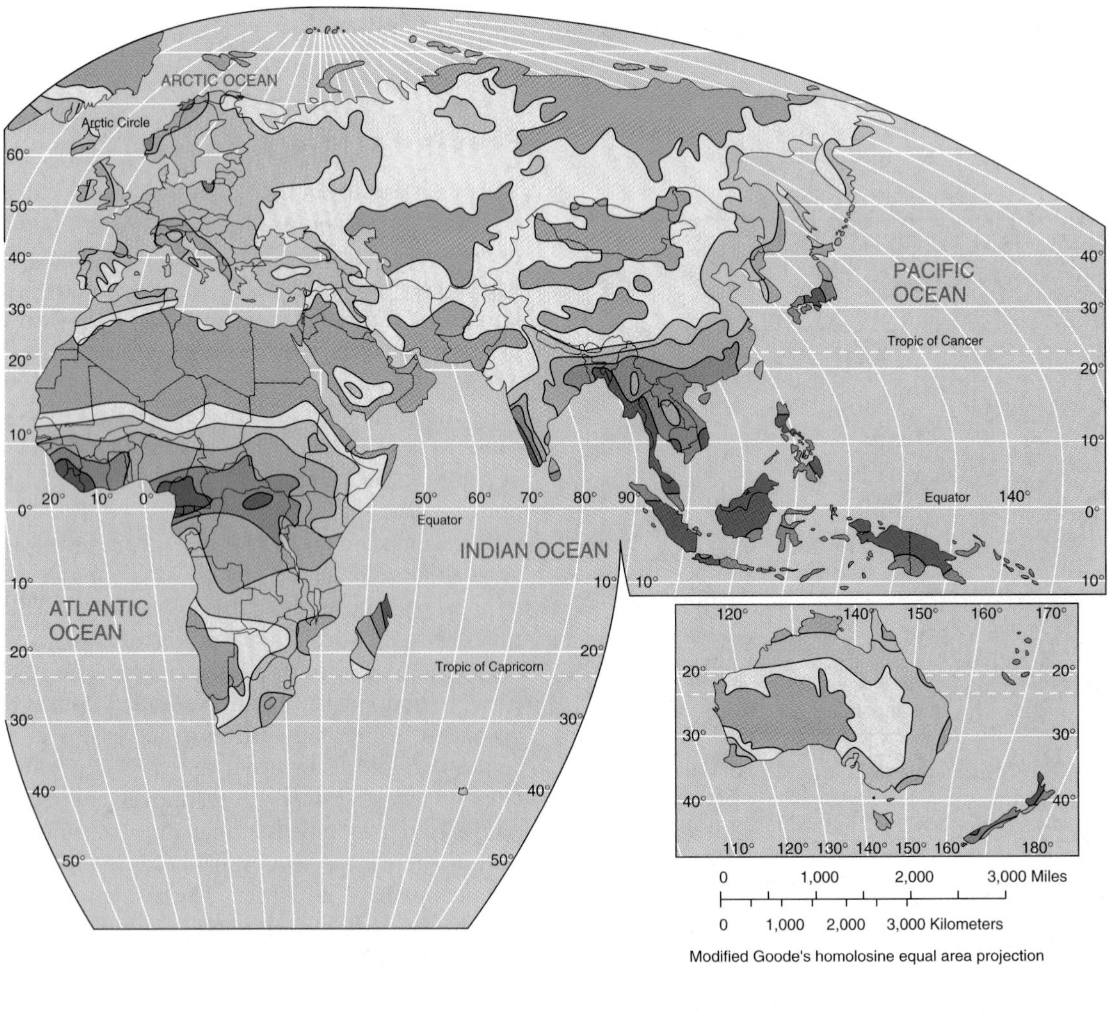

***Convection.*** Two factors—global *convection currents* and the rain shadow—may cause more or less continuously rising or falling air currents over particular regions, with major effects on precipitation. Global convection currents occur because the sun heats the Earth most intensely over and near the equator, where rays of sunlight are almost perpendicular to Earth's surface. As the air at the equator is heated, it expands, rises, and cools; condensation and precipitation occur. The constant intense heat in these equatorial areas ensures that this process is repeated often, thus causing high amounts of rainfall, which, along with continuous warmth, in turn supports tropical rain forests.

Rising air over the equator is just half of the convection current, however. The air must come down again, too. Pushed from beneath by more rising air, it literally "spills over" to the north and south of the equator and descends over subtropical regions (25° to 35° north and south of the equator), resulting in subtropical deserts. The Sahara of Africa is the prime example. The two halves of the system composed of the rising and falling air make up a **Hadley cell** (Figs. 7–6a and b). Because of Earth's rotation, winds are deflected from the strictly vertical and horizontal paths indicated by a Hadley cell and tend to flow in easterly and westerly directions—the *trade winds* (Fig. 7–6c), which blow almost continuously from the same direction.

***Rain Shadow.*** The second situation that causes continually rising and falling air occurs when moisture-laden trade winds encounter mountain ranges. The air is deflected upward, causing cooling and high precipitation on the windward side of the range. As the air crosses the range and descends on the other side, it becomes warmer and increases its capacity to pick up moisture. Hence, deserts occur on the leeward sides of mountain ranges. The dry region downwind of a mountain range is referred to as a **rain shadow** (Fig. 7–7). The severest deserts in the world are caused by the rain-shadow effect. For example, the westerly trade winds, full of moisture from the Pacific Ocean, strike the Sierra Nevada mountains in California. As the winds rise over the mountains, large amounts of water precipitate out, supporting the lush forests on the western slopes. Immediately east of the southern Sierra Nevada, however, lies Death Valley, one of the driest regions of North America.

Figure 7–5 shows general precipitation patterns for the land; the general atmospheric circulation (Fig. 7–6) and the effects of mountain ranges (Fig. 7–7) combine to give great variation in the precipitation reaching the land. The distribution of precipitation, in turn, greatly determines the ecosystems found in a given region.

## Groundwater

As precipitation hits the ground, it may either soak into the ground (**infiltration**) or run off the surface (**runoff**, a blue water flow). The amount that soaks in compared

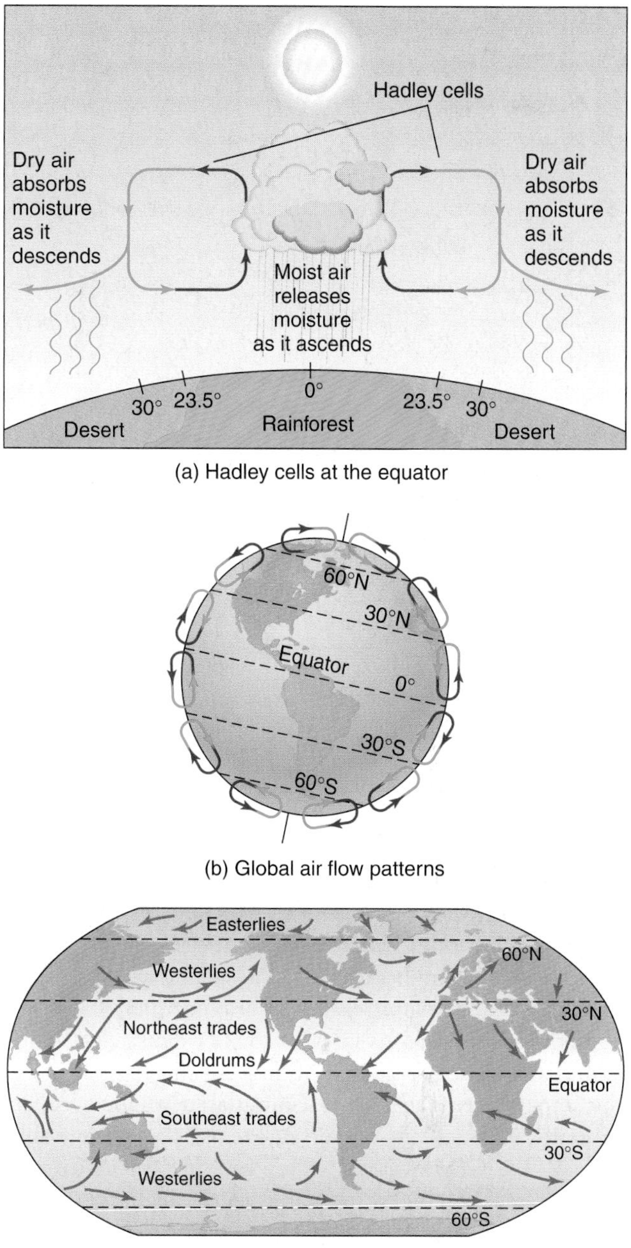

(a) Hadley cells at the equator

(b) Global air flow patterns

(c) Global trade winds

**Figure 7–6 Global air circulation.** (a) The two Hadley cells at the equator. (b) The six Hadley cells on either side of the equator, indicating general vertical airflow patterns. (c) Global trade-wind patterns, formed as a result of Earth's rotation.

with the amount that runs off is called the **infiltration–runoff ratio.**

Runoff flows over the surface of the ground into streams and rivers, which make their way to the ocean or to inland seas. All the land area that contributes water to a particular stream or river is referred to as the **watershed** for that stream or river. All ponds, lakes, streams, rivers, and other waters on the surface of Earth are called **surface waters.**

Water that infiltrates has two more alternatives (Fig. 7–3). The water may be held in the soil, in an amount that depends on the water-holding capacity of the soil.

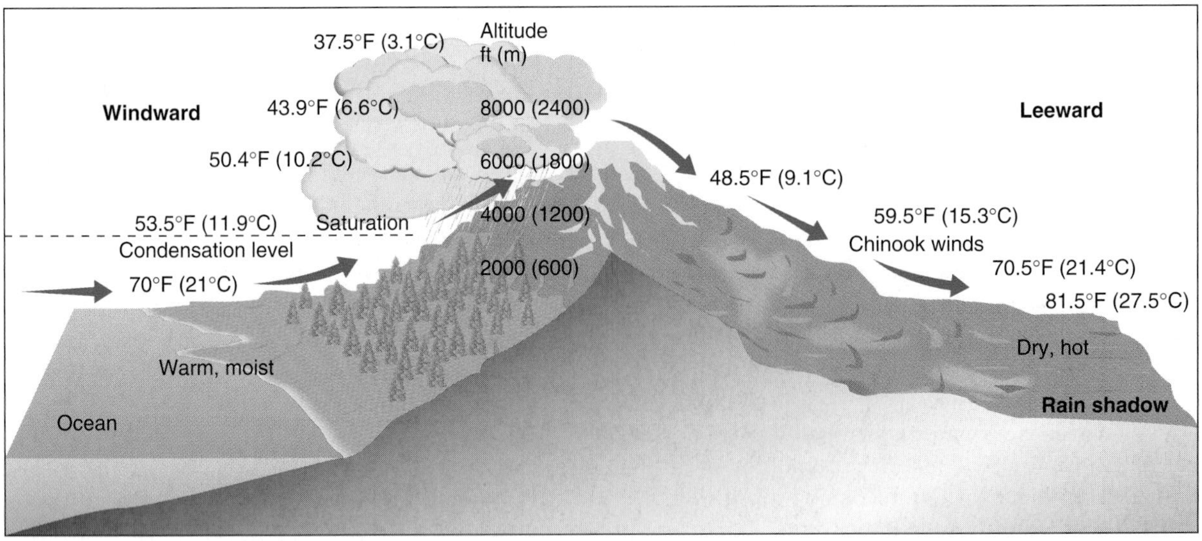


**Figure 7–7 Rain shadow.** Moisture-laden air in a trade wind cools as it rises over a mountain range, resulting in high precipitation on the windward slopes. Desert conditions result on the leeward side as the descending air warms and tends to evaporate water from the soil.

This water, called **capillary water**, returns to the atmosphere either by way of evaporation from the soil or by transpiration through plants (a green water flow). The combination of evaporation and transpiration is referred to as **evapotranspiration.**

The second alternative is **percolation** (a blue water flow). Infiltrating water that is not held in the soil is called **gravitational water**, because it trickles, or percolates, down through pores or cracks under the pull of gravity. Sooner or later, however, gravitational water encounters an impervious layer of rock or dense clay. It accumulates there, completely filling all the spaces above the impervious layer. This accumulated water is called **groundwater**, and its upper surface is the **water table** (Fig. 7–3). Gravitational water becomes groundwater when it reaches the water table in the same way that rainwater is called lake water once it hits the surface of a lake. Wells must be dug to depths below the water table. Groundwater, which is free to move, then seeps into the well and fills it to the level of the water table.

***Recharge.*** Groundwater will seep laterally as it seeks its lowest level. Where a highway has been cut through rock layers, you can frequently observe groundwater seeping out. Layers of porous material through which groundwater moves are called **aquifers**. It is often difficult to determine the location of an aquifer. Many times, layers of porous rock are found between layers of impervious material, and the entire formation may be folded or fractured in various ways. Thus, groundwater in aquifers may be found at various depths between layers of impervious rock. Also, the **recharge area**—the area where water enters an aquifer—may be many miles away from where the water leaves the aquifer. Underground aquifers hold some 99% of all *liquid* fresh water; the rest is in lakes, wetlands, and rivers.

***Purification.*** As water percolates through the soil, debris and bacteria from the surface are generally filtered out. However, water may dissolve and leach out certain minerals. Underground caverns, for example, are the result of the gradual leaching away of limestone (calcium carbonate). In most natural situations, the minerals that leach into groundwater are harmless. Indeed, calcium from limestone is considered beneficial to health. Thus, groundwater is generally high-quality fresh water that is safe for drinking. A few exceptions occur in which the groundwater leaches minerals containing sulfide, arsenic, or other poisonous elements that make the water unsafe to drink.

Drawn by gravity, groundwater may move through aquifers until it finds some opening to the surface. These natural exits may be seeps or springs. In a **seep**, water flows out over a relatively wide area; in a **spring**, water exits the ground as a significant flow from a relatively small opening. As seeps and springs feed streams, lakes, and rivers, groundwater joins and becomes part of surface water. A spring will flow, however, only if it is *lower* than the water table. Whenever the water table drops below the level of the spring, the spring will dry up.

## Pools and Fluxes in the Cycle

The hydrologic cycle consists of four physical processes: evaporation, condensation, precipitation, and gravitational flow. There are three principal loops in the cycle: (1) In the *evapotranspiration loop* (consisting of *green water*), the water evaporates and is returned by precipitation. On land, this water, the main source for natural ecosystems and rain-fed agriculture, is held as capillary water and then returns to the atmosphere by way of evapotranspiration. (2) In the *surface runoff loop* (containing blue water), the water runs across the ground surface and becomes part of the surface water system. (3) In the *groundwater loop* (also containing blue water), the water infiltrates, percolates down to join the groundwater, and then moves through aquifers, finally exiting through seeps, springs, or wells, where it rejoins the surface water. The surface runoff and groundwater loops are the usual focus for human water resource management.

In the hydrologic cycle, there are substantial exchanges of water between the land, the atmosphere, and the oceans; such exchanges are the cycle's "fluxes." Figure 7–8 shows estimates of these fluxes and of the "pools" that hold water globally. Every year, the hydrologic cycle circulates approximately 500,000 $km^3$ of water, the equivalent of 22 times all the water in the Great Lakes.

## Human Impacts on the Hydrologic Cycle

A large share of the environmental problems we face today stem from direct or indirect impacts on the water cycle. These impacts can be classified into four categories: (1) changes to Earth's surface, (2) changes to Earth's climate, (3) atmospheric pollution, and (4) withdrawals for human use.

**Changes to the Surface of Earth.** Recall from previous chapters that the direct loss of forests and other ecosystems to various human enterprises diminishes the goods and services these systems have been providing. However, the effects of human-caused changes on the water cycle may be even more profound. In most natural ecosystems, precipitation is intercepted by vegetation and infiltrates into porous topsoil. From there, water provides the lifeblood of natural and human-created ecosystems. The evapotranspiration that takes place in these ecosystems—the green water flow—not only sustains the ecosystems, but also recycles the water to local rainfall. Some of the water that infiltrates percolates down to recharge the groundwater reservoir. Then its gradual release through springs and seeps maintains the flow of streams and rivers at relatively uniform rates. The reservoir of groundwater may be sufficient to maintain a flow even during a prolonged drought. In addition, dirt, detritus, and microorganisms are filtered out as the

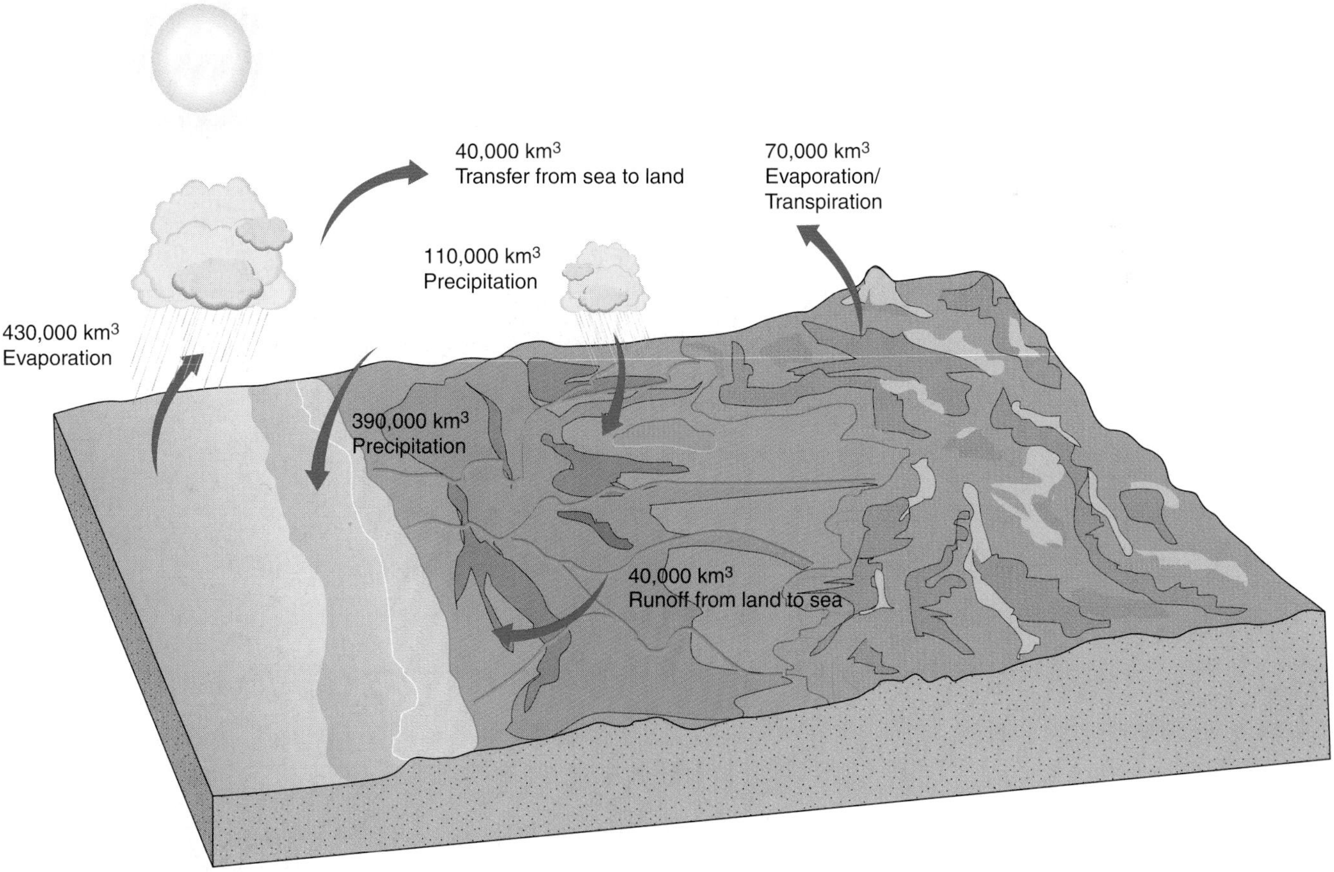


**Figure 7–8 Water balance in the hydrologic cycle.** The data show (1) the contribution of water from the oceans to the land via evaporation and then precipitation, (2) the movement of water from the land to the oceans via runoff and seepage, and (3) the net balance of water movement between terrestrial and oceanic regions of Earth. (Figure 4.2 from *Earth Science*, 8th ed., by Tarbuck and Lutgens. Copyright © 1997 by the authors. Reprinted by permission of Pearson Education, Inc., Upper Saddle River, NJ 07458.)

water percolates through soil and porous rock, resulting in groundwater that is drinkable in most cases. Similarly, streams and rivers fed by springs contain high-quality water.

As forests are cleared or land is overgrazed, the pathway of the water cycle is shifted from infiltration and groundwater recharge to runoff, so the water runs into streams or rivers almost immediately. This sudden influx of water into waterways may not only cause a flood, but also bring along sediments and other pollutants via surface erosion.

*Floods.* Floods have always been common. In many parts of the world, however, the frequency and severity of flooding are increasing—not because precipitation is greater, but because both deforestation and cultivation have increased, and they cause erosion and reduce infiltration. For example, extreme flooding in Bangladesh (a very flat country only a few feet above sea level) is now common because Himalayan foothills in India and Nepal have been deforested. Due to sediment deposited from upriver erosion, the Ganges river basin has risen 15–22 feet (5–7 m) in recent years! The 1998 flood inundated 60% of the country and caused an estimated 918 deaths (Fig. 7–9).

Increased runoff necessarily means less infiltration and therefore less evapotranspiration and groundwater recharge. Lowered evapotranspiration means less moisture for local rainfall. Groundwater may be insufficient to keep springs flowing during dry periods. Dry, barren, and lifeless streambeds are typical of deforested regions—a tragedy for both the ecosystems and the humans who are dependent on the flow. Wetlands function to store and release water in a manner similar to the way the groundwater reservoir does. Therefore, the destruction of wetlands has the same impact as deforestation: Flooding is exacerbated, and waterways are polluted during wet periods and dry up during droughts.

**Climate Change.** Groundwater in many parts of the world is actually "fossil water"; that is, it is a relic of wetter and cooler climates. There is now unmistakable evidence that Earth's climate is warming because of the rise in greenhouse gases (see Chapter 20), and as this occurs, the water cycle will be altered. A warmer climate means more evaporation from land surfaces, plants, and water bodies, because evaporation increases exponentially with temperature. A wetter atmosphere means more and, frequently, heavier precipitation and more flood events. Already, the United States and Canada have experienced a 10–15% increase in precipitation over the past 50 years (the same time frame for significant global warming to be felt). Regional and local changes are difficult to project; different computer models predict different outcomes for a given region. However, a warmer climate will likely generate more hurricanes and more droughts, which will challenge agricultural enterprises that are already under pressure to produce more food for the growing human population.

**Atmospheric Pollution.** Aerosol particles form nuclei for condensation. The more such particles there are, the greater is the tendency for clouds to form. Anthropogenic aerosols are on the increase, primarily in the form of sulfates (from sulfur dioxide in coal),

**Figure 7–9 Flooding in Bangladesh.** A flooded street in Dhaka, the capital of Bangladesh. Floods in 1998 in India and Bangladesh claimed more than 900 lives and displaced millions of people.

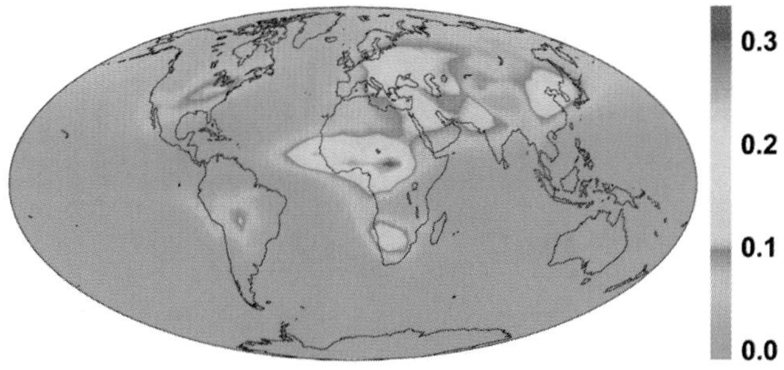


**Figure 7–10 Anthropogenic aerosols.** The global distribution of anthropogenic aerosols; where they occur, they have a cooling effect and they suppress rainfall. (Source: V. Ramathan et al., Science 294: 2120 (2001). Copyright © by AAAS. Reprinted with permission.

**table 7-2 U.S. Demands on Fresh Water**

| Use | Gallons (Liters) Used Per Person Per Day |
|---|---|
| **Consumptive** | |
| Irrigation and other agricultural use | 700 (2,800) |
| **Nonconsumptive** | |
| Electric power production | 600 (2,400) |
| Industrial use | 370 (1,500) |
| Residential use | 100 (400) |

carbon (as soot), and dust. They form a brownish haze that is associated with industrial areas, tropical burning, and dust storms. Figure 7–10 shows the global distribution of anthropogenic aerosols. Their impact on cloud formation is substantial, and where these aerosols occur, solar radiation to the earth's surface is reduced (so they have a cooling effect). Their most significant impact, however, is on the hydrologic cycle. The unique size spectrum of the anthropogenic aerosols causes them actually to suppress rainfall where they occur in abundance, even though they encourage cloud formation. As they do so, the atmospheric cleansing that would normally clear the aerosols is suppressed, and they remain in the atmosphere longer than usual. With suppressed rainfall come drier conditions, so more dust and smoke (and more aerosols) are the result. Note that this works just opposite to the impact of the greenhouse gases on climate and on the hydrologic cycle. Again, there are significant regional differences; the aerosol impact is more local, whereas the impact of greenhouse gases on climate and the hydrologic cycle is more global. In addition, the greenhouse gases are gradually accumulating in the atmosphere, while the aerosols have a lifetime measured in days.

**Withdrawing Water Supplies.** Finally, there are many problems related to withdrawals of water for human use, as we make use of surface waters and groundwater. Humans are major players in the hydrologic cycle, as is discussed in the next section.

# 7.3 Water: A Resource to Manage, a Threat to Control

## Uses and Sources

**Uses.** The major uses of fresh water in the United States are listed in Table 7–2. Most of the water used in homes and industries is for washing and flushing away unwanted materials, and the water used in electric power production is for taking away waste heat and increasing the efficiency of the process. These are **nonconsumptive** uses, because the water, though now contaminated with the wastes, remains available to humans for the same or other uses if its quality is adequate or if it can be treated to remove undesirable materials. In contrast, irrigation is a **consumptive** use, because the applied water does not return to the water resource. It can only percolate into the ground or return to the atmosphere through evapotranspiration. In either case, the water does reenter the overall water cycle, but is gone from human control.

Worldwide, the largest use of water is for irrigation (70%); second is for industry (20%) and third is for direct human use (10%). These percentages vary greatly from one region to another, depending on natural precipitation and the degree to which the region is developed (Fig. 7–11). In the United States, irrigation accounts for 85% of freshwater consumption. The total global water withdrawal in 2000 was 4,430 $km^3$, a ninefold increase since 1900. Annual global withdrawal is expected to rise by about 10% every decade. Most increases in withdrawal are due to increases in irrigated lands.

**Sources.** In the United States, about half (53%) of domestic water comes from groundwater sources, and half (47%) comes from surface waters (rivers, lakes, reservoirs). Before municipal water supplies were developed, people drew their water from whatever source they could, such as wells, rivers, lakes, or rainwater. This approach is still used in rural areas in the developing countries. Women in many of these countries walk long distances each day to fetch water. Because surface waters and shallow wells often receive runoff, they frequently are polluted with various wastes, including animal excrement and human sewage likely to contain pathogens (disease-causing organisms). In fact, an estimated 90% of wastewater in developing countries is released to surface waters

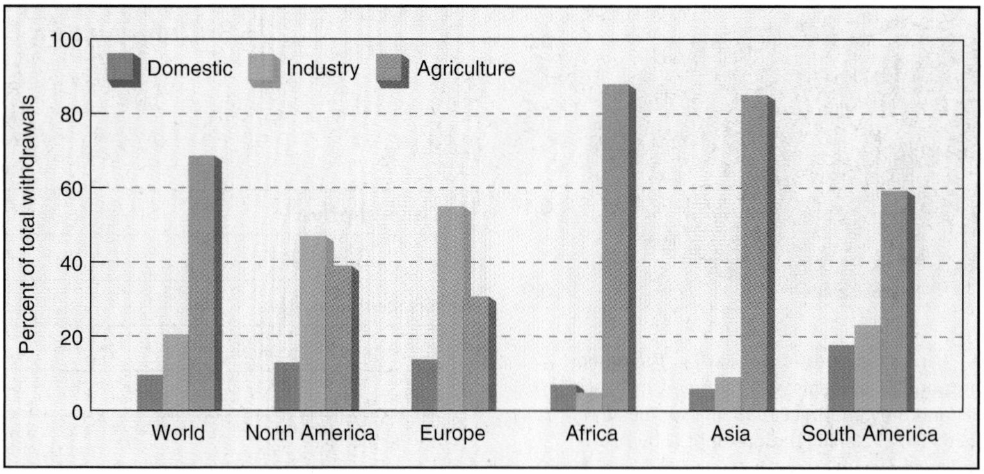


**Figure 7–11 Human usage of water.** The percentage used in each category varies with climate and relative development of the country. A less developed region with a dry climate (e.g., Africa) uses most of its water for irrigation, whereas an industrialized region (e.g., Europe) requires the largest percentage for industry.

without any treatment. Yet, unsafe as it is, this polluted water is the only water available for an estimated 1.1 billion people in less developed countries (Fig. 7–12). It is commonly consumed without purification, but not without consequences: According to the World Health Organization's (WHO's) *World Health Report 2002*, contaminated water is responsible for the deaths of over 3 million people per year, 90% of whom are children. Recall from Chapter 6 that Millennium Development Goal #7 is to reduce by half the proportion of people without access to safe drinking water.

**Figure 7–12 Water in the developing world.** In many villages and cities of the developing world, people withdraw water from rivers, streams, and ponds. These sources are often contaminated with pathogens and other pollutants. This woman in Burkina Faso is collecting water from a pond used by people and animals.

***Technologies.*** In the industrialized countries, the collection, treatment, and distribution of water is highly developed. Larger municipalities rely primarily on surface-water sources, while smaller water systems tend to use groundwater. In the former, dams are built across rivers to create reservoirs, which hold water in times of excess flow and can be drawn down at times of lower flow. In addition, dams and reservoirs may offer power generation, recreation, and flood control. Water for municipal use is piped from the reservoir to a treatment plant, where it is treated to kill pathogens and remove undesirable materials, as shown in Fig. 7–13 (see "Earth Watch" essay, p. 192). After treatment, the water is distributed through the water system to homes, schools, and industries. Wastewater, collected by the sewage system, is carried to a sewage-treatment plant, where it is treated before being discharged into a natural waterway. (See Chapter 17.) Often, wastewater is discharged into the same river from which it was withdrawn, but farther downstream.

Whenever possible, both water and sewage systems are laid out so that gravity maintains the flow through the system. This arrangement minimizes pumping costs and increases reliability. On major rivers, such as the Mississippi, water is reused many times. Each city along the river takes water, treats it, uses it, and then returns it to the river. In developing nations, the wastewater frequently is discharged with minimal or no treatment. Thus, as the water moves downstream, each city has a higher load of pollutants to contend with than the previous city had, and ecosystems at the end of the line may be severely affected by the pollution. Pollutants include industrial wastes, as well as pollutants from households, because industries and residences generally utilize the same water and sewer systems.

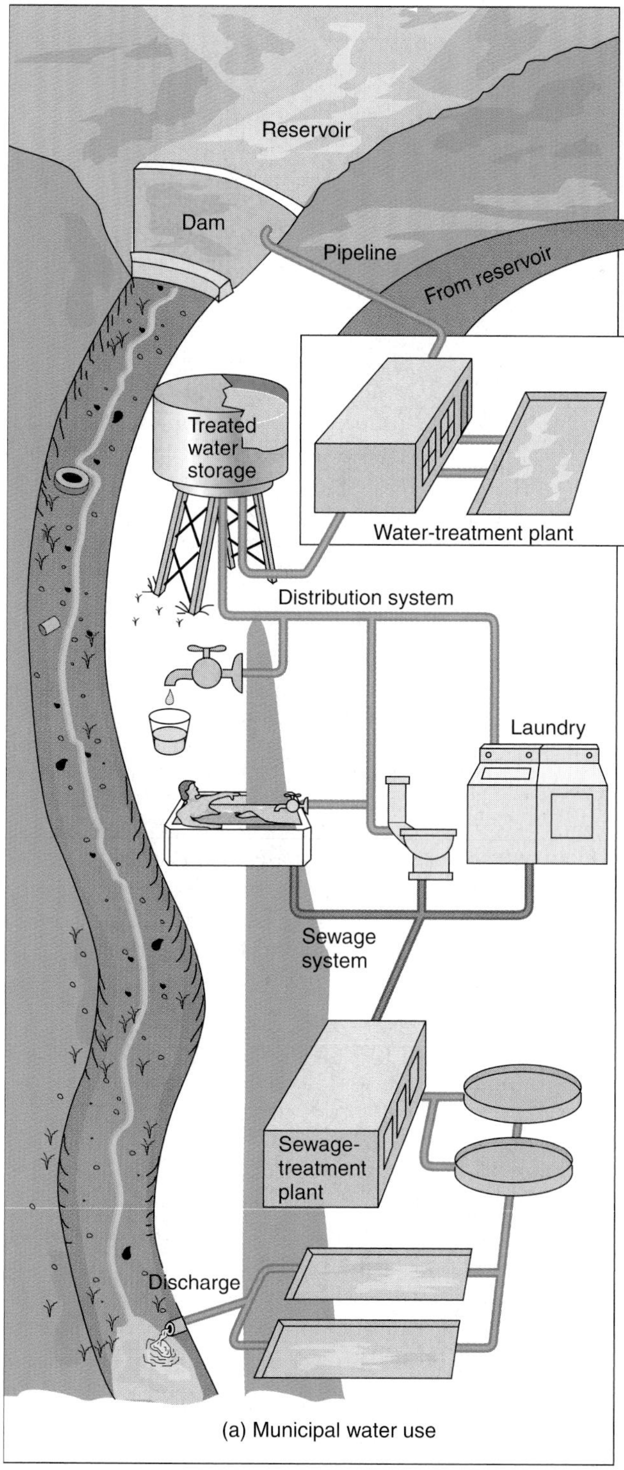


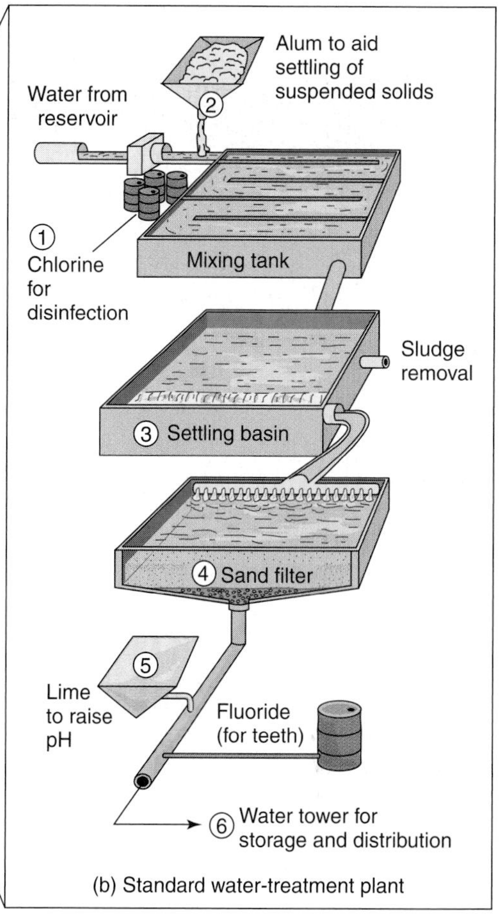


**Figure 7–13 Municipal water use and treatment.** (a) Water is often taken from a river or reservoir, piped to the treatment plant, treated, used, and then returned to the source. (b) At the treatment plant, (1) chlorine is added to kill bacteria, (2) alum (aluminum sulfate) is introduced to coagulate organic particles, and (3) the water is put into a settling basin for several hours to allow the coagulated particles to settle. The water is then (4) filtered through sand, (5) treated with lime to adjust the pH, and (6) put into a storage water tower or reservoir for distribution to your home.

Reservoirs created by dams on rivers are also major sources of water for irrigation. In this case, no treatment is required.

Smaller public drinking-water systems frequently rely on groundwater, which often may be used with minimal treatment because it has already been purified by percolation through soil. More than $1\frac{1}{2}$ billion people around the world rely on groundwater for their domestic use. Both surface water and groundwater are replenished through the water cycle (Figure 7–3). Therefore, in theory at least, these two sources of water represent a sustainable or renewable (self-replenishing) resource, but they are not inexhaustible. Over the last 50 years, groundwater use has outpaced the rate of aquifer recharge in many areas of the world. Also, the death of the Aral Sea (discussed at the beginning of the chapter) demonstrates that surface waters are being overdrawn as well. The consequences of human use of both surface and groundwater resources are discussed in the next subsection.

**earth watch**

## Water Purification

*Polluted water* contains one or more materials that make it unsuitable for a given use. *Water purification* is any method that will remove one or more of the materials. Several methods may be used in combination to obtain water that is sufficiently pure for a given use. The following are methods that are commonly used to purify water:

1. **Settling.** Soil particles and other solid material carried by flowing water may be removed by holding the water still and allowing the solids to settle. Clarified water is removed from the top. Settling may be aided by the addition of alum (aluminum sulfate). The +3 charge on aluminum ions pulls clay and other particles, which are negatively charged, into clumps that settle more readily than do the individual particles.
2. **Filtration.** Filtration is the passage of water through a porous medium. Any materials larger than the pores will be filtered out. A bed of sand is often used for this purpose.
3. **Adsorption.** Passing water through an adsorbing material that binds and holds other materials on its surface will remove certain pollutants. Activated charcoal (carbon) is commonly used in this way to remove organic contaminants from water or air.
4. **Biological oxidation.** Organic material (detritus and organisms) is fed upon by detritus feeders and decomposers, broken down in cell respiration, and thus removed. The passage of water through systems supporting the growth of such organisms removes organic material. (See Chapter 17.)
5. **Distillation.** Distillation is the evaporation and condensation of water. All materials present in the water before evaporation remain behind in the holding tank and are therefore not present when the water vapor condenses.
6. **Disinfection.** Water is treated with chlorine or other agents that kill disease-causing organisms.

The natural water cycle includes all of these purification methods except disinfection. Sitting in lakes, ponds, or the oceans, water is subject to settling. As it percolates through soil or porous rock, it is filtered. Soil and humus are good chemical adsorbents. As water flows down streams and rivers, detritus is removed by biological oxidation. As water evaporates and condenses, it is distilled.

Thus, numerous sources of fresh water might be safe to drink were it not for human pollution. The most serious threat to human health is contamination with disease-causing organisms and parasites, which come from the excrement of humans, their domestic animals, and wild animals. In human settlements, these organisms get into water and are passed on to people before any of the natural purification processes have a chance to work.

## Surface Waters

Glen Canyon Dam closed its gates in 1963 and began storing excess water in Lake Powell, the second-largest reservoir in the United States (Fig. 7–14). The dam spans the Colorado River at Lee's Ferry, Arizona, just above Grand Canyon National Park. Operated by the federal Bureau of Reclamation, the dam generates hydropower, and the reservoir stores water for distribution to Arizona, California, Nevada, and Mexico. During the dam's first two decades of operation, the water released was strictly designed to maximize hydropower; it could fluctuate manyfold in flow in the course of a day.

***Doing it Right.*** An environmental impact study in the late 1980s and early 1990s concluded that the operation of the dam had seriously damaging impacts on the downstream ecology of the Colorado River and its recreational resources (hiking, river rafting). In response, in 1996, Secretary of the Interior Bruce Babbitt issued new rules that established minimum and maximum water-release rates designed to enhance the ecological and recreational values of the river, rather than maximize hydropower revenues. The Grand Canyon Monitoring and Research Center (GCMRC) now provides scientific monitoring for the dam's operations and operates within the guidelines of adaptive ecosystem management. Many natural and social science projects have been conducted subsequently, and the project includes members from six Native American tribes living in and near the Grand Canyon. The outcome has been a public-policy success, with the dam now meeting the ecological and social needs of the river ecosystem and the people who live around it or visit it.

To trap and control flowing rivers, more than 40,000 large dams have been built around the world. Some 3,000 of these dams contain reservoirs with storage volumes greater than 95 billion liters (25 billion gallons), inundating 49 million hectares (120 million acres) of land and containing more than 6,250 km$^3$ (1,500 mi$^3$) of water. Of the 40,000 km$^3$ of annual runoff in the hydrologic cycle (Fig. 7–8), only 31% is accessible for withdrawal, because many rivers are in remote locations and because much

**Figure 7–14 Glen Canyon Dam.** Glen Canyon dam blocks the Colorado River just north of Grand Canyon National Park. Built to provide hydroelectric power and water to states in the Colorado River basin, the dam holds back Lake Powell, the second-largest reservoir in the United States.

of the remaining water is required for navigation, flood control, and the generation of hydroelectric power. Calculations indicate that human use now appropriates 6,780 $km^3$ per year, which represents 54% of the accessible freshwater runoff on Earth.

*Dam Impacts.* The United States is home to some 75,000 dams at least six feet in height and an estimated 2 million smaller structures. These have been built to run mills (now an obsolete usage), control floods, and provide water for municipal and agricultural use. These dams have enormous ecological impacts. When a river is dammed, valuable freshwater habitats, such as waterfalls, rapids, and prime fish runs, are lost. When the river's flow is diverted to cities or croplands, the waterway below the diversion is deprived of that much water. The impact on fish and other aquatic organisms is obvious, but the ecological ramifications go far beyond the river. Wildlife that depends on the water or on food chains involving aquatic organisms is also adversely affected. Wetlands occupying floodplains along many rivers, no longer nourished by occasional overflows, dry up, resulting in frequent die-offs of waterfowl and other wildlife that depended on those habitats (Fig. 7–15). Fish such as salmon, which swim from the ocean far upriver to spawn, are seriously affected by the reduced water level and have problems getting around the dam, even one equipped with fish ladders (a stepwise series of pools on the side of the dam, where the water flows in small falls that fish can negotiate). If the fish do get upriver, the hatchlings have similar problems getting back to the ocean. On the Columbia and Snake Rivers, juvenile salmon suffer a 95% mortality in their journey to the sea as a result of negotiating the dams and reservoirs that block their way.

*San Francisco Bay.* The problems extend to **estuaries**—bays in which fresh water from a river mixes with seawater. Estuaries are among the most productive ecosystems on Earth; they are rich breeding grounds for many species of fish, shellfish, and waterfowl. As a river's flow is diverted to, say, irrigated fields, less fresh water enters and flushes the estuary. Consequently, the

(a)

(b)

**Figure 7–15 Ecological effects of diversion.** (a) Ducks and geese at Upper Klamath Lake, Oregon. Vast flocks of waterfowl such as these are supported by wetlands. (b) What happens to these birds when wetlands dry up because of water diversion or a falling water table?

salt concentration increases, profoundly affecting the estuary's ecology. The San Francisco Bay is a prime example. Over 60% of the fresh water that once flowed from rivers into the bay has been diverted for irrigation in the Central Valley (4.5 million acres irrigated) and for municipal use in Southern California (22 million people served). Without the freshwater flows, salt water from the Pacific has intruded into the bay, with devastating consequences. Chinook salmon runs, for example, have almost disappeared. Sturgeon, Dungeness crab, and striped bass populations are greatly reduced. Exotic species of plants and invertebrates have replaced many native species, and the tidal wetlands have been reduced to only 8% of their former extent. The CALFED-Bay Delta Program has been established to "develop and implement a long-term comprehensive plan that will restore ecological health and improve water management for beneficial uses of the Bay-Delta System," as its mission statement reads. The estuary is huge and the CALFED mission is commendable. The question now is, Is restoration possible, or will CALFED have to settle for some degree of "rehabilitation"?

The problem is not limited to the United States. The southeastern end of the Mediterranean Sea was formerly flushed by water from the Nile River. Because this water is now held back and diverted for irrigation by the Aswan High Dam in Egypt, that part of the Mediterranean is suffering severe ecological consequences. Recall, too, the profound consequences of diverting the waters that once fed the Aral Sea.

## Groundwater

Within the seven-state High Plains region of the United States, the Ogallala aquifer supplies irrigation water to 10.4 million acres (4.2 million hectares), one-fifth of the irrigated land in the nation (Fig. 7–16). This aquifer is probably the largest in the world, but it is mostly "fossil water," recharged during the last ice age. The recent withdrawal rate has been about 23 million acre-feet (28 billion cubic meters) per year, two orders of magnitude higher than the recharge rate. Water tables have dropped 100–200 ft (30–60 meters) and are lowering at 6 ft per year. Irrigated farming has already come to a halt in some sections, and it is predicted that over the next 20 years another 3 million acres (1.2 million hectares) in this region will be abandoned or converted to dryland farming (ranching and the production of forage crops) because of water depletion.

As mentioned earlier, some 99% of all liquid fresh water is in underground aquifers. Of this groundwater, more than three-fourths is like the Ogallala aquifer—with a recharge rate of centuries or more. Such groundwater is considered *nonrenewable. Renewable groundwater* is replenished by the percolation of precipitation water, so it is vulnerable to variations in precipitation. In tapping groundwater, we are tapping a large, but not unlimited, natural reservoir. Its sustainability ultimately depends on balancing withdrawals with rates of recharge. For most of the groundwater in arid regions, there is essentially no recharge. The resource must be considered nonrenewable. Like oil, it can be removed, but any water removed today means that it will be unavailable for the future. Groundwater depletion has several undesirable consequences.

**Falling Water Tables.** Rates of groundwater recharge aside, the simplest indication that groundwater withdrawals are exceeding recharge is a falling water table, a situation that is common throughout the world. Because irrigation consumes far and away the largest amount of fresh water, depleting water resources will ultimately have its most significant impact on crop production. Although running out of water is the obvious eventual conclusion of overdrawing groundwater, falling water tables have other consequences.

**Diminishing Surface Water.** Surface waters are also affected by falling water tables. In various wetlands, for instance, the water table is essentially at or slightly above the ground surface. When water tables drop, these wetlands dry up, with the ecological results described earlier. Further, as water tables drop, springs and seeps dry up as well, diminishing even streams and rivers to the point of dryness. Thus, excessive groundwater removal creates the same results as the diversion of surface water.

**Land Subsidence.** Over the ages, groundwater has leached cavities in the ground. Where these spaces are filled with water, the water helps support the overlying rock and soil. As the water table drops, however, this support is lost. Then there may be a gradual settling of the land, a phenomenon known as **land subsidence**. The rate of sinking may be 6–12 inches (10–15 cm) per year. In some areas of the San Joaquin Valley in California, land has settled as much as 29 feet (9 m) because of groundwater removal. Land subsidence causes building foundations, roadways, and water and sewer lines to crack. In coastal areas, subsidence causes flooding, unless levees are built for protection. For example, where a 4,000-square-mile (10,000-km$^2$) area in the Houston–Galveston Bay region of Texas is gradually sinking because of groundwater removal, coastal properties are being abandoned as they are gradually being inundated by the sea. Land subsidence is also a serious problem in New Orleans, sections of Arizona, Mexico City, and many other places throughout the world.

***Oops!*** Another kind of land subsidence, a **sinkhole**, may develop suddenly and dramatically (Fig. 7–17). A sinkhole results when an underground cavern, drained of its supporting groundwater, suddenly collapses. Sinkholes may be at least 300 feet (91 m) across and as much as 150 feet deep. The problem of sinkholes is particularly severe in the southeastern United States, where

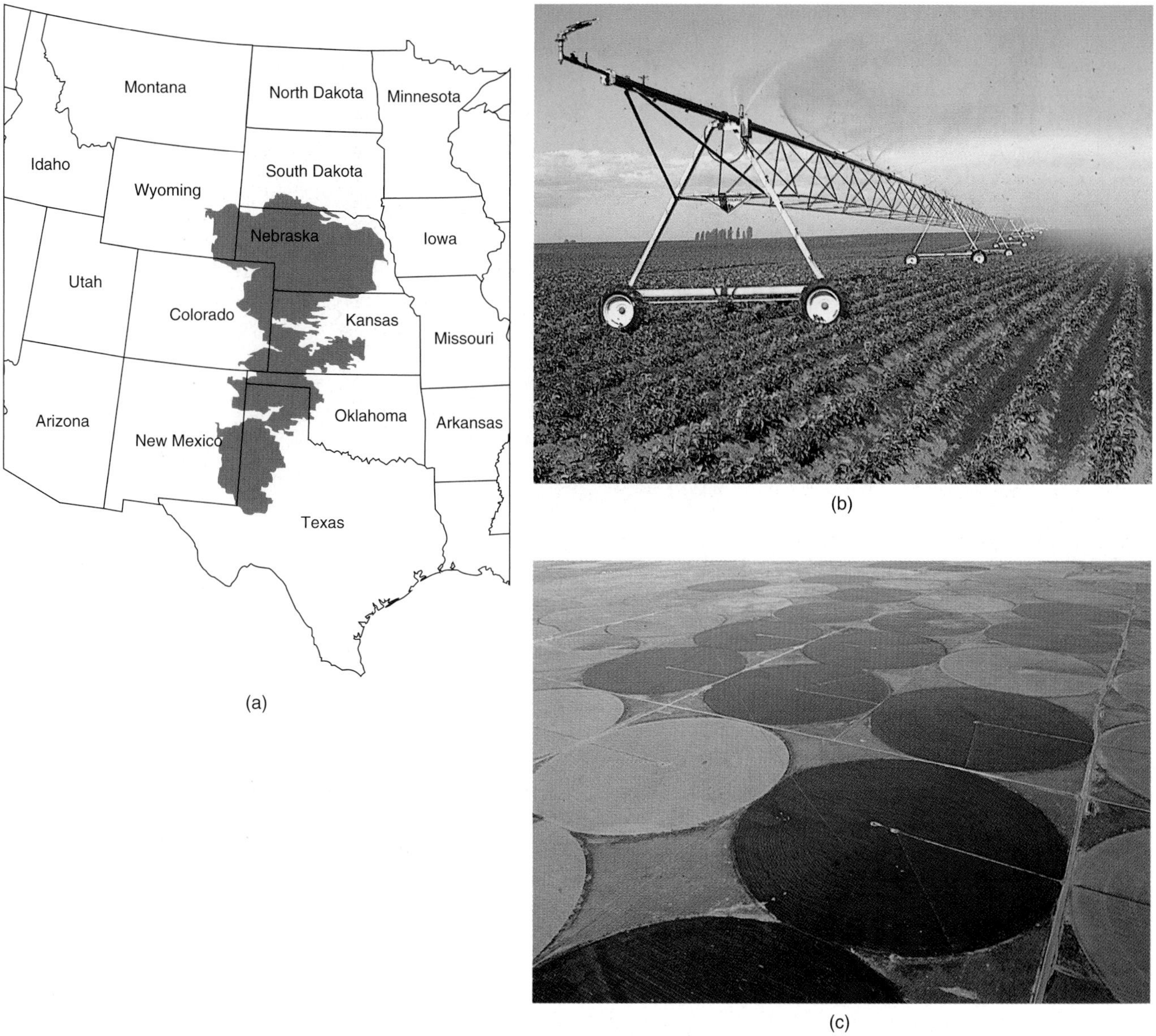


**Figure 7–16 Exploitation of an aquifer.** (a) Pumping up water from the Ogallala aquifer has made this arid region of the United States into some of the most productive farmland in the country. (b) Water is applied by means of center-pivot irrigation, in which the water is pumped from a central well to a self-powered boom that rotates around the well, spraying water as it goes. (c) An aerial photograph shows the extent of center-pivot irrigation throughout the region. Groundwater depletion will bring an end to this kind of farming.

groundwater has leached numerous passageways and caverns through ancient beds of underlying limestone. An estimated 4,000 sinkholes have formed in Alabama alone, some of which have "consumed" buildings, livestock, and sections of highways.

**Saltwater Intrusion.** Another problem resulting from dropping water tables is **saltwater intrusion**. In coastal regions, springs of outflowing groundwater may lie under the ocean. As long as a high water table maintains a sufficient head of pressure in the aquifer, fresh water will flow into the ocean. Thus, wells near the ocean yield fresh water (Fig. 7–18a). However, lowering the water table or removing groundwater at a rapid rate may reduce the pressure in the aquifer, permitting salt water to flow back into the aquifer and hence into wells (Fig. 7–18b). Saltwater intrusion is a problem at many locations along U.S. coasts.

## 7.4 Water Stewardship: Public-Policy Challenges

### Obtaining More Water

The hydrologic cycle is entirely adequate to meet human needs for fresh water, because it processes several times as much water as we require today. However,

**Figure 7–17 Sinkhole.** The removal of groundwater may drain an underground cavern until the roof, no longer supported by water pressure, collapses. The result is the sudden development of a sinkhole, such as this one, which consumed a home in Frostproof, Florida, July 12, 1991.

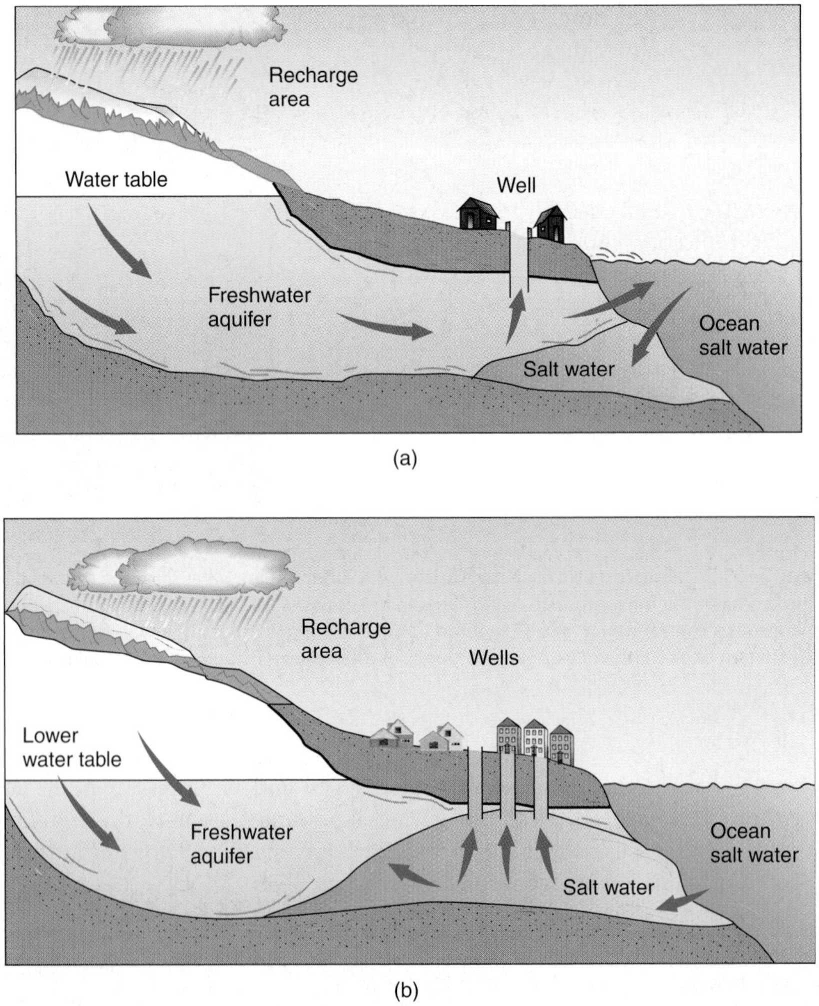


**Figure 7–18 Saltwater intrusion.** (a) Where aquifers open into the ocean, fresh water is maintained in the aquifer by the head of fresh water inland. (b) Excessive removal of water may reduce the pressure, so that salt water moves into the aquifer.

the water is often not distributed where it is most needed, and the result is persistent scarcity of water in many parts of the world. In the developing world, there is still a deficit of infrastructure, such as wells, water treatment systems, and (sometimes) large dams, for capturing and distributing safe drinking water. Despite the growing negative impacts of overdrawing water resources, expanding populations create an ever-increasing demand for additional water for irrigation, industry, and municipal use.

What are the possibilities of meeting these existing needs and growing demands in a sustainable way? Actually, there are only about four options: (1) capture more of the runoff water, (2) gain better access to existing groundwater aquifers, (3) desalt seawater, and (4) conserve present supplies by using less water.

**More Dams to Capture Runoff.** Some 260 new dams come on line each year around the world, a figure that is down from 1,000 per year in the mid 20th century. Certainly, there are opportunities to capture more of the seasonal floodwaters that rush to the sea in a short period of time, and a combination of flood control, water storage, and hydropower can be a powerful argument for dam construction.

*Three Gorges Dam.* A spectacular example currently unfolding is the construction of the Three Gorges Dam across a scenic stretch of the Yangtze River in China (Fig. 7–19). The project is the centerpiece of the Chinese government's effort to industrialize and join the modern age. When completed in 2013, the dam will be the largest in the world, generating 18,000 MW of electricity. It is also expected to provide control over the disastrous flooding of a river that took more than 300,000 lives in the 20th century. Over 1.2 million people—including entire cities, farms, homes, and factories—will be displaced and relocated to make way for the 370-mile-long (600-km-long) reservoir. Critics point to the enormous human, ecological, and aesthetic costs of the dam and claim that alternative sources of electric power are cheaper and more readily available. An international campaign to stop the dam has prevented many funding agencies, like the World Bank, from getting involved, but Chinese officials have effectively stifled internal criticism and are proceeding with work on the dam.

*Dam Problems.* Increasingly, people are recognizing the inevitable trade-offs that occur with such projects and are considering the trade-offs to be unacceptable. Dams disrupt the integrity of river systems, breaking them up into fragments of river interrupted by standing water. Increasingly, existing dams are being challenged. In fact, some 500 have already been dismantled in the United States, and others await the same fate. Removing a dam is not easy, however. Legal complexities abound where long-existing uses of a dam (to control floods or provide irrigation water or power) conflict with the expected advantages of removal (reestablishing historic fisheries such as salmon and steelhead runs and restoring the river for recreational and aesthetic use). Practical problems also exist. Often, the reservoir behind a dam has received massive amounts of sediment from upriver, and removing the dam exposes the sediments and washes them downstream. For example, a proposal to remove low dams on the Blackstone River in Massachusetts was abandoned in the early 1990s because it was found that sediments built up behind the dam were contaminated with heavy metals released from upstream factories.

*Wild Rivers.* Protection has been accorded some rivers with the passage of the Wild and Scenic Rivers Act of 1968, which keeps rivers designated as "wild and scenic" from being dammed or affected by other harmful operations. Some 11,300 miles of rivers have been protected, but more than 60,000 miles qualify for protection under the act. Those rivers designated as wild and scenic are in many ways equivalent to national parks; thus, like national parks, they need public supporters and

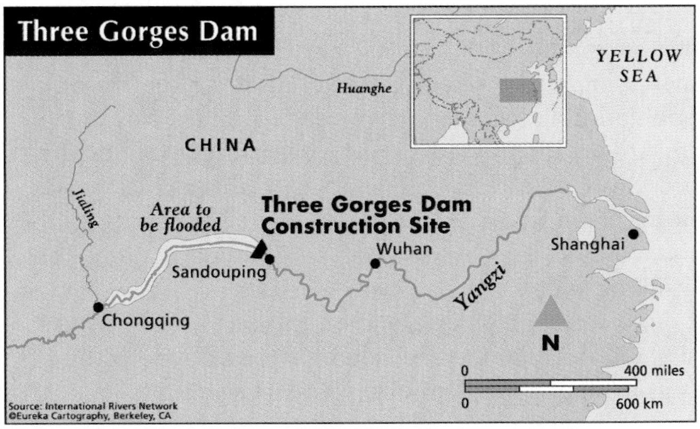


**Figure 7–19 Three Gorges Dam.** This dam, under construction in China on the upper Yangtze River, will be the world's largest when it is finished.

defenders, such as the organization American Rivers. There are consistent attempts on the part of special interests (the hydropower industry, agriculture) to undo recent work establishing wild and scenic rivers. The latest effort is in a proposed energy bill that exempts dam owners from adhering to existing environmental standards.

**Tapping More Groundwater.** Already, more than 2 billion people depend on groundwater supplies. In many areas, groundwater use exceeds aquifer recharge, leading to shortages as the water table drops below pump levels. Groundwater depletion is considered the single greatest threat to irrigated agriculture, and it is happening in many parts of the world. India, China, West Asia, countries of the former Soviet Union, the American west, and the Arabian Peninsula are all experiencing declining water tables.

Exploiting renewable groundwater will continue to be an option, but it is unlikely to provide great increases in water supply, because it is these same sustainable supplies, recharged by annual precipitation, that are being increasingly polluted. Agricultural chemicals such as fertilizers and pesticides, animal wastes, and industrial chemicals readily enter groundwater, making groundwater pollution as much of a threat to domestic water use as depletion. For example, some 60,000 $km^2$ of aquifers in western and central Europe will likely be contaminated with pesticides and fertilizers within the next 50 years. Also, India and Bangladesh are plagued with arsenic-contaminated groundwater, which has extracted a heavy toll in sickness and death for millions in those countries.

Exploiting nonrenewable groundwater carries with it all the consequences already mentioned with regard to renewable sources and is not a sustainable option.

**Desalting Seawater.** The world's oceans are an inexhaustible source of water, not only because they are vast, but also because any water removed from them will ultimately flow back in. Not many plants can be grown in full seawater, however, and although researchers are developing salt-tolerant plants via genetic modification, we are not there yet. Thus, with increasing water shortages and most of the world's population living near coasts, there is a growing trend toward **desalination** (desalting) of seawater for domestic use. More than 13,600 desalination plants already exist, primarily in Saudi Arabia, Israel, and other countries of the Middle East, providing 6.8 billion gallons of drinking water.

Two technologies—microfiltration (reverse osmosis) and distillation—are commonly used for desalination. Small desalination plants generally employ microfiltration, in which great pressure forces seawater through a membrane filter fine enough to remove the salt. Larger facilities, particularly those which can draw from a source of waste heat (for example, from electrical power plants), generally use distillation (evaporation and recondensation of vapor). Efficiency is gained by using the heat given off by condensing water to heat the incoming water. Even where waste heat is used, however, the costs of building and maintaining the plant, which is subject to corrosion from seawater, are considerable. Under the best of circumstances, the production of desalinized water costs about \$2 per 1,000 gallons (4,000 L). This is two to four times what most city dwellers in the United States currently pay, but it is still not a high price to pay for drinking water.

***Tampa Bay Leads the Way.*** In March, 2003, the Tampa Bay Seawater Desalination Plant came on line, signaling a new day for desalination (Fig. 7–20). The plant is the country's first ever built to serve as a primary water source. The plant uses salty cooling water from a power plant and reverse osmosis to produce 25 million gallons of drinking water a day, providing Pasco County, Pinellas County, St. Petersburg, and other cities with some 10% of their drinking water. The Tampa Bay region has experienced chronic water shortages and is exploiting an opportunity that has attracted worldwide attention. The water is expected to cost consumers \$2.49 per 1,000 gallons, wholesale over 30 years, more than the \$1 per 1,000 gallons for existing groundwater sources, but still a reasonable cost.

Although the higher cost might cause some people to cut back on watering lawns and to implement other conservation measures, most people in the United States could afford desalinized water without unduely altering

**Figure 7–20 Desalination plant.** The Tampa Bay Seawater Desalination Plant came on line in 2003 and now delivers 25 million gallons of drinking water a day. It is the country's first such plant built to be a city's primary water source.

their lifestyles. For irrigating croplands, however, the higher cost of desalinized seawater would probably be prohibitive.

In conclusion, it does not appear that a sustainable future can be found in desalination, large-scale groundwater exploitation, or river diversion projects. Fortunately, there are alternatives, based on the conservation and recycling of water.

## Using Less Water

A developing-nation family living where water must be carried several miles from a well finds that one gallon per person per day is sufficient to provide for all of its essential needs, including cooking and washing. Yet, a typical household in the United States consumes an average of 106 gallons (400 L) per person per day. If all indirect uses are added (including irrigation), this figure increases to 1,350 gallons (5,100 L) per person per day. Similarly, a peasant farmer may irrigate by carefully ladling water onto each plant with a dipper, while typical modern irrigation floods the whole field. The way water resource planners think is beginning to change, however: Instead of asking how much water we need and where can we get it, people are now asking how much water is available and how we can best use it. The good news is that the rate of water use per capita has actually begun to drop, which can be attributed to some well-needed water conservation strategies already put in place. Some specific measures that are being implemented to reduce water withdrawals are described in the next several subsections.

**Agriculture.** Some 40% of the world's food is grown in irrigated soils, so agriculture is far and away the largest consumer of fresh water. Most present-day irrigation wastes huge amounts of water. In fact, half of it never yields any food. Where irrigation water is applied by traditional flood or center-pivot systems, 30–50% is lost to evaporation, percolation, or runoff. Several strategies have been employed recently to cut down on this waste. One is the *surge flow* method, in which computers control the periodic release of water, in contrast to the continuous-flood method. Surge flow can cut water use in half.

*Drip, Drip.* Another water-saving method is the *drip irrigation* system, a network of plastic pipes with pinholes that literally drip water at the base of each plant (Fig. 7–21). Although such systems are costly, they waste much less water. Also, they have the added benefit of retarding salinization. (See Chapter 8.) Studies have shown that drip irrigation can reduce water use by 30 to 70%, while actually increasing crop yields, compared with traditional flooding methods. Although drip irrigation is spreading worldwide, especially in arid lands such as Israel and Australia, 97% of the irrigation in the United States and 99% throughout the world is still done by traditional flood or center-pivot methods.

**Figure 7–21 Drip irrigation.** Irrigation consumes the most water. Drip irrigation offers a conservative method of applying water.

The reason that so few farms have changed over to drip irrigation systems is that it costs about $1,000 per acre to install them. In comparison, water for irrigation is heavily subsidized by the government, so farmers pay next to nothing for it. Therefore, it makes financial sense to use the cheapest system for distributing water, even if it is wasteful. Calculations of construction costs and energy subsidies to provide irrigation water to farmers indicate an annual subsidy of $4.4 billion for the 11 million acres of irrigated land in the western United States, an average of $400 an acre. Governments collect an average of 10% or less of the actual cost in water fees charged to farmers. Reducing this subsidy would greatly encourage water conservation through the use of more efficient irrigation technologies. For example, the Broadview Water District in California has instituted a tiered pricing structure in which farmers are charged their customary fee for the first 90% of their average water use and much higher prices for the remaining 10%. Given this incentive, farmers have conserved from 9% to 31% on water use, depending on the crop, and crop yields have not been affected at all.

*Treadle Pumps.* In the developing countries, irrigation often bypasses the rural poor, who are at greatest risk of hunger and malnutrition. Affordable irrigation technologies can be remarkably successful, as in Bangladesh, where low-cost treadle pumps (Fig. 7–22) enable farmers to irrigate their rice paddies and vegetable fields at a cost of less than $35 a system. The pump works like a step exercise machine, is locally manufactured, and has been adopted by the millions.

**Figure 7–22 Treadle pump.** This treadle pump in Bangladesh is operated like an exercise machine. Over a million of these pumps now allow rural farmers to raise crops during dry seasons.

The irrigation it affords makes it possible for the farmers to irrigate small plots during dry seasons from groundwater lying just a few feet below the surface. Treadle pumps have increased the productivity of more than 600,000 acres of farmland in Bangladesh.

**Municipal Systems.** The 106 gallons of water each person consumes per day in modern homes is used mostly for washing and removing wastes—that is, for flushing toilets (3–5 gallons per flush), taking showers (2–3 gallons per minute), doing laundry (20–30 gallons per wash), and so on. Watering lawns, filling swimming pools, and other indirect consumption only add to this use.

Water conservation has long been promoted as a "save the environment" measure, though without much effect. Now numerous cities are facing the stark reality that it will be extremely expensive, and in many cases impossible, to increase supplies by the traditional means of building more reservoirs or drilling more wells. The only practical alternative, they are discovering, is to take real steps toward reducing water consumption and wastage. A considerable number of cities have programs whereby leaky faucets will be repaired and low-flow showerheads and water-displacement devices in toilets will be installed free of charge. Phoenix is paying homeowners to replace their lawns with **xeriscaping**—landscaping with desert species that require no additional watering—and the business is thriving. Many cities and towns ban certain uses of water when droughts reduce the available supply. Brown lawns and dusty automobiles are the result, signs that a city's water supply is threatened.

*A Flush World.* In 1997, the last phase of a regulation authorized by the 1992 National Energy Act took effect, and it became illegal to sell 6-gallon commodes. In their place is the new wonder of the flush world: the 1.6-gallon toilet (Fig. 7–23). When the new toilets first came into use, homeowners found that they could no longer depend on an easy flush; in fact, plumbers would usually offer a free plunger with installation of the early 1.6-gallon models. Newer versions, however, work perfectly well and save 10 gallons or more a day per person. The 50 million low-flow toilets now in place in the United States save an estimated 600 million gallons of water a day. New York City is providing significant rebates to people who replace their old toilets with the new models, and Los Angeles is offering the low-flow toilets free as part of its efforts to restore Mono Lake.

*Gray Water.* It may seem odd and wasteful to raise all domestic water to drinking-water standards and then use most of it to water lawns and flush toilets. Increasingly, *gray-water* recycling systems are being adopted in some water-short areas. **Gray water**, the slightly dirtied water from sinks, showers, bathtubs, and laundry tubs, is collected in a holding tank and used for such things as flushing toilets, watering lawns, and washing cars. Going further, a number of cities are using treated wastewater (sewage water) for irrigating golf courses and landscapes, both to conserve water and to abate the pollution of receiving waters. (See Chapter 17.) Residents of California use more than 160 billion gallons of treated wastewater for such purposes, and in Israel 70% of treated wastewater is reused for irrigation of nonfood crops. If the idea of reusing sewage water turns you off, recall that *all* water

**Figure 7–23 The 1.6-gallon commode.** Now required for all new installations, this device saves 10 gallons of water or more a day per person.

is recycled by nature. There is hardly a molecule of water you drink that has not moved through organisms—including humans—numerous times. A number of communities are already treating their wastewater so thoroughly that its quality exceeds what many cities take in from lakes and rivers.

## Public-Policy Challenges

The hydrologic cycle provides a finite flow of water through each region of Earth. When humans come on the scene, this flow of water is inevitably divided between the needs of the existing natural ecosystems and the agricultural, industrial, and domestic needs of humans, with the latter usually being met first. In fact, recent calculations indicate that humans now use 26% of total terrestrial evapotranspiration and 54% of accessible precipitation runoff. We are major players in the water cycle, and as we have seen, many facets of our water use are unsustainable.

***Water Wars.*** Maintaining a supply of safe drinking water for people is a high-priority issue. In addition, water for irrigation is vital to food production across much of the world. Often, these two demands come into conflict, generating what have been referred to as "water wars." California has been the scene of many water conflicts. For example, Southern California depends heavily on Colorado River water for irrigation of the extensive Imperial Valley farms, as well as for drinking water for its large cities. Recently, an agreement was struck that diverts 65 billion gallons of Colorado River irrigation water a year to the city of San Diego for domestic use. To purchase the water and also pay farmers to take their land out of production, the city will invest $2 billion. The deal will fallow some 30,000 acres out of a total of 450,000 acres of irrigated farmland. Normally, the irrigation water drains into the Salton Sea from the farmlands, providing the major freshwater source for the sea, which is California's largest lake and the site of a highly productive fishery. Now, many fear that the diversion will cause the sea to shrink, because the lost water represents some 25% of its intake. If the sea shrinks, the exposed seabed will dry up and will be scoured by winds, spreading salt and chemicals widely over nearby crops. The sea is also a haven for wildfowl and migrating waterbirds. If the salinity increases, it could wipe out the fishery and also threaten the birds.

The Southern California scene demonstrates that water for people can often mean less water for natural ecosystems, as rivers, wetlands, and groundwater are exploited and freshwater ecosystems are deprived of their life-sustaining water. This is not simply a matter of aesthetics, however, because these ecosystems provide a wide variety of vital goods and services. They also are host to many threatened and endangered species. A way must be found to manage water for both people and the natural world. How can this be accomplished?

**National Water Policy.** In the United States, the EPA has a mandate to provide oversight of the nation's water quality. The Clean Water Act and subsequent amendments authorize the EPA to develop programs and rules to carry out its mandate. The EPA, however, deals primarily with water quality, not water quantity. There is no federal bureaucracy to provide similar oversight for the water quantity issues discussed in this chapter.

Peter Gleick, a leading expert on global freshwater resources, addresses this public-policy need,[1] pointing out that most water policy decisions are made at local and regional levels. National policies, however, are needed to guide these lower level decisions and also to deal with interstate issues and federally managed resources. The last time a water policy report was issued was in 1950, when President Harry S. Truman established a water resources policy commission and the commission gave its report. This commission needs to be revived or a new one needs to be established, and in either case it must be given a mandate to collect data on water resources and problems and issue recommendations on how the federal government could facilitate water stewardship into the 21st century.

***Key Issues.*** The following issues need to be addressed by a water resources policy commission:

1. Water efficiency must be promoted as the primary strategy for meeting future water needs. In the United States, scarcity and rising costs have led to significant gains in efficiency. As a result, water withdrawals peaked in the early 1980s and have leveled off since then (Fig. 7–24). This change has been attributed to steps taken in all three sectors (domestic, industry, irrigation) to reduce water losses.
2. Water subsidies need to be reduced or eliminated. Sustainable water management will never happen until water is priced according to its real costs. Policy makers have persistently subsidized water resources, especially for agricultural use, and the result has been enormous wastes and inefficiencies. The removal of subsidies will also level the playing field for municipal and agricultural needs, making it possible to adjudicate those needs more fairly.
3. Polluters must be charged according to their effluents. When municipalities, farmers, and industries are allowed to pollute waterways without any accountability, the result is that everyone downstream is forced to pay more to clean up the water and aquatic ecosystems are degraded accordingly. Various approaches are possible, such as permit fees, "green taxes," direct charges, and market-based trading.

[1]Gleick, Peter H. "Global Water: Threats and Challenges Facing the United States." *Environment* 43: 18–26, March 2001.

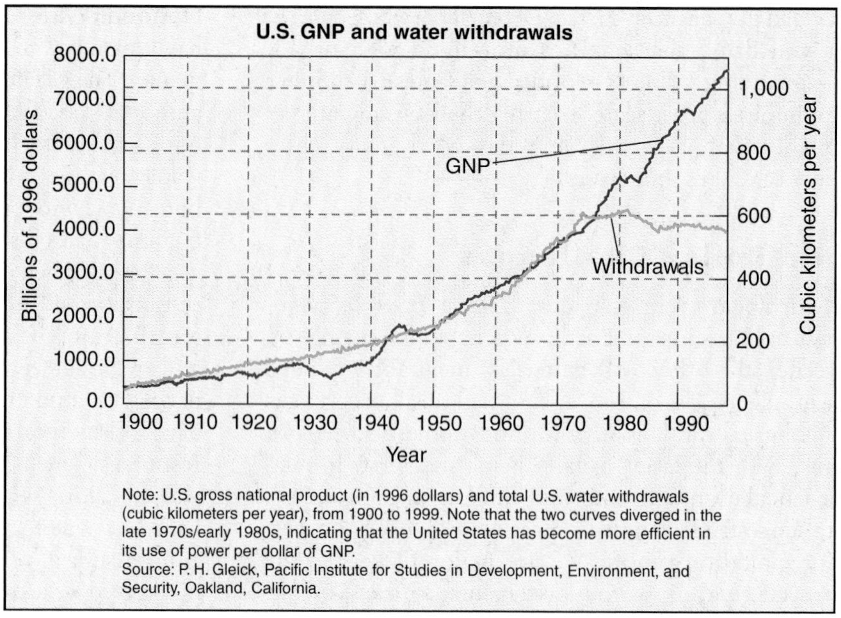
U.S. GNP and water withdrawals
8000.0
7000.0
6000.0
5000.0
4000.0
3000.0
2000.0
1000.0
0.0
Billions of 1996 dollars
1,000
800
600
400
200
0
Cubic kilometers per year
GNP
Withdrawals
1900 1910 1920 1930 1940 1950 1960 1970 1980 1990
Year
Note: U.S. gross national product (in 1996 dollars) and total U.S. water withdrawals (cubic kilometers per year), from 1900 to 1999. Note that the two curves diverged in the late 1970s/early 1980s, indicating that the United States has become more efficient in its use of power per dollar of GNP.
Source: P. H. Gleick, Pacific Institute for Studies in Development, Environment, and Security, Oakland, California.

**Figure 7–24 Water withdrawals and GNP.** Even though the U.S. GNP has continued to increase, water use has leveled off, showing the effects of conservation efforts.

4. Watershed management must be integrated into the pricing of water. Watersheds hold the key to water purity, aquifer recharge, and water storage, as well as maintaining natural wetland and riparian habitats that are vital to many plants and animals. The city of New York found out that every dollar it invested in watershed management paid off handsomely, saving as much as $200 for new treatment facilities.
5. The United States must respond to the global water crisis with adequate levels of international development aid. The most vital need is for safe drinking water. Millions of lives can be saved and the health of many more millions improved if people in the developing world have access to water that is free of pathogenic microorganisms.
6. The United States must take action to reduce the emissions of greenhouse gases that are bringing on global climate change. To date, the administration has refused to join most of the rest of the world in the reductions in emissions agreed to in the Kyoto Accord. This need is thoroughly addressed in Chapter 20.
7. Much more research and monitoring are needed to provide the basic data for making informed policy decisions. It is fair to say that if you can't measure it, you can't manage it. Essential data on the recharge and withdrawal, of water from groundwater aquifers, on stream flow, on river runoff, and on water uses are poorly collected in most developing countries and are not always well done even in the United States. The USGS National Water Use Information Program performs a minimal analysis of water use, but needs more funding to do an adequate job. The Millennium Ecosystem Assessment is providing a framework for scientists to address these issues on an international level.

**International Action.** On the international front, the World Commission on Water for the 21st Century sponsors the World Water Forum, which convened for the third time in 2003. (See "Global Perspective" essay, p. 203). The findings of the forum relate directly to solving the great water-related needs of Earth, especially in the developing countries. With a large proportion of the developing world already experiencing shortages of clean water or water for agriculture, the population increases and pressure to develop their resources that are certain to come in the next few decades will undoubtedly subject hundreds of millions to increased water stress. The problem is not that Earth contains too little fresh water; rather, it is that we have not yet learned to manage the water that our planet provides. To quote the World Water Council's "Vision" report,[2] "There is a water crisis, but it is a crisis of management. We have threatened our water resources with bad institutions, bad governance, bad incentives, and bad allocations of resources. In all this, we have a choice. We can continue with business as usual, and widen and deepen the crisis tomorrow. Or we can launch a movement to move from vision to action—by making water everybody's business."

[2]Cosgrove, William J. and Frank R. Rijsberman. "World Water Vision: Making Water Everybody's Business." World Water Council, 2000. (http://www.worldwatercouncil.org/vision.shtml)

global perspective

## The Third World Water Forum

Established in 1998, the World Commission on Water for the 21st Century was given the responsibility for developing a long-term global vision for water in the 21st century. The Commission, cosponsored by the World Bank, the FAO, the United Nations Environment Program (UNEP), and other major organizations, delivered its vision report at the second World Water Forum in The Hague, Netherlands, March 17–22, 2000. More than 4,500 people from all over the world attended the meetings. The third World Water Forum was held in Kyoto, Japan, in March of 2003. The great increase in attendance at the forum (24,000) was an indication of the world's awareness that water issues are approaching crisis levels.

At the third forum, the *World Water Development Report* was presented. This is a 600-page effort on the part of 23 U.N. agencies to produce the most comprehensive and up-to-date report on the state of the world's freshwater resources. The report, available on the Internet or in published form, monitors progress toward targets in such fields as health, food, ecosystems, cities, industries, energy, risk management, water valuation, resource sharing, knowledge-based construction of infrastructure, and governance. According to the report, by 2050 as many as 7 billion people in 60 countries could be experiencing water scarcity. Also, within 20 years, the per capita water available worldwide will drop by one-third, with the most serious impacts on the world's poorest people.

The forum itself was judged by its participants to be an exceptional success, exceeding the expectations of most. The key issues dealt with were balancing increasing human requirements for safe water, improved health, and better sanitation on the one hand, with water needs associated with food production, transportation, energy, and environmental health on the other. To facilitate this agenda, the forum addressed the need for more countries to institute effective governance and adequate financing. A particular focus was on the fundamental role of community participation in achieving the stated goals.

As is usually done at such international meetings, a statement was drafted, committing the participants to certain actions. In particular, they affirmed the Millennium Development Goals for reducing by half the world's people without access to safe drinking water or adequate sanitation. Over 100 commitments were made in which various partnerships and agencies pledged actions related to water issues. For example, the World Bank's Water and Sanitation Program agreed to fund national capacity-building projects to monitor Millennium Development Goals, and the Netherlands pledged to concentrate its support on Africa, assisting 10 countries in developing their national water plans and supporting the African Water Facility, an initiative to increase access to sources of financing related to water development on the continent.

A quote from the World Water Development Report puts it all well: "The Earth ... is facing a serious water crisis. All the signs suggest it is getting worse and will continue to do so, unless corrective action is taken. This crisis is one of water governance, essentially caused by the ways in which we mismanage water. But the real tragedy is the effect it has on the everyday lives of poor people, who are blighted by the burden of water-related disease, living in degraded and often dangerous environments, struggling to get an education for their children and to earn a living, and to get enough to eat."

# revisiting the themes

## Sustainability

The story at the beginning of the chapter showed how unsustainable practices in water diversion led to the destruction of the Aral Sea and the degradation of rivers and other aquatic ecosystems. Still, the hydrologic cycle is a remarkable global system that renews water and does so sustainably. Groundwater—especially the fraction that is nonrenewable because it was formed centuries ago—is being used in an unsustainable manner. The consequences are immediate: Water tables disappear into the depths of the earth, farms dry up because there is no water, and land subsidence ruins properties and groundwater storage itself.

## Stewardship

Exchanges between land, water, and atmosphere represent a cyclical system that has worked well for eons and that can continue to supply all our needs into the future—if we manage water wisely. However, we are changing the cycle in ways that reflect a lack of wisdom and a lack of resolve to meet people's needs in a just way. In particular, we are refusing to restrain our release of greenhouse gases, thus ensuring that global climate change will continue and intensify. Stewardship of water resources means that we consider the needs of the natural ecosystems—and especially endangered species—when we make policy decisions. It means that we address the wastefulness that characterizes so much of our water use. It means in particular that we show concern and help the millions who lack adequate clean water and are suffering because of it.

## Sound Science

What is known about the hydrologic cycle is due to sound scientific research—research that is needed to make progress in managing water resources. Hopefully, work performed under the Millennium Ecosystem Assessment will give us a better understanding of such matters as groundwater recharge, essential river flows, and the interactions between different land uses and water resources.

## Ecosystem Capital

Water in all of its manifestations—lakes, rivers, wetlands, and groundwater—represents one of the most valuable components of ecosystem capital. Life absolutely depends on it, and water provides ecosystems with its life-sustaining fluid free of charge. Water also makes possible our food production, especially when it is used in irrigation to increase the productive capacity of the land. The countries experiencing water shortages know firsthand how valuable a resource water is.

## Policy and Politics

From the opening story of the Aral Sea to the final words of the chapter, water policies have been in focus. Policy in the former Soviet Union dictated that the priority for water should be to grow cotton, rather than to sustain the Aral Sea and its fishery. Policy in the southwestern United States allocates water from the Colorado River to many states, leaving only a trickle to flow to Mexico and the sea. A policy change in the 1990s helped restore the section of the Colorado that runs from Glen Canyon Dam down through the Grand Canyon National Park. Policies and, often, local politics keep thousands of dams in place when their removal would clearly lead to environmental gains in the rivers they block. Policy in China is displacing more than 1.2 million people as the Three Gorges Dam unfolds, in exchange for hydroelectric power and flood control.

Policy made it possible for Tampa Bay to become the country's first commercially viable desalination plant. Public policy continues to favor subsidizing water costs to farmers, thus preventing market-based mechanisms from determining which needs should prevail. Policy enabled San Diego to get water at the cost of fresh water for the Salton Sea. Recall, too, from Section 7.4, that the United States lacks a coherent public policy for water resources; the last report was issued in 1950. A number of key issues were identified that should be addressed by a vitally needed new water commission.

On the political front, the Bush administration has "relaxed" (read, weakened) rules designed to protect streams, swamps, and other wetlands from being developed, dredged, or drained. This is happening in spite of a commitment to "no net loss" of wetlands, leaving many puzzled as to what that commitment means. The administration has also relinquished control of water in and under vast federal lands in the West, ceding such control to the Western states. This is supposed to be "an antidote to past federal excess," according to officials. (The Clinton administration had claimed federal rights to the water within federal lands.)

Often, private organizations play a significant role in protecting water resources. American Rivers is a private watchdog organization that publishes an annual list of the country's most endangered rivers. Heading the list in 2003 are the Klamath River in the Pacific Northwest, the Ipswich River in Massachusetts, and the Big Sunflower River in Mississippi. The Klamath and Ipswich Rivers are facing serious water shortages. The Big Sunflower River is facing a proposal to allow the U.S. Army Corps of Engineers to build a huge hydraulic system (the Yazoo Pumps) that will withdraw up to 6 million gallons of water a minute to provide irrigation water for agriculture in the region. According to American Rivers, the project could drain 200,000 acres of wetlands. Further, the Corps wants to dredge 100 miles of the riverbed, stirring up toxic deposits and essentially destroying the ecological integrity of the river. American Rivers is calling on the EPA to veto the proposal from the Corps.

## Globalization

The global hydrologic cycle represents a form of globalization that has done well for millennia. However, atmospheric pollution and changes in land use are threatening the cycle. Plainly, what the United States does about global climate change and what the tropical countries do about clearing forests eventually affect everyone.

## review questions

1. What are the lessons to be learned from the Aral Sea story?
2. Give examples of the infrastructure that has been fashioned to manage water resources.
3. What are the two processes that result in natural water purification? State the difference between them. Distinguish between green water and blue water.
4. Describe how a Hadley cell works, and explain how Earth's rotation creates the trade winds.
5. Why do different regions receive different amounts of precipitation?
6. Define *precipitation, infiltration, runoff, capillary water, transpiration, evapotranspiration, percolation, gravitational water, groundwater, water table, aquifer, recharge area, seep*, and *spring*.
7. Use the terms defined in Question 6 to give a full description of the hydrologic cycle, including each of its three loops—namely, the evaporation, surface runoff, and groundwater loops. What is the water quality (purity) at different points in the cycle? Explain the reasons for the differences.
8. How does changing Earth's surface (for example, by deforestation) change the pathway of water? How does it affect streams and rivers? Humans? Natural ecology?
9. Explain how climate change and atmospheric pollution can affect the hydrologic cycle.
10. What are the three major uses of water? What are the major sources of water to match these uses?
11. How do dams facilitate the control of surface waters? What kinds of impacts do they have?
12. Distinguish between renewable and nonrenewable groundwater resources. What are the consequences of overdrawing these two kinds of groundwater?
13. What are the four options for meeting existing water scarcity needs and growing demands?
14. Describe how water demands might be reduced in agriculture, industry, and households.
15. What is the status of water policy in the United States? Cite some key issues that should be addressed by any new initiatives to establish a national water policy.

## thinking environmentally

1. Pretend that you are a water molecule, and describe your travels through the many places you have been and might go in the future as you make your way around the hydrologic cycle time after time. Include travels through organisms.
2. Suppose some commercial interests want to create a large new development on what is presently wetlands. Have a debate between those representing the commercial interests and those representing environmentalists who are concerned about the environmental and economic costs of development. Work toward negotiating a compromise.
3. Describe how many of your everyday activities, including your demands for food and other materials, add pollution to the water cycle or alter it in other ways. How can you be more stewardly with your water consumption?
4. Describe the natural system that maintains uniform streamflow despite fluctuations in weather. How do humans upset this regulation? What are the consequences?
5. An increasing number of people are moving to the arid southwestern United States, even though water supplies are already being overdrawn. If you were the governor of one of the states in this region, what policies would you advocate to address this situation?
6. Suppose some commercial interests want to develop a golf course on what is presently forested land next to a reservoir used for city water. Describe the impacts this development might have on water quality in the reservoir.
7. Divide the class into five groups: environmentalists, farmers, city officials responsible for domestic water, industrialists, and developers. Suppose that the actual supply of water will support only 50% of prospective demands. Negotiate a compromise that will honor the interests of all parties.

# Soil: Foundation for Land Ecosystems

**Key Topics**

1. Soil and Plants
2. Soil Degradation
3. Conserving the Soil

Thirty years ago, Gabino López attended classes on sustainable agriculture in San Martín Jilotepeque, Guatemala, and then returned to his village to apply what he had learned to his own small farm. He planted vegetation barriers along the contours of his land to stop erosion and applied cow manure to enrich the soil. His maize harvest increased from 0.8 to 1.3 tons/hectare (1,050 lb/acre). López's harvest jumped to 3.4 tons/hectare when he plowed under his crop residues, as opposed to when he burned them or rotated crops. Aware of his success, his friends asked him for advice, and before long, Gabino López was an international consultant in *ecological agriculture,* ranging from Mexico to Ecuador and even India. He and his fellow farmers employ a number of practices that follow the "five golden rules of the humid tropics": (1) keeping the soil covered, (2) using minimal or even no tillage, (3) using mulch to provide nutrients for crops, (4) maximizing biomass production, and (5) maximizing biodiversity. As a result, the maize harvest has averaged 4.3 tons/hectare in a number of local villages (Fig. 8–1).

*Reaping the Harvest in Iowa.*

Recently, López reported on the concept of *farmer experimenters,* in the context of the work of COSECHA, a Honduran NGO. To overcome the problem of irregular rainfall, the farmers tested a number of different microcatchments for holding the water for small-scale irrigation. The farmers decided that 1- to 2-cubic-meter catchments were ideal and found many additional uses for the water. Water harvesting is joining green manures and natural pest control in the growing reach of agricultural development that involves villagers experimenting and teaching others what they learn. It is an ongoing process, and it is leading to sustainable agriculture for people in serious need of all the food they can raise.

***Russian Desert.*** In 1997, in Komsomolsky, southeastern Russia, Alexandra Bakhtayeva told a *Boston Globe* reporter that she had to dig herself out every day from the sand blowing in from surrounding dry land that was once prime grazing land. The windblown sand means that the area has undergone desertification—conversion from stable, rolling grasslands to drifting sand. The desert area is growing, expanding hundreds of thousands of acres annually in this Russian southland (Fig. 8–2). What used to be soil is gone and is replaced by sand,

**Figure 8–1 Farming in Guatemala.** Maize is a mainstay of nutrition for the people of Guatemala. It is possible to dramatically improve maize production with proper treatment of the soil.

and the local environmental officials lack the financial resources to reverse the process. This is a human-made desert, the result of failed Communist agricultural policies and, earlier than that, the removal of the local Kalmyk people by Stalin as he forced collectivization of the rural areas of the Soviet Union.

Before their removal by Stalin, the Kalmyks very carefully avoided plowing the land and, instead, rotated the land being grazed by their cattle and sheep. Shepherds even wore flat-bottomed shoes to avoid breaking up the topsoil, and their fat-tailed sheep were a breed with flat-bottomed hooves. The combination of plowing the thin soil in an

**Figure 8–2 Desertification in Kalmykia.** Poor agricultural policy and practice and brutal repression of the indigenous people led to widespread conversion of the fragile landscape to sandy desert.

**table 8-1 Global Degradation of Crop-, Pasture-, Forest-, and Woodlands**

| Degradation Category | Amount of Land Affected (Millions of Hectares) | Lost Production (Percent) |
|---|---|---|
| Total land | 8,735 | |
| Not degraded | 6,770 | 0 |
| Degraded | 1,965 | |
| Lightly | 650 | 5 |
| Moderately | 904 | 18 |
| Heavily | 411 | 50 |

*Source:* From Global Assessment of Soil Degradation (GLASOD), database prepared for UNEP, 1987–1990.

attempt to raise crops and grazing millions of Caucasian sheep with sharp hooves gradually broke down the soil and gave it to the winds. Now the sand has already buried 25 towns, with more on the way; nearly half of the territory of Kalmykia is severely desertified. Without soil, the region is quickly becoming a wasteland where a few people scratch out a living protected from the blowing sand by tall fences.

***Past Neglect.*** Ninety percent of the world's food comes from land-based agricultural systems, and the percentage is growing as the ocean's fish and natural ecosystems are increasingly being depleted. Protecting and nurturing agricultural soils, which are the cornerstone of food production, must be a central feature of sustainability. Yet, it is a feature that has been overlooked repeatedly in the past. The story of Easter Island in Chapter 1 is just one of many. In his book, *A Green History of the World,*[1] Clive Ponting documents how the fall of the ancient Greek, Roman, and Mayan empires was more the result of a decline in agricultural sustenance due to soil erosion than of outside forces.

As the world's population moves beyond 6 billion in the 21st century, croplands and grazing lands are being increasingly pressed to yield more crops and other products. Yet, according to the U.N.'s Global Environment Outlook 3, the poor agricultural practices of the past 50 years have led to the degradation of 23% of land used for crops, grazing, or forestry—almost 2 billion hectares, or 4.8 billion acres. (See Table 8–1.) Throughout the world, agricultural soils have been (and continue to be) degraded by erosion, the buildup of salts, and other problems that can only undermine future productivity.

In the United States, another kind of loss has also occurred: The loss of prime farmland to development averaged some 400 thousand acres per year from 1982 to 1992 and then increased to 600 thousand acres per year from 1992 to 2001. This is happening because farmlands are usually easy to develop and farmers can often sell their land to developers at huge profits.

Our objective in this chapter is to develop an understanding of the attributes of soil required to support good plant growth, of how these attributes may deteriorate under various practices, and of what is necessary to maintain a productive soil. We will then look at what is being done nationally and internationally to rescue this essential resource and to establish practices that are sustainable. In Section 8.1, we present some characteristics of soil and then examine the relationships between soil and plants that sustain productivity.

[1]Ponting, Clive. 1991. *A Green History of the World.* Penguin Books. New York, NY.

## 8.1 Soil and Plants

Soil can make the difference between harvesting an abundant crop or abandoning a field to weeds. A rich soil is much more than the dirt you might get out of any hole in the ground. Indeed, agriculturists cringe when anyone refers to soil as *dirt.* You have already learned (Chapters 2 and 3) that various detritus feeders and decomposers feed on organic detritus in an ecosystem. Nutrients from the detritus are released and reabsorbed by producers, thus recycling the nutrients. In a productive soil, the detritus feeders and decomposers constitute a biotic community of organisms that not only facilitates the transfer of nutrients, but also creates a soil environment that is most favorable to the growth of roots. In short, a productive topsoil involves dynamic interactions among the organisms, detritus, and mineral particles of the soil (Fig. 8–3).

### Soil Characteristics

Most soils are hundreds of years old and change very slowly. Soil science is an integrative science that is at the heart of agricultural and forestry practice. The results of many years of study of soils have provided a system of

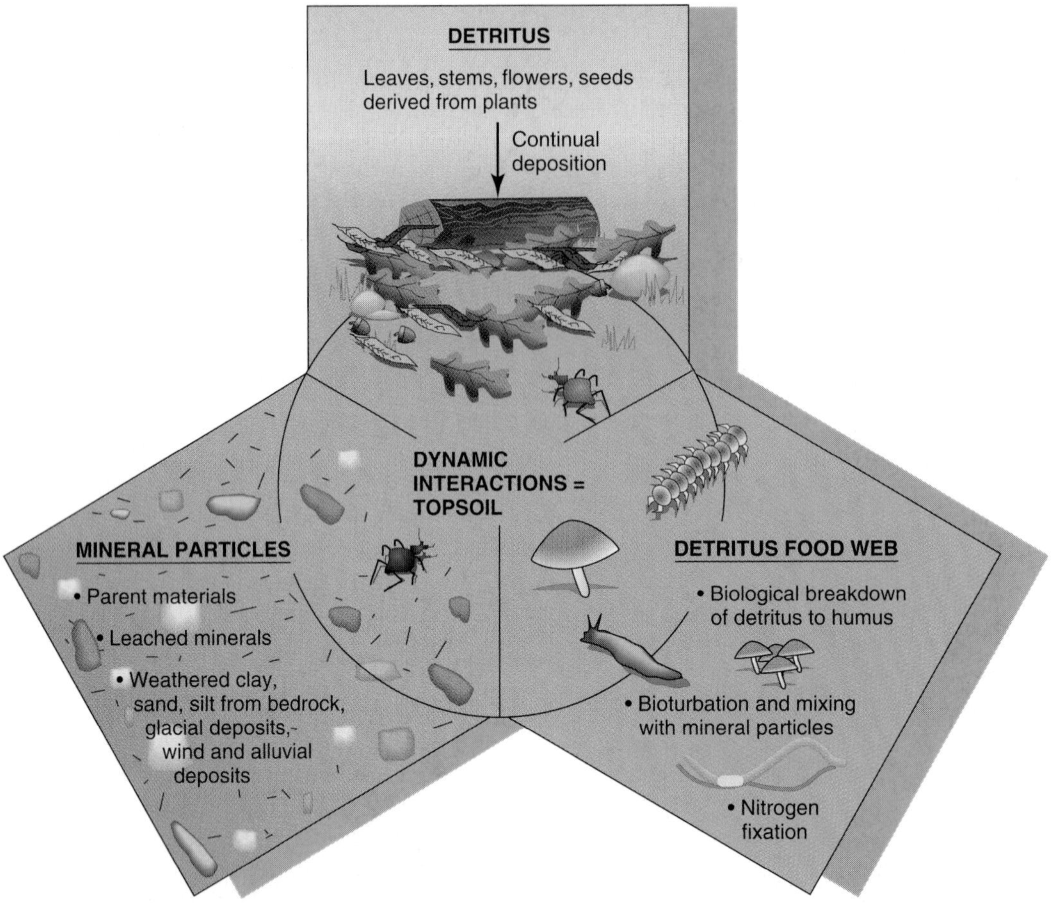


**Figure 8–3 Topsoil formation.** Soil production involves a dynamic interaction among mineral particles, detritus, and members of the detritus food web.

classification of soil profiles and soil structure and a taxonomy of soil types from all over the world—as well as a large body of scientific literature investigating relevant topics. Section 8.1 presents just enough of this information to enable you to open the door to soil science; you may even be motivated to follow up on your own in this fascinating field.

**Soil Profiles.** The processes of soil formation create a vertical gradient of layers that are often (but not always) quite distinct. These horizontal layers are known as **horizons,** and a vertical slice through the different horizons is called the **soil profile** (Fig. 8–4). The profile reveals a great deal about the factors that interact in the formation of a soil. The topmost layer, the **O horizon,** consists of dead organic matter (detritus) deposited by plants: leaves, stems, fruits, seeds, etc. Thus, the O horizon is high in organic content and is the primary source of energy for the soil community. Toward the bottom of the O horizon, the processes of decomposition are well advanced, and the original materials may be unrecognizable. At this point, the material is dark and is called **humus.**

*Below the O.* The next layer in the profile is the **A horizon,** a mixture of mineral soil from below and humus from above. The A horizon is also called **topsoil.**

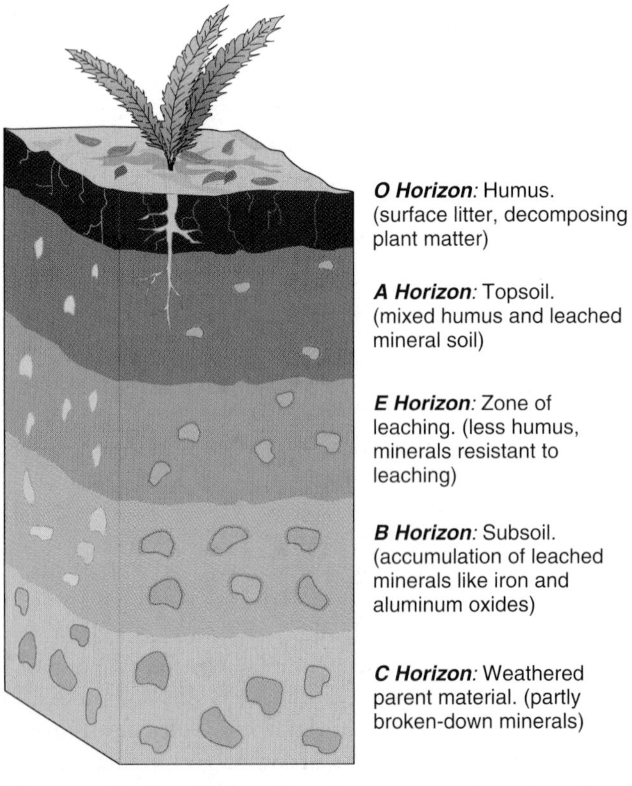


**Figure 8–4 Soil profile.** Major horizons from the surface to the parent material in an idealized soil profile.

Fine roots from the overlying vegetation cover permeate this layer. The A horizon is usually dark because of the humus that is present and may be shallow or thick, depending on the overlying ecosystem. In many soils, the next layer is the **E horizon,** where *E* stands for **eluviation**—the process of *leaching* (dissolving away) of many minerals due to the downward movement of water. This layer is often more pale in color than the two layers above it.

Below the E horizon is the **B horizon,** which is characterized by the deposition of minerals that have leached from the A and E horizons, so it is often high in iron, aluminum, calcium, and other minerals. Frequently referred to as the **subsoil,** the B horizon is often high in clay and is reddish or yellow in color. Below the B horizon is the **C horizon,** which is the parent mineral material originally occupying the site, representing weathered rock, glacial deposits, or volcanic ash and usually revealing the geological process that created the landscape. The C horizon is affected little by the biological and chemical processes that go on in the overlying layers.

Because there are innumerable soils across the diverse landscapes of the continents, soil profiles will differ in the thickness and the content of the layers. However, all soils exist in layers and can be characterized by their texture.

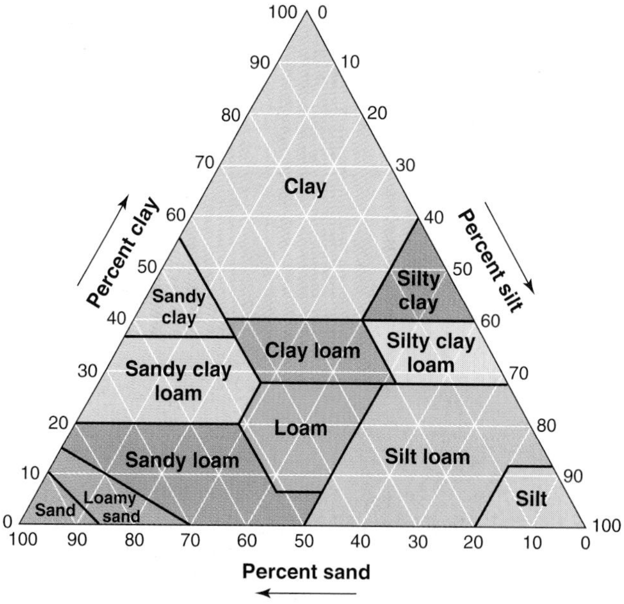


**Figure 8–5 The soil texture triangle.** Relative proportions of sand, clay, and silt are represented on each axis. Major classes of soil are indicated on the triangle. For clay, read across horizontally; for silt, read diagonally downward; for sand, read diagonally upwards to the left. The texture content of any soil should total 100% if the triangle is read properly.

**Soil Texture.** As rock weathers, it breaks down into smaller and smaller fragments. Below the size of small stones, these fragments (called *soil separates*) are classified as *sand, silt, and clay.* **Sand** is made up of particles from 2.0 to 0.02 mm in size, **silt** particles range from 0.02 down to 0.002 mm, and **clay** is anything finer than 0.002 mm. You can see the individual rock particles in sand, and you may be familiar with the finer particles called silt, but you would need a good microscope to see clay particles. If you wash clay in water, the water immediately takes on a cloudy or muddy appearance because the clay particles are suspended in it. The moldable, "gooey" quality of clay so necessary for pottery appears when just enough water is added for the particles to slide about one another on a film of water, but still cling together. On drying further, the clay particles adhere in hard clods (or elegant pots!).

***Proportions.*** The sand, silt, and clay particles constitute the mineral portion of soil. **Soil texture** refers to the relative proportions of each type of particle in a given soil. If one predominates, the soil is said to be sandy, silty, or clayey. A proportion that is commonly found in soil consists of roughly 40% sand, 40% silt, and 20% clay. A soil with these proportions is called a **loam.** You can determine the texture of a given soil by shaking a small amount of it with water in a large test tube to separate the particles and then allowing them to settle. Because particles settle according to their weight, sand particles settle first, silt second, and clay last. The rough proportion of each can then be seen. (More precise measurements require a laboratory analysis.) Soil scientists classify soil texture with the aid of a triangle that shows the relative proportions of sand, silt, and clay in a given soil (Fig. 8–5).

***Properties.*** Three basic considerations determine how several important properties of the soil are influenced by its texture:

1. Larger particles have larger spaces separating them than smaller particles have. (Visualize the difference between packing softballs and packing golf balls in the same-size containers.)
2. Smaller particles have more surface area relative to their volume than larger particles have. (Visualize cutting a block in half again and again. Each time you cut it, you create two new surfaces, but the total volume of the block remains the same.)
3. Nutrient ions and water molecules tend to cling to surfaces. (When you drain a nongreasy surface, it remains wet.)

These properties of matter profoundly affect such soil properties as infiltration, nutrient- and water-holding capacities, and aeration (Table 8–2). Note how the soil properties correspond logically to particle size (sand, silt, clay) in the table.

Soil texture also affects **workability**—the ease with which a soil can be cultivated. Workability, in turn, has a tremendous impact on agriculture. Clayey soils are very difficult to work, because, with even modest changes in moisture content, they go from being too sticky and muddy to being too hard and even bricklike. Sandy soils are very easy

**table 8-2 Relationship between Soil Texture and Soil Properties**

| Soil Texture | Water Infiltration | Water-holding Capacity | Nutrient-holding Capacity | Aeration | Workability |
|---|---|---|---|---|---|
| Sand | Good | Poor | Poor | Good | Good |
| Silt | Medium | Medium | Medium | Medium | Medium |
| Clay | Poor | Good | Good | Poor | Poor |
| Loam | Medium | Medium | Medium | Medium | Medium |

to work, because they become neither muddy when wet nor hard and bricklike when dry.

**Soil Classes.** Soils come in an almost infinite variety of vertical structures and textures. To give some order to this diversity, soil scientists have created a taxonomy of soils. It works much like the biological taxonomy with which you may be familiar. The most inclusive group in the taxonomy is the soil *order.* If you are classifying a soil, you find the order first and then work your way downward through the taxonomic categories (*suborders, groups, subgroups, families*), until you come to the soil *class* that best corresponds to the soil in question. There are literally hundreds of soil classes. In this chapter, we shall briefly examine the characteristics of four major soil orders that are most important for agriculture, animal husbandry, and forestry. (A total of 11 major orders has been established.)

*Mollisols.* Mollisols are fertile, dark soils found in temperate grassland biomes. They are the world's best agricultural soils and are encountered in the midwestern United States, across temperate Ukraine, Russia, and Mongolia, and in the pampas in Argentina. They have a deep A horizon and are rich in humus and minerals; precipitation is insufficient to leach the minerals downward.

*Oxisols.* These are soils of the tropical and subtropical rain forests. They have a layer of iron and aluminum oxides in the B horizon and have little O horizon, due to the rapid decomposition of plant matter. Most of the minerals are in the living plant matter, so oxisols are of limited fertility for agriculture. If the forests are cut, a few years of crop growth can be obtained, but in time the intense rainfall leaches the minerals downward, forming a hardpan that resists further cultivation.

*Alfisols.* Alfisols are widespread, moderately weathered forest soils. Although not deep, they have well-developed O, A, E, and B horizons. Alfisols are typical of the moist, temperate forest biome and are suitable for agriculture if they are supplemented with organic matter or mineral fertilizers to maintain soil fertility.

*Aridisols.* These are very widespread soils of drylands and deserts. The paucity of vegetation and precipitation leaves aridisols relatively unstructured vertically. They are thin and light colored and, in some regions, may support enough vegetation for rangeland animal husbandry. Irrigation used on these soils usually leads to salinization, as high evaporation rates draw salts to surface horizons, where they accumulate to toxic levels.

The descriptions of these four classes of soil demonstrate that different soils vary significantly in their properties. Soil scientists speak of "soil constraints," meaning the various characteristics of different soils that may or may not lead to successful agriculture. What are the major characteristics that enable soils to support plant growth, especially agricultural crops?

## Soil and Plant Growth

For their best growth, plants need a root environment that supplies optimal amounts of mineral nutrients, water, and air (oxygen). The pH (relative acidity) and salinity (salt concentration) of the soil are also critically important. **Soil fertility,** the soil's ability to support plant growth, often refers specifically to the presence of proper amounts of nutrients. But the soil's ability to meet all the other needs of plants is another component of soil fertility. Farmers speak of a given soil's ability to support plant growth as the *tilth* of the soil.

**Mineral Nutrients and Nutrient-holding Capacity** Mineral nutrients—phosphate ($PO_4^{3-}$), potassium ($K^+$), calcium ($Ca^{2+}$), and other ions—are present in various rocks, along with nonnutrient elements. Minerals initially become available to roots through the gradual chemical and physical breakdown of rock—the processes collectively referred to as **weathering.** Weathering, however, is much too slow to support anything approaching normal plant growth. The nutrients that support plant growth in natural ecosystems are supplied mostly through the breakdown and release (recycling) of nutrients from detritus. (See Fig. 3–22.)

*Leaching.* Nutrients may literally be washed from the soil as water moves through it, a process called **leaching.** Leaching not only lessens soil fertility, but also contributes to pollution when materials removed from the soil enter waterways. Consequently, the soil's capacity to bind and hold nutrient ions until they are absorbed by roots is just as important as the initial supply of those ions. This property is referred to as the soil's **nutrient-holding capacity** or its **ion-exchange capacity.**

*Fertilizer.* In agricultural systems, there is an unavoidable removal of nutrients from the soil with each crop, because nutrients absorbed by plants are contained in the harvested material. Therefore, agricultural systems require inputs of nutrients to replace those removed with the harvest. Nutrients are replenished with applications of **fertilizer**—material that contains one or more of the necessary nutrients. Fertilizer may be organic or inorganic. **Organic fertilizer** includes plant or animal wastes or both; manure and compost (rotted organic material) are two examples. **Inorganic fertilizers** are chemical formulations of required nutrients, without any organic matter included. Inorganic fertilizers are much more prone to leaching than are organic fertilizers.

**Water and Water-holding Capacity.** Water is constantly being absorbed by the roots of plants, passing up through the plant and exiting as water vapor through microscopic pores in the leaves—a process called **transpiration** (Fig. 8–6). The pores, called *stomata* (singular, *stoma*), are essential to permit the entry of carbon dioxide and the exit of oxygen in photosynthesis; however, the plant's loss of water via transpiration through the stomata is dramatic. A field of corn, for example, transpires an equivalent of a field-sized layer of water 17 inches (43 cm) deep in a single growing season. Inadequate water results in *wilting*, a condition that conserves water, but also shuts off photosynthesis by closing the stomata and preventing gas from being exchanged. If the wilted condition is too severe or too prolonged, the plants die.

*Infiltration.* Water is resupplied to the soil naturally by rainfall or artificially by irrigation. Three attributes of the soil are significant in this respect. First is the soil's ability to allow water to **infiltrate**, or soak in. If water runs off the ground surface, it won't be useful. Worse, it may cause erosion, which is discussed shortly.

*Water-holding Capacity.* Second is the soil's ability to hold water after it infiltrates, a property called **water-holding capacity.** Poor water-holding capacity implies that most of the infiltrating water percolates on down below the reach of roots—not very far in the case of seedlings and small plants—again becoming useless. What is desired is a good water-holding capacity—the ability to hold a large amount of water like a sponge—providing a reservoir on which plants can draw between rains. If the soil does not have such water-holding capacity, the plants will have to depend on frequent rains or irrigation or suffer the consequences of drought. Sandy soils are notorious for their poor water-holding capacity, whereas clayey (or silty) soils are good at holding water.

*Evaporation.* The third critical attribute of the soil is **evaporative water loss** from the soil surface. This kind of evaporation depletes the soil's water reservoir without serving the needs of plants. The O horizon functions well to reduce evaporative water loss by covering the soil. These aspects of the soil–water relationship are summarized in Fig. 8–7.

**Aeration.** Novice gardeners commonly kill plants by overwatering, or "drowning," them. Roots need to "breathe." Basically, they are living organs and need a constant supply of oxygen for energy via metabolism. Land plants depend on the soil being loose and porous

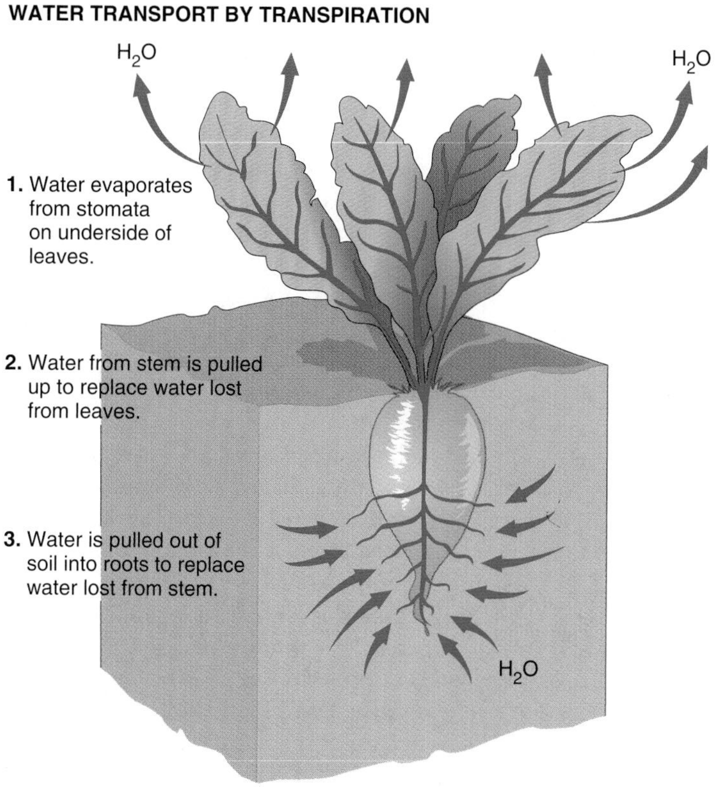
WATER TRANSPORT BY TRANSPIRATION
$H_2O$
$H_2O$
1. Water evaporates from stomata on underside of leaves.
2. Water from stem is pulled up to replace water lost from leaves.
3. Water is pulled out of soil into roots to replace water lost from stem.
$H_2O$

**Figure 8–6 Transpiration.** When water evaporates from the leaves of a plant, a vacuum is created that pulls the water up through the plant tissues. The roots draw water from the soil to replenish the evaporated water. The sun's energy drives transpiration via the initial evaporation process.

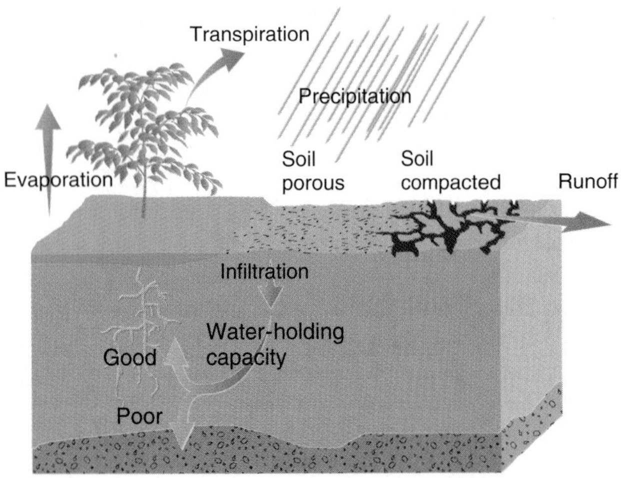


**Figure 8–7 Plant–soil–water relationships.** Water lost from the plant by transpiration must be replaced from a reservoir of water held in the soil. In addition to the amount and frequency of precipitation, the size of this reservoir depends on the soil's ability to allow water to infiltrate, to hold water, and to minimize direct evaporation.

enough to allow the diffusion of oxygen into, and carbon dioxide out of, the soil, a property called **soil aeration.** Overwatering fills the air spaces in the soil, preventing adequate aeration. So does **compaction,** or packing of the soil, which occurs with excessive foot or vehicular traffic. Compaction also reduces infiltration and increases runoff. Again, soil texture strongly influences this property, as indicated in Table 8–2.

**Relative Acidity (pH).** The term **pH** refers to the acidity or alkalinity (basicity) of any solution. A solution that is neither acidic nor alkaline is said to be neutral and has a pH of 7. The pH scale, which runs from 1 to 14, is discussed more fully in Chapter 21. For now, it is important to know that different plants are adapted to different pH ranges. Most plants (as well as animals) do best with a pH near neutral.

**Salt and Water Uptake.** A buildup of salts in the soil makes it impossible for the roots of a plant to take in water. Indeed, if salt levels in the soil get high enough, water can be drawn *out* of the plant (by osmosis), resulting in dehydration and death. Only plants with special adaptations can survive saline soils, and none of those are crop plants. The importance of the problem is explained in Section 8.2, as part of the discussion of how irrigation may lead to the accumulation of salts in soil (salinization).

## The Soil Community

In summary, to support a good crop, the soil must (1) have a good supply of nutrients and a good nutrient-holding capacity; (2) allow infiltration, have a good water-holding capacity, and resist evaporative water loss; (3) have a porous structure that permits good aeration; (4) have a pH near neutral; and (5) have a low salt content. Moreover, these attributes must be sustained. How does a soil provide and sustain such attributes?

A soil's mineral attributes—in particular, its texture—are crucial to its ability to support plant growth. Which is the best soil? Recall the principle of limiting factors (Chapter 2): The poorest attribute is the limiting factor. The poor water-holding capacity of sandy soil, for example, may preclude agriculture altogether because the soil dries out so quickly. The best textures are silts and loams, because limiting factors are moderated in these two types of soil. The good qualities are also moderated, however, so this "best" is really only "medium." It turns out that the organic parts of the soil ecosystem—the detritus and soil organisms—are necessary to optimize all attributes.

**Detritus, Soil Organisms, Humus, and Topsoil.** The dead leaves, roots, and other detritus accumulated on and in the soil supports a complex food web, including numerous species of bacteria, fungi, protozoans, mites, insects, millipedes, spiders, centipedes, earthworms, snails, slugs, moles, and other burrowing animals (Fig. 8–8). The most numerous and important organisms are the smallest—the bacteria. With the use of fluorescent staining (a process that makes bacteria glow when viewed in a fluorescence microscope), literally millions of bacteria can be seen and counted in a gram of soil (Fig. 8–9).

***Humus.*** As all these organisms feed, the bulk of the detritus is consumed through their cell respiration, and carbon dioxide, water, and mineral nutrients are released as by-products (as described in Chapter 3). However, each organism leaves a certain portion undigested; that is, a portion resists breakdown by the organism's digestive enzymes. This residue of partly decomposed organic matter is **humus,** found in high concentrations at the bottom of the O horizon. A familiar example is the black or dark brown spongy material remaining in a dead log after the center has rotted out. **Composting** is the process of fostering the decay of organic wastes under more or less controlled conditions, and the resulting compost is the same as humus. (Composting as a means of processing and recycling various wastes is discussed further in Chapter 18.)

***Soil Structure.*** As animals feed on detritus on or in the soil, they often ingest mineral soil particles as well. For example, it is estimated that as much as 15 tons per acre (37 tons per hectare) of soil pass through earthworms each year in the course of their feeding. As the mineral particles go through the worm's gut, they actually become "glued" together by the indigestible humus compounds. Thus, earthworm excrements—or **castings,** as they are called—are relatively stable clumps of inorganic particles plus humus. The burrowing activity of organisms keeps the clumps loose. This loose, clumpy characteristic, which you may experience as softness underfoot as you walk through a woods, is referred to as **soil structure** (Fig. 8–10). Whereas soil *texture* describes the *size* of soil particles, soil *structure* refers to their *arrangement.* A loose soil structure is ideal for infiltration, aeration, and workability. In

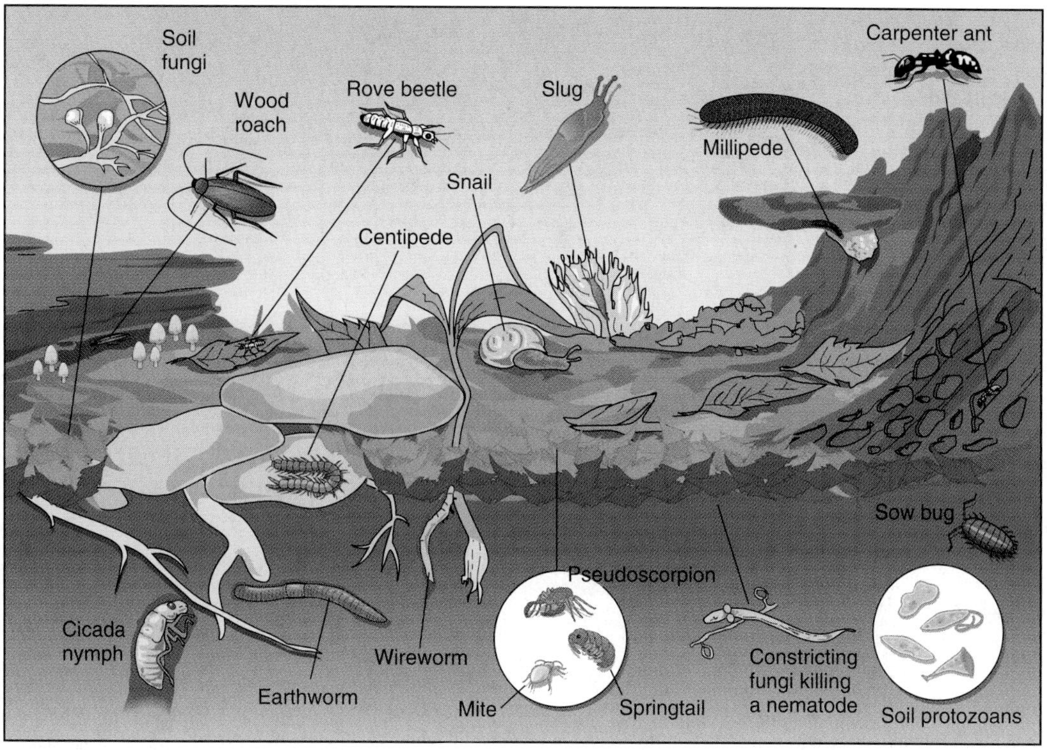


**Figure 8–8 Soil as a detritus-based ecosystem.** A host of organisms, major examples of which are shown here, feed on detritus and burrow through the soil, forming a humus-rich topsoil with a loose, clumpy structure.

addition, humus has an extraordinary capacity for holding both water and nutrients—as much as one hundredfold greater than the capacity of clay, on the basis of weight. The clumpy, loose, humus-rich soil that is best for supporting plant growth is the topsoil, represented in a soil profile as the A horizon (Fig. 8–4).

If plants are grown on adjacent plots, one of which has had all its topsoil removed, the results are striking: The yield from plants grown on subsoil is only 10 to 15% of that from plants grown on topsoil. In other words, a loss of all the topsoil would result in an 85 to 90% decline in productivity. The development of topsoil from subsoil or parent material is a process that takes hundreds of years or more. Figure 8–11 shows a site where the topsoil was removed 50 years ago. Although a lichen cover and some stunted trees have developed on the exposed gravel, there is still no soil there.

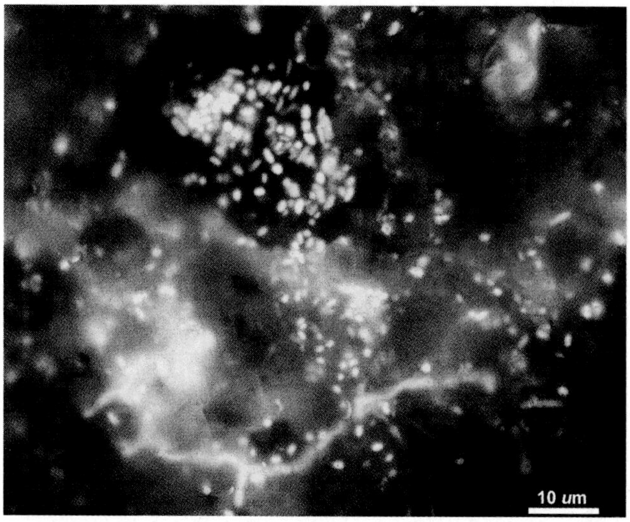


**Figure 8–9 Soil bacteria.** This photomicrograph shows soil bacteria (the bright spots) stained with acridine orange, a fluorescent dye. The bar scale indicates a length of 10 microns.

***Interactions.*** There are some important interactions between plants and soil biota. A highly significant one is the symbiotic relationship between the roots of some plants and certain fungi called **mycorrhizae.** Drawing some nourishment from the roots, mycorrhizae penetrate the detritus, absorb nutrients, and transfer them directly to the plant. Thus, there is no loss of nutrients to leaching. Another important relationship is had by certain bacteria that add nitrogen to the soil, as discussed in Chapter 3. Not all soil organisms are beneficial to plants, however. For example, *nematodes,* small worms that feed on living roots, are highly destructive to some agricultural crops. In a flourishing soil ecosystem, however, nematode populations may be controlled by other soil organisms, such as a fungus that forms little snares to catch and feed on the worms (Fig. 8–12).

**Soil Enrichment or Mineralization.** You can now see how the aboveground portion and the soil portion of an ecosystem support each other. The bulk of the detritus, which supports the soil organisms, is from green-plant producers, so green plants support the soil organisms.

**Figure 8–10 Humus and the development of soil structure.** On the left is a humus-poor sample of loam. Note that it is a relatively uniform, dense "clod." On the right is a sample of the same loam, but rich in humus. Note that it has a very loose structure, composed of numerous aggregates of various sizes.

By feeding on detritus, however, the soil organisms create the chemical and physical soil environment that is most beneficial to the growth of producers.

Green plants protect the soil and, consequently, themselves in two other important ways as well: The cover of living plants and detritus (1) protects the soil from erosion and (2) reduces evaporative water loss. Thus, it is desirable to maintain an organic mulch around those of your garden vegetables which do not maintain a complete cover themselves.

Unfortunately, the mutually supportive relationship between plants and soil can be broken all too easily. The maintenance of topsoil depends on additions of detritus in sufficient quantity to balance losses. Without continual additions of detritus, soil organisms will starve, and their benefit in keeping the soil loose will be lost. However, additional consequences occur as well. Although resistant to digestion, humus does decompose at the rate of about 2% to 5% of its volume per year, depending on the surrounding conditions. (The rate is most rapid in the tropics.) As the soil's humus content declines, the clumpy aggregate structure created by soil particles glued together with the humus breaks down. Water- and nutrient-holding capacities, infiltration, and aeration decline correspondingly. This loss of humus and the consequent collapse of topsoil is referred to as the **mineralization** of the soil, because what is left is just the gritty mineral content—sand, silt, and clay—devoid of humus.

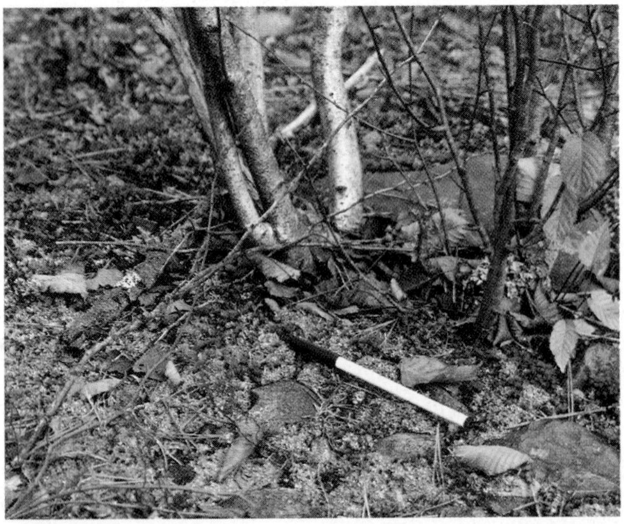

**Figure 8–11 The results of removing topsoil.** This "soil" is nothing but gravel with topsoil removed. After 50 years, only lichens and a few stunted trees have developed. The white object is a ballpoint pen.

Thus, topsoil must be seen as the result of a dynamic balance between detritus additions and humus-forming processes, on the one hand, and the breakdown and loss of detritus and humus (Fig. 8–13), on the other. If additions of detritus are insufficient, there will be a gradual deterioration of the soil. Conversely, mineralized soils can be revitalized through generous additions of compost or other organic matter.

## 8.2 Soil Degradation

In natural ecosystems, there is always a turnover of plant material, so new detritus is continuously supplied. However, when humans come on the scene and cut forests, graze livestock, or grow crops, the soil is at the mercy of our management or mismanagement. When key soil attributes required for plant growth or for other ecosystem services deteriorate over time, the soil is considered *degraded*. How serious is this problem?

*GLASOD.* The data on degraded land shown in Table 8–1 have been reproduced in many recent references, including the Pilot Analysis of Global Ecosystems (PAGE; see Chapter 1), the International Food Policy Research Institute, and Global Environment Outlook 3.

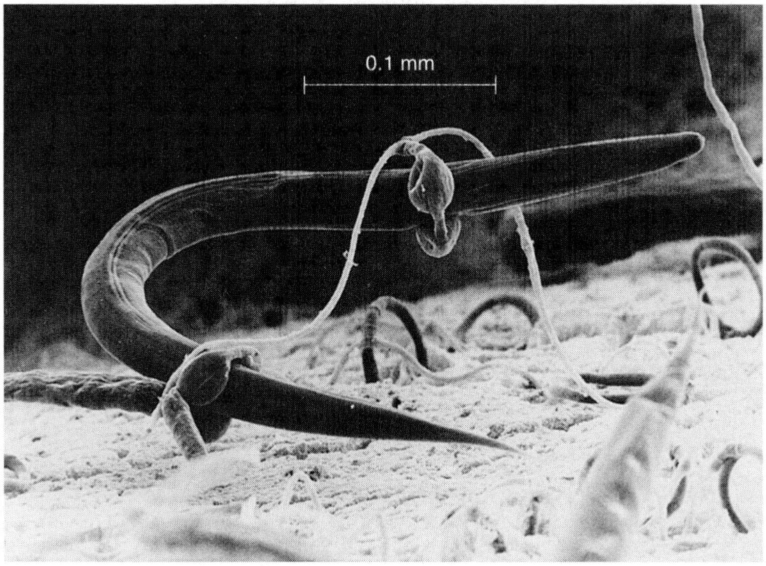


**Figure 8-12 Predatory fungus.** Soil nematode (roundworm), a root parasite, captured by the constricting rings of the predatory fungus *Arthrobotrys anchonia.*

These data are all based on one map: the Global Assessment of Soil Degradation map (GLASOD), prepared between 1987 and 1990. The data for the map came from questionnaires sent to soil experts around the world. Very little of the information has been validated by collecting data on actual soil conditions or information on crop productivity. It is the only data available, however, so it is cited again and again.

***Burkina Faso.*** Nevertheless, the GLASOD map is beginning to be challenged in places like Burkina Faso, one of the poorest and most densely populated countries in the dryland region of West Africa. The 2001 World Economic Forum calculated that that nation ranked the highest of any African dryland country in soil degradation (based on GLASOD). Other reports indicated that up to 75% of the land was degraded. In their recent

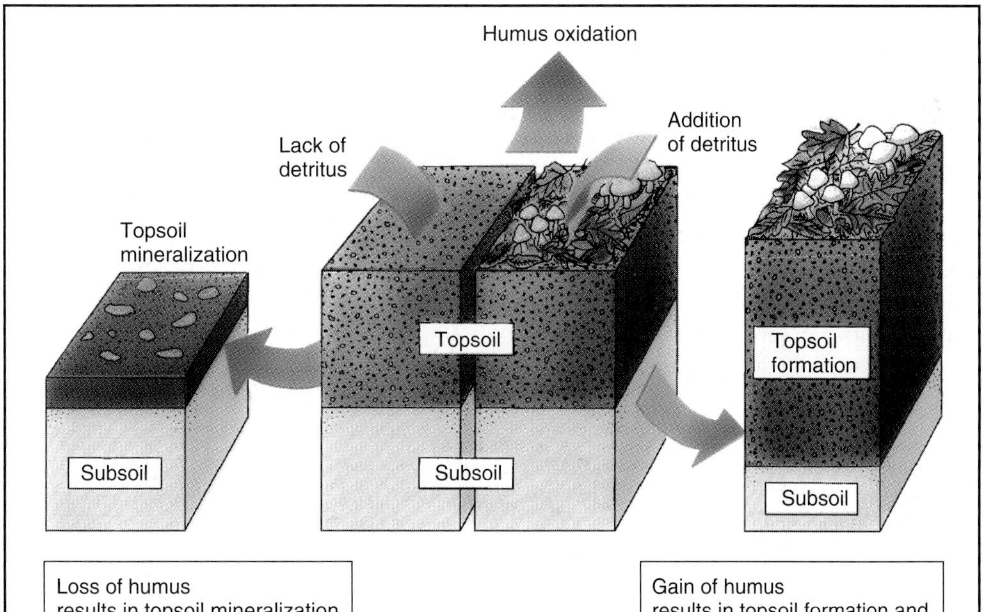

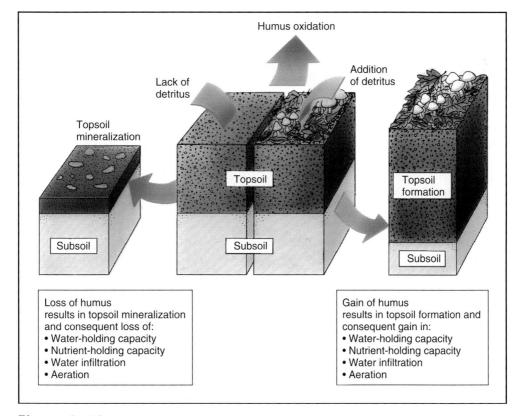

**Figure 8-13 The importance of humus to topsoil.** Topsoil is the result of a balance between detritus additions and humus-forming processes, and their breakdown and loss. If additions of detritus are insufficient, the soil will gradually deteriorate.

**Figure 8–14** **Erosion.** Severe erosion due to poor farming practices, near Bosencheve, Mexico.

paper, *Soil Degradation in the West African Sahel: How Serious Is It?*[2] however, David Niemeijer and Valentina Mazzucato reported that agricultural yields for virtually all crops have increased in the last 40 years, during which time the population more than doubled. Using subsistence agriculture, the local farmers showed a remarkable ability to manage their land by employing soil and water conservation practices appropriate to drylands. As a result of this and other recent reports, the 2002 World Economic Forum declared that the GLASOD data are suspect and certainly out of date; thus, participants in the forum did not use those data in their evaluation of environmental sustainability.

The preceding discussion is not to minimize the problem of soil degradation; it certainly exists and is a huge problem in many agricultural locations. Unfortunately, however, the frequently quoted data on worldwide soil degradation are probably quite inaccurate. More validation of actual erosion and crop production is needed. This is one of the stated tasks of the Millenium Ecosystem Assessment.

## Erosion

How is topsoil lost? The most pervasive and damaging force is **erosion,** the process of soil and humus particles being picked up and carried away by water or wind. Erosion follows anytime soil is bared and exposed to the elements. The removal may be slow and subtle, as when soil is gradually blown away by wind, or it may be dramatic, as when gullies are washed out in a single storm.

In natural terrestrial ecosystems other than deserts, a vegetative cover protects against erosion. The energy of falling raindrops is dissipated against the vegetation, and the water infiltrates gently into the loose topsoil without disturbing its structure. With good infiltration, runoff is minimal. Any runoff that does occur is slowed as the water moves through the vegetative or litter mat, so the water has too little energy to pick up soil particles. Grass is particularly good for erosion control, because when runoff volume and velocity increase, well-anchored grass simply lies down, forming a smooth mat over which the water can flow without disturbing the soil underneath. Similarly, vegetation slows the velocity of wind and holds soil particles.

***Splash, Sheet, and Gully Erosion.*** When soil is left bare and unprotected, however, it is easily eroded. Water erosion starts with what is called **splash erosion** as the impact of falling raindrops breaks up the clumpy structure of the topsoil. The dislodged particles wash into spaces between other aggregates, clogging the pores and thereby decreasing infiltration and aeration. The decreased infiltration results in more water running off and carrying away the fine particles from the surface, a phenomenon called **sheet erosion.** As further runoff occurs, the water converges into rivulets and streams, which have greater volume, velocity, and energy, and hence greater capacity to pick up and remove soil. The result is the erosion into gullies, or **gully erosion** (Fig. 8–14). Once started, erosion can readily turn into a vicious cycle if it is not controlled. Eroded soil is less able to support the regrowth of vegetation and is exposed to

[2]Niemeijer, David, and Valentina Mazzucato. "Soil Degradation in the West African Sahel: How Serious Is It?" *Environment* 44 (2): 20–31, 2002.

further erosion, rendering it even less able to support vegetation, and so on.

***Desert Pavement.*** Another very important and devastating feature of wind and water erosion is that both always involve the *differential* removal of soil particles. The lighter particles of humus and clay are the first to be carried away, while rocks, stones, and coarse sand remain behind. Consequently, as erosion removes the finer materials, the remaining soil becomes progressively coarser—sandy, stony, and, finally, rocky. Such coarse soils frequently reflect past or ongoing erosion. Did you ever wonder why deserts are full of sand? The sand is what remains after the finer, lighter clay and silt particles have blown away. In some deserts, the removal of fine material by wind has left a thin surface layer of stones and coarse sand called a **desert pavement,** which protects the underlying soil against further erosion (Fig. 8–15). Damaging this surface layer with vehicular traffic allows another episode of erosion to commence.

## Drylands and Desertification

Recall that clay and humus are the most important components of soil for both nutrient- and water-holding capacity. As clay and humus are removed, nutrients are removed as well, because they are bound to those particles. The loss of water-holding capacity is even more serious, however. Regions that have sparse rainfall or long, dry seasons support grass, scrub trees, or crops only insofar as soils have good water-holding and nutrient-holding capacity. As these soil properties are diminished by the erosion of topsoil, such areas become deserts, both ecologically and from the standpoint of production (Fig. 8–16). Indeed, the term **desertification** is used to denote this process. Note that desertification doesn't mean advancing deserts; rather, the term refers to *the formation and expansion of degraded areas of soil and vegetation cover in arid, semiarid, and seasonally dry areas, caused by climatic variations and human activities.* These lands are traditionally called **drylands.**

*Dryland* ecosystems cover over one-third of Earth's land area. They are defined by precipitation, not temperature. Beyond the relatively uninhabited deserts, much of the land receives only 10 to 30 inches (25 to 75 cm) of rainfall a year, a minimal amount to support rangeland or nonirrigated cropland. Droughts are common features of the climate in drylands, sometimes lasting for years. These lands, occupying some 5.2 billion hectares, are home to more than 2 billion people and are found on every continent except Antarctica. According to the **United Nations Convention to Combat Desertification** (**UNCCD**), some 70% of these dryland types of rangelands and nonirrigated croplands have been adversely affected by erosion and desertification, directly influencing the quality of life of some 250 million of the poorest people in the world by undermining their food production.

*UNCCD.* Recognizing the severity of this problem, the United Nations established the Convention to Combat Desertification, which was signed and officially ratified by over 100 nations in late 1996. The first Conference of the Parties was held in October 1997 in Rome. Since then, conferences were held annually until 2001 and then biennially. These conferences and other regular UNCCD meetings have been concerned with such issues as the funding of projects to reverse land degradation, "bottom-up" programs that enable local communities to help themselves, and the gathering and dissemination of traditional knowledge on effective drylands agricultural practices. Under the aegis of the UNCCD, affected countries have been developing National Action Programs;

(a) (b)

**Figure 8–15 Formation of desert pavement.** (a) As wind erosion removes the finer particles, the larger grains and stones are concentrated on the surface. (b) The result is desert pavement, which protects the underlying soil from further erosion. Traffic such as off-road vehicles breaks up the desert pavement and initiates further erosion.

**Figure 8–16 Desertification.** Cultivation without suitable protection against erosion allows finer components of the soil to be blown away, leaving the soil increasingly coarse and stony and diminishing its water-holding capacity. Persistence gradually renders the area an ecological desert, as seen in this photo, taken in east Africa.

in 2003, a UNCCD meeting made $500 million available for grants to fund projects aimed at arresting the desertification of drylands.

Even though variations in climate often play a role in the processes of desertification, it is human agency that is the greatest threat to the health of dryland ecosystems. Accordingly, let us now take a closer look at the practices that lead to erosion and desertification.

## Causing and Correcting Erosion

The three major practices that expose soil to erosion and lead to desertification are *overcultivation, overgrazing,* and *deforestation* (Fig. 8–17).

**Overcultivation.** Traditionally, the first step in growing crops has been (and, to a large extent, still is) plowing to control weeds. The drawback is that the soil is then

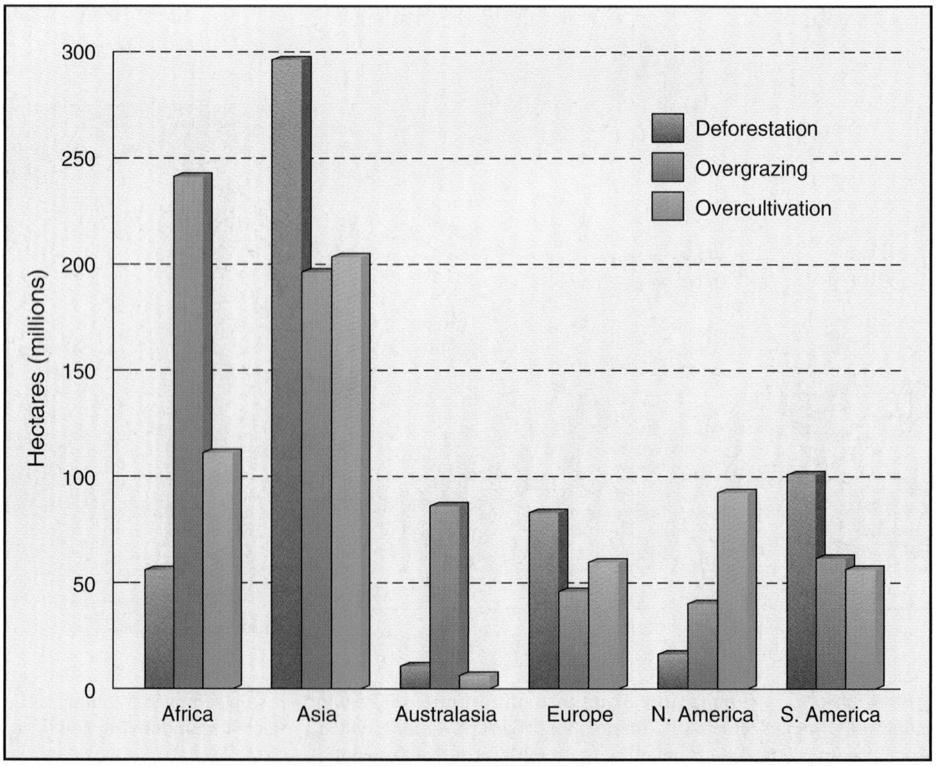


**Figure 8–17 Main causes of dryland soil degradation.** Deforestation, overgrazing, and overcultivation degrade soils in every region of the world. (*Source:* Food and Agricultural Organization of the United Nations.)

exposed to wind and water erosion. Further, it may remain bare for a considerable time before the newly planted crop forms a complete cover. After harvest, much of the soil again may be left exposed to erosion. Runoff and erosion are particularly severe on slopes, but wind erosion may extract a heavy toll on any terrain with minimal rainfall.

Plowing is frequently deemed necessary to loosen the soil to improve aeration and infiltration through it, yet all too often the effect is just the reverse. Splash erosion destroys the soil's aggregate structure and seals the surface, so that aeration and infiltration are decreased. The weight of tractors used in plowing may add to the compaction of the soil. In addition, plowing accelerates the oxidation of humus and evaporative water loss.

Despite the harmful impacts inherent in cultivation, systems of crop rotations—a cash crop such as corn every third year, with hay and clover (which fixes nitrogen as well as adding organic matter) in between—have proved sustainable. However, as food or economic demands cause farmers to abandon rotations, degradation and erosion exceed regenerative processes, and the result is a gradual decline in the quality of the soil. This is the essence of overcultivation.

*No-Till.* A technique that permits continuous cropping, yet minimizes soil erosion, is **no-till agriculture,** which is now routinely practiced over much of the United States. According to this technique, the field is first sprayed with herbicide to kill weeds, and then a planting apparatus is pulled behind a tractor to accomplish several operations at once. A steel disk cuts a furrow through the mulch of dead weeds, drops seed and fertilizer into the furrow, and then closes it (Fig. 8–18). At harvest, the process is repeated, and the waste from the previous crop becomes the detritus and mulch cover for the next. Thus, the soil is never left exposed, erosion and evaporative water loss are reduced, and there is enough detritus, including roots from the previous crop, to maintain the topsoil. A variation on this theme is **low-till farming,** which is catching on in Asia. For example, to plant wheat after a rice crop has been harvested, only one pass over a field is made, for planting seed and spreading fertilizer. Previously, farmers would plow 6 to 12 times to break up the muddy soil, dry it out for the wheat, and then irrigate the land.

*Fertilizer.* Another aspect of overcultivation involves the use of inorganic versus organic fertilizer. There is no question that optimal amounts of required nutrients can be efficiently provided by the suitable application of inorganic chemical fertilizer. The failing of chemical fertilizer is in its lack of organic matter to support soil organisms and build soil structure. Under intensive cultivation—a cash crop every year—nutrient content may be kept high with inorganic fertilizer, but mineralization, and thus soil degradation, proceed in any case. Then, with the soil's loss of nutrient-holding capacity, applied inorganic fertilizer is prone to simply leach into waterways, causing pollution.

This is not to say that chemical fertilizers do not have a valuable place in agriculture. The exclusive use of organic material may not provide enough of certain nutrients required to support optimal plant growth. What is required is for growers to understand the different roles played by organic material and inorganic nutrients and then to use each as necessary.

*NRCS.* Regardless of cropping procedures and the use of fertilizer, a number of other techniques are widely employed to reduce erosion. Among the most conspicuous are **contour strip cropping** and **shelterbelts**

**Figure 8–18 Apparatus for no-till planting.** A furrow is opened in the soil by the "wheel" (a), fertilizer is dispensed from the tank (b), seeds are dropped in from the hopper (c), and the furrow is closed by the trailing wheels (d). Weeds are controlled with herbicides. Thus, all operations of field preparation, planting, and cultivation are accomplished without disturbing the protective mulch layer on the surface of the soil.

(a)

(b)

**Figure 8–19 Contour farming and shelterbelts.** (a) Cultivation up and down a slope encourages water to run down furrows and may lead to severe erosion. The problem is reduced by plowing and cultivating along the contours at a right angle to the slope. In this photograph, strip cropping is also being utilized: The light green bands are one type of crop, the dark bands another. (b) Shelterbelts—belts of trees around farm fields—break the wind and protect the soil from erosion.

(Fig. 8–19). The **U.S. Natural Resource Conservation Service** (NRCS), formerly and still widely known as the Soil Conservation Service (SCS), was established in response to the Dust Bowl tragedy of the early 1930s. Through a nationwide network of regional offices, the NRCS provides information to farmers or other interested persons regarding soil or water conservation practices. (The regional offices can be found through county government telephone listings. Federal extension service offices will test soil samples, provide an analysis thereof, and make recommendations.) The NRCS performs a regular inventory of erosion losses in the United States. (See "Ethics essay," p. 223.) Estimates indicate that soil erosion on croplands has lessened in recent years, from 2.1 billion tons in 1992 to 1.8 billion tons in 2001. This is likely a consequence of improved conservation practices, such as windbreaks, grassed waterways, and field border strips.

**Overgrazing.** Grasslands that receive too little rainfall to support cultivated crops have traditionally been used for grazing livestock. Likewise, forested slopes that are too steep for cropping are commonly cleared and put into grass for grazing. Unfortunately, such lands are too often overgrazed. As grass production fails to keep up with consumption, the land becomes barren. Wind and water erosion follows, and desertification results. According to the UNEP's Global Environment Outlook 3, overgrazing during the past 40 years has degraded 680 million hectares.

**ethics**

## Erosion by Equation

By any name or measure, soil erosion is a highly undesirable process. It is a highly visible one, too. Ugly gashes, for example, can be seen in fields and slopes that are not well vegetated (Fig. 8–14), streams and rivers run brown with sediment after a heavy rain, deltas form where rivers empty into larger bodies of water, and reservoirs fill up with sediment over time. Eroded soil represents lost agricultural productivity, and the process can certainly lead to desertification if it is left unchecked. Erosion is also a natural geological process, wherein land surfaces undergo weathering and sediments form over geological time, eventually becoming the thick layers that form sedimentary rock. Thus, some erosion is bound to occur, no matter what farming methods are practiced.

The National Resources Inventory reports erosion rates on an annual basis. For 2001, they amounted to about 1.8 billion tons of eroded material. Much erosion is judged to be "excessive," originating from 101 million acres of cropland deemed vulnerable. How is erosion measured, especially if measurements are taken for the entire United States? In most cases, it is via map-reading, aerial photography, and the judicious use of a "universal soil loss equation and a "wind erosion equation." As a result, the figures that are obtained are *estimates* rather than *measurements*. The actual rates are not measured and, indeed, are probably not measurable for the whole country.

What happens when someone actually measures erosion and compares it with estimated rates? UCLA geographer Stanley W. Trimble asked this question in a study of the Coon Creek Basin in Wisconsin, published in 1999.[1] He had the good fortune of having available a previous historical study of the basin by the SCS, which left many markers in the sediments of the basin, and a huge amount of archived data. Trimble took soil profiles all around the basin, dated the layers of sediment, and was able to work back to the original surface of organic soil of the prairie, which dated to about 1850, when the basin was first farmed. He found rates of erosion that were high in the late 19th century, went even higher in the 1920s and 1930s, and then began to decline as farmers adopted conservation practices. Most remarkably, the rates from the 1970s to the 1990s were only 6% of their peak rate.

Although the nationally measured erosion rates showed a slight decline during the period of the 1970s to the 1990s, compared with earlier rates, the low rates found by what can be called "ground truth" in Trimble's work do not come close to the much higher estimates. Trimble's work has come under some criticism, but his response is that "the burden of proof is on those who have been making these pronouncements about big erosion numbers.... They owe us physical evidence. For one big basin, I've measured the sediment and I'm saying, I don't see it." At the very least, this kind of work suggests that more such actual measurements should be made and that the published erosion estimates should be taken with, well, a grain of silt. What do you think?

[1]From "Decreased Rates of Alluvial Sediment Storage in the Coon Creek Basin, Wisconsin, 1975–1999," by Stanley Trimble, in *Science*, 285, August 20, 1999. Copyright © by American Assn. for the Advancement of Science. Reprinted by permission of the Amerian Assn. for the Advancement of Science.

Overgrazing is not a new problem. In the United States in the 1800s, the American buffalo (bison) was slaughtered to starve out Native Americans and stock the rangelands with cattle. Overgrazing was rampant, leading to desertification and encroachment by hardy desert plants such as sagebrush, mesquite, and juniper, which are not palatable to cattle. Western rangelands now produce less than 50% of the livestock forage they produced before the advent of commercial grazing. Yet, according to a U.S. General Accounting Office (GAO) study, 20% of the rangelands remain overstocked.

The broader ecological impact of overgrazing should not go unnoticed either. The World Resources Institute reports, "Overgrazing has profoundly upset the dynamics of many range ecosystems, reducing biodiversity and altering the feeding and breeding patterns of birds, small mammals, reptiles, and insects." Further, about one-third of the endangered species in the United States are in jeopardy because of overgrazing or other practices associated with raising cattle, such as predator-control programs and the suppression of fire. Particularly hard hit are the wooded zones along streams and rivers, zones that are trampled by cattle seeking water. The resulting water pollution by sediments and cattle waste is high on the list of factors making fish species the fastest-disappearing wildlife group in the United States.

***Public Lands.*** In many cases, overgrazing occurs because the rangelands are public lands not owned by the people who own the animals. Where this is the case, herders who choose to withdraw their livestock from the range sacrifice income, while others continue to overgraze the range. Therefore, the incentive is for all to keep grazing, even though the range is being overgrazed. This problem, known as the "tragedy of the commons," is discussed further in Chapter 11.

In a variation of the tragedy of the commons, the U.S. Bureau of Land Management (BLM) and the Forest Service lease two million km$^2$ of federal (i.e., taxpayer-owned) lands for grazing rights at a nominal fee of $1.35 per animal unit (one cow-and-calf pair or five sheep) per month, about one-fourth of what a private owner would be paid for the same rights. The fee structure was established by the Taylor Grazing Act of 1934, which effectively prohibited any reduction in grazing levels and kept the fees well below market value. The income to the government is far less than the costs of administering the program (the grazing program loses $124 million a year), and the amount of livestock permitted by the BLM is considered too high by nearly all range experts. Attempts in Congress or the Department of the Interior (which houses the BLM) to raise grazing fees are consistently met with opposition from western congressmen, who threaten to cut the BLM budget.

One solution lies in better management. Because rangelands are now producing at only a 50% level or less, their restoration could benefit both wildlife and cattle production. The NRCS has a project that provides information and support to enable ranchers who own their lands to burn unwanted woody plants, reseed the land with perennial grass varieties that hold water, and manage cattle so that herds are moved to a new location before overgrazing occurs. Another solution gaining favor among ranchers and environmental groups is for the government to buy up grazing permits and "retire" rangelands. The ranchers would be offered a generous payment for the permits they hold, and the land would be used for wildlife, recreation, and watershed protection.

**Deforestation.** Forest ecosystems are extremely efficient systems both for holding and recycling nutrients and for absorbing and holding water, because they maintain and protect a very porous, humus-rich topsoil. Investigators at Hubbard Brook Forest in New Hampshire found that converting a hillside from forest to grass doubled the amount of runoff and increased the leaching of nutrients manyfold.

Much worse is what occurs if the forest is simply cut and the soil is left exposed. Pounding raindrops quickly seal the soil. The topsoil becomes saturated with water and slides off the slope in a muddy mass into waterways, leaving barren subsoil that continues to erode.

The problem is particularly acute when tropical rain forests are cut. The soils of equatorial regions have been subjected to heavy rains and leaching for millennia. Parent materials are already maximally weathered, and any free nutrients have long since been leached away. Consequently, tropical soils (oxisols) are notoriously lacking in nutrients. When the forests are cleared, the thin layer of humus with nutrients readily washes away. Only the nutrient-poor clayey subsoil, which is very poor for agriculture, is left.

*Global Loss.* Forests continue to be cleared at an alarming rate—about 23.2 million acres (9.39 million hectares) per year in the 1990s, most of which is in the developing countries, according to the 2003 *State of the World's Forests* report from the U.N. Food and Agriculture Organization (FAO). Over the past decade, an area almost three times the size of France has been lost, and much more has been degraded by fragmentation (partial clearing and thinning). About 80% of the deforestation is for agricultural purposes, despite the poor soil, and much of the agriculture is directed toward the production of cash crops or the conversion to grass to grow beef. Very little is managed with a clear aim of sustainable food production.

In all, less than 2% of tropical forests are under any kind of management plan for the protection or harvesting of forest products on a sustainable basis, although NGOs such as Conservation International are active in this regard. Also, the FAO has developed several bodies designed to address and coordinate management of the tropical forests: The Committee on Forestry and six regional forestry commissions provide a forum for member countries to discuss and develop policies that will lead to sustainability.

**The Other End of the Erosion Problem.** Water that is unable to infiltrate flows over the surface immediately into streams and rivers, overfilling them and causing flooding. Eroding soil, called **sediments**, is carried into streams and rivers, to clog channels and exacerbate flooding, fill reservoirs, kill fish, and generally upset the ecosystems of streams, rivers, bays, and estuaries. Around the world, coral reefs are dying because of sediments and other pollutants carried with them. Indeed, excess sediments and nutrients resulting from erosion are recognized as the greatest pollution problem of surface waters in many regions of the world. Meanwhile, groundwater resources are also depleted because the rainfall runs off, rather than refilling the reservoir of water held in the soil or recharging groundwater.

**Summary.** Lands suffering from, or prone to, soil degradation cover much of the globe (Fig. 8–20). In order to save this soil—the foundation of all terrestrial productivity—we must save the natural ecosystems that build, maintain, and nurture that soil, or we must at least provide mechanisms that duplicate those ecosystems. Erosion is a particularly insidious phenomenon, because the first 20 to 30% of the topsoil may be lost with only marginal declines in productivity, a loss that may be compensated for by additional fertilizer and a favorable distribution of rains. As the loss of topsoil continues, however, the decrease in productivity and increase in vulnerability to drought also continues.

## Irrigation and Salinization.

**Irrigation**—supplying water to croplands by artificial means—has dramatically increased crop production in regions that typically receive inadequate rainfall. Traditionally, water has been diverted from rivers through canals and flooded through furrows in fields, a technique known as **flood irrigation** (Fig. 8–21). In recent years, **center-pivot irrigation,** in which water is pumped from a central well through a gigantic sprinkler that slowly pivots itself around the well, has become much more popular. [See Fig. 7–16(b) and (c).] In the United States, the Bureau of Reclamation is the major governmental agency

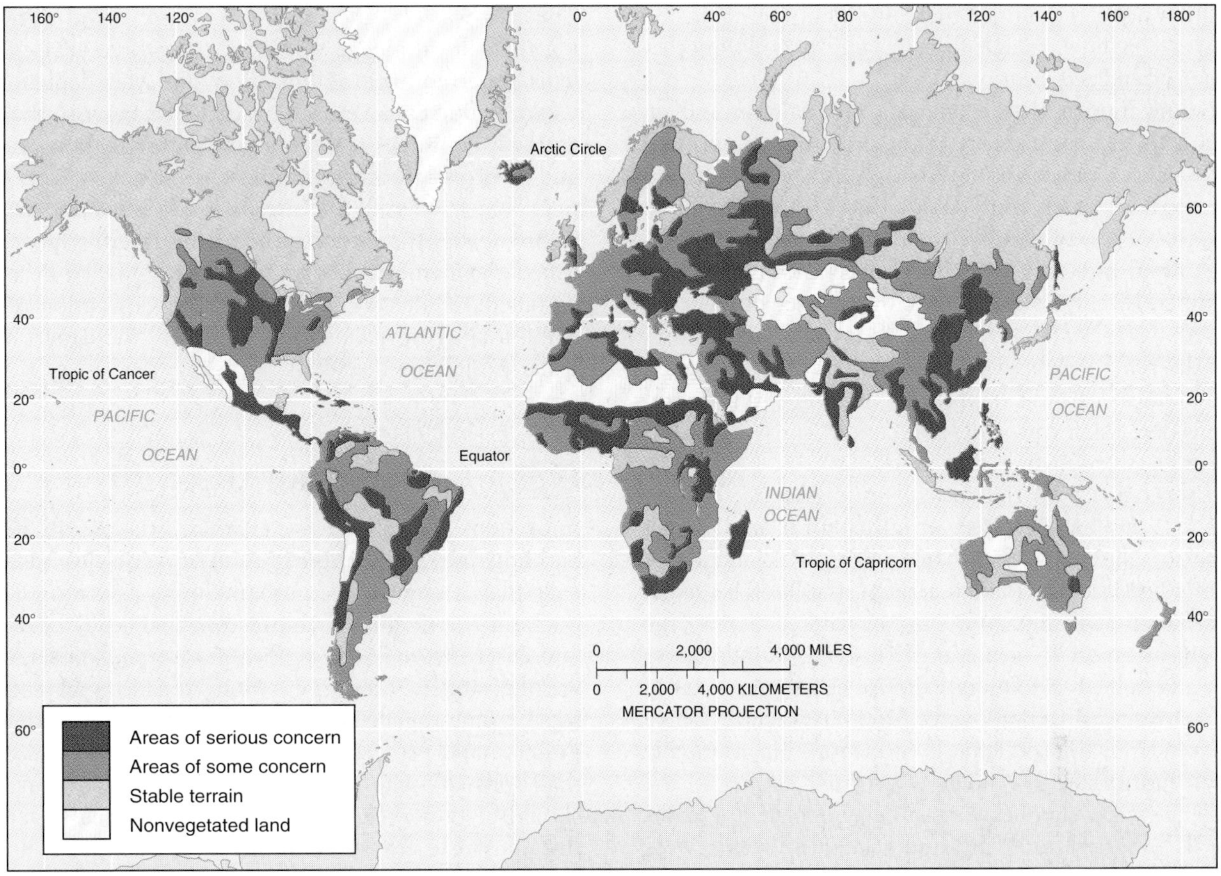


**Figure 8-20 Areas subject to soil degradation.** Throughout the world, overcultivation, overgrazing, and deforestation are degrading soil in vast areas.

involved with ongoing supplies of irrigation water to regions of the west. More than 600 dams and reservoirs have been constructed by the Bureau since its inception in 1902, and it now provides irrigation water for 10 million acres (4 million hectares) of farmland. Total irrigated land acreage in the United States is 67 million acres (27 million hectares), one-fifth of all cultivated cropland, according to the NRCS.

Worldwide, irrigated acreage has risen dramatically in the last few decades. The FAO estimates that total irrigated land now amounts to 667 million acres (270 million hectares), a 35% increase over the past 30 years. New

**Figure 8-21 Flood irrigation.** The traditional method of irrigation is to flood furrows between rows with water from an irrigation canal. This method is extremely wasteful, because most water evaporates or percolates beyond the root zone or the water table rises to waterlog the soil and prevent the crop from growing.

irrigation projects continue to be built, but the expansion in acreage is now being offset in large part by another ominous trend: *salinization*.

**Salinization** is the accumulation of salts in and on the soil to the point where plant growth is suppressed (Fig. 8–22). Salinization occurs because even the freshest irrigation water contains at least 200 to 500 ppm (0.02 to 0.05%) of dissolved salts. Adding water to dryland soils also dissolves the high concentrations of soluble minerals that are often present in those soils. As the applied water leaves by evaporation or transpiration, the salts remain behind and gradually accumulate as a precipitate. Because it happens in drylands, salinization is considered a form of desertification, since it renders the land less productive or even useless.

***Lost Land.*** Worldwide, an estimated 3.7 million acres (1.5 million hectares) of agricultural land is lost yearly to salinization and waterlogging. Asia alone has lost an estimated 320 million acres (130 million hectares) of farmland to salinization. In the United States, the problem is especially acute in the lower Colorado River basin and in the San Joaquin Valley of California, areas in which a total of 400,000 acres (160,000 hectares) have been rendered unproductive, representing an economic loss of more than $30 million per year. Adding to the problems, water supplies are fast being depleted by withdrawals for irrigation. This and related problems were discussed in Chapter 7.

Salinization can be avoided, and even reversed, if sufficient water is applied to leach the salts down through the soil. Unless there is suitable drainage, however, the soil will become a waterlogged quagmire in addition to being salinized. Drainage pipe may be installed 10–13 feet (3–4 meters) under the surface, but only at great expense, and then attention must be paid to where the salt-laden water is drained. The Kesterson National Wildlife Refuge in California was destroyed by pollution from irrigation drainage from selenium-enriched soils. The area was classified as a toxic-waste site after receiving the salty drainage water for only three years. Problems incurred from salinization and the depletion of water resources have been exacerbated because the government continues to build dams and irrigation projects and to provide water to farmers at far below cost, practices that encourage the use of excessive water. On the positive side, the Department of the Interior has established the National Irrigation Water Quality Program, designed to identify and address the irrigation-related water-quality problems associated with the department's water projects in the West.

## 8.3 Conserving the Soil

In Sections 8.1 and 8.2, we examined the nature of soils and how they support the growth of plants. We saw that soils play a crucial role in maintaining agricultural production. We examined the factors that lead to erosion and desertification—in particular, overcultivation, overgrazing, and deforestation. Along the way, we encountered some agencies and organizations working to preserve and protect the soils. There is broad consensus that soils are vital to human societies and that they should be preserved. This is encouraging, and so long as this consensus translates into action, we will be moving in the direction of sustainability.

Soil conservation must be practiced at two levels. The most important is the level of the *individual landholder*. As the chapter-opening story of Gabino López demonstrated, those working on the land are best situated to put into practice the conservation strategies that not only enhance their soils, but also bring better harvests and improve their way of life. A great deal of traditional knowledge exists with landholders, who routinely practice soil conservation techniques. The increased production in Burkina Faso is a prime example. This is stewardship in action.

**Figure 8–22 Salinization.** Millions of acres of irrigated land are now worthless because of the accumulation of salts left behind as water evaporates.

The second level is the level of *public policy*. In particular, farm policies play a crucial role in determining how soils are conserved. Policies governing the grazing of public lands and forestry practices also lead to either good or poor stewardship of the soils that underlie the resources being exploited. The story of Komsomolsky, at the beginning of the chapter, demonstrated how agricultural and human-resource policies can lead to a soil-related disaster on an epic scale. The two levels must work in harmony to bring about effective, sustainable stewardship of soil resources. Let us examine briefly what is being done in these two areas, beginning with public policy.

## Public Policy and Soils

Despite the crucial importance of saving agricultural land, forests, and rangeland, soil degradation in the form of erosion, desertification, and salinization continues to occur because of human activities. The losses are on a collision course with sustainability. Who is responsible for making changes?

***Subsidies.*** Farm policy in the United States was originally focused solely on increasing production. This goal has been achieved, as is demonstrated by continuing high yields and surpluses. In more recent years, farm policy has emphasized maintaining farm income and support of farm commodities—in other words, subsidies. In their recent book, *Perverse Subsidies,* Norman Myers and Jennifer Kent[3] point out how subsidies are bad news for the economy. Because of government subsidies, U.S. taxpayers support agriculture to the tune of $74 billion per year; artificially maintained prices for food add another $25 billion in support. Myers and Kent also argue that subsidies are bad news for the environment, because they encourage excessive use of pesticides and fertilizers, they reduce crop rotation by locking farmers into annual crop support subsidies, and they promote the continued drawdown of groundwater aquifers through irrigation. Recall from Section 8.2 how similar subsidies to ranchers for grazing on federal lands have undercut the objectives of soil conservation for those lands.

***Sustainable Agriculture.*** In comparison, the goals of sustainable agriculture are to (1) maintain a productive topsoil, (2) keep food safe and wholesome, (3) reduce the use of chemical fertilizers and pesticides, and, last, but far from least, (4) keep farms economically viable. Professional agriculturists are becoming increasingly aware of the shortfalls of modern farming and are at work developing alternatives. Many options actually mimic practices of the past, such as contouring, crop rotation, terracing, the use of smaller equipment, and the application of a reduced amount of chemicals or no chemicals at all. In 1988, the U.S. Department of Agriculture started the **Sustainable Agriculture Research and Education (SARE)** program, which provides funding for investigating ways to accomplish all of the goals of sustainable agriculture just listed. The program has continued to receive funding of $5 to $12 million annually, a small, but encouraging, amount earmarked for building and disseminating knowledge about sustainable agricultural systems.

***Conservation Reserve Program.*** Other winds of change have been blowing across the U.S. farm scene as well. In 1985, Congress passed the **Conservation Reserve Program (CRP),** under whose authority 40 million acres (16 million hectares) of highly erodible cropland were established as a "conservation reserve" of forest and grass. Farmers are paid about $50 per acre ($125 per hectare) per year for land placed in the reserve. As of 2002, 33.9 million acres (13.7 million hectares) had been placed, saving an estimated 700 million tons of topsoil per year from erosion—at a cost of $1.58 billion in "rental payments" to the 355,000 farms enrolled in the program. Under an associated bill, the **Food Security Act of 1985,** farmers are required to develop and implement soil-conservation programs in order to remain eligible for price supports and other benefits provided by the government.

***Farm Bills.*** The federal government supports U.S. agriculture through a number of programs; the farm lobby is a powerful one that has successfully maintained its support through the years. The **Federal Agricultural Improvement and Reform Act (FAIR) of 1996** (also referred to as the "Freedom to Farm" Act) attempted to bring needed reforms. Subsidies and controls over many farm commodities were reduced or eliminated, giving farmers greater flexibility in deciding what to plant, but also forcing them to rely more heavily on the market to guide their decisions (and earn their income!). Unfortunately, declining commodity prices prompted four years of emergency farm aid packages ($23 billion in additional funds), essentially maintaining the subsidies and controls targeted by FAIR. One new and encouraging program in the bill was the **Wetlands Reserve Program (WRP),** which, like the CRP, pays farmers to set aside and restore wetlands on their property.

The **2002 Farm Bill** succeeded the FAIR Act, many of whose provisions expired in 2002. The new bill continues to subsidize a host of farm products, maintaining price supports and farm income for American farmers. However, the new bill increased conservation funding for a number of programs. In 1985, for example, conservation activities supported by the Department of Agriculture received $1 billion, whereas these programs now receive more than $3 billion annually, mostly for the CRP and WRP. The new bill significantly increased funding for two newer programs, the **Wildlife Habitat Incentives Program (WHIP)** and the **Environmental Quality Incentives Program (EQIP),** both of which encourage

[3]Myers, Norman, and Jennifer Kent. 2001. *Perverse Subsidies: How Tax Dollars Can Undercut the Environment and the Economy.* Island Press. Washington, DC.

conservation-minded landowners to set aside portions of their land or address pollution problems.

Public policies aside, farmers should have incentives to nurture the soil on lands under cultivation. The more farmers incorporate conservation practices in their operations, the more productive their farms will become in the long run. The same applies to those who care for livestock that graze open rangelands. Most observers agree that the most significant obstacle to establishing sustainable soil conservation is a lack of knowledge about what such conservation can do. In particular, this applies to the developing countries, where agricultural policies often are poorly developed and the majority of landholders have little access to agricultural extension services.

## Helping Individual Landholders

In the developing world as well as the developed world, it is the individual landholders, farmers, and herders who hold the key to sustainable soil stewardship. They must be convinced that what they do will work, that it is affordable, and that, in the long run, their own well-being will improve. Frequently, this requires taking a number of small, realistic steps. It may also require help in the way of microlending (see Chapter 6), sound advice (often best coming from their peers), or simply encouragement to experiment (see the "Global Perspective" essay, this page). Toward those ends, the U.N. Convention to Combat Desertification is a large step in the right direction. Other efforts include the World Bank-coordinated *Soil Fertility Initiative for Africa* and the *Soil, Water and Nutrient Management Program* of the Consultative Group on International Agricultural Research (CGIAR). Reforms are often needed in agricultural policies in the affected countries so that the landholders and herders will be encouraged and enabled to initiate changes in their practices. Development specialists agree that the last thing small farmers in the developing countries need is a transfer of the high-technology agricultural practices employed in the industrialized countries. Most of the farmers that need to be reached are subsistence farmers working on small plots of marginal land.

***FARM.*** Various U.N. agencies and NGOs are working in the developing world to foster sustainable agriculture for subsistence farmers. One example is the **Farmer-centered Agriculture Resource Management (FARM)** program.

### global perspective

### Three-Strata Forage System for Mountainous Drylands

The Hindu Kush Himalayas (HKH) consist of a mountain chain straddling eight Asian countries from China and Myanmar in the east to Afghanistan and Pakistan in the west. The world's highest mountain region, it is also the most populous. The HKH occupies 1.65 million square miles (4.29 million km$^2$) and is home to more than 155 million people. Farming, forestry, and livestock sustain most of the people in the region. The land is fragile and is now threatened with an imbalance between population and available productive land. Rapid population growth and intensifying agriculture have led to, and are still producing, great increases in erosion and landslides as people clear forested slopes in a desperate attempt to meet their basic needs for fuelwood, timber, animal fodder, and cultivation. Aware of this intensifying problem, the **International Centre for Integrated Mountain Development** has promoted studies on how such a system can be brought into sustainable use. What is needed for the HKH region is an approach that will rehabilitate degraded lands, provide a sustainable source of fodder for animals, and also furnish fuelwood for domestic use. The *three-strata forage system* (TSFS) can accomplish all of these aims and has been shown to work in other dryland environments.

The three strata are (1) grass and legumes, (2) shrubs, and (3) fodder trees. Given the right combination of plants, this vegetation will prevent soil erosion, yield both fodder and human food and medicines, and create a pleasant environment where some greenery will persist all year. Such a system has worked well in the dry regions of Bali and is in use by almost all of the farmers there. In the HKH region, researcher Shaheena Hafeez has recommended a number of plants (grasses, legumes, shrubs, and trees) to use in a TSFS application for the mountainous terrain. One of the most interesting varieties is a tall shrub commonly found throughout the region: sea buckthorn (*Hippophae* spp.). This plant can root and thrive in poor soils, because it is a nitrogen fixer. Its root system is highly developed and holds the soil well. Sea buckthorn supplies palatable forage for animals all year and produces up to 1,300 pounds of berries per acre. The berries are a rich source of vitamins and can be made into jams, juices, and beverages. Oil from the berries is valued for its medicinal properties. A six-year-old sea buckthorn forest can provide over 7 tons of fuelwood per acre on a sustainable basis.

The TSFS is only one of a number of strategies needed for meeting what are called the "silent emergencies" of the remote HKH region (as opposed to the "loud emergencies" of global climate change, loss of biodiversity, etc.): poverty, energy and water shortages, scarcity of fuelwood and fodder, and poor access to sanitation, health, and education. The TSFS is an important strategy because it promises to provide most of the basic goods and services essential to life in these arid mountain lands.

FARM is a cooperative venture of eight Asian countries: China, India, Indonesia, Nepal, Philippines, Sri Lanka, Thailand, and Vietnam. The mission of FARM is to "support improved sustainable agricultural resource management and the attainment of household security through innovative approaches in rainfed areas of Asia." FARM's strategy is to find successful cases of sustainable agriculture used in the member countries and then create the conditions for replicating these successes. This is done by describing the effective technologies (both traditional and modern), communicating the information to professionals and policymakers in the various countries, and training farmers and development workers in these proven technologies. FARM has identified one or two field sites in each member country—small villages located in fragile ecosystems characterized by significant poverty. Site working groups composed of local farmers, community organizations, and government representatives are established at each site. As the groups develop solutions to problems, their work is communicated at regional and national levels by other FARM teams, with the objective of developing national farm strategies that incorporate the solutions found.

*Farmer Field Schools.* Under the FARM program, four farmer field schools have been established to teach integrated soil management. A typical school program involves local farmers and begins with an examination, usually conducted by a qualified soil specialist, of basic soil characteristics and typical soil-management problems that occur in the area. Then the farmers select several solutions to the problems and test the solutions in their own fields. Many of the ideas they test originate in traditions not well known by others in the group. At the end of a growing season, the results are measured and evaluated. The farmers make all the decisions and do all the fieldwork.

On a smaller scale than FARM, but using many similar methods, **Kenya Agricultural Research Institute (KARI)** has successfully spread sustainable technologies among subsistence farmers in that country. After six years of work in identifying useful approaches, KARI initiated five farmer field schools for the dissemination of the approaches. Some examples are the following:

1. Improved management and use of organic manures to improve soil fertility
2. Low-cost soil conservation structures
3. Suitable forages for waterlogged soils
4. Soil-improving green-manure legumes

By early 2002, some 475 farmers had participated in the five schools, which took place in the field and enlisted the full cooperation of the farmers through a full growing cycle.

These programs are still in their youth, but with their emphasis on the empowerment of local people, the use of traditional knowledge blended with modern agricultural science, and the strategy of communicating results to the farmers that need help the most, they promise to deliver just what sustainable development needs. To quote James Gustave Speth, former chief administrator of the UNDP and now Dean of the Yale School of Forestry and Environmental Studies, "Development will bring food security only if it is people centered, if it is environmentally sound, if it is participatory, and if it builds local and national capacity of self reliance."

# revisiting the themes

## Sustainability

Topsoils take hundreds or thousands of years to form and represent the most vital component of land ecosystems. Under most conditions, they are highly sustainable. Because of heavy human use of lands for agriculture, grazing, and forestry, however, topsoil is gradually being lost, causing the productivity of those ecosystems to decline. The key to sustainability in agriculture is soil conservation. Farmer experimenters in the developing countries are developing new approaches to farming, and farmer field schools are spreading the word on these new approaches, as well as many sound traditional approaches that are not widely known. The three-strata forage system of the mountainous drylands of Asia represents an amazing strategy for restoring soils and providing animal fodder and fuelwood that, if practiced, will promote long-term sustainability in those lands. (See "Global Perspective" essay, p. 228.)

## Sound Science

Soil science uses the principles of sound science to understand how soils are formed, how they are structured, and how they can be sustained. As with any science, however, it is possible to draw erroneous conclusions from the data, as seen with the GLASOD data on world soils and soil degradation. The same tendency can be seen in the data on soil erosion in the United States, where the measurements (or rather, estimates) are made by applying standard equations to information gathered from maps and aerial photographs. Actual "ground truth" in both cases has cast serious doubt on the accuracy of the claims made about erosion and degradation rates. The truth is that more field measurements are desperately needed, because losing soil and degrading the land bear long-term and devastating consequences.

## Stewardship

Those who work the land or graze their livestock are stewards of the soil. They are temporary tenants whose lives will be far shorter than the time it took to form the soil. Their stewardship can lead to sustainable soil health, or it can lead to further erosion and degradation of the soil. A few short years of poor management can lead to the loss of hundreds of years of topsoil. Soil is cared for at the local level. Global figures may frighten or impress, but soil stewardship takes place one field at a time, by the people who are most familiar with the land.

## Ecosystem Capital

Soil represents one of the most vital and irreplaceable components of ecosystem capital. Soils support the plants (and therefore the animals) in ecosystems, promote nutrient cycling, treat wastes, store and retain water, and are the premium resource for producing food to support the 6 billion humans on Earth.

## Policy and Politics

The plight of the drylands has caught the world's attention. Virtually all the countries in the world have ratified the U.N. CCD; many have developed National Action Programs, and the numerous meetings of the Conferences of the Parties have helped to disseminate practical information on accomplishments in many developing countries. Programs such as the farmer field schools are bringing soil science and new low-tech approaches to soil conservation to the rural farmers who need it most.

In the United States, the NRCS functions as an effective agency promoting soil conservation down to the local farm level. Funded by Congress and administered within the U.S. Department of Agriculture, the NRCS oversees the various land and soil restoration and conservation programs. Most of the $3 billion budget goes to the CRP and WRP programs to pay landowners for retiring erosion-prone croplands and wetlands.

On the negative side, however, the 2002 Farm Bill maintains an enormous budget subsidizing a host of farm commodities. The subsidies the bill authorizes are bad news for the economy, for the international market, and for the environment. The farm lobby is so strong that Congress appears to be unable to shed these subsidies, which often reward the wealthiest and largest farmers only and lock everyone into perpetual crop support. A similar subsidy rules over federal grazing lands, where ranchers (who also have a strong lobby) pay well below market price to graze their animals and the income from the fees is far below what it costs to administer the program.

The movement to promote a federal buyout of rancher's grazing permits is likely to lead to legislation; look for a bill called the Voluntary Grazing Permit Buyout Act. Although ranching associations oppose the proposal, many ranchers are in favor of it.

# review questions

1. What are the different layers, or horizons, that compose the soil profile? What makes each one distinct from the others?
2. Name and describe the properties of the three main soil *textures*. Do the same for the four major soil *orders* listed in this chapter.
3. What are the main things that plant roots must obtain from the soil? Name and describe a process (natural or unnatural) that can keep plants from obtaining the amounts of each of these things they need for survival.
4. Describe, in terms of both content and physical structure, the soil environment that will best fulfill the needs of plants.
5. State and define the relationships among soil structure, humus, detritus, and soil organisms. Describe the role each of these factors plays in creating a soil environment that best supports the needs of plants.
6. What is meant by mineralization? What causes it? What are its consequences for soil?
7. How does a vegetative cover protect and nurture the soil in a way that supports a plant's growth?
8. What is the impact of water and wind on bare soil? Define and describe the process of erosion in detail.
9. What is meant by desertification? Describe how the process of erosion leads to a loss of water-holding capacity and, hence, to desertification.
10. What are some problems with the sweeping claims made about soil erosion and soil degradation? How have these claims been challenged?
11. What are drylands? Where are they found, and how important are they for human habitation and agriculture?
12. What are the three major cultural practices that expose soil to the weather?
13. What is meant by salinization, and what are its consequences? How does salinization result from irrigation?
14. What are the two levels at which soil conservation must be practiced? What is being done at each level?

# thinking environmentally

1. Why is soil considered to be a detritus-based ecosystem? Describe how the aboveground portion and the belowground portion of an ecosystem act as two interrelated and interdependent communities.
2. Suppose you have uncovered an excellent new technique for keeping dryland soils moist and fertile. How would you get your message to the subsistence farmers?
3. Evaluate the following argument: Erosion is always with us; mountains are formed and then erode, and rivers erode canyons. You can't ever eliminate soil erosion, so it is foolish to ask farmers to do so.
4. Suppose you were going into farming. Describe the type of operation you would create, and give a rationale for each measure you would take to farm sustainably.
5. Success stories in land degradation and desertification have been solicited by the UNEP and presented on the Web site www.unep.org/unep/envpolimp/techcoop/1.htm. Consult some of these stories, and comment on what is common to, and what is different about, them.

chapter 9

# The Production and Distribution of Food

**Key Topics**

1. Crops and Animals: Major Patterns of Food Production
2. New Patterns: Genetically Modified Foods
3. Food Distribution and Trade
4. Hunger, Malnutrition, and Famine

Half the people in the world—most of them in the developing countries—eat rice daily. Although an excellent food for energy, rice is an insufficient source of vitamins and other nutrients; hence, it must be supplemented with other foodstuffs for a balanced diet. Unfortunately, millions cannot afford other foods and depend heavily on rice for their nutrition. The result is two of the most common nutritional deficiencies: vitamin A deficiency, which leads to blindness and immune-system failures, and iron deficiency, which ultimately causes anemia and also, immune-system problems. According to the WHO, over 800,000 children die each year and 140 million children are affected by vitamin A deficiency alone. If rice could somehow be supplemented with iron and vitamin A, countless millions of children could be helped.

Biochemist Ingo Potrykus, working at the Swiss Federal Institute of Technology in Zurich, believed that it could be done and, in the early 1990s, began a program of research that would lead to what many now believe may be the most remarkable accomplishment of plant biotechnology: golden rice (Fig. 9–1). Over a period of seven years, using the techniques of gene splicing, Potrykus, colleague Peter Beyer of the University of Freiburg in Germany, and their coworkers incorporated seven genes into a rice genome that would enable the plant to synthesize beta carotene (which the body can use, in turn, to synthesize vitamin A) and iron in the body of the grain. (Both beta carotene and iron are normally present in the leaves of the plant, but when the rice is milled, only the grain is saved.)

Although the level of beta carotene in golden rice is low, it is sufficient to prevent death and blindness if as little as a half pound of golden rice is consumed daily. The inventors have donated the rice to the *International Rice Research Institute* (IRRI) in the Philippines. The institute is testing the rice for environmental safety, food safety, susceptibility to diseases, digestibility, and palatability. The testing will take several years. The IRRI will make the new rice available free of charge to farmers in developing countries, once the testing has proved satisfactory.

***A Terraced Rice Plantation in Bali, Indonesia.***

**Figure 9–1 Golden rice.** The yellow rice kernels contain beta carotene, inserted into them with biotechnology to enable the rice to help meet nutritional needs for vitamin A.

***"Farmageddon."*** Golden rice has met with a less-than-universal welcome, because it is a genetically modified organism, containing genes from daffodils and beans. Some environmental organizations called it "a cruel hoax" or "grains of delusion." Genetically modified organisms are called "Frankenfoods" by many, because they contain foreign genes and were created in a laboratory. The international debate over these foods has injected a new set of fears into the food production and distribution industry, and the debate is far from settled. If society is to meet the challenge of feeding another 3 or 4 billion people in this century, however, and if we are to do so with better nutrition, genetic-engineering technology is likely to play a key role.

Throughout the world, three decades of rapid population growth have left hundreds of millions dependent on food aid, while one in five in the developing world remains undernourished. As the world population continues its relentless rise (Fig. 5–3), no resource is more vital than food. Can the world's farmers and herders produce food swiftly enough to keep up with population growth? After all, Earth holds only a finite amount of the resources needed for food production—namely, suitable land, water, energy, and fertilizer.

***Successes.*** By many measures, human societies have done very well at putting food on the table. More people are being fed than ever before, with more nutritional food. In the last 25 years, world food production has more than doubled, rising even more rapidly than population. In addition, the daily amount of food available in the developing countries has increased by 25%. A lively world trade in foodstuffs forms the bulk of economic production for many nations. The number of hungry people has declined by 40 million in the past decade. Optimists would say that, although regions of chronic hunger and malnutrition—and occasional famines—exist, these are the exceptions to an otherwise remarkable accomplishment. Those viewing the situation through rose-colored glasses are convinced that when the world population levels off (at 10 billion?), enough food will still be produced to sustain it. Others question whether that is at all possible. There is real concern whether it will even be possible to meet the Millennium Development Goal of reducing by half the proportion of people suffering from hunger.

This chapter examines food-production systems, problems surrounding the lack of food, and the global effort being made to reduce hunger and malnutrition.

## 9.1 Crops and Animals: Major Patterns of Food Production

Some 12,000 years ago, the "Neolithic Revolution" saw the introduction of agriculture and animal husbandry, which probably did more than anything else to foster the development of human civilization. Virtually all of the major crop plants and domestic animals were established in the first thousand years of agriculture. Between 1450 and 1700, world exploration and discovery led to an exchange of foods that greatly influenced agriculture and nutrition in the Western world. From the New World came potatoes, maize (corn), beans, squash, tomatoes, pineapples, and cocoa. Rice came from the Orient. In exchange, Europeans brought to the New World wheat, onions, sugar cane, and a host of domestic animals—horses, pigs, cattle, sheep, and goats. With the advancing capabilities of science and technology in the 1800s, the stage was set for a remarkable change in agriculture.

### The Development of Modern Industrialized Agriculture

Until 150 years ago, the majority of people in the United States lived and worked on small farms. Human and animal labor turned former forests and grasslands into systems that produced enough food to supply a robust and growing nation (Fig. 9–2a). Farmers used traditional approaches to combat pests and soil erosion. Thus, crops were rotated regularly, many different crops were grown, and animal wastes were returned to the soil. The land

(a) (b)

**Figure 9–2 Traditional vs. modern farming.** (a) Traditional farming practiced in Arkansas. For hundreds of years, traditional practices on American farms supported the growing U.S. population. (b) Modern agricultural practice, illustrated by a combine harvesting wheat in America's Midwest.

was good, and farming was efficient enough to allow a substantial segment of the population to leave the farm and join the growing ranks of merchants and workers living in cities and towns. Then, in the mid-1800s, the Industrial Revolution came to the United States, and it had a major impact on farming.

**The Transformation of Traditional Agriculture.** The Industrial Revolution revolutionized agriculture so profoundly that today, in the United States, some 3 million farmers and farmworkers produce enough food for all the nation's needs, plus a substantial amount for trade on world markets (Fig. 9–2b). This revolution increased the efficiency of farming remarkably. Since the mid-1930s, therefore, the number of farms has decreased by two-thirds (from 6.8 million to 2.3 million), while the size of farms has grown fourfold. (They now average 400 acres). Farming is a large and important business in the United States. All farm and farm-related employment (all jobs involved in producing the goods and services needed to satisfy the demand for farm products) accounts for 15% of the total U.S. workforce. Indeed, the revolution has achieved such gains in production that the United States routinely has to cope with surpluses of many crops.

Virtually every industrialized nation has experienced this agricultural revolution. The pattern of developments in U.S. agriculture could just as well describe those in France, Australia, or Japan. Crop production has been raised to new heights, doubling or tripling yields per acre (Fig. 9–3). To understand how this tremendous increase in production was made possible, you need to understand the components of the agricultural revolution. These, in turn, reveal some of the problems brought on by the revolution.

*Machinery.* An incredible array of farm machinery handles virtually every imaginable need for working the soil, seeding, irrigating, weeding, and harvesting. The list is impressive and includes tractors, combines, handlers, harrows, mowers, toppers, tedders, balers, spreaders, foragers, plows, tillers, disc harrows, scufflers, strippers, seed drills, swathers, and more. This machinery has enabled farmers to cultivate far more land than ever. However, the shift from animal labor to machinery has created a dependency on fossil-fuel energy. As the price of oil rises, so does the cost of food production.

*Land under Cultivation.* In the United States, cropland and pastureland comprise some 487 million acres (197 hectares). This is about one-fourth of the total land area of the lower 48 states. Agriculture, therefore, has an enormous impact on the landscape. Before 1960, much of the increased production in the United States came from bringing new land into production. Since then, increases in crop yields and consistent crop surpluses have taken the pressure off of additional conversions to cropland, enabling farmers to be selective in the land they cultivate. The surpluses have also given farmers the opportunity to take erosion-prone land out of production. Under current farm policy, the **Conservation Reserve Program** reimburses farmers for "retiring" erosion-prone land and planting it with trees or grasses. The law allows up to 39 million acres to be placed in this program, and land in the program currently totals 33.9 million acres (13.7 million hectares).

Essentially all of the good cropland in the United States is now under cultivation or held in short-term reserve. Globally, agriculture occupies 38% of the land, and the net rate of growth of this land has been constant at around 0.3% per year over the past 30 years. The expansion in cropland comes at the expense of forests and wetlands, however, which are both economically important and ecologically fragile.

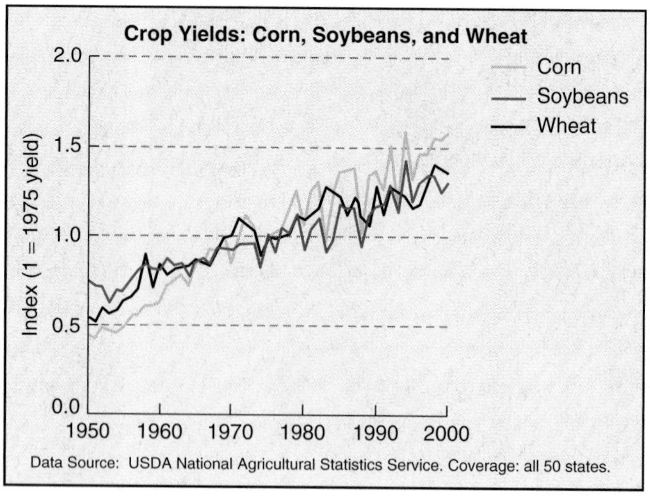


**Figure 9–3 U.S. crop yields.** Corn, wheat, and soybean yields (measured as tons or bushels per acre), indexed to 1975 as the base year. A value above 1.0 indicates a higher yield than in 1975, below 1.0 a lower yield than in 1975. (*Source:* USDA National Agricultural Statistics Service.)

*Fertilizers and Pesticides.* Farmers long ago learned that animal manure and other organic supplements to soil could increase their crop yields. When chemical fertilizers first became available and began being used, however, farmers discovered that they could achieve greater yields. Moreover, the fertilizers were more convenient to use and more available than manure. When fertilizers were first employed, 15 to 20 additional tons of grain were gained from each ton of fertilizer used. Between 1950 and 1990, worldwide fertilizer use rose tenfold. Now, however, farmers in developed countries are applying near-optimal levels of fertilizer. When levels of fertilizer are too high, the excess is washed away, resulting in groundwater and surface-water pollution. (See the "dead zone" story in Chapter 17.) The worldwide use of fertilizer resumed its rise in 1995, following a six-year decline traced to economic conditions in the countries of the former Soviet Union, and now approaches the highest level ever (146 million tons). Most of the current increase is in "underfertilized" countries such as China, India, and Brazil.

Chemical pesticides have provided significant control over insect and plant pests, but, due to natural selection, the pests have become resistant to most of the pesticides. As a result, pesticide use has tripled since 1970, but the percentage of crops lost to pests has remained constant. Also, efforts to reduce the use of pesticides have begun because of side effects to human and environmental health. Many of the chemicals that are in use have not been adequately tested for human concerns and their potential for causing genetic defects. As is discussed in Chapter 16, progress is being made toward developing natural means of control that are environmentally safe, but these new methods will be unlikely to increase yields.

*Irrigation.* Worldwide, irrigated acreage increased about 2.6 times from 1950 to 1980. By 2000, irrigated acreage represented 17% of all cropland—some 677 million acres (274 million hectares)—and produced 40% of the world's food. Irrigation is still expanding, but at a much slower pace because of limits on water resources. (Irrigation represents 70% of all water use.) Recall from Chapter 8, that much current irrigation is unsustainable, because groundwater resources are being depleted. In addition, production is being adversely affected on as much as one-third of the world's irrigated land, because of waterlogging and the accumulation of salts in the soil—consequences of irrigating where there is poor drainage.

*High-yielding Varieties of Plants.* Several decades ago, plant geneticists developed new varieties of wheat, corn, and rice that gave yields double to triple those of traditional varieties. This feat was accomplished by selecting strains that diverted more of the plant's photosynthate (photosynthetic product) to the seed and away from the stems, leaves, and roots. However, the seeds of modern wheat, rice, and corn now receive more than 50% of the photosynthate, close to the calculated physiological limit of 60%. As these new varieties were grown in the developed countries and then introduced throughout the world, production soared, and the Green Revolution was born.

## The Green Revolution

The same technologies that gave rise to the agricultural revolution in the industrialized countries were eventually introduced into the developing world. There, they gave birth to the remarkable increases in crop production called the **Green Revolution.**

In 1943, the Rockefeller Foundation sent agricultural expert Norman Borlaug and three other U.S. agricultural scientists to Mexico, with the objective of exporting U.S. agricultural technology to a less developed nation that had serious food problems. Their aim was to improve the traditional crops grown in Mexico, especially wheat. Mexican wheat was well adapted to the subtropical climate, but it gave low yields and responded to fertilization by growing very tall stalks that were easily blown over. Using wheat from other areas of the world, Borlaug and his coworkers bred a dwarf hybrid with a large head and

(a) (b)

**Figure 9–4 Traditional vs. high-yielding wheat.** Comparison of (a) an old variety of wheat, shown growing in Rwanda, with (b) a new, high-yielding variety of dwarf wheat growing in Mexico.

a thick stalk. The hybrid did well in warm weather when provided with fertilizer and sufficient water (Fig. 9–4). The program was highly successful: By the 1960s, Mexico had closed the gap between food production and food needs, wheat production had tripled, and Mexican wheat appeared on the export market.

***Nobel Effort.*** Research workers with CGIAR extended the work done in Mexico, introducing modern varieties of high-yielding wheat and rice to other developing countries. In one success story, Borlaug induced India to import hybrid Mexican wheat seed in the mid-1960s, and in six years India's wheat production tripled. Within a few years, many of the world's most populous countries turned the corner from being grain importers to achieving stability and, in some cases, even becoming grain exporters. Thus, while world population was increasing at its highest rate (2% per year), rice and wheat production increased 4% or more per year. In 35 years, global food production doubled. The Green Revolution has probably done more than any other single scientific or other achievement to prevent hunger and malnutrition. Borlaug was awarded the Nobel peace prize in 1970 in recognition of his contribution.

***Panacea?*** The high-yielding modern varieties are now cultivated throughout the world and have become the basis of food production in China, Latin America, the Middle East, southern Asia, and the industrialized nations. Because the technology raises yields without requiring new agricultural lands, the Green Revolution has also held back a significant amount of deforestation in the developing world. But as remarkable as it was, the Green Revolution is not a panacea for all of the world's food-population difficulties. More water is required to raise the high-yielding grains; thus, these grains depend critically on irrigation. Water shortages have begun to occur as a result of this dependence (Chapter 7). Also, the modern varieties require constant inputs of fertilizer, pesticides, and energy-using mechanized labor, all of which are often in short supply in developing countries. Further, because it is patterned after agriculture in the developed world, Green Revolution agriculture tends to benefit larger landholders. More food is thus raised by a smaller farm workforce, causing many farm laborers and small landholders to become displaced and migrate to the cities, joining the ranks of the unemployed.

***Greener Later.*** The CGIAR recently sponsored a study of the impact of the Green Revolution in the developing world between 1960 and 2000.[1] The study's findings have shed new light on that important agricultural development:

- The early years of the Green Revolution (1960–1980) were only the beginning; research on high-yielding crops has continued, and more varieties continue to be developed, released, and adopted by farmers.
- More recent research has featured resistance to diseases, pests, and climatic stresses. This work has been the key to the increased adoption of modern varieties.
- The early Green Revolution contributed greatly to expanded food production in Asia and Latin America; the later years of the Green Revolution have benefited Africa and the Middle East, as a greater variety of crops was developed.

[1]Evenson, R. E., and D. Gollin. "Assessing the Impact of the Green Revolution, 1960 to 2000." *Science* 300: 758–762 (May 2, 2003).

- If there had been no Green Revolution, crop yields in developing countries would certainly have been lower, leading to higher food prices and an increase in lands brought under cultivation everywhere. Less food would have been available, leading to higher levels of hunger and malnutrition and higher infant mortality.

Even with the modern varieties, agricultural production in Sub-Saharan Africa has lagged behind the rest of the developing world. The reason for this gap is the dominance of subsistence agriculture in that region.

## Subsistence Agriculture in the Developing World

In most of the developing world, plants and animals continue to be raised for food by *subsistence farmers*, using traditional agricultural methods. These farmers represent the great majority of rural populations. **Subsistence farmers** live on small parcels of land that provide them with the food for their households and, it is hoped, a small cash crop. From the point of view of the modern world, such farmers are very poor, although many do not consider themselves to be so. Like past agricultural practice in the United States, subsistence farming is labor intensive and lacks practically all of the inputs of industrialized agriculture. Also, it is often practiced on marginally productive land (Fig. 9–5).

Typically, a family owns a small parcel of land for growing food and maintains a few goats, chickens, or cows. The system is often quite sustainable, because crop residues are fed to livestock, livestock manure is used as a fertilizer, and the family's nutrition is adequate. Such a family is making the best use of very limited resources. Keep in mind, however, that subsistence agriculture is practiced in regions experiencing the most rapid population growth, even though that kind of agriculture is best suited for low population densities. An estimated 1.4 billion people in Latin America, Asia, and Africa—over one-third of the people there—depend on subsistence agriculture.

***Problems.*** The pressures of population and the diversion of better land to industrialized agriculture lead to practices that are often unsustainable and sometimes ecologically devastating. In many regions in developing countries, woodlands and forests are cleared for agriculture or removed for firewood and animal fodder, forcing the gatherers to travel farther and farther from their homes and leaving the soil susceptible to erosion. The ensuing scarcity of firewood leads the residents to burn animal dung for cooking and heat, thus diverting nutrients from the land. Erosion-prone land suited only to growing grass or trees is planted to produce annual crops. Good land is forced to produce multiple crops instead of being left fallow to recover nutrients. Growing populations compel the continued subdivision of land, which diminishes the land's ability to support each household. All these factors tend to increase the poverty that is characteristic of populations supported by subsistence agriculture, and, in a relentless cycle (see Fig. 6–7), the added poverty in turn puts increased pressures on the land to produce food and income.

***Successes.*** Because subsistence agricultural practice varies with the local climate and with local knowledge, it is difficult to draw sweeping generalities. In some areas, subsistence agriculture involves shifting cultivation within tropical forests—often called slash-and-burn agriculture.

**Figure 9–5 Subsistence farming.** Mexican farmer plowing marginally productive land with a team of oxen. (Note that he is plowing uphill, inviting erosion.) Subsistence farming feeds more than 1.4 billion people in the developing world.

**Figure 9–6 Slash-and-burn agriculture.** Countryside in Mindanao, Philippine Islands, showing the results of sustainable slash-and-burn agriculture. Cultivated fields are interspersed with trees and natural areas in a diverse ecosystem.

Research has shown that this practice can be sustainable (Fig. 9–6). The cultivators create highly diverse ecosystems in which the cleared land supports a few years of crops and gradually shifts into agroforestry—a system of tree plantations with different ground crops employed as the trees grow.

In other areas, subsistence farmers are showing remarkable success in adapting to the changing needs of local societies as they are forced to support expanding populations on the same land. In Kenya, for example, land in the Machakos region was seriously degraded during the 1930s. Now the land has recovered and supports a population six times larger, because the 1.5 million people there have diversified their agriculture and practiced soil and water conservation.

## Animal Farming and Its Consequences

Foot-and-mouth disease and mad-cow disease have plagued the livestock industry from Europe to South America. Australia, however, has been free of these diseases, and the cattle business there is booming. Australia is home to 28 million cattle, making it the world's top beef exporter. These cattle are ranch raised, grazing over vast areas of the Australian rangelands (Fig. 9–7). One cattle company alone manages a herd of 400,000 animals and would like to expand it to a million. The country's reputation as a provider of "disease-free" beef has brought in good times for Australian cattlemen.

Raising livestock—sheep, goats, cattle, buffalo, and poultry—has many parallels to raising crops (Table 9–1), and there are many connections between the two. Fully one-fourth of the world's croplands are used to feed domestic animals. In the United States alone, 70% of the grain crop goes to animals—on half the cultivated acreage. The care, feeding, and "harvesting" of the estimated 17 billion domestic animals constitute one of the most important economic activities on the planet. The primary force driving this livestock economy is the large number of the world's people who enjoy eating meat and dairy products—primarily, most of the developed world and growing numbers of people in less developed nations. As with crop farming, however, there are two patterns. On the one hand, in the developed world and on ranches in the developing world, livestock is raised in large herds and often under factorylike conditions. On the other hand, in rural societies in the developing world, livestock is raised on family farms or by pastoralists who are subsistence farmers.

***Factory Animals.*** Industrial-style animal farming can affect the environment in a host of nonsustainable ways. Because so much of the plant crop is fed to animals, all of the problems of industrialized agriculture apply to

**Figure 9–7 Cattle Ranching in Australia.** Jackaroos (Australian cowboys) drive a herd of cattle in the Injune area, Queensland.

animal farming. In addition, rangelands are susceptible to overgrazing, either because of mismanagement of prime grazing land or because the land on which the animals graze is marginal dry grasslands, used in that manner because the better lands have been converted to producing crops. Much of the western rangeland in the United States is public land leased at subsidized fees that easily lead to overgrazing (Chapter 8). Another serious problem is the management of animal manure. In developing countries, manure is a precious resource that is used to renew the fertility of the soil, build shelters, and provide fuel. In developed countries, it is a wasted resource. Close to 1.3 billion tons of animal waste is produced each year in the United States, some of which leaks into surface waters and contributes to die-offs of fish, contamination with pathogens, and a proliferation of algae. With factory farms on the sharp increase, wastes from manure and meat processing are either bypassing or overwhelming the often inadequate treatment systems available and polluting the nation's waterways. The EPA asserts that animal-based agriculture is the most widespread source of pollution in the nation's rivers.

*Rain Forest Crunched.* In Latin America, more than 49 million acres (20 million ha) of tropical rain forests have been converted to cattle pasture. Even though most of this land is best suited for growing rain-forest trees, some of it does support a rural population of subsistence farmers producing a diversity of crops. Much of the land, however, is held by relatively few ranchers who own huge spreads. The so-called hamburger connection between rain forests and fast-food hamburgers is an oversimplification, however. This conversion of land from

**table 9-1 Parallels between Plant and Animal Farming**

| | Plant | Animal |
|---|---|---|
| Major products | Grains, fruits, and vegetables for food | Meat, dairy products, and eggs for food |
| Other important products | Oils, fabrics, rubber, specialty crops (spices, nuts, etc.) | Labor, leather, wool, manure, lanolin |
| Modern practices | Industrialized agriculture on former grasslands and forests | Ranching, dairy farming, and stall-feeding |
| Traditional practices | Subsistence agriculture on marginally productive lands | Pastoral herding on nonagricultural lands |
| Current global land use | 3.7 billion acres (1.5 billion ha, or 11% of land surface) | 8.4 billion acres (3.4 billion ha, or 26% of land surface) |

forest to pasture has been the outcome of government policies meant to encourage the colonization of unoccupied land and only secondarily to develop overseas markets for agricultural products.

*Climate Change.* According to the IPCC, deforestation and other changes in land use in the tropics release an estimated 1.6 billion tons of carbon to the atmosphere annually, contributing a significant amount of carbon dioxide to the greenhouse effect. Also, because their digestive process is anaerobic, cows and other ruminant animals annually eliminate some 100 million tons of methane, another greenhouse gas, through belching and flatulence. The anaerobic decomposition of manure leads to an additional 30 million tons of methane per year. All this methane released by livestock makes up about 3% of the gases causing global warming (Chapter 20).

*Good Cow.* Even though their animals also contribute to the methane problem, it is callous to fault the subsistence farmers whose domestic animals enhance their diet and improve their quality of life. In fact, one of the most important kinds of sustainable development aid brought to rural families is the gift of a cow or a few goats or rabbits, as is carried out by Heifer Project International (Fig. 9–8). Working in 115 countries, the project has distributed large animals, beehives, fowl, and fish fingerlings to families. The organization works with grassroots groups of local people who oversee a given project. Each project is chosen to meet a genuine need and to improve the environment. Those receiving the animals must agree to be trained in how to raise them properly and must be committed to passing on the gift in the form of donating offspring of their livestock to other needy people.

The lives of millions in the developing world are tied directly to the animals they raise, and their impact on the environment is often sustainable. Animals that are well managed can enhance the soil and enable rural farmers to maintain a balanced farm ecosystem. These farmers own two-thirds of the world's domestic animals, so the rise in animal husbandry in the developing world has made an important contribution to nutrition, especially for women and small children. In sum, animal farming is far more likely to be sustainable in the rural farms and pastoral herds of the developing world than in the beef ranches and hogpens of the developed world, where there is a pressing need to address problems of pollution, overgrazing, and deforestation.

## Prospects for Increasing Food Production

The global trends in food production look rather rosy. Over the last 30 years, grain harvests and meat production have outpaced population growth (Fig. 9–9). Food production has not only kept up, but is even surpassing, population growth. In particular, rising incomes in the developing world have stimulated the rapid rise in meat consumption seen in Figure 9–9. The real concern, then, is whether these trends can continue. Looking ahead to 2020, will we be able to feed the additional $1\frac{1}{2}$ billion people and also make significant progress on meeting the Millennium Development Goal of reducing by half existing hunger and malnutrition in the world?

*IFPRI View.* A team from the International Food Policy Research Institute (IFPRI) recently looked at the prospects, asking those very questions. As a baseline, the IFPRI team expects the demand for grains to be much as it has been, slightly outpacing population increase. But it also expects a big jump in the world's consumption of meat and projects a 50% rise in consumption by 2020. Again, this increase is traced to rising demands in the developing world. Even with that projected rise, however, per capita consumption of meat in the developing world will still be far below that of the developed world. The team foresees a distinct shift away from red meat and towards poultry. The major consequence of this rise in

**Figure 9–8 Heifer Project International.** An East African woman gives a goat to a neighbor as part of the project's "passing on the gift" ceremony. Participants agree to pass on offspring of their livestock to others.

**Figure 9–9 Global population and grain and meat production.** Grain production and meat production not only have kept up with population increase, but even have outpaced it over the last 30 years. (*Source:* FAO FAOSTAT database collections.)

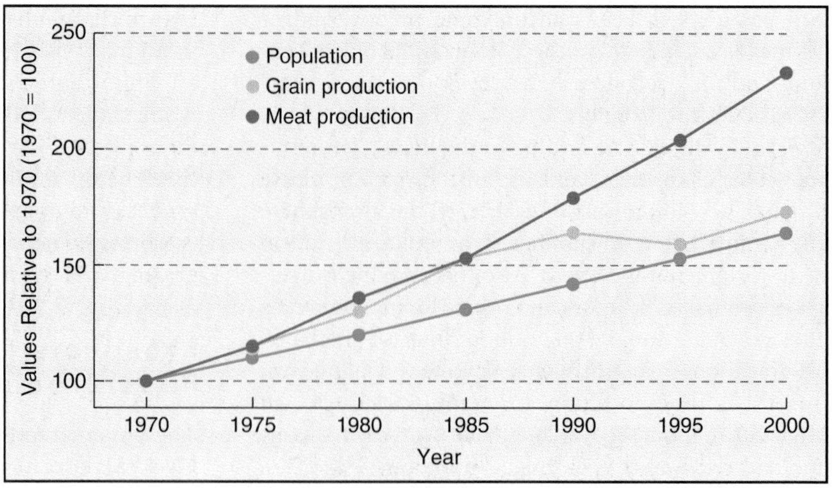


demand for meat is a shift from grains for human consumption to the production of feed grains.

The prospects for developing countries to meet the rise in demand for grains are not good. New agricultural land is scarce, and much of what will undoubtedly be brought into production is marginal land that is highly subject to erosion and soil degradation. In the opinion of agricultural experts, the opposing trends of degrading agricultural lands and opening up new lands are likely to offset each other. However, more lands probably will be irrigated, putting yet more pressure on groundwater supplies. Recall from Chapter 7 that water tables are dropping, so it is likely that, very shortly, the newly irrigated lands will be offset by abandoned irrigated lands. The pressure to increase productivity on existing land will be great, but increases in yields for grains have been slowing down. Some progress will undoubtedly be made, but clearly, not nearly enough.

The rising demand for grains in the developing world will have to be met by trade, and the major producers in the developed world will have to be willing and able to meet that demand. Net grain imports by developing countries will likely double by 2020, but only those countries making significant progress in economic development will be able to afford the imports. Poorer countries may be left behind. Because poorer countries are where much of the malnutrition and hunger already occur, the prospects for a great reduction in these conditions are not good, the team concluded. In particular, it focused on sub-Saharan Africa, which the IFPRI believes will be a "hot spot" for hunger and malnutrition for years to come.

***Perspectives.*** Is the IFPRI scenario realistic? Consider, first, the team's point relating to bringing more land into cultivation. Where this is done, it means the loss of grasslands, forests, or wetlands and the ecosystem goods and services they provide. Between 1990 and 2000, there was a net global loss of 232 million acres (94 million hectares) of forests, almost all in the developing countries and to agriculture. This loss will undoubtedly continue as population increases in the developing countries. But the IFPRI team pointed out that only about 40 million more hectares are likely to be converted to crops, due to major constraints on suitable land. Because most of the prime arable land in the world is now in cultivation, their conclusion appears valid.

***Increase Yields?*** Thus, there seem to be only two prospects for increasing food production: (1) Continue to increase crop yields and (2) begin growing food crops on land that is now used for feedstock crops or cash crops. As mentioned earlier, a dramatic rise in grain yields in the developing countries was the major accomplishment of the Green Revolution. Indeed, grain yields have continued to rise in some of the *developed* countries. In France, for example, wheat yields have quadrupled since 1950, reaching over 7 tons per hectare. Rice yields in Japan have risen 67% in the same period. Actually, the genetic potential exists for yields as high as 14 tons per hectare or higher for wheat. Can we expect yields to continue to increase up to their genetic potential?

It turns out that the great differences in yields of grain between regions (Table 9–2) have little to do with the genetic strains used. Egypt and Mexico, for example, achieve much higher yields than the United States because most of their wheat is irrigated, while U.S. wheat is rain fed and grown on former grasslands. Australian wheat yields are even lower, the result of the country's sparse rainfall. Once the agricultural land is planted with high-yielding strains and fertilized to the maximum, other factors—soil, rainfall, and available sunlight—limit productivity. These environmental limits are a reminder that agricultural sustainability is highly dependent on soil and water conservation (Chapter 8) and on the weather. Variations in climate, especially those associated with El Niño and La Niña (Chapter 20), affect harvests locally and globally, injecting significant instability in food production that makes future planning extremely chancy. Another imponderable in future food production is global climate change. Thus, as the predicted warming of the 21st century occurs, it is impossible to project how rainfall patterns will change.

***Africa.*** In the developing world, sub-Saharan Africa has the greatest need for increasing crop yields. With

**table 9-2 Annual Wheat Yield, Tons per Hectare, in Various Countries, 2002**

| Country | Yield |
|---|---|
| United Kingdom | 7.9 |
| France | 7.4 |
| Egypt | 6.0 |
| Mexico | 5.2 |
| Zimbabwe | 4.1 |
| China | 3.8 |
| Poland | 3.6 |
| Ukraine | 3.0 |
| India | 2.7 |
| United States | 2.4 |
| Pakistan | 2.3 |
| Russian Federation | 2.2 |
| Argentina | 2.2 |
| Canada | 1.7 |
| Kenya | 1.4 |
| Kazakhstan | 1.1 |
| Australia | 1.0 |

*Source:* FAOSTAT Agricultural Data, FAO, 2003.)

much of its land arid and populations continuing to grow at exponential rates, the region continues to fall behind in per capita food production and is becoming increasingly dependent on food imports. Yields would have to increase 2.5% annually just to keep up with population growth. In the face of the devastating AIDS epidemic in the region, and with chronic political upheaval, this seems unlikely. The potential is there, however, to make great progress with some very basic steps. For example, with the aid of an extension service promoted by former President Jimmy Carter, Ethiopia harvested record crops in 1995–96, showing a 32% increase in production and a 15% increase in yield in one year. The key to this growth was the use of a simple fertilizer providing nitrogen and phosphorus to the nutrient-starved soils. Poor agricultural support and a lack of capital to purchase fertilizer and new seeds are preventing many other African countries from making similar improvements.

*Less Meat?* Now, what of the possibility of switching from the production of feed grain and cash crops to food for people? Fully 70% of domestic grain in the United States is used to feed livestock. The percentages drop over other regions of the world, in proportion to the economic level of the region. Sub-Saharan Africa and India, for example, use only 2% of their grain to feed livestock. Feed grain can be considered a buffer against world hunger; if the food supply becomes critical, it might force more of the world's people to eat lower on the food chain (less meat, more grain). The trend, however, is in exactly the opposite direction. Of course, converting land use from cash crops to food crops also is possible, but it is a complex undertaking, because it involves such issues as land reform and maintaining a balance of trade.

In the face of these constraints, Gordon Conway, president of the Rockefeller Foundation, has called for a new Green Revolution. He has dubbed it a **Doubly Green Revolution**—"a revolution that is even more productive than the first Green Revolution and even more 'green' in terms of conserving natural resources and the environment. During the next three decades, it must aim to repeat the successes of the Green Revolution on a global scale in many diverse localities and be equitable, sustainable, and environmentally friendly." Conway believes that this new revolution is possible, in large part because of the emergence of biotechnology and the kinds of genetic manipulations demonstrated by the development of golden rice.

## 9.2 New Patterns: Genetically Modified Foods

Genetic engineering makes it possible to crossbreed genetically different plants and to incorporate desired traits into crop lines and animals, producing so-called transgenic breeds. Genetic engineering can also be used to clone domestic animals such as cows and goats. This is a radically different technology from the genetic research that produced the Green Revolution: Researchers are no longer limited to genes that already existed within a species or that could arise via mutation; it is now possible to exchange genes among bacteria, animals, and plants.

This kind of technology can help the developing world to produce more food. Genetically modified organisms are under intense research and development, with new products announced almost daily. Still, in spite of the obvious potential of these organisms, serious objections continue to arise to their development and use. We need to look at both sides of this controversy.

### The Promise

The first genetically altered products to be marketed were (1) the Flavr Savr™, a tomato that can be vine ripened and subsequently brought to market and kept fresh much longer than locally produced "ordinary" tomatoes; (2) cotton plants with built-in resistance to insects that comes from genes taken from a bacterium (so-called Bt, standing for *Bacillus thuringiensis,* the name of the bacterium); and (3) numerous crop plants resistant to the herbicide Roundup®, allowing farmers to employ no-till techniques. More recently, *genomics*—as the technique has come to be

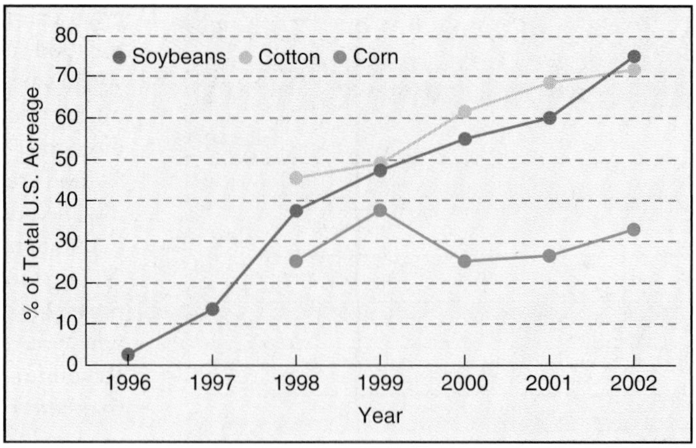


**Figure 9–10 Biotech crops in the United States.** The plot shows the rapid increase in percentage of three crops planted to produce biotech varieties. (*Source:* Biotechnology Industry Organization and USDA.)

known—has developed sorghum (an important African crop) resistant to a parasitic plant known as witchweed, which infests many crops in Africa; corn, potatoes, and cotton resistant to insects; rice resistant to bacterial blight disease; trees and salmon that grow very rapidly; and goats that produce spider silk, to name a few organisms. In just seven years since bioengineered seeds became commercially available, three-quarters of all soybeans and cotton and one-third of the corn acreage in the United States are planted with transgenic breeds (Fig. 9–10). Worldwide, more than 165 million acres (67 million hectares) are now planted with bioengineered crops, and one-fourth of this area is in developing countries.

Biotech crop research that can benefit the developing countries is proceeding at a rapid pace, often in those very countries. China, in particular, has emerged as a leader in plant biotechnology, having recently sequenced the rice genome. Among other aims, the objectives of genomics are (1) to engender resistance to diseases and pests that attack important tropical plants, (2) to increase tolerance to environmental conditions, such as drought and a high salt level, which stress most plants, (3) to improve the nutritional value of commonly eaten crops (as has been done with golden rice), and (4) to incorporate vaccines against major human diseases into the cells of commonly eaten plants such as bananas. Thus, products under development include virus-resistant cassavas, sweet potatoes, melons, squash, tomatoes, and papayas (Fig. 9–11); protein-enhanced corn and soybeans; fungus-resistant bananas; bananas and tomatoes that contain an antidiarrheal vaccine; and drought-tolerant sorghum and corn. The potential for *transgenic* crops and animals seems almost unlimited.

Among the important environmental benefits of bioengineered crops are reductions in the use of pesticides, because the crops are already resistant to pests; less erosion, because no-till cropping is facilitated by the use of herbicide-resistant crops; and less environmental damage associated with bringing more land into production, because existing agricultural lands will produce more food. When Chinese cotton farmers began turning to Bt cotton, for example, they found that they used much less pesticide, saving money, time, and their own health in the process.

## The Problems

**Environmental Concerns** Concerns about genetic engineering technology involve three considerations: environmental problems, food safety, and access to the new techniques. A major environmental concern focuses on the pest-resistant properties of the transgenic crops. With such a broad exposure to the toxin or some other resistance incorporated into the plant, it is possible that the pest will develop its own resistance to the toxin more rapidly and thus render it ineffective as an independent pesticide. The transgenic crop then loses its advantage. Another concern is the ecological impact of the crops. For example, pollen from Bt corn (resistant to the corn borer) disperses in the wind and can spread to adjacent natural areas, where beneficial insects may pick it up and be killed by the toxin. This effect was shown to occur in laboratory tests with monarch butterflies. Recent studies indicate, however, that the Bt corn poses no risk to the butterflies,

**Figure 9–11a Papaya plants in Hawaii.** The healthy plants have been genetically altered to prevent ring spot, a viral disease of papayas; the plants on the right are traditional ones.

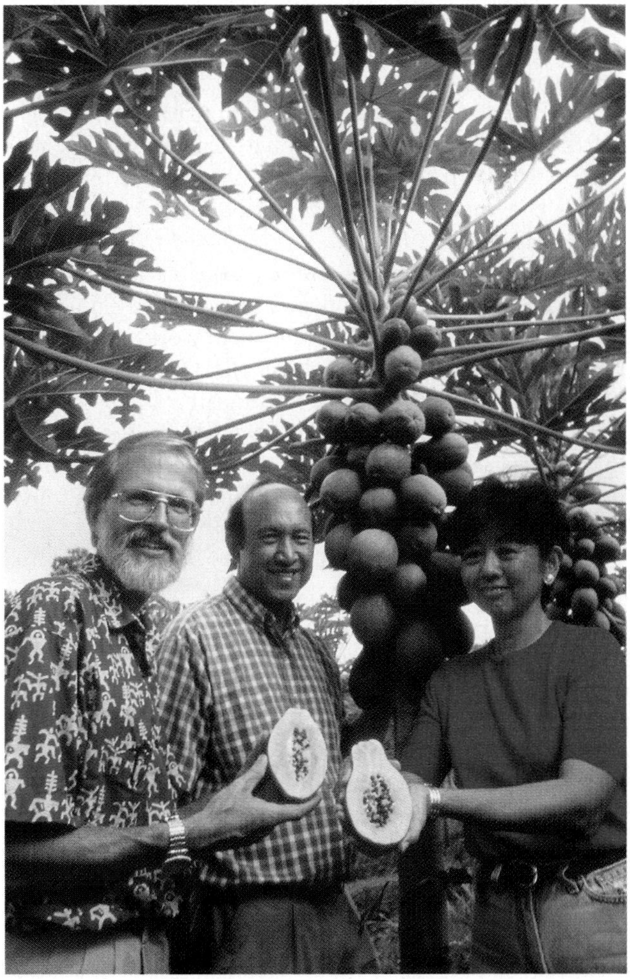

**Figure 9–11b** The developers of a papaya that is resistant to a deadly virus; from left, Dr. Richard Manshardt, Dr. Dennis Gonsalves, and Maureen Fitch, a Hawaiian graduate student.

and the EPA has cleared the corn for continued planting. However, farmers are required to plant at least 20% of their corn acreage with conventional varieties, in order to slow the development of pest resistance.

*Ask Schmeiser.* A third environmental concern arises because genes for herbicide resistance or for tolerance to drought and other environmental conditions can also spread by pollen to wild relatives of crop plants, possibly creating new or "super" weeds. Field studies indicate that this concern is valid, although the actual gene flow was not great. Canadian farmer Percy Schmeiser learned the hard way about this possibility. He was taken to court by agrochemical giant Monsanto, who accused him of "stealing" their patented Roundup-ready canola plants and profiting from the technology. His crime? His own canola crops had become "contaminated" via pollen blown in from neighboring high-tech canola fields. Schmeiser became aware of the contamination, but did nothing to profit from the presence of the hybrid plants. Nevertheless, Monsanto took him to court and won a verdict ordering Schmeiser to pay Monsanto $125,000 in fines. The court ruled that Schmeiser should have notified Monsanto of the problem in his fields, and Monsanto would have helped him to get rid of the GM canola. Schmeiser lost an appeal in 2002 and is now appealing to the Canadian Supreme Court.

**Safety Issues.** Food safety issues arise because transgenic crops contain proteins from different organisms and might trigger an unexpected allergic response in people who consume the food. Tests have shown that a Brazil nut gene incorporated into soybeans was able to express a protein in the soybeans that induced an allergic response in individuals already allergic to Brazil nuts. Also, antibiotic-resistance genes are often incorporated into transgenic organisms in order to provide a way to trace cells that have been transformed. The transgenic product could then convey the resistance to antibiotics to pathogens in human systems by a separate process of gene transfer, or the product could prevent the antibiotic from being used effectively in an individual who has consumed the transgenic food. Another concern relates to the possibility that a plant might produce new, toxic substances in its tissues in response to the presence of the foreign genes. Critics also point out that the possibility of unknown and unanticipated harm is always present. To date, none of these concerns has become evident in countries where GM crops have been grown over the past seven years.

*Corn for Aid.* In the summer of 2002, Zambia turned down a shipment of corn from the United States that was destined to provide famine relief to the country, on the grounds that the corn was likely contaminated with some Bt corn. (It had been stored in the same warehouse in the United States as Bt corn, Fig. 9–12.) Zambian scientists urged the government to reject the corn because its health risks were uncertain. The debate quickly spread to the Zambian countryside, where confusion reigned as people became convinced that the food was "poisoned" and not only would make people sick,

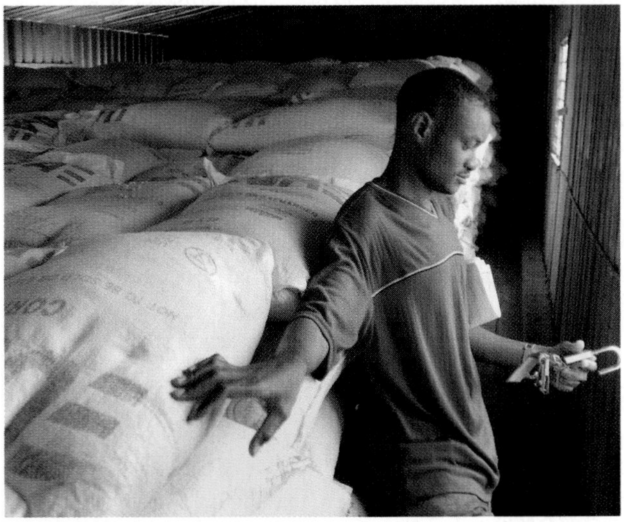

**Figure 9–12 Emergency food in Zambia.** A Zambian worker checks the lock on a warehouse full of U.S.-donated corn suspected of being genetically modified.

but also could lead to infertility and even AIDS. Many other African countries accepted the corn, however, with the stipulation that it be milled so that farmers could not grow genetically modified crops from it.

**Developing-World Access** The problems concerning access to the new technologies relate to the developing world. For the first few years, almost all genetically modified organisms were developed by large agricultural–industrial firms, with profit as the primary motive. Accordingly, farmers have been forbidden by contract from simply propagating the seeds themselves and must purchase seeds annually. Farmers in the developing countries are far less able to afford the higher costs of the new seeds, which must be paid up front each year. Aware of the potential for farmers simply to begin to produce their own seeds, researchers developed what is called the *terminator technology*, a transgenic technique that renders any seeds from the crops sterile. Faced with a firestorm of criticism, Monsanto backed down on a decision to adapt terminator technology to its transgenic crops. Still, the problem of cost remains. Fortunately, some noncommercial and donor-funded laboratories are taking aim at this problem, as is demonstrated by the development of golden rice. Biotech research is booming in China, India, and the Philippines, where most of the fruits of the research will be made available to developing-world farmers. Further, the genetically modified seeds are spreading rapidly through seed "piracy."

All of these concerns, combined with the general fear of what unknown technologies can bring, have resulted in a huge controversy over the spreading use of genetically modified organisms by farmers and the foods developed from such organisms. Activists around the world have rallied to protest the development of such organisms and the "Frankenfoods" derived from them. Concern has gone beyond protest: Agricultural test plots in the United States, Europe, and New Zealand have been vandalized. The protests have been strongest in Europe, where concerns over food have been exacerbated by other recent food scares, such as mad-cow disease. Far less concern has arisen so far in the United States. Observers believe that the fact that U.S. companies have been in the forefront of this technology might have something to do with the European response. In light of the controversy, major food-producing companies are reducing their use of genetically modified organisms in foodstuffs, suggesting that there may begin to be a reduction in the use of this technology in common food crops such as corn and soybeans. Unfortunately, farmers remain caught in the middle and are scrambling to maintain a balance between conventional and genetically modified versions of their crops in order to satisfy an unpredictable market.

## Policies

Biotechnology and its application to food crops do not exist in a regulatory vacuum. In the United States, the EPA, the U.S. Department of Agriculture (USDA), and the Food and Drug Administration (FDA) all have had regulatory oversight of different elements of the application of biotechnology to food crops. Concern over their oversight led to a report from the National Academy of Science's National Research Council, a prestigious and influential body. The report, *Genetically Modified Pest-Protected Plants: Science and Regulation,* concluded that transgenic crops have been adequately tested for environmental and health effects, but that the three agencies involved need to do a better job of coordinating their work and sharing their information more with the public. Overall, the report endorsed the science, technology, and regulation, stating, "The committee is not aware of any evidence that foods on the market are unsafe to eat as a result of genetic modification." At the same time, the report called for more research on the environmental and safety issues.

***Cartagena Protocol.*** On the international level, the U.N. Convention on Biodiversity sponsored a key conference in Montreal, Canada, in January 2000, to deal with trade in genetically modified organisms. After heated negotiations, the conference reached an agreement, called the **Cartagena Protocol on Biosafety.** Surprisingly, the agreement was welcomed by governments, the private sector, and environmental groups alike. At issue was a fundamental philosophy about how technologies, particularly new ones, should be regulated. Those skeptical of the claims of biotech companies about the safety of their products wanted *proof that the genetically modified organisms were safe* before they were allowed into their countries. Those advocating the use of such organisms wanted to see the *suspected dangers proven* before genetically modified organisms could be denied access to overseas markets.

***Precautionary Principle.*** The Cartagena Protocol states that the "lack of scientific certainty due to insufficient relevant scientific information and knowledge... shall not prevent [a country] from taking a decision" on the import of genetically modified organisms. The Protocol puts the right to deny entry of any of these organisms in the hands of the importing country, but its decision must be based on sound science (involving an assessment of the risks involved) and the broad sharing of information about the products. Thus, the agreement makes operational, for the first time in international environmental affairs, the so-called **precautionary principle,** which states that *where there are threats of serious or irreversible damage, lack of scientific certainty should not be used as a reason for failing to take measures to prevent potential damage.* According to another plank in the Protocol, shipments that contain food commodities made with genetically modified organisms must be labeled to say that they *may* contain them. This wording was a compromise that enabled opposing parties to reach a consensus on the entire Protocol. As with other agreements from U.N. conventions, participating countries must ratify the Protocol; on September 11, 2003, the Cartgena Protocol on Biosafety "entered into force."

A regulatory framework is being erected around the use of this new technology. The public response is still highly negative in Europe and positive in the United States, although activists have been quite vocal on both sides of the Atlantic and are starting to be heard in the developing countries. The European Union (EU) has lifted a ban on genetically modified foods and is requiring clear labeling on any such food or animal feed containing more than 0.9% of genetically modified ingredients. Countries everywhere are cautiously examining the technology, and the result has been a slowdown in the development and adoption of genetically modified food crops in particular. However, the future almost certainly will have to include major advances in food production from biotechnology, and many of these advances are likely to benefit the developing countries. A new plant revolution is underway, and it is unlikely to be stopped. Former U.S. President Jimmy Carter put it well: "Responsible biotechnology is not the enemy; starvation is." If food production is to keep pace with population growth, biotechnological advances will be essential, in the view of most observers.

## 9.3 Food Distribution and Trade

For centuries, the general rule for basic foodstuffs—grains, vegetables, meat, and dairy products—was *self-sufficiency.* Whenever climate, blight (as in the 19th-century Irish potato famine), or war interrupted the agricultural production of a nation or region, the inevitable result was famine and death, sometimes on the scale of millions. Once colonies were established in the New World, timber, furs, tobacco, fish, and, later, sugar, coffee, cotton, and other raw materials began to flow back to the Old World. In turn, the Old World exported manufactured goods, which helped transform the colonies into societies much like the European ones that had given birth to them. With the Industrial Revolution, trade between nations intensified, and soon it became economically feasible to ship basic foodstuffs from one part of the world to another. In time, a lively and important world trade in foodstuffs arose, and as it did, the need for self-sufficiency in food diminished. Like other sectors of the economy, food has become globalized. World trade in agricultural products was worth $412 billion in 2001.

### Patterns in Food Trade

Given this global pattern, agricultural production systems do much more than supply a country's internal food needs. For some nations (such as the United States and Canada), the capacity to produce more basic foodstuffs than the home population needs represents an extremely important entry into the international market. For many other countries (especially those of the developing world), special commodities such as coffee, fruit, sugar, spices, cocoa, and nuts provide their only significant export product. This trade clearly helps the exporter, and it allows importing nations to enjoy foods that they are unable to raise themselves or that are out of season. Given the realities of a market economy, the exchange works well only as long as the importing nation can pay cash for the food. Cash is earned by exporting raw materials, fuel, manufactured goods, or special commodities. In this way, for example, Japan imports $35 billion worth of food and livestock feed, but more than makes up for it by exporting $400 billion in manufactured goods (cars, electronic equipment, and so on) annually.

***Grain on the Move.*** The most important foodstuff on the world market is grain—namely, wheat, rice, corn, barley, rye, and sorghum. Much of this is feed grain imported by some high- and middle-income countries to satisfy the rising demand for animal protein. It is instructive to examine the pattern of global trade in grain over the past six decades (Table 9–3). In 1935, only Western Europe was importing grain; Asia, Africa, and Latin America were self-sufficient. By 1950, new patterns were emerging, and today the trade in grains—as well as other basic foodstuffs—represents a development with great economic and political implications. As the table shows, North America has become the major source of exportable grains—the world's "breadbasket" or, in another sense, the world's "meat market." Over the past 40 years, Asia, Latin America, and Africa have shown an increasing dependence on imported grain. These three regions have also experienced 40 years of continued population growth. For example, Mexico, birthplace of the Green Revolution, imported 16 million metric tons of grain in 2001 because population growth has eaten up all the gains of the Green Revolution. Although much of the food needs of these regions are met by internal production (Mexico, for example, produced 31 million tons of grain in 2001), the trend toward greater dependency is an ominous signal. If, as IFPRI projects, the developing countries import twice as much grain by 2020, the grain-exporting countries will have to either greatly increase domestic production or use less grain (not likely).

At no time in recent history has the world grain supply run out. On the average, carryover stocks (the amount held in storage by exporters—206 million tons in 2002) are enough to supply a year's worth of international trade and aid. In other words, enough food is produced to satisfy the world market—enough to feed millions of animals and keep a healthy surplus in storage. Why, then, are there people in every nation who are hungry and malnourished? Shouldn't each nation make an all-out effort to provide food for its people, and if it is unable to, shouldn't the rest of the world assume some of the responsibility for providing food to a hungry nation? Where does the responsibility lie for meeting the need for this most basic resource?

**table 9-3 World Grain Trade Since 1935**

| | Amount Exported or Imported (Million Metric Tons)[1] | | | | | | | | |
|---|---|---|---|---|---|---|---|---|---|
| Region | 1935 | 1950 | 1960 | 1970 | 1980 | 1990 | 1995 | 2000 | (2000 Production)[2] |
| North America | 5 | 23 | 39 | 56 | 131 | 123 | 108 | 98 | 396 |
| Latin America | 9 | 1 | 0 | 4 | −10 | −12 | −4 | −16 | 137 |
| Western Europe | −24 | −22 | −25 | −30 | −16 | 28 | 26 | 19 | 217 |
| Eastern Europe and Russian Federation | 5 | 0 | 0 | 0 | −46 | −38 | −5 | −6 | 169 |
| Africa | 1 | 0 | −2 | −5 | −15 | −31 | −31 | −44 | 113 |
| Asia | 2 | −6 | −17 | −37 | −63 | −82 | −78 | −74 | 996 |
| Australia and New Zealand | 3 | 3 | 6 | 12 | 19 | 15 | 13 | 21 | 35 |

[1] A minus sign in front of a figure indicates a net import; no sign indicates a net export.
(*Source:* FAOSTAT Food Balance Sheets, Food and Agricultural Organization.)
[2] For comparison, grain production figures for each region are given.

## Food Security

The Bread for the World's Hunger Institute defines **food security** as "assured access for every person to enough nutritious food to sustain an active and healthy life." To begin to answer the question of who is responsible for food security, it is helpful to examine Figure 9–13. The figure displays *three major levels of responsibility for food security:* the **family,** the **nation,** and the **global community.** At each level, the players are part of a market economy as well as a sociopolitical system. In a market economy, *food flows in the direction of economic demand. Need* is not taken into consideration. Suppose, for example, that there are hungry cats and hungry children. The food will go to the cats if the owners of the cats have money and the children's parents don't. In other words, the cash economy, following the rules of the market, provides the *opportunity* to purchase food, but not the food itself. Where the economic status of the player (a jobless parent, a poor country) is very low, the next level may be willing and able to provide the needed purchasing power or the food (but maybe not).

***Family Level.*** The most important level of responsibility is the family. The *goal* at this level is to meet the nutritional needs of everyone in the family to an extent that provides freedom from hunger and malnutrition. For an individual, there are four legitimate options for attaining food security: You can purchase the food, raise the food, gather it from natural ecosystems, or have it provided by someone, usually in the context of the family. In the event of economic or agricultural failure at the family level, the fourth option implies that there is an effective *safety net*—that, at some level, national, state, or local, there exist policies or programs whose objective is to meet the food security needs of all individuals in the society. These policies and programs can be of two types: (1) official policies, represented by a variety of welfare measures, such as the Food Stamp Program and the Supplemental Security Income program; and (2) voluntary aid through hunger-relief organizations.

***National Level.*** The policy safety net was radically transformed by the *Personal Responsibility and Work Opportunity Reconciliation Act of 1996 (PRWOR).* The act is intended to move welfare recipients from public assistance to independence by requiring that they join the workforce or perform "workfare" and by limiting the duration during which they are publicly supported. Eligibility for food stamps for adults also has work requirements. By most measures, the law was successful. Welfare caseloads in the majority of states dropped sharply, an average of 50% across the country (but this happened during economic good times). During the economic recession of the early 2000s, however, there was a rise in caseloads in most states, a dramatic increase in the number of people served by food banks, and a rise in family homelessness. PRWOR is scheduled for renewal, and an amended act is in the works in Congress.

America's Second Harvest, a network of food banks throughout the country, forms part of the voluntary safety net. The network reported a 25% increase in requests for emergency food in 2002 compared with 2001. Many more of the hungry are working poor, often living in the suburbs, who must choose between paying for food and paying for heat in their homes; or they may be the elderly, who must choose between paying for their prescription drugs and paying for food. These people often depend on one of the more than 50,000 local food charities for at least one good meal a day. America's Second Harvest retrieves food from food processors and suppliers and distributes it to the organization's network of food banks. A new law, the *Charity, Aid, Recovery and Empowerment Act of 2003* (CARE Act) allows family farmers, ranchers, and restaurant owners to deduct the costs of food donated

| Family | Country | Global community |
|---|---|---|
| *Goal:* Personal and family food security<br>*Policies:*<br>*—Employment security*<br>*—Adequate land or livestock*<br>*—Good health and nutrition*<br>*—Adequate housing*<br>*—Effective family planning*<br>*—Access to food* | *Goal:* Self-sufficiency in food and nutrition<br>*Policies:*<br>*—Just land distribution*<br>*—Support of sustainable agriculture*<br>*—Effective family planning*<br>*—Promotion of market economy*<br>*—Avoidance of militarization*<br>*—Effective safety net* | *Goal:* Sustainable food and nutrition for all countries<br>*Policies:*<br>*—Food aid for famine relief*<br>*—Appropriate technology in development aid*<br>*—Aid for sustainable agricultural development*<br>*—Debt relief*<br>*—Fair trade*<br>*—Disarmament*<br>*—Family planning assistance* |

**Figure 9–13 Responsibility for food security.** Goals and strategies for meeting food needs at three levels of responsibility: the family, the country, and the globe.

to agencies such as America's Second Harvest as a charitable contribution.

***Developing Countries.*** For developing countries, an appropriate goal at the national level would be *self-sufficiency in food*—enough food to satisfy the nutritional needs of all of a country's people, thereby providing an *effective safety net.* To meet this goal, a nation can either produce all the food its people needs or buy the food on the world market. This goal implies the existence of policies to eliminate chronic hunger and malnutrition among its people, such as a just land distribution policy, and a functioning market economy. Many nations, though, are not self-sufficient in food, and those that cannot afford to buy all they need must turn to the global community for *food aid.* (See Section 9.4.)

***Global Level.*** The WTO is the body that governs international trade. Meeting every few years, the WTO brings together developed and developing member countries for negotiations and policy decisions. The wealthy developed countries dominate the proceedings because of their economic strength. In particular, these countries have maintained policies that are highly protective of their agricultural sectors, while calling on the developing countries to liberalize their trade policies. This hypocrisy works itself out in the *tariffs* the developed countries set on agricultural products and in the *subsidies* given to the agricultural sector (Fig. 9–14). By subsidizing agricultural export products, the developed countries are able to undercut farmers in the developing countries, where such subsidies are simply not affordable. For example, Haiti cut its tariff on rice from 50% to 3% in 1995, opening the country's markets to subsidized rice from the United States. By 2000, rice production in Haiti was cut in half, leaving thousands of small rice farmers without their means of livelihood. Recall from Section 8.3 that the latest U.S. farm bill continued the subsidies of most agricultural products, even adding some new ones. The EU, notorious for its high tariff barriers and heavy subsidies to its agricultural sector (to the tune of $5 billion per year), has finally enacted agricultural reforms that change the rules for subsidies. Farmers will simply receive a one-time annual payment based on previous subsidies and the size of their farm. It is hoped that farmers will either grow less food or else grow what the market wants, eliminating the huge surpluses that have occurred in recent decades.

***Other Needs.*** There are other initiatives that could help the poorer countries to become self-sufficient in food. One of the most important is relieving the debt crisis in developing countries, an issue addressed in Chapter 6. Another factor in meeting global nutrition needs is the *trade imbalance* between the industrial and the developing countries. The developing countries typically export commodities such as cash crops, mineral ores, and petroleum, and import more sophisticated manufactured products such as aircraft, computers, machinery, and the like. Prices for the latter have risen, while those for the commodities from developing countries often decline. To some extent,

**Figure 9–14 Sugar cane harvest in Venezuela.** Developing country sugar crops are greatly disadvantaged in world markets because of subsidies that make EU sugar much cheaper.

the increasing use of labor in the developing countries for electronic assembly and in the manufacture of clothing has offset this imbalance. Indeed, the globalization of markets has begun working to the advantage of the developing countries, which have a surplus of labor and low labor costs. As a result, many industrial corporations have outsourced their manufacturing to developing countries. What the poor need is jobs, and even if the pay is far below what workers in industrial countries receive, it is often enough to help workers in developing countries improve their living conditions and their diets.

Another consequence of the trade imbalance is the tendency of the developing countries to "mine" their natural resources for the international market. Thus, lumber, fish products, minerals, and other resources are exploited in order to generate a cash flow. In effect, these countries are spending down their ecosystem capital, an unsustainable process that eventually will leave them environmentally bankrupt.

Appropriate global food security goals for the wealthy nations are listed in Fig. 9–13. These are policies that promote self-sufficiency in food production and sustainable relationships between the poorer nations and their environments, as well as addressing the trade imbalance and human-exploitation problems.

## 9.4 Hunger, Malnutrition, and Famine

At a U.N. World Food Conference in 1974, delegates from all nations subscribed to the objective "that within a decade no child will go to bed hungry, that no family will fear for its next day's bread, and that no human being's future and capacities will be stunted by malnutrition." In 1996, 22 years later, the United Nations held the World Food Summit to address *continuing* hunger, malnutrition, and famine in the world. (See the "Global Perspective" essay, p. 251.) In 2002, the FAO published an update showing that 840 million are still suffering from hunger. With the Millennium Development Goals, a new, more modest target has been proposed: *to cut world hunger in half by 2015.* In light of the evident availability of enough food to feed everyone, and in view of the global market in foodstuffs, why is there still so much hunger and malnutrition in the world?

### Nutrition vs. Hunger

**Hunger** is the general term referring to a lack of basic food required for energy and for meeting nutritional needs such that the individual is unable to lead a normal, healthy life. **Malnutrition** is the lack of essential nutrients, such as specific amino acids, vitamins, and minerals, and **undernourishment** is the lack of adequate food energy (usually measured in Calories). Hunger, malnutrition, and undernourishment exist in the United States, but one-fourth of the U.S. population is clinically obese (more than 30 pounds overweight). Worldwide, at least 1.2 billion people overeat and are overweight. To address this and other nutritional problems, the USDA established six food groups and arranged them in a pyramid—the **food guide pyramid** (Fig. 9–15)—to indicate the relative proportions of each group that should be consumed.

Following the pyramid's guidelines should facilitate weight control and cut down on the chances of developing diabetes, heart disease, high blood pressure, and some cancers, typical problems of the overnourished. Recently, the USDA pyramid has come under fire because it recommends lots of carbohydrates but few fats. Researchers have shown that some fats (plant oils) are highly protective against heart disease, and refined carbohydrates (white bread, white rice, white pasta) raise the risks of heart disease and diabetes. The best way to combat obesity is to increase the amount of exercise you get and to lower your total calorie intake.

## global perspective

### World Food Summit

Under the leadership of the Food and Agricultural Organization (FAO) of the United Nations, representatives and heads of state from 100 countries met during November 13–17, 1996. Called the World Food Summit, the meeting was the result of two years of planning and negotiations. The objective was to bring about a renewed commitment around the world to eradicate hunger and malnutrition and to promote conditions leading to food security for individuals, families, and countries everywhere. The need for this high-level attention was emphasized by new data from the FAO indicating that 840 million people, or 18% of the population of the developing world, are malnourished or hungry. In light of continuing increases in population, rising costs of grain, and declining per capita grain production, the summit was a unique opportunity for world leaders to take a new look at the meaning of sustainability.

At the start of the meeting, the delegates adopted, by acclamation, two major documents, called the Rome Declaration on World Food Security and the World Food Summit Plan of Action. These documents, in the words of the FAO, "set forth a seven-point plan stipulating concrete, political actions to ensure:

1. conditions conducive to food security;
2. the right of access to food by all;
3. sustainable increases in food production;
4. trade's contribution to food security;
5. emergency relief when and where needed;
6. the required investments to accomplish the plan;
7. concerted efforts to achieve results by countries and organizations."

One concrete objective of the plan was a 50% reduction in the number of hungry people by the year 2015. Eventually, this objective was incorporated into the Millennium Development Goals initiative. Some participants criticized the objective as being too timid, because it still leaves 420 million people malnourished or hungry. The plan emphasizes the responsibilities of countries (especially developing countries) to enact reforms that would promote food security—policies that (1) do not discriminate against agriculture or small farmers, (2) open trade, (3) foster greater efforts to bring down population growth, and (4) encourage investing in the country's infrastructure.

A five-year follow-up of the World Food Summit was scheduled to meet in November 2001, but was postponed to June 2002 because of events related to the September 11 terrorist attack in the United States. The meeting was held to assess progress in achieving the goals of the 1996 summit. The meeting in June was a disappointment to many, because very few heads of state attended and the information presented by FAO made it clear that the world was not on track to achieve the Millennium Goal of halving hunger by 2015. Achieving that goal would require a reduction of 22 million hungry per year. The actual reduction was between 6 and 8 million a year, almost all of whom were in China! The FAO proposed an antihunger program that would cost $24 billion a year and would make reaching the Millennium Goal possible. The developed countries spend about $360 billion a year on food and agricultural subsidies to their own producers, so the antihunger program would be less than one-tenth of that amount. Unless something happens to accelerate action, the basic target of reducing hunger by one-half will be 45 years behind schedule, and there is even the possibility that the number of hungry could grow during that time.

Disorders due to **overnourishment** are not very common in most parts of the developing world. There, the major nutritional problems are a lack of proteins and some vitamins (malnutrition), especially in children, and a lack of Calories for food energy (undernourishment) for all ages.

## Extent and Consequences of Hunger

Accurate, reliable figures on the worldwide extent of hunger are unavailable, mainly because few governments make any effort to document such figures. On the basis of surveys of available food in the society, with accessibility to food factored in, the FAO estimates that 799 million people in the developing world and an additional 41 million in the rest of the world are underfed and undernourished. The WHO estimates that half the people in the world, 3 billion, suffer from some form of malnutrition.

***Where?*** Almost two-thirds of the undernourished (508 million) live in Asia and the Pacific. The regions most seriously affected are India, Bangladesh, Pakistan, and Southeast Asia. Sub-Saharan Africa has the highest percentage of undernourished, with one-third of the population, some 196 million people, afflicted. Hunger is a massive problem, but it affects individual people with unique faces and personalities. Consider the following quotes from the World Bank's *Voices of the Poor:* from Senegal, "Your hunger is never satisfied, your thirst is never quenched; you can never sleep until you are no longer tired"; from a 10-year-old child from Gabon, "When I leave for school in the mornings, I don't have any breakfast. At noon there is no lunch, in the evening I get a little supper, and that is not enough. So when I see another child eating, I watch him, and if he doesn't give me something, I think I'm going to die of hunger"; and from Ukraine, "In the evenings, eat sweet potatoes, sleep; in the mornings, eat sweet potatoes, work; at lunch, go without."

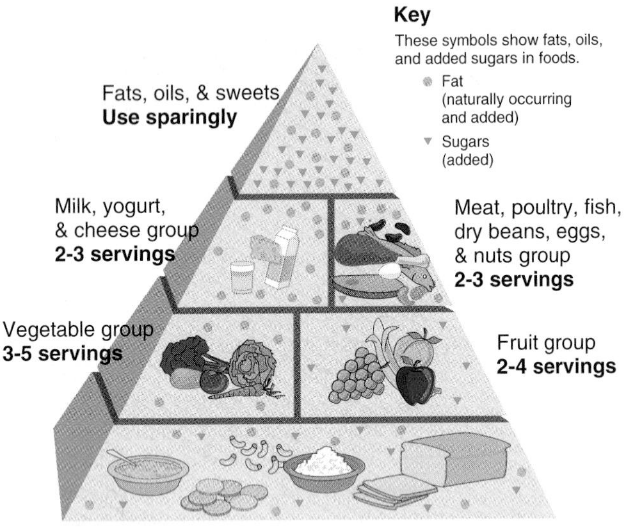


**Figure 9–15 The Food Guide Pyramid.** The pyramid is a guide produced by the U.S. Department of Agriculture to help people evaluate their food needs and food intake so as to provide adequate nutrition and keep their weight under control. Suggested numbers of servings are given as a range, because energy requirements vary for people, depending on their size, age, and level of activity. Thus, older adults and many women should use the minimum number of servings (they might need only 1,600 calories per day), while teenage boys and active men might need the maximum number of servings to get the 2,800 or so calories they need daily. This guide has recently come under criticism and may be revised soon because of its emphasis on carbohydrates and insufficient attention to beneficial fats.

**Figure 9–16 Malnourished children.** These orphaned children are often left to their own devices in this makeshift refugee camp in Sudan.

*Consequences.* The consequences of malnutrition and undernutrition vary. The effects are greatest in children and next greatest in women. Hunger can prevent normal growth in children, leaving them thin, stunted, and often mentally and physically impaired (Fig. 9–16), a condition known as protein-energy malnutrition. The U.N. Subcommittee on Nutrition has identified malnutrition and hunger in children of southern Asia as the world's most serious nutritional problem, involving at least 100 million children who are underweight because of lack of food. Research has shown that undernutrition in early childhood can seriously limit growth and intellectual development throughout life. Women in the developing world are especially vulnerable, because of widespread patterns of low status and because they are expected to perform heavy labor.

Sickness and death are companions of hunger. Because poor nutrition lowers a person's resistance to disease, measles, malaria, and diarrheal diseases become common and are major causes of death in the malnourished and undernourished. Eleven million children under the age of five die each year in the developing countries; at least half of these deaths are caused either directly or indirectly by hunger and malnutrition. Rarely do the children actually starve to death; instead, death comes from the inadequate food that leaves them weak, underweight, and vulnerable to disease. Hunger is often a seasonal phenomenon in rural areas that are supported by subsistence agriculture, as people are forced to ration their stored food in order to survive until the beginning of the next harvest. Anyone who travels in the developing world cannot help but notice that few people in the rural areas look well fed. Most are thin and spare from a lifetime of hard work and limited access to food.

## Root Cause of Hunger

In the view of most observers, *the root cause of hunger is poverty.* Our planet produces enough food for everyone alive today. However, hungry and malnourished people lack either the money to buy food or enough land to raise their own. FAO's latest report on *The State of Food Insecurity in the World* states: "Most of the widespread hunger in a world of plenty results from grinding, deeply rooted poverty."

Inadequate food is only one of many consequences of poverty. As mentioned in Chapter 6, 1.2 billion people live in absolute poverty today, with incomes of less than $1 a day. (See Fig. 6–8.) Most of these people reside in rural villages, are illiterate, spend half or more of their income on food, and represent races, tribes, or religions that suffer discrimination. Millions are trapped in a cycle of poverty that leads to the degradation of resources and perpetuates the poverty—all made worse by high fertility. It is to address this kind of poverty and its consequences that the Millennium Development Goals were articulated. It can be argued that addressing hunger and the poverty that causes it requires making progress toward all of these goals simultaneously.

Hunger and poverty do not always go from bad to worse. A number of Asian countries, including China, Indonesia, and Thailand, significantly reduced the extent of poverty and hunger in them during the 1980s and 1990s. Deliberate public policies and social services have greatly improved the welfare of millions in China, where food security is a matter of high national priority. Indonesia has benefited from oil exports and Green Revolution technology, and it continues to put a major emphasis on rural development and the country's social infrastructure. It is possible, therefore, for societies to address the needs of the hungry poor with appropriate public policies and to make progress in reducing the extent of absolute poverty and hunger. In contrast, in any given year there are countries and sometimes whole regions that face exceptional food emergencies. It is here that international responsibility comes most sharply into focus.

## Famine

A **famine** is a severe shortage of food accompanied by a significant increase in the death rate. Famine is a clear signal that a society is either unable or unwilling to distribute food to all segments of its population. Two factors—*drought* and *conflict*—have been immediate causes of famines in recent years.

***Drought.*** Drought is blamed for famines that occurred in 1968–74 and again in 1984–85 in the Sahel region of Africa. The Sahel, a broad belt of drylands south of the Sahara desert, is occupied by 70 million people who practice subsistence agriculture or tend cattle, sheep, and goats. (Such people are called *pastoralists.*) Although the region normally has enough rainfall to support dry grasslands or savanna ecosystems, the rainfall is seasonal, undependable, and prone to failure. Making matters worse, population increases in the region have led to unsound agricultural practices and overgrazing by the expanding herds of livestock. Beginning in 1965, the region experienced 20 years of subnormal rainfall, with tragic results. Crops withered, forage for livestock declined, watering places dried up, and livestock died. Both farmers and pastoralists began abandoning their land and migrating toward urban centers, where they were often herded into refugee camps. Unsanitary conditions in the camps and the already weakened condition of the refugees led to the spread of infectious diseases such as dysentery and cholera, and many thousands died before effective aid could be organized. The 1984–85 Sahelian famine is thought to have been responsible for almost a million deaths in Ethiopia alone. The number would have been higher if not for aid extended by Africans and numerous international agencies.

In 1999 and 2000, drought returned to the Sahel, once again bringing on massive food shortages and potential famine. The drought extended further south this time, affecting almost all of east and south Africa as well. Fortunately, favorable rains returned in 2001, leading to record crops. Then in 2002, rain deficits occurred once again, driving food prices up. However, 2003 saw a return of the normal rainy season, with prospects for a good agricultural season.

***Warning Systems.*** In order to prevent droughts from leading to famines, two major efforts provide crucial warnings about food insecurity in different regions of the world. One, the FAO's *Global Information and Early Warning System (GIEWS),* monitors food supply and demand all over the world, watching especially for those places where food supply difficulties are expected or are happening. The GIEWS issues frequent reports and alerts that are usually the result of its field missions or on-the-ground reports from other U.N. agencies and NGOs. The agency also relies on satellite imagery for crop-monitoring activities.

The second major effort is the *Famine Early Warning System (FEWS)* Network, funded by the U.S. Agency for International Development (USAID). FEWS is a partnership of agencies involved in satellite operations focusing on sub-Saharan Africa. The network continually measures rainfall and agricultural conditions in the region and, like the GIEWS, issues regular bulletins and special reports on the Internet, giving governments and relief agencies accurate and timely assessments of the status of food security in African countries.

***Conflict.*** Famines in which the causative factor was not drought, but war, threatened Ethiopia, Eritrea, Somalia, Rwanda, Sudan, Mozambique, Angola, and Congo in the 1990s. Devastating and prolonged civil warfare put millions of Africans at risk of famine. The civil wars disrupted the farmers' normal planting and harvesting and displaced millions from their homes and food sources. Governments in power maintained control over food and relief supplies; relief agencies operated under dangerous conditions and frequently incurred casualties. In some areas, the problem was made worse by persistent droughts. In Mozambique alone, 900,000 people died from direct military action or from indirect effects of the war there. At least 15 countries in sub-Saharan Africa were experiencing food emergencies in 2003 because of internal conflicts and their effects on neighboring countries. The recent U.S.-led wars in Afghanistan and Iraq have also triggered food shortages, requiring ongoing food assistance to Afghanistan and emergency food assistance to Iraq, where the war interrupted food production.

Famines from drought and war are preventable. India, Brazil, Kenya, and southern Africa have coped with droughts in recent years by mobilizing effective relief in the form of food, clothing, and medical assistance. Indeed, the drought of the early 1990s accelerated the peace process in Mozambique and helped change the political landscape in southern Africa. Cooperation between South Africa and the 10 nations of the Southern Africa Development Community to prevent famine lowered barriers between South Africa and its neighbors prior to the coming of democracy to that troubled nation.

### Hunger Hot Spots

***Sub-Saharan Africa.*** Much of Africa has experienced long-term and severe droughts, as well as widespread civil conflict. Erratic weather patterns and, especially, a general warming trend are thought by some scientists to be the outcome of global climate change. Regardless of the causes, the impact has been devastating to African food

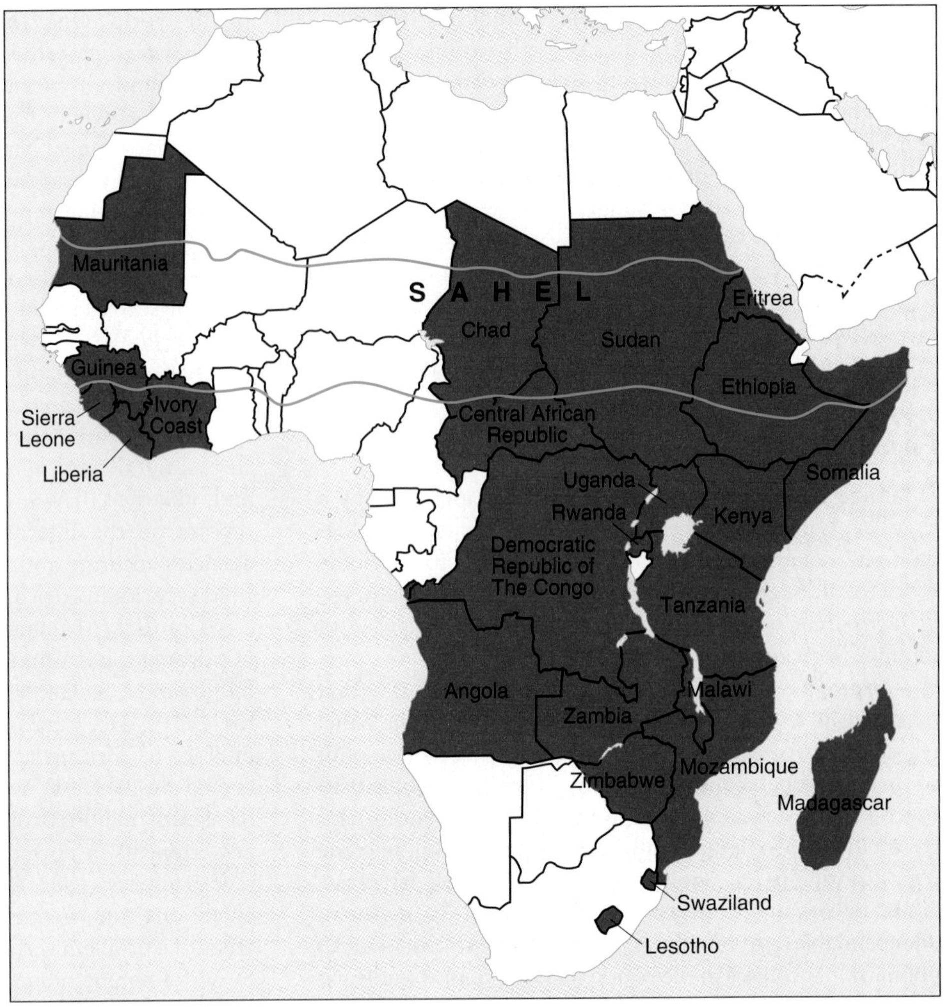


**Figure 9–17 The geography of famine.** Famines and food emergencies have occurred repeatedly in sub-Saharan Africa, especially in the Sahel (a band of dry grasslands that stretches across the continent). The map shows the countries where civil wars and droughts have recently brought on serious food shortages.

production. FEWS identifies 25 African countries facing food emergencies (Fig. 9–17). In all, over 40 million people in sub-Saharan Africa need food assistance. The following countries are especially vulnerable:

1. **Zimbabwe.** Once an agricultural breadbasket, this country is in economic chaos due to political violence and the government's land redistribution efforts. In essence, large white-owned farms have been seized and given to blacks, many of whom were connected to the government and had no intention of farming. Those opposed to Robert Mugabe's rule have been denied food aid. Over 6 million face hunger and starvation unless such aid is forthcoming.
2. **Ethiopia.** Drought and the lingering effects of war with Eritrea and refugees from neighboring countries have once again brought Ethiopia to the brink of a food crisis. Food aid has been requested, but the pledges from donors fall short. Some 11 million need aid here and in neighboring Eritrea in order to prevent famine. Government land policies are a problem: The government owns all of the land, yet 85% of the people in Ethiopia are farmers. History has shown that if farmers own their own land, they will care for it much more effectively.
3. **Sudan.** In Sudan, civil war and drought combined to put almost 1 million people in peril, triggering a massive airdrop by the World Food Program to provide almost 10,000 tons of food a month to the southern parts of Sudan (Fig. 9–18). Many humanitarian aid organizations, such as Doctors Without Borders and World Vision, are doing what they can to help, but the situation will not get better until there is peace in the region.

***North Korea.*** Another set of circumstances has brought famine to North Korea. The primary cause is the failure of centrally planned agricultural policies in a country that is one of the last bastions of communism. Floods and droughts alternatively have made conditions worse, to the point where millions are believed to be close to starvation and as many as two million have already perished. For years, North Korea refused to admit that there was a problem, but recently the country has permitted the delivery of food aid, and currently 8 million a year are being given food assistance. Most of this aid is coming from the United States.

Famine is being averted today only because of food aid. What, then, is the function of food aid and just how effective is it?

**Figure 9–18 Food Aid.** Women receive aid from the World Food Program in Sudan.

## Food Aid

The World Food Program of the United Nations coordinates global food aid, receiving most of its donations from the United States, Canada, Australia, and the countries of the European Union. In addition, many nations conduct their own programs of bilateral food aid to needy countries. Although international food aid was cut in half (to 6.2 million tons) between 1993 and 1996, it has been on the increase since then and reached 15 million tons in 1999 (Fig. 9–19). Much of the increase in 1999 was the result of a major shortfall in grain production in the Russian Federation and a consequent extension of food aid primarily from the United States (admittedly, to boost U.S. farm exports, which had been falling).

Food aid is distributed to countries all over the world, not just where famines are threatened. This raises the question, What is the proper role of food aid? Aid is vital in saving lives where food emergencies occur, but what about the people who suffer from chronic hunger and malnutrition? When should food be given to those in need, instead of being distributed according to market economics?

***When Aid Doesn't Help.*** Numerous humanitarian campaigns to end world hunger have been mounted in the last 50 years. The United States and Canada have been world leaders in giving away food (which is first purchased from farmers and therefore represents a subsidy). As mentioned earlier, a number of serious famines have been moderated or averted by these efforts, and certainly the need for ongoing emergency food aid is undeniable. However, routinely supplying food aid in an attempt to alleviate chronic hunger in developing countries may be the worst thing to do.

The problem is, people will not pay more than they have to for food. Therefore, free or very cheap foreign food undercuts the local market. In effect, local farmers must compete economically with free or low-cost imported food. When they cannot earn a profit, they stop producing and eventually enter the ranks of the poor. The cycle continues, with people who sell goods to the farmer also suffering when the farmer loses buying power. In the long run, the entire local economy deteriorates. Hence, the donation of food, while well intended, often aggravates the very conditions that it is meant to alleviate. Meanwhile, population pressures continue to build, and the magnitude of the problem increases. (See the "Ethics" essay, p. 256.)

Some food aid is strictly humanitarian. Bangladesh, for example, receives over 1 million tons per year, about half as much as it purchases on the market. The African continent receives 3–4 million tons a year. All the signs indicate that this figure will increase. A significant amount of politically expedient food aid is flowing to Eastern Europe and the countries of the former Soviet Union, largely to bolster their fragile economies.

Food aid will undoubtedly continue to be an international responsibility. It is at best a buffer against famine, and it will probably continue to be awarded to some countries for political reasons. As part of the solution to the chronic hunger and malnutrition among the poor, however, simple food aid is counterproductive. Much more good will be accomplished by restructuring the economic arrangements between rich and poor nations and by extending development aid aimed at fostering self-sufficiency in food and sustainable interactions with the environment.

## Closing Thoughts on Hunger

In 1980, a Presidential Commission on World Hunger delivered its report to President Jimmy Carter, with the major recommendation "that the United States make the elimination of hunger the primary focus of its relations with the developing world." The thrust of the report was that if we respect human dignity and have a sense of social justice, we must agree that hunger is an affront to both.

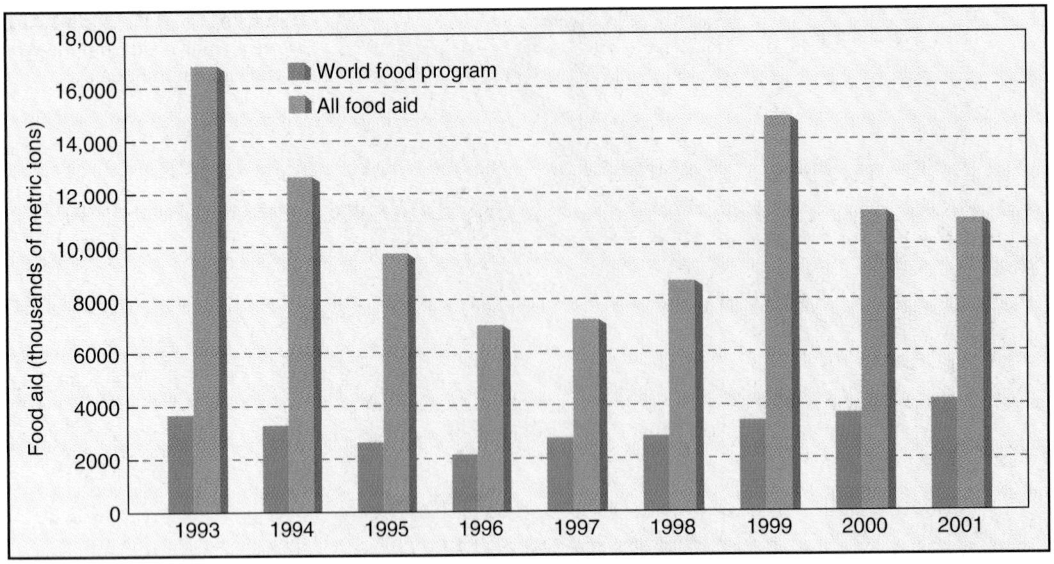


**Figure 9–19** **Global food aid.** Global food aid from 1993 to 2001. Blue indicates food aid delivered by the World Food Program; food aid from all sources is indicated in red. (*Source:* World Food Program statistics, UNFAO.)

**ethics**

## The Lifeboat Ethic of Garret Hardin

The late biologist Garret Hardin published several provocative essays addressing the worldwide food-vs.-population issue. This box outlines some of his thoughts and gives you an opportunity to respond to them.

*Carrying capacity* is the number of a species that can be supported indefinitely without degrading the environment. For human societies, the carrying capacity depends on our ability to meet food needs over the long term—that is, sustainably. If ecology had a decalogue, Hardin says, the first commandment would be "Thou shalt not transgress the carrying capacity." Around the world, numerous countries are pressing against the limits of, or have exceeded, their carrying capacity. This, says Hardin, is their problem, not ours, and he uses a lifeboat metaphor to explain why.

Picture a number of lifeboats in the sea after a ship has sunk—some crowded with people and some in which people are riding in relatively uncrowded luxury. Each lifeboat has a limited capacity. The people in the crowded boats are continually falling overboard, leaving the people on the uncrowded boats to decide whether to take them on board or not. Imagine an uncrowded boat with 50 on board and room for 10 more, with 100 people swimming in the water and begging to be taken on board. There are several options: (1) Assume that all people have an equal right to survival, and take everyone on board. This, says Hardin, would lead to catastrophe for all. (2) Admit only 10, filling all the space on the boat. Two problems with this option are that you lose your margin of safety and that you must find a basis on which to discriminate among all the people in the water. (3) Admit no more to the boat. This preserves your margin of safety and guarantees the long-term survival of the people on your boat; it is the rational solution to the lifeboat problem.

The metaphor is supposed to be applied to the problem of food aid. Some people would argue that if less grain went to feeding animals, there would be more available to feed the hungry in poor countries; or, because there are often agricultural surpluses in the rich nations, those surpluses could be used to feed the hungry. The real problem, then, is a problem of food *distribution*, not of the quantity produced. Food should be given to the hungry wherever they are.

These arguments, says Hardin, are foolish. In giving away food, we would only be encouraging the "population escalator": A population that is growing rapidly reaches the limits of its food capacity and is supplied with food from abroad, encouraging still further growth and necessitating still further food aid, and so on. Our hearts, says Hardin, tell us to send food—to let them all on board, but our heads should tell us not to. We only postpone the day of reckoning, and in the end, the amount of suffering will be greater. Overpopulated, food-poor countries have transgressed the first commandment of ecology. If we want to help, we should direct our aid toward limiting population growth, according to Hardin. What do you think? (Sample the opinion of those to whom you speak.)

The right to food must be considered a basic human right. It follows, then, that we as a nation have a moral obligation to respond to world hunger, the report concluded. Has that moral obligation made its way into public policy in the ensuing years?

Although food is our most vital resource, we do not treat it as a commons (free to all that need it). Indeed, the production and distribution of food constitute one of the most important economic enterprises on Earth. Alleviating hunger is primarily a matter of addressing the absolute poverty that afflicts one of every five people on the planet. To treat food as a commons would be to treat wealth as a commons. One of the tenets of socialist systems, this notion has proved to be unworkable. We are part of the world economy now, and it is a market economy—for the perfectly good reason that nothing else seems to work, given the realities of human nature.

What has not been done, however, is to bring this market economy under the discipline of sustainability. Short-term profit crowds out long-term sustainable management of natural resources. National self-interest in the rich countries prevents them from leveling the playing field for agricultural commodities. Why are the rich nations content to maintain the current high subsidies to agriculture that allows them to flood markets with underpriced products? Why is hunger not at the top of the policy agenda of the United States? Has "donor fatigue" dulled the consciences of leaders and the general public?

In the developing countries, there is a critical need for enlightened leadership. Sub-Saharan Africa has been plagued with corrupt and inept elites who are more interested in staying in power than in reducing poverty in their countries. Tribal conflicts have taken a terrible toll in lives and hunger. The AIDS epidemic is ravaging many of the countries, and in some the leaders have been reluctant to admit that they have a problem.

We understand the situation quite well. No new science or technology is needed to alleviate hunger and at the same time promote sustainability as we grow our food. The solutions lie in the realm of political and social action at all levels of responsibility. Given the current groundswell of concern about the environment and the end of the Cold War between capitalism and communism, there may never be a better time to turn things around and take more seriously our responsibilities as stewards of the planet and as our brothers' keepers.

# revisiting the themes

## Sustainability

Subsistence agriculture has the potential to operate sustainably, as pastoralists and small farmers tend their crops and flocks, recycling manure and feeding crop residues to their animals. Even slash-and-burn cultivation can be practiced sustainably. In contrast, factory animal farming as currently practiced wastes manure and pollutes the air and water in the name of economic efficiency. Ranching often leads to deforestation in the tropics and overgrazing on the rangelands, because the current policy highly undervalues these lands.

## Stewardship

The remarkable progress made in keeping the world's people fed is certainly good stewardship in action. This is especially true where efforts are being made, such as those by the World Food Program, FEWS, the GIEWS, and the many NGOs involved in hunger and development aid, to identify those suffering from food insecurity and to ensure that they have sufficient food for their needs. Thousands of organizations and people show that they care as this work goes on. The work of Borlaug and his team, of the CGIAR, and of the IRRI is particularly noteworthy, because they have sought to improve the yields of the more important grains and make the new high-yielding varieties available to the developing countries. America's Second Harvest and its many affiliated food banks and food pantries play an important role in the United States in keeping people from going hungry in a land of plenty.

## Sound Science

The new high-yield grains that led to the Green Revolution are sound science in practice. Agricultural research that continues to seek new ways to improve crop yields, also is sound science at its best. Although many might disagree, the new biotech work all over the world producing genetically modified crops—especially the work that is designed to help farmers in developing countries improve their ability to grow food—is sound science in action. Sound science is also needed to provide an accurate evaluation of all genetically modified crops and foods before they are implemented. The satellite imagery used by the GIEWS and FEWS is also a commendable application of sound science to human needs.

## Ecosystem Capital

Global land that is in use for agricultural crops and animal husbandry represents one of the most crucial and irreplaceable components of ecosystem capital. Those countries, such as the United States, which have an abundance of this land can expect to meet their country's food needs well into the future and even generate significant economic gain by selling surpluses to more populous countries or those less well endowed with good land. However, this ecosystem capital must be nourished; it is a renewable resource, but it can be drawn down by unsustainable practices that lead to erosion and other forms of land degradation. As populations increase, pressure will mount to turn more forests and wetlands into croplands, a transformation with trade-offs that are not usually beneficial, due to the loss of the goods and services of the natural lands.

### Policy and Politics

Public policy makes a huge difference in whether people's food needs are met and how ecosystem capital is sustained. On the negative side, some Latin American countries converted rain forests to ranches that enriched a few, but gave little assistance to peasant farmers. In a mixed situation, the new welfare act in the United States has changed the realities of the safety net. Again on the negative side, the new U.S. farm bill has extended the farm subsidies that continue to discriminate against developing countries. High tariffs and subsidies in the European Union do the same thing. Politics has turned ugly in sub-Saharan Africa, where rulers are denying food aid to those who do not support them, and land redistribution has led to the collapse of whole economies. On the positive side, Malawi has teamed up with the UNFAO to provide 200,000 treadle pumps (see Chapter 7) to poor rural families, to enable them to raise a winter crop by tapping the groundwater just a few feet below the surface.

Public policies have also had considerable difficulty in adapting to the new realities of the biotech industry. Where is the justice in fining a farmer for raising crops that were pollinated by plants in neighboring fields? The EU's use of the precautionary principle and the Cartagena Protocol are good examples of policies evolving as new technologies were developed.

In the United States, the Bush administration has responded to the food crisis in Africa with additional funds for famine relief and emergency food aid. President Bush's Millennium Challenge Account is an opportunity to fund effective development aid that will target poverty and hunger, but it needs to be funded by Congress.

### Globalization

The global food market is a $400 billion reality that can bring both benefit and hardship to those engaged in raising food crops and animals. When the rules governing trade are set by those with the most power, as has been the case in the WTO, bringing justice to the less powerful developing countries has proven elusive, with the rich countries continuing to protect their farmers with high tariffs and subsidies. However, the global market also gives developing countries a market for their commodities, which is beneficial if the trading rules are fair. The global movement of foodstuffs has facilitated hunger and famine relief, with shiploads of grains ferried from donor countries to recipients.

## review questions

1. Describe the five components of the agricultural revolution. What are the environmental costs of each?
2. What is the Green Revolution? What have been its limitations and its gains?
3. Describe subsistence agriculture, and discuss its relationship to sustainability.
4. How do sustainable animal farming and industrial-style animal farming affect the environment on different scales?
5. What are our major prospects for increasing food production in the future?
6. What are the promise and problems associated with using biotechnology in food production?
7. What is the Cartagena Protocol, and how does it embody the precautionary principle?
8. Trace the patterns in grain trade between different world regions over the last 65 years. (See Table 9–3.) Explore the consequences of these patterns.
9. Describe the three levels of responsibility for meeting food needs. At each level, list several ways food security can be improved.
10. Define hunger, malnutrition, and undernourishment. What is the extent of these problems in the world?
11. How are hunger and poverty related?
12. Discuss the causes of famine, and name the geographical areas most threatened by it.
13. How is famine relief accomplished? Why does food aid sometimes aggravate poverty and hunger?

# thinking environmentally

1. Imagine that you have been sent as a Peace Corps volunteer to a poor African nation experiencing widespread hunger. Design a strategy for assessing the needs of the people and for contacting appropriate sources for help.

2. Record your food intake over a two- to three-day period. Analyze the nutritional value of your diet. (*Hint*: Consult www.nat.uiuc.edu/ for a good tool for this kind of analysis.) Which nutrients are lacking? Which are in excess? What changes in your diet would reconcile these differences?

3. Of the following methods for increasing food production, which do you feel are viable options and why?
   Clearing forests
   Increasing yield on farmland already in production
   Irrigating arid lands
   Developing transgenic crops with biotechnology
   Aquaculture

4. Use the Web to evaluate the work of FEWS (www.fews.net) and the GIEWS (http://geoweb.fao.org/). How do these organizations function in bringing aid to people with the greatest need?

# Wild Species and Biodiversity

### Key Topics

1. The Value of Wild Species
2. Saving Wild Species
3. Biodiversity and Its Decline
4. Protecting Biodiversity

Puffins are roly-poly little seabirds that live in cold coastal waters on both sides of the North American continent. The coast of Maine once enjoyed flourishing puffin colonies on a number of offshore islands, but from colonial times the birds were hunted for their eggs, meat, and feathers, until, by 1900, only one very small colony remained. Protective laws were passed, but the islands once inhabited by the puffins began to host large colonies of predatory gulls; the puffins never returned. National Audubon Society biologist Steven Kress was teaching ornithology to summer visitors on the Maine coast in the early 1970s and began wondering whether it might be possible to bring the puffins back to Maine. Against the advice of many biologists, Kress launched a program to accomplish his dream—Project Puffin, as it is called. His plan was to obtain puffins from Newfoundland, where they were still numerous, and install them on Eastern Egg Rock, a 5 acre island 7 miles off the Maine coast. This was no easy task.

First, the predatory gulls had to be removed, with the approval of the Fish and Wildlife Service. Then, starting with four chicks in 1973, Kress and a team of volunteers and interns installed the chicks in burrows and fed them daily until they were ready to go to sea. (Puffins spend most of their time at sea.) Each year, more chicks were brought to the island. Kress' team had to convince the puffins that the island was a thriving colony, so they installed numerous painted wooden puffin decoys, played tape-recorded puffin noises, and installed bogus eggs until the first few birds began to return. Finally, in 1981, four pairs of puffins set up housekeeping on the island and raised the first Maine-bred chicks. The number of nesting pairs has grown steadily, and the island now has a thriving colony of 52 pairs.

**The Atlantic Puffin** **Dedicated researchers are helping to return puffins to the northeastern United States.**

***Social Attraction.*** The method used by Kress and his coworkers (called "social attraction") was innovative, and his success has blossomed into an enterprise now known as the Seabird Restoration Project of the Audubon Society, with an annual budget of $400,000. Using similar methods, Kress has presided over a number of reintroductions of seabirds, including several species of terns on Egg Rock and other islands in the Gulf of Maine, petrels in the Galápagos Islands, and, recently, Laysan albatrosses in Hawaii. The Fish and Wildlife Service has availed itself of Kress' methods to introduce seabirds to other islands where they once bred; many budding biologists got their life's inspiration from working with Kress. Today, children young and old can cruise out in special touring boats from nearby Boothbay Harbor and New Harbor to see puffins buzzing around Eastern Egg Rock and a number of other islands to which they have also been restored. Kress says, "Sometimes it's not enough to pass protective laws or buy land to create new sanctuaries. Every case is different, so first you try to understand the species' needs, and then you go in to re-create communities. We can either document the loss of a population or design techniques to try to save it."

The focus in this chapter is on *wild species.* We consider their importance to humans, examine the concept of *biodiversity*—the variety of life on earth—and look at the public policies that seek to protect wild species and biodiversity.

## 10.1 The Value of Wild Species

Recall from Chapter 1 that *ecosystem capital* is the sum of all the goods and services provided to human enterprises by natural systems. This capital is enormously valuable to humankind, conservatively estimated in Chapter 3 to be worth $38 trillion a year. It is so essential to human life, though, that it is priceless. The basis of most of the natural capital is ecosystems, and the basis of ecosystems is the plants, animals, and microbes—the wild species—that make them work. To maintain the sustainability of these ecosystems, their *integrity* must be preserved; this means maintaining their *resilience* and *biodiversity,* among other things. It is crucial, therefore, to put a high priority on protecting wild species and the ecosystems in which they live.

Even those who agree on the need to protect wild species, however, may disagree on the kind of protection that should be afforded to them. Some want wildlife protected so as to provide recreational hunting. Others feel strongly that hunting for sport should be banned. Many have a deeper concern for the ecological importance of wild species, and they view the ongoing loss of biodiversity as an impending tragedy. Many in the developing world, however, depend on collecting or killing wild species in order to eat or make money, so their personal survival is at stake. All of these attitudes stem from the values we place on wild species. Can the differing values be reconciled in a way that still leads to sustainable management of these important natural resources?

### Biological Wealth

About 1.75 million species of plants, animals, and microbes have been examined, named, and classified, but many more species have not yet been found or catalogued. On the basis of ratios of known to unknown species in different groups, Australian taxonomist Nigel Stork believes that there are 13 million species as yet undiscovered. This figure compares well with the estimate of 14 million in the 2001 *Global Biodiversity Outlook* produced by UNEP.

These natural species of living things, collectively referred to as **biota,** are responsible for the structure and maintenance of all ecosystems. They and the ecosystems they form represent wealth—the **biological wealth**—that comprises most of the ecosystem capital that sustains human life and economic activity with goods and services. From this perspective, the biota found in each country represents a major component of the country's wealth. More broadly, this richness of living species is one of the dimensions of Earth's **biodiversity.**

***Spending the Wealth.*** Humans have always depended on Earth's biological wealth for food and materials. As animals, we have always exploited wild species for food. Some 12,000 years ago, however, our ancestors began learning to select certain plant and animal species from the natural biota and to propagate them, and the natural world has never been the same. Forests, savannas, and plains were converted to fields and pastures as the human population grew and human culture developed. In time, many living species were exploited to extinction, and others disappeared as their habitats were developed. At least five hundred plant and animal species have become extinct in the United States alone, and thousands more are at risk. We have been drawing down our biological wealth, with unknown consequences.

Living in cities and suburbs in the industrialized world and getting all our food from supermarkets, we seem only remotely connected to nature. That is an illusion, however: Our interactions with the natural world may have changed, but we still depend on biological wealth. Moreover, millions of our neighbors in the developing world are not so insulated from the natural world; their dependence is much more immediate, because they draw sustenance and income directly from forests, grasslands, and fisheries. Due to overwhelming economic pressures, these people, too, are often engaged in practices that draw down their biological wealth, with consequences that are obvious and grave. A root source of this problem is the way we regard and value wild nature.

### Two Kinds of Value

It was not so long ago that hunters on horseback would ride out to the vast herds of bison roaming the North American prairies and shoot them by the thousands, often taking only the tongues for markets back in the east. The passenger pigeons that once darkened the skies in huge flocks were ruthlessly killed at their roosts to fill a lively

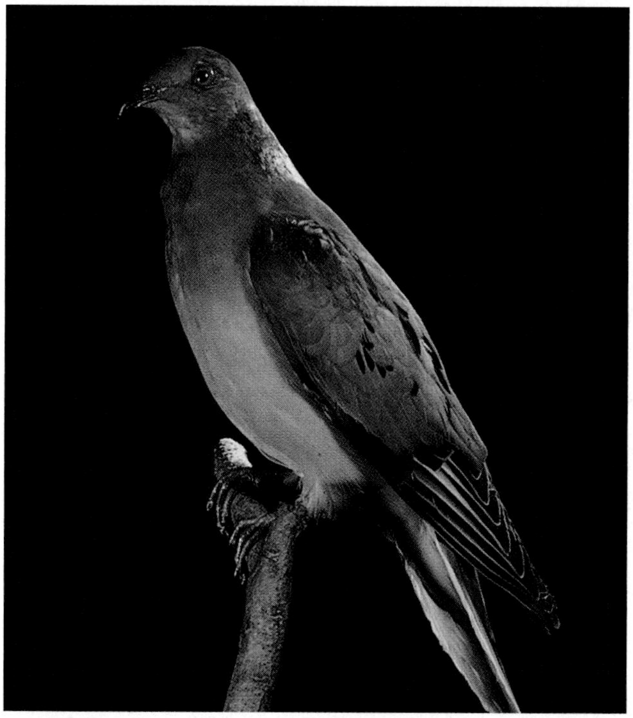

**Figure 10–1 Extinct passenger pigeon.** Clouds of passenger pigeons darkened American skies during the 18th and 19th centuries, but relentless hunting extinguished the species by the early 20th century.

demand for their meat, until the species was gone (Fig. 10–1). Plume hunters decimated egrets and other shorebirds to satisfy the demands of fashion in the late 1800s.

Appalled at this wanton destruction, 19th-century naturalists called for an end to the slaughter, and the U.S. public began to be sensitized to the losses. People began to see natural species as worthy of preservation, and naturalists began to look for ways to justify their calls to conserve nature. Thus, there was an emerging sense that species should not be hunted to extinction. But why not? Were these early conservationists just concerned that there might not be any animals left to hunt or trees left to cut down? Their problem then, and ours now, is to establish the thesis that wild species have some *value* that makes it essential to preserve them. If we can identify that value, then we will be able to assess what our duties are in relation to wild species.

***Instrumental Value.*** Philosophers who have addressed this problem identify two kinds of value that should be considered. The first is **instrumental value.** A species or individual organism has instrumental value if its existence or use benefits some other entity. This kind of value is usually *anthropocentric;* that is, the beneficiaries are human beings. Many species of plants and animals have instrumental value to humans and will tend to be preserved (conserved, that is) so that we can continue to enjoy the benefits derived from them.

***Intrinsic Value.*** The second kind of value that must be considered is **intrinsic value.** Something has intrinsic value when it has value *for its own sake;* that is, it does not have to be useful to us to possess value. How do we know that something has intrinsic value? That is a philosophical question, and it comes down to a matter of moral reasoning. People often disagree about intrinsic value, as illustrated by the animal-rights controversy. In this regard, some people argue that animals have certain rights and should be protected, for example, from being used for food, fur and hides, or experimentation. On the other side, most people agree that cruelty to animals is wrong, but they do not believe that any animals should be given rights approaching those afforded to humans.

Many people believe that no species on Earth except *Homo sapiens* has any intrinsic value. If no other species have intrinsic value, then it is difficult to justify preserving many that are apparently insignificant or very local in distribution. (The more common species are worthy of preservation because they have instrumental value of some kind or other.) Still, in spite of the problems of establishing intrinsic value for species, support is growing in favor of preserving species that not only may be useless to humans, but also may never be seen by anyone except a few naturalists or systematists (biologists who are experts on classifying organisms).

The value of natural species can be categorized as follows:

- value as sources for agriculture, forestry, aquaculture, and animal husbandry;
- value as sources for medicines and pharmaceuticals;
- recreational, aesthetic, and scientific value; and
- value for their own sake (intrinsic value).

The first three categories reflect mostly instrumental value. Aesthetic and scientific value sometimes represents intrinsic value.

## Sources for Agriculture, Forestry, Aquaculture, and Animal Husbandry

Most of our food comes from agriculture, so we tend to believe that it is independent of natural biota. This is not true. In nature, both plants and animals are continuously subjected to the rigors of natural selection. Only the fittest survive. Consequently, wild populations have numerous traits for competitiveness, resistance to parasites, tolerance to adverse conditions, and other aspects of vigor.

In contrast, populations grown for many generations under the "pampered" conditions of agriculture tend to lose these traits, because they are selected for production, not resilience. For example, a high-producing plant that lacks resistance to drought is irrigated, and the resistance to drought is ignored. Also, in the process of breeding plants for maximum production, virtually all genetic variation is eliminated. Indeed, the cultivated population is commonly called a *cultivar* (for *culti*vated *var*iety), indicating that it is a highly selected strain of the original species, with a *minimum* of genetic variation. When provided with optimal water and fertilizer, cultivars produce outstanding yields under the specific climatic conditions to which they are adapted. With their minimum genetic variation, however, they have virtually no capacity to adapt to any other conditions.

*Wild Genes.* To maintain vigor in cultivars and to adapt them to various climatic conditions, plant breeders comb wild populations of related species for the desired traits. When found, these traits are introduced into the cultivar through crossbreeding or, more recently, biotechnology (Chapter 9). For example, in the 1970s, the U.S. corn crop was saved from blight by genes from a wild strain of maize. The point is, this trait came from a related *wild* population—that is, from natural biota. If natural biota with wild populations are lost, the options for continued improvements in food plants will be greatly reduced.

*New Food Plants.* Also, the potential for developing *new* agricultural cultivars will be lost. From the hundreds of thousands of plant species existing in nature, humans have used perhaps 7,000 in all, and modern agriculture has tended to focus on only about 30. Of these, three species—wheat, maize (corn), and rice—fulfill about 50% of global food demands. This limited diversity in agriculture makes it ill suited to production under many different environmental conditions. For example, arid regions are generally considered to be unproductive without irrigation. However, many wild species of the bean family produce abundantly under dry conditions.

Scientists estimate that 30,000 plant species with edible parts could be brought into cultivation. Many of these could increase production in environments that are less than ideal. For example, every part of the winged bean, a native of New Guinea (Fig. 10–2), is edible: pods, flowers, stems, roots, and leaves. Recently introduced into many developing countries, this legume has already contributed significantly to improving nutrition. Loss of biological diversity undercuts similar future opportunities.

*Those Pests.* Another area in which wild species have instrumental value to humans is pest control. In Chapter 16, we will discuss the tremendous and invaluable opportunities there are to control pests by introducing natural enemies and increasing genetic resistance. Natural enemies and genes for increasing resistance can come only from natural biota. Destroying natural biota may destroy such opportunities.

Because species are selected from nature for animal husbandry, forestry, and aquaculture, essentially all the same arguments apply to those important enterprises, too.

Natural biota can be thought of as a bank in which the gene pools of all the species involved are deposited. As long as natural biota are preserved, there is a rich endowment of genes in the bank that can be drawn upon as needed. Thus, natural biota are frequently referred to as a **genetic bank.** Depleting this bank depletes our future.

## Sources for Medicine

Earth's genetic bank also serves medicine. For thousands of years, the indigenous people of the island of Madagascar used an obscure plant, the rosy periwinkle, in their folk medicine (Fig. 10–3). If this plant, which grows only on Madagascar, had become extinct before 1960, hardly anyone outside Madagascar would have cared. In the 1960s, however, scientists extracted two chemicals with medicinal properties—vincristine and vinblastine—from the plant. These chemicals have revolutionized the treatment of childhood leukemia and Hodgkin's disease. Before their discovery, leukemia was almost always fatal in children; today, with vincristine treatment, there is a

**Figure 10–2 The winged bean.** A climbing legume with edible pods, seeds, leaves, and roots, this tropical species demonstrates the great potential of wild species for human use.

**Figure 10–3 The rosy periwinkle.** This native to Madagascar is a source of two anticancer agents that are highly successful in treating childhood leukemia and Hodgkin's disease.

99% chance of remission. These two drugs now represent a $100-million-a-year industry.

The story of the rosy periwinkle is just one of hundreds. The venom from a Brazilian pit viper (a poisonous snake) led to the development of the drug Capoten®, used to control high blood pressure. Paclitaxel (trade name Taxol®), an extract from the bark of the Pacific yew, has proven valuable for treating ovarian, breast, and small-cell cancers. For a time, this use threatened to decimate the Pacific yew, because six trees were required to treat one patient for a year. Now the substance is extracted from the leaves of the English yew tree, a horticultural plant that is easy to maintain. Today, Taxol® is the most prescribed antitumor drug (and has earned billions of dollars for Bristol-Myers Squibb, which developed it!).

***Ethnobotany.*** Stories like these have created a new appreciation for the field of *ethnobotany*, the study of the relationships between plants and people. To date, some 3,000 plants have been identified as having anticancer properties. Drug companies have been financing field studies of the medicinal use of plants by indigenous peoples (called *bioprospecting*) for more than 10 years. Unfortunately, none have yielded any major new drugs, although there may be some in the pipeline that the drug companies will not reveal until they are ready to submit them for FDA approval. Nevertheless, the search for drugs in the tropics has led to the creation of parks and reserves to promote the preservation of natural ecosystems that are home to both the indigenous people and the plants they use in traditional healing. In Surinam, for example, a consortium made up of drug companies, the U.S. government, and several conservationist and educational groups has created the Central Surinam Reserve, a 4-million-acre area now protected.

According to the WHO, 80% of the world's people depend on non-Western medicine that in turn depends on natural products. In fact, 25% of all pharmaceuticals in the United States contain ingredients originally derived from native plants, representing $8 billion of annual revenue for drug companies and better health and longevity for countless people. Table 10–1 lists a number of well-established drugs that were discovered as a result of analyzing the chemical properties of plants used by traditional healers. It is likely that the search for such chemicals has barely scratched the surface.

### Recreational, Aesthetic, and Scientific Value

The species in natural ecosystems also provide the foundation for numerous recreational and aesthetic interests, ranging from sportfishing and hunting to hiking, camping, bird-watching, photography, and more (Fig. 10–4). Periodically, the U.S. government conducts a survey on outdoor recreation; the latest was carried out from 2000 to 2001. Of all people age 16 and older who were surveyed, an estimated 202 million were involved in some form of outdoor recreation (Table 10–2). The survey included opinions on environmental issues, and people's attitudes clearly reflected both interest and concern about the environment, an indication that their contact with the outdoors had sensitized them. For example, 82% of respondents agreed or strongly agreed with the statement "humans are severely abusing the environment," and 80% agreed or strongly agreed that "when humans interfere with nature, it often produces disastrous consequences." Very likely, the broad public support for preserving wild species and habitats stems from the aesthetic and recreational enjoyment people derive from them.

Interests under this head may range from casual aesthetic enjoyment to serious scientific study. Virtually all our knowledge and understanding of evolution and ecology have come from studying wild species and the ecosystems in which they live. Pleasure and satisfaction may even be indirect. For instance, you may never see a whale in person,

**table 10-1 Modern Drugs from Traditional Medicines**

| Drug | Medical Use | Source | Common Name |
|---|---|---|---|
| Aspirin | Reduces pain and inflammation | *Filipendula ulmaria* | Queen of the meadow |
| Codeine | Eases pain; suppresses coughing | *Papaver somniferum* | Opium poppy |
| Ipecac | Induces vomiting | *Psychotria ipecacuanha* | Ipecac |
| Pilocarpine | Reduces pressure in the eye | *Pilocarpus jaborandi* | Jaborandi plant |
| Pseudoephedrine | Reduces nasal congestion | *Ephedra sinica* | *Ma-huang* shrub |
| Quinine | Combats malaria | *Cinchona pubescens* | Cinchona tree |
| Reserpine | Lowers blood pressure | *Rauwolfia serpentina* | Rauwolfia |
| Scopolamine | Eases motion sickness | *Datura stramonium* | Jimsonweed |
| Theophylline | Opens bronchial passages | *Camellia sinensis* | Tea |
| Tubocurarine | Relaxes muscles during surgery | *Chondrodendron tomentosum* | Curare vine |
| Vinblastine | Combats Hodgkin's disease | Catharanthus roseus | Rosy periwinkle |

**Figure 10–4 Recreational, aesthetic, and scientific uses.** Natural biota provide numerous values, a few of which are depicted here.

**table 10-2 Recreational Value of the U.S. Natural Environment**

| Activity | Number of People Participating (millions) |
|---|---|
| Viewing natural scenery | 125.2 |
| Swimming in lake, river, or ocean | 87.3 |
| Boating | 75.5 |
| Fishing | 70.9 |
| Hiking | 68.8 |
| Visit wilderness or primitive area | 68.0 |
| Bird-watching | 67.4 |
| Gathering mushrooms, berries, etc. | 59.1 |
| Mountain biking | 44.6 |
| Primitive camping | 32.9 |
| Hunting | 23.6 |
| Backpacking | 22.2 |

(*Source:* National Survey on Recreation and the Environment, 2000–01)

but knowing that whales and similar exciting animals exist provides a certain aesthetic pleasure. The great popularity of nature films attests to this phenomenon.

Recreational and aesthetic values constitute a very important source of support for maintaining wild species. Recreational and aesthetic activities support commercial interests. **Ecotourism**—whereby tourists visit a place in order to observe wild species or unique ecological sites—represents the largest foreign-exchange-generating enterprise in many developing countries. As the amount of leisure time available to people increases, more and more money is spent on recreation. Because some percentage of these recreational dollars will be spent on activities related to the natural environment, any degradation of that environment affects commercial interests. As the national survey indicated, recreational, aesthetic, and scientific activities involve a great number of people and represent a huge economic enterprise.

## Value for Their Own Sake

The usefulness (instrumental value) of many wild species is apparent. But what about those other species which have no obvious value to anyone—probably the majority of plant and animal species, many of which are rare or inconspicuous in the environment? Some observers believe that the most important strategy for preserving all wild species is to emphasize the *intrinsic* value of species, rather than the unknown or uncertain ecological and economic instrumental values. Thus, the extinction of any species is an irretrievable loss of something of value.

Environmental ethicists argue that long-established existence carries with it a right to continued existence. They hold that humans have no right to terminate a species that has existed for thousands or millions of years and that represents a unique set of biological characteristics. Environmental philosopher Holmes Rolston, III, puts it this way:[1] "Destroying species is like tearing pages out of an unread book, written in a language humans hardly know how to read, about the place where they live." Some support this view by arguing that there is value in every living thing and that one kind of living thing (e.g., humans) has no greater value than any other. This argument, however, can lead to some difficulties, such as having to defend the rights of pathogens and parasites. A more common viewpoint held by ethicists is that, because humans have the ability to make moral judgments, they also have a special responsibility toward the natural world, and that responsibility includes concern for other species. Until recently, Western philosophers argued that only humans were worthy of ethical consideration; that is, the Western philosophical tradition has been strongly anthropocentric.

***Religous Support.*** Many ethicists find their basis for intrinsic value in religous thought. For example, in the Jewish and Christian traditions, Old Testament writings express God's concern for wild species when He created them. Jewish and Christian scholars alike maintain that by declaring His creation good and giving it His blessing, God was saying that all wild things have intrinsic value and therefore deserve moral consideration and stewardly care. Similarly, in the Islamic tradition, the Quran (Koran) proclaims that the environment is the creation of Allah and should be protected because it praises the Creator. Many Native American religions have a strong environmental ethic; for example, the Lakota hold that humans and other life-forms should interact like members of a large, healthy family. Thus, ethical concern for wild species underlies many religious traditions and represents a potentially powerful force for preserving biodiversity.

In sum, even if species have no demonstrable use to humans, it can still be argued that they have a right to continue to exist. Only rarely (as in the case of human parasites and pathogens) can we claim that there is any moral justification for driving other species to extinction.

[1]Rolston, Holmes III. *Environmental Ethics: Duties to and Values in The Natural World.* Philadelphia: Temple University Press, 1988, p. 129.

## 10.2 Saving Wild Species

When wild species are valued, we are saying that we want to have them around for some reason. One compelling reason is that they provide recreation and food for people who hunt them.

### Game Animals in the United States

Game animals are those animals traditionally hunted for sport, meat, or pelts. In the early days of the United States, there were no restrictions on hunting, and a number of species were hunted to extinction (great auk, heath hen, passenger pigeon) or near extinction (bison, wild turkey). As game animals became scarce in the face of increasing pressure from hunting, regulations were enacted. State governments, backed up by the federal government, enacted laws establishing hunting seasons and bag limits and hired wardens to enforce the laws. Some species were given complete protection in order to allow their populations to build up to numbers that would once again allow hunting.

***Turkey Boom.*** One success story is the wild turkey (Fig. 10–5). A favorite game species, this bird was hunted to the brink of extinction, but was making a slow comeback by the 1930s as a result of hunting restrictions. At that time, there was a total population of about 30,000 individuals in a few scattered states. After World War II, state and federal programs addressed the need for protecting turkey habitats. The birds were reintroduced into areas they had once inhabited, and hunting quotas were strictly limited. The turkey is now found in 49 states, and its population has risen to more than 5 million as a result of these measures.

Using hunting and trapping fees as a source of revenue, state wildlife managers enhance the habitats supporting important game species and provide special areas for hunting. They monitor game populations and adjust seasons and bag limits accordingly. Another source of revenue for wildlife conservation comes from federal excise

**Figure 10–5 Wild turkey.** Restoration efforts have so successfully reestablished the wild turkey throughout the United States that hunting within carefully controlled limits now takes place.

taxes placed on hunting, fishing, shooting, and boating equipment. In 2003, this source provided $478 million in matching funds (75% federal, 25% state) to state fish and wildlife conservation programs. Game preserves, parks, and other areas where hunting and fishing are prohibited are maintained to protect habitats, as well as breeding populations that can disperse to other habitats.

In spite of what seems on the surface to be a destruction of wildlife, hunting has many positive aspects. Besides the fees hunters pay, many hunters belong to organizations dedicated to the game they are interested in hunting. Organizations such as Ducks Unlimited, the National Wild Turkey Federation, and Pheasants Forever raise funds that are used for the restoration and maintenance of natural ecosystems vital to the game they are interested in hunting. For example, Ducks Unlimited recently announced that more than 10 million acres of vital wetland habitat for wildfowl have been conserved in North America. The organization's activities have helped stem the tide of loss that has reduced North America's wetlands by more than 50% over the years. The activities are in great measure responsible for the remarkable yearly hunting harvest of ducks and geese in the United States—some 16 million ducks and more than 3 million geese in 2002.

***Two-legged Predators.*** Defenders of hunting and trapping argue that their prey are often animals that lack natural predators and would increase to the point of destroying their own habitat. This is indeed frequently the case with larger animals, such as deer and elk in the United States. For example, there are so many deer in Pennsylvania that they cause some $300 million in damage annually; they eat crops, damage cars in collisions, and seriously harm forest understories through extensive browsing. In Australia, kangaroos have become such a problem that professional sharpshooters now "cull" more than 3 million kangaroos a year (out of an estimated 60 million), mostly for the pet food and leather industry. However, many members of the nonhunting public object to the killing of wildlife, and some groups, such as the Humane Society of the United States and People for the Ethical Treatment of Animals (PETA), actively campaign to limit or end hunting and trapping. Some practices that are regarded as especially cruel, such as the use of leghold steel traps, have been banned in several states as a result of ballot initiatives.

***Backyard Menagerie.*** Common game animals, such as deer, rabbits, doves, and squirrels, are well adapted to rural and suburban environments. Thus adapted and protected from overhunting, viable populations of these animals are being maintained. Some predatory animals are also on the increase. The following serious problems, however, have emerged as well:

1. The number of animals killed on roadways now far exceeds the number killed by hunters. As rural areas are developed, increasing numbers of animals found on roadways are a serious hazard to motorists. With 3.9 million miles of roadways and 225 million vehicles in the United States, over a million animals a day become roadkill. More than 200 human fatalities occur as a consequence of these collisions. In New England states, more than 1,500 moose are killed annually in accidents that often also kill or seriously injure motorists because of the size of the animals. Overpasses and tunnels are increasingly being built to provide wildlife with safe corridors (Fig. 10–6).
2. Many nuisance animals are thriving in highly urbanized areas, creating various health hazards. Opossums, skunks, and deer are attracted to urban areas by opportunities for food; unsecured garbage cans and pet food left outside will ensure visits from raccoons. In 2001, the Center for Disease Control reported 7,437 cases of rabid animals in the United States, 68% of which occurred in raccoons and skunks. A number of European countries have eradicated rabies in wild animals by spreading bait laced with rabies vaccine; several states are now employing this strategy, which shows great promise.
3. Some game animals have no predators except hunters and tend to reach population densities that push them into suburban habitats, where they cannot be hunted effectively. The white-tailed deer, for example, has become a pest to gardeners and fruit nurseries; it also poses a public health risk because it is often infested with ticks that carry Lyme disease. (See Chapter 15.) Many public parks, in which hunting is unfeasible, are home to high densities of deer. Sharon Woods, for instance, a 1-square-mile park near Columbus, Ohio, had a deer population of 500 and became an ecological disaster. Deer browsed virtually every tree from six feet down, stripped bark from many, and removed the understory. In such areas, municipalities have resorted to hiring sharpshooters to cull the deer herds. In wilder areas, rising deer populations have reduced populations of many endangered or threatened plant species by their grazing.

**Figure 10–6 Bear underpass.** To reduce black-bear roadkill, Florida has established this corridor under State Route 46 to connect important bear habitats transected by the highway. Twelve other species use the corridor, too.

**Figure 10–7 Cougar on the roof.** As suburbs encroach on cougar habitats in the western United States, attacks on humans have become more numerous. This particular cougar on the roof of a home in a California suburb has been shot with a tranquilizer dart.

4. In recent years, suburbanites have been increasingly attacked by cougars, bears, and alligators as urbanization encroaches on the wild (Fig. 10–7). California and Oregon have cougar populations numbering in the thousands due to diminished hunting, with resulting increases in contact between people and these large predators. Over the past one-hundred years, large carnivores have killed 142 people, and half of these deaths have occurred in the last decade alone.
5. Coyotes, which once roamed only in the Midwest and western states, are now found in every state and are increasing in numbers. A highly adaptable predator, the coyote will eat almost anything. In the Adirondacks, they eat mostly deer; in suburban neighborhoods, they often dine on cats, small dogs, and human garbage. (One particularly adaptive pair had a den on the median strip of Boston's beltway, Route 128, where they lived off roadkill.) Very difficult to control, the 30–50-lb animals occasionally attack small children.
6. Suburban parks and lawns, college campuses, and golf courses have become home to exploding flocks of Canada geese. These large birds are, ecologically speaking, grazing herbivores, able to consume large quantities of grass and defecating as often as every eight minutes. Protected by wildlife laws, the geese present a challenge to efforts to control them. Few predators are able to tackle these large birds, which originated from geese captured and bred to act as decoys to wild migrating flocks. One imaginative solution used around golf courses and colleges is to "hire" border collies (and their trainers), which are taught to harass the geese and keep them from comfortably grazing the lush grass they relish.

*Wildlife Services?* One highly controversial effort for controlling unwanted animals is carried out by an agency of the U.S. Department of Agriculture. In 1998, the agency changed its name from **Animal Damage Control** to **Wildlife Services** in an effort to soften its image. This agency responds to requests from livestock owners, farmers, homeowners, and others concerned with economic damage, human health, and safety to remove nuisance animals and birds. "Removal" virtually always means killing, and the Wildlife Services agency routinely uses poisons, traps, and other devices to kill almost 2 million animals and birds yearly. In 2001, for example, the agency killed 89,000 coyotes, 30,000 beavers, 2,400 bobcats, 5,300 feral hogs, 7,300 skunks, and 9,700 raccoons, among many other animals and birds. In spite of attempts by conservation organizations to limit the budget and activities of the Wildlife Services, it is still drawing \$50 million to \$65 million of taxpayers' money annually and enjoys strong support from Congress. Interestingly, this is the same Congress that for several years has been unable to agree to renew the major piece of legislation that provides protection to wild animals: the Endangered Species Act.

## Acts Protecting Endangered Species

In colonial days, huge flocks of snowy egrets inhabited coastal wetlands and marshes of the southeastern United States. In the 1800s, when fashion dictated fancy hats adorned with feathers, egrets and other birds were hunted for their plumage. By the late 1800s, egrets were almost extinct. In 1886, the newly formed Audubon Society began a press campaign to shame "feather wearers" and end the "terrible folly." The campaign caught on, and gradually attitudes changed and new laws followed.

*Lacey.* Florida and Texas were first to pass laws protecting plumed birds. Then, in 1900, Congress passed the **Lacey Act,** forbidding interstate commerce in illegally killed wildlife, making it more difficult for hunters to sell their kill. Since then, numerous wildlife refuges have been established to protect the birds' breeding habitats. With millions of people visiting these refuges and seeing the birds in their natural locales, attitudes have changed significantly. Today, the thought of hunting these birds would be abhorrent to most of us, even if official protection were removed. Thus protected, egret populations were able to recover substantially. In the meantime, the Lacey Act and its amendments have become the most important piece of legislation protecting wildlife from illegal killing or smuggling. Under the act, the U.S. Fish and Wildlife Service (FWS) can bring federal charges against anyone violating a number of wildlife laws.

*Stung.* In 2000, the president and owner of a New York–based caviar company was convicted under the Lacey Act of conspiring to smuggle protected sturgeon caviar and passing off illegally caught paddlefish caviar as Russian sturgeon caviar. He got two years in prison, and his company paid a criminal fine of $110,000. The man was caught in "Operation Malassol," a covert operation in which a special agent posed as a buyer. FWS special agents worked on over 8,700 violations of the Lacey act and other wildlife-related laws in 2001. Their efforts resulted in fines of over $13 million and 41 years in jail time for those convicted.

*Endangered Species Act.* Congress took another major step when it passed a series of acts to protect endangered species. The most comprehensive and recent of these acts is the **Endangered Species Act (ESA)** of 1973 (reauthorized in 1988). Recall from Chapter 4 that an **endangered species** is a species that has been reduced to the point where it is in imminent danger of becoming extinct if protection is not provided (Fig. 10–8). The act also provides for the protection of **threatened species,** which are judged to be in jeopardy, but not on the brink of extinction. When a species is officially recognized as being either endangered or threatened, the law specifies substantial fines for killing, trapping, uprooting (in the case of plants), modifying significant habitat of, or engaging in commerce in the species or its parts. The legislation forbidding commerce includes wildlife threatened with extinction anywhere in the world.

The ESA is administered by the *FWS* for terrestrial and freshwater species (and a few marine mammals) and by the National Oceanic and Atmospheric Administration (*NOAA*) *Fisheries* (formerly the *National Marine Fisheries Service*) for marine and anadromous species (those which migrate into freshwater). There are three crucial elements in the process of designating a species as endangered or threatened:

1. **Listing.** Species may be listed by the appropriate agency or by petition from individuals, groups, or state agencies. Listing must be based on the best available information and must not take into consideration any economic impact the listing might have. Listed species are summarized in Table 10–3.
2. **Critical Habitat.** When a species is listed, the agency must also designate as critical habitat the areas where the species is currently found or where it could likely spread as it recovers. A 1995 Supreme Court decision made it clear that federal authority to conserve critical habitats extended to privately held lands.
3. **Recovery Plans.** The agency is required to develop recovery plans that are designed to allow listed species to survive and thrive.

As Table 10–3 indicates, 1,263 U.S. species are currently listed for protection under the act; recovery plans are in place for 999 of them, and 426 species have designated critical habitat.

The ESA was scheduled for reauthorization in 1992, but because of political battles, it has been operating on year-to-year budget extensions. It has been a political lightning rod since its enactment. Opposition to the act comes from development, timber, and mineral interests, which claim that the act is costly and favors the rights of "bugs and bushes over humans and economic growth." Organizations such as the American Land Rights Association and the American Forest Resource Council continually lobby for weakening or abolishing the ESA, and they have staunch allies in the Congress—especially Rep. James Hansen of Utah, who chairs the House Resources Committee. According to Hansen, "The ESA has become a wrecking ball in this country, devastating personal finances and regional economies."

*Lawsuits.* Frustrated with the slow pace of listing species and designations of critical habitat, environmental groups like the Defenders of Wildlife and the Center for Biological Diversity have turned to the courts to force the FWS and NOAA to act on species in danger. These groups have won almost every court challenge to FWS actions. Angry with the challenges, the FWS declared a moratorium on listing activities in November 2000, claiming that the agency was spending most of its time and budget on court-related activities. The real problem, according to the environmental groups, is that the budget is far too small to accomplish the objectives of the ESA. Their claim is supported by the FWS itself, which stated that it would need some $120 million for five years to catch up with the listing backlog. The budget for listing has been $8 to $9 million per year, although the 2004 budget request has increased this line item to $12 million.

Not to be outdone, the opponents of the ESA have also taken their cause to the courts, challenging designations of critical habitat and of basic listings. Recently,

**Figure 10–8 Endangered species.** Some examples of endangered species—species whose populations in nature have dropped so low that they are in imminent danger of becoming extinct unless protection is provided.

the FWS has been settling cases before going to court. For example, the Bush administration asked a federal judge to invalidate critical habitat protection for the California gnatcatcher (a small bird designated as threatened), thus removing 500,000 acres of protective habitat. This was in response to a suit from developers. The judge ruled that the FWS did not carry out an adequate economic analysis of the critical habitat designation, as required by law.

***Impact on Species.*** Some critics of the ESA believe that the act does not go far enough. A major shortcoming is that protection is not provided until a species is officially listed by the FWS as endangered or threatened and a recovery plan is established. Species usually will not

**table 10-3 Federal Listings of Threatened and Endangered U.S. Plant and Animal Species, 2003**

| Category | Endangered | Threatened | Total Listings* | Species with Recovery Plans† |
|---|---|---|---|---|
| Mammals | 65 | 9 | 342 | 52 |
| Birds | 78 | 14 | 273 | 77 |
| Reptiles | 14 | 22 | 115 | 32 |
| Amphibians | 12 | 9 | 30 | 14 |
| Fishes | 71 | 44 | 126 | 96 |
| Snails | 21 | 11 | 33 | 22 |
| Clams | 62 | 8 | 72 | 57 |
| Crustaceans | 18 | 3 | 21 | 13 |
| Insects | 35 | 9 | 48 | 29 |
| Arachnids | 12 | 0 | 12 | 5 |
| Plants | 599 | 147 | 749 | 602 |
| Total | 987 | 276 | 1,821 | 999 |
| Total endangered U.S. species: | | 987 | (388 animals, 599 plants) | |
| Total threatened U.S. species: | | 276 | (129 animals, 147 plants) | |
| Total listed U.S. species: | | 1,263 | (517 animals, 746 plants) | |
| Total listed non-U.S. species | | 558 | (555 animals, 3 plants) | |

*Source:* Department of the Interior, U.S. Fish and Wildlife Service, Division of Endangered Species, June 9, 2003.
*Total listings refer to both U.S. and non-U.S. species.
†Recovery plans pertain only to U.S.-listed species.

make the list until their populations have become dangerously low. Currently, the backlog on listing includes some 249 "candidate species," all of which are acknowledged to be in need of listing. The FWS also claims that only about half of the species on the current list are stable or improving. Over the years, only 13 have come off the list because they recovered, while 7 have been removed because they became extinct. One of the species recently removed from the list because its numbers have recovered signifcantly is the American peregrine falcon (Fig. 10–9).

***Falcon.*** Both the peregrine falcon and the bald eagle (still a threatened species) were driven to extremely low numbers because of the use of DDT as a pesticide from the 1940s through the 1960s. Carried up to these predators through the food chain, DDT caused a serious thinning of the birds' eggshells that led to nesting failures in the two species and in numerous other predatory birds (Chapter 16). A survey indicated that by 1975 there were only 324 pairs of nesting peregrines in North America. DDT use was banned in both the United States and Canada in the early 1970s, and the stage was set for the bird to recover. Working with several nonprofit captive-breeding institutions such as the Peregrine Fund, the FWS sponsored efforts that resulted in the release of some 6,000 captive-bred young falcons in 34 states over a period of 23 years. There are now more than 1,600 known breeding pairs in the United States and Canada, well above the targeted recovery population of 631 pairs.

***Fly Away Home.*** A few species have gained exceptional public attention, and heroic efforts have been mounted to save them. Efforts to save the whooping crane, for example, include virtually full-time monitoring and protection of the single remaining flock, which for years numbered only in the teens. The historic migratory wild flock is increasing steadily and has reached a population of 184 individuals from a low of 14 cranes in 1939. This flock migrates from wintering grounds on the Texas Gulf Coast to Wood Buffalo National Park in Canada. The migration is hazardous, and accidents along the way take a toll on the birds. A nonmigratory flock is taking hold in central Florida, using eggs from captive-breeding flocks. Over 90 birds now make up the flock, and in 2003, 15 pairs formed bonds and are producing chicks.

A spectacular effort has established a new migratory flock of 21 that summers in Wisconsin and migrates to Florida. This flock was first "taught" its migratory path by following researchers in ultralight aircraft (Fig. 10–10). This eastern flock is being established by the Whooping Crane Eastern Partnership, a cooperative effort of federal, state, and nonprofit groups.

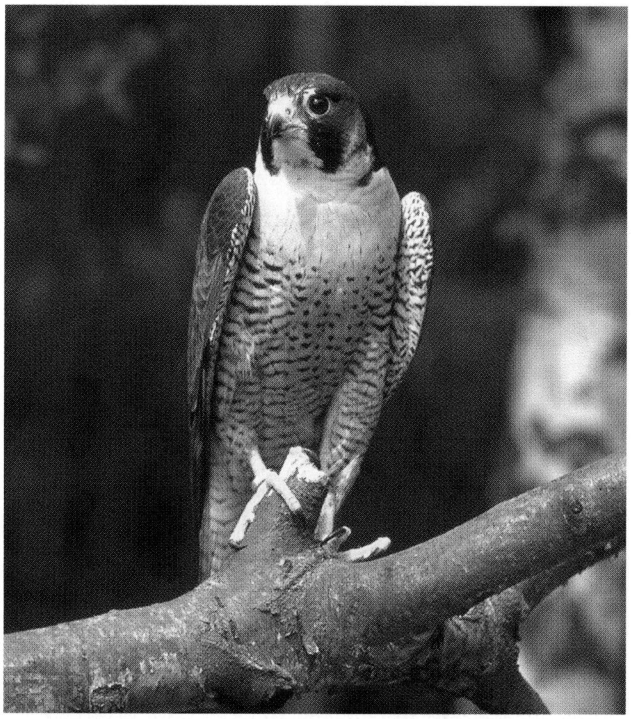

**Figure 10–9 The peregrine falcon.** The Endangered Species Act worked well in bringing this magnificent falcon to the point where it was dropped from the list.

*Spotted Owl.* Some critics of the ESA claim that it goes too far. A high-profile case in point is the northern spotted owl. This species became a focal point in the battle to save some of the remaining old-growth forests of the Pacific Northwest. The owl is found only in these forests and had dwindled to a population between 4,000 and 6,000. In June 1990, the FWS listed the owl as a threatened species. Years of bitter controversy followed, pitting environmentalists against the timber industry, until a compromise (the **Northwest Forest Plan**) was worked out in 1994. The plan, employing the ecosystem management approach now adopted by the Forest Service (Chapters 3 and 4), set aside 7.4 million acres of federal land in California, Oregon, and Washington where logging is prohibited in stands of trees older than 80 years. The latest chapter in this controversy saw the Bush administration settling a lawsuit by the American Forest Resource Council (a timber industry group) in 2003 by agreeing to review protection of the spotted owl and the amount of land that needs to be left unlogged. The FWS defended its move by pointing out that status reviews are required every five years anyway, although very few had been done in the past. In British Columbia, only 30 breeding pairs of owls may remain, so International Forest Products, a major timber company, has voluntarily stopped all logging in spotted-owl territory in an effort to save the population.

**Figure 10–10 Whooping cranes and pilot.** A pilot in this ultralight aircraft leads a flock of whooping cranes in their migration from Wisconsin to Florida.

*Klamath.* The latest controversy in the Pacific Northwest involves the Klamath River, along the Oregon–California border. This river supplies water to Upper Klamath Lake and is the source of irrigation water for some 250,000 acres of farmland. In 2001, due to an extended drought, federal authorities allocated almost all of the water to protect threatened salmon in the Klamath River and two endangered fish species in the lake, as well as many migratory species and bald eagles that depend on nearby wildlife refuges. A public outcry resulted ("fish versus people") as farms were deprived of the water they needed. In early 2002, the National Research Council (NRC) reported that there was insufficient scientific evidence to make the decision in either direction—for the lake or for the farmers. In the light of that report, the Bureau of Reclamation, which maintains federal irrigation projects, allowed most of the water to be delivered to the upstream farmers in 2002. Apparently as a result of the low water flow in the river, many thousands of coho salmon died that year. The Bureau's action was taken in spite of existing law, supported by court decisions, that requires the agency to provide the water necessary to preserve species in the Klamath basin. This is not a simple case of environmental versus agricultural concerns, because economic interests exist on both sides—harvesting salmon is a valuable industry. The problem is far from resolved: President Bush established the Klamath River Basin Federal Working Group in 2002 to investigate all the issues and make recommendations within 18 months.

*New Challenge.* The ESA requires that listing and critical-habitat decisions be made "solely on the basis of the best scientific and commercial data available." Very often, not much data *are* available; most rare species are poorly known or little studied. There is a move afoot by critics of the ESA to amend the act by calling for more peer-reviewed sound science to support a listing. This sounds like a good idea, but defenders of the act reply that it is simply a smokescreen for weakening the ESA. They maintain that often there are no funds to conduct such studies and that by the time the mandated information would be gathered, many species in decline might well become extinct. They point to the basic intent of the ESA—to save and protect wild species—and argue that all declining species should be given "the benefit of the doubt." Indeed, the *FWS Handbook* uses this very phrase in its applications of the ESA. The FWS is obligated to collect all available information on a species and to evaluate the quality of the science in that information, but to require that the listing be determined by the "best science possible" would, in the opinion of many, contradict the original intent of Congress in establishing the ESA.

In the final analysis, the ESA is a formal recognition of the importance of preserving wild species, regardless of whether they are economically important. Species listed as endangered or threatened have legal rights to protection under the law. The act is something of a last resort for wild species, but it embodies an encouraging attitude toward nature that has now become public policy. Actions taken under auspices of the act have demonstrated a unique commitment to preserve wild species. The reauthorization of a strong ESA will be a major signal of our continuing commitment to consider the value of wild species on grounds other than economic or political. In all likelihood, judicial review will continue to play an important role in determining the fairness of the many applications of the act.

In the next section, you will study the broader topic of biodiversity—the sum of richness of species and habitats on Earth—and examine reasons for its decline and strategies for its preservation.

## 10.3 Biodiversity and Its Decline

In the introduction to this chapter, biodiversity was defined as simply *the variety of life on earth*. The dimensions of biodiversity include the genetic diversity within species, as well as the diversity of ecosystems and habitats. The main focus of biodiversity, however, is the species. Recall from Chapter 4 that, over time, natural selection leads to speciation—the creation of new species—as well as extinction—the disappearance of species. Over geological time, the net balance of these processes has favored the gradual accumulation of more and more species—in other words, an increase in biodiversity.

***How Many Species?*** How much biodiversity of species is there? Estimates range widely, and the fact is, no one knows. Today, the only two certainties are that 1.75 million species have been described and many more than that exist. Most people are completely unaware of the great diversity of species within any given taxonomic category. Groups especially rich in species are the flowering plants (270,000 species) and the insects (950,000 species), but even less diverse groups, such as birds or ferns, are rich with species that are unknown to most people (Table 10–4). Taxonomists are aware that their work in finding and describing new species is incomplete. Groups that are conspicuous or commercially important, such as birds, mammals, fish, and trees, are much more fully explored and described than, say, insects, very small invertebrates like mites, and soil nematodes, fungi, and bacteria. Trained researchers who can identify major groups of organisms are few, and the task of fully exploring the diversity of life would require a major sustained effort in systematic biology.

Estimates of the number of species on Earth today are based on recent work in the tropical rain forests, which hold more living species than all other habitats combined. Costa Rica, less than half the size of New York State, is estimated to contain at least 5% of all living species. The upper boundary of the estimate keeps rising as taxonomists explore the rain forests more and more. Some (enthusiastic?) estimates place the high end at 112 million species. Whatever the number, the planet's biodiversity represents an amazing and diverse storehouse of biological wealth.

### The Decline of Biodiversity

***North America.*** The biota of the United States is as well known as any. At least 500 species native to the United States are known to have become extinct since the early days of colonization. Over 100 of these are vertebrates. At least 200,000 species of plants, animals, and microbes are estimated to live in the nation, but not much is known about most of them. An inventory of 20,897 wild plant and animal species in the 50 states, carried out in the late 1990s by a team of biologists from the Nature Conservancy and the National Heritage Network and more recently amplified by a new conservation network, NatureServe, concluded that one-third are vulnerable, imperiled, or already extinct (Fig. 10–11). Mussels, crayfish, fishes, and amphibians—all species that depend on freshwater habitats—are most at risk. Flowering plants are also of great concern, with one-third of their numbers in trouble.

It is fair to say that species *populations* are a more important element of biodiversity than just the species' *existence*. It is the populations that occupy different habitats and ranges and that contribute to biological wealth as they provide goods and services important in ecosystems. Across North America, general population declines of well-studied wild species are also occurring. Commercial landings of many species of fish are down, puzzling reductions in amphibian populations are occurring in North America and other parts of the world, and many North American songbird species, such as the Kentucky warbler, wood thrush, and scarlet tanager, have been declining and have disappeared entirely from some local regions. These birds are Neotropical migrants; that is, they winter in the neotropics of Central and South America and breed in temperate North America. According to a report from the National Audubon Society, more than one-fourth of North American birds are already declining or are in danger of doing so.

***Global Outlook.*** Worldwide, the loss of biodiversity is even more disturbing. The known causes of animal extinctions, recently assessed by the World Conservation Union, are shown in Figure 10–12. At least 726 animal species and 90 plant species have become extinct since

**table 10-4 Known and Estimated Species on Earth**

| Taxonomic Group | Number of Known Species | Estimated Numbers High | Estimated Numbers Low | Estimated Numbers Working | Accuracy |
|---|---|---|---|---|---|
| Viruses | 4,000 | 1,000,000 | 50,000 | 400,000 | Very poor |
| Bacteria | 4,000 | 3,000,000 | 50,000 | 1,000,000 | Very poor |
| Fungi | 72,000 | 2,700,000 | 200,000 | 1,500,000 | Moderate |
| Protozoa | 40,000 | 200,000 | 60,000 | 200,000 | Very poor |
| Algae | 40,000 | 1,000,000 | 150,000 | 400,000 | Very poor |
| Plants | 270,000 | 500,000 | 300,000 | 320,000 | Good |
| Nematodes | 25,000 | 1,000,000 | 100,000 | 400,000 | Poor |
| Arthropods: | | | | | |
| Crustaceans | 40,000 | 200,000 | 75,000 | 150,000 | Moderate |
| Arachnids | 75,000 | 1,000,000 | 300,000 | 750,000 | Moderate |
| Insects | 950,000 | 100,000,000 | 2,000,000 | 8,000,000 | Moderate |
| Mollusks | 70,000 | 200,000 | 100,000 | 200,000 | Moderate |
| Chordates | 45,000 | 55,000 | 50,000 | 50,000 | Good |
| Others | 115,000 | 800,000 | 200,000 | 250,000 | Moderate |
| **Total** | **1,750,000** | **111,655,000** | **3,635,000** | **13,620,000** | **Very poor** |

*Source:* United Nations Environment Programme, "Global Biodiversity Assessment" (Cambridge, U.K.: Cambridge University Press, 1995): 118, Table 3.1–1.

1500. Most of the known extinctions of the past several hundred years have occurred on oceanic islands, where small landmasses limit the size of populations and human intrusions are most severe. The extinction rate for mammals and birds has been 20 to 25 species per hundred years. This rate can be compared with the "background extinction rate" seen in the fossil record of one bird or mammal species every 400 and 800 years, respectively.

The *Global Biodiversity Outlook*, published in 2001 by UNEP to provide information for the Convention on Biological Diversity, estimates that some 24% of mammal (1,130) and 12% of bird species (1,183) are globally threatened. These two groups have been thoroughly studied. The status of all other groups of organisms is much less well known. A recent study compared historic and present-day ranges for 173 larger terrestrial mammal species.

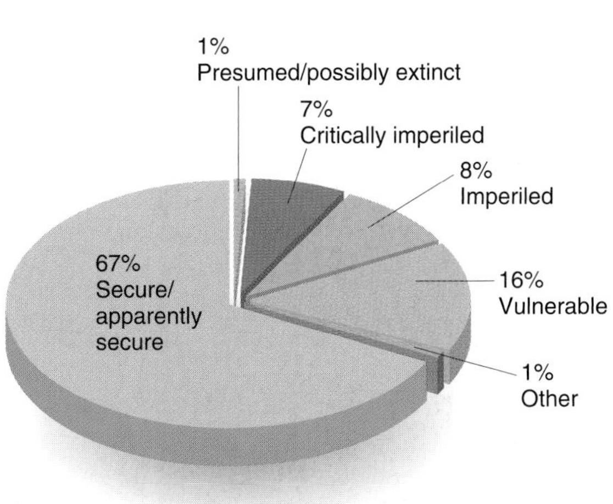


**Figure 10–11 The state of U.S. species.** Fully one-third of over 20,000 species of plants and animals surveyed were found by biologists to be at risk of extinction. (*Source:* Data from the Nature Conservancy and NatureServe.)

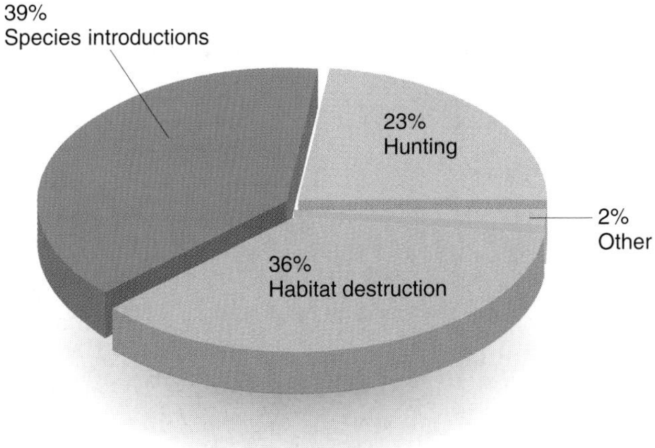


**Figure 10–12 Known causes of animal extinctions since 1600.** The introduction of foreign species, the destruction of habitats, and hunting have brought about the majority of known animal extinctions since 1600. (*Source:* Data from World Conservation Monitoring Centre.)

The research showed that the species collectively have lost two-thirds of their historic ranges due to human activities. Again, this represents a significant loss of ecosystem capital, as well as a loss of biodiversity.

Biodiversity is richest in the tropical forests—a richness that is almost unimaginable. Biologist E. O. Wilson identified 43 species of ants on a single tree in a Peruvian rain forest, a level of diversity equal to the entire ant fauna of the British Isles. Other scientists found 300 species of trees in a single 2.5-acre (1-ha) plot and as many as 10,000 species of insects on a single tree in Peru. Unfortunately, the same tropical forests that hold such high biodiversity are also experiencing the highest rate of deforestation. Because the inventory of species in these ecosystems is so poorly known, it is virtually impossible to assess extinction rates in such forests. Some scientists have tried, though, with estimates ranging from 4,000 to 17,000 species lost per year. Not all scientists agree with these calculations, however, because the estimates are based on many assumptions. Regardless of the exact rates, numerous species are in decline and some are becoming extinct. No one knows how many, but the loss is real, and it represents a continuing depletion of the biological wealth of our planet.

## Reasons for the Decline

**Physical Alteration of Habitats.** Although multiple causes are usually the rule in losses of biodiversity, one of the greatest sources of loss is the physical alteration of habitats through the processes of conversion, fragmentation, and simplification. Habitat destruction has already been responsible for 36% of the known extinctions (Fig. 10–12) and is the key factor in the currently observed population declines. Natural species are adapted to specific habitats, so if the habitat changes or is eliminated, the species go with it.

***Conversion.*** Natural areas are converted to farms, housing subdivisions, shopping malls, marinas, and industrial centers. When a forest is cleared, for example, it is not just the trees that are destroyed: Every other plant or animal that occupies the destroyed ecosystem, either permanently or temporarily (e.g., migrating birds), also suffers. The idea that this wildlife simply will move "next door" and continue to live in an undisturbed section is erroneous. Any loss of natural habitat can result in only one thing: a proportional reduction in all populations that require that habitat. Thus, the decline in North American songbird populations has been traced to the loss of winter forest habitat in Central and South America *and* the increasing fragmentation of summer forest habitat in North America. Habitats like croplands that replace natural habitats such as grasslands are quite inhospitable to all but a few species that tend to be well adapted to the new managed landscapes.

***Fragmentation.*** Natural landscapes generally have large patches of habitat that are well connected to other, similar patches. Human-dominated landscapes, however, consist of a mosaic of different land uses, resulting in small, often geometrically configured, patches that frequently contrast highly with neighboring patches (Fig. 10–13). Whether it be plowed fields or house lots, the tendency is to shrink the size of forested parcels. For the continued survival of any natural population, the number of individuals must never fall below a *critical number,* and that requires a certain minimum area that must be large enough to compensate for years of adverse weather. For many species, more area will be required during a dry year than during a normal year. If development reduces the habitat to a point where it cannot support the critical

**Figure 10–13 Fragmentation.** Development usually leads to the breaking up of natural areas, resulting in a mosaic of habitats that may not support local populations of species.

number during an adverse year, the entire population will perish. Similarly, development (such as a highway) that fragments a territory and prevents migration between the two fragments will cause a population to perish if neither area can adequately support the critical number. Also, reducing the size of a habitat creates a greater proportion of edges, a situation that favors some species, but may be detrimental to others.

For example, the Kirtland's warbler is an endangered species that is highly dependent on large patches of second-growth jack pines. The species is endangered because its habitat has been greatly fragmented, creating edges that favor the brown-headed cowbird, a nest parasite that can invade the forest and lay its eggs in the nest of the rare warbler. Edges also favor species, such as crows, magpies, and jays, that prey on nests, and these predators are thought to be partly responsible for declines in many neotropical bird populations.

***Simplification.*** Human use of habitats often simplifies them. Removing fallen logs and dead trees from woodlands for firewood, for example, diminishes an important microhabitat on which several species depend. When a forest is managed for the production of a few or one species of tree, tree diversity declines, and with it so does the diversity of the cluster of plant and animal species that depend on the less favored trees. Streams are sometimes "channelized"—their beds are cleared of fallen trees and riffles, and sometimes the stream is straightened out by dredging. Such alterations inevitably reduce the diversity of fish and invertebrates that live in the stream.

***Intrusion.*** Birds use the airways as highways, especially as they migrate north and south in the spring and fall. Recently, telecommunications towers have presented a new hazard to birds. Although television towers have been around for decades, new towers are sprouting up on hilltops and countrysides in profusion. The lights often placed on these towers can attract birds, which usually migrate at night, and the birds simply collide with the towers and supporting wires. Some 45,000 towers over 200 feet high are now in place, and the number is expected to double within a decade. The FWS has estimated that the towers kill somewhere between 5 and 50 million birds a year. No one yet knows how to prevent this slaughter, but studies are needed to determine what kinds of lighting are least attractive to the migrants.

**The Population Connection.** Past losses of biodiversity can be attributed to the expansion of the human population over the globe. Continuing human population growth will further alter natural ecosystems, resulting in the inevitable loss of more wild species and additional declines in populations. The losses will be greatest in the developing world, where biodiversity is greatest and human population growth is highest. Africa and Asia have lost almost two-thirds of their original natural habitat. People's desire for a better way of life, the desperate poverty of rural populations, and the global market for timber and other natural resources are powerful forces that will continue to draw down biological wealth on those continents.

In East Africa, where human population growth has been explosive for several decades, the conversion of savanna and woodlands to cultivation or intensive grazing by goats and cattle has driven most of the African elephant population into the existing wildlife reserves, greatly reducing their numbers. The other large African mammals have experienced similar reductions, as the needs of the rural population inevitably conflict with those of the large wild animals of East Africa. Similar losses in large-mammal populations (wolves, bison, elk, bear, cougar, etc.) occurred as North America was gradually transformed into a great agricultural and industrial continent.

One key to holding down the loss in biodiversity lies in bringing human population growth down. If the human population increases to 10 billion, as some demographers believe that it will, the consequences for the natural world will be frightening.

**Pollution.** Another major factor that decreases biodiversity is pollution, which can directly kill many kinds of plants and animals, seriously reducing their populations. For example, nutrients (phosphorus and nitrogen) from America's agricultural heartland traveling down the Mississippi River have created a 7,700-square-mile "dead zone" in the Gulf of Mexico where oxygen completely disappears from depths below 20 meters every summer. (See Chapter 17.) Shrimp, fish, crabs, and other commercially valuable sea life are either killed or forced to migrate away from this huge area along the Mississippi and Louisiana coastline.

Every oil spill kills seabirds and, often, sea mammals, sometimes by the thousands. As mentioned, DDT can devastate wild bird populations. Human wastes can spread pathogenic microorganisms to wild species, a threat called "pathogen pollution." For example, manatees (endangered aquatic mammals) have been infected by human papillomavirus, cryptosporidium, and microsporidium, with fatal outcomes. Canine distemper virus, likely spread from local dogs, killed more than 20,000 seals in the Caspian Sea in 2000. The cause of a mysterious rash of deformities in many amphibian species has finally been tracked down (Fig. 10–14). Although pesticides or other pollutants were suspected, the actual cause was a larval stage of a flatworm that invades the tadpoles and induces the deformities. The recent and rapid rise in the incidence of these deformities is traced to habitats that have been altered by human use; high nutrient pollution leads to large snail populations, which are necessary intermediate hosts of the flatworm.

***Climate Change.*** Pollution destroys or alters habitats, with consequences just as severe as those caused by deliberate conversions. Acid deposition and air pollution kill forests; sediments and nutrients kill species in lakes, rivers, and bays; and the depletion of the stratospheric ozone layer increases the impact of ultraviolet light on wild species. Global climate change, brought on by

**Figure 10–14 Deformed frogs.** Mysterious outbreaks of deformed amphibians have now been traced to a flatworm parasite that derails normal development.

anthropogenic greenhouse gas emissions, is already having an impact on many species. In the polar north, polar bears are in decline because the sea ice they depend on for hunting seals is melting at a rapid rate (up to 9% per decade). In the polar south, many penguin species are in decline as waters undergo unusual warming that disrupts the fish and invertebrate species the penguins depend on. Rising land temperatures are already encouraging the spread of vector-borne diseases (diseases carried by ticks, flies, mosquitoes, etc.) such as Lyme disease, malaria, dengue, and West Nile disease. Some of these are pathogens for both humans and wildlife. The future impact of these and other diseases of wildlife is impossible to predict, but as climate continues to change, there will certainly be impacts, and some could be devastating to wild species.

Most of the global pollution problems can be traced to the industrialized world, where energy-generating and other technologies continue to pour pollutants into the air and water at increasing rates. In this case, the developing world, where population growth is such a problem, is not to blame. Instead, global climate change and pollution from toxic substances are the legacy of the already developed nations.

**Exotic Species.** An exotic species is a species introduced into an area from somewhere else, often a different continent. Because the species is not native to the new area, it is frequently unsuccessful in establishing a viable population and quietly disappears. This is the fate of many pet birds, reptiles, and fish that escape or are deliberately released from their native habitats. Occasionally, however, an introduced species finds the new environment very much to its liking and can become an *invasive species,* thriving, spreading out, and perhaps eliminating native species by predation or competition for space or food. In Chapter 16, you will learn that most of the insect pests and plant parasites that plague agricultural production were accidentally introduced. Exotic species are major agents in driving native species to extinction and are responsible for an estimated 39% of all animal extinctions since 1600. Chapter 4 describes a number of introduced species and their ecological impact.

The transplantation of species by humans has occurred throughout history, to the point where most people are unable to distinguish between the native and exotic species living in their lands. European colonists brought literally hundreds of weeds and forage plants to the Americas, so that now most of the common field, lawn, and roadside plant species in eastern North America are exotics. Fully one-third of the plants in Massachusetts are alien or introduced—about 900 species! Among the animals, the most notorious have been the house mouse and the Norway rat; others include the wild boar, the donkey, the horse, the nutria, and even the red fox, which was brought to provide better fox hunting for the early colonial equestrians. One of the most destructive exotics is the house cat: The 100 million domestic and feral cats in the United States are highly efficient at catching small mammals and birds. They kill an estimated 1 billion small mammals (chipmunks, deer mice, and ground squirrels, as well as rats) and hundreds of millions of birds annually.

***May I Introduce...*** Deliberate introductions sanctioned by the U.S. Natural Resources Conservation Service (in the interest of "reclaiming" eroded or degraded lands) have brought us kudzu and other runaway plants, such as autumn olive, multiflora rose, and Amur honeysuckle. Many other exotic plants are introduced as horticultural desirables. One notorious example is purple loosestrife, which infests wetland habitats and replaces many plants that are valuable to wildlife. (Figure 4–13 shows photos of purple loosestrife, kudzu, and other exotic plants.) Another is Oriental bittersweet, an aggressive vine that chokes and shades small shrubs and trees. Often, it smothers young trees and prevents the forest from regrowing. In Oregon and California, the introduced star thistle invades and ruins pastureland, grows wildly, and is a fire hazard. One report calculates the annual cost of invasive species in the United States to be $137 billion.

You might think that introduced species would add to the biodiversity of a region, but many introductions have exactly the opposite effect. Foreign plants crowd out native plants. New animal species are often successful predators that eliminate native species not adapted to their presence. For example, the brown tree snake was introduced onto the island of Guam as a stowaway on cargo ships during World War II (Fig. 10–15). This snake is mildly poisonous, grows to a length of 8 feet, and eats almost anything smaller than itself. In a little over 50 years, the snakes have eliminated most of the island's birds. In fact, 9 of Guam's 12 native species of birds have become extinct, and populations of the remaining birds

**Figure 10–15 The brown tree snake.** This snake, accidentally introduced onto Guam, has decimated bird life on the island. Often lacking in natural predators, islands are especially vulnerable to the harmful effects of exotic species.

are so decimated that survivors are rarely seen or heard. Efforts at controlling the snakes have proved futile. They have no natural predators on the island, and the snake population has reached such a high density that snakes have invaded houses in search of prey (such as puppies and cats). Currently, wildlife officials are concentrating their efforts on preventing the snakes from spreading to other islands in the Marianas chain.

*Brazilian Pepper.* Introduced exotic species may also drive out native species by competing with them for resources, as described in Chapter 4. Brazilian pepper, an ornamental shrub resembling holly, has become the most widespread exotic in Florida. It is an aggressive colonizer, taking over many native plant communities and fundamentally changing ecosystems like the Everglades (Fig. 10–16). The plant now occupies more than 700,000 acres in central and southern Florida. A task force recently concluded, "The uncontrolled expansion of Brazilian pepper constitutes one of the most serious ecological threats to the biological integrity of Florida's natural systems."

*Aquaculture.* One-third of all seafood consumed worldwide is produced by aquaculture—the farming of shellfish, seaweed, and fish. Over 100 species of aquatic plants and animals are raised in aquaculture in the United States, most of which are not native to the farming locations (often a coastal embayment or freshwater pond). Parasites, seaweeds, invertebrates, and pathogens have been introduced together with the desired aquaculture species, often escaping from the pens or ponds and entering the sea or nearby river system. For example, Asian black carp are used in raising catfish because the carp eat snails that harbor parasites capable of ruining the catfish meat. So far, none are known to have escaped in the United States, but everywhere else they have been used they have escaped and spread. Sometimes the aquaculture species itself is a problem. Atlantic salmon, for example, are now raised along the Pacific coast, and many have escaped from the pens and have been breeding in some west-coast rivers.

**Overuse.** Removing whales, fish, or trees faster than they can reproduce will lead to their ultimate extinction. Nevertheless, overuse is another major assault against wild species, responsible for 23% of recent extinctions. Overuse is driven by a combination of greed, ignorance, and desperation. The plight of birds in Europe is a good example. It is said, with some justification, that a line can be drawn across the continent, north of which people watch birds and south of which they eat them. In Greece, for example, some 700,000 "protected" birds are shot

(a) (b)

**Figure 10–16 Brazilian pepper bush.** (a) A typical everglades landscape, with wooded hammocks surrounded by marshes; (b) a virtual monoculture of Brazilian pepper covers this everglades marsh area, obliterating the native biota.

each year. In Malta about 3 million a year are killed, and in Italy 50 million birds are killed and eaten each year! Most of these are small songbirds.

***Trade in Exotics.*** Another prominent form of overuse is the trafficking in wildlife and in products derived from wild species. Much of this "trade" is illegal; worldwide, Interpol estimates it to be at least $12 billion a year. It flourishes because some consumers are willing to pay exorbitant prices for such things as furniture made from tropical hardwoods (like teak), exotic pets, furs from wild animals, traditional medicines from animal parts, and innumerable other "luxuries," including polar-bear rugs, baskets made from elephant and rhinoceros feet, ivory-handled knives, and reptile-skin shoes and handbags. For example, some Indonesian and South American parrots sell for up to $10,000 in the United States, a panda-skin rug can bring in $25,000, and the gallbladder of the North American black bear, which is valued as folk medicine in Korea, can fetch as much as $2,000. A *shahtoosh* shawl can bring thousands of dollars and can be purchased in many large cities in India and elsewhere in spite of a trade ban since 1975. The shawls, extremely soft and warm, are made of the wool of the Tibetan antelope, or chiru (Fig. 10–17). They have become a "must-have" luxury item for the wealthy elite of London, New York, Hong Kong, and other large cities. The animals are killed for their wool, and three antelopes are needed to make one shawl. Some 20,000 chirus are killed each year, and the population of the species has declined to less than 75,000.

***Greed.*** The long-term prospect of extinction does not curtail the activities of exploiters, because, to them, the prospect of a huge immediate profit outweighs it. Even when the species is protected, the economic incentive is such that poaching and black-market trade continue. Rustling in the wild west has taken on a new shape: Nowadays, outlaws roam the southwestern national parks and desertlands digging up wild cacti for the lucrative landscaping trade in cities such as Phoenix, Tucson, and Las Vegas. Tiger and rhinoceros populations in the wild have declined drastically in the last two decades because of the widespread, but unfounded, belief in Far Eastern countries that parts from these animals have medicinal or aphrodisiac properties. Toward that end, there is evidence that the drug Viagra may take the pressure off some of the animals traditionally killed for the folk medicinal trade.

***eBay Sting.*** The FWS is the agency with jurisdiction over the illegal trade in wildlife in the United States. Its Special Operations force of 235 agents actively investigates cases of wildlife smuggling and poaching. Recently, an agent "purchased" an African leopard skin and a frozen stillborn tiger cub for $1,500 on eBay and nabbed the seller as a result. Agents now monitor a number of Internet on-line auction sites where collectors and traffickers often do business in endangered species and their parts.

Particularly severe is the growing fad for exotic pets—fish, reptiles, birds, and houseplants. In many cases, these plants and animals are taken from the wild; often, only a small fraction even survives the transition. When they are removed from the natural breeding populations, they are no better than dead when it comes to maintaining the species. At least 40 of the world's 330 known species of parrots face extinction because of this fad.

Poor management represents another form of overuse leading to a loss of biodiversity. Forests and woodlands are overcut for firewood, grasslands are overgrazed, game species are overhunted, fisheries are overexploited, and croplands are overcultivated. These practices not only deplete the resource in question, but often set into motion a cycle of erosion and desertification, with effects far beyond the exploited area. This problem is explored in depth in Chapter 11.

**Figure 10–17 Tibetan antelope (chiru).** This species is hunted for its delicate fur, which is used to make *shahtoosh* shawls, prized items for the wealthy elite.

## global perspective

### Biodiversity: Essential or Not?

A fierce debate is raging within the ecological academy, focused on a very important question that has serious political and economic ramifications: Will the loss of species to an ecosystem bring it closer to collapse? A corollary question asks just what the role of biodiversity is in promoting normal functioning, such as productivity, in an ecosystem. On one side of this controversy are some of the leading ecologists in the United States (especially David Tilman) and the United Kingdom (John Lawton), who have published experimental work indicating that biodiversity is essential to the healthy functioning of an ecosystem.

For example, one line of experimental work—the Ecotron Experiment—was carried out in the United Kingdom. Researchers added various plants, insects, and worms to replicate plots in enclosed chambers. They found that the more species they added to a plot, the more biomass the plot produced and thus the greater the productivity of the plot. Other work carried out in Minnesota showed that plots with a greater diversity of species were more resistant to drought. More important, Tilman and Lawton were among the coauthors of "Biodiversity and Ecosystem Functioning: Maintaining Natural Life Support Processes," an important *Issues in Ecology* paper published by the Ecological Society of America in 1999 and widely distributed on the Internet and to members of Congress. Citing their experimental work, the authors concluded, "Given its importance to human welfare, the maintenance of ecosystem functioning should be included as an integral part of national and international policies designed to conserve local and global biodiversity."

But wait a minute, say Michael Huston of Oak Ridge National Laboratory in Tennessee, Phil Grime of the University of Sheffield in the United Kingdom, and others. There is a good chance that these experimental studies and the conclusions drawn from them were "politically manipulated." Not only that, but in some important cases the experiments were flawed and therefore were inconclusive. Years ago, theoretical ecologist Robert May concluded, on the basis of mathematical modeling, that biodiversity has no consistent effect on the functioning of an ecosystem. Rather, what is more important, the critics say, is how individual organisms respond to environmental change. Citing experimental evidence in support of this view, these researchers note that if particular species are present, they will often dominate ecosystem functioning because of their superior adaptations to a given environment; the presence of many other species can be irrelevant.

This is not to say that biodiversity is unimportant; in fact, greater biodiversity will make the presence of key species more likely in an ecosystem. Thus, biodiversity is like an insurance policy against environmental variability. But to say that biodiversity is *the* key to ecosystem functioning and then to transfer this conclusion to policy recommendations goes beyond what the science supports, according to the critics. They believe that other arguments should be used to support biodiversity as an important property of natural systems and that to tie support of biodiversity to normal ecosystem functioning, as Tilman and others do, is to erect a "house of cards" which could collapse in the face of new experimental evidence.

## Consequences of Losing Biodiversity

Biodiversity almost always refers to the diversity of species found in a given ecosystem. These are the *structural* components of ecosystems. Identifying the structural components is important, because function flows from structure, and it is the *functioning* ecosystem that provides the goods and services sustaining human life and enterprises. Can ecosystems lose some species and still retain their functional integrity? Research evidence indicates that it is the dominant plants and animals that determine major ecosystem processes such as energy flow and nutrient cycling. (See "Global Perspective," this page.) Thus, simplifying ecosystems by driving rarer species to extinction is not necessarily expected to lead to ecosystem decline. Accordingly, it is best not to justify preserving wild species and biodiversity with predictions of ecological doom.

It is possible, however, to lose what ecologists call **keystone species**—species whose role is absolutely vital to the survival of many other species in an ecosystem. For example, through felling trees and building dams, beavers significantly alter forest and stream ecology and increase ecosystem diversity in a region. Alternatively, the keystone species in an ecosystem may be a predator that keeps herbivore populations under control. Sometimes, keystone species are the largest animals in the ecosystem, and because of their size, they have the greatest demands for unspoiled habitat. Thus, wolves, elephants, tigers, moose, and other large animals are considered "umbrella" species for whole ecosystems. If they can survive, there will likely still be enough undisturbed habitat for the other living species. (See Earth Watch, p. 282, for a perspective of the gray wolf as a keystone species in Yellowstone National Park).

It is also possible to introduce species (for example, Brazilian pepper) that can become new dominants in ecosystems. They have an impact on both biodiversity and functionality as they crowd out existing species, and the results are very likely to be undesirable. It is quite likely that wetlands with Brazilian pepper produce more

## earth watch

### Return of the Gray Wolf

Alaska, with 5,000 to 7,000 gray wolves, is the only state in which the wolf is not endangered and therefore not protected by the Endangered Species Act (ESA). Prior to the act, wolves had been exterminated from all but two of the lower 48 states. Currently, gray-wolf populations are on the increase. In fact, more than 3,700 of them are now found in 9 states. There has been a profound shift in public attitude toward wolves: Whereas states once poisoned and trapped wolves and paid bounties for them, Defenders of Wildlife now pays ranchers for any losses they suffer from wolf predation. (In the last 16 years, Defenders has compensated ranchers $279,000 for such losses.)

In a highly publicized program, the U.S. Fish and Wildlife Service began releasing captured Canadian wolves in Wyoming's Yellowstone National Park and in central Idaho. The 14 wolves released in Yellowstone in 1995 demonstrated their approval of their new home by producing nine pups, successfully preying on the abundant elk and bison in the park, aggressively harassing the numerous coyotes there, and rewarding visitors with views of their activities. Encouraged by their success, the wildlife biologists added 17 more wolves to Yellowstone in 1996 and then halted the reintroductions. As of mid-2003, the Yellowstone wolf population numbered 163, signaling the wolves' successful adaptation to the national park. This is bad news for the coyote population, which wolves have reduced by 50% in some areas as they regain their status as top dogs in the park. Recent observations suggest that the wolves are bringing greater ecological health to areas of Yellowstone, largely through their impact on the overabundant elk herds in the park.

A subspecies, the Mexican gray wolf, recently was reintroduced into the Southwest, where some 21 wolves are hanging on in an area with a heavy presence of range cattle. Serious proposals have been made to reintroduce the wolf into the Northeast—in particular, northern Maine, where habitat and prey are believed to be excellent.

Wolf populations in the Great Lakes and Rocky Mountain areas have, in fact, achieved population goals set by the FWS for recovery. In 2003, these populations were "downlisted" from *endangered* to *threatened*. The lower status allows for some "removal" of wolves that cause problems to livestock or people. The wolves in the Southwest will remain on the endangered list, however.

Research biologist David Mech, probably the leading authority on wolves, believes that the key to continuing the wolf's comeback is to stop short of complete control. The conflicting interests of protectionists, on the one hand, and farmers and ranchers, on the other, will have to be worked out through a careful program of regulation of wolf populations—just as many other large animals, such as bears, cougars, and coyotes, are kept under control by regulations and hunting quotas. Now that the wolf has been downlisted, in the words of David Mech, it "can be accepted as a regular member of our environment, rather than as a special saint or sinner, [and] this will go a long way toward ensuring that the howl of the wolf will always be heard throughout the wild areas of the northern world."

organic matter than wetlands without the alien species, but productivity is only one of many goods and services that result from ecosystem function. How well does the alien species fit into the complex food web of a given ecosystem? How many other species does it nourish? Is it as pleasing aesthetically? What ecosystem functions are significantly changed by the introduced species?

*If They Go?* So, what are the consequences of losing some of the amazing biodiversity Earth now holds? What will happen as more and more rare or unknown species pass from the Earth because of human activities? A few people will mourn their passing, but, as one observer put it, the Sun will continue to come up every morning, and the newspaper will appear on your doorstep. Currently, we seem to be getting away with it, because many species have already become extinct due to human activities, yet life goes on. We do not really know, however, what we are losing when species become extinct. One thing is certain, though: The natural world is less beautiful and less interesting as a result.

Extinction is serious, but so is the general decline of populations of wild species. This, too, is a loss of biodiversity, and its primary cause is the loss of habitat. Biodiversity will continue to decline as long as we continue to remove and constrict the natural habitats in which wild species live. This loss is bound to be costly, because natural ecosystems provide vital services to human societies. Recreational, aesthetic, and commercial losses will be inevitable. The loss of wild species, therefore, will bring certain and unwelcome consequences, because it is linked directly to the degradation or disappearance of ecosystems (the focus of Chapter 11).

## 10.4 Protecting Biodiversity

### International Developments

*IUCN.* Serious efforts are being made to preserve biodiversity around the world, especially in the tropics, where so much of the world's biodiversity exists. One of these efforts is the work of the *World Conservation Union* (IUCN, standing for the organization's former name, the International Union for the Conservation of Nature), which maintains a "Red List of Threatened Species." Similar to the U.S. endangered species list, the Red List uses a set of criteria to evaluate the risk of extinction to thousands of species throughout the world. The list is updated annually, and there are currently 11,167 species of plants and animals on it. Once published in book form, the Red List is now in electronic format on a searchable Web site, www.redlist.org. Each species on the list is given its proper classification, and its distribution, documentation of the listing, habitat and ecology, conservation measures, and data sources are described. The World Conservation Union is not actively engaged in preserving endangered species in the field, but its findings are often the basis of conservation activities throughout the world, and it provides crucial leadership to the world community on issues involving biodiversity. One of its efforts was to help establish the Trade Records Analysis for Flora and Fauna in International Commerce (TRAFFIC) network, which monitors wildlife trade and helps to implement CITES' decisions and provisions. (See next.)

*CITES.* Endangered species outside the United States are only tangentially addressed by the ESA. However, under the leadership of the United States, the *Convention on Trade in Endangered Species of Wild Fauna and Flora,* or CITES, was established in the early 1970s. CITES is not specifically a device to protect rare species; instead, it is an international agreement (signed by 118 nations) that focuses on trade in wildlife and wildlife parts. The treaty recognizes three levels of vulnerability of species, the highest being species threatened with extinction. Covering some 30,000 species, restrictive trade permits and agreements between exporting and importing countries are applied to those species, sometimes resulting in a complete ban on trade if the CITES nations agree. Every two or three years, the signatory countries meet at a Conference of Parties (COP). The latest (the 12th) COP was held in Santiago, Chile, in November 2002.

Perhaps the best-known act of CITES was to ban the international trade in ivory in 1990 in order to stop the rapid decline of the African elephant (from 2.5 million animals in 1950 to about 600,000 today). In response to requests from African nations, the 10th COP, in 1997, voted to lift the ban to allow a one-time ivory sale for three African countries (Namibia, Botswana, and Zimbabwe) that had demonstrated the ability to maintain sustainable populations of their elephants. As many predicted, lifting the ban led to a resumption in poaching: Over 400 elephants were killed by gangs in Zimbabwe alone. As a result, the 11th COP refused to allow any new ivory sales, despite a proposal from the three countries to expand their trade in ivory. Under pressure from the African countries, the 12th COP voted to allow the countries to sell $5 million worth of tusks. Again, poaching resumed; seven elephants were killed in Uganda in April 2003.

***Convention on Biological Diversity.*** Although CITES provides some protection for species that might be involved in international trade, it is inadequate to address broader issues pertaining to the loss of biodiversity. With the support of UNEP, an ad hoc working group proposed an international treaty that would form the basis of action to conserve biological diversity worldwide. After several years of negotiation, the **Convention on Biological Diversity** (CBD) was drafted and became one of the pillars of the 1992 Earth Summit in Rio de Janeiro. The *Biodiversity Treaty,* as the convention is called, was ratified in December 1993 and is now in force. The treaty establishes a COP as the agency that will provide oversight and report on its task during periodic meetings. Six COP meetings have been held, the latest in The Hague in April 2002. At that meeting, agreement was reached on (1) stepping up the war against invasive species; (2) guidelines for giving international companies and organizations access to genetic resources (in diversity-rich developing countries), in return for a fair share of any profits and benefits; (3) creating ways to stem the tide of deforestation and thereby foster sustainable forest management; and (4) formulating a strategic plan to coordinate the activities under the convention through 2010.

The preamble sets forth the following *basic guidelines* for the Biodiversity Treaty:

1. A concern for the intrinsic value of biodiversity.
2. Its significance for human welfare.
3. The sovereignty of a nation over its biodiversity.
4. The nation's obligations to protect and conserve biodiversity.

The signatory countries, in turn, must

- Adopt specific national biodiversity action plans and strategies
- Establish a system of protected areas and ecosystems within the country
- Establish policies that provide incentives to promote the sustainable use of biological resources
- Restore habitats that have been degraded
- Protect threatened species
- Respect and preserve the knowledge and practices of indigenous peoples
- Respect the ownership of genetic resources by countries and share the technologies developed from those resources

The secretariat of the Convention on Biological Diversity and the UNEP jointly published *Global Biodiversity Outlook* in 2001. This publication, available on the Internet, lays out in detail the work of the convention and its implementation.

Concern about access to genetic resources and funding mechanisms led President George H. W. Bush to refuse to sign the treaty in Rio in 1992. President Clinton signed the treaty in June 1993, but in order to become binding, it must be ratified by a two-thirds majority vote of the U.S. Senate.

Continued objections led the Republican-dominated Senate to shelve ratification, leaving both the treaty and U.S. involvement in it in a state of limbo, where it is likely to remain. However, many of the provisions of the treaty are being carried out by different U.S. agencies. The National Biological Service, for example, is working on several of the treaty's provisions. The United States continues to send delegations as observers to the COPs and participates in other CBD meetings, such as those dealing with genetically modified organisms. (See Chapter 9.)

*Critical Ecosystem Partnership Fund.* A new player on the international scene, the Critical Ecosystem Partnership Fund, emerged in August 2000. Jointly sponsored by the World Bank, Conservation International, and the Global Environment Facility, the fund provides grants to NGOs and community-based groups for conservation activities in biodiversity "hot spots"—25 regions in which 60% of the biodiversity has been located on just 1.4% of Earth's land surface (Fig. 10–18). The fund is providing $125 million over a five-year period.

## Stewardship Concerns

What we do about wild species and biodiversity reflects on our *wisdom* and our *values*. *Wisdom* dictates that we take steps to protect the biological wealth that sustains so many of our needs and economic activities. The scientific team that put together the U.N. Global Biodiversity Assessment focused on four themes in its recommendations:

1. **Reforming policies that often lead to declines in biodiversity.** Many governments subsidize the exploitation of natural resources and agricultural activities that consume natural habitats. These policies are frequently counterproductive, leading to the decline of important natural goods and services, which tend to be undervalued economically because they do not enter the market.
2. **Addressing the needs of people who live adjacent to or in high-biodiversity areas or whose livelihood is derived from exploiting wild species.** Although

People and Biological Diversity: Population in the Biodiversity Hotspots, 1995

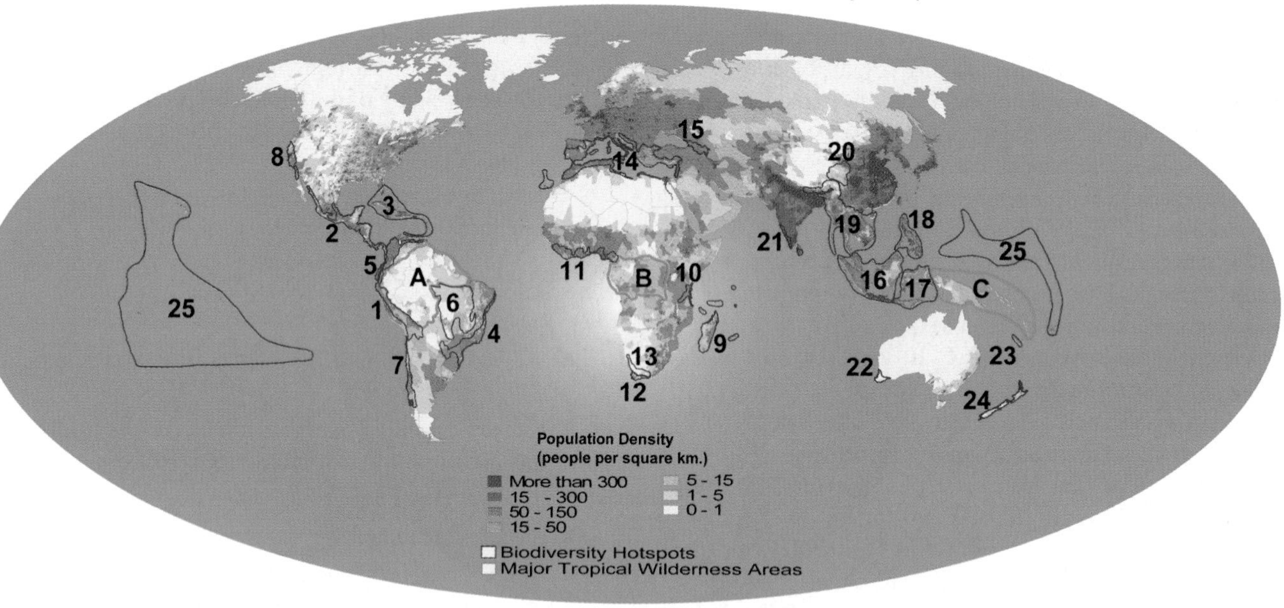


Source: Conservation International, 1999; NCGIA/CIESIN, 1995.

1. Tropical Andes
2. Mesoamerica
3. Caribbean
4. Brazil's Atlantic Forest
5. Choco/Darien/Western Ecuador
6. Brazil's Cerrado
7. Central Chile
8. California Floristic Province
9. Madagascar
10. Eastern Arc and Central Forests of Tanzania/Kenya
11. West African Forests
12. Cape Floristic Province
13. Succulent Karoo
14. Mediterranean Basin
15. Caucasus
16. Sundaland
17. Wallacea
18. Philippines
19. Indo-Burma
20. South-Central China
21. Western Ghats/Sri Lanka
22. SW Australia
23. New Caledonia
24. New Zealand
25. Polynesia/Micronesia

A, B and C Designate Major Tropical Wilderness Areas

**Figure 10–18 Biodiversity "Hot Spots."** Some 25 regions in which 60% of the Earth's biodiversity has been located on just 1.4% of the globe's land surface. (*Source:* Conservation International Web site, October 2003) protecting biodiversity benefits entire societies, the people who are closest to the protected areas are key to the success of efforts to maintain sustainability of natural resources. These people must be given ownership rights or at least communal-use rights to natural resources, and they must be involved in the protection and management of wildlife resources. Experience has shown that when local people benefit from the protection of wildlife, they will more readily take up stewardship responsibility for the wildlife. Recently, community-based natural resource management (CBNRM) emerged as a strategy for wildlife-rich African countries to preserve their biodiversity.

3. **Practicing conservation at the landscape level.** Too often, we tend to believe that a given refuge or park will protect the biodiversity of a region when, in reality, many wild species require larger contiguous habitats or a variety of habitats. For example, to address the needs of migratory birds, summer and winter ranges and flyways between them must be considered. Buffer zones should be created between major population centers and wildlife preserves. This approach requires cooperation between governments, private-property owners, and corporations.
4. **Promoting more research on biodiversity.** Far too little is known about how to manage diverse ecosystems and landscapes, and much remains to be learned about the species within the boundaries of different countries. For example, Costa Rica is engaged in a scientific inventory of its biodiversity, with an eye especially on the development of new pharmaceutical or horticultural products. Toward that end, a new field called **biodiversity informatics** has been invented. Its goal is to put taxonomic data and other information on species online—that is, to "turn the Internet into a giant global biodiversity information system," as one observer put it. Species 2000 is one such effort (www.sp2000.org), a program with the "objective of enumerating all known species of organisms on Earth." Another is the Global Biodiversity Information Facility (www.gbif.org), whose express purpose "is to make the world's biodiversity data freely and universally available."

***Values.*** Our *values* also are demonstrated by our approach to wild species. Is the natural world simply grist for our increasing consumption, or is nature something to be managed sustainably and in a stewardly fashion and then passed on to our children and their children? Because different people answer these questions differently and for various reasons, some people feel a need for alternative sets of values other than, or in addition to, simply survival or utilitarianism, to guide our course into the future.

What each individual turns to will depend on his or her beliefs, needs, education, and religious or ethnic background. Some environmentalists have embraced Deep Ecology, a movement that reminds humans that we are a part of Earth and not separate from it, thus placing nature at the forefront of human awareness. For others, religious faith forms the basis for their interaction with nature. Significantly, Harvard University's Center for the Study of World Religions has held a series of conferences exploring each of the major world faiths and its ecological views. The American Academy of Arts and Sciences, the United Nations, and the American Museum of Natural History have held interdisciplinary gatherings on faith and the environment.

Further strategies to address the loss of biodiversity must focus on preserving the natural ecosystems that sustain wild species. Ecosystems are discussed in detail in Chapter 11, including how we depend on them, what we are doing to them, and what we must do to resist the forces that seem to be leading to a future in which all that is left of wild nature is what we have managed to protect in parks, preserves, and zoos.

# revisiting the themes

## Sustainability

Because ecosystem capital is of infinite value to human enterprises, the pathway to a sustainable future depends on protecting the wild species that make ecosystems work. Protection is tied to the ways people value wild species, however, and some ways of valuing are more conducive to sustainability than others. Sustainability is diminished when species are driven to extinction. It is enhanced when we take steps to encourage healthy populations of game animals; if they thrive, so do their natural ecosystems. Sometimes, however, it is necessary to hunt or otherwise remove game animals (such as deer) from populations when their numbers increase too much in human landscapes.

In the past, egrets and other shorebirds were hunted down for their plumes and feathers, and other species were driven to extinction. Currently, many animal and plant populations are in an unsustainable decline because of habitat loss, the introduction of exotic species, pollution, and overuse. It is particularly crucial to save keystone species and to stop the spread of invasive species in order to maintain an ecosystem's integrity. The general, continued decline of populations points to an impoverished future if we do not act in time.

## Stewardship

The reestablishment of puffins off the coast of Maine is a splendid example of good stewardship at work. Kress and coworkers really cared about the missing seabirds and took effective action to bring them back. Stewardship places a high value on wild species. Although bearing instrumental value is sufficient for preserving many species, having intrinsic value might be the nearest to a pure stewardship ethic, because it works out primarily for the sake of the species. Of all the policies dealing with wild species, the ESA comes closest to supporting the intrinsic value of living things. Many of the causes of population declines and extinction demonstrate an appalling lack of concern for stewardship. Among these causal factors are the eating of millions of songbirds, the destruction of birds and small animals by domestic cats, and the cruel trade in exotic species and their parts. In the end, what we do about wild species reflects directly on our wisdom and our values.

## Sound Science

Sound science is essential for managing game animals and for carrying out the objectives of the ESA. Fortunately, many universities provide excellent programs in wildlife management, and the professionals who work with wild species are often highly motivated and skilled because of their training. However, the concept of peer-reviewed sound science is being promoted as a way to weaken the ESA, by raising the bar well beyond the act's intent of making decisions on the basis of the best scientific data available to employing the best possible science in determining which species are to be listed as endangered or threatened. Some exciting new applications of science, such as biodiversity informatics, have emerged as a result of the amazing capabilities of the Internet.

## Ecosystem Capital

The discussion of wild species and biodiversity in this chapter is all about ecosystem capital. Wild species and biodiversity represent biological wealth, providing the basis for the goods and services coming from natural ecosystems and sustaining human life and economies. They can also be thought of as a genetic bank, capable of being drawn on when new traits are needed. Ecosystem capital is diminished when more than the "interest" is drawn from the natural world as we make use of it. Drawing down ecosystem capital eventually leads to poverty and a biologically impoverished future, unless we stem the losses that are now occurring.

## Policy and Politics

The free-for-all that characterized the treatment of many wild species in the 1800s eventually led to the establishment of game laws and laws prohibiting the killing of nongame species. The ESA represents a triumph of concern for wildlife over merely stocking our freezers with game. At the same time, it has become a lightning rod, drawing attacks from property-rights advocates, developers, and industries and activities accustomed to the relatively uncontrolled use of public lands. The fortunes of the ESA wax and wane with different administrations and Congresses. Right now, the Bush administration appears to be very quick to bow to the concerns of the opponents of the ESA, making concessions in court cases that seem to go beyond the intent of the act.

Public policy operates at the international level as well, in the form of international treaties and conventions that regulate trade in wildlife and wildlife products and that focus international attention on protecting biodiversity in all of its forms. In particular, the Convention on Biological Diversity has mobilized the world community to consider many issues relating to biological diversity. Unfortunately, the United States has not yet ratified this important treaty, even though we have a lot at stake in the issues it considers.

Many nongovernmental organizations (NGOs) (some not mentioned in the chapter) serve as watchdog agencies at home and abroad, focusing attention on issues and policies related to biodiversity and wildlife. Some of the most prominent are the Defenders of Wildlife, the World Wildlife Fund, the Center for Biological Diversity, and the National Wildlife Federation.

## Globalization

The global view of biodiversity is pretty discouraging. Not only have many species become extinct, but many thousands more are threatened with extinction if current trends continue. Because of global travel, humans have introduced their plants and animals to islands and continents where these exotic species often disrupt ecosystems, causing some native species to become extinct and others to suffer population declines. Another manifestation of globalization is the enormous trade in wildlife and wildlife products, much of which is illegal and another cause of population declines. Much of this trade is indefensible, based as it is on unsupported medicinal properties or simply high fashion. The Internet, by contrast, is one form of globalization that works in favor of wild species and biodiversity. Indeed, it has become an invaluable mine of information, with organizations like the IUCN and many taxonomy advocates putting their work on-line.

## review questions

1. Define biological wealth and apply the concept to the human use of that wealth.
2. Compare instrumental value and intrinsic value as they relate to determining the worth of natural species.
3. What are the four categories into which the human valuing of natural species can be divided? Give examples of each one.
4. What means are used to preserve game species, and what are some problems emerging from the adaptations of many game species to the humanized environment?
5. What is the Lacey Act, and how is it used to protect wild species?
6. How does the Endangered Species Act preserve threatened and endangered species in the United States? What are the shortcomings of, and controversies surrounding, the act? Cite specific examples of each.
7. What is biodiversity? What are the best current estimates of its extent?
8. Describe both the known and the estimated losses of biodiversity.
9. How do changes in habitat affect biodiversity? Give examples.
10. What is an invasive species? Give examples of several and their impact on native species.
11. Discuss the idea that ecosystems may collapse if they lose some of their species.
12. What is CITES, and what are its limitations? Give three requirements of the Convention on Biological Diversity.
13. What are four recommendations for protecting biodiversity into the future?

## thinking environmentally

1. Choose an endangered or threatened animal or plant species, and use the Internet to research what is currently being done to preserve it. What dangers is your chosen species subject to? Write a protection plan for the species.
2. You have been given the task of maintaining or increasing biodiversity on a small island frequently visited in the past by mariners. What problems do you anticipate? What steps will you take to deal with them?
3. Some people argue that each individual animal has an intrinsic right to survival. Should this right extend to plants and microorganisms? What about the *Anopheles* mosquito, which transmits malaria? Tigers that sometimes kill people in India? Bacteria that cause typhoid fever? Defend your position.
4. Monies for the FWS to administer the Endangered Species Act are limited. How should we decide which species are the most important to save?
5. Is it right to use animals for teaching and research? Justify your answer.

# Ecosystem Capital: Use and Restoration

**Key Topics**

1. Global Perspective on Biological Systems
2. Conservation, Preservation, Restoration
3. Biomes and Ecosystems Under Pressure
4. Public and Private Lands in the United States

The Pacific halibut is a large flatfish (up to 8 ft in length and 800 lb in weight) found across the northern parts of that ocean in bottom waters of the continental shelf. It is a prized table fish and has supported an active commercial fishery since the late 1800s. (See the opposite page.) The eastern Pacific fishery has yielded annual catches as high as 50,000 tons and has been heavily regulated since the 1920s by the International Pacific Halibut Commission. Typically, the commission would set an annual **total allowable catch** (TAC), based on its assessment of stocks, and then the fishing fleet would fish until the TAC was reached. Soon this took on all the aspects of a fishing derby, with increasingly shorter seasons, until, in 1990, the season lasted only six days. Almost all of the fish caught in that short time had to be frozen, which lowers the quality of the fish and the price received by the fishers. The intensive fishing led to lost and damaged gear and endangered the crew and vessels as they fished in spite of adverse weather conditions.

*Halibut Fishing* **Fishermen harvest Pacific halibut on board the *Mar del Norte*, off Kodiak Island, Alaska.**

***Halibut IQ.*** Canada was the first country to move away from this fishing derby approach, in 1991, followed by the United States in 1995. Both countries adopted a management strategy called the **individual quota** (IQ) system, whereby owners of fishing vessels would be allocated a percentage of the annual TAC based on the size of their vessel and their recent fishing performance. Then they could decide when to fish within a season that lasted as long as 8 months. Entry into the fishery was strictly limited. The shift proved highly successful. The fishers saw an increase in their income, as they could sell their fish in the more lucrative fresh-fish market. Also, they could fish in better weather, and they lost less fishing gear. Indeed, to everyone's surprise, fish stocks improved so well that, from 1995 to 1999, the TAC increased by 55% and has been level since then. The change in management strategy took place with the full involvement of the fishing community, and even though the fishery is now more thoroughly regulated than ever (fishers must report when they intend to go out, fish catches are carefully monitored, and observers are present on many vessels), the fishery is healthy and the fishers are better off.

The Pacific halibut story is unusual; more often, natural ecosystems and the species in them are in a losing battle. Yet there are many notable exceptions—places around the world where people are successfully managing natural resources in sustainable ways.

Recall from Chapter 10 that the key to saving wild species is to preserve their habitats—the ecosystems in which they are found. Populations of wild species continue to decline, however, because ecosystems are also in decline throughout the world. Beyond the loss of species is the degradation or loss of the natural goods and services ecosystems provide.

The focus in this chapter is on the major ecosystems that sustain human life and economy, including the gains achieved and losses incurred as these ecosystems are converted to other uses. Particular attention is paid to those ecosystems which are in the deepest trouble. These ecosystems, the species in them, and the goods and services they generate are the wealth of nations, their *ecosystem capital.* To begin, we will examine the approaches to natural resource use that lead to trouble and contrast them with sustainable approaches.

## 11.1 Global Perspective on Biological Systems

### Major Systems and Their Goods and Services

Earth consists of ecosystems that vary greatly in their species makeup, but that exhibit common functions, such as energy flow and the cycling of matter. The major types of terrestrial and aquatic ecosystems—the biomes—reflect the response of biotas to different climatic conditions. Earth's land area can be divided into several broad categories for the purposes of this chapter (Table 11–1): **forests and woodlands, grasslands and savannas, croplands, wetlands,** and **desert lands and tundra.** (See Fig. 2–19.) The oceanic ecosystems can be categorized as **coastal ocean and bays, coral reefs,** and **open ocean.** These eight terrestrial and aquatic systems provide all of our food; much of our fuel; wood for lumber and paper; leather, furs, and raw materials for fabrics; oils and alcohols; and much more. The world economy directly depends on the exploitation of the **natural goods** (referred to in Fig. 1–13 as *provisioning services*) that can be extracted from ecosystems.

Ecosystems also perform a number of other valuable **natural services** (*regulating* and *supporting services* in Fig. 1–13) as they process energy and circulate matter in the normal course of their functioning (Fig. 11–1). Some of the services are general and pertain to essentially all ecosystems, whereas others are more specific. A thorough listing of them can be found in Table 3–2. In their normal functioning, all natural and altered ecosystems perform some or all of these natural services, free of charge. They also do it repeatedly, year after year. These services are valuable. Recall from Chapter 3 that a team of natural scientists and economists calculated the total value of a year's global ecosystem goods and services to be $38 trillion!

***Wetland to Development.*** To illustrate the value of an ecosystem, scientists have calculated that it would cost more than $100,000 a year to artificially duplicate the water purification and fish propagation capacity provided by a single acre of natural tidal wetland (Fig. 11–2). However, that wetland can be drained, filled, and converted into land suitable for vacation homes, making the acre of tidal wetland quite valuable to a developer. The benefit is local and specific (to the developer), while the loss of services is more regional and diffuse (to everyone living near the coast). If you fail to recognize and value the natural services provided by the wetland, it's a lot easier to convert it for vacation homes.

**table 11-1 Ecosystems on the Earth's Surface**

| Ecosystem | Area (million km$^2$) | Area (million mi$^2$) | Percent of Total Land Area |
|---|---|---|---|
| Forests and woodlands | 47.3 | 18.2 | 31.7 |
| Grasslands and savannas | 30.2 | 11.6 | 20.2 |
| Croplands | 16.1 | 6.2 | 10.7 |
| Wetlands | 5.2 | 2.0 | 3.5 |
| Desert lands and tundra | 51.0 | 9.6 | 34.0 |
| **Total land area** | **149.8** | **57.6** | 100.0 |
| Coastal ocean and bays | 21.8 | 8.4 | |
| Coral reefs | 0.52 | 0.2 | |
| Open ocean | 426.4 | 164.0 | |
| **Total ocean area** | **448.7** | **172.6** | |

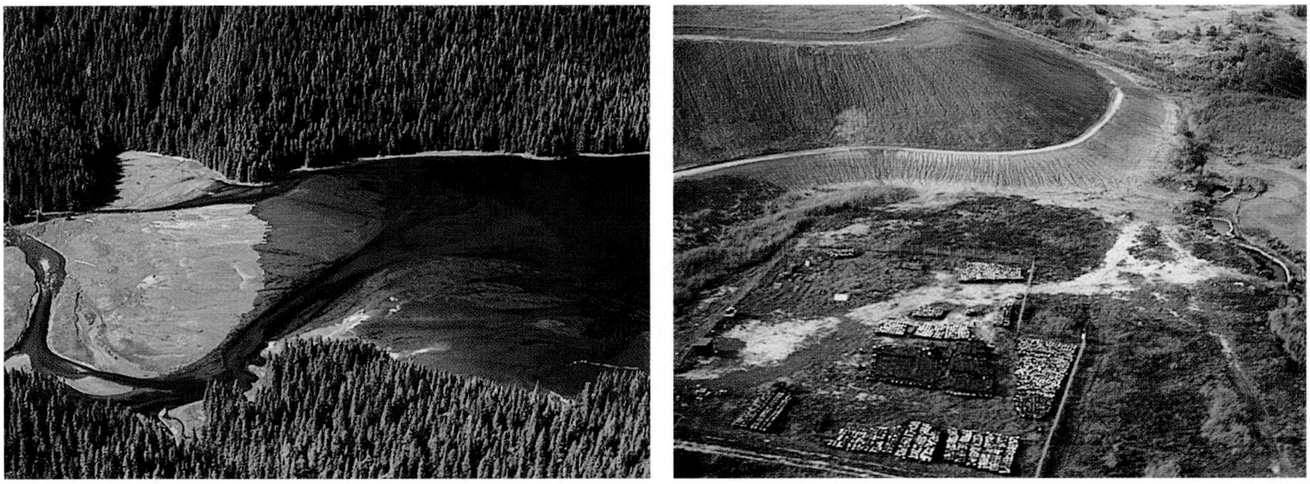

**Figure 11–1 Ecosystem services.** Natural ecosystems perform invaluable natural services if they are treated well. On the left is an aerial view of Baranof Island, Alaska, showing a salmon-spawning stream, meadows, and a coniferous forest. At the right is a view of Kin-Buc landfill in New Jersey, a Superfund site where hazardous-waste disposal has damaged vegetation and interrupted the normal functioning of the area's biota.

## Ecosystems as Natural Resources

If the natural services performed by ecosystems are so valuable, why are we still draining wetlands and removing forests? The answer is clear: *A natural area will receive protection only if the value a society assigns to services provided in its natural state is higher than the value the society assigns to converting it to a more direct human use.*

Natural ecosystems and the biota in them are commonly referred to as **natural resources.** As resources, they are expected to produce something of value, and

**Figure 11–2 Wetland services.** Tidal wetlands provide a number of valuable services, worth more than $100,000 a year for just 1 acre. Especially important are water purification and fish propagation, services that are lost when wetlands are bulkheaded and converted for vacation homes.

the most commonly understood value is *economic* value. By referring to natural systems as *resources,* we are deliberately placing them in an *economic* setting, and it becomes easy to lose sight of their *ecological* value. The concept of *ecosystem capital* is used in this text in order to avoid making this mistake.

***Valuing.*** Markets provide the mechanism for assessing economic value, but do poorly at placing monetary values on ecosystem services. People and institutions may express their preferences for different ecosystem services (resources, if you will) via the market. (Recall from Chapter 10 that this expression of value is *instrumental* in nature.) Instrumental valuation works fairly well for the *provisioning* services provided by ecosystems (goods such as timber, fish, crops, etc.). The Millennium Ecosystem Assessment (MEA) framework refers to these services as having *direct-use value.* However, the *regulating* and *supporting services* provided by natural ecosystems—the more ecological values—are far more difficult to put prices on. These services are *public goods,* not usually provided by markets, but contributing greatly to people's well-being. In fact, they are absolutely essential. In the MEA scheme, these services have *indirect-use value.*

Regulating and supporting services are also less local. Large forested areas can modify the climate in a region and absorb and hold water in the hydrologic cycle. A small forest can be cut, and the timber sold, with no noticeable impact on the weather, but with a definite economic value in a market setting, providing income to the owners. If the larger area is also cut, it will have both an immediate economic benefit and a long-term loss in regulating and supporting services. Thus, it will have an impact on people's well-being in the formerly forested area and well beyond. As discussed in Chapter 3, when the benefits and losses in such conversions are reasonably evaluated economically, losses inevitably outweigh benefits.

***Private Lands.*** Some natural ecosystems are maintained in a natural or seminatural state because that is how they provide the greatest economic (direct-use) value for their owners. For example, most of the state of Maine is owned by private corporations that exploit the forests for lumber and for paper manufacturing. However, if Maine experienced a population explosion and corporations could sell their land to developers for more money than could be gained from harvesting timber, forested lands would quickly become lots for houses. Other similar conversions can be imagined: tropical forests to grasslands for grazing, grasslands to wheat fields, agricultural fields to suburbs, and wetlands to land for vacation homes.

***Public Lands.*** Finally, some natural ecosystems either are publicly owned (state and federal lands) or cannot be owned (ocean ecosystems). These ecosystems are still considered natural resources and may be subject to economically motivated exploitation. *Sustainable* exploitation of such systems will maintain the natural services they perform. Fortunately, ecosystems not only can sustain continued exploitation, but also can be restored from destructive uses that have degraded them and removed their natural biota.

***Future Pressures.*** As population continues its relentless rise, pressures on ecosystems to provide resources will increase: More food, more wood, more water, and more fisheries products will be needed. Maintaining sustainable exploitation of ecosystem capital will become more difficult, in view of the already intense pressure being put on those ecosystems. Trade-offs among goods and services will become more commonplace. Indeed, the next 50 years is expected to see the last major expansion of agriculture. A team of analysts headed by David Tilman at the University of Minnesota has examined the past patterns of human population growth and ecosystem conversions and has calculated that croplands and grazing lands would have to expand to at least 18% larger than their present area to meet the needs of the human population by 2050.[1] Because land use and ecosystem conversion is a zero-sum game, this means that about 1 billion hectares (2.47 billion acres) of forests and grasslands will become the new grazing lands and croplands, with the loss of most of the ecosystem goods and services they once provided.

## 11.2 Conservation, Preservation, Restoration

### Conservation Versus Preservation

An ecosystem's remarkable natural capacity to regenerate makes it (and its biota) a **renewable resource.** In other words, an ecosystem has the capacity to replenish itself through reproduction despite certain quantities of organisms being taken from it, and this renewal can go on indefinitely—it is sustainable. Recall from Chapter 4 that every species has the biotic potential to increase its numbers and that, in a balanced population, the excess numbers fall prey to parasites, predators, and other factors of environmental resistance. It is difficult to find fault with activities that effectively put some of this excess population to human use. The trouble occurs when users (hunters, fishers, loggers, whalers) take more than the excess and deplete the breeding population.

**Conservation** of natural biotas and ecosystems does not—or at least should not—imply *no* use by humans whatsoever, although this may sometimes be temporarily expedient in a management program to allow a certain species to recover its numbers. Rather, the aim of conservation is to *manage or regulate use* so that it does not exceed the capacity of the species or system to renew itself. Conservation is capable of being carried out sustainably, and when sustainability is adopted in principle, conservation has a well-defined goal.

[1]Tilman, David et al. "Forecasting Agriculturally Driven Global Environmental Change." *Science* 292:281–284 (April 13, 2001).

**Figure 11–3 Muriqui monkey.** Protection of this species of monkey in Brazil involves allowing some cutting of the forest in order to provide second-growth forests.

**Preservation** is often confused with conservation. The objective of the preservation of species and ecosystems is to *ensure their continuity, regardless of their potential utility.* Effective preservation often precludes making use of the species or ecosystems in question. For example, it is impossible to maintain old-growth (virgin) forests and at the same time harvest the trees. Thus, a second-growth forest can be *conserved* (trees can be cut, but at a rate that allows the forest to recover), but an old-growth forest must be *preserved* (it must not be cut down at all).

There are times when conservation and preservation come into conflict. The Muriqui monkey of Brazil was once thought to require virgin forests, leading to a concern for protecting such forests for the sake of the species (Fig. 11–3). Recent research has shown, however, that the monkeys actually do better in second-growth forests, which support a greater range of vegetation on which the monkeys feed. Indefinite preservation of the virgin forests would lead to a decline in the population of the Muriqui monkey, a seriously endangered species. Thus, *conservation* of the forests is essential for *preservation* of this New World monkey.

## Patterns of Use of Natural Ecosystems

Before examining the specific biomes and ecosystems that are under the highest pressure from human use, you need to study a few general patterns of such use that will help you understand what is happening in these natural systems and why it is happening.

**Consumptive Use Versus Productive Use.** People in more rural areas make use of natural lands or aquatic systems that are in close proximity to them. When people harvest natural resources in order to provide for their needs for food, shelter, tools, fuel, and clothing, they are engaged in **consumptive use.** This kind of exploitation usually does not appear in the calculations of the market economy of a country. People may barter or sell such goods in exchange for other goods or services in local markets, but most consumptive use involves family members engaged in hunting and gathering to meet their own needs. Thus, people are hunting for game (Fig. 11–4), fishing, or gathering fruits and nuts in order to meet their food needs, or else they are gathering natural products like firewood, forage for animals, or wood and palm leaves to construct shelters or to use as traditional medicines.

***Bush Meat.*** Wild game, or "bush meat," is harvested in many parts of Africa and provides a large proportion of the protein needs of people. In Congo, for example, it is as high as 75% of a person's protein intake. The estimated take in the Congo basin alone is 1 million metric tons annually. Without access to these resources, people's living standard would decline and their very existence could be threatened. Unfortunately, this is a largely unregulated practice and, according to CITES, has contributed to the decimation of 30 endangered species. Commercial hunters have recently turned what was once a matter of local people using local wildlife for food into an unsustainable industry that supplies cities with tons of "fashionable" meat.

Dependence on consumptive use is most commonly associated with the developing world, but may also exist in the more rural areas of the developed world, where people may equally depend on wood for fuel or wild game for food. Their lives would be seriously affected if they were denied access to these "free" resources.

**Productive use** is the exploitation of ecosystem resources for economic gain. Thus, products such as timber and fish (and now, bush meat) are harvested and sold for national or international markets (Fig. 11–5). Productive use is an enormously important source of revenue and employment for people in every country. The commercial trade in wood products for pulp, lumber, and fuel amounts to $142 billion per year; nonwood forest products, such as rattan, cork, or Brazil nuts, are worth an additional $11 billion per year. Another productive use is the collection of wild species of plants and animals for cultivation or domestication. The wild species may provide the initial breeding stock for commercial plantations or ranches, or they may be used as sources of genes to be introduced into existing crop plants or animals to improve their resistance or to add other desirable genetic traits. Finally, as discussed in Chapter 10, wild species continue to provide sources of new medicines.

Consumptive and productive use of natural ecosystem products is strongly influenced by the political and legal institutions of a country or region. For example,

**Figure 11–4 Consumptive use.** A hunter from the Asmat tribe, in Irian Jaya, Indonesia, brings home game. Wild game is an important food source for millions in many parts of the world.

private ownership restricts access to natural areas, whereas community ownership permits their use by members of the community. State ownership of natural resources implies regulated use; here, public policy is the key to accessing these resources. In all of these arrangements—private, communal, or state—there is the potential either for sustainable use of the resources or for unsustainable use leading to the loss of the resources, as well as the services provided by the natural ecosystem being exploited. As a result, the perspective of stewardship is vital. Do the resource managers and harvesters see themselves as caretakers, as well as exploiters, of the natural resource? Are public policies that encourage stewardship and sustainability in place? Are sound scientific principles employed in management strategies? Keep these questions in mind as you next examine two patterns that are commonly seen when people exploit natural resources.

**Figure 11–5 Productive use.** Loggers load poplar logs to be transported to a lumber mill, where they will be made into wafer board for the building industry.

**Maximum Sustainable Yield.** The central question in managing a renewable natural resource is, How much continual use can be sustained without undercutting the capacity of the species or system to renew itself? The model employed to describe this amount of use is the **maximum sustainable yield (MSY),** *the highest possible rate of use that the system can match with its own rate of replacement or maintenance.* Besides pertaining to the preservation of natural biotas, MSY applies to the maintenance of parks, air quality, water quality and quantity, and soils—indeed, the entire biosphere. *Use* can refer to the cutting of timber, hunting, fishing, the number of park visitations, the discharge of pollutants into air or water, and so on. Natural systems can withstand a certain amount of use (or abuse, in terms of pollution) and still remain viable. However, a point exists at which increasing use begins to destroy the system's regenerative capacity. Just short of that point is the MSY.

***Optimal Population.*** An important consideration in understanding how the MSY works is the **carrying capacity** of the ecosystem—the maximum population the ecosystem can support on a sustainable basis. If a population is well below the carrying capacity of the ecosystem (Fig. 11–6a), then allowing that population to grow will

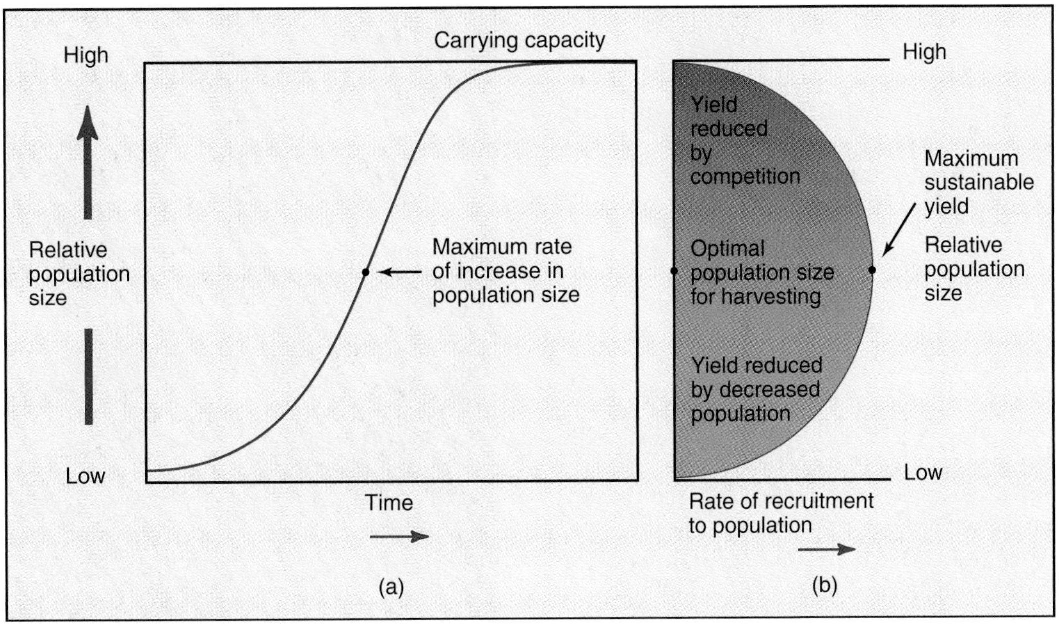


**Figure 11–6 Maximum sustainable yield.** The maximum sustainable yield occurs when the population is at the optimal level. (The rate of increase in population is at a maximum.) (a) The logistic curve of population size in relation to carrying capacity. (b) Recruitment plotted against population size, showing the effects of competition and decreased population levels.

increase the number of reproductive individuals and thus the yield that can be harvested. However, as the population approaches the carrying capacity of the ecosystem, new individuals must compete with older individuals for food and living space. As a result, recruitment may fall drastically (Fig. 11–6b). When a population is at or near the ecosystem's carrying capacity, production—and hence sustainable yield—can be increased by thinning the population so that competition is reduced and optimal growth and reproductive rates are achieved. Thus, the MSY cannot be obtained with a population that is at the carrying capacity. Theoretically, the **optimal population** is just half the population at the carrying capacity (Fig. 11–6).

The practical use of the model is complicated by the fact that the carrying capacity and, therefore, the optimal population is variable. Carrying capacity may vary from year to year as the weather fluctuates. Replacement may also vary from year to year, because some years are particularly favorable to reproduction and recruitment, while others are not. Human impacts, such as pollution and other forms of altering habitats, adversely affect reproductive rates, recruitment, carrying capacity, and, consequently, sustainable yields. For these reasons alone, managing natural populations to achieve the MSY is fraught with difficulties. Furthermore, accurate estimates of the size of the population and the recruitment rate must be made continually, and these data are hard to come by for many species.

***Precautionary Principle.*** The conventional approach has been to use the estimated MSY to set a fixed quota—in fishery management, the **total allowable catch (TAC).** If the data on population and recruitment are inaccurate, it is easy to overestimate the TAC, especially when there are pressures from fishers (and the politicians who represent them) to maintain a high harvest quota. In the face of such uncertainties, and also in response to repeated cases of overuse and depletion of resources, resource managers have been turning to the **precautionary principle.** That is, where there is uncertainty, resource managers must favor the protection of the living resource. Thus, the exploitation limits must be set well enough below the MSY to allow for uncertainties. Those using the resource ordinarily want to push the limits higher, however, so the result is frequent conflicts between users and managers. These conflicts are addressed in greater detail when the problems of fisheries are discussed later in this section.

**Using the Commons.** Where a resource is owned by many people in common or by no one, it is known as a **common pool resource,** or a **commons.** Examples of natural resource commons include federal grasslands, on which private ranchers graze their livestock; coastal and open-ocean fisheries used by commercial fishers; groundwater drawn for private estates and farms; nationally owned woodlands and forests harvested for fuel in the developing countries; and the atmosphere, which is polluted by private industry and traffic.

The exploitation of such common pool resources presents some serious problems and can lead to the eventual ruin of the resource—a phenomenon called the *tragedy of the commons*, after the late biologist Garrett Hardin's 1968 essay by that title.[2] Sustainability requires that common pool resources be maintained so as to

[2]Hardin, Garrett. "The Tragedy of the Commons." Science 162:1243–1247, 1962.

continue to yield benefits, not just for the present, but also for future users.

***Parable of the Commons.*** As described by Hardin, the original "commons" were pastures around villages in England. These pastures were provided free by the king to anyone who wished to graze cattle. In the parable, herders were quick to realize that whoever grazed the most cattle stood to benefit the most. Even if they realized that a particular commons was being overgrazed, those who withdrew their cattle simply sacrificed personal profits, while others went on using that commons. One herder's loss became another's gain, and the commons was overgrazed in any case. Consequently, herders would add to their herds until the commons was totally destroyed. They were locked into a system that led to their ruin.

Hardin's parable applies to a limited, but significant, set of problems in which there is open access to the commons, but where there is no regulating authority (or there is one, but it is ineffective) and no functioning community. Exploitation of the commons then becomes a free-for-all in which profit is the only motive. Coastal and offshore fisheries have consistently demonstrated the reality of the tragedy of the commons, with stocks of desirable fish declining all over the world. The tragedy can be avoided by limiting freedom of access.

***Limiting Freedom.*** One arrangement that can mitigate the tragedy is *private ownership*. When a renewable natural resource is privately owned, access to it is restricted, and, in theory, it will be exploited in a manner that guarantees a continuing harvest for its owner(s). This theory does not hold, however, when an owner maximizes immediate profit and then moves on. Most owners, though, will be in for the long run and manage the resources more responsibly, because it is in their best interests to do so.

Where private ownership is unworkable, the alternative is to *regulate access to the commons*. Regulation should allow for (1) protection, so that the benefits derived from the commons can be sustained, (2) fairness in access rights, and (3) mutual consent of the regulated. Such regulation can be the responsibility of the state, but it does not have to be. In fact, the most sustainable approach to maintaining the commons may be local community control, wherein the power to manage the commons resides with those who directly benefit most from their use and there are strong social ties and customs that can function well over time to protect the commons. By contrast, in many situations, state control of a commons has accelerated its ruin and led to an associated social breakdown and the impoverishment of people.

***Clamming.*** Harvesting soft-shelled clams in New England provides a modern example (Fig. 11–7). The clam flats in any coastal town are effectively a commons belonging to the town. Commercial access to them is limited to town residents and policed by the local clam warden. Clamming is a way of life for some residents, but unfortunately, it is not always a dependable enterprise. Although clammers may recognize that the clams are being overharvested, if they curtail their own takes, they diminish their personal income, while competitors do not. A clammer's loss becomes his or her competitors' gain, and the clams are depleted in any case. Thus, while the size and number of the clams diminish, indicating overharvesting, digging continues, because the clammers have to make a living. Higher prices brought on by the shortage of clams continue to make the digging profitable despite declining harvests. In most cases, the clam warden steps in and closes a number of flats to allow them to recover before local clamming is totally ruined. Coastal Massachusetts towns experience regular cycles of scarcity and recovery in the soft-shelled clam industry, an indication that the commons are often difficult to maintain. The clam warden has the power to make the difference between a full-blown tragedy of the commons and a system in which regulation leads to a sustainable harvest—not an easy or a popular task!

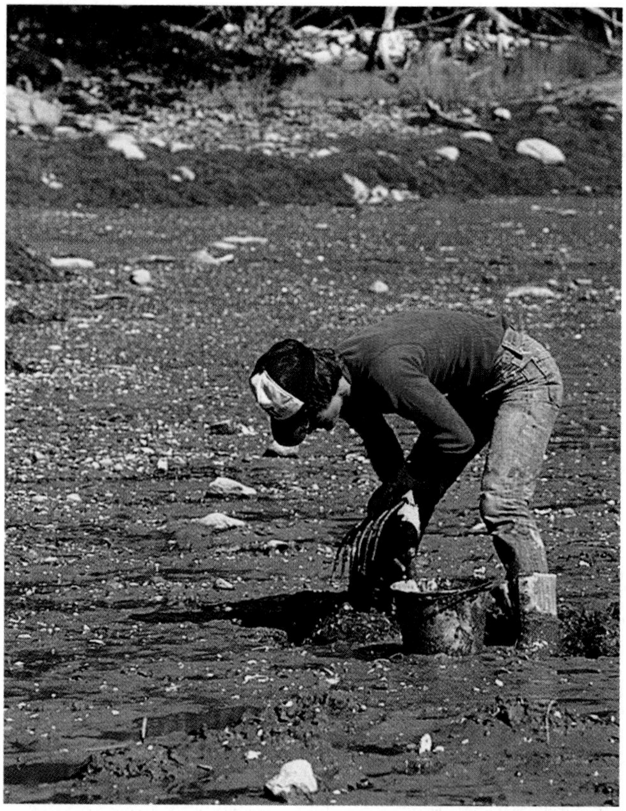

**Figure 11–7 Exploiting a commons.** Clam digger harvesting the soft-shelled clam, *Mya arenaria,* from a coastal Massachusetts clam flat.

***Public Policies.*** In sum, to achieve the objectives of conservation in industries using living resources, it is important to consider both the concept and limitations of the MSY and the social and economic factors causing overuse and other forms of environmental degradation that diminish the sustainable yield. Then, public policies can be established and enforced that protect natural resources effectively. Some principles that should be embodied in these policies are presented in Table 11–2. When policies based on principles like these are in place,

**table 11-2 Principles That Should Be Incorporated into Public Policies to Protect Natural Resources**

1. Natural resources cannot be treated as open commons. Habitats and species should be put under an authority that is responsible for their sustainability and that can regulate their use.
2. Sound science should be employed to assess the health of the resource and to set sustainable limits on its use.
3. To accommodate uncertainty, the precautionary principle should be used in setting limits for exploitation.
4. Regulations should be enforced.
5. Economic incentives that encourage the violation of regulations should be eliminated.
6. Subsidies that support exploitation of the resource should be removed.
7. Suitable habitats for the resource should be preserved and protected from pollution.
8. The sustenance needs of people living close to the resource should be met.

it is possible for people to continue to put natural resources to sustainable consumptive and productive uses. The vast harvest taken from the natural world each year is proof that sustainable use not only is possible, but also actually occurs with many resources and in many locations around the world.

In some situations, however, exploitation and degradation have gone too far, and the resilience of an ecosystem to disturbance has been overcome. In those cases, natural services and uses can be restored only if the habitats are deliberately restored. In recent years, the need for the restoration of damaged ecosystems has become clear, and a new subdiscipline with that objective—called **restoration ecology**—has been developed.

## Restoration

A great increase in restoration activity has occurred during the last 30 years, spurred on by federal and state programs and the growing science of restoration ecology. The intent of ecosystem restoration is to repair the damage to specific lands and waters so that normal ecosystem integrity, resilience, and productivity returns. Because of the complexity of natural ecosystems, however, the practical task of restoration is often difficult. Frequently, soils have been disturbed, pollutants have accumulated, important species have disappeared, and other—often exotic—species have achieved dominance.

For these reasons, a thorough knowledge of ecosystem and species ecology is essential to the success of restoration efforts. The ecological problems that can be ameliorated by restoration include those resulting from soil erosion, surface strip mining, draining wetlands, coastal damage, agricultural use, deforestation, overgrazing, desertification, and the eutrophication of lakes. In Chapter 2, the process involved in restoring a prairie ecosystem was described. Every case is different, and the largest restoration project ever conceived is under way in the (formerly) vast Everglades of southern Florida.

***Everglades Restoration.*** A much-publicized plan to restore the Everglades received congressional approval in late 2000. The plan, the **Comprehensive Everglades Restoration Plan (CERP)**, is expected to take 36 years and at least $7.8 billion to complete (Figure 11–8). The state of Florida will provide half of the funds, the federal government the other half. CERP will be managed by the U.S. Army Corps of Engineers, the federal agency that was actually responsible for creating much of the system of water control that made the restoration necessary.

The Everglades is a network of wetland landscapes once occupying the southern half of Florida. Now reduced to half its original size through development and wetlands draining, it is still an amazing and unique ecosystem, harboring a great diversity of species and many endangered and threatened species. It is home to 16 wildlife refuges and four national parks, the crown jewel being Everglades National Park.

***In Bondage.*** Because Florida is so flat, the general topography promoted a slow movement of water south from Lake Okeechobee by way of a 40-mile-wide 100-mile-long shallow "river of grass" ending in Florida Bay. Floridians once viewed it as an unproductive and forbidding swamp, so they brought the water flow under human control with a system of 1,000 miles of canals, 720 miles of levees, and hundreds of water control structures (locks, dams, spillways). The system is bounded on the east by a string of urban cities and suburban developments. In its center, 550,000 acres of croplands were created, which form a lucrative sugar industry. Now, much of the water that originally made its way through the "river of grass" is diverted to irrigate the croplands and to feed the thirsty suburbs and cities. Water shortages in the winter leave too little for the natural systems, and in the summer rainy season, too much water is diverted to the Everglades. The water quality, once pristine, is now degraded because of nutrients (especially phosphorus) from agricultural runoff. The changed nutrient regime promotes the growth of invasive species and has caused a decline in the native vegetation.

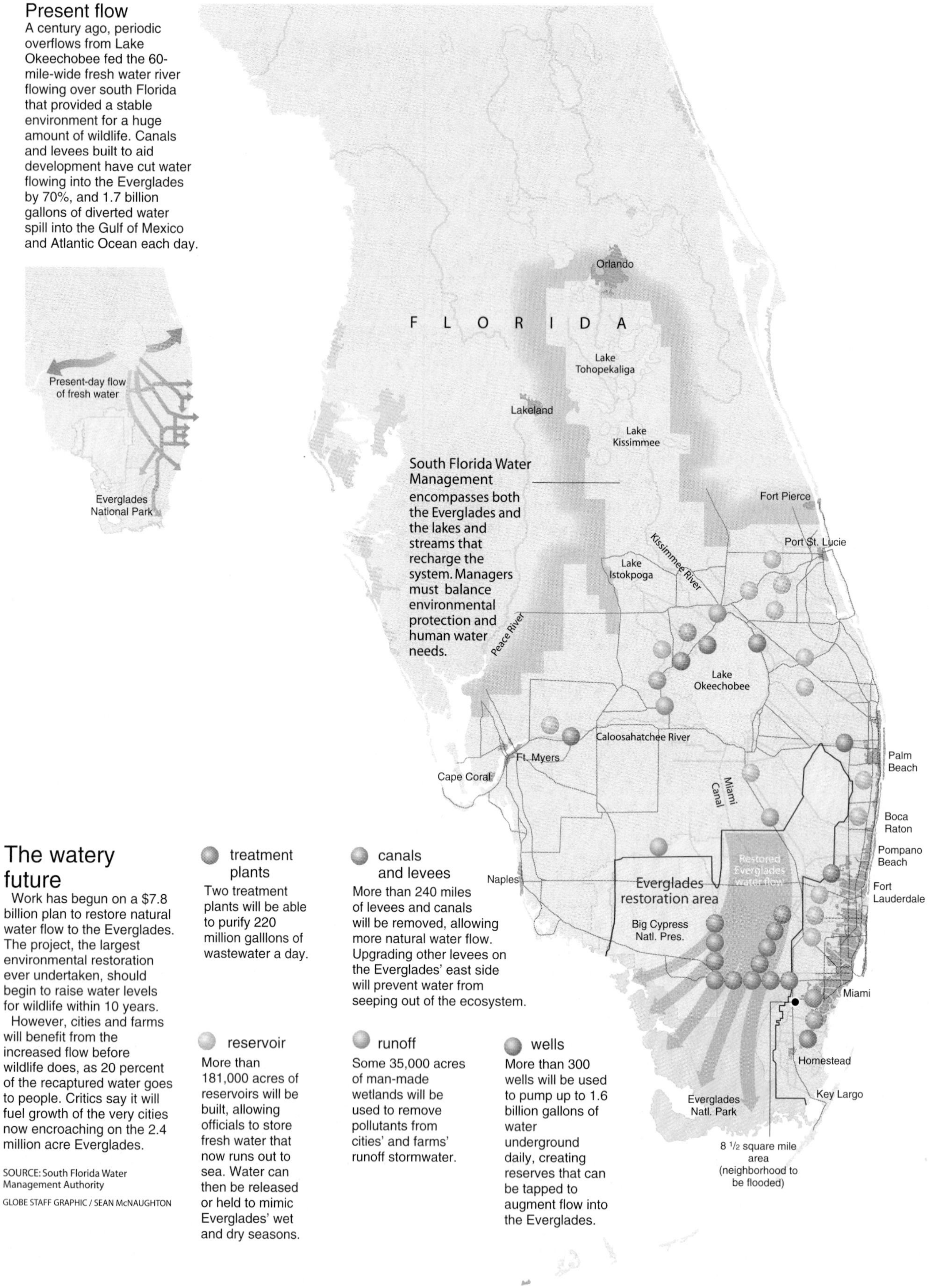
Present flow
A century ago, periodic overflows from Lake Okeechobee fed the 60-mile-wide fresh water river flowing over south Florida that provided a stable environment for a huge amount of wildlife. Canals and levees built to aid development have cut water flowing into the Everglades by 70%, and 1.7 billion gallons of diverted water spill into the Gulf of Mexico and Atlantic Ocean each day.
Present-day flow of fresh water
Everglades National Park
FLORIDA
Orlando
Lake Tohopekaliga
Lakeland
Lake Kissimmee
South Florida Water Management encompasses both the Everglades and the lakes and streams that recharge the system. Managers must balance environmental protection and human water needs.
Fort Pierce
Port St. Lucie
Kissimmee River
Lake Istokpoga
Peace River
Lake Okeechobee
Caloosahatchee River
Ft. Myers
Cape Coral
Miami Canal
Palm Beach
Boca Raton
Pompano Beach
Fort Lauderdale
Naples
Restored Everglades water flow
Everglades restoration area
Big Cypress Natl. Pres.
Miami
Homestead
Key Largo
Everglades Natl. Park
8 1/2 square mile area (neighborhood to be flooded)
The watery future
Work has begun on a $7.8 billion plan to restore natural water flow to the Everglades. The project, the largest environmental restoration ever undertaken, should begin to raise water levels for wildlife within 10 years.
However, cities and farms will benefit from the increased flow before wildlife does, as 20 percent of the recaptured water goes to people. Critics say it will fuel growth of the very cities now encroaching on the 2.4 million acre Everglades.
SOURCE: South Florida Water Management Authority
GLOBE STAFF GRAPHIC / SEAN McNAUGHTON
treatment plants
Two treatment plants will be able to purify 220 million galllons of wastewater a day.
canals and levees
More than 240 miles of levees and canals will be removed, allowing more natural water flow. Upgrading other levees on the Everglades' east side will prevent water from seeping out of the ecosystem.
reservoir
More than 181,000 acres of reservoirs will be built, allowing officials to store fresh water that now runs out to sea. Water can then be released or held to mimic Everglades' wet and dry seasons.
runoff
Some 35,000 acres of man-made wetlands will be used to remove pollutants from cities' and farms' runoff stormwater.
wells
More than 300 wells will be used to pump up to 1.6 billion gallons of water underground daily, creating reserves that can be tapped to augment flow into the Everglades.

**Figure 11–8 The Everglades restoration plan.** The goal of this plan is to restore historical water flow patterns so that ecological restoration can occur.

***Release.*** A broader restoration program was begun in 1996, coordinated by the South Florida Ecosystem Restoration Task Force. (CERP is now a major element of the task force's plans.) The task force has the following goals: (1) "getting the water right," (2) restoring, protecting, and preserving natural habitats and species, and (3) promoting the compatibility of the natural and built (human) systems. The total project is expected to cost close to $15 billion. CERP's role is to "get the water right." The plan calls for removing 240 miles of levees and canals and creating a system of reservoirs and underground wells to capture water for release during the dry season. The new flowage is designed to restore the river of grass, thereby restoring the 2.4 million acres of Everglades not to its original state, but at least to a healthy system that promotes goal 2.

***Trouble.*** If CERP does its job, close to 2 billion gallons of water that flows out of the Everglades each day will be redirected to flow southward. The plan calls for 80% of this water to feed the Everglades and 20% to feed the thirst of South Florida's cities and farms. The neighboring land will likely welcome some 2 million more residents over the life of the project, and many wonder whether one of the real reasons for the restoration project was to provide water that would allow this growth. Skeptics also wonder whether the Corps of Engineers is sufficiently committed to restoring wild areas, given its reputation for damming rivers and building canals. Problems have already surfaced with the plan to reduce nutrients: The sugar industry has won a 20-year delay from the Florida legislature in the programmed reduction of phosphorus content of waters drained from the industry's irrigated fields. If the restored flow into Florida Bay has a high nutrient content, it will likely promote undesirable algal blooms and make a bad situation there even worse.

There is no blueprint for a project like the Everglades restoration. The long time frame, the need for a continued infusion of federal and state money, and the great difficulties that will arise in carrying out the restoration guarantee that the project will attract skepticism as well as effusive praise. It is going to be in the news for a long time.

***Pending.*** As the values of natural ecosystems become more recognized, efforts to restore damaged or lost ones will become increasingly important. Many other large systems, such as San Francisco Bay, Chesapeake Bay, and the Upper Mississippi River, have suffered from human abuse and ignorance and are in need of a similarly aggressive restoration. Restoration is under way at many ecological gems, including the Galápagos Islands, under siege from many exotic species; the Illinois River, confined by levees; the Brazilian Atlantic forest, severely deforested; and Tampa Bay, affected by its urban surroundings.

The next section discusses a number of still-functioning biomes and ecosystems that are in trouble because of their great economic importance.

## 11.3 Biomes and Ecosystems under Pressure

The biomes and ecosystems of the world provide us with priceless services and economically valuable goods. Although human activities affect virtually all biomes and ecosystems, some are under more pressure than others. This section begins with a look at the forest biomes, the most important of all types of biome in terms of their economic significance and potential for active human management.

### Forest Biomes

Forests are the normal ecosystems in regions with year-round rainfall that is adequate to sustain tree growth. Forests are also the most productive systems the land can support, and they are self-sustaining. Forests and woodlands (ecosystems with mixed trees and grasses) perform a number of vital natural services. Among other things, they conserve biodiversity, moderate regional climates, prevent erosion, store carbon and nutrients, and provide recreational opportunities. They provide a number of vital goods, too, such as lumber, the raw material for making paper, fodder for domestic animals, fibers, gums, latex, fruit, berries, nuts, and fuel for cooking and heating. They provide employment for 60 million people. The total monetary value (see Chapter 3) assigned to global forest goods and services amounts to some $5.4 trillion a year. Of this amount, $153 billion is related directly to commercial exploitation of the forests. In spite of this value, the major threat to the world's forests is not simply their *exploitation,* but rather, their *total destruction.*

***FRA 2000.*** The U.N.'s FAO releases a global assessment of forest resources every two years and compares the data with previous information in order to evaluate deforestation trends. The agency published its most comprehensive analysis ever in 2001, the *Global Forest Resources Assessment 2000* (FRA 2000). One of its products is a series of global forest maps, based on satellite images with 1-kilometer resolution. Figure 11–9 shows the locations of the major world forests, which correspond with the biomes described in Chapter 2. Major findings of the new assessment (and its 2003 update) are as follows:

1. In 2000, the world's forest cover was 3.86 billion hectares (9.53 billion acres). This is some 400 million hectares *more* than was reported for 1995, the last global assessment. However, the increase is not because of major reforestation; it results, instead, from both a more accurate assessment of tropical forests and a redefinition of forests in the developed countries—from a 20% canopy cover threshold to 10%!
2. Deforestation continues to occur, primarily in the developing countries. Forest cover in the developed countries has remained stable. The FAO defines **deforestation** as *the removal of forest and replacement*

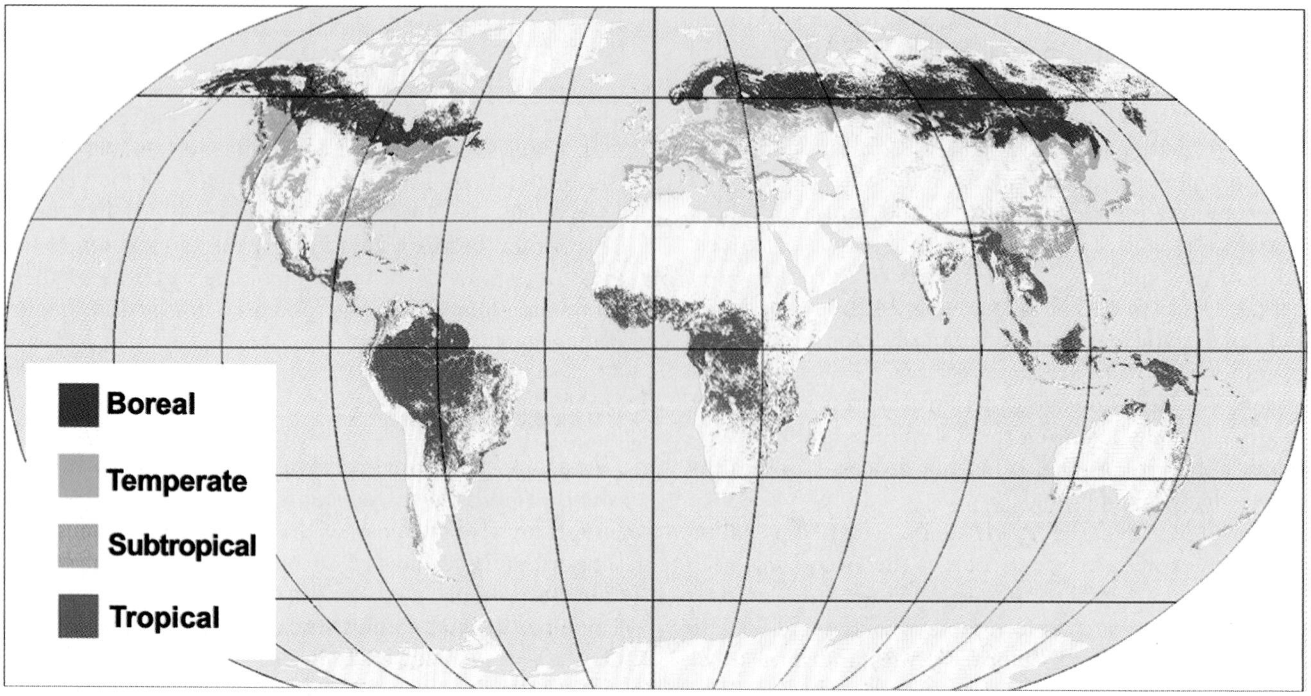


**Figure 11–9 World forest biomes.** The FAO Global Forest Resources Assessment 2000 has produced this map showing the locations of the major world forests.

*by another land use.* Thus, logging per se does not count as deforestation if the forests are allowed to regenerate. The global rate of net deforestation is estimated at 9.4 million hectares (23 million acres) per year, and the gross global deforestation is approximately 14.6 million hectares (36 million acres) per year. Reforestation is largely an outcome of an increased investment in forest plantations (4.5 million hectares per year), which account for most of the difference between net and gross deforestation.

3. Because of El Niño, intense droughts occurred in many parts of the world in the late 1990s. The droughts led inevitably to fires, which burned many millions of hectares of forests. The year 1998 was known as "the year the earth caught fire." Indonesia, the Russian Federation, Brazil, Ethiopia, and the United States were especially hard hit. Fires to clear land for agriculture continued in many tropical areas.
4. Worldwide, about 10% of the forests are protected as national parks or reserves. North, Central, and South America had the highest percentage of protected forests (16–17%), while Europe had the least (5%).
5. The role of forests in climate change was formally acknowledged in November 2001 at a meeting of the signers of the Kyoto Protocol in Marrakech, Morocco. The resulting Marrakech Accord recognized four major roles for forests: as a source of carbon dioxide when they are burned or degraded; as a sensitive indicator of changing climate; as a renewable energy source to replace fossil fuels; and as a carbon sink, by removing carbon dioxide from the atmosphere and storing it in biomass and the soils. The outcome of this accord is a much more thorough inventory of forests in the countries that have signed the Kyoto Protocol. This inventory is important because those countries can now use their forests to help mitigate their greenhouse gas emissions as they seek to meet the requirements of the Protocol. (See Chapter 20.)

***Forests as Obstacles.*** Even though the rate of deforestation is lower than in the previous decade, it is still a major cause for concern. Why are forests being cleared when they could be managed for wood production on a sustainable basis? Even though forests are highly productive systems, it has always been difficult for humans to exploit them for food. According to the FRA 2000 data, most of the deforestation was due to conversion into pastures and agricultural land. Unlike forests, grasslands have short food chains in which herbaceous growth supports large herbivores that can yield meat and other animal products. Alternatively, forests can be replaced by cultivated plants that are used directly for food. Thus, the forests have always been an obstacle to conventional animal husbandry and agriculture. Indeed, the first task the early European colonists faced when they came to the Western Hemisphere was to clear the forests so they could raise crops. Once the forests were cleared, continual grazing or plowing effectively prevented the trees from regrowing.

*Consequences.* Clearing a forest has the following immediate consequences for the land and its people:

1. The overall productivity of the area is reduced.
2. The standing stock of nutrients and biomass, once stored in the trees and leaf litter, is enormously reduced.
3. Biodiversity is greatly diminished.
4. The soil is more prone to erosion and drying.
5. The hydrologic cycle is changed, as water drains off the land instead of being released by transpiration through the leaves of trees or percolating into groundwater.
6. A major carbon dioxide sink (removal of $CO_2$ from the air) is lost.
7. The land no longer yields forest products.
8. People who depend on harvesting forest products lose their livelihood.

Unless forests are cleared and converted to other uses, they can yield a harvest of wood for fuel, paper, and building. Approximately 4.5 billion cubic feet of wood are harvested annually from the world's forests for fuel, wood, and paper (Fig. 11–10). It is unreasonable to expect the countries of the world—especially the developing countries—to forgo making use of their forests. However, forest-management practices vary greatly in their impact on the forest itself and on surrounding ecosystems.

**Types of Forest Management.** The practice of forest management, usually with the objective of producing a specific crop (hardwood, pulp, softwood, wood chips, etc.),

**Figure 11–10 Logging operations.** A forest in New Guinea is logged for its tropical wood.

is called **silviculture.** Several practices are employed, with quite different impacts on the total forest ecosystem involved. Because trees take from 25 to 100 years to mature to harvestable size, the normal process of management is a *rotation,* which is a cycle of decisions about a particular stand of trees from its early growth to the point of harvest. Two options are open to the forester: *even-aged management* and *uneven-aged management.* With even-aged management, trees of a fairly uniform age are managed until the point of harvest, cut down, and then replanted, with the objective of continuing the cycle in a dependable sequence.

*Clear-Cutting.* Typically, fast-growing, but economically valuable, trees are favored. (Others are culled from the stand.) In the Pacific Northwest, Douglas fir is frequently managed this way. Harvesting is often accomplished by *clear-cutting*—removing an entire stand at one time. This management strategy has been under severe criticism, because it creates a fragmented landscape with serious impacts on biodiversity and adjacent ecosystems; also, a clear-cut leaves an ugly, unnatural-looking site for many years. (See the "Earth Watch" essay, p. 302.) Large timber companies often employ the practice because it is efficient at the time of harvest and it does not involve much active management and planning.

*Other Methods.* Uneven-aged management of a stand of trees can result in a more diverse forest and lends itself to different harvesting strategies. For example, in *selective cutting,* some mature trees are removed in small groups or singly, leaving behind a forest that continues to maintain a diversity of biota and normal ecosystem functions. Replanting is usually unnecessary, as the remaining trees provide seeds. Another strategy is *shelter-wood cutting,* which involves cutting the mature trees in groups over a period of, say, 10 or 20 years, such that at any time there are enough trees both to provide seeds and to give shelter to growing seedlings. Like selective cutting, this method takes more active management and skill, but unlike clear-cutting, it leaves a functional ecosystem standing. In effect, it should lead to a sustainable forest, an objective that the entire wood and paper industry is currently focusing on—at least in principle.

**Sustainable Forestry.** In forestry, sustainability must be carefully defined. A common objective of forestry management is **sustained yield,** which means that the production of wood is the primary goal and the forest is managed to harvest wood continuously without being destroyed. **Sustainable forest management,** by contrast, means that forests are managed as ecosystems and that maintaining the biodiversity and function of the ecosystem is the primary objective. Thus, the forests are managed and trees are harvested in ways that will preserve most or all of the vital ecological services they perform. It is encouraging that the American Forest and

**earth watch**

## Nature's Corporations

The spotted-owl controversy pitted jobs against the preservation of the old-growth forests of the West. Timber interests maintained that preservation would result in 20,000 lost jobs. In their view, the controversy came down to the following: We should continue to cut the old-growth forests for the sake of keeping loggers employed. This view is not only shortsighted, but also illogical.

General Motors and other major businesses facing hard times lay off thousands of workers, and no one questions their need or right to do so. The survival of General Motors is more fundamental to the economy than keeping all of the company's workers employed. That is, it is assumed that those laid off will find other employment. Thus, according to conventional wisdom, corporate America must survive if the economy is to have a chance of recovering from economic bad times.

**Forest ecosystems** can be viewed as nature's corporations, providing lumber and firewood for the economy.

In a real sense, ecosystems are the corporations that sustain the economy of the biosphere. (See the inset.) Forested ecosystems can be viewed as *nature's corporations*, providing lumber and firewood for the economy. Coastal fisheries provide seafood for millions. If we want these ecosystems to survive and recover, we have to tighten our belts and withdraw some of the workforce engaged in exploiting them. The maintenance of these systems is more important than some jobs. Shouldn't out-of-work loggers and fishers seek other employment the way laid-off autoworkers, steelworkers, or computer engineers do? Why should a natural ecosystem be "bankrupted" just to maintain the *temporary* employment of a few? The work is temporary because the loggers will be out of a job in a few years, when the old-growth forests are finally all cut, and the fishers will have so depleted the fisheries that the fishers won't be able to afford to operate their boats.

Logging cutbacks due to wildlife protection were not the only reason jobs were lost. A depressed timber market and increasingly automated sawmills were also responsible. Although some 30,000 wood-related jobs were lost, there has been a net gain in employment in the region due to rapid growth in high-technology industries and federal retraining aid to many towns and workers. Industries locate in the region because of the impact of the "second paycheck," which refers to the high quality of life that attracts workers, even at lower pay scales. The same high quality of life—the forests, streams, scenic vistas, hunting, and fishing—requires preservation of the ecosystems that were being logged over.

It is shortsighted to assume that ecosystems are there just to provide jobs and that the jobs are more important than the ecosystems. When the old-growth forests are gone, we shall have lost more than just loggers' jobs: We shall have lost a priceless heritage and a major part of the natural world that provides us with vital services.

Paper Association, whose membership of privately owned forest and paper companies controls 90% of the industrial forestland in the United States, has adopted the *Sustainable Forestry Initiative*, consisting of the following principles:

1. **Sustainable forestry**—practicing a land stewardship ethic that integrates managing the reforestation, growing, nurturing, and harvesting of trees for useful products with the conservation of soil, air and water quality, wildlife and fish habitats, and aesthetics.
2. **Responsible practices**—to use in its own forests, and to promote among other forest landowners, sustainable forestry practices that are economically and environmentally responsible.
3. **Forest health and productivity**—to protect forests from wildfire, pests, diseases, and other damaging agents in order to maintain and improve long-term forest health and productivity.
4. **Protecting special sites**—to manage its forests and lands that are of special significance (e.g., biologically, geologically, or historically) in a manner that takes into account their unique qualities.

5. **Continuous improvement**—to continuously improve the practice of forest management and to monitor, measure, and report the performance of the association's members in achieving their commitment to sustainable forestry.

Later in the chapter (Section 11.4), federal forest-management policy in the United States is described as part of a general discussion on the use of public lands.

**Tropical Forests.** Because of their continued deforestation, tropical forests are of greatest concern. They are the habitat for millions of plant and animal species, a vast number of which are still unidentified. Climatologists reason that these forests are also crucial in maintaining Earth's climate, serving as a major sink for carbon and restraining the buildup of global carbon dioxide. Nevertheless, tropical forests continue to be removed at a rapid rate. Between 1960 and 1990, 20% of the tropical forests (over 445 million hectares, or 1.1 billion acres)—equivalent to two-fifths of the land area of the United States—were converted to other uses. According to FRA 2000, about 8.6 million hectares of tropical forests were lost each year during the 1990s. Although this is a disturbing rate, it is lower than the rate of deforestation in the previous decade.

*More Losses.* Unfortunately, however, the loss of tropical forests continues. Tanzania is losing up to 500,000 hectares of forest annually, and Indonesia's annual loss is at least 1.7 million hectares. Commercial logging in the Philippines has virtually disappeared because there are no forests left, and Mexico lost almost 1.2 million hectares of forest a year between 1993 and 2000.

*Reasons.* Typically, an area to be cleared is cut and then allowed to dry for a few weeks before being burned. The unmistakable signal of tropical deforestation is the plumes of smoke from burning vegetation that can often be detected from satellites or the space shuttles. Deforestation is caused by a number of factors, all of which come down to the fact that the countries involved are in need of greater economic development and have rapid population growth. The FRA 2000 study concluded that the current major cause of deforestation is *conversion to pastures and agriculture.* Many governments in the developing world are promoting deforestation by encouraging the colonization of forested lands. Indonesia, for example, has embarked on a program of resettlement and intends to convert 20% of its remaining forests to agricultural production. Thus, an area of 1 million hectares is to be converted to rice paddies. The clearing of forests for subsistence agriculture is especially intense in Africa and Asia, where population growth is unrelenting. With no end to population growth in sight for these regions, clearing is expected to account for increasing deforestation in the coming decades as people with no other options do what they have to do to attain food security.

The forests of the developing world are an important resource that can generate much needed revenue. Developing-world countries account for 20% of the international trade in forest products. Often, concessions are sold to multinational logging companies, which reap high profits by exploiting the economic desperation of the developing countries and harvest the timber with little regard for regenerating the forest. Chinese and other Asian companies are logging millions of acres of tropical forest in smaller countries like Belize, Suriname, and Equatorial Guinea, where regulation is weak and corruption is strong. More important is the consumptive use of forests by millions of people living in or on their edges. Owning few private possessions and often little land, they absolutely depend on foraging in, and extracting goods from, the forests. Among the encouraging trends in forest management in the developing world—trends which indicate that those in power are paying more attention to the needs of indigenous people and the importance of forest goods and services (other than just wood)—are the following:

1. **Sustainable forest management.** A growing number of developing countries are joining the developed countries in promoting sustainable forest management. An estimated 6% of the tropical forests are now under a formal forest management plan.
2. **Plantations of trees for wood or other products (cacao, rubber, etc.).** Forest plantations are being planted at a rate of 4.5 million hectares per year (Fig. 11–11). Although biodiversity is lower than in natural forests, plantations can continue to recycle

**Figure 11–11 Plantation forest.** This is a rubber tree plantation in Vietnam.

nutrients, hold soil, and recycle water. Plantations are much better than typical agricultural uses in preserving the natural functions of forests, and they can be managed sustainably.

3. **Extractive reserves that yield nontimber goods.** Among these goods are latex, nuts, fibers, and fruits. Recent calculations show that some forests are worth much more as extractive reserves than as sources of timber.
4. **Preserving forests as part of a national heritage and putting them to use as tourist attractions.** This practice can often generate much more income than logging can.
5. **Putting forests under the control of indigenous villagers.** The villagers can then collectively use the forest products in traditional ways. Given tenure over the land, the villagers tend to exercise stewardship over their forests in a way that is sustainable. Where this practice has been implemented, the forests have fared better than where they have been placed under state control.
6. **Remote sensing.** A new $1.4 billion Brazilian project called the System for the Vigilance of the Amazon, or SIVAM, and built with technical assistance from Raytheon Corporation monitors the Amazon with radar and satellite imagery to help Brazil protect its forests and indigenous people. Recall that FRA 2000 employed satellite data to measure forest cover throughout the world.
7. **World Bank.** Recently, the World Bank revised its forest policy. The bank will now work towards expanding the protected forest areas in developing countries, while improving the livelihoods of the millions of poor who live in forests outside the protected areas and who depend on forest products. The bank will also finance commercial logging only where sustainable forest management is taking place.

***Certification.*** A recent initiative directed toward the certification of wood products is the creation of the *Forest Stewardship Council.* This alliance of nongovernmental organizations such as the National Wildlife Federation, industry representatives, and forest scientists has developed into a major international organization with the mission of promoting sustainable forestry by certifying forest products for the consumer market. By 2001, over 24 million hectares in 47 countries had been certified.

All of these developments suggest that global awareness of the importance of the tropical forests has reached the level of serious concern. In some cases, this concern is being translated into action that is slowing the high rate of deforestation experienced in the 1980s. In others, the outcome is uncertain. The demand for tropical wood has not diminished, and the economic rewards of exploitation will undoubtedly continue to promote unsustainable uses of the forests in many tropical countries.

## Ocean Ecosystems

**Marine Fisheries.** Fisheries provide employment for at least 200 million people and account for more than 15% of the total human consumption of protein. (The term **fishery** refers either to a limited marine area or to a group of fish or shellfish species being exploited.) For years, the oceans beyond a 12-mile limit were considered international commons. By the end of the 1960s, however, numerous regions of the sea were being seriously depleted of many species by overfishing on the part of international fleets equipped with factory ships and modern fish-finding technology. In the mid-1970s, as a result of agreements forged at a series of U.N. Conferences on the Law of the Sea, nations extended their limits of jurisdiction to 200 miles offshore. The United States accomplished this with the Magnuson Act of 1976. Since many prime fishing grounds are located between 12 and 200 miles from shore, this action effectively removed most fisheries from the international commons and placed them under the authority of particular nations. As a result, some fishing areas recovered, while nationally based fishing fleets expanded to exploit the fisheries.

***The Catch.*** The total recorded harvest from marine fisheries and fish farming is reported annually by the FAO. The harvest has increased remarkably since 1950, when it was just 20 million metric tons (Fig. 11–12). By 2001, it had reached 139 million metric tons. Aquaculture accounted for 48.2 million tons, or 35% of world fish supplies that year. Based on the trends shown in the figure, the "capture" fisheries leveled off in the mid-1980s, and the continued rise in fish production is due to aquaculture. (See "Earth Watch" essay, p. 306.) Ninety-two million metric tons of the total of 139 million metric tons was the freshwater and marine fisheries catch, to which must be added some 20 million tons of bycatch (fish caught and discarded). The capture fisheries are pursued by some 4 million vessels, many equipped with fish-finding sonar and underwater television cameras and aided by spotter planes and helicopters. Because of heavy government subsidies, the fleet is able to land $81 billion worth of fish at a cost of $124 billion. It is estimated that the current fleet now has 50% more capacity for catching fish than is necessary.

***The Limits.*** The world fish catch may appear stable, but many species and areas are overfished. The FAO has concluded that 47% of fish stocks are fully exploited (at their maximum sustainable limit), 18% are overexploited and will likely decline further, and 10% are depleted and are much less productive than they once were. Much of the catch (one-third) is of low commercial value, so it is used for fish meal and oil production. Aquaculture is growing at a rate of 10% per year in the developing countries, helping to alleviate poverty and improve the well-being of rural people (Fig. 11–13).

The FAO has also concluded that the capture fisheries are at their upper limit, as indicated by the long-term

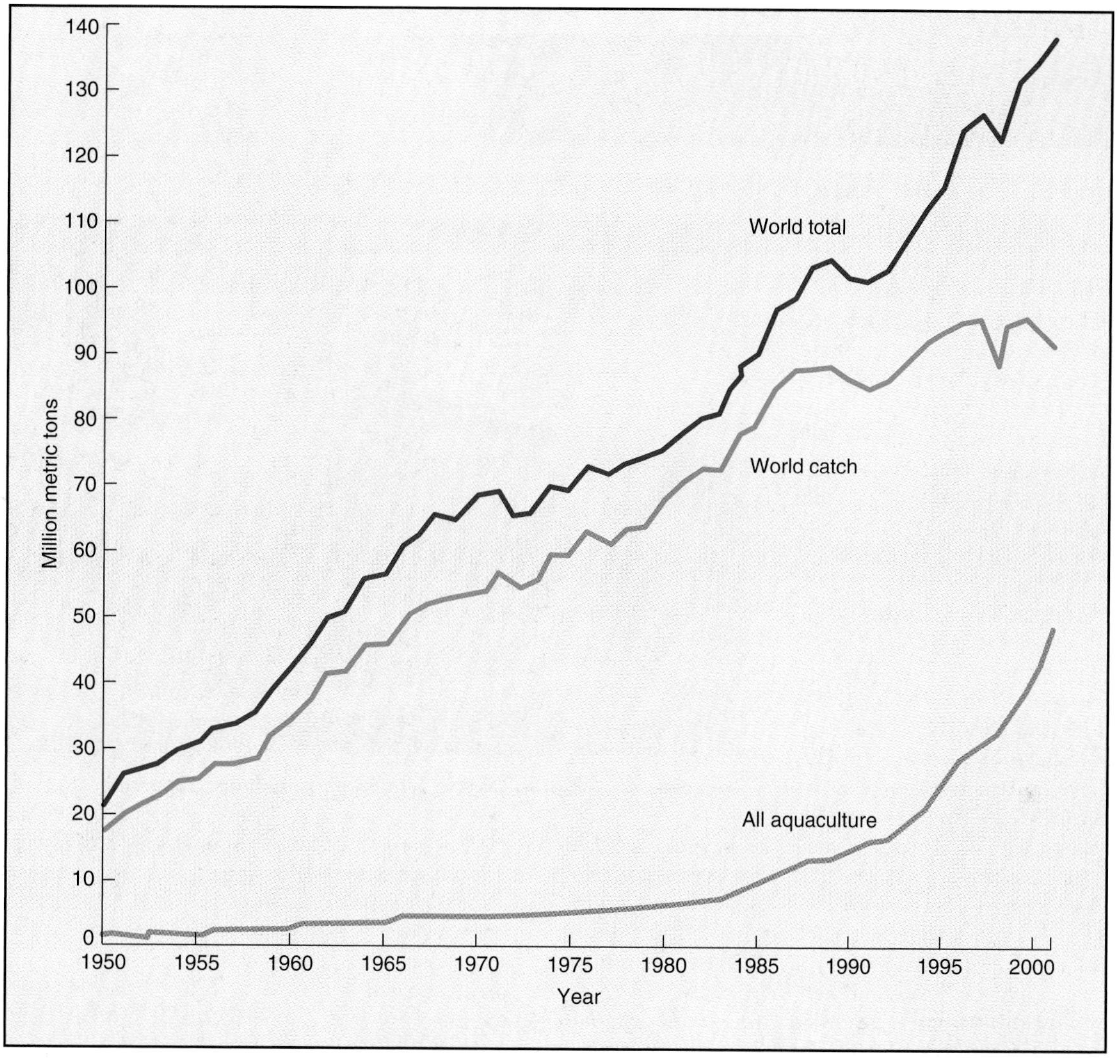


**Figure 11–12 The global fish harvest.** The global fish catch plus aquaculture equals the world total. Shown are data for 1950–2001. (*Source:* Data from FAO FAOSTAT and State of the World Fisheries 2002.)

catch data. Fisheries scientists reason that if overexploited and depleted fisheries were restored, the catch could increase by at least 10 million metric tons. That could happen, however, only if fishing for many species was temporarily eliminated or greatly reduced and the fisheries were managed on a more sustainable basis. At the 2002 Johannesburg World Summit, delegates agreed (one of the few agreements to come out of the meeting) to restore depleted global fisheries to achieve an MSY "on an urgent basis" by 2015 and to establish a network of protected marine areas by 2012. Recent studies have shown that marine reserves provide sheltered nurseries for surrounding areas and can supply world-record-size fish to adjacent fisheries.

*Georges Bank.* Events on Georges Bank, New England's richest fishing ground, illustrate the fisheries' problems well. Cod, haddock, and flounder (called groundfish, because they feed on or near the bottom and are caught by bottom trawling) were the mainstay of the fishing industry for centuries. In the early 1960s, these species accounted for two-thirds of the fish population on the bank. After 1976, fishing on Georges Bank doubled in intensity, resulting in a decline in the desirable species and a rise in the so-called rough species—dogfish (sharks) and skates. As the rough species increased their numbers, fishers focused their attention on them, even though they were less valuable. Predictably, dogfish and skate populations have also been overfished and are now subject to regulation.

*Management Councils.* The prized species declined on Georges Bank because the fishery was poorly managed. The Magnuson Act established eight regional management councils made up of government officials and industry representatives. The councils are responsible for setting management plans for their regions, while the National Marine Fisheries Service (NMFS) provides

earth watch

## Will Aquaculture Be Able to Fill the Gap?

Modern aquaculture, which is the farming of aquatic organisms, includes species ranging from seaweed to bivalve mollusks to freshwater and marine fish. Globally, aquaculture production—particularly the farming of marine and estuarine (coastal) species—has increased dramatically in the past 15 years. (See Fig. 11–12.) Without a doubt, this production is offsetting the plateau in the capture fisheries, but coastal aquaculture is not without its problems and critics. For example, shrimp farming in many tropical areas has been criticized for destroying mangrove habitats. In this case, shrimp aquaculture is being blamed, at least in part, for accelerating the decline of fisheries for some species that depend on the mangroves for a part of their life cycle. Another concern is that farming carnivorous species such as shrimp and salmon puts an even greater demand on the capture-fish species that are used to make the feed for the farmed species (although less valuable species like anchovies and herring are used for the purpose). Atlantic salmon are farmed worldwide, and there is concern that, as they escape from the net pens (which they frequently do), they will spread diseases and hybridize with the wild populations, which are often highly threatened already.

**Aquaculture. Costal salt water salmon pens in Lubec, Maine.**

Coastal aquaculture sites are typically in estuaries and lagoons or in ocean waters very close to the shoreline. (See the accompanying illustration.) These are the waters of the coastal zone that are the most used by humans for many other purposes, such as sportfishing, boating, waterskiing, etc. The nearshore coastal waters are also the most likely to be affected by sewage effluents and land runoff carrying many kinds of pollutants. The result in many cases is conflict among the various users, and the long-term outlook for coastal aquaculture is that nearshore waters are becoming less and less available for farming. There does not seem to be much argument about the need for expanded farming of coastal species to increase world fish and shellfish supplies, but the industry appears to be facing more problems than ever before. So, what can be done to make aquaculture more sustainable in the long term?

One approach is open-ocean aquaculture, which is the farming of coastal species in ocean waters several kilometers from the shoreline. This is not a new practice in some areas, but it is in North America, and even in Europe and other areas where it has been long practiced, total production levels are far lower than they could be. The most polluted areas in many coastal zones are the estuaries and embayments very close to shore. Hence, open-ocean waters are relatively clean. High water quality is important for aquaculture, but it is especially critical for species such as bivalve molluscs, which concentrate pathogens and other pollutants that can be passed on to human consumers. Also, new satellite-based remote-sensing efforts are revealing previously unknown nutrient-rich upwelling zones. In some areas in the Gulf of Maine, for example, phytoplankton concentrations can exceed the levels thought only to occur in enriched estuarine waters. This means that open-ocean waters may be able to support a high productivity of farmed species such as the blue mussel.

Mussels and other bivalve mollusks are herbivorous animals, feeding lower on the food web. A long-term sustainable approach to aquaculture would see more use of herbivorous species, reducing the impact on the wild fish stocks. Indeed, herbivorous fish like carp, tilapia, and milkfish, together with the marine mollusks, already make up 80% of the global output of aquaculture.

The problems facing aquaculture today are not unusual, because aquaculture is yet another complex practice that consists of competing demands on our environment. The problems can be solved only by approaches that are truly sustainable in the long term. In this case, it means bringing policymakers and the aquaculture industry together to work towards reducing the environmental costs of aquaculture.

---

(Contributed in part by Ray Grizzle, Jackson Estuarine Laboratory, University of New Hampshire.)

**Figure 11–13 Aquaculture in the developing world.** A tiger shrimp farm in Indonesia.

advice and scientifically based information on assessing stocks. The NMFS also has the authority to reject elements of the plans submitted to it by the councils. The New England Fishery Management Council (NEFMC) began its task by setting TAC quotas, but fishers claimed that the TACs were set too low and successfully argued for an indirect approach that employed mesh with openings of a size that allowed smaller fish to escape. In less than 10 years, the number of boats fishing Georges Bank doubled, with disastrous results, as indicated by the cod-landing data (Fig. 11–14).

With collapse of the fishery imminent, the council took a more drastic approach: In 1997, all vessels were restricted to 50% of their normal days at sea, almost 13,000 km$^2$ (about one-third of the fishing area) was totally closed to fishing, and target TACs were set at levels designed to reduce fishing mortality to 15% of the stock per year. (In 1998, mortality was still 22%.) In addition, no new vessels were allowed to enter the fishery, and the NMFS began a buyout program that paid the fishers to scrap their boats and surrender their fishing licenses. The boat buyout backfired, however, because many owners bought new boats with the money. The NMFS then turned to buying back fishing permits and recently spent $20 million to retire permits. Although the harvest is still very low, haddock and flounder stocks have rebounded well and are close to a sustainable level.

*Other Cod Fisheries.* The Grand Banks off Newfoundland were once the richest fishing grounds in the world. The cod were decimated by Canadian fishers, however, who were encouraged by their government to exploit the fishery and were given little warning of its imminent collapse. By 1991, the cod population had dropped to one-hundredth of its former size. In 1992, the fishery was closed indefinitely, costing the jobs of 35,000 fishers. So far the cod population has not rebounded, and cod are at their lowest levels in the century. Across the

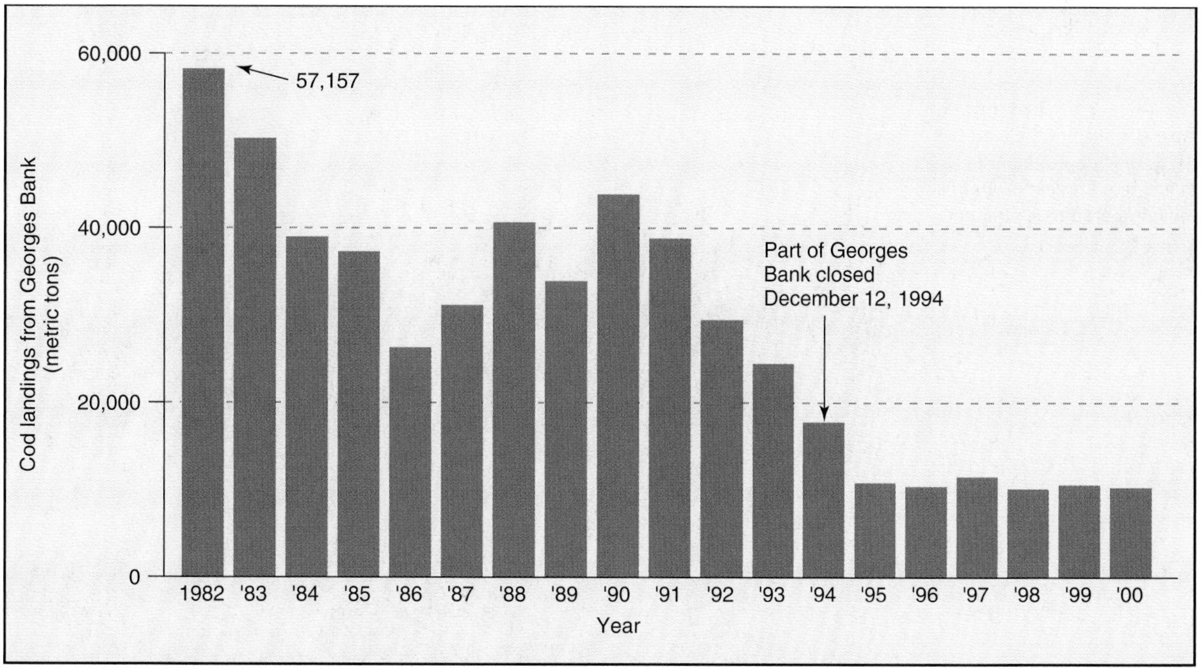


**Figure 11–14 Fishery in distress.** Data show cod landings from Georges Bank, 1982–2000. (*Source:* Northeast Fisheries Science Center, Woods Hole Laboratory, 2002.)

water, Europe's cod fishery is experiencing the same decline: The North Sea's cod stocks are now only 15% of what they were in the 1970s. Proposals to reduce harvests have met with fierce resistance from the European Community fishing nations, who argue that the reductions would destroy them.

The basic problem of the fisheries is that there are too many boats, rigged with high technology that gives the fish little chance to escape, chasing too few fish. Equipped with sophisticated electronic equipment that can find fish in the depths and with trawlers with 200-foot-wide openings that scoop up everything on and close to the bottom, the fishing industry has become too efficient (Fig. 11–15). A better system of management is needed, and the size of the fishing fleet must be reduced so that the remaining fishers are better able to make a living without overfishing. A limited system of access is needed, too, such as the IQ system used to manage the Pacific halibut fishery. The system effectively gives fishers the equivalent of property rights in the fishery. The rights are transferable and allow the fishers to determine how and when they will harvest their catch. Coupled with accurate population assessments by fishery scientists, the system has produced an economically stable and environmentally sustainable Pacific halibut fishery, and there is no reason that it should not be applied to fisheries everywhere.

*Sustainable Fisheries Act.* In 1996, the Magnuson Act was reauthorized by the Sustainable Fisheries Act. In doing so, the structure of the regional councils was retained, but they were given a clear mandate for their fishery management plans: Depleted fish stocks must be rebuilt and maintained at biologically sustainable levels, using IQs, buying out fishing vessels, and requiring that scientific information be employed in setting yields. The bill also requires steps to be taken to assess and minimize the by-catch.

The management councils are in the process of grappling with these new regulations. In 2000, for example, the NEFMC estimated that a 67% reduction in fishing mortality will be required to rebuild all the species of groundfish on Georges Bank and in the Gulf of Maine. Accordingly, the organization set TACs for 11 species at 36% lower than the 1998 landings. The measures have apparently been successful: As of 2002, a number of New England fish stocks have rebounded and are deemed no longer subject to overfishing.

*Other Issues.* Many other issues pertaining to fisheries need to be addressed, too, including banning shark finning, a practice in which just the fins of sharks are taken to satisfy the demand for shark-fin soup; the certification of "dolphin-safe" tuna (to discourage the practice of fishing that sets nets on dolphin schools because tuna are usually feeding below them); shrimping in the Gulf of Mexico that kills endangered sea turtle species; declining stocks of swordfish and tuna even as fishing intensity increases; and aquaculture demands placed on ocean fish because fish meal is fed to pen-raised fish. Another "fishery" of great concern to the international community is the great whales.

**International Whaling.** Whales, found in the open ocean and in coastal waters, were heavily depleted by overexploitation until the late 1980s. Whales once were harvested for their oil, but are now prized for their meat, considered a delicacy in Japan and a few other countries. In 1974, the International Whaling Commission (IWC), an organization of nations with whaling interests, decided to regulate whaling according to the principle of MSY. Whenever a species of whale dropped below the optimal

**Figure 11–15 Bottom trawling.** Because of the way the ocean bottom is degraded, this method, used to harvest groundfish, has been compared to clear-cutting forests.

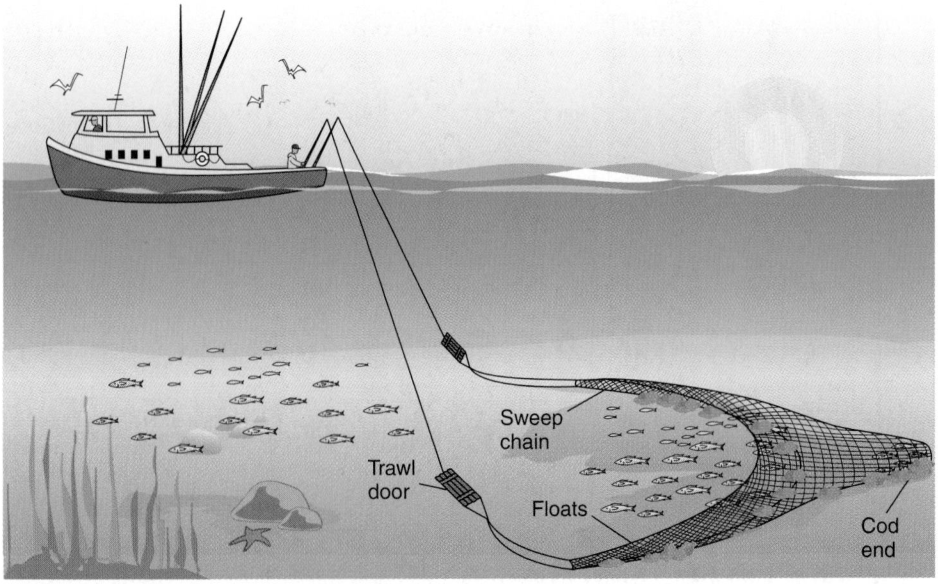

**Figure 11–16 Current whaling operation.** Butchers slice into a Baird's beaked whale at a pierside slaughterhouse in Wada, Japan.

population for such a yield, the IWC instituted a ban on hunting that species in order to allow the population to recover. At that time, three species (the right whale, bowhead whale, and blue whale) were at very low levels and were immediately protected. Because of difficulties in obtaining reliable data on and enforcing catch limits, the IWC took more drastic action and placed a moratorium on the harvesting of all whales beginning in 1986. The moratorium has never been lifted; however, some limited whaling by Japan and Norway continues, as well as harvesting by indigenous people in Canada, Alaska, and Greenland (Fig. 11–16).

Table 11–3 lists the status of 13 whale species that are of commercial interest. Estimates of the numbers remaining range from 300 northern Atlantic right whales to 1 million to 2 million sperm whales. Their main threat today is entanglement with fishing gear and collisions with ships. Although reliable data are still hard to acquire, many of the whale species appear to be recovering. The bowhead whale, for example, recently was upgraded from endangered to "conservation dependent." In the time between the IWC moratorium and the present, however, the basis of the ethical controversy over whaling has shifted from conservation to animal rights. Many people worldwide simply believe that it is wrong to kill and eat such large and unique mammals. People from whaling nations counter with the argument that their culture includes eating whales, just as other cultures include eating cows or turkeys.

*Whale Stakes.* There has been heavy pressure from two members of the IWC—Japan and Norway—to reopen whaling. The interests of these countries focus on the minke whale. In 1993, Norway resumed whaling for minkes, citing the country's right to refuse to accept specific IWC rulings. Norway based its decision on information submitted by the Scientific Committee of the IWC, which judged the minke population to be numerous enough to absorb a sustainable exploitation. Environmental scientists maintain, however, that the data on whale populations are far too uncertain to support such a judgment. Under continued pressure from antiwhaling nations (the United States, France, Australia, and others) the IWC has thus far refused to set a whaling quota on the minke. Currently, Japan and Norway each take up to 500 minke whales per year, the Japanese for "scientific purposes" and the Norwegians for human consumption. It is no secret that the Japanese "scientific" harvest ends up in commercial markets and commands a healthy price as a gourmet meat; the industry generates $27–36 million in revenue a year. In 2002, Japanese whaling vessels, in the name of science, killed almost 700 whales of several species. The newest Japanese rationale for whaling is that the whales eat too many fish that humans should be catching!

The IWC is caught in the middle of this controversy. At a recent meeting, the commission adopted a revised management procedure for setting quotas that could lead to a resumption of whaling. In subsequent annual meetings, however, IWC member countries

**table 11–3 Remaining Numbers of Large Whales**

| Endangered | Vulnerable or Conservation Dependent | Lower Risk or Insufficient Data |
|---|---|---|
| Blue whale (400–1,400) | Bowhead whale (8,500) | Minke whale (935,000) |
| N. Atlantic right whale (300) | Sperm whale (1 million–2 million) | Bryde's whale (40,000–80,000) |
| N. Pacific right whale (under 1,000) | Southern right whale (7,000) | |
| Fin whale (50,000–90,000) | Humpback whale (28,000) | |
| Sei whale (50,000) | Gray whale (27,000)<br>Southern right whale (2,900) | |

*Sources:* WWF, IUCN *Red Data List* (2000), and International Whaling Commission.

**Figure 11–17 Whale watching.** A humpback whale entertains a boatload of whale watchers in the Gulf of Maine off the Massachusetts coast. Whale watching has replaced whaling in New England waters that were once famous as whaling centers.

have refused to authorize quotas on any whales until effective inspection and observer systems are put in place. At its latest meeting in Shimonoseki, Japan, in 2002, the IWC maintained the moratorium (a three-fourths majority vote is needed to lift it), but moved closer to an agreement on monitoring, raising the possibility that limited whaling sanctioned by the IWC might soon resume. IWC officials believe that it would be advisable to bring existing whaling under international oversight.

***Whale Watching.*** One reason for the rising interest in protecting whales is the opportunity many people have had to observe them firsthand. Indeed, whale watching has become an important tourist enterprise in coastal areas (Fig. 11–17). Stellwagen Bank, within easy reach of boats from Boston, Cape Ann, and Cape Cod, Massachusetts, has become the center of a whale-watching industry estimated to be worth more than $78 million annually. From spring through fall, scores of boats venture offshore daily to watch the whales that congregate over the bank. Many of the humpback whales seem to enjoy entertaining the visitors and often frolic alongside the boats for hours. Recently, the International Fund for Animal Welfare reported that whale watching now flourishes in 87 countries and generates more than $1 billion in business annually.

Besides having aesthetic and entertainment value, whale watching is of scientific value. Whale-watching tour boats usually carry a biologist along who identifies the whale species and interprets the experience for the visitors. The biologists are often associated with groups such as the Cetacean Research Unit and the Whale Conservation Institute, which have studied the whales of Stellwagen Bank since 1979 and have published many papers on the humpback whale. Largely in response to its importance for whales, Stellwagen Bank was designated a National Marine Sanctuary in 1993.

**Coral Reefs and Mangroves.** In a band from 30° north to 30° south of the equator, coral reefs occupy shallow coastal areas and form atolls, or islands, in the ocean. Coral reefs are among the most diverse and biologically productive ecosystems in the world. The reef-building coral animals live in a symbiotic relationship with photosynthetic algae called **zooxanthellae** and therefore are found only in water shallower than 75 meters. The corals build and protect the land shoreward of the reefs and attract tourists, who enjoy the warm, shallow waters. In addition, because the reefs attract a great variety of fish and shellfish, they are important sources of food and trade for local people, generating an estimated $375 billion per year in tourism and fishing revenues. In recent years, coral reefs have been the subject of concern because they have shown signs of deterioration in areas close to human population centers. The year 1998 was a watershed: Before then, an estimated 11% of coral reefs had been destroyed by human activities.

***Bleaching.*** The 1997–98 El Niño (see Chapter 20) brought record-high sea-surface temperatures to many tropical coasts. The changing intensity and frequency of El Niño is thought to be one consequence of global warming. Extensive *coral bleaching* (wherein the coral animals lose their symbiotic algae) severely damaged an estimated 16% of the world's reefs (Fig. 11–18), according to the Global Coral Reef Monitoring Network (an affiliation of marine scientists and government agencies). Some coral bleaching is an annual phenomenon

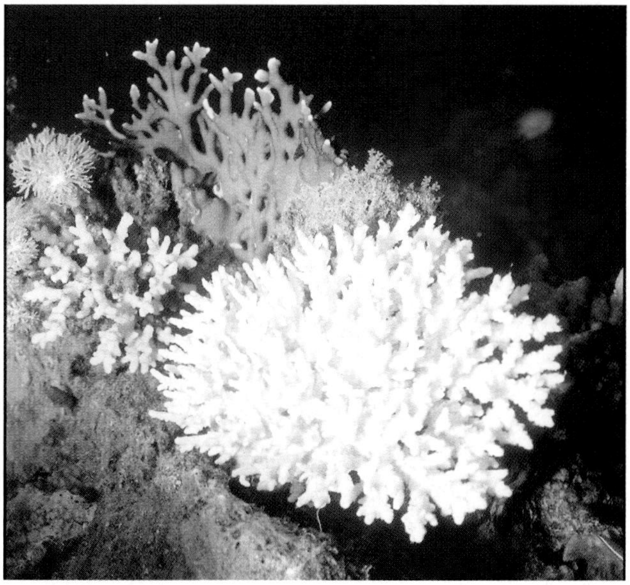

**Figure 11–18 Bleached Coral.** This coral in the Red Sea shows the effects of excessively warm temperature.

related to the normal high temperatures and light intensities of summer, but the most recent incidents are unprecedented in scope. Belize, which has the largest barrier reef in the northern hemisphere, sweltered under sustained temperatures of 31.5°C (89°F), killing off most of the coral. The high temperatures killed 90% of the coral in the central Indian Ocean. Although many of the devastated reefs are recovering, coral reproduction is unusually low in the reefs. Researchers point out that repeated occurrences of El Niño (forecast in global-warming scenarios) could permanently wipe out coral reefs over vast areas of the shallow tropical oceans, which would be an enormous loss of biodiversity and economic value.

***Exploitation.*** Another source of coral damage stems from poverty and greed. Islanders and coastal people in the tropics often scour the reefs for fish, shellfish, and other edible sea life. Some exploitation of the reefs, however, is driven by the lucrative trade in tropical fish. Local residents sometimes use dynamite and cyanide to flush the fish out of hiding in the coral, a practice that is not only destructive to the coral, but also dangerous to the perpetrators. "Fishing" with cyanide is a growing (illegal) practice that has been encouraged by exporters in the Indo-Pacific region. The intention is to stun the fish with cyanide squirted into coral crevices, but the fish may die in the process. Meanwhile, the deadly poison damages the coral, often severely. Then divers frequently use crowbars to pry apart corals sheltering stunned fish. Halting this practice will take a concerted effort on the part of the Asian countries involved to police the reefs and work toward retraining the cyanide fishers to use nondestructive fishing methods. It will also require countries that import tropical fish to monitor and certify cyanide-free imports.

***Tropical Fish?*** Diners in Hong Kong pay handsome sums for live reef fish, which are kept in aquariums and then are cooked and consumed on the spot. Other fish are destined for the aquarium trade, which, in Southeast Asia, is worth $1.2 billion, with a large percentage of the fish being captured by means of cyanide poisoning. U.S. aquarium owners who purchase tropical fish, live coral, and "rock" or reef base are inadvertently contributing to the decline of coral reefs. Many consumers admire the colorful imported tropical fish, unaware of how they were caught.

***Solutions.*** One solution to this destructive trade is to foster effective, sustainable management of the coral reef resources. Ideally, such management would be community based, so that local people benefit from the exploitation. Often, though, communities in the developing world lack the capacity to carry out this kind of management, in which case international assistance is required. The USAID is actively working towards this goal in 20 countries. The World Bank is also involved. The Coral Reef Rehabilitation and Management Project, funded at $30 million, is helping Indonesia through the bank's IBRD/IDA programs. (See Chapter 6.) The TRAFFIC network has been working to draw attention to the fish and coral species being decimated by coral reef exploitation and to bring the reefs under CITES regulation. (See Chapter 10.) Recently, the Marine Aquarium Council, an international nonprofit organization, has been working toward a method of certification which would guarantee that the marine fish sold in aquarium supply stores were collected in an "eco-friendly" manner.

***Mangroves.*** Often, just inland of the coral reefs is a fringe of mangrove trees. These trees have the unique ability to take root and grow in shallow marine sediments. There, they protect the coasts from damage due to storms and erosion and form a rich refuge and nursery for many marine fish. Despite these benefits, mangroves are under assault from coastal development, logging, and shrimp aquaculture. Between 1983 and the present, half of the world's 45 million acres of mangroves were cut down, with percentages ranging from 40% (in Cameroon and Indonesia, for example) to nearly 80% (in Bangladesh and Philippines, for instance).

Although logging (for paper pulp and chipboard) and coastal land development are believed to be responsible for most of the mangrove deforestation, the development of ponds for raising shrimp is proving destructive, too. In this practice, the mangrove forests are cleared and the land is bulldozed into shallow ponds, where black tiger prawns are fed with processed food. The aquaculture of prawns is now a multibillion-dollar industry, satisfying the great demand for shrimp in the developed countries. Unfortunately, in many cases, the ponds become polluted and no longer support shrimp, so new ponds are then created by clearing more mangroves. The regular flushing of the ponds required to remove excess nutrients and wastes often degrades adjacent coastal environments. The massive removal of mangroves has brought on the destabilization of entire coastal areas, causing erosion, siltation of sea grasses and coral reefs, and the ruin of local fisheries. Developing nations that once considered the mangroves to be useless swampland are now beginning to recognize the natural services they performed (see Global Perspective, this page). Local and international pressure to stop the destruction of mangroves is growing.

In the next section, the focus is on the connection between public policy and natural-resource management as it is reflected in public and private lands in the United States.

## 11.4 Public and Private Lands in the United States

Recall from Chapter 10 that the way to save wild species is to protect their habitats. The time has passed when the loss of ecosystems could be justified on the grounds that there were suitable substitute habitats just over the hill. With the rising human population, industrial expansion, and pressure to convert natural resources to economic gain, there will always be reasons to exploit natural ecosystems. The last resort for many species and ecosystems is protection by law in the form of national parks, wildlife refuges, and reserves. Worldwide, some 6,930 areas, representing 4.8% of the national land area on the planet, have received this kind of protection. Three hundred areas are designated more restrictive *biosphere reserves*. In the developing world, however, many of these are "paper parks," where exploitation continues and protection is given only lip service.

The United States is unique among the countries of the world in having set aside a major proportion of its landmass for public ownership. Nearly 40% of the country's land is publicly owned and is managed by state and federal agencies for a variety of purposes, excluding development. The distribution of public lands, shown in Fig. 11–19, is greatly skewed toward Alaska and other western states, a consequence of historical settlement and

### global perspective

#### The Mangrove Man

With an annual per capita income of $200, Eritrea, on the shores of the Red Sea, is one of the poorest countries in the world. Its poverty was made worse by a war with Ethiopia, which led to independence for Eritrea in 1993. The country's plight attracted the attention of biologist Gordon Sato, who visited Eritrea during the civil war and eventually spent half of his time there after peace came. Sato, a cell biologist with extensive publications and a member of the U.S. National Academy of Sciences, was concerned about the widespread hunger in Eritrea and experimented with several schemes to raise food in the hot coastal fringe. He eventually focused his attention on mangroves, tree species that grow on the coasts of many tropical and subtropical countries. Some 15% of Eritrea's 1,000 km coastline is occupied by mangroves, but the remainder is barren intertidal land. Sato observed that mangroves grow only where they receive nutrients brought to the coast by the sparse seasonal rains. He reasoned that it should be possible to greatly expand the mangrove forests if nutrients could be provided to the barren regions. The mangroves, in turn, could be used to provide fodder for goats, sheep and cattle; these in turn would form the basis of a subsistence economy for local people.

Sato proposed a unique method for growing the mangroves. First, seedlings were grown at nurseries, a relatively easy task. Then, local Eritrean workers planted the seedlings in the muddy intertidal zone and buried a plastic bag of ammonium phosphate next to each one to provide the essential nitrogen and phosphorus needed by the plant. The nutrients slowly leaked out of the bags through small holes poked in the plastic. In addition, Sato's experiments showed that iron was still a limiting nutrient, so a piece of scrap iron was thrust into the mud close to each planted seedling. To date, village workers have planted some 600,000 mangrove trees, and are now harvesting leaves and seeds from the growing trees to feed domestic animals. Sato has named the work "the Manzanar project," in a cogent reminder of his family's past history where they were interned in Manzanar, California, during World War II because of their Japanese origins.

Eritreans are now carrying out the work on their own, having seen the success of Sato's project and its remarkable use of low-technology aquaculture. Sato pictures vast forests of mangrove trees all along the coast, providing not only a base for a local subsistence economy but also establishing a rich food web that would stimulate local fisheries. His work earned him the 2002 Rolex award for Enterprise, and brought him international recognition as "the mangrove man." It also attracted some attention from coral reef scientists, who were concerned that nutrients flushing from the mangroves would pollute sensitive offshore coral reefs. In response, Sato has pointed out that there is no evidence for a negative impact of mangrove plantations on coral reefs anywhere, and no measurable increase in nutrient levels in the coastal regions around the mangrove plantations. His is a remarkable story of the application of sound science to make a difference in the lives of countless people well into the future.

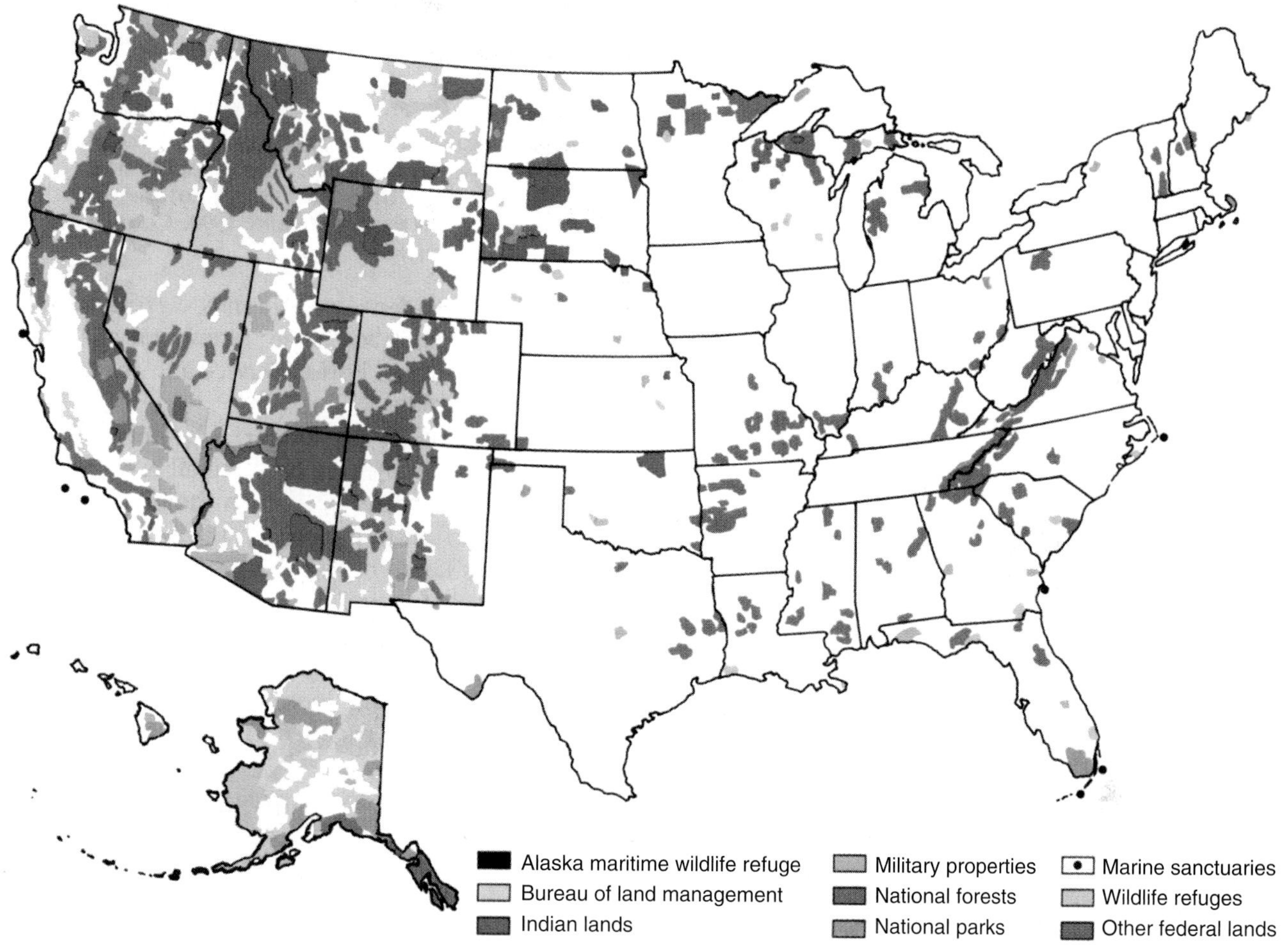


**Figure 11–19 Distribution of federal lands in the United States.** Because the east and Midwest were settled first, federally owned lands are concentrated in the west and Alaska. (*Source:* Council on Environmental Quality, *Environmental Trends,* 1989.)

land-distribution policies. Nonetheless, although most of the East and Midwest are in private hands, forests and other natural ecosystems still function on many of those regions' lands.

*Wilderness.* Land given the greatest protection (preservation) is designated **wilderness.** Authorized by the Wilderness Act of 1964, this land includes over 104 million acres in 466 locations—almost 4% of the land area of the United States. The act provides for the permanent protection of these undeveloped and unexploited areas so that natural ecological processes can operate freely. Permanent structures, roads, motor vehicles, and other mechanized transport are prohibited. Timber harvesting is excluded. Some livestock grazing and mineral development are allowed where such use existed previously; hiking and other similar activities are also allowed. Proposals by several presidents to increase wilderness protection to 5 million acres of national parklands have received no action by Congress for over 30 years. Opposition to these proposals has been mounted by various recreational organizations, which fear that the move would lock up lands they are accustomed to enjoying.

## National Parks and National Wildlife Refuges

The **national parks,** administered by the National Park Service (NPS), and **national wildlife refuges,** administered by the FWS, provide the next level of protection to 78 million and 94 million acres, respectively. Here, the intent is to protect areas of great scenic or unique ecological significance, protect important wildlife species, and provide public access for recreation and other uses. The dual goals of protection and providing public access often conflict with each other, because the parks and refuges are extremely popular, drawing so many visitors (over 280 million visits a year) that protection can be threatened by those who want to see and experience the natural sites. At one of the most popular parks—Zion National Park in Utah—auto traffic has been replaced by shuttle vehicles because of the problems of parking and habitat destruction accompanying the use of motor vehicles within the parks. Other parks will soon follow suit. All of the federal agencies involved are trying to cope with the rising tide of recreational vehicles like snowmobiles and off-road vehicles used on most federal lands. The Blue Ribbon Coalition, a lobbying group representing owners and manufacturers of off-road vehicles, fiercely defends keeping federal lands accessible to its clients.

Increasingly, agencies, environmental groups, and private individuals are working together to manage natural sites as part of larger ecosystems. For example, the Greater Yellowstone Coalition has been formed to conserve the larger ecosystem that surrounds Yellowstone National Park (Fig. 11–20). Yellowstone and Grand Teton National Parks form the core of the ecosystem, to which seven national forests, three wildlife refuges, and private lands are added to form the greater ecosystem of 18 million acres (7.2 million hectares). The coalition acts to restrain forces threatening the ecosystem. Thus, it has challenged the Forest Service's logging activities, addressed residential sprawl and road building on the private lands, and acted to protect crucial grizzly bear habitat and to curtail the senseless shooting of bison that wander off the national parklands in search of winter grazing.

This cooperative approach is important for the continued maintenance of biodiversity, because so much of the nation's natural lands remain outside of protected areas. Cooperation may also help restrict development up to the borders of the parks and refuges, keeping the refuge from becoming a fairly small natural island in a sea of developed landscape.

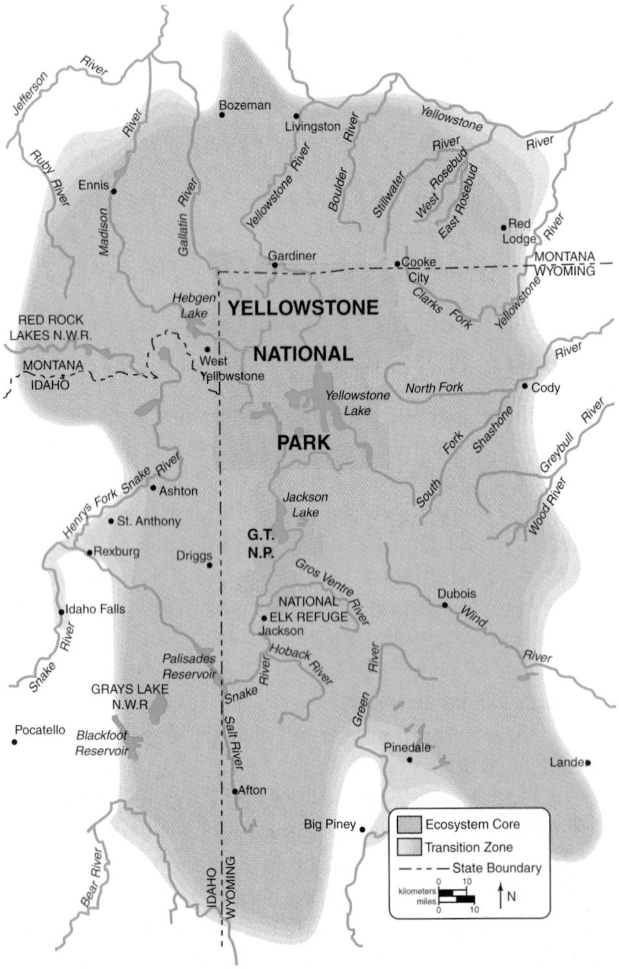


**Figure 11–20 Greater Yellowstone ecosystem.** The national park is the center of a much larger ecosystem, which has received attention from the Greater Yellowstone Coalition.

## National Forests

Forests in the United States represent an enormously important natural resource, providing habitat for countless wild species, as well as supplying natural services and products. U.S. forests range over 740 million acres, of which about two-thirds are managed for commercial timber harvest. Almost three-fourths of the managed commercial forestland is in the East and privately owned; the remainder is mainly in the West and is administered by a number of governmental agencies, primarily the National Forest Service (NFS) and the Bureau of Land Management (BLM).

Deforestation is no longer a problem in the United States. Although we have cut all but 5% of the forests that were here when the colonists first arrived, second-growth forests have regenerated wherever forestlands have been protected from conversion to croplands or house lots. There are more trees in the United States today than there were in 1920. In the East, some second-growth forests have aged to the point where they look almost as they did before their first cutting, except for the absence of the American chestnut.

The Forest Service is responsible for managing 192 million acres (78 million hectares) of national forests. BLM land, some 270 million acres (109 million hectares), is a great mixture of prairies, deserts, forests, mountains, wetlands, and tundra. Some 6% of federal timber production comes from BLM land, the remainder from National Forest lands.

***Multiple Use.*** The Forest Service's management principle in the 1950s and 1960s was *multiple use,* which meant a combination of extracting resources (grazing, logging, and mining), using the forest for recreation, and protecting watersheds and wildlife. Although the intent was to achieve a balance among these uses, multiple use actually emphasized the extractive uses; that is, it was output oriented and served to justify the ongoing exploitation of public lands by private, often favored, interest groups (ranchers, miners, and, especially, the timber industry).

To harvest timber, government foresters select tracts of forest they judge ready for harvesting and then lease the tracts to private companies, which log and sell the timber. The timber harvest from national forests has been a controversial issue. Historic data on timber sales (Fig. 11–21) reflect major shifts in logging. Note the rapid rise following World War II to satisfy the housing boom that was underway at the time, the drop reflecting rising environmental concern in the 1970s, increased logging during the Reagan–Bush years, and a steep drop with the Clinton administration as a result of renewed environmental

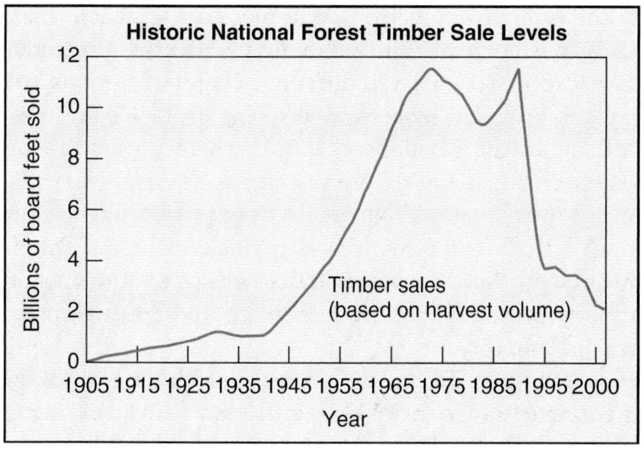


**Figure 11–21 National forest timber sales.** Timber sales over the years reflect changes in resource needs and political philosophy.

concern. According to a recent NFS analysis, growth of forests has exceeded harvests for several decades in both national and private forests, indicating that the forests are being managed sustainably.

A forestry-management strategy was introduced in the late 1980s called **New Forestry.** This practice of forestry is directed more toward protecting the ecological health and diversity of forests than toward producing a maximum harvest of logs. Thus, New Forestry involves cutting trees less frequently (at 350-year intervals instead of every 60 to 80 years); leaving wider buffer zones along streams, to reduce erosion and protect fish habitats; leaving dead logs and debris in the forests, to replenish the soil; and protecting broad landscapes—up to 1 million acres—across private and public boundaries, while involving private landowners in management decisions. The Forest Service began adopting some of these management principles in the early 1990s, and they formed the core of what is now the official management paradigm of the Forest Service: **ecosystem management.** As noted in Chapter 4, this paradigm has been adopted by all federal agencies managing public lands. It is a major philosophical shift from maximizing timber profits to the long-term stewardship of forest ecosystems. A select committee of scientists convened by the Clinton administration to recommend policy for the Forest Service reported, "The first priority for management is to retain and restore the ecological sustainability of... watersheds, forests, and rangelands for present and future generations." This, said the report, is the foundation for stewardship of the national lands the NFS manages.

***Roadless.*** The Clinton administration's Forest Service chief, Mike Dombeck, embraced the new paradigm, and one of his first acts was to declare a moratorium on building new logging roads. The Forest Service had been criticized for its road building, because the existing roads had contributed to the destruction of forests and streams. The timber industry and many members of Congress from states with national forests objected to the moratorium. They believed that it would halt timber harvests and prevent access by off-road vehicles in roadless areas. Some 58 million acres of forestland are affected by the moratorium. Court challenges to the rule by timber industries have failed, but the Bush administration has issued management directives that are circumventing the roadless-area rule. If the rule is to stand, it will require congressional action, and there is significant support for that.

***Fires.*** The years 2000 and 2002 set new records for forest fires, with 8.4 million acres burning the former year and 6.9 million acres the latter. In response, the Bush administration announced a "Healthy Forests Initiative," which is intended to prevent future catastrophic forest fires by reducing the "fuel load" (meaning trees and shrubs). The plan provides incentives to logging companies to thin fire-prone forests, especially near populated areas. In return for this work, the loggers would get to harvest larger, commercially valuable trees. In addition, arguing that citizens' appeals were stalling forest-thinning projects, President Bush ruled that environmental reviews and judicial oversight would be restricted. The General Accounting Office reported, however, that, to the contrary, over 99% of Forest Service thinning projects had proceeded without any challenge. Critics of the plan see it as a thinly veiled attempt to reward timber corporations for their political support. They maintain that fires are normal events in forests and that populated areas could be protected very well by focusing the forest thinning on areas within a half-mile of homes.

## Protecting Nonfederal Lands

In the November 2002 election, voters in 22 states passed ballot initiatives to spend $2.6 billion to protect land for open space and parks. The largest measure was in California, where voters approved $2 billion for land acquisition. In Maine, the Nature Conservancy bought 185,000 acres along the St. John River for $35 million, and the New England Forestry Foundation paid Pingree Associates $28 million for development rights on 760,000 acres of forests. Pingree will continue to harvest trees, but only in a manner that is certified by the Forest Sustainability Council. These acquisitions are indicative of a groundswell of public and private support for maintaining open space. Outright purchase by municipalities is one option, but not the only one.

***Land Trusts.*** With so much land in private hands, the future of natural ecosystems in the United States is always uncertain, particularly when the land has special appeal for recreation and aesthetic enjoyment, because there the potential for development is great. Often, landowners and townspeople want to protect natural areas from development, but are wary of turning the land over to a governmental authority. One creative option is the **private land trust,** a nonprofit organization that will accept either outright gifts of land or **easements**—arrangements in which the landowner gives up development rights into the future, but retains ownership of the parcel. The land trust may also purchase land to protect it from development. The land trust movement is growing considerably: There were 429 land trusts in the United States in 1980 and over 1,220 in 2000, as reported by the Land Trust Alliance, an umbrella organization in Washington serving local and regional trusts. The latter protect over 6.2 million acres (2.5 million hectares) of land. The Nature Conservancy, a large international land trust agency, has helped protect an additional 15 million acres (6 million hectares) in the United States and 101 million acres (41 million hectares) in other countries.

Land trusts are proving to be a vital link in the preservation of ecosystems. The oldest trust is the Trustees of Reservations in Massachusetts, founded in 1891 and now guardian of 49,000 acres throughout the state. The organization has received some ecologically prime and scenic land through the years and now maintains many of its properties for public uses that are compatible with preservation. The land trusts are serving the common desires of landowners and rural dwellers to preserve the sense of place that links the present to the past. At the same time, the undeveloped land remains in its natural state, sustaining natural populations and promising to do so into the future.

## Conclusion

Ecosystems everywhere are being exploited for human needs and profit. In addition to all the examples discussed in this chapter, other areas that are in trouble are wetlands drained for agriculture and recreation, overgrazed rangelands, rivers that are overdrawn for irrigation water, and more. The purpose of this chapter has been to highlight some of the most critical problems and to indicate steps being taken to correct them. It is certain that greater pressures will be put on natural ecosystems as the human population continues to rise. These pressures must be met with increasingly effective protective measures if we want to continue to enjoy the goods and services provided by ecosystems. The problems are most difficult in the developing world, where poverty forces people to take from nature in order to survive. If natural areas are to be preserved, the needs of people must be met in ways that do not involve destroying ecosystems. People must be provided with alternatives to exploitation, a situation requiring both wise leadership and effective international aid.

As nations formulate policies for dealing with the environment in sustainable ways, they will need better information on what is happening and what the consequences of those policies will be. Sound scientific information is essential, and in many areas there are large gaps in our knowledge. The Pilot Analysis of Global Ecosystems (PAGE) has made a good start toward closing those gaps (Chapter 1). Building on this scheme is the MEA, a four-year, $20 million effort also described in Chapter 1. The objective of this effort is to find out what is happening to the world's ecosystems and to help countries both deal with ecosystems that are in trouble and meet the goals of the numerous international environmental conventions now in place. Carried out properly, this project will yield the most accessible and rigorous data set on world ecosystems ever assembled.

U.N. Secretary-General Kofi Annan presented his millennium report to the General Assembly in April 2000 with a stern warning that far too little concern was being given to the sustainability of our planet: "If I could sum it up in one sentence, I should say that we are plundering our children's heritage to pay for our present unsustainable practices.... We must preserve our forests, fisheries, and the diversity of living species, all of which are close to collapsing under the pressure of human consumption and destruction." "In short," Annan stated, "We need a new ethic of stewardship."

# revisiting the themes

## Sustainability

The management of the Pacific halibut industry demonstrates how to achieve a sustainable harvest of a desirable natural resource. Sustainable exploitation of natural resources makes it possible for them to continue to produce the essential goods and services that support human well-being. Unfortunately, there are many cases of unsustainable use: the taking of "bush meat" by commercial hunters, the cyanide poisoning of coral reefs, the exploitation of a commons where there is open access to all, and deforestation in tropical rain forests, to name a few. The principle of sustainability is built into the model of the MSY, but for fisheries it turns out to be harder to accomplish in practice than in theory. For this reason, it is wise to use the precautionary principle in attempting to set such a threshold for a resource. Sustainable forestry, however, is not so difficult, as long as it is focused on managing the forests as functioning ecosystems and as long as biodiversity is respected. Where resources are being exploited in much of the developing world, sustainability is unworkable without taking into account the needs of local people and involving them in management decisions.

## Stewardship

Stewardly care of natural resources means consciously managing them so as to benefit both present and future generations. It is encouraging to see a stewardship ethic expressed in statements from private (the Sustainable Forestry Initiative) and public (Forest Service principles) resource managers. It is also good to see how much land (and sometimes water) is set aside for protection in perpetuity. Finally, it is heartening to see the U.N. secretary-general endorsing a stewardship ethic as an important step in preserving the natural resources that are under tremendous pressure from present generations.

## Sound Science

The MSY model reflects solid science, but it requires solid data to set accurate sustainable limits for harvest. These limits are difficult, if not impossible, to accomplish for such mobile creatures as fish and whales. Fisheries scientists are constantly challenged to assess fish stocks accurately, and their recommendations can often be trumped by political and economic decisions, as is well illustrated in the plight of the New England fisheries. Sound science by the IWC also recommends a carefully controlled whale harvest, but political opposition has prevented this from happening so far. The restoration of ecosystems requires a great deal of sound science in the form of species and ecosystem ecology. It also requires the ability to make midcourse corrections, and the Everglades restoration project is going to need a lot of these. Finally, the MEA now under way is an excellent example of enlisting the world's best scientists in an attempt to gather vital information on the state of the world's ecosystems.

## Ecosystem Capital

Like Chapter 10, the current chapter is entirely about ecosystem capital as an essential part of the wealth of nations: what it is, what it provides to the humans who exploit it, and what is happening to some of the most vital ecosystems that we depend on. The viewpoint of *ecosystem capital* is preferable to the simple approach of *natural resources,* because it keeps the ecological picture in sharp focus. In the future, it is very likely that human use will significantly disrupt the goods and services provided by ecosystem capital. Land conversions in particular (for example, from forests to croplands) will probably result in net losses of services.

## Policy and Politics

Public policy and politics were brought into focus right from the beginning of the chapter, with the recounting of the Pacific halibut fishery, which came under the management of an international agreement for sustainable exploitation. *Public policy* is the key to any system of resource management, especially one that is publicly owned. Effective management can come only from effective policies, and these are developed within a political setting. Where regulation is called for, it must be transparent and done with the mutual consent of the regulated. Table 11–2 summarizes the principles that should be incorporated into policies to protect resources. This said, *politics* rears its ugly head, again and again. Examples include the Everglades restoration, the Kyoto Protocol, the New England cod fishery, the whaling situation, wilderness protection, the roadless-area moratorium in forests, and the Healthy Forests Initiative (now signed into law—the Healthy Forests Restoration Act of 2003). Changes in administration and in the control of Congress have resulted in many serious reversals of environmental policy decisions.

## Globalization

The global trade in forest products and fisheries puts pressure on natural resources as they are exploited for their economic value. In developing countries, these resources earn important revenues that help millions to improve their economic well-being. The markets, however, can put such demand on some products, like mahogany, tuna, sharks, and tropical fish, that they are being taken well beyond a sustainable harvest. Often, the developing countries are exploited by multinational extractive corporations that leave the country much poorer in natural resources when they are finished and that frequently take advantage of corrupt politicians to gain access to the resources. Sometimes, the developing countries are willing to sacrifice some ecosystems (for example, mangrove swamps) in order to satisfy the high global demand for a product like shrimp.

# review questions

1. What were the problems in the Pacific halibut fishery, and how were they resolved?
2. Under what conditions will a natural area receive protection?
3. Compare the concept of *ecosystem capital* with that of *natural resources*. What do the two reveal about values?
4. Compare and contrast the terms *conservation* and *preservation*.

5. Differentiate between consumptive use and productive use. Give examples of each.
6. What does *maximum sustainable yield* mean? What factors complicate its application?
7. What is the tragedy of the commons? Give an example of a common pool resource and how it can be mistreated. How can such resources be protected?
8. When are restoration efforts needed? Describe efforts under way to restore the Everglades.
9. Describe some of the findings of the most recent FAO Global Forest Resources Assessment. In particular, what roles do forests play in climate change?
10. Compare and contrast even-age and uneven-age management. What are several principles of the Sustainable Forestry Initiative?
11. What is deforestation, and what factors are primarily responsible for deforestation of the tropics? Describe some encouraging trends in forest management in the developing countries.
12. What is the global pattern of exploitation of fisheries? Compare the capture fisheries with the aquaculture yield.
13. Explain the objectives of the Magnuson Act and the Sustainable Fisheries Act.
14. How does the New England fishery exhibit the worst problems of exploitation? What could fix this situation?
15. What two countries are pressuring the IWC to reopen commercial whaling, and what is their rationale for resuming the killing?
16. How are coral reefs and mangroves being threatened, and how is this destruction linked to other environmental problems?
17. Contrast the different levels of ecosystem protection given to the different categories of federal lands in the United States.
18. Describe the progression of the management of our national forests during the last half century. What are two current issues, and how are they being resolved?
19. How do land trusts work, and what roles do they play in preserving natural lands?

## thinking environmentally

1. Identify and study an area in your community where the destruction or degradation of natural land or wetland is an issue. What, if anything, is being done to protect this land? Propose a program for restoring the degraded land.
2. What incentives and assistance could the United States offer Brazil or African countries to keep their tropical rain forests from further harm due to conversion?
3. Imagine that you are the clam warden for a New England town. What policies would you put in place to prevent a tragedy of the commons from occurring?
4. What natural-resource commons do you share with those around you? Are you exploiting or conserving those commons?
5. Download the article on historical overfishing by Jeremy Jackson et al. (pangea.stanford.edu/GES56Q/Course-Materials/Jackson2001.pdf). How does this information relate to the current exploitation of marine systems?
6. Investigate the restoration of the Everglades. What political battles are going on, and what is your assessment of the potential success of this project?
7. Research the preservation efforts and effectiveness of one of the following conservation groups: the Sierra Club, the Appalachian Mountain Club, the Nature Conservancy, and the Audubon Society.

## making a difference part three: chapters 7, 8, 9, 10, 11

1. Demonstrate your concern for the hungry and homeless in your area by getting involved in local soup kitchens and by collecting goods for the needy as organized by Second Harvest.
2. Stay informed about food- and hunger-related issues. Join Bread for the World, a nationwide organization that seeks justice for the world's hungry by lobbying our nation's decision makers. Write your congressperson, asking for his or her support of legislation that benefits the hungry.
3. Produce some of your own food by planting a vegetable garden and raising animals. Practice sustainable agriculture: Avoid the use of herbicides and pesticides, compost kitchen wastes and yard wastes, and build up a humus-rich soil. Any space—even a rooftop—with six hours of sunlight is enough.
4. Contact your county soil-conservation or farm-extension agent, and ask for help in making known the work being done to prevent the erosion and runoff of farm chemicals.

5. Investigate the following issues pertaining to your community or city: Where does the water come from and how is it treated? Are water supplies adequate, or are they being overdrawn? How is storm water managed in your area? What policies or actions are being considered to meet future needs?

6. Observe where storm water from your own home and yard goes. (Does it mostly run off, or does it infiltrate?) Install a system—perhaps just a barrel at a downspout—to capture and use storm water for watering your lawn or garden and for washing your car.

7. Examine your own eating habits, and opt for a nutritionally balanced diet that includes less fat and red meat and more fruits and vegetables. Use the USDA food pyramid to guide your diet.

8. If you hunt or fish, observe all pertinent laws and kill only what you can use. Alternatively, become a photographer of nature; your pictures can be a source of enjoyment to many.

9. When you visit parks and other natural areas, stay on the designated trails. Avoid making new trails that might lead to erosion, and never disturb nesting birds and wild animals raising their young.

10. Purchase food products that are derived from the extractive use of tropical forests rather than the degradative use thereof (for example, "Rainforest Crunch" ice cream and Brazil nuts, as opposed to tropical wood products).

12. Avoid high-impact activities that cause heavy environmental damage, such as recreation involving jet skis, off-road vehicles, and snowmobiles.

# part four

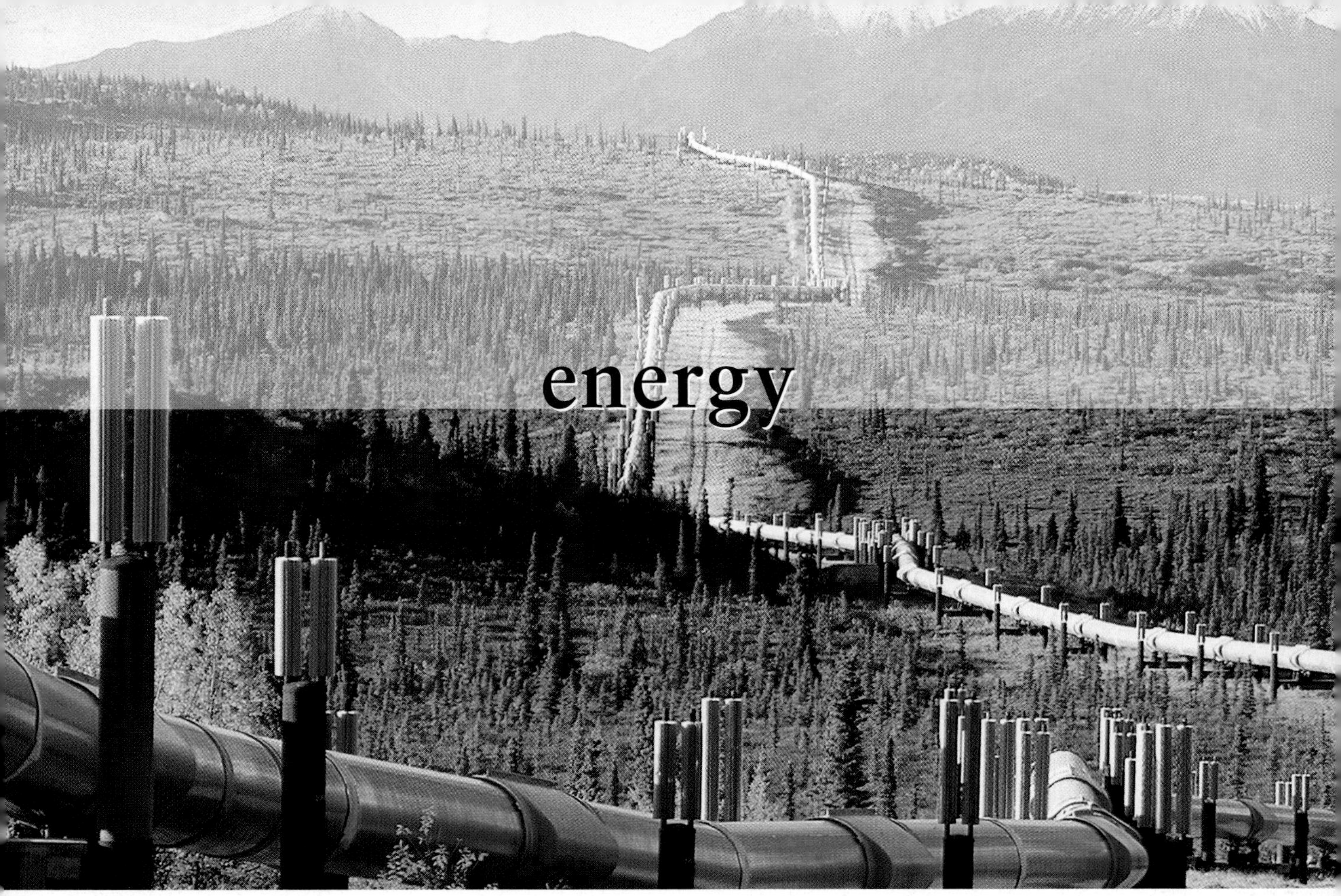

# energy

The Alaska oil pipeline snakes its way down through tundra and forest to the port of Valdez, an intrusion into a beautiful and fragile wilderness that is symbolic of the dilemma facing our society. The industrial countries are so deeply committed to fossil fuel and nuclear energy that we cannot imagine an economy apart from these energy sources. Developing countries are following the same example, hoping to enhance the well-being of their people. A recent analysis of the world's oil reserves suggests that oil production may peak within the next decade and then begin a decline that will drive prices up to new heights. Will we invade wildlife refuges and fragile coastal waters to withdraw every last drop of oil from the ground? Looming in the background is the specter of global climate change and the struggle of dealing with this threat that is tied so closely to the use of fossil fuels. Beside this is the horror of terrorism, which has the world's attention because of unprecedented episodes of brutal and deadly attacks on unsuspecting victims. Power plants, pipelines, tankers, and dams all present inviting targets, and the threat to energy security has to take terrorism into account as nations plan their energy futures.

The United States has only 2% of the world's oil reserves, but consumes 25% of its oil, leaving us at the mercy of a global oil market. The OPEC cartel and the unpredictable Persian Gulf countries possess over 80% of the planet's oil reserves, so we will have to do business with them for the foreseeable future. The intensity of our interest is reflected by two wars since 1991 and a constant military presence in the Persian Gulf countries and their surrounding waters. How shall we chart a path to a secure energy future? One approach is to keep doing what we do now, only more so. That is, pump oil, drive gas-guzzling vehicles, gouge mountains for coal, and lace the countryside with natural-gas lines and wells. An alternative, the sustainable and more secure future, is to focus our energy policy on greater energy efficiency and a move toward renewable energy sources. The challenge is there, and the choice is ours to make. Chapters 12–14 describe what's going on in this important area.

**◄ Trans-Alaska Pipeline System. Constructed in the 1970s at a cost of $8 billion, the 800-mile pipeline transports 47,000 gallons of oil per month from the Alaskan North Slope to the ice-free port of Valdez.**

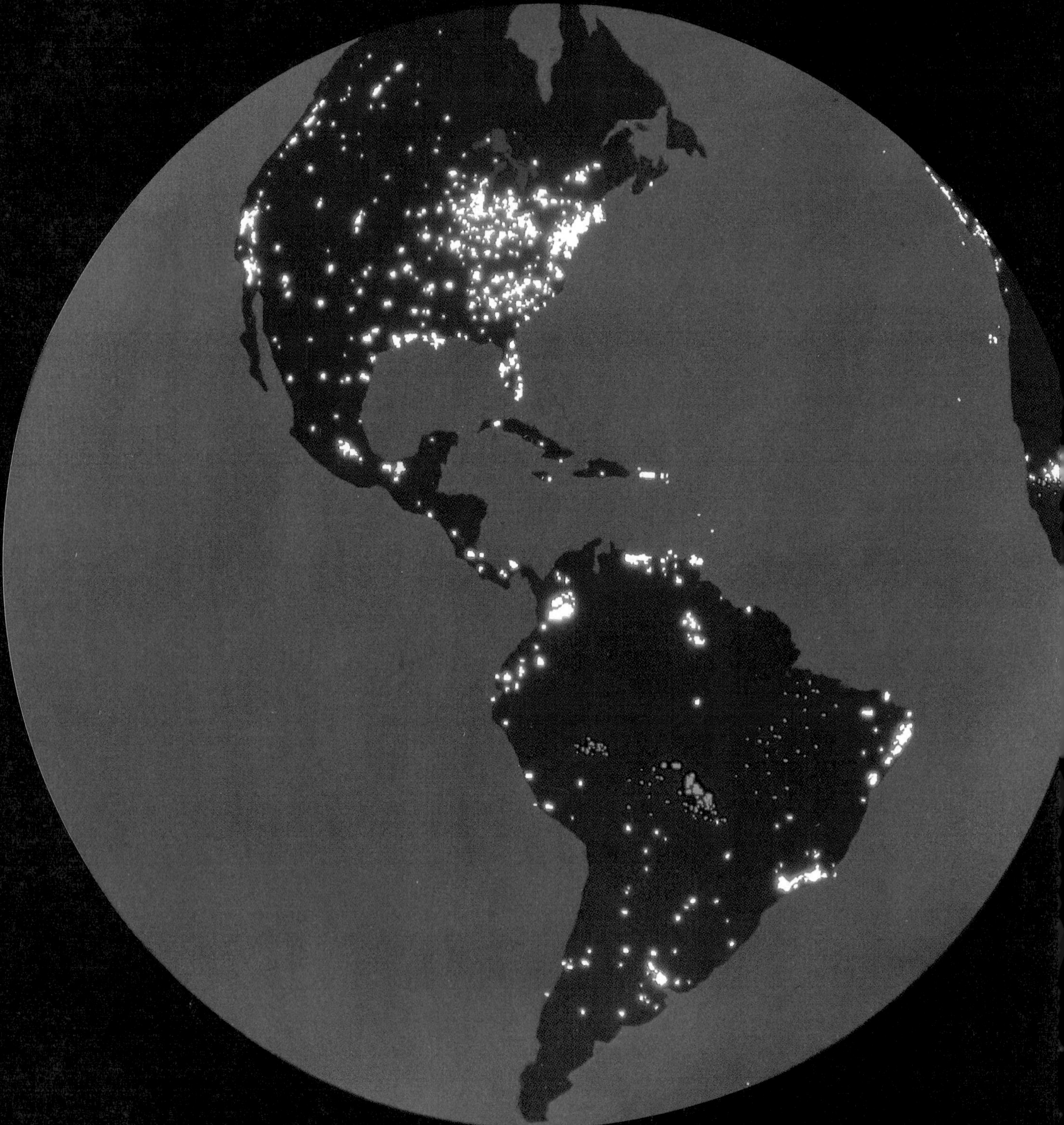

chapter 12

# Energy from Fossil Fuels

**Key Topics**

1. Energy Sources and Uses
2. Exploiting Crude Oil
3. Other Fossil Fuels
4. Fossil Fuels and Energy Security

In 1960, President Dwight Eisenhower proclaimed the creation of the Arctic National Wildlife Refuge (ANWR), 19.5 million acres (7.9 million hectares) of wilderness in the northeast corner of Alaska. Eight years later, the nation's largest-ever oil field was discovered farther to the west in Alaska, at Prudhoe Bay. The Prudhoe Bay oil stimulated the building of the Trans-Alaska oil pipeline, which will have carried some 13 billion barrels of Prudhoe oil to the pipeline terminus in ice-free Valdez, Alaska, before the field runs out. Other oil fields have been found and linked to the pipeline (Kuparuk, Alpine), but one unexploited field, the 1002 Area, sits below the ANWR. Seismic explorations by geologists have produced a broad range of estimates of the amount of oil that may be there. The U.S. Geological Service's (USGS's) best estimate puts technically recoverable oil at 7 billion barrels, about half of Prudhoe Bay's oil. If the ANWR is opened to oil development, some 7 years will elapse before the oil will actually begin to flow into the pipeline. At its peak, the flow will be about 700,000 barrels a day, exactly the amount the United States was importing from Iraq before the U.S.–Iraqi war in 2003. That amount will represent just 4% of the anticipated *daily* oil consumption in the nation by that time (around 2030). The refuge is also estimated to hold about 4 trillion cubic feet of natural gas (equal to two month's supply at current rates of use).

The ANWR also holds caribou. The Porcupine River herd of 130,000 arrives at the 1002 Area every summer to feed and produce calves (Fig. 12–1). The area also supports a herd of 250 musk oxen, as well as wolves, moose, grizzly bear, polar bear, lynx, many other mammals, and 95 species of migratory birds. Although oil development would affect a limited area of the refuge, there is concern that the impact would be especially hard on the Porcupine herd and the musk oxen. Oil development could push these animals closer to the mountains, where forage is poorer and predators are more numerous. The ANWR is pristine wilderness, and the intrusion of oil rigs, roads, pipelines, and villages (Fig. 12–2) would certainly clash with the attractions and aesthetic value found in the landscape.

***Western Hemisphere at Night.*** **Satellite photos show the widespread delivery of electricity to cities, towns, and villages throughout the hemisphere. In the United States, demand is growing faster than supply, threatening power disruptions. (White patches are city lights, and red spots in South America and Africa show large-scale burning of forests.)**

**Figure 12–1 Caribou in the Arctic National Wildlife Refuge.** The Porcupine River herd, 130,000 strong, in migration.

***Battle Lines.*** Developing the oil found in the 1002 Area has become a political firefight. On one side is the Bush administration and most Republicans in Congress. The administration has made opening up the ANWR to oil development a major piece of its national energy policy, and the Republican-controlled Congress has considered various ways to get this recommendation approved in the face of opposition. On the other side are most of the Democratic members of Congress and a host of environmental organizations. The American public either is split on the issue or is opposed to oil development in the ANWR, depending on which poll is consulted. The Inupiat Eskimos believe that the oil exploration is vital to their survival, and the Gwich'in Indians claim that it threatens to destroy their subsistence way of life, which depends on the Porcupine caribou herd. The Interior Department's Geological Survey issued a report in 2002 that summarized the scientific literature of the past 12 years and concluded that oil drilling could harm the caribou and other wildlife. One week later, the agency reversed itself after Interior Secretary Gale Norton saw the initial report, stating that the most likely drilling scenarios would have no impact on the caribou.

**Figure 12–2 Oil drilling on the North Slope.** The Alpine "village," operated by Phillips of Alaska, on the Colville River delta, a critical wildlife area.

The battle has taken on patriotic and international political dimensions, as developing the oil in ANWR has been justified in the light of unrest and war in the Mideast. President Bush presents the issue as a matter of national security. That is, developing the oil in the ANWR would help the United States reduce its dependence on foreign oil. The oil industry is predictably very interested in developing the ANWR oil, because U.S. oil reserves are continuing to decline and developing the refuge's oil would provide thousands of jobs and a great deal of profit for the industry.

This is just one of a host of energy–environment issues, albeit a highly sensitive and politically charged one. Perhaps with some reason, many critics of the U.S.-led 2003 Iraqi war are convinced that the war was much more about oil than about Saddam Hussein or the weapons of mass destruction that have yet to be found. Certainly, the 1991 Gulf War to push Iraq out of Kuwait had significant oil overtones.

***Oil Spills.*** Another major environmental issue is the oil spills that seem to happen on a regular basis. Oil for the global economy is moved from oil fields and ports and transported by oil tankers to refineries around the world. Millions of gallons are in transit at any given time, and it is inevitable that oil spills will occur. In 1989, the *Exxon Valdez* spilled 11 million gallons of crude oil into pristine Prince Edward Sound, and the oil is still leaching into the food chain of Alaskan waters. Spills occur every year. In 2002, the tanker *Prestige* sank off the Spanish coast with 21 million gallons of fuel oil, which continues to leak out of the sunken ship. The environmental consequences of oil spills are chalked up by some as one of the many costs of doing business in a fossil-fuel-driven economy. In future chapters, you will learn some of the other costs, such as air pollution and its impact on human and environmental health and rising greenhouse gases leading to global climate change.

The objective in this chapter is to gain an overall understanding of the sources of fossil-fuel energy that support transportation, homes, and industries. The consequences of our dependence on fossil fuels are discussed, as are some options for the future.

# 12.1 Energy Sources and Uses

## Harnessing Energy Sources: An Overview

Throughout human history, the advance of technological civilization has been tied to the development of energy sources. In early times (and even today in less developed regions), the major energy source was muscle power. Some people lived in relative luxury by exploiting the labor of others—slaves, indentured servants, and minimally paid workers. Human labor was supplemented by the use of domestic animals for agriculture and transportation, by water- or wind power for milling grain, and by the Sun.

By the early 1700s, inventors had already designed many kinds of machinery. The limiting factor was a continual power source to run them. The breakthrough that launched the Industrial Revolution was the development of the steam engine in the late 1700s. In a steam engine, water is boiled in a closed vessel to produce high-pressure steam, which pushes a piston back and forth in a cylinder. Through its connection to a crankshaft, the piston turns the drive wheel of the machinery. Steam engines rapidly became the power source for steamships, steam shovels, steam tractors, steam locomotives, and stationary engines to run sawmills, textile mills, and virtually all other industrial plants.

*Old King Coal.* At first, the major fuel for steam engines was firewood. Then, as demands for energy increased and firewood around industrial centers became scarce, coal was substituted. By the end of the 1800s, coal had become the dominant fuel, and it remained so into the 1940s. In addition to being used as fuel for steam engines, coal was widely used for heating, cooking, and industrial processes. In the 1920s, coal provided 80% of all energy used in the United States. By 2002, the figure had dropped to 22%.

Although coal and steam engines powered the Industrial Revolution, which greatly improved life for most people, those two sources of energy had many drawbacks. The smoke and fumes from the numerous coal fires made air pollution in cities far worse than anything seen today. Writers of the time recorded that they often could not see as far as a city block away because of the smoke. Coal is also notoriously hazardous to mine and dirty to handle, and burning coal results in large quantities of ashes that must be removed. As for steam engines, because of the size and bulk of the boiler, the engines were heavy and awkward to operate (Fig. 12–3). Frequently, the fire had to be started several hours before the engine was put into operation in order to heat the boiler sufficiently.

*Oil Rules.* In the late 1800s, the simultaneous development of three technologies—the internal combustion engine, oil-well drilling, and the refinement of crude oil into gasoline and other liquid fuels—combined to provide an alternative to steam power. The replacement of coal-fired steam engines and furnaces with petroleum-fueled engines and oil furnaces was an immense step forward in convenience. Also, the air quality improved greatly as cities were gradually rid of the smoke and soot from burning coal. (It was only in the 1960s, with the tremendous proliferation of cars, that pollution from gasoline engines became a problem.) Further, the gasoline-driven internal combustion engine provided a valuable power-to-weight advantage that allowed rapid advances in technology. A 100-horsepower gasoline engine weighs only a tiny fraction of what a 100-horsepower steam engine and its boiler weigh, and jet engines have an even greater power-to-weight ratio. Automobiles and other forms of ground transportation would be cumbersome, to say the least, without this power-to-weight advantage, and airplanes would be impossible.

Economic development throughout the world, so far as it has occurred since the 1940s, has depended largely on technologies that consume gasoline and other fuels refined from crude oil. By 1951, crude oil became the dominant energy source for the nation. Since then, it has become the mainstay not only of the United States (making up 38% of total U.S. energy), but also of most other

**Figure 12–3 Steam power.** Steam-driven tractors marked the beginning of the industrialization of agriculture. Today, tractors half the size do 10 times the work and are much easier and safer to operate.

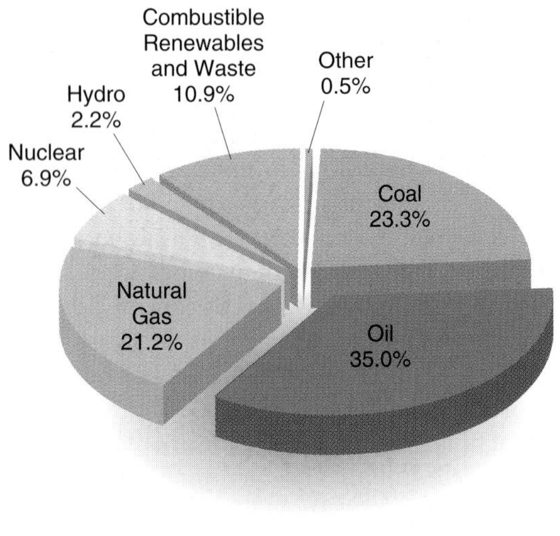


**Figure 12–4 Global primary energy supply.** The world consumes some 10,029 million tons of oil equivalent in energy every year. The sources of the total primary energy supply are shown here. (*Source:* International Energy Agency, 2003.)

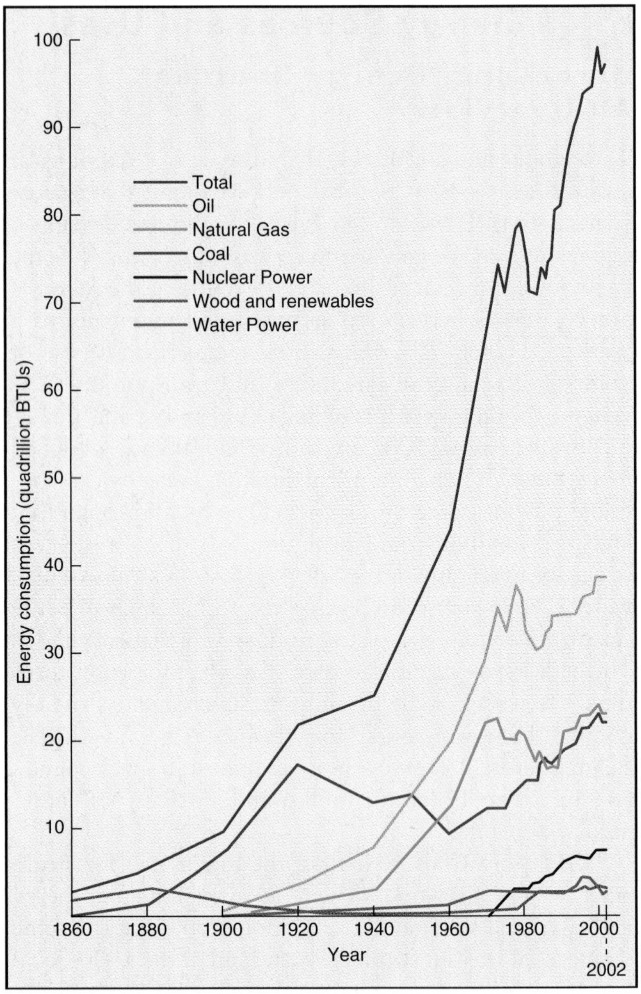


**Figure 12–5 Energy consumption in the United States.** Total consumption and major primary sources from 1860 to 2002 are shown here. Note how the mix of primary sources has changed over the years and how the total amount of energy consumed has continued to grow. Note also the skyrocketing increase in the use of oil after World War II (1945) as private cars and car-dependent commuting became common. (*Source:* Data from Energy Information Administration, U.S. Department of Energy, *Annual Energy Review 2002,* October 2003.)

countries, both developed and developing. Crude oil currently provides about 35% of the total annual global energy production of 10 billion tons of oil equivalent (Fig. 12–4). In the countries of Eastern Europe and in China, coal is still the dominant fuel. Coal ranks second globally, at 23.3%, and provides 22% of U.S. demand.

*Gas, Too.* Natural gas, the third primary fossil fuel, is found in association with oil or during the search for oil. Largely methane, which produces only carbon dioxide and water as it combusts, natural gas burns more cleanly than coal or oil; thus, in terms of pollution, it is a more desirable fuel. Despite the obvious fuel potential of natural gas, at first there was no practical way to transport it from wells to consumers. Any gas released from oil fields was (and in many parts of the world still is) flared—that is, vented and burned in the atmosphere, a tremendous waste of valuable fuel. Gradually, the United States constructed a network of pipelines connecting wells with consumers. With the completion of these pipelines, the use of natural gas for heating, cooking, and industrial processes escalated rapidly because gas is clean, convenient (no storage bins or tanks are required on the premises), and relatively inexpensive. Currently, natural gas satisfies about 23% of U.S. energy demand and 21% of the global demand.

Thus, three fossil fuels—crude oil, coal, and natural gas—provide 83.5% of U.S. energy consumption and 79.5% of the world's consumption. The remaining percentages are furnished by nuclear power, hydropower, and renewable sources. The changing pattern of energy sources in the United States is shown in Fig. 12–5. The picture is similar for other developed countries of the world, although the percentages differ somewhat, depending on the energy resources of the country. With fossil-fuel energy, people and goods have become so mobile that the human community has in fact become a global community. Adding to this linkage in a powerful way is the explosion in information storage and flow, made possible by digital devices and now the Internet—all powered by electricity.

## Electrical Power Production

A considerable portion of the total energy used is electrical power, called a **secondary energy source** because it depends on a **primary energy source** (coal or waterpower, for instance) to turn a *generator.* Electricity now defines our modern technological society, powering appliances and computers in our homes, lighting up the night (see the chapter opening photo), and enabling the global Internet and countless other vital processes in business and industry to operate. Over 33% of fossil fuel production is now used to generate electricity; in 1950, the figure was only 10%.

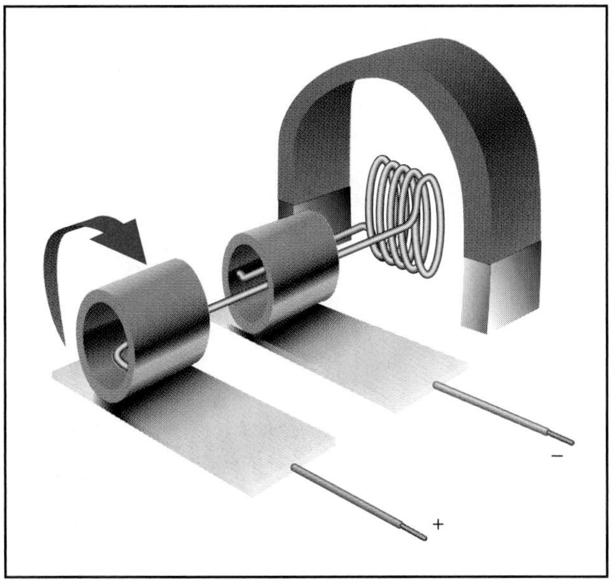


**Figure 12–6 Principle of an electric generator.** Rotating a coil of wire in a magnetic field induces a flow of electricity in the wire.

*Generators.* Electric generators were invented in the 19th century. In 1831, the English scientist Michael Faraday discovered that passing a coil of wire through a magnetic field causes a flow of electrons—an electrical current—in the wire. An **electric generator** is basically a coil of wire that rotates in a magnetic field or that remains stationary while a magnetic field is rotated around it (Fig. 12–6). The process converts mechanical energy into electrical energy. If you do just the reverse—supply electric power to a generator—it becomes a motor. As discussed in Chapter 3, anytime energy is converted from one form to another, some is lost through resistance and heat (in accordance with the second law of thermodynamics). In a generator, as the current flows in the wire, it creates a new magnetic field that is in opposition to the first and thus resists the flow of current. Some energy is lost in transmitting the electricity over wires from the generating plants to the end users. In the end, it takes three units of primary energy to create one unit of electrical energy that actually is put to use. This energy-losing proposition is justified because electrical power is so useful and, now, indispensable.

*Turbogenerators.* In the most widely used technique for generating electrical power, a primary energy source is employed to boil water, creating high-pressure steam that drives a **turbine**—a sophisticated paddle wheel—coupled to the generator (Fig. 12–7a). The combined turbine and generator are called a **turbogenerator.** Any primary energy source can be harnessed for boiling the water. Coal, oil, and nuclear energy are most commonly used at present, but burning refuse, solar energy, and geothermal energy (heat from Earth's interior) may be more widely used in the future.

In addition to steam-driven turbines, gas- and water- (hydro-) driven turbines are also used. In a gas turbogenerator, the high-pressure gases produced by the combustion of a fuel (usually natural gas) drive the turbine directly (Fig. 12–7b). For hydroelectric power, water under high pressure—at the base of a dam or at the bottom of a pipe beginning at the top of a waterfall—is used to drive a hydroturbogenerator (Fig. 12–7c). Wind turbines are also coming into use. (See Fig. 14–18.) A utility company supplying electricity to its customers will employ a mix of power sources, depending on the availability of each and the demand cycle.

**Fluctuations in Demand.** Where does your electricity come from? It depends. Most utility companies are linked together in what are called *pools.* In New England, for example, ISO [Independent System Operator] New England ties together some 350 electric power-generating plants through 8,000 miles of transmission lines. The utility is responsible for balancing electricity supply and demand, regardless of daily or seasonal fluctuations. The generating capacity of the ISO New England utilities is 31,000 megawatts (MW), produced by a variety of energy sources, as shown in Table 12–1. The greatest demands on electrical power in New England are in the summer months, when air-conditioning is at its highest use. The peak load (highest demand) for 2002 was 25,348 MW, during a hot day in August. A power pool must accommodate daily and weekly electricity demand, as well as seasonal fluctuations.

*Demand Cycle.* The typical pattern of daily and weekly demand is shown in Fig. 12–8. The base load represents the constant supply of power provided by the large coal-burning and nuclear plants with capacities as high as 1,000 MW. As demand rises during the day, the utility draws on additional plants that can be turned on and off—the intermediate and peak-load power sources that represent the utility's reserve capacity. These are the utility's gas turbines and diesel plants, or else the utility uses pumped-storage hydroelectric power (water is pumped to a reservoir using off-peak power and is allowed to flow through turbines when needed). Occasionally, a deficiency in available power will prompt a brownout (a reduction in voltage) or blackout (a total loss of power), disrupting whole regions temporarily. Such events occur most likely during times of peak power demand and may be precipitated by a sudden loss of power accompanying the shutdown of a base-load plant.

In our current economy, demand for electricity has been rising faster than supply, resulting in a decline in reserve capacity from 25% in the 1980s to the current level of 15%. Summer heat waves represent the greatest source of sudden pulses in demand, and as the climate continues to grow warmer in many parts of the United States, more and more regional utilities are being pushed to the edge of their ability to satisfy summer electricity demand. Power brownouts and blackouts are a serious threat to an economy that now depends highly on computers and other electronic devices; on August 14, 2003, the largest blackout in U.S. history hit eight eastern states and parts of Canada, imposing a $30 billion cost on the

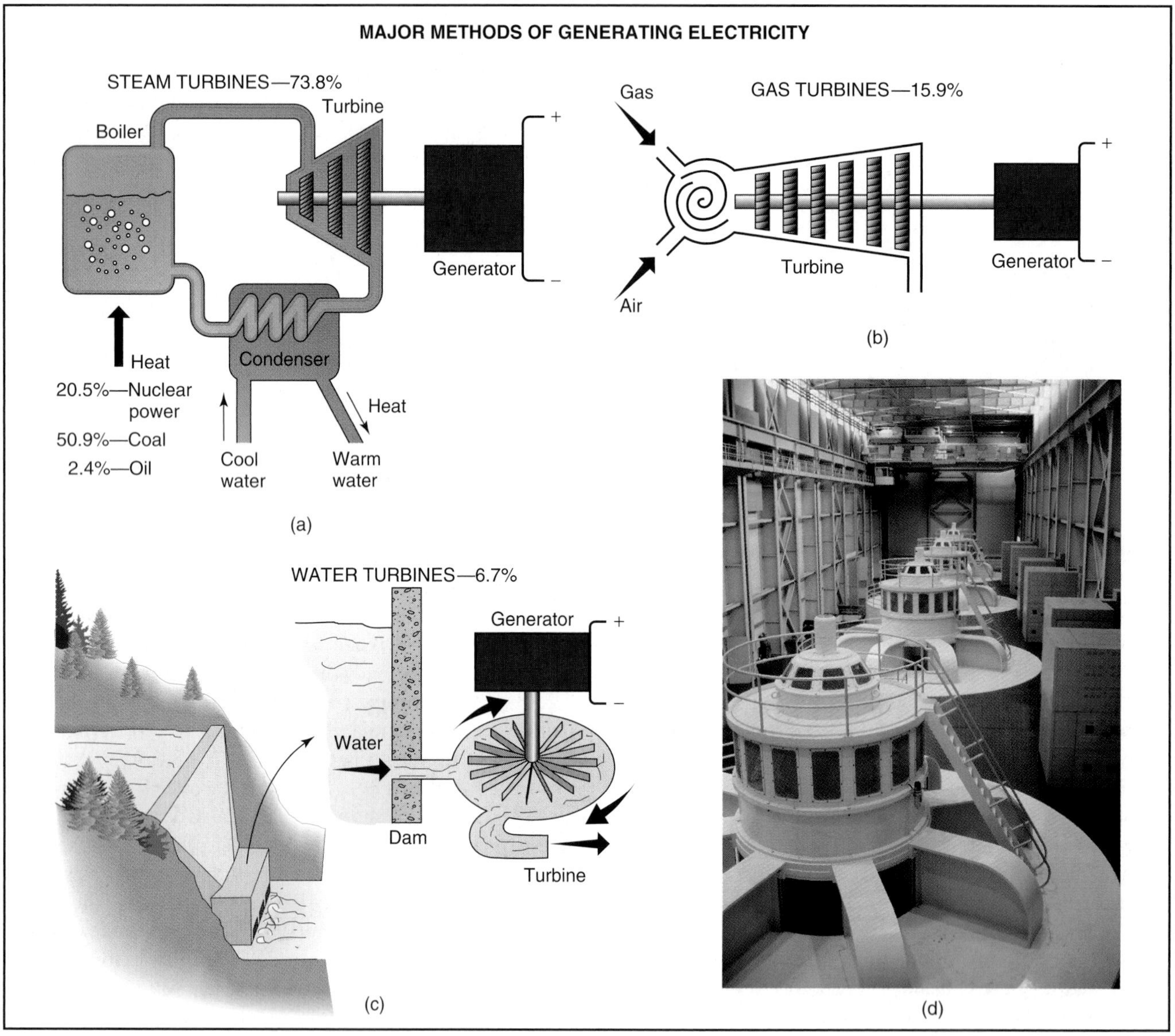


**Figure 12–7 Electrical power generation.** Electricity is produced commercially by driving generators with (a) steam turbines, (b) gas turbines, and (c) water turbines. The percentage of the U.S. electricity supply derived from each source (2002) is indicated. Small amounts of power are now also coming from solar wind energy, solar thermal energy, and photovoltaics (solar cells). (d) Each generator in this array in a hydroelectric plant is attached by a shaft to a turbine propelled by water.

**table 12-1 Energy Sources for Generating Electrical Power (% of total)**

| Source | New England* | United States | World |
|---|---|---|---|
| Nuclear power | 29.1 | 20.7 | 16.9 |
| Natural gas | 29.4 | 16.8 | 17.4 |
| Coal | 15.3 | 51.2 | 39.1 |
| Oil | 14.2 | 3.4 | 7.9 |
| Hydro | 5.0 | 5.8 | 17.1 |
| Other (wood, refuse, renewables) | 7.0 | 2.1 | 1.6 |

*Does not include an additional 11.5% from national energy grid.
(*Sources:* ISO New England, U.S. Energy Information Administration, International Energy Agency.)

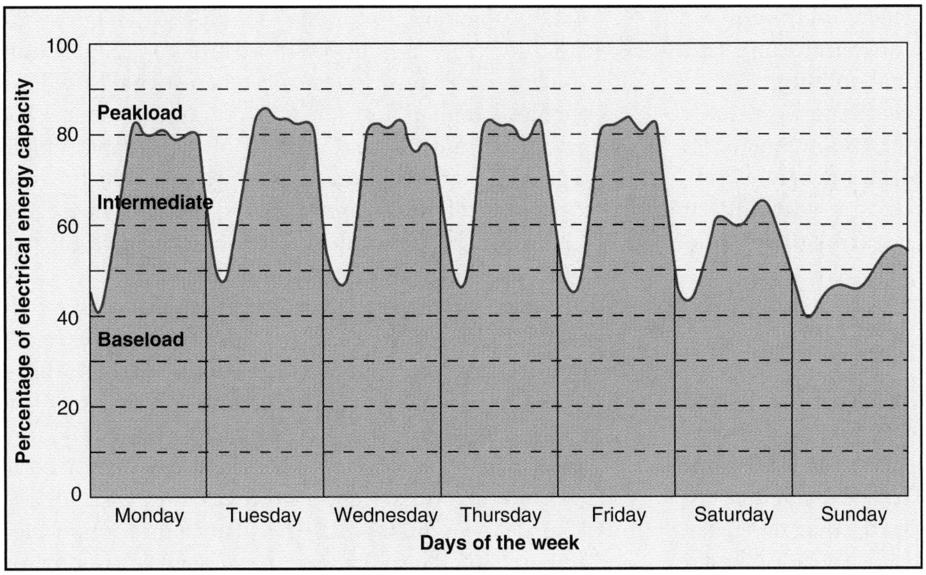


**Figure 12–8 Weekly electrical demand cycle.** Electrical demand fluctuates daily, weekly, and seasonally. A variety of generators must be employed to meet base, intermediate, and peak electricity needs.

economies of the two countries. The power industry's response to this threat is to continue building more small gas-fired turbines that can be easily turned on at times of peak demand. However, with recent surges in the price of gas, this option looks less attractive.

*Clean Energy?* Electrical power is often promoted as being the ultimate clean, nonpolluting energy source. This is true at the point of its use: Using electricity creates essentially no pollution. Hidden, however, is the fact that pollution is simply transferred from one part of the environment to another. Thus, switching to "clean" electric heat implies more demand for electricity, which must be generated from coal, hydropower, nuclear energy, or alternative energy sources. Coal-burning power plants are the major source of electricity in the United States, as well as the major source of acid deposition. They also contribute substantially to other air-pollution problems (Chapter 21). If hydroelectric power is used, creating a dam and reservoir involves displacing people, farmland, and wildlife and disrupts the migration of fish, such as salmon. Finally, nuclear power, discussed in Chapter 13, is a much-mistrusted technology because of the potential for accidents and the problem of disposal of the nuclear waste products. Additional adverse effects of using electrical energy apply to other parts of the fuel cycle, such as mining coal or processing uranium ore.

*Efficiency.* The problem of transferring pollution from one place to another becomes even more pronounced when you consider efficiency, because the thermal production of electricity has an *efficiency of only 30%*. The 70% loss is accounted for partly by some heat energy from the firebox going up the chimney and partly by a large amount of heat energy remaining in the spent steam as it comes out the end of the turbine. These unavoidable energy losses, called **conversion losses,** are a consequence of the need to maintain a high heat differential between the incoming steam and the receiving turbine, so as to maximize efficiency. Heat energy will go only toward a cooler place, so it cannot be recycled into the turbine. The most common practice is to dissipate it into the environment by means of a condenser. In fact, the most conspicuous features of coal-burning or nuclear power plants are the huge cooling towers used for that purpose (Fig. 12–9). Another important source of energy loss occurs as the electricity is transmitted and distributed to end users by connecting wires.

An alternative to cooling towers is to pass water from a river, a lake, or an ocean over the condensing system (Fig. 12–7a). With this approach, however, the waste heat energy is transferred to the body of water, with the result that all the small planktonic organisms, both plant and animal, drawn through the condensing system with the water are cooked and the warm water added back to

**Figure 12–9 Cooling towers.** These structures, the most conspicuous feature of both coal-fired and nuclear power plants, are necessary to cool and recondense the steam from the turbines before returning it to the boiler. Air is drawn in at the bottom of the tower, over the condensing apparatus, and out the top by natural convection. Cooling is aided by the evaporation of water that condenses in a plume of vapor as the moist air exits the top of the tower, as seen in this photo.

the waterway may have deleterious effects on the aquatic ecosystem. Thus, waste heat energy discharged into natural waterways is referred to as **thermal pollution.**

Electricity is a form of energy that is indispensable to modern societies as they are currently structured. Worldwide, the electrical power industry generates revenues valued at more than $800 billion a year and, at the same time, generates a challenging array of environmental problems.

## Matching Sources to Uses

In addressing whether energy resources are sufficient and where additional energy can be obtained, you need to consider more than just the source of energy, because some forms of energy lend themselves well to one use, but not to others. For example, the cars and trucks that travel the U.S. system of highways depend virtually 100% on liquid fuels. Practically all other machines used for moving, such as tractors, airplanes, locomotives, and bulldozers, depend likewise on liquid fuels. Reserves of crude oil are needed to support this transportation system. The nuclear industry has promoted nuclear power as a solution to the energy shortage. Nuclear power can do little to mitigate the demand for crude oil, however, because it is suitable only for boiling water to drive turbogenerators, which will do essentially nothing for our transportation system. The same can be said for coal. Although this situation might change with more widespread use of electric vehicles, their practicality for long distances still requires a technological breakthrough in the form of a low-cost, lightweight battery that can store large amounts of power.

***Energy Flow.*** Primary energy use is commonly divided into four categories: (1) transportation, (2) industrial processes, (3) commercial and residential use (heating, cooling, lighting, and appliances), and (4) the generation of electrical power, which goes into categories (2) and (3) as a secondary energy use. The major pathways from the primary energy sources to the various end uses are shown in Fig. 12–10. Transportation, which represents 26.5% of energy use, depends entirely on petroleum, whereas nuclear power, coal, and waterpower are limited to the production of electricity. Natural gas and oil are more versatile energy sources, and significant shifts have occurred in the use of these sources in recent years. For example, today less than 1 one-hundredth of

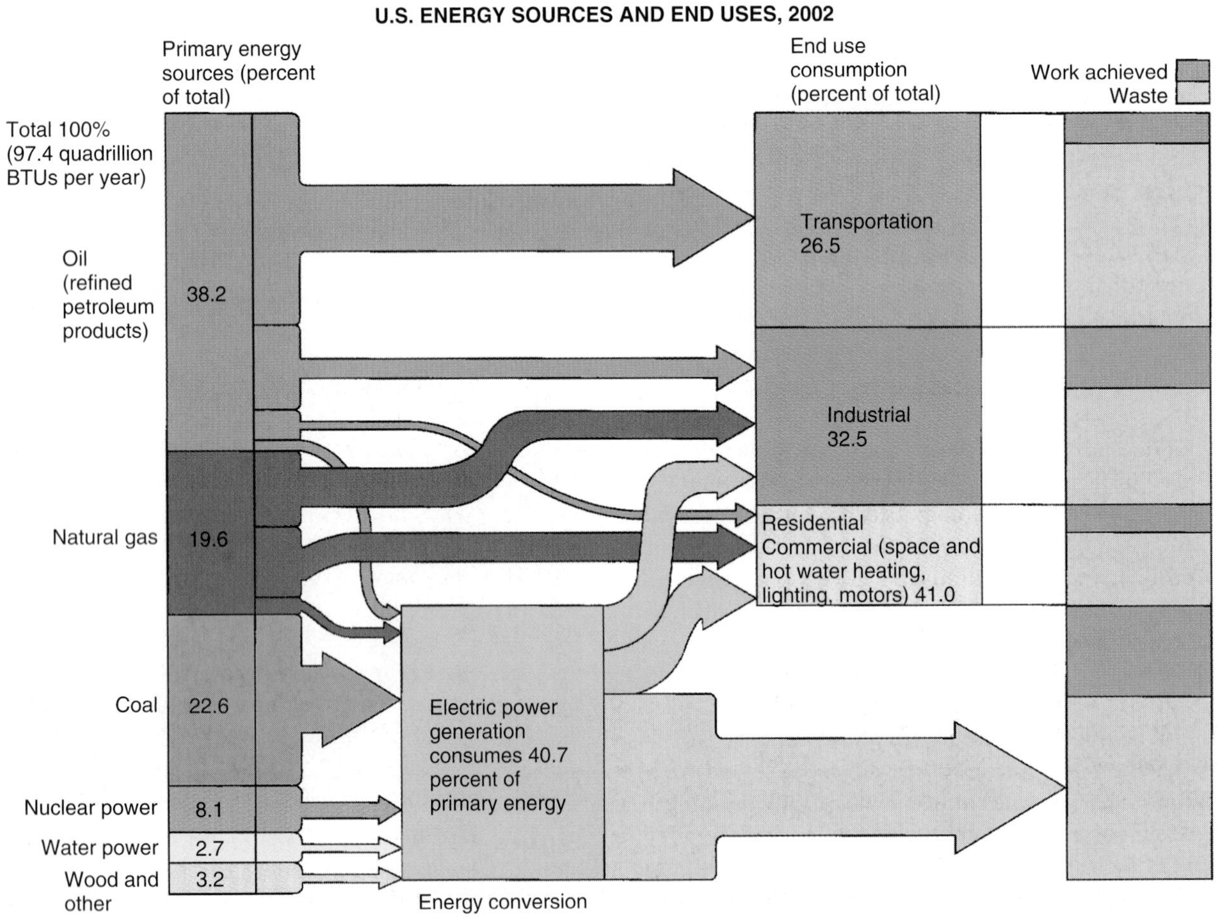


**Figure 12–10 Pathways from primary energy sources to end uses in the United States.** Only major pathways are shown. Note that end uses are connected to primary sources in specific ways. Note also the large percentage of energy that is wasted as a large portion of the energy consumed is converted to heat and lost. (*Source:* Data from Energy Information Administration, U.S. Department of Energy, *Annual Energy Review 2002,* October 2003.)

U.S. electricity is generated by oil, whereas 30 years ago it was about one-fifth.

Figure 12–10 also shows the proportion of consumed energy that goes directly to waste heat rather than for its intended purpose. Some waste is inevitable, as dictated by the second law of thermodynamics, but the current losses are much greater than necessary. The efficiency of energy use can be at least doubled for some uses, including cars that get twice as many miles per gallon and appliances that consume half as much power. For uses like electric power generation, however, somewhat lower increases in efficiency can be expected. In general, increasing the efficiency of energy use would have the same effect as increasing the energy supply, at far less expense. Such increases in efficiency are the essence of energy conservation and are discussed in more detail in Section 12.4.

In contemplating the energy future, you need to consider more than gross supplies and demands. That is, you must also examine how particular primary sources can be matched to particular end uses. Furthermore, what are the most cost-effective ways of balancing supplies and demands—by increasing supplies or by decreasing consumption through *conservation, efficiency,* and *demand management?*

## 12.2 Exploiting Crude Oil

The United States has abundant reserves of coal, adequate reserves of natural gas (at least for the time being), and the potential for the expansion of nuclear power. Therefore, no shortages loom in these areas now. (Note that the environmental concerns regarding coal and nuclear power are being set aside for the moment.) Petroleum to fuel our transportation system, however, is another matter. U.S. crude-oil supplies fall far short of meeting demand; we now depend on foreign sources for over 60% of our crude oil, and this dependency is steadily growing as U.S. reserves diminish. Foreign sources come at increasing costs, such as trade imbalances, military actions, economic disruptions, and coastal oil spills. Let us examine the situation, starting with where crude oil comes from and how it is exploited.

### How Fossil Fuels Are Formed

The reason crude oil, coal, and natural gas are called *fossil fuels* is that all three are derived from the remains of living organisms (Fig. 12–11). In the Paleozoic and Mesozoic geological eras, 100 to 500 million years ago, freshwater swamps and shallow seas, which supported abundant vegetation and phytoplankton, covered vast areas of the globe. Anaerobic conditions in the lowest layers of these bodies of water impeded the respiration of decomposers and hence the breakdown of detritus. As a result, massive quantities of dead organic matter accumulated. Over millions of years, this organic matter was gradually buried under layers of sediment and converted by pressure and heat to coal, crude oil, and natural gas. Which fuel was formed depended on the nature of the organic material buried, the specific environmental conditions, and the time involved. Just as anaerobic treatment (digestion) of sewage sludge yields methane gas and a residual organic sludge, a similar anaerobic process occurred with the buried vegetation. Natural gas is the trapped methane, and crude oil represents the residual sludge. Coal is highly compressed organic matter—mostly leafy material from swamp vegetation—that decomposed relatively little.

***Renewal.*** Additional fossil fuels may still be forming through natural processes to this day. They cannot be considered renewable resources, however, because we are using fossil fuels far faster than they ever formed. In fact, it takes 1,000 *years* to accumulate the amount of organic matter that the world now consumes each *day.* Because fossil fuels are a finite resource, the pertinent question is not *whether* they will run out, but *when* they will?

### Crude-Oil Reserves Versus Production

The total amount of crude oil remaining represents Earth's oil resources. Finding and exploiting these resources is the problem. The science of geology provides information about the probable locations and extents of ancient shallow seas. On the basis of their knowledge and their field experience, geologists make educated guesses as to where oil or natural gas may be located and how much may be found. These educated guesses are the world's **estimated reserves.** The estimates may be far off the mark, because there is no way to determine whether estimated reserves actually exist, except by exploratory drilling.

If exploratory drilling strikes oil, further drilling is conducted to determine the extent and depth of the **oil field.** From that information, a fairly accurate estimate can be made of how much oil can be economically obtained from the field. That amount then becomes **proven reserves.** Amounts are given in barrels of oil (1 barrel = 42 gallons), and the content of each reserve field is expressed in probabilities. Thus, a 50% probability, or P50, of 700 million barrels for a field means that a total of 700 million barrels is just as likely as not to come from the field. Proven reserve estimates can range from P05 (a 5% probability) to P90. A P90 estimate is more reliable, but oil producers are often vague about probabilities and might prefer to use a P05 or a P10 estimate in order to give the impression of a large reserve for political or economic reasons. Also, proven reserves hinge on the economics of extraction. Thus, such reserves may actually increase or decrease with the price of oil, because higher prices justify exploiting resources that would not be worth extracting for lower prices. The final step, the withdrawal of oil or gas from the field, is called **production** in the oil business. *Production,* as used here, is a euphemism. "It is production," said ecologist Barry Commoner, "only in the sense that a boy robbing his mother's pantry can be said to be producing the jam supply." In reality, this step is *extraction* from Earth.

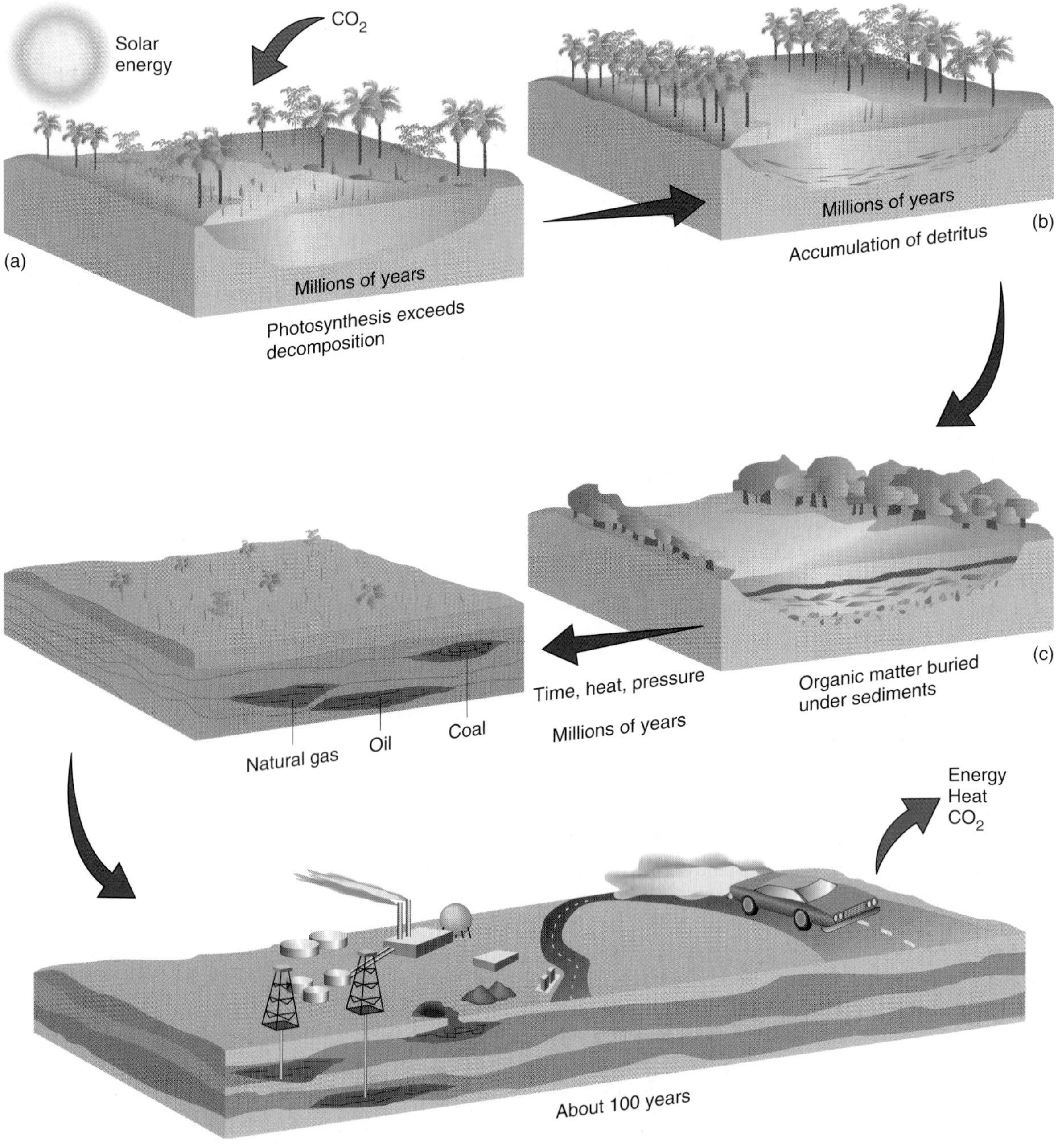


**Figure 12–11 Energy flow through fossil fuels.** Coal, oil, and natural gas are derived from biomass that was produced many millions of years ago. Deposits are finite, and because formative processes require millions of years, they are nonrenewable.

***Recovery.*** Production from a given field cannot proceed at a constant rate, because crude oil is a viscous fluid held in spaces in sedimentary rock, just as water is held in a sponge. Initially, the field may be under so much pressure that the first penetration produces a gusher. Gushers, however, are generally short lived, and the oil then seeps slowly from the sedimentary rock into a well from which it is pumped out. Ordinarily, conventional pumping (**primary recovery**) can remove only 25% of the oil in an oil field. Further removal, of up to 50 or 60% or more of the oil in the field, is often possible, but it is more costly, because it involves manipulating pressure in the oil reservoir by injecting brine, steam, or other substances—so-called **secondary** or **tertiary recovery**—that force the oil into the wells.

Economics—the price of a barrel of oil—determines the extent to which reserves are exploited. No oil company will spend more money to extract oil than it expects to make selling the oil. At $10 per barrel of crude oil, the price in the late 1990s, it was economical to extract no more than 25–35% of the oil in a given field. Thus, an increase in price makes more reserves available. In the 1970s and early 1980s, increasing oil prices justified reopening old oil fields and created an economic boom in Texas and Louisiana. Then an economic bust occurred in the late 1980s, as a drop in oil prices caused the fields to be shut down again. Now that oil prices have risen into the $20–30 range, oil exploration and recovery has resumed and the U.S. oil industry has experienced a rebirth.

## Declining U.S. Reserves and Increasing Importation

Understanding the relationships between production, price, and the exploitation of reserves can help clarify the U.S. energy dilemma. Up to 1970, the United States was largely oil independent. That is, exploration was leading to increasing proven reserves, such that production could basically keep pace with growing consumption (Fig. 12–12). In 1970, however, new discoveries fell short, so production decreased, while consumption continued to grow rapidly. This downturn vindicated the predictions of the late geologist M. King Hubbert, who proposed that oil exploitation in a region would follow a bell-shaped curve. Hubbert observed the pattern of U.S. exploitation and predicted that U.S. production would peak between 1965 and 1970. At that point, about half of the available oil would have been withdrawn, and production would decline gradually as reserves were exploited. Figure 12–12 shows a "Hubbert curve" in progress for U.S. oil production.

To fill the energy gap between rising consumption and falling production, the United States depended increasingly on imported oil, primarily from the Arab countries of the Middle East. European countries and Japan did likewise. Because imported oil cost only about $2.30 per barrel ($9.18 in 2002 dollars) in the early 1970s and Middle Eastern reserves were more than adequate to meet the demand, this course seemed to present few problems.

**The Oil Crisis of the 1970s.** Because of the low price of oil from Middle Eastern countries, the United States and most other highly industrialized nations were becoming more and more dependent on oil imports. A group of predominantly Arab countries

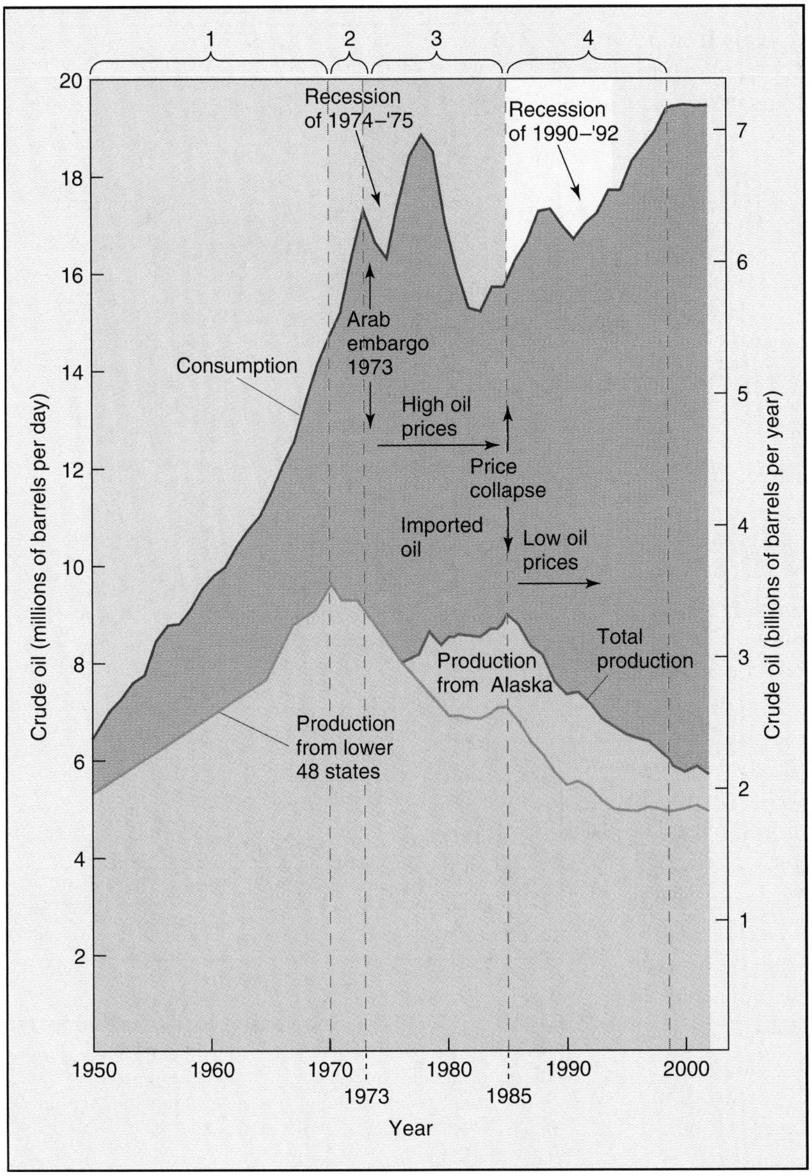


**Figure 12–12 Oil production and consumption in the United States.** Four stages can be seen: (1) Up to 1970, the discovery of new reserves allowed production to parallel increasing consumption. (2) From 1970 to 1973, a lack of new oil discoveries caused production to decline, while consumption continued to climb, resulting in a vast increase in oil imports. (3) In the late 1970s to early 1980s, high oil prices promoted both lowered consumption and increased production—which included bringing the Alaskan oil field into production. (4) From 1986, when oil prices declined sharply, until the present, consumption has again increased while production has resumed its decline, making us increasingly dependent on foreign oil. (*Source:* Energy Information Administration, *Annual Energy Review 2002,* October 2003.)

known as the Organization of Petroleum Exporting Countries (OPEC) formed a cartel and agreed to restrain production in order to get higher prices. Imported oil began to cost more in the early 1970s, and then, in conjunction with an Arab–Israeli war in 1973, OPEC initiated an embargo of oil sales to countries, like the United States, that gave military and economic support to Israel. The effect was almost instantaneous, because we depend on a fairly continuous flow from wells to points of consumption. (Compared with total use, little oil was held in storage.) Spot shortages occurred, which quickly escalated into widespread panic because of our widespread dependence on cars and trucks. We were willing to pay almost anything to have oil shipments resumed—and pay we did: OPEC resumed shipments at a price of \$10.50 a barrel (\$29 in 2002 dollars), almost four times the previous price.

Then, by continuing to limit oil production all through the 1970s, OPEC was able to keep supplies tight enough to force prices higher and higher. In the early 1980s, a barrel of oil delivered to the United States cost \$36 (\$64 in 2002 dollars), and the United States was paying accordingly. The high energy costs were devastating to the economy, as inflation and unemployment rose and the country experienced several recessions. The purchase of foreign oil became, and remains, the single largest factor in our balance-of-trade deficit (Fig. 12–13).

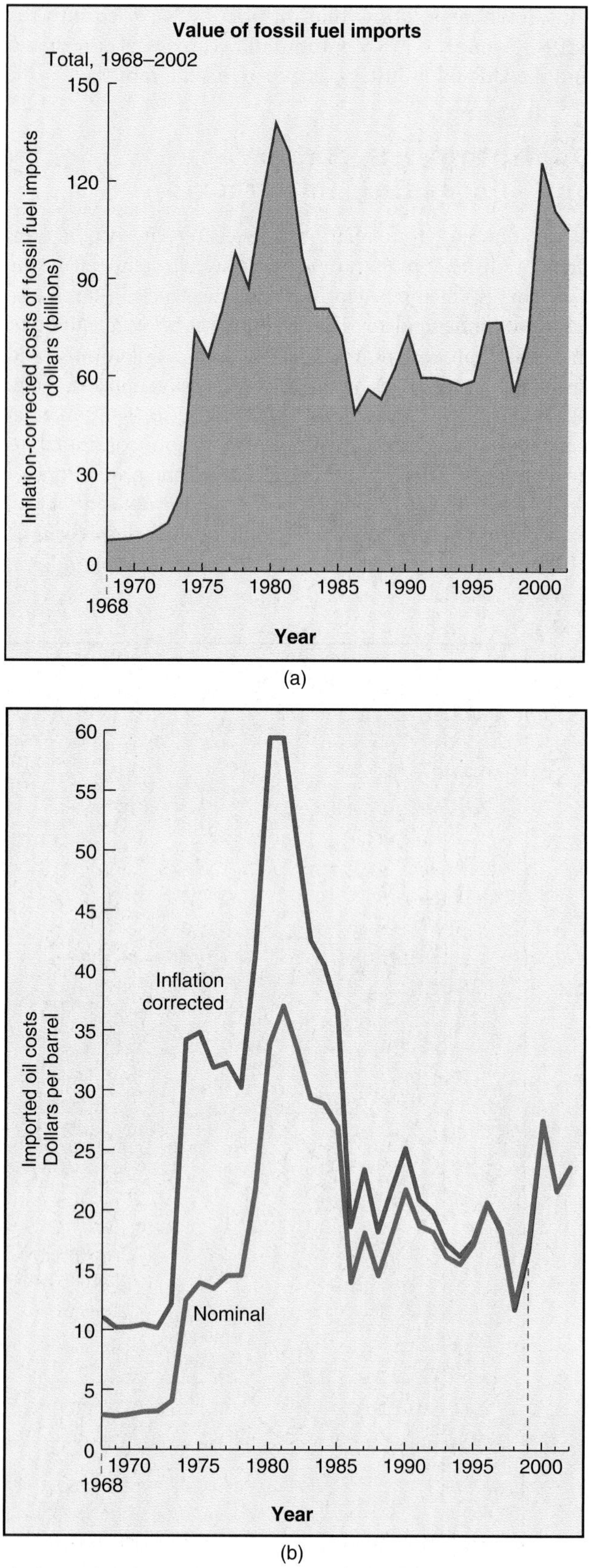


**Figure 12–13 The cost of fossil-fuel imports.** (a) The inflation-corrected cost of total oil imported to the U.S. economy and (b) the cost in dollars per barrel. (*Source:* Energy Information Administration, *Annual Energy Review 2002,* October 2003.)

**Adjusting to Higher Prices.** In response to the higher prices, the United States and other industrialized countries quickly made a number of adjustments during the 1970s. The following are the most significant changes made by the United States:

*To increase domestic production of crude oil,*

- Exploratory drilling was stepped up.
- The Alaska pipeline was constructed. (The Alaskan oil field had actually been discovered in 1968, but it remained out of production because it was so remote.)
- Fields that had been closed down as uneconomical when oil was only \$2.30 a barrel were reopened (for example, in Texas and Louisiana).

*To decrease consumption of crude oil,*

- Standards were set for automobile fuel efficiency. In 1973, cars averaged 13 miles per gallon (mpg). The government mandated stepped increases for new cars, to average 27.5 mpg by 1985. Speed limits were set at 55 mph for all U.S. highways.
- Other conservation goals were promoted for such things as insulation in buildings and efficiency of appliances.
- The development of alternative energy sources was begun. The government supported research-and-development efforts and gave tax breaks to people for installing alternative energy systems.

*To protect against another OPEC boycott,*

- A strategic oil reserve was created. As of 2002, the United States had "stockpiled" 599 million barrels of oil (equivalent to 53 days' worth of imports) in underground caverns in Louisiana.

With the exception of the stockpile, all of these steps are economically appealing only when the price of oil is high. If the price is low, the economic returns or savings may not justify the necessary investments (even though the steps are environmentally desirable).

The results of the new actions were not immediate, nor could quick returns have been expected. For example, it takes at least three to five years to design a new car model and start production. Then another six to eight years pass before enough new models are sold and old ones retired to affect the overall highway "fleet."

***Recovery.*** As we entered the 1980s, however, the efforts were definitely having an impact. The United States continued to depend substantially on foreign oil, but consumption was declining, and production, up after the Alaska pipeline was completed, was holding its own. (See Fig. 12–12.) Elsewhere in the world, significant discoveries in Mexico, Africa, and the North Sea made the world less dependent on OPEC oil. As a result of all these factors, plus OPEC's inability to restrain its own production, world oil production exceeded consumption in the mid-1980s, and there was an oil "glut."

What happens when supply exceeds demand? In 1986, world oil prices crashed from the high \$20s to \$14 per barrel. From that time until late 1999, oil prices fluctuated in the range from \$10 to \$20 per barrel and, due to inflation, were almost as low as they were in the early 1970s (Fig. 12–13). The effect of the lower oil prices was to the apparent benefit of consumers, industry, and oil-dependent developing nations, but it hardly solved the underlying problem. Indeed, it set the stage for another developing crisis: the steep rise in oil prices that began in late 1999. OPEC was concerned about the continuing oil glut and its low prices, so it decided to cut production just as East Asia was coming out of a serious recession. The rapid rise in demand brought on a shortfall in the oil supply, and the price of crude oil rose. Amid rising international concern about the impact of higher prices on the world economy, OPEC promised to increase production and has indicated that it will aim at a long-term price range of \$22–\$28 per barrel. Since then, the price has fluctuated between \$20 and \$32 per barrel, into 2004.

**Victims of Our Success.** Whereas high oil prices stimulated constructive responses, the collapse in oil prices undercut those responses:

- Exploration, which becomes more costly as wells are drilled deeper and in more remote locations, was sharply curtailed. In 1982, 3,105 drilling rigs were operating in the United States; in 1992, there were 660.
- Production from older oil fields, which was costing around \$10–\$15 per barrel to pump, was terminated. The resulting drop in production devastated the economies of Texas and Louisiana.
- Conservation efforts and incentives were abandoned. Government standards for automobile fuel efficiency were frozen at 27.5 mpg, with no intention to increase them further. Speed limits were set to 65 and even 75 mph.
- Tax incentives and other subsidies for the development or installation of alternative energy sources were terminated, destroying many new businesses engaged in solar and wind energy. Grants for research and development of alternative energy sources (except nuclear power) were cut sharply.
- The need for conservation and for the development of alternative ways to provide transportation seems to have largely passed from the mind of the public.

U.S. production of crude oil continued to drop as proven reserves were drawn down. The declining production now includes Alaska, which reached its peak production in 1988. At the same time, consumption of crude oil rose again, because both the number of cars on the highways and the average number of miles per year that each car is driven increased. Adding to that consumption, consumers began buying vehicles that were less fuel efficient. Minivans, four-wheel-drive sport-utility vehicles, light trucks, and other less fuel-efficient models now constitute over 50% of the new-car market.

Altogether, following considerable improvement in our energy situation in the late 1970s and early 1980s, the United States has reverted to old patterns. Our dependence on foreign oil has grown steadily since the mid-1980s. At the time of the first oil crisis, in 1973, the United States depended on foreign sources for about 35% of its crude oil. In 1993, it passed the 50% mark, and it is now at 60% (Fig. 12–14).

The United States is not alone in this situation. Much of the rest of the world, including both developed and developing countries, has a growing dependency on oil from those relatively few nations with surplus production. What problems does this growing dependency on foreign oil present?

## Problems of Growing U.S. Dependency on Foreign Oil

As the U.S. dependency on foreign oil grows again, we are faced with problems at three levels: (1) the costs of purchasing oil, (2) the risk of supply disruptions due to political instability in the Middle East, and (3) ultimate resource limitations in any case.

**Costs of Purchase.** The yearly cost to the United States of purchasing foreign oil and gas is shown in Fig. 12–13. The cost dropped considerably from its high of over \$130

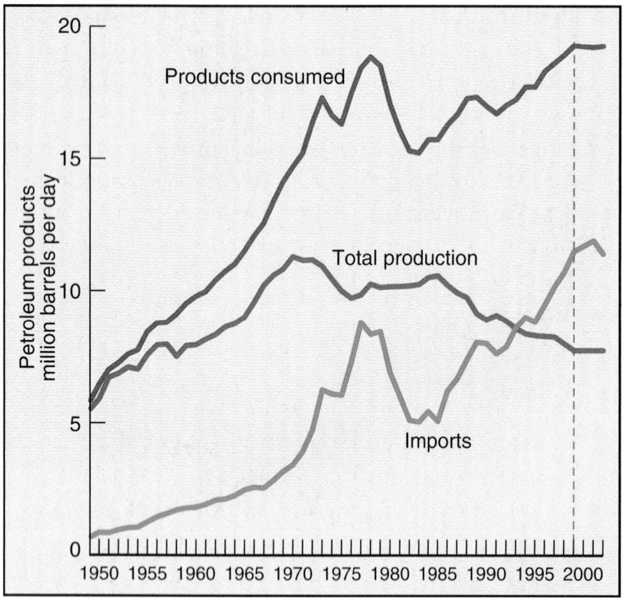


**Figure 12–14 Consumption, domestic production, and imports of petroleum products.** After the oil crises of the 1970s, the United States became less reliant on foreign oil. Since the mid-1980s, however, consumption of foreign oil has for the most part been rising steadily. (*Source:* Energy Information Administration, *Annual Energy Review 2002*, October 2003.)

billion (corrected for inflation) in the early 1980s, due to falling prices and conservation efforts. It rose rapidly in 2000, however, and is poised to climb higher when the U.S. economy recovers from the recent recession.

This cost for crude oil represents about 28% of our current balance-of-trade deficit ($418 billion in 2002), which has significant consequences for our economy. The price we pay at the pump is basically the same whether the oil is produced here or abroad. If the oil is produced at home, however, the money we pay for it circulates within our own economy, providing jobs and producing other goods and services. When we buy oil from a foreign country, we are effectively providing "foreign aid" to that country and thus draining our own economy of an equal amount.

Finally, there are the ecological costs of oil spills such as the *Exxon Valdez* incident (estimated at $15 billion) and of other forms of pollution from the refining and consumption of fuels—although estimating total damage in dollars is sometimes difficult. There are also military costs, as described in the next section.

**Persian Gulf Oil.** The Middle East is a politically unstable, unpredictable region of the world. As noted earlier, it was the unexpected Arab boycott that plunged us into the first oil crisis in 1973. Maintaining access to Middle Eastern oil is also a major reason for our efforts toward negotiating peace agreements between various parties in that region. Recognizing this political instability, the United States maintains a military capability to ensure our access to Persian Gulf oil. This capability was tested in the fall of 1990 when Saddam Hussein of Iraq invaded Kuwait, then producing 6 million barrels per day. The U.S.-led Persian Gulf War of early 1991 threw Hussein's armies out of Kuwait and gave the United States an ongoing presence in the Persian Gulf region. It was this presence, however, that eventually angered the radical Islamic terrorist Al Qaeda organization and led to the September 11, 2001, attack on the World Trade Center and the Pentagon.

***Blood for Oil?*** In 2003, a U.S.–British coalition invaded Iraq in order to overthrow Saddam Hussein's rule and eliminate suspected weapons of mass destruction (Fig. 12–15). As the momentum was building for the invasion, protesters around the world rallied to the cry "No blood for oil," implying that the real motivation behind the pending war was to gain access to Iraq's rich oil reserves. With the war now over, some hindsight suggests that oil was part of the picture. For example, coalition forces were prepared to protect Iraq's oil fields and did so immediately after the invasion. However, there were no plans to protect the Iraqi National Museum, which was looted of priceless historical artifacts. Although it is clear that the United States and Britain will not simply commandeer the Iraqi oil, U.S. and British oil companies are likely to be part of the anticipated effort to stimulate a renewed Iraqi oil industry—and the United States and other oil-hungry Western countries will be ready customers for the oil once it begins to flow again—at acceptable prices. Those countries rely on Persian Gulf oil for half of their oil imports.

***Subsidy.*** Military costs associated with maintaining access to Persian Gulf oil represent a U.S. government subsidy of our oil consumption. Before the 2003 Iraqi war, our military presence in the region was costing the

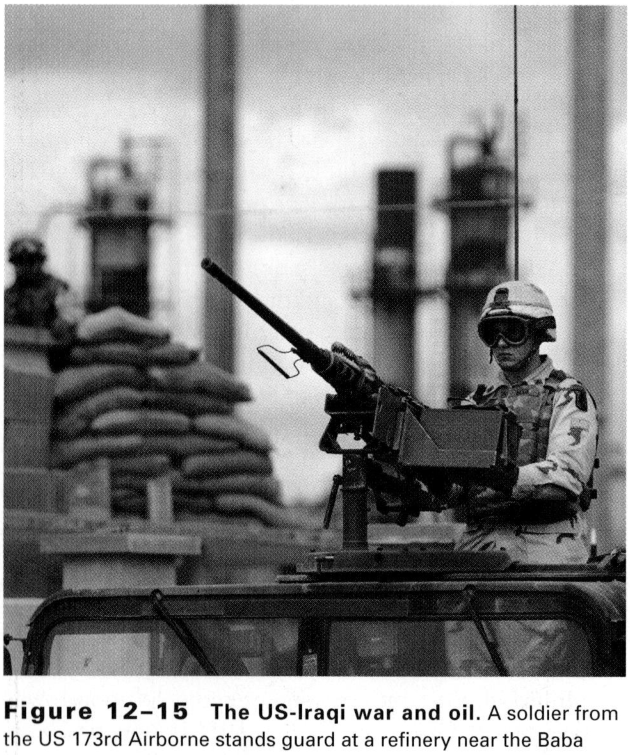

**Figure 12–15 The US-Iraqi war and oil.** A soldier from the US 173rd Airborne stands guard at a refinery near the Baba Gurgur oil field in Kirkuk, Iraq.

United States $50 billion per year. On the basis of importing about 820 million barrels a year from Persian Gulf sources (in 2002), the dollar amount of this "subsidy" comes to $61 per barrel. If this is added to the 2002 market price of $30 per barrel, then we were actually paying $91 per barrel. Now that the 2003 war has come and gone, and the nation is heavily involved in rebuilding Iraq, the real price of Persian Gulf oil has escalated and could easily be well over $100 a barrel!

*Other Sources.* The United States has turned more and more to non–Middle Eastern oil suppliers. Recently, Canada replaced Saudi Arabia as the number-one exporter of oil to the United States, with about 15% of the market. Venezuela, Columbia, and Mexico are also major sources of oil. Western Hemisphere sources now provide about 43% of our imports. In August 2002, the first tanker of Russian oil arrived at a U.S. port. Russia, a number of former Soviet republics, and several African countries are important new potential sources of oil. The relative political stability and proximity of these sources are good reasons for the United States to turn to them and away from the Middle East. However, we will continue to depend on OPEC and the Persian Gulf for a substantial share of our oil imports for the present and even more so in the future.

**Resource Limitations.** U.S. crude-oil production is decreasing because of diminishing domestic reserves. Moreover, on the basis of the "Easter-egg hypothesis," there is dim hope for major new U.S. finds. According to this hypothesis, as your searching turns up fewer and fewer eggs, you draw the logical conclusion that nearly all the eggs have been found. In terms of oil, North America is already the most intensively explored of any continental landmass. The last major find was the Alaskan oil field in 1968. Discoveries since then have been small in comparison, with increasing numbers of dry holes in between. Indeed, most of the "new" oil in the United States is coming from innovative computer mapping of geological structures and horizontal drilling technology that enables oil to be identified and withdrawn from small isolated pockets that were previously missed within old fields. Thus, an increasing percentage of our needs must continue to be met by imported oil.

How much oil is still available for future generations? About 900 billion barrels (BB) of oil has already been used, almost all in the 20th century. Currently, the nations of the world are using about 75 million barrels per day, or some 27 BBs a year. Rising demand by the developing countries and the continued rise in oil demand by the developed countries point to an increase of about 2% per year, so that demand will reach 40 BBs a year by 2020. How long can the existing oil reserves provide that much oil?

*Hubbert's Peak.* As with many issues, it depends on whom you believe. Recently, two oil geologists, Colin Campbell and Jean Laherrère, provided an analysis that puts the world's proven reserves at about 850 BBs. This is less than the 1,050 BBs claimed by oil industry reports (Table 12–2). Campbell and Laherrère derived their estimate from the P50 values of *proven* reserves and discounted some reports of reserves that were inflated for political reasons (for example, those of OPEC countries in the mid-1980s). On the basis of their estimates and the known amount already used, the two geologists calculated a Hubbert curve indicating that peak oil production will occur sometime in the current decade (Fig. 12–16). When that happens, production will begin to taper off in spite of rising demand, and the world will enter a new era of rising oil prices. Princeton professor and former oil geologist Kenneth Deffeyes uses the Hubbert technique in his book *Hubbert's Peak: The Impending World Oil Shortage*[1] to draw the same conclusions. Even if these scientists are wrong and the oil industry estimates are right, it would only delay the peak by a decade at most. To quote Deffeyes, "Global oil production will probably reach a peak sometime during this decade. After the peak, the world's production of crude oil will fall, never to rise again. . . . Ignoring the problem is equivalent to wagering that world oil production will continue to increase forever. My recommendation is for us to bet that the prediction is roughly correct."[2]

*USGS.* In 2000, on the basis of extensive geological studies, the USGS released an assessment of world oil and gas in the form of "*undiscovered* reserves"—that is, reserves from fields that have not yet been found. According to this assessment, there are undiscovered reserves of 732 BBs of oil remaining from unknown fields and an additional 688 BBs in known fields that could be unaccounted for by conventional analyses. Campbell believes that the USGS estimates are overly optimistic, based as they are on oil yet to be found (not *proven* reserves!). Even if much of this oil is found, it will at best move peak oil production back a few years or a decade, given the rising consumption rates. At the current rate of use, the proven reserves can supply only 26 years of oil demand. Thus, the 21st century will become known as an era of declining oil use, just as the 20th century saw continuously rising use.

As oil becomes more scarce, the OPEC nations will once again dominate the world oil market. The Middle East possesses 65% of the world's proven reserves, and eventually the developed world—in particular, the United States—will have to rely more and more on that region for its oil imports (Table 12–2). In light of this fact, and in light of the reality of terrorist activities, it would seem wise to begin to prepare for that day by reducing our dependency on foreign oil now. We can become more independent in three ways: (1) We can increase the fuel efficiency of our transportation system, which uses most of the oil; (2) We can use the other fossil-fuel resources we have to make fuel for vehicles; and (3) we can develop alternatives to fossil fuels, many of

---

[1]Princeton, NJ: Princeton University Press, 2001.
[2]*Ibid.*, p. 1

**table 12-2 Proven World Reserves and Annual Use of Fossil-Fuel Reserves, 2001**

| Region | Reserves: Petroleum (billion barrels)* | Reserves: Natural Gas (trillion ft$^3$)* | Reserves: Coal (billion tons)* |
|---|---|---|---|
| North America | 63.9 | 266.7 | 257.8 |
| South and Central America | 96.0 | 253.0 | 21.6 |
| Europe | 18.7 | 171.7 | 125.4 |
| Former U.S.S.R | 65.4 | 1,982.6 | 230.0 |
| Middle East | 685.6 | 1,974.6 | 1.7 |
| Africa | 76.7 | 394.8 | 55.4 |
| Far East and Oceania | 43.8 | 433.3 | 292.5 |
| Total | 1,050.0 | 5,476.7 | 984.5 |
| **Use** | **Petroleum** | **Natural Gas** | **Coal** |
| World use, 2001* | 27.5 billion barrels | 84.9 trillion ft$^3$ | 5.26 billion tons |
| World use, 2001 (quads Btus) | 157 quads | 87 quads | 93 quads |

* The fuels may be compared by calculating their energy content in British thermal units (Btus). One billion barrels of petroleum yields ca. $5.7 \times 10^{15}$ Btus, or 5.7 quadrillion (quads) Btus. One trillion cubic feet of natural gas yields ca. 1.02 quadrillion Btus, and 1 billion tons of coal yields ca. 17.65 quadrillion Btus.
*Source: BP Statistical Review of World Energy, 2002, International Energy Annual, 2001.*

which are ready now. The best answer probably lies in a suitable mix of all three choices.

## 12.3 Other Fossil Fuels

Unlike crude oil, considerable reserves of natural gas and comparatively vast reserves of coal and oil sands are still available in the United States and elsewhere in the world. Natural gas is gradually coming into use as a fuel for vehicles, and there is some potential in this respect for coal and oil sands.

### Natural Gas

In the United States, natural gas has been rising as a fossil fuel of choice—so much so, that 16% of our annual use now comes from imports, almost all by pipeline from Canada. Although most of this fuel is used for space heating and cooking, natural gas is increasingly being

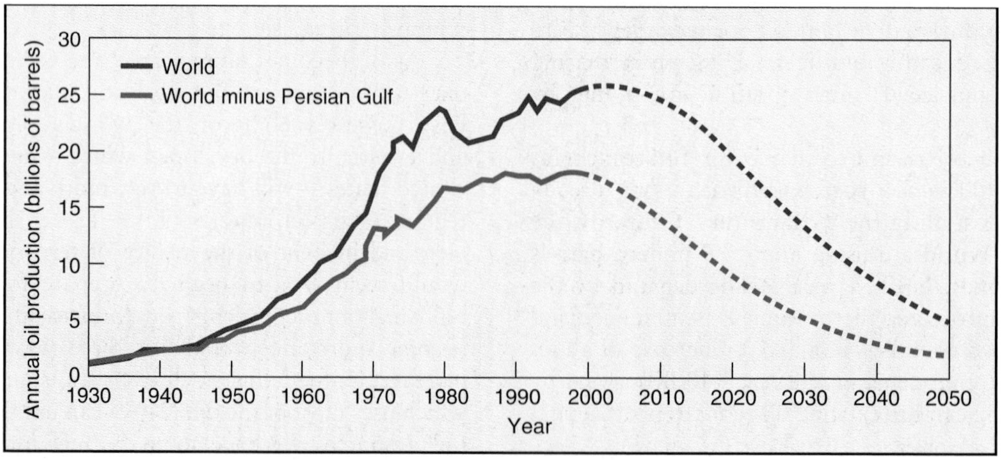


**Figure 12-16 Hubbert curves of oil production.** Oil production in a region follows a bell-shaped curve. The evidence indicates that world oil production will peak sometime in the current decade. (*Source:* Data from Richard A. Kerr, "The Next Oil Crisis Looms Large—and Perhaps Close," *Science* 281 [1998]:1128–1131. Reprinted by permission of American Assn. for the Advancement of Science.)

employed to generate electrical power. As with oil, the cost of natural gas is subject to market fluctuations in supply and demand, and they often rise and fall with the price of oil. (Between January 2002 and January 2003, gas prices tripled.) At the current rate of use, proven reserves in the United States hold only an 11-year supply of gas. This is somewhat deceiving, however, because natural gas is continually emitted from oil and gas-bearing geological deposits. The expected long-term recovery of natural gas in the United States is significantly greater than the proven reserves, putting the supply closer to 50 years.

Worldwide, the estimated natural-gas resource base is even more plentiful (Table 12–2). Almost four times as much equivalent gas (that is, equivalent in energy) is likely to be available as oil. Much of it, however, is relatively inaccessible. Gas must either be pumped through a pipeline (a process limited by the length of the line) or be subjected to high pressures so that it remains a liquid at room temperature. Nevertheless, liquefied natural gas is increasingly being produced and used.

*Gas-run Cars.* Natural gas can also meet some of the fuel needs for transportation. With the installation of a tank for compressed gas in the trunk and some modifications of the engine fuel-intake system—costing about a thousand dollars—a car will run perfectly well on natural gas, although trunk space is a problem. Such cars are in widespread use in Buenos Aires, where service stations are equipped with compressed gas to refill the tank. Compressed-gas cars now make up 15% of the personal vehicles in Argentina. Natural gas is a clean-burning fuel, producing carbon dioxide and water, but virtually no hydrocarbons or sulfur oxides. Indeed, natural gas is beginning to be used in the United States by many buses and a number of private and government car fleets. In a development that will make gas-powered cars much more feasible, the Canadian FuelMaker Corporation has developed a system that will connect to a home's natural gas line and allow homeowners to fill their own tanks.

*Synthetic Oil.* With the aid of the Fischer–Tropsch process, known since the 1920s, natural gas can be converted to a hydrocarbon that is liquid at room temperature and pressure—basically, a synthetic oil. Researchers are hotly pursuing ways to convert natural gas to synthetic oil more efficiently. Presently, the fuel produced is only about 10% more expensive than oil. Several major oil companies are spending billions on gearing up refineries to convert gas to liquid diesel and gasoline fuels, and researchers are seeking an appropriate catalyst that can convert natural gas (methane) to methanol (methyl alcohol). These options promise to make the abundant, but inaccessible, natural-gas deposits suddenly very valuable.

*The Costs.* Developing natural gas has definite trade-offs. For example, to develop the vast Northern Canadian and Alaskan gas deposits, a 3,600-mile gas pipeline will be needed to conduct the gas to the lower 48 states. Current gas prices already justify building the pipeline, which could cost $20 billion. Needless to say, such a pipeline would carry an enormous environmental cost. A substantial amount of gas reserves exists in the lower 48, especially in the Gulf Coast and the Rocky Mountain states. As these are developed, they pollute water and disturb land. Some are in sensitive wildlife areas, such as the Padre Island National Seashore.

Applying natural gas to the transportation sector—both directly as a fuel and indirectly as a synthetic oil—may extend the oil phase of our economy, but it will not be a sustainable solution, because the natural-gas reserves are also limited. What about coal, then, the most abundant fossil fuel?

## Coal

In the United States, 51% of electricity comes from coal-fired power plants. Current reserves are calculated to be 250 billion tons, and at the 2002 rate of production (1.0 billion tons/yr), the supply should last 250 years. Unlike oil and natural gas, we produce more coal each year than we use, and we export about 4% of coal production annually.

Coal can be obtained by surface mining or underground mining. The mining work can be hazardous, and injuries and fatalities are not uncommon. Both methods have substantial environmental impacts. In underground mining, at least 50% of the coal must be left in place to support the roof of the mine. Land subsidence and underground fires often occur in conjunction with underground mines. The town of Centralia, Pennsylvania, for example, has been abandoned and bought by the government because of a coal fire that started 40 years ago and could burn for another 100 years (Fig. 12–17). Coal fires around the world produce almost as much carbon dioxide as do all the cars and trucks in the United States, contributing significantly to the world's greenhouse gas emissions.

*Strip Mining.* In the more common surface, or strip, mining, gigantic power shovels turn aside the rock

**Figure 12–17 Coal fire.** Smoke seeps from an underground coal mine fire in Centralia, Pennsylvania. The town has been abandoned because of the fire.

**Figure 12–18 Strip mining of coal.** It is economical to exploit most coal deposits only by strip mining, whereby gigantic power shovels remove as much as 100 feet (32 m) of rock and soil from the surface to get at the coal seam.

and soil above the coal seam and then remove the coal (Fig. 12–18). This method totally destroys the ecology of the region, as forests are removed and streams are buried by the coal wastes. Although federal regulations require strip-mined areas to be reclaimed—that is, graded and replanted—it takes many decades before an ecosystem resembling the original one can develop. In arid areas of the west, water limits may prevent the ecosystem from ever being reestablished. Consequently, surface-mined areas may be turned into permanent deserts. Furthermore, erosion and acid leaching from the disturbed ground have numerous adverse effects on waterways and groundwater in the surrounding area.

*Coal Power.* Ninety-one percent of the coal the United States produces is used to generate electricity. Once it is mined, coal is transported to large power plants. A typical 1,000-MW plant burns 8,000 tons of coal a day—a mile-long train's worth of coal every day. Combustion of this much coal releases 20,000 tons of carbon dioxide and 800 tons of sulfur dioxide (most of which is removed if the power plant is well provided with scrubbers—see Chapter 21) into the atmosphere. Then there is the waste: 800 tons of fly ash and 800 tons of boiler residue ash daily, requiring a special landfill.

The government and industry-sponsored Clean Coal Technology (CCT) program has the mission of providing a "secure and reliable energy system that is environmentally and economically sustainable." The program has supported a spectrum of technologies applicable to existing power plants and to a new generation of coal-fired power plants. The objectives have been to remove pollutants from the coal before or after burning and to achieve higher efficiencies so that greenhouse gas emissions are reduced. To date, some of the projects have helped industry meet the sulfur dioxide ($SO_2$) emission limits of the Clean Air Act Amendments of 1990. Other technologies developed under the program are experimental, and although they are working as demonstration facilities, full implementation of the CCT portfolio will depend on policy changes that require the technologies to be employed because of stricter environmental considerations. Clean-coal technologies would be an improvement over the current ones, but they still involve substantial greenhouse gas emissions and all of the drawbacks of mining the coal.

*Synfuels.* Coal may be chemically converted to a liquid or gas fuel. Such coal-derived fuels are referred to as synthetic fuels, or **synfuels.** With both government and corporate support, a great deal of research went into upscaling these chemical processes to commercial production in the late 1970s and early 1980s. The projects were all abandoned, however, because they proved too expensive, at least for the present. The production of coal-based synthetic fuels would be profitable only if oil prices rose to \$60 to \$70 per barrel. Mining and burning coal already involves substantial hazards and pollutants, and the thought of doubling or tripling this impact to operate vehicles on coal-derived fuel runs contrary to our progress in environmental protection. In addition, converting coal to liquid fuel creates by-products that pollute the environment and that would be difficult and expensive to control.

Unfortunately, Congress failed to cancel the tax credits enacted to stimulate synfuel production; the law allows the credits to any production enterprise that changes the composition of coal for further use in power plants. More than 50 "synthetic fuel plants" have sprung up to take advantage of the tax credits, which are worth \$1 billion a year. Often, the plants do no more than spray the coal with pine tar or diesel fuel and then send the "product" off to a power plant. The Internal Revenue Service is conducting a review of the program, but lobbying by the industry has put pressure on the IRS to back off. The program is due to expire in 2007, but until it does, a number of companies, such as Marriott International and Rex Stores Corp., stand to harvest a windfall from their ownership of a few "synfuel plants."

## Oil Shales and Oil Sands

The United States has extensive deposits of oil shales in Colorado, Utah, and Wyoming. **Oil shale** is a fine sedimentary rock containing a mixture of solid, waxlike hydrocarbons called *kerogen.* When shale is heated to about 1,100°F (600°C), the kerogen releases hydrocarbon vapors that can be recondensed to form a black, viscous substance similar to crude oil, which can then be refined into gasoline and other petroleum products. However, about a ton of oil shale yields little more than half a barrel of oil. The mining, transportation, and disposal of wastes necessitated by an operation producing, for instance, a million barrels a day (just 5% of U.S. demand) would be a Herculean task, to say nothing of its environmental impact. Fortunately for the environment, though, oil shale, like oil from coal, has proved economically impractical for the present. Also, the production of oil from shale requires large amounts of water, something in short supply in the west.

**Oil sands** are a sedimentary material containing bitumen, an extremely viscous, tarlike hydrocarbon. When oil sands are heated, the bitumen can be "melted out" and refined in the same way as crude oil can. Northern Alberta, in Canada, has the world's largest oil-sand deposits (300 billion barrels), and Canada is commercially exploiting them, because the cost is competitive with current oil prices. Suncor Energy of Canada sells 88 million barrels of this unconventional oil a year and is constructing facilities to expand to 180 million barrels a year by 2012. The United States has smaller, poorer oil-sand deposits, mostly in Utah, that might produce oil for $50–$60 per barrel.

Worldwide, the combination of oil shales and oil sands has been estimated to hold several trillion barrels of oil. Large-scale commercial development of these resources will not occur until oil prices rise dramatically—something that may be underway already. Both sources involve substantial environmental impacts.

## 12.4 Fossil Fuels and Energy Security

According to the Department of Energy, the period from 1973 to 1995 saw a reduction in energy growth by 18%, saving the economy $150 billion a year in total energy expenditures. This was due to the steps taken in response to the oil crisis of the 1970s. Thus, public policy can bring about major improvements in energy use. Unfortunately, we also slipped back into old ways when the crisis faded into history. However, the events of September 11, 2001, and the subsequent war on terrorism have brought energy policy back into sharp focus. Given the absolute dependence of our economy, our homes, and our travel on fossil fuels, the United States is vulnerable to disruptions in oil supplies and also to terrorist attacks on our energy infrastructure. To address these concerns, the Union of Concerned Scientists (UCS) has produced a report titled *Energy Security: Solutions to Protect America's Power Supply and Reduce Oil Dependence*. In this section, the major themes of this report are discussed and then compared with current practices and energy policies.

### Security Threats

***Oil Dependence.*** The United States imported $105 billion worth of oil in 2002, representing 28% of our balance-of-payment deficit and 60% of our oil consumption. Most of this oil fuels the transportation sector of our economy, and there is no substitute for it. The global oil market is basically beyond our control, although we continue to invest vast sums to maintain stability in the Persian Gulf region, where 65% of the world's proven reserves are found. The U.S. (and global) economy demonstrated its sensitivity to three sudden oil price shocks over the past 30 years. Each time, although the supply dropped only 5%, the price rose rapidly and brought on inflation, a drop in the gross national product (GNP), and a recession. Relying increasingly on oil from the OPEC cartel and the volatile Persian Gulf states is simply asking for economic and political trouble in the future.

***Energy Infrastructure.*** Nuclear power plants, hydropower dams, oil and gas pipelines, refineries, tankers, and the electrical grid—all of these present attractive targets for terrorists intent on doing damage to people and the economy in the United States. Our continued involvement in the Middle East has done much to inflame the anger and frustration that can provide sufficient motivation for terrorist acts, as was demonstrated on September 11. Nuclear power plants are particularly vulnerable to a terrorist attack, and the consequences of crashing an airliner into one could be devastating. It would be relatively easy to disrupt energy flow to major areas with an attack on the infrastructure for the distribution of oil, gas, and electricity. The Trans-Alaska pipeline, for example, was shut down for three days after a hunter shot a hole in it (Fig. 12–19).

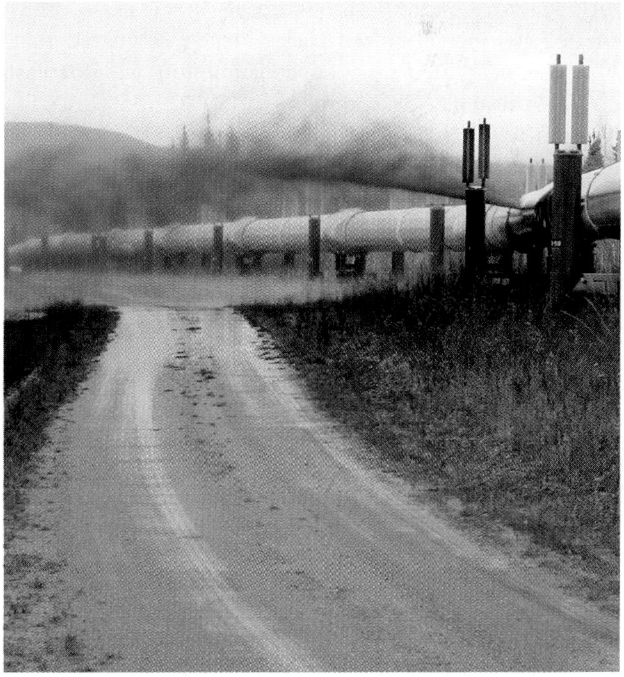

**Figure 12–19 Hunter bags pipeline.** The Trans-Alaska pipeline was shut down for three days when a hunter punctured it with a bullet hole.

*Global Climate Change.* The long-term prospect of using fossil fuels to meet our energy needs poses a huge risk to our economic, environmental, and national security because of global climate change. Recall that all fossil fuels are carbon-based compounds and none can be burned without releasing carbon dioxide ($CO_2$) as a by-product. Of the top three fossil fuels—coal, oil, and natural gas—coal produces the most, and natural gas produces the least, $CO_2$ (Table 12–3). There is overwhelming evidence that the greenhouse gas emissions of the 20th century, especially $CO_2$, have brought on a rise in global temperatures, among other observable impacts. (See Chapter 20.) Most of these emissions are brought on by using fossil fuels. If greenhouse gases continue their steady rise in the atmosphere, climate changes will melt more of the polar ice caps, causing sea level to rise. In addition, there will be major shifts in precipitation patterns, more extreme storms, droughts, heat waves, ecosystem instability, and disruptions in agriculture. Because the United States is the world's leading producer of greenhouse gases and also far outpaces other countries in per capita consumption (Fig. 12–20), what we do will profoundly affect not only our own security, but that of the entire planet.

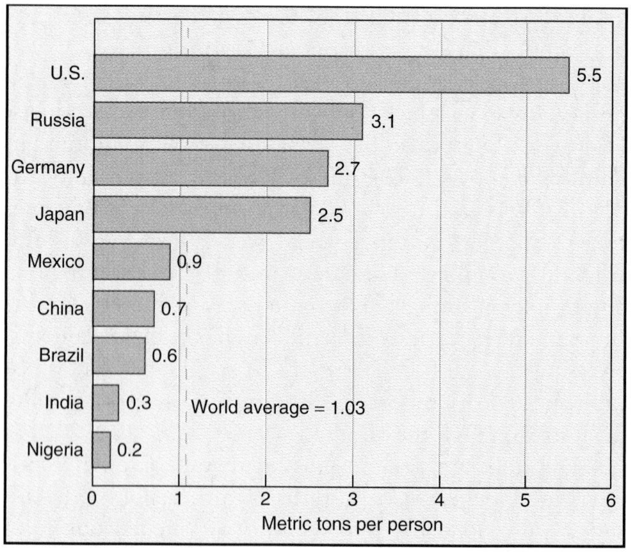


**Figure 12–20 Annual carbon dioxide emissions per capita from burning fossil fuels, selected countries, 2001.** (*Source:* Energy Information Agency, *Annual Energy Review 2002.*)

## Supply-side Policies

Both *supply-side* and *demand-side* options are possible. Because energy supply and demand is such a nationwide matter, only public policy has the sweeping power to effect changes that will make a difference.

Oil and natural-gas consumption are outpacing U.S. production even now. **The National Energy Policy Report** (*Reliable, Affordable, and Environmentally Sound Energy for America's Future*), produced in 2001 by a group chaired by Vice-President Dick Cheney, estimated that, over the next 20 years, oil consumption will increase 30%, natural-gas consumption by 50%, and electricity demand will rise 45%. The Cheney report, the basis for the Bush administration's energy policy, recommended ways to meet these rising demands with fossil fuels, including the following:

- Opening the ANWR and offshore locations to oil and gas exploration and production.
- Adding between 1,300 and 1,900 new coal-fired electric power plants in the next 20 years.
- Providing tax incentives to encourage energy production from fossil and nuclear fuels.
- Streamlining the permitting process for drilling and hydropower licensing and employing eminent domain to establish new electrical transmission lines.
- Expediting the planning and construction of a natural-gas pipeline to bring Alaskan natural gas to the lower 48 states.
- Providing support for the efforts of other countries to develop their oil and gas resources and pipelines.
- Requiring, by executive order, that all federal agencies issuing any regulations which could adversely affect energy supplies or use to document the energy impacts of the regulations and propose alternatives to the action.

**table 12-3 $CO_2$ Emissions per Unit Energy for Fossil Fuels (Natural Gas = 100%)**

| | |
|---|---|
| Natural Gas | 100% |
| Gasoline | 134% |
| Crude Oil | 138% |
| Coal | 178% |

*Source:* Data from Energy Information Administration, U.S. Department of Energy.

Policies dealing with nuclear and renewable energy will be dealt with in Chapters 13 and 14. Briefly, the Cheney report recommended building more nuclear power plants, resolving the nuclear waste repository problem, and increasing alternative energy sources. Supply-side policies, however, especially those dealing with fossil-fuel energy, are "business-as-usual" solutions to our energy security issues. They do little or nothing to reduce our vulnerability to oil price disruptions, terrorist actions, and global climate change. Demand-side policies, by contrast, have the advantage of reducing our energy needs and making it more possible to move to a future in which renewable energy can become the ruling technology. At the same time, these policies will decrease our vulnerability to terrorists and to disruptions of the oil market. They will also save money and reduce pollution from burning fossil fuels.

## Demand-Side Policies

In rethinking our energy strategies, we need to realize that energy has no real meaning or value by itself. Its only worth is in the work it can do. Thus, what we really want is not energy, but comfortable homes, transportation to and from where we need to go, manufactured goods, computers that work, and so forth. As a result, many energy analysts argue that we should stop thinking in terms of where we can get additional supplies of the old fuels to keep things running as they are. Instead, we should be thinking in terms of how we can satisfy these needs with a minimum expenditure of energy and the least environmental impact.

**Conservation.** Imagine discovering a new oil field having a production potential of at least 6 million barrels a day—three times the capacity of the Alaskan field even at its height. Furthermore, assume that this newfound field is inexhaustible and that exploiting it will not adversely affect the environment. The oil field does not exist, but it demonstrates what can be achieved through energy conservation.

*CAFE.* As noted in Section 12.2, this "conservation reserve" has already been tapped to some extent, especially with the doubling of the average fuel efficiency of cars from 13 to 27.5 mpg. Thus, drivers of those autos are satisfying their transportation needs with only half the earlier expenditure of energy and dollars. A number of energy experts believe that most of the "conservation reserve" remains to be tapped. Cars are available that get at least double the current Corporate Average Fuel Economy (CAFE) of 27.5 mpg. Light trucks, SUVs, and minivans meet a lower CAFE of 20.7 mpg. This difference is significant, because these heavier vehicles now dominate sales, with the result that the fuel economy of new vehicles has dropped. To put it bluntly, *raising the CAFE requirements for all vehicles is the key to greater oil independence and energy security.* The UCS report stated that a 40-mpg CAFE by 2012 would save 1.9 million barrels of oil a day, a 20% reduction in demand. The CAFE threshold could go higher, to 55 mpg by 2020, saving 4.8 million barrels a day. The Bush energy policy has raised the CAFE for light trucks, SUVs, and minivans to 22.2 mpg, effective in 2008, but has not touched the CAFE for cars. The Cheney report recommended that the issue receive more study. This delay has been hailed as a victory for car manufacturers and the steel and oil industries and a defeat for those concerned about future energy security. Critics point out that the Cheney group received most of its advice from those industries.

*CHPs.* A technique known as **combined heat and power (CHP)**, or cogeneration (see the "Earth Watch" essay, p. 344), is another way of gaining large amounts of energy from the conservation reserve. According to the CHP methodology, a factory or large building installs a small power plant that produces electricity and heats the building with the "waste" heat. Such systems can achieve an efficiency of 80%! In another energy-saving step, a new technology—the combined-cycle natural-gas unit—has been increasingly employed to generate electricity. Two turbines are employed in this unit, the first burning natural gas in a conventional gas turbine (Fig. 12–7b), the second in a steam turbine (Fig. 12–7a) that runs on the excess heat from the gas turbine. The use of this technology boosts the efficiency of conversion of fuel energy to electrical energy from around 30% to as high as 50%. The cost of combined-cycle systems is half that of conventional coal-burning plants, and the pollution from such systems is much lower. A large proportion of new units being built and planned for the future consists of combined-cycle systems.

*Deregulation.* Other major changes are under way for the entire electric power industry as an increasing number of states adopt deregulation, which requires the utilities to maintain distribution services (transmission lines, telephone poles, etc.), but to divest themselves of their power-generating facilities. In some states, consumers will be able to choose energy suppliers as the utilities lose their monopolies on electrical power supply. In Massachusetts, deregulation has spawned a number of new power plants, all natural-gas fired. In California, however, deregulation led to a severe energy crisis in 2000, mainly because the state allowed wholesale electricity prices to float while keeping a cap on retail prices. The final outcome of deregulation is still uncertain, but energy conservation and renewable energy appear to be much more achievable as a consequence.

*Appliances.* Many conservation measures have been put in place and are reducing energy costs. An example of how important such measures are can be seen in the standards for refrigerators and freezers, which consume energy in every home. New efficiency standards put in place (accompanied by changes in technology) have reduced the energy demand per unit to one-quarter of what it was in 1974. The net impact is remarkable: By 2001, homeowners saved $10 billion, and forty 1,000-watt power plants and 32 million tons of carbon emissions per year were avoided.

*Seeing the Light.* New building codes require improved insulation and double-pane windows for homes and buildings. Some options are underutilized, however, such as substituting fluorescent lights for conventional incandescent lightbulbs. Fluorescent lights are between 20% and 25% efficient (Fig. 12–21), whereas incandescent light bulbs are only 5% efficient, so this change would cut back on our electricity demands for lighting by 75 to 80%. Energy-conserving lightbulbs and appliances are more expensive than conventional models, but calculations show that they more than pay for themselves in energy savings. Interestingly, some utility companies are subsidizing the installation of fluorescent lightbulbs and energy-efficient appliances for consumers as a cost-effective alternative to building additional power plants.

*Internet.* Energy savings are coming from an unexpected source: the computer. Online shopping, working

earth watch

## CHP: Industrial Common Sense

In many situations, a single energy source can be put to use to produce both electrical and heat energy. This process, called *combined heat and power* (CHP), or *cogeneration*, can be employed to conserve energy in two fundamental ways: (1) A power plant whose primary function is to generate electricity also can be used to heat surrounding homes and buildings. As one example, the Consolidated Edison electric utility has been using steam from waste heat in New York City for years. (2) A facility whose primary function is to produce heat or mechanical energy can also generate electricity, which is used by the facility and can also be sold to the local utility. The system can work for all sorts of fuels and heat engines. This is by far the most common application of CHP.

The arrangement makes perfect sense. Every building and factory requires both electricity and heat. Traditionally, these two forms of energy come from separate sources—the electricity from power lines brought in from some central power station and the heat from an on-site heating system, usually fueled by natural gas or oil. Ordinary power plants are, at best, 30% efficient in their use of energy. Combining the two sources in CHP can provide an overall efficiency of fuel use as high as 80%. Besides providing savings on fuel, the cogenerating systems have at least 50% lower capital costs per kilowatt generated than conventional utilities have. This cost savings makes CHP quite competitive with utility-produced power.

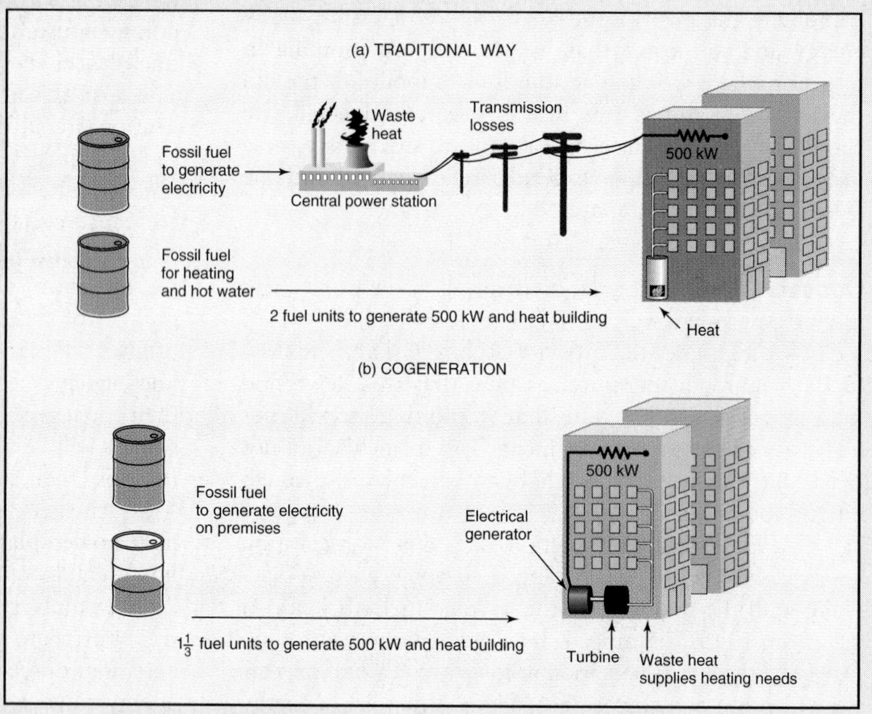


**Combined Heat and Power facility.** (a) The traditional method of providing the energy needed for buildings. (b) In cogeneration, the heat lost from an on-site power-generating system provides heating needs.

### PURPA

The Santa Barbara (California) County Hospital recently installed a cogeneration system based on natural gas as the fuel source. Natural gas drives a turbine with the capacity to generate 7 MW of electricity, much of which is used by the hospital. Boilers use the waste heat from the turbine to produce high-pressure steam, which provides all of the hospital's heating and cooling needs. Any steam not required by the hospital is directed back to the turbine to increase its power. Electrical power not used by the hospital is sold to the local utility. The system has the added advantage of protecting the hospital against wide-scale blackouts or brownouts that may occur in the region's centralized utility system.

The enactment of the **Public Utility Regulatory Policies Act (PURPA)** in 1978 was essential for stimulating the use of CHP. Under this act, a facility with cogenerative capacity (1) is entitled to use fuels that are not usually available to electricity-generating plants (for example, natural gas), (2) must produce above-standard efficiency of fuel use, and (3) is entitled to sell cogenerated electricity to a utility at a fair price. After many challenges by the utilities (which considered the procedures an interference with their business), the Supreme Court decided in favor of PURPA.

Cogeneration is becoming increasingly popular in U.S. industries as a major strategy for cutting energy costs. Given the likelihood of major price increases for petroleum and of reductions in $CO_2$ emissions, pressures favoring energy conservation will increase, and cogeneration will continue to grow. Its impact can be seen in the electric power sector of the economy: CHP electricity contributed 9.2% of the total electricity generated in the United States in 2002 and has been increasing annually.

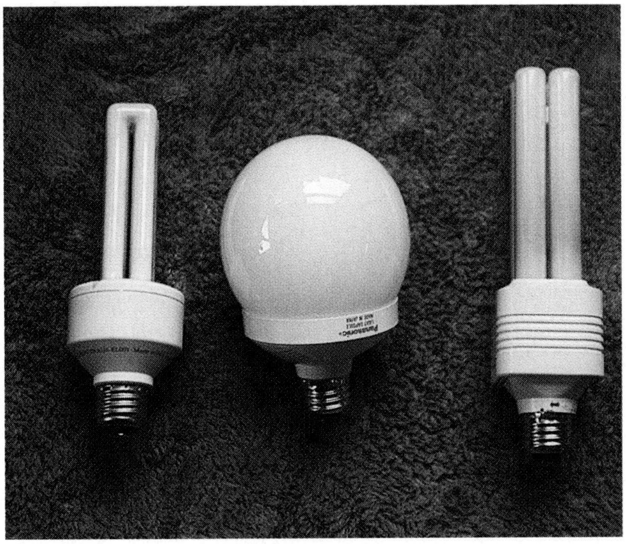

**Figure 12–21 Energy-efficient lightbulbs.** Replacing standard incandescent lightbulbs with screw-in fluorescent bulbs shown here can cut the energy demand for lighting by 75–80%. The higher cost of the fluorescent bulbs is more than offset by the greater efficiency and much longer lifetime of the bulb.

from home computers (telecommuting), business-to-business Web sites, and e-mailing are reducing energy costs substantially. With fewer catalogs and much less retail and warehouse space needed, there will be savings on paper costs, fuel for driving, and construction and heating costs for businesses, to name a few. A study by the Center for Energy and Climate Solutions (*The Internet Economy and Global Warming: A Scenario of the Impact of E-Commerce on Energy and the Environment*) suggests that the Internet will eventually bring about a 4.5 billion-square-foot reduction in commercial spaces, an energy savings worth 21 power plants per year and, through paper savings alone, a reduction of 10 million tons of $CO_2$ per year.

### Development of Non-Fossil-Fuel Energy Sources

Conserving energy is extremely important, but keep in mind that *reducing* our use of fossil fuels is not *eliminating* it. There are two major pathways for developing non-fossil-fuel energy alternatives: pursuing nuclear power and promoting renewable-energy applications. Nuclear power has been in use for 45 years, and various renewable-energy systems have also been developed. Thus, both nuclear and renewable alternatives are at a stage where they could be expanded to provide further energy needs if suitable technological solutions and public acceptance (in the case of nuclear power) or pressure and government support (of renewable energy) were provided. Chapter 13 focuses on the nuclear option, and Chapter 14 focuses on renewable options. Public policy needs are summarized at the end of Chapter 14.

## revisiting the themes

### Sustainability

The 19th, 20th, and 21st centuries have benefited immensely from the fossil fuels buried in the earth over past eons. These are, however, nonrenewable resources. They will not be available for very long, as human history goes. We are currently traveling on an unsustainable track, powered by fuels that are running out. Sustainable options are available, but we are not putting our best efforts there. Instead, we are doing our best to maintain the status quo. The most serious consequence of our current track is that we continue to pump greenhouse gases, especially $CO_2$, into the atmosphere, where they are already affecting the global climate and promise to do so at an accelerating rate as we burn ever more fossil fuels. Eventually, human civilization will have to make a transition to a future powered by renewable energy. The sooner this becomes a reality, the better off the entire planet will be.

### Stewardship

What concern, if any, should we have for future generations? During the 20th century, we used half of the available oil, and our use is accelerating. What will we leave for our grandchildren and great-grandchildren? Making the transition to a renewable energy future will be difficult, and it cannot happen overnight. A thoughtful stewardship of available resources should motivate us to refrain from pumping all the oil and natural gas to satisfy present demands and to plan for the future generations that will be facing far greater energy difficulties than we are as they struggle to complete the transition to renewable energy.

### Sound Science

Science and technology have flourished in the age of fossil fuels, not least because they have devoted much research towards developing new and often ingenious ways of extracting the energy from those fuels. The need today, however, is for scientists and engineers to point the way towards using fossil fuels with greater efficiency. A good example is the CHP technology that is being increasingly employed by factories and institutions. Another key research effort is being aimed at finding a catalyst that will convert natural gas to a liquid so that remote gas reserves can be accessed and substituted for the more polluting coal and oil resources. The greatest need is for renewable energy technologies, a subject addressed in Chapter 14.

## Ecosystem Capital

In our efforts to drill for more oil and gas and scrape away more mountains to get at coal seams, we are at the same time destroying some ecosystems and threatening the wilderness values of such jewels as the ANWR. Ecosystems everywhere are going to be affected by the rising greenhouse gases that have already had an impact on our climate. These are the trade-offs that result from our addiction to fossil fuels. The sooner we can cure our civilization of this addiction, the better off our ecosystem capital will be, and the better off all humankind will be.

## Policy and Politics

Energy policy and political battles were constantly in view in this chapter. The Bush administration has formulated an energy policy, and that's a positive development, because we have had a poorly articulated energy policy in the past. The opening case study, however, reveals the first disturbing thrust of the new policy. It is to develop the oil in a wilderness refuge, the ANWR, and the political justification for this move has tapped into America's patriotic well: It is a matter of national security.

The geopolitics of the Middle East shocked the United States in the 1970s, and we responded with a succession of policy moves that, in time, greatly increased our energy efficiency and reduced our oil consumption (and built the Alaska pipeline!). For a while, the corporate fossil-fuel interests and their allies in the automobile industry were in retreat. When oil prices plunged, however, policies to encourage renewable energy and conservation were either halted in place or terminated. The Congress has refused to move on increasing the efficiency of motor vehicles since 1985, even for a time prohibiting the Department of Transportation from studying such a move.

**Security.** Once again, Middle Eastern geopolitics brought the United States to action, when Iraq invaded Kuwait and we led a coalition of forces to rescue that oil-rich country in 1991. When we stayed on, we provided the stimulus for Osama Bin Laden and his terrorist organization to carry out the terrible attacks on America on September 11, 2001. The subsequent war on terrorism brought the U.S. to Afghanistan and then, in 2003, to Iraq to oust Saddam Hussein. The campaign was costly to the Iraqis, who took thousands of civilian casualties, not to mention the military slaughter, and costly to the United States as well, as a drain on an already faltering economy. Clearly, part of U.S. energy policy is to maintain a military capability in the Persian Gulf to ensure the continued flow of inexpensive oil.

Other elements of the Bush energy policy demonstrate that the administration is putting its greatest emphasis on the supply side. Citing the anticipated increases in energy consumption for the next 20 years, Vice President Cheney's National Energy Policy report recommended a number of steps to meet the increases in the use of fossil fuels. If energy security is the objective, however, the best way to achieve such security is to reduce future energy use, weaning us from our great dependence on imported oil and paving the way to a sustainable energy system. There are many steps that could be taken to conserve energy and move away from fossil fuels. Some of these were mentioned in the Cheney report, to be studied further (a good way to put off decisions), but no substantive new policies were articulated. The most effective move would be to increase the CAFE standards for automobiles and for the pickup trucks, minivans, and SUVs that Americans are now buying in ever greater numbers.

## Globalization

Fossil-fuel energy has linked the world together by making rapid global transportation so readily available and by facilitating communication and the instant flow of information. The global trade in oil has become one of the most important factors in the world economy: If you have oil, you can sell it, because there are many countries that need it. In particular, the United States now depends on foreign oil for 60% of its use. If there is a major disruption in this international market, the world economy will be shaken once again.

# review questions

1. How have the fuels to power homes, industry, and transportation changed from the beginning of the Industrial Revolution to the present?
2. What are the three primary fossil fuels, and what percentage does each contribute to the U.S. energy supply?
3. Electricity is a secondary energy source. How is it generated, and how efficient is its generation?
4. Name the four major categories of primary energy use in an industrial society, and match the energy sources that correspond to each.

5. What is the distinction between estimated reserves and proven reserves of crude oil, and what factors cause the amounts of each to change?
6. How does the price of oil influence the amount produced?
7. What were the trends in oil consumption, in the discovery of new reserves, and in U.S. production before 1970? After 1970? What is a Hubbert curve?
8. What did the United States do in the early 1970s to resolve the disparity between oil production and consumption? What events caused the sudden oil shortages of the mid-1970s and then the return to abundant, but more expensive, supplies?
9. How did the United States and other industrialized nations respond to the oil crisis of the 1970s, and what were the impacts of those responses on production and consumption?
10. Why did an oil glut occur in the mid-1980s, and what were its consequences in terms of oil prices and the continuing pursuit of the responses listed in Question 9?
11. What have been the directions of U.S. production, consumption, and importation of crude oil since the mid-1980s?
12. Describe the situation with respect to natural gas and coal for the United States in terms of domestic supplies, the consequences of developing and using the fuels, and the ability of those sources to supplement oil.
13. What are the security risks and consequences of our growing dependence on foreign oil? How do they relate to the Persian Gulf War of 1991? to terrorism? to the 2003 U.S.–Iraqi war?
14. What phenomenon may restrict the consumption of fossil fuels before resources are depleted?
15. Describe the Bush administration's energy policy as it develops supply- and demand-side strategies.
16. To what degree has energy conservation served to mitigate energy dependency? What are some prime examples of energy conservation? To what degree may conservation alleviate future energy shortages and the cost of developing alternative sources of energy?

## thinking environmentally

1. Statistics show that less developed countries use far less energy per capita than developed countries. Explain why this is so.
2. Between 1980 and 1999, gasoline prices in the United States declined considerably relative to inflation. Predict what they will do in the next 10 years, and rationalize your prediction.
3. Suppose your region is facing power shortages (brownouts). How would you propose solving the problem? Defend your proposed solution on both economic and environmental grounds.
4. List all the environmental impacts that occur during the "life span" of gasoline—that is, from exploratory drilling to the consumption of gasoline in your car.
5. Explore the positions of the following regarding an increase in CAFE requirements: the fossil-fuel industry, the automobile-manufacturing companies, the steel industry, Congress, the UCS, and the National Academy of Sciences. How do they differ, and why?

chapter 13

# Energy from Nuclear Power

**Key Topics**

1. Nuclear Energy in Perspective
2. How Nuclear Power Works
3. The Hazards and Costs of Nuclear Power Facilities
4. More Advanced Reactors
5. The Future of Nuclear Power

Tokaimura, a village 120 km northeast of Tokyo, is home to several nuclear installations, including a nuclear power plant, the Japan Atomic Energy Research Institute, and a fuel-reprocessing plant (Fig. 13–1). On September 30, 1999, three workers in the reprocessing plant were preparing enriched uranium for a research reactor. They mixed nitric acid with 35 lb of highly enriched uranium, adding the chemicals to a tank in buckets instead of using the plant's mechanized equipment. The amount added was well above the approved levels, and the mixture reached critical mass and began a nuclear-fission chain reaction. The reaction was not intense enough to produce an explosion, but it emitted high levels of gamma rays and neutrons, triggering alarms that sent the three workers running out of the building. It took an emergency response team almost 24 hours to shut down the reaction. By that time, enough gamma radiation had escaped to expose 439 townspeople and workers to varying levels of radiation. The three workers received very high doses, and two of the workers subsequently died of radiation poisoning. The third worker received a lower dose and survived the accident. Fortunately, only trace amounts of radioactive gases were emitted, and the impact of the accident was limited to a few hundred meters of the plant. However, the town was partially evacuated as a safety precaution until the extent of the accident became clear.

An investigation revealed that workers frequently violated regulations, and to save time, the company (JCO) had illegally revised a government-approved manual to allow workers to mix uranium fuels in buckets. Subsequently, six former JCO officials were convicted of "professional negligence" and the company was fined 1 million yen. Nevertheless, even with its violent introduction into the nuclear age via the bombing of Hiroshima and Nagasaki in 1945, Japan is determined to increase its nuclear power capacity. Nuclear power now supplies 35% of electrical power in Japan, and the goal is to reach 43% by 2010. The accident at Tokaimura, however, created a groundswell of public protest in Japan against any expansion of nuclear power and embarrassed the Japanese government. Revelations of cracks in many of its reactors and

***Trojan Nuclear Power Plant in Rainier, Oregon.***

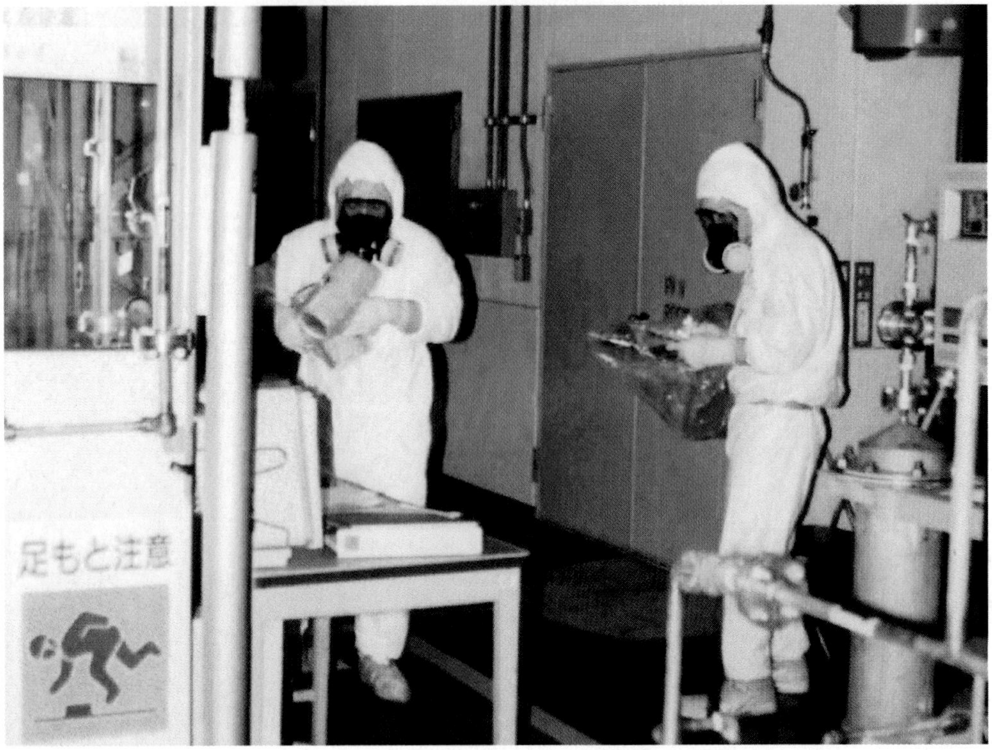


**Figure 13–1 Nuclear accident at Tokaimura, Japan.** Weeks after the accident, workers wearing protective gear test for remaining radiation in a technical room at the uranium-processing plant.

radiation leakage in 2002 have caused new waves of alarm and shaken the country's nuclear industry.

The controversy brought on by Tokaimura is only one of many in the troubled history of nuclear power. In this chapter, you delve into that history through the lenses of science and public policy as the many concerns about nuclear power are raised and discussed.

## 13.1 Nuclear Energy in Perspective

Power shortages in California, rising natural-gas prices, continued emissions of sulfur dioxide and nitrogen oxides from coal, and the long-term threat of climate change all suggest that an energy future based on fossil fuels will be difficult at best. In light of these serious energy-related problems, it seems prudent to do everything possible to develop non-fossil-fuel sources of energy. Nuclear power is an alternative that does not contribute to global warming, and there is sufficient uranium to fuel nuclear reactors well into the 21st century, with the possibility of extending the nuclear fuel supply indefinitely through reprocessing technologies. Thirty-one nations now have nuclear power plants either in place or under construction, and in some of these countries nuclear power generates more electricity than any other source. Is nuclear energy the key to the future of global energy—the best route to a sustainable relationship with our environment?

*The Nuclear Age.* For years, geologists have known that fossil fuels would not last forever. Sooner or later, other energy sources would be needed. It was anticipated that nuclear power could produce electricity in such large amounts and so cheaply that we would phase into an economy in which electricity would take over virtually all functions, including the generation of other fuels, at nominal costs. Following World War II, determined to show the world that the power of the atom could benefit humankind, the U.S. government embarked on a course to lead the world into the Nuclear Age.

Thus did the government move into the research, development, and promotion of commercial nuclear power plants (along with the continuing development of nuclear weaponry). Utilizing this research, companies such as General Electric and Westinghouse constructed nuclear power plants that were ordered and paid for by utility companies, with assurances from the federal government, via the Price–Anderson Act of 1957, that these corporations and utilities would be exempt from any legal liabilities incurred. The Nuclear Regulatory Commission (NRC), formerly the Atomic Energy Commission, an agency in the Department of Energy (DOE), set and enforced safety standards for the operation and maintenance of the new plants, as it does today.

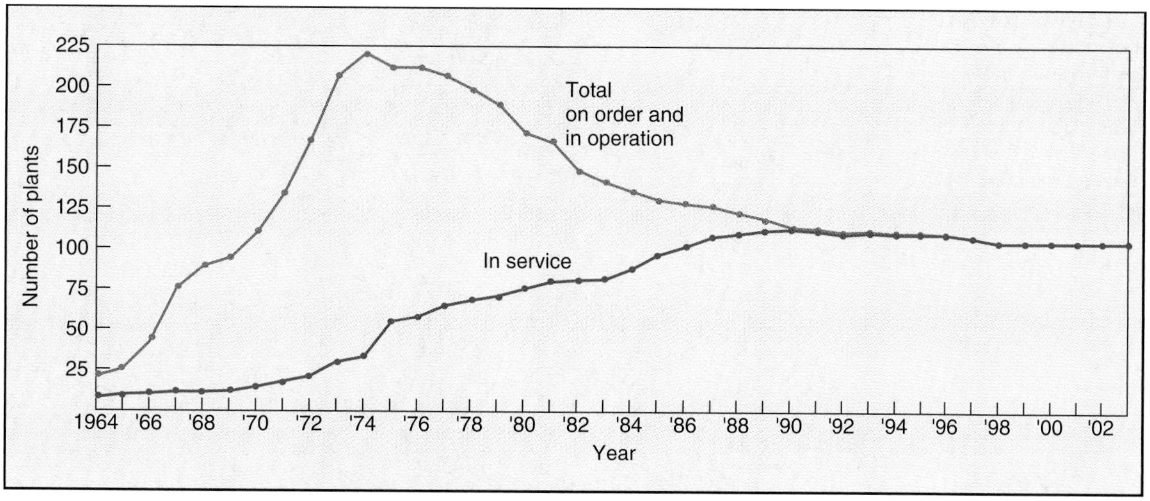


**Figure 13–2 Nuclear power in the United States.** Since the early 1970s, when orders for plants reached a peak, few utilities called for new plants and many canceled earlier orders. Nevertheless, the number of plants in service increased steadily as plants under construction were completed. The number of operating plants peaked at 112 and is now declining. (*Source:* Data from U.S. Department of Energy.)

In the 1960s and early 1970s, utility companies moved ahead with plans for numerous nuclear power plants (Fig. 13–2). By 1975, 53 plants were operating in the United States, producing about 9% of the nation's electricity, and another 170 plants were in various stages of planning or construction. Officials estimated that by 1990 several hundred plants would be on-line and by the turn of the century as many as a thousand would be operating. A number of other industrialized countries got in step with their own programs, and some less developed nations were going nuclear by purchasing plants from industrialized nations.

*Curtailed.* Since 1975, however, the picture has changed dramatically. Utilities stopped ordering nuclear plants, and numerous existing orders were canceled. The last order for a plant in the United States that was not subsequently canceled was placed in 1978. In some cases, construction was terminated even after billions of dollars had already been invested. Most strikingly, the Shoreham Nuclear Power Plant on Long Island, New York, after being completed and licensed at a cost of $5.5 billion, was turned over to the state in the summer of 1989, to be dismantled after generating electricity for only 32 hours (Fig. 13–3). The reason was that citizens and the state deemed that there was no way to evacuate people from the area should an accident occur. Similarly, just a few weeks earlier, California citizens voted to shut down the Rancho Seco Nuclear Power Plant, located near Sacramento, after a 15-year history of troubled operation. In total, 28 units (separate reactors) have been shut down permanently, for a variety of reasons.

At the end of 2003, 104 nuclear power plants were operating in the United States. When the Watts Bar nuclear plant in Tennessee came on line in February 1996, no more plants were left under construction. The 103 existing plants generate 21% of U.S. electrical power. Thus, U.S. nuclear power, which leads the world, peaked at 112 plants and is destined to head downward as more and more plants are decommissioned. The US Energy Information Agency predicts that 27% of current U.S. nuclear generating capacity will be out of commission by 2020.

*Global Picture.* Elsewhere, plants continue to be built. Including those in the United States, the world has a total of 441 operating nuclear plants, with an additional 32 under construction. Global nuclear-energy-generating electrical power has increased at a rate of 2.5% per year for the last decade. Nuclear power now generates about 17% of the world's electricity (23% in those countries with nuclear units), but dependence on nuclear energy varies greatly among the 31 countries operating nuclear

**Figure 13–3 The Shoreham Nuclear Plant on Long Island, New York.** With little electricity produced, this plant was closed because of concerns about whether surrounding areas could be evacuated in case of an accident. The plant is now being dismantled.

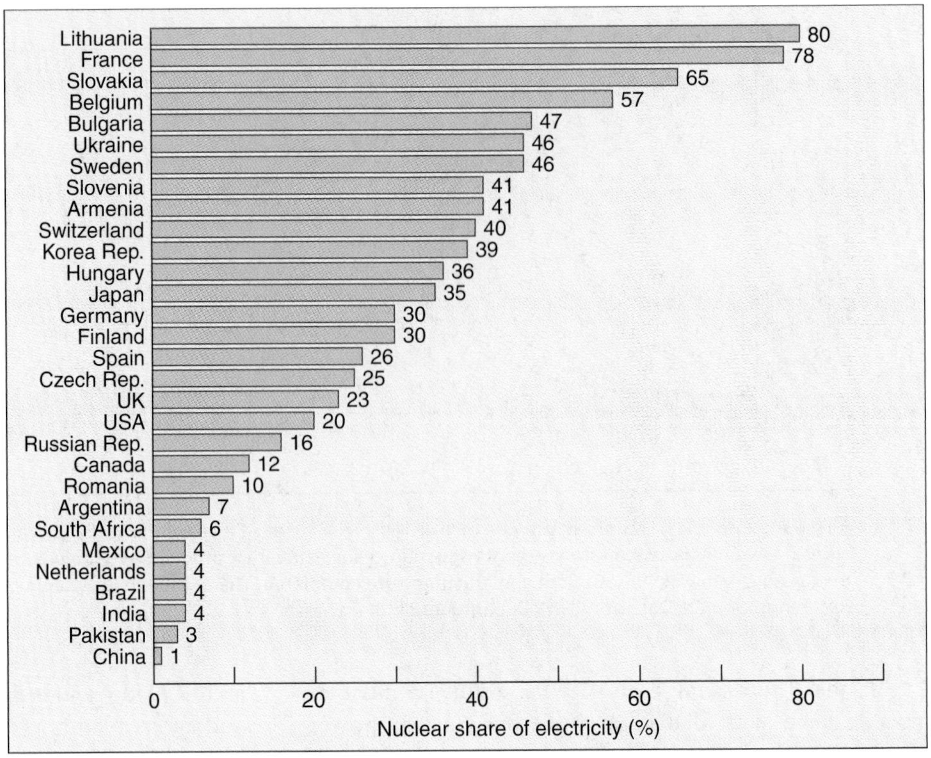


**Figure 13–4 Nuclear share of electrical power generation.** In 2002, those countries lacking fossil-fuel reserves tended to be the most eager to use nuclear power. (*Source:* Data from International Atomic Energy Agency.)

power plants (Fig. 13–4). Of the major industrial nations, only France and Japan remain fully committed to pushing forward with nuclear programs. Indeed, France now produces 78% of its electricity with nuclear power and has plans to go to more than 80%. China and India, the world's population giants, also are investing heavily into nuclear energy, with new units coming on line and many under construction. In many European countries, however, nuclear energy is facing major political opposition, and in some (Germany, Sweden), it is being phased out.

After the catastrophic accident at Chernobyl in April 1986, nuclear power is being rethought in many countries. Yet, the demand for electricity is more robust than ever, and the other means of generating electricity have their own problems. Coal-powered electricity produces more greenhouse gases and other pollutants than does any other form of energy. Oil supplies are more limited, and oil is vital to transportation and home heating. Hydroelectric power is already heavily developed. Solar-power technology is still lagging. If the technological and economic issues can be resolved, can nuclear power supply future energy needs? In order to react intelligently to this issue, you need a clear understanding of what nuclear power is and what are its pros and cons.

## 13.2 How Nuclear Power Works

The objective of nuclear power technology is to control nuclear reactions so that energy is released gradually as heat. As with plants powered by fossil fuels, the heat energy produced by a nuclear plant is used to boil water and produce steam, which then drives conventional turbogenerators. Nuclear power plants are base-load plants (see Chapter 12); they are always operating unless they are being refueled, and they are large plants, generating up to 1,400 megawatts.

### From Mass to Energy

The release of nuclear energy is completely different from the burning of fuels or any other chemical reactions discussed so far. To begin with, materials involved in chemical reactions remain unchanged at the atomic level, although the visible forms of these materials undergo great transformation as atoms are rearranged to form different compounds. Nuclear energy, however, involves changes at the atomic level through one of two basic processes: fission and fusion. In **fission,** a large atom of one element is split to produce two smaller atoms of different elements (Fig. 13–5a). In **fusion,** two small atoms combine to form a larger atom of a different element (Fig. 13–5b). In both fission and fusion, the mass of the product(s) is less than the mass of the starting material, and the lost mass is converted to energy in accordance with the law of mass–energy equivalence ($E = mc^2$), first described by Albert Einstein. The amount of energy released by this mass-to-energy conversion is tremendous: The sudden fission or fusion of a mere 1 kg of material, for example, releases the devastatingly explosive energy of a nuclear bomb.

**The Fuel for Nuclear Power Plants.** All current nuclear power plants employ the fission (splitting) of uranium-235. The element uranium, which occurs naturally in various

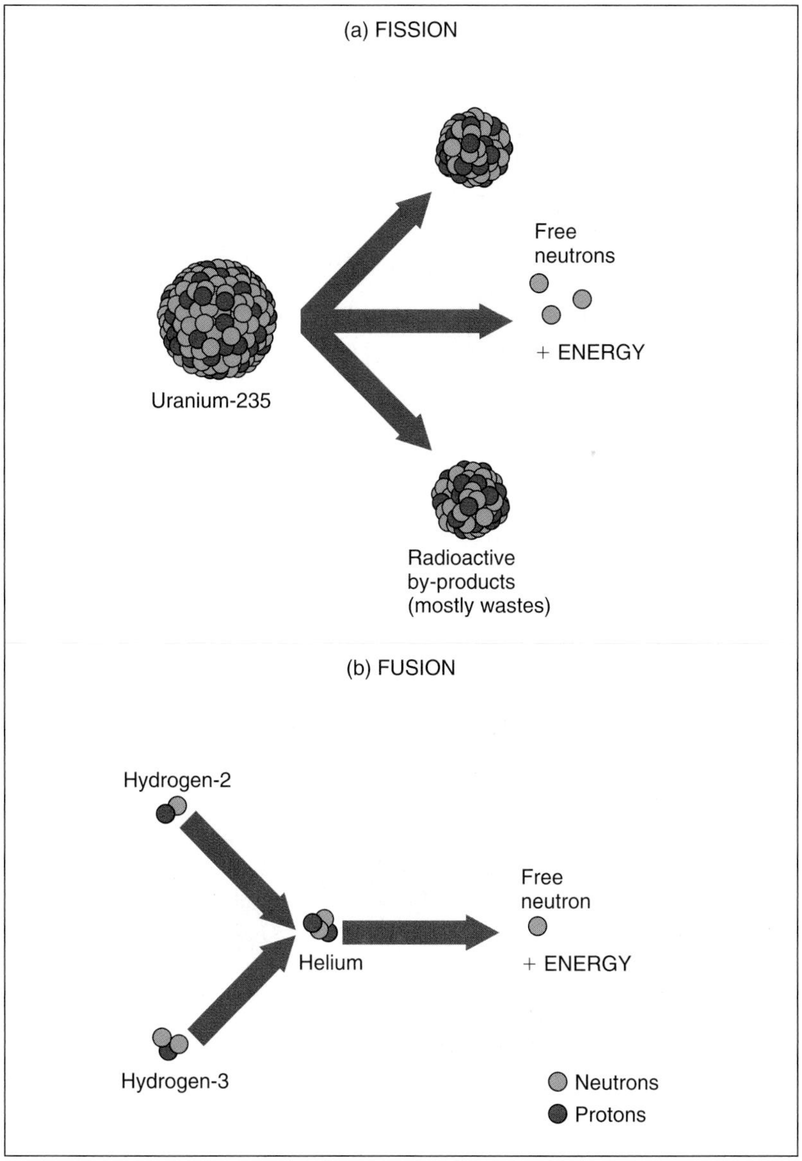


**Figure 13–5 Nuclear fission and fusion.** Nuclear energy is released from either (a) fission, the splitting of certain large atoms into smaller atoms, or (b) fusion, the fusing together of small atoms to form a larger atom. In both cases, some of the mass of the starting atom(s) is converted to energy.

minerals in Earth's crust, exists in two primary forms, or isotopes: uranium-238 ($^{238}U$) and uranium-235 ($^{235}U$). **Isotopes** of a given element contain different numbers of neutrons, but the same number of protons and electrons.

The number 238 or 235 that accompanies the chemical name or symbol for uranium is called the **mass number** of the element. It is the sum of the number of neutrons and the number of protons in the nucleus of the atom. $^{238}U$ contains 92 protons and 146 neutrons; $^{235}U$ contains 92 protons and 143 neutrons. All atoms of any given element must contain the same number of protons, so variations in mass number represent variations in numbers of neutrons. Therefore, whereas $^{238}U$ contains three more neutrons than $^{235}U$, both isotopes contain the 92 protons that *define* the element uranium. Although all isotopes of a given element behave the same chemically, their other characteristics may differ profoundly. In the case of uranium, $^{235}U$ atoms will readily undergo fission, but $^{238}U$ atoms will not. Like uranium, many other elements exist in more than one isotopic form.

It takes a neutron hitting the nucleus at just the right speed to cause $^{235}U$ to undergo fission. Because $^{235}U$ is an unstable isotope, a small, but predictable, number of the $^{235}U$ atoms in any sample of the element undergo radioactive decay and release neutrons, among other things. If one released neutron moving at just the right speed hits another $^{235}U$ atom, the latter becomes $^{236}U$, which is highly unstable and undergoes fission immediately into lighter atoms (**fission products**). The fission reaction gives off several more neutrons and releases a great deal of energy (Fig. 13–5a). Ordinarily, these neutrons are traveling too fast to cause fission, but if they are slowed down and then strike another $^{235}U$ atom, they can cause fission to occur again. In this way, more neutrons and more energy are released, with the potential to repeat the process. A domino effect, known as a *chain reaction,* may thus occur (Fig. 13–6a).

***Nuclear Fuel.*** To make nuclear "fuel," uranium ore is mined, purified into uranium dioxide ($UO_2$), and enriched. Because 99.3% of all uranium found in nature

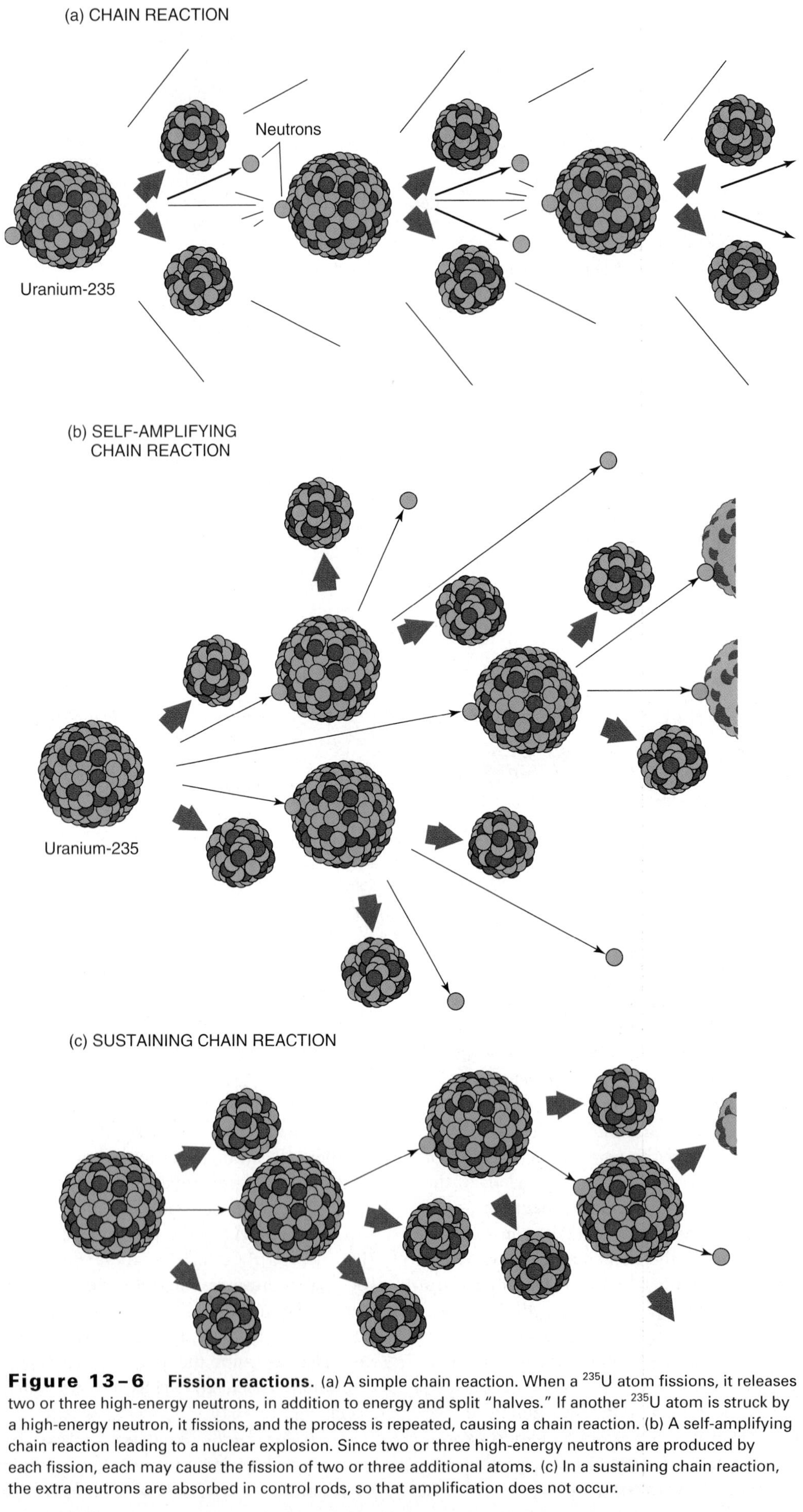


**Figure 13–6 Fission reactions.** (a) A simple chain reaction. When a $^{235}U$ atom fissions, it releases two or three high-energy neutrons, in addition to energy and split "halves." If another $^{235}U$ atom is struck by a high-energy neutron, it fissions, and the process is repeated, causing a chain reaction. (b) A self-amplifying chain reaction leading to a nuclear explosion. Since two or three high-energy neutrons are produced by each fission, each may cause the fission of two or three additional atoms. (c) In a sustaining chain reaction, the extra neutrons are absorbed in control rods, so that amplification does not occur.

is $^{238}U$ (thus, only 0.7% is $^{235}U$) **enrichment** involves separating $^{235}U$ from $^{238}U$ to produce a material containing a higher concentration of $^{235}U$. Because $^{238}U$ and $^{235}U$ are chemically identical, enrichment is based on their slight difference in mass. The technical difficulty of enrichment is the major hurdle that prevents less developed countries from advancing their own nuclear capabilities.

*Nuclear Bomb.* When $^{235}U$ is *highly* enriched, the spontaneous fission of an atom can trigger a chain reaction. In nuclear weapons, small masses of virtually pure $^{235}U$ or other fissionable material are forced together so that the two or three neutrons from a spontaneous fission cause two or three more atoms to undergo fission; each of these in turn triggers two or three more fissions, and so on. The whole mass undergoes fission in a fraction of a second, releasing all the energy in one huge explosion (Fig. 13–6b).

**The Nuclear Reactor.** A nuclear reactor for a power plant is designed to sustain a continuous chain reaction (Fig. 13–6c), but not allow it to amplify into a nuclear explosion. Control is achieved by enriching the uranium to only 4% $^{235}U$ and 96% $^{238}U$. This modest enrichment will not support the amplification of a chain reaction into a nuclear explosion. However, in the process of fission, some of the faster neutrons are absorbed by $^{238}U$ atoms, converting them into $^{239}Pu$ (plutonium), which then also undergoes fission when hit by another neutron. So much plutonium is produced in this way that at least one-third of the energy of a nuclear reactor comes from plutonium fission.

*Moderator.* A chain reaction can be achieved in a nuclear reactor only if a sufficient mass of enriched uranium is arranged in a suitable geometric pattern and is surrounded by a material called a *moderator.* The **moderator** slows down the neutrons that produce fission, so that they are traveling at the right speed to trigger another fission. In slowing down the neutrons, the moderator gains heat. In nuclear plants in the United States, the moderator is near-pure water, and the reactors are called light-water reactors (LWRs). [The term "light water" denotes ordinary water ($H_2O$), as opposed to the substance known as heavy water, or deuterium oxide ($D_2O$).] Other moderators employed in reactors of different design are graphite and deuterium oxide.

*Fuel Rods.* To achieve the geometric pattern necessary for fission, the enriched uranium dioxide is made into pellets that are loaded into long metal tubes. The loaded tubes are called **fuel elements** or **fuel rods** (Fig. 13–7a). Many fuel rods are placed close together to form a reactor core inside a strong reactor vessel that holds the water, which serves as both moderator and heat-exchange fluid (coolant) (Fig. 13–7b). Over time, daughter products that also absorb neutrons accumulate in the fuel rods and slow down the rate of fission and heat production. The highly radioactive spent-fuel elements are then removed and replaced with new ones.

*Control Rods.* The chain reaction in the reactor core is controlled by rods of neutron-absorbing material, referred to as **control rods,** inserted between the fuel elements. The chain reaction is started and controlled by withdrawing and inserting the control rods as necessary. Hence, a nuclear reactor is simply an assembly of fuel elements, moderator–coolant, and movable control rods (Fig. 13–7). As the control rods are removed and a chain reaction is initiated, the fuel rods and the moderator become intensely hot.

**The Nuclear Power Plant.** In a nuclear power plant, heat from the reactor is used to boil water to provide steam for driving conventional turbogenerators. One way to boil the water is to circulate it through the reactor. In most power plants in the United States, however, a double loop is employed. The moderator–coolant water is heated to over 600°F by circulating it through the reactor, but it does not boil, because the system is under very high pressure (155 atmospheres). The superheated water is circulated through a heat exchanger, boiling other, unpressurized water that flows past the heat exchanger tubes. This action produces the steam used to drive the turbogenerator.

*LOCA.* The double-loop design of the primary and secondary systems isolates hazardous materials in the reactor from the rest of the power plant. However, it has one serious drawback: If the reactor vessel should break, the sudden loss of water from around the reactor, called a "loss-of-coolant accident" (LOCA), could result in the core's overheating. The sudden loss of the moderator–coolant water would cause fission to cease, since the moderator would no longer be present. Nevertheless, the fuel core can still overheat, because 7% of the reactor's heat comes from radioactive decay in the newly formed fission products. In time, the uncontrolled decay would release enough heat energy to melt the materials in the core, a situation called a **meltdown.** Then, the molten material falling into the remaining water could cause a steam explosion. To guard against all this, backup cooling systems keep the reactor immersed in water should leaks occur, and the entire assembly is housed in a thick concrete containment building (Fig. 13–8).

## Comparing Nuclear Power with Coal Power

Because nuclear plants are base-load plants that provide the foundation for meeting the daily and weekly electrical demand cycle, they must be replaced with other base-load plants. With the cancellations and continuing shutdowns of nuclear power plants, the sole option is replacing such large plants with coal-burning plants. The United States has abundant coal reserves, but is burning coal the course of action we wish to pursue? Nuclear power has some decided environmental advantages over

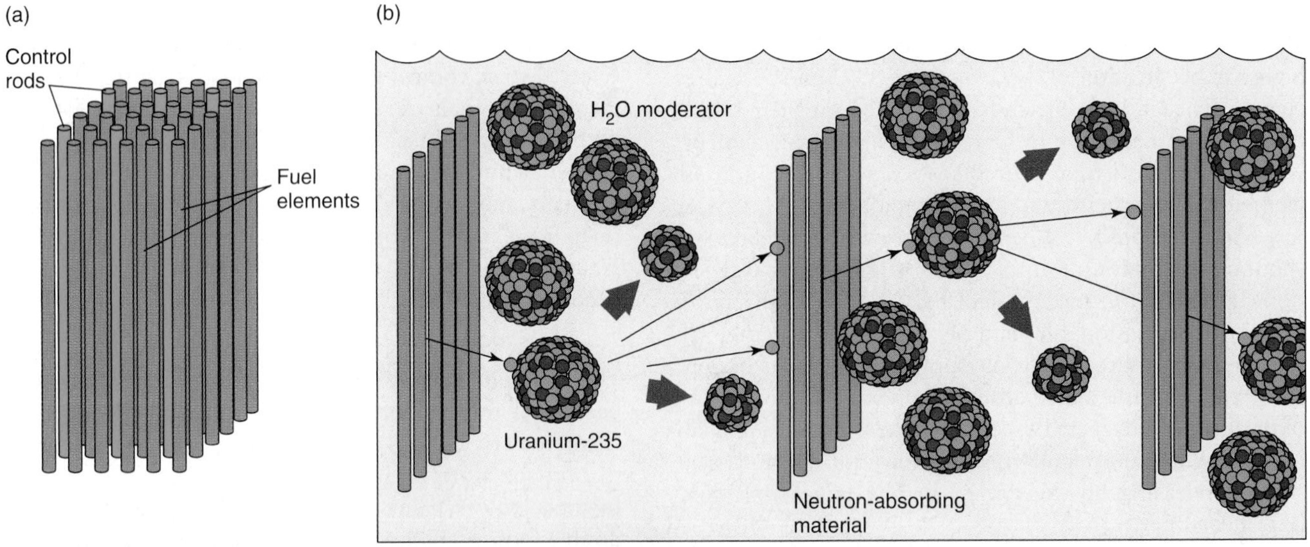

**Figure 13–7 A nuclear reactor.** (a) In the core of a nuclear reactor, a large mass of uranium is created by placing uranium in adjacent tubes, called fuel elements. The rate of the chain reaction is moderated by inserting or removing rods of neutron-absorbing material (control rods) between the fuel elements. (b) The fuel and rods are surrounded by the moderator fluid, near-pure water. (c) Technicians ready the core housing to receive uranium fuel elements in this nuclear reactor.

coal-fired power (Fig. 13–9). Comparing a 1,000-megawatt nuclear plant with a coal-fired plant of the same capacity, each operating for one year, we find the following characteristics:

- **Fuel needed.** The coal plant consumes 2–3 million tons of coal. If this amount is obtained by strip mining, some environmental destruction and acid leaching will result. If the coal comes from deep mines, there will be human costs in the form of accidental deaths and impaired health. The nuclear plant requires about 30 tons of enriched uranium, obtained from mining 75,000 tons of ore, with much less harm to humans and the environment. The fission of about 1 pound (0.5 kg) of uranium fuel releases energy equivalent to burning 50 tons of coal. Thus, one fueling of the reactor with about 60 tons is sufficient to run the nuclear power plant for as long as two years.
- **Carbon dioxide emissions.** The coal plant emits over 7 million tons of carbon dioxide into the atmosphere, contributing to global climate change. The nuclear plant emits none.
- **Sulfur dioxide and other emissions.** The coal plant emits over 300,000 tons of sulfur dioxide, particulates, and other pollutants, leading to acid rain and

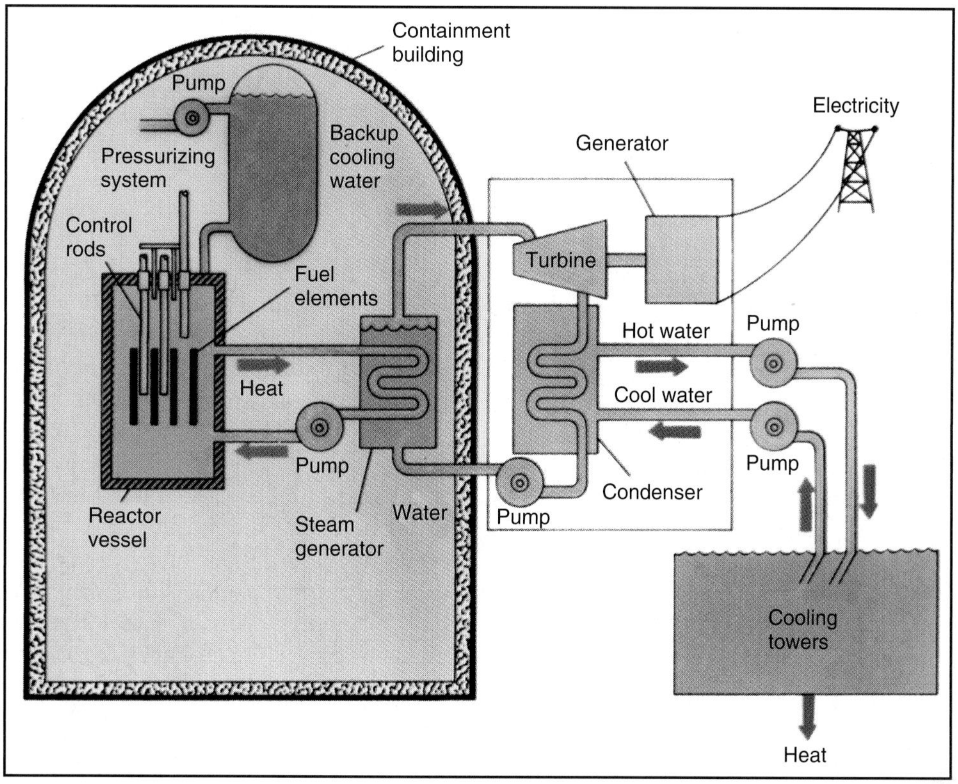


**Figure 13–8 Pressurized nuclear power plant.** The double-loop design isolates the pressurized water from the steam-generating loop that drives the turbogenerator.

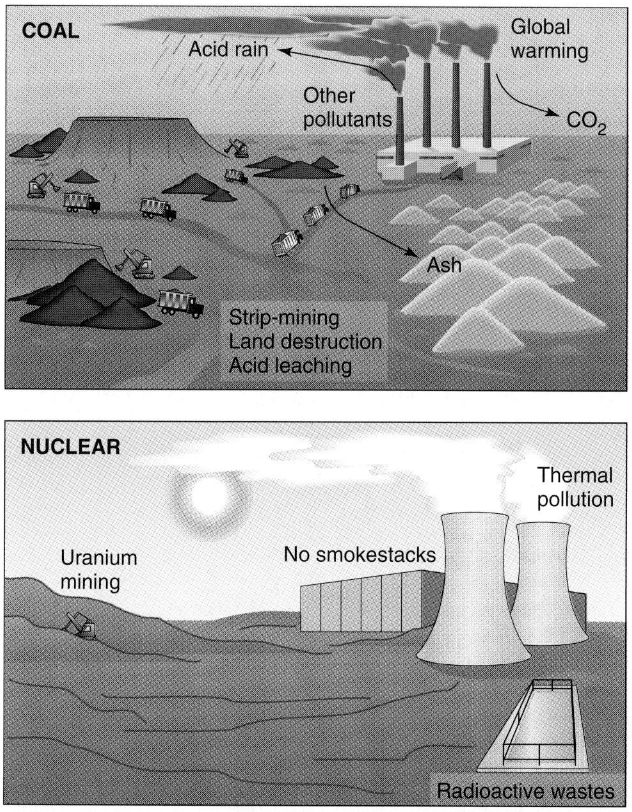


**Figure 13–9 Nuclear power vs. coal-fired power.** Both methods of generating electricity have their environmental advantages and disadvantages. The nuclear power option assumes perfect containment of radioactivity and the availability of some method for waste storage and disposal.

health-threatening air pollution. The nuclear power plant produces no acid-forming pollutants or particulates.

- **Radioactivity.** A coal plant releases 100 times more radioactivity than a nuclear power plant because of the natural presence of radioactive compounds (uranium, thorium) in the coal. The nuclear plant releases low levels of radioactive waste gases.
- **Solid wastes.** The coal plant produces about 600,000 tons of ash requiring land disposal. The nuclear plant produces about 250 tons of highly radioactive wastes requiring safe storage and ultimate safe disposal. (The handling of these radioactive wastes remains an unresolved problem.)
- **Accidents.** A worst-case accident in the coal plant could result in fatalities to workers and a destructive fire, a situation common to many industries. Accidents in a nuclear plant can range from minor emissions of radioactivity to catastrophic releases that can lead to widespread radiation sickness, scores of human deaths, untold numbers of cancers, and widespread, long-lasting environmental contamination.

It is the handling of radioactive wastes and releases and the potential for accidents that have led to public skepticism about nuclear power. Can these problems be overcome, or are other energy options less problematic and less costly than nuclear power?

## 13.3 The Hazards and Costs of Nuclear Power Facilities

Assessing the hazards of nuclear power requires an understanding of radioactive substances and their danger.

### Radioactive Emissions

When uranium or any other element undergoes fission, the split "halves" are atoms of lighter elements, such as iodine, cesium, strontium, cobalt, or any of some 30 other elements. These newly formed atoms—called the *direct products* of the fission—are generally unstable isotopes of their respective elements. Unstable isotopes (called **radioisotopes**) become stable by spontaneously ejecting subatomic particles (alpha particles, beta particles, and neutrons), high-energy radiation (gamma rays), or both. Radioactivity is measured in **curies;** one gram of pure radium-226 gives off 1 curie per second, which is approximately 37 billion spontaneous disintegrations into particles and radiation. The particles and radiation are collectively referred to as **radioactive emissions.** Any materials in and around the reactor may also be converted to unstable isotopes and become radioactive by absorbing neutrons from the fission process (Fig. 13–10). These indirect products of fission, along with the direct products, are the **radioactive wastes** of nuclear power. (Radioactive fallout from nuclear explosions also consists of direct and indirect fission products.)

**Biological Effects.** A major concern regarding nuclear power is that large numbers of the public may be exposed to low levels of radiation, thus elevating their risk of cancer and other disorders. Is this a valid concern? Radioactive emissions can penetrate biological tissue; their ability to do damage is measured in units called **sieverts** (Sv; 1 sievert = 100 rem, an older unit of measurement). The emissions leave no visible mark, nor are they felt, but they are capable of dislodging electrons from molecules or atoms that they strike.

***High Dose.*** Left behind are *ions* (charged particles), so the emissions are called *ionizing radiation.* The process of ionization may involve breaking chemical bonds or changing the structure of molecules in ways that impair their normal functions. In high doses, radiation may cause enough damage to prevent cell division. Thus, in medical applications, radiation can be focused on a cancerous tumor to destroy it. However, if the whole body is exposed to such levels of radiation (over 1 sievert is considered a high dose), a generalized blockage of cell division occurs that prevents the normal replacement or repair of blood, skin, and other tissues. This result is called *radiation sickness* and may lead to death a few days or months after exposure. The two stricken Japanese workers in the Tokaimura accident received from 10 to 17 sieverts of radiation, ultimately a lethal dose. Very high levels of radiation may totally destroy cells, causing immediate death.

***Low Dose.*** In lower doses, radiation may damage DNA, the genetic material inside the cell. Cells with damaged (mutated) DNA may then begin dividing and growing out of control, forming malignant tumors or leukemia. If the damaged DNA is in an egg or a sperm cell, the result may be birth defects (mutations) in offspring. The effects of exposure to radiation may go unseen until many years after the event (10 to 40 years is typical). Other effects include a weakening of the immune system, mental retardation, and the development of cataracts.

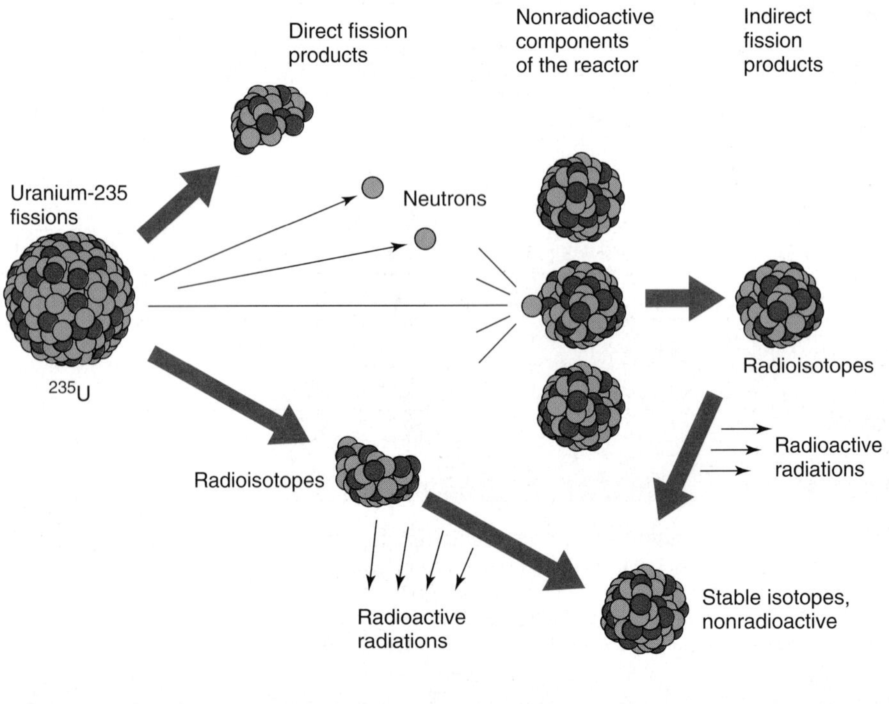


**Figure 13–10**
**Radioactive wastes and radioactive emissions.** Nuclear fission results in the production of numerous unstable isotopes, designated radioactive waste. These isotopes give off potentially damaging radiation until they regain a stable structure.

*Exposure.* Health effects are directly related to the level of exposure. There is broad agreement that doses between 100 and 500 millisieverts (10 and 50 rem; 1 millisievert = a thousandth of a sievert) result in an increased risk of developing cancer. Evidence for this hypothesis comes from studies of patients with various illnesses who were exposed to high levels of X rays in the 1930s, before the potential harm of such radiation was understood. People in these groups subsequently developed higher-than-normal rates of cancer and leukemia. Many scientists believe that no dose is without some harm; others point to the ability of living cells to repair small amounts of damage to DNA and believe that there is a threshold of radiation below which no biological effects occur. If there is such a threshold, it is virtually impossible to demonstrate, because of the many different causes of cancer, the long time it takes for cancer to develop, and the variety of sources of radiation to which people are exposed. Federal agencies employ the concept of no threshold and a direct relationship between the amount of radiation and the incidence of cancer. Thus, federal standards are set at 1.7 mSv/yr (170 mrem/yr) as the maximum exposure permitted for the general populace per year, except for medical X rays.

**Sources of Radiation.** Nuclear power is by no means the only source of radiation: There is also normal background radiation from radioactive materials, such as the uranium and radon gas that occur naturally in the Earth's crust, and cosmic rays from outer space. For most people, background radiation is the major source of radiation exposure. In addition, we deliberately expose ourselves to radiation from medical and dental X rays, by far the largest source of human-induced exposure and, for the average person, equal to one-fifth the exposure from background sources. The average person in the United States receives a dose of about 3.6 millisieverts (mSv) per year (Table 13–1). Thus, the argument becomes one of relative hazards. That is, does or will the radiation from nuclear power significantly raise radiation levels and elevate the risk of developing cancer?

*Up Close and...?* During normal operation of a nuclear power plant, the direct fission products remain within the fuel elements, and the indirect products are maintained within the containment building that houses the reactor. No routine discharges of radioactive materials (other than routine stack emissions) into the environment occur. Even very close to an operating nuclear power plant, radiation levels from the plant are lower than normal background levels. A radiation detector will pick up more radiation from the ground or concrete on a basement floor than it will when held within 150 yards of a nuclear power plant. Careful measurements have shown that public exposure to radiation from normal operations of a power plant is less than 1% of natural background radiation.

The main concern about nuclear power, therefore, should not focus on a plant's normal operation. Instead, the real problems arise from the storage and disposal of radioactive wastes and the potential for accidents.

## Radioactive Wastes

To understand the problems surrounding nuclear waste disposal, you must understand the concept of **radioactive decay,** a process in which, as unstable isotopes

**table 13-1 Relative Doses from Radiation Sources**

| Source | Dose |
|---|---|
| All sources (avg.) | 3.6 mSv/year |
| Cosmic radiation at sea level* | 0.26 mSv/year |
| Terrestrial radiation (elements in soil) | 0.26 mSv/year |
| Radon in average house | 2 mSv/year |
| X rays and nuclear medicine (avg.) | 0.50 mSv/year |
| Consumer products | 0.11 mSv/year |
| Natural radioactivity in the body | 0.39 mSv/year |
| Mammogram | 0.30 mSv |
| Chest X ray | 0.10 mSv |
| Gastrointestinal series of X rays | 14 mSv |
| Continued fallout from nuclear testing | 0.01 mSv/year |
| Living near a nuclear power station (assuming perfect containment) | <0.01 mSv/year |
| Federal standards for the general populace | 1.7 mSv/year |

*Increases 0.005 mSv for every 100 ft of elevation (1 Sv = 100 rem).
*Source:* Data from U.S. Nuclear Regulatory Commission.

eject particles and radiation, they become stable and cease to be radioactive. As long as radioactive materials are kept isolated from humans and other organisms, the decay proceeds harmlessly.

***Get a Half-Life.*** The rate of radioactive decay is such that half of the starting amount of a given isotope will decay within a certain period. In the next equal period, half of the remainder (half of a half, which equals one-fourth of the original) decays, and so on, as shown in Fig. 13–11. The time for half of the amount of a radioactive isotope to decay is known as the isotope's **half-life.** The half-life of an isotope is always the same, regardless of the starting amount.

Each particular radioactive isotope has a characteristic half-life. The half-lives of various isotopes range from a fraction of a second to many thousands of years. Uranium fissioning results in a heterogeneous mixture of radioisotopes, the most common of which are listed in Table 13–2. Some of this material—in particular, the remaining $^{235}U$ and plutonium ($^{239}Pu$) that has been created by the neutron bombardment of $^{238}U$—can be recovered and recycled for use as nuclear fuel in an operation called **reprocessing.** Although Britain and France have a large reprocessing operation, U.S. policy has prohibited the practice because of concerns over plutonium and atomic weapons. The Cheney energy report (see Chapter 12) has recommended reopening the matter, however, citing the economic advantages of reusing the spent fuel.

## Disposal of Radioactive Wastes

The development of, and commitment to, nuclear power went ahead without ever fully addressing the issue of ultimate long-term containment. Proponents of nuclear power generally assumed that the long-lived wastes could be solidified, placed in sealed containers, and buried in deep, stable rock formations (geologic burial) as the need for such containment became necessary. However, this has not yet happened. Thus, the current problem of nuclear waste disposal is twofold:

- **Short-term containment.** Allows the radioactive decay of short-lived isotopes (Table 13–2). In 10 years, fission wastes lose more than 97% of their radioactivity. Wastes can be handled much more easily and safely after this loss occurs.
- **Long-term containment.** EPA recommends a 10,000-year minimum, and the National Research Council opted for 100,000 years, to provide protection from the long-lived isotopes. Government standards require isolation for 20 half-lives. (Plutonium has a half-life of 24,000 years.)

***Tanks and Casks.*** For short-term containment, spent fuel is first stored in deep swimming pool–like tanks on the sites of nuclear power plants. The water in these tanks dissipates waste heat (which is still generated to some degree) and acts as a shield against the

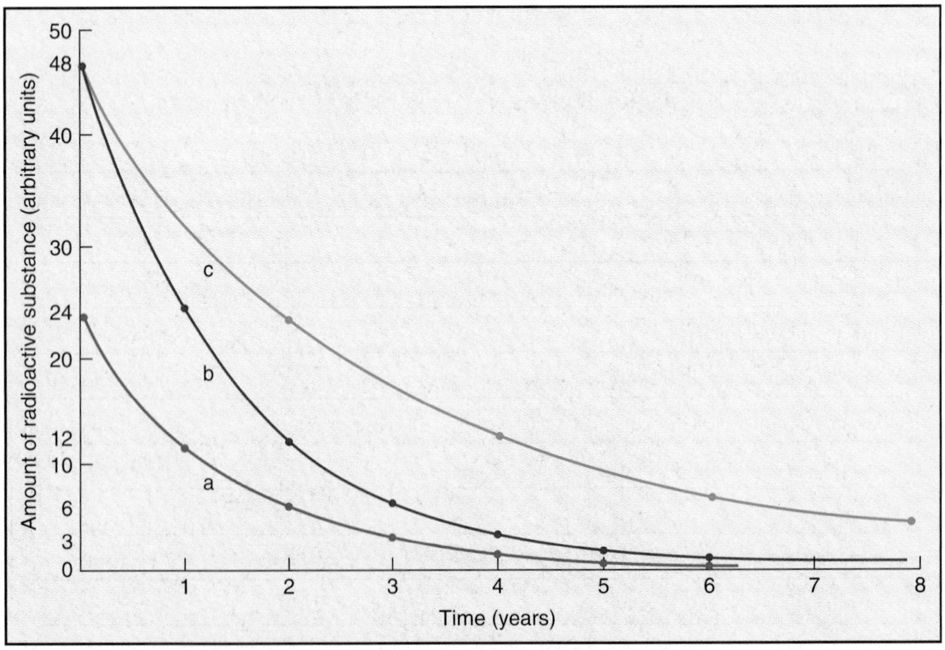


**Figure 13–11 Radioactive decay.** Regardless of the starting amount, one-half of a radioactive substance decays during each successive half-life. (a) A substance with a half-life of one year, starting with 24 units; (b) the same substance starting with 48 units; (c) a substance with a half-life of two years. (Note that the decay of a radioisotope never equals 100%; radioactivity is reduced by only half in each half-life, and there is always an undecayed portion remaining. However, it is generally held that radiation is reduced to insignificant levels after 10 half-lives.)

**table 13-2 Common Radioactive Isotopes Resulting from Uranium Fission and Their Half-Lives**

| Short-Lived Fission Products | Half-Life (days) |
|---|---|
| Strontium-89 | 50.5 |
| Yttrium-91 | 58.5 |
| Zirconium-95 | 64.0 |
| Niobium-95 | 35.0 |
| Molybdenum-99 | 2.8 |
| Ruthenium-103 | 39.4 |
| Iodine-131 | 8.1 |
| Xenon-133 | 5.3 |
| Barium-140 | 12.8 |
| Cerium-141 | 32.5 |
| Praseodymium-143 | 13.6 |
| Neodymium-147 | 11.1 |
| **Long-Lived Fission Products** | **Half-Life (years)** |
| Krypton-85 | 10.7 |
| Strontium-90 | 28.0 |
| Ruthenium-106 | 1.0 |
| Cesium-137 | 30.0 |
| Cerium-144 | 0.8 |
| Promethium-147 | 2.6 |
| **Additional Product of Neutron Bombardment** | **Half-Life (years)** |
| Plutonium-239 | 24,000 |

escape of radiation. The storage pools can typically accommodate 10 to 20 years of spent fuel. However, the capacity of storage pools at U.S. nuclear plants reached 50% by 2004 and will be at 100% by 2015. After a few years of decay, the spent fuel may be placed in air-cooled dry casks for interim storage until long-term storage becomes available. The casks are engineered to resist floods, tornadoes, and extremes of temperature. Currently, 27 nuclear plants are employing dry storage, with the prospect of many more as pool storage capacity is exhausted.

In the meantime, the wastes from the world's commercial reactors have been accumulating at a rate of 10,000 tons a year, reaching 200,000 tons at the end of 2002, all stored on-site at the power plants. Of these wastes, 47,000 tons are in the United States. Furthermore, because of neutron bombardment of the reactor walls, all nuclear power plants will eventually add to the stockpile of radioactive wastes. Scientists estimate that dismantling a decommissioned power plant will generate more nuclear waste than the plant produced during its active life!

**Military Radioactive Wastes.** Some of the worst failures in handling wastes have occurred at military facilities in the United States and in the former Soviet Union, in connection with the manufacture of nuclear weapons. Liquid high-level wastes stored at many U.S. facilities have leaked into the environment and contaminated wildlife, sediments, groundwater, and soil. Activities at these sites have been shrouded in secrecy; only recently have documents revealing past accidents and radioactive releases been declassified and made available to the public. Deliberate releases of uranium dust, xenon-133, iodine-131, and tritium gas have been documented at Hanford, Washington; Fernald, Ohio; Oak Ridge, Tennessee; and Savannah River, South Carolina. Cleaning up these sites is now the responsibility of the DOE, which has already spent $50 billion on this problem and estimates that the final cost could run as high as $250 billion, probably the largest environmental project the country will ever undertake.

***From Russia, with....*** Russian military weapons facilities have been even more irresponsible. The worst is a giant complex called Chelyabinsk-65, located in the Ural

Mountains. For at least 20 years, nuclear wastes were discharged into the Techa River and then into Lake Karachay. At least 1,000 cases of leukemia have been traced to radioactive contamination from the Chelyabinsk facility. Even today, standing for one hour on the shore of Lake Karachay will cause radiation poisoning and death within a week. The lake dried up in the summer of 1967, and winds blew radioactive dust with 5 million curies across the countryside, contaminating hundreds of thousands of people. This lake is considered to be the most polluted lake on earth, a legacy of the Cold War and an enormous continuing source of radioactive contamination. Recognizing the dangers of spreading contamination, Russian authorities have filled the lake with hollow concrete blocks, rocks, and soil, and plants now spread their greenery over this permanently contaminated site.

*Megatons to Megawatts.* The end of the Cold War has brought the welcome dismantling of nuclear weapons by the United States and the nations of the former U.S.S.R. Thousands of nuclear weapons are being dismantled, many as a result of earlier agreements between President Clinton and Russian President Boris Yeltsin. The agreements include a joint closure of all remaining plutonium weapons production reactors, funds to aid Russia in destroying missile silos and dismantling nuclear submarines and bombs, and vast cuts in the nuclear arsenals of the two former enemies. The radioactive components of the weapons—at least 100 tons of weapons-grade plutonium—must be handled with great care and disposed of where they are safe from illegal access and where they will pose no danger to human health for centuries to come—a daunting task.

One element of the U.S.–Russian agreements, the *Megatons to Megawatts* program, involves a government–industry partnership wherein a private U.S. company oversees the dilution of weapons-grade uranium to the much lower power plant uranium, sells it to U.S. nuclear power plants, and pays the Russian government a market price (to date, $2.5 billion). As a result, 7,000 nuclear warheads have been eliminated, and 175 tons of highly enriched uranium diluted to 5,124 tons of power plant fuel (Fig. 13–12). The closing down of much of the nuclear warfare enterprise is certainly a welcome development, but the problems of security and of disposal of the wastes will remain as a long-term legacy of the Cold War.

**Figure 13–12 Megatons to Megawatts.** Cylinders containing nuclear warhead–derived fuel from Russia are unloaded in the United States. This program has eliminated 7,000 nuclear warheads and currently supplies a large proportion of the fuel for U.S. nuclear power plants.

**High-Level Nuclear Waste Disposal.** The United States and most other countries that use nuclear power have decided on geologic burial for the ultimate consignment of nuclear wastes, but no nation has developed plans to the point of actually carrying out the burial. Many nuclear nations have not even been able to find a site that may be suitable for receiving wastes. Where sites have been selected, many questions about their safety have surfaced. The basic problem is that no rock formation can be guaranteed to remain stable and dry for tens of thousands of years. Everywhere scientists look, there is evidence of volcanic or earthquake activity or groundwater leaching within the last 10,000 years or so, which is to say that it may occur again in a similar period. If such events did occur, the still-radioactive wastes could escape into the environment and contaminate water, air, or soil, with consequent effects on humans and wildlife.

***Yucca Mountain.*** In the United States, efforts to locate a long-term containment facility, which have been going on for about 30 years, have been hampered by a severe *not in my backyard* (*NIMBY*) syndrome (see Chapter 18.) A number of states, under pressure from citizens, have passed legislation categorically prohibiting the disposal of nuclear wastes within their boundaries. In the meantime, the need to select and develop a long-term repository has become increasingly critical, given the buildup of spent fuel in every nuclear power plant. The Nuclear Waste Policy Act of 1982 committed the federal government to begin receiving nuclear waste from commercial power plants in 1998. At the end of 1987, Congress called a halt to the debate and selected a remote site, Yucca Mountain, in southwestern Nevada (Fig. 13–13), to be the nation's civilian nuclear waste disposal site. No other sites are being considered. Not surprisingly, Nevadans fought this selection, passing a law in 1989 that prohibits anyone from storing high-level radioactive waste in the state. The federal government has the power to override state prohibitions, though, and the Nevada site has undergone intensive study over the past 20 years, at a cost of $4 billion.

To date, the studies have shown that the storerooms, 1,000 feet (300 m) above current groundwater levels, will presumably be safe from invasion by groundwater. However, critics of the site suggest that the rock formations above the storerooms will allow a much more rapid

**Figure 13–13 Yucca Mountain, Nevada.** A shift changes at the tunnel entrance to the facility, which has been chosen as the nation's sole high-level nuclear waste depository.

permeation of rainwater than had previously been estimated. Water would speed up corrosion of the waste containers and subsequent leakage of radioactivity into groundwater. Furthermore, earthquakes or volcanic activity could change the current conditions in a short time, and the site lies in a zone of fault lines and has a history of volcanic activity only 5,000 years ago.

*Yucca–Yes.* The political wrangling came to a halt in July 2002, when President George W. Bush signed a resolution (passed by Congress) voiding a veto by Nevada's Governor Kenny Guinn that had attempted to block further development of the site. Earlier, President Bush had accepted the recommendation of Energy Secretary Spencer Abraham to officially designate the Yucca Mountain facility to be the nation's nuclear waste repository. The site could begin receiving wastes from commercial facilities all over the country in the year 2010, pending a final licensing approval. Nevada's governor and congressional delegation vow to continue their fight in the courts. In the meantime, an underground storage facility in New Mexico, the Waste Isolation Pilot Plant, has been receiving shipments of highly radioactive wastes from nuclear weapons facilities since 1999, with no apparent problems. (See "Ethics" essay, p. 364.)

## Nuclear Power Accidents

**Three Mile Island.** On March 28, 1979, the Three Mile Island nuclear power plant near Harrisburg, Pennsylvania, suffered a partial meltdown as a result of a series of human and equipment failures and a flawed design. The steam generator (Fig. 13–8) shut down automatically because of a lack of power in its feedwater pumps, and eventually a valve on top of the generator opened in response to the gradual buildup of pressure. Unfortunately, the valve remained stuck in the open position and drained coolant water from the reactor vessel. There were no sensors to indicate that this pressure-operated relief valve was open. Operators responded poorly to the emergency, shutting down the emergency cooling system at one point and shutting down the pumps in the reactor vessel. One instrument error compounded the problem: Gauges told operators that the reactor was full of water when, actually, it needed water badly. The core was uncovered for a time and suffered a partial meltdown. Ten million curies of radioactive gas were released into the atmosphere.

The drama held the whole nation—particularly the 300,000 residents of metropolitan Harrisburg, who were poised for evacuation—in suspense for several days. The situation was eventually brought under control, and no injuries or deaths occurred, but it would have been much worse if the meltdown were complete. The reactor was so badly damaged, and so much radioactive contamination occurred inside the containment building, that the continuing cleanup is proving to be as costly as building a new power plant. There are no plans to restart the reactor. GPU Nuclear, operators of the plant, have since paid out $30 million to settle claims from the accident, although the company has never admitted any radiation-caused illnesses.

**Chernobyl.** Prior to 1986, the scenario for a worst-case nuclear power plant disaster was a matter of speculation. Then, at 1:24 A.M. local time on April 26, 1986, events at a nuclear power plant in Ukraine (then a part of the Soviet Union) made such speculation irrelevant (Fig. 13–14). Since that day, Chernobyl has served as a horrible example of nuclear energy gone awry.

**ethics**

## Showdown in the New West

In a scenario reminiscent of the Civil War, two western states have become the focus of a controversy over states' rights. On one side is the federal government, especially the DOE, which is trying to solve the problem of what to do with nuclear wastes and which thinks it has a reasonable solution. On the other side are politicians in the states designated as repository sites, who are responding to the voices of citizens of those states. At stake may be the future of nuclear energy in the United States.

*WIPP.* The DOE has constructed a $1.5 billion Waste Isolation Pilot Plant (WIPP) in salt caves 2,150 feet beneath the desert in southeastern New Mexico, 26 miles from the city of Carlsbad. The plant is on federal land and has passed all necessary safety reviews. It was ready to open for business in October 1991. "Business," in this case, was for the plant to become the repository for up to a million barrels of plutonium wastes from nuclear weapons plants and laboratories around the country, as well as for the plutonium from dismantled nuclear warheads. (No commercial nuclear wastes will go to the plant.)

New Mexico sued the DOE on the grounds that it was in violation of federal law prohibiting the opening of such a repository without congressional approval. The state of Texas and four environmental groups joined New Mexico in bringing suit. In October 1992, Congress gave approval to the department to begin testing the facility's ability to store the plutonium wastes—with one significant catch: The EPA was ordered to strengthen the radiation standards protecting the public from the stored wastes. In 1998, after a lengthy review, the EPA gave the WIPP certification to begin receiving radioactive materials from nuclear weapons facilities around the country, and in 1999 the site began to receive wastes from a number of facilities. To date (2003), the WIPP has received over 500 shipments without incident. More than 14,000 drums of plutonium wastes now rest in the salt caves.

*Yucca Mountain.* Opposition to the long-term storage of nuclear wastes is even stronger in Nevada, which has the dubious distinction of having been selected for the nation's only repository for commercial nuclear wastes, at Yucca Mountain. The site is a barren ridge in the desert about 100 miles northwest of Las Vegas (Fig. 13–13). In 1989, the Nevada legislature passed a bill making it unlawful for any agent or agency to store high-level radioactive waste in the state. All sorts of surveys have been carried out to test the mood of Nevadans regarding the long-term storage of wastes in their state. The responses have been uniformly negative: Three-fourths of Nevadans agreed that the state should continue to do all it can to prevent the establishment of the Yucca Mountain site. Clearly, ordinary Nevadans perceive the risks of such a site as enormous and unacceptable, in contrast to the opinions of many scientists who have performed risk calculations regarding the site. Perhaps the most disturbing issue is the transport of the wastes through Nevada (and many other states, too) to reach the repository. What happens if there is an accident? In this case, however, the state's attempts to block further development of the site have failed, because the Supreme Court has ruled that Nevada must process applications for permits to continue work there. The federal government, moreover, is moving ahead with the site.

What's next? The DOE has submitted a request to the NRC for a construction license, and the NRC must then rule on the issue. Many years of work and billions of dollars have already been spent on the site, but independent studies suggest that there are still many issues to be resolved. While it is certainly true that, for Nevadans, the issue is a present one, the deciding concern is the safety of long-term storage of the nuclear wastes. How much risk is there that the storage containers will be corroded and leak some time in the next 10,000 years? This is an entirely unprecedented situation, so a decision must employ the highest possible standards. People in New Mexico and Nevada are indignant at their states' becoming the dumping ground for nuclear wastes from around the nation, but, in a sense, their concern is a proxy for what will be an enduring concern for generations yet unborn. We need to do something with the legacy of nuclear energy other than delay a decision. In a real sense, the nation's security is also at stake, because the stored wastes on the grounds of the existing and shut power plants are prime targets for terrorists. What do you think should be done?

Here's how it happened: While conducting a test of standby diesel generators, engineers disabled the power plant's safety systems, withdrew the control rods, shut off the flow of steam to the generators, and decreased the flow of coolant water in the reactor. However, they did not allow for the radioactive heat energy generated by the fuel core, and, lacking coolant, the reactor began to heat up. The extra steam that was produced could not escape and had the effect of rapidly boosting the energy production of the reaction. In an attempt to quell the reactor, the engineers quickly inserted the carbon-tipped control rods. The carbon tips acted as moderators, slowing down the neutrons that were produced in the reaction. The neutrons, however, were still speedy enough to trigger more fission reactions, and the result was a split-second power surge to 100 times the maximum allowed level. Steam explosions then blew the 2,000-ton top off the reactor, the reactor melted down, and a fire was ignited in graphite, burning for days. At least 50 tons of dust and debris bearing 100–200 million curies of radioactivity in the form of fission products were released in a plume that rained radioactive particles over thousands of square miles, one hundred times the radiation fallout from the bombs dropped on Hiroshima and Nagasaki in 1945.

**Figure 13–14 Chernobyl.** An aerial view of the Chernobyl reactor on April 29, 1986, three days after the explosion. The disaster was the worst nuclear power accident, directly killing at least 31 people and putting countless others in the surrounding countryside at risk for future cancer deaths.

*Consequences.* As the radioactive fallout settled, 135,000 people were evacuated and relocated. The reactor was eventually sealed in a sarcophagus of concrete and steel. A barbed-wire fence now surrounds a 1,000-square-mile exclusion zone around the reactor site. The soil remains contaminated with radioactive compounds, yet for years 2,000 Ukrainian workers were bused in daily to work at the remaining reactor in the Chernobyl complex. In December 2000, the last reactor was permanently shut down.

Only two engineers were directly killed by the explosion, but 29 of the personnel brought in to contain the reactor in the aftermath of the explosion died of radiation within a few months. Over a broad area downwind of the disaster, buildings and roadways were washed down to flush away radioactive dust. Even with these precautions, many people in or near the evacuation zone were exposed to dangerous levels of radiation, especially the short-lived radioisotope iodine-131. Because iodine collects in the thyroid gland, the radioactive iodine was responsible for great increases in the incidence of thyroid cancer in Ukraine and neighboring Belarus. (Some 3,000 cases were registered in 2001 alone.) Soviet authorities were slow to admit the seriousness of the accident and waited a week to distribute iodine pills to those affected—too late to do any good. According to the Ukrainian government, more than 4,000 Ukrainians who participated in the cleanup subsequently died, and 70,000 have become disabled. The long-term effects are estimated to range from 140,000 to 475,000 cancer deaths worldwide from the accident.

*Could It Happen Here?* Are we in the United States in danger of such an explosion occurring at a nuclear power plant? Nuclear scientists argue that the answer is no, because U.S. power plants have a number of design features that should make a repeat of Chernobyl impossible. For one thing, the Chernobyl reactor used graphite as a moderator, rather than water, as in LWRs. Further, LWRs are incapable of developing a power surge more than twice their normal power, well within the designed containment capacity of the reactor vessel. Finally, LWRs have more backup systems to prevent the core from overheating, and the reactors are housed in a thick concrete-walled containment building designed to withstand explosions such as the one that occurred in the Chernobyl reactor, which had no containment building. LWRs are not immune to accidents, however, the most serious being a complete core meltdown as a result of a total loss of coolant. This has never happened, but Three Mile Island was a close call.

The real cost of the accidents at Three Mile Island and Chernobyl must be reckoned in terms of public trust. Public confidence in nuclear energy already was declining, but it plummeted after the two accidents, which pointed to human error as a highly significant factor in nuclear safety—and human error is something the public understands well. Nuclear proponents suffered a serious loss of credibility with Three Mile Island, but Chernobyl was their worst nightmare come true—a full catastrophe, just as the antinuclear movement had predicted might someday occur.

## Safety and Nuclear Power

As a result of Three Mile Island and other, lesser incidents in the United States, the Nuclear Regulatory Commission has upgraded safety standards not only in the technical design of nuclear power plants, but also in maintenance procedures and in the training of operators. Thus, proponents contend, nuclear plants were designed to be safe in the beginning, and now they are safer than ever. Some proponents of nuclear power claim that we now have the technology to build *inherently safe* nuclear reactors, designed in such ways that any accident would result in an automatic quenching of the chain reaction and suppression of heat from nuclear decay. In reality, however, there is no such thing as an inherently safe reactor, since the concept implies no release of radioactivity under any circumstances—an impossible expectation.

*Passive Safety.* Instead, nuclear scientists are proposing a new generation of nuclear reactors with built-in *passive safety* features, rather than the *active safety* features found in current reactors. **Active safety** relies on operator-controlled actions, external power, electrical signals, and so forth. As accidents have shown, operators may override such safety factors, and electricity, valves, and meters can fail or give false information. **Passive safety,** by contrast, involves engineering devices and structures that make it virtually impossible for the reactor to go beyond acceptable levels of power, temperature, and radioactive emissions; their operation depends only on standard physical phenomena, such as gravity and resistance to high temperatures.

**New Generations of Reactors.** Nuclear scientists distinguish four generations of nuclear reactors. *Generation I* reactors are the earliest, developed in the 1950s and 1960s; few are still operating. The majority of today's reactors are *Generation II* vintage, the large baseline power plants, of several designs. *Generation III* refers to newer designs with passive safety features and much simpler, smaller power plants—among them the so-called advanced light-water reactors (ALWRs). For example, one passive safety feature would be to position a cold-water reservoir so that, in the event of a LOCA, the water would drain by gravity to the reactor core (Fig. 13–15). In addition, the design will make it impossible for operators to inactivate the passive safety systems. Another feature being planned is to build reactors as modular units small enough to conduct heat from nuclear decay outward into the soil; in this manner, the reactors could not incur a core meltdown. One such modular reactor, built in Germany, suffered no damage to its core when tested against a LOCA.

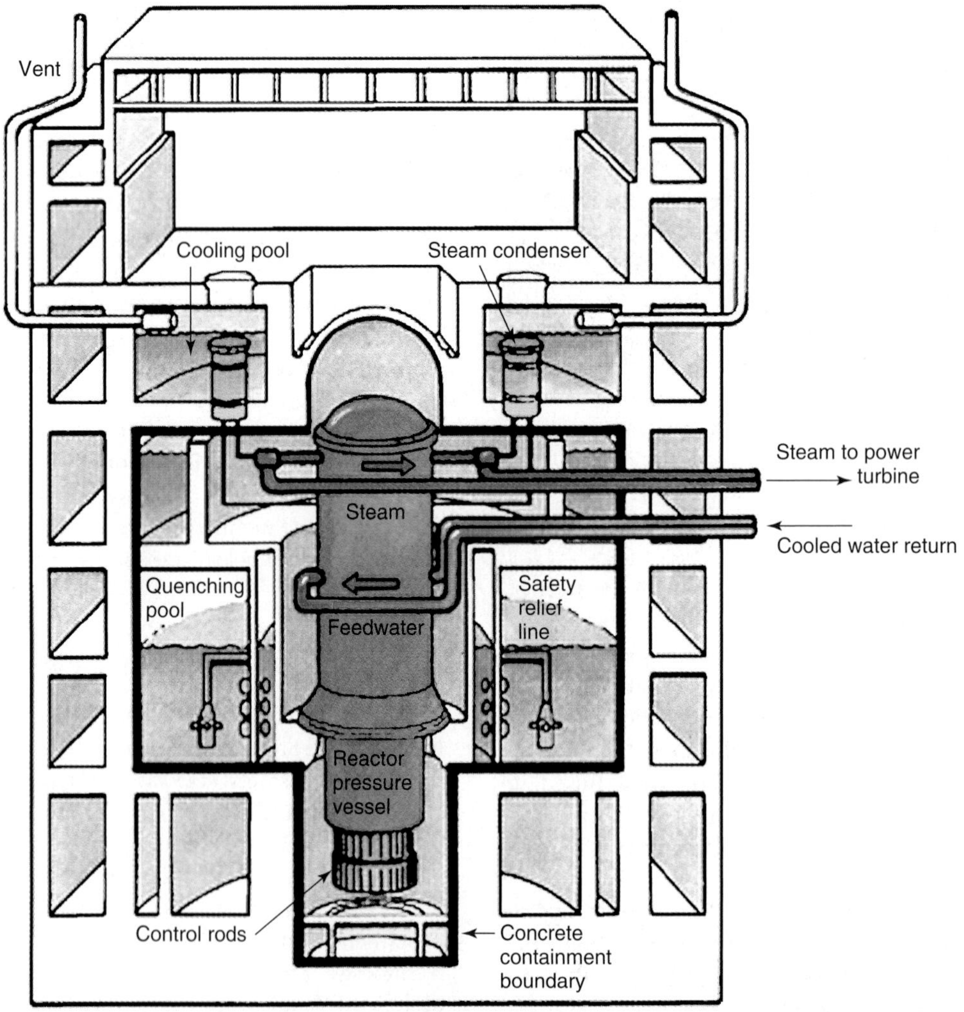


**Figure 13–15 Advanced light-water reactor.** The core is surrounded by three concentric structures: a reactor pressure vessel, in which heat from the reactor boils water directly into steam; a concrete chamber (outlined with heavy black line) and water pool, which together contain and quench steam vented from the reactor in an emergency; and a concrete building, which acts as a secondary containment vessel and shield. Any excessive pressure in the reactor will automatically open valves that release steam into a quenching pool, reducing the pressure. Water from the quenching pool can, if necessary, flow downward to cool the core. Evaporation from a pool on top of the containment building limits the buildup of containment pressure. (From "Advanced Light-Water Reactors," by M. W. Golay and N. E. Todreas. Copyright © 1990 by Scientific American, Inc. All rights reserved.)

In East Asia, the nuclear reactor design of choice is now the ALWR. Japan, with 54 nuclear plants, has had 2 of these in operation since 1996; several more are under construction in Japan and Taiwan, and there are plans for many more. The new plants are designed to last for 60 years, can be constructed within 5 years, and are simple enough that a single operator can control them under normal conditions. They generate electricity at a cost competitive with fossil-fuel options.

Generation IV plants, initiated by the U.S. Department of Energy, are now being designed and will likely be built within the next 20 years. One example of a Generation IV reactor is the pebble-bed modular reactor (PBMR), which will feed spherical carbon-coated uranium fuel pebbles gradually through the reactor vessel, like a gum-ball machine. The PBMR will be cooled with fluidized helium, an inert gas, which will also spin the turbines. These reactors will be small, will produce about 160 MW of power, and are expected to be cheap to build, passively safe, and inexpensive to operate. The modules can be built in a factory and shipped to the location of the power plant.

One motive behind the new designs is to restore the public's confidence in nuclear energy. Previously, nuclear proponents had emphasized the very low probabilities of accidents. As we have seen, though, improbable events can happen, and when they happen to nuclear power plants, the consequences can be dreadful. It may be that the public was on the way to a renewed appreciation of nuclear power, but all this changed on September 11, 2001.

**Terrorism and Nuclear Power.** What would have happened if one of the terrorist suicide groups that took over four airliners on September 11, 2001, had targeted a nuclear power plant instead of the World Trade Center? This question has raised great concern in many venues, from the NRC, to Congress, to local hearings in municipalities that host nuclear plants. After some uncertainty following September 11, the general consensus is that a jetliner could not penetrate the very thick walls of the containment vessel protecting the reactor. It could, however, destroy the control building and bring on a LOCA. Even worse, every power plant has a spent-fuel storage pool containing at least as much highly radioactive material as a nuclear core. These pools are not at all as heavily protected as the reactor vessels. (The pool walls may be only 18 inches thick.) If a pool lost its water, the fuel could heat up and cause a fire that could release greater amounts of cesium-137 than the Chernobyl accident, according to an NRC report.

Another terrorist method might be to attack the nuclear plant with a dedicated strike force, overcome the guards, and bring on a LOCA and a core meltdown by manipulating the controls. If this happened, the reactor vessel would likely be breached and a large release of radiation would occur. (This did not happen at Three Mile Island because the operators cooled the core sufficiently before it could melt completely.) Although few, if any, immediate civilian fatalities would result, there would be the long-term effects of radiation on the incidence of cancer. In another scenario, terrorists could obtain spent-fuel rods that might be incorporated into a "dirty bomb" which would rain radioactivity over a vast area. Over the years, the NRC has conducted mock terrorist attacks on nuclear plants, and a relatively high number of these attacks have penetrated the plant's security.

***Response.*** Following September 11, the NRC immediately stepped up the requirements for security around every power plant, adding more guards, more physical barriers, and vehicle checks at greater distances from the plant, as well as keeping the shoreline near the plants off limits. To date, Congress has not passed any special legislation to beef up security measures. The NRC has placed all nuclear plants at their highest level of security alert and is thoroughly reviewing its security requirements. The Project on Government Oversight (POGO), a private watchdog agency, has challenged the NRC's response, charging that the guards at many power plants, when interviewed, were highly skeptical that they could hold off a terrorist attack. According to POGO, the guards are "undermanned, underequipped, under-trained, and underpaid." The POGO report *Nuclear Power Plant Security: Voices from Inside the Fences* contains many recommendations that, if followed, would address most of the shortcomings of the security system.

## Economic Problems with Nuclear Power

Figure 13–2 shows that utilities in the United States were already turning away from nuclear power considerably before the disaster at Chernobyl. The reasons were mainly economic.

First, projected future energy demands were overly ambitious, so a slower growth rate in the demand for electricity postponed orders for all types of power plants. Second, increasing safety standards for the construction and operation of nuclear power plants caused the costs of plants to increase at least fivefold, even after inflation is considered. Adding to the rise in costs is the withdrawal of government subsidies to the nuclear industry. Third, public protests frequently delayed the construction or start-up of a new power plant. Such delays increased costs still more, because the utility was paying interest on its investment of several billion dollars even when the plant was not producing power. As these costs are passed on, consumers become yet more disillusioned with nuclear power. Finally, safety systems may protect the public, but they do not prevent an accident from financially ruining the utility. Since radioactivity prevents straightforward cleanup and repair, an accident can convert a multibillion-dollar asset into a multibillion-dollar liability in a matter of minutes, as Three Mile Island demonstrated. Thus, nuclear power involves a financial risk that utility executives are reluctant to take.

***Stranded Costs.*** A major problem facing nuclear power is the rising tide of deregulation of the U.S. electric industry. (See Chapter 12.) Still undecided is what will be done about "stranded cost recovery," meaning the repayment of unpaid costs, like that connected with the construction of power plants, to investors. This could be a real concern to nuclear power plant owners if customers choose other suppliers of electricity because of cost or philosophical considerations. The Federal Energy Regulatory Commission has ruled that these unpaid costs may be recovered in order to achieve a successful transition to the competitive environment that deregulation is designed to create. Utilities customers living within the power pool area served by a problematic nuclear plant are well aware of the stranded-cost problem.

***Half-Life?*** Another factor that promises to increase the cost of nuclear-generated electricity is a shorter-than-expected lifetime for nuclear power plants. Originally, it was thought that nuclear plants would have a lifetime of about 40 years. It now appears that lifetimes will be considerably less for many plants. Worldwide, more than 100 nuclear plants have been shut down after an average operating lifetime of 17 years. This shorter lifetime substantially increases the cost of the power produced, because the cost of the plant must be repaid over a shorter period.

The lifetimes of nuclear power plants are shorter than originally expected due to *embrittlement* and *corrosion*. **Embrittlement** occurs as neutrons from fission bombard the reactor vessel and other hardware. Gradually, this neutron bombardment causes the metals to become brittle enough that they may crack under thermal stress—for example, when emergency coolant waters are introduced in the event of a LOCA. When the reactor vessel becomes too brittle to be considered safe, the plant must either be shut down or be repaired at great cost.

**Corrosion** is a normal consequence of steam generation. Very hot, pressurized water flows from the core into the steam generator through thousands of 3/4-inch-diameter pipes immersed in water (Fig. 13–8). The water inside and outside these pipes contains corrosive chemicals that, over time, cause cracks to develop in some of the pipes. If the main line conveying steam from the generator to the turbine were to rupture, the sudden increase in pressure in the generator could cause several cracked pipes to break at once. In that case, radioactive moderator–coolant water would be released and would overload safety systems, forcing the plant to vent radioactive gas to the outside. Cracked pipes are "repaired" basically by plugging them, cutting their overall power. In March 1995, using a new high-tech probe, officials discovered that up to half of the steam generator pipes in the Maine Yankee plant had developed cracks, some penetrating 80% of the pipe's thickness. As a result, the plant was shut down after only 24 years of operation, and the owners of 70 other pressurized-water nuclear reactors were notified by the Nuclear Regulatory Commission that they, too, could be facing similar cracking problems. Recent problems of corrosion in reactor vessels in two power plants have both surprised and alarmed NRC officials; apparently the cooling water can be far more corrosive than originally expected.

***Decommissioning.*** Closing down, or decommissioning, a power plant can be extremely costly. Decommissioning Maine Yankee (Fig. 13–16) is expected to cost $635 million. (By comparison, the plant cost just $231 million to build in the late 1960s.) The owners of the plant are allowed to bill former customers to recover the costs—a "stranded cost" nobody will like. Faced with these costs, some utilities are opting to repair older plants in spite of the high costs of such repairs. Other closed reactors are on sites currently occupied by active units. The old, closed-down reactors are simply defueled and allowed to sit, until some day when all the reactors are closed and the entire site must be decommissioned. In two closed reactors, only the nuclear components were removed and the plants were converted to conventional natural-gas plants.

Overseas, nuclear power is also facing hard economic times in some regions. The most recent nuclear reactor to be built in the United Kingdom (and possibly the last for decades), Sizewell B, cost almost $3,000 per kilowatt of capacity, almost 10 times as much as a gas-fired plant. Government subsidies keep the French and Japanese from paying what would be very high prices for electricity. Yet, 31 countries are still committed to maintaining current nuclear power plants, and many are building new plants.

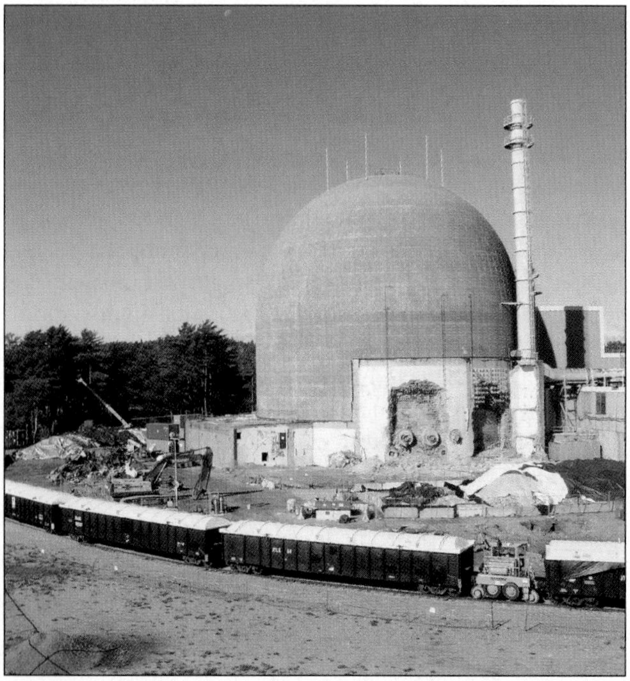

**Figure 13–16 Maine Yankee.** The Maine Yankee nuclear power plant is being decommissioned, at an anticipated cost of $635 million.

## 13.4 More Advanced Reactors

Uranium—especially $^{235}U$—is not a highly abundant mineral on Earth. At the height of optimism about nuclear energy in the 1960s, when as many as 1,000 plants were envisioned by the turn of the century, it was forecast that shortages of $^{235}U$ would develop. Breeder reactors, which utilize chain reactions, were seen as the solution to this problem.

### Breeder Reactors

Recall from Section 13.2 that when a $^{235}U$ atom fissions, two or three neutrons are ejected. Only one of these neutrons needs to hit another $^{235}U$ atom to sustain a chain reaction; the others are absorbed by something else. The breeder reactor is designed so that (nonfissionable) $^{238}U$ absorbs the extra neutrons. When this occurs, the $^{238}U$ is converted to plutonium ($^{239}Pu$), which can then be purified and used as a nuclear fuel, just like $^{235}U$. Thus, the breeder converts nonfissionable $^{238}U$ into fissionable $^{239}Pu$, and because the fission of $^{235}U$ generally produces two neutrons in addition to the one needed to sustain the chain reaction, the breeder may produce more fuel than it consumes. Furthermore, 99.7% of all uranium is $^{238}U$, so converting that to $^{239}Pu$ through breeder reactors effectively increases nuclear fuel reserves more than a hundredfold.

Breeder reactors present all of the problems and hazards of standard fission reactors, plus a few more. If a meltdown occurred in a breeder, the consequences would be much more serious than in an ordinary fission reactor, due to the large amounts of $^{239}Pu$, which has an exceedingly long half-life of 24,000 years. In addition, because plutonium can be purified and fabricated into nuclear weapons more easily than $^{235}U$ can, the potential for the diversion of breeder fuel to weapons production is greater. Hence, the safety and security precautions needed for breeder reactors are greater. Finally, breeder reactors are more expensive to build and operate.

With its scaled-down nuclear program, the United States currently has enough uranium stockpiled. Thus, there is no urgency for the nation to develop breeders. However, in the United States and elsewhere, small breeder reactors are operated for military purposes. France, Russia, and Japan are currently the only nations with commercial breeder reactors, and these have not been in continuous use. The global supply of uranium is not currently a problem, and a number of countries are reprocessing spent fuel, recovering plutonium and mixing it (at about 5% plutonium) with $^{238}U$ to produce mixed-oxide (MOX) fuel, which is suitable for further use in nuclear power plants.

### Fusion Reactors

The vast energy emitted by the Sun and other stars comes from fusion (Fig. 13–5b). The Sun, as well as other stars, is composed mostly of hydrogen. Solar energy is the result of the fusion of hydrogen nuclei (protons) into larger atoms, such as helium. Scientists have duplicated the process in the hydrogen bomb, but hydrogen bombs hardly constitute a useful release of energy. The aim of fusion technology is to carry out fusion in a controlled manner in order to provide a practical source of heat for boiling water to power steam turbogenerators.

***d–t Reaction.*** Since hydrogen is an abundant element on Earth (there are two atoms of it in every molecule of water) and helium is an inert, nonpolluting, nonradioactive gas, hydrogen fusion is promoted as the ultimate solution to all our energy problems; that is, fusion affords pollution-free energy from a virtually inexhaustible resource: water. However, most current designs do not use regular hydrogen ($^{1}H$), but rather employ the isotopes deuterium ($^{2}H$) and tritium ($^{3}H$) in what is called the *d–t reaction*. Deuterium is a naturally occurring nonradioactive isotope that can be extracted in almost any desired amount from the hydrogen in seawater. Tritium, by contrast, is an unstable gaseous radioactive isotope that must be produced artifically. Radioactive and difficult to contain, tritium is a hazardous substance. As a result, fusion reactors could easily become a source of radioactive tritium leaking into the environment, unless effective (and costly) designs were used to prevent that.

In the present state of the art, fusion power is still an energy consumer rather than a producer. The problem is that it takes an extremely high temperature—some 3 million degrees Celsius—and pressure to get hydrogen atoms to fuse. In the hydrogen bomb, the temperature and pressure are achieved by using a fission bomb as an igniter–an unthinkable way to initiate a sustained, controlled fusion reaction.

***Tokamak.*** A major technical problem is how to contain the hydrogen while it is being heated to the tremendously high temperatures required for fusion into helium. No known material can withstand these temperatures without vaporizing, although three techniques are being tested. One is the Tokamak design, in which ionized hydrogen is contained within a magnetic field while being heated to the necessary temperature. The second is laser fusion, in which a tiny pellet of frozen hydrogen is dropped into a bull's-eye, where it is hit simultaneously from all sides by powerful laser beams (Fig. 13–17). The beams heat and pressurize the pellet to the point of fusion. A third design is a device called the Z machine, which delivers very short, but powerful, pulses of current through an array of wires. This concentrated electrical energy results in a blast of X rays that recently reached 1.8 million degrees Celsius, close to the 2 to 3 million degrees it would take to initiate fusion.

Some fusion has been achieved in the Tokamak and laser fusion devices, and in late 1994 a Tokamak facility at the Princeton Plasma Physics Laboratory in Princeton,

**Figure 13–17** **Laser fusion.** In this experimental instrument at Lawrence Livermore Laboratory, 30 trillion watts of optical power are focused onto a tiny pellet of hydrogen smaller than a grain of sand and located in the center of a vacuum chamber. For less than a billionth of a second, the fusion fuel is heated and compressed to temperatures and densities near those found in the Sun.

New Jersey, achieved 10.7 megawatts of fusion power in 0.27 second of reaction. As yet, however, the break-even point has not been reached: More energy is required to run the magnets or lasers than is obtained by fusion. (It took 39.5 megawatts of energy to sustain the reaction at Princeton.) In the United States, fusion energy research is proceeding, with funding in the billions, an indication that this technology is considered an important part of the future energy scene.

*Pot of Gold?* Recently, Japanese and European researchers presented a design for an International Thermonuclear Experimental Reactor (ITER), a prototype fusion reactor of the Tokamak variety that is still in the planning stage. The design is expected to cost $5 billion, and the reactor will produce 10 times as much energy as it consumes. Recently, the United States rejoined the project, some 10 years after withdrawing because of concerns about costs and the science itself. The most optimistic view sees the ITER coming on line by 2013. As this plan indicates, developing, building, and testing a fusion power plant will require at least another 10 to 20 years and many billions of dollars. Additional plants would require more years. Thus, fusion is, at best, a very long term option. Many scientists believe that fusion power will always be the elusive pot of gold at the end of the rainbow. The standard joke about fusion power is that *it is the energy source of the future and always will be*!

## 13.5 The Future of Nuclear Power

With fossil fuels contributing to global climate change and alternatives such as solar power (Chapter 14) only in their infancy, the long-term outlook for energy is discouraging. Using nuclear power seems to be an obvious choice; however, many are opposed to nuclear energy, and it is presently the most costly option. Indeed, worldwide, public opposition to nuclear energy is higher now than it ever has been and raises the question of whether nuclear energy has any future once the currently operating plants have exhausted their life spans. In many countries of the European Union, nuclear power has a dim future. Sweden and Germany have agreed to phase out their nuclear power plants, even though nuclear power provides 46% and 30% of their electricity, respectively. Russia and other eastern European countries are holding on to their nuclear power as a present economic necessity. Only in Asia is nuclear power still being developed, with China, Japan, South Korea, and India seemingly intent on using it to fulfill a greater share of their energy needs.

## Opposition

Opposition to nuclear power is based on several premises:

- People have a general distrust of technology they do not understand, especially when that technology carries with it the potential for catastrophic accidents or the hidden, but real, capacity to induce cancer.
- Many observers are critical of the way nuclear technology is being managed. They are aware that the same agency (the Nuclear Regulatory Commission) responsible for licensing and safety regulations is also a strong supporter of the commercial nuclear industry.
- Problems involving lax safety, operator failures, and cover-ups by nuclear plants and their regulatory agencies have occurred in the United States, Canada, and Japan.
- The problems of high costs of construction and unexpectedly short operational lifetimes have already been mentioned. Thus, the economic argument can also be used to oppose nuclear power.
- The nuclear industry has repeatedly presented nuclear energy as extremely safe, arguing that the probabilities of accidents occurring are very low. However, when accidents do occur, probabilities become realities, and the arguments are moot.
- Nuclear power plants are viewed as prime targets for terrorist attacks, which could bring about a potentially devastating release of radioactivity. Critics argue that, therefore, it only makes sense to reduce the number of such targets, not add more of them.
- There remains the crucial problem of disposing of nuclear waste. All parties agree that the waste must be placed somewhere safe, but there the agreement ends, and siting a long-term nuclear waste repository has been a difficult political as well as technological problem in country after country.

Completely aside from these objections, there is the basic mismatch between nuclear power and the energy problem. The main energy problem for the United States is an eventual shortage of crude oil for transportation purposes, yet nuclear power produces electricity, which is not used for transportation. If we were moving toward a totally electric economy that included even electric cars, nuclear-generated electricity could be substituted for oil-based fuels. Unfortunately, electric cars have not yet proved practical, and the outlook for them in the near future remains uncertain. Consequently, nuclear power simply competes with coal-fired power in meeting the demands for base-load electrical power. Given the high costs and additional financial risks of nuclear power plants, coal is cheaper, and the United States does have abundant coal reserves.

Nonetheless, there are still the environmental problems of mining and burning coal, including acid precipitation and global climate change. As discussed in Chapter 12, burning coal emits more $CO_2$ per unit of energy produced than any other form of energy does. If the long-term environmental costs were factored into the price of coal, that price would be considerably higher than it is now. At the same time, the costs of the long-term confinement of nuclear wastes and the decommissioning of power plants have yet to be factored into the cost of nuclear power.

In the final analysis, nuclear power may be a unique energy option, but it is also a controversial technology. Interestingly, public opinion about nuclear power seems to be shifting. In a 2001 Associated Press poll, a majority of those polled supported nuclear power and agreed that nuclear power plants were safer now than they were 10 years ago. Opposition continues, however, because of the perceived risks to both human health and the environment and the mismatch to our most critical energy problem: an impending shortage of fuel for transportation. Opponents don't believe that the benefits are worth the risks.

## Rebirth of Nuclear Power?

What would it take to revitalize the nuclear power option? If nuclear energy is to have a brighter future, it may be because we have found the continued use of fossil fuels to be so damaging to the atmosphere that we have placed limits on their use, but have not been able to develop adequate alternative energy sources. Nuclear supporters point out that U.S. nuclear power plants prevent the annual release of 164 million tons of carbon, 5.1 million tons of sulfur dioxide, and 2.4 million tons of nitrogen oxides. Observers agree that if the rebirth of nuclear power is to come, a number of changes will have to be made:

- Reactor safety concerns will have to be addressed, perhaps through the use of smaller, advanced light-water reactor designs with built-in passive safety features.
- The potential for terrorist attacks and sabotage will also have to be addressed. The public must be convinced that every possible preventative measure has been taken.
- The industry's manufacturing philosophy will have to change to favor standard designs and factory production of the smaller reactors, instead of the custom-built reactors presently in use in the United States. Recently, the DOE selected three designs for future development, one of which—the Westinghouse advanced pressurized-water reactor—is a 600-megawatt unit similar to the one pictured in Fig. 13–15. The unit, designed for modular construction, is based primarily on passive safety. Westinghouse claims that it would take no more than three years to build such a unit.
- The framework for licensing and monitoring reactors must be streamlined, but without sacrificing safety concerns. This has largely been accomplished with the National Energy Policy Act of 1992. The rules

allow the NRC to approve a plant design prior to the selection of a specific site.

- The political problems of siting new reactors must be resolved. Perhaps the best sites will be on the grounds of present or closed reactors, taking advantage of the existing infrastructure and the familiarity with (and acceptance of) a nuclear facility.
- The waste dilemma must be resolved. Now that the Yucca Mountain waste repository has been approved by the president and the Congress, resolving the dilemma is becoming more likely; however, legal challenges to the Nevada repository are still pending.
- Political leadership will be required to accomplish all these developments, and this seems to be happening. President George W. Bush has made expanding nuclear energy a major component of his energy policy, as recommended by Vice President Cheney's *National Energy Policy* report, by taking the following steps: (1) proposing the Nuclear Power 2010 program, with plans to identify new sites for nuclear power plants, to streamline regulations associated with site licensing, and to encourage industry to build new plants that could become operational by 2010; (2) moving toward a resolution of the nuclear waste problem by approving the Yucca Mountain site for a repository; and (3) backing a congressional bill that would give federal loan guarantees for building up to six new nuclear power plants and that would authorize the construction of a special plant that would produce hydrogen from water, a key part of a coming fuel cell technology for automobiles (see Chapter 14).

# revisiting the themes

## Sustainability

Looking toward a sustainable energy future, many believe that nuclear power has a role to play in getting us there. Because uranium must be mined and is not an abundant mineral, uranium supplies will not last indefinitely. Reprocessing fuel can extend the life of nuclear power, but it carries risks that may be unacceptable. Still, nuclear fission may be a way of bridging the gap between fossil-fuel energy and renewable energy. The most important aspect of fission is that nuclear energy does not release greenhouse gases and would be capable of providing most of our electrical power, thus enabling us to avoid relying on coal.

The current dilemma is how to deal with the buildup of the highly radioactive spent-fuel wastes. Storing them in pools and casks on the grounds of the power plant is unsustainable. Consequently, countries using nuclear power must find ways to dispose of the wastes safely, which means making sure that they are contained for thousands of years.

The prospect of fusion energy continues to be intriguing, as fuel sources are abundant. The technology is extremely daunting, however, and expensive. At present, fusion looks like a dubious option.

## Sound Science

The science behind nuclear power was developed during World War II, and it was used to produce the two atomic bombs that ended the war with Japan in 1945. Nuclear physicists were eager to show that atomic energy could be put to beneficial uses, however, and with that in mind, the technology for nuclear power plants was developed. Sound science was, and continues to be, essential for this enterprise, especially in regard to improvements in safety and efficiency. Nevertheless, scientific uncertainties remain. For example, what is a safe level of exposure to radioactivity, how can the wastes be stored safely for thousands of years to come, can nuclear reactors be designed to be accident proof, and can fusion energy ever become a commercial reality?

## Policy and Politics

It was a major political decision to develop nuclear power for generating electricity. The technology is so expensive and fraught with risk that only government-level agencies can manage the development and ongoing regulation of nuclear power. There is a danger in this, however, in that one agency sets and enforces standards and at the same time is seen as promoting the technology. Some separation of responsibilities is in force today, with the EPA now overseeing the environmental standards for exposure to radiation.

Government subsidies and guarantees remain in place for this enterprise, and given its potential for massive consequences if things go wrong, that may continue to be necessary. International agreements, such as those between the United States and Russia to dismantle nuclear warheads, point to the importance of maintaining close political ties among all countries with nuclear weapons capabilities or even with nuclear power plants (since these can be a source of weapons-grade nuclear material).

The Bush administration has made nuclear energy an important element of its energy policy. This is a departure from the previous administration and represents a calculated political risk—one that the president seems to have weathered successfully so far. The decision to move forward with Yucca Mountain is part of this policy. Bush may have lost Nevada to the Democratic party in doing so, but he certainly has gained support from states with nuclear power plants burdened with wastes that have no place to go.

### Globalization

The global war against terrorism must take the threat of an attack on nuclear facilities seriously. The United States has stepped up its security measures at power plants, although some fear that those measures are inadequate for the task. Another serious security issue is the existence of highly enriched weapons material, some now in the breakaway countries of the former Soviet Union. Attempts have already been made to steal some of that material.

The global context of nuclear power was made terribly evident when the Chernobyl plant exploded and emitted radioactive particles that spread for thousands of miles. The consequences of Chernobyl continue, with elevated rates of some cancers in areas hundreds of miles from the plant. Estimations of the global toll of Chernobyl range from 140,000 to 475,000 cancer deaths, most of which will never be traceable to their origin.

## review questions

1. Compare the outlook regarding the use of nuclear power in the United States and globally in each of the following decades: 1960s, 1970s, 1980s, 1990s, and early 21st century.
2. Describe how energy is produced in a nuclear reaction. Distinguish between fission and fusion.
3. How do nuclear reactors and nuclear power plants work?
4. What environmental advantages does nuclear power offer over coal energy?
5. How are radioactive wastes produced, and what are the associated hazards?
6. Describe the two stages of nuclear waste disposal.
7. What problems are associated with the long-term containment of nuclear waste? What is the current status of the disposal situation?
8. Describe what happened at Chernobyl and Three Mile Island.
9. What features might make a nuclear power plant safer? What about terrorism?
10. What are the four generations of nuclear reactors? Describe the advantages of Generations III and IV.
11. Discuss economic reasons that have caused many utilities to opt for coal-burning, rather than nuclear-powered, plants.
12. How do breeder and fusion reactors work? Does either one offer promise for alleviating our energy shortage?
13. Explain why nuclear power does little to address our shortfall in crude-oil production.
14. Discuss changes in nuclear power that might brighten its future.

## thinking environmentally

1. Discuss the risks of nuclear power. Are we overly concerned or not concerned enough about nuclear accidents? Could an accident like Chernobyl happen in the United States?
2. Would you rather live next door to a coal-burning plant or a nuclear power plant? Defend your choice.
3. Global climate change has been cited by nuclear power proponents as one of the most important justifications for further development of the nuclear power option. Give reasons for their belief, and then cite some reasons that would counter their argument.
4. Go to the Web site www.pbmr.com and investigate the reactor described there. Where is it being designed, and where will it likely be used first? What are its features?

# Renewable Energy

### Key Topics

1. Putting Solar Energy to Work
2. Indirect Solar Energy
3. Renewable Energy for Transportation
4. Additional Renewable-Energy Options
5. Policy for a Sustainable-Energy Future

The Hummingbird Highway, one of the main roads in Belize, Central America, runs southeasterly from the capital of Belmopan, through second-growth rain forests and orange orchards, to Dandriga on the coast. About 15 miles down the highway from Belmopan, a small dirt road leads into Jaguar Creek, an environmental research and education center whose mission is to serve the Earth and the poor through sustainable development strategies. With 14 buildings occupying nearly 10,000 square feet, Jaguar Creek accommodates up to 40 people who might be there for short-term conferences or semester-long college courses. Belize is a developing country, and rural electrification is one facet that is still developing.

Many miles from any electrical power, Jaguar Creek carries on with refrigeration, lighting, washing machines, and fans in every living space. This "off-the-grid" facility has its own solar electrical system, which covers the roof of the operations building (Fig. 14–1) and which is capable of generating 50,000 watts (50 KW) of power. Backup for rainy periods is provided by a propane-powered generator. The total cost for the system, including installation, was $100,000, which breaks down to about $5.00 per watt, and the system has a life expectancy of at least 25 years. Compared with the cost of a utility system with poles, wires, transformers, and power supply, the price is a bargain. Local utility service is estimated to cost at least $25,000 per mile, and Jaguar Creek is 10 miles from the nearest power line. This sustainable-energy system is consistent with Jaguar Creek's Earth-friendly approach, which also includes composting toilets and gray-water leaching fields.

***Wind Power in Vermont*** **Green Mountain Power Corporation installed 11 wind turbines in Searsburg, Vermont, in 1997, making Searsburg the site of the largest wind power–generating station in the eastern United States. On moderately windy days, the turbines provide enough power to satisfy the needs of 2,000 homes.**

***Global Scene.*** Similar panels of solar cells are providing electricity in both developed and developing countries around the world. Throughout Israel and other countries in warm climates, it is now commonplace to have water heated by the Sun (Fig. 14–2). Even in temperate climates, many people have discovered that proper building design, insulation, and simple solar collection devices can reduce energy bills 75% or more. In the desert northeast

**Figure 14–1 Jaguar Creek.** Panels of photovoltaic cells power the Jaguar Creek facility operated by Target Earth in Belize. The system generates 50 KW of energy, enough to power this facility, which houses up to 40 people.

of Los Angeles are "farms" with rows of trough-shaped mirrors tipped toward the Sun (Fig. 14–3). These reflectors are focusing the Sun's rays to boil water or synthetic oil and drive turbogenerators. In the hills east of San Francisco, regiments of *wind turbines* standing in rows up the slopes and over the crests of the hills are producing electrical power equivalent to that produced by a large coal-fired power plant. Wind turbines are also becoming commonplace across northern Europe and are sprouting up as well in India, Mexico, Argentina, New Zealand, and other countries around the world.

**Figure 14–2 Rooftop hot water.** In warm climates, solar hot-water heaters are becoming commonplace. Note the solar water-heating panels on each of the homes in this development in southern California.

**Figure 14–3 Solar thermal power in southern California.** A solar-trough power plant. Sunlight striking the parabolic-shaped mirrors is reflected onto the central pipe, where it heats a fluid that is used, in turn, to boil water and drive turbogenerators.

The inertia of "business as usual" has kept the world on a track of growing dependence on fossil fuels, with nuclear power retaining its share of energy production for the present. In the meantime, as the preceding examples demonstrate, the use of energy from the Sun and wind has been making quiet, but steady, progress, to the point where these renewable sources of energy are becoming cost competitive with traditional energy sources—and are far more practical in many situations. Given the unsustainability of our present course—the limited supplies of fossil fuels and their polluting impacts on global climate and the air we breathe—a renewable-energy future is absolutely essential. Thus, the world has at its disposal the possibility of moving toward a nonpolluting, inexhaustible energy economy based on using the Sun's current energy output.

The objective in this chapter is to give you a greater understanding of the potential for sustainable energy from sunlight, wind, biomass, and other sources. Currently, renewable energy provides 6% of U.S. primary energy (Fig. 14–4); worldwide, renewable energy provides 14% of energy use.

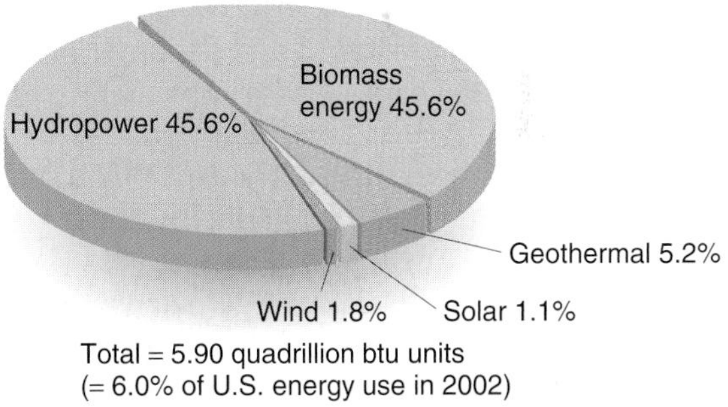


**Figure 14–4 Renewable-energy use in the United States.** A mix of sources of renewable energy provided 6% of the nation's energy use in 2002. (*Source:* Data from Department of Energy, Energy Information Agency, *Renewable Energy Annual 2002,* November 2003.)

## 14.1 Putting Solar Energy to Work

### Principles of Solar Energy

Before turning our attention to the practical ways that solar energy is captured and used, let us consider some general solar-energy concepts.

Solar energy originates with thermonuclear fusion in the Sun. (Importantly, all the chemical and radioactive polluting by-products of the reactions remain behind on the Sun.) The solar energy reaching Earth is radiant energy, entering at the top of the atmosphere at 1,370 watts per square meter, the **solar constant.** This energy ranges from ultraviolet light to visible light

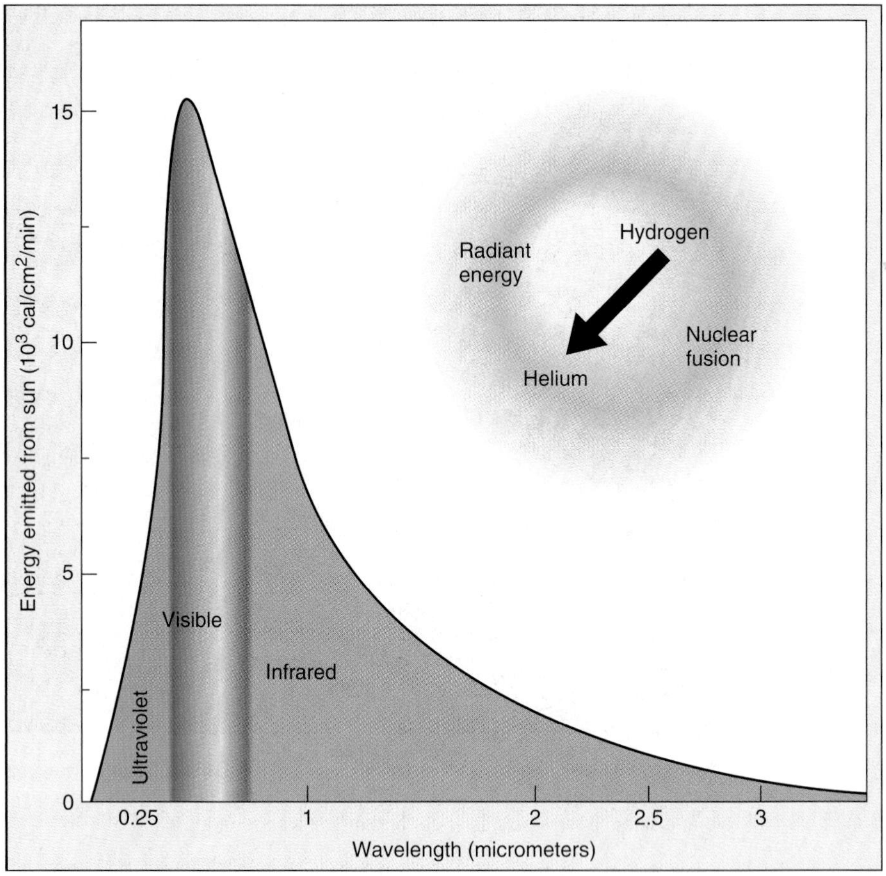


**Figure 14–5 The solar-energy spectrum.** Equal amounts of energy are found in the visible-light and the infrared regions of the spectrum.

and infrared (heat energy) (Fig. 14–5). About half of this energy actually makes it to Earth's surface; 30% is reflected, and 20% is absorbed by the atmosphere. Thus, full sunlight can deliver about 700 watts per square meter to Earth's surface when the Sun is directly overhead. At that rate, the Sun can deliver 700 megawatts of power (the output of a large power plant) to an area of 1,000 square kilometers (390 square miles).

The total amount of solar energy reaching Earth is vast almost beyond belief. Just 40 minutes of sunlight striking the land surface of the United States yields the equivalent energy of a year's expenditure of fossil fuel. The Sun delivers 10,000 times the energy used by humans.

Moreover, using some of this solar energy will not change the basic energy balance of the biosphere. Solar energy absorbed by water or land surfaces is converted to heat energy and eventually lost to outer space. Even the fraction that is absorbed by vegetation and used in photosynthesis is ultimately given off again in the form of heat energy as various consumers break down food (Chapter 3). Similarly, if humans were to capture and obtain useful work from solar energy, it would still ultimately be converted to heat and lost in accordance with the second law of thermodynamics. (See page 62.) The overall energy balance would not change.

*A Diffuse Source.* Although solar energy is an abundant source, it is also diffuse (widely scattered), varying with the season, latitude, and atmospheric conditions. The main problem associated with using solar energy is one of taking a diffuse and intermittent source and concentrating it into an amount and form, such as fuel or electricity, that can be used as heat and to run vehicles, appliances, and other machinery. In addition, what do you do when the Sun is not shining? These problems involve the *collection, conversion,* and *storage* of solar energy. Also, in the final analysis, overcoming such hurdles must be cost effective.

Consider how natural ecosystems address these three problems. Plant leaves are excellent solar *collectors,* gathering light over a wide area. With the aid of chlorophyll, photosynthesis *converts* and *stores* light energy as chemical energy—namely, starch and other forms of organic matter—in the plant. The plant biomass then "fuels" the rest of the ecosystem. In the sections that follow, you will see how we can overcome or, even better, sidestep these three problems in various ways to meet our needs from solar energy in a cost-effective manner.

## Solar Heating of Water

Solar hot-water heating is already popular in warm, sunny climates. A solar collector for heating water consists of a thin, broad box with a glass or clear plastic top and a black

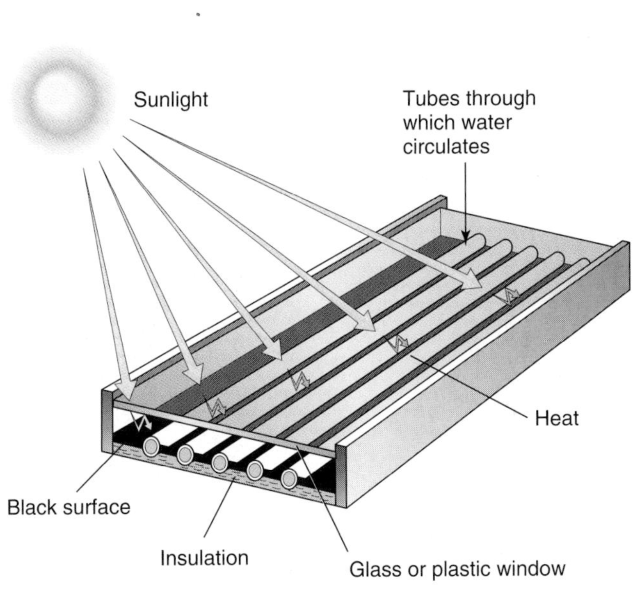


**Figure 14–6 The principle of a flat-plate solar collector.** As it is absorbed by a black surface, sunlight is converted to heat. A clear glass or plastic window over the surface allows the sunlight to enter the collector, which traps the heat. Air or water is heated as it passes over and through tubes embedded in the black surface.

bottom with water tubes embedded within (Fig. 14–6). Such collectors are called **flat-plate collectors.** Faced toward the Sun, the black bottom gets hot as it absorbs sunlight—similar to how black pavement heats up in the Sun—and the clear cover prevents the heat from escaping, as in a greenhouse. Water circulating through the tubes is thus heated and is conveyed to a tank where it is stored.

In an *active system,* the heated water is moved by means of a pump. In a *passive system,* natural convection currents are used. In a passive solar water-heating system, the system must be mounted so that the collector is lower than the tank. Thus, heated water from the collector rises by natural convection into the tank, while cooler water from the tank descends into the collector (Fig. 14–7). There are no pumps to buy, maintain, or remember to turn on and off. The size and number of collectors are adjusted to the need. The greatest economy is gained by carefully managing the requirements for hot water, so that the sizes of the tank and collectors can be minimized.

In temperate climates, where water in the system might freeze, the system may be adapted to include a heat-exchange coil within the hot-water tank. Then, antifreeze fluid is circulated between the collector and the tank. In the United States, over 1 million solar hot-water systems have been sold, but this is still only a small fraction of the total number of all hot-water heaters. The reason is the initial cost, which is 5 to 10 times as much as gas or electric heaters. However, over time, the solar system costs less to operate than an electric or gas system.

In recent years, the most common end use of solar thermal systems has been for heating swimming pools. Systems used for this purpose are inexpensive and consist of black plastic or rubberlike sheets with plastic tubing that circulates water through the system internals.

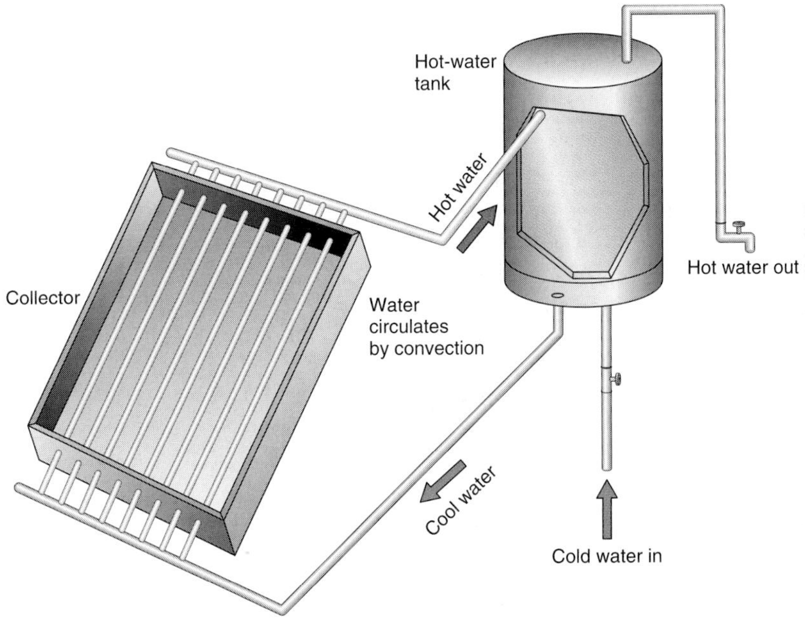


(a)

(b)

**Figure 14–7 Solar water heaters.** (a) In nonfreezing climates, simple water-convection systems may suffice. In freezing climates, an antifreeze fluid is circulated. (b) Solar panels (flat black collectors) and hot-water tanks on the roof of a hotel in Barbados.

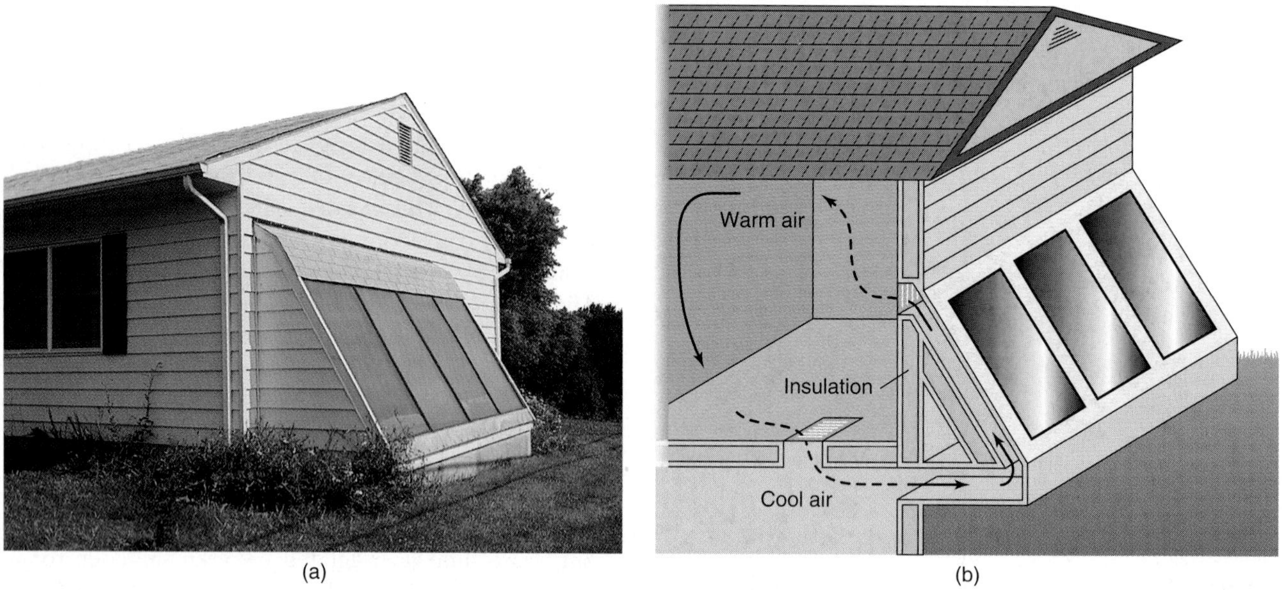


**Figure 14–8 Passive hot-air solar heating.** (a) Many homeowners could save on fuel bills by adding homemade solar collectors such as that shown here. (b) Air heated in the collector moves into the room by passive convection.

## Solar Space Heating

The same concept for heating water with the Sun can be applied to heat spaces. Flat-plate collectors such as those used in water heating can be used for space heating. Indeed, the collectors for space heating may even be less expensive, homemade devices, because it is necessary only to have air circulate through the collector box. Again, efficiency is gained if the collectors are mounted to allow natural convection to circulate the heated air into the space to be heated (Fig. 14–8).

***Building = Collector.*** The greatest efficiency in solar space heating, however, is gained by designing a building to act as its own collector. The basic principle is to have windows facing the sun. In the winter, because of the Sun's angle of incidence, sunlight can come in and heat the interior of the building (Fig. 14–9a). At night, insulated drapes or shades can be pulled down to trap the heat inside. The well-insulated building, with appropriately made doors and windows, would act as its own best heat-storage unit. Beyond good insulation, other systems for storing heat, such as tanks of water or masses of rocks, have not proved cost effective. Excessive heat load in the summer can be avoided by using an awning or overhang to shield the window from the high summer Sun (Fig. 14–9b).

***Landscaping.*** Along with design, positioning, and improved insulation, appropriate landscaping can contribute to the heating and cooling efficiency of both solar and nonsolar space-heating designs. In particular, deciduous trees or vines on the sunny side of a building can block much of the excessive summer heat while letting

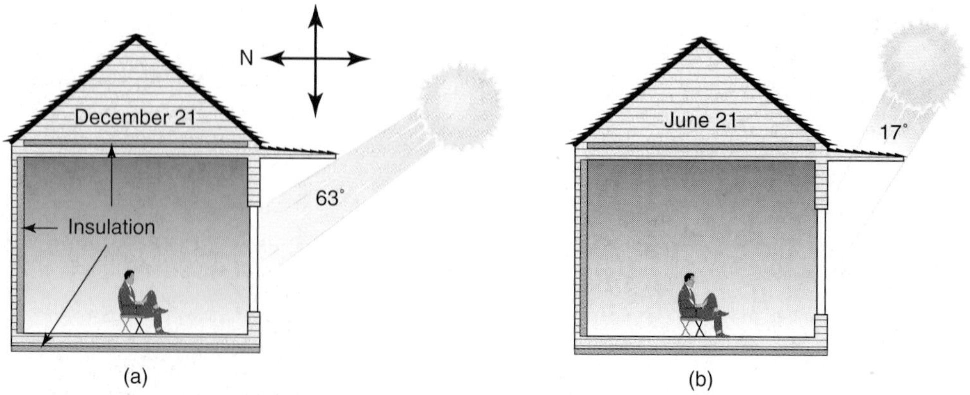


**Figure 14–9 Solar building siting.** In contrast to utilizing expensive and complex active solar systems, solar heating *can* be achieved by suitable architecture and orientation of the home at little or no additional cost. (a) The fundamental feature is large, Sun-facing windows that permit sunlight to enter during the winter months. Insulating drapes or shades are drawn to hold in the heat when the Sun is not shining. (b) Suitable overhangs, awnings, and deciduous plantings will prevent excessive heating in the summer.

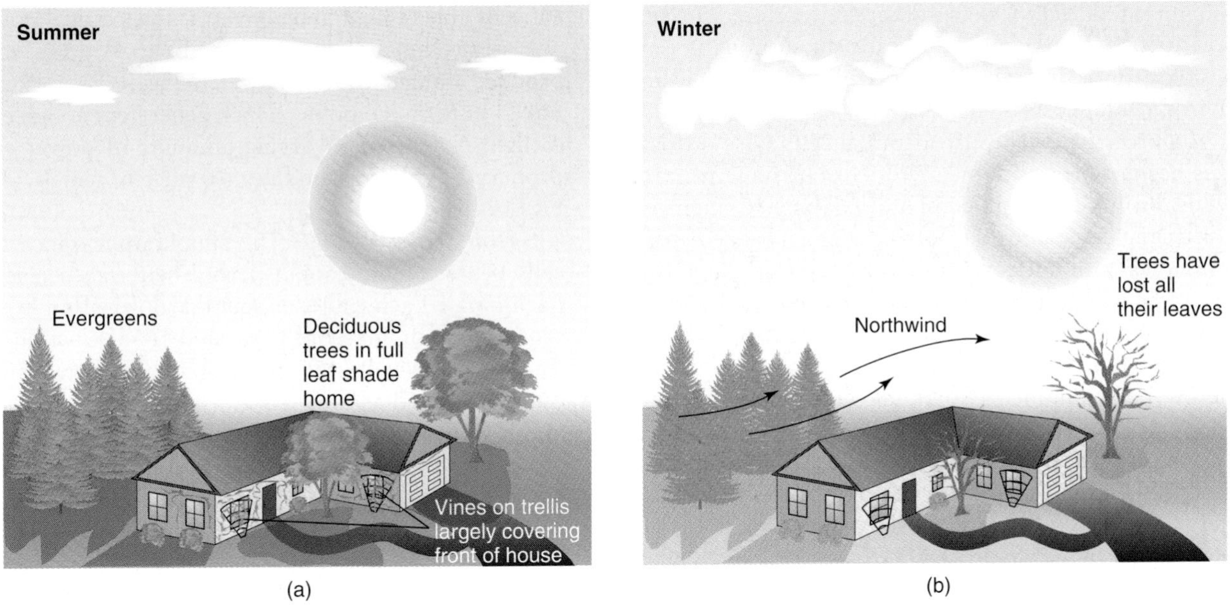


**Figure 14–10 Landscaping in solar heating and cooling.** (a) In summer, the house may be shaded with deciduous trees or vines. (b) In winter, leaves drop, and the bare trees allow the house to benefit from sunlight. Evergreen trees on the opposite side protect against, and provide insulation from, cold winds.

the desired winter sunlight pass through. An evergreen hedge on the shady side can provide protection from the cold (Fig. 14–10).

***Earth-sheltered Housing.*** One approach that combines insulation and solar heating is **earth-sheltered housing** (Fig. 14–11). The principle is to use earth as a form of insulation and orient the building for passive solar energy. Two strategies are employed. One is to build up earth against the building's walls (earth berms) in a more conventional house design; the other is to cover the building to varying degrees with earth, exposing the interior to the outside with windows facing southward. Earth has a high capacity for heat storage, and since the walls and floors of an earth-sheltered home are usually built with poured concrete or masonry, at night the building will radiate the heat stored during the day and warm the house. In the summer, the building is kept cool by its contact with earth, although moisture often will have to be controlled with dehumidifiers.

**Figure 14–11 Earth Hall.** Earth-sheltered building at Au Sable Institute of Environmental Studies in Mancelona, Michigan, houses offices, classrooms, and laboratories for the institute.

***Energy Stars.*** In almost any climate, a well-designed passive solar-energy building can reduce energy bills substantially with an added construction cost of only 5% to 10%. Since about 25% of our primary energy is used for space and water heating, proper solar design, broadly adopted, would create enormous savings on oil, natural gas, and electrical power. In 2001, the EPA extended its **Energy Star** program to buildings and began awarding the Energy Star label to public and corporate edifices. Buildings awarded the label use at least 40% less energy than others in their class and must pass a number of other criteria to qualify. In two years, over 1,000 buildings have earned the label, saving $130 million and reducing carbon dioxide emissions by 2.6 billion pounds.

A common criticism of solar heating is that a backup heating system is still required for periods of inclement weather. Good insulation is a major part of the answer to this criticism. People with well-insulated solar homes find that they have minimal need for backup heating. When backup is needed, a small wood stove or gas heater suffices. In any case, the criticism concerning the need for backup heating misses the point: The objective of solar heating is to reduce our dependency on conventional fuels. Even if solar heating and improved insulation reduced the demand for conventional fuels by a mere 20%, that would still represent sustainable savings of 20% of the traditional fuel and its economic and environmental costs.

***Opposition.*** The effort to reduce the use of fossil fuels has not pleased some sectors of our society. In the 1980s, utility and oil companies launched intensive advertising campaigns purporting that solar energy was impractical and not cost effective. These campaigns did much to curtail the growth of the solar-energy industry. In fact, fossil-fuel interests successfully lobbied the government to end incentive programs for renewable energy. For example, a solar tax credit program that had stimulated a boom in solar hot-water heating systems expired in 1985. In the 1990s, the major factor in discouraging the use of solar heating was relatively low energy prices, which were a disincentive to making any change. Given an opportunity to set targets for the worldwide use of renewable energy, the 2002 World Summit gave in to pressures from the fossil-fuel industries and governments with heavy investments in fossil fuels, ending with a watered-down call to increase renewable energy.

## Solar Production of Electricity

Solar energy can also be used to produce electrical power, thus providing an alternative to coal and nuclear power. Currently, a few methods are feasible, and two in particular, *photovoltaic cells* and *solar-trough collectors,* are proving to be economically viable.

**Photovoltaic Cells.** A solar cell, more properly called a **photovoltaic,** or **PV, cell,** looks like a simple wafer of material with one wire attached to the top and one to the bottom (Fig. 14–12). As sunlight shines on the wafer, it puts out an amount of electrical current roughly equivalent to that emitted by a flashlight battery. Thus, PV cells collect light and convert it to electrical power in one step. The cells, measuring about 4 inches square, produce about 1 watt of power. Some 40 PV cells can be linked to form a module, which generates enough energy to light a lightbulb. Varying amounts of power can be produced by wiring modules together in panels. (Recall Fig. 14–1.)

**Figure 14–12 Photovoltaic cell.** Converting light to electrical energy, this cell provides enough energy to run the small electric motor needed to turn the fan blades.

***How They Work.*** The simple appearance of PV cells belies a highly sophisticated materials science and technology. Each cell consists of two very thin layers of semiconductor material separated by a junction layer. The lower layer has atoms with single electrons in the outer orbital that are easily lost. The upper layer has atoms lacking electrons in their outer orbital; these atoms readily gain electrons. (See Appendix C for some basic chemical principles.) The kinetic energy of light photons striking the two-layer "sandwich" dislodges electrons from the lower layer, creating an electrical potential between the layers. The potential provides the energy for an electrical current to flow through the rest of the circuit. Electrons from the lower side flow through a motor or some other electrical device back to the upper side. Thus, with no moving parts, solar cells convert light energy directly to electrical power with an efficiency of about 20%.

Because they have no moving parts, solar cells do not wear out. However, deterioration due to exposure to the weather limits their life span to about 20 years. The major material used in PV cells is silicon, one of the most abundant elements on Earth, so there is little danger that the production of PV cells will ever suffer because of limited resources. The cost of these cells lies mainly in their sophisticated design and construction.

*Uses.* PV cells are already in common use in pocket calculators, watches, and numerous toys. Panels of PV cells provide power for rural homes, irrigation pumps, traffic signals, radio transmitters, lighthouses, offshore oil-drilling platforms, and other installations that are distant from power lines. It is not hard to imagine a future in which every home and building has its own source of pollution-free, sustainable electrical power from an array of PV panels on the roof. Current installations in many countries and in over half of the states in the United States involve "net metering," where the rooftop electrical output is subtracted from the customer's use of power from the power grid. More than 200,000 homes in the United States now obtain some of their electric power from solar panels.

*Cost.* The cost of PV power (cents per kilowatt-hour) is the cost of the PV cells, divided by the total amount of power they may be expected to produce over their lifetime (currently from 25 cents to a dollar per kilowatt-hour). This cost must be compared with that of other power alternatives (8–10 cents per kilowatt-hour for residential electricity). The first PV cells had a cost factor several hundred times that of electricity from conventional power stations transmitted through the power grid, so those cells were used mainly in areas far from the

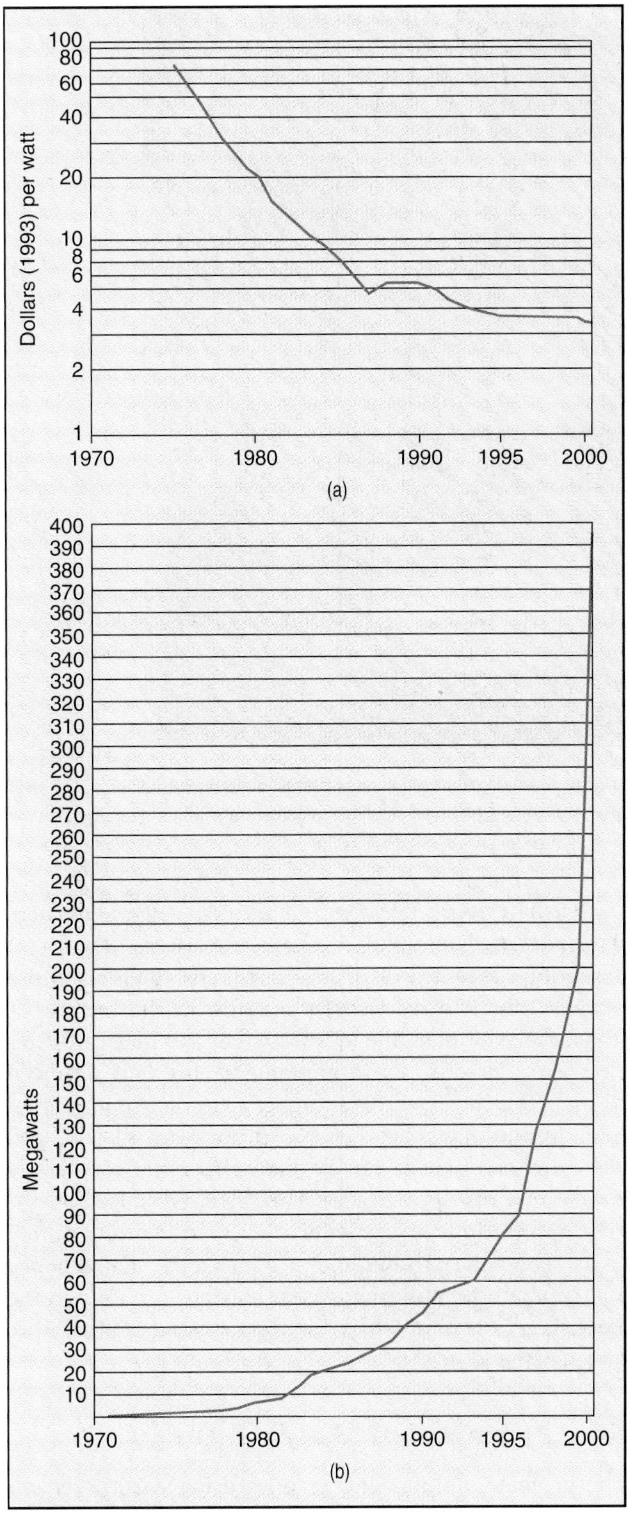


**Figure 14–13 The market for PV cells.**
(a) Through research and development, the price of photovoltaic cells has come down about twentyfold in the past 25 years.
(b) As prices have come down, sales have increased dramatically. Note the surge in sales for 2001—to 395 MW.
(*Source: Renewable Energy Annual 2002,* Energy Information Agency, Department of Energy.)

grid. PV power had its first significant application in the 1950s, in the solar panels of space satellites. This application, in fact, started the development cycle rolling. As more efficient cells and less expensive production techniques evolved, costs came down. In turn, applications, sales, and potential markets expanded, creating the incentive for further development (Fig. 14–13). Recently, the price of PV power dropped to $3,400 per kilowatt, continuing a steady decline. In response to this trend, the industry is growing rapidly. In 2002, the industry shipped solar panels worth $3 billion.

***Utilities.*** What will utility companies do as PV power becomes competitive? A demonstration PV power plant has been built in southern California (Fig. 14–14). However, the more promising future for PV power would appear to be in the installation of home-sized systems on rooftops, thus avoiding additional land and transmission costs. Toward this end, a local utility in Sacramento, California, has launched its PV Pioneers program. The utility installs the systems and pays homeowners $4 per month to maintain a 2- to 4-kw PV system on their roofs. The systems feed into the power grid and produce 3,600 kwhr/year; their cost, including installation, is expected to continue to drop until the power they generate will cost much the same as peak-load power. To date, 550 systems have been installed. A similar program in Japan has led to the installation of more than 30,000 systems in homes.

***Million Solar Roofs.*** A new federal program, the Million Solar Roofs Initiative, aims at encouraging the installation of solar energy units on a million residential and commercial rooftops by 2010, but the initiative provides few financial incentives, focusing instead on "partnering" with other organizations. The program seeks to remove market barriers and strengthen grassroots demand for solar technologies. Before such systems are widespread in the United States, policy changes will be needed to provide substantial incentives through tax credits and other means.

***New PV Technologies.*** To accelerate the market share of solar power, the cost of solar cells needs to drop dramatically. At least two new technologies promise to accomplish this: (1) thin-film PV cells, in which cheap amorphous silicon is used instead of expensive silicon crystals and the film can be applied as a coatings on roofing tiles or glass; and (2) electrically conductive plastics that are fashioned into inexpensive solar cells. Currently, the materials are too sensitive to weather conditions, but research and investment in the field are pushing forward at a rapid pace.

**Concentrating Solar Power.** With government funding, several technologies have been developed that convert solar energy into electricity by using reflectors (or concentrators) such as mirrors to focus concentrated sunlight onto a receiver that transfers the heat to a conventional turbogenerator. Unfortunately, these devices work well only in regions with abundant sunlight.

**Figure 14–14** **PV power plant.** The world's first photovoltaic (solar cell) power plant, located near Bakersfield, California. The array of 220 thirty-four-foot (eleven-m) panels produces 6.5 MW at peak, enough energy for 6,500 homes.

*Solar Trough.* One method that is proving to be cost effective is the **solar-trough** collector system, so named because the collectors are long, trough-shaped reflectors tilted toward the Sun (Fig. 14–3). The curvature of the trough is such that all of the sunlight hitting the collector is reflected onto a pipe running down the center of the system. Oil or some other heat-absorbing fluid circulating through the pipe is thus heated to very high temperatures. The heated fluid is passed through a heat exchanger to boil water and produce steam for driving a turbogenerator.

Nine solar-trough facilities in the Mojave Desert of California are now connected to the Southern California Edison utility grid. With a combined capacity of 350 MW, these facilities have about one-third the capacity of a large nuclear power plant and are converting a remarkable 22% of incoming sunlight to electrical power at a cost of 10 cents per kilowatt-hour, barely more than the cost from coal-fired facilities.

*Power Tower.* A "power tower" is an array of Sun-tracking mirrors that focuses the sunlight falling on several acres of land onto a receiver mounted on a tower in the center of the area (Fig. 14–15). The receiver transfers the heat energy collected to a molten-salt liquid, which then flows either to a heat exchanger to drive a conventional turbogenerator or to a tank at the bottom of the tower to store the heat for later use. The latest pilot plant model of this technology, Solar Two, is now connected to the region's utility grid and is generating 10 MW of electricity, enough for 10,000 homes.

*Dish–Engine System.* The dish–engine system (Fig. 14–16) is a smaller system, consisting of a set of parabolic concentrator dishes that focus sunlight onto a receiver. Fluid (often molten sodium) in the receiver is transferred to an engine (a Stirling engine) that generates electricity directly. Dish–engine systems have demonstrated efficiencies of 30%, higher than that of any other solar technology. They generate from 5 to 40 KW, and any number of them can be linked together to provide larger amounts of electricity. They are well suited for off-grid applications to provide power to remote, sunny areas. The newer technologies require little maintenance and generate no polluting gases, but they are only in the early stages of commercial development and are therefore quite costly.

## The Promise of Solar Energy

The promise of solar energy does bring with it certain disadvantages. First, the available technologies are still more expensive than conventional energy sources, although the costs are continuing to come down. Second, solar energy works only during the day, so it requires a backup energy source or a storage battery for nighttime use. Also, the climate is not sunny enough to use solar energy in the winter in many parts of the world. Against all these drawbacks should be weighed the hidden costs that are not included in the cost of power from traditional sources: air pollution, strip

**Figure 14–15 Power tower Solar Two.** Sun-tracking mirrors are used to focus a broad area of sunlight onto a molten-salt receiver mounted on the tower in the center. The hot salt is stored or pumped through a steam generator that drives a conventional turbogenerator.

mining, greenhouse gas emissions, and nuclear waste disposal, among others.

***Matching Demand.*** The Sun provides power only during the day, but 70% of electrical demand occurs during daytime hours, when industries, offices, and stores are in operation. Thus, proponents of solar energy argue that savings still can be achieved by using solar panels just for daytime needs and continuing to rely on conventional sources at night. All the solar power that can conceivably be built in the next 20 years or more could be used to supply the daytime demand, leaving traditional fuels to satisfy nighttime demand. In particular, the demand for air-conditioning, which, after refrigeration, is the second largest power consumer, is well matched to energy from PV cells. In addition, solar-powered air conditioners could operate independently from the rest of the electrical system, thus avoiding the costs of connecting the two systems. (Such air conditioners may be expected to come onto the market within the next few years.) In the long run, the nighttime load might be carried by forms of indirect solar energy, such as wind power and hydropower (discussed later in the chapter).

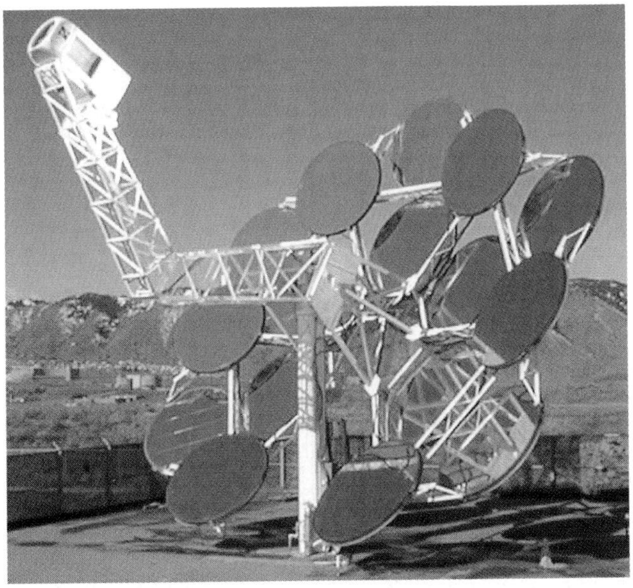

**Figure 14–16 Solar dish–engine system.** An array of circular mirrors focuses solar energy on a receiver, which transmits the energy to an engine that generates electric power.

According to Table 12–1, about 72% of our electrical power (56% worldwide) is currently generated by coal-burning and nuclear power plants. Therefore, the development of solar electrical power can be seen as gradually reducing the need for coal and nuclear power. (See "Earth Watch" essay, p. 386.) Solar power has also proved suitable for meeting the needs of electrifying villages and towns in the developing world, where, because centralized power is unavailable, rural electrification projects based on PV cells are already beginning to spread throughout the land. (See "Ethics" essay, p. 387.) The cost of the alternative—putting roughly 2 billion people who are still without electricity on the **power grid** (and thus requiring central power plants, high-voltage transmission lines, poles, wires, and transformers)—is staggering!

The promise of solar energy has not yet been realized. Although solar technology has met every expectation in terms of declining cost over time, it has not yet reached the level of market penetration that was anticipated during the years of high fossil-fuel costs, due to the declines in the cost of conventional power generation in the 1990s. Right now, natural gas and coal have the price advantage over solar energy for generating electricity; however, as discussed in Chapter 12, natural gas has limited reserves and is subject to erratic price fluctuations. (Indeed, natural-gas prices have

**earth watch**

### Economic Payoff of Solar Energy

In addition to being a nonpolluting source of renewable energy, photovoltaic, solar-trough, and wind-powered energy-generating facilities can be installed quickly and added to a utility system in relatively small increments. To understand this advantage, compare these technologies with a nuclear power plant.

A nuclear power plant can be cost effective only if it is large—about 1,000 MW in capacity, enough to power a million average homes. Such a plant, assuming that it is accepted by the public, would require 10 to 15 years from the time of the decision to build it until it came into operation and would cost around $5 billion to $10 billion. Therefore, a utility taking this route must borrow billions of dollars and keep the money tied up for 10 to 15 years before a return on the investment is realized by selling power from the new plant. Even then, if the 1,000 MW of additional power are not needed when the new plant comes on-line, the return on the investment may be meager indeed. In the late 1970s and early 1980s, a number of utility companies went bankrupt (or nearly did so) because they had huge amounts of money tied up in nuclear power plants, while demand for the power the plants produced slackened. Clearly, building large power plants (nuclear or coal) involves considerable financial risk on the part of both utilities and consumers—the ones who ultimately pay. This is why Congress is considering a federal loan guarantee as a stimulus to the nuclear industry.

By contrast, solar or wind facilities can become operational within a few months of the decision to build them, they can be installed in small increments, and additional modules can be added as demand requires. The utility does not have to guess what power demands will be in 10 or 15 years. With solar or wind power, a utility can add capacity as it is needed, make relatively small investments at any given time, and have those investments start paying back almost immediately. Thus, this approach involves much less financial risk both for the utility and for its consumers.

One further advantage of solar and wind facilities is their relative invulnerability to terrorist attack, compared with oil and gas pipelines and large power plants, especially nuclear facilities. Further, renewables produce no hazardous wastes and are geographically dispersed.

tripled since 1998!) The gap between conventional energy sources and solar sources is narrowing, and, with some encouragement from public-policy changes, solar energy will become the major source many have envisioned.

## 14.2 Indirect Solar Energy

Water, fire, and wind have provided energy for humans throughout history. Dams, firewood, windmills, and sails represent "indirect solar energy," because energy from the Sun is the driving force behind each. What is the potential for expanding these options from the past into major sources of sustainable energy for the future?

### Hydropower

Early in technological history, it was discovered that the force of falling water could be used to turn paddle wheels, which in turn would drive machinery to grind grain, saw logs into lumber, and do other laborious tasks. The modern culmination of this use of water power, or hydropower, is huge hydroelectric dams, where water under high pressure flows through channels, driving turbogenerators (Fig. 14–17). The amount of power generated is proportional to both the height of the water behind the dam—that is what provides the pressure—and the volume of water that flows through. Chapter 7 describes the role played by dams in trapping and controlling rivers for flood control, irrigation, and hydropower.

About 6.7% of the electrical power generated in the United States currently comes from hydroelectric dams, most of it from about 300 large dams concentrated in the Northwest and Southeast. Worldwide, hydroelectric dams have a generating capacity of 713,000 MW, with an additional 50,000 MW from dams in the construction or planning stage. Hydropower generates 17% of electrical power throughout the world and is by far the most common form of renewable energy in use.

*Trade-offs.* Whereas waterpower is basically a nonpolluting, renewable energy source, harnessing it by means of hydroelectric dams still involves significant ecological, social, and cultural trade-offs:

1. The reservoir created behind the dam inevitably drowns farmland or wildlife habitats and perhaps towns or land of historical, archaeological, or cultural value. Glen Canyon Dam (on the border between

**ethics**

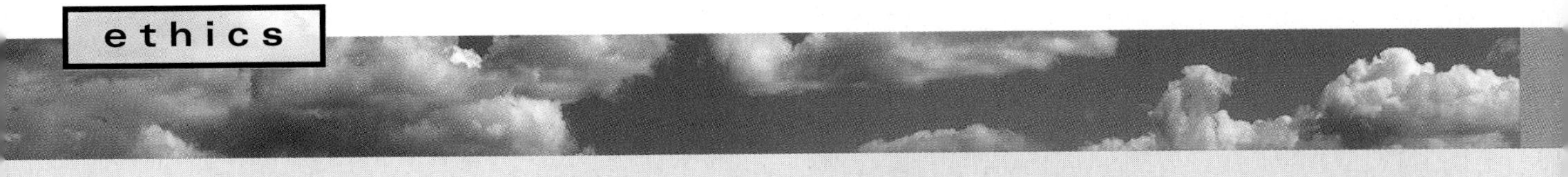

### Transfer of Energy Technology to the Developing World

The nations of the industrialized Northern Hemisphere have achieved their level of development by using energy technologies based largely on fossil fuels (and, to a lesser extent, nuclear energy). These nations' development took place during a time when fossil fuels were inexpensive. Only as those fuels became more expensive did the nations of the North begin to get serious about technologies that would make energy use more efficient. The specter of global climate change has brought a new perspective to the need to wean the industrialized North away from fossil-fuel energy in the 21st century.

The same traditional fossil-fuel-based technologies have been adopted by the developing nations of the Southern Hemisphere, except that these nations are lagging behind in their ability to implement the technology on a large scale. As discussed in Chapter 12, however, the fossil-fuel-based energy pattern of the 20th century may have to be replaced largely with renewable-energy systems in the 21st century. If the United States and other developed nations are willing to play a significant role in helping the developing nations, it is questionable whether the same developmental path the North took should be promoted. Instead, there is the opportunity to engage in some "leapfrogging" technology transfer: The North can put its development dollars into such technologies as photovoltaics, wind turbines, and efficient public transport systems for the cities.

The solar route is especially attractive for many of the climates in the developing world, where an estimated 2 billion people lack electricity. For example, some 80,000 PV systems installed in Kenya over a 10-year period now provide power to over 1% of the rural Kenyan population. The systems, with costs ranging from $300 to $1,500, are successfully marketed to people with incomes averaging less than $100 per month. Only the solar panels are imported; local companies provide the batteries and other system components. People use the energy for lighting, television, and radio. Annual sales in East Africa exceed 20,000 systems, a sign that the technology is in great demand and is fostering appropriate development in the area. Credit and financing extended to both companies and users have been key to this success. Some of the financing originates with development agencies from the North, but the majority is Kenyan. Building the infrastructure to market and service the PV systems has also been an essential part of the success story, and that can be accomplished by a relatively small number of organizations and individuals.

It is entirely possible that if this strategy of development aid is followed, the nations of the South may end up pointing the way to a sustainable-energy future for the rest of the world. What do you think?

Arizona and Utah), for example, drowned one of the world's most spectacular canyons. A number of dams have obliterated sacred lands of Native Americans.

2. Dams and the large reservoirs created behind them often displace rural populations. In the last 50 years, some 40 to 80 million people have been forced to move to accommodate the rising waters of reservoirs.
3. Dams impede or prevent the migration of fish, even when fish ladders are provided. Federal surveys show that fish habitats are suffering in the majority of the nation's rivers because of damming. Salmon fishing, one of the great industries of the Northwest, has been heavily affected.
4. Changing from a cold-flowing river to a warm-water reservoir can have unforeseen ecological consequences. The reservoir behind the Aswan High Dam in Egypt has fostered the spread of a parasitic worm that causes a debilitating disease. The reservoir has also increased humidity over a widespread area, and the higher levels of humidity are now accelerating the deterioration of ancient monuments and artifacts that have stood virtually unchanged for many centuries.
5. Ecological consequences occur below the dam as well. Because the flow of water is regulated according to the need for power, dams wreak havoc downstream. Water may go from near flood levels to virtual dryness and back to flood levels in a single day. Other ecological factors are also affected, because sediments with nutrients settle in the reservoir, so that smaller-than-normal amounts reach the river's mouth.

***More Dams?*** Even if such trade-offs were not enough, thoughts of greatly expanding water power in the United States are nullified by the fact that few sites conducive to large dams remain. Already, 75,000 dams 6 feet high or more dot U.S. rivers, with an estimated 2 million smaller structures. As a result, only 2% of the nation's rivers remain free flowing, and many of these are now protected by the Wild and Scenic Rivers Act of 1968, a law that effectively gives certain scenic rivers the status of national parks. The United States and most developed countries have already brought their hydropower to capacity, and the trend is in the direction of removing many of the dams that impede the natural flow of rivers.

**Figure 14–17 Hoover Dam.** About 5.8% of the electrical power used in the United States comes from large hydroelectric dams such as this one.

Proposals for new dams outside of the United States are embroiled in controversy over whether the projected benefits justify the ecological and sociological trade-offs. For example, the 50-meter-high Nam Theun 2 Dam, a 1,060-MW project on a tributary of the Mekong River in Laos, is expected to help that country begin a process of economic development. The project has been on hold since the mid-1990s, however, because of a number of concerns about its impact. For example, the project will displace 5,700 people and destroy 174 square miles of sensitive environments. In addition, it is basically being built to supply Thailand with power (only 75 megawatts is to be channeled to Laos at first). The World Bank is a key player because guarantees are needed for loans to a multinational consortium expecting to build and operate the dam.

***Dam Report.*** Another huge dam under construction is the Three Gorges Dam on the Yangtze River in China. (See Fig. 7–19.) When completed in 2009, it will be the largest dam ever built, displacing some 1.9 million people in order to generate 18,200 MW of electricity. Offsetting the human and economic costs is the fact that it would take more than a dozen large coal-fired power plants to produce the same amount of electricity. Recently, a special World Commission on Dams, convened to examine the impacts and controversies surrounding large dams, reported its findings. The commission began with the assertion that dams are only a means to an end, the end being "*the sustainable improvement of human welfare.*" The report gave dams a mixed blessing, concluding that large dams should be built only if no other options exist. Many guidelines were presented for assessing costs and benefits of specific projects—in particular, finding ways to incorporate participatory decision making, sustainability, and accountability in the planning and implementation of projects. Many more dams will be built, largely in the developing countries (especially China and India), and these will certainly play a role in the energy future of those countries.

## Wind Power

In 2001, the King Wind Ranch in Upton County, Texas (Fig. 14–18) became the world's largest (in keeping with the Texas tradition) wind farm. The "farm" consists of 214 wind turbines capable of generating 278 MW of power, enough to provide the electricity needs of 100,000 households. Texas established a Renewable Portfolios Standard in 1999—a mandated target for renewables requires that 2,000 MW of additional renewable capacity be added within a decade. Also aided by an energy Production Tax Credit (PTC) of 1.5 cents per kilowatt-hr

**Figure 14–18 King Mountain Wind Ranch.** Generating 278 MW of power, this wind farm in Texas is the largest in the world.

(worth 1.8 cents due to inflation) initiated by the Energy Policy Act of 1992, the King Wind Ranch wind farm and others built since 1992 brought the U.S. wind energy–generating capacity to 4,685 MW at the start of 2003. Still, the United States is only in third place, because Germany has over 12,000 MW and Spain has 4,830 MW, due to strong government encouragement and incentives. In the past five years, global wind power capacity has quadrupled, from 7,640 MW at the beginning of 1998 to 31,234 MW at the start of 2003. Wind is the fastest-growing energy source in the world, because it has become economically competitive with conventional energy sources, and countries all over the world are encouraging the installation of wind turbines by setting targets and providing subsidies.

Once a standard farm fixture, windmills fell into disuse as transmission lines brought abundant lower cost power from central generating plants. Not until the energy crisis and rising energy costs of the 1970s did wind begin to be seriously considered again as a potential source of sustainable energy. Although it is growing rapidly, wind power supplies only 0.4% of global electricity demand, but there are no technical, economic, or resource limitations that would prevent wind power from supplying at least 12% of the world's electricity by 2020, with a 1.2 million-MW generating capacity. In Denmark, which is hard pressed by a lack of fossil-fuel energy, wind now supplies over 18% of electricity.

***Design.*** Many different designs of wind machines have been proposed and tested, but the one that has proved most practical is the age-old concept of wind-driven propeller blades. The propeller shaft is geared directly to a generator. (Wind driving a generator is more properly called a **wind turbine** than a windmill.) As the reliability and efficiency of wind turbines have improved, the cost of wind-generated electricity has decreased. **Wind farms**—arrays of up to several thousand such turbines—are now producing pollution-free, sustainable power for under 5 cents per kilowatt-hour (with the PTC), a rate that is competitive with traditional sources. Moreover, the amount of wind that can be tapped is immense. The American Wind Energy Association calculates that wind farms located throughout the Midwest could meet the electrical needs for the entire country, while the land beneath the turbines could still be used for farming. Hundreds of new turbines are sprouting from midwestern farms, and the farmers are paid handsome royalties ($2,000 to $3,000 per turbine per year) for leasing their land to wind developers.

***Drawbacks.*** Still, wind power does have some drawbacks. First, it is an intermittent source, so problems of backup or storage must be considered. Second is an aesthetic consideration: One or two windmills can be charming, but a landscape covered with them can be visually tiresome. Third, windmills are a hazard to birds. Locating wind farms on migratory routes or near critical habitats for endangered species such as condors could be a problem for soaring birds in particular.

***Offshore Wind.*** Nantucket Sound lies south of Cape Cod and is a shallow body of water that rarely sees a windless day. Indeed, coastal locations are well known for their constant and often cooling winds, attracting vacationers and year-round residents who also enjoy the beaches and scenery. So, when Cape Wind Associates proposed an offshore wind farm in Nantucket Sound, the company ignited a firestorm of controversy from some very unexpected quarters: the environmental community. The proposed wind farm is not small. Some 130 turbines, each 417 feet tall, would be spread over 24 square miles of the sound. Concerns raised range from "visual pollution" to threats to the area's tourism, navigation, fishing industry, and migrating birds. Proponents of the project, however, view these concerns as a classic case of NIMBY, and point out that many who oppose the wind farm are wealthy Nantucket Islanders and prominent Massachusetts politicians who are in favor of renewable energy, except "not here." Maryland, Virginia, New York, and New Jersey have also received proposals for new offshore wind turbines, which have gotten the same mixed reception as the Nantucket Sound project.

Nevertheless, plans for offshore wind farms are putting a new spin on wind energy. Ireland has approved the world's largest offshore wind farm, with 200 turbines that will satisfy 10% of the country's electricity needs. Denmark already has many offshore turbines, as do the Netherlands and Sweden. Belgium and Germany are also in the proposal stage. The primary lure of offshore wind is the dependability and strength of the maritime winds. Siting is perhaps easier, too, as the turbines will be far enough offshore to have much less of a visual impact than onshore sites, and land does not have to be purchased.

## Biomass Energy

Burning firewood for heat is perhaps one of the oldest forms of energy that humans have used throughout history. However, there is nothing like a new name to put life into an old concept. Thus, "burning firewood" has become "utilizing biomass energy." "Biomass energy" means "energy derived from present-day photosynthesis." As Figure 14–4 indicates, biomass energy leads hydropower in renewable energy production in the United States. (Most uses of biomass energy are for heat.)

In addition to burning wood in a stove, major means of producing biomass energy include burning municipal wastepaper and other organic waste, generating methane from the anaerobic digestion of manure and sewage sludges, and producing alcohol from fermenting grains and other starchy materials.

***Burning Firewood.*** Wherever forests are ample relative to the human population, firewood, or fuelwood, can be a sustainable energy resource, and indeed, wood has been a main energy resource over much of human history. In the United States, wood stoves have enjoyed a tremendous resurgence in recent years: About 5 million homes rely entirely on wood for heating, and another 20 million use wood for some heating.

The most recent development in wood stoves is the *pellet stove,* a device that burns compressed wood pellets made from wood wastes (Fig. 14–19). The pellets are loaded into a hopper (controlled by computer chips) that feeds fuel when it is needed. The stove requires very little attention and burns efficiently, relying on one fan to draw air into it and another fan to distribute heated air to the room.

***The Fuelwood Crisis.*** In developing countries, firewood is the only source of fuel for cooking for over a billion people. In fact, 90% of the world's fuelwood is produced and used in the developing countries. Recall from Chapter 11 that two patterns of use determine how forest resources will be exploited: consumptive use and productive use. Most fuelwood use in developing countries fits the *consumptive* pattern: People simply forage for their daily needs in local woodlands and forests. Unfortunately, consumptive use often results in a deforestation that spreads outward from population centers. PROLENA, a citizen's group in the Central American country of Honduras, estimates that Hondurans cut 7 million cubic meters of firewood every year to use in cooking their meals, resulting in a high rate of deforestation. Honduras is in desperate need of a fuelwood policy that would encourage replanting and the sustainable use of wood in a land that is ecologically ideal for growing trees.

**Figure 14–19 Pellet stove.** Pellet stoves use pelletized wood waste to achieve an efficient burn that requires much less care than burns from conventional woodstoves.

The story is similar in many other parts of the developing world: Growing numbers of people go out from villages and towns daily to gather wood. The solution to the problem is to develop a market in wood, which would encourage people with land to plant trees as a cash crop. When people have to pay for fuelwood, trees become an important resource that can be put under the stewardship of local communities or private landowners, and a sustainable use of forests can result. Then, all of the other benefits of goods and services provided by forests are preserved or restored. Indeed, this is the trend in many parts of the developing world, but it often requires encouragement from governments or NGOs. (See Chapter 11.) Toward that end, the U.N.-sponsored Forest Trees and People Program (FTPP) works with local institutions and communities to encourage the management of forest resources for both conservation and sustainable development.

***Burning Wastes.*** Facilities that generate electrical power from the burning of municipal wastes (waste-to-energy conversion) are discussed in Chapter 18. Many sawmills and woodworking companies are now burning wood wastes, and a number of sugar refineries are burning cane wastes to supply all or most of their power. In Spain, wastes from olive oil production are powering three electrical facilities, providing over 32 MW of power, enough to supply 100,000 households. Power from these sources may not meet more than a few percent of a country's total electrical needs, but it represents a productive and relatively inexpensive way to dispose of biological wastes.

***Producing Methane.*** Chapter 17 examines how the anaerobic digestion of sewage sludge yields *biogas,* which is two-thirds methane, plus a nutrient-rich treated sludge that is a good organic fertilizer. Animal manure can be digested likewise. When the disposal of manure, the production of energy, and the creation of fertilizer can be combined in an efficient cycle, great economic benefit can be achieved. In China, millions of small farmers maintain a simple digester in the form of an underground brick masonry fermentation chamber with a fixed dome on top for storing gas (Fig. 14–20). Agricultural wastes, such as pig manure and cattle dung, are put into the chamber and diluted 1:1 with water, and the anaerobic digestion produces the biogas, which is used for cooking and heating. The residue slurry makes an excellent fertilizer. The concept has been expanded throughout the developing world to provide an alternative to fuelwood.

## 14.3 Renewable Energy for Transportation

Declining oil reserves and the enormous impact of transportation's demands for oil on our economy and on global climate suggest that the most critical need for a sustainable

**Figure 14–20 Biogas power.** Animal wastes are introduced to this unit in rural India and mixed with water. The wastes then decompose under anaerobic conditions, producing biogas.

energy future is a new way to fuel our vehicles. Renewable energy is already answering a small part of this need and may just have the answer for the long term in the form of hydrogen and fuel cells.

## Biofuels

Ethyl alcohol (ethanol) is produced by the fermentation of starches or sugars. The usual starting material is grain, sugar cane, or various sugary fruits, such as grapes. Fermentation is the process used in the production of alcoholic beverages. The only new part of the concept is that, instead of drinking the brew, we distill it and put it in our cars. However, production costs make ethanol almost twice as expensive as gasoline. To stimulate the industry in the United States, federal tax credits have been extended to ethanol-based fuels. The credits bring the cost down considerably. *Gasohol,* a 10% mixture of ethanol and gasoline, has been promoted and marketed in the Midwest since the late 1970s. The enthusiastic support of gasohol in that region is not motivated purely by the desire to save fossil fuel: Gasohol also provides another market for corn crops and thereby improves the economy of the region.

***Farm Products.*** In 2001, 1.8 billion gallons of ethanol for fuel, the equivalent of 42 million barrels of oil, were used in the United States. Some 10% of the nation's corn crop is dedicated to this use; an ethanol factory also produces corn oil and livestock feed. To stimulate its sugar cane industry, Brazil has also invested heavily in fuel ethanol; at one point, the country was producing 2.6 billion gallons per year, but a decision to remove subsidies has cut consumption in half. Nevertheless, ethanol, which is added at a level of 24% of fuel in Brazil, is deemed competitive with gasoline and is holding its own there.

***Air Quality?*** Besides substituting for gasoline, ethanol is a suitable substitute for methyl tertiary butyl ether (MTBE), a petroleum derivative that has been used as a fuel additive to make the burning of the gasoline cleaner. Both MTBE and ethanol improve air quality when they are made part of **reformulated gasoline,** and a number of metropolitan areas depend on this type of gasoline to improve their air quality in order to meet Clean Air Act requirements. MTBE, which is a carcinogen in animals, has been showing up in water supplies and may soon be phased out as an additive. If that happens, the use of ethanol could expand greatly, unless Congress decides to do away with the reformulated-gasoline requirement. Given the political clout of the farm lobby, this seems unlikely. Indeed, the new omnibus energy bill contains a requirement that targets a fuel ethanol increase to 5 billion gallons per year by 2012, a threefold increase over the present amount.

***Biodiesel.*** Imagine a truck exhaust that smells like French fries. Not only is that an improvement over diesel exhaust, but it is an indication that the truck is burning vegetable fuel, called biodiesel. Sales are estimated to grow to over 20 million gallons of the fuel, which is used by many government and municipal vehicles to demonstrate its potential. Standard biodiesel is a 20% mix of soybean oil in normal diesel fuel, but practically any natural oil or fat can be mixed with an alcohol to make the fuel; one excellent source is recycled vegetable oil from frying. In the United Kingdom, enterprising biodiesel users are running afoul of the law when they use vegetable oil in their diesel cars; there, it is a requirement that if a fuel is used as a substitute for one that is taxed (gasoline or diesel fuel), a tax must be paid at the rate of the substance being substituted. Police find the perpetrators by sniffing the exhaust coming from the "funny fuel."

## Hydrogen: The Fuel of the Future

Conventional cars with internal combustion engines can be run on hydrogen gas ($H_2$) as a fuel in the same manner as they are now able to run on natural gas (methane, or $CH_4$). (See Chapter 12.) Furthermore, neither carbon dioxide nor hydrocarbon pollutants are produced during the burning of hydrogen. Indeed, the only major by-product of $H_2$ combustion is water vapor:

$$2\,H_2 + O_2 \rightarrow 2\,H_2O + \text{energy}$$

(Some nitrogen oxides will be produced, too, because the burning still uses air, which is nearly four-fifths nitrogen.)

***A Good Idea, But....*** If hydrogen gas is such a great fuel, why aren't we using it? The answer is that there is virtually no hydrogen *gas* on Earth. Any hydrogen gas in the atmosphere has long since been ignited by lightning and burned to form water. And although many bacteria in the soil produce hydrogen in fermentation reactions, other bacteria are quick to use the hydrogen because it is an excellent source of energy. Thus, there are abundant amounts of the *element* hydrogen, but it is all combined with oxygen in the form of water ($H_2O$) or other compounds.

Although hydrogen can be extracted from water via **electrolysis** (the reverse of the preceding reaction), electrolysis requires an *input* of energy. Hydrogen can also be derived chemically from hydrocarbon fuels such as methane or oil, or methyl alcohol, but these processes are also energy losers; that is, more energy is put into the process than is contained in the hydrogen. Moreover, because these compounds contain carbon, it would be better to use them directly than to obtain hydrogen energy from them and use it. Therefore, the use of hydrogen as a truly clean fuel must await an energy source that is itself suitably cheap, abundant, and nonpolluting, such as solar energy.

***Plants Do It.*** For eons, the natural world has used photosynthesis to split water into hydrogen and oxygen, using light energy. (Recall that the oxygen from photosynthesis is released into the atmosphere and the hydrogen is attached to carbon dioxide to form sugars; see page 65.) To date, two ways have been proposed for mimicking this process. The first is to employ a catalyst to do what the photosynthetic pathway does. Recently, researchers from Duquesne University demonstrated that titanium dioxide ($TiO_2$), supplemented with some carbon, could absorb visible light energy and split water with an efficiency of 8.5%. Improving the efficiency to 10%, something the researchers are sure they can accomplish in time, would satisfy the Department of Energy's benchmark for a commercially useful catalyst. Stay tuned!

***Electrolysis.*** The second approach is simply to pass an electrical current through water and cause the water molecules to dissociate. Hydrogen bubbles come off at the negative electrode (the cathode), while oxygen bubbles come off at the positive electrode (the anode). (You can demonstrate this process for yourself with a battery, some wire, and a dish of water. A small amount of salt placed in the water will facilitate the conduction of electricity and the release of hydrogen.) The hydrogen is collected, compressed, and stored in gas cylinders, which can then be transported or incorporated into a vehicle's chassis. One important drawback to this technology is that it is difficult to store enough hydrogen in a vehicle to allow it to operate over practical distances. Compressing the hydrogen into a liquid would require energy, thus reducing the energy efficiency involved. As an alternative, the hydrogen can be combined with metal hydrides, which can absorb the gas and release it when needed.

***Solar Energy to Fuel.*** A rather elaborate plan has been suggested were we to take the concept of using solar energy–produced hydrogen for vehicular fuel to its logical conclusion. We would start by building arrays of solar-trough or photovoltaic generating facilities in the deserts of the southwestern United States, where land is cheap and sunlight plentiful. The electrical power produced by these methods would then be used to produce hydrogen gas by electrolysis. The most efficient way to move the hydrogen is through underground pipelines; hundreds of miles of such pipelines already exist where hydrogen is transported for use in chemical industries. Although this will be an expensive part of a hydrogen infrastructure, the technology for it is already well known. The next step is to develop vehicles that run on hydrogen.

***Model U.*** The Ford Motor Company has unveiled a "concept car" in the United States it has dubbed the Model U, since 2003 was the 100-year anniversary of the Model T (Fig. 14–21). The Model U, which made its debut at the 2003 Detroit Auto Show, sports a 2.3-liter internal combustion engine that runs on hydrogen and is equipped with a hybrid electric drive. The car uses four 10,000-psi tanks to store hydrogen, enough to range 300 miles before a fill-up. Although a working version of the drive train has been installed in a Ford Focus, this is not yet a production car. Ford is considering the car as a potential transition to a completely new way of moving a vehicle: by means of fuel cells.

**Fuel Cells.** An alternative to burning hydrogen in conventional internal combustion engines uses hydrogen in *fuel cells* to produce electricity and power the vehicle with an electric motor. **Fuel cells** are devices in which hydrogen or some other fuel is chemically recombined with oxygen in a manner that produces an electrical potential rather than initiating burning (Fig. 14–22a).

**Figure 14–21 Model U Ford.** This is a hybrid electric internal combustion vehicle powered by hydrogen and batteries. A "concept car," the Model U may or may not move to production.

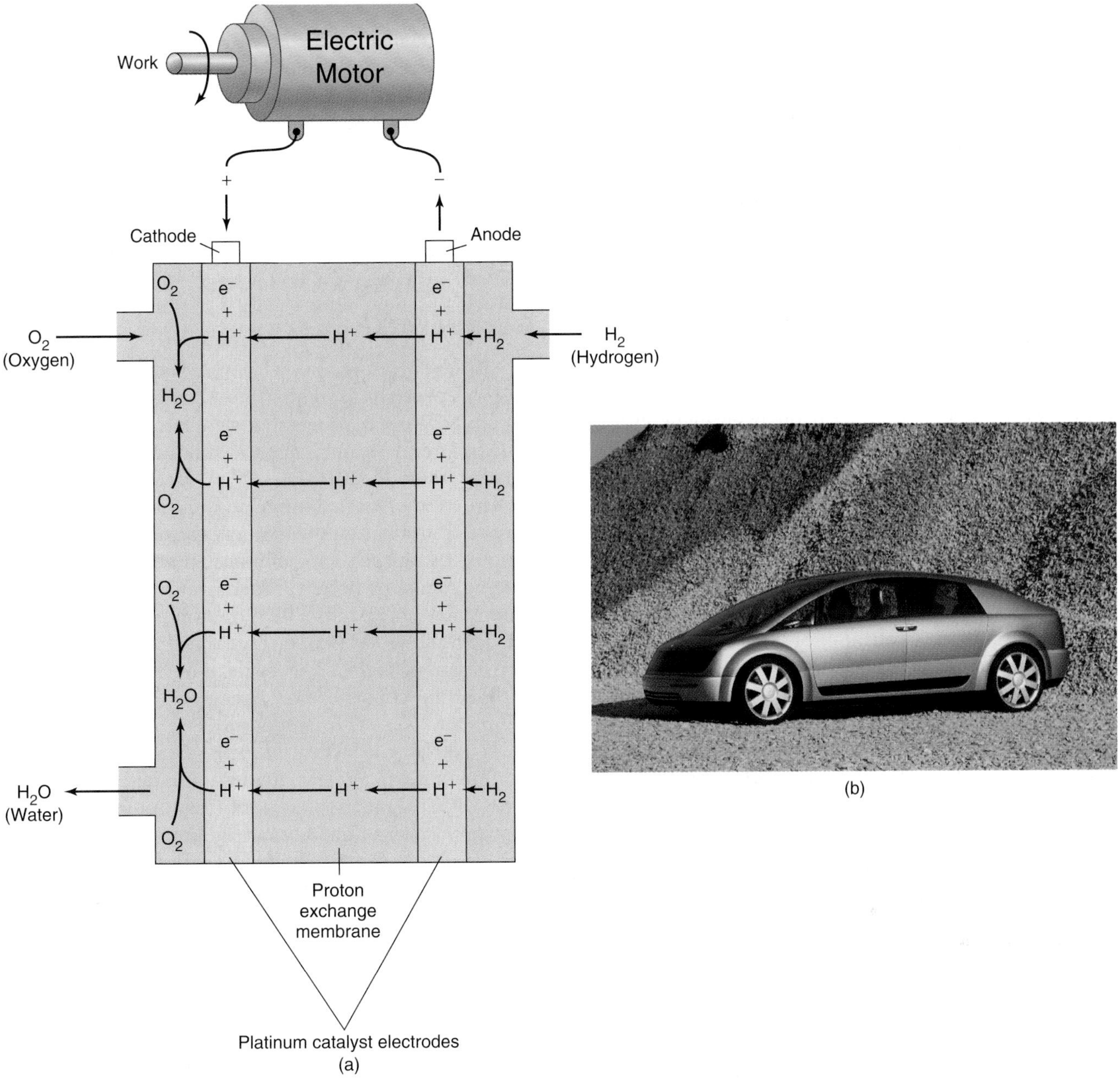


**Figure 14–22 Hydrogen–oxygen fuel cell.** (a) Hydrogen gas flows into the cell and meets a catalytic electrode, where electrons are stripped from the hydrogen, leaving protons (hydrogen ions) behind. The protons diffuse through a proton exchange membrane, while electrons flow from the anode to the motor. As the electrons then flow from the motor to the cathode of the fuel cell, they combine with the protons and oxygen to produce water. Thus, the chemical reaction pulls electrons through the circuit, producing electricity that runs the motor. Hydrogen is continuously supplied from an external tank, oxygen is drawn from the air, and the only by-product is water. **GM Hy-wire.** (b) The Hy-wire is driven by fuel cells supplied with hydrogen gas. The cells produce 94 KW of electric power. All of the propulsion and control systems are packed into an 11-inch thick "skateboard" chassis, which can support any of a number of desired body styles.

Emissions consist solely of water and heat! Because fuel cells create much less waste heat than conventional engines do, energy is transferred more efficiently from the hydrogen to the vehicle—at a rate of 45 to 60%, versus the 20% for current combustion engine vehicles. Fuel cells are now powering buses in Vancouver and Chicago, and DaimlerChrysler delivered urban buses with fuel cell drives to 10 European cities in 2003. The vehicles are based on Mercedes-Benz Citaro urban buses and will be offered for sale to transportation companies in Europe. The fuel cells develop more than 200 kilowatts of power, from hydrogen in tanks mounted on the roof of the bus.

***Hy-wire.*** Not to be outdone by Ford, GM presented a driveable car based on the fuel cell concept—the Hy-wire—at the Paris Motor Show in September 2002 (Fig. 14–22b). The vehicle has no engine compartment, because the fuel cells, hydrogen tanks, and electric motor are all part of a "skateboard" chassis. Steering, braking, and other mechanical elements of a conventional vehicle are all replaced with electronically controlled units, called by-wire systems, that connect at a single docking port to the chassis. Thus, various bodies can be mounted on the chassis and even exchanged if desired. Other vehicles operating on fuel cells are now on the road. Toyota, for example, is marketing a fuel cell version of its small Highlander SUV, making it available by lease to selected customers with access to the necessary hydrogen fuel and after-sales service. Honda has leased several fuel-cell-powered cars, the FCX, to the city of Los Angeles.

***FreedomCAR.*** Major obstacles to the widespread application of fuel-cell-driven vehicles are their high cost and the lack of an infrastructure to provide hydrogen for refueling. Still, the technology is improving rapidly and is on its way to commercial development. It received a major boost when President Bush launched his "FreedomCAR" (Cooperative Automotive Research) and "Freedom Fuel" initiatives in 2003. FreedomCAR is the government's partnering effort with the auto companies to move toward fuel-cell-powered vehicles. The president proposed $1.7 billion to support research and development over the next five years. His stated objective was to take the cars "from laboratory to showroom so that the first car driven by a child born today could be powered by hydrogen, and pollution free." It is hoped that the fuel-cell-driven cars will be competitive with cars with internal combustion engines by 2010. The research will work toward developing the cars, fuel cell technology, and the necessary hydrogen infrastructure.

Once a hydrogen infrastructure is in place, and vehicles powered by fuel cells are available, we will have entered a **hydrogen economy**—one in which hydrogen becomes a major energy carrier. With solar- and wind-powered electricity also in place, we will no longer be tied to nonrenewable, polluting energy sources and will have made the transition to a sustainable-energy system.

## 14.4 Additional Renewable-Energy Options

### Geothermal Energy

In various locations in the world, such as the northwestern corner of Wyoming (now Yellowstone National Park), there are springs that yield hot, almost boiling, water. Natural steam vents and other thermal features are also found in such areas. They occur where the hot molten rock of Earth's interior is close enough to the surface to heat groundwater, as may occur in volcanic regions. Using such naturally heated water or steam to heat buildings or drive turbogenerators is the basis of **geothermal energy.** In 2000, geothermal energy provided over 8,000 MW of electrical power (equivalent to the output of eight large nuclear or coal-fired power stations) in countries as diverse as Nicaragua, the Philippines, Kenya, Iceland, and New Zealand. Today, the largest single facility is in the United States, at a location known as The Geysers, 70 miles (110 km) north of San Francisco (Fig. 14–23). As impressive as this application of geothermal energy is, nearly double the amount is being used to directly heat homes and buildings, largely in Japan and China.

Some scientists feel that the potential for geothermal energy has barely been tapped. Just with today's technology, they predict, an additional 80,000 MW of

**Figure 14–23**
**Geothermal energy.** One of the 11 geothermal units operated by the Pacific Gas & Electric Company at The Geysers in Sonoma and Lake counties, California. California geothermal energy produces 1,600 MW of power, roughly 7% of California's energy needs.

electricity could be generated. Taking into account technologies under development as well, the predicted amount rises to an impressive 138,000 MW, which would satisfy about 4% of total world electricity production. Recently, the DOE launched a research-based initiative, called "GeoPowering the West," that aims to provide 10% of the West's energy needs with geothermal sources by 2020. (The western United States was chosen because of its unique geology.)

*Heat Pumps.* A less spectacular, but far more abundant, energy source than the large geothermal power plants exists anywhere pipes can be drilled into the Earth. Because the ground temperature below 6 feet or so remains constant, the Earth can be used as part of a heat exchange system that extracts heat in the winter and uses the ground as a heat sink in the summer. Thus, the system can be used for heating and cooling and does away with the need for separate furnace and air-conditioning systems. As Figure 14–24 shows, a geothermal heat pump (GHP) system involves loops of buried pipes filled with an antifreeze solution, circulated by a pump connected to a box (called the air handler) containing a blower and a filter. The blower moves house air through the heat pump and distributes it throughout the house via ductwork.

Four elementary schools in Lincoln, Nebraska, installed GHP systems for their heating and cooling. Their energy cost savings were 57%, compared with the cost of conventional heating and cooling systems installed in two similar schools. Taxpayers will save an estimated $3.8 million over the next 20 years with the GHP systems. According to the EPA, these systems are by far the most cost-effective energy-saving systems available. Although they are significantly more expensive to install than conventional heating and air-conditioning systems, they are trouble free and will save money over the long run.

## Tidal Power

A phenomenal amount of energy is inherent in the twice-daily rise and fall of the ocean tides, brought about by the gravitational pull of the Moon and the Sun. Many imaginative schemes have been proposed for capturing this limitless, pollution-free source of energy. The most straightforward idea is the **tidal barrage,** in which a dam is built across the mouth of a bay and turbines are mounted in the structure. The incoming tide flowing through the turbines would generate power. As the tide shifted, the blades would be reversed, so that the outflowing water would continue to generate power.

In about 30 locations in the world, the shoreline topography generates tides high enough—6 m (20 ft) or more—for this kind of use. Large tidal power plants already exist in two of those places: France and Canada. The only suitable location in North America is the Bay of Fundy, where the Annapolis Tidal Generating Station has operated since 1984, at 20-MW capacity. Thus, this application of tidal power has potential only in certain localities, and even there it would not be without adverse environmental impacts. The barrage would trap sediments, impede the migration of fish and other organisms, prevent navigation, alter the circulation and mixing of saltwater and freshwater in estuaries, and perhaps have other, unforeseen, ecological effects.

*Other Schemes.* A new application of tidal power that promises to avert the problems of the tidal barrage has been proposed by Tidal Electric, Inc. With this technology, the tidal generator is located offshore and consists of a concrete impoundment enclosure that sits on the ocean bottom, together with an adjacent turbine housing. The amount of power generated is closely related to the vertical tidal range, as the output varies with the square of the range. The impoundment is filled with water as the tide rises, creating a "head" that is gradually released through turbines when the tide is low, generating electric power. Feasibility studies are underway for two locations, one in Alaska and one in India. Other techniques are being explored which harness the currents that flow with the tides. For example, San Francisco is investigating a turbine-based system that would tap the energy of the 400 billion gallons of water that flow beneath the Golden Gate Bridge with each tide.

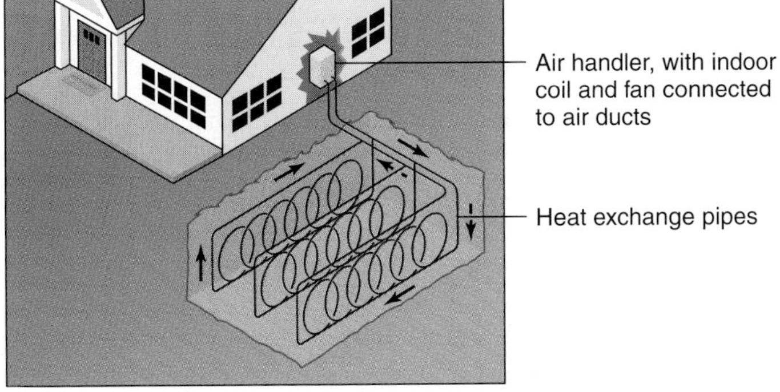


**Figure 14–24**
**Geothermal heat pump system.** The pipes buried underground facilitate heat exchange between the house and the earth. This system can either cool or heat a house and can be installed almost anywhere, although at a higher initial cost than a conventional heating, ventilation, and air-conditioning (HVAC) system.

### Ocean Thermal-Energy Conversion

Over much of the world's oceans, a thermal gradient of about 20°C (36°F) exists between surface water heated by the Sun and the colder deep water. **Ocean thermal-energy conversion (OTEC)** is the name of an experimental technology that uses this temperature difference to produce power. The technology involves using the warm surface water to heat and vaporize a low-boiling-point liquid such as ammonia. The increased pressure of the vaporized liquid would drive turbogenerators. The ammonia vapor leaving the turbines would then be recondensed by cold water pumped up from as much as 300 feet (100 m) deep and returned to the start of the cycle.

Various studies indicate that OTEC power plants show little economic promise—unless, perhaps, they can be coupled with other, cost-effective operations. For example, in Hawaii, a shore-based OTEC plant uses the cold, nutrient-rich water pumped from the ocean bottom to cool buildings and supply nutrients for vegetables in an aquaculture operation, in addition to cooling the condensers in the power cycle. Even so, interest in duplicating such operations is minimal at present.

Figure 14–25 presents a diagrammatic picture of the various renewable-energy alternatives discussed in this chapter.

## 14.5 Policy for a Sustainable-Energy Future

*Global Issues.* In Chapters 12–14, we have examined our planet's energy resources, requirements, and management options. The global issues are clear: Fossil fuels, especially oil and natural gas, will not last long at the current (and increasing) rate at which they are being consumed. Even more important, every use of fossil fuels produces unhealthy pollutants and increases the atmospheric burden of carbon dioxide, accelerating the already serious climate changes brought on by rising greenhouse gases. Renewable energy means sustainable energy, and a sustainable energy future does not include significant fossil-fuel energy. Global targets, therefore, are (1) stable atmospheric levels of greenhouse gases, especially carbon dioxide, and (2) an energy development and consumption pattern that is based on renewable energy.

### National Energy Policy

Nationally, the United States has a serious *oil–transportation problem:* We depend on importing over half of our oil, all of which is consumed in transportation. This problem is an enormous drag on our economy. In fact, more than $200,000 leaves the United States *every minute* to pay for the imported oil. *Natural-gas resources* are also being used increasingly to generate electrical power, prompting major price increases. (Prices *tripled* in 2002!) Finally, we have an *energy security problem:* Not only are we subject to the whims of OPEC and other oil producers, but we are also vulnerable to terrorist attacks on various facets of our energy infrastructure.

These national issues, as well as other, global ones, have prompted a policy response from the Bush administration (although it must be said that the administration is in denial about global climate change). In the previous two chapters, we examined policy choices and highlighted portions of the administration's National Energy Policy pertaining to *fossil fuels* and *nuclear power.* The administration is promoting a broad-based supply-side approach to fossil fuel, nuclear energy, and renewables, while also advocating demand-side savings on energy. Additional subsidies are extended to the fossil-fuel industry, while a few subsidies are given to renewables.

For *renewable energy,* **key recommendations of the National Energy Policy** would

- Facilitate access to national lands for renewable-energy production, including biomass, wind, geothermal, and solar energy.
- Increase funding for research and development on renewable energy.
- Extend the production tax credit for electricity generated with wind and biomass for another three years.
- Extend the Energy Star building energy efficiency program to include schools, retail establishments, private homes, and health care facilities.
- Develop efficiency standards for many more appliances (but not air conditioners!).
- Direct the Department of Transportation to provide recommendations for increasing the CAFE standards for a variety of vehicles.
- Initiate the FreedomCAR and Freedom Fuel projects with the necessary funding.
- Provide for a temporary income tax credit applicable to the purchase of new hybrid or fuel-cell-powered vehicles.
- Expand the ethanol tax credit program.

Notably missing from this list are targets for increasing the percentage contributions of renewable-energy sources and definite increases in the CAFE standards. However, the list indicates a serious effort to address the need to develop renewable-energy sources and, in particular, to make an effort to start the move toward a hydrogen energy economy. Because of numerous controversial energy issues, Congress was unable to reach agreement on an energy bill at the end of 2003.

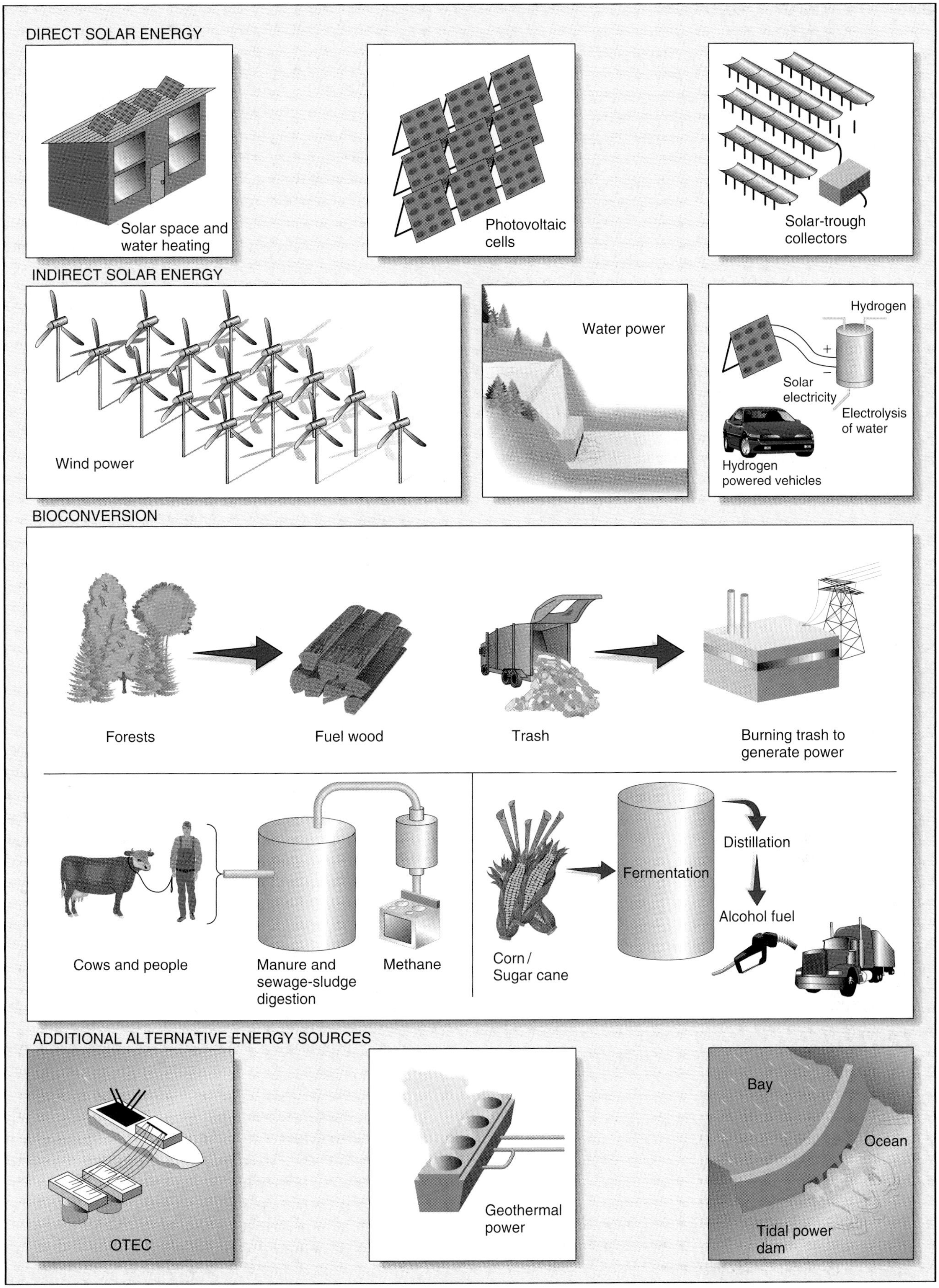


**Figure 14–25 Renewable-energy resource alternatives.** At the present time, direct solar heating of space and water, photovoltaic cells, solar-trough collectors, wind power, the production of hydrogen from solar or wind power, fuel cells, and the production of methane from animal manure and sewage sludges seem to offer the greatest potential for supplying sustainable energy with a minimal environmental impact.

## A Clean Energy Blueprint

Critics of the administration's National Energy Policy have been encouraged (by DOE Secretary Spencer Abraham) to develop an alternative energy plan. In response, the Union of Concerned Scientists, with several collaborating groups, has produced a plan (the *Clean Energy Blueprint*) that is designed to help the United States meet 20% of its electricity needs with renewables by 2020. At the same time, the plan would save hundreds of billions of dollars for consumers; reduce natural-gas and coal use; avoid the need to build hundreds of new power plants, major pipelines, and transmission lines; and reduce carbon dioxide emissions by two-thirds from the "business as usual" approach, as well as halving emissions of sulfur dioxide and nitrogen oxides.

In addition, the economic and security costs of maintaining nuclear power and oil would be reduced, and since renewables use domestically available resources, major construction and long-term job opportunities would also open up. The UCS blueprint proposes the following policy steps:

### On the Supply Side

- Establish a **Renewable Portfolio Standard (RPS)** that would require utilities to provide an increasing percentage of power from nonhydroelectric renewable sources (wind, solar, biomass). This standard would be phased in gradually, reaching 10% by 2010 and 20% by 2020.
- Maintain the **production tax credits** of 1.8 cents per kilowatt-hr for wind power and biomass, and extend the credits to all renewable-energy sources (geothermal, solar).
- Extend **net metering** so that it covers all states. Net metering is the practice of allowing people on the power grid to subtract the cost of the energy they contribute via wind turbines or photovoltaics from their electric bills.
- Substantially increase **research-and-development** spending on renewable energy and efficiency (e.g., for the National Renewable Energy Laboratory, Energy Star program).

### On the Demand Side

- Improve **efficiency standards** for appliances and other energy-consuming products, especially air conditioners.
- Enhance **building codes** so as to meet improved energy efficiency criteria, and provide **tax incentives** for going beyond the standards.
- Provide incentives for, and remove barriers to, erecting **combined heat and power (CHP)** facilities.
- Raise **fuel economy (CAFE) standards** to 40 miles per gallon by 2012 and 55 miles per gallon by 2020.
- Increase research funding for the **development of high-efficiency vehicles,** and provide tax incentives for consumers to purchase **hybrid electric and fuel cell vehicles.**

***Savings.*** The UCS analysis employed the National Energy Modeling System, a computer model maintained by the U.S. Energy Information Administration, to calculate the costs and benefits of the organization's blueprint. The savings are impressive. Figure 14–26a indicates the savings on oil resulting from the savings in efficiency, reflecting in particular the shift over to fuel-cell-powered cars. Figure 14–26b shows the impact of the Clean Energy Blueprint on electricity generation. Besides the obvious security benefits, such a program would reduce carbon emissions from power plants by two-thirds and from passenger vehicles by 40%, substantially lowering the threat to global climate change coming from the buildup of atmospheric greenhouse gases. Further, the program would save consumers by way of lower energy bills—a net savings of $640 billion by 2020.

***Congressional Action.*** Many of the elements of the Clean Energy Blueprint were incorporated into a bill (*S. 1333: Renewable Energy and Energy Efficiency Investment Act of 2001*) introduced in the Senate in 2001. The bill never made it out of the Senate Committee on Energy and Natural Resources. Similar bills have been introduced in the House (*H.R. 1343: Renewable Energy and Efficiency Act of 2003*) and the Senate (*S. 944: Renewable Energy Investment Act of 2003*) for the 108th Congress. Their fate is uncertain at this time. So far, no congressional action has been taken on increasing the CAFE standards.

***Final Thoughts.*** Renewable-energy technologies are being widely adopted in the developing world. To a great extent, this development represents a "leapfrogging" over the conventional fossil-fuel-based energy technology of the developed world. (See the "Ethics" essay, p. 387.) The United States is the only industrialized country that has seen fit to keep gasoline prices remarkably low. Fuel in all other highly developed countries is so heavily taxed that it costs consumers $3 to $5 per gallon—unlike U.S. prices, which were below $1.50 for the last 15 years. Thus, one future development that a number of environmental groups promote is the implementation of a **carbon tax**—a tax levied on all fuels according to the amount of carbon dioxide that they produce when consumed. Such a tax, proponents believe, would provide both incentives to use solar sources, which would not be taxed, and disincentives to consume fossil fuels. It is hard to imagine any step that would be more effective (or more controversial) in reducing greenhouse gas emissions in the United States. A number of European countries have already adopted such a tax.

Are the preceding developments, even the suggested Clean Energy Blueprint, enough to enable us to achieve a sustainable-energy system and to mitigate global climate change? Very likely, no. Yet, many of them are moving us in the stewardly direction that is vital to the future of the global environment.

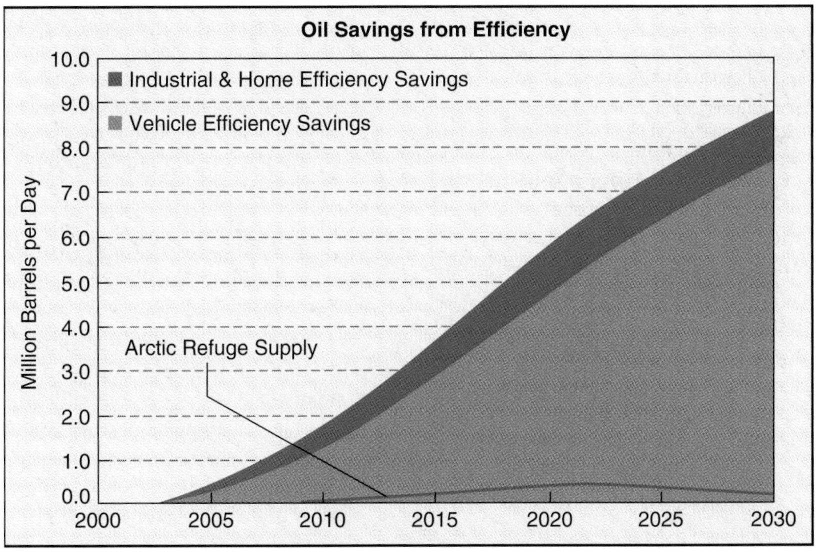


(a)

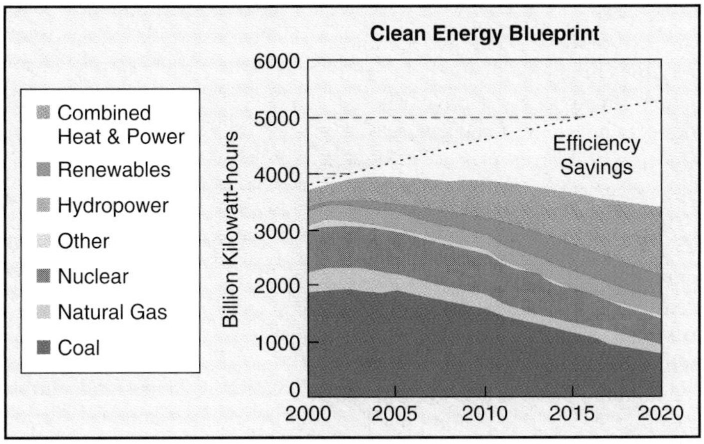


(b)

**Figure 14–26 Benefits of the UCS Clean Energy Blueprint.** (a) Oil savings from increasing the CAFE standards and switching to a fuel cell economy by 2030. (b) Savings in electricity generation from switching to renewable sources and improving efficiency. (*Source: Energy Security: Solutions to Protect America's Power Supply and Reduce Oil Dependence.* Union of Concerned Scientists, January 2002.)

## guest essay

### Caring for Planet Earth Through the Proper Use of Our Energy Resources

*Kenell J. Touryan*

*Dr. Ken Touryan is Chief Technology Analyst, assisting NREL (National Renewable Energy Laboratory) in commercializing various renewable energy technologies. Touryan received a Ph.D. from Princeton University in mechanical and aerospace sciences. Touryan also manages collaborative renewable energy technology projects between NREL and a dozen institutes in the former Soviet Union.*

Energy is central to achieving the interrelated economic, social, and environmental aims of sustainable human development. But if we are to realize this important goal, the kinds of energy we produce and the ways we use them will have to change. Otherwise, environmental damage will accelerate, inequity will increase, and global economic growth will be jeopardized. In fact, the availability of affordable energy resides at the very core of solving other critical issues such as the availability of food and fresh water, a clean environment, and poverty.

As good stewards of planet Earth, we must believe that solutions to these urgent problems are possible—that the future is much more a matter of choice than destiny. Outstanding among these choices are (1) efficient use of our depleting energy resources, (2) new emphasis on renewable energy technologies, and (3) a move toward a hydrogen economy.

#### Five Trends

Five significant trends are converging to shape the energy future of this country and the entire world. In the process, they are making renewable energy technologies more relevant than ever.

1. **World energy demand growth:** Energy demand continues to grow 4% per year in less-developed countries, as they try to catch up economically with developed countries. By 2050, energy demand is expected to grow from 14 terawatts (210 million barrels of oil equivalent/day) to 30–60 terawatts (450–900 MBOE/day).
2. **Global environmental awareness:** Ecological disasters in the 1980s, such as the widespread loss of forests in central Europe due to acid rain (caused mainly by coal-burning power plants), the explosion at the Chernobyl nuclear power plant in the Ukraine, the extensive oil slick resulting from the grounding of the Exxon *Valdez*, the growth of the ozone hole, and global warming, raised global environmental awareness. By the early 1990s, energy markets were being shaped more by environmental issues than by the economics and politics of oil.
3. **Energy security:** Fossil fuels are concentrated in Eastern Europe, the former Soviet Union, the Middle East, and many developing countries (such as Venezuela and Nigeria). As a result, the world's supply of fossil fuels is vulnerable to terrorism and/or the whims of regimes that may or may not be friendly. Renewable energy (such as wind, solar, biomass, and geothermal), on the other hand, is much more equitably distributed, so energy security is less of an issue.
4. **New technology options:** The energy crisis of the 1970s led to the development of new technologies, such as efficient gas turbines, better insulation for buildings, and energy-saving devices and appliances. Additionally, research into alternate energy resources has led to the development of solar-, wind-, biomass-, and geothermal-based sources of energy.
5. **Increasing business interest:** There are approximately 2 billion people in the world today who do not have a light bulb, and many of these people cannot be reached by traditional power lines. As a result, there is an enormous market for small-scale energy systems.

There are many reasons why the expanded use of renewables makes sense for the world's energy budget in the 21st century: proven effectiveness and reliability of advanced wind and solar technologies that are cost competitive in many applications; suitability for rural, off-grid locations; potential for significant economic growth and job creation; fuel cost is zero or low with no risk of escalation; renewable energy is environmentally responsible; and renewable resources enhance energy self-sufficiency.

## Available Now

A number of renewable energy technologies are now commercially available and several are already cost-effective. These include solar hot water for domestic consumption, industrial process heat using hot water and steam from parabolic trough concentrators, horizontal axis wind turbines, direct combustion of biomass (wood waste), geothermal power (direct steam), and small hydro (less than 10 MW).

Energy from wind farms worldwide today generates 34,000 $MW_e$ of power. In the United States alone, wind power produces more than 4 billion kWh of electricity, enough to provide electricity for a city the size of San Francisco, at competitive rates of 5¢/kWh. Additionally, worldwide sales of photovoltaic cells are increasing 20% per year, having exceeded 400 MW in 2002.

Finally, the production of ethanol from the cellulose fermentation of agricultural wastes should become commercially viable over the next decade. When it does, cropland currently idled by government programs could be used to grow enough biomass to provide more than 50% of U.S. gasoline consumption per year by 2020.

## Hydrogen Use and Hybrid Systems

A renewable energy system will only be sustainable if it includes an energy storage scheme (some way to store the renewable energy for times when it is not being generated) and an energy carrier (something to replace gasoline and other fossil fuel derived carriers). Current energy storage technologies include hydrogen ($H_2$), batteries, flywheels, superconductivity, ultracapacitors, pumped hydro, and compressed gas. Of these, the most versatile energy storage system and the best energy carrier is $H_2$.

$H_2$ can replace fossil fuels in transportation and electrical generation when renewable energy is unavailable. Transportable via gas pipelines or generated on site, any system that requires an energy carrier can use $H_2$. Moreover, the conversion of the chemical energy of $H_2$ to electrical energy via a fuel cell produces only water as waste.

Many rural villages have limited electrical service, typically provided by a diesel generator. For villages connected to the grid, power is often sporadically available and of poor quality. Hybrid systems, which use wind, photovoltaics, or biomass as the primary source (whichever is available at a given site), coupled with a diesel generator, could offset the cost of the diesel and provide more reliable service. These systems are small (several hundred kW), but sufficient for small villages and remote sites, where it is expensive to transport diesel fuel.

## Conclusion

The five trends, the more mature renewable energy technologies, and new business opportunities are converging to give renewable technologies a real boost in the energy market. Rural and remote applications using hybrid systems have opened vast new markets, where renewable technologies could make a significant impact. Finally, opening the power generation market to competition has led independent power producers to favor smaller, more innovative power projects where renewable technologies do better than most conventional fuels. Although the wind and solar energy industries are still fragile, renewable energy sources could account for 20% of the world's energy budget by 2020.

# revisiting the themes

## Sustainability

At the risk of repetition, global civilization is on an unsustainable energy course. Our intensive use of fossil fuels has created a remarkable chapter in human history, but we are now terribly dependent on maintaining and increasing our use of oil, gas, and coal. The results have already been seen in rising atmospheric greenhouse gases and the climate changes they are producing. Future fossil-fuel use will bring on more intense climate changes, and long before we burn all the available oil, gas, and coal, we will have been forced to turn to other energy sources. Renewable-energy sources derived from the Sun can provide a sustainable, inexhaustible supply of energy that is virtually free from harmful environmental impacts. It is imperative that the United States, as well as other countries, embark on a course to bring about a transition to a renewable-energy economy, most likely based on hydrogen and electricity as energy carriers, to be produced from renewable sources.

## Stewardship

Target Earth International is an NGO in Belize that has made a deliberate choice to use earth-friendly technologies whenever possible. This decision is based on the group's conviction that stewardship is essential to a sustainable-energy future. Theirs is an example that other organizations could follow, with far-reaching consequences if they did. Picture solar panels on every church, hospital, and school roof, wind turbines on college campuses, programs that distributed PV and biogas systems to people in poor, rural developing-country communities. The impact would be literally felt around the globe.

## Sound Science

Some elegant technologies have been discussed in this chapter, developed by scientists in research laboratories and then applied to solving some very practical problems. PV cells, for example, are being intensely researched in order to bring down the costs and make solar energy competitive with conventional energy sources. The National Renewable Energy Laboratory in Golden, Colorado, was established by Congress in 1977 and has developed into a cutting-edge research institute that produced the solar trough and solar dish systems. It is now exploring hydrogen technologies and improved PV cells, as well as a number of biomass technologies and low-wind-speed turbines and other renewable-energy options. This is sound science at its best, with public funding put to work to develop new ideas that may revolutionize the energy scene. (See "Guest Essay," p. 399.)

## Ecosystem Capital

Although hydroelectric power is a renewable-energy option, it has some serious drawbacks. Not only does it displace people (by the millions) as reservoirs fill behind huge dams, but also, the reservoirs drown major ecosystems and eliminate the ecosystem goods and services they were providing. Dams also prevent the natural flow of rivers and the fish living in them, often leading to downstream devastation as releases of water fluctuate due to power demands. These are costly trade-offs, and many people who promote renewable energy exclude hydropower as a technology to be encouraged.

## Policy and Politics

Renewable energy has been victimized in the past by the fossil-fuel and utilities industries, neither of which has been very interested in seeing this newcomer elbow its way into the national energy picture. Norman Myers and Jennifer Kent have singled out the fossil-fuel and nuclear energy industries as prime recipients of what they have called "perverse subsidies" in their book of the same name.[1] Because governments play such an important role in the production and distribution of energy, they also can influence the economies of different energy sources by providing or removing subsidies to the industries that deal in those sources. Myers and Kent state that the subsidy ratio for renewables versus nonrenewables is, at best, 1:10. Given the major problems associated with fossil-fuel and nuclear energy use, it is questionable whether the relevant industries should receive any subsidies at all, according to Myers and Kent.

Opposition to renewable energy led to a withdrawal of the solar tax credit program of the early 1980s. Often, the utilities are opposed to net metering, but this is one policy that deserves to be in force nationwide. Some government programs, such as the wind energy production tax credit, the Texas Renewable Portfolios Standard, and the EPA's Energy Star program, have been spectacularly successful in generating energy efficiency and encouraging renewables. Much more though, could and should be done.
The last key topic examined the Bush administration's National Energy Policy as it pertains to renewables and then the UCS Clean Energy Blueprint, which contains a number of supply-side and demand-side

[1]Norman Myers and Jennifer Kent. "*Perverse Subsidies: How Tax Dollars Can Undercut the Environment and the Economy.*" Island Press, Washington. 2001.

policy options that would move us more definitely toward a sustainable energy system. Energy is a hot topic in Congress; look for developments related to drilling in the ANWR, CAFE standards, and, less likely, a national Renewable Portfolio Standard as part of a renewable-energy bill.

### Globalization

Solar PV systems and wind turbines are now part of a global system of production and distribution. These are highly competitive and robust industries, and unfortunately, the United States is falling behind such countries as Denmark and Germany in wind turbine production and Japan in PV cell production. Manufacturing bases are growing in developing countries such as China and India. The greatest encouragement to a U.S. industrial resurgence in renewable-energy technology would be to establish a more consistent and favorable renewable-energy policy environment here in the United States and let the market provide the materials.

## review questions

1. How does solar energy compare with fossil-fuel energy in satisfying basic energy needs? What are the three fundamental problems in harnessing solar energy?
2. How do active and passive solar hot-water heaters work?
3. How can a building best be designed to become a passive solar collector for heat? What are the barriers to more widespread adoption of this alternative?
4. How does a PV cell work, and what are some present applications of such cells? What is the potential for providing more energy from PV cells in the near future?
5. Describe the solar-trough system, how it works, and its potential for providing power.
6. What is the potential for developing more hydroelectric power in North America vs. developing countries, and what would be the environmental impact of such development?
7. Where is wind power being harvested, and what is the future potential for wind farms?
8. What are four ways of converting biomass to useful energy, and what is the potential environmental impact of each?
9. What biofuels are used for transportation, and what are their limitations?
10. How can hydrogen gas be produced via the use of solar energy? How might hydrogen be collected and stored to meet the need for fuel for transportation in the future?
11. How do fuel cells work, and what is being done to adapt them to power vehicles?
12. What is geothermal energy, and what are two ways it is being harnessed?
13. What is the potential for developing tidal power in the United States?
14. How does renewable energy answer the key global and national energy issues?
15. Describe the Bush administration's approach to renewable energy as presented in its National Energy Policy.
16. What are the major policy recommendations of the UCS Clean Energy Blueprint? What would be the outcome if these recommendations were adopted?

## thinking environmentally

1. From what source(s) does a sustainable society gain its energy needs? Consider all energy sources, both conventional and alternative, and discuss each in terms of long-term sustainability. What conclusions do you reach? Defend your conclusions.
2. State what you feel should be the energy policy of the United States, based on the total range of energy options, including fossil fuels and nuclear power as well as various solar alternatives. Which energy sources should be promoted, and which should be discouraged? Suggest laws, taxes, subsidies, and so forth that might be used to bring your policy to fruition. Give a rationale for each of your recommendations.
3. Various solar-energy alternatives seem to promise a sustainable-energy future. Yet, the United States seems locked into a policy of remaining dependent on fossil fuels and nuclear power. Does this mean that solar advocates are wrong? Discuss the economic and political forces that might be at work in maintaining the status quo.
4. Audit all of the energy needs in your daily life. How much energy do you consume? Describe how each need might be satisfied by one or another renewable-energy option. (*Hint:* Check the Web for tools for this exercise.)

# making a difference part four: chapters 12, 13, and 14

1. Examine your uses of transportation. Make a car that gets high mileage per gallon a top priority, get behind car pools, use alternative transportation, keep your car tuned up and your tires inflated and balanced properly, and choose a place to live that reduces your need to drive.
2. Take advantage of the following services from your electric company: energy conservation, home-energy audits, and subsidized compact fluorescent lighting.
3. Keep informed about any nuclear power plants in your vicinity, and lobby for their strict adherence to NRC guidelines. If you do not support nuclear power, join an organization that lobbies against it.
4. Study your own home (or plans for a future one) with regard to its thermal efficiency (insulation, appliances, etc.) and the potential for using solar water, geothermal, or space heating. Improve your home's thermal efficiency, and add solar power in whatever ways you can.
5. Adopt ways to use less fuel and electricity both at home and where you work. (Turn thermostats back, turn off lights when you leave a room, wear sweaters, use energy-efficient lightbulbs, etc.)
6. Choose electricity suppliers that offer renewable energy; purchase energy-efficient appliances.
7. Using the Internet, research the EPA's Green Lights and Energy Star programs, and promote their use in an institution with which you are familiar—for example, your school, college, church, or business.

# part five

# pollution and prevention

An industrial plant sends a steady stream of exhaust gases into the air, and the wind carries the pollutants away. If it didn't, life in the vicinity of the plant would be impossible. But where is "away"? Is it the next county, state, or country? The upper atmosphere? In reality, there is no "away" for the pollution of land, water, and air resulting from the activities of the $6\frac{1}{2}$ billion people inhabiting Earth. As the human population continues to grow and develop economically, we are increasingly constrained by the wastes we produce.

In this section of the book, Chapter 15 begins with a look at the links between environmental hazards and human health. The chapters that follow examine the major forms of pollution, and you will discover that there are sustainable ways of dealing with the problems, be they water pollution, hazardous wastes and household wastes, or air pollution. It is even possible to have a sustainable system in which people and economic enterprises do not overburden the land, air, and water with wastes. You will also learn that dealing responsibly with our wastes may sometimes call for international action, as in the case of global climate change and ozone depletion. In every chapter, the public policies that have emerged to deal with each of the problems raised by pollution are discussed. Without exaggeration, the problems of pollution pose some of the most difficult and important issues facing human society. One of the essential transitions of the 21st century will be to move from pollution-intensive economic production to environmentally benign processes as we work toward a sustainable future.

◀ **Paper mill at sunset, Georgetown, South Carolina.**

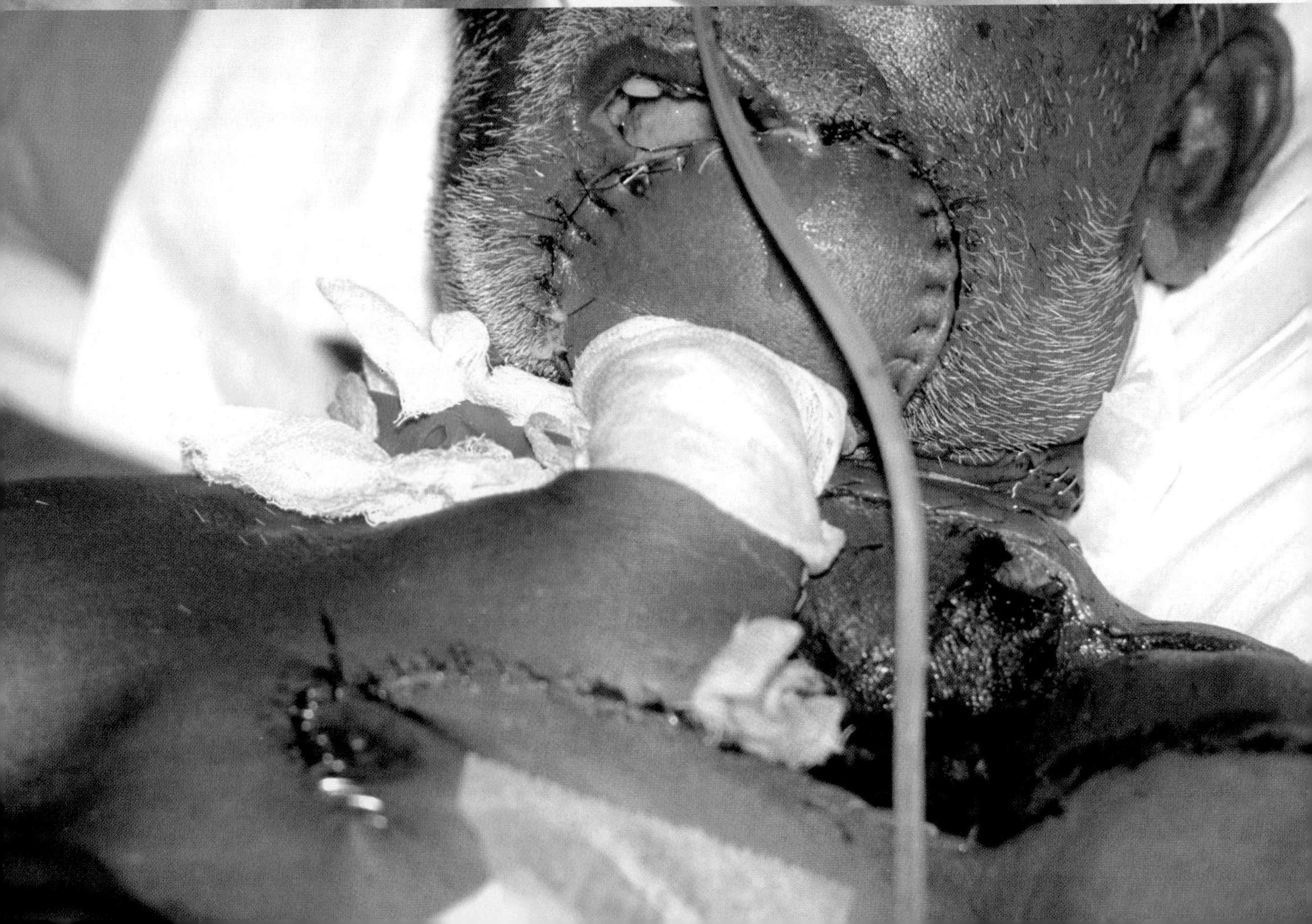

# Environmental Hazards and Human Health

**Key Topics**

1. Links between Human Health and the Environment
2. Pathways of Risk
3. Risk Assessment

No larger than a pinhead, with a light brown abdomen and a darker brown head and thorax, the black-legged tick, *Ixodes scapularis*, is easy to overlook (Fig. 15–1a). These eight-legged agile creatures can be picked up on a casual walk through the brush in many parts of the United States where deer and white-footed mice are common (Fig. 15–1b). If a tick has fed on a white-footed mouse infected with the spirochaete bacterium *Borrelia burgdorferi*, the tick can become the **vector,** or carrying agent, for Lyme disease in humans, just as it is in deer. A new bite from an infected tick usually leaves a circular red rash, that may be easy to miss. In the next few weeks, the victim develops fever, chills, fatigue, and headache, general symptoms that may be mild at their onset and hard to diagnose. If treatment is delayed, Lyme disease can become chronic and lead to long-term arthritis, memory impairment, numbness, and pain. First discovered in Lyme, Connecticut, the disease is on the increase in many parts of the United States. In 2002, 21,730 cases were reported; between 1991 and 2000, the reported incidence of Lyme disease nearly doubled.

**Smokers** **Two Indian smokers, one enjoying a cigarette, the other after being operated on for cancer. (Copyright WHP/P. Virot)**

***Of Mice and Moths.*** A recent study by a team from the Institute of Ecosystem Studies in Connecticut, led by Clive Jones, has uncovered an intriguing relationship between oak trees, gypsy moths, and Lyme disease. The gypsy moth is an introduced pest that can so ravage a forest that it looks like winter in the middle of summer. The study indicated that when the forest thrives because of natural control of the gypsy moth by mouse predation, it also presents a much greater risk of Lyme disease to people in the vicinity. If an oak forest produces a bumper crop of acorns, which happens every few years, the white-footed mouse population will thrive on the acorns. Unfortunately, the acorns also attract deer, and as a result, all the ingredients of a Lyme disease cycle are present. However, the mice also eat the overwintering pupae left behind by the gypsy moths and thereby protect the forest from being attacked the following summer, allowing it to thrive. Paradoxically, defoliation of the forest by gypsy moths, leads to poor acorn crops and declines in mouse populations, therefore greatly lowering the risk of Lyme disease, which is good for the human population in the vicinity. The dilemma seems

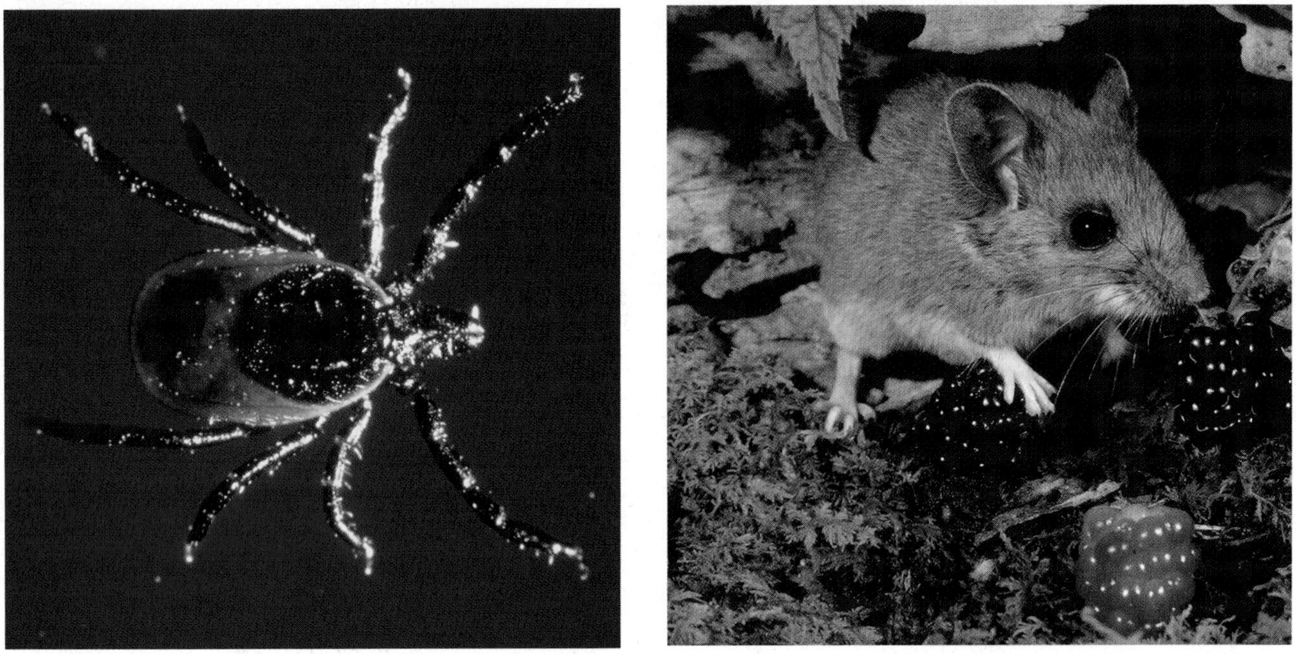

(a) (b)

**Figure 15–1 Agents in Lyme disease.** (a) The deer tick, *Ixodes scapularis,* carries the spirochaete bacteria that cause Lyme disease. (b) The white-footed mouse is host to the nymph stage of the tick. The mice and deer in woodlands in the eastern United States provide suitable conditions for the continued presence of Lyme disease.

to be a choice between a healthy forest and healthy people, although as yet, no attempt has been made to intervene in this cycle.

Lyme disease is only one of a number of emerging new diseases which continually remind us that we are not always able to control the hazards posed by our inevitable contacts with the natural environment. Severe acute respiratory syndrome (SARS), Lassa fever, the Ebola virus, the hantavirus, and the West Nile virus all have taken human lives in recent years, and there is every reason to expect more new diseases to appear as we continue to manipulate the natural environment and change ecological relationships.

***The Old Enemies.*** However, to put these diseases in perspective, it is not the new, emerging diseases that represent the greatest threat to us; rather, it is the common, familiar ones that take the greatest toll of human life and health—diseases such as malaria, diarrhea, respiratory viruses, and worm infestations, which have been kept at bay in the developed countries, but continue to ravage the developing countries. In the developed countries, cancer is the killer that is most closely linked to the environment, leaving us with questions about the chemicals we are exposed to on a daily basis.

The study of the connections between hazards in the environment, on the one hand, and human disease and death, on the other, is often called **environmental health,** but it is *human* health that is the issue here, not the health of the natural environment. In this chapter, we will examine the nature of environmental hazards and the consequences of exposure to those hazards. Then we will consider the pathways whereby humans encounter the hazards, some interventions that can alleviate the risks to health, and how public policy can address the need for continued surveillance and improvement in our management of environmental hazards.

## 15.1 Links Between Human Health and the Environment

With a focus on human health, a more precise definition of "**environment**" is *the whole context of human life—the physical, chemical, and biological setting of where and how people live.* Thus, the home, air, water, food, neighborhood, workplace, and even climate constitute elements of the human environment. Within human environments, there are *hazards* that can make us sick, cut our lives short, or contribute in other ways to human misfortune. In the context of environmental health, a **hazard** is anything that can cause (1) injury, disease, or death to humans, (2) damage to personal or public property, or (3) deterioration or destruction of environmental components.

The existence of a hazard does not mean that undesirable consequences inevitably follow. Instead, we speak of the connection between a hazard and something happening because of that hazard as a **risk,** defined here as the *probability* of suffering injury, disease, death, or some other loss as a result of exposure to a hazard. Because risks are expressed as probabilities, the analysis of the probability of suffering some harm from a hazard becomes a problem for scientists or other experts to address, and we will deal with the assessment of risk as it becomes the basis of policy decisions. For now, however, our focus is on the nature of environmental hazards.

## The Picture of Health

Health has many dimensions—physical, mental, spiritual, and emotional. The WHO defines **health** as "a state of complete mental, physical and social well-being, and not merely the absence of disease or infirmity."[1] Unfortunately, measuring all these dimensions of health for a society is virtually impossible. Thus, to study environmental health, we will focus on *disease* and consider health to be simply *the absence of disease.* Two measures are used in studies of disease in societies: morbidity and mortality. **Morbidity** is the incidence of disease in a population and is commonly used to trace the presence of a particular type of illness, such as influenza or diarrheal disease. **Mortality** is the incidence of death in a population. Records are usually kept of the cause of death, making it possible to analyze the relative roles played by infectious diseases and factors such as cancer and heart disease. **Epidemiology** is the study of the presence, distribution, and control of disease in populations.

### Public Health

***The CDC.*** Protecting the health of its people has become one of the most important facets of government in modern societies. In the United States, the lead federal agency for protecting the health and safety of the public is the *Centers for Disease Control and Prevention (CDC),* an agency under the Department of Health and Human Services. The CDC's mission is "to promote health and quality of life by preventing and controlling disease, injury, and disability." Each state also has a health department, and most municipalities have health agents. In addition, there is a huge health care industry in the United States, with federal programs such as Medicare and Medicaid, hospitals, health maintenance organizations, and local physicians and other health professionals. The delivery of health care is spread around all of these entities, but the primary responsibility for health risk management and prevention resides with the CDC and the state public-health agencies. These agencies, for example, may require public-health measures such as immunizations and quarantines, and they are responsible for monitoring certain diseases and environmental hazards and for controlling epidemics. They also gather data and provide information on health issues to state and local health care providers.

***Other Countries.*** Virtually every country has a similar *ministry of health* that acts on behalf of its people to manage and minimize health risks. The health policies that are put into place, however, are subject to limitations of information and funding. Ideally, a health ministry should direct its limited resources toward strategies that will accomplish the greatest risk prevention. Fortunately, all countries have access to the programs and information of the WHO, a United Nations agency established in 1948 with the mission of enabling "all peoples to attain the highest possible level of health." The WHO is centered at Geneva and maintains six regional offices around the world. The agency is staffed by health professionals and other experts and is governed by the U.N. member states through the World Health Assembly.

***Life Expectancy.*** One universal indicator of health is human life expectancy. In 1955, average life expectancy globally was 48 years. Today, it is 67 years and rising, and it is expected to reach 73 years by 2025. This progress is the result of social, medical, and economic advances in the latter half of the 20th century that have substantially increased the well-being of large segments of the human population. Recall from Chapter 5 the *epidemiologic transition:* the trend of decreasing death rates that is seen in countries as they modernize. Recall also that, as this transition occurred in the developed countries, it involved a shift from high mortality (due to infectious diseases) toward the present low mortality rates (due primarily to diseases of aging, such as cancer and cardiovascular diseases).

***Two Worlds.*** The countries of the world have undergone the epidemiologic transition to different degrees, with very different consequences. Despite the progress indicated by the general rise in life expectancy, one-fifth (11 million) of the annual deaths in the world are children under the age of five in the developing world. The discrepancy between the most developed and least developed countries can be seen quite clearly in the distribution of deaths by main causes, shown for 2001 in Figure 15–2. Infectious diseases were responsible for 54% of mortality in the least developed countries, compared with only 7% in the highly developed countries.

An entirely different perspective is brought into focus as we consider some of the environmental hazards that accompany industrial growth and intensive agriculture. In addition, some of the most lethal hazards in the developed world are the outcome of purely voluntary behavior—in particular, smoking tobacco and engaging in sexual activity. Let us now examine the various categories of environmental hazards.

[1]WHO Terminology Information System: www.who.int/health_system_performance/docs/glossary.htm#health (Dec. 26, 2003).

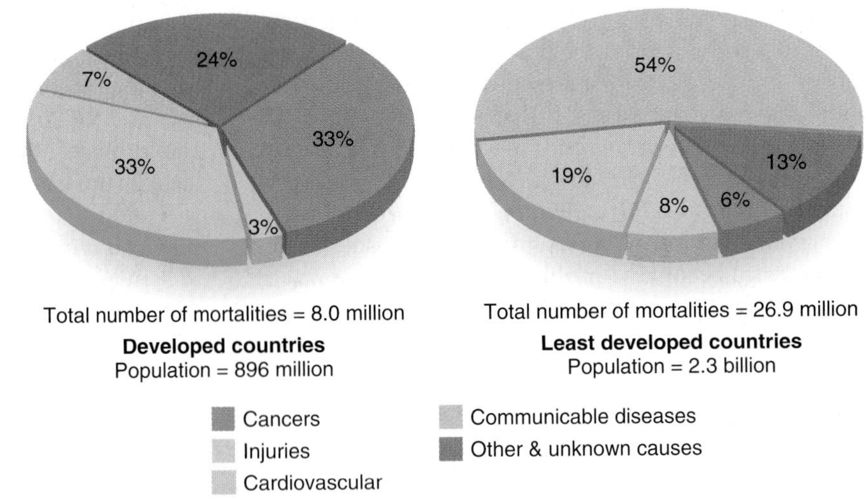


**Figure 15–2 Leading causes of mortality in the most developed and least developed countries in 2001.** Infectious diseases are predominant in the least developed countries, while cancer and diseases of the cardiovascular system predominate in the highly developed countries. (*Source:* Data from World Health Organization, *World Health Report 2002.*)

## Environmental Hazards

There are two fundamental ways to consider hazards to human health. One is to regard the lack of access to necessary resources as a hazard. For example, lack of access to clean water and nourishing food is generally harmful to a person. Investigating hazards from this perspective means considering the social, economic, and political factors that prevent a person from having access to such basic needs. Although this is a fundamentally important perspective, much of it is outside the scope of this chapter (and has already been discussed in Chapters 6 and 9). Instead, the focus will be primarily on the *exposure to hazards in the environment.* What is it in the environment that brings the risk of injury, disease, or death to people? To answer this question, you must understand the four classes of hazards: cultural, biological, physical, and chemical.

**Cultural Hazards.** Many of the factors that contribute to mortality and disability are a matter of choice or at least can be influenced by choice. People engage in risky behavior and subject themselves to hazards. Thus, they may eat too much, drive too fast, use addictive and harmful drugs, consume alcoholic beverages, smoke, sunbathe, hang glide, engage in risky sexual practices, get too little exercise, or choose hazardous occupations. People generally subject themselves to these hazards because they derive some pleasure or other benefit from them. Wanting the benefit, they are willing to take the risk that the hazard will not harm them. Factors such as living in inner cities, engaging in criminal activities, and so on are cultural sources of mortality, too. As Figure 15–3 indicates, close to half of all deaths in the United States can be traced to cultural hazards, and, in most cases,

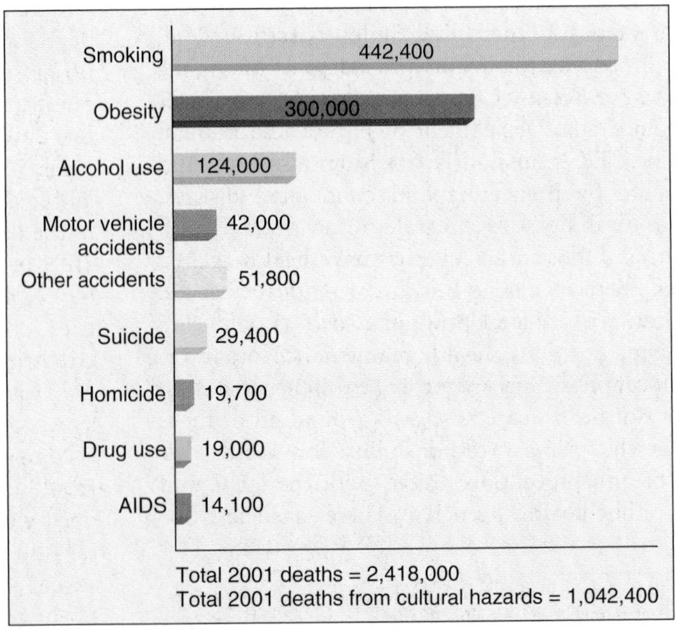


**Figure 15–3 Deaths from various cultural hazards in the United States, 2001.** Many hazards in the environment are a matter of personal choice and lifestyle. Other hazards are due to accidents and the lethal behavior of other people. (*Source:* Data from the Centers for Disease Control, National Center for Health Statistics, 2003; smoking deaths = average from 1995 to 1999; alcohol-use deaths from the National Council on Alcoholism and Drug Dependency; obesity deaths from the U.S. surgeon general.)

death from cultural hazards are preventable (if people are willing to give up their risky behaviors).

The connection between behavior and risk is especially lethal in the case of acquired immune deficiency syndrome (AIDS), caused by the human immunodeficiency virus (HIV). AIDS is one of a group of sexually transmitted diseases (STDs) that includes gonorrhea, syphilis, and genital herpes. Recall from Chapter 6 that AIDS is taking a terrible toll in the developing world, where over 90% of HIV-infected people live. Wherever it occurs, AIDS is largely the consequence of high-risk sexual behavior. At present, there is no cure or vaccine for the HIV virus.

**Biological Hazards.** Human history can be told from the perspective of the battle with pathogenic bacteria and viruses. It is a story of epidemics such as the black plague and typhus, which ravaged Europe in the Middle Ages, killing millions in every city, and of smallpox, which swept through the New World. The story continues in the 19th century with the first vaccinations and the "golden age" of bacteriology (a span of only 30 years), when bacteriologists discovered most of the major bacterial diseases and brought bacteria into laboratory culture. The 20th century saw the advent of virology (the study of viruses and the treatment of the diseases they cause); the great discovery of antibiotics; immunizations that led to the global eradication of smallpox, the defeat of polio, and the victory over many childhood diseases; and the growing influence of molecular biology as a powerful tool in the battle against disease.

The battle is not over, however, and never will be. Pathogenic bacteria, fungi, viruses, protozoans, and worms continue to plague every society and, indeed, every person. They are inevitable components of our environment. Many are there regardless of our human presence, and others are uniquely human pathogens whose access to new, susceptible hosts is mediated by the environment. Table 15–1 shows the leading global killers and the annual number of deaths they cause.

**table 15-1 Mortality from Major Infectious Diseases, 2002**

| Cause of Death | Estimated Yearly Deaths* |
|---|---|
| Acute respiratory infections | 3,845,000 |
| HIV/AIDS | 2,821,000 |
| Diarrheal diseases | 1,767,000 |
| Tuberculosis | 1,605,000 |
| Malaria | 1,222,000 |
| Measles | 760,000 |
| Pertussis (whooping cough) | 301,000 |
| Tetanus | 292,000 |
| Meningitis, bacterial | 173,000 |
| Syphilis | 157,000 |
| Hepatitis | 156,000 |
| Leishmaniasis | 51,000 |
| Trypanosomiasis (sleeping sickness) | 48,000 |
| Dengue | 19,000 |
| Schistosomiasis | 15,000 |
| Intestinal roundworms | 12,000 |

*Total deaths from infectious diseases in 2002 are estimated at 15.0 million by the WHO. (*Source:* Data from World Health Organization, *World Health Report 2003*.)

Approximately 26% of the 57 million deaths in 2002 were due to infectious and parasitic diseases. The leading causes of death in this category are the acute respiratory infections (for example, pneumonia, diphtheria, tuberculosis, whooping cough, influenza, and streptococcal infections), both bacterial and viral. Pneumonia is by far the most deadly of these (Fig. 15–4). Other bacterial and viral respiratory infections represent the most common reasons for visits to the pediatrician in the developed countries, but it is primarily in the developing countries that the respiratory infections lead to death, mostly in children who are already weakened by malnourishment or other diseases.

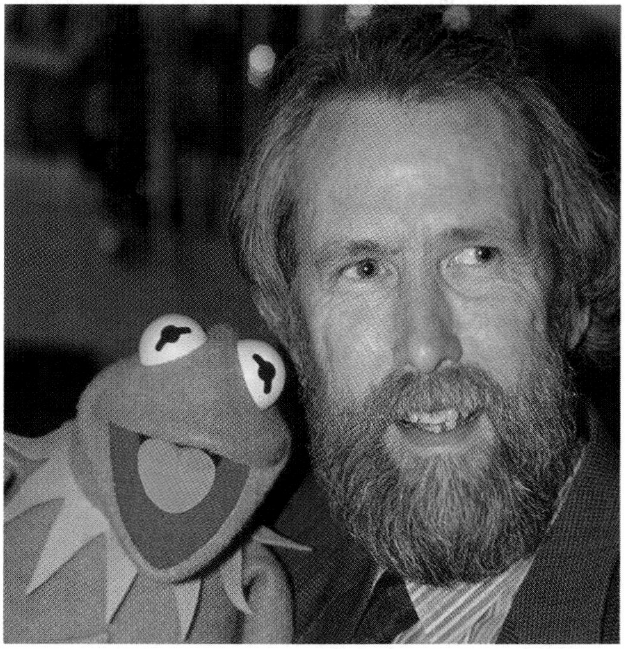

**Figure 15–4** **Jim Henson.** The creator of the Muppets succumbed in 1990 to a virulent pneumonia caused by the common bacterium *Streptococcus pneumoniae.*

***Tuberculosis.*** Although AIDS has overtaken tuberculosis as the largest cause of adult deaths from a single disease, tuberculosis continues to be a major killer. Close to one-third of the world's people are infected with the microbe *Mycobacterium tuberculosis* (Fig. 15–5), and perhaps 10% of these will develop a life-threatening case

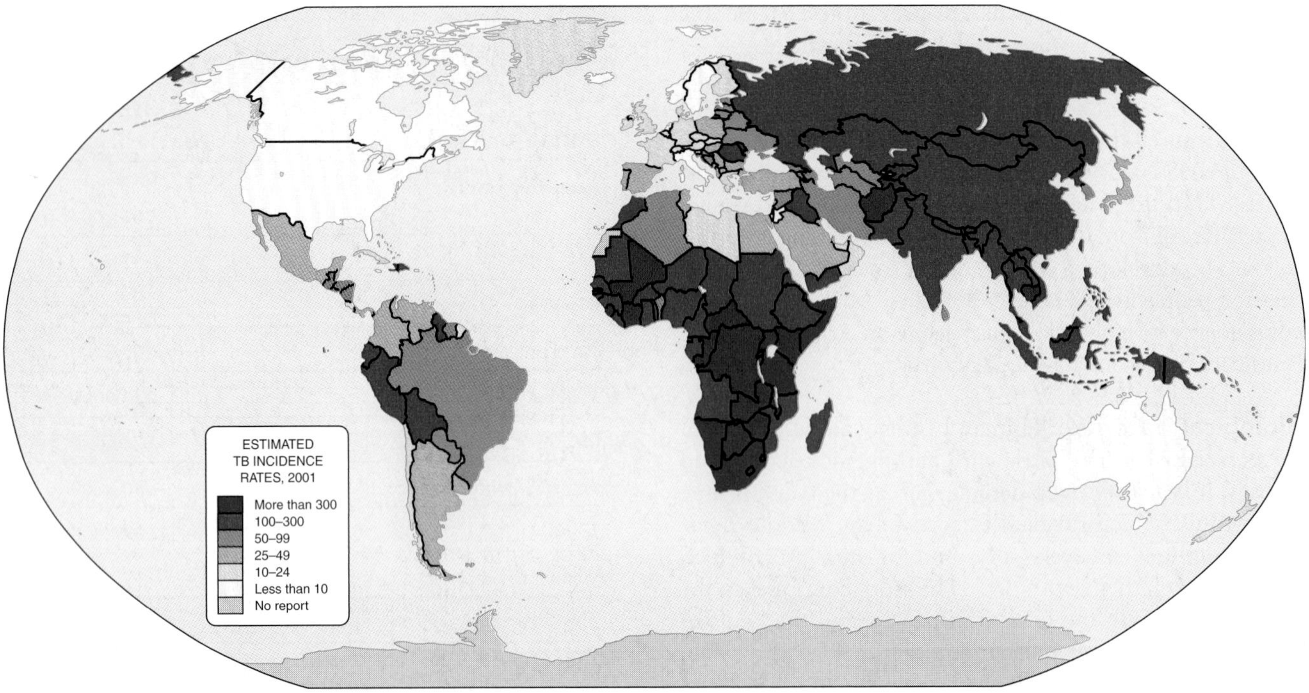


**Figure 15–5 Global map of tuberculosis, 2001.** This disease, caused by the bacterium *Mycobacterium tuberculosis,* is far from controlled. Close to one-third of the world's people are likely infected by the pathogen. TB incidence rates refer to numbers per 100,000 in the population. (*Source:* The World Health Organization)

during their lifetime. Complacency about treatment and prevention in recent years has led to a resurgence of tuberculosis, even though the bacterial pathogen that causes it has been known for over 100 years and its genome has been sequenced. Some of the resurgence can be traced to the AIDS epidemic, because HIV infection eventually leads to a compromised immune system, which allows diseases like tuberculosis to flourish. Much of the disease's comeback, however, is the result of multi-drug-resistant strains of the microbe.

*Diarrheal diseases* (for example, cholera, dysentery, salmonellosis, giardiasis, and many viral and bacterial infections), responsible for 1.7 million deaths in 2002, have the most obvious link to the environment. Most serious cases are the consequence of ingesting food or water contaminated with pathogens, such as *Salmonella, Campylobacter,* and *E. coli,* from human wastes. Infection in small children often leads to dehydration, which can quickly be life threatening. Thus, most deaths from diarrhea occur to children under five.

***Malaria.*** Of the infectious diseases present in the tropics, malaria is by far the most serious, accounting for an estimated 300 million cases each year and over a million deaths. Caused by protozoan parasites of the genus *Plasmodium,* malaria is propagated by mosquitos of the genus *Anopheles.* The life cycle of the protozoan begins with a mosquito biting an infected person and then incubating the parasite within itself (Fig. 15–6). The infected mosquito can then transmit the disease to human hosts with every bite. Within humans, the parasite invades the liver, where it multiplies and eventually breaks out to invade red blood cells. After multiplying in the red blood cells, the parasites are released, destroying the cells. The parasites then invade other red blood cells. This entire chain of events occurs in synchronized cycles, leading to the periodic episodes of fever, chills, and malaise that are typical of malaria. While in the liver and the red blood cells (most of the time), the parasites are hidden from immune system cells and antibodies. With the most virulent parasite, *P. falciparum,* 60% of the red blood cells may be destroyed in one episode, producing rapid and severe anemia. The waste products of the parasites, when they break out of the cells, produce high fever, headaches, and vomiting in the victim. If the parasites invade the brain (cerebral malaria), death occurs unless medical help is obtained quickly.

Most of the diseases responsible for mortality are also leading causes of debilitation in humans of all ages. Over 4 billion episodes of diarrhea occur to the human population in the course of a year, and, at any given time, more than 3.5 billion people suffer from the chronic effects of parasitic worms such as hookworms and schistosomes.

**Physical Hazards.** Natural disasters, including hurricanes, tornadoes, floods, forest fires, earthquakes, landslides, and volcanic eruptions (Fig. 15–7), take a toll of human life and property every year. Unfortunately, much of the loss brought on by natural disasters is a consequence of poor environmental stewardship. Hillsides and mountains are deforested, leaving the soil unprotected; people build homes and towns on floodplains; villages are nestled up to volcanic mountains; and cities

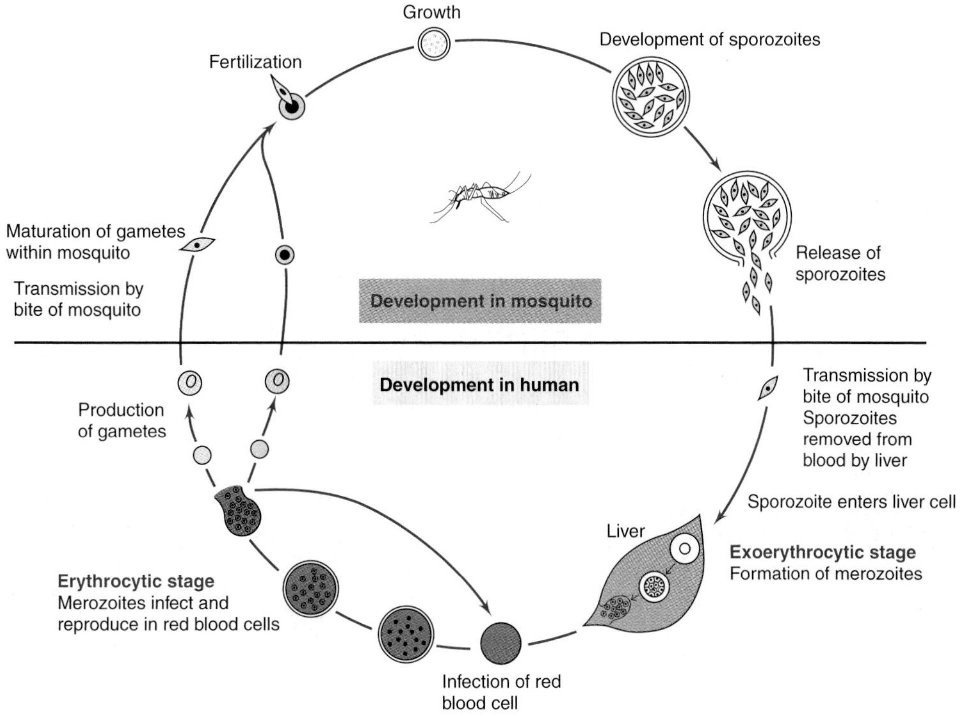


**Figure 15–6 Life cycle of malarial parasite.** Part of the life cycle of the protozoan parasite *Plasmodium,* which causes malaria in humans, is in the anopheles mosquito. In the mosquito, the protozoan develops from gametes (sex cells) into small, elongated cells called sporozoites, which multiply and migrate to the mosquito's salivary glands. The sporozoite stage can then be transmitted to humans through a mosquito bite. Sporozoites migrate to the liver and develop into small cells called merozoites, which multiply and then are released into the blood. There they enter red blood cells (erythrocytes) and multiply further. The parasite harms the human host by destroying red blood cells and by releasing fever-inducing substances into the bloodstream. Protozoal cells in the blood can then infect another anopheles mosquito when it bites the infected person, thus beginning the cycle again. (After Madigan, Michael T., Martinko, John M., and Parker, Jack, *Brock Biology of Microorganisms,* 8th ed., 1997. Reprinted by permission of Prentice Hall, Inc., Upper Saddle River, NJ.)

are constructed on known geological fault lines. A general human tendency is to assume that disasters happen only to other people in other places, and even if there is a risk, say, of a hurricane striking a coastal island, many people are willing to take that calculated risk in order to enjoy life on the water's edge. One writer suggested that town zoning boards should create a *stupid zone* for such people. Other stupid zone locations would be in the middle of 10-year floodplains, in highly flammable coniferous forests, on top of earthquake faults, and in valleys below dormant volcanoes.

*Tornadoes.* Some disasters, however, such as tornadoes, are impossible to anticipate. Each year in the United States, on average, 780 tornadoes strike, spawned from the severe weather accompanying thunderstorms. Because of the tendency for cold, dry air masses from the north to mix with warm, humid air from the Gulf of Mexico, the central United States generates more tornadoes than anywhere else on Earth. Of short duration, they develop into some of the most destructive forces known in nature, with winds reaching as high as 300 miles per hour. The most intense tornadoes have killed hundreds of people. On April 3, 1974, for example, a line of tornadoes from Canada to Georgia took more than 300 lives and caused untold property damage.

The most devastating consequences of natural hazards usually occur to those who are least capable of anticipating them and dealing with their effects. For example, in October 1998, Hurricane Mitch developed in the Caribbean Sea and slowly made its way across Central America. The combination of storm surge, winds, and rain devastated the developing countries of Honduras, El Salvador, Nicaragua, and Guatemala, killing over 10,000 people. Some locations received a year's worth of rain in one day, and the result was mud slides that obliterated whole villages. An earthquake in Gujarat, India, in January 2002 killed over 19,000 (Fig. 15–7b). Orissa state in India was hit with a massive cyclone in November 1999, killing at least 3,000 and displacing no fewer than 2 million people. In December 1999, Venezuela's Caribbean coast was lashed by devastating rains, triggering massive mud slides and floods that killed more than 30,000. Mozambique and South Africa were inundated by floodwaters in early 2000, leaving over a million homeless. Every year brings its list of disasters and the toll in deaths and in the miseries of people left homeless.

**Figure 15–7 Dangerous physical hazards.** Natural disasters bring death and destruction in their wake. (a) Effects of Hurricane Mitch in Tegucigalpa, Honduras, in 1998; (b) aftermath of an earthquake in Gujarat province, India, in January of 2001; (c) a tornado as it swept through Cantrall, Illinois, in early May 1995; and (d) Popocatépetl, a volcano located 50 miles southeast of Mexico City, erupted again on November 27, 1998.

*ProVention.* In response to this rising toll of disasters, in 2000 the World Bank launched an international partnership, the ProVention Consortium, to help equip the developing countries to cope with natural disasters. The consortium of agencies, governments, and universities addresses the link between poverty and vulnerability to disaster and works to protect the natural environments that can shield human settlements from weather-related disasters. For instance, $704 million in no-interest loans were made to India to provide relief and promote reconstruction after the Gujarat earthquake left 600,000 homeless. The consortium emphasizes an owner-driven housing program and features construction that will reduce vulnerability to future disasters.

*Climate Change.* As atmospheric greenhouse gases continue to rise during the 21st century (see Chapter 20), extreme weather events are expected to increase in frequency and intensity. Following record high heat waves in Europe and southeast Asia, an unusual outbreak of tornadoes in the United States, and a pattern of record droughts and high rainfalls, the World Meteorological Organization issued a warning in July 2003 stating that these were just the forerunners of even more intense and frequent weather and climatic events expected to occur as temperatures continue to rise due to global warming. Because these calamities will be superimposed on existing vulnerability to environmental hazards in many developing countries, the developed countries have established new funds through the Framework Convention on Climate Change (the **Least Developed Countries Fund** and the **Special Climate Change Fund**) to help the developing countries adapt to the hazards brought on by climate change.

**Chemical Hazards.** Industrialization has brought with it a host of technologies that employ chemicals such as cleaning agents, pesticides, fuels, paints, medicines, and many directly used in industrial processes (Fig. 15–8). The manufacture, use, and disposal of these chemicals often bring humans into contact with them. Exposure is either through the ingestion of contaminated food and drink, through the breathing of contaminated air, by absorption through the skin, by direct use, or by accident (discussed at length in Chapter 19). Toxicity (the condition of being harmful, deadly, or poisonous) depends not only on

(a) (b) (c) (d)

**Figure 15–8 Industrial processes and hazards.** Many industries produce chemicals and pollutants that contribute to the impact of toxic substances on human health. Some examples of hazardous workshops are (a) a paper mill in Port Angeles, Washington, (b) a tannery in Marrakech, Morocco, (c) an oil refinery in Philadelphia, Pennsylvania, and (d) a plastics factory.

*exposure,* but on the *dose* of a toxic substance—the amount actually absorbed by a sensitive organ. Further, for most substances, there is a threshold below which no toxicity can be detected. (See Fig. 19–1.)

Making things even more difficult, different people have different thresholds of toxicity for given substances. For example, children are at greater risk than adults, because children are growing rapidly and are incorporating more of their food (and any contaminants that are in it) into new tissue. Embryos are even more sensitive. Substances transmitted across the mother's placenta can have grave impacts on the embryo's development, especially in the early stages.

*Carcinogens.* Many chemicals, such as heavy metals, organic solvents, and pesticides, are hazardous to human health even at very low concentrations (Table 15–2). Episodes of acute poisoning are easy to understand and are clearly preventable, but the long-term exposure to low levels of many of these substances presents the greatest challenge to our understanding. Some chemical hazards are known *carcinogens* (cancer-causing agents). As Figure 15–2 indicates, cancer takes a very large toll in the developed countries, accounting for one-third of mortality. Because cancer often develops over a period of 10 to 40 years, it is frequently hard to connect the cause with the effect. The *Tenth Report on Carcinogens* (2002), from the Department of Health and Human Services, lists 48 chemicals now known to be human carcinogens and 171 more that are "reasonably anticipated to be human carcinogens." Many are substances used in both developed and developing countries. Other potential effects of toxic chemicals are impairment of the immune system, brain impairment, infertility, and birth defects.

Many developing countries are in double jeopardy because of both their high level of infectious disease and the rising exposure to toxic chemicals due to industrial processes and home uses of such chemicals. In both cases, the risks are largely preventable. In places where industrial growth is especially rapid, insufficient attention is often given to the pollutants which accompany that growth. For example, 13 of the 15 cities with the worst air pollution are found in Asia, where economic growth has risen rapidly.

**Cancer.** In the United States, 23% of all deaths—553,000 in 2001—were traced to cancer, and even though it is largely a disease of older adults, cancer can strike people of all ages. How is it that a simple chemical like benzene can cause leukemia? How can aflatoxin cause liver cancer? The answer to both of these questions requires an understanding of cellular

**table 15-2 Hazardous Chemicals in the Environment**

| | |
|---|---|
| **Chemicals found in the air in urban environments** | Lead (from vehicle exhausts and industrial processes)<br>Hydrocarbons and other volatile organic compounds (from vehicle exhausts and industrial processes)<br>Nitrogen oxides and sulfur oxides (from vehicle exhausts and fixed combustion processes like those carried out in power plants)<br>Particulate matter (dust, soot, and metal particles from vehicle exhausts, industrial processes, and construction)<br>Ozone (a secondary pollutant produced from the interaction of hydrocarbons and nitrogen oxides in sunlight)<br>Carbon monoxide (from vehicle exhausts and other combustion processes) |
| **Chemicals found in food and water** | Pesticides and herbicides (agricultural applications to food crops)<br>Heavy metals such as arsenic, aluminum, and mercury (in soil and water contaminated with industrial chemicals)<br>Lead (from water pipes)<br>Aflatoxins and other natural toxins (in foods contaminated by fungal growth)<br>Nitrates (in drinking water with a high nitrogen content) |
| **Chemicals found in the home and workplace (indoors)** | Lead (in household paints)<br>Smoke and particles (from the combustion of biofuels, coal, and kerosene)<br>Carbon monoxide (from the incomplete combustion of fuels)<br>Asbestos (in insulation)<br>Tobacco smoke<br>Household products (causing poisonings and exposure from their improper use) |
| **Chemicals contaminating some land sites** | Cadmium, chromium, and other heavy metals (in industrial wastes)<br>Dioxins, polychlorinated biphenyls, and other persistent organic chemicals (in industrial wastes) |

growth and metabolism. During normal growth, cells develop into a variety of specific types that perform unique functions in the body: blood cells, heart muscle cells, liver cells, absorptive cells that line the small intestine, and so forth. Some cells, like the stem cells that form the cells of the blood, keep on dividing until death. Other cells, however, are programmed to stop dividing and simply perform their functions, often throughout life. A cancer is a cell line that has lost its normal control over growth. Recent research has revealed the presence of as many as three dozen genes that can bring about a malignant (out-of-control) cancer; these genes are like ticking time bombs that may or may not go off.

*Carcinogenesis.* Carcinogenesis (the development of a cancer) is now known to be a process with several steps, often with long periods of time in between. In most cases, a sequence of five or more mutations must occur in order to initiate a cancer. (A mutation is a change in the informational content of the DNA that controls processes in the cell.) *Environmental* carcinogens are either chemicals that are able to bind to DNA and prevent it from functioning properly or agents, such as radiation, that can strike the DNA and disrupt it. When such an agent initiates a mutation in a cell, the cell may go for years before the next step—another mutation—occurs. Thus, it can take upwards of 40 years for some cells to accumulate the sequence of mutations that leads to a malignancy. In a malignancy, the cells grow out of control and form tumors that may spread (metastasize) to other parts of the body. Early detection of cancers has improved greatly over the years, and many people survive cancer who would have succumbed to it a decade ago. The best strategy, however, is prevention, so great attention is now given to the environmental carcinogens and habits, like tobacco use, that are clearly related to cancer development. The WHO estimates that 25% of cancers can be traced to environmental causes.

## 15.2 Pathways of Risk

The environmental hazards identified in Section 15.1 are responsible for untold human misery and death, but if many of the consequences of exposure to the different hazards are preventable, how is it that the hazards turn into mortality statistics? In other words, what are the pathways that lead from risks of infection, exposure to chemicals, and vulnerability to physical hazards to death? If prevention is the goal—and it should be—this knowledge is crucial. In 2002, the WHO annual report[2] was entitled *Reducing Risks, Promoting Healthy Life*. The document is a thoroughly researched presentation of the major risk factors "for which the means to reduce them are known," as the report puts it. The report shows that a remarkably small number of risk factors is responsible for a vast proportion of premature deaths and disease. The top 10 factors (Fig. 15–9) are responsible for more than one-third of all deaths and much of the global disease burden. We will consider only a few examples of these risk factors, because a comprehensive examination of this question would take up volumes.

### The Risks of Being Poor

One major pathway for hazards is *poverty*. Indeed, the WHO has stated that poverty is the world's biggest killer. Environmental risks are borne more often by the

[2]World Health Organization. "The World Health Report 2002: Reducing Risks, Promoting Healthy Life." World Health Organization, 2002. www.who.int/whr/2002/en/ (December 26, 2003).

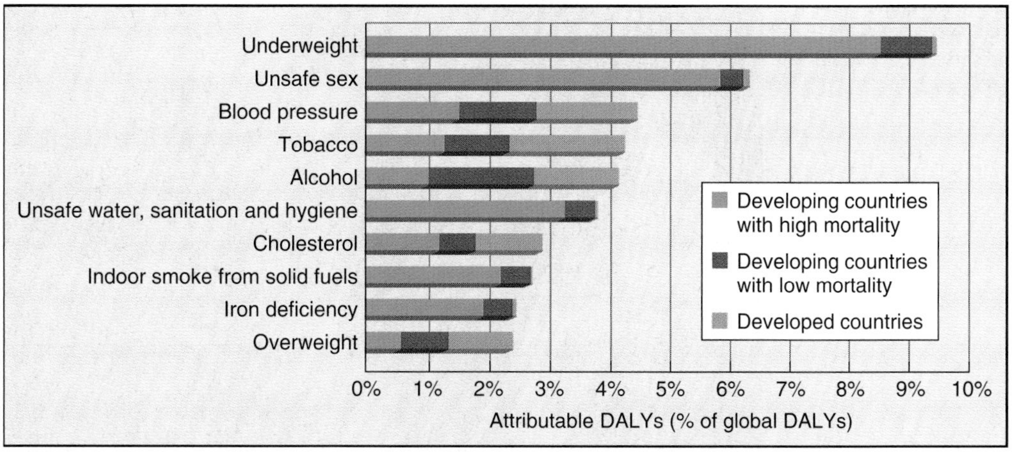


**Figure 15–9 The 10 leading global risk factors.** These 10 factors are responsible for more than one-third of all deaths and much of the global burden of disease. They are reported as a percentage of global DALYs. A DALY is a *d*isability-*a*djusted *l*ife *y*ear; one DALY equals the loss of one healthy life year. (*Source:* The World Health Organization, *World Health Report 2002*)

poor, not only in the developing countries, but also in the developed countries of the world. The people who are dying from infectious diseases are those who lack access to adequate health care, clean water, nutritious food, healthy air, sanitation, and shelter. Because of this, they are exposed to more environmental hazards and thus encounter greater risks. The world's number one risk factor (Fig. 15–9) is **underweight,** a problem of undernutrition that strikes primarily children under five years of age. (See Chapter 9.) This factor is strongly related to increasing poverty: Those households living on less than $1 per day (one-fifth of the world's population) show the highest prevalence of underweight in children. The WHO reported that this one factor affects 170 million children and was responsible for 3.4 million deaths in 2000.

***Wealth.*** Overall, the wealthier a country becomes, the healthier is its population. However, in virtually all countries, those with wealth are able to protect themselves from many environmental hazards. People in developed countries tend to live longer, and when they die, the chief causes of death are diseases of old age. Two-thirds of the death toll in the developed countries can be traced to the so-called degenerative diseases, such as cancers and diseases of the circulatory system (heart disease, strokes, etc.). By contrast, only one-third of the deaths in the developing countries are indicative of people who are living out their life span. Another indicator is age at death; people under the age of 50 account for two-fifths (more than 20 million) of the annual deaths in the world. Most of these deaths occur in the developing world, are due to infectious disease, and are preventable.

There are many reasons for this discrepancy between developed and developing nations. One is education. Advances in socioeconomic development and advances in education are strongly correlated. As people—and especially women—become more educated, they act on their knowledge to secure relief from hazards to health. They may improve their hygiene, for example, immunize their children (Fig. 15–10), or recognize dangerous symptoms such as dehydration and seek oral rehydration therapy (the replacement of lost fluids with balanced saline solutions). Other reasons for the discrepancy between health in the developed versus the developing world are the unavailability of safe drinking water and sanitation and the high indoor air pollution associated with the use of solid fuels such as wood, dung, and coal in simple cooking stoves. This latter risk factor (number 8 in Fig. 15–9) leads to a high incidence of respiratory tract infections and chronic lung disease.

***Priorities.*** Education, nutrition, and, the general level of wealth in a society do not tell the whole story, however. A nation may make a deliberate policy choice to put its resources into improving the health of its population above, say, militarization or the development of modern power sources. Countries such as Costa Rica, China, and Sri Lanka have a much longer life expectancy and lower infant mortality than their gross domestic product would predict. They have accomplished this enviable record by focusing public resources on such concerns as immunization, the upgrading of sewer and water systems, and land reform.

Another factor in the equation of environmental health is the relative distribution of wealth in a society. Where the gap between the rich and the poor is lower and the general well-being of the society is higher, the health status of the society is also higher. For example, both Japan and the United States have a similarly high

**Figure 15–10**
**Public-health clinic.** This outreach immunization center in Bangladesh provides basic health information and services to people in the surrounding area.

economic status. Japan, however, now has the highest life expectancy in the world (at 81 years) and the most equitable income distribution. The United States ranks 21st out of 157 countries in life expectancy and has great inequities between the wealthy and the poor.

## The Cultural Risk of Tobacco Use

Lifestyle choices such as refraining from exercise, overeating, driving fast, imbibing alcohol, climbing mountains, and so forth carry with them a significant risk of accident and death. However, one cultural hazard, tobacco use, is the leading cause of death in the United States (Fig. 15–11) and ranks fourth in the global tally of lethal risk factors (Fig. 15–9). Although tobacco use is declining in some developed countries, it is on the increase in most developing countries and remains high in the former socialist countries of eastern Europe. Unfortunately, recent trends show that people are likely to start smoking at a younger age. Lured by the image of smoking as "cool" or by peer pressure, youngsters are soon addicted to the nicotine in cigarettes, an addiction that is difficult to break. Recent research indicates that nicotine and cocaine share the same neural pathways, putting the two substances on a par physiologically. According to the CDC, some 46 million adults in the United States smoke cigarettes, and half of these will die or become disabled because of their habit.

***Marlboro Country?*** Smoking amounts to portable air pollution. The smoker is the primary recipient of the pollution, but the effects spread to all who breathe the smoke-clouded air and, in the case of pregnant women, to the unborn fetus. Cigarette smoking has been clearly and indisputably correlated with cancer and other lung diseases. Smoking has been shown to be responsible for at least 28% of cancer deaths in the United States. The WHO puts the number of smokers worldwide at 1.1 billion, with 4.9 million deaths per year due to cigarette smoking. Eighty percent of smokers are in the developing world, where the habit is increasing at a rate of 2% per year. In China, where smoking is practically epidemic, 63% of men are smokers.

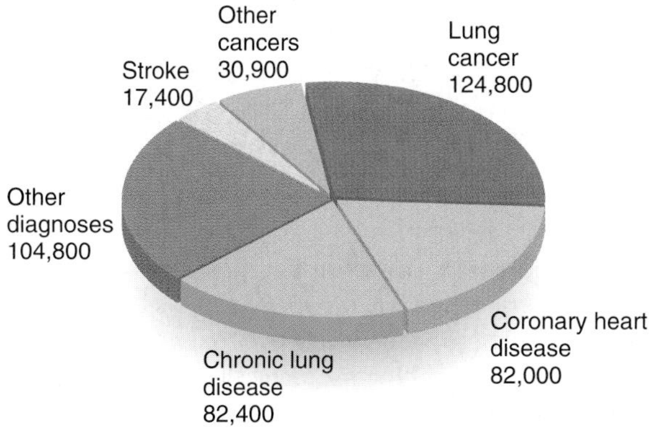


**Figure 15–11 Deaths caused by smoking.** Approximately 442,300 deaths in the United States are attributed to smoking. (*Source:* Data from the Centers for Disease Control, MMWR, 2002; average annual number of deaths, 1995–1999.)

Studies have shown that smokers living in polluted air experience a much higher incidence of lung disease than smokers living in clean air—a synergistic effect. Certain diseases typically associated with occupational air pollution show the same synergistic relationship with smoking. For example, black-lung disease is seen almost exclusively among coal miners who are also smokers, and lung disease predominates in smokers who are exposed to asbestos. Smoking is linked to half of the tuberculosis deaths in India, which leads the world in that category. Smoking also kills by impairing respiratory and cardiovascular functioning. The CDC estimates that smoking costs the nation $158 billion a year in health costs and lost job productivity because of premature death and disability.

***In the United States.*** Several measures have been taken to regulate this cultural hazard. Tobacco products are highly taxed, for example, providing states with substantial revenues ($8.2 billion in 2001). Raising cigarette taxes has been shown to be one of the most effective ways to reduce smoking, especially in young people. Surgeons general of the United States have issued repeated warnings against smoking since 1964, and public policy has taken them seriously by requiring warning labels on smoking materials, banning cigarette advertising on television, promoting smoke-free workplaces, requiring nonsmokers' areas in restaurants, and banning smoking on all domestic airline flights. Since the warnings began, the U.S. smoking population has gradually dropped from 40% to 24%.

***Secondhand Smoke.*** In January 1993, the EPA classified environmental tobacco smoke (ETS), meaning secondhand, or sidestream, smoke, as a Class A (known human) carcinogen. In the words of then EPA administrator Carol Browner, "Widespread exposure to secondhand smoke in the United States presents a serious and substantial public health risk." The EPA's action was unanimously supported by its science advisory board. Fallout from this action has been substantial: Smoke-free zones in public areas such as shopping malls, bars, restaurants, and workplaces are now common. In particular, the EPA has taken specific steps to protect children from ETS in all public places, and the Occupational Safety and Health Administration (OSHA) is promoting policies to protect workers from involuntary exposure to ETS in the workplace. The impact of these actions on public health should be substantial. One assessment indicated that just a 2% decrease in ETS would be the equivalent, in terms of human exposure to harmful particulates, of eliminating all the coal-fired power plants in the country. As public policy moves toward banning smoking in more and more locations, however, the tobacco industry is fighting back with advertisements emphasizing smoking as an important freedom. (See the "Ethics" essay, p. 420.)

ethics

## The Rights of Smokers?

Smoking in public is becoming socially unacceptable in many parts of the United States, forcing many smokers to retreat to rest rooms, specially designated smoking areas, or the sidewalks outside. Almost all states have laws restricting smoking in public places, and federal law now prohibits smoking on all domestic airline flights.

Although it was once considered simply a nuisance to have to breathe secondhand smoke, evidence has shown that consistent breathing of sidestream smoke can lead to some of the same consequences experienced by smokers. One study showed that nonsmoking wives of men who smoked were twice as likely to die of lung cancer as were nonsmoking wives of men who did not smoke. Some workers (bartenders, food servers, and musicians, for instance) involuntarily inhale the equivalent of 10 or more cigarettes per day. All of this new knowledge has put smokers on the defensive, as nonsmokers adopt slogans such as "Your right to smoke stops where my nose begins" and "If you smoke, don't exhale!"

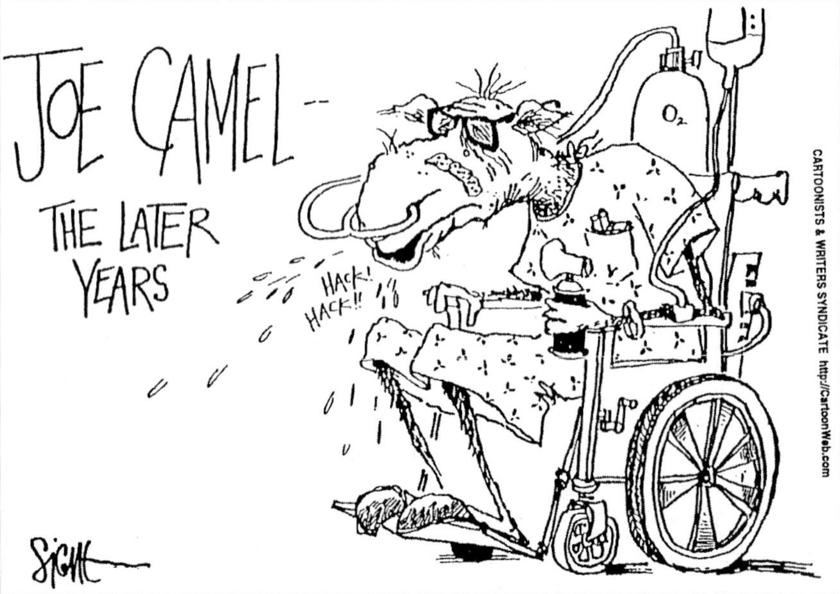


**The rights of smokers.** Confined to a wheelchair and breathing oxygen, Old Joe Camel displays the growing public awareness of the health hazards of many lifetime smokers, as well as of those who are exposed to sidestream smoke. The medical costs associated with this level of health care have prompted the many lawsuits mounted against the tobacco companies.

Smokers, however, feel that they have a right to indulge in their habit and would like nonsmokers to understand that smokers have a physical addiction that is quite difficult to break. Some smokers have become belligerent about the restrictions and are willing to take their rights into court. The newest strategy from the smoking lobby emphasizes smoking as an issue of personal freedom, like the right to use alcohol or caffeine. However, court challenges regarding the restrictions on smoking have not gone well for smokers. For example, a federal court upheld a ruling that prohibited a firefighter from smoking, whether on or off the job, because it was determined that firefighters must be in top physical condition and are subjected to hazards that smoking could aggravate.

The most controversial smoking battleground is the workplace. Many states, local communities, and private employers are promoting a smoke-free workplace. According to the American Lung Association, over 70% of U.S. workers now work in businesses that have a smoke-free policy. But some unions and businesses have been reluctant to support the ban on workplace smoking, making things difficult for workers who object to the smoke-filled air they must breathe. A ruling from OSHA could ban workplace smoking nationwide, but so far, the agency has been reluctant to make this move, arguing that it is unnecessary because of the recent positive actions on the part of the states and employers.

In the United States, nonsmokers are in the majority, and they seem to be gaining the upper hand in the controversy over public smoking. Have we gone too far, or should we do even more to curb sidestream smoke? What do you think?

*Goodbye, Joe.* Another significant development came from the FDA. In February 1994, citing the addictive properties of nicotine, then FDA administrator David Kessler proposed that the agency have the authority to regulate cigarettes as a drug. At congressional hearings following the FDA's initiative, tobacco executives denied that cigarette smoking was addictive. The hearings included allegations of a cover-up of the tobacco industry's own research that revealed the addictive properties of nicotine. The FDA launched efforts to curb tobacco sales to the underaged. One early casualty was "Joe Camel," a cartoon character that seemed to be targeted at the young (Fig. 15–12). In March 2000, however, the U.S. Supreme Court ruled that the FDA did not have the authority to regulate tobacco. That authority, said the Court, belonged to

**Figure 15–12** **Joe Camel.** This cartoon character was featured for years in ads by R. J. Reynolds. It developed a very high recognition rate among children, and because of this, the FDA pressured the manufacturer to stop using its Joe Camel ads.

Congress. In another blow to anti-smoking forces, in 2001 the Supreme Court ruled that states could not ban cigarette advertising near schools, citing an existing federal law.

*Legally Speaking.* On the legal front, there has been more success. The attorneys general of a number of states brought suit to recover some of their Medicaid expenses brought about by smoking-related illnesses. In late 1998, 46 states reached a $246 billion settlement with the top four tobacco companies wherein the companies would remunerate the states over a period of years, while the states would agree to drop lawsuits asking the companies to pay for their Medicaid expenses. The money is being used not only to reimburse the states, but also to help finance programs to discourage people—especially youths—from smoking. One predicted benefit of the settlement was an almost twofold rise in the wholesale price of cigarettes by the manufacturers. The settlement does not preclude other suits against the tobacco companies, and these are increasingly meeting with success after decades in which juries refused to award judgments to litigants, believing that smokers should be held responsible for their actions. In July 2000, a two-year class-action case in Florida was finally decided, and the litigants (thousands of smokers and former smokers) were awarded $149 billion! Two cases in Oregon in 2002 have led to verdicts against Phillip Morris, one for $150 million and another for $79.5 million.

*FCTC.* The battle against smoking is rapidly spreading worldwide. After several years of efforts, the WHO *Framework Convention on Tobacco Control (FCTC)* was adopted in May, 2003, by the World Health Assembly. This convention is a treaty that would reduce the spread of smoking by requiring signatory countries to raise taxes, restrict advertising, require larger health warnings on tobacco materials, and reduce ETS (steps already taken in the United States). As of 2003, there were 77 signatory countries. The next step is for the treaty to be ratified by the signatories. Once 40 countries ratify, it will go into effect.

## Risk and Infectious Diseases

Epidemiology has been described as "medical ecology," because the epidemiologist traces a disease as it occurs in geographic locations, as well as tracing the mode of transmission and consequences of the disease. Epidemiologists point to many reasons infectious diseases and parasites are more prevalent in the developing countries. One major pathway of risk is contamination of food and water due to inadequate hygiene and inferior sewage treatment. If drinking water is not treated thoroughly, it becomes a vehicle for transmitting human pathogens from feces. Some of the risk can be traced to the failure to educate people about diseases and how they may be prevented. Much, however, is simply a consequence of the general lack of resources needed to build effective public-health-related infrastructure in cities, towns, and villages of the developing world. Globally, an estimated 2.5 billion people are without adequate sanitation, and 1.2 billion lack access to safe drinking water. Many diarrheal diseases are transmitted in this way, and some of these, such as cholera and typhoid fever, are deadly and capable of spreading to epidemic proportions. Any diarrheal infection in small children can be fatal if it is accompanied by malnutrition or dehydration. By making oral rehydration therapy kits readily available throughout the country, Mexico reduced mortality rates in children by 70% in less than a decade.

***Even Here?*** These problems are not unique to the developing world. Inadequate surveillance in developed countries sometimes leads to major outbreaks of diarrheal diseases, with a few fatal outcomes. For instance, in 1993, the water supply of Milwaukee, Wisconsin, became contaminated with *Cryptosporidium parvum*, a protozoal parasite originating in farm animal wastes. Some 370,000 people developed diarrheal illnesses from the outbreak, and over 4,000 were hospitalized. The factorylike conditions under which food processing in the developed countries takes place often give pathogens an opportunity to contaminate food on a wide scale, resulting in diarrheal outbreaks and major recalls of food products. In continental Europe and the United Kingdom, mad-cow disease led to the deaths of about 100 people and caused hundreds of thousands of cattle to be slaughtered to contain the disease's spread. The occurrence of such outbreaks, the resurgence of known diseases, and the emergence of new ones like SARS, AIDS, and the Ebola virus are the concern of public-health agents and epidemiologists in organizations such as WHO and the CDC. (For a brief description of their work, see the "Global Perspective" essay, this page.)

***Tropical Diseases.*** The tropics, where most of the developing countries are found, have climates ideally suited for the year-round spread of disease by insects. Mosquitoes are *vectors* for several deadly and debilitating diseases, such as yellow fever, dengue fever, elephantiasis, Japanese encephalitis, West Nile virus, and malaria. Of these, malaria is by far the most serious. Public-health attempts to control malaria have been aimed at eradicating the anopheles mosquito with the use of insecticides. Malaria was once prevalent in the southern United States, but an aggressive campaign to

## global perspective

### An Unwelcome Globalization

We live in a time when globalization is rapidly encompassing travel, information, trade, and investment. The Internet ties people together in ways unimagined a few years ago. The globalization of health, however, remains an elusive goal, similar to the globalization of economic well-being. Laurie Garrett, in *The Coming Plague*, describes an unwelcome form of globalization: the globalization of disease. Garrett examines the recent history of emerging diseases such as AIDS, Ebola, hantavirus, Rift Valley fever, Legionnaires' disease, and others. She also explains the resurgence of familiar diseases like tuberculosis, cholera, and pneumonia as a consequence of the widespread and unwise use of antibiotics. Many of the new diseases are clearly linked to changes in land use, which bring humans into close contact with rodents or other animals that harbor viruses previously unknown to medicine and often deadly to humans. Resurgent diseases, by contrast, are a creation of our medical practice. By treating people with antibiotics without restraint, we unknowingly select strains that are immune to the antibiotics and that pass on their resistant genes to unrelated bacteria by way of plasmid transfer. (Plasmids are small circular pieces of DNA that can move from one bacterium to another in a semisexual process.)

The heroes of the book are the women and men on the front lines of epidemiology, people such as Joe McCormick of the CDC, who doggedly pursued lethal viruses in Africa, or Patricia Webb, who worked on the deadly viruses in maximum-containment laboratories back in the United States. Garrett makes a plea for a greater commitment from our universities, medical schools, and government agencies to train workers who will be capable of recognizing new diseases and who will be able to move about equally well in the laboratory, the hospital, and the field in pursuit of knowledge and public-health intervention around the world.

The importance of Garrett's perspective was recently emphasized with the appearance of several new and deadly viral diseases. Wood rats in California were found to be the source of the Whitewater Arroyo virus, which caused several deaths in 2000. The West Nile virus, which is carried by birds (and is fatal to them), first appeared in the United States in 1999 and in four years has spread to almost every state. In 2002, over 4,000 cases were recorded, with 284 deaths. This virus, spread by mosquitoes, may have traveled to the United States in a bird or an infected human. A particularly deadly virus causing SARS appeared in China in late 2002 and spread rapidly to several countries. SARS infected more than 8,400 people and killed over 900 before it was contained in mid-2003. The epidemic is believed to have originated in civets, animals considered a delicacy in southern China. A recent count indicated that three-fourths of emerging diseases are diseases of wild animals (*zoonoses*) and that their appearance is increasing in frequency. The potential of these diseases for causing sickness and death has been proven; the one sure fact is that we have not seen the last of them.

eliminate anopheles mosquitoes and identify and treat all human cases of malaria in the 1950s led to the complete eradication of the disease there.

Unfortunately, in the tropics, the mosquitoes have developed resistance to all of the pesticides employed, and eradication has remained an elusive goal. DDT is still used in some developing countries to spray the walls of huts and houses, but its use in this manner is highly controversial because of the pesticide's well-known harmful environmental and health impacts. (See "Ethics" essay, Chapter 16, p. 439.) However, the spraying has proven to be highly effective. Resuming their use of DDT after several years with another pesticide, South Africa and Zambia were able to dramatically reduce their malaria caseloads. Further complicating control of the disease, the *Plasmodium* protozoans have also developed resistance to one treatment drug after another. Until recently, chloroquine was quite effective against malaria in Africa. Now it is ineffective, and mefloquine is the drug of choice for treating cases of malaria and for prophylactic protection.

*Good News.* In 2002, molecular biologists successfully sequenced the genomes of the *Anopheles* mosquito and the most lethal malaria parasite, *P. falciparum.* Armed with this information, researchers are now able to target potential weak points in both organisms in the search for new vaccines and drugs. Although this is indeed good news, the fact is that effective treatments for malaria are already available, but either they are too expensive or those needing them most live in remote regions.

One highly promising effort funded by WHO's Tropical Disease Research program found that children provided with insecticide-treated nets over their beds (bed nets) experienced a substantial reduction in mortality from all causes (Fig. 15–13). This thrust is being recommended for large-scale intervention throughout Africa as a cost-effective way of reducing the high mortality from malaria in African children. Bed nets are especially effective against anopheles mosquitoes, which emerge from hiding and feed primarily at night. In Vietnam, a combination of bed nets and more effective drugs reduced malaria deaths 98% in seven years. Elsewhere, WHO has documented increases in the incidence of malaria in association with land-use changes such as deforestation, irrigation, and the creation of dams (Fig. 15–14). In semiarid areas of Africa, a seventeenfold increase in malaria resulted from development projects involving irrigation and other high-intensity agriculture.

Research continues on the development of new, more effective antimalarial drugs and on the development of an effective vaccine. Because the drug companies have been reluctant to fund this research (it is not likely to be profitable), the Bill and Melinda Gates Foundation has provided $150 million funding for the Malaria Vaccine Initiative. (Altogether, the Gates Foundation has given $288 million to fight malaria.) However, the parasite's complicated life cycle and the fact that malaria does not readily induce long-term immunity have made a vaccine an elusive goal, so the current public-health strategy is to inhibit access of mosquitoes to hosts (with bed nets and house spraying) and provide prompt treatment and prophylaxis to a greater proportion of the affected populations. Recently, WHO initiated the "Roll Back Malaria" campaign, designed not to eradicate the disease, but to reduce the malaria-caused mortality by half by 2010 and then to halve mortality again by 2015. The emphasis of this program is on getting research findings into policy and practice.

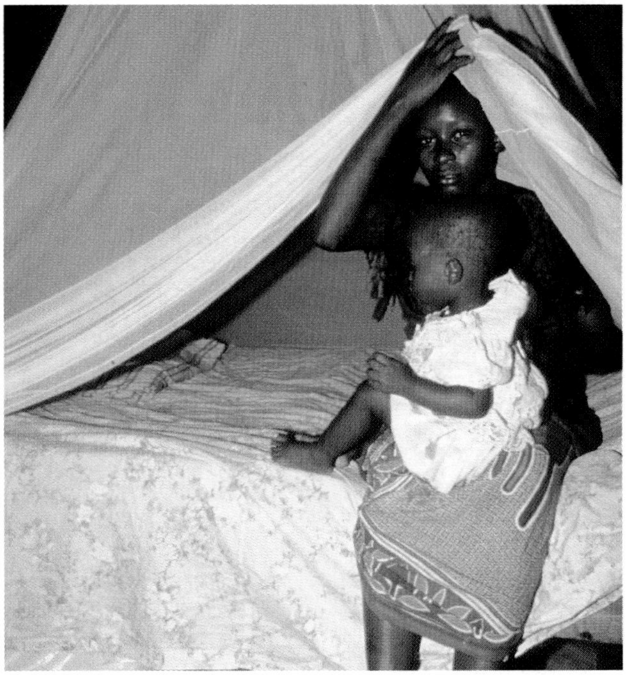

**Figure 15–13 Insecticide-treated bed net.** This Kenyan woman and her child were protected at night from malaria-bearing mosquitoes. Bed nets have been shown to prevent infection.

## Toxic Risk Pathways

How is it that people are exposed to chemical substances that can bring them harm? Airborne pollutants represent a particularly difficult set of chemical hazards to control, as they are difficult to measure and difficult to avoid. Humans breathe 30 lb (14 kg) of air into their lungs each day. Although some of the symptoms of pollution that people suffer involve the moist surfaces of the eyes, nose, and throat, the major site of impact is the respiratory system. Three categories of impact can be distinguished:

1. **Chronic.** Pollutants cause the gradual deterioration of a variety of physiological functions over a period of years.
2. **Acute.** Pollutants bring on life-threatening reactions within a period of hours or days.

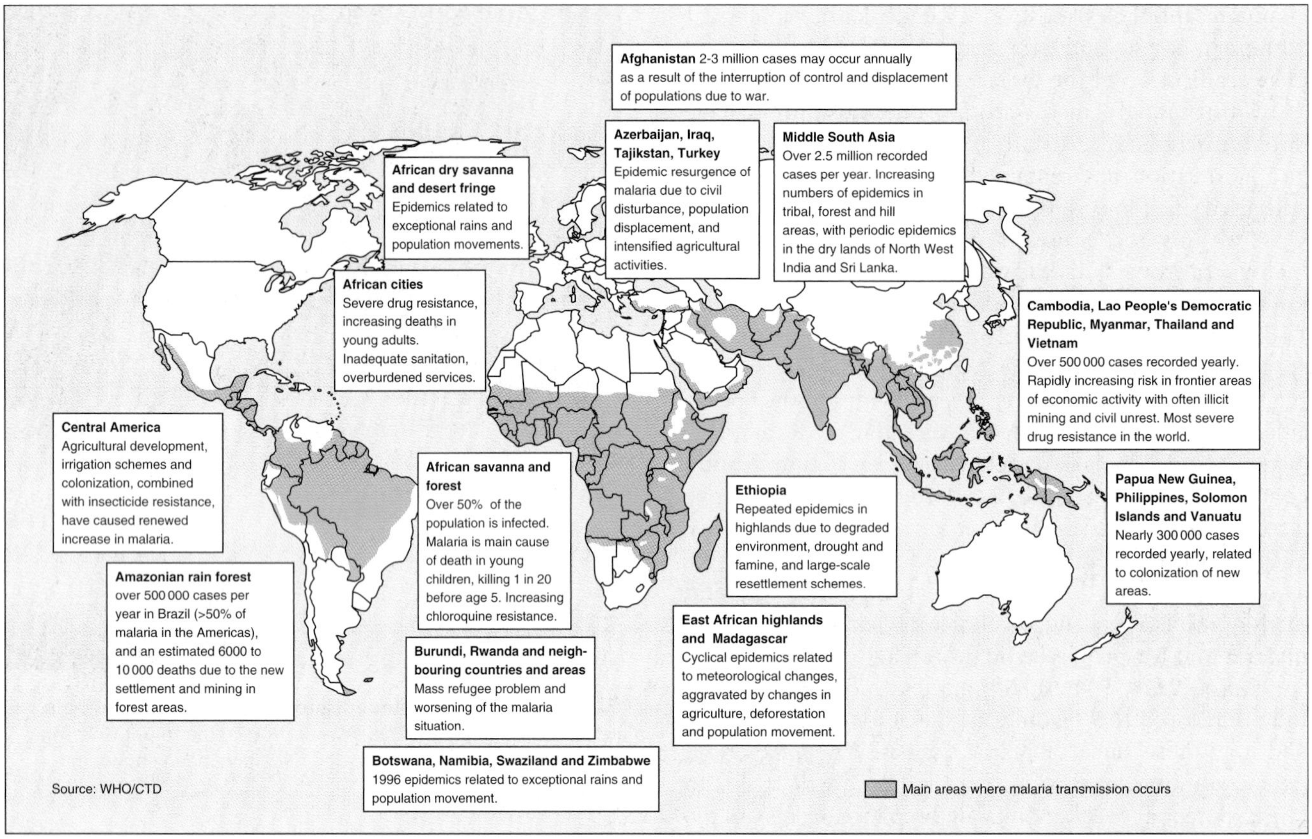


**Figure 15–14 Malaria distribution and problem areas.** Malaria is found in the tropics and subtropics around the world. Drug resistance and environmental disruptions are contributing to the strong resurgence of the disease in the last two decades. (*Source:* World Health Organization, *Tropical Disease Research: Progress 1995–96: Thirteenth Programme Report* (Berne, Switzerland: Berteli Printers, 1997, p. 42).)

3. **Carcinogenic.** Pollutants initiate changes within cells that lead to uncontrolled growth and division (cancer), which is frequently fatal.

***Indoor Air Pollution.*** Much attention is given to air pollutants released to the atmosphere, and rightly so. (Chapter 21 discusses these pollutants.) However, *indoor air pollution* can pose an even greater risk to health. The air inside the home and workplace often contains much higher levels of hazardous pollutants than outdoor air does! In the developed countries, the overall indoor air pollution problem is threefold. First, increasing numbers and types of products and equipment used in homes and offices give off hazardous fumes. Second, buildings have become increasingly well insulated and sealed; hence, pollutants are trapped inside, where they may accumulate to dangerous levels. Third, people are exposed more to indoor pollution than to outdoor pollution. The average person spends 90% of his or her time indoors, and the people who spend the most time inside are those most vulnerable to the harmful effects of pollution, namely, small children, pregnant women, the elderly, and the chronically ill.

The sources of indoor air pollution are quite varied, as evidenced in Figure 15–15. One of the least excusable of these is smoking, which carries a much higher health risk than any of the sources shown in Figure 15–15. ETS is known to contain thousands of substances, including 40 that are known carcinogens and many others that are potent respiratory irritants.

***Developing Countries.*** The most serious indoor air pollution threat is found in the developing world, where at least 3 billion people continue to rely on biofuels like wood and animal dung for cooking and heat. This threat was identified as one of the top 10 health risks by the WHO (Fig. 15–9). Fireplaces and stoves are often improperly ventilated (if at all), exposing the inhabitants to very high levels of pollutants, such as carbon monoxide, nitrogen and sulfur oxides, soot, and benzene (the last two being known carcinogens). Four major problems have been associated with indoor air pollutants in the developing world: acute respiratory infections in children; chronic lung diseases, including asthma and bronchitis; lung cancer; and birth-related problems. One study in Mexico showed that women who were continually exposed to indoor smoke had 75 times the risk of developing a chronic lung disease than women who weren't exposed. Public-policy solutions to this problem include ventilated stoves that burn more efficiently and the conversion from biofuels to bottled gas or liquid fuels such as kerosene.

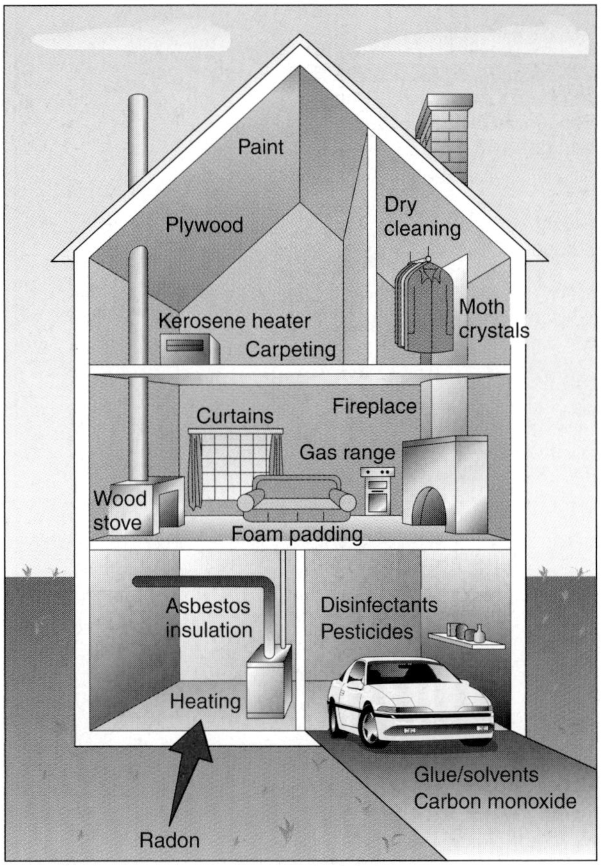


**Figure 15–15 Indoor air pollution.** Pollutants originate from many sources and can accumulate to unhealthy levels, leading to asthma and "sick-building syndrome" (characterized by eye, nose, and throat irritations, nausea, irritability, and fatigue), as well as other serious problems.

*Asthma.* One common consequence of indoor air pollution is asthma, which is currently at epidemic proportions in the United States, afflicting over 20 million people. An allergic disease, asthma attacks the respiratory system so that air passages tighten and constrict, causing wheezing, tightness in the chest, coughing, and shortness of breath. In acute attacks, the onset is sudden and sometimes life threatening. Substances that can trigger asthma attacks include dust, animal dander, mold, secondhand smoke, and various gases and particles. Asthma is expensive, too, with direct and indirect costs totaling more than $11 billion a year in the United States. Over 5,000 die from the condition annually. Through its Indoor Environments Division Web site (www.epa.gov/iaq/), the EPA provides helpful guidelines for controlling many of the substances that trigger asthma in schools, homes, and other indoor environments.

*Worms, Anyone?* Oddly, asthma is much more common in developed countries, in spite of the greater exposure to indoor air pollution in developing countries. Recent research has suggested that frequent parasitic worm infections stimulate a well-regulated anti-inflammatory network in immune systems of most people in the developing countries, protecting them against many allergic diseases like asthma. In the developed countries, where such parasitic infections are uncommon, this protection is lacking, so allergic disorders are much more common. Although this disparity suggests that it might be beneficial to play host to a few worms, it would be preferable to identify the parasite molecules that induce the protective immune response and to employ those molecules as vaccines.

*Toxicology.* Even though the substances found in indoor air are hazardous, the link between their presence in the air and the development of a health problem is harder to establish than with the infectious diseases. As with cancer, often the best that can be done is to establish a statistical correlation between exposure levels and the development of an adverse effect. The science of **toxicology** studies the impacts of toxic substances on human health and investigates the relationships between the presence of such substances in the environment and health problems like cancer. Chapter 19 discusses toxic substances and toxicology in greater detail.

In the meantime, we give our attention to *risk assessment,* one of the most important tools of the toxicologist in the developed countries. Risk assessment is an approach to the problems of environmental health that is now a major element of public policy, but also the subject of some controversy.

## 15.3 Risk Assessment

Although, in the developed countries, our lives are much more free of hazards and risks than ever before, our society still faces hazards and the risks that they present to our health. For our own self-interest, we might like to know about and evaluate the risks to which we are subjected. Such knowledge would enable us to make informed choices as we consider the benefits and risks of the hazards around us. In the developing countries, the high burden of disease so common in many of the poorer regions calls for a medical response out of humanitarian concern. But treating the sick, as helpful and appropriate as it is, will go on indefinitely if prevention is ignored. According to the WHO 2002 report, "Focusing on risks to health is the key to preventing disease and injury."[3] When the risks are thoroughly understood, strategies for reducing disease and death are far easier to identify and prioritize. If governments are to act as stewards of the health of their people, they desperately need the kind of information a thorough evaluation of health risks can bring them.

**Risk assessment** is *the process of evaluating the risks associated with a particular hazard before taking some action in a situation in which the hazard is present.* Figure 15–16 shows a number of risks, expressed as days of lost life expectancy, for people in a typical developed country. Alternatively, risks may be expressed as the probability of dying from a given hazard. According to

[3] Ibid, p.9.

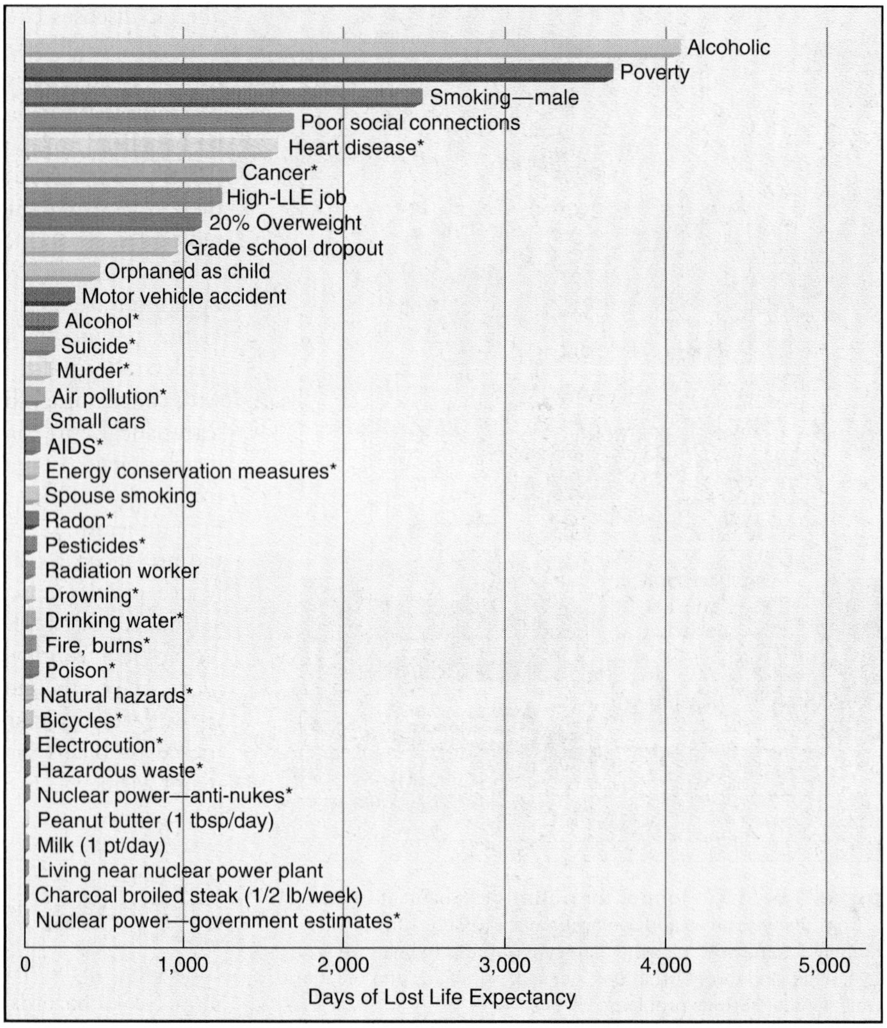


**Figure 15–16 Loss of life expectancy from various risks.** One way to illustrate the relative risks posed by different hazards is to rank them in terms of average days of lost life associated with each risk. (*Source:* R. Wilson and E. A. C. Crouch, *Risk-Benefit Analysis,* 2d ed., 2001, Harvard University Press)

Table 15–3, for example, the annual risk of dying from a motor vehicle accident is 1.6 per 10,000. That is, in the course of a year, 1.6 out of every 10,000 who ride in or drive a motor vehicle will die as the result of an accident. Figure 15–16 suggests that this is equivalent to about 300 days of lost life expectancy. In other words, people who never enter a motor vehicle will live an average of 300 days longer than those who do.

Truthfully, not very many people actually make choices about the hazards in their lives on the basis of such risk assessments. However, risk assessment has become an important process in the development of public policy and has long been a major way of applying sound science to the hard problems of environmental regulation. The WHO, however, is now recommending risk assessment as an ideal way to promote the health of people everywhere.

## Environmental Risk Assessment by the EPA

Risk assessment began at the EPA in the mid-1970s as a way of addressing the cancer risks associated with pesticides and toxic chemicals. As currently performed at the EPA, risk assessment has four steps: hazard assessment, dose-response assessment, exposure assessment, and risk characterization. The description that follows includes elements of new guidelines for risk assessment that are currently being evaluated by the EPA.

**Hazard Assessment: Which Chemicals Cause Cancer?** **Hazard assessment** is the process of examining evidence linking a potential hazard to its harmful effects. In the case of accidents, the link is obvious. The use of cars, for example, involves a certain number of crashes and deaths. In these cases, *historical data,* such as the annual highway death toll, are useful for calculating risks.

In other cases the link is not so clear, because there is a time delay between the first exposure and the final outcome. For example, establishing a link between exposure to certain chemicals and the development of cancer some years later is often very difficult. In cases where linkage is not obvious, the data may come from epidemiological studies or animal tests. An **epidemiological study** tracks how a sickness spreads through a community. Thus, to find a link between cancer and exposure to some chemical, an epidemiological study

**table 15-3 Some Commonplace Hazards, Ranked According to Their Degree of Risk**

| Hazardous Action | Annual Risk* | | |
|---|---|---|---|
| Cigarette smokers | 10 | per | 1,000 |
| All cancers | 2.0 | per | 1,000 |
| Firefighters | 4.0 | per | 10,000 |
| Police killed in line of duty | 2.9 | per | 10,000 |
| Hang gliding | 2.6 | per | 10,000 |
| Air pollution | 2.5 | per | 10,000 |
| Motor vehicle accident | 1.6 | per | 10,000 |
| Snowmobiling | 1.3 | per | 10,000 |
| Home accidents | 1.1 | per | 10,000 |
| Firearms | 1.1 | pre | 10,000 |
| Airline pilot | 10 | per | 100,000 |
| Mountain hiking | 6.4 | per | 100,000 |
| Alcohol consumption | 6.4 | per | 100,000 |
| Boating | 5 | per | 100,000 |
| Swimming | 3 | per | 100,000 |
| Four tablespoons peanut butter per day (cancer risk from aflatoxin) | 1.3 | per | 100,000 |
| Drinking water containing EPA limit of chloroform | 7 | per | 1,000,000 |
| Chest X ray | 3 | per | 1,000,000 |
| Electrocution | 1.8 | per | 1,000,000 |

* Probability of dying.
*Source:* R. Wilson and E. A. C. Crouch, *Risk-Benefit Analysis*, 2d ed., 2001, Harvard University Press. Copyright © 2001 by Richard Wilson and Edmund Crouch. Reprinted by permission of the authors.

would examine all the people exposed to the chemical and determine whether that population has more cancer than the general population. These findings are considered to be the best data for risk assessment and have resulted in scores of chemicals being labeled as known human carcinogens.

***Ask the Mice.*** We do not want to wait 20 years to find out that a new food additive causes cancer, so we turn to **animal testing** (Fig. 15–17) to find out now what *might* happen in the future. A test involving several hundred animals (usually mice) takes about three years and costs more than $250,000. If a significant number of the animals develop tumors after having been fed the substance being tested, then the substance is either a possible or probable human carcinogen, depending on the strength of the results.

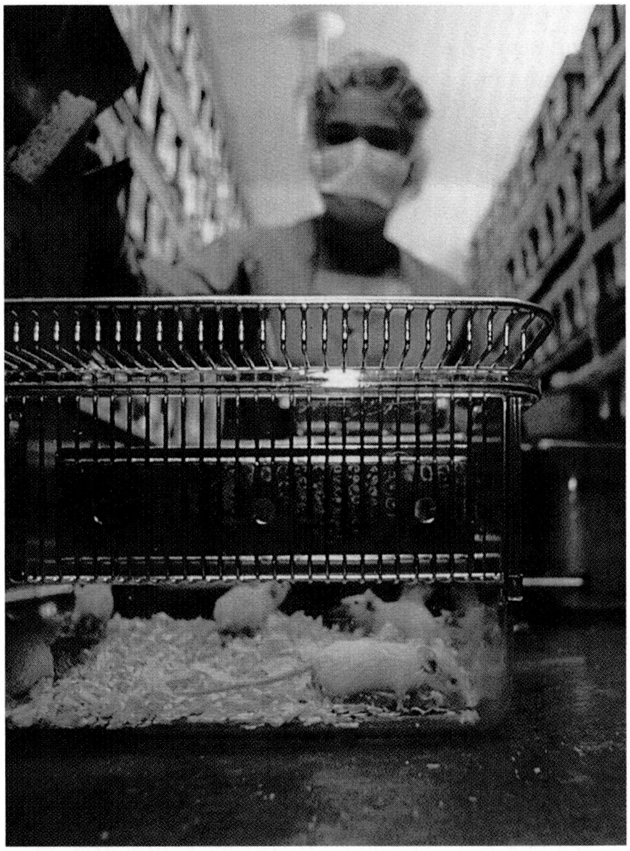

**Figure 15–17 Animal testing.** Laboratory mice are routinely used to test the potential of a chemical to cause cancer. The mice are an important source of information helping to assess the presence of a hazard in food, cosmetics, or the workplace.

Three objections to animal testing have been raised: (1) Rodents and humans may have very different responses to a given chemical; (2) the doses used on the animals are often much higher than those to which humans may be exposed; and (3) some people are opposed on ethical grounds to the use of animals for such purposes. Although there are obvious differences between rodents and humans, all chemicals shown by epidemiological studies to be human carcinogens are also carcinogenic to test animals, suggesting that animal tests have some predictive value for humans.

***Why Not Ask People?*** In the past, chemical manufacturers that were interested in assessing the safety of pesticides (in order to have their products approved for use) paid human subjects to test the products. Such tests are far less expensive than the usual animal testing. (Volunteers may be paid only a few hundred dollars; rats charge much higher fees!). A National Academy of Sciences panel addressed this topic in 1998, stating that the use of human volunteers "to facilitate the interests of industry or of agriculture" is unjustifiable. The EPA agreed and banned any use of such tests. However, recently the Bush administration, responding to requests from manufacturers, has pressed the EPA to reevaluate its policy and allow human testing. Fortunately, in

July 2003, the House of Representatives passed an amendment (the Bishop amendment) to HR 2681 (an appropriation bill for 2004) that reinstates EPA's moratorium on human testing of pesticides.

***Or the Chemical?*** Another source of information is the chemical or process itself: What are its physical and chemical properties, and what mode of action might the chemical or process take in inducing cancer?

Hazard assessment takes a "weight-of-evidence" approach to determining the carcinogenic potential of a chemical or process. Standard descriptors are used in its conclusions, such as "*likely to be carcinogenic to humans.*" In the end, hazard assessment tells us that we *may* have a problem.

**Dose–Response and Exposure Assessment: How Much for How Long?** When animal tests or human studies show a link between exposure to a chemical and an ill effect, the next step is to analyze the relationship between the concentrations of chemicals in the test (the **dose**) and both the incidence and the severity of the response. From this information, projections are made about the number of cancers that may develop in humans who are exposed to different doses of the chemical. Unless there is reason to assume otherwise, a linear response is used to determine an acceptable level of exposure. This process is called **dose–response assessment.**

Following dose–response assessment is **exposure assessment.** This procedure involves identifying human groups already exposed to the chemical, learning how their exposure came about, and calculating the doses and length of time of their exposure.

**Risk Characterization: What Does the Science Say?** The final step, **risk characterization,** is to pull together all the information gathered in the first three steps in order to determine the risk and its accompanying uncertainties. Here is where sound science is able to inform the risk manager. The characterization summary should provide information about key supportive data and methods, their limitations, and, especially, the estimates of risk and the uncertainties involved. Commonly, risk is expressed as the probability of a fatal outcome due to the hazard, as in Table 15–3 for common, everyday risks. For example, the EPA expresses cancer risk as an "upper-bound, lifetime risk," meaning the highest value in the range of probabilities of the risk, calculated over a lifetime. The Clean Air Act of 1990 directs the EPA to regulate chemicals that have a cancer risk of greater than one in a million ($>1 \times 10^6$) for the people who are subject to the highest doses. That is the same standard employed by the FDA in regulating chemicals in food, drugs, and cosmetics. One important limitation of risk characterization is that the analysts must work with the available information, which never seems to be sufficient. The final estimate of risk therefore includes a measure of the imprecision that always accompanies the use of imperfect information.

## Public-Health Risk Assessment

In its task of composing the *2002 World Health Report,* the WHO was breaking new ground in the area of risk assessment. Until then, most risk assessment took place as a function of environmental regulation. The new effort began by looking at all of the risk factors commonly known to be responsible for poor health and mortality and then asking the question, "Of all the disease burden in this population, how much could be caused by this [particular] risk?" Then certain risk factors were chosen for special attention, on the basis of the following criteria:

- **Potential global impact.** The risk is likely to be among the leading causes of poor health and of mortality. For example, the risks of being overweight are now well documented in many of the developed countries, where the condition is a serious problem.
- **High likelihood of causality.** It should be possible to trace the cause–effect relationships between poor health and the factor in question. Thus, overweight persons have a greater tendency to develop diabetes, high blood pressure, and high cholesterol; these, in turn, lead to greater risks of heart disease, stroke, and many forms of cancer.
- **Modifiability.** It should be possible to develop countrywide risk reduction policies that would substantially lessen the impact of the factor. For instance, overeating is a matter of individual behavior, which can be difficult to change, but a society can draw greater attention to the epidemic of obesity, its health risks, and the steps that can be taken to prevent overeating.
- **Availability of data.** There must be reliable data on the prevalence of the risk factor and its relationship to disease and mortality. With regard to overeating, the CDC coordinates a state-based monitoring system (the Behavioral Risk Factor Surveillance System) that collects information via telephone interviews. The CDC reports information on overeating and other, similar behavior-related health risks (insufficient physical activity, failing to eat fruits and vegetables, using tobacco and alcohol) to the state and local health departments.

***The DALY.*** The data collected must be assessed with the use of a "common currency," one that facilitates a comparison of different types of risks. The WHO report uses the measurement known as the *DALY—the "disability-adjusted life year."* One DALY represents the loss of one healthy year of a person's life. This measure, now in common usage in the global health community, assesses the burden of a disease in terms of life lost due to premature mortality and time lived with a disability.

The WHO report uncovered some remarkable differences in risk factors in different groups of countries (Table 15–4). Each factor calls for a different suite of

**table 15-4 Ten Leading Selected Risk Factors as Percentage Causes of Disease Burden, Measured in DALYs**

| | |
|---|---|
| **Developing countries** | |
| **High-mortality countries** | |
| Underweight | 14.9% |
| Unsafe sex | 10.2% |
| Unsafe water, sanitation, and hygiene | 5.5% |
| Indoor smoke from solid fuels | 3.7% |
| Zinc deficiency | 3.2% |
| Iron deficiency | 3.1% |
| Vitamin A deficiency | 3.0% |
| Blood pressure | 2.5% |
| Tobacco | 2.0% |
| Cholesterol | 1.9% |
| **Low-mortality countries** | |
| Alcohol | 6.2% |
| Blood pressure | 5.0% |
| Tobacco | 4.0% |
| Underweight | 3.1% |
| Overweight | 2.7% |
| Cholesterol | 2.1% |
| Indoor smoke from solid fuels | 1.9% |
| Low fruit and vegetable intake | 1.9% |
| Iron deficiency | 1.8% |
| Unsafe water, sanitation, and hygiene | 1.7% |
| **Developed countries** | |
| Tobacco | 12.2% |
| Blood pressure | 10.9% |
| Alcohol | 9.2% |
| Cholesterol | 7.6% |
| Overweight | 7.4% |
| Low fruit and vegetable intake | 3.9% |
| Physical inactivity | 3.3% |
| Illicit drugs | 1.8% |
| Unsafe sex | 0.8% |
| Iron deficiency | 0.7% |

(*Source:* World Health Report, 2002)

interventions. As a country evaluates its highest health risks, it will often have to prioritize its risk prevention strategies. This is especially true of the high-mortality developing countries, where public health resources are severely limited. The work done by the WHO in the 2002 report should be an invaluable resource in prompting governments to take more seriously their important role as the stewards of their country's health.

It is the opinion of the WHO that the scientific basis for the main risk factors presented in the 2002 report is well understood. Also, the strategies for cost-effective risk prevention for the main risks are straightforward. There is broad agreement about the need for action, for example, to address the increase in tobacco consumption or the role of unsafe sex in the HIV/AIDS epidemic. The WHO has an important role to play in coordinating efforts to translate scientific information into action, and its 2002 report is a major step in carrying out that role.

## Risk Management

**Regulatory Decisions.** In environmental risk assessment, the analysis of hazards and risks is a task that falls primarily to the scientific community. The task is incomplete, however, if no further consideration is given to the information gathered. Risk management naturally follows and is the responsibility of lawmakers and administrators. EPA policy separates risk assessment and risk management within the agency.

**Risk management** involves *(1) a thorough review of the information available pertaining to the hazard in question and the risk characterization of that hazard* and *(2) a decision as to whether the weight of the evidence justifies a regulatory action.* Without doubt, public opinion can play a powerful role in both processes. In general, however, a regulatory decision will hinge on one or more of the following considerations:

1. **Cost–benefit analysis.** This type of analysis compares costs and benefits relative to a technology or proposed chemical product; if done fairly, the analysis can make a regulatory decision very clear cut.
2. **Risk–benefit analysis.** A decision can be made with regard to the benefits versus the risks of being subjected to a particular hazard, especially when those benefits cannot be easily expressed in monetary values. The use of medical X rays is a good example. X rays carry a calculable risk of cancer, but the benefit derived when you X ray a broken bone is much greater than the small cancer risk involved.
3. **Public preferences.** As we will see, people have a much greater tolerance for risks that they feel are under their own control or are voluntarily accepted.

Risk management has been thoroughly incorporated into the EPA's policymaking process for at least 25 years.

Its most common use has been in the design of regulations. The process has enabled the agency to target appropriate hazards for regulation and to determine where to aim the regulations—that is, at the source, at the point of use, or at the point of disposal of the hazardous agent. As was done for toxic chemicals and cancer, a given risk level may be adopted as a standard against which policymakers can measure new risks.

## Risk Perception

The U.S. public's concern about environmental problems can be traced to the fear of hazards that pose a risk to human life and health. People perceive that their lives are more hazardous than ever before, but that is not true. In fact, our society is freer from hazards than it ever has been, as is evidenced by increased longevity. Why, then, do people protest against nuclear power plants, waste sites, and pesticide residues in food when, according to experts, these hazards pose extremely small risks (Fig. 15–16)? The answer lies in people's **risk perceptions**—their intuitive judgments about risks. In short, people's perceptions are not consistent with the results coming from a scientific analysis of risk. There are some good reasons for this discrepancy—and some lessons to be learned from it.

*Hazard versus Outrage.* The reason for the inconsistency between public perception and actual risk calculations, according to Peter Sandman of Rutgers University, is that the public perception of risks is often more a matter of *outrage* than *hazard.* Sandman holds that, while the term *hazard* expresses primarily a concern for fatalities only, the term *outrage* expresses a number of additional concerns:[4]

1. **Lack of familiarity with a technology.** Examples include how nuclear power is produced and how toxic chemicals are handled.
2. **Extent to which the risk is voluntary.** Research has shown that people who have a choice in the matter will accept risks roughly 1,000 times as great as when they have no choice.
3. **Public impression of hazards.** Accidents involving many deaths (as in Bhopal) or a failure of technology (for example, at Three Mile Island) are thoroughly imprinted into public awareness by media coverage and are not quickly forgotten.
4. **Overselling of safety.** The public becomes suspicious when scientists or public-relations people play up the benefits of a technology and play down the hazards.
5. **Morality.** Some risks have the appearance of being morally wrong. If it is wrong to foul a river, the notion that the benefits of cleaning it are not worth the costs is unacceptable. You should obey a moral imperative regardless of the costs.
6. **Control.** People are much more accepting of a risk if they are in control of the elements of that risk, as in driving an automobile, swimming, and overeating.
7. **Fairness.** The benefits and the risks should be connected. If the benefits go to someone else, why should you accept any risk?

*Media's Role.* The public perception of risk is strongly influenced by the media, which are far better at communicating the outrage elements of a risk than they are at communicating the hazard elements. Public concern over oil spills rose precipitously following the *Exxon Valdez* oil spill in Alaska, an accident that received extensive media coverage and had a high "outrage quotient." Cigarette smoking, however, which causes 442,000 deaths a year in the United States alone, receives minimal media attention, because it is not "news." Hence, there is no outrage factor. Indeed, it has been suggested that if all the year's smoking fatalities occurred on one day, the media would go ballistic, and smoking would be banned the next day!

*Public Concern and Public Policy.* Generally speaking, *public concern,* rather than cost–benefit analysis or risk analysis conducted by scientists, drives public policy. The EPA's funding priorities are set largely by Congress, which reflects public concern. Is this a problem?

If public outrage is the primary impetus for public policy, some serious risks may get less attention than they deserve. In particular, risks to the environment are commonly perceived as much less important than they really are, because of the public's preoccupation with risks to human health. This difference underscores the importance of *risk communication,* a task that should not be left to the media. For example, the value of ecosystems and their connections to human health and welfare need far greater emphasis in the public consciousness, and this responsibility falls to the scientific and educational communities, as well as to governmental agencies. Studies have shown that the most effective risk communication occurs by starting with what people already know and what they need to know and then tailoring the message so that it provides people with the knowledge that helps them make a more informed evaluation of risks. The process should not stop there, though: To be successful, risk communication should also involve dialogue among all the parties involved, namely, policymakers, the public, experts, and other interested groups.

The public's concern for more than the probabilities of fatalities has merit. Public concern must be heard, understood, and given a reasonable response. It may not be the best source of public policy, but it reflects certain values and concerns that could easily be omitted by an "objective" risk assessment. The fact is, subjective judgments are going to play a role at every step in the risk-assessment

[4]Covello, V. and P. Sandiman. "Risk Communication, Evolution and Revolution." in Wolbarst, A. (ed.). Solutions to an Environment in Peril. Baltimore: Johns Hopkins University Press, 164–178, 2001.

process, from hazard assessment to risk perception to risk management. The uncertainties involved in risk assessment should remind us that the process is only a tool—and an imperfect one at best.

***Precautionary Principle.*** The use of risk assessment has come under attack in recent years, with three major concerns: (1) It does not adequately reflect the uncertainty inherent in much of the assessment process; (2) it has been largely employed in situations where a chemical or process is already in use; and (3) the burden of proof of harm falls largely on the regulators.

In recognition of these problems, there is a strong movement toward employing the **precautionary principle** in formulating public policy to protect the environment and human health. The Wingspread Conference of 1998—a meeting of 35 academic scientists, government researchers, labor representatives, and grassroots environmentalists from the United States, Canada, and Europe—stated the following *consensus definition* of the principle: "Where an activity raises threats of harm to human health or the environment, precautionary measures should be taken even if some cause-and-effect relationships are not fully established scientifically. In this context, the proponent of an activity, rather than the public, should bear the burden of proof."

The precautionary principle is well established in Europe, where it originated, and has become the basis for European environmental law. For example, the principle has been employed to restrict the import of genetically modified foods (Chapter 9). It is also an essential component of many U.N. environmental treaties. How it might relate to U.S. risk-based environmental policy is still under debate. The principle operates in some measure in our approach to pharmaceuticals and pesticides, according to which a substance is presumed guilty until proven innocent; that is, the substance is not permitted to be put on the market until tests show that it is safe to use.

The precautionary principle represents a potent tool for implementing stewardship in the environmental health arena. Although risk assessment is useful in many situations, it should not be used as a blanket policy. The precautionary principle enables us to act to prevent some potential environmental health problems, even at the risk of being wrong or at the risk of spending more than a problem technically deserves. The old saying, "An ounce of prevention is worth a pound of cure" captures the sense of this approach. Where uncertainty is substantial, and especially where the penalty for being wrong is great, it would seem wise to make the precautionary principle a guiding principle overseeing the entire risk assessment process.

**Stewards of Health.** In a very real sense, public-health ministries and agencies are the primary stewards of the health and welfare of a country's people. Public policy should clearly reflect stewardship principles that put the health of people and the environment above the economic bottom line that often seems to drive our decisions. To quote the *2002 World Health Report,* "Governments are the stewards of health resources. This stewardship has been defined as 'a function of a government responsible for the welfare of the population and concerned about the trust and legitimacy with which its activities are viewed by the citizenry.'" Governments can accomplish much in the arena of preventive health, but they must nurture their scarce resources carefully. It is in that sense that the WHO 2002 report focuses on reducing risks in order to promote health and provides a valuable tool for accomplishing this to all countries. Beyond the government, health care providers are also stewards whose responsibility is vitally important to environmental health.

# revisiting the themes

## Sustainability

A foundational strategy of sustainable development is to reduce poverty, in part because being poor makes one susceptible to many hazards. Numerous risk factors are a consequence of the lack of access to clean water, healthy air, sanitation, nutritious food, and health care. The infectious diseases in particular are a tremendous impediment to the economic progress in poorer countries. It is hard to see how those countries can move toward sustainable societies when HIV/AIDS, malaria, and tuberculosis take such a heavy toll on their children and young adults. The first seven Millennium Development Goals all target environmental health problems. (See Table 6–2.) If the world community takes these goals seriously, the progress that is made will bring about a revolution in environmental health and move scores of countries much farther along in their quest for sustainable development.

## Stewardship

Governments are the stewards of their country's health resources. Health ministries have a difficult task if they are not given the funding and personnel to carry out their mission. Fortunately, organizations such as the WHO can provide the vital information governments need to address their countries' greatest health problems. The WHO exemplifies the stewardship concept of a community of nations acting together to bring care and information to those whose health is at risk. Those who work alongside of the WHO—the health ministries, health-related NGOs, and health care providers—are the hands and hearts of the world community.

### Sound Science

The continuing battle against pathogens and against human disorders like cancer and heart disease requires the sustained attention of dedicated research laboratories around the world. This is sound science in the direct interest of humankind. Risk assessment utilizes the scientific community in the final step of the process, at which risk characterization takes place. At that point, the scientists have evaluated the risk as best they can, so it is now up to those responsible for risk management. The WHO states that the basis for the major risk factors plaguing most societies has been scientifically established. There is more work to do, of course, but the failure to prove scientific causality cannot be an excuse for inaction by those responsible for their people's health.

### Policy and Politics

The U.S. CDC is an excellent example of government policy put into action to protect people from health risks. The CDC is a competent agency, and it carries out its mandate aggressively as it pursues new diseases and keeps the public health community informed about important issues. In developing countries, ministries of health are responsible for addressing the major health risks confronting the people of their countries. Theirs is a difficult task in many of the poorer countries. Not only are the ministries likely to have severely limited resources, but the health challenges they must confront are often enormous. Some countries have placed higher priorities on their health needs than on military armament or other sectors and, as a result, are experiencing significant progress in public health.

The international community has responded to global health issues on several fronts. Chief among these are the Millennium Development Goals, which have focused attention on poverty, mortality, and diseases in the developing countries. The World Bank maintains a Web site, www.developmentgoals.org, that tracks indicators for monitoring progress towards these goals in the different regions of the world.

The developed countries can contribute much, especially in official development assistance, debt forgiveness, and the reduction of tariffs on developing-country exports. Some of the reticence of the rich countries in extending this aid has been countered by the generosity of many NGOs and wealthy individuals (such as the Bill and Melinda Gates Foundation and the Turner Foundation). Another significant effort has been the Framework Convention on Tobacco Control.

The WHO has provided excellent world leadership in promoting awareness and action in combating the health risks people face in developed and developing countries alike. In focusing on risk reduction, the WHO's 2002 report has done the world health community a great service. The entire process of risk assessment highlights the need for political action as the focus moves from assessment to management of risks.

### Globalization

As an "Unwelcome Globalization" (Global Perspective, p. 422), the rapid movement of disease poses a frightening threat to people everywhere. Viruses and other pathogens can often be spread in a matter of days from an origin to locations on the other side of the world, as the SARS epidemic illustrated so graphically in 2003. This rapid transmission points to the important role of scientists in the front line of epidemiology; at times, these scientists must even take great risks to track down pathogens.

## review questions

1. Differentiate between a hazard and a risk.
2. Define morbidity, mortality, and epidemiology.
3. Describe the public-health roles of the CDC and the WHO.
4. What are the four categories of human environmental hazards? Give examples of each.
5. List as many as you can of the top 10 risk factors that are responsible for global mortality and disease.
6. Document the struggles to control tobacco use both in the United States and globally.
7. What is the significance of malaria worldwide, and what are some recent developments in the battle against this disease?
8. Discuss the four steps of risk assessment used by the EPA.
9. Describe the process of risk assessment for public health as it was recently carried out by the WHO. What is a DALY?

10. What concerns about risks and hazards tend to generate public outrage?

11. Discuss the relationship between public risk perception and assessment, on the one hand, and public policy, on the other.

12. What is the precautionary principle, and how would employing the principle better our approach to environmental public policy?

## thinking environmentally

1. Consider Table 15–4. Discuss each group of countries' risk factors that are primarily a consequence of human choice. Pick one of these factors, and describe how you might proceed to bring it under control in an appropriate country.

2. Imagine that you have been appointed to a risk–benefit analysis board. Explain why you approve or disapprove of the widespread use of
   - genetically modified foods
   - drugs to slow the aging process
   - nuclear power plants
   - asbestos

3. Suppose your town wanted to spray trees to rid them of a deadly pest. How would you use cost–benefit analysis to determine whether spraying is a good idea? Is there a better approach?

4. Use the Internet to investigate your state's (or country's) department of public health. Describe ways in which this agency interacts with the CDC or the WHO (or both).

# Pests and Pest Control

**Key Topics**

1. The Need for Pest Control
2. Promises and Problems of the Chemical Approach
3. Alternative Pest Control Methods
4. Socioeconomic Issues in Pest Management
5. Pesticides and Policy
6. Revisiting the Themes

The French Quarter of New Orleans is a world-famous tourist destination. In its 10 square blocks it has great jazz, colonial architecture, Bourbon Street, and outstanding restaurants. It also has an overseas invader called the Formosan subterranean termite (Fig. 16–1) that is eating away at its buildings. The termites, natives to China, are believed to have arrived in the United States on military ships returning from the Pacific theater after World War II. More robust and voracious than native species, the Formosan termites build large nests above and below ground, with millions in each nest. They infest trees as well as buildings and have damaged more than 30% of the historic live oaks of New Orleans. The termites have expanded their range from a few coastal cities to major areas of 10 states in the South and West. Currently, they cost property owners more than $1 billion a year for damage control and repair.

***Full Stop.*** The Agricultural Research Service (ARS) of the U.S. Department of Agriculture has launched a $5 million-a-year program called Operation Full Stop against the Formosan subterranean termite. Ordinary treatment with approved pesticides can hold the insects at bay for a time, but they seem very adept at exploiting the smallest contact between soil and house. Once inside, they can eat rafters, studs, door frames, and flooring, often remaining undetected until something collapses. The ARS has been employing a system of baiting the termites by putting out blocks of wood to monitor for their presence. (See opposite page.) Once termites are found in a block, a toxic baited block is switched with it, and

*French Quarter, New Orleans*
**The French Quarter, 10 square blocks of residences, offices, restaurants, and jazz emporiums, is infested with the Formosan subterranean termite. These workers are baiting a trap in one of many monitoring stations maintained by Operation Full Stop.**

**Figure 16–1 Formosan subterranean termites.** These termites are feeding on Sudan-red impregnated paper, which allows researchers to trace their numbers and foraging range.

the termites carry off the toxic bait to share with their unsuspecting fellows back in the nest. By fully enlisting all the households in the French Quarter, Full Stop has reduced the number of termites significantly.

The ARS recently announced the discovery of a cottony mold that is able to kill 100% of the termites contacting it within a week; Full Stop team members are now working on a method for producing the mold and employing it as part of the baiting approach. The researchers envision infected termites picking up the fungal spores and carrying them back to their colony, initiating a lethal epidemic that wipes out millions and eradicates the colony. If this tactic works, it may be only jazz bands that bring down the house in the French Quarter of New Orleans.

## 16.1 The Need for Pest Control

Since earliest times, humans have suffered frustration and food losses brought on by destructive pests. To this day, farmers, pastoralists, and homeowners wage a constant battle against the insects, plant pathogens, and unwanted plants that compete with their human rivals for the biological use of crops, animals, and homes.

**Defining Pests.** One dictionary defines a **pest** as "any organism that is noxious, destructive, or troublesome." This definition includes a broad variety of organisms that interfere with humans or with our social or economic endeavors: pathogens, nuisance wild animals, annoying insects, molds, etc. Earlier chapters considered nuisance wild animals (Chapter 10) and pathogens (Chapter 15); the emphasis in this chapter will be on those pests which interfere with crops, grasses, domestic animals, and structures—primarily the agricultural pests and weeds.

**Agricultural pests** are organisms that feed on ornamental plants or agricultural crops or animals. The most notorious of these organisms are various insects, but certain fungi, viruses, worms, snails, slugs, rats, mice, and birds also fit into the category (Fig. 16–2(a)–(f)). **Weeds** are plants that compete with agricultural crops, forests, and forage grasses for light and nutrients (Fig. 16–2(g)–(j)). Some unwanted plants poison cattle or have other serious effects; others simply detract from the appearance of lawns and gardens. Bringing these pests under control has three main purposes: to protect our food, to protect our health, and for convenience.

**The Importance of Pest Control.** Part of the credit for modern human prosperity can be attributed to pest control. For example, pesticides are vital elements in the prevention of the diseases that kill and incapacitate humans. In addition to having an agricultural use, pesticides have become important public-health tools used to combat diseases such as malaria and sleeping sickness.

Insects, plant pathogens, and weeds destroy an estimated 37% (before and after harvest) of potential agricultural production in the United States, at a yearly loss of $122 billion to consumers and producers. Elsewhere, pests also destroy important crops: (1) The fungus causing late blight in potatoes, *Phythophtora infestans,* is considered to be global agriculture's worst crop disease. (Yields in Russia suffered a 70% loss in the late 1990s.) (2) The "vampire weed," *Striga hermonthica,* is a root parasite that causes several billion dollars in losses each year in East Africa. (3) Desert locusts form swarms periodically that migrate thousands of miles and are such a threat in Africa and Asia that the U.N.'s FAO maintains a special program to monitor and predict coming locust plagues. This crop destruction threatens food security in many developing countries, perpetuating hunger and poverty among the poorest people.

Efforts to control these losses in the United States in 1999 involved the use of 912 million pounds (415 thousand metric tons) of **herbicides** (chemicals that kill plants) and **pesticides** (chemicals that kill animals and insects considered to be pests) annually (Fig. 16–3), at a direct cost of $11.1 billion. Worldwide, 2.84 million tons of pesticides were used in 1999, costing $34 billion. Many of the changes in agricultural technology, such as monoculture and the widespread use of genetically identical crops, which have boosted yields, have also brought on an increase in the proportion of crops lost to pests—from 31% in the 1950s to 37% today. During the last half-century, the use of herbicides and pesticides multiplied manyfold, leading to a disturbing and unsustainable dependency on them.

**Different Philosophies of Pest Control.** Medical practice employs two basic means of treating infectious diseases. One approach is to give the patient a massive

**Figure 16–2** **Important pests.** (a) Fire ants. (b) Cotton boll weevil. (c) Aedes aegypti mosquito. (d) male medfly. (e) Colorado potato beetle. (f) migratory locust. (g) vampire weed. (h) crabgrass.

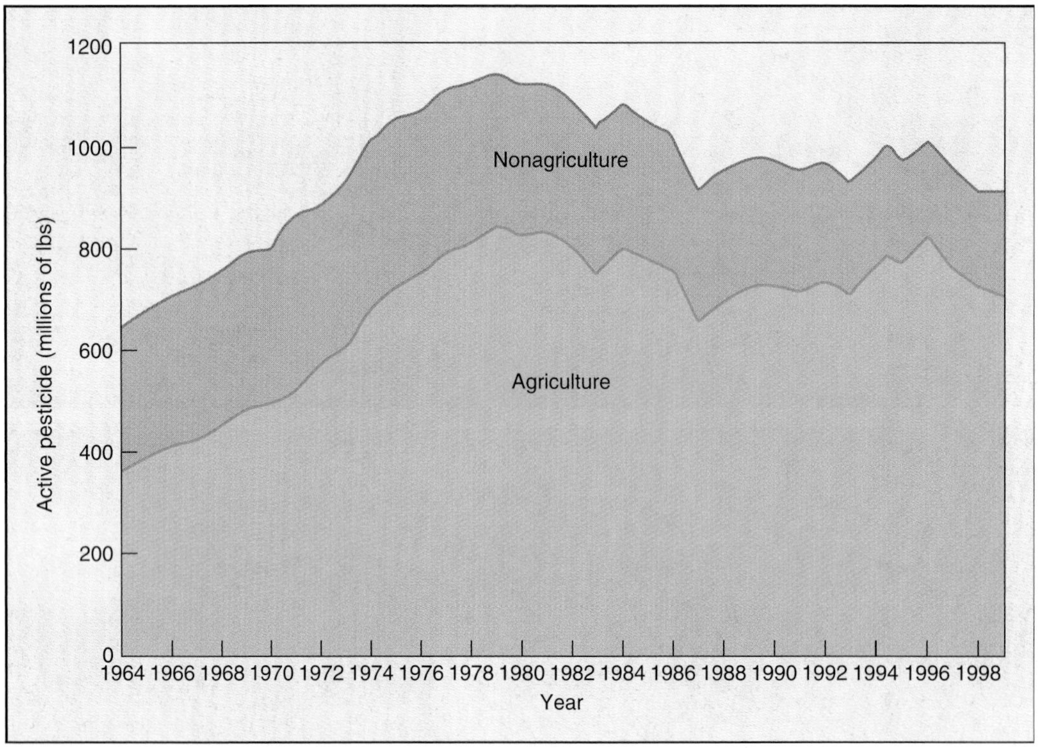


**Figure 16–3 Pesticide use in the United States.** Nonagricultural and agricultural use of pesticides (active ingredients) is shown for the period 1964–1999. (*Source:* Data from the Environmental Protection Agency, Office of Pesticide Programs.)

dose of antibiotics, hoping to eliminate the pathogen causing the problem or to stop the pathogen before it can get established. The other approach is to stimulate the patient's immune system with a vaccine to produce long-lasting protection against any future invasion. In practice, both means are often used to keep a particular pathogen under control.

The same basic philosophies govern the control of agricultural pests. The first is **chemical treatment.** Like the use of antibiotics, chemical treatment seeks a "magic bullet" that will eradicate or greatly lessen the numbers of the pest organism. Although it has had much success, this approach gives only short-term protection. Furthermore, the chemical often has side effects that are highly damaging to other organisms.

The second philosophy is **ecological control.** Like stimulating the body's immune system, this approach seeks to give long-lasting protection by developing control agents on the basis of knowledge of the pest's life cycle and of ecological relationships. Such agents, which may be other organisms or chemicals, work in one of two ways: Either they are highly specific for the pest species being fought, or they manipulate one or more aspects of the ecosystem. Ecological control emphasizes the protection of people and domestic plants and animals from damage from pests, rather than eradication of the pest organism. Thus, the benefits of pest control can be obtained while maintaining the integrity of the ecosystem.

These two philosophies are combined in the approach called **integrated pest management (IPM).** IPM is *an approach to controlling pest populations by using all suitable methods—chemical and ecological—in a way that brings about long-term management of pest populations and also has minimal environmental impact.* This approach is increasing in usage, especially where pesticides are seen as undesirable because of health risks and in developing countries, where the cost of pesticides is often prohibitive.

## 16.2 Promises and Problems of the Chemical Approach

Pesticides are categorized according to the group of organisms they kill. There are insecticides (for insects), rodenticides (for mice and rats), fungicides (for fungi), herbicides (for plants), and so on. None of these chemicals, however, is entirely specific to the organisms it is designed to control; each poses hazards to other organisms, including humans.

### Development of Chemical Pesticides and Their Successes

Finding effective materials to combat pests is an ongoing endeavor. The early substances used (frequently referred to as **first-generation pesticides**) included toxic heavy metals such as lead, arsenic, and mercury. Scientists now recognize that these substances may accumulate in soils, inhibit plant growth, and poison animals and humans. In addition, toxic heavy metals lost their effectiveness as pests became increasingly resistant to them. For example, in the early 1900s, citrus growers were able to kill 90% of injurious **scale insects** (minute insects that suck the juices from plant cells) by placing a tent over an infested tree and piping in deadly cyanide gas for a short time. By 1930, this same technique killed as little as 3% of these pests.

The next step had its origins in the science of organic chemistry in the early 1800s. During the 19th century, chemists synthesized thousands of organic compounds, but for the most part, these compounds sat on shelves because uses for them had not yet been found. By the 1930s, however, with agriculture expanding to meet the needs of a rapidly increasing population and with first-generation (inorganic) pesticides failing, farmers needed something new. In time, **second-generation pesticides,** as they came to be called, were developed as a result of synthetic organic chemistry.

**The DDT Story.** In the 1930s, a Swiss chemist, Paul Müller, began systematically testing some organic chemicals for their effect on insects. In 1938, he hit upon the chemical dichlorodiphenyltrichloroethane (DDT), a chlorinated hydrocarbon that had first been synthesized some 50 years earlier. Just traces of DDT killed flies in Müller's laboratory.

DDT appeared to be nothing less than the long-sought "magic bullet," a chemical that was extremely toxic to insects and yet seemed nontoxic to humans and other mammals. It was very inexpensive to produce. At the height of its use in the early 1960s, DDT cost no more than about 20 cents a pound. It was **broad spectrum,** meaning that it was effective against a multitude of insect pests. It was also **persistent,** meaning that it did not break down readily in the environment and hence provided lasting protection. This last attribute provided additional economy by eliminating both the material and labor costs of repeated treatments.

***In War. . . .*** DDT quickly proved successful in controlling important insect disease carriers. During World War II, for example, the military used DDT to control body lice, which spread typhus fever among the men living in dirty battlefield conditions. As a result, World War II was one of the first wars in which fewer men died of typhus than of battle wounds. DDT sprayed over the island of Saipan enabled the U.S. marines to defeat both dengue fever and the Japanese. The WHO of the United Nations used DDT throughout the tropical world to control mosquitoes and greatly reduced the number of deaths caused by malaria. (See the "Ethics" essay, this page.) There is little question that DDT saved millions of lives. In fact, the virtues of DDT seemed so outstanding that Müller was awarded the Nobel prize for medicine in 1948 for his discovery.

**ethics**

## DDT for Malaria Control: Hero or Villain?

Malaria exacts a terrible toll in death and disease: 300 million bouts of sickness and a million deaths a year. Fifty years ago, malaria seemed to be on the way out, as new synthetic drugs killed the protozoan parasites that caused the disease and DDT wiped out mosquitoes everywhere. But resistance emerged with a vengeance, and now the parasites are highly resistant to most common malaria drugs, and the mosquitoes are resistant to DDT. Although DDT has been banned in developed countries, because of its human health and environmental dangers, it is still used in many developing countries, largely because it is so inexpensive. DDT is one of a group of 12 persistent organic pollutants targeted for phasing out by the UNEP, with the deadline set for 2007. This is the story from the developed world.

In malarial regions in the developing world, there is a different story. Here, villagers are exposed to mosquitoes because they cannot afford screens on their homes, mosquito nets to sleep under, or repellent to spray on their children, four of whom die from malaria every minute. The anopheles mosquito, which carries malaria, is a night hunter, biting people as they rest or sleep. In 22 countries, DDT is still the main line of defense against the mosquito. The typical application involves spraying inside houses and eaves at a concentration of $2\ g/m^2$, once or twice a year. Even where the mosquitoes are resistant to its killing effects, the DDT acts as a repellent or irritant that keeps the mosquitoes from staying around. Scientists from the WHO have referred to indoor DDT spraying as the most cost-effective and easily applied control measure for malaria. To quote one angry malaria program chief from a developing country, "Banning DDT is eco-imperialism. It's a model of putting the interests of the environment ahead of human lives."

Those in favor of pushing forward with the ban cite the problems of keeping DDT out of agricultural fields, where it can enter food chains. (Once the pesticide is made available, controlling its use is difficult.) Alternative pesticides, such as synthetic pyrethroids, are already available (but resistance to these is appearing, too). An integrated public-health approach is needed wherein mosquito breeding places are eliminated, bed nets are employed, and malaria cases are treated promptly to intercept the life cycle of the parasite. However, the public-health sector is unconvinced, and organizations such as the Malaria Foundation International and the Malaria Project are actively campaigning against the proposed action for a worldwide ban on DDT. To date, these organizations have obtained a special exemption for countries to continue to use DDT if they wish. What do you think should be done?

**Figure 16–4 Aerial spraying.** A crop is being sprayed with pesticides to keep insects under control. Spraying is the basic technique still used in most control programs, although nowadays the pesticides in use do not persist in the environment for more than a week or two.

*And In Peace....* Postwar uses of DDT expanded dramatically. The chemical was sprayed on forests to control defoliating insects such as the spruce budworm. It was routinely sprayed on salt marshes to kill nuisance mosquitoes. It was sprayed on suburbs to control the beetles that spread Dutch elm disease. DDT even proved highly effective in controlling agricultural insect pests. (Fig. 16–4). Indeed, DDT was so effective, at least in the short run, that many crop yields increased dramatically. Growers could ignore other, more painstaking methods of pest control such as crop rotation and the destruction of old crop residues. They could grow varieties that were less resistant, but more productive. They could grow certain crops in warmer or moister regions, where formerly the damage from pests had been devastating. In short, DDT gave growers more options for growing the most economically productive crop.

The success of DDT led to the development of a great variety of synthetic organic pesticides. Currently, the EPA Pesticide Program regulates more than 18,000 pesticide products, most of which are synthetic organic pesticides. Examples and some characteristics of these insecticides and herbicides are listed in Table 16–1.

## Problems Stemming from Chemical Pesticide Use

The problems associated with the use of synthetic organic pesticides can be categorized as follows:

- development of resistance by pests
- resurgences and secondary-pest outbreaks
- adverse environmental and human health effects

**Development of Resistance by Pests.** The most fundamental problem for growers is that chemical pesticides gradually lose their effectiveness. Over the years, it becomes necessary to use larger and larger quantities, to try new and more potent pesticides, or to do both to obtain the same degree of control. Synthetic organic pesticides fared no better than first-generation pesticides in this respect. For example, in 1946, 1 kg

**table 16-1 Characteristics of Pesticides**

**Insecticides**

| Type | Examples | Toxicity to Mammals | Persistence |
|---|---|---|---|
| Organophosphates | Parathion, malathion, phorate, chloropyrifos | High | Moderate (weeks) |
| Carbamates | Carbaryl, methomyl, aldicarb, aminocarb | Moderate | Low (days) |
| Chlorinated hydrocarbons | DDT, toxaphene, dieldrin, chlordane, lindane | Relatively low | High (years) |
| Pyrethroids | Permethrin, bifenthrin, esfenvalerate, decamethrin | Low | Low (days) |

**Herbicides**

| Type | Examples | Effects on Plants |
|---|---|---|
| Triazines | Atrazine, simazine, cyanizine | Interfere with photosynthesis, especially in broadleaf plants |
| Phenoxy | 2, 4-D, 2, 4, 5, -T, methylchlorophenoxybutyrate (MCPB) | Cause hormonelike effects in actively growing tissue |
| Acidamine | Alachlor, Propachlor | Inhibit germination and early seedling growth |
| Dinitroaniline | Trifluralin, oryzalin | Inhibit cells in roots and shoots; prevent germination |
| Thiocarbamate | ethylpropylthiolcarbamate (EPTC), cycloate, butylate | Inhibit germination, especially in grasses |

*Source:* U.S. Environmental Protection Agency, *Private Pesticide Applicator Training Manual.*

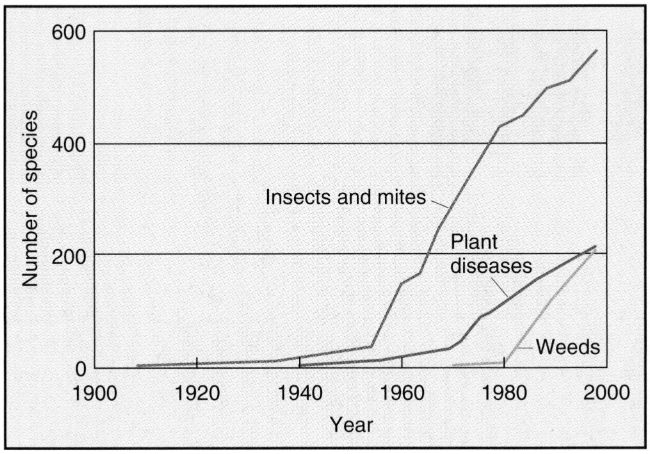


**Figure 16–5 Number of species resistant to pesticides, 1908–1998.** Using pesticides causes selection (survival of the fittest) for individuals that are resistant. As we continue to use pesticides, we breed insects and other pests that are increasingly resistant to those pesticides. [*Source:* Lester Brown et al., *Vital Signs 1999* (Washington, DC: Worldwatch Institute, 1999).

(2.2 lb) of pesticides provided enough protection to produce about 60,000 bushels of corn. By 1971, it took 64 kg (141 lb) to produce the same amount, and losses due to pests actually *increased* during the intervening years.

***Evolution at Work.*** Resistance builds up because pesticides destroy the sensitive individuals of a pest population, leaving behind only those few that already have some resistance to the pesticide. Resistant insect populations develop rapidly because insects have a phenomenal reproductive capacity. A single pair of houseflies, for example, can produce several hundred offspring that may mature and reproduce themselves only two weeks later. Consequently, repeated pesticide applications result in the unwitting selection and breeding of genetic lines that are highly, if not totally, resistant to the chemicals that were designed to eliminate them. Cases have been recorded in which the resistance of a pest population has increased as much as twenty-five-thousandfold. Recent work on mosquito resistance to a group of pesticides has shown that resistance genes (insecticide-detoxifying carboxylesterase genes) can be spread globally in a brief time. In short, pesticide resistance simply confirms that "natural" selection is a powerful evolutionary force.

***The last Roundup®.*** When one pesticide is used extensively, resistance is virtually inevitable. A case in point is the herbicide glyphosate (trademark Roundup®), which is used on widely grown genetically modified (Roundup Ready®) corn, soybeans, and cotton. These crops contain a gene that makes them resistant to glyphosate. Thus, use of the modified crops has led to enormous increases in the use of the herbicide. Several weeds—mare's tail, water hemp, and ryegrass—recently developed resistance to glyphosate, and there is concern that the resistance could spread because of the absolute dependency of the Roundup Ready® crops on glyphosate.

Over the years of pesticide use, the number of resistant species has climbed steadily. Many major pest species are resistant to all of the principal pesticides. Indeed, over a thousand species of insects, plant diseases, and weeds have shown resistance to pesticides (Fig. 16–5). Interestingly, as a pest population becomes resistant to one pesticide, it also may gain resistance to other, unrelated pesticides, even though it has not been exposed to the other chemicals. (See the "Earth Watch" essay, p. 442.)

**Resurgences and Secondary-Pest Outbreaks.** The second problem with the use of synthetic organic pesticides is that after a pest has been virtually eliminated with a pesticide, the pest population not only recovers, but explodes to higher and more severe levels. This phenomenon is known as a **resurgence.** To make matters worse, small populations of insects that were previously of no concern because of their low numbers suddenly start to explode, creating new problems. This phenomenon is called a **secondary-pest outbreak.** For example, with the use of synthetic organic pesticides, mites have become a serious pest problem, and the number of serious pests on cotton has increased from 6 to 16.

To illustrate the seriousness of resurgences and secondary-pest outbreaks, a recent study in California listed a series of 25 major pest outbreaks, each of which caused more than $1 million worth of damage. All but one involved resurgences or secondary-pest outbreaks. Moreover, the species appearing in secondary-pest outbreaks quickly became resistant to pesticides, thus compounding the problem. At first, pesticide proponents denied that resurgences and secondary-pest outbreaks had anything to do with the use of pesticides. However, careful investigations have shown that resurgences and secondary-pest outbreaks occur *because the insect world is part of a complex food web.*

Thus, the chemical approach fails because it ignores basic ecological principles. It assumes that the ecosystem is a static entity in which one species, the pest, can simply be eliminated. In reality, the ecosystem is a dynamic system of interactions, and a chemical assault on one species will inevitably upset the system

**earth watch**

### The Ultimate Pest?

If we were to conjure up the ultimate insect pest, we might imagine that it would

1. Attack a broad variety of plants and fruits.
2. Spread viruses that can devastate crop plants.
3. Be resistant to a host of pesticides.
4. Be highly prolific and have a rapid life cycle.
5. Thrive on roadside weeds.
6. Have few natural predators.

Such a pest, if unleashed on agriculture, could cause millions of dollars in crop losses and devastate many farmers.

Unfortunately, something very close to the ultimate pest is already with us. The silverleaf whitefly (*Bemisia argentifolii*) is a tiny white insect that emerged from Florida's poinsettia greenhouses in 1986 and has become established in California, Texas, Florida, and Arizona. It has all the characteristics listed and has been dubbed the "superbug" by farmers who have encountered it. The fly is known to eat at least 500 species of plants—just about everything except asparagus and onions—and it lives on roadside weeds when it can't get its favorite crop food. The insects swarm all over the plants, sucking them dry and leaving them withered and rotten. Swarms become so dense that they interfere with the breathing and vision of anyone caught in them. Total crop losses to this insect are greater than \$200 million each year, and it continues to cause extensive damage. A national campaign against the insect has been launched by the Agricultural Research Service of the USDA, complete with five-year plans. A few pesticides remain effective, but resistance is a lurking problem whenever these are used. So far, the judicious use of pesticides and cultural controls like plowing under a crop as soon as it is harvested are holding the bug at bay. For now, all of the farm communities in the affected areas are involved in programs to suppress the silverleaf whitefly, and research programs are frantically looking for effective biological control agents.

and produce other, undesirable effects. Populations of plant-eating insects are frequently held in check by other insects that parasitize or prey on them (Fig. 16–6). Pesticide treatments often have a greater impact on these natural enemies than on the plant-eating insects they are meant to control. Consequently, with their natural enemies suppressed, both the population of the original target pest and populations of other plant-eating insects explode. To achieve sustainability, therefore, we must understand how ecosystems work and adapt our interventions accordingly.

***Treadmill.*** The late entomologist Robert van den Bosch coined the term **pesticide treadmill** to describe attempts to eradicate pests with synthetic organic chemicals. It is an apt term, because the chemicals do not eradicate the pests. Instead, they increase resistance and secondary-pest outbreaks, which lead to the use of new and larger quantities of chemicals, which in turn lead to more resistance and more secondary-pest outbreaks, and so on. The process is an unending cycle constantly increasing the risks to human and environmental health and is clearly not sustainable (Fig. 16–7). In the words of one observer, "Modern agriculture has a serious chemical dependency—an addiction to pesticides."

**Human Health Effects.** As with all toxic substances, pesticides can be responsible for both acute and chronic health effects. According to the American Association of Poison Control Centers, over 90,000 persons suffered acute poisonings from pesticides in the United States during 2001, and 17 of those victims died. Most were farmworkers or employees of pesticide companies who came in direct contact with the chemicals. The WHO surveyed pesticide poisoning and estimated that there are between 3.5 and 5 million cases of acute occupational pesticide poisoning each year in developing countries, of which at least 20,000 result in death. The use of pesticides by untrained persons is considered to be the major cause of these poisonings, but in many cases children and families come in contact with the pesticides through aerial spraying, the dumping of pesticide wastes, or the use of pesticide containers to store drinking water. The organophosphates and carbamates were responsible for most of the poisonings.

***Chronic Effects.*** Pesticides are applied to fields and orchards to reduce pest damage to crops. They are also used to protect harvested food so that it is brought undamaged to market. Because of the wide-ranging use of pesticides by most farmers, consumers are inevitably exposed to pesticide residues on their food. Also, many occupational exposures to pesticides are subacute. The public-health concern is that the pesticides might have chronic effects, even at low levels of exposure. Among the chronic effects of pesticides is the potential for causing cancer, as indicated by animal testing. In fact, evidence of carcinogenicity has been observed for 20 pesticides. Epidemiological evidence has implicated organochlorine pesticides in various cancers, including lymphoma and breast cancer.

Other chronic effects include dermatitis, neurological disorders, birth defects, and infertility. For example, male

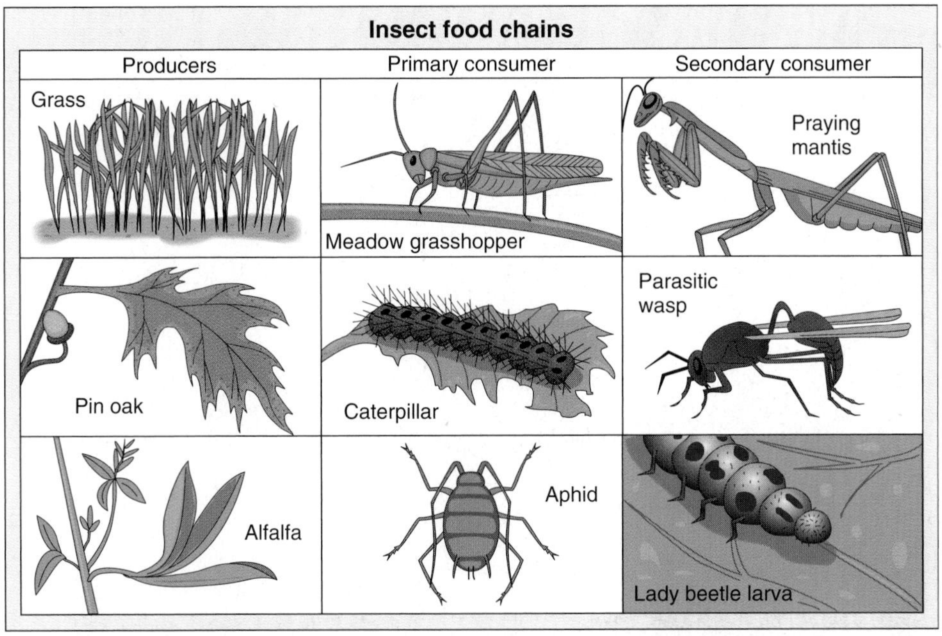


(a)

(b)

**Figure 16–6 Insect food chains.** (a) Food chains exist among insects, just as they do among higher animals. (b) Aphid lion (on right) impaling and eating a larval aphid (a small insect that sucks sap from plants).

sterility has been clearly linked to dibromochloropropane (DBCP), once used to control nematodes and now banned. Two disorders recently associated with pesticide use are suppression of the immune system and disruption of the endocrine system. Again, the evidence for an impact on humans comes from laboratory animal testing and epidemiological studies. Persistent exposure to pesticides among factory workers in India led to abnormally low white blood cell counts. (White blood cells are crucial elements of the immune system.) Similar results were found for agricultural workers in the former Soviet Union. Laboratory studies show many forms of immune suppression by different types of pesticides.

***Endocrine Disruptors.*** Laboratory tests have shown that a number of pesticides, including atrazine and alachlor (weed killers) and DDT, endosulfan, diazinon, and methoxychlor (insecticides), interfere with reproductive hormones. Both a rise in the incidence of breast cancer among humans and reports of abnormal sexual development in alligators and other animals in the wild have given credence to the possibility that very low levels of a number of chemicals are able to mimic or disrupt the effects of estrogenic hormones (sexual hormones that are highly potent at low concentrations). It is well documented that farmworkers and herbicide sprayers have defective sperm counts. A recent study confirmed that men exposed to low levels of several pesticides, probably through drinking water, were many times more likely to have defective sperm or low sperm counts than those with little or no exposure.

Concern about this mimicry has reached Congress, and the EPA was directed to develop procedures for testing chemicals for endocrine disruption activity. An EPA-established advisory committee strongly recommended proceeding with a screening program, and in 1999 the EPA established the Endocrine Disruptor Screening

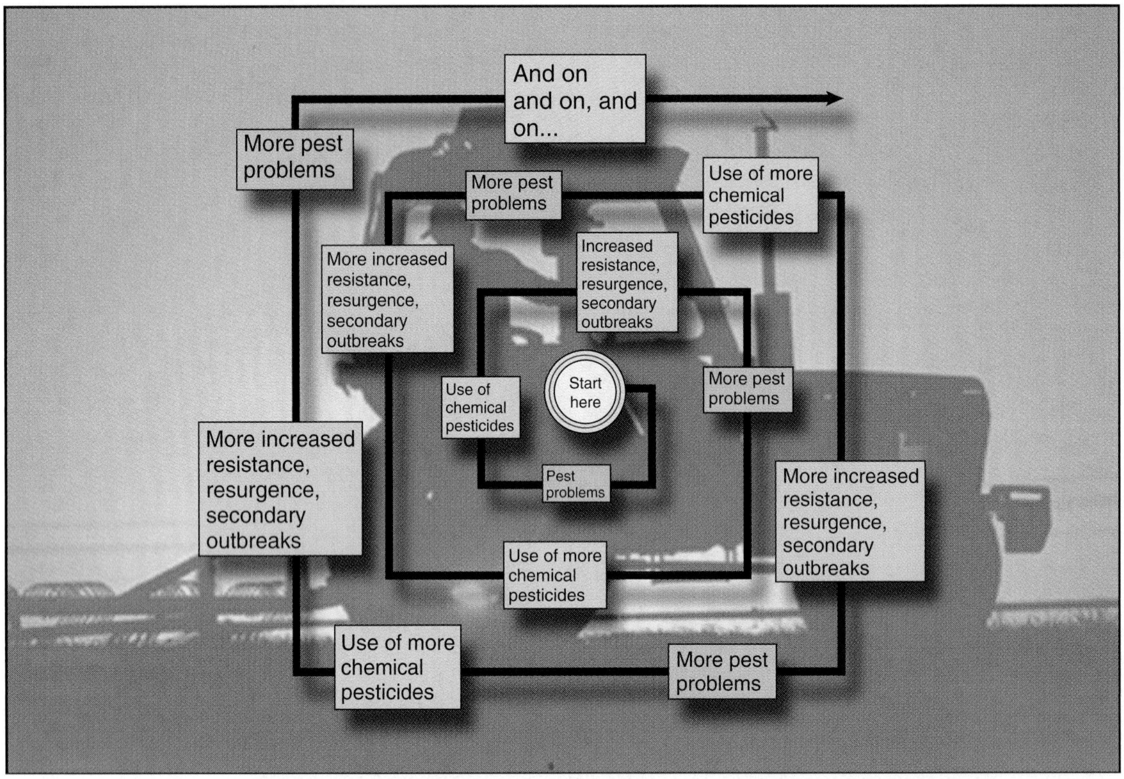


**Figure 16–7 The pesticide treadmill.** Treatment with chemical pesticides aggravates many pest problems. The continued use of these products demands ever-increasing dosages of pesticides, which further aggravate pest problems and produce more contamination of foodstuffs and ecosystems.

Program. Currently, the program is evaluating processes to be used in screening the potential endocrine disruptors. Pesticides are prime suspects, and chemical manufacturers will soon be required to test many pesticides in use in the United States.

**Environmental Effects.** Aerial spraying for insect control has pointed most clearly to the adverse environmental effects of pesticides. The story of DDT, used so widely during the 1940s, 1950s, and early 1960s (80,000 metric tons in 1962 alone), illustrates the hazards.

*DDT, Continued.* In the 1950s and 1960s, ornithologists (people who study birds) observed drastic declines in populations of many species of birds that fed at the tops of food chains. Fish-eating birds such as the bald eagle and osprey (Fig. 16–8) were so affected that their extinction seemed imminent. Investigators at the U.S. Fish and Wildlife National Research Center near Baltimore, Maryland, showed that the problem was reproductive failure: Eggs were breaking in the nest before hatching. The investigators also showed that the fragile eggs contained high concentrations of dichlorodiphenyldichloroethylene (DDE), a product of the partial breakdown of DDT by the animal's body. DDE interferes with calcium metabolism, causing birds to lay thin-shelled eggs. Further study revealed that birds were acquiring high levels of DDT and DDE by bioaccumulation and biomagnification, the

**Figure 16–8 Bald eagle.** Populations of fish-eating birds such as the osprey, brown pelican, and bald eagle were decimated in the 1950s and 1960s by the effects of widespread spraying with DDT. With the banning of the pesticide, these populations have greatly recovered. The bald eagle was taken off the endangered species list in 1994.

process of accumulating higher and higher doses through the food chain.

Because of accumulation, small, seemingly harmless amounts received over a long period of time may reach toxic levels. This phenomenon—referred to as **bioaccumulation**—can be understood as follows: Many synthetic organics are highly soluble in lipids (fats or fatty compounds), but sparingly soluble in water. As they pass through cell membranes, which are lipid, they come out of water solution and enter into the lipids of the body. Thus, traces of synthetic organics like pesticides and their breakdown products that are absorbed with food or water are trapped and held by the body's lipids, while the water and water-soluble wastes are passed in the urine. Because synthetics are unnatural compounds, the body cannot fully metabolize them and has no mechanism to excrete them. Thus, trace levels consumed over time gradually accumulate in the body and may produce toxic effects sooner or later.

*Up the Chain.* Bioaccumulation, which occurs in the individual organism, may be compounded through a food chain. Each organism accumulates the contamination in its food, so it accumulates a concentration of contaminant in its body that is many times higher than that in its food. In effect, the next organism in the food chain now has a more contaminated food and accumulates the contaminant to yet a higher level. Essentially all the contaminant accumulated by the large biomass at the bottom of the food pyramid is concentrated, through food chains, into the smaller and smaller biomass of organisms at the top of the food pyramid. This multiplying effect of bioaccumulation that occurs through a food chain is called **biomagnification.** One of the most distressing aspects of bioaccumulation and biomagnification is that there are no warning symptoms until concentrations of contaminant in the body are high enough to cause problems. Then it is too late to do much about it. As is often the case, bioaccumulation and biomagnification go unrecognized until serious problems bring the phenomena to light. Figure 16–9 shows how DDT and its metabolic products worked their way up the food chain to the top predators.

*Silent Spring.* In the 1950s, Rachel Carson, a U.S. Fish and Wildlife Service biologist and an accomplished science writer, began reading the disturbing scientific accounts of the effects of DDT and other pesticides on wildlife. Carson was finally galvanized into action by a letter from a friend distressed over the large number of

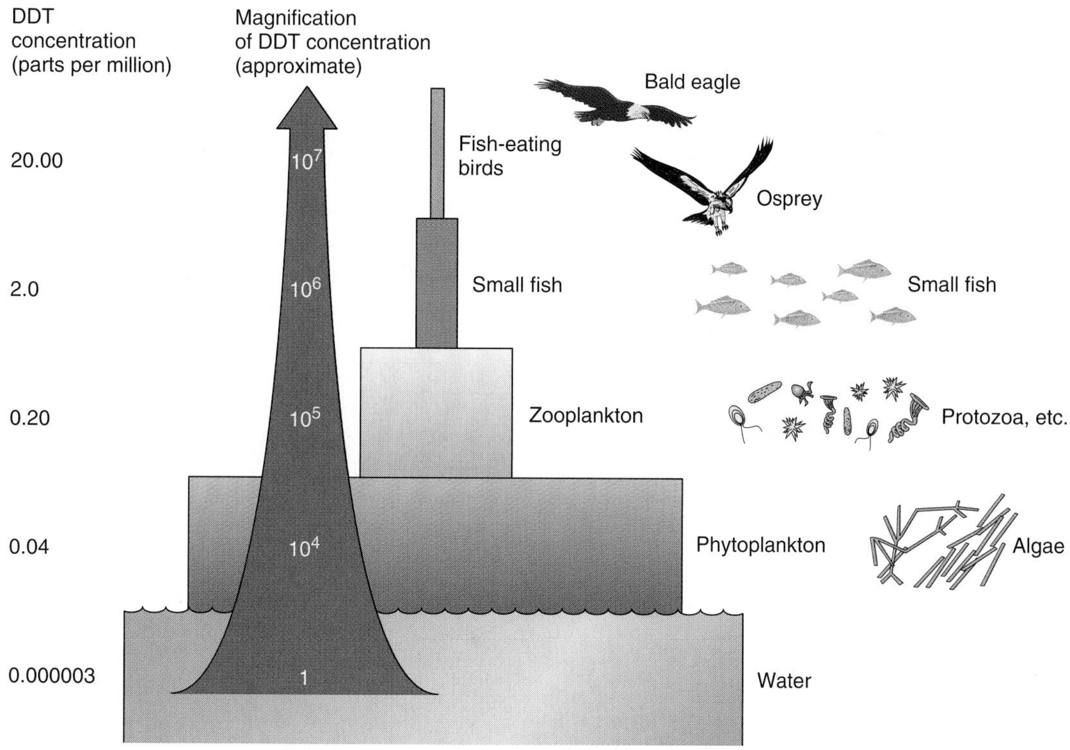


**Figure 16–9 Biomagnification.** Organisms on the first trophic level absorb the pesticide from the environment and accumulate it in their bodies (bioaccumulation). Then, each successive consumer in the food chain accumulates the contaminant to yet a higher level. Thus, the concentration of the pesticide is magnified manyfold throughout the food chain. Organisms at the top of the chain are likely to accumulate toxic levels.

birds killed when the friend's private bird sanctuary was sprayed for mosquito control. By 1962, Carson finished a book-length documentation of the effects of the almost uncontrolled use of insecticides across the United States. The book, *Silent Spring*, became an instant best-seller. Its basic message was that if insecticide use continued as usual, there might someday come a spring with no birds—and with ominous impacts on humans as well.

*Silent Spring* triggered a debate that has not ended. Representatives of the agricultural and chemical industries claimed that the book was an unreasonable and unscientific account that, if taken seriously, would halt human progress. At the same time, however, the book was hailed as an unparalleled breakthrough in environmental understanding. Eventually, concerns about environmental and long-term health effects led to the banning of DDT in the United States and most other industrialized countries in the early 1970s. Numerous other related chlorinated hydrocarbon pesticides (for example, chlordane, dieldrin, endrin, and heptachlor) were also banned because of their propensity for bioaccumulation in the environment and suspected potential for causing cancer. In the years since the banning of DDT, the bird species that were adversely affected have recovered. Today, the debate centers on the continued use of DDT against malaria in developing countries. (See the "Ethics" essay, p. 439.)

Forty years later, Rachel Carson is credited with stimulating the start of the modern environmental movement and the creation of the EPA. *Silent Spring* has become a classic, and the regulation of insecticides and other toxic chemicals is in a sense a monument to Rachel Carson, who died of cancer only two years after her book was published.

### Nonpersistent Pesticides: Are They the Answer?

A key characteristic of chlorinated hydrocarbons is their persistence; that is, they take a long time to break down. Lingering in the environment for years, they can contaminate many organisms and become biomagnified. This persistence results because these chemicals have such complex structures that soil microbes are unable to fully break them down. DDT, for example, has a half-life on the order of 20 years.

Because the persistent pesticides have been banned, the agrochemical industry has substituted *nonpersistent* pesticides for the banned compounds. The synthetic organic phosphates (for example, malathion, parathion, and chlorpyrifos), as well as carbamates such as aldicarb and carbaryl, have been used extensively in place of chlorinated hydrocarbons (see Table 16–1). These compounds are potent inhibitors of the enzyme cholinesterase, which is essential for proper functioning of the nervous system in all animals. They break down into simple nontoxic products within a few weeks after their application. Thus, there is no danger of their migrating long distances through the environment and affecting wildlife or humans long after being applied.

***Toxicity.*** For several reasons, however, nonpersistent pesticides are not as environmentally sound as they might appear. First of all, they are persistent enough to "ride the food supply from farmer to consumer," as Anne Platt McGinn put it in her WorldWatch booklet "*Why Poison Ourselves?*" The total environmental impact of a pesticide is a function not only of its persistence, but also of its toxicity, its dosage, and the location where it is applied. Many of the nonpersistent pesticides are far more toxic than DDT. This higher toxicity, combined with the frequent applications needed to maintain control, presents a significant hazard to agricultural workers and others exposed to these pesticides. For instance, the organophosphates are responsible for an estimated 70% of all pesticide poisonings. The Food Quality Protection Act of 1996 requires the EPA to develop new health-based standards that address the risk of children's exposure to such pesticides. The fallout from the new standards has led to the banning of any use of parathion on produce and of chlorpyrifos for all over-the-counter uses, such as on flea collars and in lawn products. Chlorpyrifos may no longer be used on tomatoes and is prohibited for use on apples after blooming has ceased. Diazinon, a widely used agricultural insecticide and a common household product, has been banned for many agricultural, and all indoor residential, uses. Outdoor uses are being gradually phased out.

***Hawk Kill.*** Second, nonpersistent pesticides may still have far-reaching environmental impacts. For example, to control outbreaks of grasshoppers that eat sunflowers, farmers in Argentina began spraying monocrotophos, an acutely toxic organophosphate pesticide, in 1995. Shortly afterwards, thousands of dead Swainson's hawks were seen in the sunflower fields. The hawks, which spend summers in the United States, winter in great numbers in the Argentine pampas and feed voraciously on grasshoppers there. It was estimated that 20,000 hawks died in one year, about 5% of the world's population of the species. In response, the Argentine government banned the use of monocrotophos. (It had already been withdrawn from use in the United States in 1988.)

Third, desirable insects may be just as sensitive as pest insects to nonpersistent pesticides. Bees, for example, which play an essential role in pollination, are highly sensitive to them. Thus, the use of these compounds creates an economic problem for beekeepers and jeopardizes pollination. Also, regular spraying of neighborhoods with malathion to control mosquitoes leads inevitably to great declines in butterflies and fireflies.

Finally, nonpersistent chemicals are just as likely to cause resurgences and secondary-pest outbreaks as are persistent pesticides, and pests become resistant to nonpersistent pesticides just as readily, too.

## 16.3 Alternative Pest Control Methods

Numerous ecological and biological factors affect the relationship between a pest and its host. Ecological control seeks to manipulate one or more of these natural factors so that crops are protected without upsetting the rest of the ecosystem or jeopardizing environmental and human health. Because ecological control involves working with natural factors instead of synthetic chemicals, the techniques are referred to as **natural control** or **biological control** methods. This natural approach, unlike the chemical treatment approach, depends on an understanding of the pest and its relationship with its host and with its ecosystem. The more we know about the organisms involved, the greater are our opportunities for natural control.

To illustrate, the life cycle typical of moths and butterflies is shown in Fig. 16–10. Many groups of insects have a similarly complex life cycle. The development of each stage may be influenced by numerous abiotic factors, and at each stage the insect may be vulnerable to attack by a parasite or predator. The proper completion of each stage depends on internal chemical signals provided by insect hormones. Locating mates, finding food, and other behaviors depend on external chemical signals. All these findings suggest ways in which pest populations may be controlled without resorting to synthetic chemical pesticides.

The four general categories of natural or biological pest control are (1) **cultural control,** (2) **control by natural enemies,** (3) **genetic control,** and (4) **natural chemical control.** We will consider each of these in turn.

### Cultural Control

A cultural control is a nonchemical alteration of one or more environmental factors in such a way that the pest finds the environment unsuitable or is unable to gain access to its target.

**Cultural Control of Pests Affecting Humans.** We routinely practice many forms of cultural control against diseases and parasitic organisms. Some of these practices are so familiar and well entrenched in our culture that we no longer recognize them for what they are. For instance, disposing properly of sewage and avoiding drinking water from unsafe sources are cultural practices that protect against waterborne disease-causing organisms. Combing and brushing the hair, bathing, and wearing clean clothing are cultural

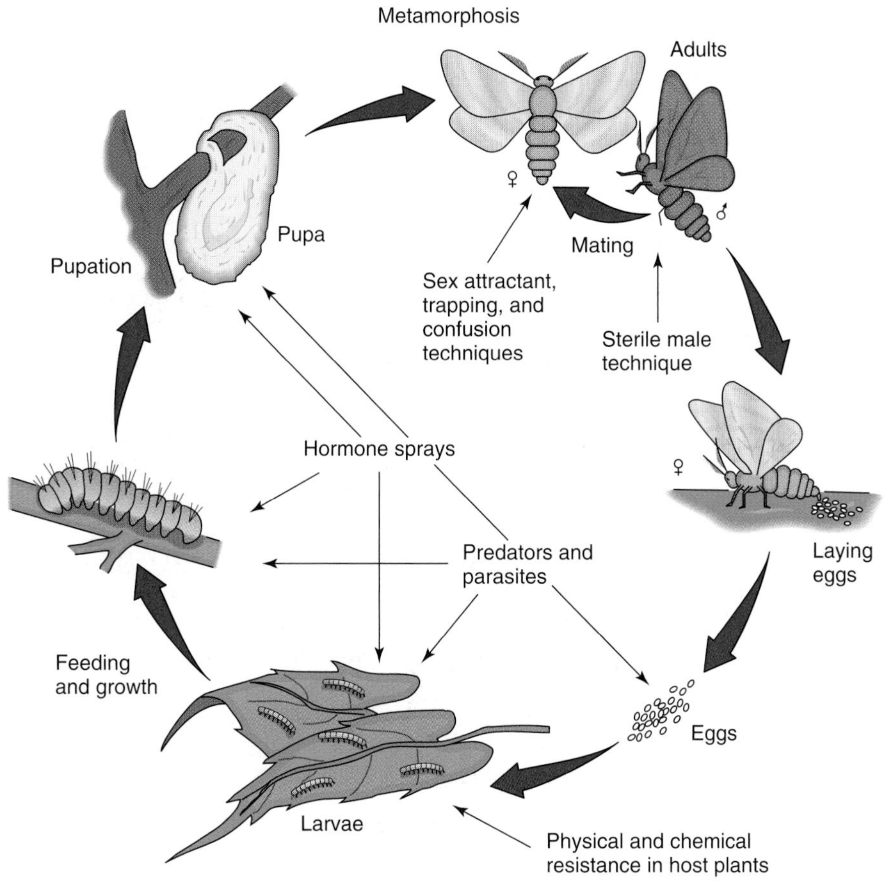


**Figure 16–10 Complex life cycle of insects.** Like the moth shown here, most insects have a complex life cycle that includes a larval stage and an adult stage. Biological control methods recognize the different stages and attack the insect, using knowledge of its needs and life cycle.

practices that eliminate head and body lice, fleas, and other parasites. Changing of bed linens regularly protects against bedbugs. Properly and systematically disposing of garbage and keeping a clean house with all cracks sealed and with good window screens are effective methods for keeping down populations of roaches, mice, flies, mosquitoes, and other pests. Sanitation requirements in handling and preparing food are cultural controls designed to prevent the spread of disease. The refrigeration, freezing, canning, and drying of foods are cultural controls that inhibit the growth of organisms that cause rotting, spoilage, and food poisoning.

If such practices of personal hygiene and sanitation are compromised, as they usually are in any major disaster, there is the very real danger of additional widespread mortality resulting from outbreaks of parasites and diseases. As discussed in Chapter 15, sickness and death are the consequences where these practices are not broadly pursued in a society, as in many less developed countries.

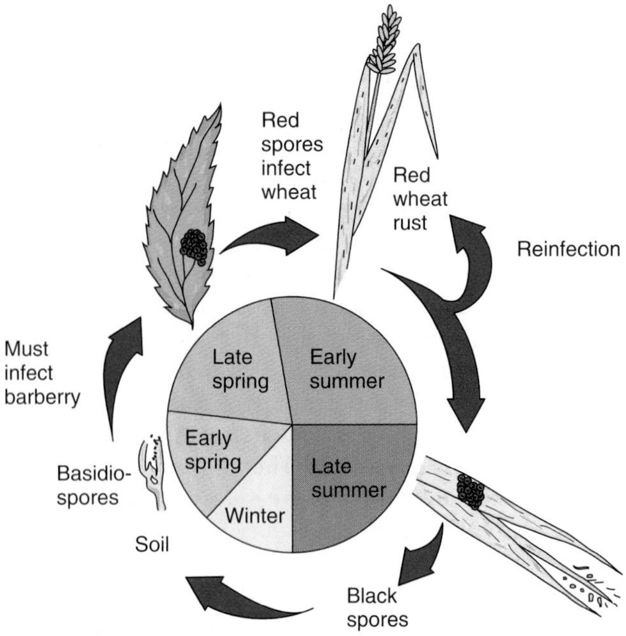


**Figure 16–11 Cultural control.** Part of the life cycle of wheat rust, a parasitic fungus that is a serious pest on wheat, requires that the rust infest barberry, an alternative host plant. The elimination of barberry in wheat-growing regions has been an important cultural control.

**Cultural Control of Pests Affecting Lawns, Gardens, and Crops.** Homeowners are prone to use excessive amounts of herbicides to maintain a weed-free lawn. Weed problems in lawns are frequently a result of cutting the grass too short. If grass is left at least 3 inches (8 cm) high, it will usually maintain a dense enough cover to keep out most crabgrass and other noxious weeds. Thus, monitoring the height of grass is a form of cultural weed control. Also, many homeowners allow a diversity of plants in their lawns, tolerating a variety of plants some people might consider weeds.

Some plants are particularly attractive to certain pests; others are especially repugnant. In either case, the effect may spill over to adjacent plants. Thus, a gardener may control many pests by paying careful attention to eliminating plants that act as attractants (for example, roses) and growing those which act as repellents. Marigolds and chrysanthemums are justly famous for being insect repellents.

Some parasites require an alternative host and so can be controlled by eliminating that host (Fig. 16–11). Also, hedgerows, fencerows, and shelterbelts can provide refuges where natural enemies of pests (birds, amphibians, praying mantises, and so on) can be maintained.

*Crops.* Managing any crop residues that are not harvested is important. Spores of plant disease organisms and insects may overwinter or complete part of their life cycle in the dead leaves, stems, or other plant residues that remain in the fields after harvesting. Plowing under or burning the material may be quite effective in keeping pest populations to a minimum. In gardens, a clean mulch of material such as grass clippings or hay will keep down the growth of weeds and protect the soil from drying and erosion.

Growing the same crop on a plot of land year after year keeps the pest's food supply continuously available (Fig. 16–12). Hence, crop rotation—the practice of changing crops from one year to the next—may provide control because pests of the first crop cannot feed on the second crop and vice versa. Crop rotation is especially effective in controlling root nematodes (roundworms that live in the soil and feed on roots)

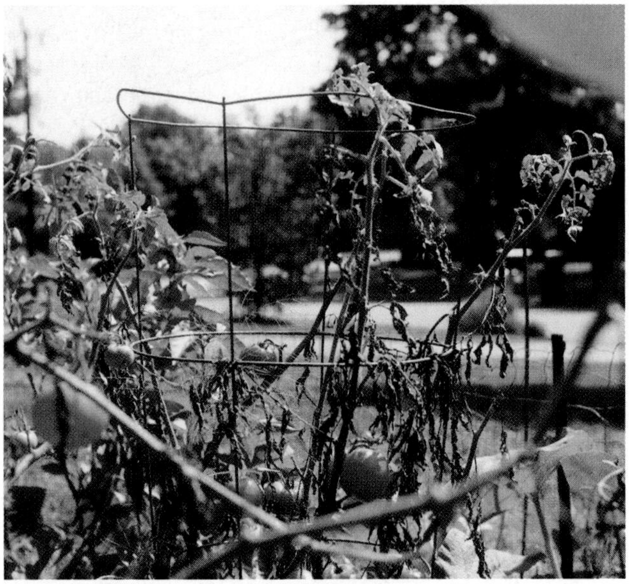

**Figure 16–12 Tomato wilt.** This tomato plant has been devastated by tomato wilt. Growing tomatoes in the same plot maintains wilt spores in the soil, guaranteeing that wilt will strike the plants every year.

and other pests that do not have the ability to migrate appreciable distances.

For economic efficiency, agriculture has moved progressively toward monoculture; witness the Corn Belt, Cotton Belt, and so on. When a pest outbreak occurs, monoculture is most conducive to its rapid multiplication and spread. The spread of a pest outbreak is impeded, and other natural controls may be more effective, if there is a mixture of crop species, some of which are not vulnerable to attack. One approach that works well in the British Isles is to intersperse cultivated with uncultivated strips that are not treated with pesticides. Natural enemies of pests are maintained in the uncultivated strips.

***Border Patrol.*** Most of the pests that are hardest to control were unwittingly imported from other parts of the world, and many other species in other regions would be serious pests if introduced into the United States. Therefore, it is important to keep would-be pests out of the country. This is a major function of the U.S. Customs Bureau and of the agriculture departments of some states. Biological materials that may carry pest insects or pathogens are either prohibited from crossing the border or subjected to quarantines, fumigation, or other treatments to ensure that they are free of pests. The cost of these procedures is small compared with the costs that could be incurred if the pests gained entry and became established in the nation.

## Control by Natural Enemies

The following examples illustrate the range of possibilities for controlling pests with natural enemies:

- Tiny predator beetles have been imported in an attempt to control the (also imported) hemlock wooly adelgid, which is killing off hemlock trees all over the eastern United States.
- Various caterpillars have been controlled by parasitic wasps (Fig. 16–13).
- The prickly pear cactus and numerous other weeds have been controlled by plant-eating insects (Fig. 16–14). (More than 30 weed species worldwide are now limited by insects introduced into their habitats.)
- Rabbits in Australia are controlled by an infectious virus.
- Water hyacinth, which has blanketed many African lakes, is coming under control following the introduction of Brazilian weevils.
- Mealybugs in Africa are controlled by a parasitic wasp. (See the "Global Perspective" essay, p. 451.)
- In a test spraying in Niger (in sub-Saharan Africa), the swarming desert locust was completely controlled with a newly developed product called "Green Muscle," a mix of dry spores of the fungus *Metarhizium anisopliae* with oil. The product is sprayed on infested fields.

***First, Protect the Natives.*** The problem with using natural enemies is finding organisms that provide control of the target species without attacking other, desirable species. Entomologists estimate that of the 50,000 known species of plant-eating insects that have the potential for being serious pests, only about 1% actually are. The populations of the other 99% are held in check by one or more natural enemies, so they do not cause significant amounts of damage. Therefore, the first step in using natural enemies for control should be *conservation*—protecting the natural enemies that already exist. Conservation means avoiding the use of broad-spectrum chemical pesticides, which may affect natural enemies of the target species even more than they do the target pests themselves. Eliminating or placing considerable restrictions on the use of broad-spectrum chemical pesticides will, in many cases, allow natural enemies to reestablish themselves and control secondary pests—those which become serious problems only after the use of pesticides.

***Import Aliens as a Last Resort.*** However, effective natural enemies are not always readily available. In many cases, the absence of a pest's natural enemies is the result of accidentally importing the pest without also importing its natural enemies. Quite often, effective natural enemies have been found by systematically combing the home region of an introduced pest and finding its various predators or parasites. To date, the U.S. Department of Agriculture (USDA) has released almost 1,000 insects to control pests, with generally favorable results. Identifying the enemy species is vital, and the species must be carefully tested before intentionally releasing it, something that has not always been done well. Some 16% of 313 parasitoid wasp species introduced into North America have attacked native species. In one recent case, a weevil (a type of beetle) imported to control an exotic musk thistle (a weed), attacked five native thistle species that pose no threat to agriculture. When effective natural enemies are found, however, they can provide control indefinitely, saving millions of dollars per year.

## Genetic Control

Most plant-eating insects and plant pathogens attack only one species or a few closely related species. This specificity implies a genetic incompatibility between the pest and species that are not attacked. Most genetic-control strategies are designed to develop genetic traits in the host species that provide the same incompatibility—that is, resistance to attack by the pest. The technique has been utilized extensively in connection with plant diseases caused by fungal, viral, and bacterial parasites. For example, in the years 1845–1847, the potato crop in Ireland was devastated by late blight, a fungal

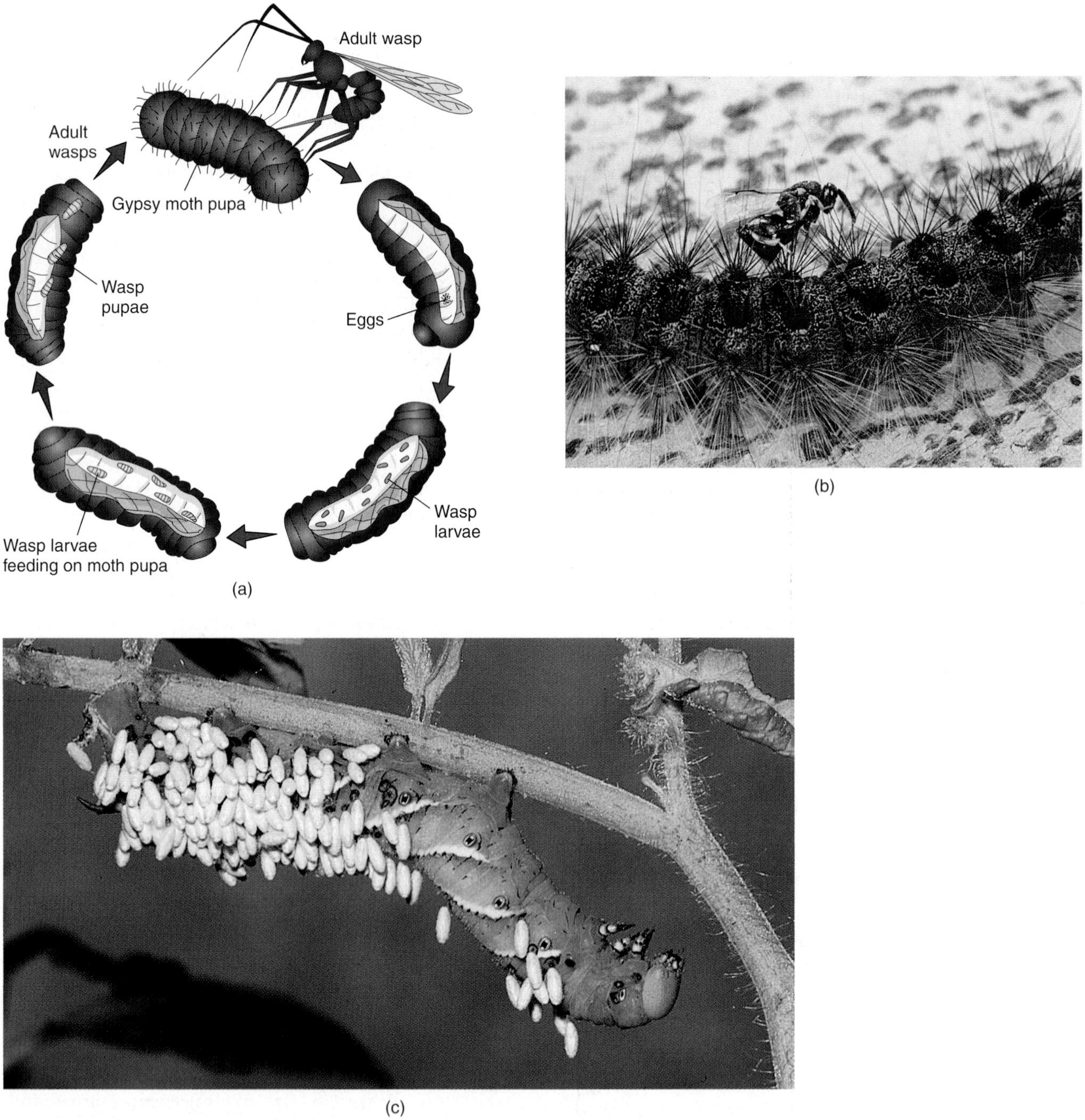


**Figure 16–13** **Parasitic wasps.** (a) The life cycle of the parasitic wasp that uses the gypsy moth as its host. (b) A wasp depositing eggs in a gypsy moth larva. (c) Another insect parasite, a braconid wasp, lays its eggs on the pest known as the tomato hornworm, which is the larva of the sphinx moth. The wasp larvae feed on the caterpillar, and shortly before the caterpillar dies, they emerge and form the cocoons seen here.

parasite. Nearly a million people starved, and another million people emigrated to escape the same fate. Nowadays, protection against such disasters is provided in large part by growing varieties of potato that are resistant to the blight. It is no overstatement to say that the world owes much of its production of potatoes, corn, wheat, and other cereal grains to the painstaking work of plant geneticists who selected and bred disease-resistant varieties.

The same potential exists for breeding plants that are resistant to insect pests. Traits that provide resistance may act as chemical or physical barriers.

**Control with Chemical Barriers.** A chemical barrier is a chemical produced by the plant we want to protect; the substance is lethal or at least repulsive to the would-be pest. Once they have identified such a barrier, plant breeders can use selection and crossbreeding to enhance

(a) (b)

**Figure 16–14 Prickly pear cactus.** Biological control of the weed known as the prickly pear cactus by a cactus-eating moth in Queensland, Australia. (a) The land shown here, once settled, had to be abandoned because of prickly pear infestation. (b) The same land was reoccupied following the destruction of the prickly pear by the moth.

the trait in the desirable plant. The relationship between wheat and the Hessian fly provides an example (Fig. 16–15). The fly lays its eggs on wheat leaves, and the larvae move down the leaves and into the main stem as they feed. The weakened stem either dies or is broken in the wind. The Hessian fly was introduced into the United States in the straw bedding of Hessian soldiers during the Revolutionary War. The fly eventually spread throughout much of the Midwest, causing widespread devastation until scientists at the University of Kansas developed a variety of wheat that produces a chemical that is toxic to the insect; the chemical kills the larvae when they feed on the leaves. Increasing resistance through breeding has reduced losses to this pest to less than 1% where resistant varieties have been planted for several years.

global perspective

## Wasps 1, Mealybugs 0

The cassava (manioc) plant originated in South America, but has been cultivated throughout the tropical world. Currently, in sub-Saharan Africa, it is the primary food for more than 200 million people, one-third of the human population in that region. Cassava is a high-yielding crop that requires no modern technology and, accordingly, is raised by subsistence farmers everywhere.

The mealybug, an insect not previously seen in Africa, appeared in Congo in the early 1970s and spread across sub-Saharan Africa, leaving a trail of ruined harvests and hunger in its wake. Zaire and Congo, unable to afford pesticides, turned for help to the International Institute of Tropical Agriculture in Nigeria. The group formed the Biological Control Program and began to look for natural enemies of the mealybug.

Returning to the land of origin of the cassava, a researcher found the mealybug in Paraguay and observed that the insect was kept under control by natural predators and parasites. After extensive testing, researchers identified a parasitic wasp, *Epidinocarsis lopezi*, as the prime candidate for controlling the bugs. No larger than a comma on this page, the female wasp seeks out mealybugs, paralyzes them with a sting, and deposits eggs that hatch inside the mealybug and eat their way through the insect.

Once this wasp–bug relationship was identified, the wasps were reared by the thousands on captive mealybugs and then spread across the continent over a period of eight years. The Biological Control Program has trained 400 workers to monitor the progress of the program, and all indications are that the battle has been won. It is estimated that every dollar invested in the control program has yielded $149 in crops saved from destruction. The best news is that the control is permanent and does not require the repeated application of expensive and environmentally damaging pesticides. In the wake of the project, national biological control programs have been established and are now focusing on pests of other crops of sub-Saharan Africa.

**Figure 16–15 Hessian fly.** This pest of wheat is controlled by maintaining wheat plants that produce a chemical that is toxic to the fly.

**Control with Physical Barriers.** Physical barriers are structural traits that impede the attack of a pest. For example, leafhoppers are significant worldwide pests of cotton, soybeans, alfalfa, clover, beans, and potatoes, but they can damage only plants that have relatively smooth leaves. Hooked hairs on the leaf surfaces of some plants tend to trap and hold immature leafhoppers until they die. Similarly, glandular hairs that exude a sticky substance fatally entrap alfalfa weevil larvae (Fig. 16–16). Such traits can be enhanced in vulnerable plants through selective breeding.

**Control with Sterile Males.** Another genetic-control strategy involves flooding a natural population with sterile males that have been reared in laboratories. Combating the screwworm fly provides a prime illustration. This fly, which is closely related to the housefly and looks much like it, lays its eggs in open wounds of cattle and other animals. The larvae feed on blood and lymph, keeping the wound open and festering. Secondary infections frequently occur and often lead to the death of the animal.

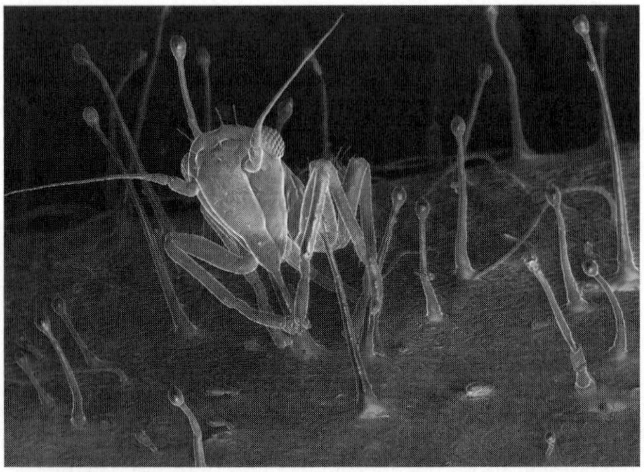

**Figure 16–16 Alfalfa glandular hairs.** This scanning electron micrograph shows an immature potato leafhopper trapped by sticky glandular hairs on the stem of a resistant alfalfa strain.

Early in the 20th century, the problem became so severe that cattle ranching from Texas to Florida and northward was becoming economically impossible. In studying the situation during the 1940s, Edward Knipling, an entomologist with the USDA, observed two essential features of screwworm flies: (1) Their populations are never very large, and (2) the female fly mates just once, lays her eggs, and then dies. Knipling reasoned that if the female mated with a sterile male, no offspring would be produced. His hypothesis was correct, and today sterile males are routinely used to control this pest. After huge numbers of screwworm larvae are grown on meat in laboratories, the resulting pupae are subjected to just enough high-energy radiation to render them sterile. These sterilized pupae are then air-dropped into the infested area. Ideally, 100 sterile males are dropped for every normal female in the natural population, giving a 99% probability that wild females will mate with one of the sterile males. The technique proved so successful that it eliminated the screwworm fly from Florida in 1958–59 and was subsequently employed to eradicate the screwworm fly from Mexico and most of Central America. The sterile-insect technique has saved billions of dollars for the cattle industry and has been used successfully on a number of agriculturally important pests.

Recently, the technique was used to eradicate the tsetse fly from the island of Zanzibar, thus eliminating the disease trypanosomiasis (sleeping sickness), which the tsetse fly carries. Indeed, now herdsmen can raise their cattle without the devastating losses the disease causes elsewhere in sub-Saharan Africa, more than 10 million square miles of which are inhabited by tsetse flies. Research efforts are under way to extend this success to other parts of the continent.

**Strategies Using Biotechnology.** Biotechnology has multiplied the potential for genetic control. More complex than basic plant breeding, genetic engineering makes it possible to introduce genes into crop plants from other plant species, bacteria, and viruses. The new transgenic crops (called genetically modified organisms) are undergoing rapid development and testing. In fact, over 40 transgenic crops have received regulatory approval already. More than 150 million acres of transgenic crops are planted globally, the most important being soybeans, cotton, and corn. This approach is not without its problems, however: Recall from Chapter 9 that public acceptance of genetically modified organisms is far behind the exploding technology, especially in Europe.

One promising strategy is to incorporate the protein coat of a plant virus into the plant itself. When the plant "expresses" (that is, manufactures) the virus's protein coat, the plant becomes resistant to infection by the real virus. In this way, crop plants have been made resistant to more than a dozen plant viruses.

**Figure 16–17 Bioengineered potatoes.** The center row shows potato plants that have the Bt gene incorporated into their genome. These plants are protected from the Colorado potato beetle, while surrounding rows show typical devastation of nonengineered potato plants by beetles.

***Bt.*** Several strategies that employ biotechnology have the potential for reducing pesticide use. Endowing plants with resistance to insects has been engineered with the incorporation of a potent protein produced by a bacterium, *Bacillus thuringiensis* (Bt). This protein kills the larvae of a number of plant-eating insects and is harmless to mammals, birds, and most other insects. Scientists have engineered the gene into a number of plants, including cotton, potatoes, and corn (Fig. 16–17). This development alone is expected to cut in half the use of pesticides on cotton—a crop that consumes more than 10% of the pesticides used throughout the world. The protection comes from a naturally occurring protein used for three decades by home gardeners, organic growers, and other farmers. Wide-scale use of the Bt cotton (Monsanto's Bollgard® cotton) revealed that it is effective against two of three pests, but farmers had to employ pesticides to control the bollworm (a kind of moth larva) in 40% of the 2 million acres planted of the new crop. Bt corn has been developed to combat the European corn borer, a pest responsible for over $1 billion in losses and control costs every year.

***Roundup® Again.*** One biotech strategy has been to give a crop species resistance to a broad-spectrum herbicide. Although this might appear to encourage the use of herbicides, it is actually expected to make it possible to use less total amounts of herbicide. The herbicide glyphosate (Roundup®), first discussed in Section 16.2, inhibits an amino acid pathway that occurs only in plants and not in animals, biodegrades rapidly, and can control both grasses and broadleaf plants. (Because it is nonspecific and will kill every plant, it must be applied and cleaned up carefully.) Genes resistant to glyphosate have been successfully introduced into cotton, canola, soybean, and corn. Farmers save money on herbicides because they need to spray Roundup® only once to control weeds in their soybean fields. (Three-quarters of the U.S. soybean crop is now planted with Roundup Ready® soybeans.)

***Not So Fast. . . .*** There is a downside, however, to these current biotech innovations. First, they are not well suited to developing countries, as they foster a dependence on a costly annual supply of seeds under current corporation practices. Also, there is the definite possibility of developing "superweeds" that are resistant to the herbicides or plant viruses, as resistance genes are shared with genetically close wild species. Furthermore, insects are likely to develop resistance to Bt plants much more readily than they would to occasional spraying of the Bt insecticide directly, because they are constantly exposed to Bt toxic effects season after season. This has already happened in the case of the herbicide Roundup®, as discussed in Section 16.2. The same potential for developing resistance applies to chemical and physical control strategies. Thus, scientists have had to develop new resistant varieties of wheat seven times in the case of wheat and the Hessian fly. Nevertheless, it is hoped that the applications of biotechnology will gradually reduce the burden of pesticides in the environment and their potential impact on human health. One sign that this strategy is working is the rapid decline in the crop-duster industry (Fig. 16–4), which is in a tailspin because of the growing use of genetically modified crops.

## Natural Chemical Control

**Hormones** are chemicals produced in humans and other animals which provide "signals" that control developmental processes and metabolic functions. In addition, insects produce many **pheromones**—chemicals secreted by one individual that influence the behavior of another individual of the same species.

The aim of natural chemical control is to isolate, identify, synthesize, and then use an insect's own hormones or pheromones to disrupt its life cycle. Two advantages of natural chemicals are that they are nontoxic and that they are highly specific to the pest in question. (They do not affect the natural enemies of the pest to any appreciable extent.) If the affected pest is eaten by another organism, it is simply digested.

Scientists have discovered that caterpillar pupation is triggered by a decrease in the level of the chemical called **juvenile hormone.** If this chemical is sprayed on caterpillars, pupation does not occur. The larvae simply continue to feed and grow, become grossly oversized, and eventually die. One newly developed insecticide, called Mimic, is a synthetic variation of **ecdysone,** the molting hormone of insects. Mimic begins the molting process in insect larvae, but doesn't complete it. As a result, the larva is trapped in its old skin and eventually starves to death. Mimic is specific to moths and butterflies and is viewed as a potent agent in the control of the spruce budworm, the gypsy moth, the beet armyworm, and the codling moth—all highly devastating pests. Mimic is now registered for use throughout North America, and has proven effective against the hemlock looper in Newfoundland and the gypsy moth in the United States.

***Perfume?*** Adult insects secrete pheromones to attract mates. Once identified and synthesized, these pheromones may be used in either the trapping technique or the confusion technique. In the trapping technique, the pheromone is used to lure males or females into traps or into eating poisoned bait (Fig. 16–18). In the confusion technique, the pheromone is dispersed over the field in such quantities that males become confused, cannot find females, and thus fail to mate.

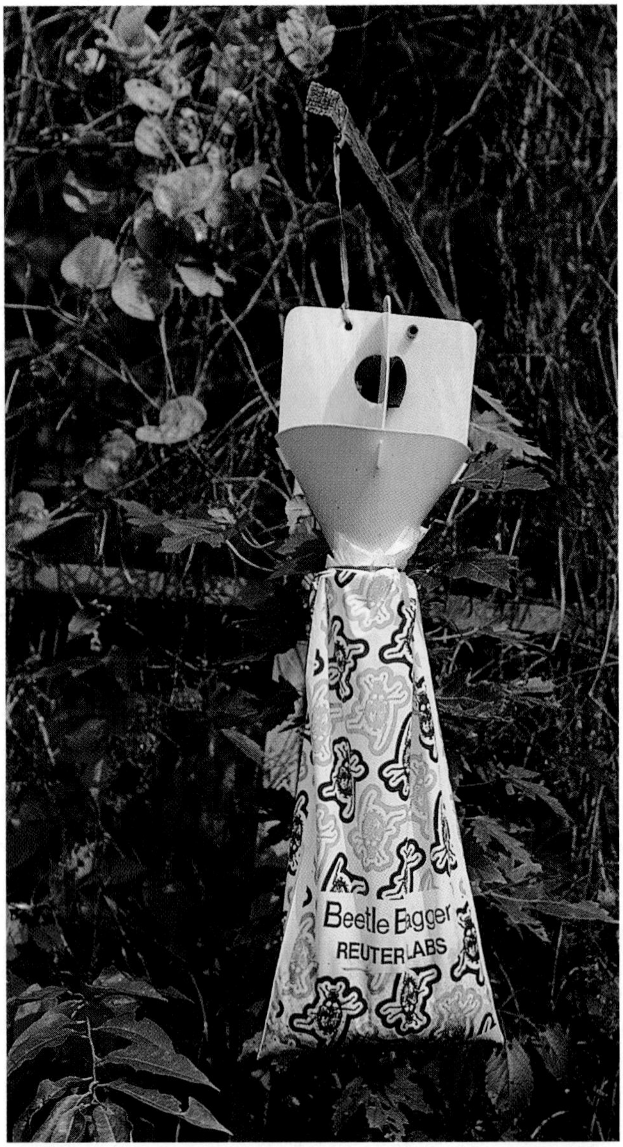


**Figure 16–18 Trapping technique.** Adult Japanese beetles are lured into a trap by scented bait they find attractive. Once inside the trap, baffles prevent the beetles from escaping. These traps are available in hardware stores and are quite effective, although some users find that they are too effective, luring beetles to their yards and gardens without actually catching them. (Japanese beetles are notorious lawn grubs as larvae; as adults, they are voracious consumers of garden flowers and vegetables.)

The enormous potential of natural chemicals for controlling insect pests without causing ecological damage or disruption has been recognized for at least 30 years. Research has identified a large number of pheromones, and these natural chemicals are now key tools employed in IPM on many insects, such as the boll weevil and the pink bollworm for cotton, the codling moth for pears and apples, the tomato pinworm for tomatoes, and several bark beetles that infest forest trees, among others.

## 16.4 Socioeconomic Issues in Pest Management

The increasing availability of alternative pest control methods and the accumulating evidence of the failures of the chemical approach to pest management have led to a growing movement toward avoiding the use of pesticides, particularly on foods. This movement must contend with a strong tendency on the part of farmers to employ pesticides.

### Pressures to Use Pesticides

A species becomes a pest only when its population multiplies to the point of causing significant damage. Natural controls are generally aimed at keeping pest populations below damaging levels, not at total eradication of the populations. By keeping pest populations down, natural controls avert significant damage while preserving the integrity of the ecosystem.

Therefore, the question to be asked when facing any pest species is, Is the species causing significant damage? Damage should be deemed significant only when the economic losses due to the damage considerably outweigh the cost of applying a pesticide. This point is called the **economic threshold** (Fig. 16–19). If significant damage is not occurring, natural controls are already operating, and the situation is probably best left as is. Spraying with synthetic chemicals at this stage is more than likely to upset the natural balance and make the situation worse through resurgences. It is also not cost effective. If, however, significant damage *is* occurring, a pesticide treatment may be in order.

It makes little difference whether the threat of loss as a result of pest infestation is real or imagined, close at hand or remote. What is important is how the grower perceives the threat. Even if there is no evidence of immediate damage from a pest, a grower who believes that his or her plantings are at risk is likely to resort to **insurance spraying**—the use of pesticides to prevent losses to pests. For example, pesticides are heavily used

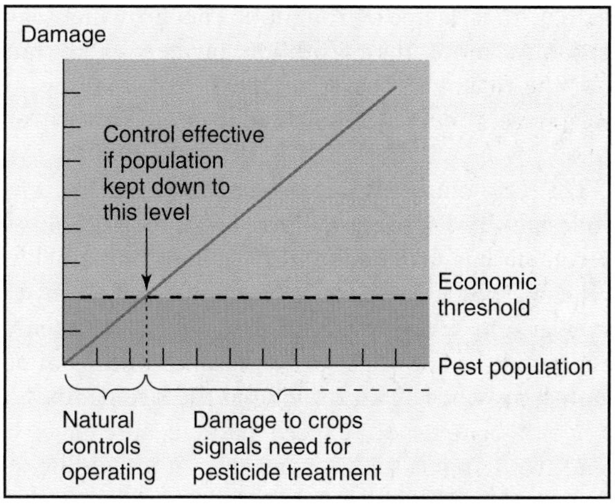


**Figure 16–19 The economic threshold.** The objective of pest control should not be to eradicate a pest totally. All that is needed is to keep population levels below the economic threshold.

in European apple crops, where the most important threats are mildew and scab—airborne diseases that, once established, cause significant economic damage. Farmers typically employ insurance spraying to control these diseases.

***The Snow White Syndrome.*** Consumers also put indirect pressure on growers to use pesticides. From customers to supermarket chains to canneries, there is a tendency to select the best-looking fruits and vegetables, leaving the remainder to be sold at lower prices or trashed. Like Snow White, we all tend to reach for the apparently perfect, unblemished apple. Growers know that blemished produce means less profit, so they indulge in **cosmetic spraying**—the use of pesticides to control pests that harm only the item's outward appearance. Cosmetic spraying accounts for a significant fraction of pesticide use, does nothing to enhance crop yield or nutritional value, and results in an increase in pesticide residues remaining on the produce.

## Integrated Pest Management

Integrated pest management aims to minimize the use of synthetic organic pesticides without jeopardizing crops. This is made possible by addressing all the interacting sociological, economic, and ecological factors involved in protecting crops. With IPM, the crop and its pests are seen as part of a dynamic ecosystem. The goal is not the eradication of pests, but maintaining crop damage below the economic threshold. The EPA refers to a four-tiered approach in describing IPM:

1. **Set action thresholds.** With the economic threshold in mind, IPM identifies a point at which pest populations or environmental conditions indicate that some control action is needed.
2. **Monitor and identify pests.** Pest populations are monitored, often by persons employed by local agricultural extension services or farm cooperatives or by persons acting as independent consultants. These **field scouts** are trained in identifying and monitoring pest populations (traps baited with pheromones are used for this purpose) and in determining whether the population exceeds the economic threshold.
3. **Prevention.** Cultural and biological control practices form the core of IPM techniques. Such practices as crop rotation, polyculture (instead of monoculture), the destruction of crop residues, the maintenance of predator populations, and carefully timed planting and fertilizing are basic to IPM. "Trap crops" are often used. That is, early strips of a crop such as cotton are planted to lure existing pests, and then the pests are destroyed by hand or by the limited use of pesticides before they can reproduce. Spraying is not performed during the growing season, in order to encourage natural enemies of the pests to flourish.
4. **Control.** If the preceding steps indicate that pest control is needed in spite of the preventive methods, measures are taken to decrease the pest population. Here, pesticides may be used, in quantities and brands that do the least damage to the natural enemies of the pest.

Making the economic benefits of natural controls known to growers is an important aspect of IPM. Because pesticides increased yields and profits when they were first used, many farmers still cling to them, believing that they offer the only way to bring in a profitable crop. In some instances, however, the rising costs of pesticides, along with their tendency to aggravate pest problems, have eliminated their economic advantage. **Pest-loss insurance,** which pays the farmer in the event of loss due to pests, has enabled insured growers to refrain from unnecessary and costly "insurance spraying."

***Brown Plant Hopper.*** Particularly in the developing world, governmental agricultural policy often determines the extent to which IPM is adopted. Governments and aid agencies usually subsidize the purchase of pesticides, thereby strongly encouraging growers to step onto the pesticide treadmill. By contrast, an Indonesian experience has provided a viable IPM model for other rice-growing countries. Faced with declining success in controlling the brown plant hopper (Fig. 16–20), Indonesian rice growers have switched from heavy pesticide spraying to a light spraying regime that preserves the natural enemies of the insect. The success of the program can be traced to close cooperation between the Indonesian government and the FAO (Fig. 16–21). FAO workers conducted training sessions for farmers (farmer field schools), weekly

**Figure 16–20 Brown plant hoppers.** Immature brown plant hoppers, shown on the stem of a rice plant.

meetings that lasted throughout the growing season. Eventually, more than 200,000 farmers were taught about the rice agroecosystem and IPM techniques, and these farmers formed a corps that could in turn teach others.

The economic and environmental benefits of the Indonesian IPM program have been remarkable. The government has saved millions of dollars annually by not purchasing pesticides, farmers have not had to invest in pesticides and spraying equipment, the environment has been spared the application of thousands of tons of pesticide, fish are once again thriving in the rice paddies, and the health benefits of reduced pesticide use have been spread from applicators to consumers and wildlife. (The FAO is sponsoring similar IPM training programs in eight other rice-growing nations and has started programs on vegetable crops in six nations.)

Recently, the FAO, the World Bank, the UNDP, and the UNEP have joined to cosponsor the *Global IPM Facility.* Its mission is to establish networks among farmers, researchers, and extension services and to foster policy development in member countries. During the facility's several-year tenure, thousands of farmers have been trained in IPM techniques, crop yields have improved, and farmers' costs have declined. Although IPM does rely on chemical pesticides when needed, if

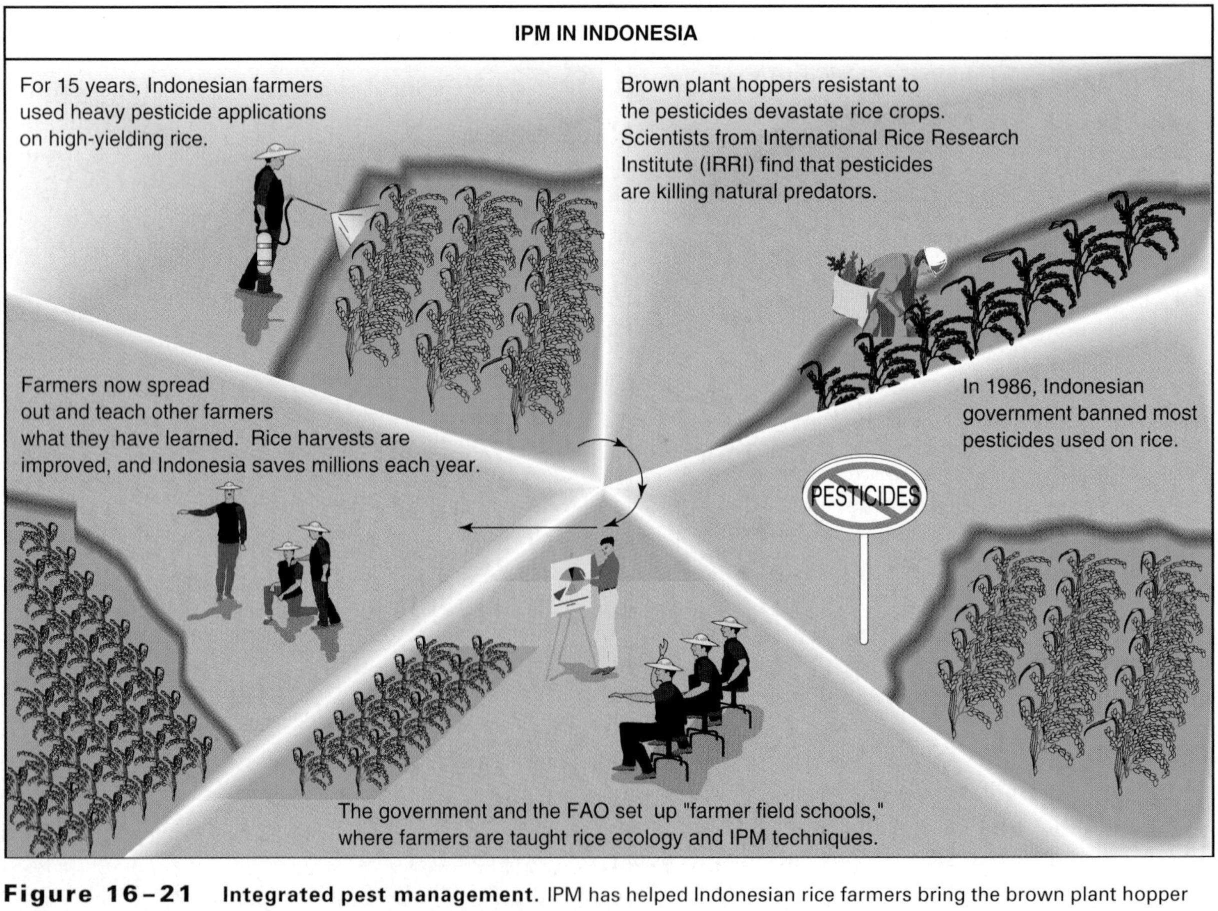

IPM IN INDONESIA
For 15 years, Indonesian farmers used heavy pesticide applications on high-yielding rice.
Brown plant hoppers resistant to the pesticides devastate rice crops. Scientists from International Rice Research Institute (IRRI) find that pesticides are killing natural predators.
In 1986, Indonesian government banned most pesticides used on rice.
PESTICIDES
The government and the FAO set up "farmer field schools," where farmers are taught rice ecology and IPM techniques.
Farmers now spread out and teach other farmers what they have learned. Rice harvests are improved, and Indonesia saves millions each year.


**Figure 16–21 Integrated pest management.** IPM has helped Indonesian rice farmers bring the brown plant hopper under control after years of frustration on the pesticide treadmill.

IPM is broadly adopted, it is a logical pathway to a sustainable future in which pesticides are not polluting groundwater, contaminating food, killing pollinating insects, and, in the end, creating new, more resistant pests.

### Organically Grown Food

For consumers concerned about pesticides in their food, **organic food** is an option. Many farmers are turning away from the use of pesticides, chemical fertilizers, antibiotics, and hormones as they raise grains, vegetables, and livestock. Typically, organic farms are small, employ traditional farming methods with diverse crops, and are tied to local economies. For the farmers, the organically raised crops often represent lower crop yields, but also lower expenses. The soil in their fields and orchards is richer, and they have a product that is increasingly in demand. Sales of organics have increased 20% a year over the past decade, and organically grown food is now an $11 billion enterprise in the United States and $25 billion worldwide. The major reason for this expanding market is concern over health and safety, although consumers also report that the organic foods are usually more tasty. They are also more costly, but recent tests have shown that organic foods are far less likely to have any pesticide residues, and those which do contain much lower amounts than conventionally grown foods. Most residues on organic foods came from older, persistent pesticides lingering in soils, such as DDT.

*USDA Organic.* The certification of organic foods has been a contentious process, and eventually the industry requested help from the government. In response, in 1990 Congress passed the Organic Foods Protection Act, which established the National Organic Standards Board (NOSB), under USDA auspices. After many years of work, proposed guidelines for certification from the NOSB were aired in early 1998. Possibly in response to lobbying from the food industry, the board allowed genetically engineered foods, irradiated foods (foods subjected to radiation to kill bacteria), and crops fertilized with sewage sludge to be included as organic. A firestorm of protest from the organic-food community ensued. The NOSB went back to work, and in late 2002 the USDA published its final standards for certifying organic foods. As a result, no product involving genetically engineered or irradiated foods or foods fertilized with sewage sludge can be certified as 100% organic. The standards also prohibit the use of conventional pesticides, antibiotics, growth hormones, and chemical fertilizers. To market certified organic foods, farmers now must have their operations scrutinized by USDA-approved inspectors. Only then may they use the USDA seal (Fig. 16–22) on their products. If the food is at least 95% organic, it may also display the seal, but may not claim to be 100% organic.

**Figure 16–22 Organic foods.** This symbol on food signals that the food or crop has been produced on a certified organic farm.

## 16.5 Pesticides and Policy

In regulating the use of pesticides, three concerns are paramount: (1) Pesticides must be evaluated both for their intended uses and for their impacts on human health and the environment; (2) those who use the pesticides—especially agricultural workers—must be appropriately trained and protected from the risks of close contact; and (3) because most of the agricultural applications involve food, the public must be protected from the risks of pesticide residues on food products.

### FIFRA

The **Federal Insecticide, Fungicide, and Rodenticide Act,** commonly known as FIFRA, addresses the first two of the preceding concerns. This law, established by Congress in 1947 and amended in 1972, is administered by the EPA, which has total jurisdiction over the manufacture, sale, use, and testing of pesticides. FIFRA requires manufacturers to register pesticides with the EPA before marketing them. The registration procedure includes testing for toxicity to animals (and, by extrapolation, to humans). From the test results, usage standards are set. For example, chlordane, which was used for 40 years as a pesticide on crops, lawns, and gardens, was tested and shown to be acutely toxic to animals in high doses. Chronic exposure led to damaging effects on most vital systems and caused liver cancer in animals. Acute and chronic exposure of humans to chlordane showed similar results, except for the development of cancer. On the basis of these results, the EPA canceled all uses of chlordane for food crops in 1978, but allowed continued use against termites until 1988, when all uses of the chemical in the United States were banned. The law also stipulates that the EPA must determine whether a product "generally causes unreasonable adverse effects on the environment," and if it so determines, the product may be restricted or banned. This was the basis for the banning of DDT.

***Priority Pesticides.*** The EPA's Pesticide Program places high priority on registering "reduced-risk" pesticides, especially those which promise to replace more toxic chemicals. Most are **biopesticides**—microbial pesticides like Bt, plant-incorporated protectants such as corn bioengineered to manufacture Bt in its tissues, or biochemical substances that are naturally occurring, like the pheromones (e.g., Mimic). These pesticides often clear the registration process within a year, compared with the typical three years a conventional pesticide might take. In 2002, for example, the EPA registered 26 new products, 15 of which were reduced-risk pesticides. Once registered, a pesticide must have an EPA-approved label that lists its active ingredients, instructs the public on its proper use, and informs users of the risks involved with its use.

***Pesticide Workers.*** In protecting workers from risks posed by occupational exposure to pesticides, the EPA works with state environmental agencies, which have primary responsibility for enforcement. Under its Worker Protection Standard program, the EPA has issued regulations involving requirements for safety training, the use of protective equipment, the provision of emergency assistance, and permitting standards. The states must then demonstrate to the EPA that, on the basis of the federal regulations, they have adequate regulation and enforcement mechanisms.

## FQPA of 1996

Protecting consumers from exposure to pesticides on food was a contentious and confusing scene for many years. Three agencies are involved: the EPA, the FDA, and the Food Safety and Inspection Service of the USDA. Basically, the EPA sets the allowable tolerances for pesticide residues, based on its assessment of toxicity, and the FDA monitors and enforces the tolerances on all foods except meat, poultry, and egg products, which are managed by the USDA.

***Delaney Clause.*** Prior to 1996, the **Federal Food, Drug, and Cosmetic Act of 1958 (FFDCA)** was the enabling legislation assigning the foregoing responsibilities. One clause of the FFDCA, the so-called **Delaney clause,** was highly controversial. The clause states, "No [food] additive shall be deemed to be safe if it is found to induce cancer when ingested by man or animal." Since pesticides represent one of the largest categories of toxic chemicals to which people are exposed, this clause was applied in prohibiting many pesticides from being used on foodstuffs when those pesticides had been found to cause cancer in laboratory tests with animals. In essence, the law states that if a given pesticide presents *any* risk of cancer, *no detectable residue* may remain on the food. As analytical chemical techniques became more and more sensitive, extremely low traces of a pesticide could be measured, and according to the Delaney clause, if that pesticide caused cancer in experimental animals, farmers could not use it.

The National Research Council examined this dilemma and recommended in 1987 that the anti-cancer clause be replaced with a "negligible risk" standard, based on risk-analysis procedures that had been developed subsequent to the Delaney clause. After years of debate, Congress passed the **Food Quality Protection Act (FQPA)** in 1996 and did away with the Delaney clause. The following are the major requirements of the act:

- The new safety standard is "a reasonable certainty of no harm" for substances applied to foods.
- Special consideration must be given to the exposure of young children to pesticide residues.
- Pesticides or other chemicals are prohibited if, when consumed at average levels over the course of a lifetime, they can be shown to carry a risk of more than one case of cancer per million people.
- All possible sources of exposure to a given pesticide must be evaluated, not just exposure from food.
- All older products on the market before 1996 must be reassessed according to the new requirements.
- A special attempt must be made to assess the potential harmful effects of the so-called hormone disrupters.
- The same standards are to be applied to raw and processed foods (previously regulated separately).

***Care for Kids.*** The focus on children is the outcome of a 1993 report by the National Research Council, *Pesticides in the Diets of Infants and Children.* The focus is on children both because they typically consume more fruits and vegetables per unit of body weight than do adults and because studies have shown that youths are frequently more susceptible than adults to carcinogens and neurotoxins. Accordingly, the FQPA requires the EPA to add a tenfold safety factor in assessing children's risks from pesticides.

The USDA Pesticide Data Program (PDP) provides annual analyses of pesticides in fresh and processed foods. Under the FQPA, the EPA is to evaluate toxicity data on individual pesticides and then, using the USDA's PDP analyses and other information on risk assessment (see Chapter 15), set *limits on the amounts of a pesticide that remain in or on foods*—called **tolerances.** To date, the EPA is on target with the directives of the FQPA, having published tolerance reassessments for about 6,400 pesticide products (out of a total of over 9,700). More than 1,900 tolerances were rejected. As mentioned earlier, two pesticides—methyl parathion and chlorpyrifos—recently were banned on the basis of these reevaluations, because they were found to pose unacceptable risks to children. For an example of a "tolerance," another, less toxic organophosphate, malathion, is listed at 8 ppm tolerance for many crops, meaning that no more than 8 ppm of malathion may be found on or in the product. For milk, however, the tolerance is set at 0.5 ppm, an indication of milk's special place in children's diets. USDA and FDA inspectors constantly monitor foods in interstate commerce to enforce these tolerance limits.

### Pesticides in Developing Countries

The United States currently exports more than 35,000 tons of pesticides (active ingredients) each year, worth $1.8 billion. Some 25% of this total consists of products banned in the United States itself. FIFRA requires "informed permission" prior to the shipment of any pesticides banned in the United States. This permission comes from the purchaser, which can often be a foreign subsidiary of the exporting company. The EPA must then notify the government as to the identity of the importing country.

*PIC.* Fortunately, the international community has erected a more effective system, through the cooperative work of two U.N. agencies: the FAO and the UNEP. There is now a process of **prior informed consent (PIC),** whereby exporting countries inform all potential importing countries of actions they have taken to ban or restrict the use of pesticides or other toxic chemicals. Governments in the importing country respond to the notifications via the U.N. agencies, which then disseminate all the information they receive to the exporting countries and follow up by monitoring the export practices of those countries. As with many other U.N. actions, the process involves establishing a secretariat and a "convention," in this case the *Rotterdam Convention on the Prior Informed Consent Procedure for Certain Hazardous Chemicals and Pesticides in International Trade.* The PIC process was officially approved at a U.N. conference in 1998 and will go into force after 50 countries have ratified it. To date, only 30 have.

*Code of Conduct.* The more serious problem of unsafe pesticide use in the developing countries has also been addressed by the FAO. In spite of early opposition from the pesticide industry and exporting countries, an international *Code of Conduct on the Distribution and Use of Pesticides* was approved by the FAO in November 2002 whereby conditions of safe pesticide use were addressed in detail. The code makes clear the responsibilities of both private companies and countries receiving pesticides in promoting their safe use. Unlike the PIC Convention, this code calls for voluntary action and so is not legally binding, but it has already proven useful in holding private industry and importing countries to standards of safe use. In the view of some observers, both PIC and the FAO Code of Conduct are not as strong as they should be. Nevertheless, their very existence represents great progress in addressing a difficult problem in the developing countries.

## revisiting the themes

### Sustainability

Ecological control is a sustainable approach to controlling pests, while the chemical treatment approach guarantees that pesticides will need to be used indefinitely. Pest resistance, resurgences, and secondary outbreaks have been the norm. This is good news for the pesticide industry, bad news for the farmers, the public, and the environment. The ecological control methods provide the means to achieve control of most pests in a sustainable way, especially when they are controlled by natural enemies. IPM presents a reasonable alternative for managing pests without the unnecessary use of pesticides. The organic food movement is proof that the old farming ways are often workable and, for consumers and the environment, a great way to keep pesticides out of the natural world. Reducing or eliminating the use of pesticides and keeping food safe and healthy are two of the pillars of the sustainable-agriculture movement. (See Chapter 8.)

### Stewardship

The widespread use of DDT during and just after World War II was an exercise in good stewardship. Millions of lives were saved. However, the continued, often indiscriminate, use of this persistent pesticide and others like it led to trouble in the environment, and the careful work of Rachel Carson in documenting this trouble in the 1960s was a landmark in environmental stewardship. Because stewardship also includes human welfare and health, the focus of the FQPA on children was a major event in stewardly care. It can be argued that the work of the EPA in following up this important legislation is stewardship in action. The agency is an aggressive promoter of reduced use of the more toxic pesticides.

### Sound Science

It was sound science that uncovered the connections between DDT and the decline of many predatory birds—biomagnification was a great surprise to everyone. Some amazing applications of sound science are seen in the imaginative work on alternative pest controls, such as Knipling's sterile-male technique, the use of pheromones to control pests, and the many studies that have uncovered natural enemies of pests, leading to their control.

### Ecosystem Capital

One of the most important ecosystem services is the regulating work of predators, parasites, and diseases that keeps most organisms from exploding and doing serious economic damage. For this reason, it is vitally important to preserve these natural control agents in and around agricultural landscapes. Minimizing or eliminating the use of broad-scale chemical pesticides is a primary strategy for accomplishing this objective. Other ways also exist, such as maintaining hedgerows, fencerows, and shelterbelts where the natural predator populations can find refuge.

### Policy and Politics

The public policy of caring for the environment was galvanized by *Silent Spring*. Soon, the EPA was created, and many landmark laws were enacted to protect the public and the environment from air, land, and water pollution. FIFRA and FQPA represent the two most important pieces of legislation dealing with pesticides. The work of the EPA in establishing tolerances sensitive to the special needs of children demonstrates a strong commitment by government to protecting its most vulnerable citizens. Finally, the organic-food movement has its certification, 12 years after passage of the Organic Foods Protection Act. On the world scene, the Global IPM Facility, the Rotterdam Convention, and the Code of Conduct represent the best of international efforts to bring pesticide use under control and stimulate sustainable agriculture in the developing countries.

Several years ago, a group of leading entomologists called for a 50% reduction in pesticide use in the United States, supporting their recommendation by detailing the economic and environmental benefits of such a reduction. This goal could become public policy if it is included in legislative or regulatory law. Shortly afterwards, the heads of the EPA, FDA, and USDA issued a joint announcement of their commitment to reducing pesticide use in the United States. Pesticide reform is in the wind. Yet, as Figure 16–3 indicates, we have a long way to go to achieve a 50% reduction from the 1979 high point of pesticide use. As with many of the issues discussed in this book, significant progress in pesticide control has benefited from grassroots action and pressure from public-interest groups and nongovernmental organizations, such as the Consumers Union and the Pesticide Action Network. These and other groups are pressing for continued movement in the direction of ecological pest management and continued progress in keeping food free of pesticide residues.

### Globalization

Because the sale and use of pesticides is now a global phenomenon, its impact is also global. Chapter 19 documents many toxic substances (such as DDT) that are spread globally via the atmosphere and then deposited on land. Resistance to pesticides also spreads globally, as insects and weeds evolve in an environment that selects for resistance. Fortunately, the global community has mobilized to act, through the FAO. This one U.N. agency fully deserves whatever financial means it takes to enable it to carry out its many missions (including the establishment and continued support of IPM field schools, locust control, PIC, and the Code of Conduct for pesticide use) to foster sustainable agriculture and the responsible use of pesticides.

## review questions

1. Define pests. Why do we control them?
2. Discuss two basic philosophies of pest control. How effective are they?
3. What were the apparent virtues of the synthetic organic pesticide DDT?
4. What adverse environmental and human health effects can occur as a result of pesticide use?
5. Define *bioaccumulation* and *biomagnification*.
6. Why are nonpersistent pesticides not as environmentally sound as first thought?
7. Describe the four categories of natural, or biological, pest control. Cite examples of each and discuss their effectiveness.

8. Define the term *economic threshold* as it relates to pest control. What are cosmetic spraying and insurance spraying?
9. What are the four steps in IPM? Explain how IPM worked in Indonesia.
10. What is organic food, and how is it now certified?
11. How does FIFRA attempt to control pesticides? What new perspective does FQPA bring to the policy scene?
12. Discuss recent policy regarding the export of pesticides to developing countries.

## thinking environmentally

1. U.S. companies export pesticides that have been banned or restricted in this country. Should the practice be allowed to continue? Support your answer.
2. Almost one-third of the chemical pesticides bought in the United States are for use in houses and on gardens and lawns. What should manufacturers and users do to ensure the limited and prudent use of pesticides?
3. Should the government give farmers economic incentives to switch from pesticide use to IPM? How might that work?
4. Investigate how bugs and weeds are controlled on your campus. Are IPM techniques being used? Organic fertilizers? If not, consider lobbying for their use.
5. Read or reread the "Ethics" essay entitled "DDT for Malaria Control: Hero or Villain?" (p. 439). How would you resolve this issue?

# Water Pollution and Its Prevention

**Key Topics**

1. Water Pollution
2. Eutrophication
3. Sewage Management and Treatment
4. Public Policy

The Mississippi River watershed encompasses 40% of the land area of the United States, and as it collects water from this huge area, it eventually delivers the water to the Gulf of Mexico. Because much of the watershed is America's agricultural heartland, the Mississippi is a reflection of what happens on the farm. As U.S. farm production experienced tremendous growth during the latter half of the 20th century, more and more fertilizers were spread on agricultural fields, and more animals were raised on feedlots. At the same time, wetlands bordering river tributaries were drained and no longer intercepted agricultural runoff. The result was a tripling of soluble nitrogen delivered to the river and, finally, to the Gulf of Mexico.

***Dead Zone.*** The impact of this major flow of nitrogen to the Gulf of Mexico was first detected by marine scientists in 1974, who found areas where oxygen had disappeared from bottom sediments and much of the water column above them. This absence of oxygen is deadly to the bottom-dwelling animals and any oxygen-breathing creatures in the water column that are not able to escape. At first, the **hypoxic area,** or "**dead zone,**" as it is called, was thought to be similar to other bodies of water, such as the lower Chesapeake Bay, that received high nutrient levels—a minor disturbance that would disappear seasonally. But then the hypoxic area suddenly doubled in size following the 1993 floods in the Midwest, and it continues to grow (Fig. 17–1). In 2002, the area extended over 8,500 square miles (22,000 $km^2$) off the Louisiana coast, an area larger than the state of Massachusetts. The shrimpers and fishers who work these waters know that something has been happening and have learned to avoid the hypoxic area.

***The Culprit.*** To the scientists working on the problem, the connection between nitrogen and hypoxia became clear. Nitrogen is a common limiting factor in coastal marine waters. Where it is abundant, it promotes the dense growth of phytoplankton—photosynthetic microorganisms that live freely suspended in the water. Zooplankton—microscopic animals in the water—rapidly consume the

*The Mississippi Delta* **The Mississippi basin consists of 40% of the landmass of the lower 48 states. Runoff from this huge area delivers enormous amounts of sediment and nutrients to the Gulf of Mexico.**

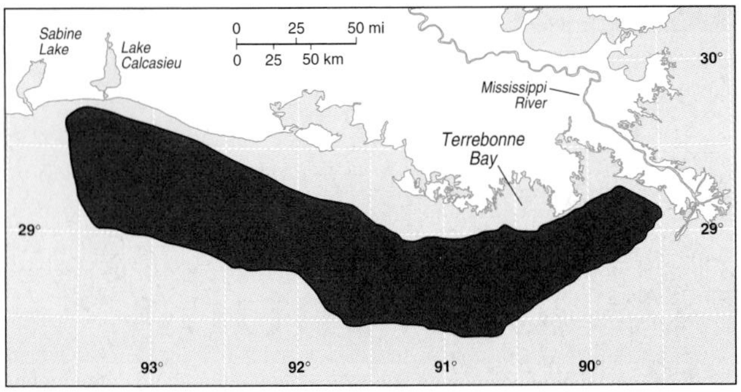


**Figure 17–1 The Dead Zone in the Gulf of Mexico.** Each year, oxygen disappears from middepths to the bottom over a vast area of the Gulf.

phytoplankton and multiply. As the abundant phytoplankton and zooplankton (and their fecal pellets) die or sink toward the bottom, they are decomposed by bacteria, a process that consumes dissolved oxygen. Eventually, the oxygen is used up, creating the dead zone. In the Gulf, the zone can extend up from the bottom of the sea to within a few meters of the surface. Typically, it stays from May through September, finally being dispersed by the mixing of the water column as colder weather sets in.

Because the Gulf fishery is a $2.8 billion enterprise, the problem eventually got national attention. In 1998, Congress passed the **Harmful Algal Bloom and Hypoxia Research and Control Act,** establishing an interagency task force of scientists charged with assessing the dead zone in the Gulf of Mexico, as well as harmful algal blooms, another water pollution problem discussed later in the chapter. The task force released its report in May 2000. The report confirmed the connection between nitrogen and the dead zone and laid out options for reducing the nitrogen load coming downriver. Two strategies were key to mitigation of the problem: (1) reducing nitrogen loads to streams and rivers in the Mississippi basin (by using less fertilizer and thereby greatly reducing farm runoff) and (2) restoring and promoting nitrogen retention and denitrification processes in the basin. The fertilizer industry and farm groups are unhappy with the findings, contending that nutrients are only one of several factors contributing to the hypoxic area. When the EPA and NOAA presented their *Action Plan to Reduce Gulf Hypoxia* in January 2001, their primary goal was to reduce the size of the hypoxic area by half by 2015, mainly by cutting nitrogen input by 30%. The plan was crafted by representatives from nine states along the Mississippi River, many federal agencies, and two Native American tribes.

The Gulf dead zone turns out to be one more costly lesson about **eutrophication** (literally, enrichment): It is unreasonable to use vital water resources to receive pollutants and then expect those resources to continue to provide us with their usual bounty of goods and services.

In this chapter, we will focus on water *quality* (Chapter 7 dealt with water *quantity*—the global water cycle and water resources). Eutrophication and other forms of water pollution are discussed, as is sewage treatment. The next four chapters focus on pollution by solid and hazardous wastes, and air pollution.

## 17.1 Water Pollution

The EPA defines pollution as "the presence of a substance in the environment that, because of its chemical composition or quantity, prevents the functioning of natural processes and produces undesirable environmental and health effects." Any material that causes the pollution is called a **pollutant.**

### Pollution Essentials

Pollution is not usually the result of deliberate mistreatment of the environment. The additions that cause pollution are almost always the *by-products* of otherwise worthy and essential activities—producing crops, creating comfortable homes, providing energy and transportation, and manufacturing products—and of our basic biological functions (excreting wastes). Pollution problems have become more pressing over the years because both growing population and expanding per capita use of materials and energy have increased the amounts of by-products that go into the environment. Also, many materials now widely used, such as aluminum cans, plastic packaging, and synthetic organic chemicals, are **nonbiodegradable.** That is, they resist attack and breakdown by detritus feeders and decomposers and consequently accumulate in the environment. An overview of the categories of pollutants that result from various activities is shown in Figure 17–2.

It is important to note the breadth and diversity of pollution. Any part of the environment may be affected, and almost anything may be a pollutant. The only criterion is that the addition of a pollutant results in undesirable changes. The impact of an undesirable change may be largely aesthetic—hazy air obscuring a distant view or the unsightliness of roadside litter, for instance. The impact may be on ecosystems as a whole—the die-off of fish or forests, for example. Or the impact may be on human health—such as human wastes contaminating water supplies. The impact also may range from very local—the contamination of an individual well, for instance—to global—adding ozone-depleting chemicals

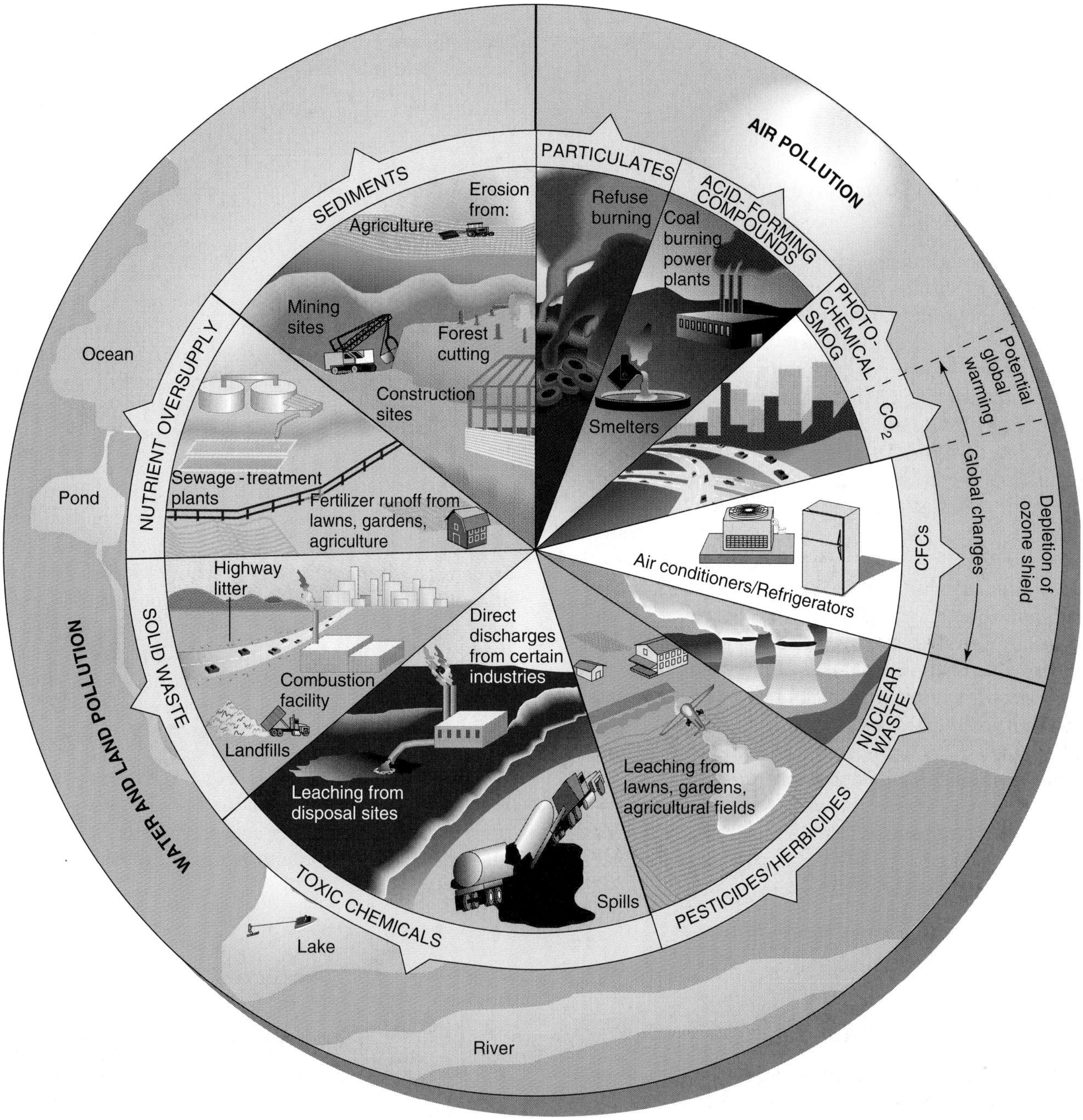


**Figure 17–2 Categories of pollution.** Pollution is an outcome of otherwise worthy human endeavors. Major categories of pollution and some activities that cause them are shown here.

to the atmosphere. We tend to think of pollution as the introduction of human-made materials into the environment, but undesirable changes may be caused by the introduction of too much of otherwise natural compounds, such as fertilizer nutrients introduced into waterways and carbon dioxide introduced into the atmosphere (Chapter 20).

Thus, the slogan "Don't pollute" is a gross oversimplification. The very nature of our existence necessitates the production of wastes. Our job in remediating present and future pollution problems is parallel to the concept of sustainable development itself. It is to adapt the means of meeting our present needs so that by-products are managed in ways that will not jeopardize present and future generations. The general strategy in each case must be to

1. Identify the material or materials that are causing the pollution—the undesirable change.
2. Identify the sources of the pollutants.
3. Develop and implement pollution control strategies to prevent the pollutants from entering the environment.
4. Develop and implement alternative means of meeting the need that do not produce the polluting by-product—in other words, avoid the pollution altogether.

Basically, the strategy for addressing pollution accomplishes one of the transitions to a sustainable society presented in Chapter 1: "A technology transition from pollution-intensive economic production to environmentally benign processes."

## Water Pollution: Sources, Types, Criteria

Water pollutants originate from a host of human activities and reach surface water or groundwater through an equally diverse host of pathways. For purposes of regulation, it is customary to distinguish between **point sources** and **nonpoint sources** of pollutants (Fig. 17–3). Point sources involve the discharge of substances from factories, sewage systems, power plants, underground coal mines, and oil wells. These sources are relatively easy to identify and therefore are easier to monitor and regulate than nonpoint sources, which are poorly defined and scattered over broad areas. Pollution occurs as rainfall and snowmelt move over and through the ground, picking up pollutants as they go. Some of the most prominent nonpoint sources of pollution are agricultural runoff (from farm animals and croplands), storm-water drainage (from streets, parking lots, and lawns), and atmospheric deposition (from air pollutants washed to earth or deposited as dry particles).

Two basic strategies are employed in attempting to bring water pollution under control: (1) Reduce or remove the sources and (2) treat the water so as to remove pollutants or convert them to harmless forms. Water treatment is the best option for point sources. Source reduction can be employed for both kinds of sources and is the best option for nonpoint sources.

**Pathogens.** The most serious water pollutants are the infectious agents that cause sickness and death. (See Chapter 15.) The excrement from humans and other animals infected with certain **pathogens** (disease-causing bacteria, viruses, and other parasitic organisms) contains large numbers of these organisms or their eggs. Table 17–1 lists the most common waterborne pathogens. Even after symptoms of disease disappear, an infected person or animal may still harbor low populations of the pathogen, thus continuing to act as a carrier of disease. If wastes from carriers contaminate drinking water, food, or water used for swimming or bathing, the pathogens can gain access to, and infect, other individuals (Fig. 17–4).

***Public Health.*** Before the connection between disease and sewage-carried pathogens was recognized in the mid-1800s, disastrous epidemics were common in cities. Epidemics of typhoid fever and cholera killed thousands of people prior to the 20th century. Today, public-health measures that prevent this disease cycle have been adopted throughout the developed world and, to a considerable extent, in the developing world. The following measures were far more important than modern medicine in controlling waterborne diseases:

1. Purification and disinfection of public water supplies with chlorine or other agents (Chapter 7).
2. Sanitary collection and treatment of sewage wastes.

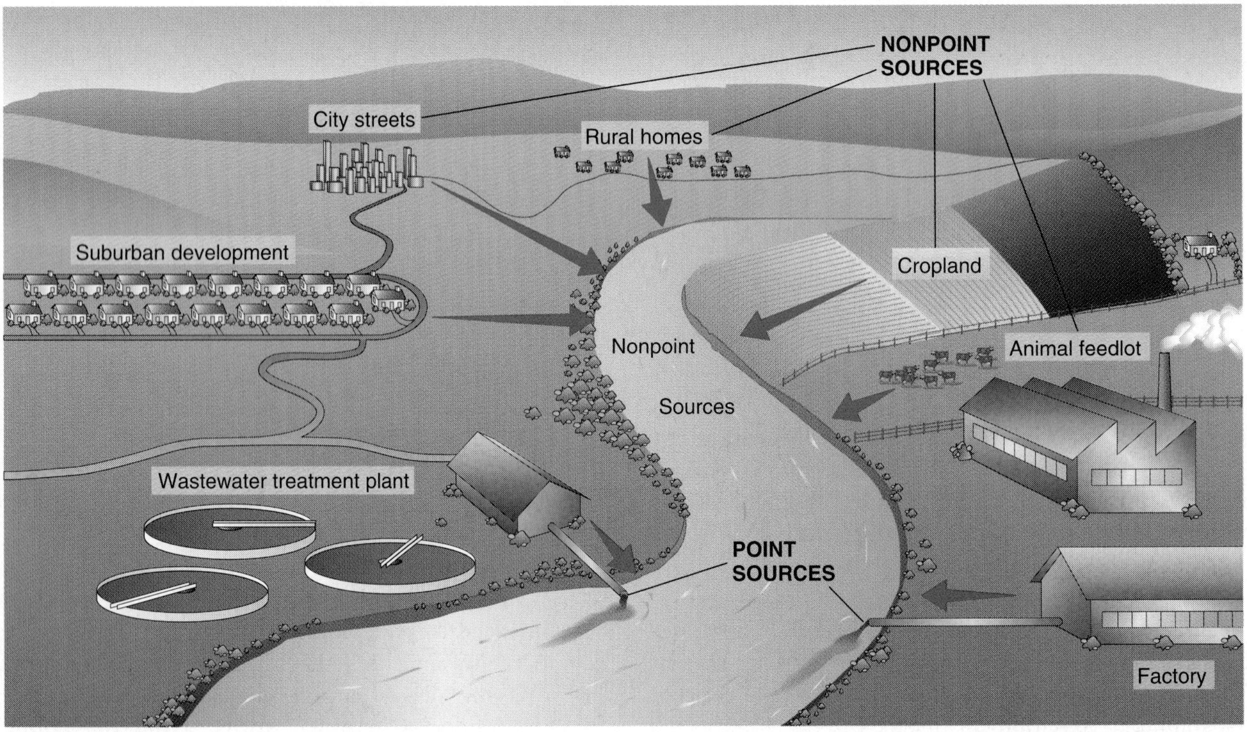


**Figure 17–3 Point and nonpoint sources.** Point sources are far easier to identify and correct than the diffuse nonpoint sources.

**table 17-1 Pathogens Carried by Sewage**

| Disease | Infectious Agent |
|---|---|
| Typhoid fever | *Salmonella typhi* (bacterium) |
| Cholera | *Vibrio cholerae* (bacterium) |
| Salmonellosis | *Salmonella* species (bacteria) |
| Diarrhea | *Escherichia coli*, *Campylobacter* species (bacteria) *Cryptosporidium parvum* (protozoan) |
| Infectious hepatitis | Hepatitis A virus |
| Poliomyelitis | Poliovirus |
| Dysentery | *Shigella* species (bacteria) *Entamoeba histolytica* (protozoan) |
| Giardiasis | *Giardia intestinalis* (protozoan) |
| Numerous parasitic diseases | (Roundworms, flatworms) |

3. Maintenance of sanitary standards in all facilities in which food is processed or prepared for public consumption.
4. Public education in personal and domestic hygiene practices (for example, washing the hands with soap).

Standards regarding these measures are set and enforced by government public-health departments. A variety of other measures are enforced as well. For example, if bathing areas are contaminated with raw sewage, health departments close them to swimming. Implicit in all measures is monitoring for sewage contamination. (See the "Earth Watch" essay, p. 468.) Our own personal hygiene, sanitation, and health precautions, such as not drinking water from untested sources and making sure that foods like pork, chicken, and hamburger are always well cooked, remain the last and most important line of defense against disease.

***Sanitation = Good Medicine.*** Many people attribute good health in a population to modern medicine, but *good health is primarily a result of the prevention of disease through public-health measures.* Over 1 billion people do not have access to safe drinking water. Some $2\frac{1}{2}$ billion people live in areas having poor (or no) sewage collection or treatment. Over 3 million deaths each year are traced to waterborne diseases (mostly in children under 5). One of the Millennium Development Goal (MDG) targets is "to reduce by half the proportion of people without sustainable access to safe drinking water." The 2002 Johannesburg World Summit added another target to the MDGs: to "halve, by the year 2015, the proportion of people who do not have access to basic sanitation."

Largely because of poor sanitation regarding water and sewage, a significant portion of the world's population is chronically infected with various pathogens. Moreover, populations in areas where there is little or no sewage treatment are extremely vulnerable to deadly epidemics of any and all diseases spread by way of sewage. This is especially true of *cholera*. Because of unsanitary conditions, an outbreak of cholera in Peru in 1990 killed several thousand people as it spread through Latin America. The civil unrest in Liberia in 2003 led to an outbreak of cholera. Thousands of cases occurred as normal water treatment and sanitary waste disposal were disrupted by the fighting. The WHO reports scores of cholera outbreaks annually, and cases frequently number in the thousands, with hundreds of deaths.

**Figure 17-4 The Ganges River in India.** In many places in the developing world, the same waterways are used simultaneously for drinking, washing, and disposal of sewage. A high incidence of disease, infant and childhood mortality, and parasites is the result.

**earth watch**

## Monitoring for Sewage Contamination

An important aspect of public health is the detection of sewage in water supplies and other bodies of water with which humans come in contact. It is worth understanding how this detection is done. It is virtually impossible to test for each specific pathogen that might be present in a body of water. Therefore, an indirect method called the **fecal coliform test** has been developed. This test is based on the fact that huge populations of a bacterium called *E. coli* (*Escherichia coli*) normally inhabit the lower intestinal tract of humans and other animals, and large numbers of the bacterium are excreted with fecal material.

In temperate regions at least, *E. coli* does not last long in the outside environment. Therefore, the presumption is that when *E. coli* is found in natural waters, it is an indication of recent and probably persisting contamination with sewage wastes. In most situations, *E. coli* is not a pathogen itself, but is referred to as an **indicator organism:** Its presence indicates that water is contaminated with fecal wastes and that sewageborne pathogens may be present. Conversely, the absence of *E. coli* is taken to mean that water is free from such pathogens.

The fecal coliform test, one technique of which is shown in the accompanying figures, detects and counts the number of *E. coli* in a sample of water. Thus, the results indicate the relative degree of contamination of the water and the relative risk of pathogens. For example, to be safe for drinking, water may not have any fecal coliform in a 100-mL (0.4-cup) sample. Water with as many as 200 *E. coli* per 100 mL is still considered safe for swimming. Beyond that level, a river may be posted as polluted, and swimming and other direct contact should be avoided. By contrast, raw sewage (99.9% water, 0.1% waste) has *E. coli* counts in the millions.

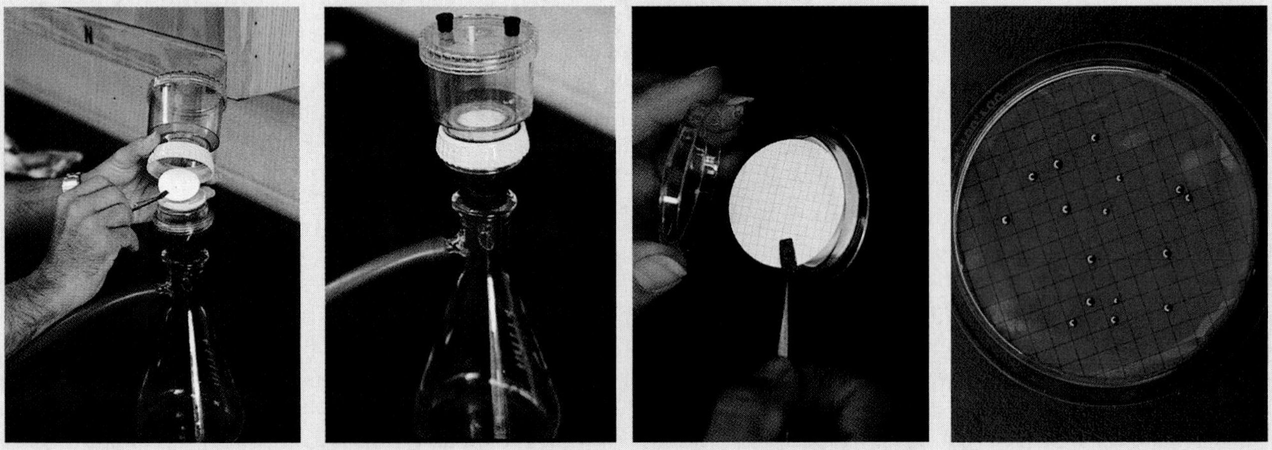

(a) (b) (c) (d)

**Testing water for sewage contamination by the Millipore technique.** (a) A Millipore filter disk is placed in the filter apparatus. (b) A sample of the water being tested is drawn through the filter, and any bacteria present are entrapped on the disk. (c) The disk is then placed in a petri dish on a special medium that supports the growth of bacteria and that will impart a particular color to fecal *E. coli* bacteria. Next, the dish is incubated for 24 hours at 38°C, during which time the bacteria on the disk will multiply to form a colony visible to the naked eye. (d) *Escherichia coli* bacteria, indicating sewage contamination, are identifiable as the colonies with a metallic green sheen.

**Organic Wastes.** Along with pathogens, human and animal wastes contain organic matter that creates serious problems if it enters bodies of water untreated. Other kinds of organic matter (leaves, grass clippings, trash, etc.) can enter bodies of water as a consequence of runoff and, in the case of excessive aquatic plant growth, can grow within the water. With the exception of plastics and some human-made chemicals, these wastes are biodegradable. As mentioned in connection with the Gulf dead zone, when bacteria and detritus feeders decompose organic matter in water, they consume oxygen gas dissolved in the water. The amount of oxygen that water can hold in solution is severely limited. In cold water, dissolved oxygen (DO) can reach concentrations up to 10 parts per million (ppm); even less can be held in warm water. Compare this ratio with that of oxygen in air, which is 200,000 ppm (20%), and you can understand why even a moderate amount of organic matter decomposing in water can deplete the water of its DO. Bacteria keep the water depleted in DO as long as there is dead organic matter to support their growth and oxygen replenishment is inadequate.

*BOD.* **Biochemical oxygen demand (BOD)** is a measure of the amount of organic material in water, in terms of how much oxygen will be required to break it down biologically, chemically, or both. The higher the BOD measure, the greater is the likelihood that dissolved oxygen will be depleted in the course of breaking it down. A high BOD causes so much oxygen depletion that animal life is severely limited or precluded, as in the bottom waters of the Gulf of Mexico. Fish and shellfish are killed when the DO drops below 2 or 3 ppm, though some are less tolerant at even higher DO levels. If the system goes anaerobic (without oxygen), only bacteria can survive, using their abilities to switch to fermentation or anaerobic respiration (metabolic pathways that do not require oxygen). A typical BOD value for raw sewage would be around 250 ppm. Even a moderate amount of sewage, added to natural waters containing at most 10 ppm DO, can deplete the water of its oxygen and produce highly undesirable consequences well beyond those caused by the introduction of pathogens.

**Chemical Pollutants.** Because water is such an excellent solvent, it is able to hold many chemical substances in solution that have undesirable effects. Water-soluble **inorganic chemicals** constitute an important class of pollutants that include heavy metals (lead, mercury, cadmium, nickel, etc.), acids from mine drainage (sulfuric acid) and acid precipitation (sulfuric and nitric acids), and road salts employed to melt snow and ice in the colder climates (sodium and calcium chlorides). The **organic chemicals** are another group of substances found in polluted waters. Petroleum products pollute many bodies of water, from the major oil spills in the ocean (Chapter 12) to small streams receiving runoff from parking lots. Other organic substances with serious impacts are the pesticides that drift down from aerial spraying or that run off from land areas (Chapter 16) and the various industrial chemicals, such as polychlorinated biphenyls (PCBs), cleaning solvents, and detergents.

Many of these pollutants are toxic even at low concentrations (Chapter 19). Some may become concentrated by passing up the food chain in a process called biomagnification (Chapter 16). Even at very low concentrations, they can render water unpalatable to humans and dangerous to aquatic life. At higher concentrations, they can change the properties of bodies of water so as to prevent them from serving any useful purpose except perhaps navigation. Acid mine drainage, for example, pollutes thousands of miles of streams in coal-bearing regions of the United States (Fig. 17–5). Stream bottoms are coated with orange deposits of iron, and the water is so acidic that only a few hardy bacteria and algae can tolerate it.

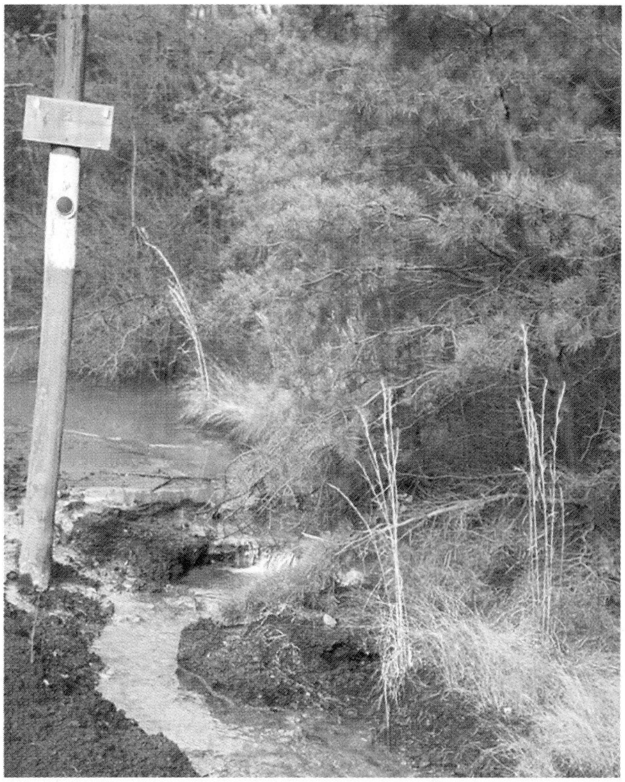

**Figure 17–5 Acid mine drainage.** Acidic water from mines contains high concentrations of sulfides and iron, which are oxidized when they come into the open, thus producing unsightly and sterile streams.

**Sediments.** As natural landforms weather, and especially during storms, a certain amount of sediment enters streams and rivers. However, erosion from farmlands, deforested slopes, overgrazed rangelands, construction sites, mining sites, stream banks, and roads can greatly increase the load of sediment entering waterways. Frequently, storm drains simply lead to the nearest depression or some natural streambed. Sediments (sand, silt, and clay) have direct and extreme physical impacts on streams and rivers. When erosion is slight, streams and rivers of the watershed run clear and support algae and other aquatic plants that attach to rocks or take root in the bottom. These producers, plus miscellaneous detritus from fallen leaves and so on, support a complex food web of bacteria, protozoa, worms, insect larvae, snails, fish, crayfish, and other organisms, which keep themselves from being carried downstream by attaching to rocks or seeking shelter behind or under rocks. Even fish that maintain their position by active swimming occasionally need such shelter to rest (Fig. 17–6a).

Sediment entering waterways in large amounts has an array of impacts. Sand, silt, clay, and organic particles (humus) are quickly separated by the agitation of flowing water and are carried at different rates. Clay and humus are carried in suspension, making the water muddy and reducing the amount of light penetrating the water and, hence, reducing photosynthesis as well. As the material settles, it coats everything and continues to block photosynthesis. It also kills the animals by clogging their gills and feeding structures. The eggs of fish and other aquatic organisms are particularly vulnerable to being smothered by sediment.

*Bed Load.* Especially destructive is the **bed load** of sand and silt, which is not readily carried in suspension, but is gradually washed along the bottom. As particles roll and tumble along, they scour organisms from the rocks.

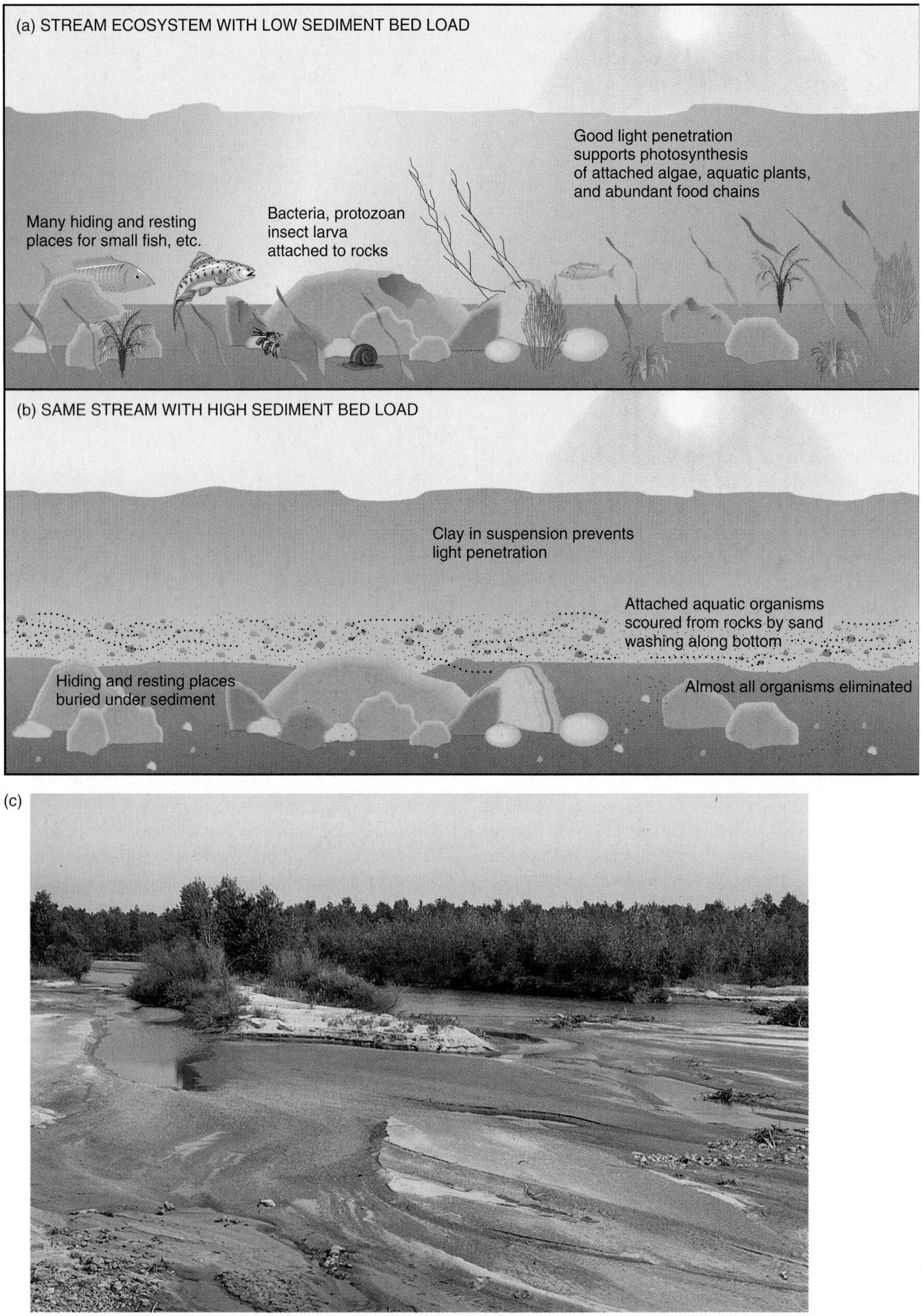

(a) STREAM ECOSYSTEM WITH LOW SEDIMENT BED LOAD
Good light penetration
supports photosynthesis
of attached algae, aquatic plants,
and abundant food chains
Many hiding and resting
places for small fish, etc.
Bacteria, protozoan
insect larva
attached to rocks
(b) SAME STREAM WITH HIGH SEDIMENT BED LOAD
Clay in suspension prevents
light penetration
Attached aquatic organisms
scoured from rocks by sand
washing along bottom
Hiding and resting places
buried under sediment
Almost all organisms eliminated
(c)

**Figure 17–6 Impact of sediment on streams and rivers.** (a) The ecosystem of a stream that is not subjected to a large sediment bed load. (b) The changes that occur with large sediment inputs. (c) Platte River at Lexington, Nebraska. The sandbars seen here constitute the bed load; they shift and move with high water, preventing the reestablishment of aquatic vegetation.

**Figure 17–7 Storm-water management.** Rather than letting excessive runoff from developed areas cause flooding and other environmental damage, runoff can be funneled into a retention pond, as shown here. Then it can drain away slowly, maintaining natural streamflow, or it can recharge groundwater. The retention pond may be designed to retain a certain amount of water and thus create a pocket of wildlife habitat in an otherwise urban or suburban setting.

They also bury and smother the bottom life and fill in the hiding and resting places of fish and crayfish. Aquatic plants and other organisms are prevented from reestablishing themselves because the bottom is a constantly shifting bed of sand (Fig. 17–6b and c). Modern storm-water management is designed to reduce the bed load, usually via storm drains that are periodically emptied of their sediment. Many housing developments include a storm-water retention reservoir, which is simply a pond that receives and holds runoff from the area during storms. Water may infiltrate into the soil or may create a pocket of natural wetland habitat supporting wildlife (Fig. 17–7).

Sediments do not receive the attention the news media give to hazardous wastes and certain other pollution problems, but erosion is so widespread throughout the world that few streams and rivers escape the harsh impact of excessive sediment loads.

**Nutrients.** Some of the inorganic chemicals carried in solution in all bodies of water are classified as nutrients—essential elements required by plants. The two most important nutrient elements for aquatic plant growth are *phosphorus* and *nitrogen,* and they are often in such low supply in water that they are the limiting factors for phytoplankton or other aquatic plants. More nutrients mean more plant growth, so nutrients become water pollutants when they are added from point or nonpoint sources and stimulate undesirable plant growth in bodies of water. The most obvious point sources of excessive nutrients are sewage outfalls. As we will see in Section 17.3, it takes a special effort to remove nutrients from sewage during the treatment process, so that is not always done. If the sewage is not treated at all, higher levels of nutrients will enter the water, which is what's happening in much of the developing world.

Agricultural runoff is the most notorious nonpoint source of nutrients. The nutrients are applied to agricultural crops as chemical fertilizers, manure, water for irrigation, sludge, and crop residues. When nutrients are applied in excess of the needs of plants, or when runoff from agricultural fields and feedlots is heavy, the nutrients are picked up by water that eventually enters streams, rivers, lakes, and the ocean. Other nonpoint sources of nutrients include lawns and gardens, golf courses, and storm drains.

**Water Quality Standards.** When is water considered polluted? Many water pollutants, like pesticides, cleaning solvents, and detergents, are substances that are found in water only because of human activities. Others, such as nutrients and sediments, are always found in natural waters, and they are a problem only under certain conditions. In both cases, "pollution" means "any quantity that is harmful to human health or the environment." The mere presence of a substance in water does not necessarily pose a problem. Rather, it is the *concentration* of the pollutant that must be of primary concern.

*Criteria Pollutants.* But what concentration is worrisome? To provide standards for assessing water pollution, the EPA has established the **National Recommended Water Quality Criteria.** On the basis of the latest scientific knowledge, the EPA has listed 158 chemicals and substances as criteria pollutants. The majority of these are toxic chemicals, but many are also natural chemicals or conditions that describe the state of water, such as nutrients, hardness (a general measure of dissolved calcium and magnesium salts), and pH (a measure of the acidity of the water). The list identifies the pollutant and then recommends concentrations for fresh water, salt water, and human consumption (usually fish and shellfish). Values are given for the **criteria maximum concentration** (CMC, the highest single concentration beyond which environmental impacts may be expected) and **criterion continuous concentration** (CCC, the highest sustained concentration beyond which undesirable impacts may be expected). The criteria are *recommendations* meant to be used by the states, which are given the primary responsibility for upholding water pollution laws. States and Native American tribes may revise these criteria, but their revisions are subject to EPA approval.

Drinking water standards are stricter. For these, the EPA has established the **Drinking Water Standards and Health Advisories,** a set of tables that are updated periodically. These standards, covering some 90 contaminants, are enforceable under the authority of the **Safe Drinking Water Act (SDWA).** They are presented as **maximum contaminant levels (MCLs).** To see how these two sets of standards work, consider the heavy metal arsenic.

***Arsenic.*** Arsenic is listed as a known human carcinogen by the Department of Health and Human Services; it occurs naturally in groundwater, often reaching high concentrations. The CMC and CCC values for arsenic are 340 and 150 $\mu$g/L (1 $\mu$g/L = 1 part per billion) for freshwater bodies, and 69 and 36 $\mu$g/L for saltwater bodies. The drinking-water MCL concentration, however, is 10 $\mu$g/L, and therein lies a story.

For years, the MCL concentration for arsenic was 50 $\mu$g/L, despite warnings by the U.S. Public Health Service and the WHO that this was much too high a level. (Both bodies recommended 10 $\mu$g/L.) The problem is the cancer risk. From standard risk assessment processes (see Chapter 15), the cancer mortality risk was assessed at 1 in 100 for people drinking water regularly at the 50 $\mu$g/L level. This ratio is more than 100 times higher than the permitted risk for any other contaminant. Before leaving office, the Clinton EPA administration determined that the MCL for arsenic should be lowered, but the pending rule was withdrawn by the Bush EPA, on the premise that sound science did not clearly support the new rule. Many states and rural areas were complaining that attaining the lower arsenic concentration would be too costly. In the end, after a National Academy of Sciences report and urgings from both the House and Senate, then EPA administrator Christine Whitman upheld the Clinton era rule and set the arsenic MCL at 10 $\mu$g/L.

***Other Applications.*** Two important applications of water quality criteria are the **National Pollution Discharge Elimination System (NPDES)** and **Total Maximum Daily Load (TMDL)** programs. The NPDES program addresses point-source pollution and issues permits that regulate discharges from wastewater treatment plants and industrial sources. The TMDL program, by contrast, evaluates all sources of pollutants entering a body of water, especially nonpoint sources, according to the water body's ability to assimilate the pollutant. For both of these programs, water quality criteria provide an essential means of assessing the condition of the receiving body of water, as well as evaluating the levels of the various pollutant discharges.

On the basis of data supplied to the EPA by the states, over 20,000 rivers, lakes, and estuaries—44% of our waters—are not meeting the recommended water quality standards. The major pollutants are sediments, excess nutrients, pathogens, and toxic substances. According to the EPA, 94% of the United States is served with drinking water that meets the drinking-water standards, although some have disputed this percentage as being unrealistic. At the end of the chapter we will address the specifics of public-policy responses to this serious national problem. We turn now to eutrophication, the most widespread water pollution problem in the nation.

## 17.2 Eutrophication

Recall from Chapter 2 that the term "trophic" refers to feeding. Literally, "eutrophic" means "well nourished." As we will see, there are some quite undesirable consequences of high levels of "nourishment" in bodies of water. Although eutrophication can be an entirely natural process, the introduction of pollutants into bodies of water has greatly increased the scope and speed of eutrophication.

### Different Kinds of Aquatic Plants

To understand eutrophication, you need to be able to distinguish between *benthic* plants and *phytoplankton.*

**Benthic plants** (from the Greek *benthos,* deep) are aquatic plants that grow attached to, or are rooted in, the bottom of a body of water. All common aquarium plants and sea grasses are benthic plants. As shown in Figure 17–8a, benthic plants may be categorized as **submerged aquatic vegetation (SAV),** which generally grows totally under water, or **emergent vegetation,** which grows with the lower parts in water, but the upper parts emerging from the water.

To thrive, SAV requires water that is clear enough to allow sufficient light to penetrate to allow photosynthesis. As water becomes more turbid, light is diminished. In extreme situations, it may be reduced to penetrating to just a few centimeters beneath the water's surface. Thus, increasing turbidity decreases the depth at which SAV can survive.

Another important feature of SAV is that it absorbs its required mineral nutrients from the bottom sediments through the roots, just as land plants do. SAV is not limited by water that is low in nutrients. Indeed, enrichment of the water with nutrients is counterproductive for SAV, because it stimulates the growth of phytoplankton.

***Phytoplankton.*** *Phytoplankton* consists of numerous species of photosynthetic algae, protists, and chlorophyll-containing bacteria (cyanobacteria, formerly referred to as blue-green algae) that grow as microscopic single cells or in small groups, or "threads," of cells. Phytoplankton lives suspended in the water and is found wherever light and nutrients are available (Fig. 17–8b). In extreme situations, water may become literally pea-soup green (or tea colored, depending on the species involved), and a scum of phytoplankton may float on the surface and

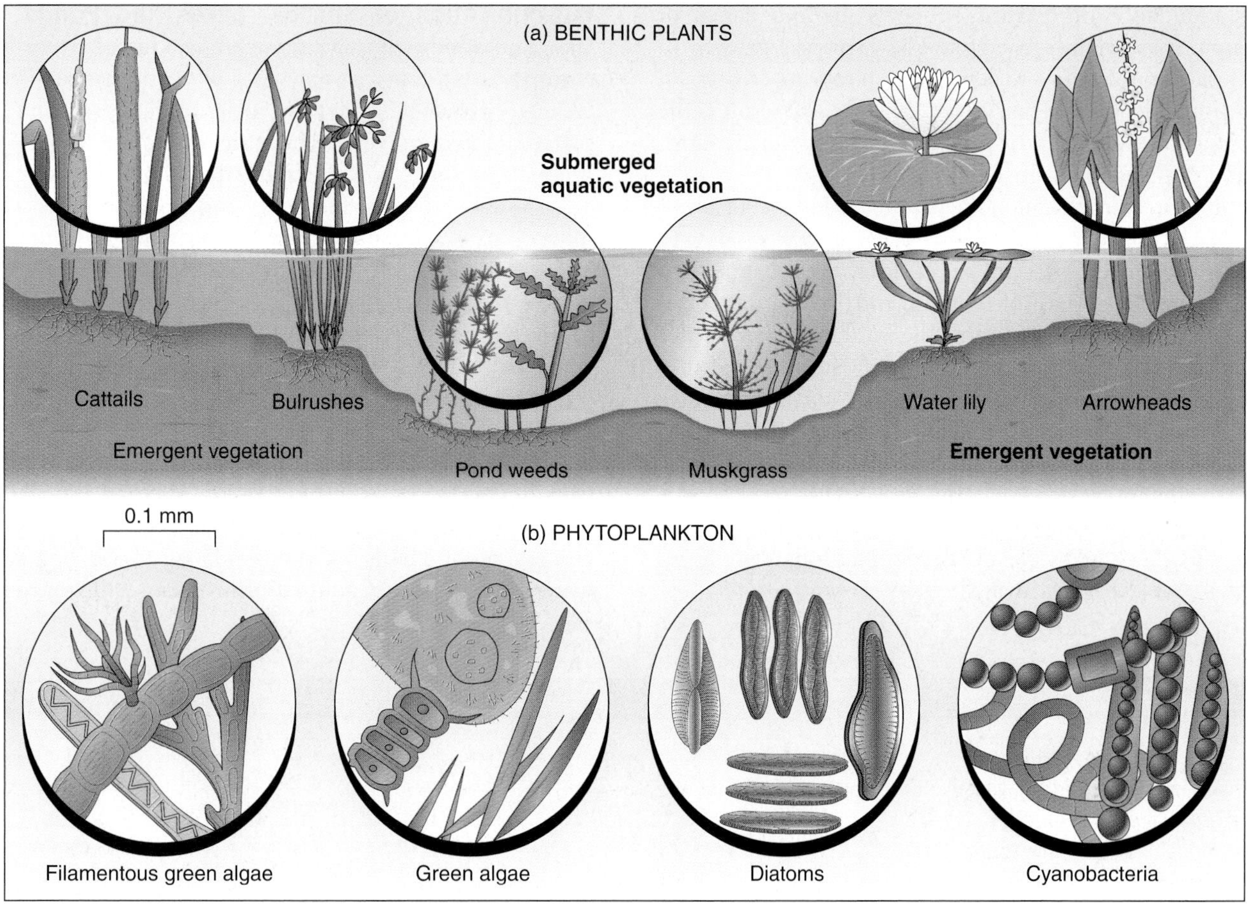


**Figure 17–8 Aquatic photosynthesizers.** (a) Benthic, or bottom-rooted, plants. These are subdivided into submerged aquatic vegetation (SAV) and emergent vegetation. (b) Phytoplankton, various photosynthetic organisms that are either single cells or small groups or filaments of cells, float freely in the water.

absorb essentially all the light. However, phytoplankton reaches such densities only in nutrient-rich water, because, not being connected to the bottom, it must absorb its nutrients from the water. A low level of nutrients in the water limits the growth of phytoplankton accordingly.

Considering the different requirements of phytoplankton and SAV, the balance between them is altered when nutrient levels in the water are changed. As long as water remains low in nutrients, populations of phytoplankton are suppressed, the water is clear, and light may penetrate to support the growth of SAV. As nutrient levels increase, phytoplankton can grow more prolifically, making the water turbid, thus shading out the SAV.

## The Impacts of Nutrient Enrichment

A lake in which light penetrates deeply—one in which the bottom is visible beyond the immediate shoreline—is **oligotrophic** (low in nutrients). Such a lake is fed by a watershed that holds its nutrients well. A forested watershed, for example, holds nitrogen and phosphorus tightly and allows very little of those elements to enter the water draining into the streams and rivers. If the streams feed into a lake, it will reflect their low nutrient content.

In an oligotrophic lake, the low nutrient levels limit the growth of phytoplankton and allow enough light to penetrate to support the growth of SAV, which draws its nutrients from the bottom sediments. In turn, the benthic plants support the rest of a diverse aquatic ecosystem by providing food, habitats, and dissolved oxygen. The oligotrophic body of water is prized for its aesthetic and recreational qualities, as well as for its production of fish and shellfish.

**Eutrophication.** As the water of an oligotrophic body becomes enriched with nutrients, numerous changes are set in motion. First, the nutrient enrichment allows the rapid growth and multiplication of phytoplankton, increasing the turbidity of the water. The increasing turbidity shades out the SAV that live in the water. With the die-off of SAV, there is a loss of food, habitats, and dissolved oxygen from their photosynthesis.

Phytoplankton has a remarkably high growth and reproduction rate. Under optimal conditions, phytoplankton biomass may double every 24 hours, a capacity far beyond that of benthic plants. Thus, phytoplankton soon reach a maximum population density, and continuing growth and reproduction are balanced by die-off. Dead phytoplankton settles out, resulting in heavy deposits of detritus on the lake or river bottom. In turn, the abundance of detritus supports an abundance of decomposers, mainly bacteria. The explosive growth of bacteria, consuming oxygen via respiration, creates an additional demand for dissolved oxygen. The result is the depletion of dissolved oxygen, with the consequent suffocation of fish and shellfish.

In sum, **eutrophication** refers to this whole sequence of events, starting with nutrient enrichment, and proceeding to the growth and die-off of phytoplankton, the accumulation of detritus, the growth of bacteria, and, finally, the depletion of dissolved oxygen and the suffocation of higher organisms (Fig. 17–9). For humans, the eutrophic system is unappealing for swimming, boating, and sportfishing. Also, if the lake is a source of drinking water, its value may be greatly impaired because phytoplankton rapidly clog water filters and may cause a foul taste. Even worse, some species of phytoplankton secrete various toxins into the water that may kill other aquatic life and be injurious to human health as well. (See the "Earth Watch" essay, p. 475.)

**Eutrophication of Shallow Lakes and Ponds.** In lakes and ponds whose water depth is 6 feet (2 m) or less, eutrophication takes a somewhat different course. There, SAV may grow to a height of a meter or more, reaching the surface. Thus, with nutrient enrichment, the SAV is not shaded out, but grows abundantly, sprawling over and often totally covering the water surface with dense mats of vegetation that make boating, fishing, or swimming impossible (Fig. 17–10). Any vegetation beneath these mats is shaded out. As the mats of vegetation die and sink to the bottom, they create a BOD that often depletes the water of dissolved oxygen, causing the death of aquatic organisms other than bacteria.

**Natural vs. Cultural Eutrophication.** In nature, apart from human impacts, eutrophication is part of the process of aquatic succession, discussed in Chapter 4. Over periods of hundreds or thousands of years, bodies of water are subject to gradual enrichment with nutrients.

**Figure 17–9**
**Eutrophication.** As nutrients are added from sources of pollution, an oligotrophic system rapidly becomes eutrophic and undesirable.

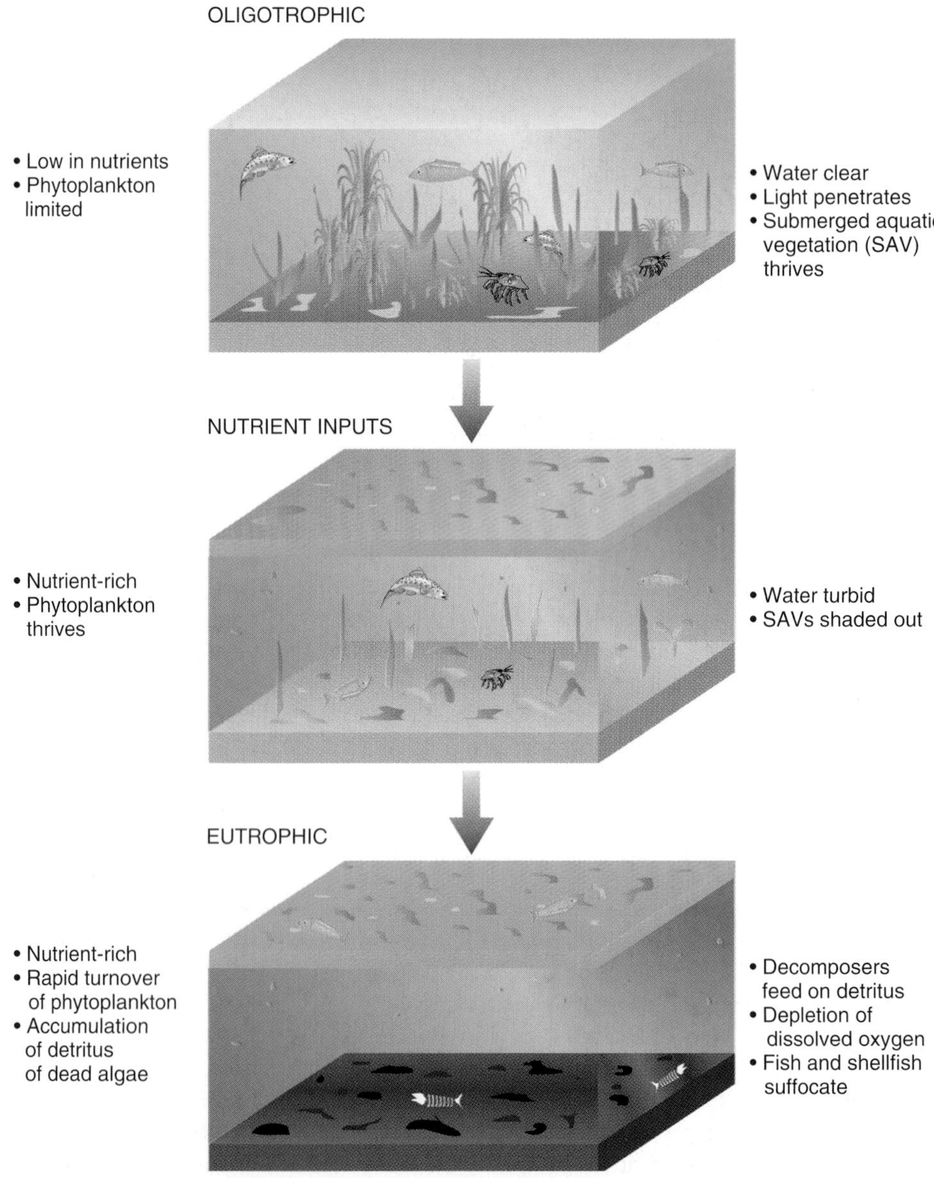
OLIGOTROPHIC
• Low in nutrients
• Phytoplankton limited
• Water clear
• Light penetrates
• Submerged aquatic vegetation (SAV) thrives
NUTRIENT INPUTS
• Nutrient-rich
• Phytoplankton thrives
• Water turbid
• SAVs shaded out
EUTROPHIC
• Nutrient-rich
• Rapid turnover of phytoplankton
• Accumulation of detritus of dead algae
• Decomposers feed on detritus
• Depletion of dissolved oxygen
• Fish and shellfish suffocate

### The Algae from Hell

Surprises dot the environmental landscape of the last 50 years, but one of the most bizarre and unexpected ones emerged when, in 1988, a North Carolina State University plankton biologist, Dr. Joanne Burkholder, examined some open sores that were found on fish. She identified a new organism, a microscopic dinoflagellate (a type of alga) she named *Pfiesteria piscicida.* (Recently, a second toxic species, *Pfiesteria shumwayae,* has been identified.) Subsequent work in laboratory cultures of *Pfiesteria* revealed that the species has no fewer than 24 different life-forms, a phenomenon unprecedented in the annals of science. Sometimes *Pfiesteria* is an amoeba, sometimes it is a flagellate, and sometimes it is in a cyst. Sometimes it secretes a toxin that stuns fish, and then it turns into a form that proceeds to feed on the fish, which have been disoriented by the toxin. Sometimes it behaves like a decent, ordinary alga and lives by photosynthesis. *Pfiesteria* has been implicated in at least 50 major fish kills in Albemarle and Pamlico Sounds, North Carolina, and has been identified in waters from Virginia, Maryland, Delaware, Florida, and even northern Europe. In the summer of 1997, fish kills struck several tributaries of Chesapeake Bay, particularly the Pocomoke River. The fish kills and catches of fish with ulcerous lesions finally led the state to close the river to all fishing and harvesting of shellfish.

*Pfiesteria* is toxic not just to fish and shellfish; it has also been implicated in skin rashes and neurological disorders in people who have encountered water containing the organism. The problems first appeared in Burkholder's laboratory. Later, watermen and others in close contact with infested waters began reporting similar symptoms: short-term memory loss, respiratory problems, fatigue, disorientation, and blurred vision. When news of the fish kills in Chesapeake Bay and the impacts on humans became public, there was a predictable panic response, and consumers all around the Chesapeake stopped eating seafood. In Maryland alone, the economic loss was estimated at between $15 and $20 million in sales during 1997.

The recent increase in *Pfiesteria* episodes joins a long list of what are called *harmful algal blooms (HABs).* Red tides, consisting of concentrations of different species of toxic dinoflagellates, have been occurring more and more frequently in coastal waters all around North America. They may be responsible for more than $1 billion in economic damage to coastal systems and fisheries. Many scientists believe that nutrients—especially nitrogen and phosphorus—are the basic culprits in feeding the growth of the algae in estuaries and coastal waters. In the Delmarva peninsula, comprising parts of Delaware, Maryland, and Virginia, some 600 million chickens are produced each year, and runoff from the poultry farms is believed to be a major source of nutrients to Chesapeake tributaries. North Carolina State University has opened a *Pfiesteria* research facility to provide Dr. Burkholder with the capability of analyzing samples from suspected infestations and to continue her work with the basic biology of the organism. Recently, five federal agencies joined to sponsor the Ecology and Oceanography of Harmful Algal Blooms (ECOHAB), a national research program studying HABs in U.S. coastal waters. There is little doubt that the blooms are increasing in frequency, and there is a good possibility that the explanation will be found in the pollutants that are still entering our waterways.

Thus, *natural eutrophication* is a normal process. Wherever nutrients come from sewage-treatment plants, poor farming practices, urban runoff, and certain other human activities, however, humans have inadvertently managed to vastly accelerate the process of nutrient enrichment. The accelerated eutrophication caused by humans is called **cultural eutrophication.**

## Combating Eutrophication

There are two approaches to combating the problem of eutrophication. One is to *attack the symptoms*—the growth of vegetation, the lack of dissolved oxygen, or both. The other is to *get at the root cause*—excessive inputs of nutrients and sediments.

**Attacking the Symptoms.** Attacking the symptoms is applicable in certain situations in which immediate remediation is the goal and costs are not prohibitive. Methods of attacking the symptoms of eutrophication include (1) chemical treatments (with herbicides), (2) aeration, (3) harvesting aquatic weeds, and (4) drawing water down.

***Herbicides.*** *Herbicides* are often applied to ponds and lakes to control the growth of nuisance plants. To control phytoplankton growth, copper sulfate and diquat are frequently used. For controlling SAV and emergent vegetation, fluridone, glyphosate, and 2,4-D are employed. However, these compounds are toxic to fish and aquatic animals, sometimes at concentrations required to bring the vegetation under control. Also, fish are often killed after herbicide is applied, because the rotting vegetation depletes the water of dissolved oxygen. In sum, herbicides provide only cosmetic treatment, and as soon as they wear off, the vegetation grows back rapidly.

***Aeration.*** The depletion of dissolved oxygen by decomposers and the consequent suffocation of other aquatic life is the final and most destructive stage of eutrophication. *Artificial aeration* of the water can avert this terminal stage. An aeration technique currently gaining

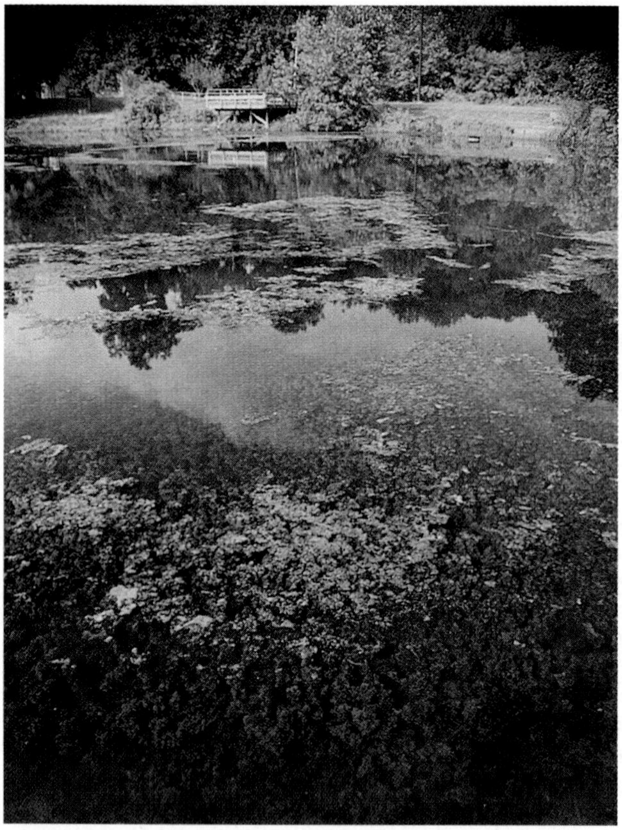

**Figure 17–10 Eutrophication in shallow lakes and ponds.** In shallow water, sufficient light for photosynthesis continues to reach the submerged aquatic vegetation. The oversupply of nutrients stimulates growth, so that vegetation reaches the surface and forms mats.

in popularity is to lay a network of plastic tubes with microscopic pores on the bottom of the waterway to be treated. High-pressure air pumps force microbubbles from the pores, and the bubbles dissolve directly into the water. The method is proving effective in speeding up the breakdown of accumulated detritus, improving water quality, and enabling the return of more desirable aquatic life. Despite its high cost, the technique is applicable to harbors, marinas, and some water-supply reservoirs, where the demand for better water quality justifies the cost.

*Harvesting.* In shallow lakes or ponds, where the problem is bottom-rooted vegetation reaching and sprawling over the surface (see Fig. 17–10), *harvesting the aquatic weeds* may be an expedient way to improve the water's recreational potential and aesthetics. Commercial mechanical harvesters are used, and nearby residents sometimes get together to remove the vegetation by hand. The harvested vegetation makes good organic fertilizer and mulch. But even harvesting has a limited effect: The vegetation soon grows back because roots are left in the nutrient-rich sediments.

*Drawdown.* Because many recreational lakes are dammed, another option for shallow-water weed control is to draw the lake down for a period each year. This process kills most of the rooted aquatic plants along the shore, although they grow back in time.

All of these approaches are only temporary fixes for the problem and have to be repeated often, at significant cost, to keep the unwanted plant growth under control.

**Getting at the Root Causes.** Controlling eutrophication requires long-term strategies for correcting the problem, which ultimately means reducing the inputs of nutrients and sediments—two of the types of pollution discussed at the beginning of this chapter. The first step is to identify the major point and nonpoint sources of nutrients and sediments. Then it is a matter of developing and implementing strategies for correction. Which source or factor is most significant will depend on the human population and the land uses within the particular watershed. Therefore, each watershed must be analyzed as a separate entity, and appropriate measures must be taken to reduce the levels of nutrients and sediments exiting from that watershed. In the sections that follow, we shall discuss major strategies used to control eutrophication.

First, recall the concept of limiting factors (Chapter 2), in which only one nutrient need be lacking to suppress growth. In natural freshwater systems, **phosphorus** is the most common limiting factor. In marine systems (like the coastal Gulf of Mexico), the limiting nutrient is most often **nitrogen.** Both in the environment and in biological systems, phosphorus (P) is present as phosphate ($PO_4^{3-}$), and nitrogen is present in a variety of compounds, most commonly as nitrate ($NO_3^-$) or the ammonium ion ($NH_4^+$). Therefore, we will focus primarily on phosphate and nitrogen compounds.

*Ecoregional Nutrient Criteria.* Beginning in 2001, the EPA began publishing water quality nutrient criteria aimed at preventing and reducing the eutrophication that impacts so many bodies of water. The agency listed its recommended criteria for *causative* factors—*nitrogen* and *phosphorus*—and criteria for *response* factors—*chlorophyll a* as a measure of phytoplankton density and a measure of water clarity. The EPA divided the country into ecoregions and determined criteria levels deemed appropriate for the specific region. For example, ecoregion VIII, which represents the northern Great Lakes and the northeastern states, has the following criteria for lakes and reservoirs: total phosphorus 8.0 $\mu$g/L, total nitrogen 0.24 mg/L, chlorophyll *a* 2.43 $\mu$g/L, and Secchi depth (a measurement of transparency) 4.93 meters. Other ecoregions often have higher criteria levels, reflecting natural levels of these criteria that are different from those of the pristine waters around the northern Great Lakes. Like the other water quality criteria, the nutrient criteria are provided as targets for the states as they address their water pollution (and especially eutrophication) problems.

*Control Strategies for Point Sources.* In heavily populated areas, discharges from sewage-treatment plants are major sources of nutrients entering waterways. The levels of phosphate in effluents from these plants are elevated even more in areas where laundry detergents containing phosphate are used. Phosphate contained in detergents for cleaning purposes goes through the system and out with the discharge. In regions where eutrophication has been identified as a problem, a key step toward prevention has been to ban the sale of phosphate detergents or at least to regulate the maximum allowable level of phosphates. Total or partial phosphate bans are now in effect in 20 states and the District of Columbia. Major detergent manufacturers are shifting to the general production of nonphosphate formulations, although the consumer should read a product's active ingredients. Also, the bans do not cover dishwashing detergents, some brands of which are high in phosphate.

Bans on detergents with phosphate and upgrades of sewage-treatment plants have brought about marked improvements in waterways that were heavily damaged by effluents from the plants. In a sense, however, these are the easy measures, in that the target for correction, the effluent, is an obvious *point source,* and methods for correcting the problem are clear cut. The NPDES permitting process is the basic regulatory tool for reducing point-source pollutants. Under the Clean Water Act, "anyone discharging pollutants from any point source into waters of the U.S. [must] obtain an NPDES permit from EPA or an authorized state." More than 400,000 facilities are now regulated by way of NPDES permits. The permit must be drawn up in the context of the total pollutant sources (point and nonpoint) impacting a watershed or body of water.

***Control Strategies for Nonpoint Sources.*** Correction becomes more difficult, but no less important, when the source is diffuse, as in the case of farm and urban runoff. Remediation of such *nonpoint sources* will involve thousands—even millions—of individual property owners adopting new practices regarding management and the use of fertilizer and other chemicals on their properties. Nevertheless, that is the challenge. When the Clean Water Act was reauthorized in 1987, a new section (Section 319) was added, requiring states to develop management programs to address nonpoint sources of pollution. Thus, there is a legal mandate to address the issues of agricultural and urban runoff.

The EPA has opted to develop regulations via the TMDL program. The concept is straightforward:

- Identify the pollutants responsible for degrading a body of water.
- Estimate the pollution coming from all sources (point and nonpoint).
- Estimate the ability of the body of water to assimilate the pollutants while remaining below the threshold designating poor or inadequate water quality.
- Determine the maximum allowable pollution load.
- Within a margin of safety, allocate the allowable level of pollution among the different sources such that the water quality standards are achieved.

As with most attempts at solving environmental problems, the devil is in the details. The states are responsible for administering TMDL pollution abatement, but the EPA maintains oversight and approval of the results. State administrators have a number of options for managing nonpoint pollution sources, including regulatory programs, technical assistance, financial assistance, education and training, and demonstration projects. By 2002, the states had listed over 25,800 water bodies as being impaired. More than 5,400 were impaired due to nutrient overload. The EPA had approved a total of 9,100 TMDLs submitted by the states to correct the impairments, and closing this workload gap is a high priority for the agency.

***BMPs.*** Reducing or eliminating pollution from nonpoint sources will involve different strategies for different sources. For example, where a significant portion of the watershed is devoted to agricultural activities, the major sources of nutrients and sediments are likely to be (1) erosion and leaching of fertilizer from croplands and (2) runoff of animal wastes from barns and feedlots (Fig. 17–11). All the practices that may be used to minimize such erosion, runoff, and leaching are lumped under a single term, **best management practices (BMPs),** which includes all the methods of soil conservation discussed in Chapter 8. Table 17–2 lists examples of BMPs for a variety of nonpoint sources.

Once control measures have been put in place, the polluted body of water must be monitored to determine whether water quality standards are being attained. Several years of data collection may be necessary to detect genuine trends in quality. Once water quality standards are reached and sustained, the body of water is removed from the list of impaired waters. If standards are not met, the TMDL process must be revisited and new pollution allowances allocated.

**Recovery.** The good news is that cultural eutrophication and other forms of water pollution can be controlled and often reversed, provided that a total watershed management approach is undertaken. For a small lake fed from a modest-sized watershed, the task may be quite straightforward because major sources of nutrients can be identified and addressed. For large bodies of water, such as the Chesapeake Bay—the watershed of which includes some 12,000 square miles (27,000 km$^2$) in five states and the District of Columbia—the task is significantly greater. The Chesapeake suffered from heavy nutrient pollution, with cultural eutrophication destroying all but a tenth of the 600,000 acres of vital sea grasses that once formed the basis of the bay's rich ecosystem. The governmental jurisdictions involved have been cooperating under a Chesapeake Bay program with the common goal of achieving federal clean water standards by 2010. The watershed is divided into the smaller watersheds of each of the tributaries, and each jurisdiction is taking responsibility for reducing the nutrients emanating from its tributaries.

***Lake Washington.*** One remarkable success story is Lake Washington, east of Seattle (Fig. 17–12). This 34-square-mile lake is at the center of a large metropolitan

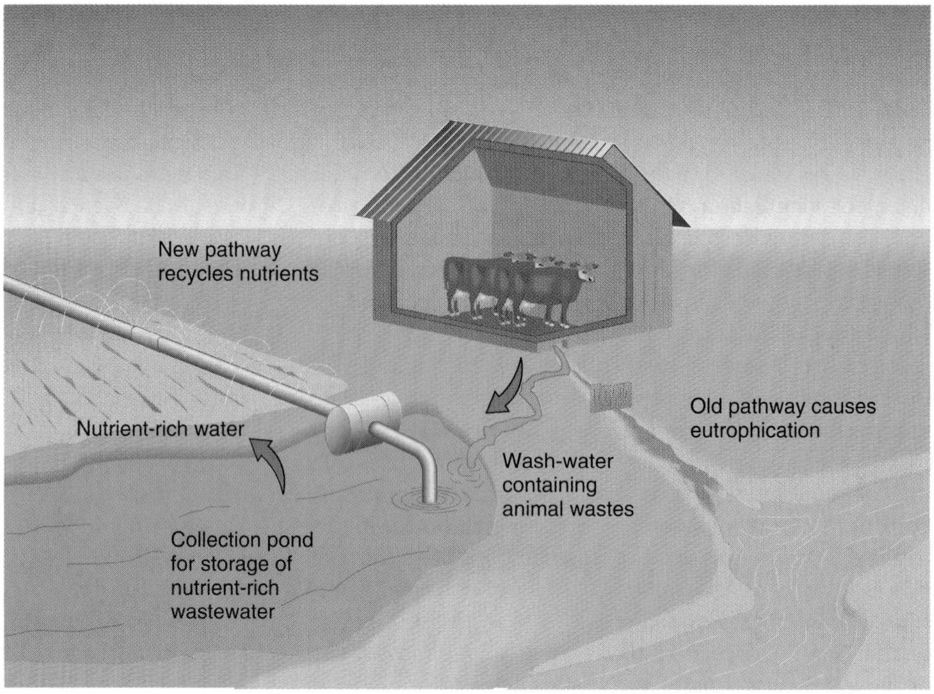


**Figure 17–11 A collection pond for dairy-barn washings.** When washings from animal facilities are flushed directly into natural waterways, they contribute significantly to eutrophication. This situation may be avoided by collecting the flushings in ponds from which both the water and the nutrients may be recycled.

**table 17-2 Best Management Practices for Reducing Pollution from Nonpoint Sources**

| **Agriculture** | **Forestry** |
|---|---|
| Animal waste management | Ground cover maintenance |
| Conservation tillage | Limiting disturbed areas |
| Contour farming | Log removal techniques |
| Strip-cropping | Pesticide and herbicide management |
| Cover crops | Proper management of roads |
| Crop rotation | Removal of debris |
| Fertilizer management | Riparian zone management |
| Integrated pest management | **Mining** |
| Range and pasture management | Underdrains |
| Terraces | Water diversion |
| **Construction** | **Multicategory** |
| Limiting disturbed areas | Buffer strips |
| Nonvegetative soil stabilization | Detention and sedimentation basins |
| Runoff detention and retention | Devices to encourage infiltration |
| **Urban** | Vegetated waterways |
| Flood storage | Interception and diversion of runoff |
| Porous pavements | Sediment traps |
| Runoff detention and retention | Streamside management zones |
| Street cleaning | Vegetative stabilization and mulching |

*Source: Guidance for Water Quality-Based Decisions: The TMDL Process* (USEPA, 1991/EPA440-4-91-001).

**Figure 17–12 Lake Washington.** A testimony to the fact that bodies of water can recover from cultural eutrophication, Lake Washington is now thriving because sources of nutrients enriching the lake have been removed.

area. During the 1940s and 1950s, 11 sewage-treatment plants were sending state-of-the-art treated water into the lake at a rate of 20 million gallons per day. At the same time, phosphate-based detergents came into wide use. The lake responded to the massive input of nutrients by developing unpleasant "blooms" of noxious algae. The water lost its clarity, the desirable fish populations declined, and masses of dead algae accumulated on the shores of the lake.

Citizen concern led to the creation of a system that diverted the treatment-plant effluents into nearby Puget Sound, where tidal flushing would mix them with open-ocean water. The diversion was complete by 1968, and the lake responded quickly. The algal blooms diminished, the water regained its clarity, and, by 1975, recovery was complete. Careful studies by a group of limnologists from the University of Washington showed that phosphate was the culprit. Before the diversion, the lake was receiving 220 tons of phosphate per year from all sources. Afterward, the total dropped to 40 tons per year, well within the lake's normal capacity to absorb phosphate by depositing it in deep bottom sediments. One clear lesson learned was that sewage-treatment effluent must never be allowed to enter a lake unless nutrients are removed as part of the treatment. The major lesson, however, was that eutrophication can be reversed by paying attention to nutrient inputs and addressing the various sources. Lake management, in other words, is a matter of controlling the phosphate loading into a lake from all sources: human and animal sewage, agricultural and yard fertilizers, street runoff, and failing septic systems. Many lakes are now protected by associations of concerned citizens who see themselves as stewards of their lake.

In recent years, the United States and other developed nations, in recognition of the problem of eutrophication, have made substantial progress in developing or modifying sewage-treatment systems that will keep nutrients in a cycle on the land and not discharge them into waterways. But, at the other extreme, particularly in developing countries, much **raw sewage**—completely untreated—is still discharged, creating not only the eutrophication of waterways, but also a significant health hazard.

## 17.3 Sewage Management and Treatment

Before the late 1800s, the general means of disposing of human excrement was the outdoor privy. Seepage from the privy frequently contaminated drinking water and caused disease, especially in places where privies and wells were located near one another. In the late 1800s, Louis Pasteur and other scientists showed that sewage-borne bacteria were responsible for many infectious diseases. This important discovery led to intensive efforts to rid cities of human excrement as expediently as possible. Cities already had drain systems for storm water, but using these systems for human wastes had been prohibited. With the urgency of the situation, however, minds quickly changed. The flush toilet was introduced, and sewers were tapped into storm drains. Thus, Western civilization initiated the one-way flow of flushing sewage wastes into natural waterways.

As a result of this practice, receiving waters that had limited capacity for dilution became open cesspools of foul odors, vermin, and filth as the overload of organic matter depleted dissolved oxygen and all aquatic life suffocated. For increasing distances around or downstream from the sewage outfall, the water became unfit to swim in because of the sewage contamination. Any further use of the water for human consumption required extensive treatment to remove pathogens.

### Development of Sewage Collection and Treatment Systems

To alleviate the problem of sewage-polluted waterways, facilities were designed and constructed to treat the outflow before it entered the receiving waterway. The first sewage-treatment plants in the United States were built around 1900. However, the combined volumes of sewage and storm water soon proved impossible to handle. During heavy rains, wastewater would overflow the treatment plant and carry raw sewage into the receiving waterway. Gradually, regulations were passed requiring developers to install separate systems—**storm drains,** for collecting and draining runoff from precipitation, and **sanitary sewers,** to receive all the wastewater from sinks, tubs, and toilets in homes and other buildings. (Note the distinction between the two terms; it is incorrect to speak of storm drains as sewers.)

The ideal modern system, then, is a system in which all sewage water is collected separately from storm water and is fully treated to remove all pollutants before the water finally is released. But progress toward this goal has been extremely uneven. Up through the 1970s, even in the United States and other developed countries, thousands of communities still discharged untreated sewage directly into waterways, and for many more the degree of treatment was minimal. Indeed, the increasing sewage pollution of waterways and beaches was the major impetus behind the passage of the Clean Water Act of 1972.

In the meantime, much of the developing world still exists in the most primitive stage of sewage treatment. Innumerable poor villages and poor areas of mushrooming cities of the developing world lack sewer systems for collecting wastewater. (Only about 10% of wastewater in the developing countries is collected.) It is not uncommon to find raw sewage littering the ground and overflowing gutters and streams. Even where flush toilets and collection systems exist, only a small percentage of the treatment plants operate effectively. Many of the people living in these regions must use these badly polluted waters for bathing, laundering, and even drinking (Fig. 17–4).

Before addressing methods of treating sewage, let us clarify which contaminants and pollutants we are talking about.

## The Pollutants in Raw Sewage

Raw sewage is not only the flushings from toilets; it is also the collection of wastewater from all other drains in homes and other buildings. A sewer system brings all tub, sink, and toilet drains from all homes and buildings together into larger and larger sewer pipes, just as the branches of a tree eventually come together into the trunk. This total mixture collected from all drains, which flows from the end of the "trunk" of the collection system, is called **raw sewage** or **raw wastewater.** Because we use such large amounts of water to flush away small amounts of dirt, especially as we stand under the shower or often just run the water, most of what goes down sewer drains is water. Raw sewage is about 1,000 parts water for every 1 part waste—99.9% water to 0.1% waste.

Given our voluminous use of water, the quantity of raw-sewage output is on the order of 150–200 gallons (600–800 liters) per person per day. That is, a community of 10,000 persons will produce on the order of 1.5 to 2.0 million gallons (6–8 million liters) of wastewater each day. With the addition of storm water, raw sewage is diluted still more. Nevertheless, the pollutants are sufficient to make the water brown and smell foul.

*Types of Pollutants.* The pollutants generally are divided into the following four categories, which correspond to the techniques used to remove them:

1. Debris and grit: rags, plastic bags, coarse sand, gravel, and other objects flushed down toilets or washed through storm drains.
2. Particulate organic material: fecal matter, food wastes from garbage-disposal units, toilet paper, and other matter that tends to settle in still water.
3. Colloidal and dissolved organic material: very fine particles of particulate organic material, bacteria, urine, soaps, detergents, and other cleaning agents.
4. Dissolved inorganic material: nitrogen, phosphorus, and other nutrients from excretory wastes and detergents.

In addition to the four categories of pollutants in "standard" raw sewage, variable amounts of pesticides, heavy metals, and other toxic compounds may be found in sewage because people pour unused portions of products containing such materials down sink, tub, or toilet drains. Also, industries may discharge various toxic wastes into sewers.

## Removing the Pollutants from Sewage

The challenge of treating sewage is more than installing a technology that will do the job; it is also finding one that will do the job *at a reasonable cost.* A diagram of wastewater treatment procedures is shown in Figure 17–13.

### Preliminary Treatment (Removal of Debris and Grit).

Because debris and grit will damage or clog pumps and later treatment processes, removing them is a necessary first step and is called **preliminary treatment.** Usually, preliminary treatment involves two steps: the screening out of debris and the settling of grit. Debris is removed by letting raw sewage flow through a **bar screen,** a row of bars mounted about 1 inch (2.5 cm) apart. Debris is mechanically raked from the screen and taken to an incinerator. After passing through the screen, the water flows through a **grit chamber**—a swimming-pool-like tank—in which its velocity is slowed just enough to permit the grit to settle. The settled grit is mechanically removed from these tanks and taken to landfills.

**Primary Treatment (Removal of Particulate Organic Material).** After preliminary treatment, the water moves on to **primary treatment,** where it flows very slowly through large tanks called **primary clarifiers.** Because its flow is slow, the water is nearly motionless for several hours. The particulate organic material, about 30% to 50% of the total organic material, settles to the bottom, where it can be removed. At the same time, fatty or oily material floats to the top, where it is skimmed from the surface. All the material that is removed, both particulate organic material and fatty material, is combined into what is referred to as **raw sludge,** which must be treated separately.

Primary treatment involves nothing more complicated than putting polluted water into a "bucket," letting

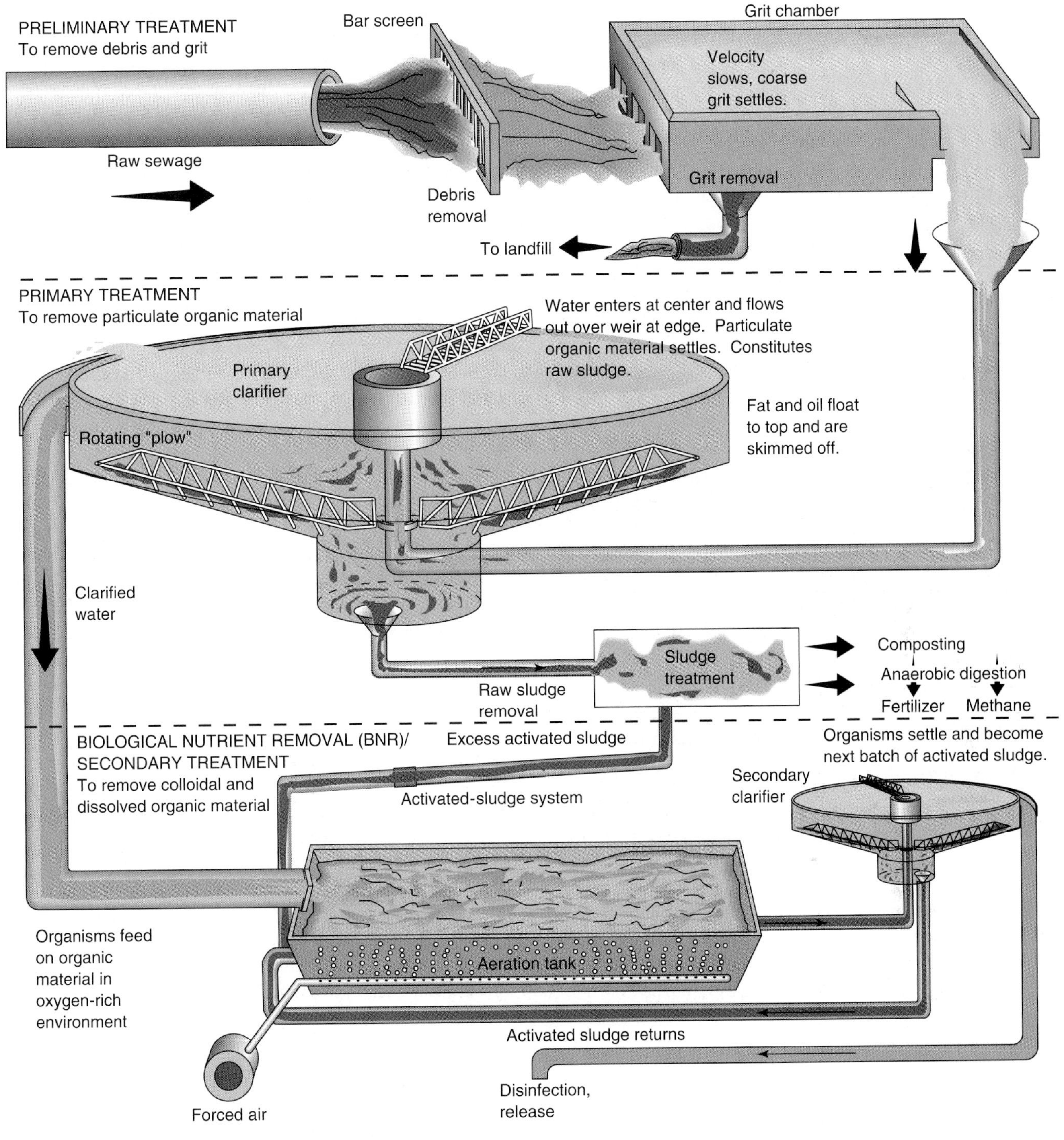


**Figure 17–13 A diagram of wastewater treatment.** Raw sewage moves from the grit chamber to undergo primary treatment, in which sludge is removed. The clarified water then proceeds to undergo secondary treatment (shown here as activated-sludge treatment).

material settle, and pouring off the water. Nevertheless, such treatment removes much of the particulate organic matter at minimal cost.

**Secondary Treatment (Removal of Colloidal and Dissolved Organic Material).** Secondary treatment is also called **biological treatment,** because it uses organisms—natural decomposers and detritus feeders. Basically, an environment is created that enables these organisms to feed on the colloidal and dissolved organic material and break it down to carbon dioxide and water via their cell respiration. The sewage water from primary treatment is a food- and water-rich medium for the decomposers and detritus feeders. The only thing that needs to be added to the water is oxygen to enhance the organisms' respiration and growth. Either of two systems may be used to add oxygen to the water: a *trickling-filter system* or an *activated-sludge system.* In the **trickling-filter**

(a)

(b)

**Figure 17–14 Trickling filters for secondary treatment.** (a) The water from primary clarifiers is sprinkled onto, and trickles through, a bed of rocks 6–8 feet deep. (b) Various bacteria and other detritus feeders adhering to the rocks consume and digest organic matter as it trickles by in the water, which is collected at the bottom of the filters and goes on for final clarification and disinfection.

**system,** the water exiting from primary treatment is sprinkled onto, and allowed to percolate through, a bed of fist-sized rocks 6–8 feet (2–3 m) deep (Fig. 17–14). The spaces between the rocks provide good aeration. Like a natural stream, this environment supports a complex food web of bacteria, protozoans, rotifers (organisms that consume protozoans), various small worms, and other detritus feeders attached to the rocks. The organic material in the water, including pathogenic organisms, is absorbed and digested by decomposers and detritus feeders as it trickles by.

The **activated-sludge system** (Fig. 17–13, bottom) is the most common secondary-treatment system. Water from primary treatment enters a large tank that is equipped with an air-bubbling system or a rapidly churning system of paddles. A mixture of detritus-feeding organisms, referred to as **activated sludge,** is added to the water as it enters the tank, and the water is vigorously aerated as it moves through the tank. Organisms in this well-aerated environment reduce the biomass of organic material, including pathogens, as they feed. As the organisms feed on each other, they tend to form into clumps, called *floc,* that settle readily when the water is stilled. Thus, from the aeration tank, the water is passed into a secondary **clarifier tank,** where the organisms settle out and the water—now with more than 90% of all the organic material removed—flows on. The settled organisms are pumped back into the aeration tank. (They are the activated sludge that is added at the beginning of the process.) Surplus amounts of activated sludge, which occur as populations of organisms grow, are removed and added to the raw sludge. The organic material is oxidized by microbes to form carbon dioxide, water, and mineral nutrients that remain in the water solution.

**Biological Nutrient Removal (Removal of Dissolved Inorganic Material).** The original secondary-treatment systems were designed and operated in a manner to simply maximize biological digestion, because the elimination of organic material and its resulting BOD was considered the prime objective. Ending up with a nutrient-rich discharge was not recognized as a problem. Today, with increased awareness of the problem of cultural eutrophication, secondary activated-sludge systems have been added and are being modified and operated in a manner that both removes nutrients and oxidizes detritus, in a process known as **biological nutrient removal (BNR).**

***Nitrogen.*** Recall that, in the natural nitrogen cycle (Fig. 3–19), various bacteria convert nutrient forms of nitrogen (ammonia and nitrate) back to nonnutritive nitrogen gas in the atmosphere through **denitrification.** For the biological removal of nitrogen, then, the activated-sludge system is partitioned into zones, and the environment in each zone is controlled in a manner that promotes the denitrifying process (Fig. 17–15).

***Phosphorus.*** In an environment that is rich in oxygen, but relatively lacking in food—the environment of zone 3—bacteria take up phosphate from solution and store it in their bodies. Thus, phosphate is removed as the excess organisms are removed from the system. These organisms, together with the phosphate they contain, are added to, and treated with, the raw sludge, ultimately producing a more nutrient-rich treated-sludge product.

Various chemical treatments are often used as an alternative to BNR. One such process is simply to pass the effluent from standard secondary treatment through a filter of lime, which causes the phosphate to precipitate out as insoluble calcium phosphate. Another is to treat the effluent with ferric chloride, which produces insoluble ferric phosphate, or with an organic polymer, which

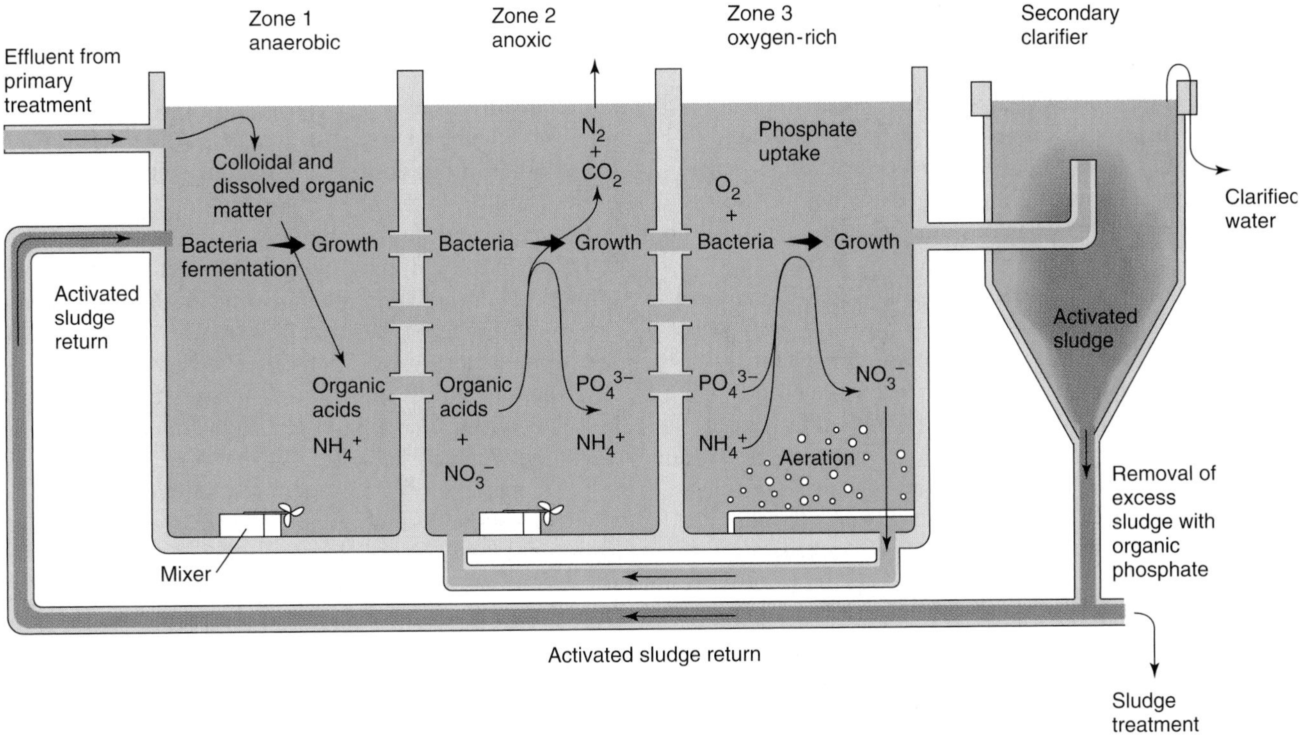


**Figure 17–15 Biological nutrient removal (BNR).** The secondary treatment—the activated-sludge process—may be modified to remove nitrogen and phosphate while at the same time breaking down organic matter. For this BNR process, the aeration tank is partitioned into three zones, only the third of which is aerated. As seen in the diagram, ammonium ($NH_4^+$;) is converted to nitrate ($NO_3^-$;) in zone 3. Recycled to zone 2, which is without oxygen gas (anoxic), the nitrate supplies the oxygen for cell respiration. In the process, the nitrate is converted to nitrogen gas and is released into the atmosphere. Phosphate is taken up by bacteria in zone 3 and removed with the excess sludge.

gives rise to a floc. With these methods, most of the phosphorus is removed along with the sludge.

**Final Cleansing and Disinfection.** With or without BNR, the wastewater is subjected to a final clarification and disinfection (Fig. 17–13, bottom). Although few pathogens survive the combined stages of treatment, public-health rigors still demand that the water be disinfected before being discharged into natural waterways. The most widely used disinfecting agent is chlorine gas, because it is both effective and relatively inexpensive. But this treatment also introduces chlorine into natural waterways, and even minute levels of chlorine can harm aquatic animals. Also, chlorine gas is dangerous to work with. More commonly, sodium hypochlorite (Clorox®) provides a safer way of adding chlorine to achieve disinfection.

Other disinfecting techniques are coming into use as well. One alternative disinfecting agent is ozone gas, which kills microorganisms and breaks down to oxygen gas, improving water quality in the process. Ozone is unstable and hence explosive, however, so it must be generated at the point of use, a step that demands considerable capital investment and energy. Another disinfection technique is to pass the effluent through an array of ultraviolet lights mounted in the water. (Standard fluorescent lights without the white coating emit ultraviolet light.) The ultraviolet radiation kills microorganisms, but does not otherwise affect the water.

***Final Effluent.*** After these treatment steps, the wastewater has a lower organic and nutrient content than many bodies of water into which it is discharged. BOD values are commonly 10 to 20 ppm, as opposed to 200 ppm in the incoming sewage. In other words, discharging the wastewater may actually contribute toward improving water quality in the receiving body. In water-short areas, there is every reason to believe that the treated water itself, with little additional treatment, could be recycled into the municipal water-supply system.

Bear in mind that the techniques we have described here represent the state of the art. Many cities, even in the developed world, are still operating with antiquated systems that provide lower quality treatment. A few coastal cities still persist in discharging sewage that has received only primary treatment directly into the ocean, although in the United States this practice has almost disappeared.

## Treatment of Sludge

Recall that the particulate organic matter that settles out or floats to the surface of sewage water in primary treatment forms the bulk of *raw sludge,* although the sewage also contains excesses from activated sludge and BNR systems. Raw sludge is a gray, foul-smelling, syrupy liquid with a water content of 97% to 98%. Pathogens are certain to be present in raw sludge because it includes material directly from toilets. Indeed, raw sludge is considered to be a biologically hazardous material. However, as nutrient-rich organic material, it has the potential to be used as organic fertilizer if it is suitably treated to kill pathogens and if it does not contain other toxic contaminants.

Three commonly used methods for treating sludge and converting it into organic fertilizer are *anaerobic digestion, composting,* and *pasteurization.* Since this is a developing industry, it is unclear which method will prove most cost effective and environmentally acceptable over the long run. Also, none of these methods is capable of removing toxic substances such as heavy metals and nonbiodegradable synthetic organic compounds. The presence of such toxins can preclude the use of sludge as fertilizer.

**Anaerobic Digestion.** **Anaerobic digestion** is a process of allowing bacteria to feed on the detritus *in the absence of oxygen.* The raw sludge is put into large airtight tanks called **sludge digesters** (Fig. 17–16). In the absence of oxygen, a consortium of anaerobic bacteria breaks down the organic matter. The end products of this decomposition are carbon dioxide, methane, and water. Thus, a major by-product of anaerobic processes is **biogas,** a gaseous mixture that is about two-thirds *methane.* The other one-third is made up of carbon dioxide and various foul-smelling organic compounds that give sewage its characteristic odor. Because of its methane content, biogas is flammable and can be used for fuel. In fact, it is commonly collected and burned to heat the sludge digesters, because the bacteria working on the sludge do best when maintained at about 104°F (38°C).

After four to six weeks, anaerobic digestion is more or less complete, and what remains is called **treated sludge** or **biosolids,** consisting of the remaining organic matter, which is now a relatively stable, nutrient-rich, humuslike material suspended in water. Pathogens have been largely, if not entirely, eliminated, so they no longer present any significant health hazard.

Such treated sludge makes an excellent organic fertilizer that may be applied directly to lawns and agricultural fields in the liquid state in which it comes from the digesters, providing the benefit of both the humus and the nutrient-rich water. Alternatively, the sludge may be dewatered by means of belt presses, whereby the sludge is passed between rollers that squeeze out most of the water (Fig. 17–17a) and leave the organic material as a semisolid **sludge cake** (Fig. 17–17b) that, after disinfection, is easy to stockpile, distribute, and spread on fields with traditional manure spreaders.

**Composting.** Another process sometimes used to treat sewage sludge is **composting.** Raw sludge is mixed with wood chips or some other water-absorbing material to reduce the water content. It is then placed in **windrows**—long, narrow piles that allow air to circulate conveniently through the material and that can be turned with machinery. Bacteria and other decomposers break down the organic material to rich humuslike material that makes an excellent treatment for soil.

**Pasteurization.** After the raw sludge is dewatered, the resulting sludge cake may be put through ovens that operate like oversized laundry dryers. In the dryers, the sludge is **pasteurized**—that is, heated sufficiently to kill any pathogens (exactly the same process that makes milk safe to drink). The product is dry, odorless organic pellets. Milwaukee, Wisconsin, which has a particularly

**Figure 17–16 Anaerobic sludge digesters.** In these tanks, the bacterial digestion of raw sludge in the absence of oxygen leads to the production of methane gas, which is tapped from the tops of the tanks, and humuslike organic matter that can be used as a soil conditioner. The egglike shape of the tanks facilitates mixing and digestion.

(a)

(b)

**Figure 17–17 Dewatering treated sludge.** (a) Sludge (98% water) may be dewatered by means of belt presses, as shown here. The liquid sludge is run between canvas belts going over and between rollers, so that much of the water is pressed out. (b) The resulting "sludge cake" is a semisolid humuslike material that may be used as an organic fertilizer.

rich sludge resulting from the brewing industry, has been using this process for over 60 years. The city bags and sells the pellets throughout the country as an organic fertilizer under the trade name Milorganite®.

## Alternative Treatment Systems

**On-site Wastewater Treatment Systems.** Despite the expansion of sewage collection systems, many homes in rural and suburban areas lie outside the reach of a municipal system. For these homes, on-site treatment systems are required. Currently, 25% of the U.S. population is served by such systems. The traditional and still most common on-site system is the *septic tank and leaching field* (Fig. 17–18). Wastewater flows into the tank, where particulate organic material settles to the bottom. The tank acts like a primary clarifier in a municipal system. Water containing colloidal and dissolved organic material, as well as dissolved nutrients, flows into the drain field and gradually percolates into the soil. Organic material that settles in the tank is digested by bacteria, but accumulations still must be pumped out every three to five years. Soil bacteria decompose the colloidal and dissolved organic material that comes through the drain field. Some people establish successful vegetable gardens over septic drain fields, thus instantiating the sound principle of recycling the nutrients. Prerequisites for this traditional system are suitable land area for the drain field and subsoil that allows sufficient percolation of water.

Unfortunately, the on-site systems frequently fail, resulting in unpleasant sewage backup into homes and pollution of groundwater and surface waters. Homeowners are often unaware of how the systems work, and most local regulatory programs do not require the homeowner to be accountable for them. Substantial nutrient and pathogen pollution of groundwater and surface waters is traced to failed on-site systems. For this reason, the EPA

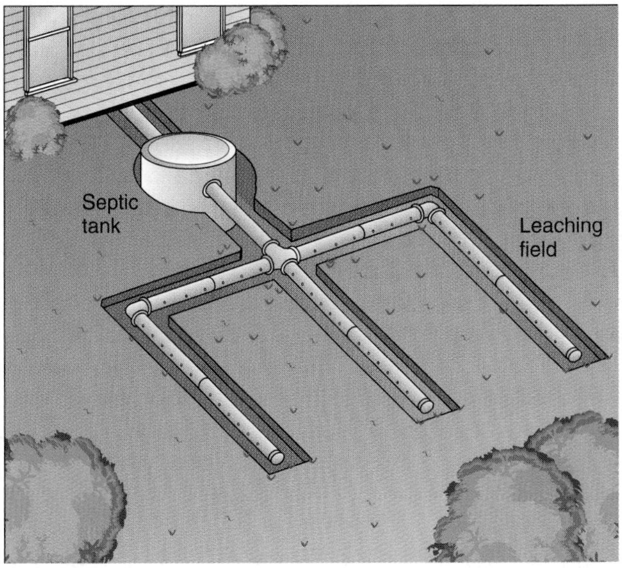


**Figure 17–18 Septic tank treatment.** Sewage treatment for a private home, using a septic tank and drain field. Normally, the pipes and the tank are buried underground. They are shown uncovered here only for illustration. (*Source:* USDA, Soil Conservation Service.)

has made several resources available to local and state agencies responsible for maintaining water quality: voluntary national guidelines, a handbook for management, and a manual for on-site wastewater systems.

*Septic System Primer.* The following are some simple suggestions for successful septic system maintenance:

1. Use caution in disposing materials down the drain. Do not dump coffee grounds, cat litter, diapers, grease, feminine hygiene products, pesticides, paint, gasoline, household chemicals, or other products that could either fill up or clog the septic tank or kill the bacteria that make the system work.

2. Maintain your system by having it inspected regularly and pumped out at least every five years and more frequently if you have a larger family. Keep records of maintenance, and have a map of the location of your septic tank and the leaching field.
3. If you have a garbage disposal unit, consider disabling it; the added materials will shorten the life of the system and force more frequent pumping out.
4. Keep heavy equipment and vehicles off your system and leaching field.

**Using Effluents for Irrigation.** The nutrient-rich water coming from the standard secondary-treatment process is beneficial for growing plants. The problem is that we don't want to put that water into waterways, where it will stimulate the growth of undesirable algae. But why not use it for irrigating plants we do want to grow? This is a way of completing the nutrient cycle. Indeed, the concept has been put into practice in a considerable number of locations as an alternative to upgrading the treatment to remove nutrients. Again, effluents must not be contaminated with toxic materials.

To illustrate, the nutrient-rich effluent from standard secondary treatment from St. Petersburg, Florida, was causing cultural eutrophication in Tampa Bay. Now St. Petersburg uses the effluent to irrigate 4,000 acres (1,600 ha) of urban open space, from parks and residential lawns to a golf course. Revenues from the water sales help offset operating costs. Similarly, Bakersfield, California, receives a $30,000 annual income from a 5,000-acre (2,000-ha) farm irrigated with its treated effluent. And Clayton County, Georgia, is irrigating 2,500 acres (1,000 ha) of woodland with partially treated sewage. Hundreds of other similar projects are underway around the country.

***Bad Idea.*** A number of developing countries irrigate croplands with raw (untreated) sewage effluents. This practice results in getting the bad with the good. The crops respond well, but parasites and disease organisms can easily be transferred to farmworkers and consumers. Therefore, it is important to emphasize that only *treated* effluents should be used for irrigation.

**Reconstructed Wetland Systems.** In treating wastewater, it is also possible to make use of the nutrient-absorbing capacity of wetlands in suitable areas and under suitable climatic conditions. The project may be part of a wetlands recovery program, or artificial wetlands may be constructed.

In the 1960s and 1970s, for example, much of the land around Orlando, Florida, which was originally wetlands, was drained and converted to cattle pasture. At the time, Orlando was discharging 13 million gallons (50 million liters) per day of nutrient-rich effluent into the James River following secondary treatment. Through the Orlando Easterly Wetlands Reclamation Project, 1,200 acres (480 ha) of pastureland was converted back to wetlands. The project involved scooping soil from, and building berms (mounds of earth) around, pastures to create a chain of shallow lakes and ponds. In addition, 1.2 million wetland plants ranging from bulrushes and cattails to various trees were planted. The effluent entering the upper end now percolates through the wetland for about 30 days before entering the James River virtually pure. Thus, the project has recreated a wildlife habitat. Wetland systems can be designed for small as well as large areas and are becoming an increasingly popular alternative for small communities. The key to success for such systems is to ensure that they are kept in balance and not loaded beyond their ability to handle inputs.

## 17.4 Public Policy

In the United States, the responsibility for overseeing the health of the nation's waters rests with the EPA, which, however, can develop regulations only if Congress gives it the authority to do so. Hence, the foundation for public policy must be the laws passed by Congress. Some major legislative milestones for protecting the nation's waters are listed in Table 17–3. The landmark legislation is the **Clean Water Act (CWA) of 1972,** which gave the EPA jurisdiction over, and for the first time required permits for, all point-source discharges of pollutants. This act and subsequent amendments provided billions of dollars to help cities build treatment plants to meet the federal requirement for secondary treatment of all sewage. As the table shows, the 1987 amendments to the CWA established a revolving loan fund—the **Clean Water State Revolving Fund (SRF)** program—to replace the direct-grants program. To build treatment facilities, local governments borrow at low interest rates, and as they repay the loans, the funds received are used for more loans. The fund may also be used to control nonpoint-source pollution. To date, over $38 billion in SRF loans have been made by the states. However, the EPA estimates that an additional $148 billion will be needed over the next 20 years to meet funding needs for all eligible municipal wastewater treatment systems. This amount represents a gap of about $6 billion between current annual expenditures and projected needs.

***Reauthorization.*** Reauthorization of the Clean Water Act is long overdue. Congress has been hung up on a debate over whether requirements should be strengthened or weakened, whether additional mandates should be subjected to a cost–benefit analysis, and how regulatory relief should be provided to industries, states, cities, and individuals required to take actions to comply with the regulations. In particular, Congress has been unsure of how to deal with the TMDL program. Some piecemeal legislation has dealt with water quality issues, reauthorizing funding for several existing CWA programs.

**table 17-3 Legislative Milestones in Protecting Natural Waters and Water Supplies**

| |
|---|
| **1899—Rivers and Harbors Act:** First federal legislation protecting the nation's waters in order to promote commerce. |
| **1948—Water Pollution Control Act:** The federal government provides technical assistance and funds to state and local governments to promote efforts to protect water quality. |
| **1972—The Clean Water Act:** Legislation establishing a comprehensive federal program to achieve the goal of protecting and restoring the physical, chemical, and biological integrity of the nation's waters. Requires permits for any discharge of pollution, strengthens water quality standards, and encourages the use of the best achievable pollution control technology. Provided billions of dollars for construction of sewage-treatment plants. |
| **1972—Marine Protection, Research, and Sanctuaries Act:** Prevents unacceptable dumping in oceans. |
| **1974—Safe Drinking Water Act:** Authorizes the EPA to regulate the quality and safety of public drinking-water supplies, maintain drinking-water standards for numerous contaminants, and set requirements for the chemical and physical treatment of drinking water. |
| **1977—Clean Water Act amendments:** Strengthen controls on toxic pollutants and allow states to assume responsibility for federal programs. |
| **1987—Water Quality Act:** These major amendments to the Clean Water Act (1) create a revolving loan fund to provide ongoing support for the construction of treatment plants, (2) address regional pollution with a watershed approach, and (3) address nonpoint source pollution in Section 319 of the act, with requirements for states to assess the problem and develop and implement plans for dealing with it. The act makes available $400 million in grants to carry out its provisions. |
| **1996—Safe Drinking Water Act amendments:** These amendments impose numerous requirements for managing water supplies, establish a revolving loan fund to help municipalities update their water system infrastructure, and provide flexibility for the EPA to base its selection of contaminants on health risk and costs and benefits. |

***Problems and Progress.*** The EPA has identified nonpoint-source pollution as the nation's number-one water pollution problem, with the construction of new wastewater treatment facilities not far behind. Other significant issues, including storm-water discharges, combined and separate sewer overflows, wetlands protection, and animal feeding operations, are also receiving the EPA's attention and are in Congress' sights for possible action. The continuing high percentage of the nation's water resources that fail to meet the water quality standards makes it clear that much remains to be accomplished. According to the U.S. Public Interest Research Group, four out of five wastewater treatment plants are polluting waterways above levels that their NPDES permits allow.

Nevertheless, much progress has been made in the 30 years since the enactment of the Clean Water Act. The number of people in the United States served by adequate sewage treatment plants has more than doubled, from 85 million to 173 million. Soil erosion has been reduced by 1 billion tons annually, and two-thirds of the nation's waterways are safe for fishing and swimming, double the number in 1972. Many of the nation's most heavily used rivers, lakes, and bays including the Androscoggin River in Maine, Boston Harbor, the upper Mississippi River, the South Platte River, the upper Arkansas River, Lake Erie, the Illinois River, and the Delaware River, have been cleaned up and restored. Fish now swim in rivers once so polluted that only bacteria and sludge worms could survive. Significantly, a 70% increase in bottom vegetation has been achieved since the mid-1980s in the Chesapeake Bay. A national sense of stewardship has been applied to the rivers, lakes, and bays that are our heritage from a previous generation, and public policy has been enacted and billions of dollars spent to bring our waters back from a polluted condition.

***Developing Countries.*** The picture in developing countries is far less encouraging. A 2003 World Bank analysis suggests that, in order to meet the MDGs, about 1.5 billion people must be provided access to safe drinking water and 2 billion people will require basic sanitation facilities. (These estimates do not even include the enormous need for treating sewage wastes adequately). *Recall that satisfying the MDGs only means reducing by half the proportion of people having these needs.* The analysis found that only about 25% of the developing countries are likely to meet these goals by 2015. An estimated $30 billion a year will be required, and that represents a doubling of current investments in these sectors. Policy needs go beyond simply financing for building treatment facilities. Since it is the poorest in the developing countries who are the majority of those unserved, policies in those countries must focus on meeting their needs within existing capacities for water and sanitation. Further, there is a substantial need for knowledge sharing, to educate consumers about good hygiene habits and to inform local authorities about providing sanitation services and clean water. Finally, the international community must scale up its support of development assistance targeted toward meeting the MDGs, as well as expanding public–private partnerships for meeting the needs.

# revisiting the themes

## Sustainability

In ecosystems, sustainability is accomplished when wastes are broken down and the nutrients they contain are recycled. Although recycling the nutrients is not always achievable in modern societies, handling wastes responsibly is one way we can preserve an environment that will meet the needs of future generations. One excellent approach is to aim at avoiding dangerous pollution, by shifting from pollution-intensive economic production to environmentally benign processes. Another approach is to aim at accomplishing the MDGs and help the developing countries bring under control the waterborne diseases that still cause much mortality. Until they do, sustainable development will be impossible.

## Stewardship

Stewardship means intentionally caring for people and the natural world. The public-health measures put in place in the last two centuries represent the stewardship of human resources. Also, as we divert public and private resources to cleaning up our waterways, we are acting as good stewards in the best sense of the concept. Lakes and watersheds often have people and associations which have provided leadership in correcting the problems that lead to eutrophication. The Lake Washington story is an excellent example. Stewardship also means caring for our neighbors' needs, and nowhere are those needs more apparent than in the need for clean water and sanitation in so many developing countries.

## Sound Science

Sound scientific research and communication was essential in establishing the basic knowledge about pollutants and their impacts on human health and the environment. It continues to be highly important as the EPA carries out its research and regulatory activities, administering the Clean Water Act and the Safe Drinking Water Act. The criteria pollutant standards for natural waters and maximum contaminant standards for drinking waters must be based on accurate risk assessments as the EPA defends its regulatory activities.

## Ecosystem Capital

Our natural waters represent an enormous bank of ecosystem capital, providing essential goods and services to people everywhere. The opening story of the dead zone demonstrates how it is possible to degrade an enormous area of the Gulf of Mexico by nutrient overload. Unfortunately, a similar situation exists in the Chesapeake Bay and many other bodies of water. The result is a significant decline in the fish and shellfish we have traditionally harvested from those waters. These are just a couple of the more spectacular examples. The more important battlegrounds are the thousands of smaller lakes, rivers, and coastal estuaries that are also suffering from human impacts.

## Policy and Politics

The EPA is able to carry out activities only as it is empowered by the Congress, which writes the basic laws. The Clean Water Act and its amendments represent the latest in a series of laws that have as their intent protecting and restoring the integrity of the nation's waters. The Safe Drinking Water Act and its amendments are meant to ensure that Americans have tap water that is safe to drink at all times. The details of these laws are voluminous. Log on to the Web site for the EPA Office of Water to see how far reaching the EPA's activities and authority are. The story of arsenic is a good example of how closely most of the water quality criteria are scrutinized and guarded by NGOs and others. Perhaps because this task is so complex, Congress seems to be paralyzed into inaction as it faces the need to reauthorize the Clean Water Act. Nevertheless, more needs to be done, especially at the state level, to improve our national water quality, and the Congress can play an important role as it establishes funding for the states to carry out its mandates. Many states are behind in assessing water quality, and the need for updating sewage treatment systems is enormous.

## Globalization

A sense of the global reach of water pollution is difficult to imagine, yet billions of people in the developing countries are held back from achieving their physical and intellectual potential because of the terrible effects of polluted water and lack of sanitation. The disastrous situation affects social, economic, and political affairs in the developing countries, and since the world is now one global economy, it also affects the developed world. At the very least, there is the continuing need for development aid, yet there is always too little aid to make the problems go away. How many people in the United States have even heard of the MDGs? There are calls for abolishing the United Nations—some from our own representatives—and the ignorance and callous disregard they reflect are part of the shame of our country's hubris as we play the role of superpower.

## review questions

1. Define *pollution, pollutant, nonbiodegradable, point source,* and *nonpoint source* of pollutants.
2. Discuss each of the following categories of water pollutants and the problems they cause: pathogens, organic wastes, chemical pollutants, and sediments.
3. How are water quality standards determined? Distinguish between water quality criteria pollutants and maximum contaminant levels.
4. Describe and compare submerged aquatic vegetation and phytoplankton. Where and how does each get nutrients and light?
5. Explain the difference between oligotrophic and eutrophic waters. Describe the sequential process of eutrophication.
6. Distinguish between natural and cultural eutrophication.
7. What is being done to establish nutrient criteria?
8. How does the NPDES program address point-source pollution by nutrients?
9. Describe the TMDL program. How does it address nonpoint-source pollution, and what role do water quality criteria play in the program?
10. Give a brief history of how humans' handling of sewage wastes has changed as the risks and potential benefits have become better understood.
11. Name and describe the facility and the process used to remove debris, grit, particulate organic matter, colloidal and dissolved organic matter, and dissolved nutrients from wastewater.
12. Why is secondary treatment also called biological treatment? What is the principle involved? What are the two alternative techniques used?
13. What are the principles involved in, and what is accomplished by the removal of biological nutrients from waste? Where do nitrogen and phosphate go in the process?
14. Name and describe three methods of treating raw sludge, and give the end product(s) that may be produced from each method.
15. How may sewage from individual homes be handled in the absence of municipal collection systems? What are some problems with these on-site systems?
16. What are some of the important issues relating to water quality in public policy?

## thinking environmentally

1. A large number of fish are suddenly found floating dead on a lake. You are called in to investigate the problem. You find an abundance of phytoplankton and no evidence of toxic dumping. Suggest a reason for the fish kill.
2. A number of regions in the United States have banned the use of phosphate-containing detergents in recent years. What harm is caused by such detergents, and what is hoped to be achieved by the bans?
3. Nitrogen pollution has damaged the San Diego Creek–Newport Bay ecosystem in California by encouraging the heavy growth of intertidal algae. Report on the TMDL for this system. Consult the EPA Web site: www.epa.gov/owow/tmdl/examples/nutrients.html.
4. Arrange a tour to the sewage-treatment plant that serves your community. Compare it with what is described in this chapter. Is the water being purified or handled in a way that will prevent cultural eutrophication? Are sludges being converted to, and used as, fertilizer? What improvements, if any, are in order? How can you help promote such improvements?
5. Suppose a new community of several thousand people is going to be built in Arizona (with a warm, dry climate). You are called in as a consultant to design a complete sewage system, including the collection, treatment, and use or disposal of by-products. Write an essay describing the system you recommend, and give a rationale for the choices involved.

chapter 18

# Municipal Solid Waste: Disposal and Recovery

**Key Topics**

1. The Solid-Waste Problem
2. Solutions to the Solid-Waste Problem
3. Public Policy and Waste Management

Danehy Park features four soccer fields, four softball fields, three basketball courts, a baseball diamond, two playgrounds, and a 2 mile jogging trail. The 55-acre park, opened in 1990, is located close to a heavily populated area of North Cambridge, Massachusetts, and is in almost constant use in good weather (Fig. 18–1). The city of Cambridge sponsors annual family days in September that draw over 4,000 people to the park, and nearby upscale Walden Park Apartments features Danehy Park as one of its selling points. An unusual feature of the park is a big red light in the public rest room that warns users to vacate the park if the light goes on. The light is tied into an elaborate venting system that prevents methane from building up below ground. The system is necessary because Danehy Park is built on the former city dump. Aside from the presence of the light and a few minor settling problems that have interfered with water drainage on the soccer fields, a newcomer would never know that this was once a blight on the neighborhood—an open, burning dump in the 1950s and then a "sanitary landfill" that was closed in 1972. The park increased Cambridge's open space by 20% when it was created.

Danehy Park is a lesson in land use that is being learned only slowly in the United States—and, for that matter, in the rest of the world. The lesson is that if we could picture more effectively what 50 years can bring in the way of change, we could solve one of the most contentious problems facing local communities: what to do with municipal waste. As old dumps and landfills were closed because of environmental concerns, they created a temporary problem identified as the "solid-waste crisis" in the 1970s and 1980s. Now many of those dumps and landfills are being converted into parks, golf courses, and nature preserves, and future generations will hardly suspect that their recreational site was once a landfill.

***Still a Crisis?*** In some ways, the solid-waste crisis is still with us, as we will see when we examine the problem of interstate trash movement. It is commonly said that we are running out of space to put our trash and garbage. Perhaps we are, but if so, it is because we are happy to purchase the goods displayed so prominently in our malls and advertised in the media, but we are reluctant to

***A Tired Landscape*** **Scrap tires as far as the eye can see, accumulated by a private contractor in Smithfield, Rhode Island.**

**Figure 18–1 From landfills to playing fields.** Thomas W. Danehy Park in Cambridge, Massachusetts, a former landfill that has been recycled into a recreational park. In addition to spotlighting playing fields, the park features a half-mile "glassphalt" pathway (built with recycled glass and asphalt), shown in the foreground.

accept the consequences of getting rid of them responsibly.

This chapter is about solid-waste issues. We examine current patterns of disposal—landfills, combustion, and recycling—and look for sustainable solutions to our solid-waste problems. The ideal would be to imitate natural ecosystems and reuse everything. Some solutions do well in conforming to this principle, whereas others do not—and possibly cannot.

## 18.1 The Solid-Waste Problem

The focus of this chapter is **municipal solid waste** (MSW), defined as the total of all the materials (commonly called trash, refuse, or garbage) thrown away from homes and commercial establishments and collected by local governments. MSW is different from hazardous waste (covered in Chapter 19) and nonhazardous industrial waste. The latter is no small matter: Industrial facilities generate and manage 7.6 billion tons of nonhazardous industrial waste annually. Included in the category are wastes from demolition and construction operations, agricultural and mining residues, combustion ash, sewage treatment sludge, and wastes generated by industrial processes. The states oversee these wastes, because Congress has not delegated any authority to the EPA to regulate them, as it has for MSW and hazardous waste.

### Disposal of Municipal Solid Waste

Over the years, the amount of MSW generated in the United States has grown steadily, in part because of a growing population, but also because of changing lifestyles and the increasing use of disposable materials and excessive packaging. In 1960, the nation generated 2.7 pounds (1.2 kg) per person per day. In 2001, we generated a total of 229 million tons (208 million metric tons) of MSW, an average of 4.4 pounds (2 kg) per person per day. With a population (2000 U.S. census figure) of 281 million, that is enough waste to fill 84,000 garbage trucks each day. The *solid-waste problem* can be stated simply: *We generate huge amounts of MSW, and it is increasingly expensive to dispose of it in ways that are environmentally responsible and protective of human health.*

The refuse generated by municipalities is a mixture of materials from households and small businesses, in the proportions shown in Figure 18–2. However, the proportions vary greatly, depending on the generator (commercial versus residential), the neighborhood (affluent versus poor), and the time of year (during certain seasons, yard wastes, such as grass clippings and raked leaves, add greatly to the solid-waste burden). Little attention is given to what people throw away in their trash. Even if there are restrictions and prohibitions, they can be bypassed with careful packing of the trash containers. Thus, many environmentally detrimental substances—paint, used motor oil, small batteries, and so on—are discarded, with the feeling that they are gone forever.

***Whose Job?*** Customarily, local governments have had the responsibility for collecting and disposing of MSW. The local jurisdiction may own trucks and employ workers, or it may contract with a private firm to provide the collection service. Traditionally, the cost of waste pickup is passed along to households via taxes. Alternatively, some municipalities have opted for a "pay-as-you-throw"

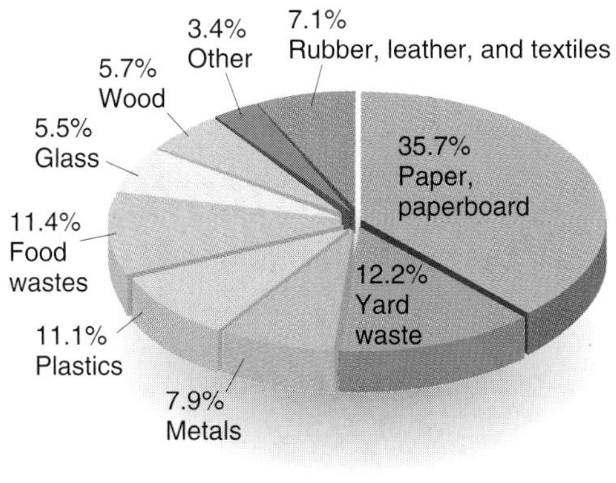


**Figure 18–2 U.S. municipal solid-waste composition.** The composition of municipal solid waste in the United States in 2001. (*Source:* Data from EPA, Office of Solid Waste, *Municipal Solid Waste in the United States: 2001 Facts and Figures.* October 2003.)

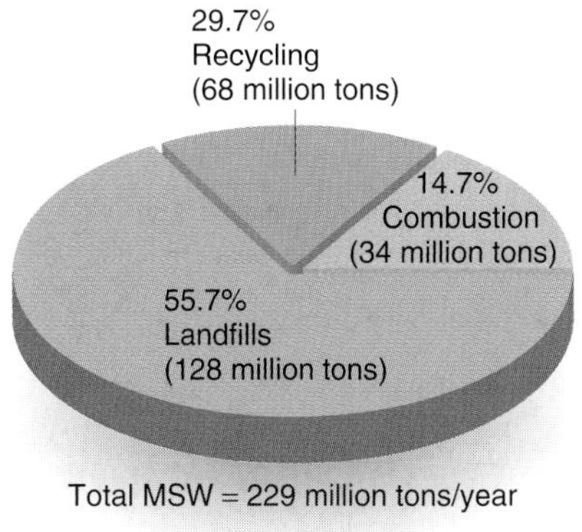


**Figure 18–3 U.S. municipal solid-waste disposal.** Disposal of solid waste to landfills, combustion, and recycling, 2001. (*Source:* Data from EPA, Office of Solid Waste, *Municipal Solid Waste in the United States: 2001 Facts and Figures.* October 2003.)

(PAYT, as it is called) system, in which households are charged for waste collection on the basis of the amount of trash they throw away. Some municipalities put all trash collection and disposal in the private sector, with the collectors billing each home by volume and weight of trash. The MSW that is collected is then disposed of in a variety of ways, and it is at the point of disposal that state and federal regulations begin to apply.

***Past Sins.*** Until the 1960s, most MSW was burned in open dumps. The waste was burned to reduce its volume and lengthen the life span of the dump site, but refuse does not burn well. Smoldering dumps produced clouds of smoke that could be seen from miles away, smelled bad, and created a breeding ground for flies and rats. Some cities turned to incinerators, or combustion facilities, as they are called today—huge furnaces in which high temperatures allow the waste to burn more completely than in open dumps. Without controls, however, incinerators were prime sources of air pollution. Public objections and air pollution laws forced the phaseout of open dumps and many incinerators during the 1960s and early 1970s. Open dumps were then converted to landfills.

***Where Does it Go?*** In the United States in 2001, 55.7% of MSW was disposed of in landfills, 29.7% was recovered for recycling and composting, and the remainder (14.7%) was combusted (Fig. 18–3). Over the last 10 years, the landfill and combustion components have been declining, while recycling has shown a steady increase. The pattern is different in countries where population densities are higher and there is less open space for landfills. High-density Japan, for instance, combusts 75% of its trash. Most Western European countries also deposit less than half of their MSW in landfills and combust most of the rest.

## Landfills

In a **landfill**, the waste is put on or in the ground and is covered with earth. Because there is no burning, and because each day's fill is covered with at least six inches of earth, air pollution and populations of vermin are kept down. Unfortunately, aside from those concerns and the minimizing of cost, no other factors were given real consideration when the first landfills were opened. Municipal waste managers generally had no understanding of, or interest in, ecology, the water cycle, or what products decomposing wastes would generate, and they had no regulations to guide them. Therefore, in general, any cheap, conveniently located piece of land on the outskirts of town became the site for a landfill. Frequently, this site was a natural gully or ravine, an abandoned stone quarry, a section of wetlands, or a previous dump (Fig. 18–4). Once the municipality acquired the land, dumping commenced without any precautions. After the site was full, it would be covered with earth and ignored. Only recently have landfills been seen as a valuable open-space resource.

**Problems of Landfills.** Landfills are subjected to biological and physical factors in the environment and will undergo change over time as a consequence of the operation of those factors on the waste that is deposited.

**Figure 18–4 New Orleans dump.** This burning dump, sited on wetlands, demonstrates the worst of the MSW disposal practices of the past.

Several of the changes are undesirable, because, if they are not dealt with effectively, they present the following problems:

- leachate generation and groundwater contamination
- methane production
- incomplete decomposition
- settling

***Leachate Generation and Groundwater Contamination.*** The most serious problem by far is groundwater contamination. As water percolates through any material, various chemicals in the material may dissolve in the water and get carried along in a process called *leaching* (Chapter 7). The water with various pollutants in it is called *leachate.* As water percolates through MSW, a noxious leachate is generated that consists of residues of decomposing organic matter combined with iron, mercury, lead, zinc, and other metals from rusting cans, discarded batteries, and appliances—all generously "spiced" with paints, pesticides, cleaning fluids, newspaper inks, and other chemicals. The nature of the landfill site and the absence of precautionary measures (noted earlier) can funnel this "witches' brew" directly into groundwater aquifers.

All states have some municipal landfills that are, or soon will be, contaminating groundwater, but Florida has some unique problems. Flat and with vast areas of wetlands, much of the state is only a few feet above sea level and rests on water-saturated limestone. No matter where Florida's landfills were located, they were either in wetlands or just a few feet above the water table. As a result, more than 150 former municipal landfill sites in Florida are now on the Superfund list. (Superfund is the federal program with the responsibility for cleaning up sites that are in imminent danger of jeopardizing human health through groundwater contamination.) All landfills in Florida are now required to have state-of-the-art landfill liners.

***Methane Production.*** Because it is about two-thirds organic material, MSW is subject to natural decomposition. Buried wastes do not have access to oxygen, however, so their decomposition is anaerobic. A major by-product of the process is *biogas,* which is about two-thirds methane and the rest hydrogen and carbon dioxide, a highly flammable mixture. (See Chapter 17.) Produced deep in a landfill, biogas may seep horizontally through the soil and rock, enter basements, and even cause explosions if it accumulates and is ignited. Homes at distances up to 1,000 feet from landfills have been destroyed, and some deaths have occurred as a result of such explosions. Also, gases seeping to the surface kill vegetation by poisoning the roots. Without vegetation, erosion occurs, exposing the unsightly waste.

A number of cities have exploited the problem by installing "gas wells" in old and existing landfills. The wells tap the landfill gas, and the methane is purified and used as fuel. There are now over 340 landfill gas energy projects in the United States. In 2002, commercial landfill gas in the nation produced 8 billion kilowatt-hours of electricity and 75 billion cubic feet of gas for direct-use applications. For example, Riverview, Michigan, collaborates with Detroit Edison to "mine" the landfill gas under the city's 212-acre landfill (locally referred to as Mount Trashmore). This gas-to-energy project provides the energy needs of 3,700 homes, and the closed landfill is a wintertime skiing and recreation area (Fig. 18–5).

The recovery of landfill gas has significant environmental benefits. It directly reduces greenhouse gas emissions; methane (a powerful greenhouse gas) released from landfills is the largest anthropogenic source of methane emissions in the United States. Methane recovery also reduces the use of coal and oil, which are nonrenewable and highly polluting energy resources.

***Incomplete Decomposition.*** The commonly used plastics in MSW resist natural decomposition because of their molecular structure. Chemically, they are polymers of petroleum-based compounds that microbes are unable to digest. Biodegradable plastic polymers have been developed from sources such as cornstarch, cellulose, lactic acid, soybean protein, and amides. For example, the German company Bayer AG recently announced plans to market a 100% biodegradable plastic structured as a polyester amide polymer. Plastic film made from the polymer degrades completely in 70 days, according to company tests. So far, however,

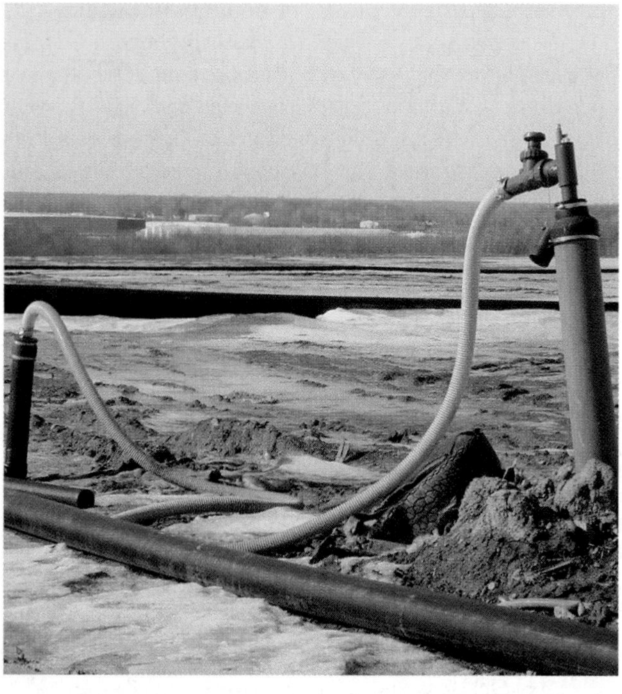

**Figure 18–5 Riverview Land Preserve.** This Michigan land preserve is sited on an old landfill. Gas recovery from the landfill provides energy for 3,700 homes, and the site is a recreational attraction.

none of the biodegradable plastics has seen common usage in consumer products.

A team of "archaeologists" from the University of Arizona, led by William Rathje, has been carrying out research on old landfills. The research has shown that even materials formerly assumed to be biodegradable—newspapers, wood, and so on—are often degraded only slowly, if at all, in landfills. In one landfill, 30-year-old newspapers were recovered in a readable state, and layers of telephone directories, practically intact, were found marking each year. Since paper materials are 37% of MSW, this is a serious matter. The reason paper and other organic materials decompose so slowly is the absence of suitable amounts of moisture. The more water percolating through a landfill, the better paper materials biodegrade. Unfortunately, the more percolation there is, the more toxic leachate is produced!

***Settling.*** Finally, waste settles as it compacts and decomposes. Luckily, this eventuality was recognized from the beginning, so buildings have never been put on landfills. Settling does present a problem in landfills that have been converted to playgrounds and golf courses, because it creates shallow depressions (and sometimes deep holes) that collect and hold water. This problem can be addressed by continual monitoring of the facility and the use of fill to restore a level surface.

**Improving Landfills.** Recognizing the foregoing problems, the EPA upgraded siting and construction requirements for new landfills. Under current regulations,

- New landfills are sited on high ground, well above the water table, not in a geologically unstable area, and away from airports (because of bird hazards).
- The floor is contoured so that water will drain into a tile leachate-collection system. The floor and sides are covered with a plastic liner and at least two feet of compacted soil. With such a design, any leachate percolating through the fill will move into the leachate-collection system. Collected leachate can be treated as necessary.
- Layer upon layer of refuse is positioned such that the fill is built up in the shape of a pyramid. Finally, it is capped with at least 18 inches of earthen material and a layer of topsoil and then seeded. The cap and the pyramidal shape help the landfill shed water. In this way, water infiltration into the fill is minimized, and less leachate is formed.
- The entire site is surrounded by a series of groundwater-monitoring wells that are checked periodically, and such checking must go on indefinitely.

These design features are summarized in Figure 18–6. Most landfills currently in operation have the improved technologies, which protect both human health and the environment.

Although the regulations protect groundwater, the landfill pyramids themselves may well last as long as the Egyptian pyramids. However, the creative siting and construction of landfills has the potential to address some highly significant future needs, because the abandoned landfill can become an attractive golf course, recreational facility, or wildlife preserve.

***Siting New Landfills.*** From 1988 to 2001, the number of municipal landfills declined from around 8,000 to 1,858. Because the size of landfills has increased, and because recycling is on the rise, the EPA does not believe that landfill capacity is a problem. However, as the agency puts it, "regional dislocations" may be expected. These are due to the one problem associated with landfills that gets more attention than any other: **siting.** It is not that landfills take up enormous amounts of land. One particular landfill occupying 121 acres (Central Landfill) serves the entire state of Rhode Island, and Fresh Kill, the largest landfill in the world, served much of New York City and its environs for many years and is only 2,200 acres. But, as old landfills are closed, it has become increasingly difficult to find land for the new ones needed to take their place.

People in residential communities (where MSW is generated) invariably reject proposals to site landfills anywhere near where they live, and those who already live close to existing landfills are anxious to close them down. Weary of the odor and heavy truck traffic, residents of Staten Island, New York, pressured the city of New York to shut down the huge Fresh Kill landfill, once the recipient of 27,000 tons of garbage per day. The city fulfilled its promise to phase out the landfill in 2001, and Staten Islanders are now anticipating future uses of the former landfill for recreation and wildlife. Gas from the huge landfill is providing the area with a new energy source. Currently, New York City exports its 11,000 tons per day of MSW to other states, at an average cost of $64 a ton.

***Anywhere But Here.*** With spreading urbanization, few suburban areas are not already dotted with residential developments. Practically any site selected, then, is met with protests and legal suits. The problem has been repeated in so many parts of the country (and globe!) that it has given rise to several inventive acronyms: **LULU** (locally unwanted land use), **NIMBY** (not in my backyard), and **NIMTOO** (not in my term of office). You can probably imagine how these attitudes apply to the landfill-siting problem.

***Outsourcing.*** The siting problem has some undesirable consequences. First, it drives up the costs of waste disposal, as alternatives to local landfills are invariably more expensive. Second, it leads to the inefficient and equally objectionable practice of the long-distance transfer of trash, as waste generators look for private landfills whose owners are anxious to receive trash. Often, this transfer occurs across state and even national lines, leading to resentment and opposition on the part of citizens of the recipient state or nation. Led by Pennsylvania,

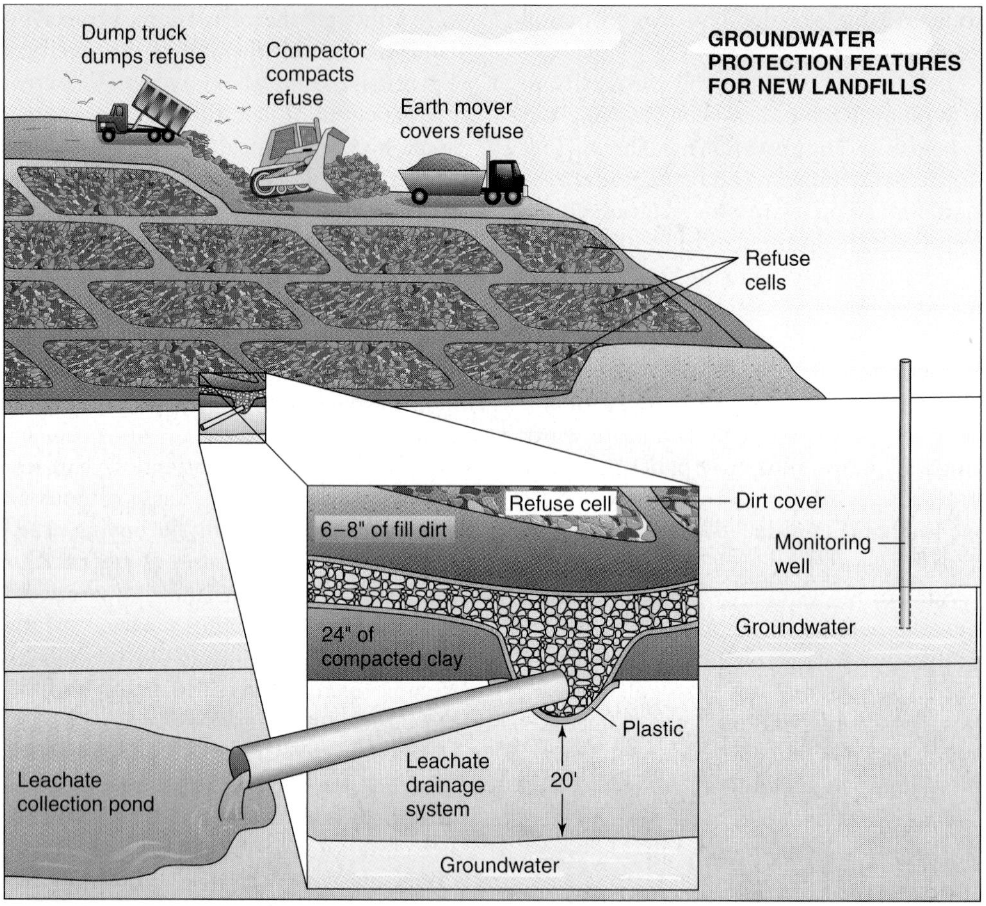


**Figure 18–6 Features of a modern landfill.** The landfill is sited on a high location, well above the water table. The bottom is sealed with compacted soil and a plastic liner, overlain by a rock or gravel layer, with pipes to drain the leachate. Refuse is built up in layers as the amount generated each day is covered with soil, so that the completed fill has a pyramidal shape that sheds water. The fill is provided with wells for monitoring groundwater.

eight states import more than 1 million tons of MSW a year (Table 18–1), while seven states export more than 1 million tons per year, led by New York. Not surprisingly, the heaviest importing states are adjacent to the largest exporting states.

One positive impact of the siting problem is that it encourages residents to reduce their waste and recycle as much as possible. (This alternative is discussed in Section 18.3.) Another effect is that the problem stimulates the use of combustion as an option for waste disposal.

**table 18-1 Interstate Transfer of MSW, 2001 (Amounts over 1 Million Tons)**

| Exporting States and Provinces | Exported MSW (Million Tons) | Importing States | Imported MSW (Million Tons) |
|---|---|---|---|
| New York | 5.6 | Pennsylvania | 9.8 |
| New Jersey | 1.8 | Virginia | 3.9 |
| Maryland | 1.5 | Michigan | 3.1 |
| Missouri | 1.7 | Illinois | 1–1.5 million |
| Massachusetts | 1.2 | Indiana | 1–1.5 million |
| Washington | 1.2 | Oregon | 1–1.5 million |
| North Carolina | 1.1 | Ohio | 1–1.5 million |
| Ontario | 1.1 | Wisconsin | 1–1.5 million |

## Combustion: Waste to Energy

Because it has a high organic content, refuse (especially the plastic portion) can be burned. Currently, some 102 combustion facilities are operating in the United States, burning about 34 million tons of waste annually—14.5% of the waste stream. This process is really waste *reduction*, not waste *disposal*, though, because, after incineration, the ash must still be disposed of.

**Advantages of Combustion.** The combustion of MSW has some advantages:

- Combustion can reduce the weight of trash by over 70% and the volume by 90%, thus greatly extending the life of a landfill (which is still required to receive the ash).
- Toxic or hazardous substances are concentrated into two streams of ash, which are easier to handle and control than the original MSW. The *fly ash* (captured from the combustion gases by air pollution control equipment) contains most of the toxic substances and can be safely put into a landfill. The *bottom ash* (from the bottom of the boiler) can be used as fill in some construction sites and roadbeds. Some combustion facilities process the bottom ash further to recover metals and then convert the remainder into concrete blocks.
- No changes are needed in trash collection procedures or people's behavior. Trash is simply hauled to a combustion facility instead of to the landfill.
- Practically all modern combustion facilities are waste-to-energy (WTE) facilities equipped with modern emission-control technology that brings the emissions into compliance with Clean Air Act regulations. When burned, unsorted MSW releases about 35% as much energy as coal, pound for pound. To their waste processing, many of these facilities add resource recovery, in which many materials are separated and recovered before (and sometimes after) combustion.

**Drawbacks of Combustion.** Combustion has some drawbacks, too:

- Adverse health effects can result, especially from older or poorly managed facilities. Particulate matter, lead, mercury, dioxins, and furans are the most serious pollutants entering the air and contributing to local and regional air pollution. Workers at such facilities are especially at risk for health problems.
- Combustion facilities are expensive to build, and their siting has the same problem as that of landfills: No one wants to live near one.
- Combustion ash is often loaded with metals and other hazardous substances and must be disposed of in secure landfills.
- To justify the cost of its operation, the combustion facility must have a continuing supply of MSW. For that reason, the facility enters into long-term agreements with municipalities, and these agreements can lessen the flexibility of the community's solid-waste management options.
- Even if the combustion facility generates electricity, the process wastes both energy and materials, unless it is augmented with recycling and recovery. However, even when it is so augmented, the facility competes directly with recycling for burnable materials such as newspapers, so such facilities represent a major impediment to recycling in some municipalities.

**An Operating Facility.** Let us look at the operation of a typical modern WTE facility, which might serve a number of communities or a large metropolitan area. Servicing a population of a million or more, the plant receives about 3,000 tons of MSW per day. The waste comes in by rail and truck, and the communities pay tipping fees (the costs assessed at the disposal site) that average $58 per ton. Waste processing is efficient, because, overall, about 80% of the MSW is burned for energy, 12% is recovered, and 8% is put in landfills. The process, pictured in Figure 18–7, is as follows:

1. Incoming waste is inspected, and obvious recyclable and bulky materials are removed.
2. Waste is then pushed onto conveyers that feed shredders capable of reducing the width of waste particles to 6 inches or less.
3. Strong magnets remove about two-thirds of ferrous metals for recycling before combustion.
4. The waste is then blown into boilers, where light materials burn in suspension and heavier materials burn on a moving grate.
5. Water circulated through the walls of the boilers produces steam, which drives turbines for generating electricity.
6. After the waste is burned, the bottom ash is conveyed to a processing facility, where further separation of metals may occur in a process that recovers brass, aluminum, gold, copper, and iron. In some facilities, this process nets $1,000 a day in coins alone!
7. Combustion gases are passed through a lime-based spray dryer–absorber to neutralize sulfur dioxide and other noxious gases. Then the gases go through electrostatic precipitators that remove particles. The resulting waste stream is significantly lower in pollutants than that which would be emitted by an energy-equivalent utility based on coal or oil combustion.

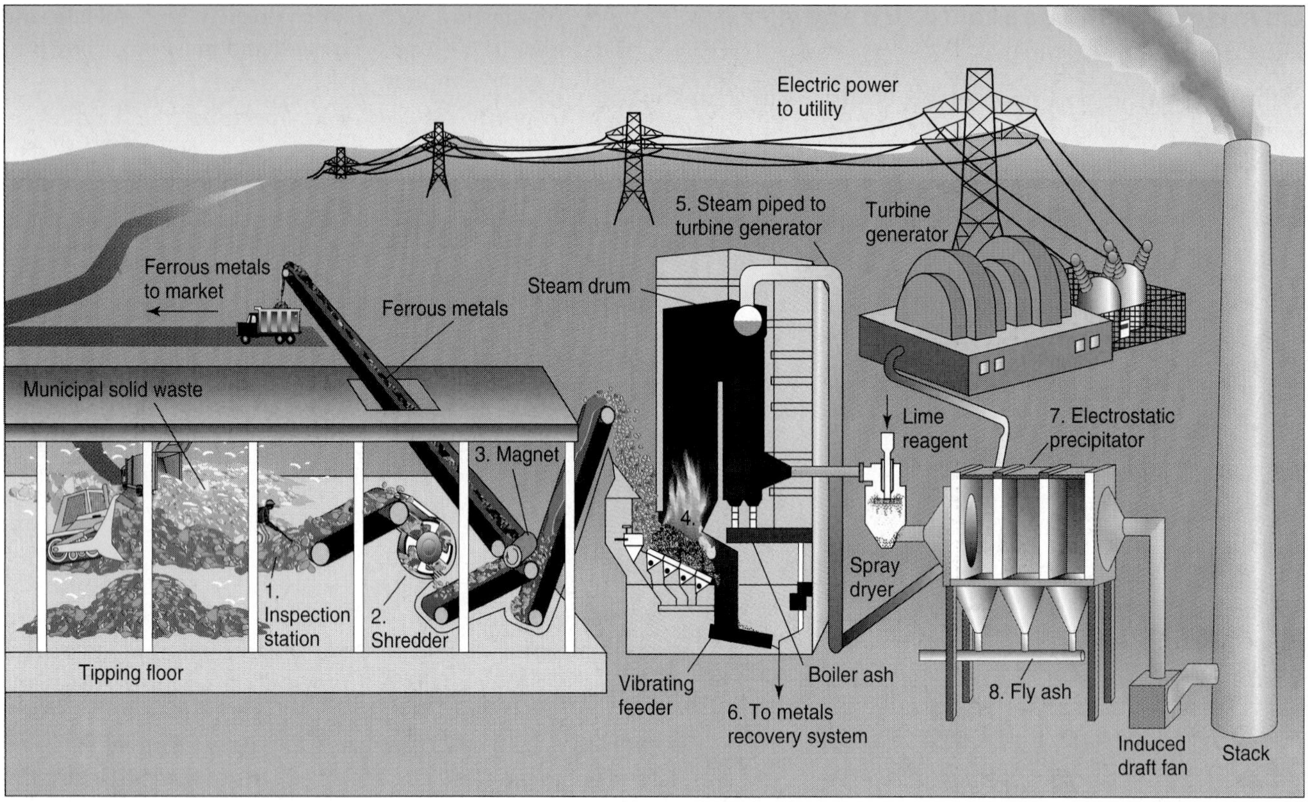


**Figure 18–7 Waste-to-energy combustion facility.** Schematic flow for the separation of materials and combustion in a typical modern waste-to-energy (WTE) combustion facility. Numbers refer to steps in the process as described in the text.

8. The fly ash and bottom ash residues are put into landfills.

An appreciation for the impact of such a facility can be gained from looking at the outcome of a year's operation. In one year, 1 million tons of MSW are processed, 40,000 tons of metal are recycled, and 570,000 megawatt-hours of electricity are generated—the equivalent of more than 60 million gallons of fuel oil and enough electricity to power 65,000 homes. All this comes from stuff that people have thrown away!

Because of the drawbacks mentioned previously, WTE facilities are strongly opposed during the siting, permitting, and construction phases. Opponents cite public-health concerns about dioxins, mercury, and other heavy metals, as well as traffic and property value concerns.

## Costs of Municipal Solid-Waste Disposal

The costs of disposing of MSW are escalating, and not just because of the new design features of landfills. More and more, they reflect the expenses of acquiring a site and providing transportation. Tipping fees average about $34 a ton, and at some landfills they exceed $100 a ton. The waste collector must recover this cost, as well as transportation costs to the site, and generate a profit (if the facility is privately operated). With the closing of the Fresh Kills landfill, the total cost of disposing of New York City's MSW rose to $263 a ton in 2002. At 11,000 tons per day, this amounts to over $1 billion a year! Nationally, all revenues from MSW disposal amounted to more than $14 billion a year in 2002.

Getting rid of *all* trash is becoming more expensive, and one sad consequence of this increasing expense is illegal dumping. Some towns are charging up to $5 a bag for MSW disposal, and it now costs $1 or more to get rid of an automobile tire and $30 or more to dispose of a refrigerator. As a result, tires, refrigerators, yard waste, car parts, and construction waste are appearing all over the landscape. Institutions and apartments that operate with dumpsters often must put padlocks on them in order to prevent their unauthorized use by people trying to avoid disposal costs. Many states have established a corps of environmental police to track down midnight dumpers and bring them to justice.

# 18.2 Solutions to the Solid-Waste Problem

Our domestic wastes and their disposal represent a huge stream of material that flows in one direction: from our resource base to disposal sites. Just as natural ecosystems depend on recycling nutrients, we can move in the direction of sustainability only if we also learn to recycle more of our wastes. There is strong evidence that we are moving in this direction, but the best strategy of all is to reduce waste at its source, called "source reduction."

## Source Reduction

Source reduction accomplishes two goals: It reduces the amount of waste that must be managed, and it conserves resources. It is noteworthy that, after rising rapidly during the latter decades of the 20th century, the amount of waste per capita in the United States peaked at 4.5 lb in 1990 and has since leveled off. (We still lead the world in this dubious statistic, however.) The leveling off is a signal that some changes in lifestyle may be occurring that have an impact on waste generation.

Source reduction is difficult to measure, because it means trying to measure something that no longer exists. The EPA measures source reduction by measuring consumer spending, which reflects the goods and products that ultimately make their way to the trash bin. After 1990, consumer spending continued to grow, but the MSW stream slowed down. If the MSW had grown at the same pace as consumer spending, some 287 million tons would have been generated in 2000 instead of 232 million. Thus, some 55 million tons never made it into the waste stream in 2000, and the EPA considers this to be due to source reduction activities.

*Examples.* Source reduction can involve a broad range of activities on the part of homeowners, businesses, communities, manufacturers, and institutions. Consider the following examples:

- Lightening the weight of many items has reduced the amount of materials used in manufacturing. Steel cans are 60% lighter than they used to be; disposable diapers contain 50% less paper pulp, due to absorbent-gel technology; and aluminum cans contain only two-thirds as much aluminum as they did 15 years ago.
- The Information Age may be having an impact on the use of paper. Electronic communication, data transfer, and advertising are increasingly performed on personal computers, and the number of homes with personal computers continues to rise. The growth of the Internet as a medium for information transfer has been phenomenal. These developments have the potential to reduce paper waste.
- Many durable goods are reusable, and indeed, the United States has a long tradition of resale of furniture, appliances, and carpets. Other goods may be donated to charities and put to good use. The growing popularity of yard sales, flea markets, consignment clothing stores, and other markets for secondhand goods (Fig. 18–8) is an encouraging development.

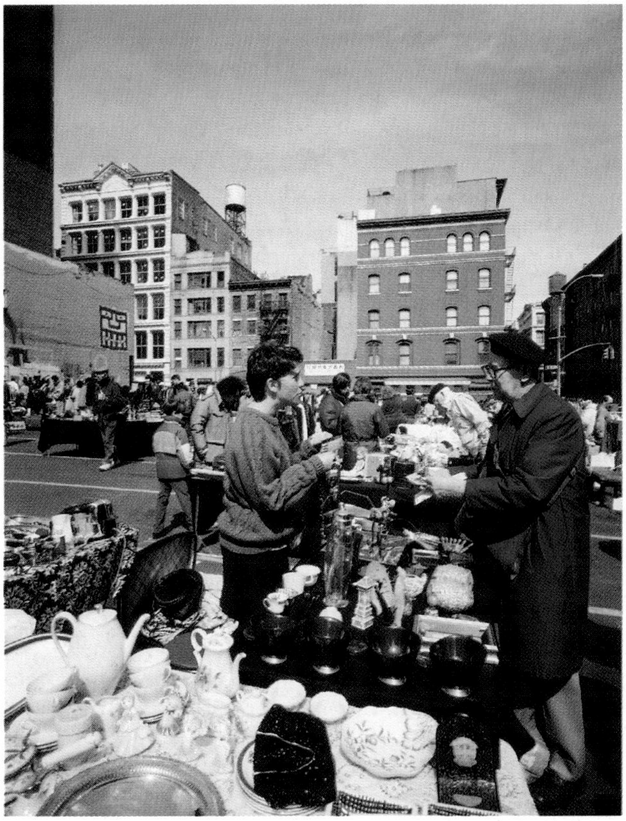

**Figure 18–8 Waste reduction by reuse.** Yard sales and flea markets are keeping many materials from becoming MSW after a single use.

- Lengthening a product's life can keep that product out of the waste stream. If products are designed to last longer and be easier to repair, consumers will learn that they are worth the extra cost. (You get what you pay for; cheap products usually take the shortest route to the waste bin.)
- We are all on bulk-mailing lists, because these lists are sold and shared widely, so we are guaranteed to receive increasing volumes of advertising. To stay off such lists, simply inform mail-order companies and other organizations involved that you do not want your name and address shared. (You can write to Mail Preference Service, Direct Marketing Association, P.O. Box 9008, Farmingdale, NY 11735-9008. The service provides, at no charge, a master list of people who do not wish to receive mass advertising. It works as long as marketers choose to consult it.)
- An increasingly popular way of treating yard waste (currently 12% of MSW) is composting, especially as more and more states are banning yard waste from MSW collections. The natural biological decomposition (rotting) of organic matter in the presence of air, composting can be carried out in the backyard by homeowners, or it can be promoted by municipalities as a way of dealing with a large fraction of MSW.

## The Recycling Solution

In addition to reuse, recycling is an obvious solution to the solid-waste problem. More than 75% of MSW is recyclable material. There are two levels of recycling: **primary** and **secondary.** *Primary recycling* is a process in which the original waste material is made back into the same material—for example, newspapers recycled to make newsprint. In *secondary recycling,* waste materials are made into different products that may or may not be recyclable—for instance, cardboard from waste newspapers.

*Why Recycle?* Recycling is a hands-down winner in terms of energy use and pollution:

- **It saves energy and resources.** One ton of recycled steel cans saves 2,500 lb of iron ore, 1,000 lb of coal, and more than 5,400 BTUs of energy. One ton of papers recycled saves 17 trees, 6,953 gallons of water, 463 gallons of oil, and 4,000 kilowatt hours of energy.
- **It decreases pollution.** Making recycled paper requires 64% less energy and generates 74% less air pollution and 35% less water pollution than does using wood from trees. For every ton of waste processed, a thorough recycling program will eliminate 620 lb of carbon dioxide, 30 lb of methane, 5 lb of carbon monoxide, 2.5 lb of particulate matter, and smaller amounts of other pollutants.

*What Gets Recycled?* The primary items from MSW currently being heavily recycled are cans (both aluminum and steel), bottles, plastic containers, newspapers, and yard wastes. Yet, there are many alternatives for reprocessing various components of refuse, and people are coming up with new ideas and techniques all the time. A few of the major established techniques, together with current percentages of their recovery by recycling, are as follows:

- Paper and paperboard (45% recovery) can be remade into pulp and reprocessed into recycled paper, cardboard, and other paper products; finely ground and sold as cellulose insulation; or shredded and composted.
- Most glass (19% recovery) that is recycled is crushed, remelted, and made into new containers; a smaller amount is used in fiberglass or "glasphalt" for highway construction.
- Some forms of plastic (5.5% recovery) can be remelted and fabricated into carpet fiber, outdoor wearing apparel, irrigation drainage tiles, building materials, and sheet plastic (Fig. 18–9).
- Metals can be remelted and refabricated. Making aluminum (25% recovery) from scrap aluminum saves up to 90% of the energy required to make aluminum from virgin ore. In addition, aluminum ore is imported and represents part of the mounting U.S. trade deficit. National recycling of aluminum saves energy, creates jobs, and reduces the trade deficit.
- Yard wastes (leaves, grass, and plant trimmings—57% recovery) can be composted to produce a humus soil conditioner.

**Figure 18–9 Plastic recycling.** Bales of plastic bottles in a recycling center.

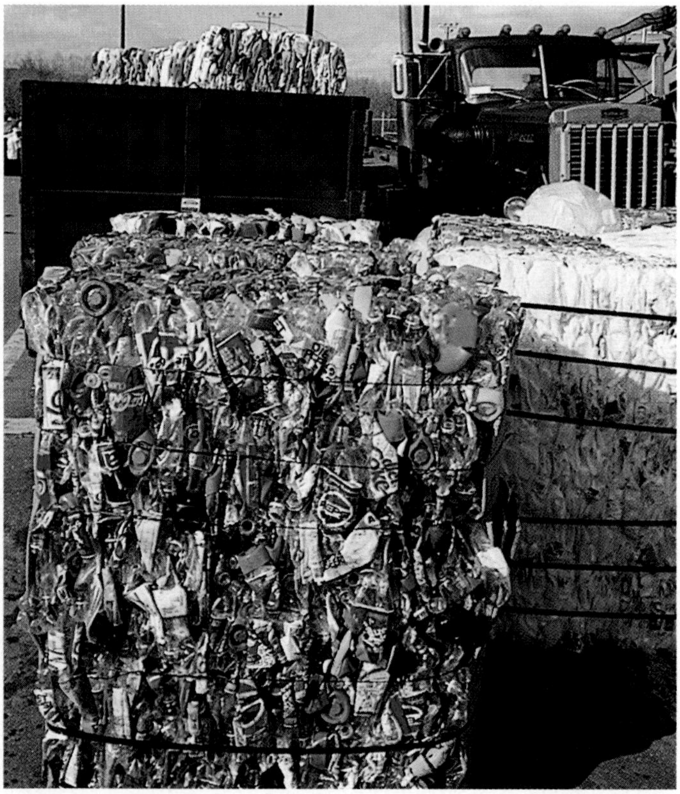

- Textiles (14.6% recovery) can be shredded and used to strengthen recycled paper products.
- Old tires (38% recovery) can be remelted or shredded and incorporated into highway asphalt. Not included in the 38% figure are some 2 million tons of tires burned in special combustion facilities.

Recycling is both an environmental and an economic issue. Many people are motivated to recycle because of environmental concern, but the use of recycled materials is also driven by economic factors. The Global Recycling Network (www.grn.com) is an electronic information exchange that promotes the trade of recyclables from MSW and the marketing of "ecofriendly" products made from recycled materials.

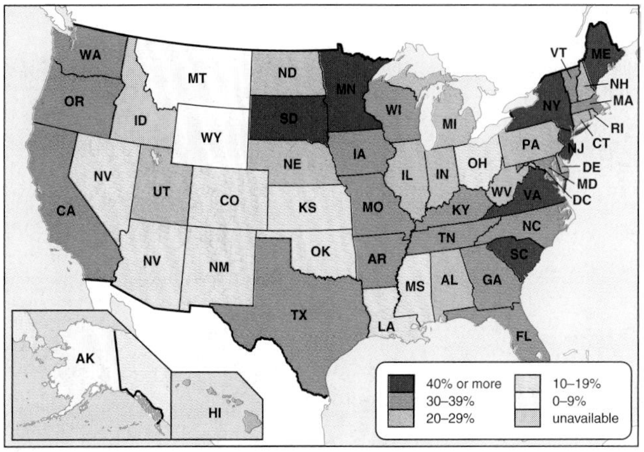


**Figure 18–10 State recycling rates.** There is great variation in recycling from state to state. (*Source:* EPA, Office of Solid Waste).

## Municipal Recycling

Recycling is probably the most direct and obvious way most people can become involved in environmental issues. If you recycle, you save some natural resources from being used (trees, in the case of paper), and you prevent landfills from becoming "landfulls."

Recycling's popularity is well established: Virtually every state has specific recycling goals, met with varying degrees of success (Fig. 18–10). EPA sources report that only 6.7% of MSW was recycled in 1960, versus 29.7% in 2001 (Fig. 18–11).There is a great diversity of approaches to recycling in municipalities, from recycling centers requiring residents to drive miles to them to curbside recycling with sophisticated separation processes. The most successful programs have the following characteristics:

1. There is a strong incentive to recycle, in the form of pay-as-you-throw charges for general trash and no charge for recycled goods.
2. Recycling is not optional; mandatory regulations are in place, with warnings and sanctions for violations.
3. Residential recycling is curbside (Fig. 18–12), with free recycling bins distributed to households. (Curbside recycling has risen rapidly. Currently, 49% of the U.S. populace is served via 9,700 curbside programs.)
4. Drop-off sites are provided for bulky goods like sofas, appliances, construction and demolition materials, and yard waste.
5. Recycling goals are ambitious, yet clear and feasible. Some percent of the waste stream is targeted, and progress is followed and communicated.
6. A concerted effort is made to involve local industries in recycling.
7. The municipality employs an experienced and committed recycling coordinator.

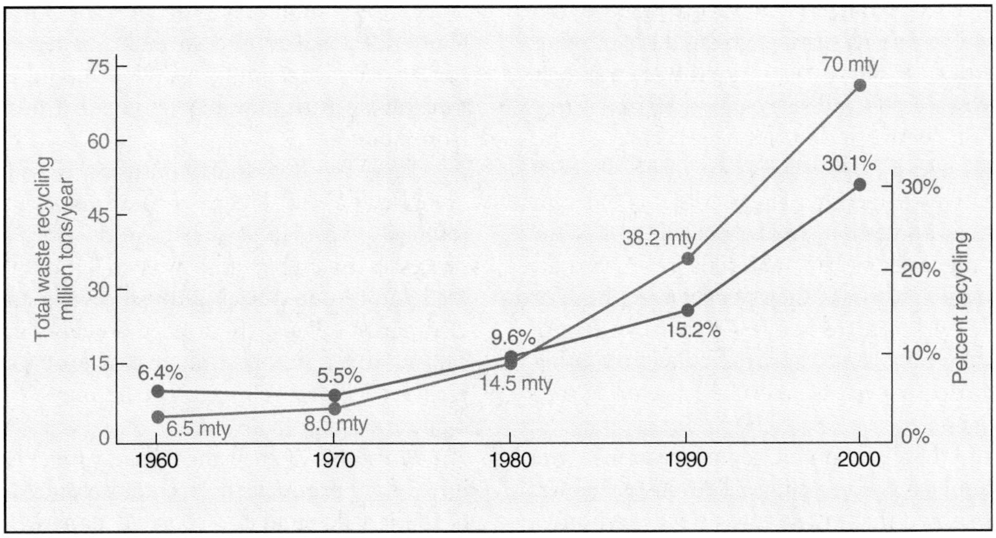


**Figure 18–11 MSW recycling in the United States.** MSW recycled from 1960 to 2001, as total waste and percentage recycled. The EPA's new goal is to reach 35% by 2005 (*Source:* EPA. Office of Solid Waste, *Municipal Solid Waste in the United States: 2001 Facts and Figures.* October 2003).

**Figure 18–12 Curbside recycling.** Curbside pickup by a waste hauler, using a special truck that presorts recyclables. Curbside recycling reaches 49% of the U.S. population.

Nevertheless, municipalities experience very different recycling rates. New York City, for example, has made great progress—from recycling only 5% to recycling 30% in four years—by mandating curbside recycling for all of its 3 million households. Boston, by contrast, with only 11% recycling, held Massachusetts back in its goal to achieve 46% by 2000. Worcester, farther to the west in the state, has achieved a commendable 48% recycling rate.

*Critics of Recycling.* Recycling has its critics, who base their arguments primarily on economics. If the costs of recycling (from pickup to disposal of recycled components) are compared with the costs of combustion or placing waste in landfills, recycling frequently comes out second best. Markets for recyclable materials fluctuate wildly, and residents often end up subsidizing the recycling effort. The shortfall between recycling costs and market value range from \$20 to \$135 a ton for some recyclables (items such as green glass and colored plastic bottles). In some cases (notably, newspaper and glass), prices are controlled by the industry of origin. Competition between landfills and combustion facilities often lowers tipping fees, creating an even greater disincentive to recycle. Thus, critics of current recycling practices argue that, unless recycling pays for itself through the sale of the materials recovered, it should not be done. However, the critics ignore the subsidies that make virgin materials less expensive to use than recycled materials. Also, garbage collection is big business, and those involved see recycling as cutting into their trash business.

In spite of these economic considerations, the demonstrated support for recycling programs is strong. Experience has shown that at least two-thirds of households will recycle if presented with a curbside pickup program. The percentage goes up when recycling is combined with PAYT programs. Dover, New Hampshire, a city of 26,000, was facing escalating costs of MSW collection and disposal after its landfill closed. The city instituted both a PAYT program and curbside recycling in 1991 and saw its per household trash plummet from 6 to 2.3 lb per day in six years. At the same time, the cost per household dropped from \$122 to \$73 per year, and the recycling rate rose quickly to over 50%. The city's solid-waste management costs dropped 27% in spite of a population increase.

**Paper Recycling.** By far the most important recycled item is newspapers, because of their predominance in the waste stream. It is a simple matter to tie up or bag household newspapers, and the amount recovered by recycling is increasing dramatically. Currently, more newspaper is being recycled (60%) than discarded. Since more than 25% of the trees harvested in the United States are used to make paper, recycling paper saves trees. Depending on the size and type of trees, a 1-meter stack of newspapers equals the amount of pulp from one tree.

There is often some confusion in what is meant by "recycled paper." The key is the amount of *postconsumer* recycled paper in a given product. In manufacturing processes, much paper is "wasted" and is routinely recovered and rerouted back into processing and called "recycled paper." Thus, the total recycled content of a paper can be 50%, although the actual postconsumer amount recovered by recycling might be only 10% of the total amount of paper.

*Markets.* After the wastepaper is incorporated into a final product, the market becomes a critical factor. Is there a demand for recycled paper? The technology for producing high-quality paper from recycled stock has improved greatly, to the point where it is virtually

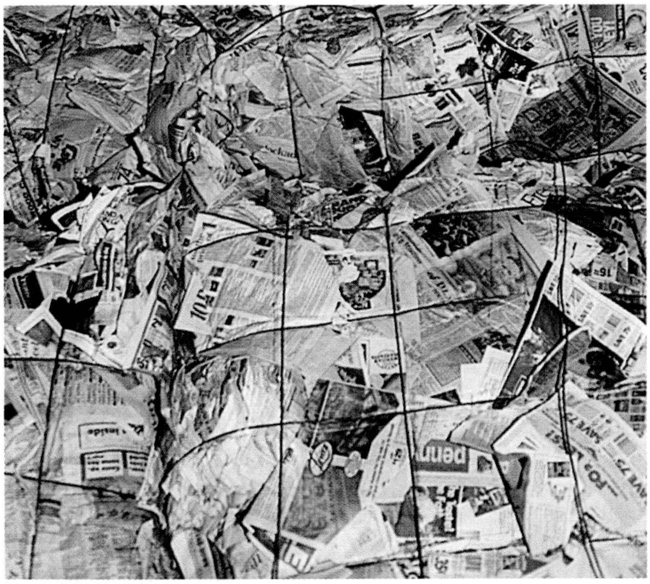

**Figure 18–13 Wastepaper exports.** Bales of wastepaper being loaded for shipment overseas.

impossible to distinguish the recycled product from "virgin" paper.

The market for used newspapers has fluctuated greatly over the past several decades. During the late 1980s, the market was saturated, and municipalities often had to pay to get rid of newspapers. In 1995, discarded papers were so valuable—up to $160 a ton—that thieves were stealing them off the sidewalks before the recycling trucks could pick them up. A year later, as more recycling programs came on line, the market collapsed again, and many cities once more paid as much as $25 a ton to have the newspapers hauled away—but this is still less expensive than the cost of landfills.

There is a lively international trade as well in used paper (Fig. 18–13). Forest-poor countries in Europe and Asia purchase wastepaper from the United States and other industrial countries in the Northern Hemisphere, where there continues to be a surplus of such paper. The largest importer is Taiwan (almost 2 million tons per year), which also boasts the highest percentage of reused content in its paper: 98%. The United States is at a much lower 33%.

**Glass Recycling and Bottle Laws.** The glass in MSW is primarily in the form of containers, most of which held beverages. The average person drinks about a quart of liquid each day. Given that there are 290 million Americans, this daily consumption amounts to some 28 billion gallons of liquid per year for the nation as a whole. Most of this volume is packaged in single-serving containers that are used once and then thrown away. Nonreturnable glass containers constitute 5.5% of the solid waste stream in the United States and about 50% of the nonburnable portion; they also make up about 90% of the nonbiodegradable portion of roadside litter. Broken bottles along roads, beaches, and parklands are responsible for innumerable cuts and other injuries, not to mention flat tires. Both the mining of the materials and the process used to manufacture the beverage containers create pollution. All of these factors produce hidden costs that do not appear on the price tag of the item; however, we pay for them with taxes to clean up litter, as well as with our injuries, flat tires, environmental degradation, and so on.

***Bottle Laws.*** In an attempt to reverse these trends, environmental and consumer groups have promoted *bottle laws* that facilitate the recycling or reuse of beverage containers. Such laws generally call for a deposit on all beverage containers, both reusable and throwaway, glass and plastic. Retailers are required to accept the used containers and pass them along for recycling or reuse. Bottle laws have been proposed in virtually all state legislatures over the last decade. Nevertheless, in every case, the proposals have met with fierce opposition from the beverage and container industries and certain other special-interest groups. The reason for their opposition is economic loss to their operations. The container industry contends that bottle laws will result in loss of jobs and higher beverage costs for the consumer.

In most cases, the industry's well-financed lobbying efforts have defeated bottle laws. However, some states—10, as of 2003—have adopted bottle laws of varying types despite industry opposition (Table 18–2). The experience of these states has proved the beverage and bottle industry's claims false. In fact, more jobs are gained than lost, costs to the consumer have not risen, a high percentage of bottles is returned, and there is a marked reduction in the can and bottle portion of litter. A final measure of the success of bottle laws is continued public approval: Despite industry

**table 18-2 States that Have Passed Bottle Laws**

| State | Year Passed | State | Year Passed |
|---|---|---|---|
| Oregon | 1972 | Iowa | 1978 |
| Vermont | 1973 | Massachusetts | 1978 |
| Maine | 1976 | Delaware | 1982 |
| Michigan | 1976 | New York | 1983 |
| Connecticut | 1972 | California | 1991 |

efforts to repeal these laws, no state that has one has repealed it.

Repeated attempts have been made to get a national bottle law through Congress—to date, unsuccessfully. Opponents (the same ones that oppose state-level bottle bills) argue that such a law will threaten the newly won successes in curbside recycling, of which beverage containers represent the most important source of revenue. However, since curbside recycling currently reaches only half of the U.S. population, a national bottle law would undoubtedly recover a greater proportion of beverage containers. (States with bottle laws report that 80% to 97% of containers are returned, a significant proportion from the curbside collections from people who don't bother to redeem the containers.) Also, a national bottle law would be labor intensive, employing tens of thousands of new workers.

In 2001, recycling and bottle laws resulted in the recovery of 22% of glass containers, 49% of aluminum containers, 59% of steel containers, and about 10% of plastic containers, with an aggregate weight of 5.8 million tons.

**Plastics Recycling.** Plastics have a bad reputation in the environmental debate, for several reasons. First, plastics have many uses that involve rapid throughput—for example, packaging, bottling, the manufacture of disposable diapers, and the incorporation of plastics into a host of cheap consumer goods. Second, plastics are conspicuous in MSW and litter. Ironically, most trash is disposed of in plastic bags manufactured just for that purpose. Finally, plastics do not decompose in the environment, because no microbes (or any other organisms) are able to digest them. (Imagine plastics in landfills delighting archaeologists hundreds of years from now.) Therefore, the possibility of recycling at least some of the plastic in products—containers of liquid, for example—interests the environmental community.

*PETE and HDPE.* If you look on the bottom of a plastic container, you will see a number (or some letters) inside the little triangle of arrows that has come to represent recycling (Fig. 18–14). The number or the letters are used to differentiate between and sort the many kinds of plastic polymers, not all of which are recyclable. The two recyclable plastics in most common use are high-density polyethylene (HDPE; code 2) and polyethylene terephthalate (PETE; code 1). In the recycling process, plastics must be melted down and poured into molds, and unfortunately, some contaminants from the original containers may carry over, making it difficult to reuse the plastic for food containers, so uses for the recycled plastic are necessarily restricted. However, some new uses for recycled plastic are appearing, with more in the development stage. PETE, for example, is turned into carpet fiber and fill for outdoor wearing apparel, and HDPE becomes irrigation drainage tiles, sheet plastic, and, appropriately, recycling bins.

Critics of the recycling of plastics point out that if the process were driven purely by the market, it wouldn't happen at all. They argue that recovering plastics is more costly than starting from scratch—that is, beginning with petroleum derivatives—and that manufacturers are getting involved mainly because environmentally concerned consumers are demanding their involvement. Industry-supported critics also point out that plastics in landfills create no toxic leachate or dangerous biogas. Moreover, plastics in combustion facilities burn wonderfully hot and leave almost no ash. Perhaps we can conclude that even though plastic recycling has its problems, continued attention to plastics in MSW will keep us moving in a sustainable direction.

## Regional Recycling Options

As landfills close down and MSW is transferred out of cities and towns, transfer stations are set up to collect the wastes and move them on to their final destination. Some 3,970 transfer stations existed in 2001, an increase of 500 over the previous year. It is an encouraging sign that a number of these transfer stations are actually **materials recovery facilities,** referred to in the trade as MRFs, or "murfs." In 2001, there were 480 MRFs operating in the United States, handling over 22 million tons of MSW a year.

*How a MRF Works.* Basic sorting takes place when waste is collected, either through curbside collection or by town recycling stations (sites to which

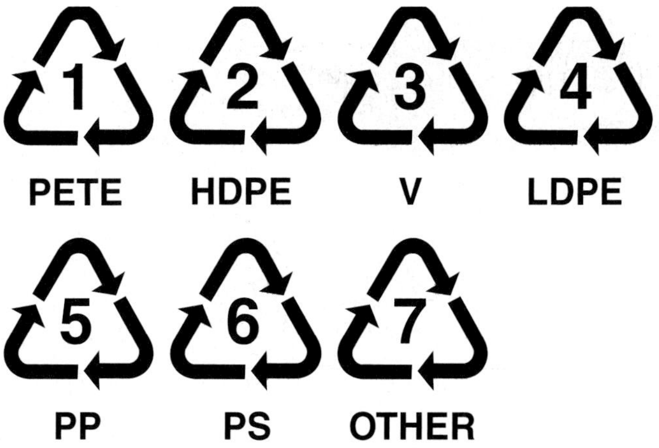


**Figure 18–14** **Recycling codes.** These symbols are marked on all solid plastic items. 1: PETE = polyethylene terephthalate; 2: HDPE = high-density polyethylene; 3: V = polyvinyl chloride; 4: LDPE = low-density polyethylene; 5: PP = polypropylene; 6: PS = polystyrene; 7: OTHER = any other plastic.

townspeople can bring wastes to be recycled). The waste is then trucked to the MRF and handled on three tracks—one for metal cans and glass containers, another for paper products, and a third for plastics. The materials are moved through the facility by escalators and conveyor belts, tended by workers who inspect and sort further. The objective of the process is to prepare materials for the recycled-goods market. Glass is sorted by color, cleaned, crushed into small pebbles, and then shipped to glass companies, where it replaces the raw materials that go into glass manufacture—sand and soda ash—and saves substantially on energy costs. Cans are sorted, flattened, and sent either to detinning plants or to aluminum-processing facilities. Paper is sorted, baled, and sent to reprocessing mills. Plastics are sorted into four categories, depending on their color and type of polymer, and then sold.

The facility's advantages are its economy of scale and its ability to produce a high-quality end product for the recycled materials market. Towns know where to bring their waste, and they quickly become familiar with the requirements for initial sorting during the collection and transfer of the waste. Some MRFs have gone to high technology, replacing the manual sorting with magnetic pulleys, optical sensors, and air sorters. As the movement grows, this is likely the direction the technology will take, to improve the efficiency of operations and cut costs.

*Mixed Waste Processing.* Less common than conventional MRFs, the mixed waste processing facility receives MSW just as if it were going to a landfill or a WTE facility. The waste is loaded on a conveyer and is sorted for recovery of recyclable materials before being landfilled or combusted. A total of 43 such facilities were operating in the United States in 2001, processing about 10.5 million tons of MSW a year.

*Mixed Waste and Yard Trimmings Composting.* Taking advantage of the fact that MSW has a high organic content, a few regional facilities compost the MSW after removing large items and metals. Often, they will mix treated sewage sludge with the MSW (called *co-composting*), which provides a rich source of bacteria and nutrients. These facilities are frequently plagued with odor problems, and some have had serious fires. Sometimes, such facilities are combined with a MRF, as in the case of Nantucket Island, Massachusetts. (See "Earth Watch," p. 506.) Much more common are the yard-trimmings composting programs—over 3,800 in 2001, handling about 16 million tons per year.

## 18.3 Public Policy and Waste Management

The management of MSW used to be entirely under the control of local governments. In recent years, however, state and federal agencies have played an increasingly important role in waste management, partly through regulation and partly through encouragement and facilitation.

### The Regulatory Perspective

At the federal level, the following legislation has been passed:

- The first attempt by Congress to address the problem was the **Solid Waste Disposal Act** of 1965. The legislation gave jurisdiction over solid waste to the Bureau of Solid Waste Management, but the agency's mandate was basically financial and technical rather than regulatory.

**earth watch**

## The Nantucket Story

Nantucket Island lies some 30 miles south of Cape Cod and is the year-round home to 10,000 residents, but hosts more than 50,000 during the summer months. Nantucket's MSW was taken to a landfill for many years, but in the late 1980s the landfill was overflowing and leaching wastes into sensitive surrounding wetlands. In 1994, the state of Massachusetts mandated that the island close its landfill, leaving islanders with few options. The least attractive one was to ship the MSW off the island, which would quadruple the island's waste management costs. In 1996, the island decided to privatize waste-handling operations and signed a 25-year agreement with Waste Options to build an integrated waste management system that encompassed recycling, composting, and cleanup of the existing landfill. After several years of operation, the island's MSW has been reduced to only 14% of what is generated, and it goes to a landfill that is actually shrinking in size.

The process begins with a MRF, which has boosted Nantucket's recycling rate from 17% to 42%. Cardboards, newspaper, magazines, cans, glass, and plastic bottles are sorted, baled, and shipped off the island. A unique feature is a Take It or Leave It exchange, where residents can drop off used clothing and other unwanted items and others can pick up what they want from the dropped-off items. Other items, such as mattresses, tires, yard wastes, and construction materials, are handled separately and are either shipped to the mainland for further processing or processed on-site.

The remaining MSW is processed together with dewatered sewage sludge from the waste treatment plant (thus, co-composting) to create a high-grade compost. The heart of this process is the Eweson Digester, a 185-foot-long steel drum that rotates continually and promotes the decomposition of much of the organic matter in the wastes. Air and water are added as needed to optimize biological decomposition, and after three days in the digester the digested material is screened, with the larger particles going to the landfill and the screenings further processed in a Maturation Building, in which, over several weeks, the screenings are repeatedly turned to reach a stable, high-quality compost that is useful for gardening and landscaping. Air in the building is controlled for odor with a biofilter.

The landfill materials represent only 14% of the original MSW and are added to the island landfill no faster than previously buried waste is taken from the landfill (called landfill mining) to the digester during the winter months, when trash and people are at a low ebb on the island. The price paid by the Nantucket islanders is $90 a ton, which is quite competitive with MSW costs in other Massachusetts communities. This public–private partnership on Nantucket Island has produced a waste treatment process that could become a model for the rest of the country.

- The **Resource Recovery Act** of 1970 gave jurisdiction over waste management to the newly created EPA and directed attention to recycling programs and other ways of recovering resources in MSW. The act also encouraged the states to develop some kind of waste management program.
- The passage of the **Resource Conservation and Recovery Act (RCRA)** of 1976 saw a more regulatory ("command and control") approach to dealing with MSW, as the EPA was given power to close local dumps and set regulations for landfills. Combustion facilities were covered by air-pollution and hazardous-waste regulations, again under the EPA's jurisdiction. The RCRA also required the states to develop comprehensive solid-waste management plans.
- The **Superfund Act** of 1980 (described in Chapter 19) addressed abandoned hazardous-waste sites throughout the country, 41% of which are old landfills.
- The **Hazardous and Solid Waste Amendments** of 1984 gave the EPA greater responsibility to set solid-waste criteria for all hazardous-waste facilities. Since even household waste must be assumed to contain some hazardous substances, this meant that the EPA had to determine and monitor all landfill and combustion criteria more closely.

Largely in response to these federal mandates, the states have been putting pressure on local governments to develop *integrated waste management plans,* with goals for recycling, source reduction, and landfill performance.

## Integrated Waste Management

It is not necessary to fasten onto just a single method of handling wastes. Source reduction, waste-to-energy combustion, recycling, materials recovery facilities, landfills, and composting all have roles to play in waste management. Different combinations of these options will work in different regions of the country. A system having several processes in operation is called *integrated waste management.* Let us examine the available MSW management options, with a view toward developing recommendations that make environmental sense—that is, which push us in a more sustainable direction.

**Waste Reduction.** The modern United States leads the world in per capita waste production and per capita energy consumption—we really are a "throwaway society" (Fig. 18–15). How can we turn this situation around? True management of MSW begins in the home (or dorm room). Materialistic lifestyles, affluence, and overconsumption can pack our homes and our trash cans with stuff we simply do not need. Indeed, some observers see this as a malady called "*affluenza.*" (See "Ethics" essay, p. 508.)

*WasteWise.* Several incentives have been put in place on the governmental level to encourage waste reduction. For example, the EPA sponsors the **WasteWise** program, targeting the reduction of MSW by establishing partnerships with not only local governments, schools, and organizations, but multinational corporations as well. WasteWise is a voluntary partnership that allows the partners to design their own solid-waste reduction programs. Almost 1,200 organizations throughout the country have partnered with the EPA to reduce and recycle waste and, in a new focus of the program, to calculate their reductions of waste-related greenhouse gas emissions. The EPA regularly honors partner organizations that have achieved exemplary waste reduction. For example, Arkansas-based Virco Mfg. Corporation, a furniture manufacturer, has recycled more than 75 million tons of materials since 1991. The company regularly purchases materials with recycled content, having spent $30 million on these materials in 2001. Indeed, company employees have been in the forefront of a volunteer effort to help the town of Conway, Arkansas, handle an overload of cardboard, collecting and selling it and giving the profits to the city.

*PAYT.* The EPA also brings attention to a national trend of "unit pricing," or charging households and other "customers" for the waste they dispose, with the agency's PAYT program. Instead of paying for trash collection and disposal through local taxes, which provides no incentive to reduce waste, communities levy curbside charges for all unsorted MSW—say, $1–5 per container (Fig. 18–16). Between 1990 and 2000, the population served by PAYT MSW collection more than doubled (to over 40 million). More than 5,000 cities and towns now employ the PAYT method. Following this highly successful strategy, PAYT communities reduce their waste from 14 to 27%, on average, and

**Figure 18–15 The throwaway society.** A view from space. (*Source:* © Tribune Media Services, Inc. All Rights Reserved. Reprinted with permission.

**ethics**

## "Affluenza": Do You Have It?

First, take this sample quiz to see if you have the bug:

1. I'm willing to pay more for a T-shirt if it has a cool corporate logo on it.
2. I'm willing to work at a job I hate so I can buy lots of stuff.
3. I usually make just the minimum payment on my credit cards.
4. When I'm feeling blue, I like to go shopping and treat myself.
5. I spend much more time shopping each month than I do being involved in my community.
6. I'd rather be shopping right now.
7. I'm running out of room to store my stuff.

Give yourself two points for true and one point for false. If you scored 10 or above, you may have a full-blown case of "affluenza."

Just what is "affluenza"? Jessie H. O'Neill, who coined the term, defines it as a "dysfunctional relationship with wealth or money." Symptoms of "affluenza" include a love of shopping, a glut of stuff in your home or dorm that you really don't need or use, and a dissatisfaction with what you have. It can strike anyone, regardless of their economic status. According to two PBS programs produced about it, "Affluenza" and "Escape from Affluenza," "affluenza" is a plague of materialism and overconsumption that is so pervasive that it actually characterizes our modern society.

The United States leads the world in per capita waste generation, a symptom of a societywide problem, according to O'Neill. It begins at an early age, as children are bombarded by TV commercials urging them to get the latest toys. Peer pressure makes it worse, with children sometimes "forced" to wear only the approved apparel. Every fad that comes along must be accommodated. Eventually, the conditioned children grow up, carry credit cards, and drive—and then "affluenza" takes on more serious consequences. The toys get more expensive. Adults caught by the disease acquire so much that they have to rent a self-storage bin to hold things they can't part with. Bankruptcy and credit card overloads are commonplace, a consequence of people's inability to control their spending.

How do you escape from "affluenza"? There are several steps you can take. First, admit that you have a problem. Then, begin to take small withdrawal steps. Before you buy something, ask yourself, Do I need it? Could I borrow it from a friend or relative? How many hours do I have to work to pay for it? Another suggested step is to avoid those recreational shopping trips to the mall. Take a walk or play ball with some kids instead. Become an advertising critic. Make a budget. These are a few of the many possible pathways for escape from "affluenza." As you do these things, you may be amazed at how challenging and rewarding it can be to live more simply—and, certainly, more sustainably.

increase their recycling from 32 to 59%. As with the WasteWise program, EPA's role is that of a facilitator, not a regulator.

***EPR.*** Another policy for bringing about waste reduction would be to establish a program of *extended product responsibility* (EPR), a concept that involves assigning some responsibility for reducing the environmental impact of a product at each stage of its "life cycle," especially the end. For example, Hewlett-Packard and Xerox make it easy for customers to return spent copier cartridges, and the two companies then recycle the components of the cartridges. EPR originated in Western Europe. Germany, the Netherlands, and Sweden have well-developed regulations that require companies to take back many items, that encourage the reuse of products, and that promote the manufacture of more durable products.

Again, the EPA is active in providing information to manufacturers and purchasing departments to help them design and buy more environmentally sound products.

**Figure 18–16 "Pay-as-you-throw" trash pickup.** Consumers purchase the empty bags for a set price, and only these bags are picked up by the trash hauler. Next to the bags are three recycling containers, which are emptied free of charge.

**Waste Disposal Issues.** No human society can avoid generating some waste. There will always be MSW, no matter how much we reduce, reuse, and recycle. Most experts see WTE combustion facilities holding their own, a decrease in the percentage of waste going to landfills, and a continued rise in recycling.

It will be important to break the gridlock at local and state levels on the siting of new landfills and WTE combustion facilities. Landfills will still be needed, although they should last longer than in the past. Policy makers have long known about the landfill shortage, but have opted for short-term solutions with the lowest political cost. One result of this choice is the long-distance hauling of MSW by truck and rail described earlier. If regions and municipalities are required to handle their own trash locally, as some states do, they will find places to site landfills, regardless of NIMBY considerations. New landfills must use the best technologies, such as proper lining, leachate collection and treatment, groundwater monitoring, biogas collecting, and final capping. If these technologies are employed, a landfill will not be a health hazard. Also, sites should be selected with a view to some future use that is attractive.

***Just Say No!*** With the closing of New York City's huge Fresh Kill landfill, the city is now sending its MSW to other states. Yet most citizens in the other states object to being New York's dumping ground, and there is the deeper stewardship issue of local responsibility for local wastes, as well as the inefficiency of long-distance waste hauling. It will take an act of Congress to give states or local jurisdictions the right to ban imports of waste, because of the interstate commerce clause in the Constitution. The **Solid Waste Interstate Transportation Act** of 2003 (H.R. 1730) has been introduced in Congress, with bipartisan support, especially from states that are currently receiving waste from other states. The legislation would give local and state governments the authority to limit or prohibit out-of-state wastes transported to landfills, unless a prior agreement existed. It also would allow states to put a cap on the amount of waste transported in, depending on the enactment of a statewide recycling program. The bill is a demand for fairness to states that are working hard to deal responsibly with their own wastes, only to see the unrestricted transport of wastes from other states.

Another policy goal should be to encourage more co-composting of MSW. Nantucket Island was forced to change its ways of managing MSW and has come up with a very attractive waste management program. (See Earth Watch, p. 506.) As on Nantucket Island, the technologies employed should be the best available, and if these are used, there should be no significant threat to human health. This option seems to be the best way to deal with mixed waste that is nonrecyclable. Recyclable materials are recovered before the composting step, the waste is greatly reduced in volume, and the compost is a highly useful product that can aid in the recovery of damaged lands in need of soil.

**Recycling and Reuse.** Recycling is certainly the wave of the future. It should not, however, be pursued in lieu of waste reduction and reuse. The move toward more durable goods is an overlooked and underutilized option. In fact, waste reduction remains the most environmentally sound and least costly goal for MSW management, because wastes that are never generated do not need to be managed.

***States.*** Many states and municipalities set ambitious recycling goals in the 1980s and early 1990s. California, for example, mandated a 50% rate of diversion from landfills (via source reduction, recycling, and composting), with heavy fines for cities that failed to meet the goal. Many California cities have reached and surpassed the goal, but others are lagging behind. The state has begun to levy fines in support of the goal. For example, the city of Gardena was fined $70,000 in 2003 for its poor showing (a 13% diversion rate). Massachusetts opted for a statewide goal of 46% by 2000 and achieved 41%. A new solid-waste master plan has raised the goal to 60% of MSW reduction through source reduction and recycling combined, to be achieved by 2010. One objective of the plan is to completely phase out the export of solid waste to other states. EPA's target goal for the country is 35% by 2005, which seems readily achievable.

Banning the disposal of recyclable items in landfills and combustion facilities makes very good sense. Many states have incorporated this regulation into their management programs. For example, Massachusetts phased out yard waste in 1991, metals and glass in 1992, and recyclable papers and plastic in 1994. Landfill- and combustion-facility operators are supposed to conduct random truck searches and are authorized to turn back trucks with significant amounts of banned items.

Enacting a national bottle-deposit law would be a giant stride toward the reuse and recycling of beverage containers and would also greatly reduce roadside litter in states lacking bottle laws.

Finally, closing the "recycling loop" remains a significant action to be taken by governments to encourage recycling. A number of states have opted for one or more of the following approaches: (1) minimum postconsumer levels of recycled content for newsprint and glass containers; (2) requirements stating that purchases of certain goods must include recycled products even if they are more expensive than "virgin" products; (3) requirements that all packaging be reusable or be made (at least partly) of recycled materials; (4) tax credits or incentives that encourage the use of recycled materials in manufacturing; and (5) assistance in the development of recycling markets.

# revisiting the themes

## Sustainability

Sustainable natural systems recycle materials, but we create so many products that are difficult to recycle that it becomes necessary to be intentional about source reduction and product responsibility. The greatest departure from sustainable living is the interstate transfer of trash, which wastes energy and allows cities and states to go for the cheapest immediate solution and avoid coming to grips with a responsible approach to waste management.

## Stewardship

Instead of allowing economic considerations to dictate MSW management choices, many municipalities and organizations are consciously choosing options that reflect (and often bear the name of) stewardship. Here is one ideal stewardly waste management plan: (1) Emphasize source reduction wherever possible; (2) employ mandatory curbside recycling and a PAYT collection program; (3) if feasible, establish a MRF for efficient handling of recyclables (and possibly MSW); (4) employ co-composting of the remaining MSW with treated sewage sludge; (5) deposit residual materials in a local landfill; and (6) prohibit all interstate transfer of MSW. This plan features local MSW management, and local or regional recycling, composting, and landfilling.

## Policy and Politics

The role of the EPA in waste management is crucial. The agency sets the rules for landfill operations, but takes a proactive role in encouraging source reduction, recycling, and state MSW management programs. WasteWise and the PAYT program are two excellent examples of EPA's facilitating role. The greatest current public policy needs are a national bottle law and a law dealing with interstate waste transport (like the Solid Waste Interstate Transportation Act currently being considered).

# review questions

1. List the major components of municipal solid waste (MSW).
2. Trace the historical development of refuse disposal. What percentage now goes to landfills, combustion, and recycling?
3. What are the major problems with placing waste in landfills? How can these problems be managed?
4. Explain the difficulties accompanying landfill siting and outsourcing. How can these processes be handled responsibly?
5. What are the advantages and disadvantages of WTE combustion?
6. What is the evidence for increasing source reduction, and what are some examples of how it is accomplished?
7. What are the environmental advantages of recycling?
8. Describe the materials that are recycled and how recycling is accomplished.
9. Discuss the attributes of successful recycling programs.
10. Discuss MRFs, mixed waste processing, and mixed waste co-composting.
11. What laws has the federal government adopted to control solid-waste disposal?
12. What is integrated waste management? What would be a stewardly and sustainable solid waste management plan?

## thinking environmentally

1. Compile a list of all the plastic items you used and threw away this week. How can you reduce the number of items on the list?
2. How and where does your school dispose of solid waste? Is a recycling program in place? How well does it work?
3. Suppose your town planned to build a combustion facility or landfill near your home. Outline your concerns, and explain your decision to be for or against the site.
4. Does your state have a bottle bill? If not, what has prevented the bill from being adopted?
5. Consider the author's "ideal stewardly waste management plan" (in the "Stewardship" section of "Revisiting the Themes"). How many of these components does your city or town employ?
6. Read the "Ethics" essay on "affluenza." Take the sample quiz. If you score 10 or above, check out the "affluenza" Web site (http://www.affluenza.com/) and get help!

chapter 19

# Hazardous Chemicals: Pollution and Prevention

**Key Topics**

1. Toxicology and Chemical Hazards
2. A History of Mismanagement
3. Cleaning Up the Mess
4. Managing Current Hazardous Wastes
5. Broader Issues

There are strange things done in the midnight sun
by the men who moil for gold;
The arctic trails have their secret tales
that would make your blood run cold.
The Northern Lights have seen queer sights, but the
queerest they ever did see
Was that night on the marge of Lake LeBarge
I cremated Sam McGee.

So begins "The Cremation of Sam McGee" by Robert Service. In the poem, Lake LeBarge, in the Canadian Yukon Territory, eventually received the cremated ashes of Sam McGee. Unfortunately, the lake has been receiving a much different burden in recent years, to the point where its fish have become hazardous to eat because of their DDT content. Much to research scientists' surprise, fish, birds, and mammals all over the Arctic are showing elevated body burdens of a number of **persistent organic pollutants** (POPs, as they are called)—in particular, DDT, toxaphene, chlordane, PCBs, and dioxins. Moreover, because the Inuit people of the far north depend on these animals for food, they, too, are carrying high loads of POPs—indeed, some of the highest in the world.

The real mystery is how these chemicals—all of which are toxic—get to such remote and pristine environments. No one is spraying pesticides there, and there are no industries to produce PCBs and dioxins. The key to the presence of such chemicals can be found in several factors that characterize POPs: persistence, bioaccumulation, and the potential for long-range transport. Added to these factors is the unique climate of the Arctic, where the extreme cold promotes a process called "cold condensation." POPs are semivolatile chemicals carried in global air patterns to the Arctic, where they condense on the snowpack and are then washed into the water with the spring thaw. Once in the lakes and coastal waters, the chemicals are picked up by plankton and passed up the food chain through bioaccumulation and biomagnification. (See Fig. 16–9.)

A recent Canadian government study found that three-fourths of Inuit women have PCB levels up to

**A Nunavut Hazard This Inuit mother is scraping the fat from a sealskin on Baffin Island, Nunavut. Seals and other animals used for food by the Inuit harbor high levels of persistent organic pollutants (POPs).**

five times above those deemed safe. Dioxins in Nunavut territory are transferred to the Inuit via caribou, a major food source, as well as through the aquatic food chain. The caribou get the dioxins from deposition on lichens and mosses. Although some of the POPs are on the decline in parts of the Arctic, some new arrivals—brominated fire retardants (polybrominated diphenyl ethers, or PDBEs)—have begun to accumulate in polar bears, seals, and foxes.

***Boon or Bane?*** The potential health effects of the chemical load carried by the Arctic natives range from immune-system disorders to disruption of their hormone systems and, over time, to cancer. (All five of the POPs mentioned are on the National Toxicology Program's list of carcinogens.) Stories like that of the Inuit, and others that can only be called disasters, have convinced the public that significant dangers are associated with the manufacture, use, and disposal of many chemicals. Very few, however, would advocate giving up the advantages of the innumerable products that derive from our modern chemical industry. Better, instead, to learn to handle and dispose of chemicals in ways that minimize the risks. Thus, over the past 25 years, regulations surrounding the production, transport, use, and disposal of chemicals have mushroomed. From having almost no controls at all, the chemical industry is now among the most thoroughly regulated of all industries, and everyone using or handling chemicals that are deemed hazardous is affected by the regulations in one way or another.

This chapter focuses on hazardous chemicals—their nature and how they are investigated. An overview is presented of laws and regulations designed to protect human and environmental health, and the strengths and weaknesses of those measures are examined.

## 19.1 Toxicology and Chemical Hazards

**Toxicology** is the study of the harmful effects of chemicals on human and environmental health. A toxicologist might, for example, investigate the potential for *phenolphthalein* (a common chemical laboratory reagent that has also been used in over-the-counter laxatives) to cause health problems. The toxicologist might study the **acute** toxicity effects that might occur upon ingestion of the chemical or upon its contact with the skin, as well as the effects of **chronic** exposure over a period of years and, finally, the **carcinogenic** potential of phenolphthalein. Indeed, such studies have shown that phenolphthalein causes tumors in mice, and because of this finding, the FDA recently cancelled all laxative uses of phenolphthalein, which is now listed as "reasonably anticipated to be a human carcinogen."

Data on toxic chemicals are made available to health practitioners and the public via a number of sources, including the National Toxicology Program (NTP) Chemical Repository and the National Institute of Environmental Health Sciences (NIEHS), the latter of which provides an annual updated *Report on Carcinogens,* with a complete list of carcinogenic agents. The EPA also makes available data on toxic chemicals in its Integrated Risk Information System (IRIS). All of these sources of information are on the Internet. Environmental Defense, a nonprofit organization linking science, economics, and law to combat environmental problems, uses these and other data sources in its "Scorecard," an exhaustive Web-based profile of 6,800 chemicals, including information on their manufacture and uses.

### Dose Response and Threshold

In our discussion of risk assessment (Chapter 15), we introduced the concepts of *dose response* and *exposure.* In investigating a suspect chemical, a toxicologist would conduct tests on animals, investigate human involvement with the chemical, and present information linking the **dose** (the level of exposure multiplied by the length of time over which exposure occurs) with the **response** (some acute or chronic effect or the development of tumors). For example, studies of phenolphthalein toxicity indicate that the $LDL_0$ (lowest dose at which death occurred in animal testing) was 500 mg/kg. This is a low toxicity, so concern about the chemical would center on *chronic* or *carcinogenic* issues.

Human **exposure** to a hazard is a vital part of its risk characterization, and such exposure can come through the workplace, food, water, or the surrounding environment. For example, information in the NTP Chemical Repository gives the various uses of phenolphthalein, followed by both precautions to take in handling it and the symptoms of exposure (there are many!). Getting an accurate determination of human exposure is often the most difficult area of risk assessment.

***Threshold Level.*** In the dose-response relationship, there may or may not be a *threshold.* Organisms are able to deal with certain levels of many substances without suffering ill effects. The level below which no ill effects are observed is called the **threshold level.** Above this level, the effect of a substance depends on both its concentration and the duration of exposure to it. Higher levels may be tolerated if the exposure time is short. Thus, the threshold level is high for short exposures, but gets lower as the exposure time increases (Fig. 19–1). In other

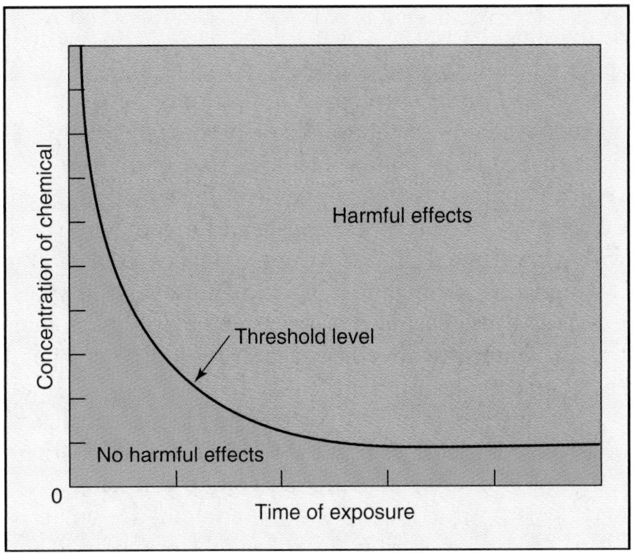


**Figure 19–1 The threshold level.** The threshold level for harmful effects of toxic pollutants is lower as the exposure time increases. It is also different for different chemicals.

words, it is not the absolute amount, but rather the *dose,* that is important.

Where carcinogens are concerned, the EPA generally takes a zero-dose, zero-response approach. That is, there is no evidence of a discrete threshold level for any carcinogenic chemicals. However, the lower the dose, the more likely it is that the response cannot be distinguished from the background level of cancers in a population. In such cases, the risk becomes very low and drops below the level for which regulatory action is needed.

These are just some basics of toxicology. The field is well established and is the most important source of sound scientific information that leads to regulation by the FDA and the EPA. The NTP was established in 1978 to provide this kind of information and has become the world's leader in assessing chemical toxicity and carcinogenicity. We turn our attention now to the chemicals themselves.

## The Nature of Chemical Hazards: HAZMATs

A chemical that presents a certain hazard or risk is known as a **hazardous material (HAZMAT).** The EPA categorizes substances on the basis of the following hazardous properties:

- **Ignitability.** Substances that catch fire readily (e.g., gasoline and alcohol)
- **Corrosivity.** Substances that corrode storage tanks and equipment (e.g., acids)
- **Reactivity.** Substances that are chemically unstable and that may explode or create toxic fumes when mixed with water [e.g., explosives, elemental phosphorus (not phosphate), and concentrated sulfuric acid]
- **Toxicity.** Substances that are injurious to health when they are ingested or inhaled (e.g., chlorine, ammonia, pesticides, and formaldehyde)

Containers in which HAZMATs are stored, and vehicles that carry HAZMATs, are required to display placards identifying the hazards (flammability, corrosiveness, the potential for poisonous fumes, and so on) of the material inside (Fig. 19–2).

*Radioactive materials,* discussed in Chapter 13, are probably the most hazardous of all and are treated as an entirely separate category.

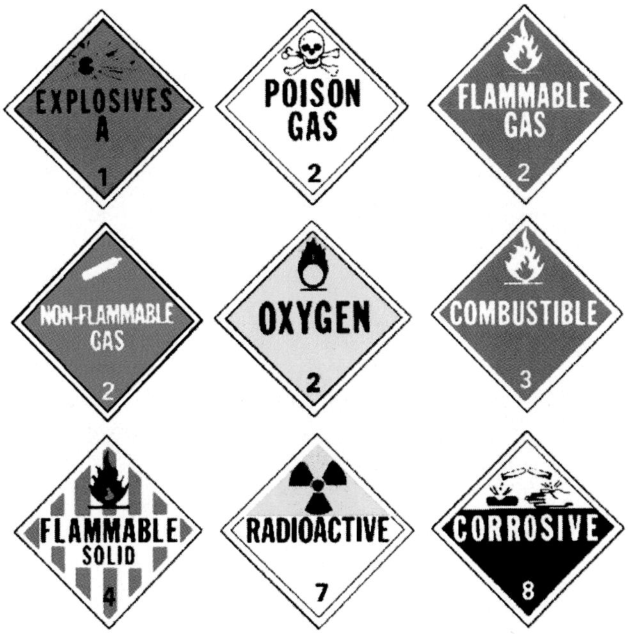


**Figure 19–2 HAZMAT placards.** These are examples of some of the placards that are mandatory on trucks and railcars carrying hazardous materials. Numbers in place of the word on the placard or on an additional orange panel will identify the specific material. Placards alert workers, police, and firefighters to the kinds of hazards they face in case an accident occurs.

## Sources of Chemicals Entering the Environment

To understand how HAZMATs enter our environment, we need to look at how people in our society live and work. First, the materials making up almost everything we use, from the shampoo and toothpaste we apply in the morning to the TV set we watch in the evening, are products of chemical technology. Our use constitutes only one step in the **total product life cycle,** a term that encompasses all steps, from obtaining raw materials to final disposal of the product. Implicit in our use of hair spray, for example, is that raw materials were obtained and various chemicals were produced to make both the spray and its container. Chemical wastes and by-products are inevitable in production processes. In addition, consider the risks of accidents or spills occurring in the manufacturing process and in the transportation of raw materials, the finished product, or wastes. Finally, what are the risks of breathing hair spray? What happens to the container you throw into the trash, which still holds some of the hair spray when the propellant is used up?

***Better Living Through . . . ?*** Multiply these steps by all the hundreds of thousands of products used by billions of people in homes, factories, and businesses, and you can appreciate the magnitude of the situation. More than 80,000 chemicals are registered for commercial use within the United States. At every stage—from mining raw materials through manufacturing, use, and final disposal—various chemical products and by-products enter the environment, with consequences for both human and environmental health (Fig. 19–3).

The use of many products—pesticides, fertilizers, and road salts, for example—directly introduces them into the environment. Alternatively, the intended use may leave a fraction of the material in the environment—the evaporation of solvents from paints and adhesives, for instance. Then there are the product life cycles of materials that are used tangentially to the

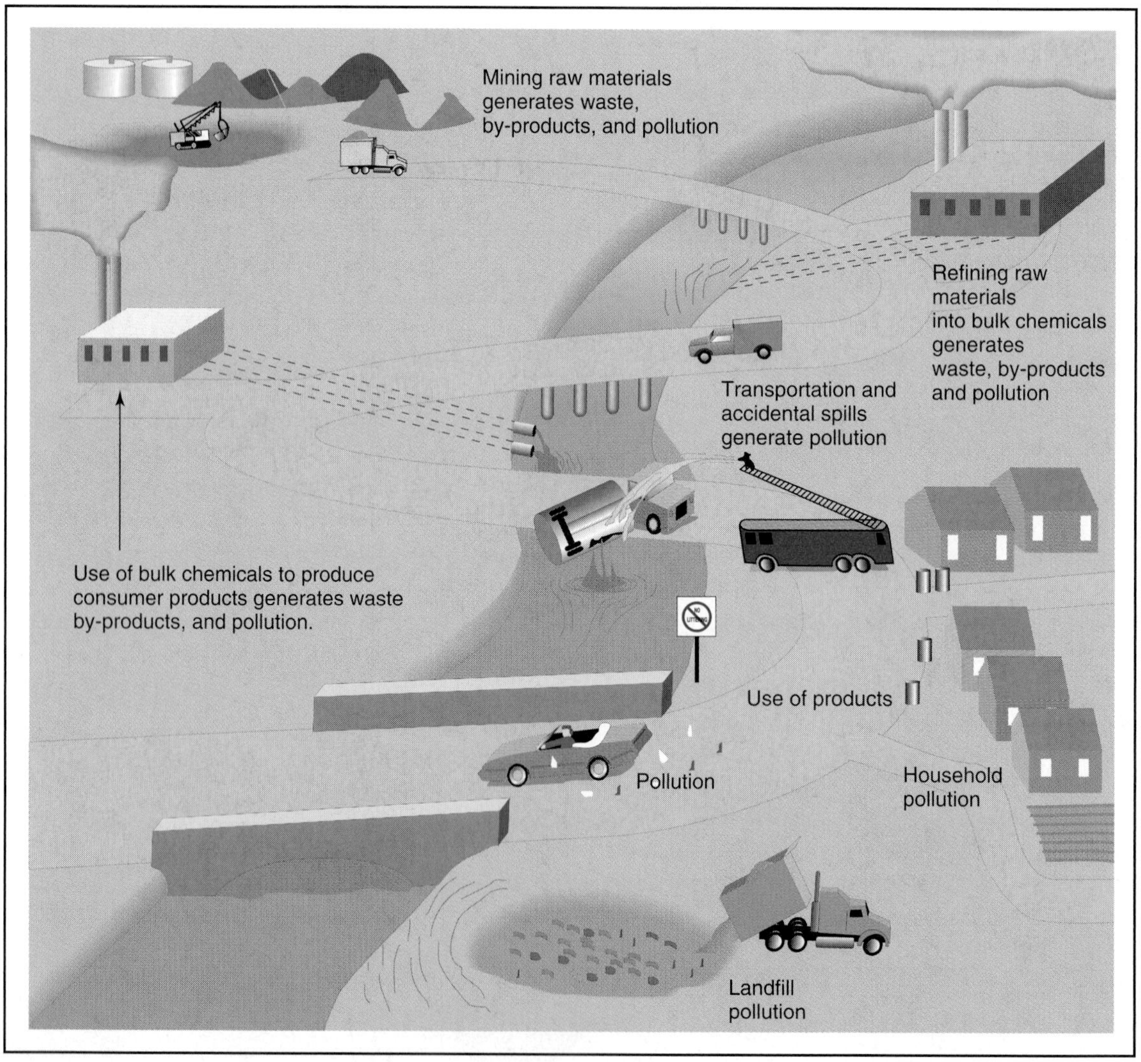


**Figure 19–3 Total product life cycle.** The life cycle of a product begins when the raw materials are obtained and ends when the used-up product is discarded. At each step—or in transport between steps—wastes, by-products, or the product itself may enter the environment, causing pollution and creating various risks to human and environmental health.

desired item. Consider lubricants, solvents, cleaning fluids, cooling fluids, and so on, with whatever contaminants they may contain. Likewise, there are the product life cycles of gasoline, coal, and other fuels that are consumed for energy. In addition to the unavoidable wastes and pollutants produced, in every case there is the potential for accidental releases, ranging from minor leaks from storage tanks to the release of toxic gas from a pesticide plant in Bhopal, India, in 1984, that killed at least 10,500 people.

**Toxics Release Inventory.** The introduction of chemicals into the environment may occur in every sector, from major industrial plants to small shops and individual homes. Whereas single events involving large amounts of one chemical may constitute a disaster and make headlines, the total amounts entering the environment from millions of homes and businesses is far greater and presents much more of a health risk to society. Some idea of the quantities involved can be obtained from the *Toxics Release Inventory (TRI)*. The **Emergency Planning and Community Right-to-Know Act (EPCRA) of 1986** requires industries to report the locations and quantities of toxic chemicals stored on each site to state and local governments and to report releases of toxic chemicals to the environment. [The TRI does not cover small businesses, such as dry-cleaning establishments and gas stations, or household hazardous waste (HHW).] The TRI is disseminated on the Internet and provides an annual record of releases of almost 650 designated chemicals. Waste not released to the environment was either recycled, combusted for energy, or treated on-site. For 2001, the TRI reveals the following information about toxic chemicals:

- Total production-related toxic wastes: 26,740 million lb.
- Releases to the air: 1,679 million lb.
- Releases to the water: 221 million lb.
- Releases to land disposal sites and underground injection: 4,258 million lb.
- Total environmental releases: 6,158 million lb (23% of wastes).

Add to the last figure the estimated 3.4 billion lb of HHW each year and an unknown quantity coming from small businesses, and you have some idea of the dimensions of the problem. The good news is that over the 13 years since the TRI has been in effect, the quantities of virtually all categories of toxic waste have kept going down (Fig. 19–4), and total releases have declined by 54%. (The data in the figure are based on a consistent set of chemicals and industries and so do not agree with the 2001 information presented previously. Many chemicals and industries have been added since the TRI began in 1988.)

## The Threat from Toxic Chemicals

All toxic chemicals, by definition, are hazards that pose a risk to humans. (See Chapter 15.) Fortunately, a large portion of the chemicals introduced into the environment are gradually broken down and assimilated by natural processes. Therefore, once these chemicals are diluted sufficiently, they pose no long-term human or environmental risk, even though they may be highly toxic in acute doses (high-level, short-term exposures).

Two major classes of chemicals do not readily degrade in the environment, however: (1) *heavy metals and their compounds* and (2) *synthetic organics*. Again, if sufficiently diluted in air or water, most of these compounds do not pose a hazard, although there are some notable exceptions.

**Heavy Metals.** The most dangerous heavy metals are lead, mercury, arsenic, cadmium, tin, chromium, zinc, and copper. These metals are used widely in industry, particularly in metalworking or metal-plating shops and

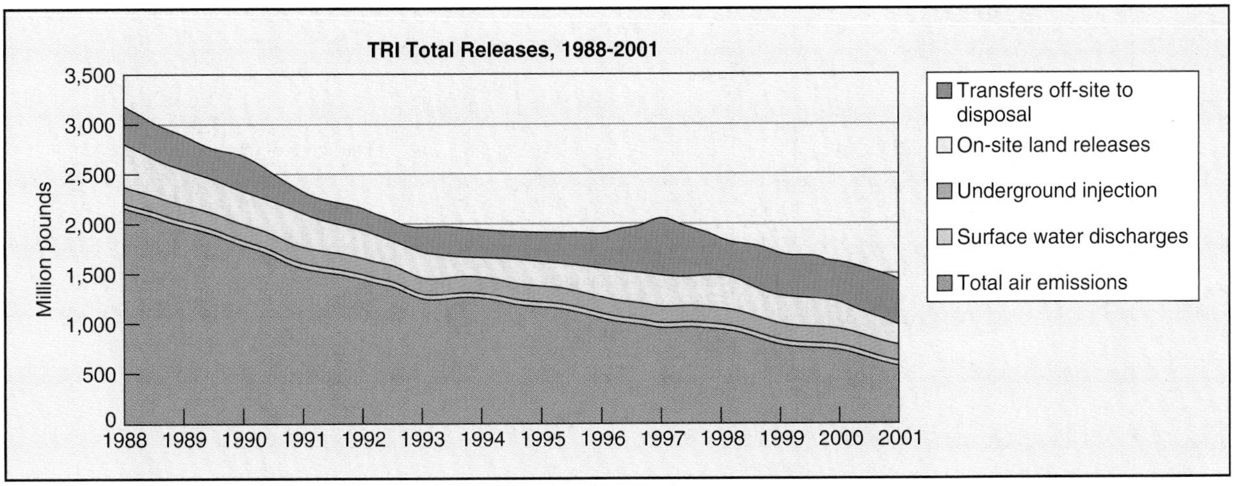


**Figure 19–4 Toxic Release Inventory Releases, 1988–2001.** The data here follow a consistent set of chemicals and industries from 1988 to 2001, showing changes in the amounts released to the air and water, disposed of on land, or transferred off-site for treatment and disposal. (*Source:* EPA, 2001 Toxics Release Inventory, July 2003)

in such products as batteries and electronics. They are also used in certain pesticides and medicines. In addition, because compounds of heavy metals can have brilliant colors, they are used in paint pigments, glazes, inks, and dyes. Thus, heavy metals may enter the environment wherever any of these products are produced, used, and ultimately discarded.

Heavy metals are extremely toxic because, as ions or in certain compounds, they are soluble in water and may be readily absorbed into the body, where they tend to combine with and inhibit the functioning of particular vital enzymes. Even very small amounts can have severe physiological or neurological consequences. The mental retardation caused by lead poisoning and the insanity and crippling birth defects caused by mercury are particularly well-known examples.

**Organic Compounds.** Petroleum-derived and *synthetic* organic compounds are the chemical basis for all plastics, synthetic fibers, synthetic rubber, modern paintlike coatings, solvents, pesticides, wood preservatives, and hundreds of other products. Because of their chemical structure, many synthetic organics are resistant to biodegradation. Ironically, this resistance is an important part of what makes many such compounds useful: We wouldn't want fungi and bacteria attacking and rotting the tires on our cars, and paints and wood preservatives function only insofar as they are both nonbiodegradable and toxic to decomposer organisms.

These compounds are toxic because they are often readily absorbed into the body, where they interact with particular enzymes, but their nonbiodegradability prevents them from being broken down or processed further. When a person ingests a sufficiently high dose, the effect may be acute poisoning and death. With low doses over extended periods, however, the effects are insidious and can be mutagenic (mutation causing), carcinogenic (cancer causing), or teratogenic (birth defect causing). Synthetic organic compounds also may cause serious liver and kidney dysfunction, sterility, and numerous other physiological and neurological problems.

***Dirty Dozen.*** A particularly troublesome class of synthetic organics is the **halogenated hydrocarbons,** organic compounds in which one or more of the hydrogen atoms have been replaced by atoms of chlorine, bromine, fluorine, or iodine. These four elements are classed as *halogens*—hence the name *halogenated hydrocarbons* (Fig. 19–5). Of the halogenated hydrocarbons, the **chlorinated hydrocarbons** (also called **organic chlorides**) are by far the most common. Organic chlorides are widely used in plastics (polyvinyl chloride), pesticides (DDT, Kepone, and mirex), solvents (carbon tetrachlorophenol and tetrachloroethylene), electrical insulation (polychlorinated biphenyls), and many other products. Most of the so-called "dirty dozen" POPs (Table 19–1) are halogenated hydrocarbons. All are toxic to varying extents, and most are known animal carcinogens. Many are also suspected endocrine disruptors at very low levels. (See Chapter 16.) These 12 POPs have been banned or highly restricted as a result of the **Stockholm Convention on Persistent Organic Pollutants,** signed in 2001 by delegates from 127 countries.

***PERC.*** Let us consider one of the halogenated hydrocarbon compounds, tetrachloroethylene (Fig. 19–5), in a simple case study. This substance, also called perchloroethylene, or PERC, is colorless and nonflammable and is the major substance in dry-cleaning fluid. It is an effective solvent and finds uses in all kinds of industrial cleaning operations. It can also be found in shoe polish and typewriter correction fluid. PERC evaporates readily

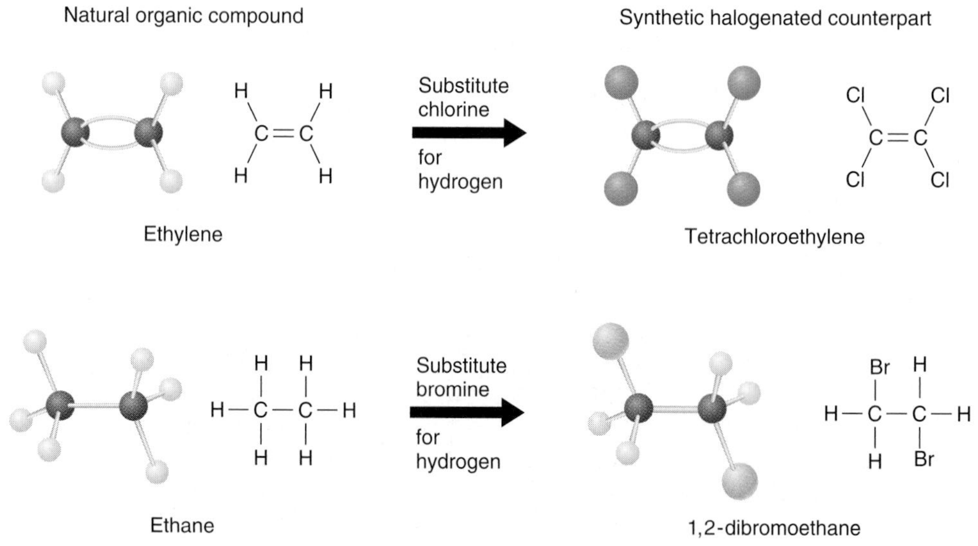


**Figure 19–5** **Halogenated hydrocarbons.** These are organic (carbon-based) compounds in which one or more hydrogen atoms have been replaced by halogen atoms (chlorine, fluorine, bromine, or iodine). Such compounds are particularly hazardous to health because they are nonbiodegradable and they tend to bioaccumulate. Shown here are tetrachloroethylene and 1,2-dibromoethane.

**table 19-1 The "Dirty Dozen" Persistent Organic Pollutants (POPs)**

| Chemical Substance | Designed Use | Major Concerns |
|---|---|---|
| Aldrin | Pesticide to control soil insects and to protect wooden structures from termites | Toxic to humans, may be carcinogenic |
| Chlordane | Broad-spectrum insecticide to protect crops | Biomagnification in food webs |
| DDT | Widely used insecticide, malaria control | Biomagnification in food webs |
| Dieldrin | Termite control, crop-pest control | Toxic, biomagnification in food webs, high persistence |
| Endrin | Insecticide and rodenticide | Toxic, especially in aquatic systems |
| Heptachlor | General insecticide | Toxic, carcinogenic |
| Hexachlorobenzene | Fungicide | Toxic, carcinogenic |
| Mirex | Insecticide against ants | High toxicity to aquatic animals, carcinogenic |
| Toxaphene | General insecticide | High toxicity to aquatic animals, carcinogenic |
| PCBs | Variety of industrial uses, especially in transformers and capacitors | Toxic, teratogenic, carcinogenic |
| Dioxins | No known use; by-products of incineration and paper bleaching | Toxic, carcinogenic, reproductive system effects |
| Furans | No known use; by-products of incineration, PCB production | Toxic, especially in aquatic systems |

*Source:* UNEP Persistent Organic Pollutants from United Nations Environment Program—Chemicals Division (http://irptc.unep.ch/pops/).

when exposed to air, but in the soil it can enter groundwater easily because it does not bind to soil particles. Human exposure can occur in the workplace, especially in connection with dry-cleaning operations. It can also occur if people use products containing PERC, such as dry-cleaned garments. PERC enters the body most readily when breathed in with contaminated air. Breathing PERC for short periods can bring on dizziness, fatigue, headaches, and unconsciousness. Over longer periods, PERC can cause liver and kidney damage.

Laboratory studies also show PERC to be carcinogenic, and it is listed in NTP's *Report on Carcinogens* as "reasonably anticipated to be a human carcinogen." Indeed, a study by the National Institute for Occupational Safety and Health revealed that dry-cleaning workers showed elevated rates of esophageal and bladder cancer. PERC is not known to cause any environmental harm at levels normally found in the environment, except that it does add to the burden of volatile organic chemicals causing photochemical smog. PERC is produced in large amounts in the United States, and according to the EPA's TRI, 2.8 million pounds of PERC were released to the environment in 2001 from industrial sources, mostly by evaporation.

*MTBE.* An especially troublesome issue emerged as officials implemented Clean Air Act regulations to create a cleaner burning gasoline. Methyl tertiary butyl ether (MTBE) is an oxygenate now added to a quarter of the gasoline sold in the United States. It is a suspected carcinogen and is being found in increasing levels in wells and surface waters throughout the country. Because it has been added to gasoline, it shows up in surface waters wherever two-stroke engines are used for watercraft and in groundwater wherever underground gasoline storage tanks develop leaks. MTBE imparts a nasty odor to water at very low concentrations. A number of states have recently moved to ban the chemical (it will take several years to do so, however), which is good news for Midwestern corn growers, as ethyl alcohol is the likely substitute for MTBE as an oxygenate. (See Chapter 14.)

## Involvement with Food Chains

The trait that makes heavy metals and nonbiodegradable synthetic organics particularly hazardous is their tendency to accumulate in organisms. We discussed the phenomena of bioaccumulation and biomagnification in connection with DDT in Chapter 16.

*Minamata.* A tragic episode in the early 1970s, known as Minamata disease, revealed the potential for biomagnification of mercury and other heavy metals. The disease is named for a small fishing village in Japan where the episode occurred. In the mid-1950s, cats in Minamata began to show spastic movements, followed by partial paralysis, coma, and death. At first, this was thought to be a syndrome peculiar to felines, and little attention was paid to it. However, when the same symptoms began to occur in people, concern escalated quickly. Additional symptoms, such as mental retardation, insanity, and birth defects were also observed. Scientists and health experts eventually diagnosed the cause as acute mercury poisoning.

A chemical company near the village was discharging wastes containing mercury into a river that flowed into the bay where the Minamata villagers fished. The mercury, which settled with detritus, was first absorbed and bioaccumulated by bacteria and then biomagnified as it passed up the food chain through fish to cats or to humans. Cats had suffered first and most severely because they fed almost exclusively on the remains of fish. By the time the situation was brought under control, some 50 persons had died and 150 had suffered serious bone and nerve damage. Even now, the tragedy lives on in the crippled bodies and retarded minds of some Minamata descendants.

## 19.2 A History of Mismanagement

In the past, chemical wastes of all kinds were disposed of as quickly and easily as possible. From the dawn of the Industrial Age to the relatively recent past, it was common practice to exhaust all combustion fumes up smokestacks, vent all evaporating materials and solvents into the air, and flush all waste liquids and contaminated wash water into sewer systems or directly into natural waterways. Many human health problems ensued, but they either were not recognized as being caused by the pollution or were accepted as the "price of progress." Indeed, much of our understanding regarding human health effects of hazardous materials is derived from those uncontrolled exposures. For example, the expression "mad as a hatter" comes from the fact that people who made hats in the 1800s frequently became insane. The insanity, it was later found, was caused by poisoning from the mercury used in the production process.

In the 1950s, as production expanded and synthetic organics came into widespread use in the developed countries, many streams and rivers essentially became open chemical sewers, as well as sewers for human waste. These waters not only were devoid of life, but were themselves hazardous. For example, in the 1960s, the Cuyahoga River, which flows through Cleveland, Ohio, carried so much flammable material that it actually caught fire and destroyed seven bridges before the fire burned itself out (Fig. 19–6). Worsening pollution (both chemical and sewage) and increasing recognition of adverse health effects finally created a degree of public outrage that pushed Congress to pass the Clean Air Act of 1970 and the Clean Water Act of 1972. These acts set standards for allowable emissions into air and water and timetables for reaching those standards.

The Clean Air and Clean Water Acts and their subsequent amendments remain cornerstones of environmental legislation. However, their passage in the early 1970s left an enormous loophole. If you can't vent wastes into the atmosphere or flush them into waterways, what do you do with them? Industry turned to *land disposal,* which was essentially unregulated at the time, as an expedient alternative. Indiscriminate air and water disposal became indiscriminate land disposal. Thus, in retrospect, we see that the Clean Air and Clean Water Acts, for all their benefits in improving air and water quality, also succeeded in transferring pollutants from one part of the environment to another.

### Methods of Land Disposal

In the early 1970s, there were three primary land-disposal methods: (1) deep-well injection, (2) surface impoundments, and (3) landfills. With the conscientious implementation

**Figure 19–6 Cuyahoga River on fire.** Prior to laws and regulations curtailing pollution, all manner of wastes were indiscriminately discharged. In the 1960s, the Cuyahoga River actually caught fire. Incidents such as this contributed to public outrage that led to the passage of the Clean Water Act of 1972.

of safeguards, each of these methods has some merit, and each is still heavily used for hazardous-waste disposal. Without adequate regulations or enforcement, however, contamination of groundwater is virtually inevitable.

**Deep-Well Injection.** Deep-well injection involves drilling a "well" thousands of feet below groundwater into a porous geological formation or brine. A well consists of concentric pipes and casings that isolate the wastes as they are injected, and the well is sealed at the bottom to prevent wastes from backing up (Fig. 19–7). Over time, the wastes often will undergo reactions with naturally occurring material that will make them less hazardous. This method is presently used for various volatile organic compounds, pesticides, fuels, and explosives. Currently, 163 wells are operated at 51 locations, mostly in the Gulf Coast region. The EPA's Underground Injection Control Program regulations require that the wells be limited to geologically stable areas, so that the hazardous-waste liquids pumped into the well remain isolated indefinitely. The total amount of deep-well injection has declined over the years, from 685 million tons in 1988 to 116 million tons in 2001 (representing 5.4% of on-land disposal). The technique remains useful because it has a greater potential than other methods for keeping toxic wastes from contaminating the hydrologic cycle and the food web.

**Surface Impoundments.** Surface impoundments are simple excavated depressions ("ponds") into which liquid wastes are drained and held. They were the least expensive, and hence most widely used, way to dispose of large amounts of water carrying relatively small amounts of chemical wastes. The impoundments continue to be widely used, often to carry out wastewater treatment prior to discharging the wastes. As waste is discharged into the pond, solid wastes settle and accumulate while water evaporates (Fig. 19–8). If the bottom of the pond is well sealed, and if the loss of water through evaporation equals the gain from input, impoundments

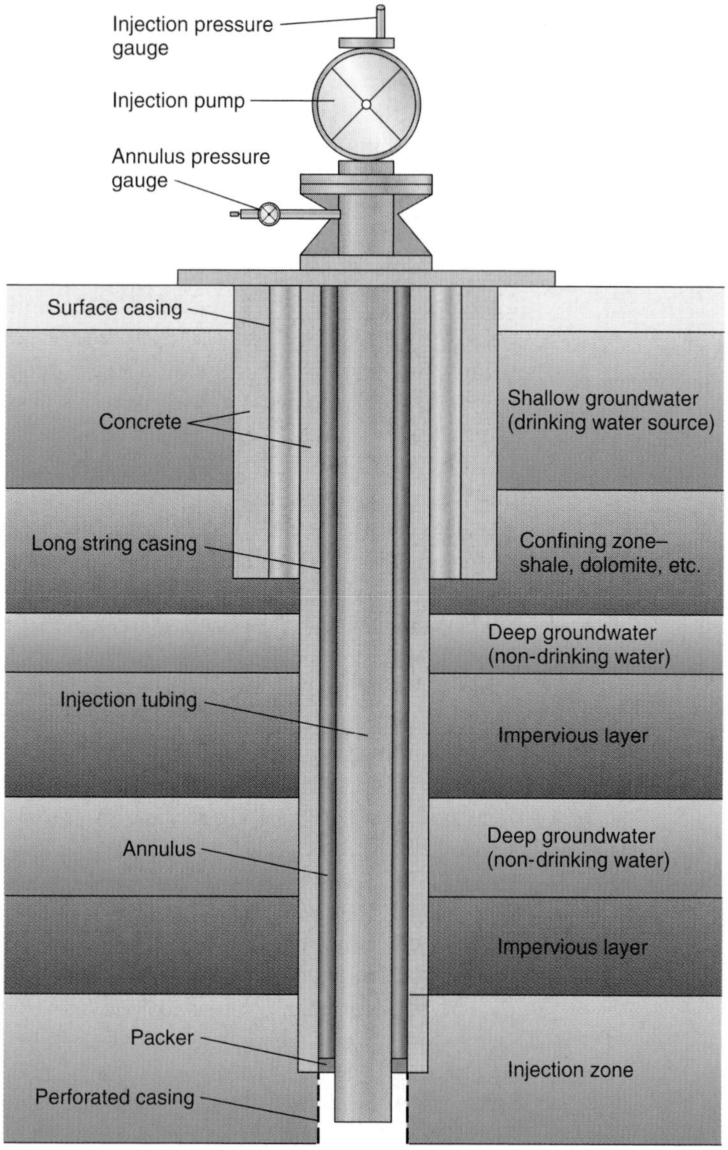


**Figure 19–7 Deep-well injection.** This technique is used for the disposal of large amounts of liquid wastes. The concept is that toxic wastes may be drained into dry, porous strata below ground, where they may reside harmlessly "forever." The injection tubing is surrounded by a space (the annulus) which is filled with an inert, pressurized fluid. The long string casing is filled with cement from the surface to the injection zone. (Adapted from the EPA Underground Injection Program Web site)

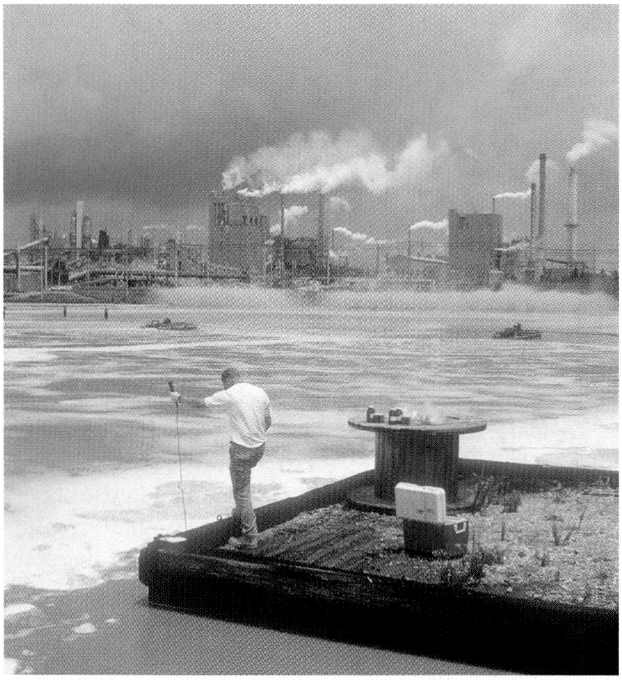

**Figure 19–8 Surface impoundment.** Surface impoundments are often used for preliminary treatment of liquid wastes. Shown here is an impoundment at a paper manufacturer in Alabama.

may receive wastes indefinitely. However, inadequate seals may allow wastes to percolate into groundwater, exceptional storms may cause overflows, and volatile materials can evaporate into the atmosphere, adding to air pollution problems.

A 1996 **Resource Conservation and Recovery Act (RCRA)** amendment prompted the EPA to conduct a study of surface impoundments. The study (*Industrial Surface Impoundments in the United States*), released in 2001, surveyed the scope of impoundments, the characteristics of the wastes they managed, and the potential risks posed by chemical constituents in existing surface impoundments. The 2001 TRI reports 965 million pounds of toxics released to on-site surface impoundments. (*On-site* refers to disposal by manufacturers or institutions on their own facilities; *off-site* means that the wastes have been transferred to a waste-treatment or disposal facility.) Some of the findings in the EPA study (which was limited to the 1990–2000 time frame) are as follows:

- Approximately 18,000 industrial surface impoundments were in use at 7,500 facilities, mostly east of the Mississippi or along the Pacific Coast.
- At least two-thirds of the impoundments contained hazardous chemicals with carcinogenic and other human health concerns, originating largely from the chemical, paper, concrete, and petroleum industries.
- Most of the impoundments were only a few meters above groundwater, and more than half lacked liner systems that would prevent the movement of chemicals into groundwater or soil.
- More than 20 million people lived within 1.2 miles (2 km) of an industrial impoundment, and some 10% of impoundments were located within 500 feet (150 meters) of a domestic drinking-water well.
- From 2 to 5% of the sites were judged to pose possible risks to human health, whereas 24% were found to pose a risk of release to the environment.

The study concluded that existing state and federal regulations should be adequate to cover most of the impoundment-related problems; however, in view of the existing risks, gaps in regulatory coverage clearly exist and may need future regulations, but the EPA has not yet determined what these might be.

**Landfills.** RCRA sets rigid standards for the disposal of hazardous wastes to landfills. When hazardous wastes are in a concentrated liquid or solid form, they are commonly put into drums, typically collected by a waste management company (for a hefty fee!), and then treated in accordance with their chemical and physical characteristics. *Treatment standards* for wastes are established by the EPA and characterized as **best-demonstrated available technologies (BDATs)**. These standards apply to all solid and liquid hazardous wastes and have the effect of reducing their toxicity and mobility. The BDAT technologies include stabilization and neutralization of sludges, soils, liquids, powders, and slurries; chemical oxidation of organic materials; and various specific techniques applied to metal-bearing wastes that convert them to an insoluble solid material for landfill disposal. Some 447 million pounds of wastes were delivered to on-site landfills in 2001, and an additional 365 million pounds were deposited in off-site landfills and surface impoundments (categories not separated by TRI). Only 23 landfills are licensed to receive off-site hazardous wastes in North America, and these are owned by just a few waste management companies (e.g., Safety-Kleen, Waste Management, and Envirosource).

If a landfill is properly lined, supplied with a system to remove leachate, provided with monitoring wells, and appropriately capped, it may be reasonably safe and is referred to as a **secure landfill** (Fig. 19–9). The various barriers are subject to damage and deterioration, however, requiring adequate surveillance and monitoring systems to prevent leakage.

Because early land disposal was not regulated, in numerous instances not even the most rudimentary precautions were taken. Indeed, in many cases, deep wells were injecting wastes directly into groundwater, abandoned quarries were sometimes used as landfills with no additional precautions being taken, and surface impoundments frequently had no seals or liners whatsoever. Even worse, considerable amounts of waste failed to get to any disposal facility at all.

**Figure 19–9 Secure landfill.** Landfills are a technique that is widely used for disposing of concentrated chemical wastes. Shown is an operating hazardous-waste landfill in Alabama. (Adapted from an illustration by Rick Farrell. All rights reserved.)

**Midnight Dumping and Orphan Sites.** The need for alternative methods to dispose of waste chemicals created an opportunity for a new enterprise: waste disposal. Many reputable businesses entered the field, but—again in the absence of regulations—there were also disreputable operators. As stacks of drums filled with hazardous wastes "mysteriously" appeared in abandoned warehouses, vacant lots, or municipal landfills, it became clear that some operators simply were pocketing the disposal fee and then unloading the wastes in any available location, frequently under cover of darkness—an activity termed **midnight dumping** (Fig. 19–10). Authorities trying to locate the individuals responsible would learn that they had gone out of business and were nowhere to be found.

Some companies or individuals simply stored wastes on their own properties and then went out of business, abandoning the property and the wastes. These locations became known as **orphan sites**—hazardous-waste sites without a responsible party to clean them up. As drums containing hazardous chemicals corroded and leaked, there was great danger of reactive chemicals combining and causing explosions and fires. One of the most famous abandoned sites is the "Valley of the Drums" (VOD), in Kentucky (Fig. 19–11).

**Figure 19–10 Midnight dumping.** Hazardous wastes were often left on remote or unoccupied properties by unscrupulous haulers.

## Scope of the Mismanagement Problem

The mounting problem of unregulated land disposal of hazardous wastes was brought vividly to public attention by the episode at Love Canal, near Niagara Falls, New York. The area was occupied by a school and a number of houses, all of which were perched on top of a chemical waste dump that had been filled over and developed. The surface of the dump began to collapse, exposing barrels of chemical wastes. Fumes and chemicals began seeping into cellars. Ominously, people began reporting serious health problems, including birth defects and miscarriages. An aroused neighborhood, led by activist Lois Gibbs, began demanding that the state do something about the problem. Following confirmation of the contamination by toxic chemicals, President Carter signed an emergency declaration in 1978 to relocate hundreds of residents. The state and federal governments closed the school and demolished a number of the homes nearest the site (Fig. 19–12). In all, more than 800 families moved out of the area because of the fear of damage to their health from the toxic chemicals.

*Occidental?* The absence of a public policy for dealing with the disposal of hazardous chemicals contributed to the situation at Love Canal. Hooker Chemical and Plastics Company had purchased an abandoned canal near Niagara Falls in 1942 and proceeded to fill it with an estimated 17,000 tons of hazardous wastes. Hooker covered the canal over with a clay cap and sold it to the school board for a small sum, reportedly after warning the board that there were chemicals buried on the property. Subsequent

**Figure 19–11 "Valley of the Drums," an orphan waste site.** Thousands of drums of waste chemicals, many of them toxic, were unloaded near Louisville, Kentucky, around 1975 and left to "rot," seriously threatening the surrounding environment, waterways, and aquifers.

construction penetrated the clay cap, and rain seeped in and leached chemicals in all directions. Hooker Chemical's parent company, Occidental Petroleum, eventually spent more than $233 million on the cleanup and subsequent lawsuits.

As both government and independent researchers began surveying the extent of the problem, bad disposal practices were found to be rampant. The World Resources Institute estimated that, in the United States in the early 1980s, there existed 75,000 active industrial

**Figure 19–12 Love Canal.** These are a few of the nearly one hundred homes that the state was forced to buy and later destroy as a result of contamination from Love Canal. The area has now been cleaned up and redeveloped with new homes.

landfill sites, along with 180,000 surface impoundments and 200 other special facilities that were or could be sources of groundwater contamination. In most cases, the contaminated area was relatively small—200 acres (80 ha) or less—but in total, the problem was immense and affected every state in the country. As studies and tests proceeded, thousands of individual wells and some major municipal wells were closed because they were contaminated with toxic chemicals.

Many cases of private-well contamination caused by careless disposal were discovered only after people experienced "unexplainable" illnesses over prolonged periods. While these incidents did not receive the media attention of Love Canal, they were nevertheless devastating to the people involved. Even more important than the damage already done was the recognition that the problem was ongoing.

The problems concerning toxic chemical wastes can be divided into three areas:

- cleaning up the "messes" already created, especially where they threaten drinking-water supplies;
- regulating the handling and disposal of wastes currently being produced, so as to protect public and environmental health; and
- reducing the quantity of hazardous waste produced.

Each of these problems is addressed in the sections that follow.

## 19.3 Cleaning Up the Mess

A major public-health threat from land-disposed toxic wastes is the contamination of groundwater that is subsequently used for drinking. The first priority is to ensure that people have safe water. The second is to clean up or isolate the source of pollution so that further contamination does not occur.

### Ensuring Safe Drinking Water

To protect the public from the risk of toxic chemicals contaminating drinking-water supplies, Congress passed the **Safe Drinking Water Act of 1974.** Under this act, the EPA sets national standards (*Drinking Water Standards and Health Advisories*) to protect the public health, including allowable levels of 83 specific contaminants. If any contaminants are found to exceed maximum contaminant levels (MCLs), the water supply should be closed until adequate purification procedures or other alternatives are adopted. The act was amended in 1986, and the EPA now has jurisdiction over groundwater and sets MCLs for 90 contaminants. States and public water agencies are required to monitor drinking water to be sure that it meets the standards. (See Chapter 17.)

Because of concerns about public drinking water, many people have turned to drinking bottled water to avoid presumed contamination of municipal water supplies. Bottled water is considered a food and, as such, is regulated by the FDA. However, FDA standards for bottled water specify only that it be as safe as tap water. Ironically, consumers pay for a product for which there are no guarantees that it is any better than tap water. Indeed, numerous samples of bottled water in one study showed higher levels of contamination than did municipal supplies.

### Groundwater Remediation

If dumps, leaking storage tanks, or spills of toxic materials have contaminated groundwater, and if such groundwater is threatening water supplies, all is not yet lost. **Groundwater remediation** is a developing and growing technology. Techniques involve drilling wells, pumping out the contaminated groundwater, purifying it, and reinjecting the purified water back into the ground or discharging it into surface waters (Fig. 19–13). Cleaning up the source of the contamination is mandatory. If, however, the contamination is extensive, remediation may not be possible, and the groundwater must be considered unfit for use as drinking water.

### Superfund for Toxic Sites

Probably the most monumental task we are facing is the cleanup of the tens of thousands of toxic sites resulting from the years of mismanaged disposal of toxic materials. Wherever facilities were still operating, pressures were brought to bear on the operators to clean up the sites. Many operators who could not afford to do so, however, simply declared bankruptcy, and the sites joined those already abandoned.

The **Comprehensive Environmental Response, Compensation, and Liability Act of 1980 (CERCLA),** popularly known as Superfund, initiated a major federal program aimed at cleaning up abandoned chemical waste sites. Through a tax on chemical raw materials (the authorization for collecting more tax monies expired in 1995), this legislation provides a trust fund for the identification of abandoned chemical waste sites, protection of groundwater near the site, remediation of groundwater if it has been contaminated, and cleanup of the site. Federal spending for the program grew from about $300 million per year in the early 1980s to $2 billion in 1995, but the fund is expected to run out of money by 2004. Congress has refused to renew the tax on industry, leaving U.S. taxpayers with the continuing cleanup bill. However, the cleanup job is nowhere near complete. A recent study commissioned by Congress (*Superfund's Future: What Will It Cost?*) estimated that it would cost between $14 and $16.4 billion to clean up the remaining nonfederal listed sites in the 2000–2010 decade. Superfund, one of the EPA's largest ongoing programs, works as described in the next several subsections.

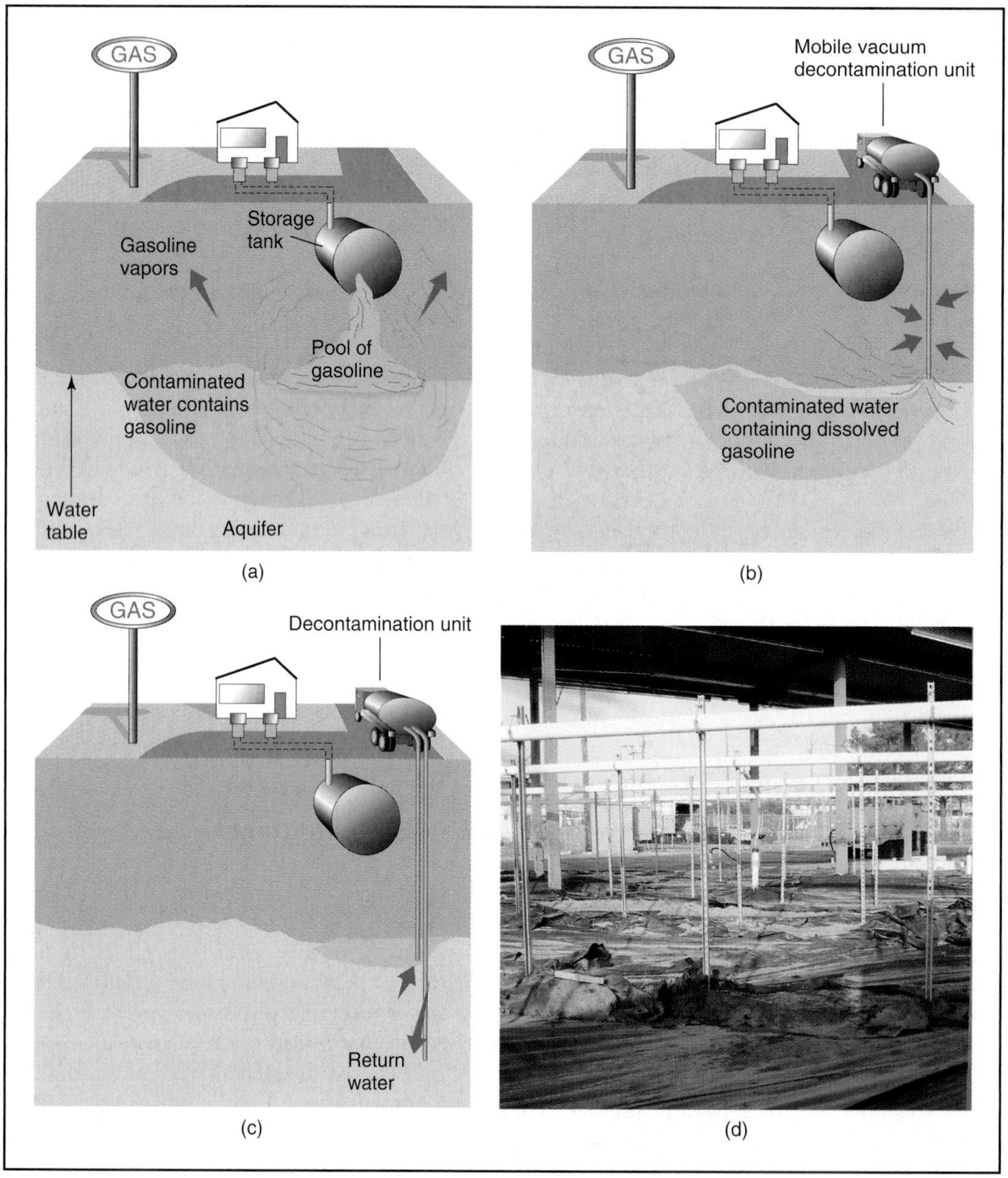


**Figure 19–13 Leaking underground storage tank remediation.** (a) Typical subsurface contamination from a leaking fuel tank at a gas station. (b) After the leak has been repaired, the vacuum-extraction process causes gasoline and residual hydrocarbons in the soil and on the water table to evaporate and then removes the vapors, preventing further contamination of the groundwater. (c) Contaminated groundwater is pumped out, treated, and returned to the ground. (d) Soil vapor extraction system removing volatile organic compounds in soil at an abandoned gasoline station.

**Setting Priorities.** Resources are insufficient to clean up all sites at once. Therefore, a system for setting priorities has been developed:

- As sites are identified—note that many abandoned sites had long since been forgotten—their current and potential threat to groundwater supplies are initially assessed by taking samples of the waste, determining its characteristics, and testing groundwater around the site for contamination. If it is determined that no immediate threat exists, nothing more may be done.
- If a threat to human health does exist, the most expedient measures are taken immediately to protect the public. These measures may include digging a deep trench, installing a concrete dike around the site, and recapping it with impervious layers of plastic and clay to prevent infiltration. Thus, the wastes are isolated, at least for the short term. If the situation has gone past the threat stage, and contaminated groundwater is or will be reaching wells, remediation procedures are begun immediately.
- The worst sites (those presenting the most immediate and severe threats) are put on a **National Priorities**

**List** (NPL) and scheduled for total cleanup. A site on this list is reevaluated to determine the most cost-effective method of cleaning it up, and finally cleanup begins with efforts to identify "responsible parties," namely, industries in which the wastes originated. The industries are "invited" to contribute to the cleanup, and they do so either by contributing financially or by participating in cleanup activities.

**Cleanup Technology.** If the chemical wastes are contained in drums, the drummed wastes can be picked up, treated, and managed in proper fashion. For the VOD site (Fig. 19–11), the EPA cleanup started by removing some 4,200 drums for off-site treatment and disposal. The bigger problem for many sites is the soil (often millions of tons) contaminated by leakage. One procedure is to excavate contaminated soil and run it through an incinerator or kiln assembled on the site to burn off chemicals (Fig. 19–14). (Synthetic organic chemicals can be broken down by incineration, and heavy metals are converted to insoluble, stable oxides.) Another method is to drill a ring of injection wells around the site and a suction well in the center. Water containing a harmless detergent is injected into the injection wells and drawn into the suction well, cleansing the soil along the way. The withdrawn water is treated to remove the pollutants and is then reused for injection.

The VOD site was situated in a poorly drained shale and limestone deposit. No local use was made of the groundwater. The EPA determined that the most feasible (BDAT) treatment was simply to contain the contaminated soil and groundwater, so the agency installed a clay cap, a perimeter drainage system, monitoring wells, and a security fence. The remedial work was completed in 1989, at a total cost of $2.5 million, and the site was deleted from the NPL in 1996. The third five-year review (in 2003) indicated that the cap was functioning well, the site was still undeveloped, and no human uses were being made of groundwater.

***Bioremediation.*** Another rapidly developing cleanup technology is *bioremediation.* In many cases, the soil is contaminated with toxic organic compounds that are biodegradable. The problem is that they do not degrade, because the soil lacks organisms, oxygen, or both. In **bioremediation,** oxygen and organisms are injected into contaminated zones. The organisms feed on and eliminate the pollutants (as in secondary sewage treatment) and then die when the pollutants are gone. Considerable research is being done to find and develop microbes that will break down certain kinds of wastes more readily. Bioremediation, which may be used in place of detergents to decontaminate soil, is a rapidly expanding and developing technology that is applicable to leaking storage tanks and spills, as well as to waste-disposal sites.

***Plant Food?*** When the soil contaminants are heavy metals and nonbiodegradable organic compounds, *phytoremediation* has been employed with some success. **Phytoremediation** uses plants to accomplish a number of desirable cleanup steps: stabilizing the soil, preventing further movement of contaminants by erosion, and extracting the contaminants by direct uptake from the soil. After full growth, the plants are removed and treated as toxic waste products. However, this is a slow process and can be used only at sites where the contaminants and their concentrations are not toxic to plants. Nevertheless, scientists have found sunflowers that will capture uranium, poplar trees that soak up dry-cleaning solvents, and ferns that thrive on arsenic. Phytoremediation is now used at a level above $100 million a year.

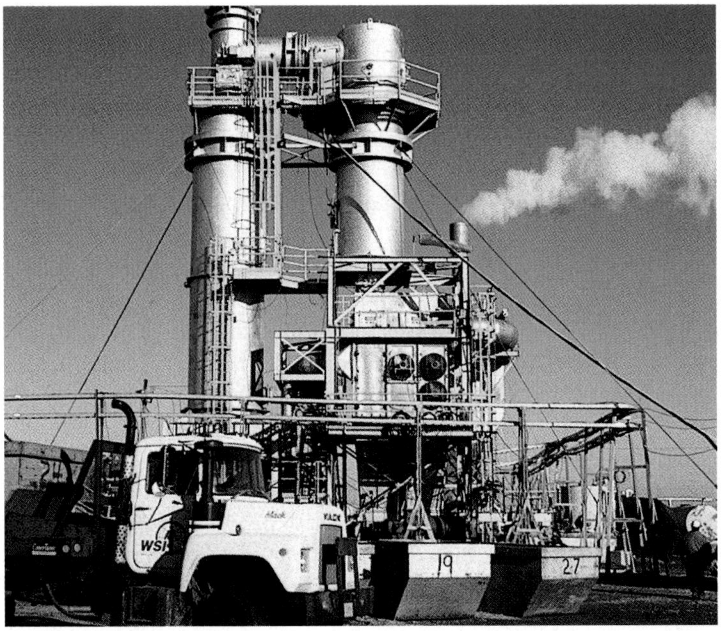

**Figure 19–14 Mobile incinerator for toxic waste.** The EPA is cleaning up some Superfund sites by running contaminated materials through this incinerator, erected at the site.

**Evaluating Superfund.** Of the literally hundreds of thousands of various waste sites, over 44,000 have been deemed serious enough to be given Superfund status. Over time, the EPA judged that about 33,000 of these sites did not pose a significant public-health or environmental threat. Such sites were accordingly assigned to the category "no further removal action planned" (NFRAP). Still, over 11,300 sites remain on the active list, a figure that, it is assumed, now includes most, if not all, of the sites in the United States that pose a significant risk. Interestingly, some of the worst that have come to light are on military bases, the result of what many characterize as the totally heedless and unconscionable discarding of toxic materials from military operations. (13% of the NPL sites are "federal facilities").

***Progress.*** As of 2003, 1,302 sites were still on the NPL. Additions are made to the list as sites from the master list are assessed, and subtractions are made when remediation on a site is complete. Since 1980, when CERCLA went into effect, about 846 sites have received all the necessary cleanup-related construction, and 267 of these have been deleted from the list. Cleanup of sites takes an average of 12 years, at a mean cost of $20 million. Total cleanup often takes that long because groundwater remediation is a slow process. The other NPL sites are in various stages of analysis, remediation, and construction. (An example of a Superfund site and its progress toward cleanup is presented in the "Earth Watch" essay, p. 529.)

***Who Pays?*** CERCLA is based on the principle that "the polluter pays." However, the history of a waste-disposal site may go back 50 years or more, and users may have included everything from schools, hospitals, and small businesses to large corporations, all of which mount legal defenses to disclaim responsibility. When liability is difficult to track down or the responsible parties are unable to pay, the Superfund trust fund kicks in. However, over 70% of the cleanup costs to date have come from polluters (so-called "responsible parties").

Over the years, the EPA has gained a great deal of experience in dealing with Superfund sites. The technology has become quite sophisticated, and many hazardous-waste remediation companies have become established. A great deal of progress has been made, and, interestingly, although hazardous waste was once a top environmental concern, the American public now places it only eighth among a host of other environmental problems, according to a recent Roper survey.

***Critics.*** Nevertheless, critics of the program are skeptical about its success, citing the high costs and slow progress. Industries claim that they are unfairly blamed for pollution that reaches back to activities that were legal before the enactment of CERCLA. Many feel that overly stringent standards of cleanup are costing large sums of money without providing any additional benefit to public health. (This problem of finding a suitable balance between costs and benefits is discussed in more depth in Chapter 22.) Many of the critics' concerns have been reflected in attempts to reauthorize CERCLA by the last three Congresses, but bipartisan support for the proposed legislation has been lacking, and the law continues to operate only on previously established regulations. Bipartisan support is also lacking for reauthorization of the Superfund trust fund; a Democratic move to revive the tax creating the fund was shot down in the Senate in 2003. With the imminent disappearance of the trust fund monies, the public will be paying the costs of Superfund cleanups that are not recovered from "responsible parties." The Bush administration's 2004 budget proposes $1.39 billion for the Superfund program, 80% to come from public funds and the remainder from the possibly nonexistent trust fund.

**Brownfields.** One highly successful recent Superfund development is the *brownfields* program. **Brownfields** are "abandoned, idled, or underused industrial and commercial facilities where expansion or redevelopment is complicated by real or perceived environmental contamination" (EPA definition). It has been estimated that environmental hazards not serious enough to be put on the Superfund NPL impair the value of $2 trillion worth of real estate in the United States. Bipartisan support led to passage of the **Brownfield Act** in 2002, which provides grants for the assessment of sites and remediation work and authorizes $200 million for the ensuing five years for the program. Previously authorized under CERCLA, the brownfields program now stands on its own. The new legislation limits liability for owners and prospective purchasers of contaminated land, thus clearing the way for more cleanup of the estimated 450,000 sites that would qualify as brownfields. States are also moving on this issue. One of the nation's most successful brownfields programs is in Massachusetts, which has provided a $30 million brownfields redevelopment fund managed by MassDevelopment, the state's economic development agency. Over 300 projects are in the development pipeline under the program.

Many brownfield sites lie in economically disadvantaged communities, and their rehabilitation contributes jobs and exchanges a functional facility for an unsightly blight on the neighborhood. For example, the city of Chicago recently acquired and cleaned up a former bus barn in west Chicago that was becoming an illegal indoor garbage dump. The Illinois EPA cleared the site for reuse, and it was acquired by Scott Peterson Meats, which erected a $5.2 million smokehouse that employs 100 workers. Frequently, the rehabilitation of brownfield sites provides industries and municipalities with centrally located, prime land for facilities that would otherwise have been carved out of suburban or "greenfield" lands (land occupied by natural ecosystems). A further advantage is that the new developments go back on the tax rolls, turning a liability into a community asset that lightens the load of residential taxpayers.

**earth watch**

### The Case of the Obee Road NPL Site

The following is but one example of the more than 1,200 sites still on the EPA's Superfund NPL:

#### Conditions at Listing (January 1987)

The Obee Road Site consisted of a plume of contaminated groundwater in the vicinity of Obee Road in the eastern section of Hutchinson, Reno County, Kansas. The Kansas Department of Health and Environment had been investigating the area since July 1983. At that time, the state detected volatile organic chemicals, including benzene, *trans*-1,2-dichloroethylene, chlorobenzene, 1,1-dichloroethylene, tetrachloroethylene, trichloroethylene, vinyl chloride, and toluene, in wells drawing on a shallow aquifer. An estimated 1,900 residents of suburban Obeeville obtained drinking water from private wells in the aquifer.

To protect public health, Hutchinson connected the homes of the Obeeville residents and a school that was drawing water from a contaminated well to the Reno County Rural Water District. Preliminary work by the state identified the source of the contamination as the former Hutchinson City landfill, which is located at the eastern edge of what is now the Hutchinson Municipal Airport. Before it was closed in 1968, the landfill had accepted unknown quantities of liquid wastes and sludges from local industries. Later, an adjacent industrial area, the Airport Road subsite, was also identified as a source of organic solvents. The Department of Defense (DOD), which owned or maintained the airport until 1963, may also have disposed of solvents in the landfill. No records have been found for the design, use, or closure of the landfill or for the wastes that were delivered there.

#### Status as of 2003

Further testing and analysis showed that the original landfill had "stabilized." That is, leaching had already run its course, and there was little likelihood of further contamination from the landfill. Therefore, excavation or other cleanup of the landfill subsite has been deemed unnecessary, but monitoring of the site to detect any further leakage will continue for the foreseeable future. According to the EPA's assessment, there are no present risks to human health or the environment from the site. Remedial action that was selected includes restricting access to, and preventing future development of, the landfill, which is now covered with vegetation. Eight wells have been installed, and groundwater monitoring is to occur for a five-year period to ascertain whether further groundwater contamination is occurring. For the Airport Road subsite, three pumping wells have been installed, and groundwater is continually being pumped and treated via air-stripping towers.

**Leaking Underground Storage Tanks (LUST).** One consequence of our automobile-based transportation system is the millions of underground fuel-storage tanks at service stations and other facilities. Putting such tanks underground greatly diminishes the risk of explosions and fires, but it also hides leaks. Underground storage tanks were traditionally made of bare steel, so they had a life expectancy of about 20 years before they began leaking and contaminating groundwater. By 2003, over 436,000 failures had been reported. Without monitoring, small leaks can go undetected until nearby residents begin to smell fuel-tainted water flowing from their faucets.

**Underground storage tank (UST)** regulations, part of RCRA, now require strict monitoring of fuel supplies, tanks, and piping so that leaks may be detected early. When leaks are detected, remediation must begin within 72 hours. Rules now require all USTs to be upgraded with interior lining and cathodic protection (to retard electrolytic corrosion of the steel), and new tanks must be provided with the same protection if they are steel. Many service stations are turning to fiberglass tanks, which do not corrode. A LUST trust fund, financed by a 0.1-cent-per-gallon tax on motor fuel, pays for federal activities involved with oversight and cleanup. States are required to have UST programs, and the states implement the federal regulations, sometimes adding more stringent requirements. By 2003, some 297,000 cleanups were completed, with 139,000 yet to go. The states have reported that leaking USTs are the most common source of groundwater contamination.

## 19.4 Managing Current Hazardous Wastes

The production of chemical wastes is, and will continue to be, an ongoing phenomenon as long as modern societies persist. Therefore, human and environmental health can be protected only if we have management procedures to handle and dispose of wastes safely so that we will not be creating more and more Superfund sites. We have already noted that the Clean Water Act and the Clean Air Act limit discharges into water and air, respectively. When problems regarding the disposal of wastes on land became evident in the mid-1970s, Congress passed the RCRA in 1976 in order to control land disposal. Thus, a company producing hazardous wastes in the United States today is under the regulations of these three major environmental acts.

## The Clean Air and Water Acts

The Clean Air Act of 1970, the Clean Water Act of 1972, and their various amendments make up the basic legislation limiting discharges into the air or water. More is said about the Clean Air Act in Chapter 21; under the Clean Water Act (see Chapter 17), any firm (including facilities such as sewage-treatment plants) discharging more than a certain volume into natural waterways must have a **discharge permit** (NPDES permit). Discharge permits are a means of monitoring who is discharging what. Establishments with discharge permits are required to report all discharges of substances covered by the TRI. The renewal of the permits is then made contingent on reducing pollutants to meet certain standards within certain periods. Standards are being made continually stricter as technologies for pollution control improve. Some manufacturing firms discharging wastewater into municipal sewer systems are required to pretreat such water to remove any pollutant that cannot be removed by sewage-treatment plants—that is, nonbiodegradable organics and heavy metals.

Nevertheless, even this restriction does not end all water pollution. Certain amounts of wastes are still legally discharged under permits, and although they may account for a low percentage of total emissions, they still add up to large numbers, as we have seen. Moreover, small firms, homes, and farms are exempt from regulation and contribute an unknown quantity of toxics to air and water. Further still, a great deal of pollution comes from nonpoint sources such as urban and farm runoff.

## The Resource Conservation and Recovery Act (RCRA)

The 1976 RCRA and its subsequent amendments are the cornerstone legislation designed to prevent unsafe or illegal disposal of all solid wastes on land. The RCRA has three main features. *First,* it requires that all disposal facilities, such as landfills, be sanctioned by *permit.* The permitting process requires that the facilities have all the safety features mentioned in Section 19.2, including monitoring wells. This requirement caused most old facilities to shut down—many subsequently became Superfund sites—and new high-quality landfills with safety measures to be constructed.

*Second,* the RCRA requires that toxic wastes destined for landfills be pretreated to convert them to forms that will not leach. Such treatment now commonly includes biodegradation or incineration in various kinds of facilities, including cement kilns (Fig. 19–15). Biodegradation involves the use of systems similar to secondary sewage treatment, as discussed in Chapter 17, and perhaps new species of bacteria capable of breaking down synthetic organics. If treatment is thorough, there may be little or nothing to put in a landfill. That is the ultimate objective.

For whatever is still going to disposal facilities, the *third* major feature of RCRA is to require "cradle-to-grave" tracking of all hazardous wastes. The generator (the company that originated the wastes) must fill out a form detailing the exact kinds and amounts of waste generated. Persons transporting the waste, who are also required to be permitted, and those operating the disposal facility must each sign the form, vouching that the amounts of waste transferred are accurate. Copies of their signed forms go to the EPA. All phases are subject to unannounced EPA inspections. The generator remains responsible for any waste "lost" along the way or any inaccuracies in reporting. This provision of the RCRA thus ensures that generators will deal only with responsible parties and curtails midnight dumping.

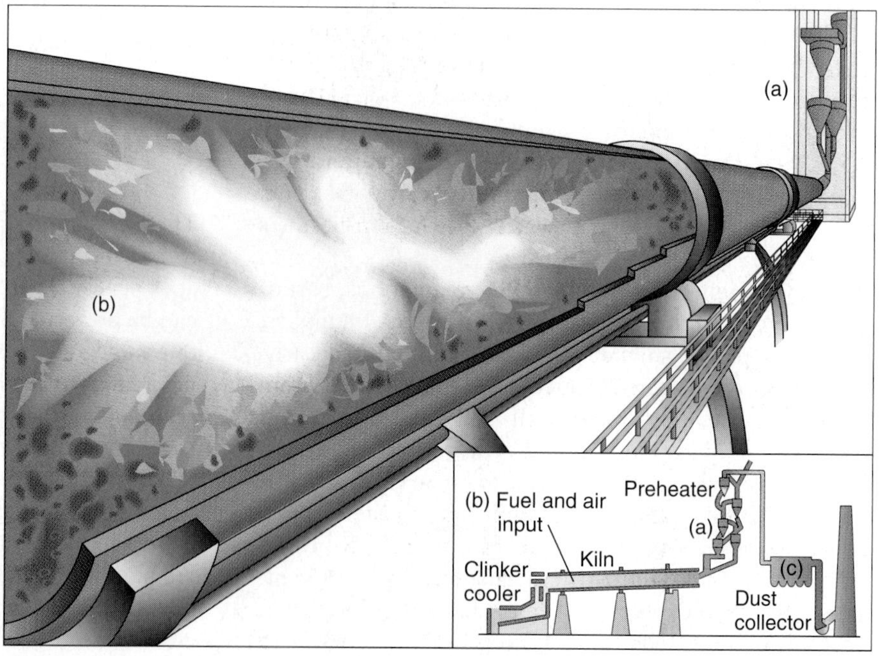


**Figure 19–15 Cement kiln to destroy hazardous wastes.** A cement kiln is a huge rotating "pipe," typically 15 feet in diameter and 230 feet long, mounted on an incline. (a) Solid wastes fed in with raw materials are fully incinerated and made to react with cement compounds as they gradually tumble toward the combustion chamber. (b) Flammable liquid wastes added with the fuel and air impart fuel value as they are burned. (c) Waste dust is trapped and recycled into the kiln. (Redrawn with permission. Southdown, Inc., Houston, TX 77002.)

## Reduction of Accidents and Accidental Exposures

A significant risk to personal and public health lies in exposures that occur as a result of leaks, accidents, and the misguided use of hazardous chemicals in the home or workplace. A considerable number of laws bear on reducing the probability of accidents and on minimizing the exposure of both workers and the public should accidents occur.

**Department of Transportation Regulations.** Transport is an area that is particularly prone to accidents. As modern society uses increasing amounts and kinds of hazardous materials, the stage is set for accidents to become widescale disasters. To reduce this risk, **Department of Transportation Regulations (DOT Regs)** specify the kinds of containers and methods of packing to be used in the transport of various hazardous materials. Such regulations are intended to reduce the risk of spills, fires, and poisonous fumes that could be generated from mixing certain chemicals in case an accident occurs.

In addition, DOT Regs require that every individual container and the outside of a truck or railcar carry a standard placard identifying the hazards (flammability, corrosiveness, the potential for poisonous fumes, and so on) of the material inside (Fig. 19–2). Such placards enable police and firefighters to identify the potential hazard and respond appropriately in case of an accident. You may also see highway HAZMAT signs restricting truckers with hazardous materials to particular routes or lanes or to using a highway only during certain hours.

**Worker Protection: OSHA Act and the "Worker's Right to Know."** In the past, it was not uncommon for industries to require workers to perform jobs that entailed exposure to hazardous materials without informing the workers of the hazards involved. This situation is now addressed by certain amendments to the **Occupational Safety and Health Act (OSH Act of 1970).** These amendments make up the **hazard communication standard** or **"worker's right to know."** Basically, the law requires businesses, industries, and laboratories to make available both information regarding hazardous materials and suitable protective equipment. One form taken by this information is the **material safety data sheet (MSDS)**, which must accompany the shipping, storage, and handling of over 600 chemicals. These sheets contain information about the reactivity and toxicity of the chemicals, with precautions to follow when one is using the chemical. Notably, however, the responsibility to read the information and exercise proper precautions remains with the worker.

**Community Protection and Emergency Preparedness: SARA, Title III.** In 1984, an accident at a Union Carbide pesticide plant in Bhopal, India, caused the release of some 30 to 40 tons of methyl isocyanate, an extremely toxic gas (Fig. 19–16). An estimated 600,000 people in communities surrounding the plant were exposed to the deadly fumes. The official death toll stands at 10,500, but unofficial estimates range much higher. At least 50,000 people suffered various degrees of visual impairment, respiratory problems, and other injuries from the exposure, and many more have experienced chronic illnesses related to the disaster. An Indian investigation of the accident found

**Figure 19–16 Toxic chemical disaster, Bhopal, India.** Officials view some of the victims of a deadly gas release at a Union Carbide plant in 1984. At least 10,500 died, and many more suffered injuries.

that the disaster likely happened because Union Carbide officials scaled back safety and alarm systems at the plant in order to cut costs. Ironically, most of the deaths and injuries could have been avoided if people had known that methyl isocyanate is highly soluble in water, that a wet towel over the head would have greatly reduced one's exposure, and that showers would have alleviated aftereffects. Unfortunately, neither the people affected nor the medical authorities involved had any idea of the chemical they were confronted with, much less of the way to protect or treat themselves.

After the Bhopal disaster, Congress passed legislation to address the problem of accidents. For expediency, the measure was added as Title III to a bill reauthorizing Superfund: the **Superfund Amendments and Reauthorization Act of 1986 (SARA)**. Title III of SARA is better known as **Emergency Planning and Community Right-to-know Act (EPCRA)**, which we have already encountered in connection with the TRI.

***EPCRA Rules.*** EPCRA requires companies that handle in excess of 5 tons of any hazardous material to provide a "complete accounting" of storage sites, feed hoppers, and so on. The information goes to a *local emergency planning committee,* which is also required in every governmental jurisdiction. The committee is made up of officials representing local fire and police departments, hospitals, and any other groups that might be involved in case of an emergency, as well as the executive officers of the companies in question.

The task of the committee is to draw up scenarios for accidents involving the chemicals on-site and to have a contingency plan for every case. This means everything from having firefighters trained and properly equipped to fight particular kinds of chemical fires, to having hospitals stocked with medicines to treat exposures to the particular chemicals, to implementing procedures. Thus, there can be an immediate and appropriate response to any kind of accident. Together with the information in the TRI, the act enables communities to draw up a chemical profile of their local area and encourages them to initiate pollution prevention and risk reduction activities.

**The Toxic Substances Control Act.** In the past, new synthetic organic compounds were introduced for a specific purpose without any testing of their potential side effects. For example, in the 1960s, a new compound known as TRIS was found to be an effective flame retardant and was widely used in children's sleepwear. It was only later discovered that TRIS was a potent carcinogen. Treated sleepwear was immediately withdrawn from the market. It is not known how many children (if any) developed cancer from TRIS, but this and other such cases revealed a hole in the systems of protection.

Congress responded by passing the **Toxic Substances Control Act of 1976 (TSCA)**, which requires that, before manufacturing a new chemical in bulk, manufacturers submit a "premanufacturing report" to the EPA in which the potential environmental and human health impacts of the substance are assessed (including those which may derive from the ultimate disposal of the chemical). Depending on the results of the assessment, the manufacturer may be required to test the effects of the product on living things. Following the results of testing, a product's uses may be restricted or the product may be kept off the market altogether. The TSCA also authorizes the EPA to develop an inventory of the chemicals already in use at the time of the legislation and to require testing wherever there are indications of potential risks. Over 80,000 chemical substances are currently on the inventory.

Laws applying to hazardous wastes are summarized graphically in Fig. 19–17. Note that nongovernmental consumer advocate groups have been a major force behind the passage of these laws, as well as of regulations requiring ingredients to be labeled on all products. Citizen action does make a difference!

## 19.5 Broader Issues

Before we leave the topic of hazardous wastes, we still need to address two important broader issues: the problem of environmental justice and the movement toward pollution prevention.

### Environmental Justice and Hazardous Wastes

- The largest commercial hazardous-waste landfill in the United States is located in Emelle, Alabama. African-Americans make up 90% of Emelle's population. This landfill receives wastes from Superfund sites and every state in the continental United States.
- A Choctaw reservation in Philadelphia, Mississippi, was targeted to become the home of a 466-acre hazardous-waste landfill. The reservation is entirely Native American.
- A recent study found that 870,000 U.S. federally subsidized housing units are within a mile of factories that reported toxic emissions to the EPA. Most of the occupants of these apartments are minorities.

The issue is *environmental justice,* introduced in Chapter 1. The EPA defines **environmental justice** as "the fair treatment and meaningful involvement of all people regardless of race, color, national origin, or income with respect to the development, implementation, and enforcement of environmental laws, regulations, and policies. Fair treatment means that no group of people, including racial, ethnic, or socioeconomic group[s], should bear a disproportionate share of the negative environmental consequences resulting from

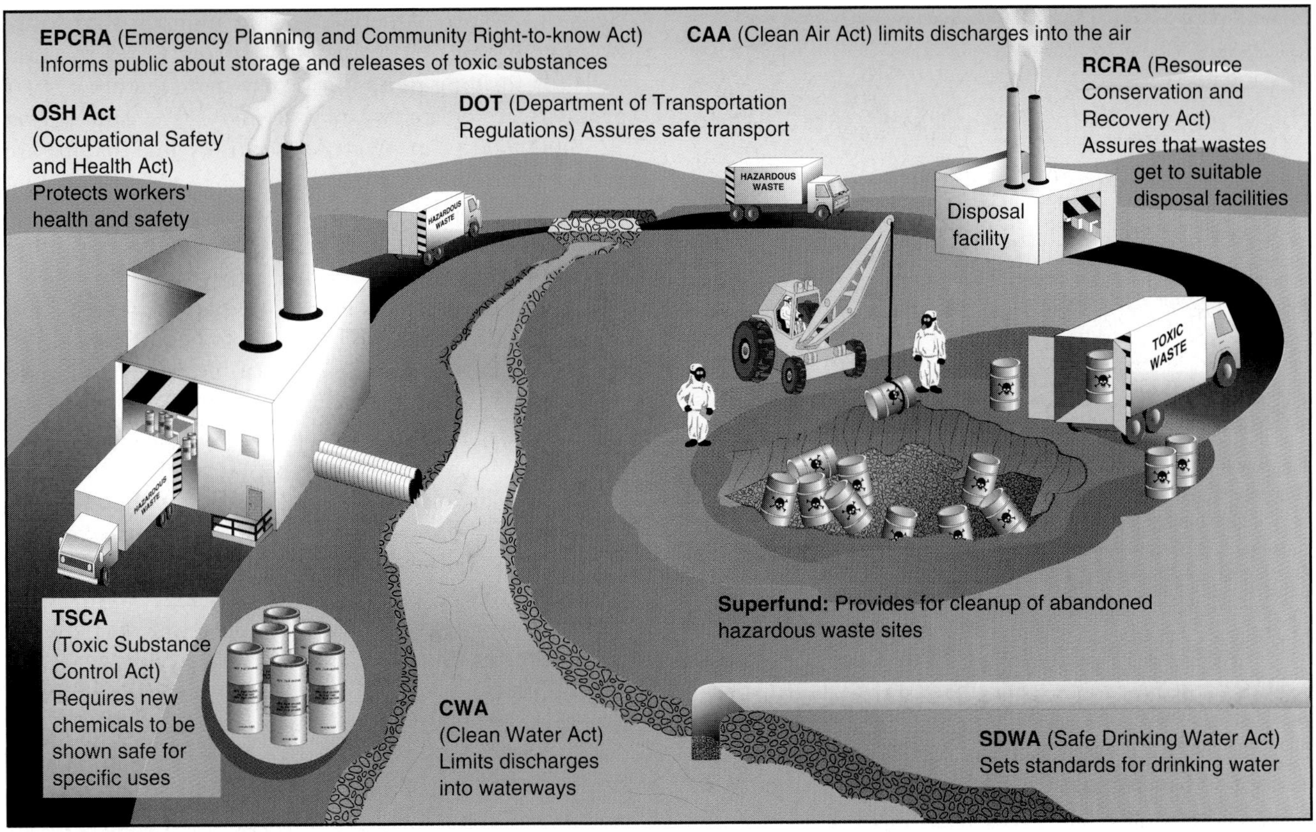


**Figure 19–17** **Major hazardous-waste laws.** Summary of the major laws pertaining to the protection of workers, the public, and the environment from hazardous materials.

industrial, municipal, and commercial operations or the execution of federal, state, local, and tribal programs and policies."

Several recent studies have shown that, all across the United States, waste sites and other hazardous facilities are more likely than not to be located in towns and neighborhoods where most of the residents are non-Caucasian. These same towns and neighborhoods are also less affluent, a further element of environmental injustice. The wastes involved are generated primarily by affluent industries and the affluent majority, but somehow the wastes tend to end up well away from where they were generated and in the backyards of people of color. It seems fair to assume that the siting of hazardous facilities is a matter of political power, and those with the power would like to have the sites well away from their own backyards.

***Federal Response.*** The federal administration has taken this problem seriously. In 1994, President Clinton issued Executive Order 12898, focusing federal agency attention on environmental justice. The EPA's response has been to establish an Environmental Justice (EJ) program, put in place early in 1998, to capture the intent of Executive Order 12898 and to further a number of strategies already established by the EPA's Office of Solid Waste and Emergency Response. For example, in connection with the EPA's Brownfields Initiative, communities are putting abandoned properties back into productive use. Many EJ efforts are directed toward addressing justice concerns *before* they become problems. In 1997, the EPA rejected a permit for a plastics plant in Convent, Louisiana, in a predominantly minority community. Subsequently, the chemical company decided to build the plant in an industrial area near Baton Rouge.

***International EJ.*** One form of trade the developing countries can do without is the international trade in toxic wastes. Some countries or local jurisdictions in search of ready cash have found a source in the form of toxic-waste shipments, which invariably come from more developed countries and are usually illegal (Fig. 19–18). The **Basel Convention** is an international agreement that places a ban on most international toxic-waste trade, while the *Basel Action Network* is a nongovernmental organization that seeks to prevent such trade when it is not in conformity with the convention. The network is vigilant in publicizing and coordinating legal challenges to incidents involving toxic-waste shipments and has made it clear that this trade is a matter of international environmental justice.

## Pollution Prevention for a Sustainable Society

*Pollution control* and *pollution prevention* are not the same. **Pollution control** involves adding a filter or some other device at the "end of the pipe" to prevent

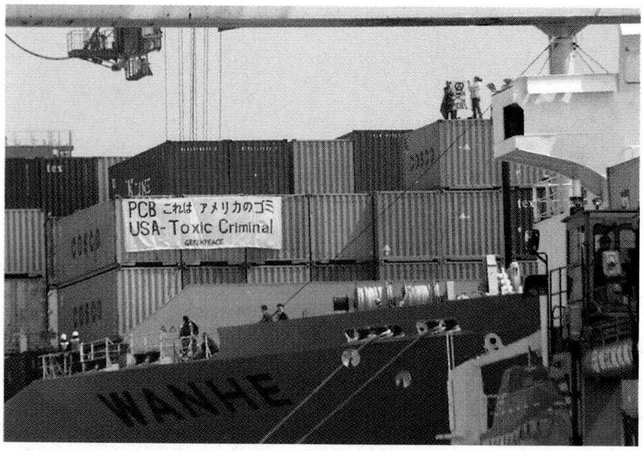


**Figure 19–18 International toxic waste.** The *Wan He* is carrying toxic waste from U.S. military bases in Japan. It was refused entry in the United States and Canada. Greenpeace protesters are shown on board in Yokohama, Japan, where the ship was docked pending a final decision on its destination.

pollutants from entering the environment. The disposal of the captured pollutants still has to be dealt with, and this entails more regulation and control. **Pollution prevention,** by contrast, involves changing the production process, the materials used, or both so that harmful pollutants won't be produced in the first place. For example, adding a catalytic converter to the exhaust pipe of your car is pollution control, whereas redesigning the engine so that less pollution is produced or switching to an electric car is pollution reduction and avoidance, respectively.

Pollution prevention often results in better product or materials management—that is, less wastage. Thus, pollution prevention frequently creates a cost savings. For example, Exxon Chemical Company added simple "floating roofs" to tanks storing its most volatile chemicals, thereby reducing evaporative emissions by 90% and gaining a savings of $200,000 per year. That paid for the cost of the roofs in six months. As another example, United Musical Instruments in Nogales, Arizona, worked with the Arizona Department of Environmental Quality's Pollution Prevention Unit to install a closed-loop system that reduced water use by 500,000 gallons per year. The system also reduced hazardous waste by 58% in three years and saved the company $127,000 in its first year alone. These are examples of the *minimization* or *elimination* of pollution.

***Green Chemistry.*** A second angle on pollution avoidance is *substitution*—that is, finding nonhazardous substitutes for hazardous materials. For example, the dry-cleaning industry, which uses large quantities of toxic organic chemicals, is exploring the use of water-based cleaning, or **wet cleaning.** Preliminary results from some entrepreneurs indicate that wet cleaning is quite comparable in cost and performance to dry cleaning. There are now more than 150 wet cleaners in North America, and the number is growing. In another case, Clairol switched from water to foam balls for flushing pipes during the manufacture of hair products, reducing wastewater by 70% and saving $250,000 per year. Water-based inks and paints are being substituted for those containing synthetic organic solvents, and inks and dyes based on biodegradable organic compounds are being substituted for those based on heavy metals.

A third approach is *reuse*—that is, cleaning up and recycling solvents and lubricants. Some military bases, for example, have been able to distill solvents and reuse them, instead of discarding them into the environment. This approach has increased greatly during the past decade, as measured by the TRI data. Indeed, all three approaches have been employed increasingly by industries handling hazardous chemicals, and all have contributed to significant reductions in hazardous waste releases into the environment. (See Fig. 19–4.) The public disclosure of TRI data is judged to have played a crucial role in these reductions. The EPA holds that the system has been "one of the most effective environmental programs ever legislated by Congress and administered by EPA."

***You, the Consumer.*** Finally, pollution avoidance can also be applied to the individual consumer. So far as you are able to reduce or avoid the use of products containing harmful chemicals, you are preventing those amounts of chemicals from going into the environment. You are also reducing the by-products resulting from producing those chemicals. The average American home contains as much as 100 pounds of HHW—substances such as paints, stains and varnishes, batteries, pesticides, motor oil, oven cleaners, and more. These materials need to be stored safely, used responsibly, and disposed of properly if they are not to pose a risk of injury or death and a threat to environmental health. (Many communities hold regular HHW collection days; if yours doesn't, you could initiate one!)

Increasingly, companies are gradually beginning to produce and market **green products,** a term used for products that are more environmentally benign than their traditional counterparts. How fast and to what degree these green products replace traditional products will in large part depend on how we behave as consumers. In other words, consumerism can be an extremely potent force.

In conclusion, there are four ways to address the problems of chemical pollution: (1) pollution prevention, (2) recycling, (3) treatment (breaking down or converting the material to harmless products), and (4) safe disposal. The first three of these methods promote a minimum of waste, in harmony with the second principle of sustainability. There is every reason to believe that we can have the benefits of modern technology without destroying the sustainability of our environment by polluting it.

## career link

### Daniel S. Granz, EPA Environmental Engineer

There are times when Dan Granz, environmental engineer in the EPA's Lexington, Massachusetts, Office of Environmental Measurement and Evaluation, has to suit up with a full-body protective white suit, complete with respirator, gloves, and booties. As uncomfortable as the suits are, they are essential when one is dealing with some of the unknown and hazardous chemicals encountered in the field. Properly suited, Granz has dealt with some serious hazards, such as mislabeled drums and tanks at a hazardous-waste transfer station, open lagoons on the site of a defunct tannery, and an aging factory whose floorboards were still saturated with PCBs from a manufacturing process that had ended 20 years ago. He regards the great variety of projects that come his way as a major attraction of his job.

Recently, Granz worked on a project involving lead sampling from vacant lots in Providence, Rhode Island. The project's objective is to rehabilitate the urban lots so that they can be put to good use and be brought back onto the city's tax rolls. The problem is contamination by the lead-based paint used virtually everywhere a few decades ago. Lead is especially hazardous to children, who might be playing on the lots. If the lead levels prove to be moderately low, the EPA may recommend that uncontaminated soil be brought in to mix with the urban soil and dilute its lead content. If the levels are high, the topsoil might have to be removed or covered to seal it in place. Granz and his team measure soil samples on the spot with portable equipment and then follow up with testing back in the laboratory for confirmation.

Daniel S. Granz

Granz brings a unique kind of preparation to his occupation. He graduated with a B.A. in biology from Gordon College in Massachusetts in 1977, but his eye was on the job market (not too promising for biologists at the time). Accordingly, he enrolled in a civil engineering program at Northeastern University in Boston and graduated in 1980 with a bachelor-of-science degree in civil engineering. One of the attractions of Northeastern was its co-op program, which places students as interns with different agencies or businesses. Granz elected a placement with the EPA. When his degree was granted, he was offered two jobs: a position with EPA in the same office where he now works and a job with the Exxon Corporation. His co-op experience had convinced him that he would like to be on the regulatory side of the ledger, working on improving the environment.

When asked about how students might get to work in a federal agency like the EPA, Granz admitted that it was not necessarily easy, but that turnover does occur and jobs do open up regularly. His advice is to have a specialty. He believes that the combination of biology and civil engineering gave him a unique profile that was attractive to prospective employers. The EPA and other agencies often hire college students as summer interns, and co-op programs are also a good idea, Granz stated. His job satisfaction comes from knowing that he can make a difference as he works with industries and the problems they have created. Quite often, he is able to work directly with polluters, convincing them to do the right thing before legal action becomes necessary. Over the years, Granz has seen great improvement in the environment and is convinced that, without the work of the EPA, our air, water, and land would be far more polluted and hazardous than it is now. That is to say, there is still work to be done as the EPA addresses some of the harder problems, such as nonpoint-source pollution, groundwater contamination, and the cleanup of brownfield sites.

# revisiting the themes

## Sustainability

In our recent past, society has been on an unsustainable course in dealing with hazardous chemicals. More and more sites were being contaminated, and we were seeing rising rates of cancer and other health problems. Although the connection between hazardous waste sites and actual cancer cases has been difficult to prove, the rising tide of hazardous chemicals in the environment was certainly a prime suspect. The legislation and programs discussed in this chapter have turned the situation around, and the move toward pollution prevention and the recycling of hazardous chemicals indicates that we are moving in a sustainable direction in our dealings with toxic chemicals.

## Stewardship

One vital task of stewardship is to protect fellow human beings from unnecessary risks. We were not doing well with this responsibility, and a few wake-up calls led to stewardly action on the part of people such as Lois Gibbs, Jan Schlichtmann,[1] and others who put their lives and reputations on the line to bring some sanity to our permissive dealings with toxic-waste dumps. As a result, our society stopped simply accepting contamination as the "price of progress," and after a great deal of public outcry, Congress responded with legislation to address the many problems. We have made real progress in cleaning up the Superfund sites, the leaking underground storage tanks, and the brownfield sites. Industries can no longer release products without testing their potential for doing harm. The TRI in particular has provided the incentive for industry to clean up its act. It is not time, however, to let down our guard. Injustice to minority groups, pressures by industries to ease up on the regulations, and the attitude that hazardous wastes no longer pose a threat indicate that this issue requires constant surveillance and repeated pressure on Congress and the EPA. New chemicals are constantly appearing, and old ones have not yet been thoroughly researched; the Superfund program is still far from completion.

## Sound Science

Toxicology is a well-established and crucial branch of science that has helped us to develop a thorough inventory of hazardous chemicals and their health effects. Federal agencies and NGOs have made certain that this information is available to the public via the Internet (through the NTP, IRIS, NIEHS, and Scorecard), an excellent wedding of sound science and social needs.

## Ecosystem Capital

Hazardous chemicals in the environment had reached the point where they could no longer be ignored, as rivers began to catch fire and fish were no longer safe food sources because they accumulated high levels of toxic substances. Many bodies of water had received toxic wastes, and the legacy of this misuse is still with us today in the form of contaminated sediments. The POPs still plague much of the Arctic regions, with toxic substances having accumulated in the tissues of the animals in some of the most remote regions of the Earth. Thankfully, many of these chemicals have now been banned, their effects are diminishing, and many of the species especially affected (e.g., by DDT) have recovered.

## Policy and Politics

This chapter is all about environmental public policy. The EPA is in the center of the hazardous-chemical picture and has been given many laws to administer as the agency encourages, enforces, cajoles, and sometimes prosecutes those who deal with hazardous chemicals. From the cradle to the grave of toxic chemicals, there are laws and regulations to prevent future disasters on the order of Superfund sites or Bhopal-like accidents. Political and ideological issues persist, however, as in the controversy over funding the Superfund trust and in battles to reauthorize many of the laws. Public policy has also gone international, in the Stockholm Convention on Persistent Organic Pollutants.

## Globalization

The unwelcome globalization of toxic substances can be seen in the global spread of POPs to remote regions via air and water currents and in the business of disposing of hazardous wastes by shipping them to developing countries eager to earn some cash. Both of these problems are being dealt with by international agreements, but again, surveillance is necessary to enforce the agreements.

[1]See Jonathan Harr, *A Civil Action* (New York: Vintage Books, 1996).

# review questions

1. How do toxicologists investigate hazardous chemicals? How is this information disseminated to health practitioners and the public?
2. What four categories are used to define hazardous chemicals?
3. Define what is meant by "total product life cycle," and describe the many stages at which pollutants may enter the environment.
4. What are the two classes of chemicals that pose the most serious long-term toxic risk, and how do they affect food chains?
5. What are the "dirty dozen" POPs? Why are they on a list?
6. How were chemical wastes generally disposed of before 1970?
7. What two laws pertaining to the disposal of hazardous wastes were passed in the early 1970s? Describe how the passage of the laws shifted pollution from one part of the environment to another.
8. Describe three methods of land disposal that were used in the 1970s. How has their use changed over time?

9. What law was passed to cope with the problem of abandoned hazardous-waste sites? What are the main features of the legislation?

10. What is being done about leaking underground storage tanks and brownfields?

11. What law was passed to ensure the safe land disposal of hazardous wastes? What are the main features of the legislation?

12. What laws exist to protect the public against exposures resulting from hazardous chemical accidents? What are the main features of the legislation?

13. What role does the Toxic Substances Control Act play in the hazardous waste arena?

14. Why does the EPA have an environmental justice program?

15. Describe the advantages of pollution prevention efforts.

## thinking environmentally

1. Select a chemical product or drug you suspect could be hazardous, and investigate it, using the resources made available on the Internet by federal agencies and Environmental Defense.

2. Before the 1970s, it was not illegal to dispose of hazardous chemicals in unlined pits, and many companies did so. Should they be held responsible today for the contamination those wastes are causing, or should the government (taxpayers) pay for the cleanup? Give a rationale for your position.

3. Use the Toxics Release Inventory on the Web to investigate the locations and amounts of toxic chemicals released in your state or region.

4. Suppose an incineration facility is to be built near your community. It is proposed that hazardous wastes currently being landfilled at the same location will be disposed of in the new facility. Would you support or oppose the proposal? Give a rationale for your position.

5. Do you favor cutting back on regulations pertaining to hazardous chemicals in order to help balance the federal budget? What laws or regulations, if any, would you cut back on? Defend your answer.

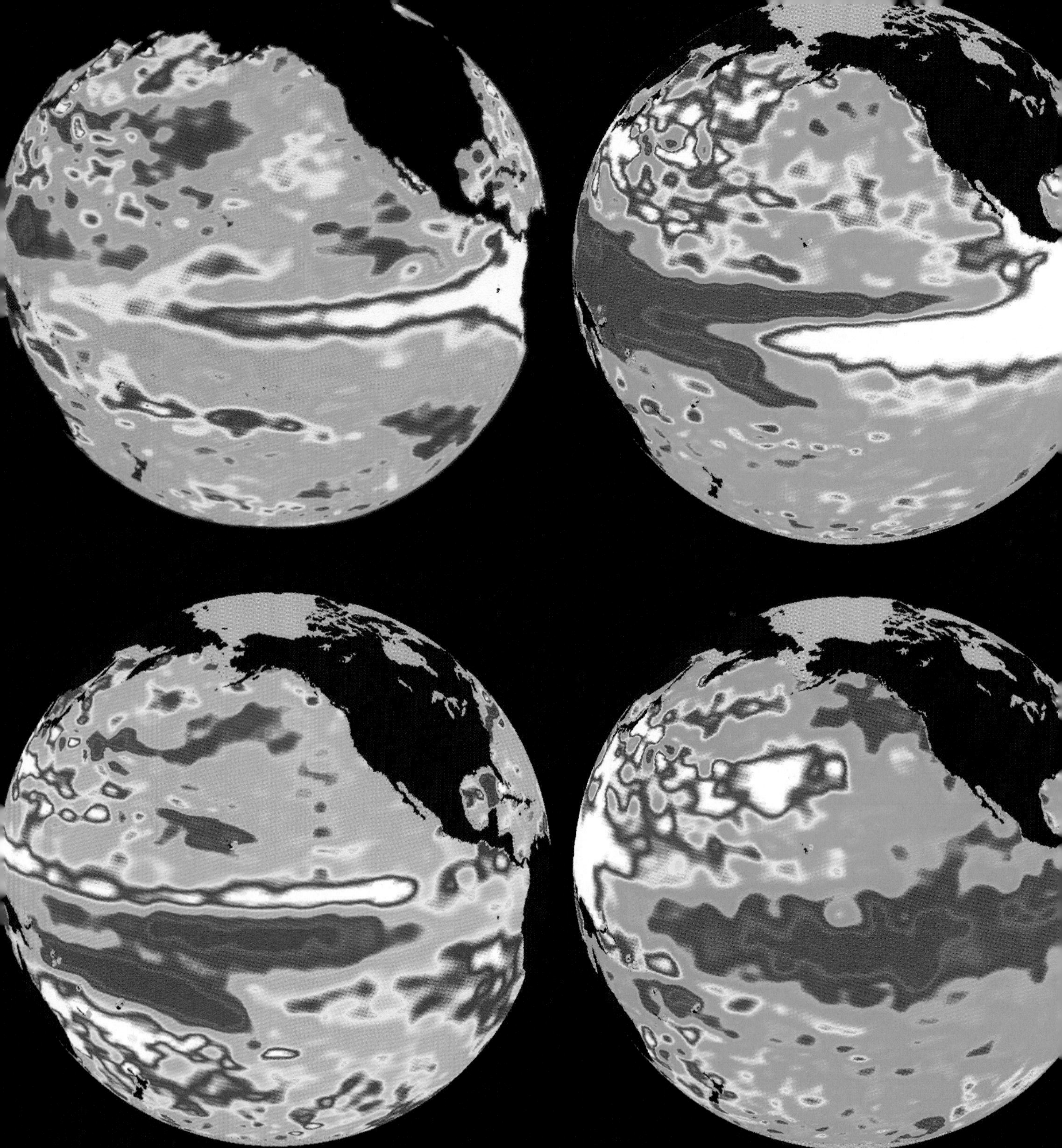

# chapter 20

# The Atmosphere: Climate, Climate Change, and Ozone Depletion

## Key Topics

1. Atmosphere and Weather
2. Climate
3. Global Climate Change
4. Response to Climate Change
5. Depletion of the Ozone Layer

El Niño has become a household name. Beginning in April 1997 and extending through the spring of 1998, an especially intense El Niño linked the world together. In California and Oregon, unusually severe storms battered the coastline, causing major coastal erosion and flooding rivers (Fig. 20–1). On the eastern seaboard and the Gulf of Mexico, residents relaxed through a hurricane season that was the mildest in many years. By contrast, rainfall at least five times the normal deluged East Africa, often a region of drought. Fires blackened 1,400 square miles of drought-affected forests in Indonesia, creating huge clouds of smoke that blanketed much of Southeast Asia. In New Guinea, the lack of rain brought crop failures, necessitating a massive effort in food relief in order to prevent famine. Record crop harvests were enjoyed in India, Australia, and Argentina. Unusual rainfall in California and Florida triggered lush growth of vegetation.

***La Niña.*** Shortly after El Niño was officially declared over in May 1998, the world began to hear about La Niña, as weather conditions shifted 180°. Florida was hit by hot, dry weather, triggering wildfires that swept through woods and suburbs, fueled by the lush vegetation nurtured by El Niño. Heavy rains returned to drought-ridden areas, and Venezuela, China, and Mozambique experienced disastrous floods. The tropical Atlantic spawned an unusual number of strong hurricanes. By the spring of 2000, La Niña had dissipated and weather conditions had returned to normal. With global damage estimated upwards of $36 billion, and with over 22,000 deaths, the 1997–2000 El Niño–La Niña has been a lesson in global climate the world will not soon forget. The atmosphere and oceans teamed up to produce a reminder that we live at the mercy of a system we neither control nor understand.

What caused these incredible changes in weather over so much of the globe? Briefly, El Niño occurs

*Development of the 1997–2000 El Niño–La Niña* **Satellite images of the central Pacific Ocean from May 25 (upper left) and December 18 (upper right), 1997, showing El Niño conditions, and from June 26, 1998 (lower left), and March 10, 1999, (lower right), showing La Niña conditions. (White is warmest, red is next, and blue is coldest.)**

(a) (b) (c) (d)

**Figure 20–1 Impacts of El Niño.** (a) Landslides on the California coast. (b) Food relief in New Guinea. (c) Smoke and fires in Indonesia. (d) Flooding in Kenya.

when a major shift in atmospheric pressure over the central equatorial Pacific Ocean leads to a reversal of the trade winds that normally blow from an easterly direction. Warm water spreads to the east, the jet streams strengthen and shift from their normal courses, patterns in precipitation and evaporation are affected, and the system is usually sustained for more than a year. La Niña conditions are just the reverse: The easterly trade winds are reestablished with even greater intensity, upwelling of colder ocean water in the eastern Pacific from the depths replaces the surface water blown westward, the jet streams are weakened, and weather patterns are again affected.

Meteorologists are quite good at explaining what El Niño and La Niña are and where and how they may influence the different continents and oceans, but they are still unable to explain why they happen. What we do know, however, is that they are occurring at an unprecedented frequency. In the past 15 years, for example, there have been six El Niños. The latest was in 2002, a short-term and less intense event. If such changes in major weather patterns persist, we will in fact be experiencing a climate change. The El Niño–La Niña phenomenon has revealed to people everywhere that the atmosphere, oceans, and land are linked together and that, when normal patterns are disrupted, the climate on the whole Earth can be affected. Could it be that the warming trend now evident in global temperatures is responsible for these changes? Are we in for a major climate change? Indeed, what does control the climate?

We will seek answers to these questions as we investigate the atmosphere, how it is structured, and how it brings us our weather and climate. Then we will consider the evidence for global climate change, finishing with a look at what is happening to ozone in the upper atmosphere.

## 20.1 Atmosphere and Weather

### Atmospheric Structure

Recall from Chapter 3 that the atmosphere is a collection of gases that gravity holds in a thin envelope around Earth. The gases within the lowest layer, the **troposphere,** are responsible for moderating the flow of energy to Earth and are involved with the biogeochemical cycling of many elements and compounds—oxygen, nitrogen, carbon, sulfur, and water, to name the most crucial ones. The troposphere ranges in thickness from 10 miles (16 km) in the tropics to 5 miles (8 km) in higher latitudes, due mainly to differences in heat energy budgets. This layer contains practically all of the water vapor and clouds in the atmosphere; it is the site and source of our weather. Except for local temperature inversions, the troposphere gets colder with altitude (Fig. 20–2). Air masses in this layer are well mixed vertically, so pollutants can reach the top within a few days. Substances entering the troposphere—including pollutants—may be changed chemically and washed back to Earth's surface by precipitation (Chapter 21). Capping the troposphere is the **tropopause.**

***Higher.*** Above the tropopause is the **stratosphere,** a layer within which temperature *increases* with altitude, up to about 40 miles above the surface of Earth. The temperature increases primarily because the stratosphere contains ozone ($O_3$), a form of oxygen that absorbs high-energy radiation emitted by the Sun. Because there is little vertical mixing of air masses in the stratosphere and no precipitation from it, substances that enter it can remain there for a long time. Beyond the stratosphere are two more layers, the *mesosphere* and the *thermosphere,* where the ozone concentration declines and only small amounts of oxygen and nitrogen are found. Because none of the reactions we are concerned with occur in the mesosphere or thermosphere, we shall not discuss those two layers. Table 20–1 summarizes the characteristics of the troposphere and stratosphere.

### Weather

The day-to-day variations in temperature, air pressure, wind, humidity, and precipitation—all mediated by the atmosphere—constitute our **weather. Climate** is the result of long-term weather patterns in a region. The scientific study of the atmosphere—both of weather and of climate—is **meteorology.** It is fair to think of the atmosphere–ocean–land system as an enormous weather engine, fueled by the Sun and strongly affected by the rotation of Earth and its tilted axis. Solar radiation enters

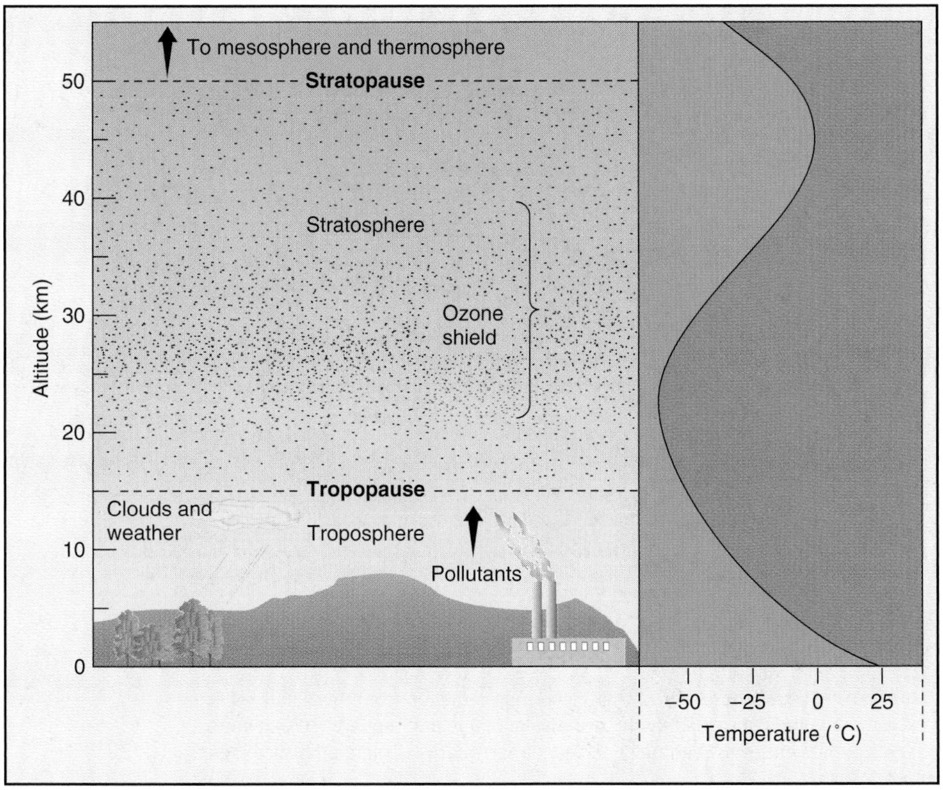


**Figure 20–2 Structure and temperature profile of the atmosphere.** The left-hand plot shows the layers of the atmosphere and the ozone shield, while the plot on the right shows the vertical temperature profile.

**table 20-1 Characteristics of Troposphere and Stratosphere**

| Troposphere | Stratosphere |
|---|---|
| Extent: Ground level to 10 miles (16 km) | Extent: 10 miles to 40 miles (16 km to 65 km) |
| Temperature normally decreases with altitude, down to −70°F (−59°C) | Temperature increases with altitude, up to +32°F (0°C) |
| Much vertical mixing, turbulent | Little vertical mixing, slow exchange of gases with troposphere, via diffusion |
| Substances entering may be washed back to Earth | Substances entering remain unless attacked by sunlight or other chemicals |
| All weather and climate take place here | Isolated from the troposphere by the tropopause |

the atmosphere and then takes a number of possible courses (Fig. 20–3). Some is reflected by clouds and Earth's surfaces, but most is absorbed by the atmosphere, oceans, and land, which are heated in the process. The land and oceans then radiate some of their heat back upward as infrared energy.

*Flowing Air.* Some of the heat that is radiated back is transferred to the atmosphere. Thus, air masses will grow warmer at the surface of Earth and will tend to expand, becoming lighter. The lighter air will then rise, creating *vertical* air currents. On a large scale, this movement creates the major *convection currents* we encountered in Chapter 7 (Fig. 7–6). Air must flow in to replace the rising warm air, and the inflow leads to *horizontal* airflows, or wind. The ultimate source of the horizontal flow is cooler air that is sinking, and the combination produces the Hadley cell (Fig. 7–6). As discussed in Chapter 7, these major flows of air create regions of high rainfall (equatorial), deserts (25° to 35° north and south of the equator), and horizontal winds (trade winds).

*Convection.* On a smaller scale, **convection currents** bring us the day-to-day changes in our weather as they move in a general pattern from west to east. Weather reports inform us of regions of high and low pressure, but

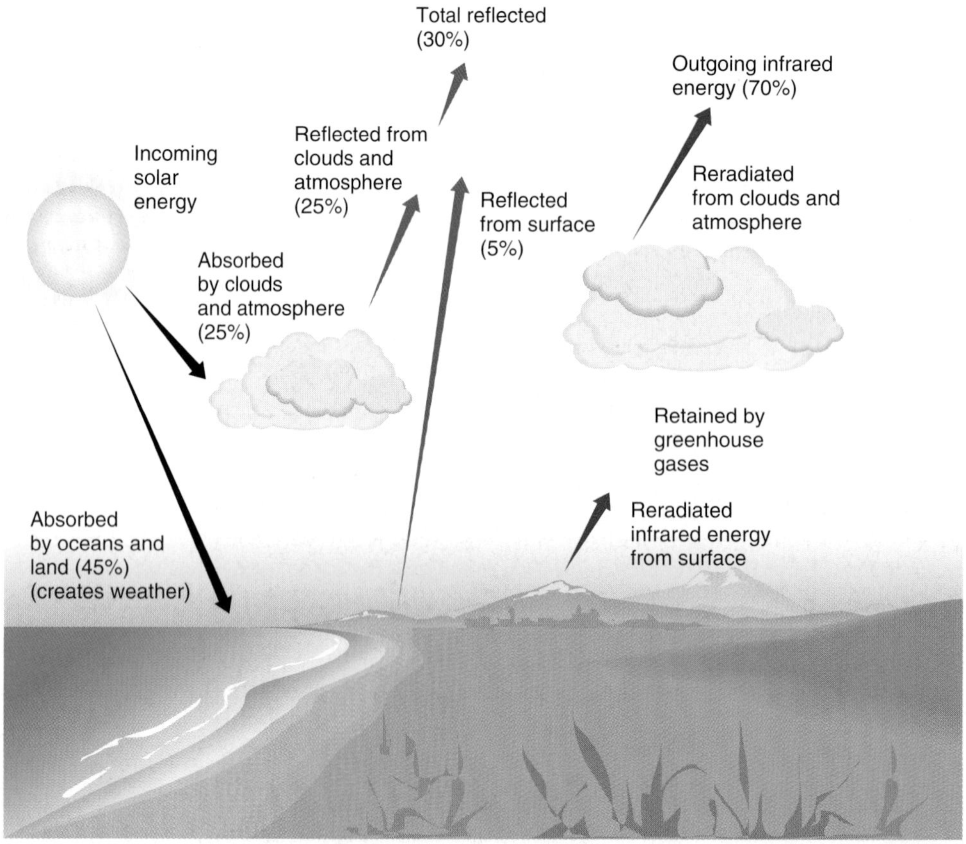


**Figure 20–3 Solar-energy balance.** Much of the incoming radiation from the Sun is reflected back to space (30%), but the remainder is absorbed by the oceans, land, and atmosphere (70%), where it creates our weather and fuels photosynthesis. Eventually, this absorbed energy is radiated back to space as infrared energy (heat).

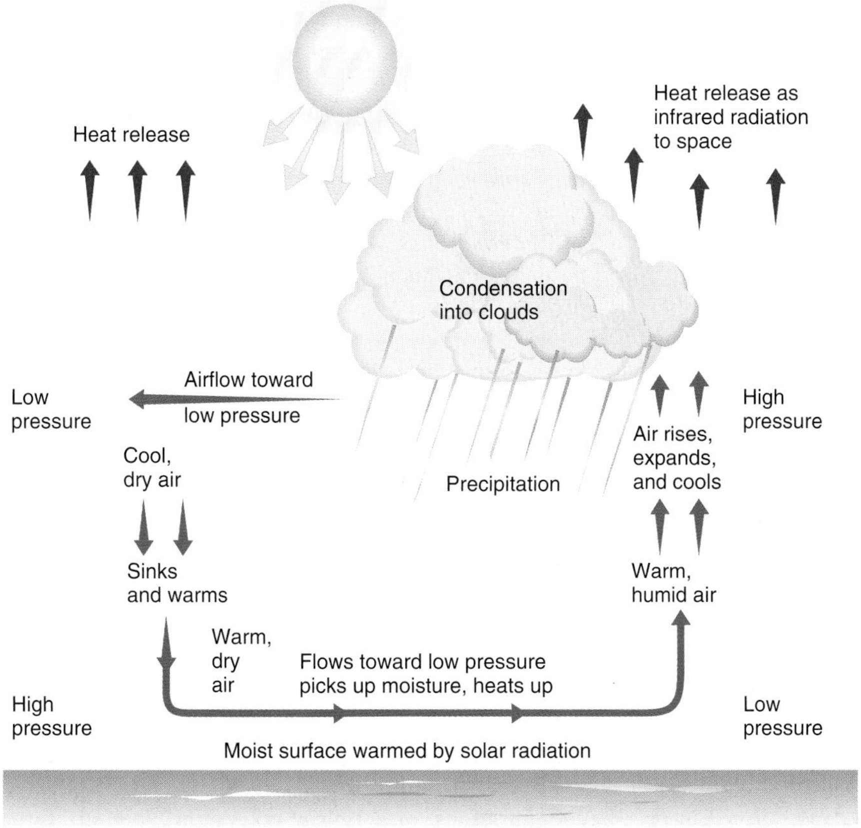


**Figure 20–4 A convection cell.** Driven by solar energy, these cells produce the main components of our weather as evaporation and condensation occur in rising air and precipitation results, followed by the sinking of dry air. Horizontal winds are generated in the process.

where do these come from? Rising air (due to solar heating) creates high pressure up in the atmosphere, leaving behind a region of lower pressure close to Earth. Conversely, once the moist, high-pressure air has cooled by radiating heat to space and losing heat through condensation (thereby generating precipitation), the air then flows horizontally toward regions of sinking cool, dry air (where the pressure is lower). There, the air is warmed at the surface and creates a region of higher pressure (Fig. 20–4). The differences in pressure lead to airflows, which are the winds we experience. As the figure shows, the winds tend to flow from high-pressure regions toward low-pressure regions.

*Jet Streams.* The larger scale air movements of Hadley cells are influenced by Earth's rotation from west to east. This creates the trade winds over the oceans and the general flow of weather from west to east. Higher in the troposphere, Earth's rotation and air-pressure gradients generate veritable rivers of air, called **jet streams,** that flow eastward at speeds of over 300 mph and that meander considerably. Jet streams are able to steer major air masses in the lower troposphere. One example is the polar jet stream, which steers cold air masses into North America when it dips downward in latitude.

*Put Together. . . .* Air masses of different temperatures and pressures meet at boundaries we call **fronts,** which are regions of rapid weather change. Other movements of air masses due to differences in pressure and temperature include hurricanes and typhoons and the local, but very destructive, tornadoes. Finally, there are also major seasonal airflows: the **monsoons,** which often represent a reversal of previous wind patterns. Monsoons are created by major differences in cooling and heating between oceans and continents. The summer monsoons of the Indian subcontinent are famous for the beneficial rains they bring and notorious for the devastating floods that can occur when the rains are heavy. Putting these movements all together—taking the general atmospheric circulation patterns and the resulting precipitation, then adding the wind and weather systems generating them, and finally mixing all this with the rotation of Earth and the tilt of the planet on its axis, which creates the seasons—yields the general patterns of weather that characterize different regions of the world. In any given region, these patterns are referred to as the region's **climate.**

## 20.2 Climate

Climate was described in Chapter 2 as the average temperature and precipitation expected throughout a typical year in a given region. Recall that the different temperature and moisture regimes in different parts of the world "created" different types of ecosystems called *biomes,* representing the adaptations of plants, animals, and microbes to the prevailing weather patterns, or climate, of a region. The temperature and precipitation patterns themselves are actually caused by other forces, namely, the major determinants of weather previously outlined. Humans can adjust to practically any climate (short of the brutal conditions on high mountains or burning deserts), but this is not true of the other inhabitants of the particular regions we occupy. If other living organisms in a region are adapted to a particular climate, then a major change in the climate represents a major threat to the structure and function of the existing ecosystems. The subject of climate change is such a burning issue today because we depend on these other organisms for a host of vital goods and services without which we could not survive (Chapter 3). If the climate changes, can these ecosystems change with it in such a way that the vital support they provide us is not interrupted? How rapidly can organisms and ecosystems adapt to changes in climate? How rapidly do climates change? One way to answer these questions is to look into the past, which may harbor "records" of climate change.

## Climates in the Past

Searching the past for evidence of climate change has become a major scientific enterprise, one that becomes more difficult the further into the past we try to search. Systematic records of the factors making up weather—temperature, precipitation, storms, and so forth—have been kept for little more than a hundred years. Nevertheless, these records already inform us that our climate is far from constant. The record of surface temperatures, from weather stations around the world and from literally millions of observations of temperatures at the surface of the sea, tells an interesting story (Fig. 20–5): Since 1855, global average temperature has shown periods of cooling and warming, but, in general, has increased 0.6°C (1°F). During the 20th century, two warming trends occurred, one from 1910 to 1945 and the latest dramatic increase from 1976 until the present.

***Further Back.*** Observations on climatic changes can be extended much further back in time with the use of **proxies**—measurable records that can provide data on factors such as temperature, ice cover, and precipitation. For example, historical accounts suggest that the Northern Hemisphere enjoyed a warming period from 1100 to 1300 A.D. This was followed by the "Little Ice Age," between 1400 and 1850 A.D. Additional proxies include tree rings, pollen deposits, changes in landscapes, marine sediments, corals, and ice cores.

Some work done quite recently on ice cores has both provided a startling view of a global climate that oscillates according to several cycles and afforded some evidence that remarkable changes in the climate can occur within as little as a few decades. Ice cores in Greenland and the Antarctic have been analyzed for thickness, gas content (specifically, carbon dioxide ($CO_2$) and methane ($CH_4$), two greenhouse gases), and **isotopes,** which are alternative chemical configurations of a given compound, due to different nuclear components. Isotopes of oxygen, as well as isotopes of hydrogen, behave differently at different temperatures when condensed in clouds and incorporated into ice.

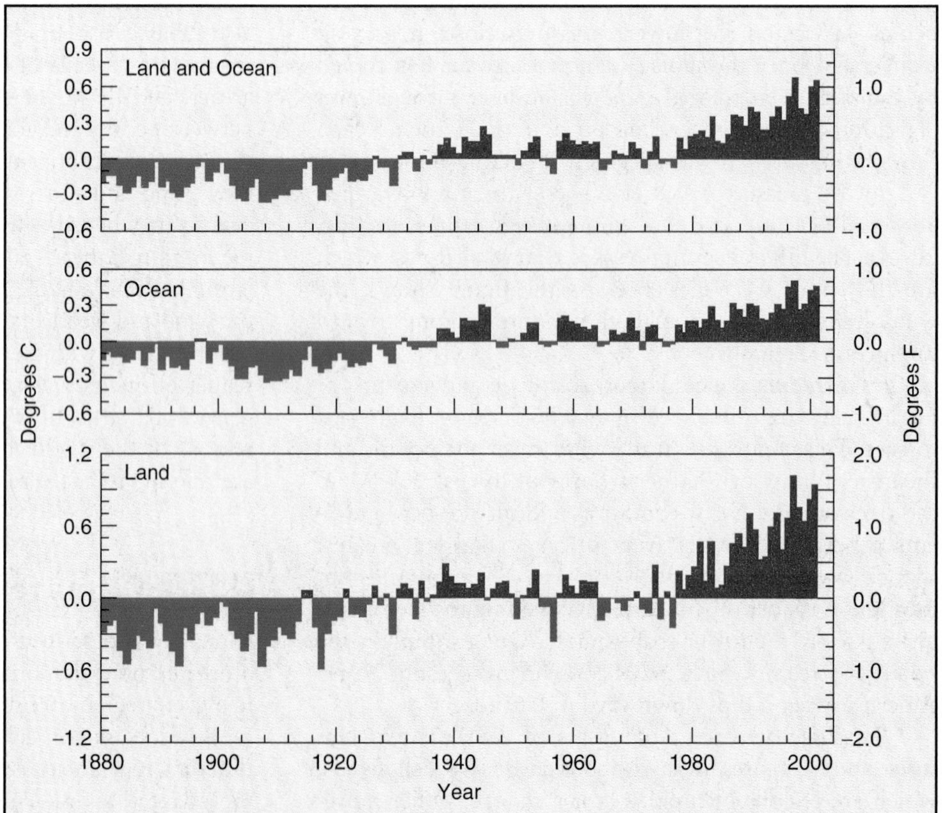


**Figure 20–5 Annual mean global surface atmospheric temperatures.** The baseline, or zero point, is the 1880–1999 long-term average temperature. The warming trend since 1970 is conspicuous. (*Source:* National Climatic Data Center, NOAA.)

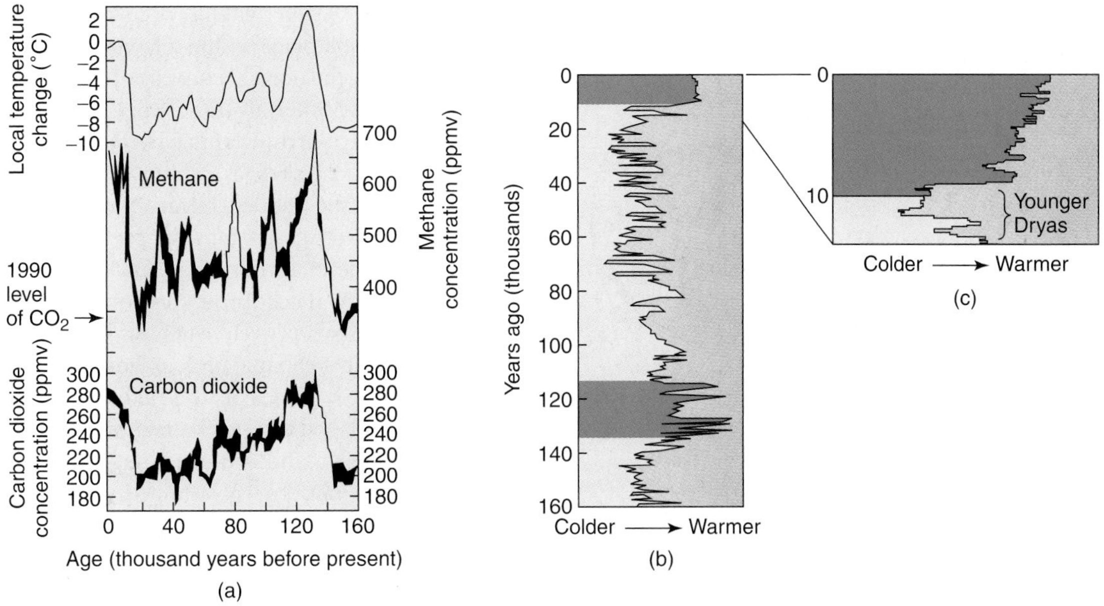


**Figure 20–6 Past climates, as determined from ice cores.** (a) Temperature, methane, and carbon dioxide data from Antarctic ice cores, covering the past 160,000 years. (b) Temperature patterns of the last 160,000 years, demonstrating climatic oscillations. (c) Higher resolution of the last 12,000 years. The "Younger Dryas" cold spell occurred at the start of this record. Note how rapidly the cold spell dissipated, at around 10,700 years before the present. [After Christopherson, Robert W., *Geosystems: An Introduction to Physical Geography,* 4th ed. (Upper Saddle River, NJ: Prentice Hall, 2000.)]

The record based on these analyses indicates that Earth's climate has oscillated between ice ages and warm periods (Fig. 20–6). During major ice ages, huge amounts of water were tied up in glaciers and ice sheets, and the sea level was lower by as much as 400 feet (120 m). The most likely explanation for these major oscillations is the existence of known variations in Earth's orbit, such that, in different modes of orbital configuration, the distribution of solar radiation over different continents and latitudes varies substantially. These oscillations take place according to several periodic time intervals, called **Milankovitch cycles** (after the Serbian scientist who first described them): 100,000, 41,000, and 23,000 years.

***Rapid Changes.*** Superimposed on the major oscillations is a record of rapid climatic fluctuations during periods of glaciation and warmer times (Fig. 20–6c). One such rapid change, called the *Younger Dryas* event (*Dryas* is a genus of Arctic flower), occurred toward the end of the last ice age. Earth had been warming up for 6,000 years and then plunged again into 1,500 years of cold weather. At the end of this event, 10,700 years ago, Arctic temperatures rose 7°C in 50 years! The impact on living systems must have been enormous. It is unlikely that this warming was due to variations in solar output: There is simply no evidence that the Sun has changed over the last million years, let alone undergone major changes in just a few decades. Instead, scientists have narrowed the field of possible explanations to the link between the atmosphere and the oceans. We return now to this link, having already seen the crucial role played by the oceans in El Niño events.

## Ocean and Atmosphere

Earth is mostly a water planet, covered more than two-thirds by oceans. Since we all live on land, it is hard for us to imagine that the oceans play a dominant role in determining our climate. Nevertheless, the oceans are the major source of water for the hydrologic cycle and the main source of heat entering the atmosphere. Recall from Chapter 9 that the *evaporation of ocean water* supplies the atmosphere with water vapor, and when water vapor condenses in the atmosphere (Fig. 7–6), it supplies the atmosphere with heat (latent heat of condensation). The oceans also play a vital role in climate because of their innate *heat capacity*—the ability to absorb energy when water is heated. Indeed, the entire heat capacity of the atmosphere is equal to that of just the top 3 m of ocean water! The well-known ameliorating effect of oceans on the climate of coastal land areas is a consequence of this property. Finally, through the movement of currents, the oceans *convey heat* throughout the globe.

***Thermohaline Circulation.*** A thermohaline circulation pattern dominates oceanic currents, where *thermohaline* refers to the effects that temperature and salinity have on the density of seawater. This **conveyor** system (Fig. 20–7) acts as a giant, complex conveyor belt, moving water masses from the surface to deep oceans and back again, according to the density of the mass. A key area is the high-latitude North Atlantic, where salty water from the Gulf Stream moves northward on the surface and is cooled by Arctic air currents. Cooling increases the density of the water, which then sinks to depths of up to 4,000 m—the *North Atlantic Deep Water* (NADW). This

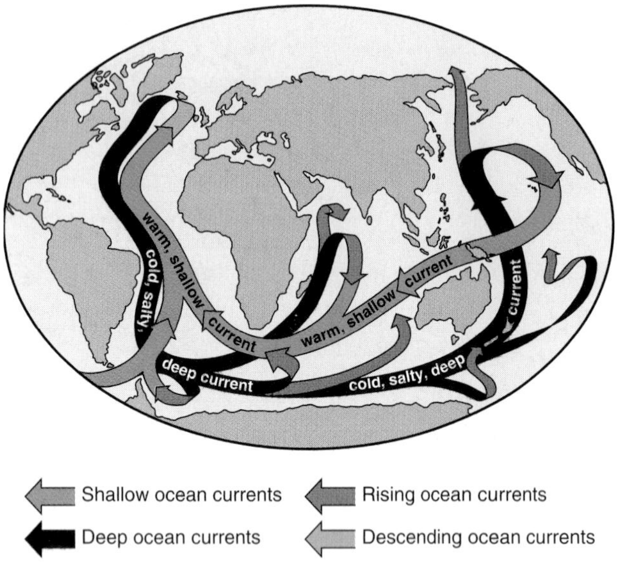


**Figure 20–7 The oceanic conveyor system.** Salty water flowing to the North Atlantic is cooled and sinks, forming the North Atlantic Deep Water system. This deep flow extends southward and is joined by Antarctic water, whereupon it extends into the Indian and Pacific oceans. Surface currents then proceed in the opposite direction, returning the water to the North Atlantic. (Adapted from Figure 3.15 from *Our Changing Planet,* 2d ed., by Fred T. MacKenzie. Copyright © 1998 by Prentice Hall, Inc. Reprinted by permission of Pearson Education, Inc., Upper Saddle River, NJ 07458.)

deep water spreads southward through the Atlantic to the southern tip of Africa, where it is joined by cold Antarctic waters. Together, the two streams spread northward into the Indian and Pacific oceans as deep currents. Gradually, the currents slow down and warm, becoming less dense and welling up to the surface, where they are further warmed and begin to move surface waters back again toward the North Atlantic. This movement transfers enormous quantities of heat toward Europe, providing a climate that is much warmer than the high latitudes there would suggest. The circulation pattern operates over a period of about 1,000 years for one complete cycle and is vital to the maintenance of current climatic conditions.

*Abrupt Change.* Recent evidence indicates that the conveyor system has been interrupted in the past, changing climate abruptly. One mechanism that could accomplish such a change is the appearance of unusually large quantities of fresh water in the North Atlantic, lowering the density of the water and therefore preventing much of the massive sinking that normally occurs there and blocking the northward movement of warmer, saltier water. North Atlantic marine sediments show evidence of the periodic invasion (six times in the last 75,000 years) of icebergs from the polar ice cap that supplied huge amounts of fresh water as they melted—called Heinrich events, after the scientist who first described them. The evidence indicates that these invasions coincided with rapid cooling, as recorded in ice cores, and suggests that the conveyor system shifted southward, with deep water forming nearer to Bermuda than Greenland. When this shift occurred, a major cooling of the climate took place within a few decades. Although it is not clear how, the normal conveyor pattern returned and brought about another abrupt change, this time warming conditions in the North Atlantic. The Younger Dryas event likely involved just such a shift in the conveyor system, brought on by the sudden release of dammed-up water from glacial Lake Agassiz into the St. Lawrence drainage. Referring to these abrupt changes in a recent article, oceanographer Wallace Broecker commented,[1] "Earth's climate system has proven itself to be an angry beast. When nudged, it is capable of a violent response."

*What if...?* One of the likely consequences of extended global warming is increased precipitation over the North Atlantic and more melting of sea ice and ice caps. If such a pattern is sustained, it could lead to a breakdown in the normal operation of the conveyor and a rapid change in climate, especially in the northern latitudes, a possibility that Broecker has called "The Achilles' heel of our climate system." Indeed, oceanographers have seen a gradual decline in the salinity of the northern seas consistent with known glacial melt and a thinning of the ice in the Arctic Ocean. Although it is too soon to know whether this freshening will affect the conveyor, the trend has raised concern over climate change another notch.

## 20.3 Global Climate Change

### The Earth as a Greenhouse

Factors that influence the climate include interactive, *internal components* (oceans, the atmosphere, snow cover, sea ice, etc.) and *external factors* (solar radiation, the Earth's rotation, slow changes in our planet's orbit, and the gaseous makeup of the atmosphere). **Radiative forcing** is the influence a particular factor has on the energy balance of the atmosphere–ocean–land system. The factors can be *positive,* leading to *warming,* or *negative,* leading to *cooling,* as they affect the energy balance. If the factors change over time, they can lead to a change in the climate.

**Warming Processes.** The interior of a car heats up when the car is sitting in the Sun with the windows closed. This heating occurs because sunlight comes in through the windows and is absorbed by the seats and other interior objects, thus converting light energy into heat energy, which is given off in the form of infrared radiation. Unlike sunlight, infrared radiation is blocked by glass and so cannot leave the car. The trapped heat energy causes the interior air temperature to rise. This is the same phenomenon that keeps a greenhouse warmer than the surrounding environment.

*Greenhouse Gases.* On a global scale, water vapor, $CO_2$, and other gases in the atmosphere play a

[1]Broecker, W. S. "Does the Trigger for Abrupt Climate Change Reside in the Ocean or in the Atmosphere?" *Science* 300 (June 6, 2003): 1519–1522.

role analogous to that of the glass in a greenhouse. Therefore, they are called **greenhouse gases (GHGs).** Light energy comes through the atmosphere and is absorbed by Earth and converted to heat energy at the planet's surface. The infrared heat energy radiates back upward through the atmosphere and into space. The GHGs that are naturally present in the troposphere absorb some of the infrared radiation and reradiate it back toward the surface; other gases ($N_2$ and $O_2$) in the troposphere do not (Fig. 20–8). This greenhouse effect was first recognized in 1827 by French scientist Jean-Baptiste Fourier and is now firmly established. The GHGs are like a heat blanket, insulating Earth and delaying the loss of infrared energy (heat) to space (Fig. 20–8). Without this insulation, average surface temperatures on Earth would be about −19°C instead of +14°C, and life as we know it would be impossible. Therefore, our global climate depends on Earth's concentrations of GHGs. If these concentrations increase or decrease significantly, their influence as positive "forcing" agents will change, and our climate will change accordingly.

**Cooling Processes.** Earth's atmosphere is also subject to negative forcing factors. For example, on average, clouds cover 50% of Earth's surface and reflect some 21% of solar radiation away to space before it ever reaches the ground. Sunlight reflected in this way is called the **planetary albedo,** and it contributes to overall cooling by preventing a certain amount of warming in the first place. The effect of the albedo is especially true for *low-lying* clouds and can be readily appreciated if you think of how you are comforted on a hot day when a large cloud passes between you and the Sun. The radiation that was heating you and your surroundings is suddenly intercepted higher up in the atmosphere, and you feel cooler. However, *high-flying,* wispy clouds have a positive forcing effect, absorbing some of the infrared radiation and emitting some infrared themselves. Overall, the current net impact of clouds is judged to be slightly negative forcing.

Snow and ice also reflect sunlight, contributing to the planetary albedo. Recent work, however, suggests that this effect is reduced greatly by the widespread soot originating from anthropogenic sources. By darkening the snow and ice, the soot promotes the absorption of radiant energy rather than reflection.

***Volcanoes.*** Volcanic activity can also lead to planetary cooling. When Mount Pinatubo in the Philippines erupted in 1991, some 20 million tons of particles and aerosols entered the atmosphere and contributed to a significant drop in global temperature as radiation was

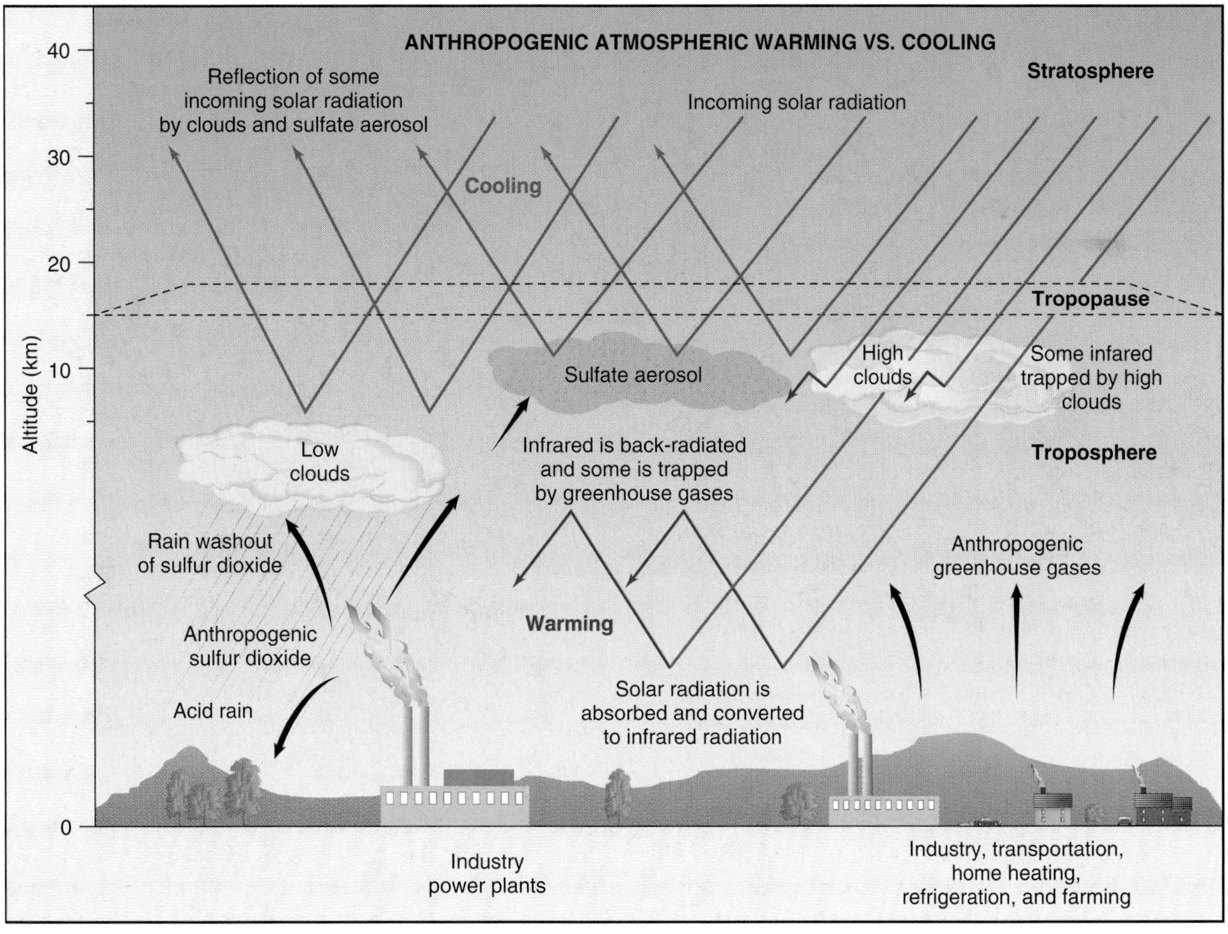


**Figure 20–8 Global warming and cooling.** Factors involved in atmospheric warming and cooling are illustrated. Greenhouse gases promote global warming; sulfur dioxide leads to cooling.

reflected and scattered away. This global cooling effect lasted until the volcanic debris was finally cleansed from the atmosphere by chemical change and deposition, a process that took several years.

*Aerosols.* Climatologists have found that anthropogenic sulfate aerosols (from ground-level pollution) play a significant role in canceling out some of the warming from GHGs. Sulfur dioxide from industrial sources enters the atmosphere and reacts with compounds there to form a high-level aerosol—a sulfate haze. This haze reflects and scatters some sunlight and also contributes to the formation of clouds, with a concomitant increase in planetary albedo. The mean residence time of the sulfates forming the aerosol is about a week, so the aerosol does not increase over time, as the GHGs do. However, because anthropogenic sulfate is more than double that coming from natural sources, its effect is substantial and persistent. Climatologists estimate that the cooling effect of these pollutants has counteracted some 20–30% of the global warming in recent years. Newer findings indicate that sulfate aerosols and other pollutants can prevent precipitation from occurring within clouds, adding substantially to their impact on climate.

*Ozone Depletion.* Paradoxically, the depletion of the stratospheric ozone layer by anthropogenic sources (Section 20.5) has led to a cooling of the lower stratosphere. In turn, this cooling offsets about 20% of the observed positive forcing traced to the increases in greenhouse gases in the atmosphere.

**Solar Variability.** Since the Sun is the source of radiant energy that heats Earth, any variability in the sun's radiation reaching Earth will likely influence the climate. Direct monitoring of solar irradiance goes back only a few decades, but it has confirmed that there is an 11-year cycle of changes involving slight increases during times of high sunspot activity. These changes in turn may affect internal components such as surface and tropospheric temperatures, ocean currents, and the position of the jet stream. Variations in solar forcing are thought to be partly responsible for the early-20th-century warming trend, but attempts to correlate other climatic changes with sunspot cycles have not worked well. The problem is that the solar variations are interacting with climate system factors which themselves have stronger impacts than the solar forcing itself. In particular, the solar variability that has been observed is not consistent with the warming of the latter part of the 20th and early 21st centuries.

*Thus....* Global atmospheric temperatures are a balance between the positive and negative forcing from natural causes (volcanoes, clouds, natural GHGs, solar irradiance) and anthropogenic causes (sulfate aerosols, soot, ozone depletion, increases in GHGs). The net result varies, depending on one's location. As we will see, this balance makes it difficult to be certain about the cause-and-effect link between forcing factors and climate parameters such as temperature, precipitation, and storm events. It also contributes much uncertainty to our predictions of what will happen in the future as GHGs continue to increase. Figure 20–8 depicts most of the natural and anthropogenic factors that interact to influence the temperature at any given location.

## The Greenhouse Gases

**Carbon Dioxide.** More than 100 years ago, Swedish scientist Svante Arrhenius reasoned that differences in $CO_2$ levels in the atmosphere could greatly affect Earth's energy budget. Arrhenius suggested that, in time, the burning of fossil fuels might change the atmospheric $CO_2$ concentration, although he believed that it would take centuries before the associated warming would be noticeable. He was not concerned about the impacts of this warming, arguing that such an increase would be beneficial. (He lived in Sweden!)

*Monitoring.* In 1958, Charles Keeling began measuring $CO_2$ levels on Mauna Loa, in Hawaii. Measurements there have been recorded continuously, and they reveal a striking increase in atmospheric levels of the gas (Fig. 20–9). The concentrations increased exponentially until the energy crisis in the mid-1970s and have been rising more or less linearly ever since. The data also reveal an annual oscillation of 5–7 ppm, which reflects seasonal changes of photosynthesis and respiration in terrestrial ecosystems in the Northern Hemisphere. When respiration predominates (late fall through spring), $CO_2$ levels rise; when photosynthesis predominates (late spring through early fall), $CO_2$ levels fall. Most salient, however, is the relentless rise in $CO_2$ levels, ranging from 0.8 to 1.9 ppm per year in recent years. As of early 2004, atmospheric $CO_2$ levels were over 375 ppm, 35% higher than they were before the Industrial Revolution and higher than they have been for over 400,000 years. Thus, our insulating blanket is thicker, and it is reasonable to expect that this will have a warming effect.

*Sources.* As Arrhenius suggested, the obvious place to look for the source of increasing $CO_2$ levels is our use of fossil fuels. Every kilogram of fossil fuel burned produces about 3 kg of $CO_2$. (The mass triples because each carbon atom in the fuel picks up two oxygen atoms in the course of burning and becoming $CO_2$.) Currently, 6.6 billion metric tons (gigatons, or Gt) of fossil-fuel carbon (GtC) are burned each year, all added to the atmosphere as $CO_2$. Most of this amount, as well as past fossil-fuel $CO_2$, comes from the industrialized countries (Fig. 20–10). It is estimated that the burning of forest trees is another anthropogenic source that adds some 1.6 GtC annually to the carbon already coming from industrial processes.

*Sinks.* Careful calculations show that if all the $CO_2$ emitted from burning fossil fuels accumulated in the atmosphere, the concentration would rise by at least 3 ppm per year, not the 1.9 ppm or less charted in Fig. 20–9. In the 1990s, this meant an addition of approximately 8 GtC per year added to the atmosphere by anthropogenic sources, yet only 3.2 GtC per year actually accumulated. Thus, there must be carbon "*sinks*" that absorb $CO_2$ and

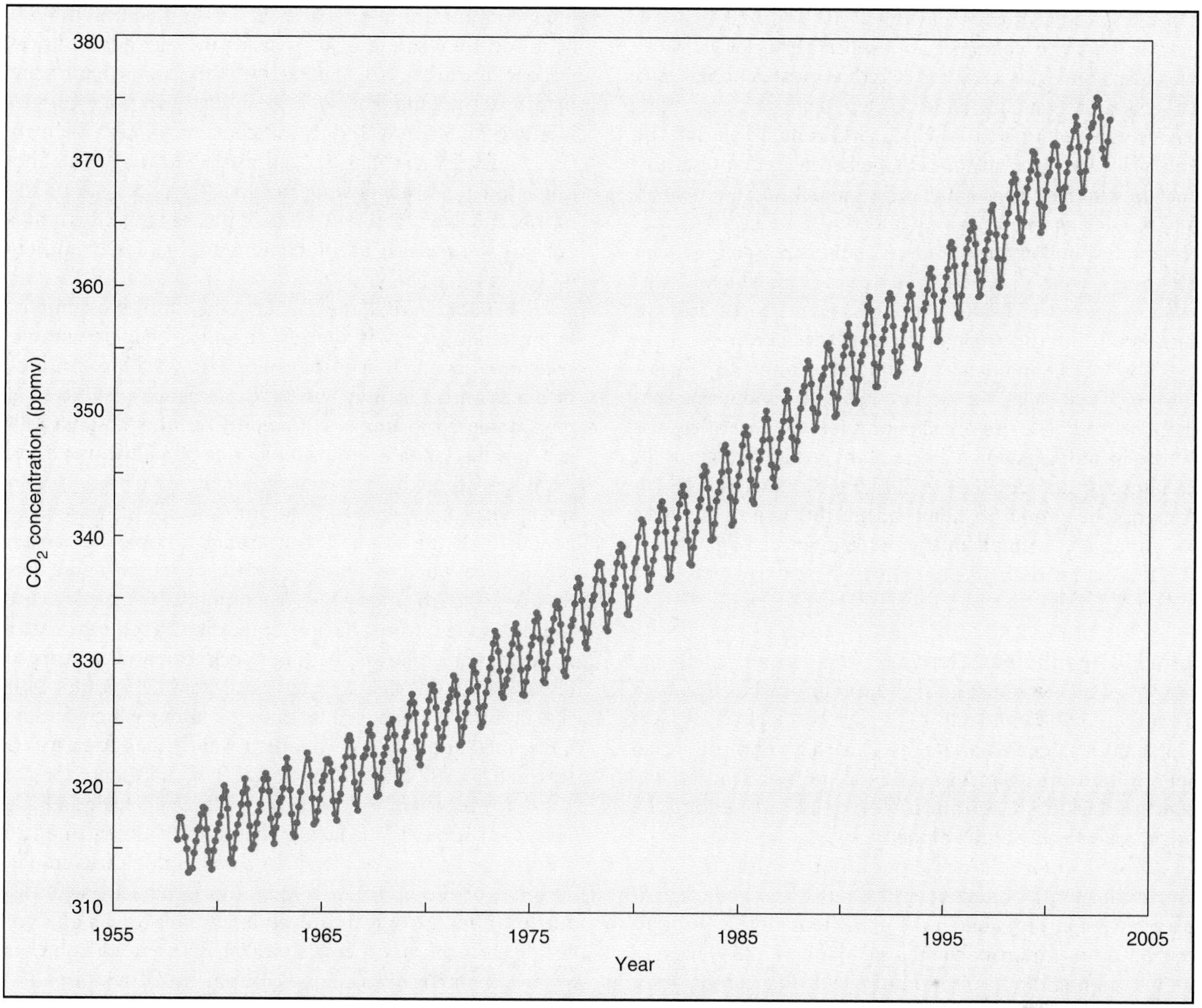


**Figure 20–9 Atmospheric carbon dioxide concentrations.** The concentration of $CO_2$ in the atmosphere fluctuates between winter and summer because of seasonal variation in photosynthesis. The average concentration is increasing owing to human activities—in particular, burning fossil fuels and deforestation. (All measurements made at Mauna Loa Observatory, Hawaii, by Dave Keeling and Tim Whorf, Scripps Institute of Oceanography.)

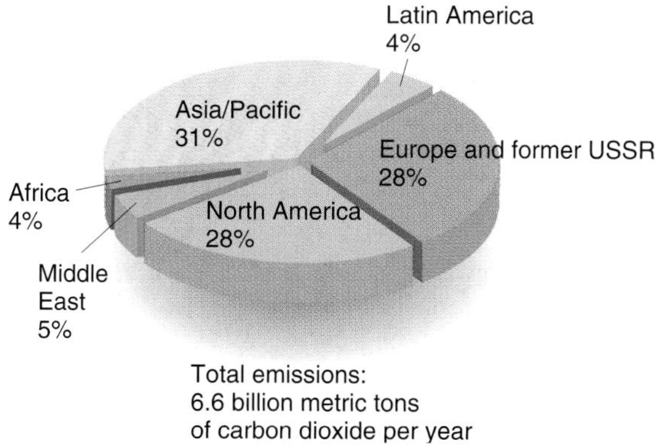


**Figure 20–10 Sources of carbon dioxide emissions from fossil-fuel burning.** Total emissions in 2001 were approximately 6.6 GtC, or 24 billion metric tons of $CO_2$. (*Source:* Data from U.S. Energy Information Agency.)

keep it from accumulating at a more rapid rate in the atmosphere. Recent work with stable carbon and oxygen isotopes and careful measurements of atmospheric $CO_2$ from 120 stations around the globe have brought us much closer to quantifying annual fluxes in $CO_2$ and identifying the missing sinks. There is broad agreement that the oceans serve as a sink for much of the $CO_2$ emitted; some of this is due to the uptake of $CO_2$ by phytoplankton and its subsequent sinking, and some is a consequence of the undersaturation of $CO_2$ in seawater. There are limitations to the ocean's ability to absorb $CO_2$, however, because only the top 300 m of the ocean is in contact with the atmosphere. (As mentioned earlier, the deep ocean layers do mix with the upper layers, but the mixing time is over a thousand years.) Calculations indicate that, in the 1990s, the ocean sink accounted for an uptake of 2.0 ($\pm$0.6) GtC annually.

The seasonal swings illustrated in Figure 20–9 show that the biota can influence atmospheric $CO_2$ levels. Measurements indicate that terrestrial ecosystems can also serve as a carbon sink. Indeed, these land ecosystems apparently stored a net 1.5 (±0.8) GtC annually during the 1990s, a figure that includes the losses from deforestation and thus implies an average gross annual uptake of some 3 GtC during those years. For this reason, terrestrial ecosystems, and especially forests, are increasingly valued because of their ability to sequester carbon. Much of this carbon uptake is being attributed to increased rainfall associated with the warming trend in temperature.

Both ocean and land sinks fluctuate in their uptake of carbon, and indeed, during the 1980s, the land was essentially neutral. The best explanation for the observed variability in the behavior of these sinks is the climate itself, perhaps as a consequence of El Niño–La Niña and North Atlantic circulation phenomena, but this hypothesis is as yet unproven. A simple model of the major dynamic pools and fluxes of carbon, as presently understood, is presented in Figure 20–11.

**Other Greenhouse Gases.** Water vapor, methane, nitrous oxide, ozone, and chlorofluorocarbons absorb infrared radiation and add to the insulating effect of $CO_2$ (Table 20–2). Some of these gases also have anthropogenic sources and are increasing in concentration, raising the concern that future warming will extend well beyond the calculated effects of $CO_2$ alone.

*Water Vapor.* Water vapor absorbs infrared energy and is the most abundant greenhouse gas. Although it plays an important role in the greenhouse effect, its concentration in the troposphere is quite variable. Through evaporation and precipitation (the hydrologic cycle—see Chapter 7), water undergoes rapid turnover in the lower atmosphere, and water vapor does not tend to accumulate over time. Water vapor does, however, appear to be a major factor in what has been called the "supergreenhouse effect" in the tropical Pacific Ocean. As it traps energy that has been radiated back to the atmosphere, the high concentration of water vapor contributes significantly to the heating of the ocean surface and the lower atmosphere in the tropical Pacific. This means that if temperatures over the land and oceans rise, evaporation will increase and the water vapor concentration will likely rise, which in turn causes even more warming. The phenomenon is called **positive feedback** and is one of the more disturbing features of future warming, because it increases the sensitivity of climate to increased anthropogenic GHGs.

*Methane.* Methane ($CH_4$), the third-most-important greenhouse gas, is a product of microbial fermentative reactions; its main natural source is wetlands. Anthropogenic sources include livestock (methane is generated in the stomachs of ruminants), landfills, coal mines, natural-gas production and transmission, rice cultivation, and manure. Although methane is gradually destroyed in reactions with other gases in the atmosphere, it was being added to the atmosphere faster than it could be broken down. The concentration of atmospheric methane has doubled since the Industrial Revolution, as revealed in core samples taken from glacial ice (Table 20–2) and, after rising at a rate of around 1.8 ppm per year, now appears to have leveled off. Measurements indicate that two-thirds of the present methane emissions are anthropogenic. Like $CO_2$, methane is more abundant than it has been for at least the last 400,000 years.

*Nitrous Oxide.* Nitrous oxide ($N_2O$) levels have increased some 15% during the last 200 years and are still rising. Sources of the gas include agriculture and the burning of biomass; lesser quantities come from fossil-fuel burning. $N_2O$ is produced in agriculture via anaerobic denitrification processes (see Chapter 3), which occur wherever nitrogen (a major component of fertilizers) is highly available in soils. The buildup of nitrous oxide is particularly unwelcome, because its long residence time (114 years) makes the gas a problem in not only the troposphere, where it contributes to warming, but also the stratosphere, where it contributes to the destruction of ozone.

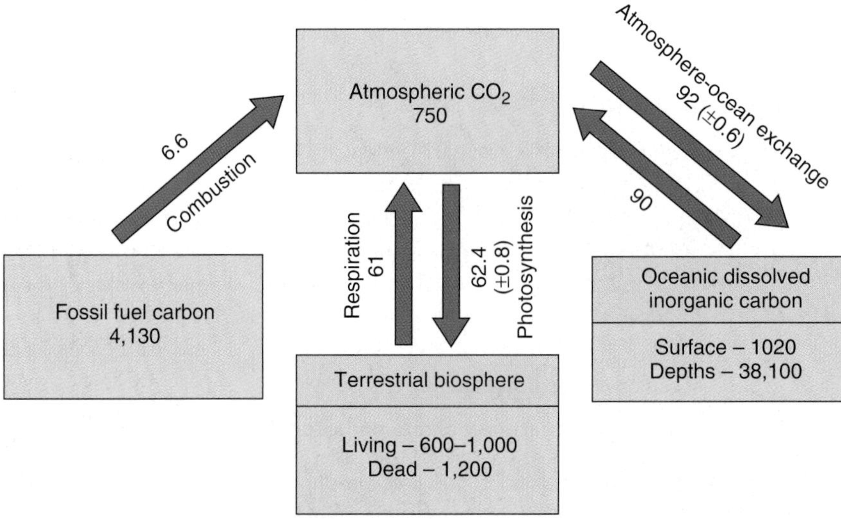


**Figure 20–11 Global carbon cycle.** Data are given in GtC (billion metric tons of carbon). Pools are in the boxes, and fluxes are indicated by the arrows. (*Data sources:* Various.)

**table 20-2 Anthropogenic Greenhouse Gases in the Atmosphere**

| Gas | Average Concentration 100 years ago (ppb)[1] | Approximate Current Concentration (ppb) | Average Residence Time in Atmosphere (years) |
|---|---|---|---|
| Carbon dioxide ($CO_2$) | 288,000 | 375,000 | 120 |
| Methane ($CH_4$) | 848 | 1,850 | 12 |
| Nitrous oxide ($N_2O$) | 285 | 316 | 114 |
| Chlorofluorocarbons and halocarbons | 0 | 1.2 | 50–100 |

[1] Parts per billion; 1,000 ppb = 1 part per million (ppm).
(*Source:* Carbon Dioxide Information Analysis Center, Oak Ridge National Laboratory.)

*Ozone.* Although ozone in the troposphere is short lived, it is a potent greenhouse gas. Some ozone from the stratosphere (where it is formed) descends into the troposphere, but the greatest source is anthropogenic, through the action of sunlight on pollutants (discussed later in the chapter and in Chapter 21). Estimates indicate that the concentration of ozone in the atmosphere has increased 36% since 1750. Major sources are automotive traffic and burning forests and agricultural wastes.

*CFCs and Other Halocarbons.* Emissions of halocarbons are entirely anthropogenic. Like nitrous oxide, halocarbons are long lived and contribute to both global warming in the troposphere and ozone destruction in the stratosphere. Used as refrigerants, solvents, and fire retardants, halocarbons have a much greater capacity (10,000 times) for absorbing infrared radiation than does $CO_2$. The rate of production of chlorofluorocarbons (CFCs) has declined since the Montreal Accord of 1987, and the concentration of CFCs in the troposphere leveled off in the late 1990s and is now slowly declining. However, these gases are highly stable and will continue to exert their warming effects for many decades.

Together, the other anthropogenic GHGs are estimated to trap as much infrared radiation as $CO_2$ does. Although the tropospheric concentrations of some of these gases are rising, it is hoped that they will gradually decline in importance because of steps being taken to reduce their levels in the atmosphere, leaving $CO_2$ as the primary greenhouse gas to cope with in the future.

## Evidence of Climate Change

**Intergovernmental Panel on Climate Change.** In 1988, the United Nations Environment Program and the World Meteorological Society established the Intergovernmental Panel on Climate Change (IPCC) in order to provide accurate and relevant information that would lead to an understanding of human-induced climate change. The IPCC has established three working groups: one to assess the scientific issues (Working Group I), another to evaluate the impact of global climate change and the prospects for adapting to it (Working Group II), and a third to investigate ways of mitigating the effects (Working Group III). The IPCC working groups consist of more than two thousand experts in the appropriate fields from over a hundred countries. These stalwarts are unpaid and participate at the cost of their own research and other professional activities. The work of the IPCC has been guided by two basic questions:

- **Risk assessment**—Is the climate system changing, and what is the impact on society and ecosystems?
- **Risk management**—How can we manage the system through adaptation and mitigation?

*Third Assessment.* In January 2001, Working Group I released its third assessment (the next is due in 2007). The major headings of the "Summary for Policymakers" provide a useful outline for examining the evidence for climate change and the IPCC's assessment of that evidence:

**1. An Increasing Body of Observations Gives a Collective Picture of a Warming World and Other Changes in the Climate System.** There is much natural variation in weather from year to year, and local temperatures do not necessarily follow globally averaged ones. It is a fact, however, that 17 of the hottest years on record have occurred since 1980 (Fig. 20–5); indeed, 1990–2000 was the hottest decade ever recorded—and that in spite of two cooler years following the eruption of Mount Pinatubo in the Philippines in 1991. After the debris from Mount Pinatubo was cleansed from the atmosphere, we returned to the warming trend that was evident in the 1980s. The years 1997, 1998, 2001, and 2002 set new records for global temperatures, adding further compelling evidence for global warming.

*Correlation?* Scientists have puzzled over the absence of a more direct correlation between $CO_2$ and global atmospheric temperatures over the last several decades. Calculations, for example, predict significantly higher increases in temperature than have actually occurred. Critics have used this discrepancy to bolster their arguments that global warming is, so to speak, a lot of "hot air." Recent work has addressed the problem. First, it is apparent that sulfate aerosol in the industrialized regions of the Northern Hemisphere appears to be canceling out much of the greenhouse warming over those regions. The cooling effects of the aerosol occur over the very regions most responsible for greenhouse gas emissions. However, scientists point out that the aerosol cooling effect is temporary. As the industrialized nations continue to reduce sulfate emissions (because of acid rain and its impacts on human health), and as GHGs continue to build up, the Northern Hemisphere will likely experience its own share of warming.

The second finding confirms what climatologists have thought, but, until now, have never been able to prove: The oceans are absorbing much of the heat, slowing down the rate at which the atmosphere is warming. Recently, oceanographers analyzed millions of oceanic temperature records from 1950 to 1995 and found that, between 1970 and 1995, the heat content of the global oceans increased dramatically. Indeed, the heat record in the oceans follows the same general trend as that in the atmosphere, having accelerated during the 1990s (Fig. 20–12).

***Satellites?*** An interesting and controversial aspect of the warming trend has arisen with measurements of tropospheric temperatures (from Earth's surface to 8 km) made by 13 different satellites over two decades. Two early analyses of the data indicated either virtually no increase (+0.01°C per decade) or a significant increase (+0.10°C per decade), the latter consistent with the +0.17°C-per-decade warming found in the surface temperature measurements. Much has been made of the first group's apparent discrepancy with the surface temperature trends, especially by skeptics eager to disprove global warming. A National Research Council panel examined this discrepancy and concluded both that the surface warming was real and that satellite records should be used with caution. A more recent analysis of the satellite data corrected for seasonal and diurnal effects, instrument biases, and orbital discrepancies and found a trend of +0.22–0.26°C per decade. If anything, the tropospheric temperatures appear to be rising even more rapidly than the surface temperatures.

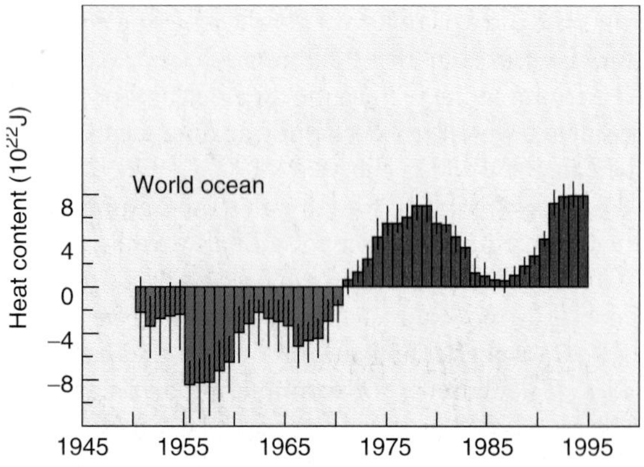


**Figure 20–12 Heat capture by the oceans.** The plot shows the changes in heat content of the upper 3,000 meters of the world's oceans, based on an analysis of millions of temperature data points. (*Source:* Reprinted with permission from *Warming of the World Ocean* by S. Levitus et al. in *Science,* 287, March 24, 2000. Copyright © 2000 by American Association for the Advancement of Science. Reprinted by permission.)

***Other Changes.*** Other significant impacts of global climate change were noted by the IPCC. One is the retreat of glaciers. The World Glacier Monitoring Service has followed the melting and slow retreat of mountain glaciers all over the world. The response time of glaciers to climatic trends is slow—10 to 50 years—but the data collected show a wide-scale recession of glaciers over the past hundred years, consistent with an increase in GHGs. Another impact is the thinning of polar ice. Recent work has documented large decreases in the thickness and extent of sea ice in the Arctic. The polar ice cap has lost nearly 20% of its volume in the last two decades, and permafrost has been melting all over the polar regions, lifting buildings, uprooting telephone poles, and breaking up roads. During 2002 and 2003, the largest ice shelf in the Arctic (150 square miles) was broken up. The Greenland Ice Sheet, a huge reservoir of water, is melting at unprecedented rates, undoubtedly a response to the rise in temperatures in the northern latitudes. While overall global temperature has increased a moderate 1°F, temperatures in Alaska, Siberia, and northwest Canada have risen 5°F in summer and 10°F in winter. Indeed, winters above 40° north latitude have been shortened by an average of two weeks.

The weather has been changing. Further impacts reported by the IPCC include a significant increase in precipitation—especially heavy precipitation; a reduction in the frequency of very low temperatures; a greater frequency of El Niño events; and more frequent and intense droughts in parts of Asia and Africa. In July 2003, the World Meteorological Society issued a bulletin to the effect that the world's weather has seen record-breaking extremes during the late spring, in the form of heat waves, rainfall, and tornadoes, and linked them to climate change.

***Sea Level.*** By far the most ominous of the impacts is the *rise in sea level*. With global warming, the sea level has been rising due to two factors: thermal expansion as ocean waters warm, and the melting of glaciers and ice fields. Sea level has risen between 0.1 and 0.2 meter during the 20th century and is continuing to rise at rate of 2 mm per year. As we have noted, the oceans are clearly warming (Fig. 20–12).

**2. Emissions of GHGs and Aerosols Due to Human Activities Continue to Alter the Atmosphere in Ways That Are Expected to Affect the Climate.** The IPCC examined long-term records of GHGs and sulfate aerosols and demonstrated that these agents were on the increase in the atmosphere (Fig. 20–13) and were exerting a predictable forcing impact on the climate system—positive for the GHGs and negative for the aerosols. All of these trends and impacts are the result of human activities, and are all currently increasing. The combined radiative forcing of these and other anthropogenic and natural factors of the climate are shown in Figure 20–14. Note that the level of scientific certainty for some of these factors is low, but it is high for the GHGs, whose positive radiative forcing is highly significant.

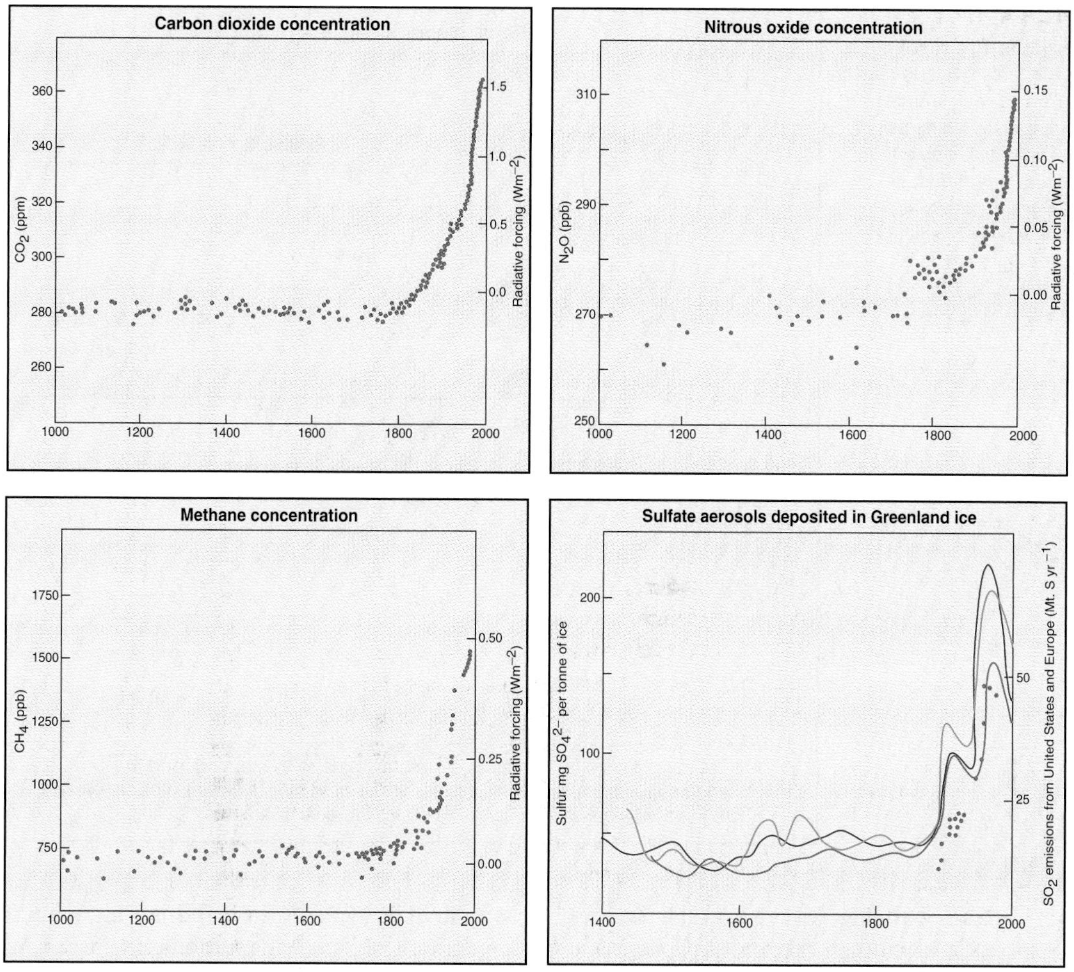


**Figure 20–13 Human influence on the atmosphere during the industrial era.** The plots show concentrations of the greenhouse gases carbon dioxide, methane, and nitrous oxide as they have risen since the early 1800s. Sulfate aerosols are also shown. The lines depict three ice core measurements (left axis), while the dots show levels of sulfate emissions from the U.S. and Europe (right axis). (*Source:* IPCC Working Group I.)

**3. Confidence in the Ability of Models to Project Future Climate Has Increased.** Weather forecasting employs powerful computers capable of handling large amounts of atmospheric data and applying appropriate mathematical equations modeling the processes taking place in the atmosphere, oceans, and land. Forecasts of conditions for 72 hours or more have become quite accurate. Even though computing power has increased greatly in recent years, the somewhat chaotic behavior of weather parameters prevents more long-term forecasting.

Modeling climate is an essential strategy for exploring the potential future impacts of rising GHGs. Climatologists employ the same powerful computers used for weather forecasting and have combined global atmospheric circulation patterns with ocean circulation and radiation feedback from clouds to produce *coupled general circulation models* (CGCMs) that are capable of simulating long-term climatic conditions. Fourteen centers around the world are now intensely engaged in exploring climate change by running models coupling the atmosphere, oceans, and land. The latest information on the cooling effects of sulfate aerosols has been incorporated into current simulations, giving modelers more confidence in their results. Figure 20–15 is taken from the IPCC assessment and indicates simulated global mean temperatures based on natural and known anthropogenic forcings. Note how well the combined model simulates the observed temperature changes. Many models have now successfully simulated past climates on the basis of known differences in Earth's orbit and ocean–atmosphere coupling. The most complex models, requiring supercomputers, recently scored a major triumph in predicting the intensity and locations of the latest El Niño. As we will see, the main purpose of the models is to project the future global climate.

**4. There Is New and Stronger Evidence That Most of the Warming Observed Over the Last 50 Years Is Attributable to Human Activities.** In 1995, in its second assessment report, the IPCC stated cautiously, "The balance of evidence suggests a discernible human influence on global climate." (See "Ethics" essay, p. 556.) In the panel's third assessment, the tone of consensus had shifted, to the more definite statement that anthropogenic GHGs have "contributed substantially to the observed warming over the last 50 years." The panel also concluded

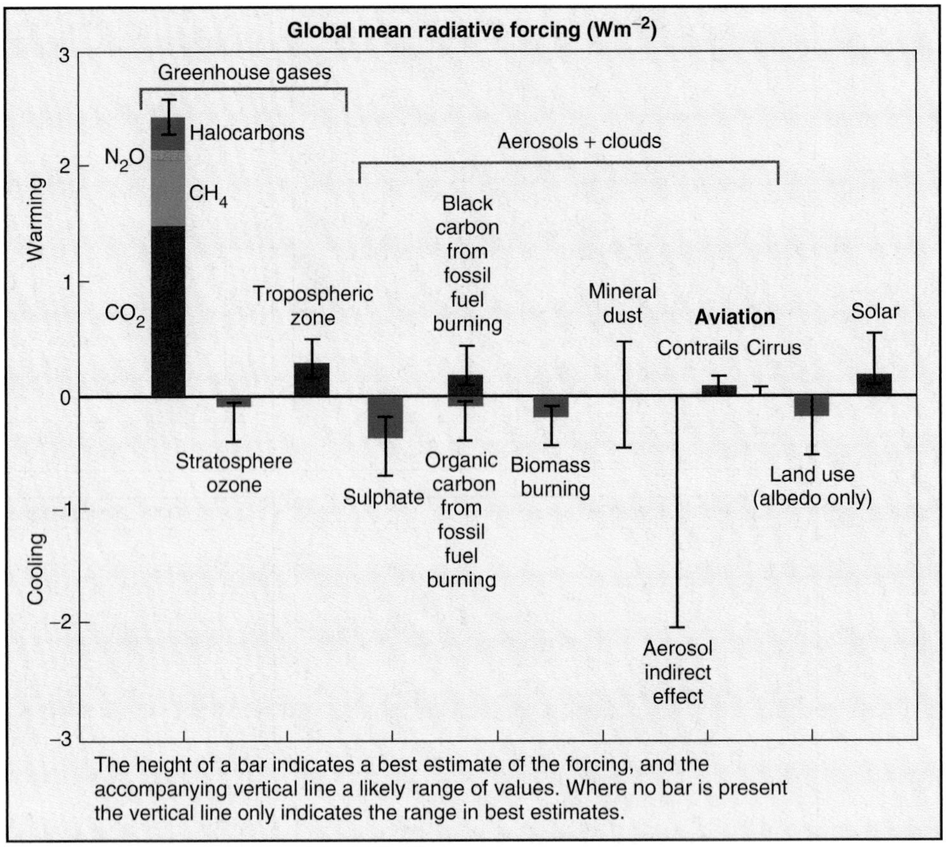


**Figure 20–14 Anthropogenic and natural climate forcing.** Positive and negative forcing from various sources are shown, with units in watts per square meter. The balance clearly indicates a net warming effect, estimated at 1.6 watts per square meter. The relative level of scientific understanding of the sources is indicated also. (*Source:* IPCC Working Group I.)

that the 20th century warming was responsible for the observed rise in sea level through thermal expansion and the extensive loss of glaciers and sea ice.

**5. Human Influences Will Continue to Change Atmospheric Composition Throughout the 21st Century.** The coupled general circulation models were employed to examine the changes that might take place over the 21st century. A number of assumptions about greenhouse gas emissions were combined with a range of responses of climate to greenhouse gas concentrations (climate sensitivity) to project some 35 scenarios of 21st-century climate changes. All scenarios project a rise in $CO_2$ as a consequence of burning fossil fuels. The IPCC expects that this one greenhouse gas will be the dominant influence on climate. Figure 20–16 shows several of the scenarios as they play out in the models; the rising $CO_2$ emissions and atmospheric concentrations are clearly indicated.

*Assumptions.* The assumptions underlying the scenarios are critical. The A1 scenarios reflect a world in which population reaches its peak in midcentury and then declines, economic growth is rapid, and new technologies are employed. A1F1 shows a fossil-fuel-intensive society, A1T reflects shifts to non-fossil-energy sources, and A1B is a balance across all energy sources. The A2 series reflects a world of independent countries going their own way, with increasing population and varied economic growth and technological change. B1 reflects a world with population developments as in A1, but with rapid changes in economies and effective global cooperation to reach sustainability in all sectors. B2 has increasing populations and more regional adaptations, as in the A2 series, but with significant efforts to achieve sustainability. IS92a is a scenario carried over from the second IPCC assessment, for comparison purposes. Notably, none of the scenarios includes specific adjustments based on the targets of the Framework Convention on Climate Change or the Kyoto Protocol (discussed shortly).

The IPCC scenarios demonstrate the crucial importance of energy choices. The great range of $CO_2$ emissions and consequent atmospheric concentrations (540–970 ppm) in the scenarios reflect the range of energy options from "business as usual" fossil-fuel use (A1F1) to a shift to renewable energy (B1). Even the most optimistic scenario pictures a doubling of atmospheric $CO_2$ over preindustrial levels. The most important issue here is that of stabilizing atmospheric $CO_2$ concentrations. It will be necessary to bring $CO_2$ emissions below 1990 levels (The Kyoto Protocol target) within a few decades, to stabilize $CO_2$ at 450 ppm, but if it took a century to achieve the 1990 target, $CO_2$ would be at 650 ppm, and if it took two centuries, 1,000 ppm would be the ambient $CO_2$ concentration.

**6. Global Average Temperature and Sea Level Are Projected to Rise Under All IPCC Scenarios.** The major consequence of rising greenhouse gas levels during the 21st century is rising temperatures—the amount of rise again depending on the energy choices and other factors, such as population growth. Rising global temperatures are linked to two major impacts: *regional climatic changes* and

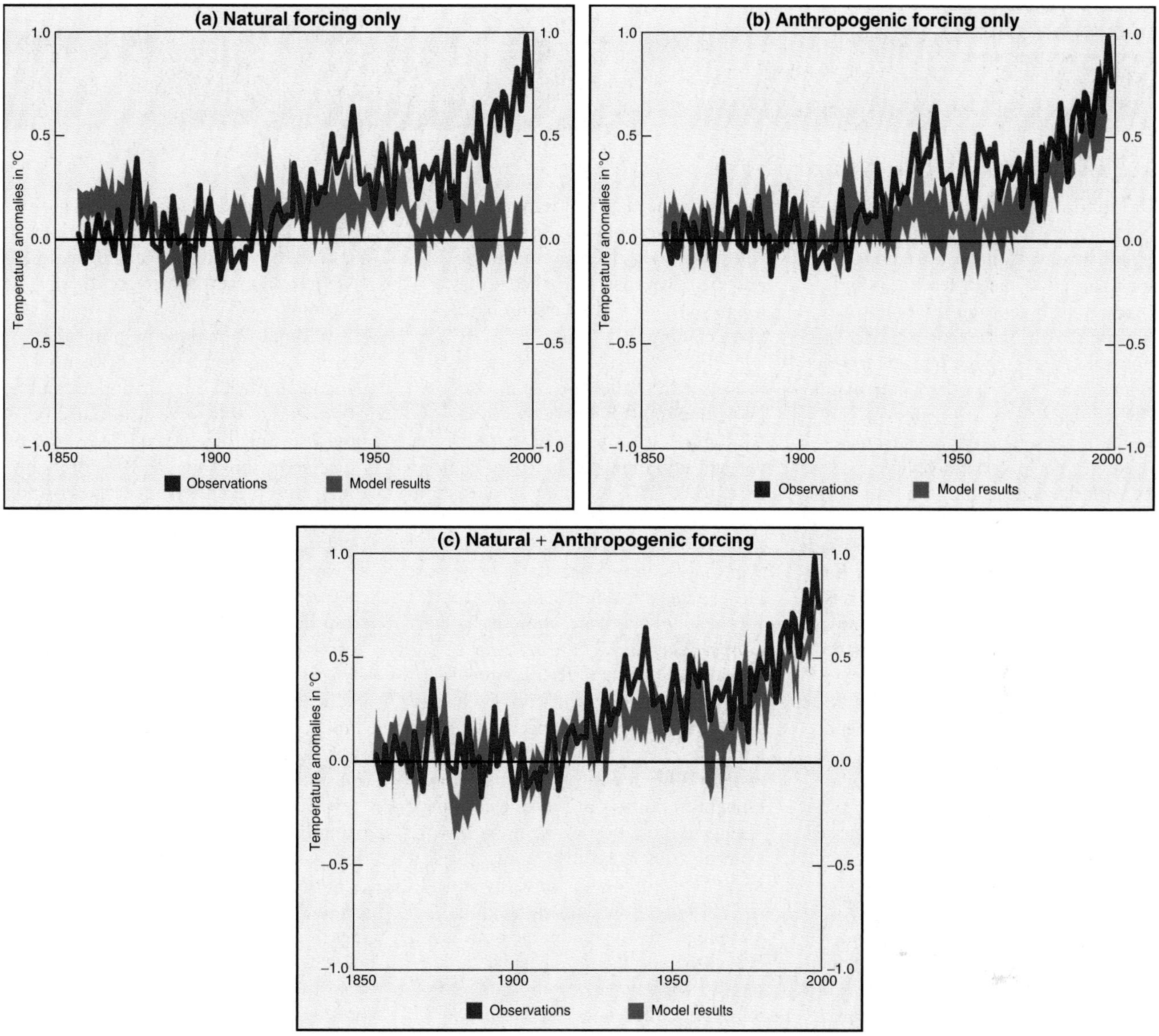


**Figure 20–15 Comparison between modeled and actual data on temperature rise since 1860.** (a) Model results with only natural forcing. (b) Model results with only anthropogenic forcing. (c) Model results combining natural and anthropogenic forcing. Note how well the latter follows the actual temperature data. (*Source:* IPCC Working Group I.)

a *rise in sea level* (Fig. 20–17a). Both of these effects show up in all of the models and are expected to become evident in a matter of a few decades. The IPCC assessment states, "The globally averaged surface temperature is projected to increase by 1.4 to 5.8°C over the period 1990 to 2100." This projection is based on the range of scenarios and numerous climate models. For doubled $CO_2$ (560 ppm), the best estimate is a temperature rise of 2.5°C. The difference in temperature between an ice age and the warm period following is only 5°C, so in a century, unless some drastic measures are taken, Earth's climate will change rapidly. Responding to these changes will likely involve unprecedented and costly adjustments. The impact on natural ecosystems could be highly destabilizing.

***Climate Changes.*** Warming will seriously affect rainfall and agriculture. The present-day difference between temperatures at the poles and those at the equator is a major force driving the atmospheric circulation. The warming associated with a doubling of greenhouse gas levels is likely to be more pronounced in polar regions (as much as 10°C) and less pronounced in equatorial regions (1 to 2°C). Greater heating at the poles than at the equator will change atmospheric circulation patterns, as well as rainfall distribution patterns. Higher temperatures promote more evaporation and a greater ability of the atmosphere to hold moisture, which can enhance precipitation. Some regions will be drier, and some will be wetter. The frequency of droughts is expected to increase, as well as the frequency and intensity of rainstorms (and, therefore, floods). Further changes in climate could be initiated by alterations in the global oceanic circulation, which could be brought on by greenhouse warming. Oceanographers fear that if the ocean's conveyor goes into a stall, sudden shifts in land temperature may occur in the northern hemisphere.

ethics

## Stewardship of the Atmosphere

The atmosphere is a global commons—a resource used by all countries, yet not owned by any. Countries are free to draw from it to meet their needs and to add to it to get rid of pollutants. Because the air in the troposphere moves so quickly, what one country does to it can easily affect others downwind. The stratosphere is less turbulent, but inevitably it is affected by many gases that enter the lower atmosphere—as we see with the depletion of the ozone layer. Managing a global commons presents a difficult problem: Treaties must be made between countries and compliance must be assured if the global commons is to be protected. Some progress has been made in forging international agreements to curb acid emissions, which represent one form of transboundary air pollution. The Montreal Protocol and its amendments represent the most far-reaching and potentially effective international effort to protect a global resource: the stratospheric ozone layer.

Taking action on ozone-depleting chemicals is child's play, however, compared with what lies ahead in dealing with global climate change. Perhaps the most challenging aspect of the problem is that what we do about it now will do little to affect the present generation. It will likely take several generations for greenhouse gas concentrations to come to a doubling of preindustrial levels, with almost certain profound impacts. Thus, our dilemma is this: Shall we take action *now* to restrain our use of fossil fuels in order to benefit *future* generations? Increasing numbers of people are answering this question with a definite *yes*! For example, more than 2,000 economists agreed that there are many policies to reduce greenhouse gas emissions for which the total benefits outweighed the costs. Over 1,500 scientists, including 104 Nobel laureates, signed a statement urging government leaders attending the 1997 Kyoto conference to "act immediately to prevent the potentially devastating consequences of human-induced global warming." In fact, these people are assuming the role of stewards of the atmosphere and are concerned about what they are passing on to their children and grandchildren.

One scientist whose efforts have been crucial to the work of the Intergovernmental Panel on Climate Change (IPCC) is British meteorologist Sir John Houghton. Houghton has been chairman of the IPCC Working Group I and chief editor of four of the group's published reports, including the 2001 Working Group I Scientific Assessment. He also has published a succinct and very readable book on climate change: *Global Warming: The Complete Briefing*, currently in its second edition. Chapter 8 of Houghton's book, entitled "Why Should We Be Concerned," addresses the question of human responsibility for caring for the planet. Houghton points out that the current human exploitation of the natural world is unsustainable and that going "back to nature" or hoping for some "technical fix" to correct the damage of global warming is not a reasonable option. Houghton speaks of "a basic instinct that we wish to see our children and our grandchildren well set up in the world and wish to pass on to them some of our most treasured possessions," in particular "an earth that is well looked after and which does not pose to them more difficult problems than those we have had to face." He states, "We have no right to act as if there is no tomorrow."

Houghton makes the point that action on environmental problems depends not only on our understanding of how the world works, but also on how we value our world. He examines our responsibility to care for all living things and presents perspectives on this view from a number of the world's religions. He speaks of the concept of stewardship—one of the continuing themes of this text—that is derived from the Judaeo-Christian Old Testament. He recommends as a model for stewardship the biblical picture of the Garden of Eden, where the first humans were placed "to work and take care of it," as Genesis puts it. He states, "At the basis of this stewardship should be a principle extending what has traditionally been considered wrong—or in religious parlance as sin—to include unwarranted pollution of the environment or lack of care for it."

Stewardship of the atmosphere suggests that the world's nations act on the precautionary principle and begin to curb greenhouse gas emissions now and more deeply in the future. What do you think, and what will you do?

Sir John Houghton (left) and text author Richard Wright (right) discuss global climate change at a conference in Cambridge, UK.

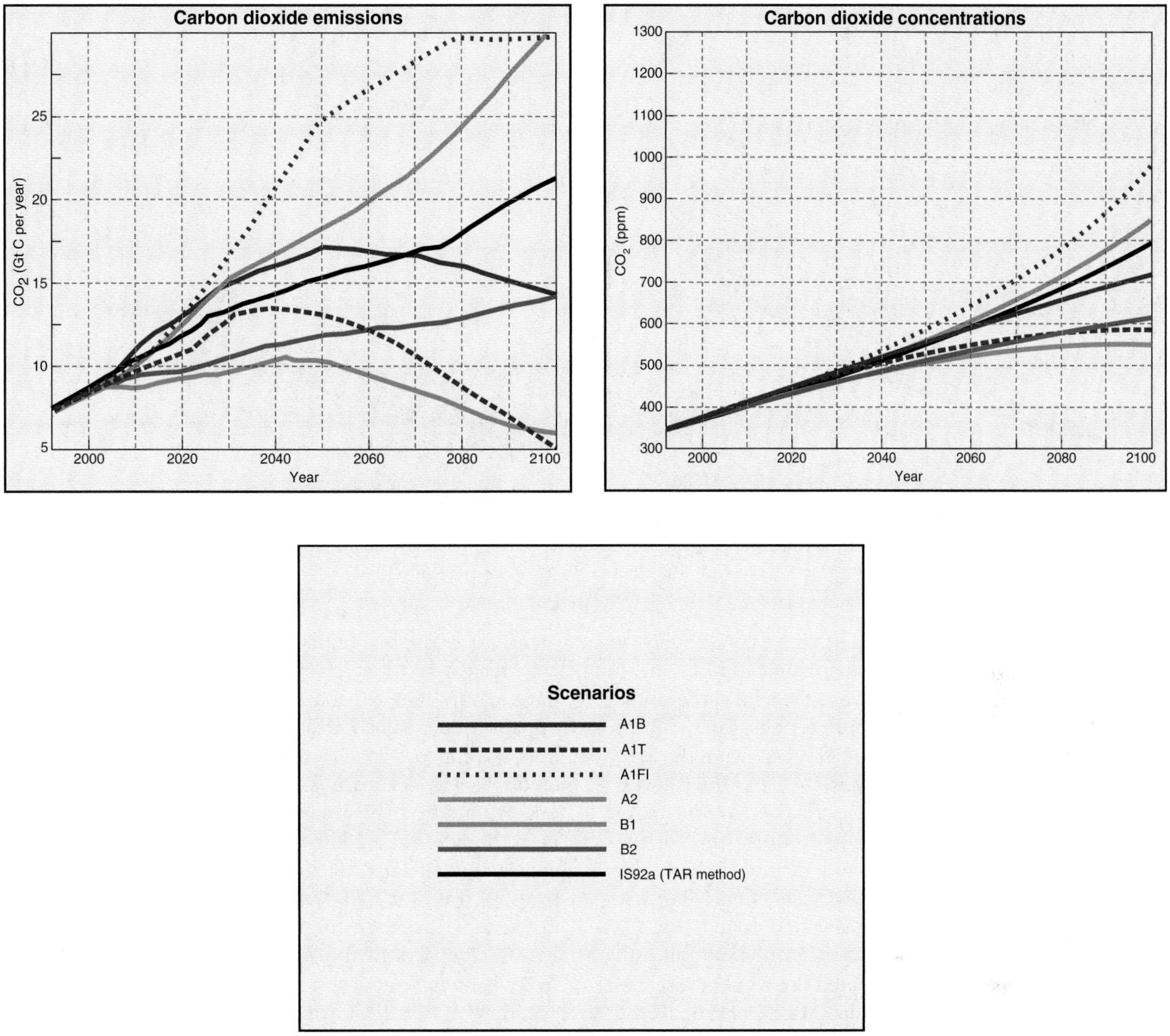


**Figure 20–16 Twenty-first century carbon dioxide changes according to model scenarios.** The projected carbon dioxide emissions and atmospheric concentrations are shown for seven computer model runs. Each is based on different assumptions. (See text.) (*Source:* IPCC Working Group I.)

For the agricultural community, the greatest difficulty in coping with climatic change is not knowing what to expect. Already, farmers lose an average of one in five crops because of unfavorable weather. As the climate shifts, the vagaries of weather will become more pronounced, and crop losses are likely to increase. Fortunately, farmers are capable of rapidly switching crops and land uses, so the impact may not be as bad as some observers think.

***Benefits?*** Humans and ecosystems will be subjected to unprecedented changes in climate under most of the scenarios presented by the IPCC. Many regions will experience hotter summer weather, but many northern regions will also enjoy warmer winters and longer growing seasons. The increased $CO_2$ in the atmosphere will likely stimulate plant growth under some conditions and in some regions, but in many regions the increased heat will offset this benefit. Because of the rapid rate of changes that are anticipated, the net harm very likely will greatly outweigh the benefits.

***Rising Sea Level.*** The warming at the poles due to rising temperatures could have an enormous effect when the ice melts, because so much water is stored in the world's remaining ice—enough to raise sea level by 75 m. The area of greatest concern is the Greenland ice sheet, where local warming is expected to be up to three times the global average. If this ice sheet melts completely, sea level could rise 7 meters! Another area of concern is the Antarctic, which holds most of the world's ice. West Antarctica's ice sheet is not greatly elevated above sea level; in fact, most of it rests on land that is below sea level. The Ross and Weddell Seas extend outward from this ice field, and the landward fringes of these seas are covered with thick floating shelves of ice. Currently, the rate of melting and of buildup seem to be about equal,

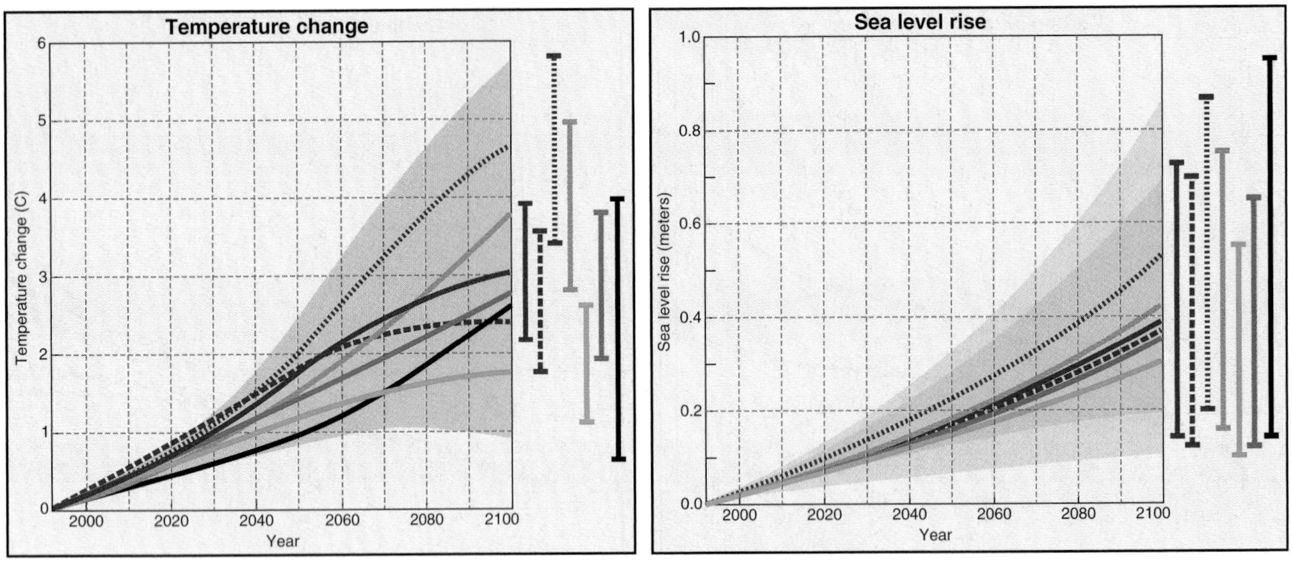


**Figure 20–17 Twenty-first century temperature change and sea-level rise.** The same model scenarios as in Figure 20–16 are used to project the range of temperature change and the rise in sea level based on the different assumptions. The vertical bars indicate the range of uncertainty for each model as it reaches 2100. (See text.) (*Source:* IPCC Working Group I.)

and many climate models indicate that ice could build up even more in the Antarctic with increasing GHGs.

There is great uncertainty about the magnitude of the rise in sea level. IPCC model projections for the next century range from 9 to 88 centimeters, with the best estimate at 50 cm (1/2 meter) (Fig. 20–17b). A half-meter rise in sea level will flood many coastal areas and make them much more prone to damage from storms, forcing people to abandon properties and migrate inland. For many of the small oceanic nations, such as the Maldives, off the coast of India, or the Marshall Islands, in the Pacific, a rise in sea level would mean obliteration, not just alteration. The highest estimated rise would cause disasters in most coastal cities, which are home to half the world's population and its business and commerce.

**7. Anthropogenic Climate Change Will Persist for Many Centuries.** The estimated rise in sea level extends only through the next century, but the impact will certainly be greater beyond the year 2100. Are inland cities and communities ready to accommodate the billions of people that will be displaced? Are we ready to build dikes or modify all ports to accommodate the higher sea level? Once atmospheric greenhouse gas levels are stabilized, temperatures and sea levels will continue to rise for hundreds of years because of the slow response time of the oceans.

**8. Further Action Is Required to Address Remaining Gaps in Information and Understanding.** There are many uncertainties in our understanding of how global climate works—especially in how it will respond to the challenges of 21st-century additions of GHGs. There also is a great need for more research and observations on current climate variables, such as the occurrence and impacts of clouds; the extent of, and changes in, glaciers and sea ice; the functioning of the carbon cycle (particularly the extent to which oceans and land sequester $CO_2$); natural variability in climate; and the direct and indirect impacts of aerosols. Further research is needed as well to improve climate models, so that they may be given greater credibility as they are used to project future climates under a range of assumptions. Especially crucial is the need to understand and better assess the regional changes expected in climate. All of these uncertainties point to the pressing need for more research, but uncertainty should not become an excuse for putting off action to prevent the harmful impacts of future climate changes. There is an overwhelming consensus among scientists regarding the accuracy of the IPCC third assessment. In May 2001, 17 academies of science from various parts of the world issued a statement in support of the IPCC third assessment. Asserting that "business as usual is no longer a viable option," they stated, "The balance of the scientific evidence demands effective steps now to avert damaging changes to Earth's climate."

## Other Assessments

***U.S. National Assessment.*** In the Global Change Research Act of 1990, Congress mandated an assessment of the findings of research on global climate change and its implications for the 21st century in the United States. The assessment, titled *Climate Change Impacts on the United States: The Potential Consequences of Climate Variability and Change,* was the work of a team of climate experts from major U.S. universities and agencies and was released in November 2000, after extensive peer reviews and numerous drafts.

The assessment team employed several CGCMs in projecting changes, but based most of its work on two: one from the Canadian Climate Centre and the other from the Hadley Centre in the United Kingdom. Simulations run on

**table 20-3 Key Findings of the 2000 U.S. Climate Change Assessment**

| | |
|---|---|
| **1. Increased warming** | Temperatures in the United States will rise 5–10°F (3–6°C) by 2100. Heat waves will become more intense and frequent in urban areas. |
| **2. Differing regional impacts** | Temperature and precipitation changes will vary from one region to the next. Instances of heavy and extreme precipitation will increase, and some regions will get drier. |
| **3. Vulnerable ecosystems** | Many ecosystems are highly vulnerable to the projected climate changes. Alpine meadows may disappear, coral reefs will be threatened by the higher sea level, and some barrier islands will disappear. Many of the valuable goods and services we expect from natural ecosystems will be affected and possibly lost. |
| **4. Widespread water concerns** | Droughts, floods, and water quality will be a concern in many regions. Changes in the snowpack will occur in the West and Alaska. |
| **5. Agriculture** | The agriculture sector is likely to survive the changes, and U.S. crop productivity may actually increase over the first half of the 21st century, although the gains will not be uniform. |
| **6. Forests** | Forest growth may increase in the first decades, but changes in species will occur, with some significant losses, such as the sugar maple forests of the Northeast. Fire, insects, drought, and diseases will likely start impairing forest growth in the long term. |
| **7. Coastlines** | Rising sea levels and increasing frequencies and intensities of storms will inundate many beaches along the east and south coasts, damaging the coastal infrastructure. |
| **8. Surprises and uncertainty** | The complexity of the atmosphere–land–ocean system will undoubtedly bring us some surprises, because of the uncertainty involved in the model's projections and in human responses to real climate changes. |

the models can make any number of assumptions about future releases of GHGs, but the team based its projections on the assumption that world greenhouse gas emissions would rise at a rate of 1% per year, because there would be no significant interventions to reduce emissions. This is a "business-as-usual" assumption. As a result, warming was projected to rise 3–6°C (5–10°F) in the United States by the end of the 21st century. Many other impacts also are projected in the assessment, which examines them for the various regions of the United States, as well as for various sectors, such as agriculture, forests, human health, and water resources. The key findings of the assessment are presented in Table 20–3. They are consistent with similar scenarios projected by the IPCC. It is clear that the consequences of global climate change will eventually be experienced by everyone, everywhere in the United States, if "business as usual" is our response.

***U.S. National Research Council.*** Following the publication of the IPCC third assessment, President George W. Bush asked the National Academy of Sciences to review the scientific assessment, give its own judgment of the report's accuracy and estimates of future changes, and answer a list of questions originating with the administration. The National Academy appointed a special committee of the National Research Council to conduct this review, and the committee released its report in June 2001, entitled *Climate Change Science: An Analysis of Some Key Questions*. The report clearly endorsed the IPCC Working Group I (WG I) assessment, stating, "The committee generally agrees with the assessment of human-caused climate change presented in the IPCC Working Group I scientific report.... Human-induced warming and associated sea level rises are expected to continue through the 21st century.... [T]he body of the WG I report is scientifically credible and is not unlike what would be produced by a comparable group of only U.S. scientists.... "

## 20.4 Response to Climate Change

The world's industries and transportation networks are so locked into the use of fossil fuels that massive emissions of $CO_2$ and other GHGs seem sure to continue for the foreseeable future, bringing on significant climate change. One possible response to this situation is **adaptation.** That is, we must anticipate some harm to natural and human systems and should plan adaptive responses to lessen the vulnerability of people, their property, and the biosphere to coming changes.

Still, we can lessen the rate at which emissions are added to the atmosphere and, eventually, bring about a sustainable balance—although no one thinks that doing so will be easy. This is the **mitigation** response: Take action to reduce emissions.

***Skeptics.*** Skeptics about global warming can be found in many sectors, from predictable sources such as the fossil-fuel industry, Rush Limbaugh, and many conservative think tanks, to members of the scientific community. They all stress the uncertainties of the global climate models and, in particular, emphasize that there is much that we do not know about the role of the oceans, the clouds, the biota, and the chemistry of the atmosphere. In the absence of "convincing evidence," many think that the threat of global warming may well be overplayed. Why take enormously costly steps, they ask, to prevent something that may have positive results or may even never happen?

Indeed, why respond at all? The FCCC has proposed **three principles** that speak to this question:

1. The first is the **precautionary principle,** as articulated in the 1992 Rio Declaration. The principle states, "In order to protect the environment, the precautionary approach shall be widely applied by States according to their capabilities. Where there are threats of serious or irreversible damage, lack of full scientific certainty shall not be used as a reason for postponing cost-effective measures to prevent environmental degradation." Employing this principle is like taking out insurance: We are taking measures to avoid a highly costly, but uncertain, situation (albeit one that is appearing more and more likely). The risk of serious harmful consequences is real; the scientific consensus is strong, and the political response, as we will see, is becoming robust.
2. The second is the **polluter pays principle.** Polluters should pay for the damage their pollution causes. This principle has long been part of environmental legislation. $CO_2$ is where it is because the developed countries have burned so much fossil fuel since the beginning of the industrial era.
3. The third is the **equity principle.** Currently, the richest 1 billion people produce 55% of $CO_2$ emissions, while the poorest 1 billion produce only 3%. International equity and intergenerational equity are ethical principles that should compel the rich and privileged to care about those generations which follow—especially those in the poor countries who are even now experiencing the consequences of global climate change. If we are able to take action that will prevent both present and future harm, it would be wrong not to do so.

## Response 1: Mitigation

A number of steps have been suggested to combat global climate change, with the goal of *stabilizing the greenhouse gas content of the atmosphere at levels and on a time scale that would prevent dangerous anthropogenic interference with the climate system and that are consistent with sustainable development.* (This is the stated objective of the Framework Convention on Climate Change—see below.) The various steps are presented in Table 20–4. All of these steps move societies in the direction of sustainability and have many benefits beyond their impacts on global climate change. For example, investing in more efficient use and production of energy reduces acid deposition, is economically sound, lowers the harmful health effects of air pollution, lowers our dependency on foreign oil and our vulnerability to terrorism—and reduces $CO_2$ emissions.

### What Has Been Done?

*Framework Convention on Climate Change.* One of the five documents signed by heads of state at the UNCED Earth Summit in Rio de Janeiro in 1992 was the **Framework Convention on Climate Change (FCCC).** This convention agreed to the goal of stabilizing greenhouse gas levels in the atmosphere, starting by reducing greenhouse gas emissions to 1990 levels by the year 2000 in all

**table 20-4 Options for Mitigation of Greenhouse Gas Emissions**

| |
|---|
| Place a worldwide cap on GHG emissions; for $CO_2$, this can be accomplished by limiting the use of fossil fuels in industry and transportation. Rights to emit GHGs would be allocated to different countries and would be tradeable in a market-based system. |
| Invest in and deploy an increasing percentage of energy in the form of renewable-energy technologies: wind power, solar collectors, solar thermal energy, photovoltaics, hydrogen-powered vehicles, and geothermal energy, among others. |
| Remove fossil-fuel subsidies like depletion allowances, tax relief to consumers, support for oil and gas exploration, and the substantial costs of maintaining access to Middle Eastern oil. |
| Encourage the development of nuclear power, but only if issues concerning cost-effectiveness, reliability, spent fuel, and high-level waste are resolved. |
| Stop the loss of tropical forests and encourage the planting of trees and other vegetation over vast areas now suffering from deforestation. |
| Make energy conservation rules much more stringent. (Tighten building codes to require more insulation, use energy-efficient lighting, and so forth.) |
| Reduce the amount of fuels used in transportation by raising mileage standards, encouraging car pooling, stimulating mass transit in urban areas, and imposing increasingly stiff carbon taxes on fuels. |
| Sequester $CO_2$ emitted from burning fossil fuels by capturing it at the site of emission, converting it into liquid $CO_2$, and pumping it into the deep oceans, where the low temperatures and high pressure would preserve it as a solid mass. |
| Make a greater effort to slow the growth of the human population; a continually growing population will inevitably produce increasing amounts of greenhouse gas emissions. |

industrialized nations. Countries were to achieve the goal by voluntary means. Five years later, it was obvious that the voluntary approach was failing. All the developed countries except those of the European Union *increased* their greenhouse gas emissions by 7 to 9% in the ensuing five years. The developing countries increased theirs by 25%!

***Kyoto Protocol.*** Prompted by a coalition of island nations (whose very existence is threatened by global climate change), the third Conference of Parties to the FCCC met in Kyoto, Japan, in December 1997 to craft a binding agreement on reducing greenhouse gas emissions. In Kyoto, 38 industrial and former Eastern bloc nations agreed to reduce emissions of six GHGs to 5.2% *below* 1990 levels, to be achieved by 2012. The signatories to this agreement are called the *Annex I parties;* all others are *non-Annex I parties*—the developing countries. The percentages of the reductions varied (Table 20–5). At the conference, the developing countries refused to agree to *any* reductions, arguing that the developed countries had created the problem and that it was only fair that the developing countries continue on their path to development as the developed countries did, energized by fossil fuels.

Each participating country must *ratify* the accord, signifying that country's compliance with its terms. The accord requires that at least 55 countries ratify it, and Annex I ratifying countries must account for at least 55% of Annex I GHG emissions. At Bonn, Germany, in July 2001, and later in Marrakesh, Morocco, in October 2001, the signers of the Kyoto Protocol met to thrash out the details of the "rule book" for carrying out the requirements of the protocol. Following these meetings, ratification proceeded. Over 110 parties have ratified (many are non-Annex I parties), thus meeting the first requirement for the protocol. However, two countries—the United States and Australia—have pulled out of the agreement, and several Annex I parties have not yet committed themselves. The most important of the latter parties is Russia. With 17% of the Annex I 1990 carbon emissions, Russia's signature could put the Kyoto Protocol over the top. (Similarly, if the United States signed, its 36% share would push the accord way over the top.)

The targeted reductions of the Kyoto Protocol will *not* stabilize atmospheric concentrations of GHGs. The IPCC calculates that it would take immediate reductions in emissions of at least 60% worldwide to stabilize greenhouse gas concentrations at today's levels. Since this has no chance of happening, the concentration of greenhouse gases in the atmosphere will continue to rise. Nevertheless, Kyoto is an important first step toward the stabilization of the atmosphere. To achieve stabilization at 450 ppm will require an energy system that is revolutionary, because current carbon emissions at 6 ½ GtC/year will have to fall to 2 GtC by 2100 and then drop even further.

The signers of the Kyoto Protocol have a great deal of latitude in deciding how they will achieve their GHG reductions. The most desirable approach would be to select from a portfolio of policy options. The list of mitigation options presented in Table 20–4 includes several that are market based. One that is already underway is **carbon credit trading.** The World Bank has established the Prototype Carbon Fund, which provides developing countries and those in transition economies (Eastern Europe, the former Soviet Union) with the opportunity to initiate clean technologies—especially renewable energy—and their reductions in emissions are credited to the Fund's contributors (developed countries) in the form of credits. In 2002, a reduction of some 67 million tons of carbon emissions was accomplished in this way. Sample projects are a wind farm in Costa Rica and the reforestation of worn-out land in Romania.

**table 20-5 Kyoto Emission Allowances and Eleven-Year Greenhouse Gas Emission Changes**

| Country | Kyoto Allowances (%) (1990–2008/12) | Observed Change (%) (1990–2001) | 2001 Greenhouse Gas Emissions ($10^6$ Metric Tons of Carbon) |
|---|---|---|---|
| United States | −7 | +15.7 | 1,510 |
| European Union | −8 | −0.2 | 829 |
| United Kingdom | −12.5 | −4.1 | 147 |
| Germany | −21 | −17.1 | 218 |
| Australia | +8 | +32.3 | 95 |
| Canada | −6 | +11.5 | 129 |
| Japan | −6 | +10.8 | 311 |
| Russian Federation | 0 | −30.5 | 403 |
| All Annex I countries | −5.2 | −1.7 | 3,649* |
| Non-Annex I countries | | | 2,621* |

*1999 figures
(*Sources:* WorldWatch Institute, 2002; Carbon Dioxide Information Analysis Center.)

*U.S. Policy.* Early in March 2001, U.S. representatives signed a statement with the Group of Eight industrial countries that reaffirmed our commitment to uphold the Kyoto Protocol and expressed concern about the serious threat of global climate change. A week later, in a letter responding to a query from four senators about his administration's views on global climate change, President George W. Bush stated that he opposed the Kyoto Protocol for two reasons:

1. It exempts the developing countries and thus is unfair.
2. It would cause serious harm to the U.S. economy.

At the same time, the president retreated from his most significant and explicit campaign promise to protect the environment: to reduce $CO_2$ emissions as part of an overall strategy to regulate four air pollutants emitted by power plants. He also cited the "incomplete state of scientific knowledge of the causes of, and solutions to, global climate change. . . . " Later in March, EPA administrator Christine Whitman officially notified the world that the United States would withdraw from the Kyoto agreement.

***Global Climate Change Initiative.*** The move away from the Kyoto Protocol does not mean that the United States is doing nothing about climate change. Instead, we are committed to going our own way. In February 2002, that way became clearer with the release of the administration's **Global Climate Change Initiative** (GCCI). The flagship policy of this initiative is an 18% cut in **emissions intensity** over the next 10 years. What is emissions intensity? It is *not* the same as emissions; it is the *ratio* of greenhouse gas emissions to economic output, the latter measured as gross domestic product (GDP). Emissions intensity in 2002 was 183 tons of $CO_2$ per million dollars of GDP, and the goal is to reduce it to 151 tons per million dollars of GDP.

Critics point out that the program is voluntary, and even if it works, it will simply continue a trend already in progress that has seen emissions intensity reduced by 16% from 1990 to 2000. While that happened, our total greenhouse gas emissions actually rose 14% (Fig. 20–18). Under the Bush plan, U.S. emissions in 2012 would be about 30% above the goal of returning emissions to their 1990 level, and since the Kyoto Protocol calls for a 7% reduction *below* 1990 levels, we are on a track of a 35% or more increase in greenhouse gas emissions, compared with our Kyoto-agreed goal!

Many of the programs of the GCCI, such as the Green Lights and Energy Star programs and the Climate Challenge and Natural Gas STAR programs, were already put in place during the Clinton administration. New initiatives include tax incentives for a number of climate-friendly energy technologies and increased funding for renewable-energy technologies and climate research. It is clear that the GCCI simply will not even come close to a reasonable reduction on the order of those being mandated by the Kyoto Protocol nations.

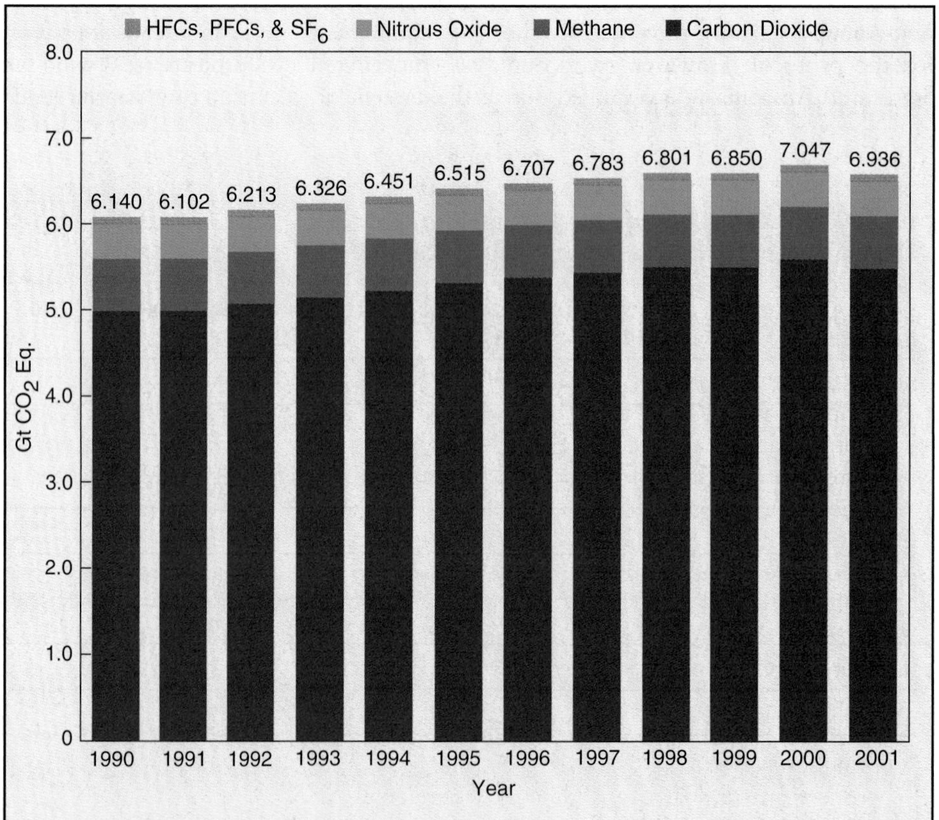


**Figure 20–18 U.S. greenhouse gas emissions, 1990–2001.** Four types of greenhouse gases are shown. Total emissions rose 14%, while greenhouse gas intensity fell 16%. Units in $GtCO_2$ equivalents. (*Source:* EPA, Inventory of U.S. Greenhouse Gas Emissions and Sinks, 1990–2001. April 15, 2003)

*U.S. Climate Change Science Program (CCSP).* In July 2003, the Bush administration released a report that provides guidelines for the government's Climate Change Science Program and puts a sharper image on the government's approach to climate research. The plan seeks to address a number of issues in climate science, such as the natural variability in climate, a quantitative approach to forces causing climate change, projections of future climate change, and the sensitivity and adaptability of human and natural ecosystems to climate change. Reports on key topics all have specific timetables, from two to four years. Critics of the plan see it as an excuse to stall action on global warming, holding that many of the questions which are addressed have already been well answered. Some scientists are cautiously optimistic that the CCSP is a serious effort to reduce the uncertainties in the scientific analysis of climate change so that improved policies will be forthcoming.

*States and Corporations.* It is encouraging that some U.S. states are taking unilateral action to address global climate change. Fourteen states are adopting renewable portfolio standards (see Chapter 14) and are mandating reporting of greenhouse gas emissions. In many states, bills have been introduced to build a framework for regulating $CO_2$, and 12 states have petitioned the U.S. Court of Appeals to force the EPA to regulate greenhouse gas emissions, using their authority under the Clean Air Act.

At a recent meeting of the World Economic Forum, the CEOs of the world's 1,000 largest corporations voted that climate change was the most urgent problem facing humanity. Some observers believe that this "bottom-up" approach to climate change action could provide the catalyst that will bring U.S. federal policy in line with that of the rest of the world.

## Response 2: Adaptation

Climate change is already happening and will likely accelerate in the future, regardless of the mitigating steps taken. *Mitigation* is the response that is essential to reducing the magnitude of future climate change, but *adaptation* will also be needed, especially in view of the potential for the following consequences, some of which are already happening:

1. Crop yields are likely to be reduced in tropical and subtropical regions as warming and droughts become more severe. Some agriculture in the north temperate regions may be enhanced.
2. Water is likely to become more scarce in many regions already suffering from water scarcity, especially the subtropics.
3. Increased heat and moisture in many regions will likely lead to an increase in infectious disease and potentially lethal heat waves.
4. Increased intensity and frequency of storm events will bring severe flooding to many regions that already are prone to natural disasters. The rise in sea level will intensify the risks in coastal areas.

*New Funds.* In light of these risks, the IPCC stated, "The effects of climate change are expected to be greatest in developing countries in terms of loss of life and relative effects on investment and the economy." As a start, the FCCC 2001 Marrakesh meeting established two new funds to address this concern: a **Least Developed Countries Fund,** to advise countries on national adaptation strategies, and a **Special Climate Change Fund,** to provide additional financial assistance to developing countries affected by climate change. The emphasis of adaptation is on the developing countries, since it is assumed that the developed countries will be far more able to afford the costs of adaptation.

*New Report.* In 2003, the World Bank and a host of U.N. and other organizations released a report that addresses the preceding concerns. Called *Poverty and Climate Change: Reducing the Vulnerability of the Poor through Adaptation,* the report makes some telling points:

1. Climate change impacts will be superimposed on existing vulnerabilities—water availability, food security, health and disease, life in low-lying regions. The poor already suffer disproportionately from these risks.
2. The countries with the fewest resources will likely bear the greatest burden of climate change, with greater loss of life and economic impacts.
3. Ecosystem goods and services are likely to be disrupted by climate change. The poor depend especially on these for survival, so they will be most affected.
4. Even today, over 96% of disaster-related deaths take place in developing countries. Major weather events can set back development for decades.
5. Because poverty is so severe in these countries, adaptation must be treated in the context of reducing poverty and achieving the Millennium Development Goals. (See Table 6–2.) Practically every one of the goals is threatened by the anticipated impacts of climate change, and the report addresses these impacts in detail.

*Strategies.* Specific adaptation strategies will vary with the different circumstances, but, in general, the World Bank report suggests that "the best way to address climate change impacts on the poor is by integrating adaptation measures into sustainable development and poverty reduction strategies." Adaptations addressed in the report include improved governance, vulnerability assessments, access to accurate information on climate change, and the integration of impacts into economic processes. Those adaptations which have "no regrets" benefits—measures that would foster desirable social benefits, regardless of the intensity or existence of climate change impacts—are especially encouraged. *General poverty reduction* is one such measure. It is no surprise that the developing countries—especially the poorest ones—will need substantial assistance from the developed countries. The two new FCCC funds are a good start, but external support also needs to increase from the many development assistance channels, including NGO assistance and intergovernmental aid.

*Conclusion.* Global climate change is perhaps the greatest challenge facing human civilization in the 21st century. Both adaptation and mitigation are needed. The two are complementary responses to the many threats posed by increasing GHG emissions. Whatever the outcome, it is certain that we are conducting an enormous global experiment, and our children and their descendants will be living with the consequences.

Let us now turn our attention to ozone depletion, an atmospheric problem the public often confuses with global warming.

## 20.5 Depletion of the Ozone Layer

The stratospheric ozone layer protects Earth from harmful ultraviolet radiation. The depletion of this layer is another major atmospheric challenge that is due to human technology. Environmental scientists have traced the problem to a widely used group of chemicals: the CFCs. As with global warming, however, there were skeptics who were unconvinced that the problem was serious. However, scientists have achieved a strong consensus on the ozone depletion problem, reflected in the most recent report of the Scientific Assessment Panel of the Montreal Protocol: **Scientific Assessment of Ozone Depletion: 2002.** This section draws from that report and gives you an opportunity to examine the evidence.

### Radiation and Importance of the Shield

Solar radiation emits electromagnetic waves with a wide range of energies and wavelengths (Fig. 20–19). Visible light is that part of the electromagnetic spectrum which can be detected by the eye's photoreceptors. Ultraviolet (UV) wavelengths are slightly shorter than the wavelengths of violet light, which are the shortest wavelengths visible to the human eye. UVB radiation consists of wavelengths that range from 280 to 320 nanometers (0.28 to 0.32 $\mu$m), whereas UVA radiation is from 320 to 400 nanometers. Since energy is inversely related to wavelength, UVB is more energetic and therefore more dangerous, but UVA can also cause damage.

On penetrating the atmosphere and being absorbed by biological tissues, UV radiation damages protein and DNA molecules at the surfaces of all living things. (This damage is what occurs when you get a sunburn.) If the full amount of ultraviolet radiation falling on the stratosphere reached Earth's surface, it is doubtful that any life could survive. We are spared more damaging effects from UV rays because most UV radiation (over 99%) is absorbed by ozone in the stratosphere. For that reason, stratospheric ozone is commonly referred to as the **ozone shield** (Fig. 20–2).

However, even the small amount (less than 1%) of UVB radiation that does reach us is responsible for sunburns and more than 700,000 cases of skin cancer and precancerous ailments per year in North America, as well as for untold damage to plant crops and other life-forms. (See "Global Perspective" essay, p. 565.)

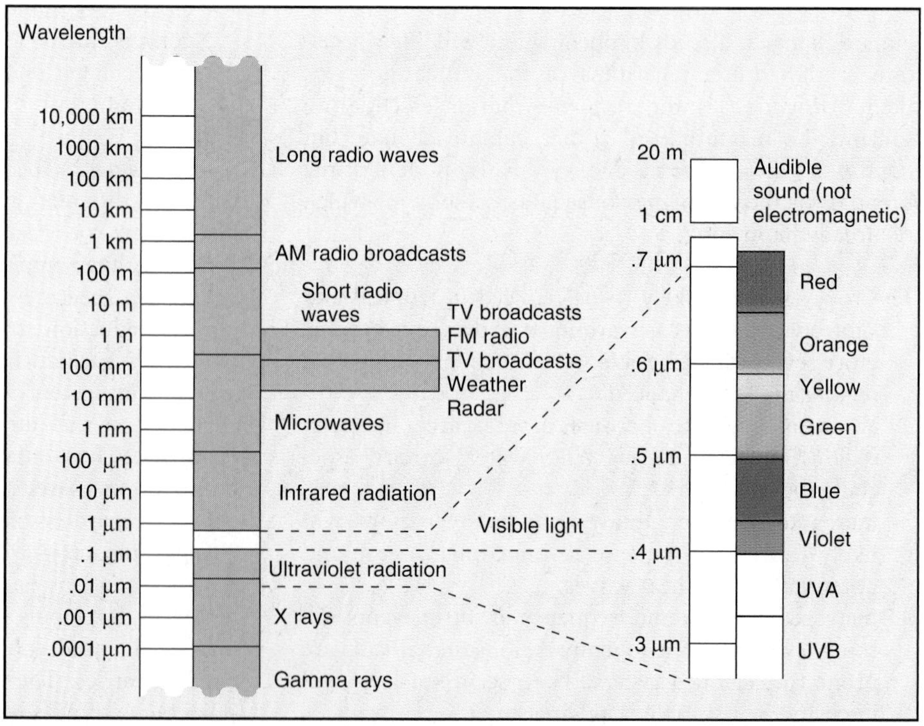


**Figure 20–19 The electromagnetic spectrum.** Ultraviolet, visible light, infrared, and many other forms of radiation are different wavelengths of the electromagnetic spectrum.

## global perspective

### Coping with UV Radiation

People living in Chile, Australia, and New Zealand are no strangers to the effects of the thinning ozone shield. UV alerts are given in those parts of the Southern Hemisphere during their spring months as lobes of ozone-depleted stratospheric air move outward from the Antarctic. At times, the ozone above those countries can be less than half its normal concentration, which means that greater intensities of UV radiation reach Earth's surface. It is now predicted that one out of every three Australians will develop serious and perhaps fatal skin cancer in his or her lifetime.

A thinning ozone layer has appeared above the Northern Hemisphere, too, and is getting serious attention. Concerned over the rising incidence of skin cancer and cataract surgery in the United States, the EPA, together with NOAA and the CDC, has initiated a new *UV index*. (See table below.) The index is in the form of daily forecasts of UV exposure, issued by the Weather Service for 58 cities. Satellite measurements of stratospheric ozone are combined with other weather patterns. The goal of the index is to remind people of the dangers of UV radiation and to prompt them to take appropriate action to avoid cancer, premature aging of the skin, eye damage, cataracts, and blindness. For at least the next two decades, the dangers will be quite noticeable, ranging from a greater incidence of sunburn to a likely epidemic of malignant melanoma.

Less obvious, but no less important, are the chronic effects of exposure to the Sun. The normal aging of skin is now known to be largely the result of damage from the Sun: coarse wrinkling, yellowing, and the development of irregular patches of heavily pigmented and unpigmented skin. This can happen even in the absence of episodes of sunburn. Further, a large proportion of the million cataract operations performed annually in the United States can be traced to UV exposure, another of the chronic effects of this kind of radiation.

The most serious impact of chronic exposure to the Sun is skin cancer, which occurs in some 700,000 new cases each year in the United States. Three types of skin cancer are traced to UV exposure: basal cell carcinoma (BCC), squamous cell carcinoma (SCC), and melanoma. Most of the skin cancers (75% to 90%) are BCCs, slow growing and therefore treatable. SCC accounts for about 20% of skin cancers and can also be cured if treated early. However, it metastasizes (spreads away from the source) more readily than BCC and thus is potentially fatal. Melanoma is the most deadly form of skin cancer, as it metastasizes easily. Melanoma is often traced to occasional sunburns during childhood or adolescence or to sunburns in people who normally stay out of the Sun.

What is an appropriate response to this disturbing information? First, know when UV intensity is greatest (three hours on either side of midday), and take precautions accordingly. Protect your eyes with sunglasses and a hat, and apply sunscreen with a protection factor (SPF) rating of 15 or higher during such times. Think twice before considering sunbathing or tanning beds, and never proceed without sunscreen that is strong enough to prevent sunburn. Always protect children with sunscreen. Their years of potential exposure are many, and their skin is easily burned. If you detect a patch of skin or a mole that is changing in size or color or that is red and fails to heal, consult a dermatologist for treatment.

**UV Index (EPA)**

| Exposure Category | Index Value | Minutes to Burn for "Never Tans" (Most Susceptible) | Minutes to Burn for "Rarely Burns" (Least Susceptible) |
|---|---|---|---|
| Minimal | 0–2 | 30 | >120 |
| Low | 4 | 15 | 75 |
| Moderate | 6 | 10 | 50 |
| High | 8 | 7.5 | 35 |
| Very high | 10 | 6 | 30 |
| Very high | 15 | <4 | 20 |

## Formation and Breakdown of the Shield

Ozone is formed in the stratosphere when UV radiation acts on oxygen ($O_2$) molecules. The high-energy UV radiation first causes some molecular oxygen ($O_2$) to split apart into free oxygen (O) atoms, and these atoms then combine with molecular oxygen to form ozone via the following reactions:

$$O_2 + \text{UVB} \rightarrow O + O \qquad (1)$$

$$O + O_2 \rightarrow O_3 \qquad (2)$$

Not all of the molecular oxygen is converted to ozone, however, because free oxygen atoms may also combine with ozone molecules to form two oxygen molecules in the following reaction:

$$O + O_3 \rightarrow O_2 + O_2 \quad (3)$$

Finally, when ozone absorbs UVB, it is converted back to free oxygen and molecular oxygen:

$$O_3 + UVB \rightarrow O + O_2 \quad (4)$$

Thus, the amount of ozone in the stratosphere is dynamic. There is an equilibrium due to the continual cycle of reactions of formation [Eqs. (1) and (2)] and reactions of destruction [Eqs. (3) and (4)]. Because of seasonal changes in solar radiation, ozone concentration in the Northern Hemisphere is highest in summer and lowest in winter. Also, in general, ozone concentrations are highest at the equator and diminish as latitude increases—again, a function of higher overall amounts of solar radiation. However, the presence of other chemicals in the stratosphere can upset the normal ozone equilibrium and promote undesirable reactions there.

**Halogens in the Atmosphere.** **Chlorofluorocarbons (CFCs)** are a type of halogenated hydrocarbon. (See Chapter 19.) CFCs are nonreactive, nonflammable, nontoxic organic molecules in which both chlorine and fluorine atoms have replaced some hydrogen atoms. At room temperature, CFCs are gases under normal (atmospheric) pressure, but they liquefy under modest pressure, giving off heat in the process and becoming cold. When they revaporize, they reabsorb the heat and become hot. These attributes led to the widespread use of CFCs (over a million tons per year in the 1980s) for the following applications:

- In refrigerators, air conditioners, and heat pumps as the heat-transfer fluid.
- In the production of plastic foams.
- By the electronics industry for cleaning computer parts, which must be meticulously purified.
- As the pressurizing agent in aerosol cans.

*Rowland and Molina.* All of the preceding uses led to the release of CFCs into the atmosphere, where they mixed with the normal atmospheric gases and eventually reached the stratosphere. In 1974, chemists Sherwood Rowland and Mario Molina published a classic paper[2] (for which they were awarded the Nobel prize in 1995) concluding that CFCs could damage the stratospheric ozone layer through the release of chlorine atoms and, as a result, would increase UV radiation and cause more skin cancer. Rowland and Molina reasoned that, although CFCs would be stable in the troposphere (they have been found to last 70 to 110 years there), in the stratosphere they would be subjected to intense UV radiation, which would break them apart, releasing free chlorine atoms via the following reaction:

$$CFCl_3 + UV \rightarrow Cl + CFCl_2 \quad (5)$$

Ultimately, all of the chlorine of a CFC molecule would be released as a result of further photochemical breakdown. The free chlorine atoms would then attack stratospheric ozone to form chlorine monoxide (ClO) and molecular oxygen:

$$Cl + O_3 \rightarrow ClO + O_2 \quad (6)$$

Furthermore, two molecules of chlorine monoxide may react to release more chlorine and an oxygen molecule:

$$ClO + ClO \rightarrow 2\ Cl + O_2 \quad (7)$$

Reactions 6 and 7 are called the **chlorine catalytic cycle,** because chlorine is continuously regenerated as it reacts with ozone. Thus, chlorine acts as a **catalyst,** a chemical that promotes a chemical reaction without itself being used up in the reaction. Because every chlorine atom in the stratosphere can last from 40 to 100 years, it has the potential to break down 100,000 molecules of ozone. Thus, CFCs are judged to be damaging because they act as transport agents that continuously move chlorine atoms into the stratosphere. The damage can persist, because the chlorine atoms are removed from the stratosphere only very slowly. Figure 20–20 shows the basic processes of ozone formation and destruction, including recent refinements to our knowledge of those processes that will be explained shortly.

*EPA Action.* After studying the evidence, the EPA became convinced that CFCs were a threat and, in 1978, banned their use in aerosol cans in the United States. Manufacturers quickly switched to non-damaging substitutes, such as butane, and things were quiet for several years. CFCs continued to be used in applications other than aerosols, however, and skeptics demanded more convincing evidence of their harmfulness.

Atmospheric scientists reason that any substance carrying reactive halogens to the stratosphere has the potential to deplete ozone. These substances include halons, methyl chloroform, carbon tetrafluoride, and methyl bromide. Chemically similar to chlorine, bromine also attacks ozone and forms a monoxide (BrO) in a catalytic cycle. Because of its extensive use as a soil fumigant and pesticide, methyl bromide is thought to cause between 30 and 60% of current stratospheric ozone loss. (Bromine is 40 times as potent as chlorine in ozone destruction.)

**The Ozone "Hole."** In the fall of 1985, British atmospheric scientists working in Antarctica reported a gaping "hole" (actually, a thinning of one area) in the stratospheric ozone layer over the South Pole (Fig. 20–21). There, in an area the size of the United States,

[2]Molina, M. J. and Rowland, F. S. "Stratospheric Sink for Chlorofluoro-methanes: Chlorine-atom Catalyzed Distribution of Ozone." *Nature* 249 (1974): 810–812.

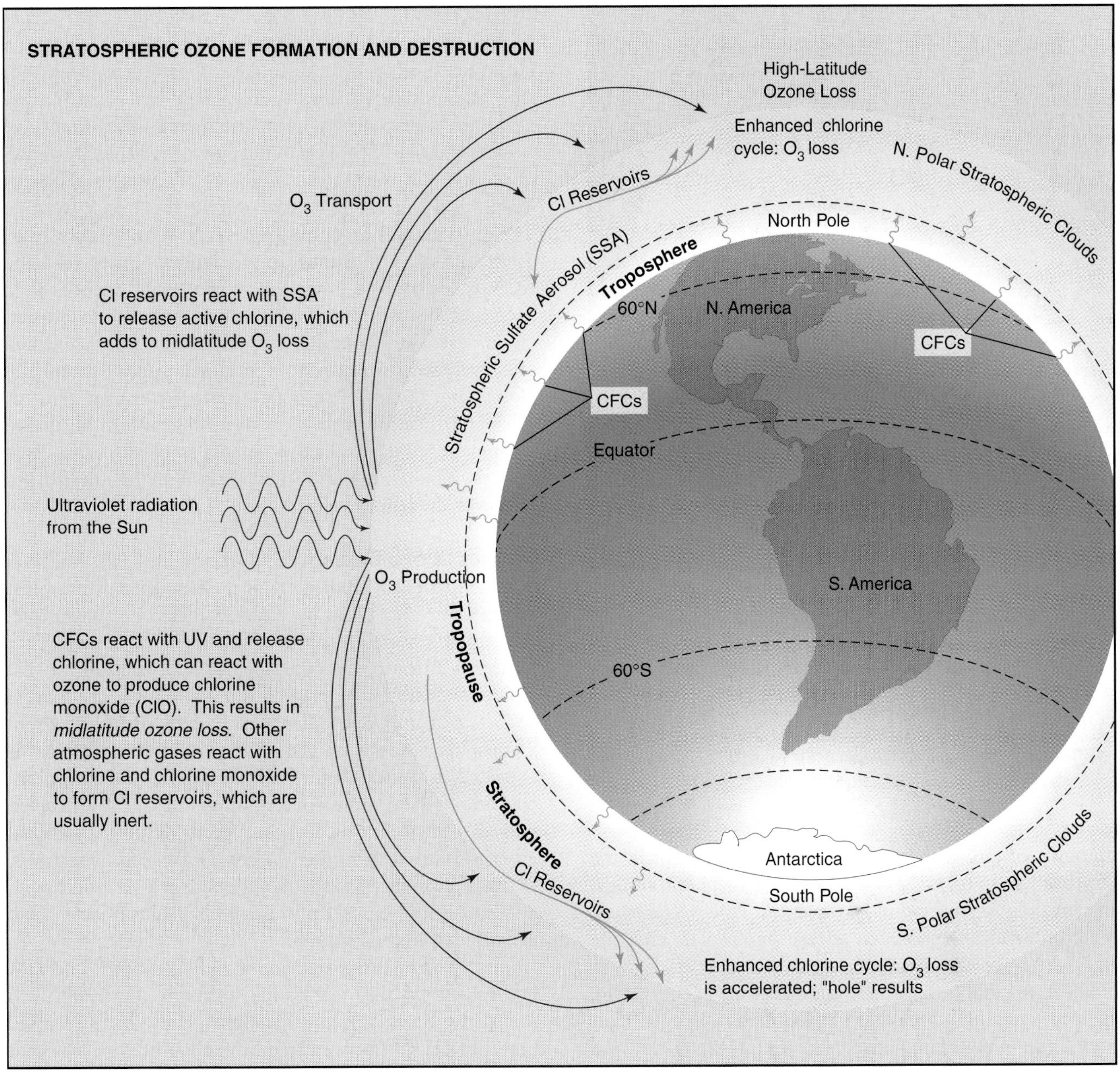


**Figure 20–20 Stratospheric ozone formation and destruction.** UV radiation stimulates ozone production at the lower latitudes, and ozone-rich air migrates to high latitudes. At the same time, CFCs and other compounds carry halogens into the stratosphere, where they are broken down by UV radiation, to release chlorine and bromine. Ozone is subject to high-latitude loss during winter, as the chlorine cycle is enhanced by the polar stratospheric clouds. Midlatitude losses occur as chlorine reservoirs are stimulated to release chlorine by reacting with stratospheric sulfate aerosol.

ozone levels were 50% lower than normal. The hole would have been discovered earlier by NASA satellites monitoring ozone levels, except that computers were programmed to reject data showing a drop as large as 30% as due to instrument anomalies. Scientists had assumed that the loss of ozone, if it occurred, would be slow, gradual, and uniform over the whole planet. The ozone hole came as a surprise, and if it had occurred anywhere but over the South Pole, the UV damage would have been extensive. As it is, the limited time and area of ozone depletion there have not apparently brought on any catastrophic ecological events so far.

News of the ozone hole stimulated an enormous scientific research effort. A unique set of conditions was found to be responsible for the hole. In the summer, gases such as nitrogen dioxide and methane react with chlorine monoxide and chlorine to trap the chlorine, forming so-called **chlorine reservoirs** (Fig. 20–20), preventing much ozone depletion.

***Polar Vortex.*** When the Antarctic winter arrives in June, it creates a vortex (like a whirlpool) in the stratosphere, which confines stratospheric gases within a ring of air circulating around the Antarctic. The extremely cold temperatures of the Antarctic winter cause the small

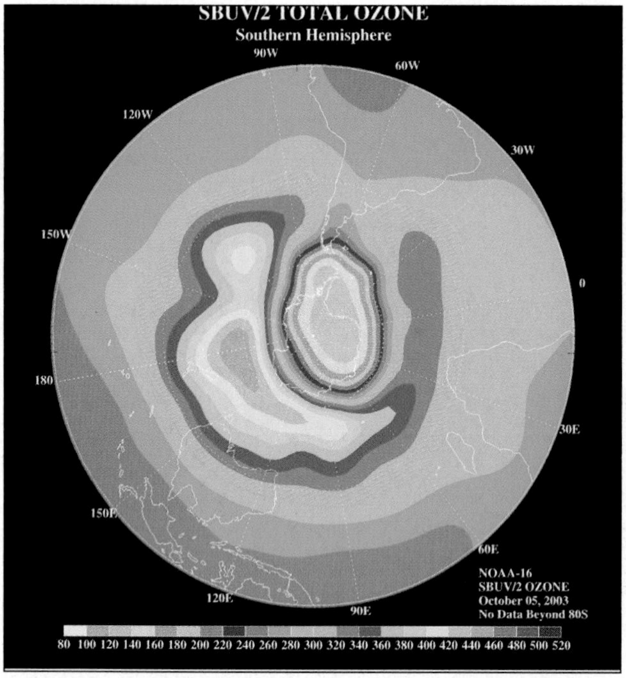


**Figure 20–21 The Antarctic ozone hole.** In 2003, Total Ozone Mapping Spectrometer (TOMS) instruments aboard a satellite recorded the second-largest-ever ozone hole over Antarctica. This image was taken on October 5, 2003, while the hole was still very large. The color scale represents ozone concentrations in Dobson units (DU). Areas enclosed by the 220-DU contour (all of the purple regions) are considered to be within the hole. (*Source:* Earth Observatory, NASA.)

amounts of moisture and other chemicals present in the stratosphere to form the south polar stratospheric clouds. During winter, the cloud particles provide surfaces on which chemical reactions release molecular chlorine ($Cl_2$) from the chlorine reservoirs.

When sunlight returns to the Antarctic in the spring, the Sun's warmth breaks up the clouds. UV light then attacks the molecular chlorine, releasing free chlorine and initiating the chlorine cycle, which rapidly destroys ozone. By November, the beginning of the Antarctic summer, the vortex breaks down and ozone-rich air returns to the area. However, by that time, ozone-poor air has spread all over the Southern Hemisphere. Shifting patches of ozone-depleted air have caused UV radiation increases of 20% above normal in Australia. Television stations there now report daily UV readings and warnings for Australians to stay out of the Sun. On the basis of current data, estimates indicate that in Queensland, where the ozone shield is thinnest, three out of four Australians are expected to develop skin cancer. The ozone hole intensified during the 1990s and has shown signs of leveling off between 9 and 10 million square miles, an area as large as North America. The 2003 hole reached 10.9 million square miles, however, the second largest on record. The 2002 hole was much smaller than usual (6 million square miles), a consequence of warmer-than-normal temperature patterns over Antarctica.

**Arctic Hole?** Because severe ozone depletion occurs under polar conditions, observers have been keeping a close watch on the Arctic. To date, no hole has developed. The higher temperatures and weaker vortex formation there are expected to prevent the severe losses that have become routine in the Antarctic. However, ozone depletion does occur in the Arctic, with ozone levels as much as 20–25% lower than normal during the Arctic winters. The loss intensifies during especially cold winters (e.g., 1999–2000), a phenomenon attributed largely to some recently discovered large particles (called stratospheric "rocks" because they are so much larger than normal cloud particles) containing nitric acid hydrate, which effectively strip nitrogen from the north polar stratospheric clouds and then sink into the troposphere. The effect of this denitrification is to allow more chlorine and bromine to remain in their reactive forms, since nitrogen compounds would ordinarily moderate the ozone-destroying effects of the halogens.

**Further Ozone Depletion.** Ozone losses have not been confined to Earth's polar regions, although they are most spectacular there. A worldwide network of ozone-measuring stations sends data to the World Ozone Data Center in Toronto, Canada. Reports from the center reveal ozone depletion levels of 3 and 6% over the period 1997–2001 in midlatitudes of the Northern and Southern Hemispheres, respectively. Ozone loss everywhere is expected to peak before 2010, when, hopefully, the chlorine and bromine concentrations in the *stratosphere* will start to decline as a consequence of the international agreements that have been forged. In fact, concentrations of these substances in the *troposphere* are now declining, and ozone loss in the *upper stratosphere* has diminished, according to recent reports.

Is the ozone loss significant to our future? The EPA has calculated that the ozone losses of the 1980s will eventually have caused 12 million people in the United States to develop skin cancers over their lifetime and that 93,000 of these cancers will be fatal. Americans are estimated to have developed more than 900,000 new cases of skin cancer a year in the 1990s. The ozone losses of that decade allowed more UVB radiation than ever to reach Earth. Studies have confirmed increased UVB levels, especially at higher latitudes.

## Coming to Grips with Ozone Depletion

The dramatic growth of the hole in the ozone layer (Fig. 20–22) has galvanized a response around the world. In spite of the skepticism in the United States, scientists and politicians here and in other countries have achieved treaties designed to avert a UV disaster.

***Montreal Protocol.*** In 1987, under the auspices of its environmental program, the United Nations convened a meeting in Montreal, Canada, to address ozone depletion. Member nations reached an agreement, known as the

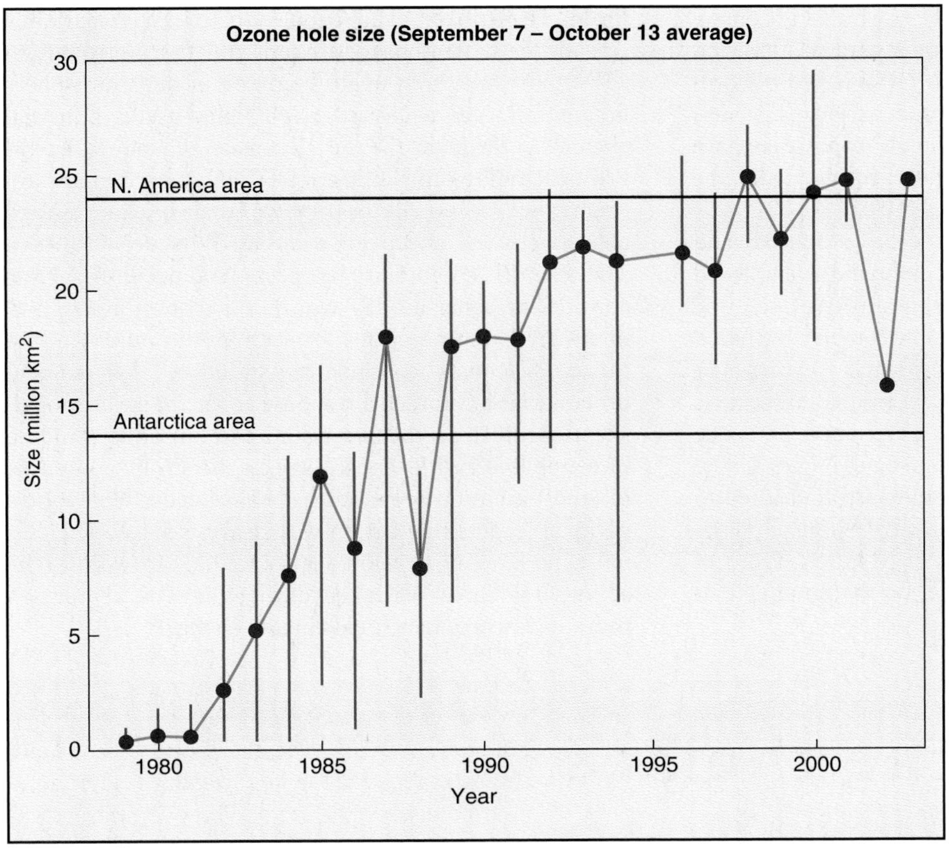


**Figure 20–22 Ozone hole size.** The size of the ozone hole is plotted from 1979 to 2003, showing the time of origin of the hole to its leveling off in the last few years. Note that the hole was as great as all of N. America for four different years. (*Source:* NASA.)

**Montreal Protocol,** to scale CFC production back 50% by 2000. To date, 184 countries (including the United States) have signed the original agreement.

The Montreal protocol was written even before CFCs were so clearly implicated in driving the destruction of ozone and before the threat to Arctic and temperate-zone ozone was recognized. Because ozone losses during the late 1980s were greater than expected, an amendment to the protocol was adopted in June 1990. The amendment requires participating nations to phase out the major chemicals destroying the ozone layer by 2000 in developed countries and by 2010 in developing countries. In the face of evidence that ozone depletion was accelerating even more, another amendment to the protocol was adopted in November 1992, moving the *target date for the complete phaseout of CFCs to January 1, 1996*. Timetables for phasing out all of the suspected ozone-depleting halogens were shortened at the 1992 meeting.

Quantities of CFCs are still being manufactured to satisfy legitimate demand in the developing countries. These legal CFCs, however, are being sidetracked into black-market trade because of hefty taxes on the legal sales in developing countries. Some of the black-market CFCs are smuggled into developed countries. By 2003, the U.S. Justice Department had charged 114 individuals and numerous businesses with smuggling CFCs into the United States, leading to hefty fines, significant jail time, and the recovery of tons of the banned chemicals.

*Action in the United States.* The United States was the leader in the production and use of CFCs and other ozone-depleting chemicals, with du Pont Chemical Company being the major producer. Following 15 years of resistance, du Pont pledged in 1988 to phase out CFC production by 2000. In late 1991, a spokesperson announced that, in response to new data on ozone loss, the company would accelerate its phaseout by three to five years. Many of the large corporate users of CFCs (AT&T, IBM, and Northern Telecom, for example) phased out their CFC use by 1994. Also, du Pont spoke in opposition to three bills introduced into the 104th Congress in September 1995. The bills were designed to terminate U.S. participation in, and compliance with, the CFC-banning protocols.

The Clean Air Act of 1990 also addresses this problem, in Title VI, "Protecting Stratospheric Ozone." Title VI is a comprehensive program that restricts the production, use, emissions, and disposal of an entire family of chemicals identified as ozone depleting. For example, the program calls for a phaseout schedule for the hydrochlorofluorocarbons (HCFCs), a family of chemicals being used as less damaging substitutes for CFCs until nonchlorine substitutes are available. Halons—used in chemical fire extinguishers—were banned in 1994. The act also regulates the servicing of refrigeration and air-conditioning units.

January 1, 1996, has come and gone, and in most of the industrialized countries CFCs are no longer being produced or used. Substitutes for CFCs are readily available and, in some cases, are even less expensive than the CFCs. The most commonly used substitutes are HCFCs, which still contain some chlorine and are scheduled for a gradual phaseout. The most *promising* substitutes are HFCs—hydrofluorocarbons, which contain no chlorine and are judged to have no ozone-depleting potential. CFCs in the troposphere have peaked and are slowly declining, and scientists predict the ozone shield will recover entirely by 2050.

Methyl bromide continues to be produced and released into the atmosphere; the compound is employed as a soil fumigant to control agricultural pests. Under the Montreal Protocol, it is scheduled to be completely phased out by 2005 in the developed countries and 10 years later in the developing countries. In the meantime, bromine is likely to contribute an increasing percentage of ozone depletion relative to chlorine. Responding to requests from the agricultural industry, the Bush administration tried to get an exemption to the scheduled methyl bromide phaseout at a November 2003 meeting of Montreal Protocol nations. The exemption was denied, on the grounds that suitable substitutes were available.

**Final Thoughts.** The ozone story is a remarkable episode in human history. From the first warnings in 1974 that something might be amiss in the stratosphere because of a practically inert and highly useful industrial chemical, through the development of the Montreal Protocol and the final steps of CFC phaseout that are still occurring, the world has shown that it can respond collectively and effectively to a clearly perceived threat. The scientific community has played a crucial role in this episode, first alerting the world and then plunging into intense research programs to ascertain the validity of the threat. Scientists continue to influence the political process that has forged a response to the threat. Although skeptics still stress the uncertainties in our understanding of ozone loss and its consequences, the strong consensus in the scientific community convinced the world's political leaders that action was clearly needed. This is a most encouraging development as we look forward to the actions that must be taken during the 21st century to prevent catastrophic global climate change.

# revisiting the themes

## Sustainability

Earth is in the midst of an unsustainable rise in atmospheric greenhouse gas levels, the result of our intense use of fossil fuels. In short, we completely depend on a technology that is threatening our future. All projections of future fossil-fuel use and greenhouse gases point to global consequences that are serious, but not inevitable. The United States and other developed countries will not escape these consequences, but the gravest of them affects the developing countries. A sustainable pathway is still open to us, and it involves a combination of steps we can take to mitigate the emissions and bring the atmospheric concentration of greenhouse gases to a stable, even declining, level. The ozone story represents such a sustainable pathway. Levels of ozone-destroying chemicals are now stabilizing and will soon decline, and it is very likely that the ozone layer will be "healthy" by midcentury.

## Stewardship

Stewardly care for Earth is not really an option. We *must* act, and several principles have been cited as arguments for effective action to prevent dangerous climate change: the precautionary principle, the polluter pays principle, and the equity principle. If we care about our neighbors and our descendants, we will take action both to mitigate the production of greenhouse gases and to enable especially the poorer countries to adapt to coming unavoidable climate changes, even if taking such action is costly. The "Ethics" essay points to the work of a caring steward: Sir John Houghton. The ozone story, again, is a model for effective, stewardly action in the face of a major threat to people yet unborn.

## Sound Science

Much of this chapter is about sound science, since our knowledge of climate change and ozone depletion depends on the solid work of thousands of scientists. It is largely the scientists who are calling attention to the perils of climate change. They were the first to discover the risks involved, and they are now the chief advocates of effective action. Nevertheless, more good science needs to be done, because there are still many unanswered questions and uncertainties. We know enough now to act, however, and the model of ozone depletion stands as proof that we can act even before every loose end is tied up.

## Ecosystem Capital

Ecosystems depend on climate, and if we change the climate too rapidly, many ecosystems are likely to suffer serious disruptions and will no longer provide the essential goods and services societies now depend on. We allow these ecosystems to change at our own peril. In the end, ecosystems will redevelop and many species will adapt or move to new locations, but there may well be major losses in biodiversity, and the ecosystem adaptations will not happen overnight. The short-term impacts will be especially hard on those societies which live close to the land (that is, the developing countries).

### Policy and Politics

The story of the global response to climate change and ozone depletion is a study in international politics. Enforceable global policies are now in place for ozone restoration and are well underway for a first pass at global climate change. The United States was a leader in the first case and is a significant holdout in the second. As the world's leader in greenhouse gas emissions, we have established ourselves also as the world's most selfish nation as we refuse to join the other nations in taking effective action to prevent climate change, arguing that it would not be good for our economy. The Bush administration may have a lot to answer for as the future unfolds.

### Globalization

The issues presented in this chapter are surely global ones, with the entire atmosphere being affected and global agreements being forged. The benefits afforded by U.N. agencies are evident to nations everywhere. Never has global leadership been more necessary.

## review questions

1. Explain the significance of the El Niño–La Niña patterns for Earth's climate and weather.
2. What are the important characteristics of the troposphere? Of the stratosphere?
3. How does solar radiation create our weather patterns?
4. What has been discovered about global climate trends in the past? What is the contemporary trend showing?
5. What is the conveyor system? How does it work? How does it connect with global warming?
6. What is radiative forcing? Describe some positive and negative forcing agents.
7. Which of the greenhouse gases are the most significant contributors to global warming? How do they work?
8. What evidence does the IPCC cite in support of its conclusion that the world is warming and the climate is changing?
9. How are models employed in climate change research? Describe several scenarios for 21st-century climate change.
10. What are the major IPCC projections for temperature and sea level for the 21st century?
11. What mitigation steps could be taken to stabilize the greenhouse gas content of the atmosphere?
12. Trace the political history of the FCCC and the Kyoto Protocol. What is the current status of the protocol?
13. Describe current U.S. climate change policy, and compare it with that of other developed countries.
14. Evaluate the importance of adaptation to climate change. Where is adaptation most critical, and why?
15. How is the ozone shield formed? What causes its breakdown?
16. How do CFCs affect the concentration of ozone in the stratosphere and contribute to the formation of the ozone hole?
17. Describe the international efforts that are currently in place to protect our ozone shield. What evidence is there that such efforts are effective?

## thinking environmentally

1. Evaluate your personal impact on climate change with the Climate Change Calculator at www.climcalc.net/. What steps could you take to lower your impact?
2. In your opinion, is global climate change occurring? What about ozone depletion? Consider the precautionary principle introduced at the 1992 Earth Summit. What might be the short- and long-term economic effects of using this approach to solve environmental problems such as climate change and ozone depletion?
3. Suggest a range of international agreements that would lead to the kind of reductions in greenhouse gas emissions called for by the Kyoto Protocol and that would satisfy the concerns of global equity between developed and developing countries.
4. Critique the decision to allow developing countries to continue using CFCs until 2010. What will happen after that? Why was this decision made? What are some consequences of the decision?

ISSSTE

# Atmospheric Pollution

**Key Topics**

1. Air-Pollution Essentials
2. Major Air Pollutants and Their Sources
3. Impacts of Air Pollutants: Health and Environment
4. Bringing Air Pollution under Control
5. Unresolved Issues

On Tuesday morning, October 26, 1948, the people of Donora, Pennsylvania (population 13,000), awoke to a dense fog (Fig. 21–1). Donora lies in an industrialized valley along the Monongahela River. On the outskirts of town, a sooty sign read, "DONORA: NEXT TO YOURS, THE BEST TOWN IN THE U.S.A." The town had a large steel mill, which used high-sulfur coke, and a zinc-reduction plant that roasted ores laden with sulfur. At the time, most homes in the area were heated with coal. At first, the fog did not seem unusual: Most of Donora's fogs lifted by noon, as the Sun warmed the upper atmosphere and then the land. This one, however, didn't lift for five days.

Through Wednesday and Thursday, the air began to smell of sulfur dioxide—an acrid, penetrating odor. By Friday morning, the town's physicians began to get calls from people in trouble. At first, the calls were from elderly citizens and those with asthmatic conditions. They were having difficulty breathing.

***Smog in Mexico City*** **Because of its geographic location and heavy automobile traffic, Mexico City has had some of the worst air pollution anywhere.**

The calls continued into Friday afternoon. People young and old were complaining of stomach pain, headaches, nausea, and choking. Work at the mills went on, however. The first deaths occurred on Saturday morning. By 10 A.M., one mortician had nine bodies; two other morticians had one each. On Sunday morning, the mills were shut down. Even so, the owners were certain that their plants had nothing to do with the trouble. Mercifully, the rain came on Sunday and cleared the air—but not before more than 6,000 townspeople were stricken and 20 elderly people had died. During the next month, 50 more people died.

Eventually, it was determined that the cause of the deaths was a combination of polluting gases and particles, a thermal inversion in the lower atmosphere, and a stagnant weather system that, together, brought home the deadly potential of *air pollution.* Today, a historical marker in the town commemorates the Donora Smog of 1948, but the lasting legacy of the smog and of those who suffered sickness and death are the state and national laws to control air pollution, culminating in the Clean Air Act of 1970. Donora was a landmark event.

**Figure 21–1 Donora, Pennsylvania.** Donora was the scene of a major air-pollution disaster in 1948.

In the previous chapter, we encountered the structure and function of the atmosphere and looked at its links to climate, climate change, and ozone depletion. Here, we will consider what happens when we add pollutants to the air.

## 21.1 Air-Pollution Essentials

### Pollutants and Atmospheric Cleansing

With the advent of the Industrial Revolution, the mixture of gases and particles in our atmosphere began to change. For a long time, we have known that the atmosphere contains numerous *gases*. The major constituents of the atmosphere are $N_2$ (nitrogen), at a level of 78.08%; $O_2$ (oxygen), at 20.95%; Ar (argon), at 0.93%; $CO_2$ (carbon dioxide), at 0.03%; and water vapor, ranging from 0 to 4%. Smaller amounts of at least 40 "trace gases" are normally present as well, including ozone, helium, hydrogen, nitrogen oxides, sulfur dioxide, and neon. In addition, *aerosols* are present in the atmosphere—microscopic liquid and solid particles such as dust, carbon particles, pollen, sea salts, and microorganisms, originating primarily from land and water surfaces and carried up into the atmosphere.

**Air pollutants** are substances in the atmosphere—certain gases and aerosols—that have harmful effects. Three factors determine the level of air pollution:

- The amount of pollutants entering the air
- The amount of space into which the pollutants are dispersed
- The mechanisms that remove pollutants from the air

*Atmospheric Cleansing.* Environmental scientists distinguish between natural and anthropogenic air pollutants. For millions of years, volcanoes, fires, and dust storms have sent smoke, gases, and particles into the atmosphere. Trees and other plants emit volatile organic compounds into the air around them as they photosynthesize. However, there are mechanisms in the biosphere that remove, assimilate, and recycle these natural pollutants. First, as shown in Figure 21–2, a naturally occurring cleanser, the **hydroxyl radical (OH)**, oxidizes many gaseous pollutants to products that are harmless or that can be brought down to the ground or water by precipitation. Sea salts picked up from sea spray as air masses move over the oceans are a second cleansing agent. These salts, now aerosols, act as excellent nuclei for the formation of raindrops. The rain then brings down many particulate pollutants (other aerosols) from the atmosphere to the ocean, cleansing pollutant-laden air coming from the land. Third, microorganisms in the soil further convert some of the gaseous pollutants into harmless compounds. These three processes hold natural pollutants below toxic levels (except in the immediate area of a source, such as around an erupting volcano).

Many of the pollutants oxidized by the hydroxyl radical are of concern because human activities have raised their concentrations far above normal levels. Recent studies have shown that the hydroxyl radical plays the key

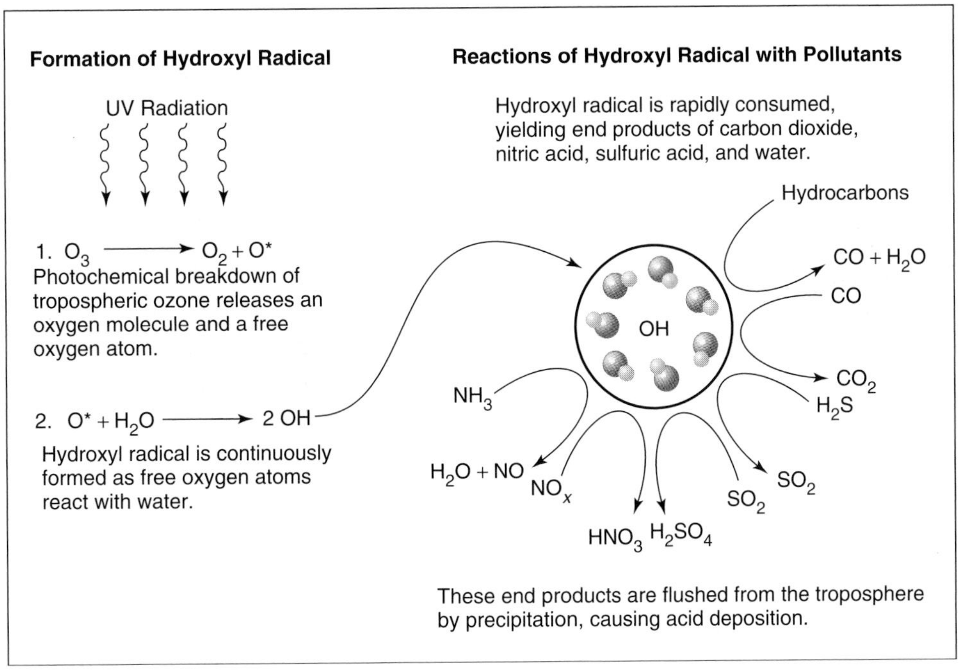


**Figure 21–2** **The hydroxyl radical.** This is a simplified model of atmospheric cleansing by the hydroxyl radical. The first step is the photochemical destruction of ozone, which is the major process leading to ozone breakdown in the troposphere. The second step produces hydroxyl radicals, which react rapidly with many pollutants, converting them to substances that are less harmful or that can be returned to Earth via precipitation.

role in the removal of anthropogenic pollutants from the atmosphere. Thus, highly reactive hydrocarbons are oxidized within an hour of their appearance in the atmosphere, and nitrogen oxides ($NO_x$) are converted to nitric acid ($HNO_3$) within a day. It takes months, however, for less reactive substances, such as carbon monoxide (CO), to be oxidized by hydroxyl. It appears that the atmospheric levels of hydroxyl are in turn determined by the levels of anthropogenic pollutant gases; thus, hydroxyl's cleansing power is "used up" when high concentrations of pollutants are oxidized, and the pollutants are subsequently able to build up to damaging levels. In contrast, as Figure 21–2 shows, the photochemical breakdown of tropospheric ozone is the major source of the hydroxyl radical, and, as we will see, higher levels of ozone result from higher concentrations of other polluting gases.

## The Appearance of Smog

***Industrial Smog.*** Down through the centuries, the practice of venting the products of combustion and other fumes into the atmosphere remained the natural way to avoid their noxious effects within buildings. With the Industrial Revolution of the 1800s came crowded cities and the use of coal for heating and energy. It was then that air pollution began in earnest. In *Hard Times,* Charles Dickens described a typical English scene: "Coketown lay shrouded in a haze of its own, which appeared impervious to the sun's rays. You only knew the town was there because there could be no such sulky blotch upon the prospect without a town." This shrouding haze became known as **industrial smog** (a combination of *sm*oke and f*og*!), an irritating, grayish mixture of soot, sulfurous compounds, and water vapor (Fig. 21–3a). This kind of smog continues to be found wherever industries are concentrated and where coal is the primary energy source. The Donora incident is a classic example. Currently, industrial smog can be found in the cities of China, Korea, and a number of Eastern European countries.

***Photochemical Smog.*** After World War II, the booming economy in the United States led to the creation of vast suburbs and a mushrooming use of cars for commuting to and from work. As other fossil fuels were used in place of coal for heating, industry, and transportation, industrial smog was largely replaced by another kind of smog. Increasingly, Los Angeles and other cities served by huge freeway systems began to be enshrouded daily in a brownish, irritating haze that was different from the more familiar industrial smog (Fig. 21–4). Weather conditions were usually warm and sunny rather than cool and foggy, and typically the new haze would arise during the morning commute and begin to dissipate only by the time of the evening commute. This **photochemical smog,** as it came to be called, is produced when several pollutants from automobile exhausts—nitrogen oxides and volatile organic carbon compounds—are acted on by sunlight (Fig. 21–3b).

***Inversions.*** Certain weather conditions intensify levels of both industrial and photochemical smog. Under normal conditions, the daytime air temperature

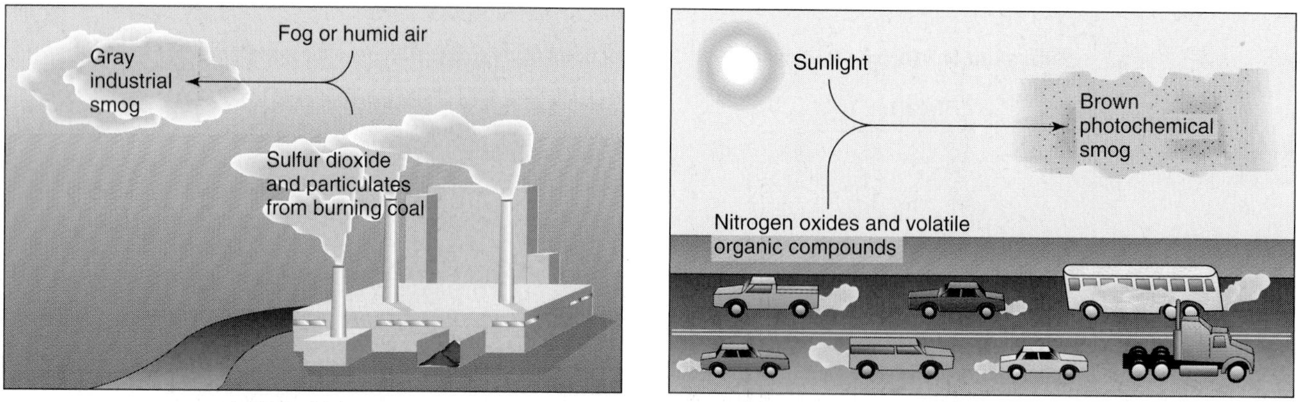


(a) Industrial smog (b) Photochemical smog

**Figure 21–3 Industrial and photochemical smog.** (a) Industrial smog, or gray smog, occurs when coal is burned and the atmosphere is humid. (b) Photochemical smog, or brown haze, occurs when sunlight acts on vehicle pollutants.

is highest near the ground, because sunlight strikes Earth and the absorbed heat radiates to the air near the surface. The warm air near the ground rises, carrying pollutants upward and dispersing them at higher altitudes (Fig. 21–5a). At times, however, a warm air layer occurs above a cooler layer. This condition of cooler air below and warmer air above is called a **temperature inversion** (Fig. 21–5b). Inversions often occur at night, when the surface air is cooled by radiative heat loss. Such inversions are usually short lived, as the next morning's Sun begins the heating process anew and any pollutants that accumulated overnight are carried up and away. During cloudy weather, however, the Sun may not be strong enough to break up the inversion for hours or even days. Or a mass of high-pressure air may move in and sit above the cool surface air, trapping it. Topography can also intensify smogs, as cool ocean air flows into valleys and is trapped there by nearby mountain ranges. Los Angeles has both topographical features!

*Impact of Smogs.* When such long-term temperature inversions occur, pollutants can build up to dangerous levels, prompting local health officials to urge people with breathing problems to stay indoors. For many people, smog causes headaches, nausea, and eye and throat irritation. (See Global Perspective, p. 577). It may aggravate preexisting respiratory conditions, such as asthma and emphysema. In some industrial cities (Donora, Pennsylvania, for example), air pollution reached lethal levels under severe temperature inversions. These cases became known as *air-pollution disasters.* London experienced repeated episodes of inversion-related disasters in the mid-1900s. One episode in 1952 resulted in 4,000 pollution-related deaths.

The effects of air pollution smogs are not limited to people. In recent years, many species of trees and other vegetation in and near cities began to die back, and farmers near cities suffered damage to, or even total destruction of, their crops because of air pollution. A conspicuous acceleration in the rate of metal corrosion and the deterioration of rubber, fabrics, and other materials was also noted.

(a) (b)

**Figure 21–4 Los Angeles air.** A typical episode of photochemical smog. (a) Early in the morning, the air is clear. (b) Midmorning of the same day, the air is hazy with smog.

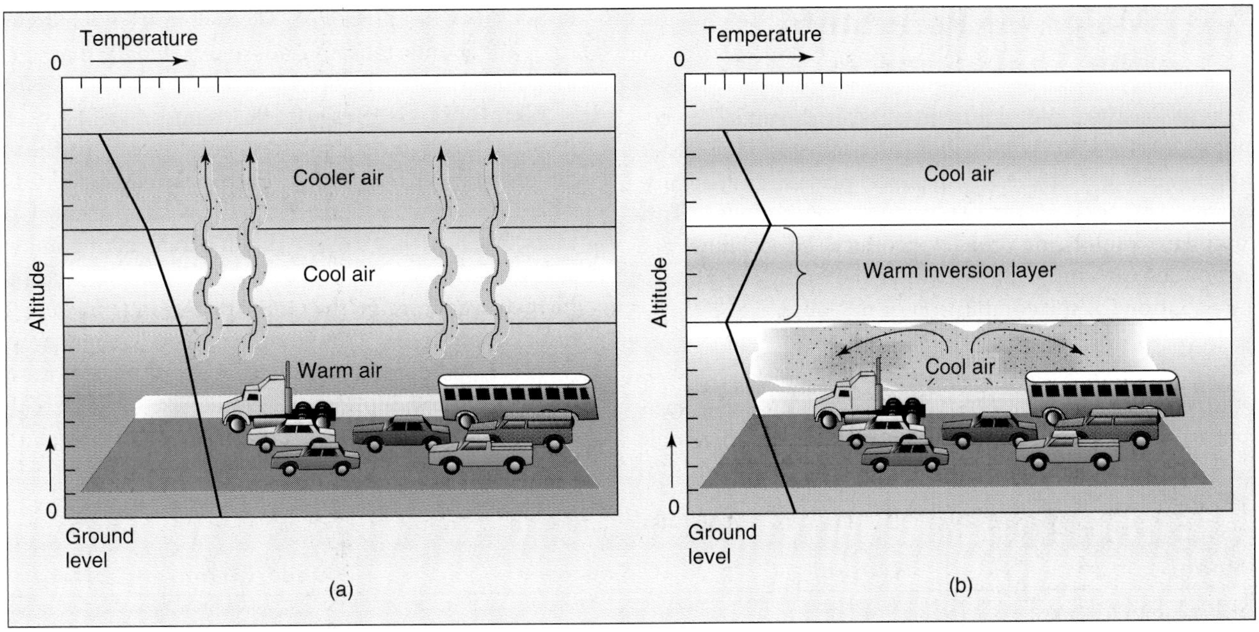


**Figure 21–5** **Temperature inversion.** A temperature inversion may cause episodes of high concentrations of air pollutants. (a) Normally, air temperatures are highest at ground level and decrease at higher elevations. (b) In a temperature inversion, a layer of warmer air overlies cooler air at ground level.

## global perspective

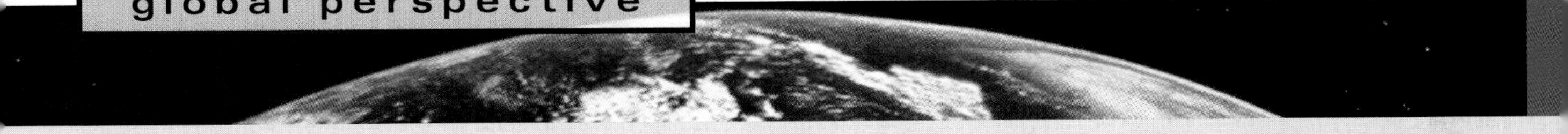

### Mexico City: Life in a Gas Chamber

"I'm getting out of here one of these days," said one resident of Mexico City. "The ecologists are right. We live in a gas chamber." For many years, Mexico City residents have had the worst air in the world. The city exceeded safety limits for ozone 300 days a year, with values sometimes almost 400% over norms set by the WHO. Smog emergencies occur frequently. The city's 3.5 million vehicles (a third of which are old and in disrepair) and 60,000 industries pour millions of tons of pollutants into the air every day. Because it is nestled in a natural bowl, Mexico City is affected by thermal inversions every day during the winter. In 1999, the World Resources Institute named Mexico City the world's most dangerous city for children in terms of air pollution. One study showed that 40% of the city's young children suffer chronic respiratory sickness during the winter months. The pollution is estimated to cost $11.5 million a day in terms of health effects and lost industrial production, and an estimated 500,000 residents suffer from some form of smog-related health problem. Writer Carlos Fuentes nicknamed the metropolis "Makesicko City"!

Mexico City's problems are indicative of a general dilemma throughout the developing world. Jobs and industrial activity, vital to a nation's economy, depend on the use of motor vehicles and on the industrial plants surrounding the city. Many automobile owners in turn depend on their cars, but can ill afford to keep them repaired in order to reduce exhaust emissions. Some people see cars as status symbols and will acquire one as soon as possible—often a "clunker" discarded from the developed world. Industries in Mexico are not regulated as they are in the United States and lack the technology and capital to control emissions.

In response to the problem, the Mexican government has initiated an air-monitoring system and mandated emissions tests and catalytic converters for automobiles. When pollution alerts are sounded, high-emission vehicles are banned from the roads. People have begun buying newer, far less polluting vehicles in order to keep driving. New air-pollution laws are beginning to be enforced: One major refinery responsible for 4% of the air-quality problem was shut down, and 80 other industries were temporarily or permanently closed for violating the new laws. New subway lines are being built, and buses that are less polluting are being added to the fleet. Lead, sulfur dioxide, and carbon monoxide levels have declined greatly, but ozone and particulates are still too high by international norms. The city had its first pollution alert in almost three years in September 2002, a sign that Mexico City's 19 million inhabitants are breathing more freely these days. Several Asian cities have replaced Mexico City as suffering from the world's unhealthiest air.

## 21.2 Major Air Pollutants and Their Sources

In large measure, air pollutants are direct and indirect by-products of the *combustion* of coal, gasoline, other liquid fuels, and refuse (wastepaper, plaster, and so on). These fuels and wastes are organic compounds. With complete combustion, the by-products of burning are carbon dioxide and water vapor, as shown by the following equation for the combustion of methane:

$$CH_4 + 2\ O_2 \rightarrow CO_2 + 2\ H_2O \qquad (1)$$

Unfortunately, oxidation is seldom complete, and substances far more complex than methane are involved. Other processes producing air pollutants are *evaporation* (of volatile substances) and *strong winds* that pick up dust and other particles.

Table 21–1 presents the names, symbols, major sources, characteristics, and general effects of the major air pollutants. The first seven (**particulates, VOCs, CO, $NO_x$, $SO_2$, lead,** and **air toxics**) are called **primary pollutants,** because they are the direct products of combustion and evaporation. Some primary pollutants may undergo further reactions in the atmosphere and produce additional undesirable compounds, called **secondary pollutants** (**$O_3$, peroxyacetyl nitrates, sulfuric acid, nitric acid**). Sources of air pollution are summarized in Figure 21–6. Notice that, while power plants are the major source of

**table 21-1 Major Air Pollutants and Their Effects**

| Name | Symbol | Source | Description and Effects |
|---|---|---|---|
| **Primary Pollutants** | | | |
| Suspended particulate matter | PM | Soot, smoke, metals, and carbon from combustion; dust, salts, metal, and dirt from wind erosion; atmospheric reactions of gases. | Complex mixture of solid particles and aerosols suspended in air. Affect respiration and can be carcinogenic; impair visibility. |
| Volatile organic compounds | VOC | Incomplete combustion of fossil fuels; evaporation of solvents and gasoline; emission from plants. | Mixture of compounds, some carcinogenic; major agent of ozone formation. |
| Carbon monoxide | CO | Incomplete combustion of fuels. | Invisible, odorless, tasteless gas; poisonous because of ability to bind to hemoglobin and block oxygen delivery to tissues. |
| Nitrogen oxides | $NO_x$ | From nitrogen gas due to high combustion temperatures when burning fuels. | Reddish-brown gas and lung irritant capable of producing acute disease; major source of acid rain when converted to nitric acid in atmosphere. |
| Sulfur oxides | $SO_x$ | Combustion of sulfur-containing fuels, especially coal. | Poisonous gas that impairs breathing; major source of acid rain when converted to sulfuric acid in atmosphere. |
| Lead | Pb | Combustion of leaded fuels and solid wastes. | Toxic at low concentrations; accumulates in body and can lead to brain damage and death. |
| Air toxics | Various | Industry and transportation | Toxic chemicals of many kinds, many of which are known human carcinogens (e.g., benzene, asbestos, vinyl chloride); Clean Air Act identifies 188 air toxics. |
| Radon | Rn | Rocks and soil; natural breakdown of radium and uranium. | Invisible and odorless radioactive gas can accumulate inside homes; second leading cause of lung cancer in United States. |
| **Secondary Pollutants** | | | |
| Ozone | $O_3$ | Photochemical reactions between VOCs and $NO_x$ | Toxic to animals and plants and highly reactive in lungs; oxidizes surfaces and rubber tires. |
| Peroxyacetyl nitrates | PAN | Photochemical reactions between VOCs and $NO_x$ | Damage to plants and forests; irritants to mucuous membranes of eyes and to lungs. |

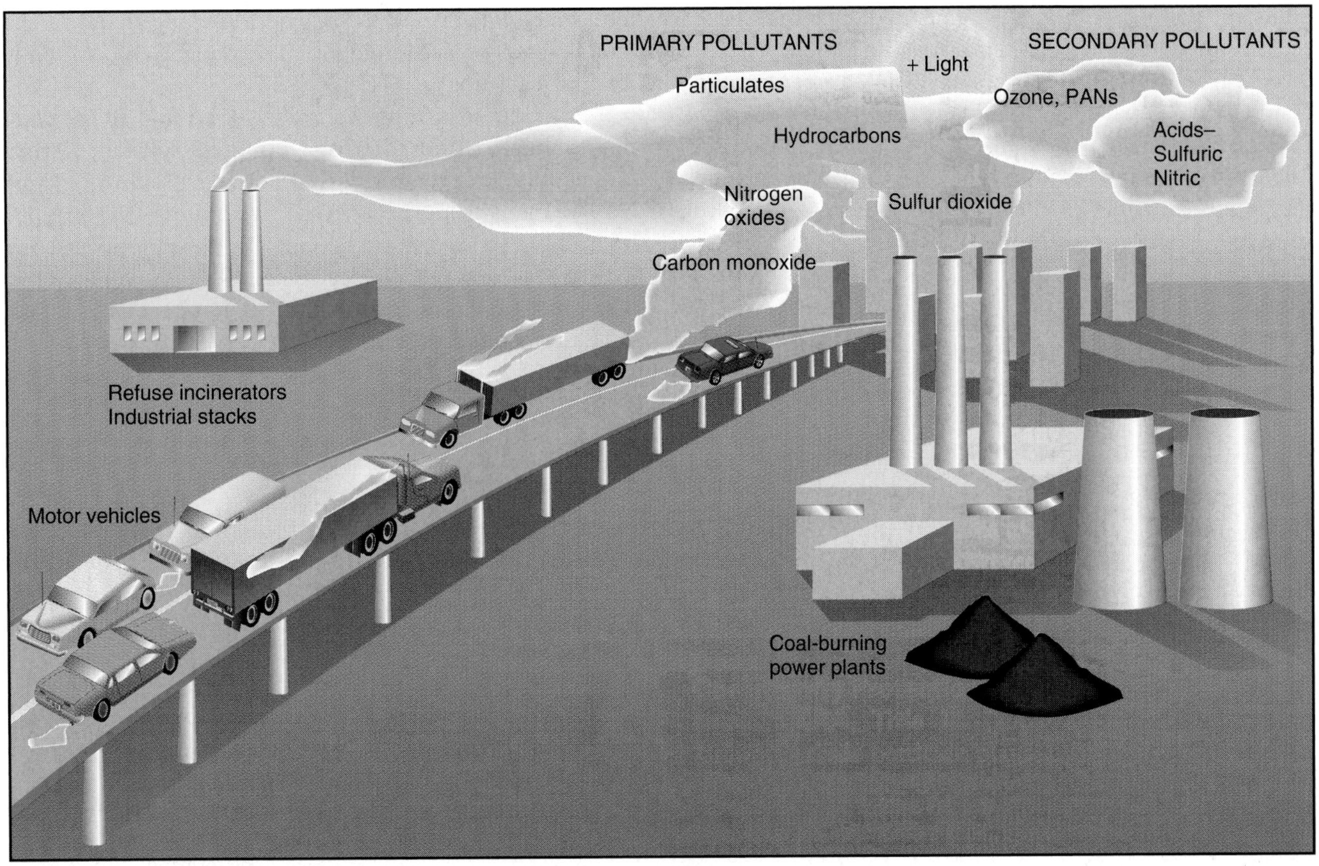


**Figure 21–6** **The prime sources of the major air pollutants.**

sulfur dioxide, industrial processes are the major source of particulates, and transportation accounts for the lion's share of carbon monoxide and nitrogen oxides. This diversity suggests that strategies for controlling air pollutants will be different for the different sources.

## Primary Pollutants

When fuels and wastes are burned, particles consisting mainly of carbon are emitted into the air; these are the particulates we see as soot and smoke (Fig. 21–7). In

**Figure 21–7** **Industrial pollution.** Because the stack of this wood-products mill in New Richmond, Quebec Province, Canada, lacks electrostatic precipitators, the plume contains substantial amounts of suspended particulate matter that can be seen from a great distance.

addition, various unburned fragments of fuel molecules remain; these are the VOC emissions. Incompletely oxidized carbon is carbon monoxide (CO), in contrast to completely oxidized carbon, which is carbon dioxide ($CO_2$). Combustion takes place in the air, a mixture of 78% nitrogen and 21% oxygen. At high combustion temperatures, some of the nitrogen gas is oxidized to form the gas nitric oxide (NO). In the air, nitric oxide immediately reacts with additional oxygen to form nitrogen dioxide ($NO_2$) and nitrogen tetroxide ($N_2O_4$). (These compounds are collectively referred to as the **nitrogen oxides,** or **$NO_x$**.) Nitrogen dioxide absorbs light and is largely responsible for the brownish color of photochemical smog.

In addition to organic matter, fuels and refuse contain impurities and additives, and these substances, too, are emitted into the air during burning. Coal, for example, contains from 0.2% to 5.5% sulfur. In combustion, this sulfur is oxidized, giving rise to several gaseous oxides, the most common being sulfur dioxide ($SO_2$). Coal may contain heavy-metal impurities, and refuse contains an endless array of "impurities."

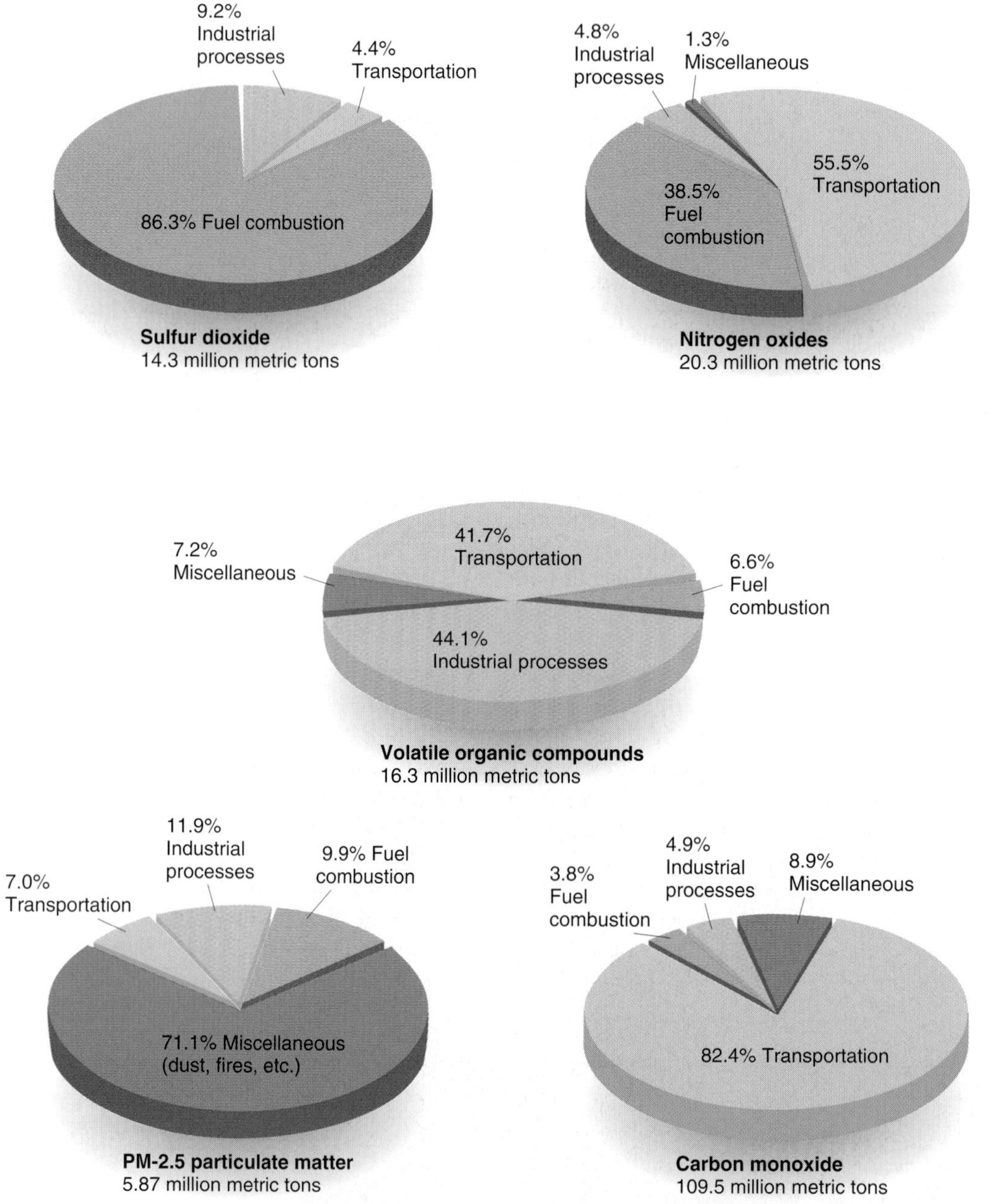


**Figure 21–8 U.S. emissions of five primary air pollutants, by source, for 2001.** *Fuel combustion* refers to fuels burned for electrical power generation and for space heating. Note especially the different contributions by transportation and fuel combustion, the two major sources of air pollutants. (*Source:* EPA Office of Air Quality, 2003.)

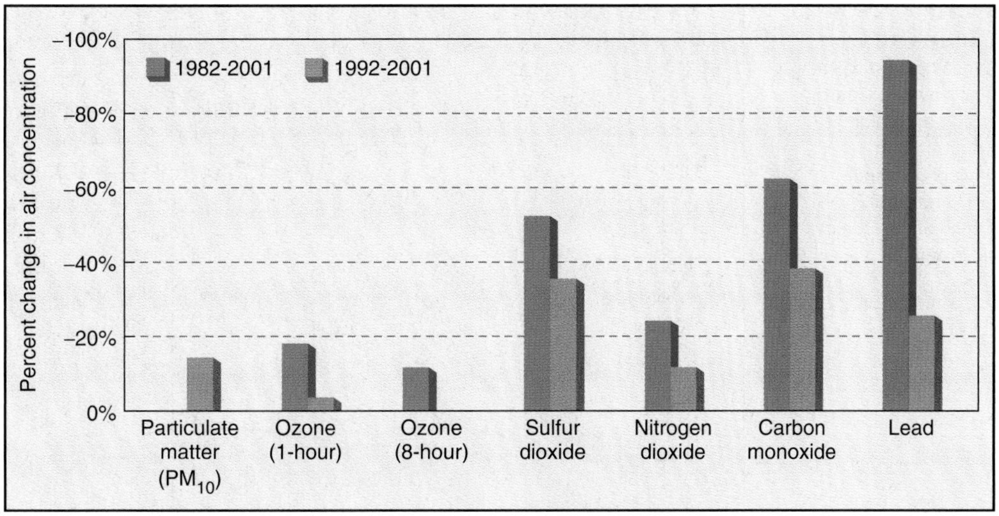


**Figure 21–9 Clean Air Act impacts.** Percent reductions in the concentration of six criteria air pollutants, 1982–2001. (*Source:* EPA Draft Report on the Environment, 2003)

*Tracking Pollutants.* The EPA Office of Air Quality tracks trends in national **emissions** of the primary pollutants from all sources. The EPA also follows air quality by measuring **ambient concentrations** of the pollutants at thousands of monitoring stations across the country. According to EPA data, in 2001 *emissions* in the United States of the first five primary pollutants amounted to 166 million metric tons. By comparison, in 1970, when the first Clean Air Act became law, these same five pollutants totaled 244 million tons. The relative amounts emitted in the United States in 2001 (the most recent data), as well as their major sources, are illustrated in Figure 21–8. The general picture shows continued improvement for most emissions over the last two decades (Fig. 21–9), reflecting the effectiveness of Clean Air Act regulations. This progressively improving trend over time has occurred in spite of the fact that, between 1970 and 2002, the economy increased by 164% and vehicular miles traveled increased by 155% (Fig. 21–10).

*Getting the Lead Out.* The sixth type of primary pollutant—lead and other heavy metals—is discussed separately because the quantities emitted are far less than the levels for the first five. Before the EPA-directed phaseout, lead was added to gasoline as an inexpensive way to prevent engine knock. Emitted with the exhaust from gasoline-burning vehicles, lead remained airborne and traveled great distances before settling as far away as the glaciers of Greenland. Since the phaseout, concentrations of lead in the air of cities in the United States have declined remarkably. Between 1983 and 2002, for example, ambient lead concentrations and lead emissions fell 94% and 93%, respectively. At the same time, levels of lead in children's blood have dropped greatly. Lead concentrations in Greenland ice have also decreased significantly. All these declines indicate that lead restrictions in the United States and a few other Northern Hemisphere nations have had a global impact. The data indicate that we have reached a steady state in lead emissions and that any new reductions must target the lead smelters and battery manufacturers, which are now by far the largest source of lead emissions.

*Toxics and Radon.* As with lead, the concentrations of toxic chemicals and radon in the air are small compared with those of the other primary pollutants. Some of the air's toxic compounds—benzene, for example—originate with transportation fuels. Most, however, are traceable to industries and small businesses. Radon, by contrast, is produced by the spontaneous decay of fissionable material in rocks and soils. Radon escapes naturally to the surface and seeps into buildings through cracks in foundations and basement floors, sometimes collecting in the structures.

## Secondary Pollutants

**Ozone** and numerous reactive organic compounds are formed as a result of chemical reactions between nitrogen oxides and volatile organic carbons. Because sunlight provides the energy necessary to propel the reactions, these products are known collectively as **photochemical oxidants.** In preindustrial times, ozone concentrations ranged from 10 to 15 parts per billion (ppb), well below harmful levels. Summer concentrations of ozone in unpolluted air in North America now range from 20 to 50 ppb. Polluted air may contain ozone concentrations of 150 ppb or more, a level considered quite unhealthy if encountered for extended periods. EPA data indicate that ambient ozone levels, monitored across the United States at some 1,100 sites, declined between 1983 and 1993, but no improvement has been seen since 1993. Ozone air quality standards are the leaders in noncompliance across the country, with over 136 million people living in counties that still do not meet standards.

*Ozone Formation.* The major reactions in the formation of ozone and other photochemical oxidants are shown in Figure 21–11. Nitrogen dioxide absorbs light

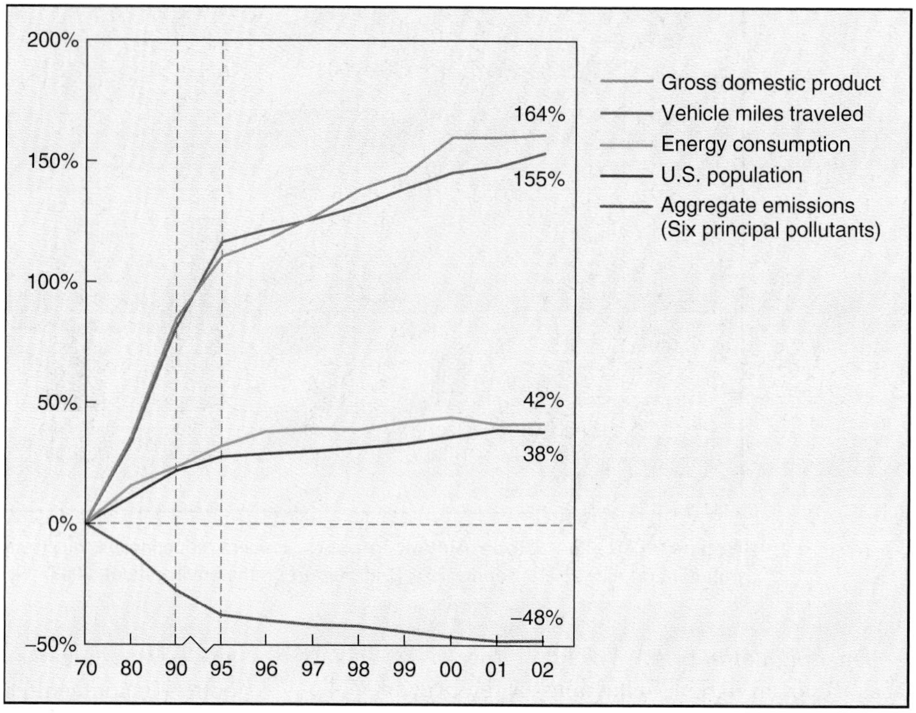


**Figure 21–10**
**Comparison of growth vs. emissions.** Gross domestic product, vehicle miles traveled, energy consumption, and population growth are compared with reductions in the six principal pollutants for 1970 and 2002. (*Source:* EPA Office of Air Quality, 2002 Status and Trends)

energy and splits to form nitric oxide and atomic oxygen, which rapidly combines with oxygen gas to form ozone. If other factors are not involved, ozone and nitric oxide then react to form nitrogen dioxide and oxygen gas. A steady-state concentration of ozone results, and there is no appreciable accumulation of the gas (Fig. 21–11a). When VOCs are present, however, the nitric oxide reacts with them instead of with the ozone, causing several serious problems. First, the reaction between nitric oxide and the VOCs leads to highly reactive and damaging compounds known as peroxyacetyl nitrates, or PANs (Fig. 21–11b). Second, numerous aldehyde and ketone compounds are produced as the VOCs are oxidized by atomic oxygen, and these compounds are also noxious. Finally, with the nitric oxide tied up in this way, the ozone tends to accumulate. Because of the complex air chemistry involved, ozone concentrations usually peak 30 to 100 miles (50 to 160 km) downwind of urban centers where the primary pollutants were generated. As a result, ozone concentrations above the health standards for the gas in the United States are often found in rural and wilderness areas.

In a sense, **sulfuric acids** and **nitric acids** can also be considered secondary pollutants, since they are products of sulfur dioxide and nitrogen oxides, respectively, reacting with atmospheric moisture and oxidants such as hydroxyl. Sulfuric and nitric acids are the acids in acid rain (technically referred to as acid deposition). We now turn our attention to this important and widespread problem.

## Acid Deposition

*Acid precipitation* refers to any precipitation—rain, fog, mist, or snow—that is more acidic than usual. Because dry acidic particles are also found in the atmosphere, the combination of precipitation and dry-particle fallout is called **acid deposition.** In the late 1960s, Swedish scientist Svante Odén first documented the acidification of lakes in Scandinavia and traced it to air pollutants originating in other parts of Europe and Great Britain. Since then, careful monitoring has shown that broad areas of North America, as well as most of Europe and other industrialized regions of the world, are regularly experiencing precipitation that is between 10 and 1,000 times more acidic than usual. This is affecting ecosystems in diverse ways, as illustrated in Figure 21–12.

To understand the full extent of the problem, first we must understand some principles about acids and how we measure their concentration.

**Acids and Bases.** Acidic properties (for example, a sour taste and corrosiveness) are due to the presence of hydrogen ions ($H^+$, a hydrogen atom without its electron), which are highly reactive. Therefore, an **acid** is any chemical that releases hydrogen ions when dissolved in water. The chemical formulas of a few common acids are shown in Table 21–2. Note that all of them ionize—that is, their components separate—to give hydrogen ions plus a negative ion. The higher the concentration of hydrogen ions in a solution, the more acidic is the solution.

A **base** is any chemical that releases hydroxide ions ($OH^-$, oxygen–hydrogen groups with an extra electron) when dissolved in water. (See Table 21–2.) The bitter taste and caustic properties of all alkaline, or basic, solutions are due to the presence of hydroxide ions.

***pH.*** The concentration of hydrogen ions is expressed as **pH.** The pH scale goes from 0 (highly acidic) through 7 (neutral) to 14 (highly basic) (Fig. 21–13). Each of the numbers on the scale represents the negative logarithm

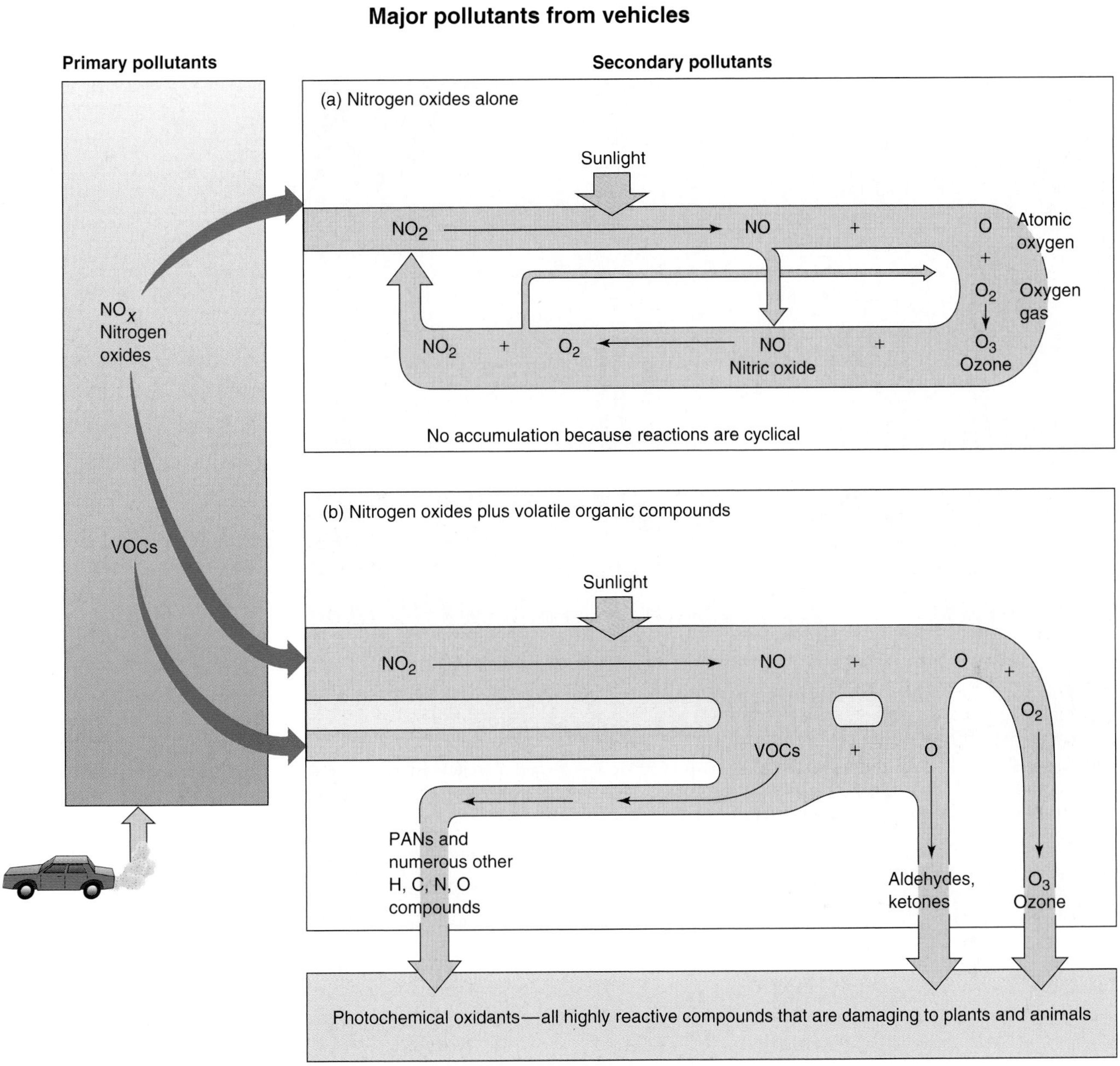


**Figure 21–11 Formation of ozone and other photochemical oxidants.** (a) Nitrogen oxides, by themselves, would not cause ozone and other oxidants to reach damaging levels, because reactions involving nitrogen oxides are cyclic. (b) When VOCs are also present, however, reactions occur that lead to the accumulation of numerous damaging compounds—most significantly, ozone, the most injurious.

(power of 10) of the hydrogen ion concentration, expressed in grams per liter. For example, to say that a solution has a pH of 1 means that the concentration of hydrogen ions in the solution is $10^{-1}$ g/L (0.1 g/L); pH = 2 means that the hydrogen ion concentration is $10^{-2}$ g/L, and so on. At pH = 7, the hydrogen ion concentration is $10^{-7}$ (0.0000001) g/L, and the hydroxide ion concentration is also $10^{-7}$ g/L. This is the neutral point, where small, but equal, amounts of hydrogen ions and hydroxide ions are present in pure water. The pH numbers above 7 continue to express the negative exponent of hydrogen ion concentration, but they also represent an increase in hydroxide ion concentration. Because solutions of pH greater than 7 contain higher concentrations of hydroxide ions than of hydrogen ions, they are called basic solutions.

Since numbers on the pH scale represent powers of 10, there is a *tenfold difference* between each unit and the next. For example, pH 5 is 10 times as acidic (has 10 times as many $H^+$ ions) as pH 6, pH 4 is 10 times as acidic as pH 5, and so on.

**Extent and Potency of Acid Precipitation.** In the absence of any pollution, rainfall is normally slightly acidic, with a pH of 5.6, because carbon dioxide in the air readily dissolves in, and combines with, water to

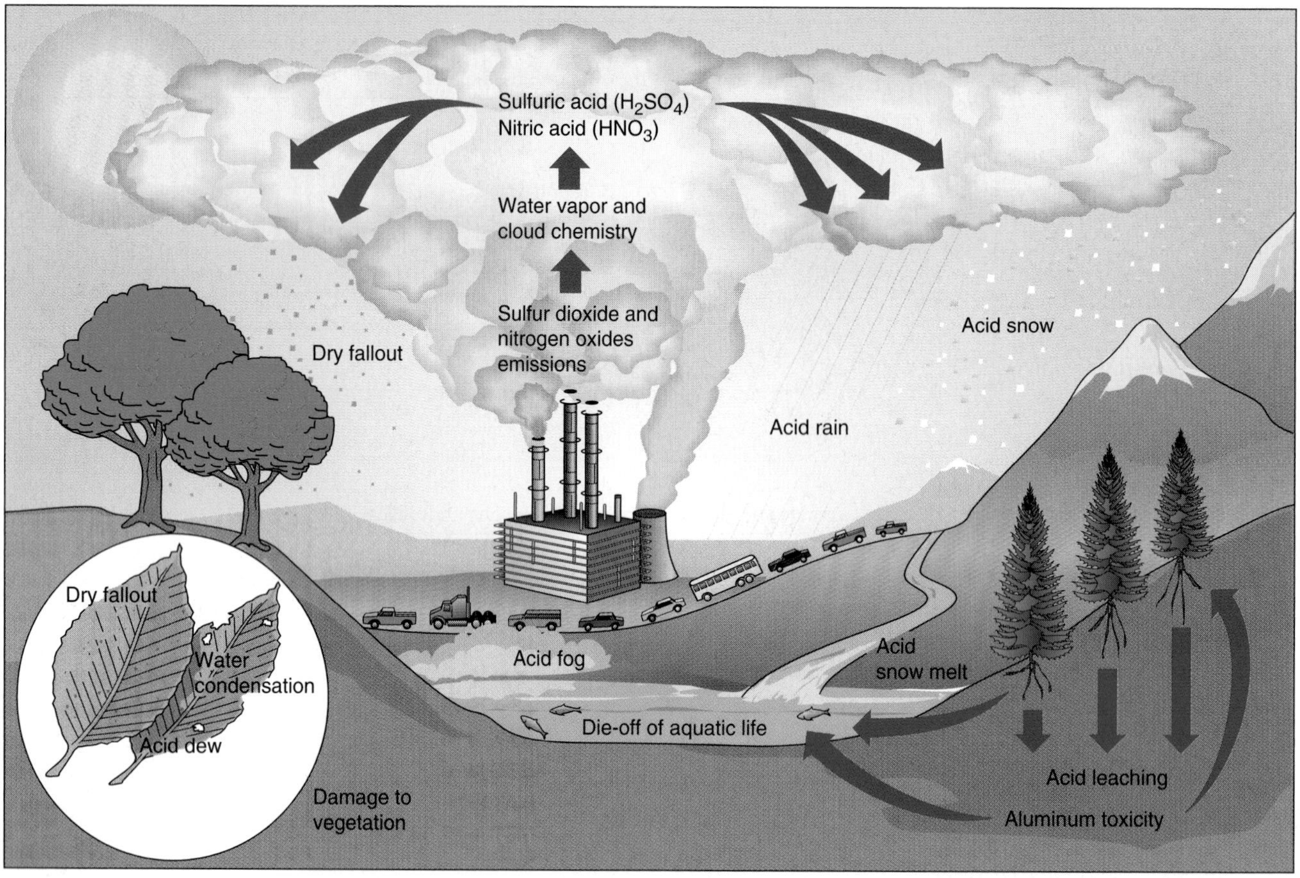


**Figure 21–12 Acid deposition.** Emissions of sulfur dioxide and nitrogen oxides react with hydroxyl radicals and water vapor in the atmosphere to form their respective acids, which return to the surface either as dry acid deposition or, mixed with water, as acid precipitation. Various effects of acid deposition are noted.

**table 21-2 Common Acids and Bases**

| | Formula | Yields | $H^+$ Ion(s) | Plus | Negative Ion |
|---|---|---|---|---|---|
| **Acid** | | | | | |
| Hydrochloric acid | $HCl$ | → | $H^+$ | + | $Cl^-$ Chloride |
| Sulfuric acid | $H_2SO_4$ | → | $2\ H^+$ | + | $SO_4^{2-}$ Sulfate |
| Nitric acid | $HNO_3$ | → | $H^+$ | + | $NO_3^-$ Nitrate |
| Phosphoric acid | $H_3PO_4$ | → | $3\ H^+$ | + | $PO_4^{3-}$ Phosphate |
| Acetic acid | $CH_3COOH$ | → | $H^+$ | + | $CH_3COO^-$ Acetate |
| Carbonic acid | $H_2CO_3$ | → | $H^+$ | + | $HCO_3^-$ Bicarbonate |
| | **Formula** | **Yields** | **$OH^-$ Ion(s)** | **Plus** | **Positive Ion** |
| **Base** | | | | | |
| Sodium hydroxide | $NaOH$ | → | $OH^-$ | + | $Na^+$ Sodium ion |
| Potassium hydroxide | $KOH$ | → | $OH^-$ | + | $K^+$ Potassium ion |
| Calcium hydroxide | $Ca(OH)_2$ | → | $2\ OH^-$ | + | $Ca^{2+}$ Calcium ion |
| Ammonium hydroxide | $NH_4OH$ | → | $OH^-$ | + | $NH_4^+$ Ammonium ion |

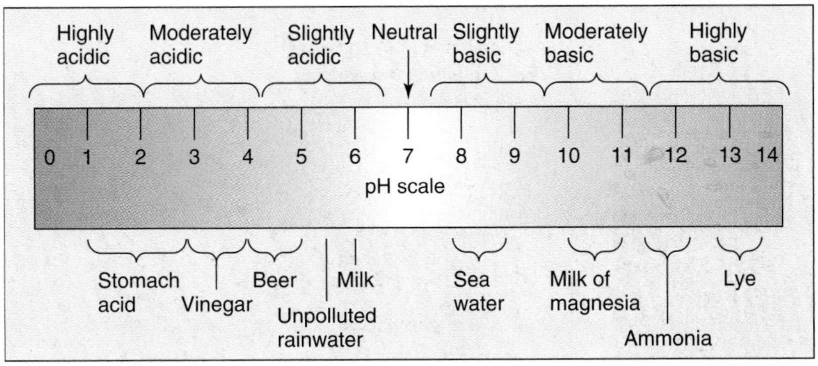


**Figure 21–13 The pH scale.** Each unit on the scale represents a tenfold difference in hydrogen ion concentration.

produce carbonic acid. **Acid precipitation**, then, is any precipitation with a pH less than 5.5.

Unfortunately, acid precipitation is now the norm over most of the industrialized world. The pH of rain and snowfall over a large portion of eastern North America is typically below 4.5, reflecting the west-to-east movement of polluted air from the Midwest (Fig. 21–14). Many areas in this region regularly receive precipitation having a pH of 4.0 and, occasionally, as low as 3.0. Fogs and dews can be even more acidic. In mountain forests east of Los Angeles, scientists found fog water of pH 2.8—almost 1,000 times more acidic than usual—dripping from pine needles. Acid precipitation also has been heavy in Europe, from the British Isles to central Russia. It is now documented in Japan as well.

**Sources of Acid Deposition.** Chemical analysis of acid precipitation in eastern North America and Europe reveals the presence of two acids, **sulfuric acid** ($H_2SO_4$) and **nitric acid** ($HNO_3$), in a ratio of about two to one. (In the western United States and in the western provinces of Canada, nitric acid predominates, principally formed from tailpipe emissions.) As we have seen, burning fuels produce sulfur dioxide and nitrogen oxides, so the source of the acid deposition problem is evident. These oxides enter the troposphere in large quantities from both anthropogenic

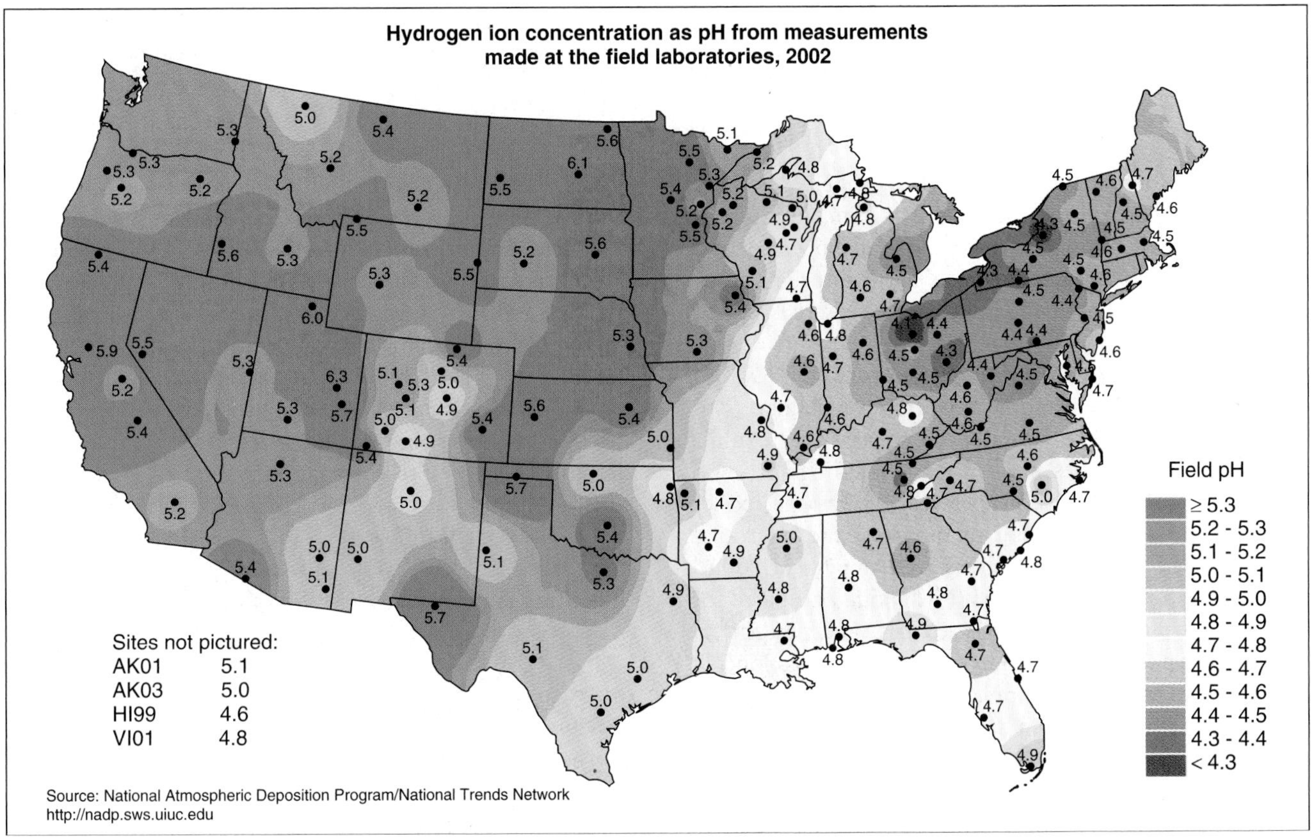


**Figure 21–14 Acid deposition in the United States.** Data from over 200 monitoring sites across the United States. Such measurements indicate that acid deposition continues to prevail throughout the East and much of the Midwest (AK sites are in Alaska, HI is the Hawaiian Islands, and VI refers to the Virgin Islands.) (*Source:* National Atmospheric Deposition Program, 2003)

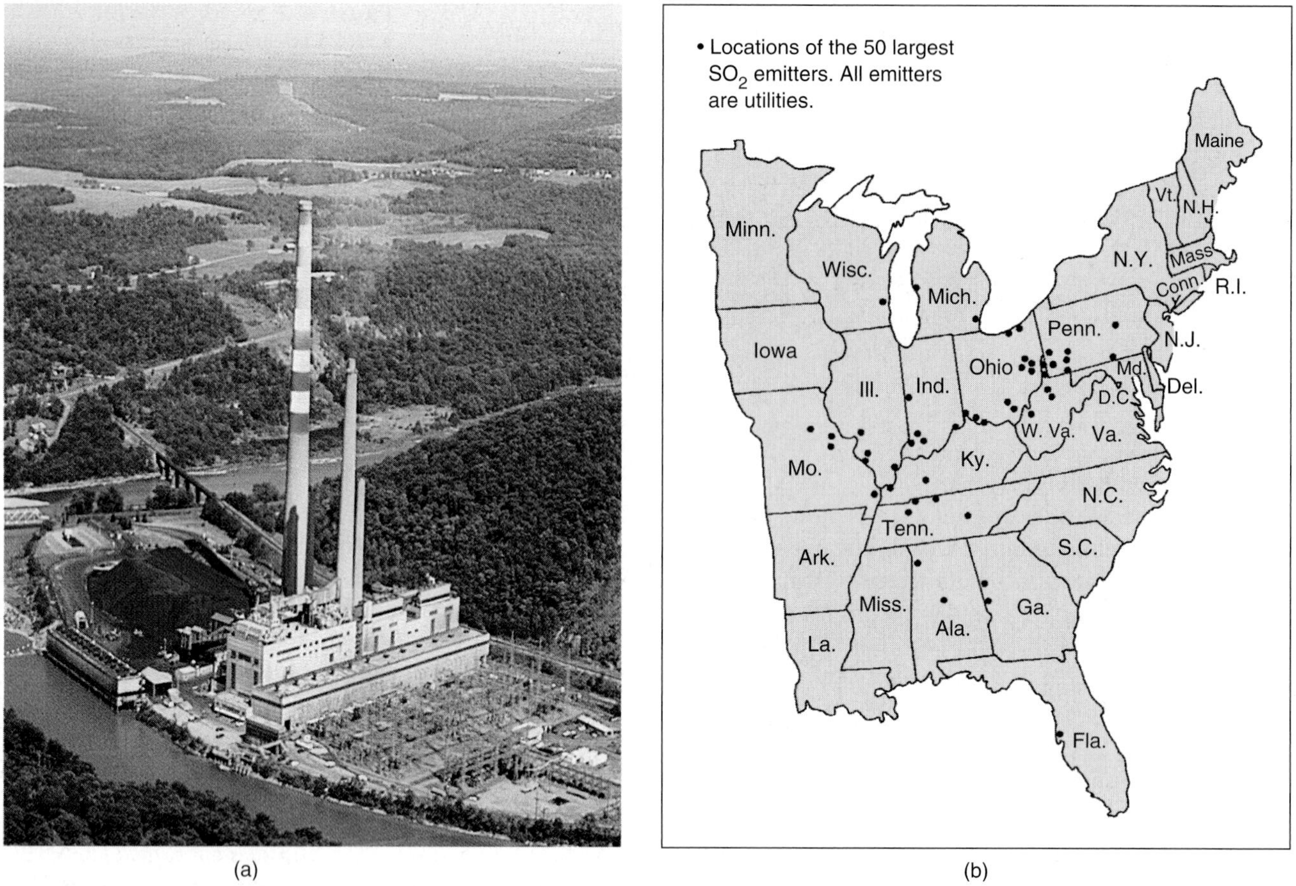


**Figure 21–15 Midwestern coal-burning power plant.** (a) Standard smokestacks of this coal-burning power plant were replaced by new 1,000-foot (330-m) stacks to aid in the dispersion of pollutants into the atmosphere. The taller stacks alleviated local air-pollution problems, but created a more widespread distribution of acid-generating pollutants. (b) Locations of the 50 largest sulfur dioxide emitters, all of which are utility coal-burning power plants.

and natural sources. Once in the troposphere, they are oxidized by hydroxyl radicals (Fig. 21–2) to sulfuric and nitric acids, which dissolve readily in water or adsorb to particles and are brought down to Earth in acid deposition. This usually occurs within a week of the oxides' entering the atmosphere.

*Natural versus Anthropogenic Sources.* *Natural sources* contribute substantial quantities of pollutants to the air, including 50 to 70 million tons per year of sulfur dioxide (from volcanoes, sea spray, and microbial processes) and 30 to 40 million tons per year of nitrogen oxides (from lightning, the burning of biomass, and microbial processes). *Anthropogenic sources* are estimated at 100 to 130 million tons per year of sulfur dioxide and 60 to 70 million tons per year of nitrogen oxides. The vital difference between these two sources is that anthropogenic oxides are strongly concentrated in industrialized regions, whereas the emissions from natural sources are spread out over the globe and are a part of the global environment. Levels of the anthropogenic oxides have increased at least fourfold since 1900, while levels of the natural emissions have remained fairly constant.

As Fig. 21–8 indicates, 14.3 million tons of sulfur dioxide were released into the air in 2001 in the United States; 86% was from fuel combustion (mostly from coal-burning power plants). Some 20.3 million tons of nitrogen oxides were released, 56% of which can be traced to transportation emissions and 39% to fuel combustion at fixed sites. In the eastern United States, the source of much of the acid deposition was identified as the tall stacks of 50 huge coal-burning power plants (Fig. 21–15). The tall stacks were built to alleviate local sulfur dioxide pollution at ground level. These same plants along with over 200 other large coal-fired power plants targeted by the EPA, are now reducing their emissions as a result of the Clean Air Act. However, 69% of $SO_2$ emissions and 22% of $NO_x$ emissions still originate from fossil-fuel-burning electric utility plants.

## 21.3 Impacts of Air Pollutants

Air pollution is an alphabet soup of gases and particles, mixed with the normal constituents of air. The amount of each pollutant present varies greatly, depending on its proximity to the source and various conditions of wind and weather. As a result, we are exposed to a mixture that varies in makeup and concentration from day to day—even from hour to hour—and from place to place. Consequently, the effects we feel or observe are rarely, if ever, the effects of a single pollutant. Instead, they are the combined impact of the whole mixture of pollutants acting over our life span up to that point, and frequently these effects are *synergistic*—that is, two or more factors combine to produce an effect greater than their simple sum.

For example, plants may be so stressed by pollution that they become more vulnerable to other environmental factors, such as drought or attack by insects. Humans may develop lung disease due to the combined effects of ozone and $NO_x$. Given the complexity of this situation, it is extremely difficult to determine the role of any particular pollutant in causing an observed result. Nevertheless, some significant progress has been made in linking cause and effect.

## Effects on Human Health

The air-pollution disasters in Donora and London have demonstrated that exposure to air pollution can be deadly. Every one of the primary and secondary air pollutants (Table 21–1) is a threat to human health, particularly the health of the respiratory system (Fig. 21–16). *Acute* exposure to some pollutants can be life threatening, but many effects are *chronic,* acting over a period of years to cause a gradual deterioration of physiological functions. Moreover, some pollutants are *carcinogenic,* adding significantly to the risk of lung cancer as they are breathed into the lungs.

**Chronic Effects.** Almost everyone living in areas of urban air pollution suffers from chronic effects. Long-term exposure to *sulfur dioxide* can lead to bronchitis (inflammation of the bronchi). Chronic inhalation of *ozone* can cause inflammation and, ultimately, fibrosis of the lungs, a scarring that permanently impairs lung function. *Carbon monoxide* reduces the capacity of the blood to carry oxygen, and extended exposure to carbon monoxide can contribute to heart disease. Chronic exposure to *nitrogen oxides* impairs lung function and is known to affect the immune system, leaving the lungs open to attack by bacteria and viruses. Exposure to airborne *particulate matter* can bring on a broad range of health problems, including respiratory and cardiovascular pathology.

***Asthma.*** Those most sensitive to air pollution are small children, asthmatics, people with chronic pulmonary or heart disease, and the elderly. Asthma, an immune system disorder characterized by impaired breathing caused by the constriction of air passageways, is brought on by contact with allergens and many of the compounds in polluted air. According to the Clean Air Task Force, ground-level ozone and the smog it generates are responsible for sending 53,000 asthmatics to the hospital each summer. For example, automobile traffic was reduced in Atlanta during the 1996 Summer Olympic Games. A 28% reduction in peak ozone concentrations was achieved, coinciding with a significantly lower rate of childhood asthma episodes. In the last decade, the incidence of asthma in the United States almost doubled, to over 20 million people (including 6 million children) affected.

***Strong Evidence.*** Two important studies recently analyzed by the Health Effects Institute have mounted strong evidence of the harmful effects of fine particles and sulfur pollution. These studies followed thousands of adult subjects living in 154 U.S. cities for as long as 16 years. In the studies, higher concentrations of fine particles were correlated with increased mortality, especially from cardiopulmonary disease and lung cancer. The more polluted the city, the higher was the mortality. The studies were used by the EPA to step up its regulatory attention to fine particles, as we will see later.

One of the heavy-metal pollutants, lead, deserves special attention. For several decades, lead poisoning has been recognized as a cause of mental retardation. Researchers once thought that eating paint chips containing lead was the main source of lead contamination in humans. In the early 1980s, however, elevated lead levels in blood were shown to be much more widespread than previously expected, and they were present in adults as well as children. Learning disabilities in children, as well as high blood pressure in adults, were correlated with levels of lead in the blood. The major source of this widespread contamination was traced to leaded gasoline. (The lead in exhaust fumes may be inhaled directly or may settle on food, water, or any number of items that are put into the mouth.) This knowledge led the EPA to mandate the elimination of leaded gasoline by the end of 1996. The result has been a dramatic reduction in lead concentrations in the environment (Fig. 21–17).

**Acute Effects.** In severe cases, air pollution reaches levels that cause death, although such deaths usually occur among people already suffering from severe respiratory or heart disease or both. The gases present in air pollution are known to be lethal in high concentrations, but such concentrations occur in cases of accidental poisoning only. Therefore, deaths attributed to air pollution are not the direct result of simple poisoning. However, intense air pollution puts an additional stress on the body, and if a person is already in a weakened condition (as for example, the elderly or asthmatics), this additional stress may be fatal. Most of the deaths in air-pollution disasters reflect the acute effects of air pollution. However, recent research shows that even moderate air pollution can cause changes in cardiac rhythms of people with heart disease, triggering fatal heart attacks.

**Carcinogenic Effects.** The heavy-metal and organic constituents of air pollution include many chemicals known to be carcinogenic in high doses. According to the industrial reporting required by the EPA, 4.7 million tons of hazardous air pollutants (air toxics) are released annually into the air in the United States. The presence of trace amounts of these chemicals in air may be responsible

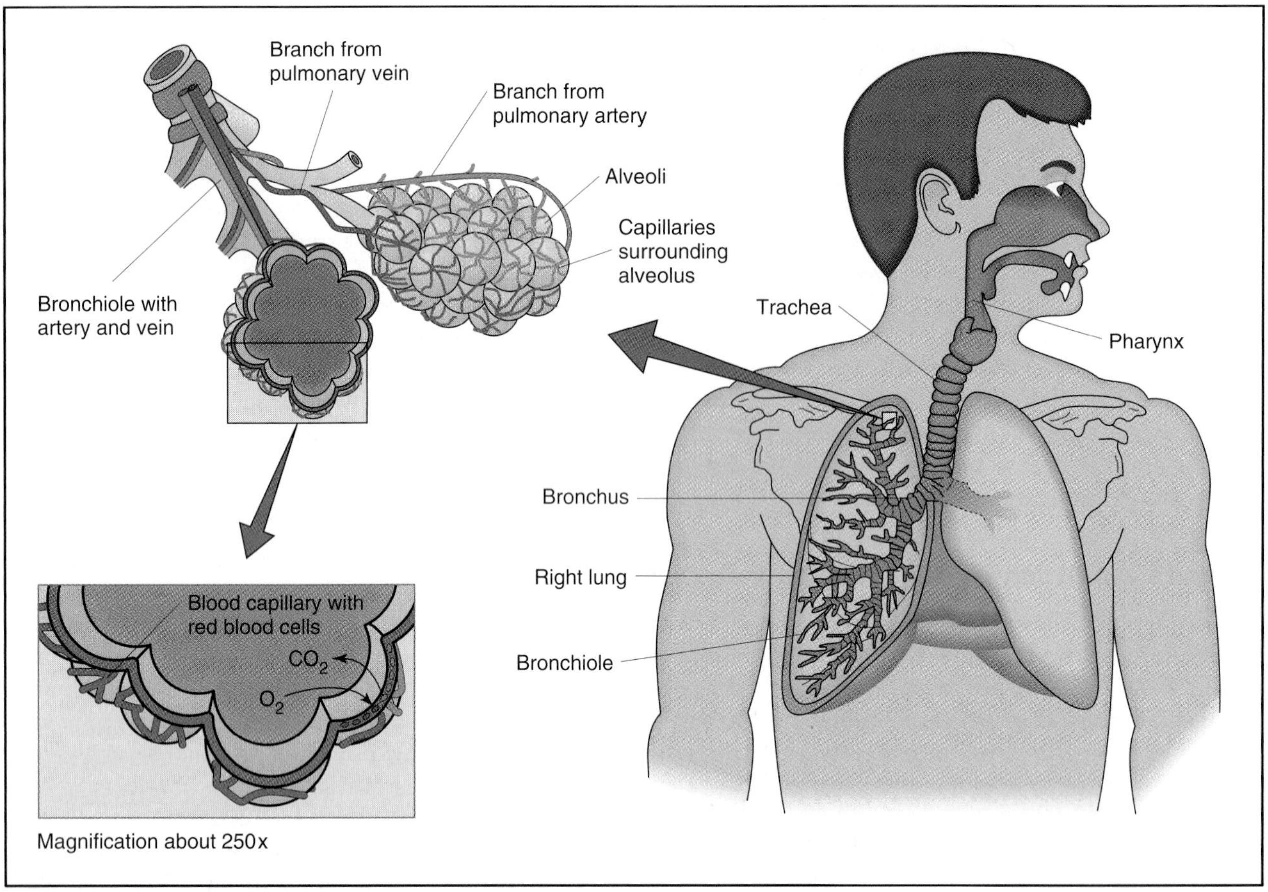


(a)

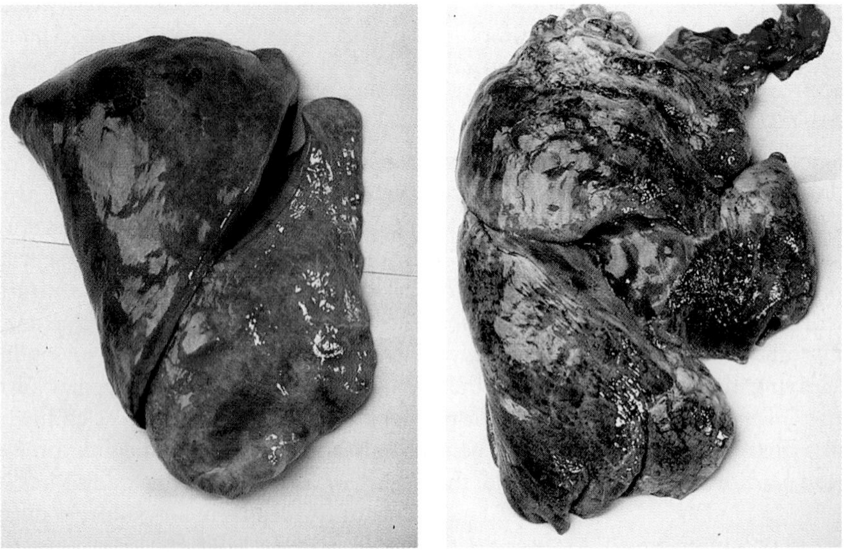

(b)

**Figure 21–16 The respiratory system.** (a) In the lungs, air passages branch and rebranch and finally end in millions of tiny sacs called alveoli. These sacs are surrounded by capillaries. As blood passes through the capillaries, oxygen from inhaled air diffuses into the blood from the alveoli. Carbon dioxide diffuses in the reverse direction and leaves the body in the exhaled air. (b) On the left is normal lung tissue, and on the right is lung tissue from a person who suffered from emphysema, a chronic lung disease in which some of the structure of the lungs has broken down. Cigarette smoking and heavy air pollution are associated with the development of emphysema and other chronic lung diseases.

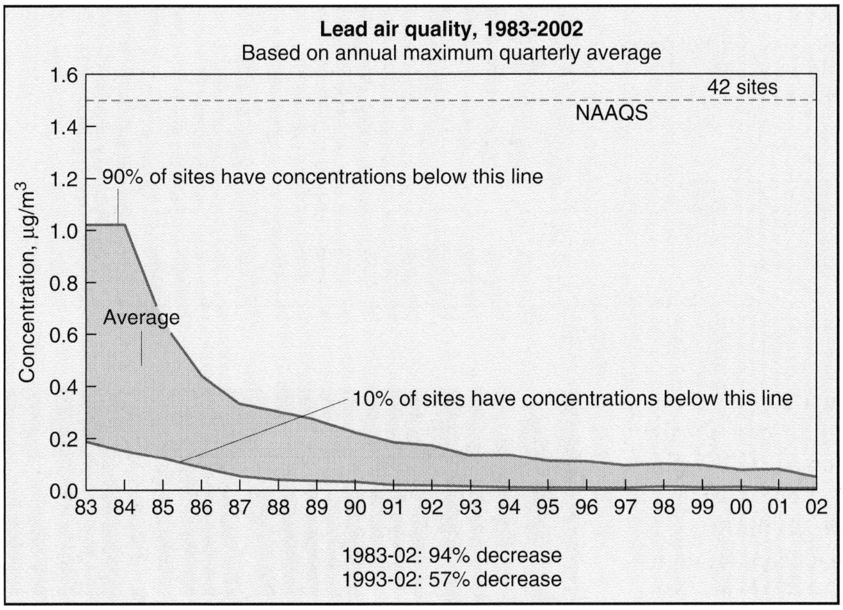


**Figure 21–17 Lead air quality.** Changes in the ambient air concentrations of lead, 1983–2002. (*Source:* EPA Office of Air Quality, *2002 Status and Trends*)

for a significant portion of the cancer observed in humans. One major source of such carcinogens is diesel exhausts. Soot is a known human carcinogen, and the EPA has just classified diesel exhaust as a likely human carcinogen and is moving ahead with regulations that require drastic reductions in the pollutants from this source.

In some cases, exposure to a pollutant can be linked directly to cancer and other health problems by way of epidemiological evidence. One pollutant that clearly and indisputably is correlated with cancer and other disorders is benzene. This organic chemical is present in motor fuels and is also used as a solvent for fatty substances and in the manufacture of detergents, explosives, and pharmaceuticals. Environmentally, benzene is found in the emissions from fossil-fuel combustion—namely, motor-vehicle exhaust and burning coal and oil. Benzene is also present in tobacco smoke, which accounts for half of the public's exposure to the chemical. The EPA has classified benzene as a known human carcinogen, linked to leukemia in persons encountering the chemical through occupational exposure. Chronic exposure to benzene can also lead to numerous blood disorders and damage to the immune system.

**Figure 21–18 Air-pollution damage.** The countryside around this Butte, Montana, smelter is devastated by toxic pollutants from the industrial processes that take place in the smelter.

## Effects on the Environment

Experiments show that plants are considerably more sensitive to gaseous air pollutants than are humans. Before emissions were controlled, it was common to see wide areas of totally barren land or severely damaged vegetation downwind from smelters or coal-burning power plants (Fig. 21–18). The pollutant responsible was usually sulfur dioxide.

*Crop Damage.* The dying off of vegetation in large urban areas and the damage to crops, orchards, and forests downwind of urban centers are caused mainly by exposure to ozone and other photochemical oxidants. Estimates of crop damage by ozone range from \$2 billion to \$6 billion per year, with an estimated \$1 billion in California alone. Crops vary in their susceptibility to ozone, but damage for many important crops (such as soybeans, corn, and wheat) is observed at common ambient levels of ozone (Fig. 21–19). Much of the world's grain production occurs in regions that receive enough ozone pollution to reduce crop yields. The same parts of the world that produce 60% of the world's food—North America, Europe, and eastern Asia—also produce 60% of the world's air pollution.

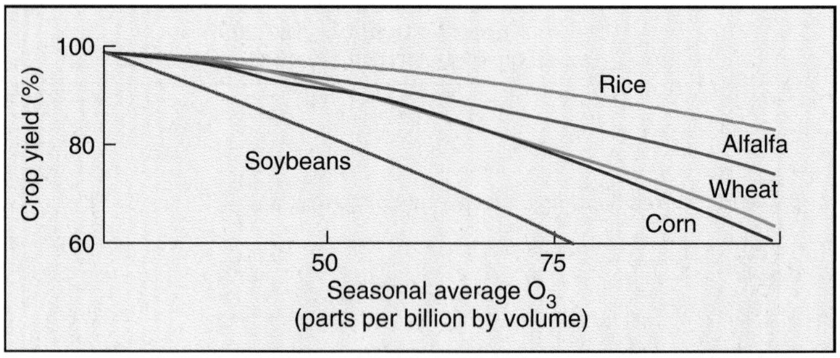


**Figure 21–19 Ozone impact on crop yields.** Some crop species react more strongly to ozone than others. (*Source:* NASA, *Earth Observatory: The Ozone We Breathe,* April 19, 2002)

*Forest Damage.* The negative impact of air pollution on wild plants and forest trees may be even greater than on agricultural crops. Significant damage to valuable ponderosa and Jeffrey pines occurred along the entire western foothills of the Sierra Nevada in California. U.S. Forest Service studies have concluded that ozone from the nearby Central Valley and San Francisco–Oakland metropolitan region was responsible for this damage. Forests under stress from pollution are more susceptible to damage by insects and other pathogens than are unstressed forests. The deaths of the ponderosa and Jeffrey pines in the Sierra Nevada were attributed to western pine beetles, which invade trees weakened by ozone.

Heavy metals, ozone, and acids carried in by clouds from the Midwest were implicated in the death of red spruce in Vermont's Green Mountains. During the 1970s in the San Bernardino Mountains of California, an area that receives air pollution from Los Angeles, 50% of the trees died in some areas. As a result of air-pollution controls, those same areas have shown significant improvement in tree growth in recent years, an encouraging sign.

*Effects on Materials and Aesthetics.* Walls, windows, and other exposed surfaces turn gray and dingy as particulates settle on them. Paints and fabrics deteriorate more rapidly, and the sidewalls of tires and other rubber products become hard and checkered with cracks because of oxidation by ozone. Metal corrosion is increased dramatically by sulfur dioxide and acids derived from sulfur and nitrogen oxides, as are weathering and the deterioration of stonework (Fig. 21–20). These and other effects of air pollutants on materials increase the costs for cleaning or replacing them by hundreds of millions of dollars a year. Many of the damaged materials are irreplaceable.

*Visibility.* A clear blue sky and good visibility are matters of health, but they also carry significant aesthetic value and can have a deep psychological impact on people. Can a value be put on these benefits? Many of us spend thousands of dollars and hundreds of hours commuting long distances to work so that we can live in a less polluted environment than the one in which we work. Ironically, the resulting traffic and congestion cause much of the very pollution we are trying to escape. Then we travel to national parks and wilderness areas and too often find that the visibility in *these* natural areas is impaired by what has been called "regional haze," coming from particulates and gases originating hundreds of miles away. In an encouraging move, the EPA established a Regional Haze Rule in 1999 aimed at improving the visibility at 156 national parks and wilderness areas. The regulations call on all 50 states to establish goals for improving visibility and to develop long-term strategies for reducing the emissions (especially particles) that cause the problem.

**Figure 21–20 Effect of pollution on monuments.** The corrosive effects of acids from air pollutants are dissolving away the features of many monuments and statues, as seen in the faces of these statues in Brooklyn, New York.

**Effects of Acid Deposition.** Acid deposition has been recognized as a problem in and around industrial centers for over 100 years. Its impact on ecosystems, however, was noted only about 40 years ago, when anglers started noticing sharp declines in fish populations in many lakes in Sweden, Ontario, and the Adirondack Mountains of upper New York State. Since that time, as ecological damage has continued to spread, studies have revealed many ways in which acid deposition alters and may destroy ecosystems.

*Impact on Aquatic Ecosystems.* The pH of an environment is extremely critical, because it affects the function of virtually all enzymes, hormones, and other proteins in the bodies of all organisms living in that environment. Ordinarily, organisms are able to regulate their internal pH within the narrow limits necessary to function properly. A consistently low environmental pH, however, often overwhelms the regulatory mechanisms in many life-forms, thus weakening or killing them. Most freshwater lakes, ponds, and streams have a natural pH in the range from 6 to 8, and organisms are adapted accordingly. The eggs, sperm, and developing young of these organisms are especially sensitive to changes in pH. Most are severely stressed, and many die, if the environmental pH shifts as little as one unit from the optimum.

As aquatic ecosystems become acidified (pH 5 and below), higher organisms die off, either because the acidified water kills them or because it keeps them from reproducing. Figure 21–12 shows that acid precipitation may leach aluminum and various heavy metals from the soil as the water percolates through it. Normally, the presence of these elements in the soil does not pose a problem, because they are bound in insoluble mineral compounds and therefore are not absorbed by organisms. As the compounds are dissolved by low-pH water, however, the metals are freed. They may then be absorbed and are highly toxic to both plants and animals. For example, mercury tends to accumulate in fish as lake waters become more acidic. Indeed, mercury levels are so high in the Great Lakes that many bordering states advise against eating fish caught in those waters.

Acid deposition hit some regions especially hard in the latter part of the 20th century. In the southern half of Norway, 19% of fish stocks have been wiped out; of 85,000 lakes in Sweden, 4,000 are severely acidified. In Ontario, Canada, approximately 1,200 lakes lost their fish life. In the Adirondacks, a favorite recreational region for New Yorkers, 346 lakes were without fish in 1990. In New England and the eastern Catskills, over 1,000 lakes have suffered from recent acidification. The physical appearance of such lakes is deceiving. From the surface, they are clear and blue, the outward signs of a healthy condition. However, the only life found below the surface is acid-loving mosses growing on the bottom.

***Buffer, the Acid Slayer.*** As Figure 21–14 indicates, wide regions of the United States receive roughly equal amounts of acid precipitation, yet not all areas have acidified lakes. Apparently, many areas remain healthy, whereas others have become acidified to the point of becoming lifeless. How is this possible? The key lies in the system's *buffering capacity*. Despite the addition of acid, a system may be protected from changes in pH by a **buffer**—a substance that, when present in a solution, has a large capacity to absorb hydrogen ions and thus maintain the pH at a relatively constant value.

Limestone ($CaCO_3$) is a natural buffer (Fig. 21–21) that protects lakes from the effects of acid precipitation in many areas of the North American continent. Lakes and streams receiving their water from rain and melted snow that has percolated through soils derived from limestone will contain dissolved limestone. The regions that are sensitive to acid precipitation are those containing much granitic rock that does not yield good buffers. For these areas, the most critical time of year is the spring thaw, when accumulated winter snow melts. If the thaw is sudden, streams, rivers, and lakes are hit with what has been called "acid shock," as accumulated acids send pH levels plummeting in a sudden burst of meltwater. Making matters worse, the acid shock often coincides with spawning and egg laying in aquatic animals, when they are at their most vulnerable.

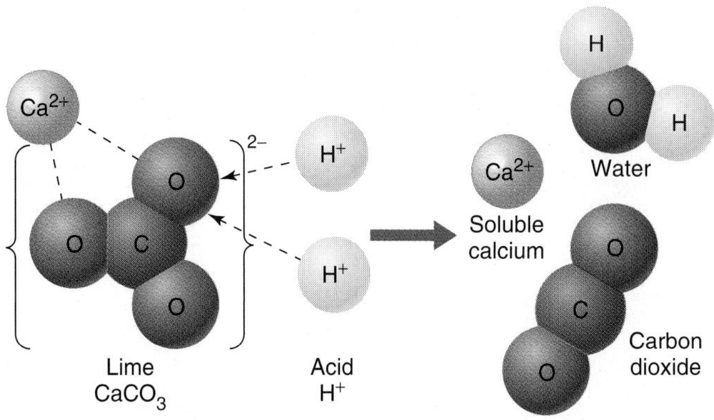


**Figure 21–21 Buffering.** Acids may be neutralized by certain nonbasic compounds called *buffers*. A buffer such as limestone (calcium carbonate) reacts with hydrogen ions as shown. Hence, the pH of a lake or river remains close to neutral despite the additional acid.

Any buffer has limited capacity. Limestone, for instance, is used up by the buffering reaction and so is no longer available to react with more added hydrogen ions. Those ecosystems which had very little buffering capacity have already acidified and collapsed. Those which have greater buffering capacity remain healthy. In recent years, Sweden embarked on a program of liming (with $CaCO_3$) to restore the buffering capacity of lakes, and some 13,500 lakes were saved in that way.

***Impact on Forests.*** From the Green Mountains of Vermont to the San Bernardino Mountains of California, the die-off of forest trees in the 1980s caused great concern. Red spruce forests are especially vulnerable. In New England, 1.3 million acres of high-elevation forests were devastated (Fig. 21–22). Commonly, the damaged trees lost needles as acidic water drew calcium from them, rendering them more susceptible to winter freezing. Sugar maples, important forest trees in the Northeast, have shown extensive mortality, ranging from 20 to 80% of all trees in some forests.

Much of the damage from acid precipitation to forests is due to chemical interactions within the forest soils. Sustained acid precipitation at first adds nitrogen and sulfur to the soils, which stimulate tree growth. In time, though, these chemicals leach out large quantities of the buffering chemicals (usually calcium and magnesium salts). When these buffering salts no longer neutralize the acid rain, aluminum ions, which are toxic, are dissolved from minerals in the soil. The combination of aluminum and the increasing scarcity of calcium, which is essential to plant growth, leads to reduced tree growth. Research at the Hubbard Brook Experimental Forest in the White Mountains of

**Figure 21–22 Effect of pollution on forests.** Heavy metals and acids from distant urban areas were responsible for the deaths of red spruce trees in Vermont's Green Mountains.

New Hampshire has shown a marked reduction in calcium and magnesium in the forest soils from the 1960s on, which is reflected in the amount of calcium in tree rings over the same period. The net result of these changes has been a severe decline in forest growth. Experimentally liming soils in forests showing maple tree dieback was able to restore the trees to health, and aluminum ion concentrations were seen to decrease greatly.

*In Europe.* Dying forest trees (**waldsterben**) are a serious problem in many parts of Europe, and the evidence indicates that the same kinds of soil–chemical exchanges are occurring there. Because of variations in the buffering capacity of soils and the differing amounts of sulfur and nitrogen brought in by acid precipitation, forests are affected to varying degrees. Some forests continue to grow, whereas many decline. One result of sustained acidification is a gradual shift toward more acid-tolerant species. For example, in New England, the balsam fir is moving in to replace the dead spruce.

**Impact on Humans and Their Artifacts.** Limestone and marble (which is a form of limestone) are favored materials for the outsides of buildings and for monuments (collectively called **artifacts**). The reaction between acid and limestone is causing these structures to erode at a tremendously accelerated pace. Monuments and buildings that have stood for hundreds or even thousands of years with little change are now dissolving and crumbling away, as seen in Fig. 21–20. The corrosion of buildings, monuments, and outdoor equipment by acid precipitation costs billions of dollars for replacement and repair each year in the United States. In an extreme case of corrosion, a 37-foot bronze Buddha in Kamakura, Japan, is slowly dissolving away as precipitation from Korea and China bathes the statue in rain that is more acidic than that found in the United States.

Although the decay of such artifacts is a tragic loss in itself, it should also stand as a grim reminder of how we are dissolving away the buffering capacity of ecosystems. In addition, some officials are concerned that acid precipitation's mobilization of aluminum and other toxic elements may result in the contamination of both surface water and groundwater. Increased acidity in water also mobilizes lead from the pipes used in some old plumbing systems and from the solder used in modern copper systems. As the song says, "What goes up must come down."[1] The sulfur and nitrogen oxides pumped into the troposphere in the United States—some 11.4 million metric tons in 1900—gradually increased and reached their peak in 1973, at 53 million metric tons. These pollutants have come back down as acid deposition, generally to the east of their origin, because of the way weather systems flow. The deposits cross national boundaries, too. As a result, Canada receives half of its acid deposition from the United States, and Scandinavia gets it mostly from Great Britain and other Western European nations. Japan receives the windborne pollution from widespread coal burning in Korea and China. There is now a broad scientific consensus that the problem of acid deposition must be addressed at national and international levels. Accordingly, we turn next to the efforts being made to curb air pollution.

## 21.4 Bringing Air Pollution under Control

By the 1960s, it was obvious that pollutants produced by humans were overloading natural cleansing processes in the atmosphere. The unrestricted discharge of pollutants into the atmosphere could no longer be tolerated.

*Clean Air Act.* Under grassroots pressure from citizens, the U.S. Congress enacted the **Clean Air Act of 1970 (CAA).** Together with amendments passed in 1977 and 1990, this law, administered by the EPA, represents the foundation of U.S. air-pollution control efforts. The act calls for identifying the most widespread pollutants, setting **ambient standards**—levels that need to be achieved to protect environmental and human health—and establishing control methods and timetables to meet the standards.

*NAAQS.* The CAA mandated the setting of standards for four of the primary pollutants—particulates, sulfur dioxide, carbon monoxide, and nitrogen oxides—and for the secondary pollutant ozone. At the time, these five pollutants were recognized as the most widespread and objectionable ones. Today, with the addition of lead, they are known as the **criteria pollutants** and are covered by the **National Ambient Air Quality Standards (NAAQS)** (Table 21–3). The **primary standard** for each

[1]Brook Benton, "It's Just a Matter of Time," Mercury Records, 1959; Blood, Sweat, and Tears, "Spinning Wheel," Columbia Records, 1969.

**table 21-3 National Ambient Air Quality Standards for Criteria Pollutants**

| Pollutant | Averaging Time[1] | Primary Standard |
|---|---|---|
| $PM_{10}$ particulates | 1 year<br>24 hours | 50 $\mu g/m^3$<br>150 $\mu g/m^3$ |
| $PM_{2.5}$ particulates[2] | 1 year<br>24 hours | 15 $\mu g/m^3$<br>65 $\mu g/m^3$ |
| Sulfur dioxide | 1 year<br>24 hours | 0.03 ppm<br>0.14 ppm |
| Carbon monoxide | 8 hours<br>1 hour | 9 ppm<br>35 ppm |
| Nitrogen oxides | 1 year | 0.053 ppm |
| Ozone | 8 hours<br>1 hour | 0.08 ppm<br>0.12 ppm |
| Lead | 3 months | 1.5 $mg/m^3$ |

[1] The averaging time is the period over which concentrations are measured and averaged.
[2] $PM_{2.5}$ is the particulate fraction having a diameter smaller than or equal to 2.5 micrometers. It supplements the existing $PM_{10}$ standard.
*Source:* EPA's Draft Report on the Environment 2003.

pollutant is based on the presumed highest level that can be tolerated by humans without noticeable ill effects, minus a 10% to 50% margin of safety. For many of the pollutants, long-term and short-term levels are set. The short-term levels are designed to protect against acute effects, while the long-term standards are designed to protect against chronic effects.

Table 21–3 reflects two standards announced by EPA in July 1997: the new $PM_{2.5}$ category and a revision of the ozone standard from 0.12 ppm for 1 hour to 0.08 ppm for 8 hours. These changes were established in response to health studies which indicated that the modifications could improve lung function substantially and prevent early death.

*NESHAPs.* In addition, **National Emission Standards for Hazardous Air Pollutants (NESHAPs)** have been issued for eight toxic substances: arsenic, asbestos, benzene, beryllium, coke-oven emissions, mercury, radionuclides, and vinyl chloride. The Clean Air Act of 1990 greatly extended this section of the EPA's regulatory work by specifically naming 188 toxic air pollutants for the agency to track and regulate.

## Control Strategies

The basic strategy of the 1970 Clean Air Act was to regulate the *emissions* of air pollutants so that the *ambient* criteria pollutants would remain below the primary standard levels. This approach is called **command and control,** because industry was given regulations to achieve a set limit on each pollutant, to be accomplished by specific control equipment. The assumption was that human and environmental health could be significantly improved by a reduction in the output of pollutants. If a particular region was in violation for a given pollutant, a local government agency would track down the source(s) and order reductions in emissions until the region came into compliance.

Unfortunately, this strategy proved difficult to implement. Most of the regulatory responsibility fell on the states and cities, which were often unable or unwilling to enforce control. In 1990, 230 areas violated the standards. Even today, after 34 years of legislated air pollution control, 124 metropolitan areas still fail to meet the standards. The good news is that total air pollutants were reduced by some 48% during a time when both population and economic activity increased substantially and even in the noncompliant regions the severity of air pollution episodes lessened.

***Reducing Particulates.*** Prior to the 1970s, the major sources of particulates were industrial stacks and the open burning of refuse. The CAA mandated the phaseout of open burning of refuse and required that particulates from industrial stacks be reduced to "no visible emissions."

The alternative generally taken to dispose of refuse was landfilling, a solution that has created its own set of environmental problems (discussed in Chapter 18). To reduce stack emissions, many industries were required to install filters, electrostatic precipitators, and other devices. Unfortunately, the solid wastes that are removed from exhaust gases frequently contain heavy metals and other toxic substances (Chapter 19). Although these measures have markedly reduced the levels of particulates since the 1970s, particulates continue to be released from steel mills, power plants, cement plants, smelters, construction sites, diesel engines, and so on. Wood-burning stoves and wood and grass fires also contribute to the particulate load, making regulation even more difficult.

The EPA added a new ambient air quality standard for particulates ($PM_{2.5}$) in 1997, on the basis of information that smaller particulate matter (less than 2.5 micrometers in diameter) has the greatest effect on health, because the finer particulate matter tends to originate from combustion processes and from atmospheric chemical reactions between pollutants. The coarser material ($PM_{10}$) often comes from windblown dust, and although it is still regulated, it is not considered to play as significant a role in health problems as do the finer particles. Because of legal challenges and other delays, this standard is expected to be implemented in 2004.

*1990 Amendments.* The **Clean Air Act Amendments (CAAA)** of 1990 target specific pollutants more directly and enforce compliance more aggressively, through such means as the imposition of sanctions. As with the earlier law, the states do much of the work in carrying out the mandates of the 1990 act. Each state must develop a *State Implementation Plan (SIP)* that is required to go through a process of public comment before being submitted to the EPA for approval. The SIP is designed to reduce emissions of every NAAQS pollutant whose control standard (Table 21–3) has not been attained. One major change is a permit application process (already in place for the release of pollutants into waterways). Polluters must apply for a permit that identifies the kinds of pollutants they release, the quantities of those pollutants, and the steps they are taking to reduce pollution. Permit fees provide funds the states can use to support their air-pollution control activities. The amendments also afford more flexibility than the earlier command-and-control approach, by allowing polluters to choose the most cost-effective way to accomplish the goals. In addition, the legislation uses a market system to allocate pollution among different utilities.

Under the CAAA, regions of the United States that have failed to attain the required levels must submit *attainment plans,* based on **reasonably available control technology (RACT)** measures. Offending regions must convince the EPA that the standards will be reached within a certain time frame.

## Limiting Pollutants from Motor Vehicles

Cars, trucks, and buses release nearly half of the pollutants that foul our air. Vehicle exhaust sends out VOCs, carbon monoxide, and nitrogen oxides that lead to ground-level ozone and PANs. Additional VOCs come from the evaporation of gasoline and oil vapors from fuel tanks and engine systems. The CAA mandated a 90% reduction in these emissions by 1975. This timing proved to be unrealistic, but enough improvements have been made over the years that a new car today emits 75% less pollution than did pre-1970 cars (Fig. 21–23). This improvement is fortunate, because driving in the United States has been increasing much more rapidly than the population has. Between 1970 and 2000, the number of vehicle miles increased from 1 trillion to 2.75 trillion miles per year, and between 1980 and 2000, the number of vehicles on the road increased over 44%. (See inset data in the figure.) It is hard to imagine what the air would be like without the improvements mandated by the CAA.

The reductions in automobile emissions have been achieved with a general reduction in the size of passenger vehicles, along with a considerable array of pollution control devices, among which is one that affords the

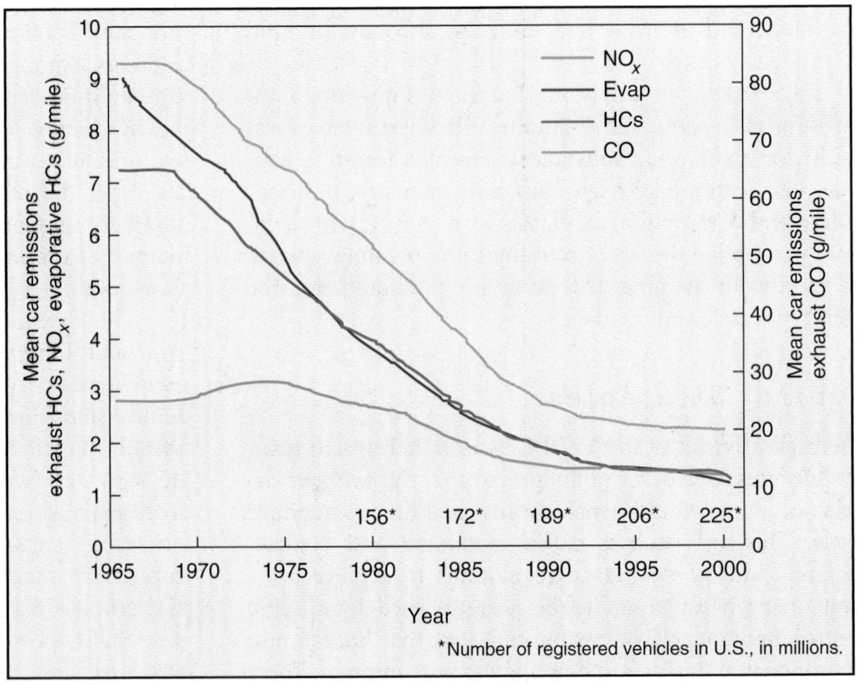


**Figure 21–23 Trends in automobile emissions.** Average car emissions from vehicles, in grams per vehicle mile traveled, from 1965 to (estimated) 2000. Note the numbers of vehicles on the road in the United States from 1980 to 2000. (*Sources: Science,* 261, p. 39. July 2, 1993; U.S. Department of Commerce Web site)

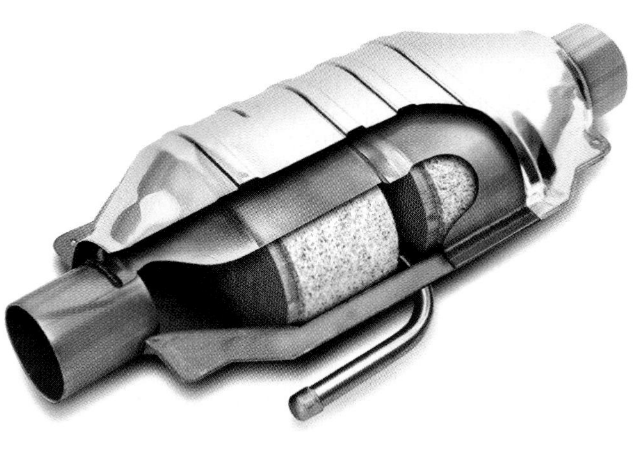

**Figure 21–24 Cutaway of a catalytic converter.** Engine gas is routed through the converter, where catalysts promote chemical reactions that change the harmful hydrocarbons, carbon monoxide, and nitrogen oxides into less harmful gases, such as carbon dioxide, water, and nitrogen.

computerized control of fuel mixture and ignition timing, allowing more complete combustion of fuel and decreasing VOC emissions. To this day, however, the most significant control device on cars is the **catalytic converter** (Fig. 21–24). As exhaust passes through this device, a chemical catalyst made of platinum-coated beads oxidizes most of the VOCs to carbon dioxide and water. The catalytic converter also oxidizes most of the carbon monoxide to carbon dioxide. Although newer converters reduce nitrogen oxides as well, the level of reduction is not impressive.

*CAAA Changes.* Despite these efforts, the continuing failure to meet standards in many regions of the United States, as well as the aesthetically poor air quality in many cities, made it clear during the 1980s that further legislation was necessary. The following are the highlights of the CAAA of 1990 on motor vehicles and fuels:

1. New cars sold in 1994 and thereafter were required to emit 30% less VOCs and 60% less nitrogen oxides than cars sold in 1990. Emission-control equipment had to function properly (as represented in warranties) to 100,000 miles. Buses and trucks were required to meet more stringent standards. Also, the EPA was given the authority to control emissions from all nonroad engines that contribute to air pollution, including lawn and garden equipment, motorboats, off-road vehicles, and farm equipment.
2. Starting in 1992 in the regions with continuing carbon monoxide problems, oxygen was to be added to gasoline in the form of MTBE or alcohols, to stimulate more complete combustion and to cut down on carbon monoxide emissions.
3. Before 1990, only a few states and cities required that vehicle inspection stations be capable of measuring emissions accurately. The new amendments required more than 40 metropolitan areas to initiate inspection and maintenance programs, while other cities were to improve their programs.

Two factors now affect fuel efficiency and consumption rates. First, the elimination of federal speed limits in 1996 reduced fuel efficiencies because of the higher speeds (and increased the rate of traffic fatalities). Second, the sale of sport utility vehicles (SUVs) and light trucks has surged since 1990. In 2001, these vehicles made up 51% of all vehicle sales (in 1980, it was only 20%). In city traffic, SUVs get 10 to 16 mpg; on the highway, they get no higher than 22 mpg. To compound the problem, as fuel efficiency decreases, tailpipe emissions increase.

*CAFE Standards.* Under the authority of the **Energy Policy and Conservation Act of 1975** and its amendments, the DOT was given the authority to set *corporate average fuel economy (CAFE)* standards for motor vehicles. The intention of the law was to conserve oil and promote energy security, but reducing gasoline consumption also means less air pollution. The current standard for passenger cars requires a fleet average of 27.5 miles per gallon (mpg). The standard for light trucks (pickups, SUVs, minivans) is set lower, at 20.7 mpg. Actual 2003 vehicle performances in both categories are several mpg lower than the CAFE standards. (For comparison, the fleet fuel economy in 1974 was only 12.9 mpg!). Intense lobbying by the automobile and petroleum industries prevented more stringent standards of fuel efficiency from being included in the CAAA of 1990. After years during which the DOT was prohibited by Congress from even studying new CAFE rulemaking, growing concern about the increasing dominance of "light trucks" in the national fleet led to a new rule raising the light-truck standard to 22.2 mpg, to be phased in by 2007.

**Managing Ozone.** Because ozone is a secondary pollutant, the only way to control ozone levels is to address the compounds that lead to its formation. For a long time, it was assumed that the best way to reduce ozone levels was simply to reduce emissions of VOCs. The steps pertaining to motor vehicles in the Clean Air Act Amendments of 1990 address the sources of about half of the VOC emissions. **Point sources** (industries) account for another 30% of such emissions, and **area sources** (numerous small emitters, such as dry cleaners, print shops, and users of household products) represent the remaining 20%. RACT measures have already been mandated for many point sources, and much progress has been made through EPA, state, and local regulatory efforts to reduce emissions from those sources. In the last 20 years, VOC emissions have declined by 35%.

However, recent understanding of the complex chemical reactions involving $NO_x$, VOCs, and oxygen has thrown some uncertainty into this strategy of emphasizing a reduction in VOCs. The problem is that *both* $NO_x$ and VOC concentrations are crucial to the generation of ozone. Either one or the other can become the rate-limiting species in the reaction that forms ozone. (See Fig. 21–11b.) Thus, as the ratio of VOCs to $NO_x$ changes, the concentration of $NO_x$ can become the controlling chemical factor because of an excess of VOCs. This happens more commonly in air-pollution episodes that occur over a period of days than in the daily photochemical smog of the urban city.

***New Ozone Standard.*** The revised ozone standard (0.12 to 0.08 ppm) announced in 1997 met with strong opposition from industry groups, which obtained a court injunction in 1999 prohibiting the EPA from enforcing the new standard. However, the U.S. Supreme Court has upheld the EPA in the dispute. This and other delays have put off implementation to 2004. Cost–benefit estimates indicated that the anticipated health benefits far outweighed the costs of compliance (estimated at $9.7 billion per year). The new ozone standards are also expected to prevent a substantial amount of damage to vegetation by ozone.

***Down with $NO_x$.*** In a response to petitions from states in the Northeast that were having trouble with ozone and smog because of out-of-state emissions, the EPA established the *Ozone Transport Rule,* which sets $NO_x$ budgets for Midwestern and Southern states (each region home to hundreds of coal-fired power plants) that, if carried out, would reduce their $NO_x$ emissions by 75 to 85%. The EPA has also issued further regulations designed to address urban and suburban smog and ozone levels. Under the new *Tier 2 Standards,* which will be phased in gradually between 2004 and 2009, emissions for all SUVs, light trucks, and passenger vans will be held to the same standards as those for passenger cars. For $NO_x$, this means a gradual reduction to 0.07 gram per mile, a 90% reduction over current passenger car standards. Because the new technologies will affect the sticker price of the already expensive SUVs, the auto industry opposes the regulations. Further, by 2004, gasoline manufacturers will have to cut the sulfur content of gasoline from 300 ppm to 30 ppm and of diesel fuel down to 15 ppm (opposed by the refinery industry, of course). The EPA believes that the combined effects of these new regulations should remove a further 2 million tons of pollutants emitted annually.

**Controlling Toxic Chemicals in the Air.** By EPA estimates, the total amount of toxic substances emitted into the air in the United States is around 4.7 million tons annually. Cancer and other adverse health effects, environmental contamination, and catastrophic chemical accidents are the major concerns associated with this category of pollutants. Under the CAAA of 1990, Congress identified 188 toxic pollutants. It then directed the EPA to identify major sources of these pollutants and to develop **maximum achievable control technology (MACT)** standards. Besides affecting control technologies, the standards include options for substituting nontoxic chemicals, giving industry some flexibility in meeting MACT goals. State and local air-pollution authorities will be responsible for seeing that industrial plants achieve the goals. In response to the CAAA, the EPA developed the NTI to track emissions of the affected substances. To reduce the contribution from vehicles, the agency is requiring cleaner burning fuels in urban areas. Industrial sources are being addressed by setting emission standards for some 82 stationary-source categories (e.g., paper mills, oil refineries). The program is expected to reduce emissions by at least 1.5 million tons annually.

## Coping with Acid Deposition

Scientists working on acid-rain issues calculated that a 50% reduction in acid-causing emissions in the United States would effectively prevent further acidification of the environment. This reduction was not expected to correct the already bad situations, but, together with natural buffering processes, it was estimated to be capable of preventing further environmental deterioration. Because we know that about 50% of acid-producing emissions come from coal-burning power plants, control strategies focused on these sources.

**Political Developments.** Although evidence of the link between power-plant emissions and acid deposition was well established by the early 1980s, no legislative action was taken until 1990. The problem was one of different regional interests. Western Pennsylvania and the states of the Ohio River valley, where older coal-burning power plants produce most of the electrical power of the region, argued that controlling their sulfur dioxide emissions would make electricity unaffordable in the region. Throughout the 1980s, a coalition of politicians from these states, fossil-fuel corporation representatives of high-sulfur coal producers, and representatives from the electric power industry effectively blocked all attempts at passing legislation that would take action on acid deposition.

On the other side of the issue were New York and the New England states, as well as most of the environmental and scientific community, which argued that it was both possible and necessary to address acid deposition and that the best way to do so was to control emissions from power plants. Also, since 70% of Canada's acid-deposition problem came from the United States, diplomatic pressure toward a resolution was applied.

***Action.*** With the passage of Title IV of the CAAA of 1990, the outcome of the controversy became history. However, two decades of action were lost because of political delays. In the wake of the passage of the act, Canada and the United States signed a treaty stipulating that Canada would cut its sulfur dioxide emissions by half and cap them at 3.2 million tons by the year 2000. (U.S. obligations from the treaty are discussed shortly.) Canada has done well; its emissions are now about 2.7 million tons and are expected to stay down. As a result of several treaties in Europe, sulfur dioxide emissions there are down 40% from 1980 levels and should decline another 18% by 2010 if the signatories fulfill their obligations. The Europeans also targeted nitrogen oxide emissions, with the goal of a 30% reduction by 1999.

The main beneficiary of these agreements is the remaining healthy aquatic and forested ecosystems, which, if all goes according to plan, will be protected from future acid deposition. In addition, it is hoped that ecosystems already harmed will be able to recover from current damage and that we will be back on the road to a sustainable interaction with the atmosphere.

**Title IV of the Clean Air Act Amendments of 1990.** Title IV of the CAAA is the first law in U.S. history to address the acid-deposition problem, by mandating reductions in both sulfur dioxide and nitrogen oxide levels. The major provisions of Title IV are as follows:

1. By the year 2010, total $SO_2$ emissions must be reduced 10 million tons below 1980 levels. This is the 50% reduction called for by scientists, and it involves the setting of a permanent *cap* of 8.9 million tons on such emissions. The reduction will be implemented in phases.
2. In a major departure from the command-and-control approach, Title IV authorizes the EPA to use a free-market approach to regulation. Each plant is granted **emission allowances** based on formulas in the legislation. One allowance permits plants to emit 1 ton of $SO_2$ each. The plants are free to choose how they will achieve their allowances. For example, they can reduce emissions, or they can buy allowances from other utilities that are under their own allowances in emissions.
3. In the future, new utilities will not receive allowances. Instead, they will need to buy into the system by purchasing existing allowances. Thus, there will be a finite number of allowances.
4. Nitrogen oxide emissions from power plants were to be reduced by 2 million tons by the year 2001. This was to be accomplished by regulating the boilers used by the utilities and by mandating the continuous monitoring of emissions.

*Accomplishments of Title IV.* The utilities industry has responded to the new law with three actions:

1. Many utilities are switching to low-sulfur coal, available in Appalachia and the western United States.
2. Many older power plants are adding scrubbers—"liquid filters" that put exhaust fumes through a spray of water containing lime. $SO_2$ reacts with the lime and is precipitated as calcium sulfate ($CaSO_4$). Since the technology for high-efficiency scrubbing is well established, more and more power plants are installing scrubbers, which have been required for all coal-burning power plants built after 1977.
3. Many utilities are trading their emission allowances. At current prices, the purchase of allowances often represents a less costly way to achieve compliance than by purchasing low-sulfur coal or adding a scrubber. A typical transaction might involve a trade in the rights to emit, say, 10,000 tons of sulfur dioxide at a cost of $200 a ton. The combination of approaches has created so many efficiencies that it has cut compliance costs to 10% of what was expected.

To carry out Title IV of the CAAA, the EPA initiated a two-phase approach. Phase I addressed 398 of the largest and highest-emitting coal-fired power plants, aiming at a reduction of 3.5 million tons of $SO_2$ by 2000. Remarkably, the goal has been met and even exceeded—at a cost far below the gloomy predictions of the industries (Fig. 21–25). Phase II began in 2000 and targets the

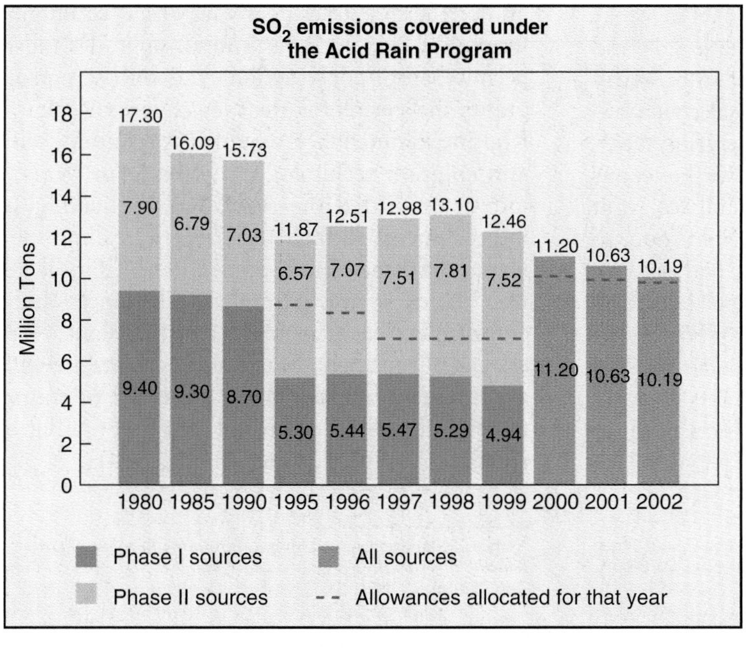


**Figure 21–25 $SO_2$ emissions covered under the Acid Rain Program.** Under Title IV of the CAAA of 1990, emissions from Phase I coal-fired power plants were 1.6 million tons below the targeted amount. Phase II sources were still above allowances, but declining. (*Source:* EPA Office of Air Quality, *2002 Status and Trends*)

remaining sources of $SO_2$ in order to reach the 8.9 million-ton cap by 2010.

***$NO_x$.*** The 2001 goal for a 2 million-ton reduction in $NO_x$ emissions has not been reached. Emissions had decreased by 1.5 million tons by 2002, and the trend for the power-plant sources is definitely downward. However, these sources account for only 22% of all $NO_x$ emissions, and the trend in total $NO_x$ emissions has been a much slower decline than that for $SO_2$.

***Field News.*** The news from the field reflects the trends. Concentrations of sulfate in rain and deposition on the land have shown a significant decline (10% to 25%) over a large part of the eastern United States in the last six years. The decline is likely due to the Phase I emission reductions. However, the forests, streams, and fish of the Adirondacks and the White Mountains of New Hampshire have shown very few signs of recovery. One reason is the continued impact of nitrogen deposition. Apparently, nitrogen plays a much larger role in acid deposition than was once believed. In passing the CAAA, Congress did not set curbs on nitrogen emissions, but simply opted to reduce them by a minimal 2 million tons. The other reason that little progress has been made in recovery is the long-term buildup of sulfur deposits in soils. Residual sulfur takes a long time to flush from natural ecosystems. Accordingly, most scientists believe that the recovery of the affected ecosystems will require further reductions in emissions of both sulfur and nitrogen. In a recent article,[2] 10 leading acid-rain researchers evaluated the current conditions of the northeastern U.S. forests, lakes, and streams and concluded that further cuts of 80% in $SO_2$ emissions would be needed in order to achieve full recovery of the affected ecosystems. Acid rain still falls on eastern forests, as Figure 21–14 indicates.

## 21.5 Unresolved Issues

We have air pollution because we want the goods and services that inevitably generate it. We want to be warm or cool, depending on the weather; we use electricity to power our homes, institutions, and manufacturing facilities; we crave freedom of movement, so we have our motor vehicles, trains, boats, and planes; and we want the array of goods produced by industries, from food, to DVDs, to computers, to toys—the list is endless. We embrace all of these because we derive some benefit from them. The benefits may be essential or trivial, but we are accustomed to having them, as long as we can pay for them. Satisfying all of these needs and wants is what drives the American economy—and it produces millions of tons of pollutants in the process.

***Costs versus Benefits.*** In recent decades, we have been learning that air pollution does not have to get worse—that, in fact, it can be remedied enough to keep most of us from getting sick and the environment from being degraded. But the remedies come with a price. Without question, measures taken to reduce air pollution carry an economic cost. Some critics have charged that air pollution controls are not cost effective; that is, the benefits are not nearly as great as the costs. They see lost opportunities for economic growth and tend to disregard the *costs avoided* (from improved health). In a surprising development, the Office of Management and Budget (OMB), a White House agency, declared that environmental regulations are good for the economy. The 2003 report to Congress[3] focused especially on new clean-air regulations and found that the benefits in terms of reduced hospital visits, fewer premature deaths, and fewer lost workdays were five to seven times greater than the costs of complying with the rules. In perhaps the most complete study ever of costs and benefits of regulatory decision making, the OMB found that the yearly benefits of environmental regulations ranged from \$121 to \$193 billion, while the costs ranged from \$37 to \$43 billion. This work confirms former analyses of the benefits of the Clean Air Act. The OMB report is a striking contrast to a perceived tendency on the part of the Bush administration to roll back environmental regulations or at least loosen them. The New Source Review controversy is a prime example of this tendency.

***New Source Review.*** The Clean Air Act explicitly requires all power plants and other polluting industrial facilities built after 1970 to incorporate "best available" pollution control technology, such as modern scrubbers. Rather than impose large and immediate costs on industry and the consumer, the CAA exempted all older facilities from having to install new pollution controls and chose rather to require them to do so only when they upgraded their facilities in any way that would increase pollution. All new facilities and all upgrades must obtain a permit before construction, and the EPA program that grants the permits is the New Source Review Program. Routine maintenance would be permitted, with no new antipollution technology required. For years, however, industries skirted the regulation by making substantial improvements to their facilities and calling the improvements "routine maintenance." The Clinton administration cracked down on this practice and brought enforcement cases against 51 power-plant operators and dozens of refineries. Many of these suits are still pending, and quite a few have been settled out of court, with the utilities agreeing to install effective pollution-control technology in their plants (Table 21–4).

---

[2]Driscoll, Charles T. et al. "Acidic Deposition in the Northeastern United States: Sources and Inputs, Ecosystem Effects, and Management Strategies." *Bioscience* 51 (March 2001): 180–198.

[3]Office of Management and Budget. *Informing Regulatory Decisions: Costs and Benefits of Federal Regulations and Unfunded Mandates on State, Local, and Tribal Entities.* OMB, 2003.

**table 21-4 Status of Some New Source Review Enforcement Cases**

| Company | Enforcement Status |
|---|---|
| American Electric Power Co. | Trial scheduled for early 2004 |
| Cinergy Corp. | Oct. 2004 deadline for settlement or trial |
| Duke Energy Corp. | Trial begun in late 2003 |
| Dynegy/Illinois Power | Awaiting court decision |
| FirstEnergy Corp./Ohio Edison Co. | Guilty verdict Aug. 7, 2003; awaiting penalty. |
| Southern Co. | Awaiting trial date |
| Vectren Corp. | $33 million settlement reached in June 2003 |
| TECO Energy Inc./Tampa Electric | $1 billion settlement reached in February 2000 |
| Tennessee Valley Authority | Justice Department is appealing a ruling dismissing the case |
| Dominion Virginia Power | $1.2 billion settlement reached in April 2003 |
| PSEG | $337 million settlement reached January 2002 |
| Wisconsin Electric Corp. | $600 million settlement reached April 2003 |

*Source:* Public Citizen; note that unsettled cases are in jeopardy because of changes in the New Source Review rules in late 2003.

Prompted by Vice President Cheney's energy task force (see Chapter 12), the EPA announced in August 2003 that it would change the rules for the New Source Review. Factories and power plants would no longer be required to update their pollution controls, unless the improvements or changes involved more than 20% of the entire facility's value. The EPA's position is that "routine maintenance" is too vague a criterion. Instead, industry needed a more precise definition of when the New Source Review process would be triggered. Many state officials and environmental groups were outraged at the change and warned that the change in regulations would allow some 17,000 facilities (including over 500 older coal-fired power plants) across the country to increase their pollution emissions simply by upgrading 20% or less at a time. They charged that the Bush administration was rewarding the utility industry for its multi-million-dollar contributions to the 2000 presidential and congressional Republican campaigns. In establishing its energy recommendations, Cheney's energy task force solicited most of its advice from the fossil-fuel and utility lobbies. The new rule was praised by the National Association of Manufacturers, the American Petroleum Institute, and other groups that lobby for the industries affected by the Clean Air Act. However, 14 states have mounted a legal challenge to the new rule, and in December, 2003, a federal court issued a "stay" preventing the rule from going into effect. At stake are billions of dollars of pollution-control upgrades that the utility industry could now avoid making. One immediate impact of the change was to put all of the Justice Department's pending enforcement cases in limbo. In fact, in November 2003, the EPA announced that, because of the changed policy, the agency would drop its investigations into 50 power plants for past violations.

***Clear Skies.*** The Bush administration has introduced legislation (the Clear Skies Act) that would change the CAA regulations for power plants by addressing three major pollutants simultaneously: $SO_2$, $NO_x$, and mercury. Clear Skies would use the same "cap-and-trade" strategy currently in place for $SO_2$ in the acid-rain legislation of the CAA. The 2003 version of the legislation would cut $SO_2$ emissions to a cap of 3 million tons, $NO_x$ emissions to a cap of 1.7 million tons, and mercury to a cap of 15 tons, all of these amounts to be phased in gradually until 2018. The proposed legislation would certainly help states achieve the NAAQS for particulates and ozone and would go far toward further addressing the continuing acid-rain problems. The EPA calculates the costs to be around $6.3 billion by 2020 and the health benefits alone to total $110 billion a year, far greater than the costs. The prospects for this legislation are uncertain.

***Getting Around.*** The attention given to pollution from power plants and industries is necessary and important, but half of the major air pollutants come from vehicles. Raising the CAFE standards would help address (1) our dependence on imported oil (and all that that implies), (2) all of the health issues from smog and particulates, and (3) carbon dioxide emissions that are bringing on global climate change. Yet, the automakers and the auto industry unions, the fossil-fuel industry, and politicians from Michigan and rural states all continue to resist raising CAFE mileage requirements. America's love affair with pickups, vans, and SUVs continues unabated. Still, there are *some* signs of change.

California, which by all measures has the greatest problem with photochemical smog and vehicular pollutants, passed a law in 1990 requiring that, by 2003, 10% of all new vehicles sold in the state be "emission free"—in other words, powered by electricity. The state has been backing off of that law and, in 2003, completely rescinded it. Problems with electric cars have soured California and other states that had been looking in that direction. Electric cars are considerably lighter than conventional cars and quite limited in range (traveling 50 to 100 miles before needing recharging). They also lack such amenities as air-conditioning and other power accessories. Moreover, switching from gasoline to electrical power simply transfers the site of pollution emission from the moving vehicle to the power plant—a trade-off with uncertain consequences, unless we make strides in renewable-energy sources.

California has two near-term choices for addressing the state's air-pollution problems. One is the hybrid electric vehicle, which is an electric car with a small internal combustion engine equipped with an electric generator to charge the batteries. Such vehicles are already being marketed by Honda and Toyota and are the most fuel-efficient cars in the country. The Toyota Prius (Fig. 21–26) averages 48 mpg, while the Honda Insight gets 56 mpg. Detroit's big three are also getting ready to market hybrid electric vehicles, SUVs and pickups, as well as passenger cars. As more and more of the new hybrids are sold, the price is expected to drop substantially. The other option is the "partial zero-emission vehicle," a super-clean gasoline car that emits only 5% of the pollutants of a standard new car. A number of models are already available, including the Toyota Camry and Sienna. Many others are being prepared for the 2004 model year. This option does not, however, do anything for carbon dioxide emissions: The gasoline engines are still large and fuel-hungry.

**Figure 21–26 Hybrid Electric car.** Toyota's Prius, a hybrid electric vehicle.

An encouraging trend is the increase in mass-transit ridership. (See Earth Watch, this page.) In the past six years, ridership on buses, subways, and commuter rail lines has increased by 24%, faster than highway or air-transport ridership. This is an option that would encourage people to live closer to their workplaces, given the appropriate public-policy commitment. (See Chapter 23.)

## earth watch

### Portland Takes a Right Turn

For many of us, the American lifestyle includes spending time in an automobile each day going to and from work. Vehicle miles per year in the United States have increased much more rapidly than population, and too often the only response of state and city governments is to build more lanes on expressways.

Portland, Oregon, had its share of expressways 20 years ago, but took a different tack when faced with the prospect of more and more commuters being on the road. In response to an Oregon land-use law, Portland threw away its plans for more expressways and instead built a light-rail system, the **Metropolitan Area Express** (MAX). This public transportation system now carries the equivalent of two lanes of traffic on all the roads feeding into downtown Portland. As a result, smoggy days have declined from 100 to 0 per year, the downtown area has added 30,000 jobs with no increase in automobile traffic, and Portland's economy has prospered.

This was clearly a situation in which everyone won. MAX has made a major contribution to the clean air and continued economic success of downtown Portland. The Portland solution seems so sensible that you have to ask why it is the exception and not the rule.

# revisiting the themes

## Sustainability

Air pollution was on an unsustainable track, with industrial smog producing health disasters, photochemical smog making city living intolerable for many, and crops and natural ecosystems deteriorating seriously. Less obvious, but just as important, chronic exposure to air pollutants was responsible for poor health and increased mortality for millions. Are we back on a more sustainable track now? The important indicators would be a long-term decline in lead in the environment, annual amounts of emissions of pollutants, acid and sulfate deposition in the Northeast, ambient levels of pollutants, asthma hospitalizations, and locations that are noncompliant. Most of these parameters are moving in the right direction, but it is far too soon to declare victory.

## Stewardship

Stewardship means doing the right thing when faced with difficult problems and choices. It is fair to say that our society has responded well in caring for children threatened by high lead levels in the blood and, in general, in taking the costly actions we have documented in this chapter. Further, in addressing acid deposition, we have even shown care for the natural world. The need for stewardship is still very strong, however, and, thankfully, there are many NGOs and other organizations, and indeed many in the EPA, that are determined to stay the course and make sure that effective action is taken. Often, though, they have to do battle with industry lobbies and special interests that are concerned with the economic bottom line and that seem unable to embrace the common good represented by effective action to curb pollution.

## Sound Science

Atmospheric science has come a long way in the last half century. Some of its triumphs include understanding causes and effects that are operative in photochemical smog, discovering the cleansing role of the hydroxyl radical, tracing the sources of lead and mercury in humans, and working out causes and effects in the acid-rain story. Sound science is increasingly important in measuring ambient pollutant levels and in devising technologies to prevent more pollution. Many scientists have put their reputations and careers on the line as they fought for policy changes to correct the human and environmental damage they were uncovering.

## Ecosystem Capital

Ecosystems have been dealing with natural air pollutants for eons, but the Industrial Revolution began to change the air, bringing new and much more concentrated anthropogenic pollutants that began to damage ecosystems that were important to human welfare. Ozone-related crop loss, forest damage, lakes without fish, soils that lost their buffering capacity, and loss of visibility were all diminishing the goods and services we both enjoy and need. It is in our best interests to restore ecosystems that have been damaged by air pollution and to preserve the ecosystem capital that supports human life and economic welfare.

## Politics and Policies

The Clean Air Act was a triumph over complacency about air pollution. Congress showed then, and continues to show now, that it can act for the common good to bring about effective changes in public policy. It is still a very contentious process to get antipollution laws passed and new regulations aired and approved. This chapter has probed the laws in some depth and, indeed, has highlighted cases in which it continues to be difficult to pass laws and regulations that make very good sense. Examples are the CAFE standards, the new ozone and particulates standards, and the great recalcitrance of the utilities to curb the pollutants coming from older power plants. Political battles continue, with moves made at high levels (for example, the Cheney energy task force) often trumping the best-laid plans of conscientious legislators. The Republican party's control of Congress and the White House makes it easier for the fossil-fuel and automobile industries and utilities to have their way with rules and new laws, as evidenced by the New Source Review controversy. Thankfully, some voices are *not* being silenced: The 2003 report of the OMB spelled out the benefits of clean-air regulations. Further, President Bush's Clear Skies Initiative is a hopeful sign that, in the end, much good will be done. It remains to be seen whether the proposal actually becomes law.

## Globalization

Acid precipitation demonstrates that the air knows no boundaries: Canada receives our polluted air, and Japan receives acids from China's heavy use of coal. Fortunately, the cleansing work of the ocean salts as they form aerosols and promote rain lessens the global reach of air pollutants. One new manifestation of globalization is the so-called Asian brown cloud, a huge blanket of pollution that frequently arises over South and central Asia, stirred up by winds that pick up dust, ash, acids, and aerosols from forest fires, industries, and deforested soils. The cloud moves eastward, fouling the air and likely contributing to countless fatalities due to respiratory disease. The brown cloud is thought to be responsible for major reductions in rainfall because of the nature of its aerosol particles. Ozone and acid rain are carried in its train, and the impact of the cloud extends from Asia itself across the Pacific to North America. The phenomenon is currently under intense study.

## review questions

1. Give three examples each of natural and anthropogenic air pollutants.
2. What naturally occurring cleanser helps to remove pollutants from the atmosphere? What molecule (that we studied in Chapter 20) is the major source of this cleanser?
3. Describe the origin of industrial smog and photochemical smog. What are the differences in the cause and appearance of each?
4. What are the major primary air pollutants and their sources?
5. What are secondary pollutants, and how are they formed?
6. Distinguish between emissions and ambient concentrations. How are they measured?
7. What is the difference between an acid and a base? What is the pH scale?
8. What two major acids are involved in acid deposition? Where does each come from?
9. What impact does air pollution have on human health? Give the three categories of impact and distinguish among them.
10. Describe the negative effects of pollutants on crops, forests, and other materials. Which pollutants are mainly responsible for these effects?
11. How can a shift in environmental pH affect aquatic ecosystems? In what other ecosystems can acid deposition be observed? What are its effects?
12. What are the National Ambient Air Quality Standards, and how are they used?
13. Discuss ways in which the Clean Air Act Amendments of 1990 address the failures of previous legislation.
14. What technological and political changes are taking place in the United States to reduce acid deposition?
15. What are some unresolved air pollution issues that deserve our attention?

## thinking environmentally

1. Motor vehicles release close to half the pollutants that dirty our air. What alternatives might be introduced to encourage a decrease in our use of automobiles and other vehicles with internal combustion engines?
2. How would our pattern of life be different in the absence of the Clean Air Act and its amendments? Write a short essay describing the possibilities.
3. What arguments might the auto and utility industries present to delay regulatory action on particulates and ozone? How would health officials respond to their arguments?
4. The Swedish government has made efforts to attack the symptoms of acid deposition directly by trying to neutralize acidic lakes with lime. Discuss the pros and cons of this method.

## making a difference part five: chapters 15 to 21

1. If you are a smoker, enlist in a smoking cessation program. Insist that others not smoke in your presence or in your home or workplace.
2. Monitor the hazards in your own life, and take steps to reduce behavior ranked as risky by the experts. Be discriminating as you hear about hazards and risks. Be especially aware of the media's tendency to exploit the outrage element of risks.
3. Survey the lakes and ponds in your area and determine whether cultural eutrophication is occurring. If you find that it is, follow up by consulting a local environmental group, and get involved in combating the problem.

4. Use phosphate-free detergents for all washing. Information regarding the use of the product is printed on the container (often in very fine print!).

5. Investigate sewage treatment in your community. To what extent is the sewage treated? Where does the effluent go? The sludge? Are there any problems? Are there alternatives that are ecologically more sound? Consult local environmental groups for help.

6. Investigate how municipal solid waste is handled in your school or community and what the future plans are for waste management. Promote curbside recycling if it is not happening, and investigate the possibility of adopting a system that charges for each container of unrecycled waste.

7. Consider your consumption patterns. Live as simply as possible so as to minimize your impact on the environment. Buy and use durable products, and minimize your use of disposables.

8. Read labels and become informed about the potentially hazardous materials you use in the home and workplace. Whenever possible, use nontoxic substitutes. If toxic substances are used, ensure that they are disposed of properly.

9. Encourage your community to set up a hazardous-waste collection center where people can take leftover hazardous materials.

10. Reduce your contribution to air pollution by amending your lifestyle to drive fewer miles: Arrange to live near your workplace, carpool, use public transportation, bicycle to work or school, and avoid unnecessary trips.

11. Maintain all fuel-burning equipment (oil burner, gas heater, lawn mower, outboard motor, automobile, etc.) so that it will burn fuel efficiently.

12. When you buy a new car, consider purchasing one of the new hybrid electric vehicles, or at least buy a car that gets good mileage.

13. Be an advocate of energy conservation on all fronts by supporting the steps we can take to combat global climate change. Insist on full U.S. participation in global efforts to curb greenhouse gas emissions, such as the Kyoto Accord of 1997.

14. Make sure that anyone servicing a refrigerator or air-conditioning unit for you will recapture and recycle the CFCs if the system needs to be opened.

# part six

# toward a sustainable future

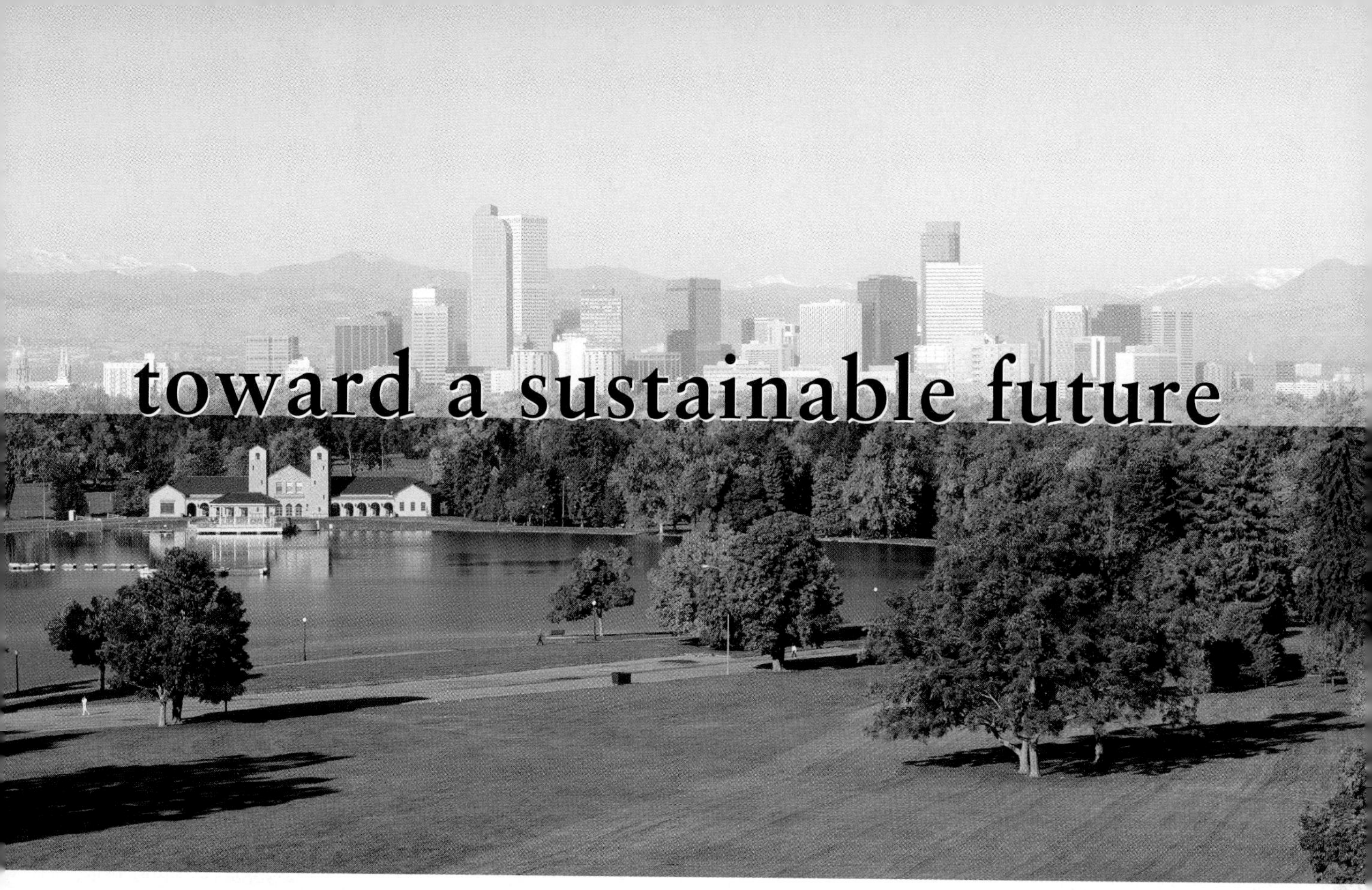

How will we shape a sustainable future? In previous chapters, we discussed many issues that need to be addressed and resolved if we are indeed to achieve a sustainable relationship with the environment. Environmental public policy and individual lifestyles are key to that goal.

In Chapter 22, we highlight economics and public policy as they relate to the environment. We examine how a nation can call on its wealth to address societal problems, and we consider the three factors that constitute the wealth of nations: produced assets, natural capital, and human resources. In Chapter 23, the last one of the book, we explore one of the most difficult problems faced by every nation today: promoting livable and sustainable communities, particularly in the many crowded cities of the developing world. We look at examples of cities moving in a sustainable direction. We also look at some of the work being done by organizations that are grappling with the question of sustainability. Finally, we look at the vital, but difficult, task of adopting lifestyles that embrace some of the stewardship values capable of shaping a sustainable future. The outcome is uncertain, but we know enough now to be able to point in a sustainable and stewardly direction.

***Denver, Colorado.*** As this view from Ferril Lake in City Park shows, Denver's setting, with the snow-capped Rockies to the west, makes it a very attractive place to live. The "mile-high" city is considered one of the most livable cities in America.

chapter 22

# Economics, Public Policy, and the Environment

**Key Topics**

1. Economics and Public Policy
2. Resources and the Wealth of Nations
3. Pollution and Public Policy
4. Benefit–Cost Analysis
5. Politics, the Public, and Public Policy

In 1997, the EPA announced a set of new air quality standards called **40 CFR Part 50** in the *Federal Register*. The new standards for particulate matter and ozone required industries and regions to take effective action to reduce the levels of those pollutants. Among the documents supporting the new standards was a regulatory impact analysis of the *costs and benefits* associated with them. The EPA estimated that, by 2010, when the regulations would be in full effect, the annual benefits would amount to some $92 billion, whereas the costs of the regulations would be some $6.4 billion. The new standards were met with strong protests from industry groups (see Chapter 21), which argued that compliance would instead cost as much as $150 billion a year. The industries and the EPA became locked in a battle over the regulations that was ultimately decided by the U.S. Supreme Court.

The EPA was required by law to perform the benefit–cost analysis. Back in the 1970s, however, most environmental policies were developed with little consideration of economics. Early policies were typical command-and-control responses to air and water pollution that focused on controlling emissions from cars and factories, pesticides being sprayed over large areas, and releases from sewage outfalls and other point sources. The policies had a significant economic impact on businesses, consumers, and the workforce.

***Wait a Minute....*** Reacting to the situation, and concerned that the U.S. economy in general and businesses in particular might be overregulated and thus unduly restricted, President Ronald Reagan issued Executive Order 12291 in February 1981. This order required all executive departments and agencies to support every new *major* regulation with **benefit–cost analysis.** To qualify as major, a regulation had to (1) impose annual costs of at least $100 million, *or* (2) cause a significant increase in costs or prices for some sector of the economy or geographic region, *or* (3) have a significant adverse effect on competition, investment, productivity, employment, innovation, or the ability of U.S. firms to compete with foreign firms.

***Capitol Building, Washington, DC***
**The Senate and House of Representatives meet and establish public policy in this center of activity of the U.S. government.**

No regulations were to be issued unless the benefits clearly outweighed the costs. It is no secret that the goal of Reagan's order was to roll back environmental regulation.

In 1993, President Bill Clinton issued Executive Order 12866, which continued most of the policies of Executive Order 12291 (with the exception of the third qualification) and added the processes of public and interagency review of proposed regulatory rules. The 1993 executive order softened the quantitative test by stating that benefits should "justify" costs and that a number of additional non-monetary factors (such as the impacts on different groups, or issues that cannot be quantified) could be considered in making a final decision on a given regulation.

President George W. Bush has sustained Executive Orders 12291 and 12866 and added Executive Order 13272, which stipulaties that agencies should give proper consideration of the impacts of their rulemaking on small businesses and other smaller entities.

***Bottom Line.*** Benefit–cost analysis is an economic measure applied to environmental policy. It is a reminder that all environmental decisions take place in the context of a society in which money and politics are the language of exchange and the bottom line is not always *what is right to do,* but *what is possible to do.* In this chapter, our major focus is on environmental public policy, but we will see that economics plays a dominant role in virtually every environmental issue, both domestically and internationally. This should come as no surprise, as we have seen economic issues in many previous chapters. In Chapter 3, for example, the value of ecosystem goods and services was discussed; in Chapter 10, value concepts were applied to wild species; in Chapter 11, the question of common property resources was addressed; in Chapter 15, benefit–cost analysis was identified as a major tool of risk management; and in Chapter 21, benefit–cost analysis was applied to air pollution laws. We have also seen the great disparity of wealth among nations and the profound consequences of the lack of wealth in many of the countries still in the early stages of economic development. Economic concerns are highly important, but even more important is the development of just and effective environmental public policies.

## 22.1 Economics and Public Policy

Environmental public policy includes all of a society's laws and agency-enforced regulations which deal with that society's interactions with the environment. Two sets of environmental issues are encompassed in environmental public policy: the prevention or reduction of air, water, and land pollution; and the use of natural resources such as forests, fisheries, oil, land, and so forth. Public policies addressing these two sets of issues are developed at all levels of government: local, state, federal, and global.

### The Need for Environmental Public Policy

Recall from Chapter 1 that the purpose of environmental public policy is *to promote the common good.* The "*Policy and Politics*" integrative theme (*the human decisions that determine what happens to the natural world, and the political processes that lead to those decisions*) is reviewed at the end of each chapter; you are encouraged to scan this section in different chapters. As you do, you will see that public policies focus both on *the improvement of human welfare* and *the protection of the natural world.* Since human welfare is strongly tied to the goods and services derived from the natural world, protecting the environment is, in the best sense, serving ourselves, even as we preserve and restore environments and the living landscape.

What are the consequences of not having an effective environmental public policy? Human societies and their economic activities have the potential for doing great damage to the environment, and that damage has a direct impact on present and future human welfare (Table 22–1). The effects of pollution and the misuse of resources are seen most clearly in those parts of the world where environmental public policy is often not well established and implemented—the developing world. As Table 22–1 shows, millions of deaths and widespread disease are directly traceable to degraded environments. The costs to human welfare are felt in the areas of health, economic productivity, and the ongoing ability of the natural environment to support human life needs. Therefore, laws to protect the environment are not luxuries to be tolerated only if they do not interfere with individual freedoms or economic development; they are an essential part of the foundation of any well-organized human society.

### Relationships Between Economic Development and the Environment

In a human society, an economy is the system of exchanges of goods and services worked out by members of the society. Goods and services are produced, distributed, and consumed as people make economic decisions about what they need and want and what they will do to become players in the system—what they might provide that others would need and want.

**table 22-1 Principal Health and Productivity Consequences of Poor Environmental Management**

| Environmental Problem | Effect on Health | Effect on Productivity |
|---|---|---|
| Water pollution and water scarcity | More than 3 million deaths and billions of illnesses a year are attributable to pollution; poor household hygiene and added health risks are caused by water scarcity. | Declining fisheries; increase in rural household time (time spent fetching water) and in municipal costs of providing safe water; depletion of aquifers, leading to irreversible compaction; constraint on economic activity because of water shortages. |
| Air pollution | Many acute and chronic health impacts: excessive levels of urban particulate matter are responsible for 300,000 to 700,000 premature deaths annually and for half of childhood chronic coughing; 400 million to 700 million people, mainly women and children in poor rural areas, are affected by smoky indoor air. | Restrictions on vehicles and industrial activity during critical episodes; effect of acid rain on forests, bodies of water, and human artifacts. |
| Solid and hazardous wastes | Diseases spread by rotting garbage and blocked drains. Risks from hazardous wastes are typically local, but often acute. | Pollution of groundwater resources. |
| Soil degradation | Reduced nutrition for poor farmers on depleted soils; greater susceptibility to drought. | Field productivity losses in the range of 0.5–1.5% of gross national product are common on tropical soils; offsite siltation of reservoirs, river-transport channels, and other hydrologic systems. |
| Deforestation | Localized flooding, leading to death and disease. | Reduced potential for sustainable logging and for prevention of erosion, increased watershed instability, and diminished carbon storage capability by forests. Loss of nontimber forest products. |
| Loss of biodiversity | Potential loss of new drugs. | Reduction in ecosystem adaptability and loss of genetic resources. |
| Atmospheric changes | Possible shifts in vectorborne diseases; risks from climatic natural disasters; diseases attributable to ozone depletion (300,000 more skin cancers per year and 1.7 million cases of cataracts per year). | Damage to coastal investments from rise in sea level; regional changes in agricultural productivity; disruption of marine food chain. |

*Source:* The World Bank, *World Development Report, 1992* (New York: Oxford University Press, 1992).

As societies develop, economic activity assumes increasingly broader dimensions, with increasingly pervasive impacts on the whole society. In previous chapters, we examined numerous instances in which economic activity can damage the environment and human health. In addition, an unregulated economy can make intolerable inroads on natural resources. For example, during the latter part of the 19th century in the United States, private enterprise had unrestrained access to forests, grazing lands, and mineral deposits. This unsustainable exploitation of resources began to be addressed at the turn of the century, as government rules and regulations imposed necessary limits.

Fortunately, economic activity in a nation also can provide the resources needed to solve the problems that that very activity creates. A strong relationship exists between the level of development of a nation and the effectiveness of its environmental public policies. Figure 22–1 shows a number of environmental indicators in relation to per capita income levels. (The data were combined from studies of different countries.) Several patterns emerge:

- Many problems decline (for example, sickness due to inadequate sanitation or water treatment) as income levels rise, because the society gains the resources available to address the problems with effective technologies.
- Some problems increase and then decline (for example, urban air pollution) when the consequences of the problem are recognized and then public policies are developed to address them.
- Increased economic activity causes some problems to increase without any clear end in sight (for example, municipal solid waste and $CO_2$ emissions).

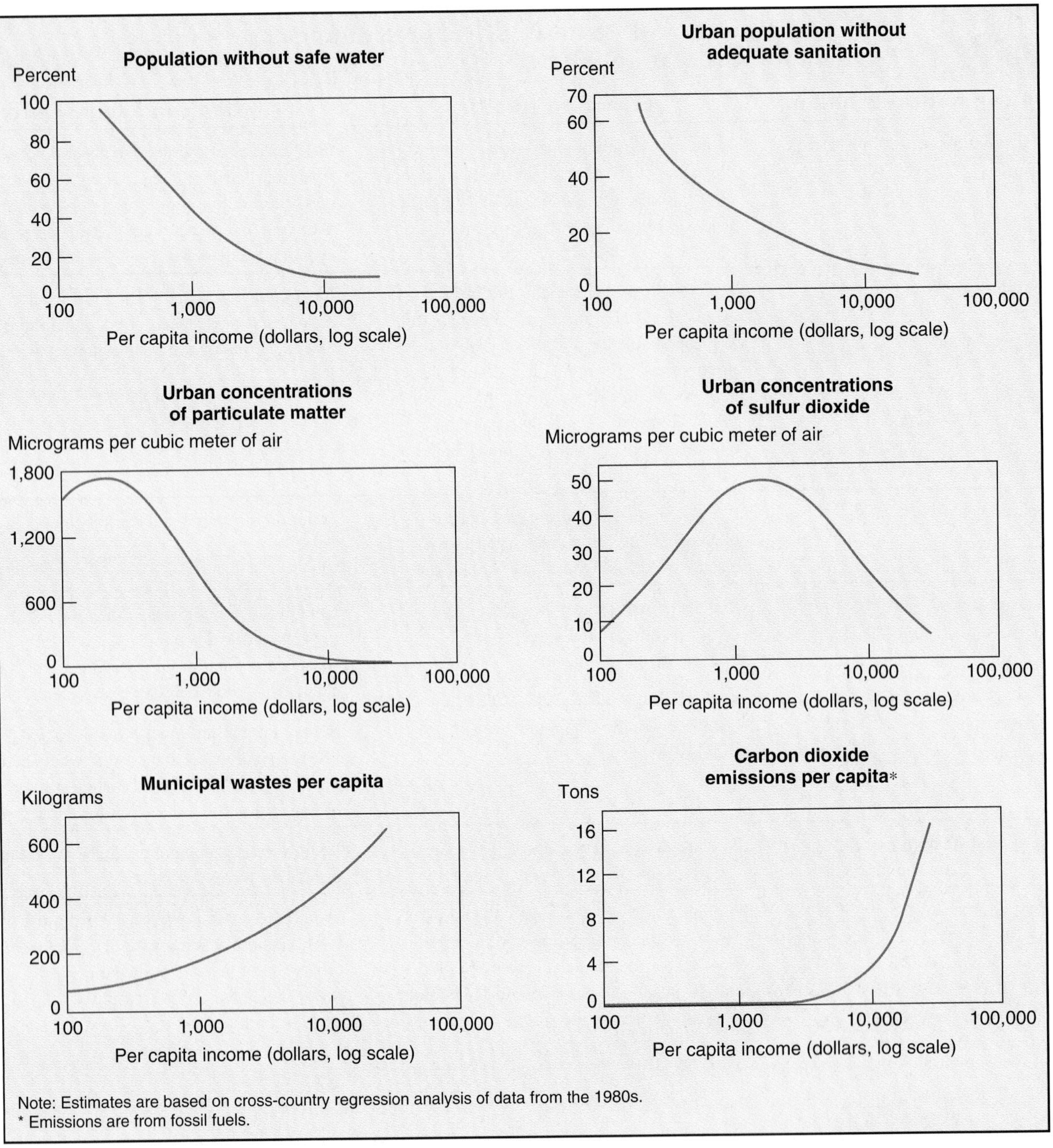


**Figure 22–1 Environmental indicators and per capita income.** Some of the most serious environmental problems can be improved with income growth; others get worse and then improve, while some problems just worsen. (*Source:* World Bank, *The World Development Report, 1992* [New York: Oxford University Press, 1992], p. 11.)

The key to solving all of these problems brought on by economic activity is the *development of effective public policies and institutions.* When this does not happen, environmental degradation and human disease and death are the inevitable outcomes. To understand this relationship better, let us look at how economic systems work.

## Economic Systems

Economic systems are social and legal arrangements people construct in order to satisfy their needs and wants and improve their well-being. Economic systems range from the communal barter systems of primitive societies to the current global economy, which is so complex that it defies description. In the organized societies of today's world, two kinds of economic systems have emerged: the **centrally planned economy,** characteristic of socialist countries, and the **free-market economy,** characteristic of the capitalist countries.

In any kind of economy, there are basic components that determine the economic flow of goods and services. Classical economic theory considers **land** (natural resources), **labor,** and **capital** as the three elements constituting the "factors of production." *Economic activity* involves the circular flow of money and products, as shown in Figure 22–2. Money flows from

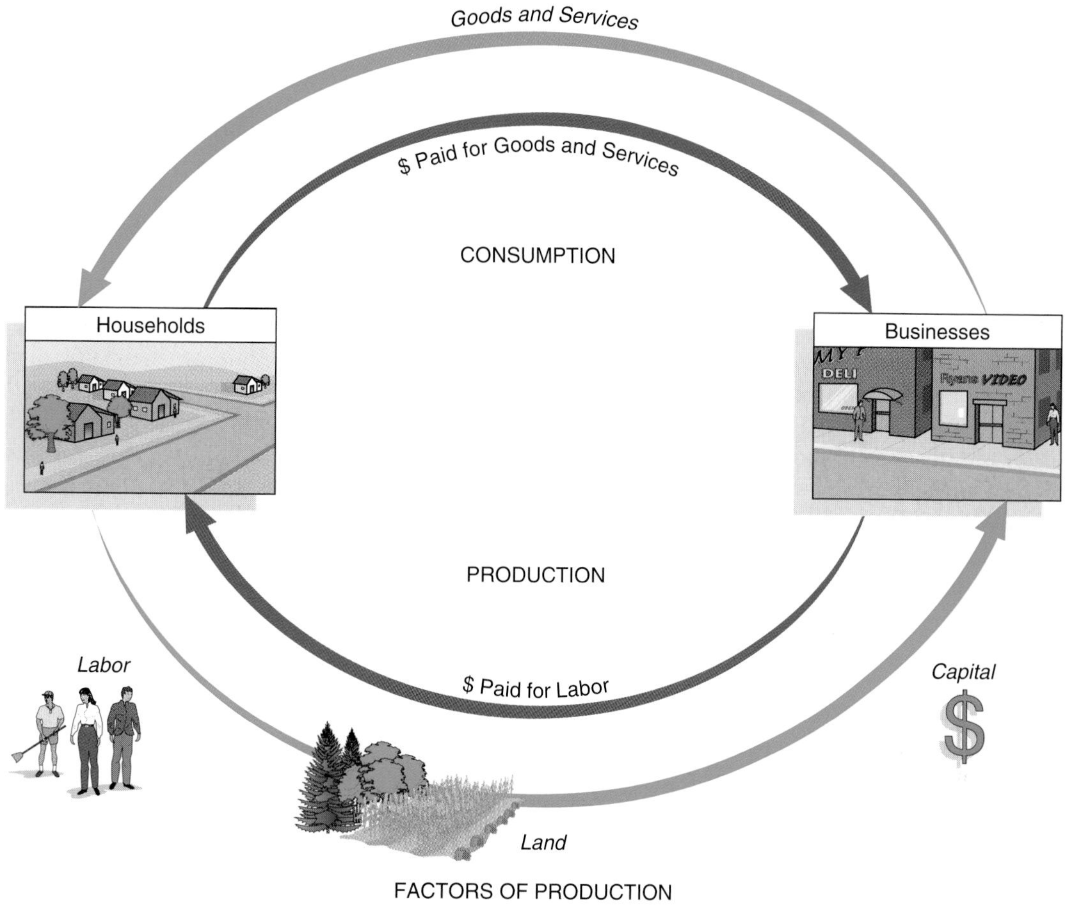


**Figure 22–2** **Classical view of economic activity.** Land (natural resources), labor, and capital are the three elements constituting the "factors of production." Economic activity involves the circular flow of money and products.

households to businesses as people pay for products and from businesses to households as people are paid for their labor. Labor, land, and capital are invested in the production of goods and services by businesses, and the products provided by businesses are "consumed" by households.

***The Party Decides.*** The cycles shown in the figure apply to both kinds of economic systems in existence today; the two systems differ mostly in how economic decisions are made. In a pure centrally planned economy, the leaders make all the basic decisions regarding what and how much will be produced where and by whom. Those who adopt such a system believe that it is a more efficient system in which the real needs of the people will be met because of the wise decisions of their rulers. It is a system in which equity and efficiency are theoretically achievable ("from each according to his abilities, to each according to his needs," as Karl Marx put it). The failure of the centrally planned economy in the former Soviet Union has led to economic chaos in the countries that it once comprised. North Korea, Cuba, a few African countries, and the People's Republic of China (although moving to a market economy) remain as the last holdouts of this economic model.

***The Market Decides.*** In the pure free-market economy, the market itself determines what will be exchanged. Goods and services are offered in a market that is free from governmental interference. The system is completely open to competition and the interplay of supply and demand. If supply is limited relative to demand, prices rise; if there is an oversupply, prices fall, because people and corporations will economize and pay the lowest price possible. The whole system is in private hands and is driven by the desire of people and businesses to acquire goods, services, and wealth as they act in their self-interest. All "players" have free access to the market. Competition spurs efficiency as inferior products and services are forced out of the market. People can make informed decisions about their purchases because there is sufficient information about the benefits and harm associated with economic goods. The free-market economic system is thought to be at its best when left completely alone.

***Government's Role.*** The preceding discussion describes only the very basics of the two economic systems. In reality, no country has a pure form of either economy. The developed countries all function with market economies, but government involvement occurs at many levels, more so in some of the "social democracies" of Western Europe than in the United States. Governments can, for example, control interest rates, determine the amount of money in circulation, and adopt policies that stimulate economic growth in times of slowdown. Governments also maintain surveillance over financial processes such as stock markets and bank operations. On the downside, powerful business interests can manipulate a market economy, and unscrupulous people can exploit the freedom of the market to defraud people (as in the recent U.S. business scandals involving WorldCom, Enron, and many mutual funds). Only governments can provide the policing that the society needs.

Many people believe that the free-market economic system lacks a "conscience." They argue that there are many workings of the market economy that create hardships for people at the bottom of the economic ladder. If population growth is rapid and there are not enough jobs to go around, for example, the result often is exploitation of workers, as we saw in the case of the factory workers in Nairobi, Kenya, in Chapter 5. A market economy only offers people *access* to goods and services; if people lack the means to pay, however, access alone will not meet their needs. Recall from Chapter 9 the importance of a "safety net" in connection with food security for impoverished people. Also, it is too easy for self-interest to lead businesses and individuals to exploit natural resources or to avoid costs by polluting the environment instead of producing a "clean" product. Over the years, the democracies have been learning what laws need to be put in place to *control access to natural resources* and to *prevent damaging pollution.* These two areas of concern are the substance of environmental public policy.

## 22.2 Resources and the Wealth of Nations

What are the resources on which a country draws to establish and maintain an economy? The classic economic paradigm would say that *land* (which represents the environment), *labor,* and *capital* are the essential resources needed for a country to be able to mount its economy. However, a new breed of economists, often called **ecological economists** for want of a better term, has emerged in recent years and has taken issue with this view. These economists point out that the classic view sees the environment simply as one set of resources within the larger sphere of the human economy. The environment's vital role in supplying the goods and services on which human activities depend is largely ignored. They argue that the classical approach looks at things backwards and that the natural environment actually *encompasses* the economy, which is constrained by the limits of resources in the environment (Fig. 22–3). Without the vital raw materials provided by the environment, and without the capacity of the environment to absorb wastes (pollution), there is no economy. Thus, the ecosystems and natural resources found in a given country provide the context for that country's economy. Similarly, the total global biosphere is the context for the global economy.

***Trees to Paper to Trash.*** As described by Thomas Prugh, Robert Costanza, and Herman E. Daly in *The Local Politics of Global Sustainability,*[1] **economic production** can thus be seen as it really is: *the process of converting the natural world to the manufactured world.* Renewable and nonrenewable resources and the ecosystems containing them are turned into cars, toys, books, food, buildings, highways, computers, etc. Thus, to quote Prugh, Costanza, and Daly, "Resources flow into the economy from the enfolding ecosystem and are transformed by labor and capital (using energy, also a resource), and then pass out of the economy and back into the ecosystem in the form of wastes."[2] (See Fig. 22–3.)

The ecological economists' view emphasizes the role of natural ecosystems as essential life-support elements—*ecosystem capital,* as we referred to it earlier. The approach brings the concepts of carrying capacity and limits into perspective and places sustainability sharply into focus. If the economy continues to grow, the natural world continues to shrink. In time, economic growth must come to a sustainable steady state, and this must happen before our natural capital is consumed beyond its ability to support an economy.

If we add to ecosystem capital the nonrenewable mineral resources, such as fossil fuels and metal ores, we can use the more inclusive term *natural capital.* Without a doubt, the ecosystems and mineral resources of a given country—its *natural capital*—are a major element in the wealth of that country. In recent years, the World Bank Environment Department has been working on ways of measuring the wealth of nations and has produced some insightful analyses, which we consider next.

### The Wealth of Nations

Environmentally sustainable development means improving human well-being over time, a process that is especially important to the world's impoverished people. It is a process that requires societies to manage a portfolio of assets that can contribute to improving human well-being. The World Bank group asked, What are the components of a nation's wealth—its assets? In a document entitled *Expanding the Measure of Wealth: Indicators of Environmentally Sustainable Development,* the World Bank indicates three major components of capital that determine a nation's wealth: *produced assets, natural capital,* and *human resources.*

[1]Washington, DC: Island Press, 2000.
[2]p. 19.

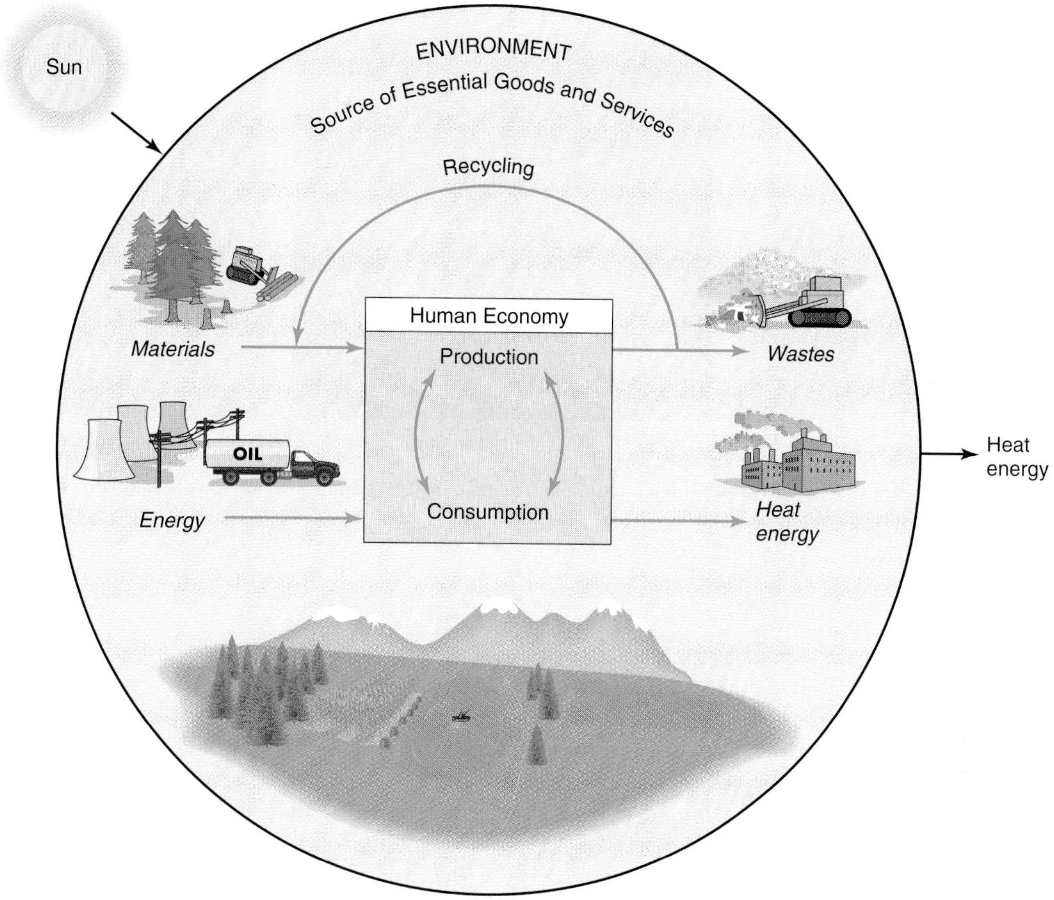


**Figure 22–3 Environmental economic view of economic activity.** The natural environment encompasses the economy, which is constrained by the resources found within the environment.

**Produced assets** (Fig. 22–4a) are the human-made buildings and structures, machinery and equipment, vehicles and ships, monetary savings and stocks, highways and power lines, etc., that are essential to the production of economic goods and services. These are often the major focus of national economic planning, and they are the most easily measured of the three components of a nation's wealth. Produced assets have very clear income-earning potential, but they may also be subject to obsolescence and must be renewed continually. For example, a clothing factory produces a flow of goods destined for consumers, but the machinery wears out over time, and the building itself ages. Thus, we have income—a flow of goods—and depletion, which is referred to as capital consumption.

The **natural capital** (Fig. 22–4b) refers to the goods and services supplied by natural ecosystems and the mineral resources in the ground. Natural capital represents an indispensable set of resources, some of which are *renewable* and some *nonrenewable*. Renewable natural capital is represented by forests, fisheries, agricultural soil, water resources, and the like, which can be employed in the production of a flow of goods (lumber, fish, corn). Often overlooked is the fact that this same natural capital also provides a flow of services in the form of waste breakdown, climate regulation, oxygen production, and so forth. (See Table 3–2 for an extensive list of the goods and services provided by natural capital.) As we have seen in earlier chapters, renewable natural capital (ecosystem capital) is subject to depletion, but also has the capacity to yield a sustained income if it is managed responsibly. Nonrenewable natural capital, such as oil and mineral deposits, also is subject to depletion and provides no services unless it is extracted and converted into some useful form. It is part of the wealth of a nation, however, as long as it is in the ground. Some components of natural capital are easier to measure than others; for instance, the lumber of a forest can be reckoned in board feet, but how does one measure the impact of a forest on the local climate and on its capacity to sustain a high level of biodiversity?

**Human resources** (Fig. 22–4c) can be divided into three elements. The first is **human capital,** which refers to the population and its physical, psychological, and cultural attributes (innate talents, competencies, abilities, etc.). To this is added the value imparted by education, which enables members of a population to acquire skills that are useful in an economy. Education can be either formal or informal; the point is that

(a) Produced assets

(b) Natural capital

(c) Human resources

**Figure 22–4 The wealth of nations.** Three major components of capital determine a nation's wealth: (a) produced assets, (b) natural capital, and (c) human resources.

people acquire a productive capacity through education. Investments in health and nutrition also contribute to increases in human capital.

The second element of human resources is what the World Bank group has called **social capital**—the social and political environment that people create for themselves in a society. Social capital includes more formal structures and relationships as defined by government, the rule of law, the court system, and civil liberties. It also includes the horizontal relationships of people as they associate into religious or ethnic affiliations or join organizations in which they have a common interest. Social capital is considered a vital element of a nation's wealth, because social relationships clearly affect, and are affected by, economic processes. Human resources are the hardest of the three components to measure.

A third element of human resources is **knowledge assets**—the codified or written fund of knowledge that can be readily transferred to others across space and time. Although these assets constantly grow, their value can be imparted only if people have access to them (hence the vital importance of libraries, schools, universities, and the Internet).

*Complementarity.* The three classes of assets usually complement each other as they are called on to improve human well-being. For instance, the natural assets of clean water and healthy food can improve human health and the capability to work on produced assets. Knowledge assets can provide the information for building effective social capital in the form of laws, government agencies, and interest groups. As these components of a nation's wealth are understood and managed, they enhance human well-being in many ways (Fig. 22–5). Their influence can be direct, as in the effects of a beautiful natural setting or a network of trusted friends, or indirect, as in the contribution they make to economic production and its eventual consumption.

*Calculations.* On the basis of these components of the wealth of nations (their assets), the World Bank Environment Department devised various means of measuring each component and then set about evaluating the wealth of different nations. Some of the results of the bank's analysis are shown in Table 22–2 and Fig. 22–6. One finding that should come as no surprise is the notable differences in per capita wealth among nations. Interestingly, the dominant source of wealth for most nations is human resources, but natural capital ranks high in a number of countries. Although natural capital often ranks third in most countries, this does not mean that it is less important than human resources or produced assets. (The World Bank authors offer a caution in interpreting their work, as they did not assess the services component of natural capital and calculated mainly the actual resource inputs into production—that is, forests as a source of timber, mineral assets, etc.) The dominance of the human resources component emphasizes the importance of investing in health, education, and nutrition in a society that needs to move further along in development. The bank has focused attention on natural capital as an essential element of economic production and has provided the basics of a tool for measuring progress in sustainability. Interestingly, the World Bank did not use the commonly employed indicator of economic status known as the gross national product (GNP).

## Shortcomings of the GNP

The *gross national product,* or GNP, refers to the sum of all goods and services produced (really, consumed) in a country in a given time frame. The GNP is the most commonly used indicator of the economic health and wealth of a country. In most countries, the comparable *gross domestic product,* or GDP, is now used. The GDP is the GNP minus net income from abroad. As a per

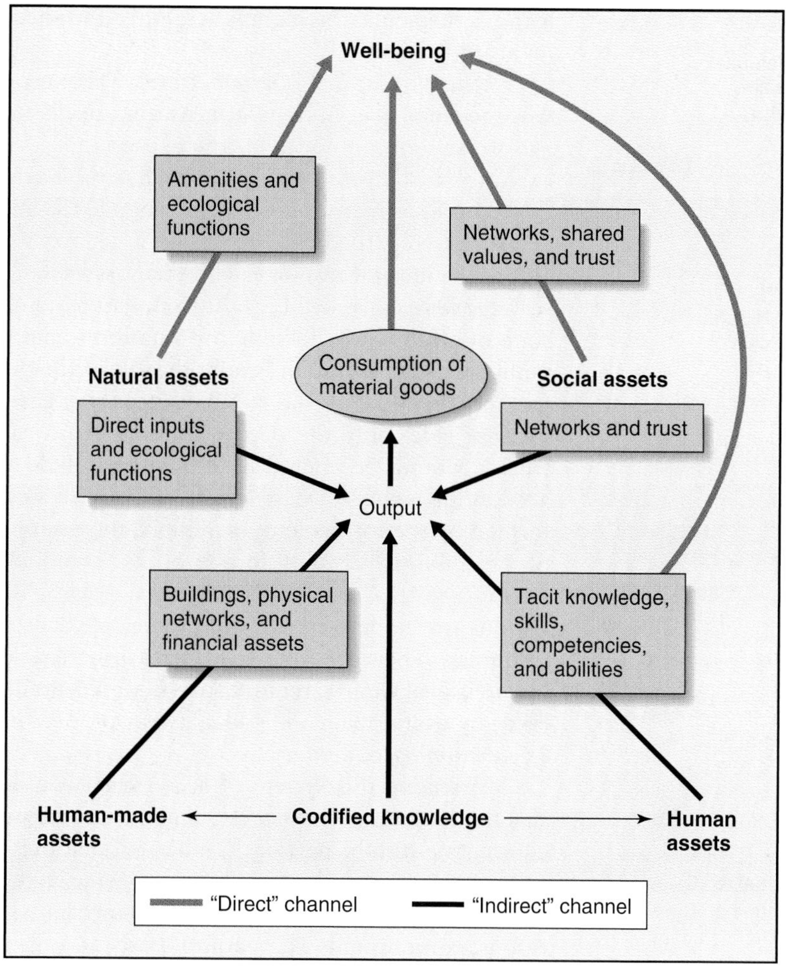


**Figure 22–5 How society's assets complement each other.** The three factors known as human resources, natural assets, and human-made assets all enhance human well-being. (*Source: World Development Report 2003,* the World Bank and Oxford University Press. By Permission)

**table 22-2 Wealth of Nations per Capita, by Region**

| | Dollars Per Capita | | | | Percent Share of Total Wealth | | |
|---|---|---|---|---|---|---|---|
| | Total Wealth | Human Resources | Produced Assets | Natural Capital | Human Resources | Produced Assets | Natural Capital |
| North America | 326,000 | 249,000 | 62,000 | 16,000 | 76 | 19 | 5 |
| Pacific members of OECD* | 302,000 | 205,000 | 90,000 | 8,000 | 68 | 30 | 2 |
| Western Europe | 237,000 | 177,000 | 55,000 | 6,000 | 74 | 23 | 2 |
| Middle East | 150,000 | 65,000 | 27,000 | 58,000 | 43 | 18 | 39 |
| South America | 95,000 | 70,000 | 16,000 | 9,000 | 74 | 17 | 9 |
| North Africa | 55,000 | 38,000 | 14,000 | 3,000 | 69 | 26 | 5 |
| Central America | 52,000 | 41,000 | 8,000 | 3,000 | 79 | 15 | 6 |
| Caribbean | 48,000 | 33,000 | 10,000 | 5,000 | 69 | 21 | 11 |
| East Asia | 47,000 | 36,000 | 7,000 | 4,000 | 77 | 15 | 8 |
| Eastern and Southern Africa | 30,000 | 20,000 | 7,000 | 3,000 | 66 | 25 | 10 |
| West Africa | 22,000 | 13,000 | 4,000 | 5,000 | 60 | 18 | 21 |
| South Asia | 22,000 | 14,000 | 4,000 | 4,000 | 65 | 19 | 16 |

*Source:* The World Bank, *Expanding the Measure of Wealth: Indicators of Environmentally Sustainable Development* (Washington, DC: International Bank for Reconstruction and Development, 1997).
*Organization for Economic Cooperation and Development; Pacific members are Japan, Australia, New Zealand, and the Republic of Korea.

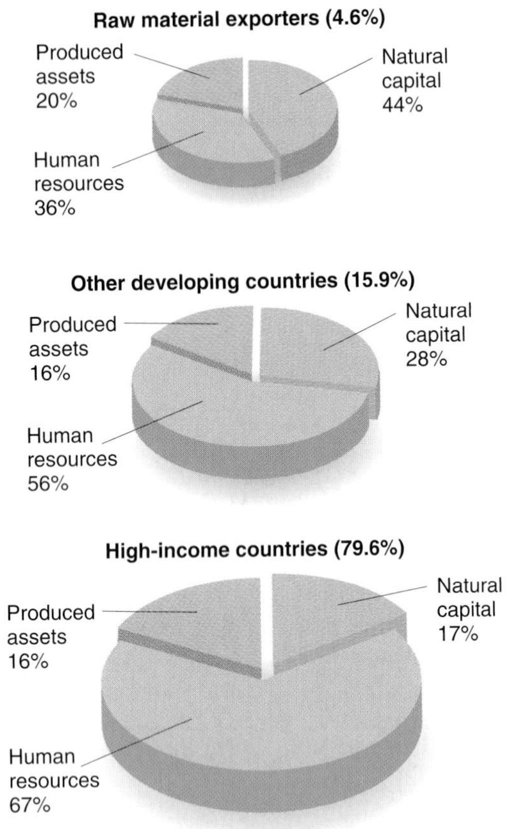


**Figure 22–6 Composition of world wealth.** Some developing countries depend on exporting raw materials. These countries account for 4.6% of the world's wealth. The remaining developing countries contain 15.9% of the world's wealth. The high-income developed countries possess 79.6% of the wealth. Relative percentages of produced assets, natural capital, and human resources are shown for each category. (*Source:* The World Bank, *Monitoring Environmental Progress: A Report on Work in Progress* [Washington, DC: International Bank for Reconstruction and Development, 1995].)

capita index, the GDP is often used to compare rich and poor countries and to assess economic progress in the developing countries. An important element in calculating a *net* GDP is accounting for the assets that are used in the production processes. Buildings and equipment, for example, are essential to production, but they gradually wear out or depreciate. Their depreciation is charged against the value of production (called capital depreciation); that is, an accounting of capital depreciation is routinely subtracted from the production of goods and services in order to generate a net GDP. (We shall call net GDP NNP—*net national product.*) This measurement is flawed, however, because it does not include goods traded in "black markets" and it does not include goods and services performed by a family for itself. Thus, subsistence farming in developing countries is not measured, and this omission surely produces a falsely lower NNP for these countries. The NNP also omits the natural services provided by ecosystems. For example, even though clean air is as important as asthma medicine for asthmatics, only the medicine is included in the NNP.

***Natural Capital Depreciation.*** The economists who invented the GNP as a measuring device 60 years ago never took into account the depreciation of natural capital—an omission that recently has received a great deal of criticism from ecological economists. Such concerns were regarded as external to a country's balance sheet. The natural resources and their associated natural services were considered a "gift from nature." Given this state of affairs, it is possible for a nation to cut down a million acres of forest and count the sale of the timber on the income side of the NNP ledger, whereas on the expense side, only the depreciation of chain saws and trucks would be included. Completely hidden from accounting is the loss of all the natural services once performed by the forests and, incredibly, the disappearance of the million acres of forests as an economic asset! As long as this discrepancy remains, it is possible for nations to underestimate the value of their natural resources. They are able to deplete fisheries, lose soil by intensive farming, remove forests, and degrade rangelands by overgrazing and still *account for these activities as economic productivity*!

Correcting this system of accounting would require calculating the current market value of natural assets and considering it to be part of the stock of a nation's wealth. To this value would be added the value due to the natural services performed by the ecosystems in which the resources are found. The natural assets and the services they perform are the nation's natural capital. When the natural resources are drawn down, the depreciation represented by the loss of natural capital would be entered into the ledger as the NNP is calculated. Similar arguments can be made for human resources as a measure of a nation's wealth. A more accurate NNP would have to take all of these depreciations into account, yielding the relationship

$$\mathrm{NNP} = \mathrm{GDP} - d_{\mathrm{P}} - d_{\mathrm{N}} - d_{\mathrm{H}} \qquad (1)$$

where $d$ represents a depreciation and the subscripts $P$, $N$, and $H$ indicate produced assets, natural capital, and human resources, respectively, all of which are components of a nation's wealth.

***Environmental Accounting.*** *Agenda 21*—one of the major documents that came out of the 1992 Earth Summit meeting in Rio—recognized the need for "a program to develop national systems of integrated environmental and economic accounting in all countries." The U.N. Statistical Division, which coordinates the accounting procedures of member countries, has been preparing a manual (*Integrated Environmental and Economic Accounting*) designed to accomplish *Agenda 21*'s objective. In the manual, countries are encouraged to perform environmental accounting by putting their environmental assets and services in monetary units and keeping a parallel account of their *net domestic product* (similar to the

NNP) as it is affected by environmental accounting. After revisions by numerous international working groups, the U.N. manual, named *SEEA-2003* (for *System for Integrated Environmental and Economic Accounting for the year 2003*), is scheduled to be published in 2004. The 600-page manual is intended to be used as a reference work showing nations how to implement environmental accounting. The document includes case studies and step-by-step instructions and strongly advocates the adoption of integrated environmental and economic accounting by U.N. member nations. In recent years, some 25 countries have experimented with different forms of environmental accounting, indicating that some form may become broadly adopted before long. Currently, SEEA is seen as an extension of the System of National Accounts (SNA), a globally coordinated accounting framework that sets standards for compiling and presenting economic data. Like the GDP, the SNA deals only with goods, services, and assets that are economic and have market value. Both the SNA and SEEA are functions of the U.N. Statistics Division.

## Resource Distribution

None of the resources that are crucial to an economy are evenly distributed among nations. Table 22–2 shows that great differences in wealth exist among the nations of the world, something we have seen many times in previous chapters. It is significant that the greatest differences occur in the two categories of assets that are most closely tied to the process of development: *human resources* and *produced assets*. With the exception of the oil-rich Middle East, the per capita differences in natural capital are unremarkable. Thus, *economic development*, which fosters growth in human resources and produced assets, is the pathway to achieving more equity in the distribution of resources.

***Essential Conditions.*** There are many reasons why some countries have undergone rapid development and others have not; population growth is one that has been cited in this text (Chapter 6). Perhaps the most important reasons relate to the major institutions that coordinate human behavior and that make up the social capital of a society (Fig. 22–7). For example, the definition of rights, the enforcement of those rights, and the facilitation of economic exchange are all considered essential to the successful development of human capital and produced assets. A well-developed body of law, an honest legal system, inclusiveness, broad civic participation, and a free press will go far in maintaining rights within a society. A well-developed market economy and free entry into and exit from markets for people and businesses are also essential for development. Functioning communication and transportation networks and viable financial markets are necessary for sustaining a market economy. History tells us that if these resources are in place, a society will make progress in the development of human resources and produced assets. In societies where such resources are not in place, corruption, inefficiency, banditry, and human rights abuses often prevail, and continuing poverty and environmental degradation are the predictable result.

**Intergenerational Equity.** What we have just discussed can be called *intragenerational equity*—the golden rule of *making possible for others what is possible for you*. All forms of current assets need to be managed well to improve the well-being of people who already exist. Sustainable development, by contrast, is more about intergenerational equity—*meeting the needs of the present without compromising the ability of future generations to meet their needs*. Thus, we can conserve resources and avoid long-term pollution partly out of a concern for future generations. Economists point to some

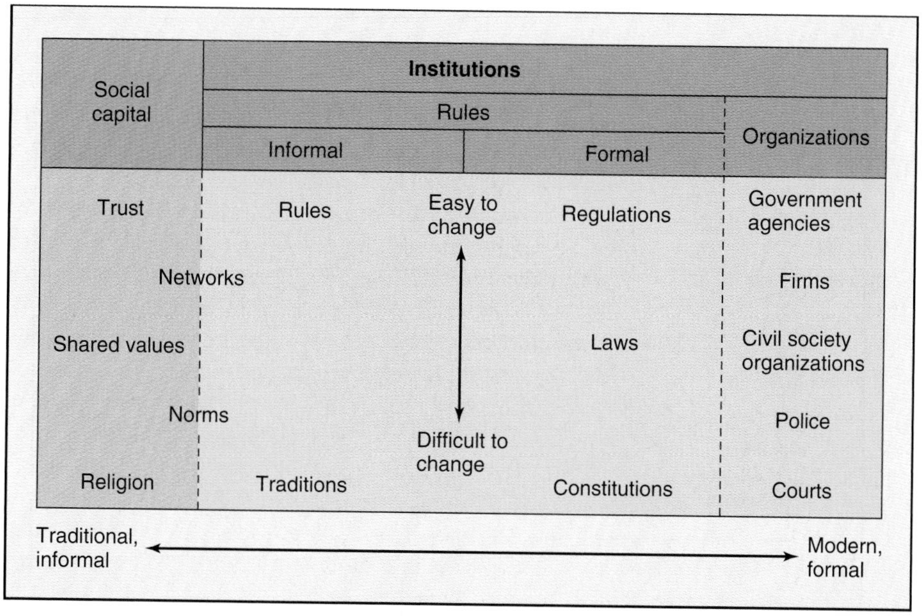


**Figure 22–7 The components of social capital.** Social norms, rules, and organizations for coordinating human behavior. (*Source:* World Development Report 2003. The World Bank. Used with permission.)

problems with this perspective, however. Economics deals with future resources by applying what is called the *discount rate*. By this, we mean the *rate* used for finding the present value of some future benefit or cost. The discount rate often approximates current interest rates. For example, a dollar is worth more to us today than it will be five years from now even without inflation, because we can put the dollar to use now and can earn interest on that dollar if we invest it. At a 5% rate of interest, in five years our dollar would be worth $1.28. Turning this idea around, a dollar five years from now is worth only $0.78 today.

***Use it Up!*** The conclusion is that some resource or benefit is worth more now than it will be in the future. This concept can be applied to a stand of trees. The owner has several options: (1) Cut all the trees and sell the timber today; (2) hold onto the trees in the hope that they will bring a better price at some time in the future; (3) harvest the trees on a sustainable basis, spreading out profits over time. If it turns out that the trees grow more slowly than current interest rates, the economically profitable decision would be to cut them now and invest the proceeds. This is a problem for natural capital, since it would appear to argue in many cases for maximizing short-term profit over sustainable, long-term use of the resource. In other words, reap the economic value now, and put off the ecological costs to the future. Similarly, if we let future generations cope with global climate change, we can continue to enjoy the benefits of the unrestrained use of fossil fuels. The same can be said about other resources, such as swordfish: Harvest them now, and when they're gone, switch to some other species of fish.

This approach presents a problem for intergenerational equity. What is at odds here is the self-interest of present individuals and generations versus the longer term (more sustainable) interests of a community or a society. To make things a bit more challenging, it might also be argued that conserving resources for the future could actually be thought of as putting the interests of future generations above those of, say, the poor today. Thus, intragenerational justice is no less important than intergenerational justice. Hopefully, the topic at least gives us an insight into how difficult it is to develop appropriate public policy that deals with the allocation of resources.

## 22.3 Pollution and Public Policy

Let us now focus more directly on environmental public policy as it is developed in order to cope with pollution. As we do, we will see additional areas of intersection between public policy and economics.

### Public-Policy Development: The Policy Life Cycle

Environmental public policy is developed in a sociopolitical context, usually in response to a problem. As we saw (Fig. 22–1), poor countries or regions face a group of poverty-related environmental issues, such as a lack of sanitation facilities or unsafe water supplies. As these countries develop, they become better able to deal with these basic issues through effective public policies. However, more affluence can bring new problems, like industrial air pollution, requiring further public-policy development, and sometimes unexpected problems, such as stratospheric ozone depletion, arise. When specific problems are addressed in a democratic society, the development of public policy often takes a predictable course, called the policy life cycle.

The typical policy life cycle has four stages: *recognition, formulation, implementation,* and *control* (Fig. 22–8). Each of the stages carries a certain amount of

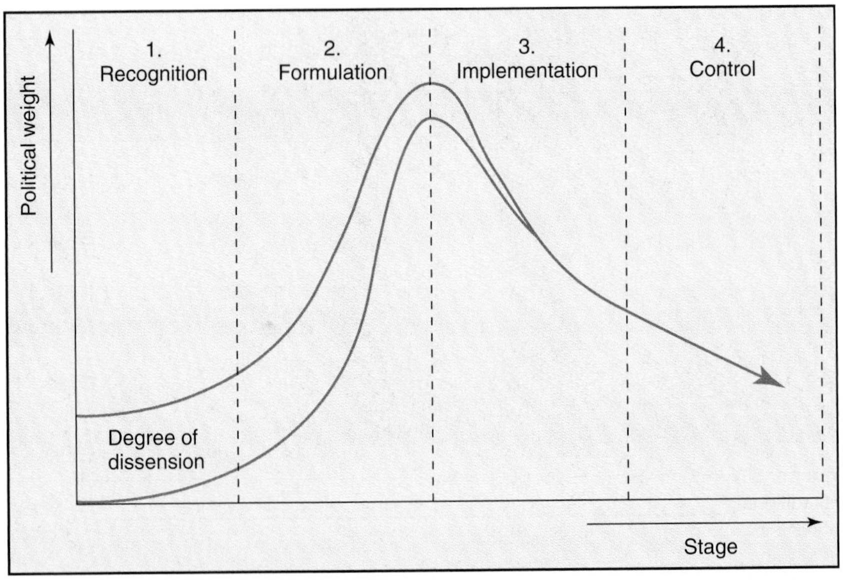


**Figure 22–8 The policy life cycle.** Most environmental issues pass through a policy life cycle in which an issue is accorded different degrees of political "weight" as it moves through the cycle. The final result is a policy that has been incorporated into the society and a problem that is under control.

political "weight," which varies over time and is represented in the figure by the course of the life-cycle line. The thickness of the line represents controversy and political uncertainty. We will use Rachel Carson and the formation of the Environmental Protection Agency as a case study to illustrate the policy life cycle.

***Recognition Stage.*** Alerted in 1957 by a close friend who had experienced the aerial spraying of her small bird sanctuary and the subsequent death of many birds, biologist Rachel Carson (Fig. 22–9) began to turn to technical and scientific literature to investigate what was known about the impacts of pesticides on the natural world. It took several years before Carson was ready to write a book on the subject. By that time, she was thoroughly aroused and angry at what she saw as a conspiracy to keep the public uninformed about pesticides. When her book *Silent Spring* was finally published in 1962, it ignited a firestorm of criticism from the chemical and agricultural establishment. However, it also caught the public's eye, and it quickly made its way to the president's Science Advisory Committee when President John F. Kennedy read a serialized version of it in the *New Yorker.* Kennedy charged the committee with studying the pesticide problem and recommending changes in public policy.

**Figure 22–9 Rachel Carson.** Marine scientist and author Rachel Carson, who wrote the book *Silent Spring,* which exposed the wide-scale use of pesticides and jump-started the environmental movement.

Illustrated by these events, the *recognition stage* of public policy is low in political weight. The stage begins with the early perceptions of an environmental problem, often coming as a result of scientific research. Scientists then publish their findings, and the media pick up the information and popularize it. When the public becomes involved, the political process is underway. During the recognition stage, dissension is high. Opposing views on the problem surface as businesses or technological industries respond to the bad news that they are at the root of some new environmental problem. Eventually, the problem gets attention from some level of the government, and the possibility of addressing it with public policy is considered.

***Formulation Stage.*** In 1963, President Kennedy's committee made recommendations that fully supported Carson's thesis in *Silent Spring.* It became clear to policymakers during the 1960s that environmental policy was weak and often compromised by ties to the very industries that were creating the problems. Public debate ensued, environmentalism in the form of new organizations and legal action developed, and the executive and legislative branches of government wrestled with what to do to establish effective environmental public policy. Finally, in 1969, Congress passed a bill known as the **Environmental Policy Act,** the first bill to recognize the interconnectedness of ecological systems and human enterprises. Shortly after that, a commission appointed by President Richard Nixon to study environmental policy recommended the creation of a new agency that would be responsible for dealing with air, water, solid waste, the use of pesticides, and radiation standards, in recognition of the interrelated web of water, soil, air, and biological life so eloquently pictured by Rachel Carson. The new agency, called the **Environmental Protection Agency** (EPA), was given a mandate to protect the environment against pressures from other governmental agencies and from industry, on behalf of the public. The year was 1970, the same year that 20 million Americans celebrated the first Earth Day.

Thus, the *formulation stage* is a stage of rapidly increasing political weight. The public is now aroused, and debate about policy options occurs in the corridors of power. The political battles may become fierce, as questions dealing with regulation and who will pay for the proposed changes are addressed. Media coverage is high, and politicians begin to hear from their constituencies. Lobbyists for special interests or environmental groups pressure legislators to soften or harden the policy under consideration. During the formulation stage, policymakers consider what may be called the *"Three E's"* of environmental public policy: **effectiveness** (whether the policy accomplishes what it intends to do in improving the environment), **efficiency** (whether the policy accomplishes its objectives at the least possible cost), and **equity** (whether the policy parcels out the financial burdens fairly among the different parties involved). Often, policymakers emphasize effectiveness over efficiency and equity at this stage of development, because they are looking for a workable solution and trying to make it into a law as soon as possible.

***Implementation Stage.*** Given its broad jurisdiction, the newly created EPA embarked idealistically on a course that would take the agency into regulatory enforcement and the environmental research needed to determine how to develop environmental protection standards. The 1970s were declared by President Nixon to be the "environmental decade." The EPA's first administrator, William Ruckelshaus, proposed that the agency's mission was, first and foremost, the "development of an environmental ethic." Within a few years, Congress enacted more than 20 major pieces of environmental legislation, giving the EPA increasing power to develop regulations. One of its most significant actions was to ban DDT and other persistent pesticides. Since then, the EPA has become the most powerful regulatory agency in the country, able to affect everything from the design of automobiles to standards for nuclear power plants and sanitary landfills.

At this point, the policy has reached the *implementation stage*, in which its real political and economic costs are exacted. The policy has been determined, and the focal point moves to regulatory agencies. During the implementation stage, public concern and political weight are declining. By now the issue is not very interesting to the media, and the emphasis shifts to the development of specific regulations and their enforcement. Industry learns how to comply with the new regulations. Over time, greater attention may be given to efficiency and equity as all the players in the process gain experience with the policy.

***Control Stage.*** In its three decades of existence, the EPA has become a mature and powerful agency as it has carried out its mandate to protect the environment and human health. (See the Guest Essay, p. 621, for a view from within of how this mandate is carried out.) We have seen many examples of the work of the EPA in previous chapters. The final stage in the policy life cycle is the *control stage*. By this point, years have passed since the early days of the recognition stage. Problems are rarely completely resolved, but the environment is improving, with things moving in the right direction. Policies (and their derived regulations) are broadly supported and often become embedded in the society, although their vulnerability to political shifts continues. Regulations may become more streamlined. The policymakers must now see that the problem is kept under control, and in due time the public often forgets that there ever was a serious problem.

The policy life cycle is a simplified view of what is frequently a highly complex and contentious political process. At any one time, different problems will be in different stages of the life cycle, as shown in Fig. 22–10 for a number of environmental problems in the industrialized countries. Also, countries in different stages of economic development will be in different stages of the policy life cycle for a given problem. For example, sickness and death due to polluted water are still serious problems in many developing countries, because public policies have not yet caught up with the need for water treatment and sewerage. In contrast, this problem is in the control stage in the industrialized societies. The reason for the discrepancy can often be traced directly to the costs that lie behind the development and implementation of public policy.

## Economic Effects of Environmental Public Policy

What are the relationships between a country's economy and its environmental public policy? Are there some policies that are too costly? How should a society parcel out its limited resources to address environmental problems? Are environmental regulations a burden on the economy? Let us begin to sort out these questions by looking at the relationship between environmental policies and costs.

As we have seen, policies do not just appear out of thin air. Instead, they are the result of some version of the policy life cycle. In countries undergoing economic development, the choice of environmental policies can often make the difference between rapid growth and stagnation. In the already industrialized countries, the best policies are those that conform to the "Three E's" (effectiveness, efficiency, and equity).

**Costs of Policies.** Some policies have relatively little or no direct *monetary cost*—they do not require major investments of administration or resources. Thus, removing subsidies to special interests and denying special access to national resources can result in a more efficient and equitable operation of the economy and can protect the environment. For example, the use of national lands for cattle grazing and timber harvesting is subsidized in the United States. As a result, the real costs—and environmental consequences—of these activities are borne not just by the special-interest groups that have access to the resources in question, but by all taxpayers. Removing such subsidies, however, can have very real *political costs*, as powerful interests do everything they can to hold onto their privileges.

Most environmental policies involve some very real costs that must be paid by some segment of society. For instance, resources are invested in policies that control pollution because their benefits are judged to outweigh their costs—that is, the public welfare is improved, and the environment is protected.

Who pays the costs? The equity principle implies that those benefiting from the policies—the customers of businesses whose activities are regulated and the people whose health is protected by laws against polluting—should pay. (When businesses are taxed or industries are forced to add new technologies, they pass their costs along to consumers, so the costs of public policies are borne by the public in one way or another.)

guest essay

## A Transformational Environmental Policy

Robert Claplanhoo Martin

*(The writer, Robert Claplanhoo Martin, is the former national ombudsman of the United States Environmental Protection Agency. He served for nearly 10 years, interceding with the government on behalf of American citizens and communities, from 1992 to 2002. He is a descendant of the traditional chief of the Makah Indian Nation in the Pacific Northwest. He received his B.A. in political science from Gordon College in 1979 and his J. D. from the George Washington University National Law Center in 1984. He now serves as the president and chief executive officer of the Legal Environmental Assistance Foundation in Tallahassee, Florida.)*

> Always do what is right. It will astonish most people and surprise the rest.
> —Mark Twain

For nearly 10 years, it was my duty and privilege to represent American citizens and communities as the national ombudsman of the United States Environmental Protection Agency (EPA). The historical purpose of the ombudsman centuries ago in Europe was to stand between the king and the people as an intercessor. The purpose of the ombudsman today is to stand between the people and the large, powerful institutions of modern-day society.

The ombudsman function established within the EPA by Congress stood as a true mediator between American communities harmed by toxic waste sites and the scientific bureaucracy responsible for protecting the health and the environment of those communities. The ombudsman function was an intercessor for the people with the senior management of the government and with the large corporations often responsible for toxic sites in communities.

On the one hand, the ombudsman functioned in a complex technical world inside an EPA full of environmental scientists and risk assessments. On the other hand, the ombudsman functioned in a world outside of the EPA, facing penultimate questions: Will citizens truly be heard? Will communities have a meaningful role in situations where their health and environment are at stake? Will justice be achieved, considering the harm done to children and the degradation of the environment? Will right be done?

Perhaps the most significant environmental policy achievement of the ombudsman function was to always strive for the government to do what was right for the people—that is, to stand firmly so that citizens might be truly heard, that communities could have a meaningful role when dealing with large corporations responsible for contaminating their neighborhoods, and that government science be thorough and sound.

As Benjamin Disraeli said long ago, "Justice is truth in action." This represents a *transformational paradigm* that can influence the making of environmental policy. The functional criteria of this paradigm are openness, transparency, and meaningful involvement. These attributes drive the percolation of the truth. This transformational paradigm operates to raise the quantum of truth in the formulation of environmental policy, much like an Archimedes screw can be turned to raise a quantum of water. And, when truth is operating between communities and the government, then environmental justice may be done. Government then can fulfill its appropriate role of assuring the common good in the making of environmental policy.

The EPA ombudsman function served to advocate for this kind of transformational paradigm for environmental policy making over and against a more rigid *transactional paradigm.* The transactional paradigm places the truth of citizens outside of the decision-making process and leaves no room for scientific anomalies that can contribute to our understanding of vexing toxic problems. In so doing, the ombudsman function questioned, often at a basic level, the standards governing permissible problems, concepts, and explanations that the EPA would allow at many toxic sites. This questioning sometimes led to valuable scientific discoveries.

It is not that we must oppose all efforts by government to manage the environment. On the contrary, we must every day support effective environmental policies, such as remedial investigations, feasibility studies, remedial designs, risk assessments, and "records of decision" (official EPA decisions leading to the resolution of specific problems). What must be opposed is when environmental policies result from the politics, power, and money that drive so much else in the government. Additionally, these policies must never become divorced from truth, justice, and the fundamental rights of citizens to due process.

Thus, by striving within the system for right, truth, and the ultimate empowerment of the American citizenry to set the course of environmental policy at both the community level and the national level, the ombudsman function may well have fundamentally changed the way government forms environmental policy. Simply put, in the purely transactional model of environmental policy making, what is possible drives what is right. In the transformational model of environmental policy making advocated by the ombudsman function, what is right drives what is possible to accomplish.

Positive outcomes of the ombudsman transformational paradigm may be seen in changed EPA decisions and approaches in communities across the nation. Examples include moving radioactive wastes out of Denver after the ombudsman questioned the risks and benefits of leaving it in place; replacing thermal incineration in Houston with a more innovative technology; and a reassessment of environmental contamination in New York City following the collapse of the World Trade Center on September 11, as a result of Ombudsman probing.

When a transformational paradigm of environmental policy making is adhered to, the public and the government engage in an upward spiral of truth, resulting in policies and programs that benefit the common good. When a transactional paradigm of environmental policy making is rigidly followed, however, the public and the government are likely to find themselves in a downward spiral of narrow-mindedness and even deception that does not allow for sound science and the assurance of the public welfare.

To the extent that the ombudsman function always provided a choice for transformational relationships between American communities and the government, it may have contributed toward a more sustainable path for the development of environmental policy and a more just future for all those who must live with the consequences of such policies.

We always have choice and are responsible for choosing our path wisely. We would do well to affirm a path that encourages transformational relationships between "we the people" and our government. We have the great fortune in our nation of being responsible for the government we have created.

**Impact on the Economy.** For years, special-interest groups have argued that environmental regulations are excessive and bad for the economy—that our concern for environmental protection costs hundreds of thousands of jobs, reduces our competitiveness in the marketplace, drives up the price of products and services, and, in general, imposes costs on the economy that are non-productive. These views are widespread and are routinely used to try to roll back the laws and regulations that represent environmental public policy.

How sound are these arguments? This concern has been addressed by many economists in recent years, and the general findings of their studies indicate that the "environment vs. economy trade-off" is a myth. Even without considering the environmental or health benefits, economists found that environmental protection has no significant adverse effect on the economy and, in fact, often has a positive effect.

***Anecdotal.*** First, we must recognize that much of the evidence used to support the charges that have been made is anecdotal. For example, according to the American Petroleum Institute, "environmental restrictions on energy extraction and production have caused the loss of 400,000 jobs," and according to Florida sugar growers, "Measures to protect the Everglades will cause the loss of 15,000 jobs." Indeed, numerous cases can be cited where *some* jobs were lost because of environmental regulations.

Anecdotal evidence can also be used in *favor* of environmental policy: according to the California Planning and Conservation League Foundation, "Recycling has created 14,000 jobs in California," and according to the EPA, "The Clean Air Act of 1990 will generate 60,000 new jobs." Indeed, *many* sectors of the U.S. economy that are most subject to environmental regulations—plastics, fabrics, etc.—have improved their efficiency and their competitiveness in the international market.

***Careful Studies.*** Is there something more substantial to help us with this important controversy? The best evidence comes from careful studies of what is actually happening. These studies reveal the following interesting findings:

- Estimated pollution control costs are only 1.72% of value added. (Value added is the increase in value from raw materials to final production.) Industries are not going to shut down, move overseas, or lay off thousands of workers for costs of this size.

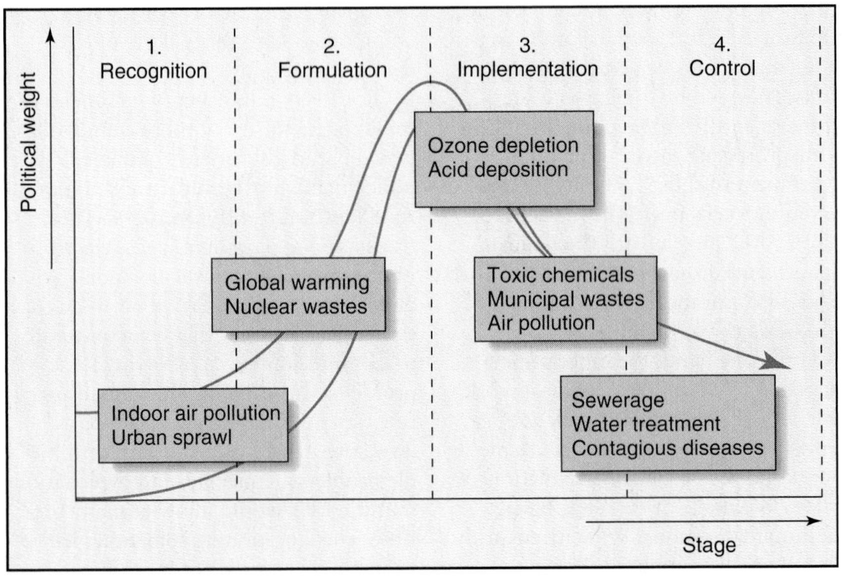


**Figure 22–10**
**Environmental problems in the policy life cycle.** Different environmental problems are in different stages of the policy life cycle in industrial societies.

- Actually, only 0.1% of job layoffs were attributed by employers to environment-related causes, according to a study by the U.S. Bureau of Labor Statistics.
- In general, states with the strictest environmental regulations also had the highest rates of job growth and economic performance.
- Nations with the highest environmental standards also had the most robust economies and rates of job creation.

In short, the results of many careful studies of the relationship between environmental protection and economic growth refute the suggestion that environmental regulation is bad for jobs and the economy. In fact, economic performance is highest where environmental public policy is most highly developed. Healthier workers are more productive and spend less on medical care. It may be no coincidence that in the former communist countries in Europe, where environmental public policy was deliberately suppressed in order to favor industry, both the economy and the environment are disaster areas. The evidence indicates that concerns over energy efficiency, pollution control, and conservation of resources have encouraged businesses to modify their technologies in ways that make them *more* competitive, not less, in the national and international marketplace. The evidence also suggests that at least as many jobs have been created by environmental protection as have been lost.

***How Much Cost?*** Taken as a whole, environmental protection is a huge industry. In a 1990 landmark study (*Environmental Investments: The Cost of a Clean Environment*), the EPA calculated that the estimated cost of complying with environmental regulations was about 2.1% of GDP, a figure that was projected to increase to 2.7% of GDP by 1997. In 1997 dollars, this is about $210 billion. Admittedly, this estimate is very rough, and some believe it is too low, while others think it is too high. To put the percentage into perspective, in 1997 the United States spent 10.6% of GDP on health care and 4.3% for national defense. Is it worth 2.7% of GDP to provide healthy air to breathe and clean water and ecosystems that can continue to provide essential goods and services if we are willing to spend 15% on our personal health and national security?

A study of total expenditures for the environmental protection industry for one year revealed that it was responsible for the following:

- $355 billion in total industry sales
- $14 billion in corporate profits
- $63 billion in federal, state, and local government revenues
- 4 million jobs

In sum, we can draw the following conclusions from our examination of the impact of environmental policy on the economy:

- Environmental public policy does not *diminish* the wealth of a nation; rather, it *transfers* wealth from polluters to pollution controllers and to less polluting companies.
- The "environmental protection industry" is a major job-creating, profit-making, sales-generating industry.
- The argument that environmental protection is bad for the economy is simply unsound. Not only is environmental protection good for the economy, but environmental public policy is responsible for a less hazardous, healthier, and more enjoyable environment.

## Policy Options: Market or Regulatory?

Suppose that the decision has been made to develop a policy to deal with an environmental problem. Now what? The objective of environmental public policy is to *change the behavior of polluters and resource users so as to benefit public welfare and the environment.* The two main ways to accomplish these behavior changes are by using a direct regulatory approach, in which standards are set and technologies are prescribed—the command-and-control strategy discussed in Chapter 21—and by using basic market approaches to set prices on pollution and the use of resources. Both approaches have their advantages and disadvantages. Many newer policies include elements of each.

***Command and Control.*** Traditionally, most of the laws that give the EPA its authority have prescribed command-and-control solutions. The laws directed the EPA to regulate pollution problems, often specifying many of the details to be used. For example, on the basis of knowing the health effects of pollutants, the legislators may choose to set standards that reflect the health of the most vulnerable members of the population. This is the case with the basic criteria covering air pollutants (Chapter 21). To meet these standards, regulations are established, and polluters are required by law to comply with the regulations. The regulatory approach also works well with land-use issues, in which certain values are upheld that will not necessarily be protected by a straightforward market approach. The regulatory approach has been used most commonly in environmental public policy. Indeed, many environmental problems are not readily amenable to market-based policies.

One of the shortcomings of the regulatory approach is that it practically guarantees a certain sustained level of pollution. If a polluter is told to use a particular technology or is given a cap on emissions, the policy gives the polluter no incentive to invest in technologies or a reduction in the source that would keep pollution at lower levels than allowed. A better way might be to set a standard (air quality or water quality criteria, for example), and let the polluter decide how best to achieve the standard. Here there is a command, but control is in the hands of the polluter. Pollution *prevention,* as discussed in Chapter 19, is the preferred course of action, and it is encouraging to see the EPA moving deliberately in that direction.

*Use the Market?* Market-based policies have the virtues of simplicity, efficiency, and (theoretically) equity. All polluters are treated equally and will choose their responses on the basis of economic principles having to do with profitability. With a market-based approach, there are strong incentives to reduce the costs of using resources or of paying for the right to a certain amount of pollution. The trading of emission allowances under the Clean Air Act of 1990 is a good example of the market approach to pollution control. User fees (pay as you throw) for the disposal of municipal solid waste is another. Several countries have restored and sustained fishery stocks by assigning individual quotas that would add up to a sustainable fishing take—the total allowable catch. (See Chapter 11.)

We turn next to the details of *benefit–cost analysis,* a specific tool that is employed in developing environmental public policy. In doing so, we will address the question of how to measure the cost-effectiveness of regulating pollution.

## 22.4 Benefit–Cost Analysis

As mentioned at the beginning of this chapter, Executive Order 12866 requires that all major regulations be supported by a benefit–cost analysis. Such an analysis begins by establishing a compelling need for the proposed regulation, proceeds to describe a range of alternative approaches, and then compares the estimated costs of the proposed action and the main alternatives with the benefits that will be achieved. All costs and benefits are given monetary values (where possible) and compared by means of what is commonly referred to as a *benefit–cost* (or *cost–benefit*) *ratio.* A favorable ratio for an action means that the benefits outweigh the costs. The action is said to be cost effective, and thus there is an economic justification for proceeding with it. The analysis may involve selecting the option with the best benefit–cost ratio, but if some of the benefits or costs cannot be expressed in monetary units, such a ratio can be misleading. If costs are projected to outweigh benefits, the project may be revised, dropped, or shelved for later consideration.

Benefit–cost analysis of environmental regulations is intended to build efficiency into policy so that society does not have to pay more than is necessary for a given level of environmental quality. If the analysis is done properly, it will consider *all* of the costs and benefits associated with a regulatory option. In so doing, it must address the problem of *externalities.*

### External and Internal Costs

In the language of economics, an external cost is an effect of a business process that is not included in the usual calculations of profit and loss. For example, when a business pollutes the air or water, the pollution imposes a cost on society in the form of poor health or the need to treat water before using it. This is an *external bad.* When workers improve their job performance as a result of experience and learning, the improvement is not credited in the company's ledgers, although it is considered an *external good.*

Amenities such as clean air, uncontaminated groundwater, and a healthy ozone layer are not privately owned. In the absence of regulatory controls, there are no direct costs to a business for degrading these amenities. (In other words, they are externalities.) Therefore, there are no incentives to refrain from polluting the air or water. By including *all* of the costs and benefits of a project or a regulation, benefit–cost analysis effectively brings the externalities into the economic accounting. One suggestion for accomplishing this is the use of green fees or taxes. (See the "Earth Watch" essay, p. 625.)

### The Costs of Environmental Regulations

The costs of pollution control include the price of purchasing, installing, operating, and maintaining pollution-control equipment and the price of implementing a control strategy. Even the banning of an offensive product costs money, because jobs are lost, new products must be developed, and machinery may have to be scrapped. In some instances, a pollution-control measure may result in the discovery of a less expensive way of doing something. In most cases, however, pollution control costs money. Thus, the effect of most regulations is to prevent an external bad by imposing economic costs that are ultimately shared by government, industry, and consumers.

*Costs Go Up.* Pollution-control costs generally increase exponentially with the level of control to be achieved (Fig. 22–11). That is, a partial reduction in pollution may be achieved by a few relatively inexpensive measures, but further reductions generally require increasingly expensive measures, and 100% control is likely to be impossible at any cost. Because of this exponential relationship, regulatory control often has to focus on stimulating new ways of reducing pollution. Indeed, the costs of pollution control constitute a powerful incentive to make substitutions, to recycle materials, or to redesign industrial processes.

## earth watch

### Green Fees and Taxes

Benefit–cost analysis works well only if all of the costs and benefits of a polluting activity are included in the calculations. Let us assume that this is possible. What would happen if everyone had to pay the true costs of their use of the environment as they used resources or degraded the air or water? Such charges are often called **green fees.** Oddly enough, two groups that frequently are battling—economists and environmentalists—agree that such a policy would go a long way toward solving many of our environmental problems.

Economists point out that, in a free-market economy, the market will guarantee that resources will be used in the most efficient way, because people will economize. This means that most businesses and people who use, say, 100 gallons of gasoline and have to pay a levy that reflects the true costs of that gasoline and the pollution it produces will be motivated to keep their use of gasoline to an absolute minimum. They will adopt more efficient vehicles or alternative modes of transportation. Including all of the external costs of the gasoline in its price would involve adding together the cost of maintaining our military presence in the Persian Gulf, the costs of air pollution, and even the expected costs of coping with global climate change. The green fee could be imposed as a tax on gasoline, and to ease its impact on the economy, the fee could be implemented gradually until it brings the price of gasoline to its true level.

Internalizing all these costs, if it were possible, would make $5 per gallon of gasoline at the pumps seem a bargain. Many observers have suggested that a tax of, say, $100 per ton of fossil-fuel carbon would go a long way toward accomplishing the objective of paying the real costs of the fuel. Great, say the environmentalists. This would mean that the people who do the most damage pay for their environmental impact, something environmental groups have been calling for all along. They insist that some of the revenues collected be used to mitigate the impacts of pollution and resource use. If that were to be accomplished, the outcome would move our society in the direction of sustainability.

Pollution fees, gasoline taxes, and user fees all have the potential for "internalizing" the external costs of environmental degradation and the use of resources. If such green fees could become public policy, other taxes (e.g., personal income taxes) might be able to be lowered, since these taxes are often the sources of revenue that are paying much of the external costs.

Would such a policy be politically feasible? Actually, it is already at work at the local level in thousands of communities with "pay as you throw" garbage collection. Communities simply charge a fixed rate per bag of trash collected at curbside, and the proceeds pay for the costs of municipal trash collection. If people are asked whether they *like* higher taxes, the answer is a highly predictable *no*. But if they are asked whether they would rather be taxed on their energy use, their trash production, and their purchases of high-impact goods than be taxed on their incomes and investments, they might well choose the former, especially if they can appreciate the environmental benefits that accrue from such a policy.

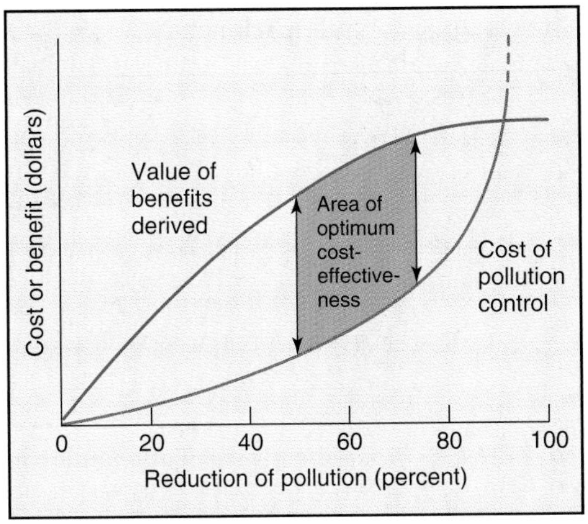


**Figure 22–11 The benefit–cost ratio for reducing pollution.** The cost of pollution control increases exponentially with the degree of control to be achieved. However, benefits to be derived from pollution control tend to level off and become negligible as pollutants are reduced to near or below threshold levels. The optimum cost-effectiveness is achieved at less than 100% control.

***Costs Go Down.*** In most cases, pollution-control technologies and strategies are understood and available. Thus, equipment, labor, and maintenance costs can be estimated fairly accurately. Unanticipated problems that increase costs may occur, but as technology advances and becomes more reliable, experience is gained, lower cost alternatives frequently emerge, and such unforeseen increases are negligible. The costs of pollution control are likely to be highest at the time they are initiated; then they decrease as time passes (Fig. 22–12a). The importance of this trend will become evident when we consider the time span over which costs and benefits are compared.

How accurate are the cost estimates? A look at some historical examples reveals that industry reports predict higher costs of regulations than government studies do, and surprisingly, even government analysts often overestimate the costs of regulations. For example, industry sources claimed that the costs of eliminating chlorofluorocarbons from automobile air conditioners would increase the price of new cars by anywhere from $650 to $1,200. In reality, the price increase turned out to be much lower—from $40 to $400. A recent study of benefit–cost analyses revealed that overestimates of costs occurred in 12 of 14 regulations. The primary reason given for the overestimates is a failure to anticipate the impact of technological innovation, admittedly a factor that is difficult to assess, although another reason might be political—to project costs so high that the regulators will back down.

The costs of improving the quality of the environment represent a major economic outlay. What benefits have we received in return?

## The Benefits of Environmental Regulation

**The Regulatory Right to Know Act** directs the OMB to submit an annual report to Congress containing, among other things, an estimate of the costs and benefits of federal rules. The 2003 report (*Informing Regulatory Decisions*) represents the most comprehensive federal study yet of costs and benefits over the 1992–2002 decade. The OMB found that the benefits (those which could be measured in monetary units) were three to six times the costs. Benefits, for example, were estimated at between $146 and 231 billion, while industries, states, and municipalities spent an estimated $37 to 43 billion to comply with the rules. Fewer hospital visits and reductions in both premature deaths and lost workdays were benefits associated with reductions in air pollutants mandated by the Clean Air Act.

**Calculating Benefits.** Such benefits of regulatory policies are seldom as easy to calculate as the costs. Estimating benefits is often a matter of estimating the costs of poor health and damages that *would have* occurred if the regulations were not imposed (Fig. 22–12b). For example, the projected environmental damage that would be brought on by a given level of sulfur dioxide emissions from a coal-fired power plant (an external bad) becomes a benefit (an external good) when a regulatory action prevents half of those emissions. Benefits include such things as improved human health, reduced corrosion and deterioration of materials, reduced damage to natural resources, the preservation of aesthetic and spiritual qualities of the environment, increased opportunities for outdoor recreation, and continued opportunities to use the environment in the future. The dollar value of these benefits is derived by estimating, for instance, the reduction in health-care costs, the reduction in maintenance and replacement costs, and the economic value generated by the enhanced recreational activity. Examples of benefits are listed in Table 22–3. The values of some benefits can be estimated fairly accurately. For example, air-pollution episodes increase the number of people seeking medical attention. Since the medical attention provided has a known dollar value, eliminating a certain number of air-pollution episodes provides a health benefit of that value.

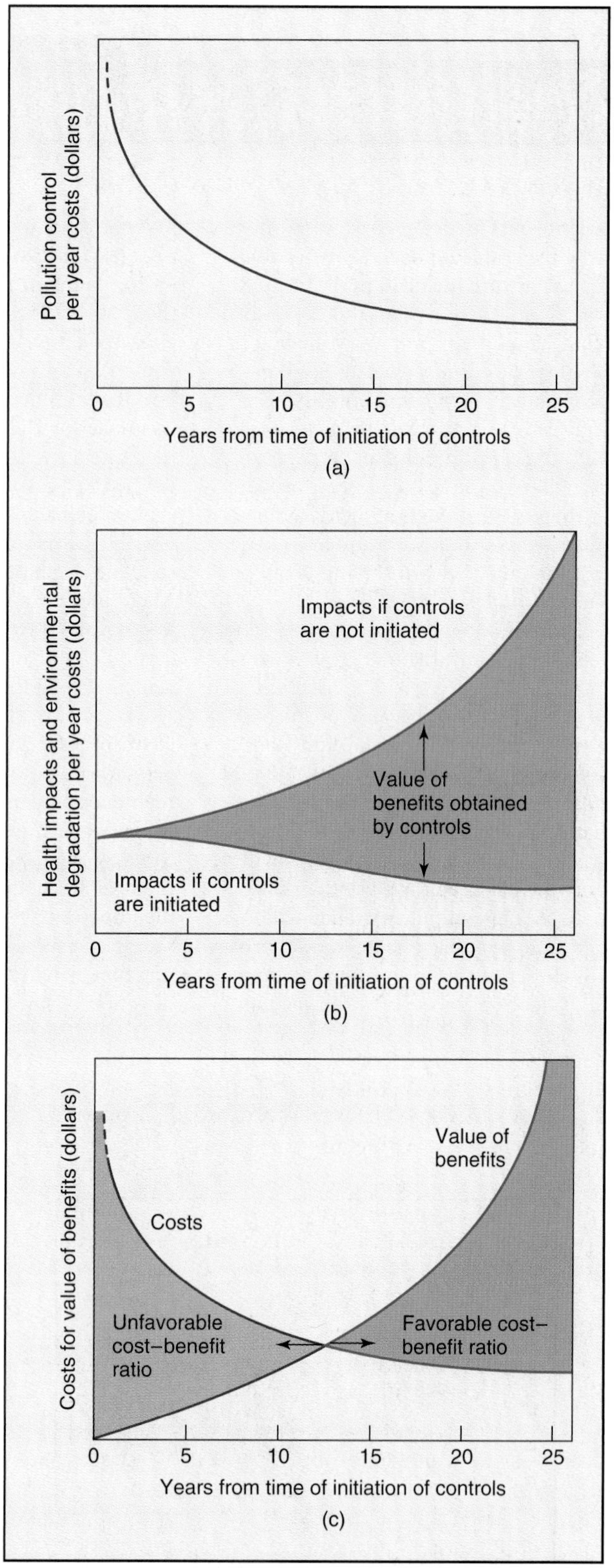


**Figure 22–12 Cost-effectiveness of pollution control over time.** (a) Pollution-control strategies generally demand high initial costs. The costs then generally decline as those strategies are absorbed into the overall economy. (b) Benefits may be negligible in the short term, but they increase as environmental and human health recover from the impacts of pollution or are spared increasing degradation. (c) When the two curves are compared, we see that what may appear as cost-ineffective expenditures in the short term (5–10 years) may, in fact, be highly cost effective in the long term.

**table 22-3 Benefits That May Be Gained by the Reduction or Prevention of Pollution**

| | |
|---|---|
| **1. Improved human health** | Reduction and prevention of pollution-related illnesses and premature deaths<br>Reduction of worker stress caused by pollution<br>Increased worker productivity |
| **2. Improved agriculture and forest production** | Reduction of pollution-related damage<br>More vigorous growth by removal of stress due to pollution<br>Higher farm profits, benefiting all agriculture-related industries |
| **3. Enhanced commercial or sport fishing** | Increased value of fish and shellfish harvests<br>Increased sales of boats, motors, tackle, and bait<br>Enhancement of businesses serving fishermen |
| **4. Enhancement of recreational opportunities** | Direct uses, such as swimming and boating<br>Indirect uses, such as observing wildlife<br>Enhancement of businesses serving vacationers |
| **5. Extended lifetime of materials and less necessity for cleaning** | Reduction of corrosive effects of pollution, extending the lifetime of metals, textiles, rubber, paint, and other coatings<br>Reduction of cleaning costs |
| **6. Enhancement of real-estate values** | Increased property values as the environment improves |

***Shadow Pricing.*** Many benefits, however, are difficult to estimate. Accurate benefit–cost analysis depends on assigning monetary values to every benefit, but how can a dollar value be put on maintaining the integrity of a coastal wetland or on the enjoyment of breathing cleaner air? The answer is to find out how much people are willing to pay to maintain these benefits. How can this be done, though, if there is no free market for the benefits? Again, economists have an answer: *Shadow pricing* involves asking people what they *might* pay for a particular benefit if it were up to them to decide. For example, to evaluate the benefits of cleaner air in the Los Angeles basin, homeowners were asked to place a value on improving their air quality from *poor* to *fair*. The average response (in 1977) was $30 a month. The total benefit was then calculated from the number of households in the basin, multiplied by the $30 average value.

**The Value of Human Life.** Shadow pricing becomes difficult when the analysis has to place a value on human life. Many of the pollutants to which people are exposed are hazardous, exacting a toll on health and life expectancy. To estimate the benefits of regulating such pollutants, it is necessary to calculate how many lives will be saved or how many people will enjoy better health. Finding a value for these benefits is fraught with ethical difficulties. For example, the EPA calculated that the new clean-air standards for ozone and particulates would prevent 15,000 premature deaths annually. Different approaches have resulted in a range of values for human life from tens of thousands of dollars to $10 million per life saved. Any benefit–cost analysis that must factor in the risk to human life requires that human lives be valued somewhere within this very broad range. For example, the OMB valued each cancer case at $5 million, each traffic fatality at $3 million, and each lost-workday injury at $50,000.

**Nonhuman Environmental Components.** How does shadow pricing work for the nonhuman components of natural environments—for a population of rare wildflowers, for example, or a wilderness site? These things depend entirely on how willing people are to pay for the preservation of those components. Again, monetary values must be assigned to their existence, and again the outcome will be strongly influenced by a highly subjective element in the analysis.

## Cost-Effectiveness Analysis

Cost-effectiveness analysis is an alternative option for evaluating the costs of regulations. Here, the merits of the goal are accepted, and the question is, How can that goal be achieved at the least cost? To find out, alternative strategies for reaching the goal are analyzed for costs, and the least costly method is adopted.

Cost-effectiveness can be applied to the desired level of pollution control. As Figure 22–11 shows, a significant benefit may be achieved by modest degrees of cleanup. Note, however, how differently the cost and benefit curves behave with increasing reduction of pollution. Little additional benefits are realized when cleanup begins to approach 100%, yet costs increase exponentially. This behavior follows from the fact that living organisms—including humans—can often tolerate a threshold level of pollution without ill effect. (See Fig. 19–1.) Therefore, reducing the level of a pollutant below threshold levels will not yield an observable improvement.

Figure 22–11 demonstrates that, with a modest degree of cleanup, benefits can outweigh costs. At some point in the cleanup effort, the lines cross and the costs exceed the benefits. Consequently, while it is tempting to argue that we should strive for 100% control, demanding more than 90% control may involve enormous costs with little or no added benefit. At the point when the control of a particular pollutant reaches 90%, it makes more sense to allocate dollars and effort to other projects in which greater benefits may be achieved for the money spent. Optimum cost-effectiveness that meets the efficiency criterion for public policy is achieved at the point where the benefit curve is the greatest distance above the cost curve.

However, the perspective of time should be considered in calculating cost-effectiveness (Fig. 22–12c). A situation that appears to be cost ineffective in the short term may prove extremely cost effective in the long term. This is particularly true for problems such as acid deposition or groundwater contamination from toxic wastes.

**Progress.** What has been the result of benefit–cost analyses to date? Pollution of air and surface water reached critical levels in many areas of the United States in the late 1960s, and since that time huge sums of money have been spent on pollution abatement. Benefit–cost analysis shows that, overall, these expenditures have more than paid for themselves in decreased health-care costs and enhanced environmental quality. Performing the analysis also forces regulators and the regulated to assess the impacts of proposed policy changes more carefully.

***The EPA Checks In.*** The costs of controlling pollution are high, so what benefits do we receive in exchange for this cost? The EPA prepared a *Draft Report on the Environment* in 2003 and listed progress made, as well as many of the challenges ahead. The following are some of the accomplishments of environmental public policy, as reported by the EPA:

- Since 1970, total emissions of six common air pollutants have decreased by an average of 25%.
- Since 1976, average levels of lead in the blood of children have declined by nearly 85%.
- Between 1988 and 2000, releases of toxic chemicals, as reported to the Toxics Release Inventory, have decreased 48%.
- In 2002, states reported that 94% of the population on community water systems had drinking water that met all of the standards, as opposed to 79% in 1993.
- Toxic air emissions have fallen by 24% since 1990.
- More than 297,000 underground storage tanks have been cleaned up since 1990.
- Since 1980, 846 out of 1,450 Superfund sites on the National Priorities List have been completely cleaned up.
- Recovery of municipal solid wastes by recycling has increased from 7% in 1970 to 30% in 2000.
- The rate of annual wetland losses has declined from 500,000 acres a year three decades ago to less than 100,000 acres a year today.

***In Spite of Growth!*** Since 1970, these improvements have occurred even as the economy grew by 161%, the population rose by 39%, and the number of motor vehicle miles driven increased by 149%. Do the benefits outweigh the costs of the improvements? To answer this question, consider the phaseout of leaded gasoline as just one example. The project has cost about $3.6 billion, according to an EPA benefit–cost report. Benefits were valued at over $50 billion, $42 billion of which were for medical costs that were avoided!

Executive Order 12866 is still in effect. A benefit–cost analysis accompanies every new regulatory rule from the EPA and other federal agencies. Even when a rule is not classified as significant, the accompanying documentation must demonstrate, by benefit–cost analysis, that the rule does not qualify as significant. In the Unfunded Mandates Act of 1995, Congress backed up Executive Order 12866 by requiring all agencies to prepare benefit–cost analyses for any regulations imposing costs of $100 million or more in a given year. Like it or not, benefit–cost analysis is now part of public policy in the United States.

## 22.5 Politics, the Public, and Public Policy

Environmental public-policy legislation is the responsibility of Congress. Once a piece of legislation is passed by the House of Representatives and the Senate and is signed into law by the president, it is then implemented by appropriate governmental agencies. This is where the EPA gets its authority to draw up regulations for setting new standards for air pollutants. One final role Congress must play is to pass appropriations for the various programs that have already been authorized by law. The executive branch (the president) draws up a budget asking for appropriations for all of the governmental agencies and programs, and Congress decides how much of what the president asks for will actually get funded. Thus, the two basic times Congress can influence a given issue are when it is formulated into law and when it receives appropriations. The appropriations must be authorized every year, and new laws can always amend or abolish existing ones. Given this basic process, there are many opportunities for partisan politics to set environmental public policy.

## Politics and the Environment

Party affiliations can make a difference in environmental public policy. The Republican party has a large conservative contingent that embraces policy reform to reduce the regulation of industry so that market forces and private enterprise can be freed up "to move the economy forward." The same contingent also defends private property rights threatened by programs like the Endangered Species Act and wetlands regulations. A very well articulated environmental backlash called the Wise Use movement emerged in the 1990s. This movement views environmentalists as the enemy of free enterprise, responsible for promoting a false view of the Earth as fragile and in crisis because of human mismanagement. The anti-environmental movement has received the support of many conservative Republican legislators.

*Party Control.* Republicans have controlled Congress since 1995 (the 104th Congress), with the exception of a Democratically controlled Senate in 2001–2002. Under President Bill Clinton, the Democrats held the White House from 1993 to 2000, while Republican President George W. Bush has been in office since 2001. During the last 10 years, there have been many struggles over environmental legislation and regulations.

The 104th Congress (1995–96) began with an effort to dismantle many key laws, especially the Clean Water Act, the Clean Air Act, and Superfund. When it became apparent what was happening, heavy grassroots work by many environmental organizations and extensive media coverage made it clear that the public did not support this anti-environmental thrust. As a result, most of the anti-environmental legislation died in the Senate or was vetoed by the president. The 105th and 106th Congresses saw some of the same anti-environmental forces attempting once more to get their legislation through. Again, President Clinton and the Democratic minority were able to hold off new attempts to weaken environmental laws. However, one effect of these partisan battles was a failure to reauthorize a number of important laws, such as the Endangered Species Act and the Superfund program. The laws continued in force nevertheless.

*Special-Interest Politics.* The role and power of special-interest groups became evident during these struggles. In the 1970s and 1980s, special interests (such as the coal industry or the fisheries industry) would hire lobbyists who would use their influence to try to convince congressional committee members to act in their favor on issues that were important to them. Environmental NGOs would then use the media to expose the issues to the public and call on their constituencies to telephone and send mail to their representatives in Congress. The anti-environmental interests caught on to the NGO strategy and began to mimic what they were doing, but more effectively, because they had deeper pockets. Soon a constant battle between special-interest groups emerged in which each would use the media and telephone calls (enhanced by phone banks, which can generate thousands of calls to legislators in a short period) to pressure members of Congress. Added to this picture is the massive explosion in campaign contributions. Because they have much more money, the anti-environmental forces have been able in many instances to overwhelm the environmental organizations in this battle of influence. For example, a huge effort was mounted by the fossil-fuel industries to discredit the scientific basis for global climate change and thereby scuttle the Kyoto accord in 1997. (The effort failed, and the United States signed the accord, although Congress did not ratify it.)

What all this means is that environmental concerns have become a battle of special-interest groups—that environmental protection is seen as one among many interests in American politics. The issues are up for grabs, and protecting the environment often depends largely on the strength of environmental-interest groups. James Skillen, executive director of the Center for Public Justice, says that this is a seriously flawed development. In this setting, ecological well-being not only must have scientific proof of its validity, but also has to intrude into the political arena in competition with opposing special interests and *win*. Skillen believes that environmental justice—meaning the protection of land, air, water, and the associated biota—should not have to depend on an interest group for its advocacy. Rather, it is a matter of the *public good* and, as such, should be seen as a basic responsibility of government. To quote Skillen, "Recognition in basic law of the necessity of ecological health as the precondition of all public and market relations should become as fundamental as the recognition that certain human rights exist as the precondition of all public rules and market regulations."[3] The NGOs stepped into the gap because they found that doing so was necessary.

***W Steps In.*** George W. Bush identified himself as an environmentalist during the 2000 presidential campaign and, at the time, promised to reduce carbon dioxide emissions in an effort to combat global warming. Shortly after his contested election (over irregularities in Florida's electoral process), President Bush began what was called by one commentator an "environmental meltdown in the White House." On his first day in office, he directed all cabinet members to freeze more than 50 regulations that had been approved toward the end of the Clinton administration. This was to be the beginning of a concerted move to use the regulatory process as the key to making changes in environmental policies. There is no doubt that regulations should be examined from time to time and, indeed, reformed. The 2003 OMB report (*Informing Regulatory Decisions*) identified over 300 regulations that were nominated by solicitation from various (regulated)

---

[3]Skillen, James, "How can we Do Justice to Both Public and Private Trusts?" October 1996. Global Stewardship Initiative National Conference, Gloucester, MA. http://cesc.montreat.edu/GSI/GSI-Conf/discussion/Skillen.html.

organizations to be reformed. The OMB then submitted the recommendations to the appropriate regulatory agencies, which would decide what to do about the suggestions. For example, the Boeing company proposed that the Federal Aviation Administration lower the minimum amount of fresh air required in airplane cabins because the current minimum is "unattainable."

***What's the Record?*** Most of the changes in regulations, however, have been initiated from within the administration (possibly influenced by recommendations from affected parties). The Natural Resources Defense Council (NRDC) has been keeping tabs on the Bush administration's environmental record. The tally of efforts to weaken regulations is already far more extensive than that of any administration (see the NRDC Web site, www.nrdc.org/bushrecord/)—even that of President Ronald Reagan, who was no friend of environmental regulation. A number of these proposed or accomplished regulatory changes have been cited in earlier chapters. Perhaps the most notorious of President Bush's decisions were his rejection of the Kyoto Protocol and his retreat from his campaign promise to reduce carbon dioxide emissions (Chapter 20). The Bush–Cheney energy plan (Chapter 12) clearly favors the petroleum and automotive industries and proposes drilling for oil on the ANWR. These and other elements of the plan are included in a major energy bill making its way through Congress. You are encouraged to consult the NRDC Web site to explore the Bush record.

***What's the Score?*** Not all Republicans are in favor of the administration's environmental record. The organization Republicans for Environmental Protection works to make "natural resource conservation and sound environmental protection fundamental elements of the Republican Party's vision for America," as the group's Web site states it. Some Republican governors, such as George Pataki of New York and Bob Taft of Ohio, are making important environmental policy in their states. On the other side of the coin, not all Democrats are pro environment. The League of Conservation Voters keeps track of how senators and representatives vote on environmental issues and rates every member of Congress on the organization's "National Environmental Scoreboard" (www.lcv.org/scoreboard/).

In the end, however, the ultimate responsibility for the nation's environmental policy should reside with the public.

### Citizen Involvement

How is the general public involved in public policy? The public is involved directly in public policy because people pay for the benefits they receive through higher taxes and higher costs on the products they consume. Another way to ask this question is, What can you do to see that environmental concerns get the attention they deserve? Here are some suggestions:

- Become involved in local environmental problems. Local jurisdictions often have the power to make crucial decisions affecting the environment, and groups of citizens working together can often make a difference in how the decisions go.
- Become a member of an environmental NGO that informs its constituencies of environmental concerns around the country.
- Contact your legislators to inform them of your support for a particular environmental public policy under consideration during the policy formulation stage.
- Take note of the viewpoint of any political candidate on matters of environmental policy. Public concern about environmental policy can play an important role in the election or defeat of political candidates. Volunteer to work for environment-friendly candidates.
- Comment on regulations when they are first proposed; in some situations, the public can bring a class action suit against polluters.
- Stay informed on environmental affairs; only when the public is informed is it likely that grassroots support will be maintained and public environmental policy will really reflect public opinion.

## revisiting the themes

### Sustainability

The ecological economists have put sustainability in the center of their economic theories. They rightly point out that the natural environment encompasses the economy. Therefore, there are limits to global economic growth that are set by the environment, and the economy must come to a steady state at some point. The World Bank's efforts to measure all of the wealth of nations also focuses on sustainability—in this case, sustainable development to improve human well-being over time. A nation must manage its assets, but it can't do that unless it knows what they are and how to measure them. Depleting assets of any kind is a clear signal of unsustainable practice, so a tool like the U.N.'s System for Integrated Environmental Accounting is a positive development. Countries now have no excuse not to do environmental accounting. One problem that crops up is that of intergenerational equity: It may be more profitable to use a resource now than to conserve it for future generations. In this case, people must be willing to put aside their own self-interests in favor of those yet unborn.

### Stewardship

Table 22–1 demonstrates the consequences of poor environmental stewardship; laws to protect the environment are not a luxury, but an essential foundation for a society. Then EPA Administrator Christine Todd Whitman introduced the agency's *Draft Report on the Environment 2003* with these words: "We are all stewards of this shared planet, responsible for protecting and preserving a precious heritage for our children and grandchildren. As long as we work together and stay firmly focused on our goals, I am confident we will make our air cleaner, our water purer, and our land better protected for future generations." Environmental public policy over the past 30 years has accomplished much in achieving a safer and more healthy environment. This represents our society's clear commitment to planetary stewardship. In light of the Bush administration's policies, however, it seems fair to ask, Where is the stewardship?

### Ecosystem Capital

It took a long time for economists to discover natural capital. The ecological economists have now made it an essential concept to be reckoned with in any economic analysis. It is now unthinkable to ignore ecosystem capital (hence, we have focused on renewable resources) in an accounting of a nation's economy. The GDP has been exposed as a poor measure of true economic activity—one that should be replaced with a measure like the NNP, which includes depreciation of all forms of a nation's wealth, especially ecosystem capital, which had long been thought of as a "gift from nature."

### Policy and Politics

Since this entire chapter is about policy and politics, it would be redundant to try to summarize this theme. Instead, we remind you that economics and public policy form a bottom line for environmental decisions, as unfortunate as this seems. It's not *what is right to do,* but *what is possible to do.*

### Globalization

The United Nations is a form of globalization, one that, through its Statistics Division and the World Bank, has helped formulate public policy and make it understandable. In contrast, the WTO has attempted to organize global trade, but has consistently demonstrated some of the more ugly effects of globalization, as the wealthy nations continue to maintain their economic interests vis-à-vis the poor nations. (See "Global Perspective," this page.) Free trade seems like a good idea, but few countries really want to engage in it with other nations. Subsidies and tariffs remain the sticking point in WTO negotiations, and only as the poor nations organize are they going to get anywhere with their protests against the wealthy nations' subsidies.

## global perspective

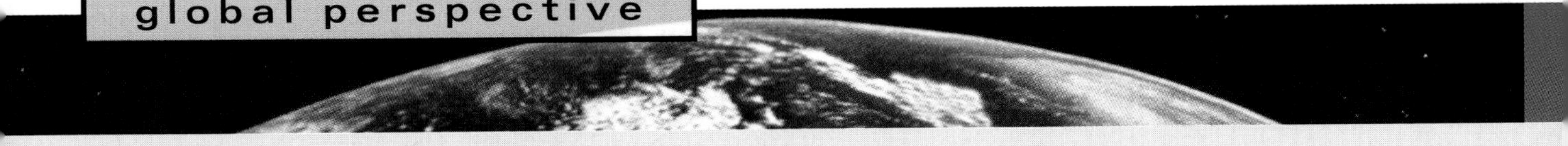

### The World Trade Organization

Trade is a fundamental economic activity. The exchange of goods drives the engine of the free-market economy. In today's world, trade has become increasingly globalized, and in the process, world trade has become one of the most dominant forces in society. Many, however, question whether this is good, and their concerns become headline news every time the World Trade Organization (WTO) meets. For example, in Seattle, Washington, in December 1999, tens of thousands of protesters (called "globophobes" by some) took to the streets to block the delegates from assembling—and were the recipients of tear gas, pepper spray, and rubber bullets as things began to turn ugly. The protesters were an unusual alliance of environmental activists and organized labor, groups that are often at odds over timber harvests and fishing. What were their issues?

The WTO was created in 1993 by trade ministers from many countries meeting to build on the foundation for global trade laid by the General Agreement on Tariffs and Trade (GATT). The WTO was given the responsibility for implementing numerous trade rules and, in doing so, was empowered to enforce the rules, with the option of imposing stiff trade penalties on violators. In effect, the WTO became a force for the globalization of trade, facilitating it by reducing or eliminating trade barriers (tariffs) between countries and subsidies to domestic industries that tended to discriminate against foreign-produced goods. Those who support the WTO claim that free trade between countries is the key to global economic growth and prosperity. This all sounds very good, but as the saying goes, "The devil is in the details."

### Free-trade Rules

Since its creation, the WTO has adjudicated a number of international trade issues and, in the process, has established a reputation for elevating free trade over substantial human rights and environmental resource concerns. Furthermore, many trade agreements and the dispute rulings are carried out in closed sessions. For example, the WTO ruled against a U.S. law requiring any imports of shrimp to be certified as turtle safe (that is, the shrimp were to be certified as harvested by trawlers employing "turtle exclusion devices," or TEDs). Even though TEDs have been proven to reduce turtle mortality by 97%, the WTO ruled that the restriction was a violation of WTO rules. Other issues that have come before the WTO are a U.S. ban on the import of wooden crates from China because they harbored the Asian long-horned beetle, a serious threat to forest trees; a European Union ban of hormone-treated beef and any products involving genetically modified organisms from the United States; and bans against developing countries' products that utilize child labor.

### Cancún

In the Seattle meetings, the talks broke down, not because of the protests, but because of a refusal of the key players—the United States, Japan, and Europe—to move from their positions. In the WTO meeting held in Cancún, Mexico, in September 2003, the talks broke down again before negotiations could proceed. This time, a coalition of 22 developing nations (G-22) representing 60% of the world's people simply walked out of the meetings. Their issue was primarily the import barriers and subsidies that distort global trade in agricultural products. The wealthy nations spend $300 billion a year to subsidize their farmers. The result is depressed world markets for such products as cotton. The United States, for example, subsidizes cotton farmers with $3.6 billion. This depresses the world cotton market prices by one-fourth, since American farmers supply some 40% of the global cotton market.

The future of the WTO is unclear. It retains the power already delegated to it, but in the absence of any new negotiations, the controversies over secrecy, labor, agriculture, and environmental issues will cloud its work. The need for reform is evident. The new clout of the G-22 nations (China, India, Mexico, Indonesia, and many others) may well bring reform and force the rich countries to come to terms.

## review questions

1. What is the overall objective of environmental public policy, and what are the objective's two most central concerns?
2. Review Table 22–1, and then describe the effects on health and productivity associated with the following environmental problems: water pollution, air pollution, hazardous wastes, soil degradation, deforestation, loss of biodiversity, and atmospheric changes.
3. Three patterns of environmental indicators are associated with differences in the level of development of a nation. What problems decline, what problems increase and then decline, and what problems increase with the level of development?
4. Name the two basic kinds of economic systems and explain how they differ.
5. Describe the three components of a nation's wealth. Evaluate the wealth of the world's nations, as measured by these three components.
6. Why is the gross national product (GNP) an inaccurate indicator of a nation's economic status? How is it corrected by the net national product (NNP)?
7. What conditions are necessary for a country to make progress in the development of human resources and produced assets?
8. Explain the importance of considering both intragenerational equity and intergenerational equity in addressing resource allocation issues.
9. List the four stages of the policy life cycle, and show how *Silent Spring* and the formation of the EPA illustrate these stages.
10. What are the "Three E's" for evaluating environmental public policy?
11. What is the conclusion of careful studies regarding the relationship between environmental policies, on the one hand, and jobs and the economy, on the other?
12. What are the advantages and disadvantages of the regulatory approach vs. a market approach to policy development?
13. Define the term *benefit–cost analysis* as it relates to environmental regulation. How does this analysis method address external costs? Distinguish it from cost-effectiveness analysis.
14. List specific costs and benefits of pollution control.
15. Discuss the cost-effectiveness of pollution control. How much progress have we made in pollution control in the last 25 years?
16. How has the political arena affected environmental public policy in the last few years? What has been the role of special-interest groups? What can your role be?

## thinking environmentally

1. Investigate the environmental public policies of a developing country, and measure them against the information presented in Table 22–1.

2. Suppose it was discovered that the bleach that is commonly used for laundry was carcinogenic. Referring to the policy life cycle, describe a predictable course of events until the problem is brought under control.

3. Stage a debate on the following resolution: "Environmental regulation is bad for the economy."

4. Consult the League of Conservation Voters' Scoreboard (www.lcv.org/scoreboard/) to find out how your state's senators and representatives vote on environmental issues.

5. Consider the complexities and limitations of benefit–cost analysis. Should we continue to support its use to determine public policy? How great a role should it play? Support your answers.

# Sustainable Communities and Lifestyles

**Key Topics**

1. Urban Sprawl
2. Urban Blight
3. Moving Toward Sustainable Communities
4. Toward the Common Good

Waitakere City (opposite) is the sixth-largest city in New Zealand, with a population of 169,000 and a land area of 367 mi$^2$ (954 km$^2$), of which 7% is designated "living" area. Waitakere City has been experiencing the typical problems of urban growth, such as urban sprawl, traffic congestion, and pressure on natural areas within the city. In addition, the city reflects a diverse ethnic makeup: a distinctly European ethnic majority and minorities of several cultural groups, but especially the native Maori. This city formally adopted *Agenda 21* in 1993 and became New Zealand's first ecocity.

***Greenprint.*** Adopting the *Agenda 21* goal of protecting the indigenous people and cultures of countries while meeting other goals of sustainable development, Waitakere City focused on Maori principles and values first and then on New Zealand legislation and Agenda 21 principles to develop the guidelines for the city's efforts. Using the Maori perspective of *tangata whenua*, or "people of the land," which incorporates ecological and stewardship principles into the relationship between people and land, the city developed a strategic plan called *Greenprint*. In this plan, the Maori are included in decision making by the city council, and Maori principles serve as guidelines for the city's ongoing growth and development. The Greenprint plan makes specific reference to the principles of *taonga* (a community's treasures, such as forests, rivers, animals, humans, and the health of the environment) and *kaitiaki* (those selected by a tribe to be stewards of the *taonga*).

***Green Network.*** With input from the Maori, the city council has developed a Green Network Plan with the objective of "protecting, restoring, and enhancing the natural environment across the whole city, while increasing human enjoyment and appreciation of that environment. This goal places human beings within the natural world, as stewards of its natural and spiritual health." The whole community is enlisted in "greening the city": restoring native plants, providing

**Waitakere City, New Zealand New Zealand's first ecocity, Waitakere City, covers a large area in Auckland metropolitan area, New Zealand. Shown here is one of the more developed areas, called Titirangi.**

corridors for wildlife, and providing access to the parks and reserves that will maintain people's safety (from crime). Because the city's population is growing, the network plan will be continually challenged by issues related to urban sprawl, such as loss of habitat and the impacts of exotic species and pets on native wildlife. Other Greenprint initiatives include improving housing, developing subdivisions, purchasing and retailing goods and services, and fostering local business ventures, all guided by stewardship principles. With these actions, Waitakere City has accepted the challenge of becoming an ecocity and has embarked on a course that has made it a world leader in the development of sustainable cities.

***Trenton, New Jersey.*** On the other side of the world, citizens of Trenton, New Jersey, have made remarkable progress in rehabilitating their city, with the aid of Isles, Inc. (from the concept of neighborhood-scale "islands of redevelopment"), an urban-planning company. Trenton, with a multicultural population of 89,000, is the capital of New Jersey. Isles, Inc., was founded in 1981 by nearby Princeton University personnel, but quickly evolved into a locally owned and controlled development organization.

A thriving industrial city a century ago, Trenton was suffering from the all-too-common problems of a decaying urban center, ethnic segregation, and urban blight. To date, Isles has fostered the following changes: Vacant land has been transformed into 15 community garden sites; almost 300 units of affordable housing have been created; numerous unemployed young people have been given job training in connection with the affordable housing projects; a career center provides academic and job training, counseling, and leadership training to at-risk, inner city youths; environmental education programs are provided yearly to 3,000 students; a recent capital campaign raised funding for an urban environmental education center and an endowment; and the city is redeveloping nearly 60 "brownfield" sites (see Chapter 19). An indication of why Isles has been so successful can be seen in one of its guiding principles: "The only sustainable solutions for urban problems are those which empower people with the knowledge and skills to help themselves."

***Showcase.*** Trenton has received over $2 million in EPA grants to renew vacant and abandoned industrial sites (Fig. 23–1). The city routinely involves local residents and businesses in the redevelopment process. As a consequence, Trenton has been selected as a "Showcase Community" by the Brownfields National Partnership and is a model for cooperative efforts to accomplish locally based initiatives. Like Waitakere City, Trenton has established a course of action and has found help in its efforts toward renewal.

This tale of two cities is tied together by a common resurgence of attention to urban living and a desire on the part of city residents to take charge of their own communities and reverse the trends that have been leading to a decline in the livability of their cities. It is significant that the direction they take is toward sustainability. These city residents recognize that things need to change, and they are becoming agents of changes that involve individual lifestyles and institutional redirection. One city is coping with urban sprawl, and the other is tackling urban blight.

**Figure 23–1 Trenton brownfield site.** An abandoned service station with contaminated soil and underground storage tanks (left) was converted to a new fire station (right) in Trenton's West Ward.

Our objective in this chapter is first to show how urban sprawl and urban blight are related and how they can be remedied. Then we shall consider the global and national efforts being undertaken to forge sustainable cities and communities. Finally, we shall look at some ethical concerns and personal lifestyles as we consider the urgent needs of people and the environment upon entering a new millennium.

## 23.1 Urban Sprawl

What is urban sprawl? It's been said that, like pornography, you know it when you see it. One thing is certain: Sprawl has been getting a lot of attention from the academic community. A team of researchers from Rutgers University, Cornell University, and the NGO Smart Growth America worked for several years on a comprehensive effort to define, measure, and evaluate sprawl and its impacts. Their work, *Measuring Sprawl and Its Impact,* was published on Smart Growth America's Web site in 2002. A 2003 study entitled *Measuring the Health Effects of Sprawl* is published on the same site and attempts to quantify the health effects of sprawl. The Federal Transit Administration sponsored a third important study, *Costs of Sprawl—2000,* published by the National Academy Press in 2002. These studies will inform our analysis of urban sprawl.

Sprawl's signature manifestation is a far-flung urban–suburban network of low-density residential areas, shopping malls, industrial parks, and other facilities loosely laced together by multilane highways. The word *sprawl* is used because the perimeters of the city have simply been extended outward into the countryside, one development after the next, with little plan as to where the expansion is going and no notion as to where it will stop. Almost everywhere we go near urban areas, we are confronted by farms and natural areas giving way to new developments, new highways being constructed, and old roadways being upgraded and expanded (Fig. 23–2). In fact, cars and sprawl are co-dependent; they need each other!

### The Origins of Urban Sprawl

Until the end of World War II, a relatively small percentage of people owned cars. Cities had developed in ways that allowed people to meet their needs by means of the transportation that was available—mainly walking. Every few blocks had a small grocery, a pharmacy, and other stores, as well as professional offices integrated with residences. Often, buildings had a store at the street level and residences above (Fig. 23–3). Schools were scattered throughout the city, as were parks for outdoor recreation or relaxation. Thus, walking distances were generally short, and bicycling made going somewhere even more

**Figure 23–2 New-highway construction.** The hallmark of development over the past several decades has been the construction of new highways, which, however, have only fed the environmentally damaging car-dependent lifestyle.

**Figure 23–3 The integrated city.** Before the widespread use of automobiles, cities had an integrated structure. A wide variety of small stores and offices on ground floors, with residences on upper floors, placed everyday needs within walking distance. This scene is from a historic section of Philadelphia.

convenient. For more specialized needs, people boarded public transportation—electric trolleys, cable cars, and buses—from neighborhoods to the "downtown" area, where big department stores, specialty shops, and offices were located. Public transportation did not change the compact structure of cities, because people still needed to walk to the transit line. At the outer ends of the transit lines, cities gave way abruptly to farms, which provided most of the food for the city, and natural areas. The small towns and villages surrounding cities—the original suburbs—were compact for the same reasons and mainly served farmers in the immediate area. At the end of World War II, this pattern began to change dramatically.

*The Dream.* Despite the many advantages of urban life, many people found cities less-than-pleasant places to live. Especially in industrial cities, poor housing, inadequate sewage systems, inadequate refuse collection, pollution from home furnaces and industry, and generally congested, noisy conditions were common. A decrease in services during World War II aggravated those problems. Hence, many people had the desire—the "American dream"—to live in their own house, on their own piece of land, away from the city.

**Suburbs.** Because of the massive production of war materiel, consumer goods were in short supply during World War II. As a result, when the war ended, there was a pent-up demand for consumer goods. When mass production of cars resumed at the end of the war, people flocked to buy them. With private cars, people were no longer restricted to living within walking distance of their workplaces or of transit lines. They could move out of their cramped city apartments and into homes of their own outside the city. Their cars would allow them to drive back and forth easily to their jobs, shopping, recreation, and so on.

*Housing Boom.* Returning veterans and the resulting baby boom created a housing demand that rapidly exhausted the existing inventory. Developers responded quickly to the demand for private homes. They bought farmlands and natural areas outside the cities and put up houses. The government aided this trend by providing low-interest mortgages through the Veterans Administration and the Federal Housing Administration, and interest payments on mortgages were made tax deductible. (Rent was not.) Property taxes in the suburbs were much lower than in the city. These financial factors meant that, for the first time in history, making monthly payments on one's own home in the suburbs was cheaper than paying rent for equivalent or less living space in the city.

Thus, the mushrooming development around cities did not proceed according to any plan; rather, it happened wherever developers could acquire land (Fig. 23–4). No governing body existed to devise—much less enforce—an overall plan. Cities were usually surrounded by a maze of more or less autonomous local jurisdictions (towns, villages, townships, and counties). Local governments were quickly thrown into the catch-up role of trying to provide schools, sewers, water systems, other public facilities, and, most of all, roads to accommodate the uncontrolled growth. Local zoning laws kept residential and commercial uses separate. It became more difficult to walk to a grocery store or an office.

**Highways.** The influx of commuters into previously rural areas soon resulted in traffic congestion, creating a need for new and larger roads. To raise money to build and expand highways, Congress passed the Highway Revenue Act of 1956, which created the **Highway Trust Fund.** This legislation placed a tax on gasoline and earmarked the revenues to be used exclusively for building new roads.

**Figure 23–4** **Urban sprawl.** Prime land is being sacrificed for development around most U.S. cities and towns. The locations of developments are determined more by the availability of land than by any overall urban planning. Here, suburban development is encroaching onto prime farmland in Hailey, Idaho.

Ironically, the Highway Trust Fund perpetuates development. A new highway not only alleviates existing congestion, but also encourages development at farther locations, because *time,* not *distance,* is the limiting factor for commuters. (When people describe how far they live from work, it is almost always put in terms of minutes, not distance.) The average person is willing to spend 20 to 40 minutes each way on a daily commute. Given this time limit, people who have to walk would have to live within 1 or 2 miles of their work. With cars and expressways, people can live 20 to 40 miles from work and still get there in the same amount of time.

***Vicious Cycle.*** Therefore, new highways that were intended to reduce congestion actually fostered the development of open land and commuting by more drivers from distant locations (Fig. 23–5). Soon, traffic conditions became as congested as ever. Average commuting distance has doubled since 1960, but average commuting time has remained about the same. The increase in commuting distance, however, requires more fuel, which generates more money for the Highway Trust Fund. Thus, the whole process is repeated in a continuous cycle (Fig. 23–6). As a result, government policy, far from curtailing urban sprawl, helped support and promote it.

Residential developments are inevitably followed (or sometimes led) by shopping malls, industrial parks, motels, and office complexes. These commercial centers are usually situated in such a way that the only possible access to them is by car. This situation, in turn, has only changed the direction of commuting: Whereas, in the early days of suburban sprawl, the major traffic flow was into and out of cities, it is now between suburban centers. Multilane highways connecting suburban centers are perpetually congested with traffic going in both directions.

***Exurbs, Too.*** In broad perspective, then, urban sprawl is a process of **exurban migration**—that is, a relocation of residences, shopping areas, and workplaces from their traditional spots in the city to outlying areas. Population growth, although it has occurred, has played a relatively minor part in the development of urban sprawl. The populations of many eastern and midwestern U.S. cities, excluding the suburbs, have been declining over the last 50 years as a result of exurban migration (Table 23–1). Exurban migration is continuing in a leapfrog fashion as people from older suburbs move to **exurbs** (communities farther from cities than suburbs are).

The "love affair" with cars is not just a U.S. phenomenon: Around the world, in both developed and developing countries, people aspire to own cars and adopt the car-dependent lifestyle. Consequently, urban sprawl is occurring around many developing-world cities as people become affluent enough to own cars. Love of the car-dependent lifestyle, however, is insufficient to make it sustainable. In fact, this lifestyle is at the crux of a number of the unsustainable trends described in earlier chapters.

(a)

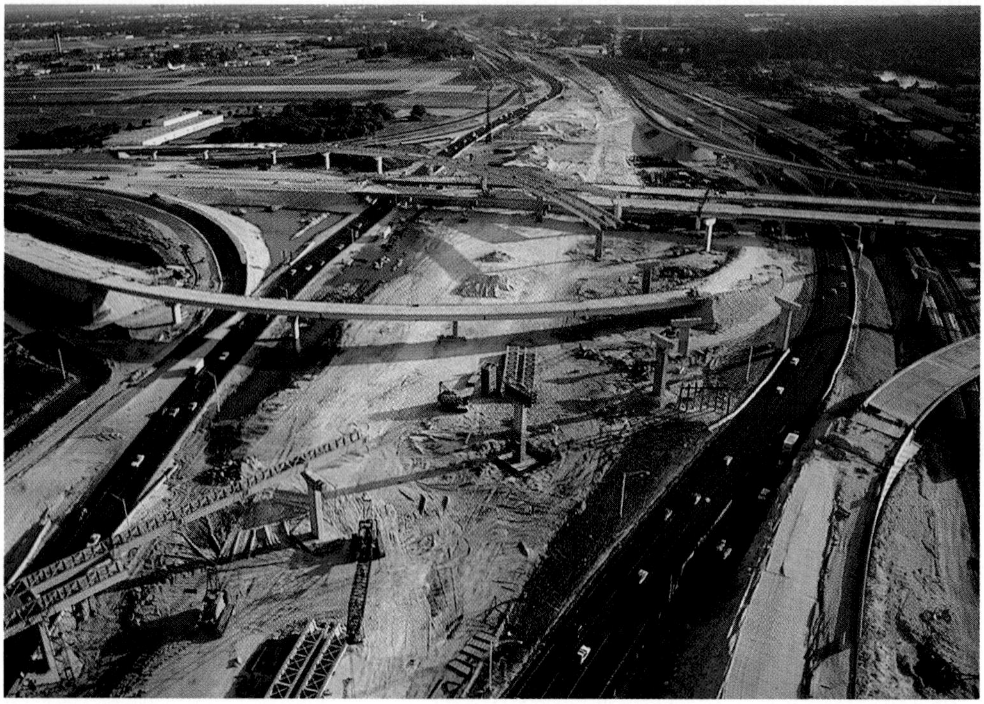

(b)

**Figure 23–5 The new highway–traffic-congestion cycle.** (a) Growing traffic congestion creates the need for new and upgraded highways. (b) However, upgraded highways encourage development at more distant locations, thus creating more traffic congestion and hence the need for more highways.

## Measuring Sprawl

The Smart Growth America research team labeled sprawl as "the process in which the spread of development across the landscape far outpaces population growth." They described four dimensions characterizing sprawl and used them to analyze metropolitan regions in order to create a sprawl index:

- Population widely dispersed in low-density development; thus, measure *residential density*.

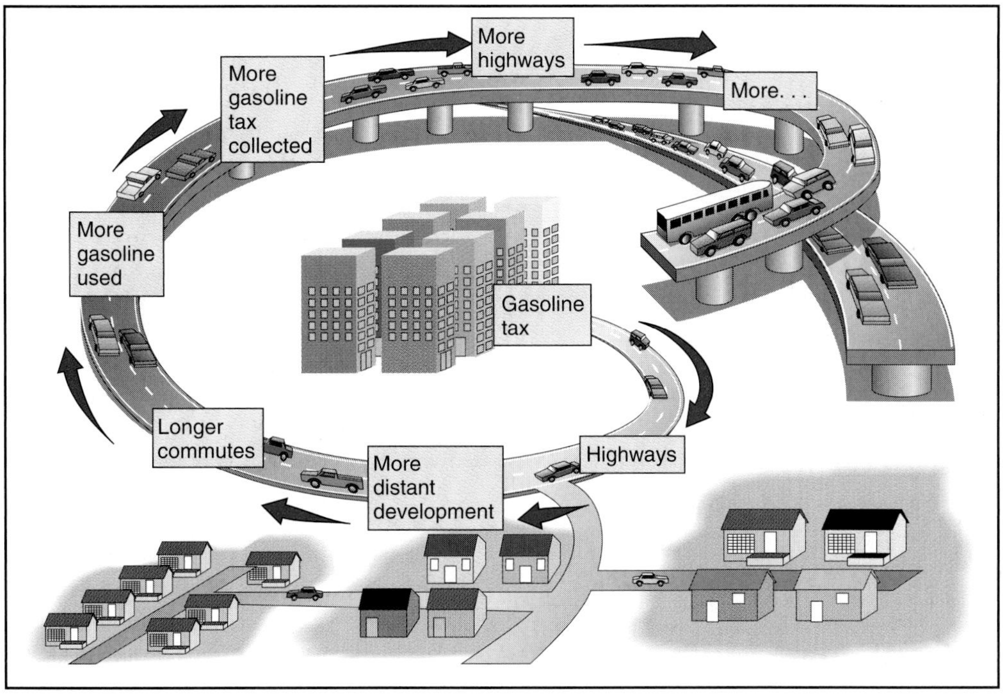


**Figure 23–6** **The development cycle spawned by the Highway Trust Fund.**

- Sharply separated homes, stores, and workplaces; thus, measure *neighborhood mix* of these three.
- A network of roads marked by poor connections and huge superblocks that concentrate traffic onto a few routes; thus, measure *accessibility of the street networks.*
- A general lack of well-defined downtowns and activity centers; thus, measure *strength of activity centers and downtowns.*

***Rankings.*** Each of the preceding four dimensions was standardized numerically so as to give 100 points as the average of the factor; higher means less sprawl, and lower means more. Scores for each factor were combined into a total *sprawl score* that also averaged 100. The team analyzed 83 metropolitan areas, representing almost half of the U.S. population. The results ranged from 14 to 178 (Fig. 23–7), with the Riverside–San Bernardino, California, metropolitan area the most sprawled and New York

**table 23-1** **Decline in Population of American Cities, 1950–2000**

| City | Population (thousands) 1950 | 1970 | 1990 | 2000 | Percent Change, 1950–2000 |
|---|---|---|---|---|---|
| Baltimore, MD | 950 | 905 | 736 | 651 | −31 |
| Boston, MA | 801 | 641 | 574 | 589 | −26 |
| Buffalo, NY | 580 | 463 | 328 | 293 | −49 |
| Cleveland, OH | 915 | 751 | 505 | 478 | −48 |
| Detroit, MI | 1,850 | 1,514 | 1,028 | 951 | −49 |
| Louisville, KY | 369 | 362 | 270 | 256 | −31 |
| Minneapolis, MN | 522 | 434 | 368 | 383 | −27 |
| Philadelphia, PA | 2,072 | 1,949 | 1,586 | 1,517 | −27 |
| St. Louis, MO | 857 | 662 | 396 | 348 | −59 |
| Washington, DC | 802 | 757 | 607 | 572 | −29 |
| U.S. population (millions) | 151 | 218 | 249 | 281 | +86 |

(*Source:* Data from Population Division, U.S. Bureau of the Census, 2003.)

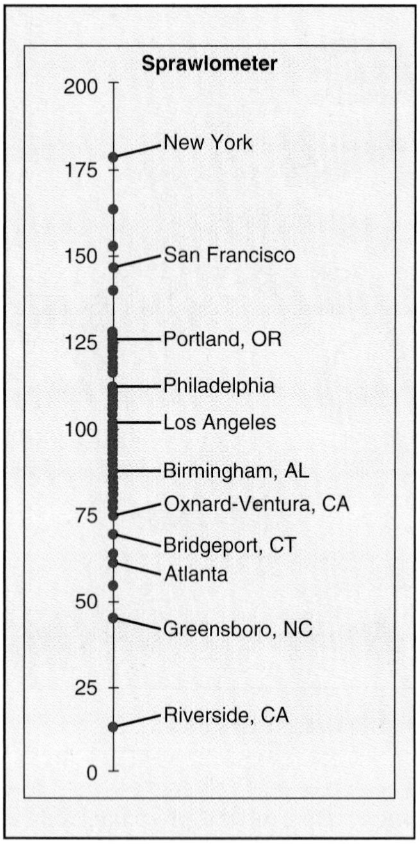


**Figure 23-7 Sprawl scores for 83 metropolitan regions.** The "sprawlometer" shows rankings of numerous metropolitan areas. Low scores mean high sprawl, and vice versa. (*Source: Measuring Sprawl and Its Impact, 2002;* Smart Growth America)

City the least. The 10 most sprawling metropolitan regions are ranked in Table 23-2. The Smart Growth authors point out that even though an area ranks lower on the sprawl score, that does not mean that sprawl is not a problem. Each of the four dimensions and the overall score were used to examine a range of travel and transportation-related outcomes as the authors examined the impacts of sprawl on daily life.

## Impacts of Urban Sprawl

**Environmental Impacts.** Urban sprawl affects the quality of life and the environment in numerous ways. The *environmental impacts* of urban sprawl are many and serious, as the following discussion indicates:

*Depletion of Energy Resources.* Shifting to a car-dependent lifestyle has entailed an ever-increasing demand for petroleum, with many consequences (Chapter 12). From 1950 to 2000, U.S. oil consumption nearly tripled, while population grew by just 86%, a nearly threefold increase in per capita consumption.

*Air Pollution.* Despite improvements in pollution control, many cities still fail to meet desired air-quality standards. (Chapter 21). Vehicles are responsible for an estimated 80% of the air pollution in metropolitan regions. The higher use of gasoline produces greater amounts of carbon dioxide, the most serious anthropogenic greenhouse gas (Chapter 20).

*Water Pollution and Degradation of Water Resources.* All the highways, parking lots, driveways, and other paved areas associated with urban sprawl lead to a substantial increase in runoff, resulting in increased flooding and erosion of stream banks. Also, water quality is degraded by the runoff of fertilizer, pesticides, crankcase oil (the oil that drips from engines), pet droppings, and so on.

*Loss of Agricultural Land.* Perhaps most serious in the long run is the loss of prime agricultural land. In the United States alone, sprawling development is eating up 590,000 acres of prime farmland a year. Most of the food for cities used to be locally grown on small, diversified family farms surrounding the city. With most of these farms turned into housing developments, it is estimated

**table 23-2 Ten Most Sprawling Metropolitan Regions**

| Metropolitan Region | Overall Sprawl Index Score | Rank |
|---|---|---|
| Riverside–San Bernardino, CA | 14.2 | 1 |
| Greensboro–Winston-Salem–High Point, NC | 46.8 | 2 |
| Raleigh–Durham, NC | 54.2 | 3 |
| Atlanta, GA | 57.7 | 4 |
| Greenville–Spartanburg, SC | 58.6 | 5 |
| West Palm Beach–Boca Raton–Delray Beach, FL | 67.7 | 6 |
| Bridgeport–Stamford–Danbury, CT | 68.4 | 7 |
| Knoxville, TN | 68.7 | 8 |
| Oxnard–Ventura, CA | 75.1 | 9 |
| Forth Worth–Arlington, TX | 77.2 | 10 |

(*Source:* Ewing, Reid, Pendall, Rolf, and Chen, Don. *Measuring Sprawl and Its Impact.* Smart Growth America, 2002) that food now travels an average of 1,000 miles (1,500 km) from where it is produced—mostly on huge commercial farms—to where it is eaten. The loss is not just the locally grown produce, but also the social interactions and ties with the farm community.

*Loss of Landscapes and Wildlife.* New developments are consuming land at an increasing pace. According to the National Resources Inventory, 2.2 million acres a year were developed from 1992 to 2001, compared with 1.4 million acres a year from 1982 to 1992 (Fig. 23–8). The result is the sacrifice of aesthetic, recreational, and wildlife values in metropolitan areas, the very places where they are most important.

The "fragmentation" of wildlife habitat due to urban sprawl has led to marked declines in many species, ranging from birds to amphibians. Also, the expanding highways lead to increasing numbers of roadkills. Today, much more wildlife is killed by vehicles than by hunters.

**Impacts on Quality of Life.** Major negative impacts of sprawl on the *quality of life,* as found by the two studies, are as follows:

*Higher Vehicle Ownership and Driving Mileages.* Cars are driven greater distances per person in high-sprawl areas. In Atlanta (sprawl score of 58), for example, daily use per person was 34 miles each day, while in Portland, Oregon (sprawl score of 126), it was only 24 miles each day. In the top 10 sprawl areas, there were 180 vehicles per 100 households, while in the lowest 10, there were 162. Chauffeuring kids to and from school or after-school activities (such as sports and music lessons) requires more driving in sprawl areas. As these values for miles driven and cars owned aggregate over many metropolitan areas, they add up to millions more miles driven and cars on the road.

*Greater Risk of Fatal Accidents.* The higher rates of vehicle use lead to higher highway fatalities. In Riverside, California, at the top of the sprawl index, 18 of every 100,000 people die each year in highway accidents. In the eight lowest areas on the index, fewer than 8 of 100,000 residents die annually on the highways.

*Lowered Rates of Walking and Lessened Use of Mass-Transit Facilities.* The percentage of commuter trips by transit in the 10 highest sprawl areas was 2%, compared with 7% in the 10 lowest sprawl areas. Similar trends are found for people walking to their workplace.

*No Change in Congestion Delays.* There was no difference in commute times between high- and low-sprawl areas. This indicates that moving out to the suburbs or exurbs to get away from traffic congestion does not work. Thus, metropolitan regions cannot sprawl their way out of congestion.

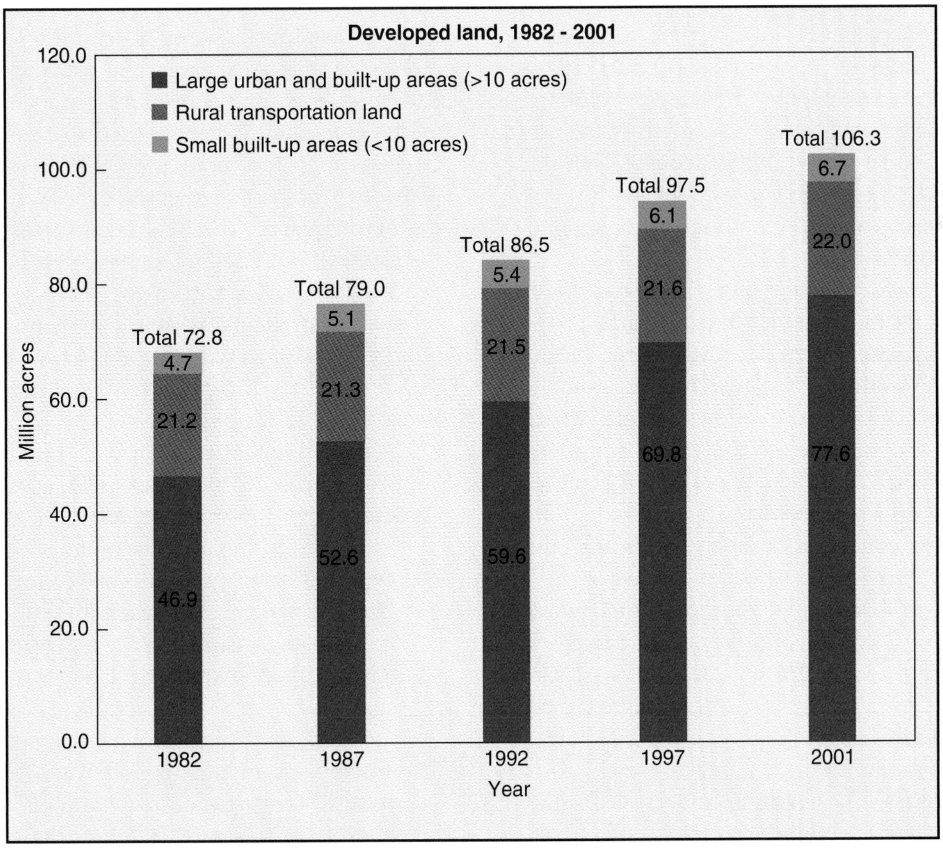


**Figure 23–8 Conversion of land to developed uses.** The buildup of developed land in the United States is shown. From 1982 to 2001, 34 million acres were converted to developed uses from natural lands and farmlands. (*Source:* National Resources Inventory, 2001 Annual NRI; 2003)

***Higher Costs for Municipal Services.*** The developments in outlying regions of metropolitan areas all have to be serviced with schools, sewers, water, electricity, roads, and other infrastructure elements, often forcing county and town budgets (and therefore taxes) to escalate. More compact forms of development are less expensive to service.

***Higher Incidence of Obesity and High Blood Pressure.*** People living in high-sprawl areas drive more, while those in more compact communities walk more. Walking and other moderate physical activity has many health benefits. The study of sprawl and its health effects found a highly significant correlation between the degree of sprawl, on the one hand, and weight, obesity, and high blood pressure, on the other.

With all of these undesirable effects of sprawl, one might question why people put up with them? The answer is that people choose to locate in low-density areas. They perceive that it is better to live in such areas, so they move there. There are benefits to sprawl, and it is important to examine them and compare them with the costs. Such a comparison has recently been made in the *Costs of Sprawl—2000* study.

**The Benefits of Urban Sprawl.** Quality-of-life issues tend to be decisive for most people, and these issues seem to be heavily weighted *in favor of sprawl.* In general, sprawl involves lower density residential living, larger lot sizes (often 2 acres or more), larger single-family homes, better quality public schools, lower crime rates, better social services, and greater opportunity for participation in local governments. As people move further out from an urban center, housing costs usually decline, and communities tend to be more homogeneous. Given these actual or perceived benefits, it is no wonder that sprawl is the dominant form of growth occurring in major metropolitan areas.

***Net Benefits?*** As we have seen, however, there is little or no advantage in commute times or traffic congestion and actually greater distances to drive for those in the outlying suburbs and exurbs. And the traffic fatality rate is worse, since speed alone allows people to drive greater distances and achieve an equivalent commuting time. Several of the perceived benefits of sprawl have serious negative side effects for society as a whole. Thus, people move to the exurbs in order to live in neighborhoods with lower crime rates, in order to have access to better schools, or in order to live in more homogeneous communities. In one way or another, these choices are often exclusionary in their effects. The results of thousands of people making these choices accentuate the concentration of poorer people in high-density, inner-city or aging suburban communities.

***Common Good?*** The environmental costs of sprawl are very real, but these are seldom perceived as decisive by those people moving to the sprawling suburbs. They are more a matter of the common good, and people tend to make choices based on *personal good* rather than the *common good.* On the whole, most observers who have studied sprawl and its impacts believe that the costs of sprawl outweigh the benefits. It is worth the effort to bring sprawl under control, they feel. The key to controlling sprawl is to provide those quality-of-life benefits that attract people, but to do it without incurring the serious environmental and social costs that are clearly associated with sprawl. This is what "smart growth" intends to do.

## Reining In Urban Sprawl: Smart Growth

The way to curtail urban sprawl might appear to be simply to pass laws that bring exurban development under control. Indeed, a number of European countries and Japan have such laws. However, laws cannot be passed without the support of the public. In the United States, the strong sentiment toward the right of a property owner to develop property as he or she feels fit has made such restrictions, with few exceptions, politically impossible to pass.

Yet, there are signs that a sea change is occurring in the attitudes of Americans toward uncontrolled growth. For example, the recently emerged concept of "*smart growth*" is inviting communities and metropolitan areas to address sprawl and purposely choose to develop in more environmentally sustainable ways. The concept recognizes that growth will occur and focuses on economic, environmental, and community values that together will lead to more sustainable communities. Table 23–3 presents the basic principles of smart growth. Zoning laws are being changed in many localities, and a new generation of architects and developers is beginning to focus on creating integrated communities, as opposed to disassociated facilities. By directing growth to particular places, smart growth provides protection for sensitive lands. Early indications in the marketplace—the final determining factor—are that the new communities are a great success.

***Initiatives.*** Smart-growth initiatives are appearing on the ballot in many states and municipalities. (Four hundred were enacted into law between 1999 and 2001.) These initiatives involve several key strategies:

1. **Setting boundaries on urban sprawl.** Oregon led the way in passing a land-use law in the 1970s requiring every city to set a boundary beyond which urban growth is prohibited. The boundary is a line permitting expected future growth—but controlled growth. Portland has had great success with this strategy, and today it is a highly desirable place to live and work.
2. **Saving open space.** The best way to preserve open space is to own it or own the development rights to the land, and states are enacting programs to acquire crucial remaining open spaces. For example, New Jersey voters overwhelmingly passed a \$1 billion bond issue to set aside half of the state's remaining 2 million acres of open space.
3. **Developing existing urban space.** In 1997, Maryland's General Assembly passed the "Smart Growth and

**table 23-3 Principles of Smart Growth**

| |
|---|
| Create a range of housing opportunities and choices for people of all income levels. |
| Create walkable neighborhoods. |
| Encourage the collaboration of citizens, corporations, businesses, and municipal bodies within communities. |
| Foster distinctive, attractive communities with a strong sense of place, setting standards for development that respond to community values. |
| Make development decisions predictable, fair, and cost effective. |
| Mix land uses to provide a more attractive and healthy place to live. |
| Preserve open space, farmland, natural beauty, and critical environmental areas. |
| Provide a variety of transportation choices. |
| Strengthen and direct development toward existing communities (called "infilling"). |
| Take advantage of compact building design instead of conventional land-hogging development. |

Neighborhood Conservation" initiative, which includes measures to channel new growth to community sites where key infrastructure is already in place. Abandoned and brownfield sites in metropolitan areas are already well served with roads, electricity, water, and sewerage and can be developed into new homes, workplaces, and shopping areas, as in Trenton, New Jersey.

4. **Creating new towns.** A key to developing sustainable communities is changing zoning laws to allow stores, light industries, professional offices, and higher density housing to be integrated together. Affordable and attractive housing can be created around industrial parks and shopping malls for the people who work and shop in those places. Safe bicycle and pedestrian access can be provided between residential areas and workplaces. Reston, Virginia; Columbia, Maryland; and Valencia, California, were established as self-contained towns along these lines in the 1960s, and they are still highly desirable as business and residential locations. Fairfax County, Virginia, by contrast, next to Washington, DC, is now almost 100% covered by sprawl-type development. It is estimated that if Fairfax had pursued smart-growth policies, the county would still be 70% open space and would be far less congested and less polluted.

***Raiding the Fund.*** In addition to the foregoing key strategies, there have been repeated attempts to break the cycle created by the Highway Trust Fund by allowing revenues to be used for purposes other than the construction of more highways. The first significant inroad was made in 1991 with the passage of the **Intermodal Surface Transportation Efficiency Act (ISTEA)**, a bill environmentalists had been promoting for many years. Under this act, almost half of the money levied by the Highway Trust Fund is now eligible to be used for other modes of transportation, including cycling, walking, and mass transit. The legislation also requires states to hire bicycle and pedestrian coordinators to oversee programs and to establish long-range pedestrian–bicycle plans. In addition, the act requires that states spend a total of $3 billion on "transportation enhancements," which include pedestrian and bicycle facilities. The objectives of ISTEA were significantly reinforced and extended with the passage of the **Transportation Equity Act for the Twenty-first Century (TEA-21)** in 1998. TEA-21 authorized a total of $218 billion for federal highway and mass-transit programs for five years and provided about $7 billion per year for transit-related programs, a major increase over ISTEA funding. TEA-21 has been given an extension into 2004, but needs to be reauthorized.

Another key strategy for reducing sprawl is to revitalize the cities and older suburbs. This means coping with urban blight.

## 23.2 Urban Blight

In the developed world, the other end of the exurban migration is the city from which people are moving. *Exurban migration* is the major factor underlying an urban decay, or blight, that has occurred over the last 50 years. A completely different set of factors is responsible for urban blight in the developing world. In that case, people are moving *to* the cities at a rate that far exceeds the capacity of the cities to assimilate them. The result is urban slums that surround virtually every city in the poorer developing countries. U.N.-Habitat estimates that 837 million people live in urban slums in the developing world, often under conditions that are so squalid that they defy description.

We look first at urban blight in the developed countries, with the United States as our focus.

### Economic and Ethnic Segregation

Historically, U.S. cities have included people with a wide diversity of economic and ethnic backgrounds. However, moving to the suburbs required some degree of affluence—at least the ability to manage a down payment and mortgage on a home and the ability to buy and drive a car. Therefore, exurban migration often excluded the poor, the elderly, and the handicapped. The poor were largely African-Americans and other minorities, largely because a history of racial discrimination had kept them from education and well-paying jobs. Discriminatory lending practices by banks and dishonest sales practices by real-estate agents (redlining) kept minorities out of the suburbs even when money was not a factor. Civil rights laws passed in the 1960s made such practices illegal, but there is much evidence that they still persist in various forms. Even without the racial segregation, economic segregation continues to exist and has even intensified.

In short, exurban migration and urban sprawl have led to segregation of the population into groups sharing common economic, social, and cultural backgrounds. Moving into the new suburban and exurban developments (often labeled "white flight") were the economically advantaged, whereas many areas of the cities and older suburbs were effectively abandoned to the economically depressed, who in large part were ethnic minorities, the handicapped, and the elderly (Fig. 23–9).

**Figure 23–9 Segregation by exurban migration.** Exurban migration, the driving force behind urban sprawl, has also led to segregation of the population along economic and racial lines. In (a) an area of suburbia and (b) an area of inner-city Baltimore, you see the contrast.

(a)

(b)

## The Vicious Cycle of Urban Blight

Affluent people moving to the suburbs set into motion a vicious cycle of exurban migration and urban blight that continues today. To understand how this downward cycle occurs, we must note several points about local governments. First, local governments (usually city or county agencies, although the particular entities vary from state to state) are responsible for providing public schools, maintaining local roads, furnishing police and fire protection, collecting and disposing of refuse, maintaining public water and sewers, and providing welfare services, libraries, and local parks. Second, a local government's major source of revenue to pay for these services is local property taxes, a yearly tax proportional to the market value of the property and home or other buildings on it. If property values increase or decrease, property taxes are adjusted accordingly. Third, in most cases, the central city is a governmental jurisdiction that is separate from the surrounding suburbs.

*Economic Dysfunction.* Because of the preceding three characteristics of local government, exurban migration has the following consequences: By the economics of supply and demand, property values in the suburbs escalate with the influx of affluent newcomers. Thus, suburban jurisdictions enjoy an increasing tax revenue with which they can improve and expand local services. At the same time, property values in the city decline because of decreasing demand. The lower property taxes create a powerful disincentive toward maintaining property. Many properties end up being abandoned by the owner for nonpayment of taxes. The city "inherits" such abandoned properties, but they are a liability rather than a source of revenue (Fig. 23–10). The declining tax revenue resulting from falling property values is referred to as an **eroding,** or **declining, tax base,** and it has been a serious handicap for most U.S. cities since the exurban migration started in the late 1940s. Adding to the problem is the fact that many immigrants tend to settle in the inner cities, drawn there by ethnic bonds and the hope of employment. The result is that many inner-city dwellers require public assistance of one sort or another. Thus, cities bear a disproportionate burden of welfare obligations, as well as of the declining tax revenues.

The eroding tax base forces city governments to cut local services, increase the tax rate, or, generally, both. Hence, the property tax on a home in a city is often two to three times greater than on a comparably priced home in the suburbs, a difference that amounts to $2,000 to $3,000 per year in the tax bill on an average-priced home. At the same time, schools, refuse collection, street repair, libraries, parks, and other services and facilities deteriorate from neglect. Increasing taxes and deteriorating services causes yet more people to leave the city and become part of urban sprawl, even if the car-dependent lifestyle is not their primary choice. However, it is still only the relatively more affluent

**Figure 23–10 Abandoned buildings.** Many cities' ills are because of too *little* population, not too much. When the exurban migration causes a city's population to drop below a certain level, businesses are no longer supported. The result is abandonment and urban blight. Shown here is a section of Baltimore only a few blocks from the redeveloped Inner Harbor seen in Fig. 23–11.

individuals who can make this move; thus, the whole process of exurban migration, a declining tax base, deteriorating real estate, and worsening schools and other services and facilities is perpetuated in a vicious cycle. This downward spiral of conditions is referred to as **urban blight** or **urban decay.**

## Economic Exclusion of the Inner City

The exurban migration includes more than just individuals and families: The flight of affluent people from the city removes the purchasing power necessary to support stores, professional establishments, and other enterprises. Faced with declining business, merchants and practitioners of all kinds are forced either to go out of business or to move to the new locations in the suburbs. Either way, vacant storefronts are the result, and people remaining in the city lose convenient access to goods, services, and, most important, jobs, because each business also represents employment opportunities.

Unemployment rates run at 50% or more in depressed areas of inner cities, and what jobs do exist are mostly at the minimum wage. By contrast, the new jobs being created are in the shopping malls and industrial parks in the outlying areas, which are largely inaccessible to inner-city residents for lack of public transportation. Therefore, not only are people who remain in the inner city poor from the outset, but the cycle of exurban migration and urban decay has now led to their economic exclusion from the mainstream. As a result, drug dealing, crime, violence, and other forms of deviant social behavior are widespread and worsening in such areas.

To be sure, in the last two decades, many cities have redeveloped core areas. Shops, restaurants, hotels, convention and entertainment centers, office buildings, residences, and places for walking and relaxing are all combining to bring a new spirit of life and hope, as well as the practical assets of taxes and employment, back into cities (Fig. 23–11). This is an encouraging new trend, so far as it goes. However, walk a few blocks in any direction from the sparkling, revitalized core of most U.S. cities, and you will find urban blight continuing unabated.

## Urban Blight in Developing Countries

More than half of the 3 million residents of Nairobi, Kenya's capital city, live in slums consisting of flimsy shacks made of scrap wood, metal, plastic sheeting, rocks, and mud (Fig. 23–12). The slum neighborhoods of Nairobi surround the outskirts of the city. Burning trash and charcoal cooking fires cloud the air, and the lack of sewers adds to the smell. People get their water from private companies, since the city does not provide water or any other public amenities. There is no land ownership, because the settlements are built on public land, or on private land where landowners exact high rents. Crime and disease are endemic, and AIDS and tuberculosis spread readily in such slums.

Similar slum neighborhoods can be found around most of the developing-world cities, and in spite of the conditions, the cities are expected to continue growing well into the middle of the century. The continued rapid growth is a function of both natural population growth (which is high in many developing countries) and

**Figure 23–11 Inner Harbor, Baltimore.** High population densities help make a city, not destroy it. Redeveloping key areas of cities with heterogeneous mixtures of office buildings, shopping, restaurants, residences, and recreational facilities, all of which attract people, has helped to bring new life back into the cities.

**Figure 23–12**
**Shantytown in Nairobi, Kenya.** More than half of Nairobi's residents live in such slums, where conditions defy description.

migration into the city from rural areas. The economies of rural areas, often based only on subsistence farming, simply do not provide the jobs needed by a growing population. So people move to the cities, where they at least have the hope of employment. The city housing is overwhelmed by the influx of migrants, who could not afford the rents even if housing were available. The slums actually represent a great deal of entrepreneurial energy, because the residents build the dwellings and erect their own infrastructure, usually including food stands, coffee shops, barber shops, and, in some cases, schools for their children.

*The Needs.* The vast slums surrounding the cities are a tremendous challenge to the institutional structure of developing countries. Because slum areas are unauthorized, cities rarely provide them with electricity, water, sanitation facilities, and other social amenities. Yet, the slums often represent an essential workforce to the city, taking low-paying jobs that keep the city and its inhabitants functioning. A great need in such neighborhoods is *home security*. People live in fear of bulldozers coming at any time and leveling the shantytown. In Brazil, governments are providing legal status to existing slums (favelas), granting lawful titles to the land. Peru has undertaken a huge titling program, giving recognition to some 1 million urban land parcels in a four-year period. Providing this level of security to the inhabitants is often the key to mobilizing further assets, such as acquiring access to credit and negotiating with the city government or utilities for additional services. People will improve their living conditions if they are given the assurance that they can stay where they are.

The people in the shantytowns also need more *jobs*, and again, local governments could provide employment at low cost to accomplish many improvements in their own neighborhoods and elsewhere—collecting trash, building sewers, composting organic wastes, and growing food. Providing *cheap transportation* is another need. People need to get to where the jobs are and could often do so if they had bicycles or could afford to ride public transportation. Perhaps the greatest need of people living in the informal neighborhoods of city slums is *government representation*. Their voices are seldom heard, because they are among the poorest and most powerless in a society. They are often plagued by government corruption, wherein nothing gets done unless accompanied by bribes (which these people can ill afford). In some regions, however, community organizations are emerging to deal with the basic issues of the city slums. People are organizing and demanding their rights as citizens. **Slum Dwellers International (SDI)** (www.sdinet.org) now represents slum people to institutions like the World Bank and city governments. Change can happen, and people are learning how to make it happen. As cities devote energy and resources to providing the essential services and infrastructure to the informal slums that surround them, they will only be strengthened. The cities, and even the slums, will become more livable.

## What Makes Cities Livable?

The environmental consequences of urban sprawl and the social consequences of urban blight are two sides of the same coin. A sustainable future will depend on both reining in urban sprawl and revitalizing cities. The urban blight of cities in the developing world requires deliberate policies to address the social needs of the people flocking to the shantytowns. The only possible way to sustain the global population is by having viable, resource-efficient cities, leaving the countryside for agriculture and natural ecosystems. The key word is *viable*, which means *livable*. No one wants to, or should, be required to live in the conditions that have come to typify urban blight.

*Livability* is a general concept based on people's response to the question "Do you like living here, or would you rather live somewhere else?" Crime, pollution, recreational, cultural, and professional opportunities, as well as many other social and environmental factors, are summed up in the subjective answer to that question. Although many people assume that the social ills of the city are an outcome of high population densities, crime rates and other problems in U.S. cities have climbed while populations dwindled as a result of the exurban migration.

Looking at livable cities around the world, we find that the common denominator is (1) maintaining a high population density; (2) preserving a heterogeneity of residences, businesses, stores, and shops; and (3) keeping layouts on a human dimension, so that people can meet, visit, or conduct business incidentally over coffee at a sidewalk café or stroll on a promenade through an open area. In short, the space is designed for, and devoted to, people (Fig. 23–13). In contrast, development of the past 50 years has focused on accommodating automobiles and traffic. Two-thirds of the land in cities that have grown up in the era of the automobile is devoted to moving, parking, or servicing cars, and such space is essentially alien to the human psyche (Fig. 23–14). William Whyte, a well-known city planner, remarked, "It is difficult to design space that will not attract people. What is remarkable is how often this has been accomplished."

***A Matter of Design.*** The world's most livable cities are not those with "perfect" auto access between all points. Instead, they are cities that have taken measures to reduce outward sprawl, diminish automobile traffic, and improve access by foot and bicycle in conjunction with mass transit. For example, Geneva, Switzerland, prohibits automobile parking at workplaces in the city's center, forcing commuters to use the excellent public transportation system. Copenhagen bans all on-street parking in the downtown core. Paris has removed 200,000 parking places in the downtown area. Curitiba, Brazil, with a population of 1.6 million, is cited as the most livable city in all of Latin America. The achievement of Curitiba is due almost entirely to the efforts of Jaime Lerner, who, serving as mayor since the early 1970s, has guided development with an emphasis on mass transit rather than cars. The space saved by not building highways and parking lots has been put into parks and shady walkways, causing the amount of green area per inhabitant to increase from 4.5 square feet in 1970 to 450 square feet today.

***Charinkos.*** In Tokyo, millions of people ride *charinkos*, or bicycles (Fig. 23–15), either all the way to work or to subway stations from which they catch fast, efficient trains, including the "bullet train," to their destination. By sharply restricting development outside certain city limits, Japan has maintained population densities within cities and along metropolitan corridors that ensure the viability of commuter trains. Japan's cities have maintained a heterogeneous urban structure that mixes small shops, professional offices, and residences in such a way that a large portion of the population meets its needs without cars. In maintaining an economically active city, it is probably no coincidence

**Figure 23–13 Livable cities.** The key to having a livable city is to provide a heterogeneity of residences, businesses, and stores and to keep layouts on a human dimension so that people can meet or conduct business incidentally over coffee in a sidewalk café or stroll on a promenade through an open area. In short, the space is designed for, and devoted to, people. Shown here is Quincy Market, Boston.

**Figure 23-14 Car-centered cities.** In contrast to what is required for livability, city development of the last several decades has focused on moving and parking cars and creating homogeneity. Note the sterility of parking lots surrounding buildings in the foreground of this photo of Los Angeles.

**Figure 23-15 Bicycles in Tokyo.** Most people in Tokyo do not own automobiles, but ride bicycles or walk to stations, from which they take fast, efficient, inexpensive subways to reach their destinations.

that street crime, vagrancy, and begging are virtually unknown in the vast expanse of Tokyo, despite the seeming congestion.

***Portland.*** Portland, Oregon, is a pioneering U.S. city that has taken giant steps to curtail automobile use. The first step was to encircle the city with an urban growth boundary, a line outside of which new development was prohibited. Thus, compact growth, rather than sprawl, was ensured. Second, an efficient light-rail and bus system was built (see the "Earth Watch" essay in Chapter 21, p. 600), which now carries 45% of all commuters to downtown jobs. (In most U.S. cities, only 10% to 25% of commuters ride public transit systems.) By reducing traffic, Portland was able to convert a former expressway and a huge parking lot into the now-renowned Tom McCall Waterfront Park (Fig. 23-16). Portland is now ranked among the world's most livable cities.

***The Big Dig.*** Other U.S. cities are discovering hidden resources in their inner core areas. Expressways, which were built with federal highway funds to speed vehicles through the cities, frequently dissected neighborhoods, separating people from waterfronts and consuming existing open space. Now, with federal help from TEA-21 programs, many cities are removing or burying the intruding expressways. Boston, for example, is in the closing stages of its "Big Dig," a $15 billion project that has moved its central freeway below ground, added parks and open space, and reconnected the city to its waterfront. The waterfronts themselves are often the neglected relics of bygone transportation

**Figure 23–16 Portland, Oregon.** By reducing traffic, Portland, Oregon, was able to convert a former expressway and a huge parking lot into the now-renowned Tom McCall Waterfront Park. Due to such planning, Portland is currently ranked among the world's most livable cities.

by water and rail. In recent years, cities such as Cleveland, Chicago, San Francisco, and Baltimore have redeveloped their waterfronts, turning rotting piers and abandoned freight yards into workplaces, high-rise residences, and public and entertainment space.

It turns out that the U.S. economic boom of the 1990s contributed immeasurably to urban renewal, filling the federal coffers and at the same time boosting employment in the inner cities. Retail businesses and manufacturers are discovering that inner-city neighborhoods can provide a valuable labor pool, as well as customers for their goods and services.

**Livable Equals Sustainable.** The same factors that underlie livability also lead to sustainability. The reduction of auto traffic and a greater reliance on foot and public transportation reduce energy consumption and pollution. Urban heterogeneity can facilitate the recycling of materials. Housing can be retrofitted with passive solar space heating and heating for hot water. Landscaping can provide cooling, as described in Chapter 14. A number of cities are developing vacant or cleared areas into garden plots (Fig. 23–17), and rooftop hydroponic gardens are becoming popular. Such gardens will not make cities agriculturally self-sufficient, but they add to urban livability, provide an avenue for recycling compost and nutrients removed from sewage, give a source of fresh vegetables, and have the potential to generate income for many unskilled workers. If urban sprawl is curbed, relatively close-in farms could provide most of the remaining food needs for the city.

***To the Point.*** The preceding discussion of urban issues may seem to have strayed far from nature and environmental science. However, there is a close connection between the two: The decay of our cities is hastening the degradation of our larger environment. As the

**Figure 23–17 Urban gardens.** Urban garden plots on formerly vacant lots in Philadelphia. Urban gardening is becoming recognized as having sociological, economic, environmental, and aesthetic benefits.

growing human population spreads outward from the old cities in developed countries, we are co-opting natural lands and prime farmlands and ratcheting up the rate of air pollution and greenhouse gases. Similarly, as increasing numbers in the developing world stream *into* the cities, they are degrading the human and social resources that are vital to all environmental issues. Thus, without the creation of sustainable human communities, there is little chance for the sustainability of the rest of the biosphere.

By making urban areas more appealing and economically stable, we not only improve the lives of those who choose to remain or must remain in cities, but also spare surrounding areas. Our parks, wildernesses, and farms will not be replaced by exurbs and shopping malls, but rather, will be saved for future generations. In the last decade, some remarkable developments have begun to accelerate the movement toward building sustainable communities. We turn to these developments now.

## 23.3 Moving Toward Sustainable Communities

Recall from Chapter 1 that the 1992 UNCED created the Commission on Sustainable Development. This new body was given the responsibility for monitoring and reporting on the implementation of the agreements that were made at UNCED—in particular, those related to sustainability. Following Agenda 21, one of the accords signed at UNCED, countries agreed to work on sustainable development strategies and action plans that would put them on a track toward sustainability. Some 100 countries have now developed national environmental action plans or sustainable development strategies that lay out public-policy priorities. Most of these countries are currently actively engaged in transforming the policies into action plans, and we will investigate two of the outcomes of these activities: the *Sustainable Cities Program* and the *Sustainable Communities movement.*

### Sustainable Cities

A joint facility of the U.N. Environment Program and the U.N. Human Settlements Program, the Sustainable Cities Program (SCP) is designed to foster the planning and management needed to move cities in the developing countries toward sustainability. The program defines a sustainable city as a city in which "achievements in social, economic and physical development are made to last." Cities are the focus of the program because they are absorbing two-thirds of the population growth in the developing countries and, in the process, are experiencing serious environmental degradation in and around their growing urban centers.

***Bottom Up.*** The SCP is viewed as a "capacity-building program," meaning that it is intended to help cities and countries mobilize resources and capabilities that are mostly available within the cities themselves (building management capacity). The approaches fostered by the program are thus "bottom up" in nature, calling for the involvement of people at all economic and social levels and reconciling their interests when conflicts are evident. Common principles seen in sustainable-development theory—principles such as social equity, economic efficiency, and environmental planning and management—are put into practice. Working groups specific to unique city issues are the core element of the program. To date, 20 SCP cities have been identified and are working on implementing the program, and more than 20 others are in the early stages of participation. Among the SCP cities are Madras (India), Dar es Salaam (Tanzania), Accra (Ghana), Shenyang (China), Concepción (Chile), and many others (Fig. 23–18). The process involves a preparatory phase and then moves into demonstration activities at the level of the whole

**Figure 23–18 Dar es Salaam, Tanzania.** The first Sustainable Cities Program demonstration city, Dar es Salaam is growing at a rate of 8% per year and is struggling to cope with deteriorating environmental conditions caused by its rapid growth.

city, finally extending to replications in other cities of the country. The following are the processes in a city's involvement:

- The preparation of an environmental profile and the identification of priority issues, involving key stakeholders from public, business, and community sectors.
- A city consultation bringing together key actors from government and the community to deliberate and agree on the issues of prime concern and to establish working groups.
- The creation of appropriate institutions for an environmental planning and management process that will continue to serve city needs after the SCP activities end.

Funding for the demonstration phase comes primarily from the city itself, but additional aid often originates in partnerships with other countries. Denmark, Canada, France, the Netherlands, the United Kingdom, and Italy have provided support and expertise. The World Bank and other U.N. agencies are frequently participants in SCP activities. Indeed, the program's approach is quite conducive to collaboration among NGOs, other countries, and the host country and city. The program is in its early growth stages, with a budget now at $30 million per year, and it is a promising approach to sustainable development in exactly the places where it is needed most.

## Sustainable Communities

Revitalizing urban economies and rehabilitating cities requires coordinated efforts on the part of all sectors of society. In fact, such efforts are beginning to occur. What is termed a "sustainable communities movement" is taking root in cities around the United States. The **Sustainable Communities Network** (www.sustainable.org/) and the **Smart Communities Network** (www.sustainable.doe.gov/) provide a cornucopia of information, help, case studies, and linkages, demonstrating that the movement is gaining ground throughout the nation. Chattanooga, Tennessee, is now regarded as a prototype of what can occur with such a movement.

***Chattanooga.*** Thirty years ago, Chattanooga, which straddles the Tennessee River, was a decaying industrial city with high levels of pollution. Employment was falling, and residents were fleeing to the suburbs, leaving abandoned properties and increasing crime. In 1969, the EPA presented the city with a special award for "the dirtiest city in America." Then, **Chattanooga Venture,** a nonprofit organization founded by community leaders, launched *Vision 2000,* the first step of which was to bring people from all walks of life together to build a consensus about what the city *could* be like. Literally thousands of ideas were gradually distilled into 223 specific projects. Then, with the cooperation of all sectors—including government, business, lending institutions, and average persons—work on the projects began, providing employment in construction for more than 7,300 people and permanent employment for 1,380 and investing more than $800 million. Among the projects were the following:

- With the support of the Clean Air and Clean Water Acts, local government clamped down hard on industries to control pollution.
- Chattanooga Neighborhood Enterprise fostered the building or renovation of more than 4,600 units of low- and moderate-income housing.
- A new industry to build pollution-control equipment was spawned, as was another industry to build electric buses, which now serve the city without noise or pollution.
- A recycling center employing mentally handicapped adults to separate materials was built.
- An urban greenway demonstration farm, which schoolchildren may visit, was created.
- A zero-emissions industrial park utilizing pollution-avoidance principles was built.
- The river was cleaned up, and a 22-mile Riverwalk reclaimed the waterfront and built parks, playgrounds, and walkways along the riverbanks.
- Theaters, museums, and a freshwater aquarium were renovated or built.
- All facilities and a renovated business district were made pedestrian friendly and accessible.

***Revision.*** With these and numerous other projects, many of them still ongoing, Chattanooga has moved its reputation from one of the worst to one of the best places to live (Fig. 23–19). Chattanooga Venture developed a step-by-step guide for community groups to assist them in similar efforts to build sustainable communities, and the city's experience is being modeled in other cities throughout the United States as well as internationally. Recently, Chattanooga sponsored *Revision 2000,* with over 2,600 participants taking up where Vision 2000 left off. Revision 2000 identified an additional 27 goals and more than 120 recommendations for further improving Chattanooga. The city's experience demonstrates that visioning and change must occur and continue to successfully create and maintain sustainable community development. Although Chattanooga Venture is no longer in operation, its past work has built an expectation throughout the city that public projects will involve the public and that the process will produce results. The latest manifestation of this expectation was *Recreate 2008,* in which hundreds of people engaged in a visioning process in 1998 that created a plan for revitalizing the city's park system.

**Figure 23–19 Chattanooga, Tennessee.** With numerous projects to reduce pollution and traffic and to provide attractions and amenities for people, Chattanooga, Tennessee, has changed its reputation from one of the worst to one of the best places to live in the United States.

**The President's Council on Sustainable Development.** Between 1993 and 1999, the **President's Council on Sustainable Development** attempted to focus the country's attention on that issue. The council, made up of members from the administration, the business community, and the environmental community, produced three major reports and coordinated a number of "national town meetings." The reports, available at http://clinton2.nara.gov/PCSD/, are well done and, if heeded, could make a valuable contribution to the country's need to think about its future. In winding up its work, the council asked President Clinton to carry on the work of the council and "assure the existence of some forum for the thoughtful consideration of sustainable development issues by high-level leaders from all sectors." This forum has not yet been established, and the work of the council has been ignored by the Bush administration.

## 23.4 Toward the Common Good

Sustainability, stewardship, and sound science—these are the three strategic themes that have kept our eyes on the basics of how we must live on our planet. As discussed in Chapter 1, sustainability is the practical goal that our interactions with the natural world should be working toward, stewardship is the ethical and moral framework that informs our public and private actions, and sound science is the basis for our understanding of how the world works and how human systems interact with it. Are we making progress in incorporating these concepts into our society?

### Our Dilemma

Our intention in this book has been to avoid dwelling on the bad news and instead to raise the hope—indeed, the certainty—that all environmental problems can be addressed successfully. Throughout the text, we have pointed to policies and possibilities that can help move human affairs in a sustainable direction. There is much at stake. When the year 2015 arrives, will we be able to say that we have met all the Millennium Development Goals, alleviating much of the developing world's grinding poverty and its consequences? Will we reduce our use of fossil fuels to halt global warming and the rising sea level? Can we hope to feed the 3 billion more who will join us by midcentury and simultaneously reduce the malnutrition that still plagues 840 million?

*Human Decisions.* As the late Nobel Laureate Henry Kendall said, "Environmental problems at root are human, not scientific or technical." Even though our scientific understanding is incomplete and our grasp of sustainable development is still tentative, we know enough to be able to act decisively in most circumstances. Therefore, it is human decisions, at both the personal and societal level, that can bring about change, and it is these decisions that define our stewardship relationship with Earth.

However, making stewardly decisions is not a simple task. Decision making is affected by our personal values and needs, and competing values and needs exist at every turn. One group wins, and another loses. If we stop all destruction of the rain forests, the peasant who needs land to farm will not eat. If we force a halt to the harvesting of shrimp because of concerns about the by-catch, thousands of shrimpers will be out of work. Making decisions means considering competing needs and values and reaching the best conclusion in the face of the numbing complexity of demands and circumstances. It is the nature of our many dilemmas that "business as usual" resolves only a very small fraction of them. Even though public policies at the national and international level are absolutely essential elements of our success in turning things around, these in the end are the outcome of decisions by very human people; and even if thoughtful decisions are made by those in power, they will fail unless they are supported by the people who are affected by them.

***The Common Good.*** In Chapter 1, we defined the common good in the context of public policy: *to improve human welfare and to protect the natural world.* What compels people to act to promote the common good? We can identify a number of *values* (indeed, *virtues*) that can be the basis of stewardship and help us in our ethical decisions involving both public policies and personal lifestyles:

- *Compassion* for those less well off—those suffering from extreme poverty or other forms of deprivation. Compassion can become a major element of our foreign policy, energizing our country's efforts in promoting sustainable development in the countries that are most in trouble. Compassion can also motivate young people to spend two years of their life in the Peace Corps or a hunger relief agency.
- *A concern for justice.* Just policies can become the norm in international economic relations, and a concern for justice can also move people to protest the placement of hazardous facilities in communities of color.
- *Honesty,* or a concern for the truth—in the sense of keeping the laws of the land and in openly examining issues from different perspectives.
- *Sufficiency*—rather than ever-increasing consumption, using no more than is necessary of Earth's resources.
- *Humility*—instead of demanding our rights, being willing to share and even defer to others.
- *Neighborliness*—being concerned for other members of our communities (even the global community), such that we do not engage in activities that are harmful to them, but instead make positive contributions to their welfare.

To achieve sustainability, we will have to couple values like these with the knowledge that can come from our understanding of how the natural world works and what is happening to it as a result of human activities. For this plan to work, people must be willing to grapple with scientific evidence and basic concepts and develop a respect for the consensus that scientists have reached in most areas of environmental concern. It has been our objective to convey that consensus to you in this text. Recall also that sustainable solutions have to be economically feasible, socially desirable, and ecologically viable (Fig. 1–9). Again, we have tried to emphasize all three of these components as we point to the political decisions that are shaping our future in so many of our environmental concerns.

The good news is that many people are engaging environmental problems—working with governments and industries, bringing relief to people in need, and demonstrating exactly the kinds of decisions that are needed for effective stewardship. (See "Ethics" essay, p. 657.) Thus, traditional environmentalist organizations, as well as people from other sectors of society, are working together on the many issues that we have highlighted in this text, bringing about policy changes as well as changes in personal lifestyles (the final topic of the text).

## Lifestyle Changes

What are some of the lifestyle changes that are needed, and are they in fact occurring? We should be encouraged and inspired by the literally millions of people in all walks of life who are acutely aware of the problems and who are making outstanding efforts to bring about solutions. Every pathway toward solutions that we have mentioned represents the work of thousands of dedicated professionals and volunteers, ranging from scientists and engineers to businesspeople, lawyers, and public servants. Indeed, we are all involved, whether we recognize it or not. Simply by our existence on the planet, everything we do—the car we drive, the products we use, the wastes we throw away, virtually every choice we make and action we take—has a certain environmental impact and a certain consequence for the future. Therefore, it is not a matter of having an effect, but of what and how great that effect will be. It is a matter of each of us asking ourselves, Will I be part of the problem or part of the solution? The outcome will depend on how each of us responds to the challenges ahead.

There are a number of levels on which we may participate to work toward a sustainable society:

- individual lifestyle changes
- political involvement
- membership and participation in nongovernmental environmental organizations

**ethics**

## The Tangier Island Covenant

Chesapeake Bay, the largest estuary in the United States, is famous for its blue crabs and the unique "watermen" who fish for the crabs and other marine animals. Watermen are independent fishers who have lived on the bay for generations and who defend their way of life fiercely. In recent years, the watermen of Tangier, a small island on the Virginia part of the eastern shore of the bay, have been in constant battle with the Chesapeake Bay Foundation and the state regulators over crab and oyster laws restricting when and where they can harvest. The watermen skirted restrictions by keeping undersized crabs and dredging clams and oysters out of season. They were also accustomed to throwing their trash overboard, causing the shores of the island to be littered with debris.

In 1997, the island community of Tangier became part of an unusual, but significant, process called "values-based stewardship," a form of conflict resolution that requires working within the value system of the community to bring about sustained changes in behavior toward the environment. Susan Drake, a doctoral candidate in environmental science at the University of Wisconsin, came to the island to carry out her thesis research on how conflicts are dealt with in homogeneous communities such as Tangier. Originally attracted to the project because of the highly religious community present on the island—the two churches are the center of life and authority there—Drake knew that the watermen considered themselves part of a rich bounty in an environment created by God. After working with watermen on their boats and spending time in their churches and with their families, Drake became concerned about the apparent lack of a connection between what the God-fearing watermen believed and what they actually did in their profession.

Convinced that she might be able to intervene in a positive way, Drake requested permission from her thesis committee to engage in "participatory action research," wherein the researcher may enter the conflict and attempt to bring about a resolution. Working within the value system of the community—in this case, a value system based on faith—Drake addressed a joint meeting of the two churches on the island one Sunday morning. Referring to an image of Jesus standing behind a young fisherman at the wheel of a boat in rough seas that almost all Tangier watermen displayed in the cabins of their vessels, Drake addressed the issues of dumping trash overboard and skirting the laws. She showed the picture to the congregation and then put paper blinders over the eyes of Jesus as she listed the ways in which the watermen were disobeying the law. The day of her sermon coincided with a very high tide that gathered the debris from around the shores of the island and washed it up into the streets and people's yards.

***The Covenant.*** The meeting ended with a call by Drake to the watermen to adopt a covenant she had drafted earlier with the help of some of the island's women. Fifty-six watermen went to the altar and promised to abide by the following covenant:

> The Watermen's Stewardship Covenant is a covenant among all watermen, regardless of their profession of religious faith. As watermen, we agree to (1) be good stewards of God's creation by setting a high standard of obedience to civil laws (fishery, boat, and pollution laws) and (2) commit to brotherly accountability. If any person who has committed to this covenant is overtaken in any trespass against this covenant, we agree to spiritually restore such a one in a spirit of gentleness. We also agree to fly a red ribbon on the antennas of our boats to signify that we are part of the covenant.

The next day, and up to the present, scores of watermen began flying red ribbons, taking trash bags out on their boats, and fishing according to the laws. Trucks began to pick up the trash on the island early in the morning. The two churches proceeded to work with the entire island community to draft a plan for the future sustainability of the island and its way of life. They formed the "Tangier Watermen's Stewardship for the Chesapeake" in order to implement the plan, which is far reaching and involves the women of the island lobbying the state legislature and opening up lines of communication with scientists and government regulators involved in overseeing the bay's health.

Values-based stewardship is a call to life-changing stewardship, put within the value system of a community. In this case, putting the call within the context of the faith of the watermen resulted in the establishment of structures within the community that show great promise in helping to bring the entire island toward a sustainable future.

- volunteer work
- career choices

*Lifestyle changes* may involve such things as switching to a more fuel-efficient car or walking or using a bicycle for short errands; recycling paper, cans, and bottles; retrofitting your home with solar energy; starting a backyard garden and composting and recycling food and garden wastes into your soil; putting in time at a soup kitchen to help the homeless; choosing low-impact recreation, such as canoeing rather than speedboating; living closer to your workplace; and any number of additional things.

*Political involvement* ranges from supporting and voting for particular candidates to expressing your support for particular legislation through letters or phone calls. *Membership in nongovernmental environmental organizations* can enhance both lifestyle changes and political involvement. As a member of an environmental organization, you will receive, and may help disseminate, information, making you and others more aware of specific environmental problems and things you can do to help. In particular, you will be informed about environmentally significant legislation so that you may focus your political efforts at the most effective time and place. Also, your membership and contribution serve to support the lobbying efforts of the organization. A lobbyist representing only him- or herself has relatively little impact on legislators. If, however, the lobbyist represents a million-member organization that can follow up with that many phone calls and letters (and, ultimately, votes), the impact is considerable. Finally, in cases where enforcement of the existing law has been the weak link, some organizations, such as the Public-Interest Research Groups, the Natural Resources Defense Council, and the Environmental Defense Fund, have been highly influential in bringing polluters or the government to court to see that the law is upheld. Again, this can be done only with the support of members.

Another form of involvement is *joining a volunteer organization*. Many effective actions that care for people and the environment are carried out by groups that depend on volunteer labor. Political organizations and virtually all NGOs depend highly on volunteers. Many helping organizations, such as those dedicated to alleviating hunger and to building homes for needy families (Fig. 23–20), are adept at mobilizing volunteers to accomplish some vital tasks that often fill in the gaps left behind by inadequate public policies.

Finally, you may choose to devote your *career* to implementing solutions to environmental problems. Environmental careers go far beyond the traditional occupation of wildlife or park management. There are any number of lawyers, journalists, teachers, research scientists, engineers, medical personnel, agricultural extension workers, and others focusing their talents and training on environmental issues or hazards. Business and job opportunities abound in pollution control, recycling, waste management, ecological restoration, city planning, environmental monitoring and analysis, nonchemical pest control, the production and marketing of organically grown produce, and so on. Some developers concentrate on rehabilitation and the reversal of urban blight, as opposed to contributing more to urban sprawl. Some engineers are working on the development of pollution-free vehicles to help solve the photochemical smog dilemma of our cities. Indeed, it is difficult to think of a vocation that cannot be directed toward promoting solutions to environmental problems.

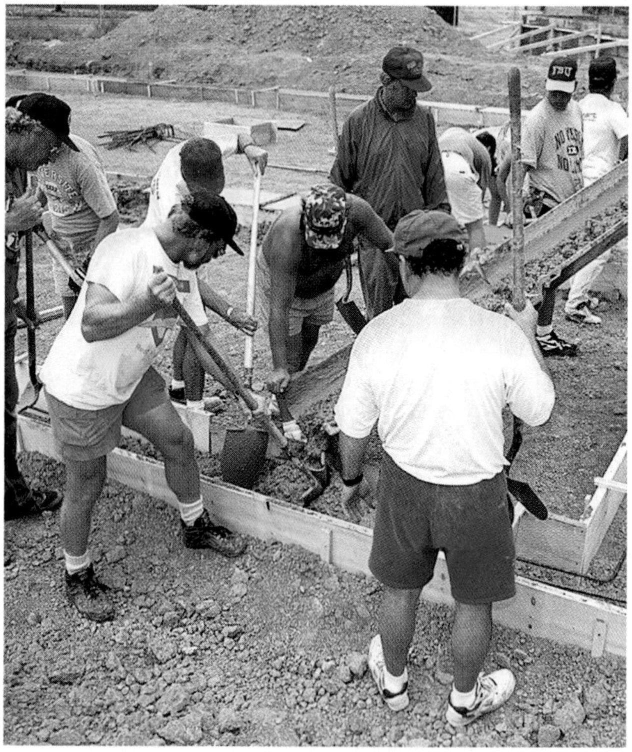

**Figure 23–20 Habitat for Humanity volunteers.** By using charitable contributions and volunteer labor, Habitat for Humanity creates quality homes—over 150,000 built throughout the world by 2003—that needy families can afford to buy with no-interest mortgages, enabling the money to be recycled to build more homes. Prospective home buyers participate in the construction and may gain valuable skills in the process.

We began this book with the story of Easter Island. The Easter Islanders failed to grasp the connections between the essential goods and services provided by their island environment, and their sustainable existence on the island. Their failure didn't lead to their extinction, but if sustainability means that people should thrive and not just survive, they certainly failed.

Human society is moving into a new millennium, and it is going to be an era of rapid changes unprecedented in human history. Will it be just survival, or will we thrive and become a sustainable global society? We are engaged in an environmental revolution—a major shift in our worldview and practice from seeing nature as resources to be exploited to seeing nature as life's supporting structure that demands our stewardship.

You are living in the early stages of this revolution, and we invite you to be one of the many who will make it happen.

> For civilization as a whole, the faith that is so essential to restore the balance now missing in our relationship to the earth is the faith that we do have a future. We can believe in that future and work to achieve it and preserve it, or we can whirl blindly on, behaving as if one day there will be no children to inherit our legacy. The choice is ours; the earth is in the balance.
>
> Al Gore, *Earth in the Balance*[1]

[1]Gore, Al. *Earth in the Balance*. Boston: Houghton Mifflin, 1992, p. 368.

# revisiting the themes

## Sustainability

Both Waitakere City and Trenton have sustainability in their sights. Waitakere City is reigning in urban sprawl, while Trenton is dealing with urban blight. These two widespread social developments are unsustainable, and to bring them under control is to move toward sustainability. They both lead to the loss of prime agricultural land and natural areas, and they consume other resources and produce pollution. The sustainable answers to these two trends are smart growth and the transformation of cities into livable habitats for people. The global Sustainable Cities Program and the American Sustainable Communities movement demonstrate the attractiveness of sustainability as a goal for sustainable development.

## Stewardship

The stewardship ethic includes a concern for justice. "White flight" into the suburbs has left minorities trapped in the inner cities, held down by racial and economic discrimination. Equally important is the migration of rural people into the cities of the developing world, where they are forced to live in slums and are unjustly denied ordinary social and environmental services. Both of these developments have energized many individuals and numerous NGOs to make common cause with those who are suffering the injustice and help them mobilize their human and social resources to bring about change. A similar motivation to bring about change led Chattanooga Venture to begin the visioning process that eventually turned that city completely around. This is exactly what stewards should be doing. Stewards are those who put their lives and energies on the line for the common good, enlisting those virtues which are the basis of stewardly care: compassion, a concern for justice, honesty, sufficiency, humility, and neighborliness. It is a fact that every one of us is given many opportunities to work together to create a truly sustainable society, and stewardship is the ethic that is there to guide our actions.

## Ecosystem Capital

Ecosystem capital is expended far more rapidly when urban sprawl is left unchecked. In the United States, some 2.2 million acres a year are converted from natural areas and farmlands to developments. This transformation depletes ecosystem capital in the very places where they are most needed: close to the metropolitan areas.

## Policy and Politics

The story of urban sprawl is a story of poor or nonexistent public policy. Land use is traditionally under local control (if it is under any control at all), and absent any public policy dealing with land use, sprawl happens. In fact, public policy fed sprawl, as the Highway Trust Fund taxed gasoline sales in order to build roads and highways. Crafting effective policy to deal with sprawl means recognizing why people choose to live on the suburban fringes in low-density houses and changing the way we organize our communities. It requires changing public policy to do this, but smart-growth initiatives are taking root in state after state. This is a completely nonpartisan issue, as Democratic and Republican governors in equal numbers supported smart growth initiatives in their states.

Urban blight, too, was the outcome of poor or nonexistent public policy. Cities stood by helplessly and watched the flight of wealthier people and businesses out to the suburbs. In a similar way, developing-world cities are simply watching the poor flock in, and those cities refuse to devote their resources to helping the poor meet their needs. Politics and policies rule in these situations, and it often takes people with compassion and a vision to bring about change. Frequently, change can be brought about at the grassroots level, but it helps immensely to have leaders who care about the poor in their midst.

### Globalization

The global economy continues to favor the wealthy, developed countries, but as we saw in Chapter 22, the developing countries are beginning to make progress in bringing their issues forward. If the country's economy moves forward, the cities will, too, because they are the engines of the economy.

## review questions

1. What problems do Trenton and Waitakere City illustrate, and how are they organizing to deal with those problems?
2. How did the structure of cities begin to change after World War II? What factors were responsible for the change?
3. Why are the terms "car dependent" and "urban sprawl" used to describe our current suburban lifestyle and urban layout?
4. What federal laws and policies tend to support urban sprawl?
5. Compare the costs and benefits of urban sprawl, looking at the environmental impacts and the impacts on quality of life.
6. What is "smart growth"? What are four smart-growth strategies that address urban sprawl?
7. What services do local governments provide, and what is their prime source of revenue for these services?
8. What is meant by "erosion of a city's tax base"? Why does it follow from exurban migration? What are the results?
9. How do exurban migration, urban sprawl, and urban decay become a vicious cycle?
10. What is meant by "economic exclusion"? How is it related to the problems of crime and poverty in cities?
11. What are the reasons for urban blight in the cities of many developing countries? What do the people in the shantytowns need most?
12. What are some characteristics of livable cities?
13. How does the U.N. Sustainable Cities Program work?
14. What is the Sustainable Communities Program? How does Chattanooga illustrate the program?
15. What are some stewardship virtues, and what role could they play in fostering sustainability?
16. State five levels on which people can participate in working toward a sustainable future.

## thinking environmentally

1. Interview an older person (say, 65 years or older) about what his or her city or community was like before 1950 in terms of meeting people's needs for shopping, recreation, getting to school, work, and so on. Was it necessary to have a car?
2. Do a study of your region. What aspects of urban sprawl and urban blight are evident? What environmental and social problems are evident? Are they still going on? Are efforts being made to correct them?
3. Identify nongovernmental organizations in your region that are working toward reining in urban sprawl or toward preserving or bettering the city. What specific projects are underway in each of these areas? What roles are local governments playing in the process? How can you become involved?
4. Suppose you are a planner or developer. Design a community that is people oriented and that integrates the principles of sustainability and self-sufficiency. Consider all the fundamental aspects: food, water, energy, sewage, solid waste, and how people get to school, shopping, work, and recreation.

## making a difference part six: chapters 22, 23

1. Analyze your environmental impacts as a consumer. What modes of transportation do you use, what foods do you consume or throw away, what appliances do you own, how do you heat or cool your home, and how much water do you use? Consider especially your impact on the four leading consumption-related environmental problems: air pollution, global climate change, changes in natural habitats, and water pollution.

2. Use the available Web resources to locate specific regulations developed by the EPA or other regulatory agencies, and research the details of how the agency conforms to Executive Order 12866 in developing the economic impact of the regulations.
3. Investigate land-use planning and policies in your region. Is urban sprawl or urban blight a problem? Become involved in zoning issues, and support zoning and policies that will promote, rather than hinder, ecologically sound land use ("smart growth").
4. Find out about the President's Council on Sustainable Development, and investigate what is happening in your state or city to promote sustainable development.
5. Contact local conservation organizations or land trusts, and offer your help in protecting and preserving open land in your region.
6. Find out if there is a Habitat for Humanity project or group in your area, and if there is, volunteer some time in support of their work.

# Environmental Organizations

This appendix presents a list of selected nongovernmental organizations that are active in environmental matters. Listed are national organizations, as well as some smaller, specialized ones. These organizations offer a variety of fact sheets, brochures, newsletters, publications, educational materials, and annual reports, many of which are available on the Internet. Web addresses are provided in the list. Requests for information should be specific and are best made by e-mail; Web sites have e-mail addresses and query options. Some organizations have internship positions available for those wishing to do work for an environmental group. An exhaustive listing of over 3,000 environmental, governmental, and educational organizations can be found in the *Conservation Directory* (2003), available from the National Wildlife Federation ($70.00). **www.nwf.org/conservationDirectory/print.cfm** (January 28, 2004).

AMERICAN FARMLAND TRUST. Focuses on the preservation of American farmland. 1200 18th Street, N.W., Suite 800, Washington, DC 20036. **www.farmland.org**

AMERICAN LUNG ASSOCIATION. Research, education, legislation, lobbying, advocacy: indoor and outdoor air pollution effects, smoking. 1740 Broadway, New York, NY 10019. **www.lungusa.org**

AMERICAN RIVERS, INC. Mission is "to preserve and restore America's rivers' systems and to foster a river stewardship ethic." Education, litigation, lobbying: wild and scenic rivers, hydropower relicensing. Information on specific rivers available. 1025 Vermont Ave., N.W., Suite 720, Washington, DC 20005. **www.amrivers.org**

BREAD FOR THE WORLD. National advocacy group that lobbies for legislation dealing with hunger. 50 F Street, N.W., Suite 500, Washington, DC 20001. **www.bread.org**

CENTER FOR SCIENCE IN THE PUBLIC INTEREST. Research and education: alcohol policies, food safety, health, nutrition, organic agriculture. 1875 Connecticut Avenue, N.W., Suite 300, Washington, DC 20009. **www.cspinet.org**

CHESAPEAKE BAY FOUNDATION. Research, education, and litigation: environmental defense and management of Chesapeake Bay and surrounding area. Philip Merrill Environmental Center, 6 Herndon Avenue, Annapolis, MD 21403. **www.cbf.org**

CLEAN WATER ACTION PROJECT. Lobbying, education, research: water quality. 4455 Connecticut Ave., N.W., Suite A300, Washington, DC 20008-2328. **www.cleanwateraction.org**

COMMON CAUSE. Lobbying: government reform, energy reorganization, clean air. 1250 Connecticut Ave., N.W., Washington, DC 20036. **www.commoncause.org**

COMMUNITY TRANSPORTATION ASSOCIATION OF AMERICA. Provides technical assistance for rural and specialized transportation systems. 1341 G Street, N.W., Suite 600, Washington, DC 20005. **www.ctaa.org**

CONGRESS WATCH. Lobbying: consumer health and safety, pesticides (a function of Public Citizen). 215 Pennsylvania Avenue, S.E., Washington, DC 20003. **www.citizen.org/congress**

CONSERVATION INTERNATIONAL. Education and research to preserve and promote awareness about the world's most endangered biodiversity. 1919 M Street, N.W., Suite 600, Washington, DC 20036. **www.conservation.org**

CORAL REEF ALLIANCE. Works with the diving community and others to promote coral reef conservation around the world. 2014 Shattuck Avenue, Berkeley, CA 94704-1117. **www.coral.org**

CRITICAL MASS ENERGY AND ENVIRONMENT PROJECT. Research and education: alternative energy and nuclear power (a function of Public Citizen). 215 Pennsylvania Avenue, S.E., Washington, DC 20003. **www.citizen.org/cmep**

DEFENDERS OF WILDLIFE. Research, education, and lobbying: endangered species. 1101 14th Street, N.W., #1400, Washington, DC 20005. **www.defenders.org**

DUCKS UNLIMITED, INC. Hunters working to fulfill the annual life cycle needs of North American waterfowl by protecting, enhancing, restoring, and managing important wetlands and associated uplands. One Waterfowl Way, Memphis, TN 38120. **www.ducks.org/**

EARTHWATCH INSTITUTE. Environmental research is encouraged by organizing teams to make expeditions to various locations all over the world. Team members contribute to the expenses and spend from several weeks to several months on the site. 3 Clock Tower Place, Suite 100, Box 75, Maynard, MA 01754. **www.earthwatch.org**

ENVIRONMENTAL DEFENSE FUND. Research, litigation, and lobbying: cosmetics safety, drinking water, energy, transportation, pesticides, wildlife, air pollution, cancer prevention, radiation. 1875 Connecticut Ave. N.W., Washington, DC 20009. **www.edf.org**

ENVIRONMENTAL LAW INSTITUTE. Training, educational workshops, and seminars for environmental professionals, lawyers, and judges on institutional and legal issues affecting the environment. 1616 P Street, N.W., Suite 200, Washington, DC 20036. **www.eli.org**

FOREST STEWARDSHIP COUNCIL. Promotes sustainable forest management and a system of certification of sustainable forest products. 1134 29th Street, N.W., Washington, DC 20007. **www.fscus.org**

FREEDOM FROM HUNGER. Develops programs for the elimination of hunger worldwide. 1644 DaVinci Court, Davis, CA 95617. **www.freefromhunger.org**

FRIENDS OF THE EARTH. Research, lobbying: all aspects of energy development, preservation, restoration, and rational use of the earth. 1025 Vermont Ave., N.W., Washington, DC 20005. **www.foe.org**

GLOBAL ACTION PLAN. An international program enlisting families and organizations in sustainable lifestyle changes, with special programs for children (now part of

Empowerment Institute). P.O. Box 428, Woodstock, NY 12498. **http://empowermentinstitute.net/**

Greenpeace, USA, Inc. An international environmental organization dedicated to protecting the planet through nonviolent, direct action providing public education, scientific research, and legislative lobbying. 702 H Street, N.W., Washington, DC 20001. **www.greenpeace.org**

Habitat for Humanity International. Fosters homebuilding and ownership for the poor. Education on housing issues. 121 Habitat Street, Americus, GA 31709-3498. **www.habitat.org**

Heifer Project International. Works throughout the developing world to bring domestic animals to poor families. P.O. Box 8058, Little Rock, AR 72203. **www.heifer.org**

Institute for Local Self-Reliance. Research and education: appropriate technology for community development. 2425 18th Street, N.W., Washington, DC 20009. **www.ilsr.org**

International Food Policy Research Institute. Mission is to identify and analyze policies for sustainably meeting the food needs of the developing world. IFPRI seeks to make its research results available to all those in a position to use them and to strengthen institutions in developing countries that conduct research relevant to its mandate. **www.ifpri.org/**

Izaak Walton League of America, Inc. Research, education, endowment grants: conservation, air and water quality, streams. 707 Conservation Lane, Gaithersburg, MD 20878. **www.iwla.org**

Land Trust Alliance. Works with local and regional land trusts to enhance their work. 1331 H Street, N.W., Suite 400, Washington, DC 20005. **www.lta.org**

League of Conservation Voters. Political arm of the environmental community. Works to elect candidates to the U.S. House and Senate who will vote to protect the nation's environment, and holds them accountable by publishing the National Environmental Scorecard each year, available from their Web site. 1920 L Street, N.W., Suite 800, Washington, DC 20036. **www.lcv.org**

League of Women Voters of the U.S. Education and lobbying, general environmental issues. Publications on groundwater, agriculture, and farm policy, with additional topics available. 1730 M Street, N.W., Suite 1000, Washington, DC 20036. **www.lwv.org**

Millennium Ecosystem Assessment. "An international work program designed to meet the needs of decision makers and the public for scientific information concerning the consequences of ecosystem change for human well-being and options for responding to those changes." **www.millennium-assessment.org**

Millennium Development Goals. The World Bank site for information and monitoring of progress towards the Millennium Development Goals. **www.developmentgoals.org/**

National Audubon Society. Research, lobbying, education, litigation, and citizen action: broad-based environmental issues. 700 Broadway, New York, NY 10003. **www.audubon.org**

National Council for Science and the Environment. Mission is to improve the scientific basis for making decisions about environmental issues. An umbrella organization, one important affiliate, The National Library for the Environment, provides briefings on environmental issues and the legislation addressing them. 1707 H St. N.W., Suite 200, Washington, D.C. 20006. **www.ncseonline.org/**

National Park Foundation. Education, land acquisition, management of endowments, grant making. National Park Foundation, 1101 17th Street, N.W., Suite 1102, Washington, DC 20036. **www.nationalparks.org**

National Parks Conservation Association. Research and education: parks, wildlife, forestry, general environmental quality. 1300 19th St., N.W., Suite 300, Washington, DC 20036. **www.npca.org**

National Wildlife Federation. Research, education, lobbying: general environmental quality, wilderness, and wildlife. 8925 Leesburg Pike, Vienna, VA 22184. **www.nwf.org**

Natural Resources Defense Council. Research and litigation: water and air quality, land use, energy, pesticides, toxic waste. 40 West 20th Street, New York, NY 10011. **www.nrdc.org**

Nature Conservancy. "Preserve plants, animals, and natural communities that represent the diversity of life on Earth by protecting the land and the water they need to survive." 4525 North Fairfax Drive, Suite 100, Arlington, VA 22203. **http://nature.org/**

Oxfam America. Funding agency for projects to benefit the "poorest of the poor" in South America, Africa, India, Central America, the Caribbean, and the Philippines. Provides whatever resources are needed. 733 15th Street, N.W., Suite 340, Washington, DC 20005. **www.oxfamamerica.org**

Pew Center on Global Climate Change. The center's mission "is to provide credible information, straight answers, and innovative solutions in the effort to address global climate change." 2101 Wilson Blvd., Suite 550, Arlington, VA 22201. **www.pewclimate.org/**

Planned Parenthood Federation of America. Education, services, and research: fertility control, family planning. 810 7th Avenue, New York, NY 10019. **www.plannedparenthood.org**

The Population Institute. Education, research, and speaking engagements: population control. 107 2nd Street, N.E., Washington, DC 20002. **www.populationinstitute.org**

Population Connection. Public education, lobbying, and research toward stable populations. 1400 16th Street, N.W., Suite 320, Washington, DC 20036. **www.populationconnection.org**

Population Reference Bureau, Inc. Organization engaged in collection and dissemination of objective population information. Excellent publications. 1875 Connecticut Avenue, N.W., Suite 520, Washington, DC 20009. **www.prb.org**

Rachel Carson Council, Inc. Publication and distribution of information on pesticides and toxic substances, educational conferences, and seminars. 8940 Jones Mill Road, Chevy Chase, MD 20815. **http://members.aol.com/rccouncil/ourpage/rcc_page.htm**

Rain Forest Action Network. Information and educational resources: world's rain forests. 221 Pine St., Suite 500, San Francisco, CA 94104. **www.ran.org**

Rainforest Alliance. Education, medicinal plants project, timber project to certify "smart wood," and news bureau in Costa Rica. 65 Bleeker St., New York, NY 10012. **www.rainforest-alliance.org**

Renew America. "A nationwide clearinghouse for environmental solutions by seeking out and promoting successful programs. We offer positive, constructive models to help communities meet environmental challenges." Many different environmental categories are addressed. 1200 18th Street, N.W., Suite 1100, Washington, DC 20036. **http://sol.crest.org/environment/renew_america/**

RESOURCES FOR THE FUTURE. Research and education think tank: connects economics and conservation of natural resources, environmental quality. 1616 P Street, N.W., Washington, DC 20036. **www.rff.org**

SHACK/SLUM DWELLERS INTERNATIONAL. Promoting awareness and support for the urban poor in developing countries. **www.sdinet.org/**

SIERRA CLUB. Education and lobbying: broad-based environmental issues. 85 Second Street, 2nd Floor, San Francisco, CA 94105. **www.sierraclub.org**

SPRAWL WATCH CLEARINGHOUSE. A resource center which provides information on sprawl, smart growth and livable communities. **www.sprawlwatch.org/**

TRUST FOR PUBLIC LAND. Works with citizen groups and government agencies to acquire and preserve open space. 116 New Montgomery, 4th Floor, San Francisco, CA 94105. **www.tpl.org/**

UNION OF CONCERNED SCIENTISTS. Connecting citizens and scientists, augments scientific analysis with innovative thinking and committed citizen advocacy to build a cleaner, healthier environment and a safer world. 2 Brattle Square, Cambridge, MA 02238. **www.ucsusa.org/**

U.S. PUBLIC INTEREST RESEARCH GROUP. Research, education, and lobbying, working through state chapters: alternative energy, consumer protection, utilities regulation, public interest. 218 D Street, S.E., Washington, DC 20003. **www.pirg.org**

WATER ENVIRONMENT FEDERATION. Research, education, and lobbying. 601 Wythe Street, Alexandria, VA 22314. **www.wef.org**

WILDERNESS SOCIETY. Research, education, and lobbying: wilderness, public lands. 1615 M St., N.W., Washington, DC 20036. **www.tws.org**

WORLD RESOURCES INSTITUTE. Conducts research and publishes reports for educators, policymakers, and organizations on environmental issues. 10 G Street, N.E., Suite 800, Washington, DC 20002. **www.wri.org**

WORLD WILDLIFE FUND. Preservation of wildlife habitats and protection of endangered species. 1250 Twenty-Fourth Street, N.W., P.O. Box 97180, Washington, DC 20037. **www.worldwildlife.org**

WORLDWATCH INSTITUTE. Research and education: energy, food, population, health, women's issues, technology, the environment. 1776 Massachusetts Avenue, N.W., Washington, DC 20036. **www.worldwatch.org**

# Units of Measure

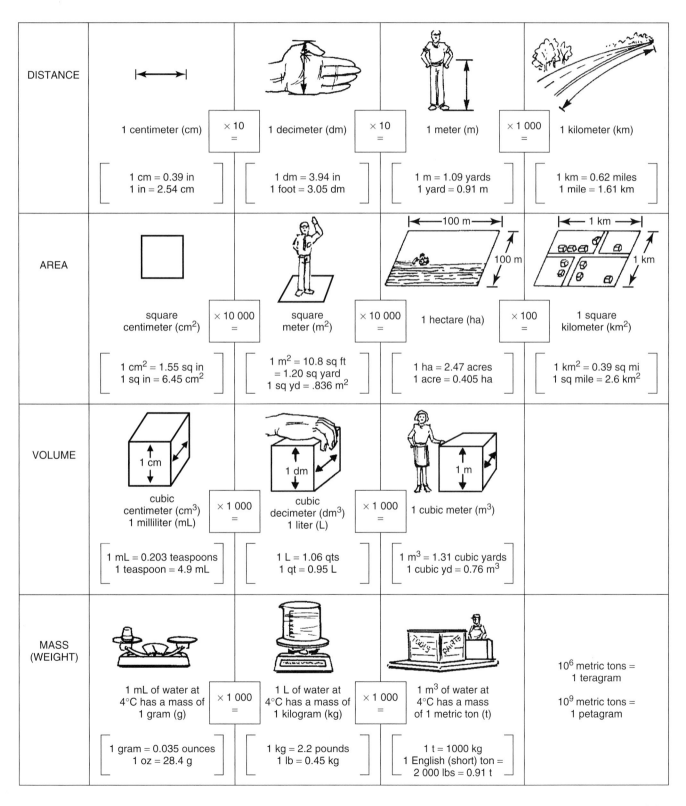

| | | | | | | | |
|---|---|---|---|---|---|---|---|
| DISTANCE | <br>1 centimeter (cm)<br>1 cm = 0.39 in<br>1 in = 2.54 cm | × 10<br>= | <br>1 decimeter (dm)<br>1 dm = 3.94 in<br>1 foot = 3.05 dm | × 10<br>= | 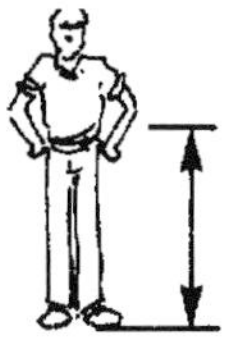<br>1 meter (m)<br>1 m = 1.09 yards<br>1 yard = 0.91 m | × 1 000<br>= | 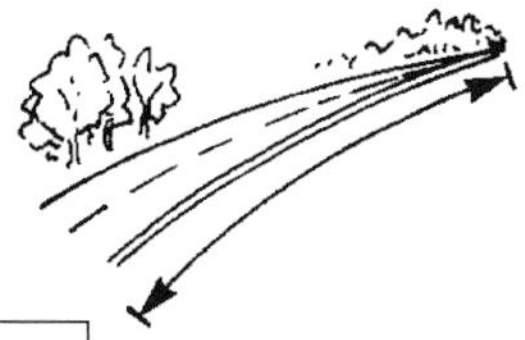<br>1 kilometer (km)<br>1 km = 0.62 miles<br>1 mile = 1.61 km |
| AREA | <br>square centimeter ($cm^2$)<br>1 $cm^2$ = 1.55 sq in<br>1 sq in = 6.45 $cm^2$ | × 10 000<br>= | <br>square meter ($m^2$)<br>1 $m^2$ = 10.8 sq ft<br>= 1.20 sq yard<br>1 sq yd = .836 $m^2$ | × 10 000<br>= | 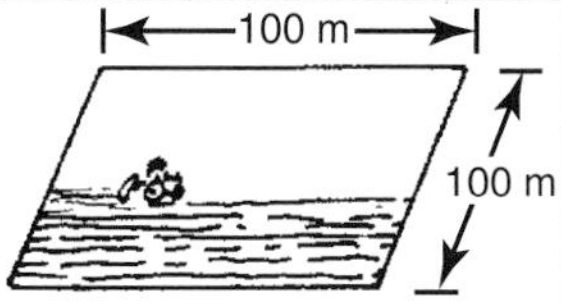<br>1 hectare (ha)<br>1 ha = 2.47 acres<br>1 acre = 0.405 ha | × 100<br>= | 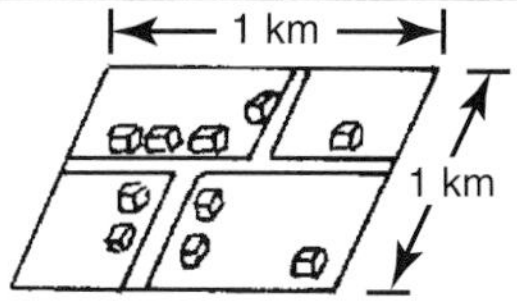<br>1 square kilometer ($km^2$)<br>1 $km^2$ = 0.39 sq mi<br>1 sq mile = 2.6 $km^2$ |
| VOLUME | 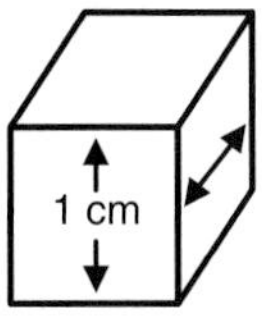<br>cubic centimeter ($cm^3$)<br>1 milliliter (mL)<br>1 mL = 0.203 teaspoons<br>1 teaspoon = 4.9 mL | × 1 000<br>= | 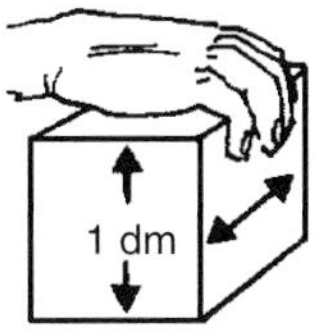<br>cubic decimeter ($dm^3$)<br>1 liter (L)<br>1 L = 1.06 qts<br>1 qt = 0.95 L | × 1 000<br>= | 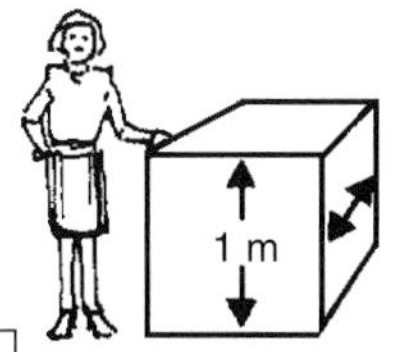<br>1 cubic meter ($m^3$)<br>1 $m^3$ = 1.31 cubic yards<br>1 cubic yd = 0.76 $m^3$ | | |
| MASS (WEIGHT) | 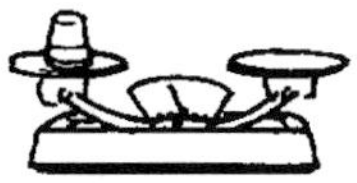<br>1 mL of water at 4°C has a mass of 1 gram (g)<br>1 gram = 0.035 ounces<br>1 oz = 28.4 g | × 1 000<br>= | <br>1 L of water at 4°C has a mass of 1 kilogram (kg)<br>1 kg = 2.2 pounds<br>1 lb = 0.45 kg | × 1 000<br>= | <br>1 $m^3$ of water at 4°C has a mass of 1 metric ton (t)<br>1 t = 1000 kg<br>1 English (short) ton = 2 000 lbs = 0.91 t | | $10^6$ metric tons = 1 teragram<br>$10^9$ metric tons = 1 petagram |

**Energy Units and Equivalents**

1 Calorie, food calorie, or kilocalorie—The amount of heat required to raise the temperature of one kilogram of water one degree Celsius (1.8°F).

1 BTU (British Thermal Unit)—The amount of heat required to raise the temperature of one pound of water one degree Fahrenheit.

1 joule—a force of one newton applied over a distance of one meter (A newton is the force needed to produce an acceleration of 1 m per sec per sec to a mass of one kg)

1 Calorie = 3.968 BTUs = 4,186 joules
1 BTU = 0.252 Calories = 1,055 joules

1 therm = 100,000 BTUs
1 quad = 1 quadrillion BTUs

1 watt = standard unit of electrical power

1 watt = 1 joule per second

1 watt-hour (wh) = 1 watt for 1 hr. = 3.413 BTUs

1 kilowatt (kw) = 1000 watts

1 kilowatt-hour (kwh) = 1 kilowatt for 1 hr. = 3413 BTUs

1 megawatt (Mw) = 1,000,000 watts

1 megawatt-hour (Mwh) = 1 Mw for 1 hr. = 34.13 therms

1 gigawatt (Gw) = 1,000,000,000 watts or 1,000 megawatts

1 gigawatt-hour (Gwh) = 1 Gw for 1 hr. = 34,130 therms

1 horsepower = 0.7457 kilowatts; 1 horsepower-hour = 2545 BTUs

1 cubic foot of natural gas (methane) at atmospheric pressure = 1031 BTUs

1 gallon gasoline = 125,000 BTUs

1 gallon No. 2 fuel oil = 140,000 BTUs

1 short ton coal = 25,000,000 BTUs

1 barrel (oil) = 42 gallons

# Some Basic Chemical Concepts

## Atoms, Elements, and Compounds

All matter, whether gas, liquid, or solid, living or nonliving, or organic or inorganic, is composed of fundamental units called **atoms.** Atoms are extremely tiny. If all the world's people, more than 6 billion of us, were reduced to the size of atoms, there would be room for all of us to dance on the head of a pin. In fact, we would only occupy a tiny fraction (about 1/10,000) of the pin's head. Given the incredibly tiny size of atoms, even the smallest particle that can be seen with the naked eye consists of billions of atoms.

The atoms making up a substance may be all of one kind, or they may be of two or more kinds. If the atoms are all of one kind, the substance is called an **element.** If the atoms are of two or more kinds bonded together, the substance is called a **compound.**

Through countless experiments, chemists have ascertained that there are only 93 distinct kinds of atoms that occur in nature. (An additional 21 have been synthesized.) These natural and most of the synthetic elements are listed in Table C–1, together with their chemical symbols. By scanning the table, you can see that a number of familiar substances, such as aluminum, calcium, carbon, oxygen, and iron, are elements; that is, they are a single, distinct kind of atom. However, most of the substances with which we interact in everyday life, such as water, stone, wood, protein, and sugar, are not on the list. Their absence from the list is indicative that they are not elements; rather, they are compounds, which means that they are actually composed of two or more different kinds of atoms bonded together.

## Atoms, Bonds, and Chemical Reactions

In chemical reactions, atoms are neither created nor destroyed, nor is one kind of atom changed into another. What occurs in chemical reactions, whether mild or explosive, is simply a rearrangement of the ways in which the atoms involved are bonded together. An oxygen atom, for example, may be combined and recombined with different atoms to form any number of different compounds, but a given oxygen atom always has been, and always will be, an oxygen atom. The same can be said for all the other kinds of atoms. To understand how atoms may bond and undergo rearrangement to form different compounds, it is necessary to examine several concepts concerning the structure of atoms.

## Structure of Atoms

Every atom consists of a central core called the **nucleus** (not to be confused with a cell nucleus). The nucleus of an atom contains one or more **protons** and, except for hydrogen, one or more **neutrons.** Surrounding the nucleus are particles called **electrons.** Each proton has a positive (+) electric charge, and each electron has an equal and opposite negative (−) electric charge. Thus, in any atom the charge of the protons may be balanced by an equal number of electrons, making the whole atom neutral. Neutrons have no charge.

Atoms of all elements have this same basic structure, consisting of protons, electrons, and neutrons. The distinction between atoms of different elements is in the number of protons. The atoms of each element have a characteristic number of protons that is known as the **atomic number** of the element. (See Table C–1.) The number of electrons characteristic of the atoms of each element also differs, in a manner corresponding to the number of protons. The combined total of protons and neutrons in the nucleus of an element is the **mass number,** or **atomic weight.** The general structure of the atoms of several elements is shown in Figure C–1.

The number of protons and electrons in an atom of an element (i.e., the atomic number of the element) determines the chemical properties of the element. The number of neutrons may vary. For example, most carbon atoms have six neutrons in addition to the six protons, as indicated in Figure C–1. But some carbon atoms have eight neutrons. Atoms of the same element that have different numbers of neutrons are known as **isotopes** of the element. The total number of protons plus neutrons is used to define different isotopes. For example, the usual isotope of carbon is referred to as carbon-12, while the isotope with eight neutrons is referred to as carbon-14. The chemical reactivities of different isotopes of the same element are identical. (However, certain other properties may differ.) Many isotopes of various elements prove to be radioactive, such as carbon-14. The atomic weights in Table C–1 are expressed as the average of the isotopes of an element as they occur in nature. The weights listed in parentheses are those of the most stable known isotope.

### Bonding of Atoms

The chemical properties of an element are defined by the ways in which its atoms will react and form bonds with other atoms. By examining how atoms form bonds, we shall see how the number of electrons and protons determines these properties. There are two basic kinds of bonding: (1) **covalent bonding** and (2) **ionic bonding.**

In both kinds of bonding, it is important to recognize that electrons are not randomly distributed around the atom's nucleus. Rather, there are, in effect, specific spaces in a series of layers, or **orbitals,** around the nucleus. If an orbital is occupied by one or more electrons, but is not filled, the atom is unstable; it will then tend to react and form bonds with other atoms to achieve greater stability. A stable state is achieved by having all the spaces in the orbital filled with electrons. It is important, however, to keep the charge neutral (i.e., the total number of electrons should be equal to that of the protons).

## table C-1 The Elements

| Element | Symbol | Atomic Number | Atomic Weight |
|---|---|---|---|
| Actinium | Ac | 89 | (227) |
| Aluminum | Al | 13 | 27.0 |
| *Americium | Am | 95 | (243) |
| Antimony | Sb | 51 | 121.8 |
| Argon | Ar | 18 | 39.9 |
| Arsenic | As | 33 | 74.9 |
| Astatine | At | 85 | (210) |
| Barium | Ba | 56 | 137.3 |
| *Berkelium | Bk | 97 | (245) |
| Beryllium | Be | 4 | 9.01 |
| Bismuth | Bi | 83 | 209.0 |
| *Bohrium | Bh | 107 | (262) |
| Boron | B | 5 | 10.8 |
| Bromine | Br | 35 | 79.9 |
| Cadmium | Cd | 48 | 112.4 |
| Calcium | Ca | 20 | 40.1 |
| *Californium | Cf | 98 | (251) |
| Carbon | C | 6 | 12.0 |
| Cerium | Ce | 58 | 140.1 |
| Cesium | Cs | 55 | 132.9 |
| Chlorine | Cl | 17 | 35.5 |
| Chromium | Cr | 24 | 52.0 |
| Cobalt | Co | 27 | 58.9 |
| Copper | Cu | 29 | 63.5 |
| *Curium | Cm | 96 | (245) |
| *Dubnium | Db | 105 | (262) |
| Dysprosium | Dy | 66 | 162.5 |
| *Einsteinium | Es | 99 | (254) |
| Erbium | Er | 68 | 167.3 |
| Europium | Eu | 63 | 152.0 |
| *Fermium | Fm | 100 | (254) |
| Fluorine | F | 9 | 19.0 |
| Francium | Fr | 87 | (223) |
| Gadolinium | Gd | 64 | 157.3 |
| Gallium | Ga | 31 | 69.7 |
| Germanium | Ge | 32 | 72.6 |
| Gold | Au | 79 | 197.0 |
| Hafnium | Hf | 72 | 178.5 |
| *Hassium | Hs | 108 | (265) |
| Helium | He | 2 | 4.00 |
| Holmium | Ho | 67 | 164.9 |
| Hydrogen | H | 1 | 1.01 |
| Indium | In | 19 | 114.8 |
| Iodine | I | 53 | 126.9 |
| Iridium | Ir | 77 | 192.2 |
| Iron | Fe | 26 | 55.8 |
| Krypton | Kr | 36 | 83.8 |
| Lanthanum | La | 57 | 138.9 |
| *Lawrencium | Lr | 103 | (257) |
| Lead | Pb | 82 | 207.2 |
| Lithium | Li | 3 | 6.94 |
| Lutetium | Lu | 71 | 175.0 |
| Magnesium | Mg | 12 | 24.3 |
| Manganese | Mn | 25 | 54.9 |
| *Meitnerium | Mt | 109 | (265) |
| *Mendelevium | Md | 101 | (256) |
| Mercury | Hg | 80 | 200.6 |
| Molybdenum | Mo | 42 | 95.9 |
| Neodymium | Nd | 60 | 44.2 |
| Neon | Ne | 10 | 20.2 |
| *Neptunium | Np | 93 | 237.0 |
| Nickel | Ni | 28 | 58.7 |
| Niobium | Nb | 41 | 92.9 |
| Nitrogen | N | 7 | 14.0 |
| *Nobelium | No | 102 | (254) |
| Osmium | Os | 76 | 190.2 |
| Oxygen | O | 8 | 16.0 |
| Palladium | Pd | 46 | 106.4 |
| Phosphorus | P | 15 | 31.0 |
| Platinum | Pt | 78 | 195.1 |
| Plutonium | Pu | 94 | (242) |
| Polonium | Po | 84 | (210) |
| Potassium | K | 19 | 39.1 |
| Praseodymium | Pr | 59 | 140.9 |



*continued*

**table C-1 The Elements (Continued)**

| Element | Symbol | Atomic Number | Atomic Weight |
|---|---|---|---|
| Promethium | Pm | 61 | (145) |
| Protoactinium | Pa | 91 | 231.0 |
| Radium | Ra | 88 | 226.0 |
| Radon | Rn | 86 | (222) |
| Rhenium | Re | 75 | 186.2 |
| Rhodium | Rh | 45 | 102.9 |
| Ruthenium | Ru | 44 | 101.1 |
| *Rutherfordium | Rf | 104 | (261) |
| Samarium | Sm | 62 | 150.4 |
| Scandium | Sc | 21 | 45.0 |
| *Seaborgium | Sg | 106 | (263) |
| Selenium | Se | 34 | 79.0 |
| Silicon | Si | 14 | 28.1 |
| Silver | Ag | 47 | 107.9 |
| Sodium | Na | 11 | 23.0 |
| Strontium | Sr | 38 | 87.6 |
| Sulfur | S | 16 | 32.1 |
| Tantalum | Ta | 73 | 180.9 |
| Technetium | Tc | 43 | 98.9 |
| Tellurium | Te | 52 | 127.6 |
| Terbium | Tb | 65 | 158.9 |
| Thallium | Tl | 81 | 204.4 |
| Thorium | Th | 90 | 232.0 |
| Thulium | Tm | 69 | 168.9 |
| Tin | Sn | 50 | 118.7 |
| Titanium | Ti | 22 | 47.9 |
| Tungsten | W | 74 | 183.8 |
| Uranium | U | 92 | 238.0 |
| Vanadium | V | 23 | 50.9 |
| Xenon | Xe | 54 | 131.3 |
| Ytterbium | Yb | 70 | 173.0 |
| Yttrium | Y | 39 | 88.9 |
| Zinc | Zn | 30 | 65.4 |
| Zirconium | Zr | 40 | 91.2 |

*Elements that do not occur in nature.
Parentheses indicate the most stable isotope of the element.

## Covalent Bonding

These two requirements—filling all the spaces and keeping the charge neutral—may be satisfied by adjacent atoms sharing one or more pairs of electrons, as shown in Figure C–2. The sharing of a pair of electrons holds the atoms together in what it is called a **covalent bond.**

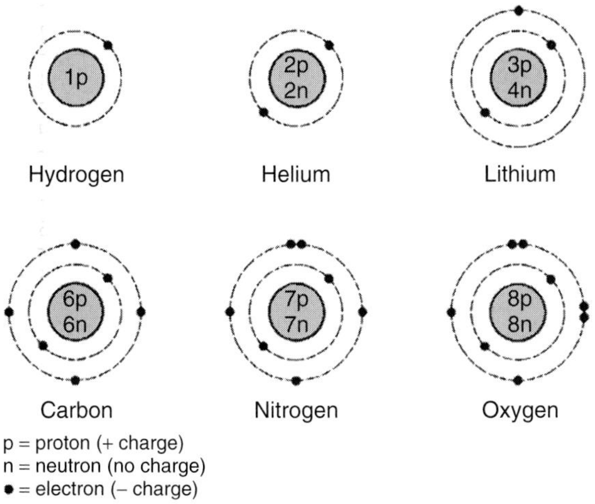


**Figure C–1 Structure of atoms.** All atoms consist of fundamental particles: protons (*p*), which have a positive electric charge, neutrons (*n*), which have no charge, and electrons, which have a negative charge.

Covalent bonding, by satisfying the charge–orbital requirements, leads to discrete units of two or more atoms bonded together. Units of two or more covalently bonded atoms are called **molecules.** A few simple, but important, examples are shown in Figure C–2.

A chemical formula is simply a shorthand description of the number of each kind of atom in a given molecule. The element is given by the chemical symbol, and a subscript following the symbol gives the number of atoms of that element present in the molecule, no subscript being understood as one atom. A molecule with two or more different kinds of atoms may be called a compound, but a molecule composed of a single kind of atom—oxygen ($O_2$), for example—is still defined as an element.

Only a few elements—carbon, hydrogen, oxygen, nitrogen, phosphorus, and sulfur—have configurations of electrons that lend themselves readily to the formation of covalent bonds. Carbon in particular, with its ability to form four covalent bonds, can produce long, straight, or branched chains or rings (Fig. C–3). Thus, an infinite array of molecules can be formed by using covalently bonded carbon atoms as a "backbone" and filling in the sides with atoms of hydrogen or other elements. Accordingly, it is covalent bonding among atoms of carbon and these few other elements that produces all natural organic molecules, those molecules which make up all the tissues of living things, and also synthetic organic compounds such as plastics.

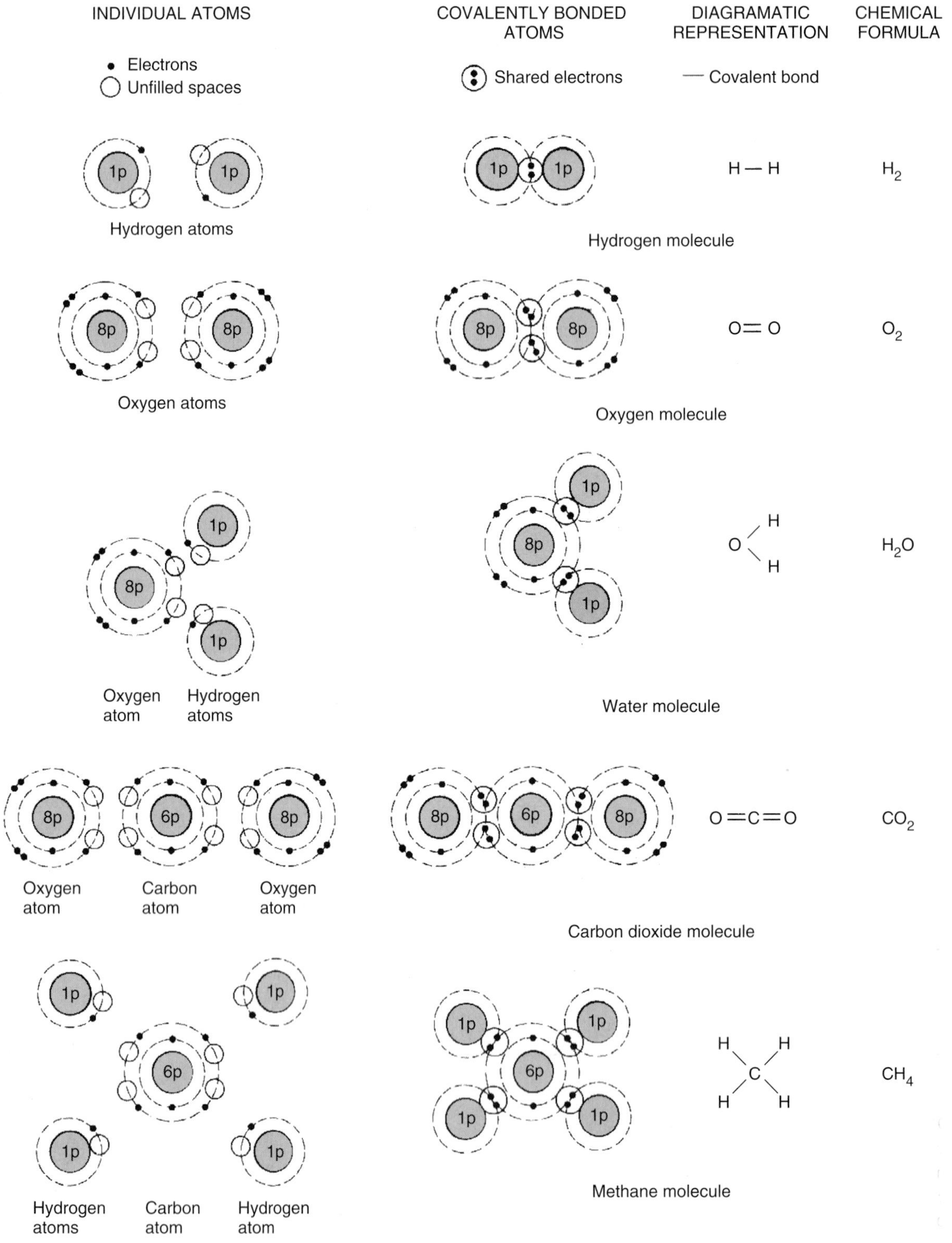


**Figure C–2 Atoms to molecules.** Molecules are composed of atoms covalently bonded to each other in stable configurations in which electrons are shared.

## Ionic Bonding

Another way in which atoms may achieve a stable electron configuration is to gain additional electrons to complete the filling of an orbital or to lose excess electrons in an incompleted orbital. In general, the maximum number of electrons that can be gained or lost by an atom is three. Therefore, an element's atomic number determines whether one or more electrons will be lost or gained. If an atom's outer orbital is one to three electrons short of being filled, it will always tend to gain additional electrons. If an atom has one to three electrons in excess of its last complete orbital, it will always tend to lose those electrons.

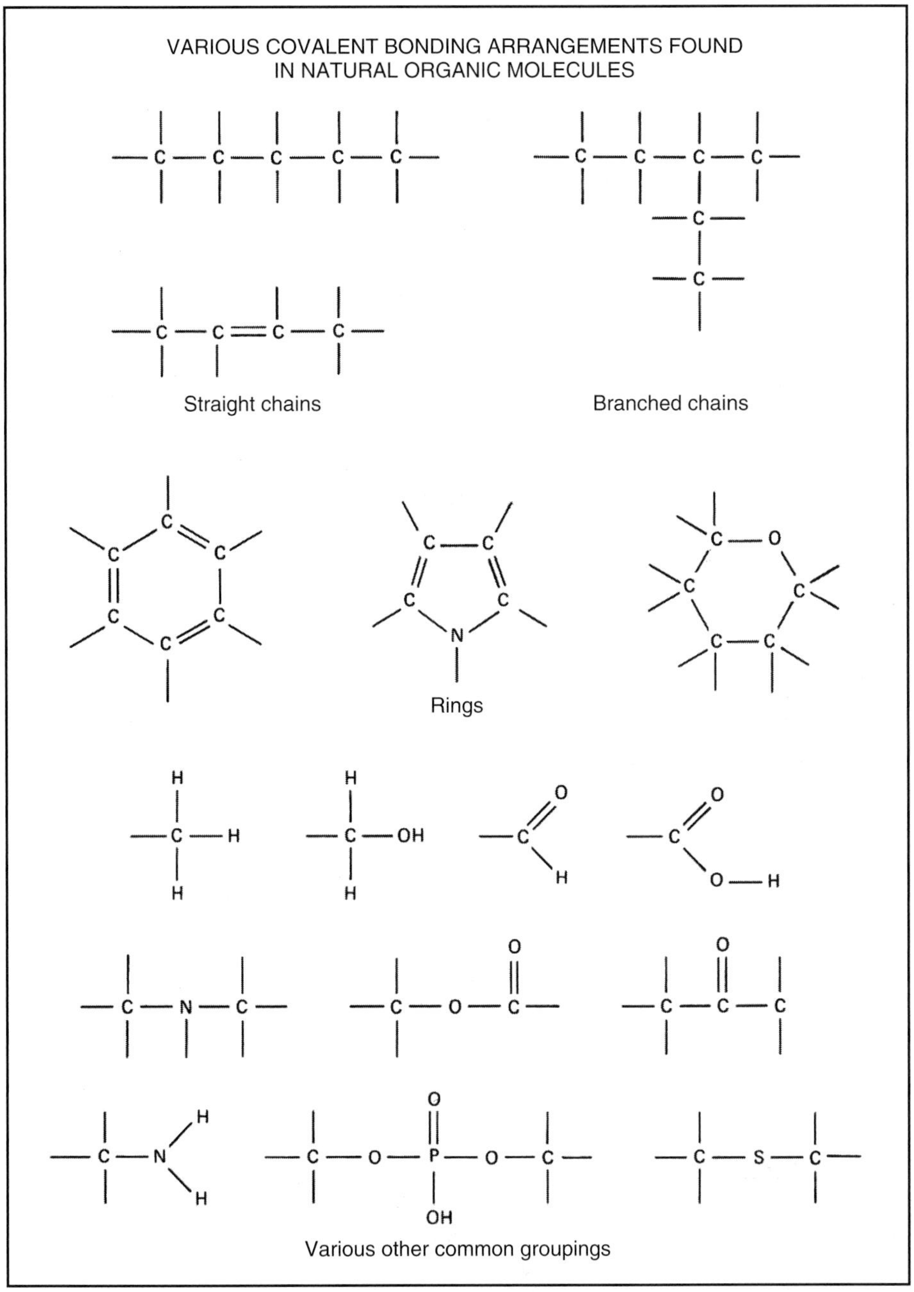


**Figure C–3 Covalent bonding and organic molecules.** The ability of carbon and a few other elements to readily form covalent bonds leads to an infinite array of complex organic molecules, which constitute all living things. A few major kinds of groupings are shown here. Note that each element forms a characteristic number of bonds: carbon, 4; nitrogen, 3; oxygen, 2; hydrogen, 1; sulfur, 2; and phosphorus, 5. Bonds left hanging (lines without atoms attached) indicate attachments to other atoms or groups of atoms.

Of course, gaining or losing electrons results in the number of electrons in an atom being greater or less than the number of protons; the atom consequently has an electric charge. The charge will be one negative unit for each electron gained and one positive unit for each electron lost (Fig. C–4). A covalently bonded group of atoms may acquire an electric charge in the same way. An atom or group of atoms that has acquired an electric charge in this way is called an **ion,** positive or negative. Ions are designated by a superscript following the chemical symbol, which denotes the number of positive or negative charges. The absence of superscripts indicates that the atom or molecule is neutral. Some important ions are listed in Table C–2.

Since unlike charges attract, positive and negative ions tend to join and pack together in dense clusters in such a way as to neutralize an atom's overall electric charge. This joining together of ions through the attraction of their opposite charges is called **ionic bonding.** The result is the formation of hard, brittle, more or less crystalline substances of which all rocks and minerals are examples (Fig. C–5).

It is significant to note that whereas covalent bonding leads to discrete molecules, ionic bonding does not. Any number and combination of positive and negative ions may enter into an ionic bond to produce crystals of almost any size. The only restriction is that the overall charge of positive ions be balanced by that of negative ions. Thus, substances bonded in an ionic fashion are properly called compounds, but not molecules. When chemical formulas are used to describe such compounds, they define the ratio of various elements involved, not specific molecules.

## Chemical Reactions and Energy

While atoms themselves do not change, the bonds between atoms may be broken and re-formed, with different atoms producing different compounds or molecules. This is essentially what occurs

**Figure C-4 Formation of ions.** Many atoms will tend to gain or lose one or more electrons in order to achieve a state of complete (electron-filled) orbitals. In doing so, these atoms become positively or negatively charged ions, as indicated.

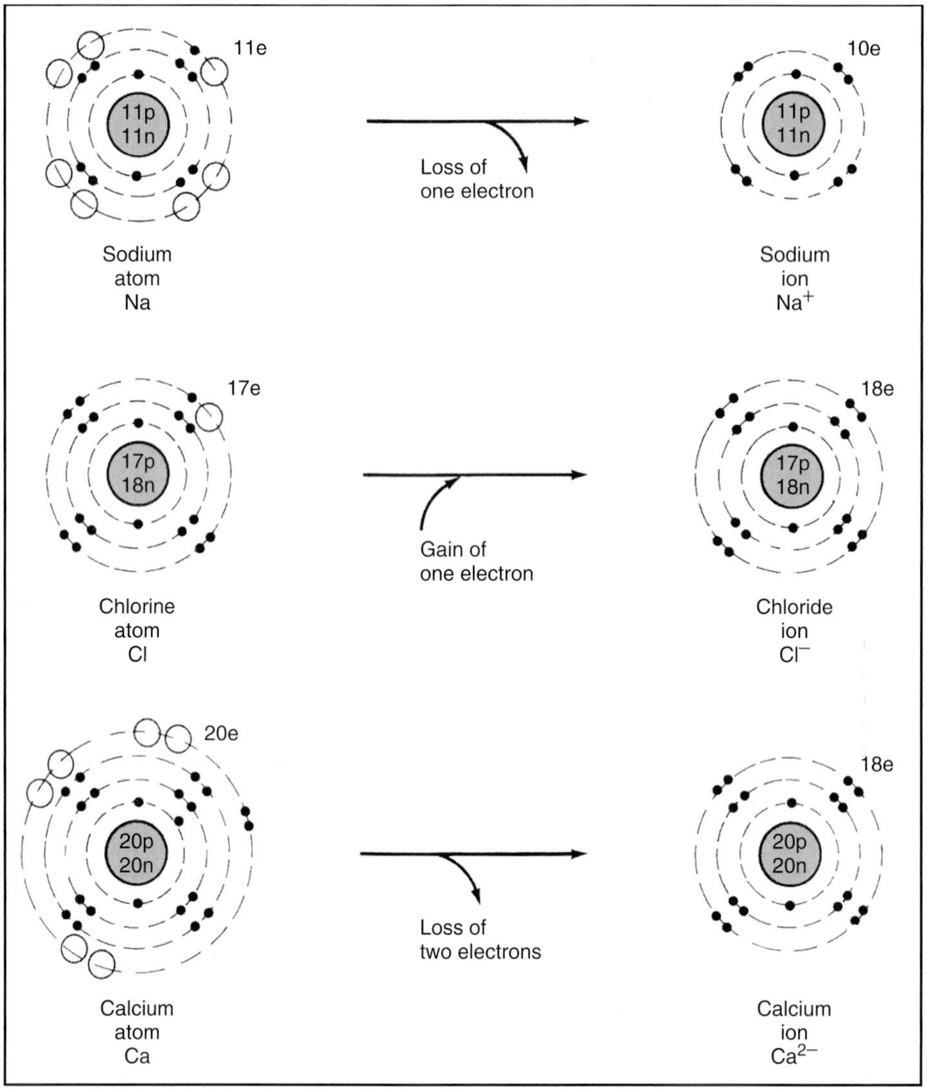


in all chemical reactions. What determines whether a given chemical reaction will occur? Earlier, we noted that atoms form bonds because they achieve a greater stability by doing so. However, some bonding arrangements may provide greater overall stability than others. Consequently, substances with relatively unstable

**table C-2 Ions of Particular Importance to Biological Systems**

| Negative (−) Ions | | Positive (+) Ions | |
|---|---|---|---|
| Phosphate | $PO_4^{3-}$ | Potassium | $K^+$ |
| Sulfate | $SO_4^{2-}$ | Calcium | $Ca^{2+}$ |
| Nitrate | $NO_3^-$ | Magnesium | $Mg^{2+}$ |
| Hydroxyl | $OH^-$ | Iron | $Fe^{2+}$, $Fe^{3+}$ |
| Chloride | $Cl^-$ | Hydrogen | $H^+$ |
| Bicarbonate | $HCO_3^-$ | Ammonium | $NH_4^+$ |
| Carbonate | $CO_3^{2-}$ | Sodium | $Na^+$ |

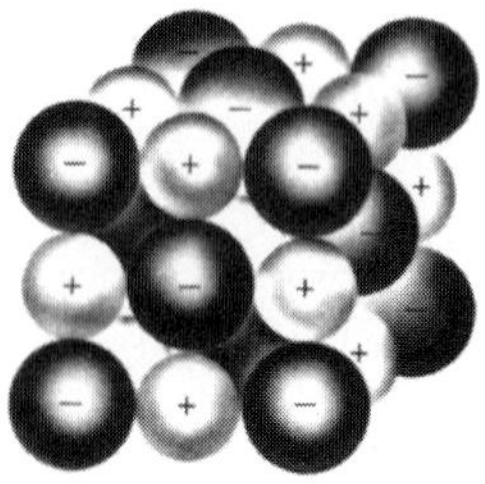

**Figure C-5 Crystals.** Positive and negative ions bond together by their mutual attraction and form crystals.

bonding arrangements will tend to react to form one or more different compounds that have more stable bonding arrangements. Common examples are the reaction between hydrogen and oxygen to produce water and the reaction between carbon and oxygen to produce carbon dioxide (Fig. C-6).

Energy is always released when the constituents of a reaction gain greater overall stability, as indicated in the figure. Thus, when energy is released in a chemical reaction, the atoms achieve more stable bonding arrangements. Therefore, it may be said that chemical reactions always tend to go in a direction that releases energy, as well as giving greater stability.

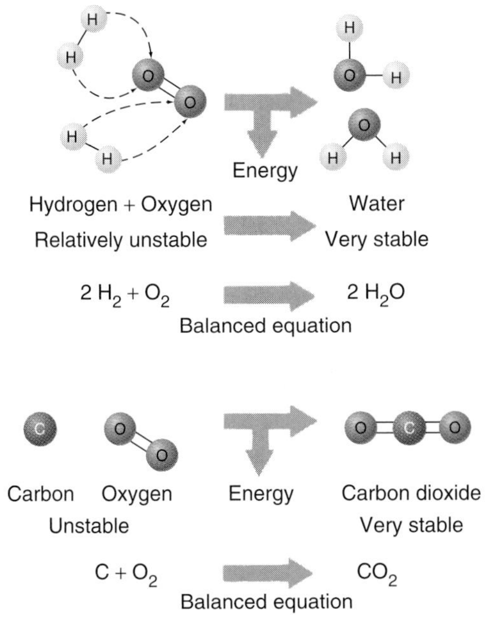


**Figure C–6 Stable bonding arrangements.** Some bonding arrangements are more stable than others. Chemical reactions will go spontaneously toward more stable arrangements, releasing energy in the process. However, reactions may be driven in the opposite direction with suitable energy inputs.

However, chemical reactions can be made to go in a reverse direction. With suitable energy inputs and under suitable conditions, stable bonding arrangements may be broken and less stable arrangements formed. As described in Chapter 2 of the text, this is the basis of the photosynthesis that occurs in green plants. Light energy causes the highly stable hydrogen–oxygen bonds of water to split and form less stable carbon–hydrogen bonds, thus creating high-energy organic compounds.

Obviously, much more could be said about the involvement of chemistry in environmental science. We would highly recommend a course (or several courses) in chemistry to the student who wants to gain a deeper understanding of how the subject is involved in environmental concepts and issues.

# photo credits

**Table of Contents**
TOC–1, Michael Melford, Getty Images, Inc.–Image Bank; TOC–3, Michael P. Gadomski, Animals Animals Earth Scenes; TOC–5, Jeremy Walker, Getty Images, Inc.–Stone Allstock; TOC–6, Jeff Greenberg, Index Stock Imagery, Inc.

**Chapter 1**
CO-1, © Don and Elaine Dvorak, Kite Aerial Photography; Figure 1–1, Jan Halaska/Photo Researchers, Inc.; Figure 1–2, Fred J. Eckert/Eckert Images; Figure 1–7, Malcolm S. Kirk/Peter Arnold, Inc.; Figure 1–10, AP/Wide World Photos; Figure 1–11a, Reuters/Aladin /Getty Images Inc.–Hulton Archive Photos; Figure 1–11b, Maya Vidon/Agence France Presse/Getty Images; Figure 1–11c, Alain Evrard/Photo Researchers, Inc.; Figure 1–14, Wally McNamee/Corbis; Figure 1–15, Susan Jones/SIPA Press; Figure 1–16, AFP/Corbis

**Chapter 2**
PO-01, Michael Melford/Getty Images, Inc./The Image Bank; CO-2 and inset, Richard T. Wright; Figure 2–1, Richard T. Wright; Figure 2–2, Richard T. Wright; Figure 2–7, Tom McHugh/Photo Researchers, Inc.; Figure 2–8a, Runk/Schoenberger/Grant Heilman Photography, Inc.; Figure 2–8b, Sinclair Stammers/Science Photo Library/Photo Researchers, Inc.; Figure 2–8c, Brian Parker/Tom Stack & Associates, Inc.; Figure 2–15, Stephen P. Parker/Photo Researchers, Inc.; Figure 2–16, Bernard J. Nebel; Figure 2–23, Anthony Bannister/Animals Animals/Earth Scenes; Figure 2–24, Archiv/Photo Researchers, Inc.

**Chapter 3**
CO-3, M.P. Kahl/Photo Researchers, Inc.; Figure 3–14, Reproduced from Transmission electron micrograph by Breznak & Pankratz (1977) with permission of the American Society for Microbiology; Figure 3–12UN, NASA Goddard Laboratory for Atmospheres; Figure 3–20, U.S. Department of Agriculture; Figure 3–23UN, Gonzalo Arcila/Columbia University Biosphere 2 Center

**Chapter 4**
CO-4, David Hosking/Photo Researchers, Inc.; Figure 4–1a & b, Renee Lynn/Photo Researchers, Inc.; Figure 4–3, John Burnley/Photo Researchers, Inc.; Figure 4–6b, Fran Allan/Animals Animals/ Earth Scenes; Figure 4–7, Jim Zipp/Photo Researchers, Inc.; Figure 4–8, Johann Schumacher/Peter Arnold, Inc.; Figure 4–9, Grant Heilman/Grant Heilman Photography, Inc.; Figure 4–10, Bernard J. Nebel; Figure 4–11, Australian Consulate General; Figure 4–12a, Ron Peplowski/The Detroit Edison Company; Figure 4–12b, Peter Parks/Oxford Scientific Films/Animals Animals/Earth Scenes; Figure 4–13a, John Pontier/Animals Animals/Earth Scenes; Figure 4–13b, Leonard Lee Rue III/Photo Researchers, Inc.; Figure 4–13c, Robert Bornemann/Photo Researchers, Inc.; Figure 4–13d, John Mitchell/Photo Researchers, Inc.; Figure 4–14a, Nuridsany et Perennou/Photo Researchers, Inc.; Figure 4–14b, J.H. Robinson/Photo Researchers, Inc.; Figure 4–14c, Michael Fogden/Animals Animals/ Earth Scenes; Figure 4–14d, R.H. Armstrong/Animals Animals/ Earth Scenes; Figure 4–17b, Jeff Lepore/Photo Researchers, Inc.; Figure 4–17c, C.K. Lorenz/Photo Researchers, Inc.; Figure 4–18b, E.C. Williams/Visuals Unlimited; Figure 4–18c, Norman D. Tomalin/Bruce Coleman Inc.; Figure 4–19, Bernard J. Nebel; Figure 4–20b, Bernard J. Nebel; Figure 4–21b, Bernard J. Nebel; Figure 4–22, Francis Lepine/Biosphere Photographie/Animals Animals/Earth Scenes; Figure 4–23, USDA/NRCS/Natural Resources Conservation Service; Figure 4–24b, Michael Giannechini/Photo Researchers, Inc.; Figure 4–21c, Jacques Jangoux/Photo Researchers, Inc.; Figure 4–25, Mitch Reardon/Photo Researchers, Inc.

**Chapter 5**
PO-2, AP/Wide World Photos; CO-5, Betty Press/Woodfin Camp & Associates; Figure 5–8, Asa C. Thoresen/Photo Researchers, Inc.; Figure 5–9b, Nickelsberg/Liaison Agency, Inc.; Figure 5–10, Jerry Cooke/Photo Researchers, Inc.

**Chapter 6**
CO-6, Reuters/Kamal Kishore/Getty Images, Inc./Hulton Archive Photos; Figure 6–5, Bruno J. Zehner/Peter Arnold, Inc.; Figure 6–14-02UN, © Darrell Plowes; Figure 6–9, Nigel J.H. Smith/Animals Animals/Earth Scenes; Figure 6–11, Jagdish Agarwal/Pictor/ImageState/International Stock Photography Ltd.; Figure 6–12, Claude Poulet/ Figaro/Getty Images–Liaison; Figure 6–14, John Van Hasselt/Corbis; Figure 6–15, William Campbell/Peter Arnold, Inc.

**Chapter 7**
PO-3, Michael P. Gadomski/Animals Animals/Earth Scenes; CO-7, Image courtesy USGS Eros Data Center; Figure 7–2a, Pete Saloutos/ Corbis/Stock Market; Figure 7–2b, Scott Peterson/Getty Images, Inc.–Liaison; Figure 7–9, AP Wide World Photos; Figure 7–10, Reproduced with permission from Fig 2C from V. Ramanathan et al., SCIENCE 294:2119–2124(2001), American Association for the Advancement of Science; Figure 7–12, Mark Edwards/Still Pictures/Peter Arnold, Inc.; Figure 7–14, Michael Collier; Figure 7–15a, Margot Conte/Animals Animals/Earth Scenes; Figure 7–15b, Dick Rowan/ Photo Researchers, Inc.; Figure 7–16b, Earl Roberge/Photo Researchers, Inc.; Figure 7–16c, Francois Gohier/ Photo Researchers, Inc.; Figure 7–17, R. Veronica Decker/The Orlando Sentinel/Getty Images, Inc.–Liaison; Figure 7–19a, Liu Liqun/ChinaStock Photo Library; Figure 7–20, Courtesy Tampa Bay Water; Figure 7–21, Lowell Georgia/Photo Researchers, Inc.; Figure 7–22, © International Development Enterprises; Figure 7–23, TOTO Inc.

**Chapter 8**
CO-8, Martin Bond/Science Photo Library/Photo Researchers, Inc.; Figure 8–1, Photo by Gene Hettel/Courtesy of CIMMYT; Figure 8–2, Ruben Mangasaryan/Patker Photo Agency; Figure 8–9, © Frank Dazzo/Michigan State University; Figure 8–10, Bernard J. Nebel; Figure 8–11, Richard T. Wright; Figure 8–12, N. Allen and G. Barron, University of Guelph, Guelph, Ontario; Figure 8–14, Charles R. Belinky/Photo Researchers, Inc.; Figure 8–15b, Steve Kaufman/Peter Arnold, Inc.; Figure 8–16, Rene Grosjean/Getty Images, Inc.–Liaison; Figure 8–18, Larry Lefever/Grant Heilman Photography, Inc.; Figure 8–19a, U.S. Department of Agriculture; Figure 8–19b, Dan McCoy/Corbis/Stock Market; Figure 8–21, Betty Derig/Photo Researchers, Inc.; Figure 8–22, John M. White/Texas Agricultural Extension Services.

**Chapter 9**
CO-9, Denis Waugh/Getty Images Inc./Stone Allstock; Figure 9–1, Jon Christensen; Figure 9–2a, Garry D. McMichael/Photo Researchers, Inc.; Figure 9–2b, E.R. Degginger/Color–Pic/Animals Animals/Earth Scenes; Figure 9–4a, 9–4b, Dr. Nigel Smith/Animals Animals/Earth Scenes; Figure 9–5, Dick Davis/Photo Researchers, Inc.; Figure 9–6, Roland Seitre/Peter Arnold, Inc.; Figure 9–7, Robin Smith/Photo-Library.com; Figure 9–8, © Heifer Project International; Figure 9–11a, Stephen A. Ferreira, University of Hawaii at Manoa; Figure 9–11b, Richard M. Manshardt, University of Hawaii at Manoa; Figure 9–12, Andy Nelson/2002 © The Christian Science Monitor; Figure 9–13L,

Susan Kuklin/Photo Researchers, Inc.; Figure 9–13M, John Spragens, Jr./Photo Researchers, Inc.; Figure 9–13R, European Space Agency/Science Photo Library/Photo Researchers, Inc.; Figure 9–14, Eugene Gordon/Pearson Education/PH College; Figure 9–16, © Robert Patrick/Corbis Sygma; Figure 9–18, Scott Peterson/Liaison Agency, Inc.

**Chapter 10**

CO-10, Richard T. Wright; Figure 10–1, Gary G. Gibson/National Audobon Society/Photo Researchers, Inc.; Figure 10–2, ECHO: Educational Concerns for Hunger Organization; Figure 10–3, Richard T. Wright; Figure 10–4TL, Lowell Georgia/Photo Researchers, Inc.; Figure 10–4BL, Explorer/Yves Gladu/Photo Researchers, Inc.; Figure 10–4TM, Mark Burnett/Photo Researchers, Inc.; Figure 10–4BM, Chris Rogers/Corbis/Stock Market; Figure 10–4TR, Jack Wilburn/Animals Animals/Earth Scenes; Figure 10–4BR, Leonard L.T. Rhodes/Animals Animals/Earth Scenes; Figure 10–5, Tom McHugh/Photo Researchers, Inc.; Figure 10–6, Steve Homan/Federal Highway Administration; Figure 10–7, AP/Wide World Photos; Figure 10–8a, C. Allan Morgan/Peter Arnold, Inc.; Figure 10–8b, Steve Kaufman/Peter Arnold, Inc.; Figure 10–8c, Julia Sims/Peter Arnold, Inc.; Figure 10–8d, A.H. Rider/Photo Researchers, Inc.; Figure 10–8e, Jeff Lepore/Photo Researchers, Inc.; Figure 10–8f, C.C. Lockwood/Animals Animals/Earth Scenes; Figure 10–8g, Tom McHugh/Steinhart Aquarium/Photo Researchers, Inc.; Figure 10–8h, Mickey Gibson/Animals Animals/Earth Scenes; Figure 10–8i, Douglas Faulkner/Photo Researchers, Inc.; Figure 10–9, Darek Karp/Animals Animals/Earth Scenes; Figure 10–10, © Operation Migration, Inc.; Figure 10–13, John Colwell/ Grant Heilman Photography, Inc.; Figure 10–14, Allen Blake Sheldon/Animals Animals/Earth Scenes; Figure 10–15, Klaus Uhlenhut/Animals Animals/Earth Scenes; Figure 10–16a, Richard N. Mack, Washington State University; Figure 10–16b, Richard N. Mack, Washington State University; Figure 10–17a, © Alain Dragesco-Joffe/Animals Animals/Earth Scenes; Figure 10–17b, © Dinodia Picture Agency

**Chapter 11**

CO-11, Paul A. Souders/Corbis/Bettmann; Figure 11–1L, Bud Lehnhausen/Photo Researchers, Inc.; Figure 11–1R, Ray Pfortner/Peter Arnold, Inc.; Figure 11–3, Andrew L. Young/Photo Researchers, Inc.; Figure 11–4, BIOS (A. Compost)/Peter Arnold, Inc.; Figure 11–5, Grant Heilman/Grant Heilman Photography, Inc.; Figure 11–7, Ted Levin/Animals Animals/Earth Scenes; Figure 11–8, Republished with permission of Globe Newspaper Company, Inc., from an issue of the Boston Globe; Figure 11–9, Courtesy Food & Agriculture Organization of the United Nations; Figure 11–10, David Austen/Woodfin Camp & Associates; Figure 11–10UN, William Hubbel/Woodfin Camp Associates; Figure 11–11, Tim Hall/Robert Harding World Imagery; Figure 11–12UN, © Mark C. Burnett/Photo Researchers, Inc.; Figure 11–13, © C. Ruoso/BIOS/Peter Arnold, Inc.; Figure 11–16, © AP/Wide World Photos; Figure 11–17, Jan Robert Factor/Photo Researchers, Inc.; Figure 11–18, Klaus Jost/Peter Arnold, Inc.

**Chapter 12**

PO-4, Paul McCormick/Getty Images Inc./Image Bank; CO-12, Mark Marten/NASA/Photo Researchers, Inc.; Figure 12–1, © Michio Hishino/Minden Pictures; Figure 12–2, © Gary Schultz; Figure 12–3, Tony Wells/Mason Dixon Historical Society, Inc.; Figure 12–7d, C.P. Hickman/Visuals Unlimited; Figure 12–9, Neal Palumbo/Getty Images, Inc.–Liaison.; Figure 12–15, © AP/Wide World Photos; Figure 12–17, © Walter Choroszewski; Figure 12–18, Chris Jones/Corbis/Stock Market; Figure 12–19, Courtesy Alyeska Pipeline Service Company; Figure 12–21, Bernard J. Nebel

**Chapter 13**

CO-13, Pete Saloutos/Corbis/Stock Market; Figure 13–1, AP Wide World Photos; Figure 13–3, DiMaggio/Kalish/Corbis/Stock Market; Figure 13–7c, Erich Hartmann/Magnum Photos, Inc.; Figure 13–12, United States Enrichment Corporation; Figure 13–13, AP Wide World Photos; Figure 13–14, Vlastimir Shone-Zoufarov/Getty Images, Inc.–Liaison; Figure 13–16, Courtesy Maine Yankee; Figure 13–17, Alexander Tsiaras/Science Source/Photo Researchers, Inc.

**Chapter 14**

CO-15, Republished with permission of Globe Newspaper Company, Inc., from the July 21, 1997 issue of The Boston Globe, © 1997; Figure 14–1, James D. Beard; Figure 14–2, Tom McHugh/Photo Researchers, Inc.; Figure 14–3, Gabe Palmer/Corbis/Stock Market; Figure 14–7b, Bernard J. Nebel; Figure 14–8a, Bernard J. Nebel; Figure 14–11, Au Sable Institute of Environmental Studies; Figure 14–12, Bernard J. Nebel; Figure 14–14, T.J. Florian/Rainbow; Figure 14–15, Sandia National Laboratories; Figure 14–16, Sandia National Laboratories; Figure 14–17, Lowell Georgia/Photo Researchers, Inc.; Figure 14–18, © Renewable Energy Systems, Inc.; Figure 14–19, Creative Energy Corporation/Avanti Pellet Stove; Figure 14–20, Glenn Edwards/Panos Pictures; Figure 14–21, Ford World Headquarters; Figure 14–22b, Copyright 2004 General Motors Corp. Used with permission of GM Media Archives; Figure 14–23, Pacific Gas [Electric Company]

**Chapter 15**

PO-5, Jeremy Walker/Getty Images Inc./Stone Allstock; CO-15a,b, P. Virot/World Health Organization; Figure 15–1a, E.R. Degginger/Animals Animals/Earth Scenes; Figure 15–1b, Zig Leszcynski/Animals Animals/Earth Scenes; Figure 15–4, AP/Wide World Photos; Figure 15–7a, Gregory Bull/AP/Wide World Photos; Figure 15–7b, Reuters/Corbis/Bettman; Figure 15–7c, Rich Saal/State Journal Register/AP/Wide World Photos; Figure 15–7d, Blanca Gutierrez/AP/Wide World Photos; Figure 15–8a, Calvin Larsen/Photo Researchers, Inc.; Figure 15–8b, Gilles Mingasson/Getty Images, Inc.–Liaison; Figure 15–8c, Doug Wechsler/Animals Animals/Earth Scenes; Figure 15–8d, Deep Light Productions/Science Photo Library/Photo Researchers, Inc.; Figure 15–10, Shehzad Noorani/ Woodfin Camp [Associates]; Figure 15–11UN, EB, Signe Wilkinson/Cartoonists & Writers Syndicate; Figure 15–12, A. Ramey/ Woodfin Camp [Associates]; Figure 15–13, Copyright WHO/Andy Crump; Figure 15–17, Slide Works/Jackson Laboratory

**Chapter 16**

CO-16, Richard T. Wright; Figure 16–1, USDA/ARS/Agricultural Research Service; Figure 16–2a, Richard T. Wright; Figure 16–2b, Richard T. Wright; Figure 16–2c, Richard T. Wright; Figure 16–2d, Richard T. Wright; Figure 16–2e, Richard T. Wright; Figure 16–2f, Christine Pemberton, Omni–Photo Communications, Inc.; Figure 16–2g, © Dan Nickrent, Southern Illinois University; Figure 16–2h, Scott Camazine/Photo Researchers, Inc.; Figure 16–4, U.S. Department of Agriculture; Figure 16–6b, David Pimentel, U.S. Department of Agriculture; Figure 16–7 bkgd, Getty Images, Inc.–PhotoDisc; Figure 16–8, Johnny Johnson/Animals Animals/ Earth Scenes; Figure 16–12, Richard T. Wright; Figure 16–13b, Ed Degginger/Color-Pic, Inc.; Figure 16–13c, Harry Rogers/Photo Researchers, Inc.; Figure 16–14a, Department of Lands, Queensland, Australia; Figure 16–14b, Department of Lands, Queensland, Australia; Figure 16–15, © Louis Quitt/Photo Researchers, Inc.; Figure 16–16, Dr. Christopher M. Ranger and Dr. Arthur A. Hower; Figure 16–17, NatureMark Potatoes, a unit of Monsanto Company; Figure 16–18, Jack Dermid/Photo Researchers, Inc.; Figure 16–20, Nigel Cattlin/Holt Studios International/Photo Researchers, Inc.; Figure 16–22, Richart T. Wright.

**Chapter 17**

CO-18, NASA/Mark Marten/Science Source/Photo Researchers, Inc.; Figure 17–4, Tom Stoddart/Katz Pictures/Woodfin Camp [Associates]; Figure 17–4UNa–d, Bob Hudson/George Waclawin/Bernard J. Nebel; Figure 17–5, Thomas D. Brock; Figure 17–6c, Charles R. Belinky/Photo Researchers, Inc.; Figure 17–7, Bernard J. Nebel; Figure 17–10, Bernard J. Nebel; Figure 17–12, Loren Callahan/AP Wide World Photos; Figure 17–14a, 17–14b, Bernard J. Nebel; Figure 17–16, Bernard J. Nebel; Figure 17–17a, 17–17b, Bernard J. Nebel

**Chapter 18**

CO-18, Matt York/AP/Wide World Photos; Figure 18–1, Photo © Cymie Payne, courtesy of Cambridge Arts Council; Figure 18–4, Van

Bucher/Photo Researchers, Inc.; Figure 18–5, City of Riverview; Figure 18–8, Rafael Macia/Photo Researchers, Inc.; Figure 18–9, Alan L. Detrick/Photo Researchers, Inc.; Figure 18–12, Susan Greenwood/Liaison Agency, Inc.; Figure 18–13, John Griffin/The Image Works; Figure 18–15, © Tribune Media Services, Inc. All Rights Reserved. Reprinted with permission; Figure 18–16, Robert J. Fiore, City of Worcester, Massachusetts

**Chapter 19**

CO-19, Gordon Wiltsie/Peter Arnold, Inc.; Figure 19–6, Corbis/Bettmann; Figure 19–8, U.S. Environmental Protection Agency Headquarters; Figure 19–9, Alon Reininger/Contact Press/Woodfin Camp [Associates]; Figure 19–10, Nancy J. Pierce/Photo Researchers, Inc.; Figure 19–11, Van Bucher/Photo Researchers, Inc.; Figure 19–12, Joe Traver, Getty Images, Inc.–Liaison; Figure 19–13d, Tetra Tech EM Inc.; Figure 19–14, Tim Lynch, Getty Images, Inc.–Liaison; Figure 19–16, Sondeep/AP/Wide World Photos; Figure 19–18, Greenpeace, Inc.

**Chapter 20**

CO-20a–d, Jet Propulsion Laboratory/NASA/CNES; Figure 20–1a, AP/Wide World Photos; Figure 20–1b, David Gray/AP/Wide World Photos; Figure 20–1c, Tantyo Bangun/Still Pictures/Peter Arnold, Inc.; Figure 20–1d, Reuters/George Mulala, Getty Images, Inc.–Hulton Archive Photos; Figure 20–15UN, Russ Camp/Richard T. Wright; Figure 20–21, National Weather Service

**Chapter 21**

CO-21, Steven D. Elmore/Corbis/Stock Market; Figure 21–1, Copyright Pittsburgh Post-Gazette, 1999. All rights reserved. Reprinted with permission. Photo originally from Pittsburgh Sun-Telegraph.; Figure 21–4a, Joe Sohm/Chromosohm/The Image Works; Figure 21–4b, Frank Hanna/Visuals Unlimited; Figure 21–7, Richard T. Wright; Figure 21–15, Grapes/Michaud/Photo Researchers, Inc.; Figure 21–16b1, b2, Matt Meadows/Peter Arnold, Inc.; Figure 21–18, Joern Gerdis/Photo Researchers, Inc.; Figure 21–20, Ray Pfortner/Peter Arnold, Inc.; Figure 21–22, Ray Portner/Peter Arnold, Inc.; Figure 21–24, Courtesy Car Sound Exhaust Systems; Figure 21–26, © Toyota Motor North America, Inc.; Figure 21–26UN, Tri-Met, Portland, Oregon

**Chapter 22**

PO-6a & b, Jeff Greenberg/Index Stock Imagery, Inc.; CO-22, Wes Thompson/Corbis/Stock Market; Figure 22–4a, Larry Lefever/Grant Heilman Photography, Inc.; Figure 22–4b, Noah Poritz/MacroWorld/Photo Researchers, Inc.; Figure 22–4c, David Weintraub/Photo Researchers, Inc.; Figure 22–9, Bettmann/Corbis/Bettmann.

**Chapter 23**

CO-23, G.R. "Dick" Roberts Photo Library; Figure 23–1a,b, City of Trenton; Figure 23–2, David Pollack/Corbis/Stock Market; Figure 23–3, Esbin-Anderson/The Image Works; Figure 23–4, Michael Wickes/The Image Works; Figure 23–5a, Steve Elmore/Corbis/Stock Market; Figure 23–5b, David Lawrence/Corbis/Stock Market; Figure 23–9a, b, Bernard J. Nebel; Figure 23–10, Bernard J. Nebel; Figure 23–11, Bernard J. Nebel; Figure 23–12, Sean Sprague/Peter Arnold, Inc.; Figure 23–13, Geri Engberg/Corbis/Stock Market; Figure 23–14, Pete Saloutos/Corbis/Stock Market; Figure 23–15, Kim Newton/ Woodfin Camp [Associates]; Figure 23–16, © Mr. Janis Miglavs 1995; Figure 23–17, 18th [Glenwood Garden, photo courtesy of the Pennsylvania Horticultural Society]; Figure 23–18, Bill Cardoni/Liaison Agency, Inc.; Figure 23–19, Chattanooga Area Convention and Visitors Bureau; Figure 23–19UN, © Jeffrey Pohorski; Figure 23–20, Karl Gehring/Liaison Agency, Inc.

**abiotic.** Pertaining to factors or things that are separate and independent from living things; nonliving.
**abortion.** The termination of a pregnancy by some form of surgical or medicinal intervention.
**absolute poverty.** The lack of sufficient income in cash or exchange items for meeting the most basic human needs for food, clothing, and shelter.
**acid.** Any compound that releases hydrogen ions when dissolved in water. Also, a water solution that contains a surplus of hydrogen ions.
**acid deposition.** Any form of acid precipitation and also fallout of dry acid particles. (See **acid precipitation.**)
**acid precipitation.** Includes acid rain, acid fog, acid snow, and any other form of precipitation that is more acidic than normal (i.e., less than pH 5.6). Excess acidity is derived from certain air pollutants, namely, sulfur dioxide and oxides of nitrogen.
**activated sludge.** Sludge made up of clumps of living organisms feeding on detritus that settles out and is recycled in the process of secondary wastewater treatment.
**activated sludge system.** A system for removing organic wastes from water. The system uses microorganisms and active aeration to decompose such wastes. The system is used most as a means of secondary sewage treatment following the primary settling of materials.
**active safety features.** Those safety features of nuclear reactors that rely on operator-controlled reactions, external power sources, and other features that are capable of failing. (See **passive safety features.**)
**adaptation.** An ecological or evolutionary change in structure or function that enables an organism to adjust better to its environment and hence enhances the organism's ability to survive and reproduce.
**adiabatic cooling.** The cooling that occurs when warm air rises and encounters lower atmospheric pressure. Adiabatic *warming* is the opposite process, whereby cool air descends and encounters higher pressure.
**adsorption.** The process whereby chemicals (ions or molecules) stick to the surface of other materials.
**aeration.** *Soil*: The exchange within the soil of oxygen and carbon dioxide necessary for the respiration of roots. *Water*: The bubbling of air or oxygen through water to increase the amount of oxygen dissolved in the water.
**aerosols.** Microscopic liquid and solid particles originating from land and water surfaces and carried up into the atmosphere.
**affluenza.** A term used to describe a dysfunctional relationship with wealth or money.
**Agenda 21.** A definitive program to promote sustainable development that was produced by the 1992 Earth Summit in Rio de Janeiro, Brazil, and adopted by the U.N. General Assembly.
**age structure.** Within a population, the different proportions of people who are old, middle aged, young adults, and children.
**AID.** The U.S. Agency for International Development, the development arm of the U.S. State Department responsible for administering international aid from the United States.
**AIDS.** Acquired immune deficiency syndrome, a fatal disease caused by the human immunodeficiency virus (HIV) and transmitted by sexual contact or the use of nonsterile needles (as in drug addiction).
**air pollution disaster.** Short-term situation in industrial cities in which intense industrial smog brings about a significant increase in human mortality.
**air toxics.** A category of air pollutants that includes radioactive materials and other toxic chemicals which are present at low concentrations, but are of concern because they often are carcinogenic.
**alga, pl. algae.** Any of numerous kinds of photosynthetic plants that live and reproduce while entirely immersed in water. Many species (the planktonic forms) exist as single cells or small groups of cells that float freely in the water. Other species (the seaweeds) may be large and attached.
**algal bloom.** A relatively sudden development of a heavy growth of algae, especially planktonic forms. Algal blooms generally result from additions of nutrients, whose scarcity is normally limiting.
**alkaline, alkalinity.** A *basic* substance; chemically, a substance that absorbs hydrogen ions or releases hydroxyl ions; in reference to natural water, a measure of the base content of the water.
**alley cropping.** Agricultural cropping in which rows of shade-producing trees are alternated with rows of food crops to promote growth in dry climates.
**ambient standards.** Air-quality standards (set by the EPA) determining certain levels of pollutants that should not be exceeded in order to maintain environmental and human health.
**anaerobic.** Oxygen free.
**anaerobic digestion.** The breakdown of organic material by microorganisms in the absence of oxygen. The process results in the release of methane gas as a waste product.
**anaerobic respiration.** Respiration carried on by certain bacteria in the absence of oxygen. Methane, which can be used as fuel gas (it is the same as natural gas), may be a by-product of the process.
**animal testing.** Procedures used to assess the toxicity of chemical substances using rats, mice, and guinea pigs as surrogates for humans who might be exposed to the substances.
**anthropogenic.** Referring to pollutants and other forms of impacts on natural environments that can be traced to human activities.
**appropriate technology.** Technology that seeks to increase the efficiency and productivity of hand labor without displacing workers. That is, it seeks to enable people to improve their well-being without disrupting the existing social and economic system.
**aquaculture.** A propagation or rearing of any aquatic (water) organism in a more or less artificial system.
**aquifer.** An underground layer of porous rock, sand, or other material that allows the movement of water between layers of nonporous rock or clay. Aquifers are frequently tapped for wells.

**artificial selection.** Plant and animal breeders' practice of selecting individuals with the greatest expression of desired traits to be the parents of the next generation.

**asbestos fibers.** Crystals of asbestos, a natural mineral, that have the form of minute strands; asbestos is a serious health hazard in indoor spaces.

**association.** A unique combination of plants on a given site; the fundamental level of classification used in the United States National Vegetation Classification system.

**asthma.** Chronic disease of the respiratory system in which the air passages tighten and constrict, causing breathing difficulties; acute attacks may be life threatening.

**atom.** The fundamental unit of all elements.

**autotroph.** Any organism that can synthesize all its organic substances from inorganic nutrients, using light or certain inorganic chemicals as a source of energy. Green plants are the principal autotrophs.

**background radiation.** Radiation that comes from natural sources apart from any human activity. We are all exposed to such radiation.

**bacterium, pl. bacteria.** Any of numerous kinds of prokaryotic microscopic organisms that exist as simple single cells that multiply by simple division. Along with fungi, bacteria constitute the decomposer component of ecosystems. A few species cause disease.

**base.** Any compound that releases hydroxyl ($OH^-$) ions when dissolved in water. A solution that contains a surplus of hydroxyl ions.

**base-load power plant.** A power plant, usually large and fueled by coal or nuclear energy, that is kept operating continuously to supply electricity.

**bed load.** The load of coarse sediment, mostly coarse silt and sand, that is gradually moved along the bottom of a riverbed by flowing water rather than being carried in suspension.

**benefit–cost analysis.** An analysis or a comparison of the benefits in contrast to the costs of any particular action or project. (See **cost–benefit ratio.**)

**benthic plants.** Plants that grow under water and that are attached to or rooted in the bottom of the body of water. For photosynthesis, these plants depend on light penetrating the water.

**best management practice.** Farm management practice that serves best to reduce soil and nutrient runoff and subsequent pollution.

**bioaccumulation.** The accumulation of higher and higher concentrations of potentially toxic chemicals in organisms. Bioaccumulation occurs in the case of substances that are ingested, but that cannot be excreted or broken down (nonbiodegradable substances).

**biochemical oxygen demand (BOD).** The amount of oxygen that will be absorbed or "demanded" as wastes are being digested or oxidized in both biological and chemical processes. Potential impacts of wastes are commonly measured in terms of the BOD.

**biocide.** Pesticide or other chemical that is toxic to many, if not all, kinds of living organisms.

**bioconversion.** The use of biomass as fuel. Burning materials such as wood, paper, and plant wastes directly to produce energy, or converting such materials into fuels such as alcohol and methane.

**biodegradable.** Able to be consumed and broken down to natural substances, such as carbon dioxide and water, by biological organisms, particularly decomposers. Opposite: **nonbiodegradable.**

**biodiesel.** Diesel fuel made from a mixture of vegetable oil and regular diesel oil.

**biodiversity.** The diversity of living things found in the natural world. The concept usually refers to the different species, but also includes ecosystems and the genetic diversity within a given species.

**biodiversity hot spots.** 25 locations around the world that are home to some 60% of the world's known biodiversity

**biodiversity informatics.** A new discipline with the goal to put on the Internet all available taxonomic data and collection information on species

**biogas.** The mixture of gases—about two-thirds methane, one-third carbon dioxide, and small portions of foul-smelling compounds—resulting from the anaerobic digestion of organic matter. The methane content enables biogas to be used as a fuel.

**biological control.** Control of a pest population by the introduction of predatory, parasitic, or disease-causing organisms.

**biological nutrient removal.** A process employed in sewage treatment to remove nitrogen and phosphorus from the effluent coming from secondary treatment.

**biological species concept.** The concept that defines a species as a group of interbreeding natural populations that does not breed with other groups because of various barriers.

**biological wealth.** The life-sustaining combination of commercial, scientific, and aesthetic values imparted to a region by its biota.

**biomagnification.** Bioaccumulation occurring through several levels of a food chain.

**biomass.** Mass of biological material. Usually the total mass of a particular group or category; for example, biomass of producers.

**biomass energy.** Energy produced by burning plant-related materials such as firewood. (See **bioconversion.**)

**biomass fuels.** Fuels in the form of plant-related material (e.g., firewood). (See **bioconversion.**)

**biomass pyramid.** The structure that is obtained when the respective biomasses of producers, herbivores, and carnivores in an ecosystem are compared. Producers have the largest biomass, followed by herbivores and then carnivores.

**biome.** A group of ecosystems that are related by having a similar type of vegetation governed by similar climatic conditions. Examples include prairies, deciduous forests, arctic tundra, deserts, and tropical rain forests.

**bioremediation.** The use of microorganisms for the decontamination of soil or groundwater. Usually involves injecting organisms or oxygen into contaminated zones.

**biosolids.** Organic material removed from sewage effluents in the course of treatment. Formerly referred to as sludge.

**biosphere.** The overall ecosystem of Earth. The sum total of all the biomes and smaller ecosystems, which ultimately are all interconnected and interdependent through global processes such as the water cycle and the atmospheric cycle.

**biosphere reserves.** Natural areas in various parts of the world set aside to protect genetic biodiversity.

**biota.** The sum total of all living organisms. The term usually is applied to the setting of natural ecosystems.

**biotechnology.** The use of genetic-engineering techniques, enabling researchers to introduce new genes into food crops and domestic animals and thereby produce valuable products and more nutritious crops.

**biotic.** Living or derived from living things.

**biotic community.** All the living organisms (plants, animals, and microorganisms) that live in a particular area.

**biotic potential.** Reproductive capacity. The potential of a species for increasing its population and/or distribution. The biotic potential of every species is such that, given optimum conditions, its population will increase. (Contrast environmental resistance).

**biotic structure.** The organization of living organisms in an ecosystem into groups such as producers, consumers, detritus feeders, and decomposers.

**birth control.** Any means, natural or artificial, that may be used to reduce the number of live births.

**blue water flow.** In the hydrologic cycle, precipitation and its further movement in infiltration, runoff, surface water, and groundwater.

**BOD.** See **biochemical oxygen demand.**

**bottle law (bottle bill).** A law that provides for the recycling or reuse of beverage containers, usually by requiring a returnable deposit upon purchase of the item.

**bottom-up regulation.** Basic control of a natural population occurs as a result of the scarcity of some resource.

**breeder reactor.** A nuclear reactor that, in the course of producing energy, also converts nonfissionable uranium-238 into fissionable plutonium-239, which can be used as fuel. Hence, a reactor that produces at least as much nuclear fuel as it consumes.

**broad-spectrum pesticides.** Chemical pesticides that kill a wide range of pests. These pesticides also kill a wide range of nonpest and beneficial species; therefore, they may lead to environmental disturbances and resurgences. The opposite of narrow-spectrum pesticides and biorational pesticides.

**brownfields.** Abandoned, idled, or underused industrial and commercial facilities whose further development is inhibited because of real or perceived chemical contamination.

**BTU (British thermal unit).** A fundamental unit of energy in the English system. The amount of heat required to raise the temperature of 1 pound of water 1 degree Fahrenheit.

**buffer.** A substance that will maintain the pH of a solution by reacting with the excess acid in the solution. Limestone is a natural buffer that helps to maintain water and soil at a near-neutral pH.

**buffering capacity.** The amount of acid that may be neutralized by a given amount of buffer.

**calorie.** A fundamental unit of energy. The amount of heat required to raise the temperature of 1 gram of water 1 degree Celsius. All forms of energy can be converted to heat and measured in calories. Calories used in connection with food are kilocalories, or "big" calories, the amount of heat required to raise the temperature of 1 liter of water 1 degree Celsius.

**cancer.** Any of a number of cellular changes that result in uncontrolled cellular growth.

**capillary water.** Water that clings in small pores, cracks, and spaces against the pull of gravity (e.g., water held in a sponge).

**carbon monoxide.** A highly poisonous gas, the molecules of which consist of a carbon atom with one oxygen attached. Not to be confused with carbon dioxide, a natural gas in the atmosphere.

**carbon tax.** A tax levied on all fossil fuels in proportion to the amount of carbon dioxide that is released as they burn.

**carcinogenic.** Having the property of causing cancer, at least in animals and, by implication, in humans.

**carnivore.** An animal that feeds more or less exclusively on other animals.

**carrying capacity.** The maximum population of a given species that an ecosystem can support without being degraded or destroyed in the long run. The carrying capacity may be exceeded, but not without lessening the system's ability to support life in the long term.

**Cartagena Protocol.** An international agreement governing trade in genetically modified organisms. The Cartagena Protocol was signed in January 2000.

**castings.** The humus-rich pellets resulting from earthworm activity.

**catalyst.** A substance that promotes a given chemical reaction without itself being consumed or changed by the reaction. Enzymes are catalysts for biological reactions. Also, catalysts are used in some pollution control devices (e.g., the catalytic converter).

**catalytic converter.** The device used by U.S. automobile manufacturers to reduce the amount of carbon monoxide and hydrocarbons in a car's exhaust. The converter contains a catalyst that oxidizes these compounds to carbon dioxide and water as the exhaust passes through.

**cell.** The basic unit of life; the smallest unit that still maintains all the attributes of life. Many microscopic organisms consist of a single cell. Large organisms consist of trillions of specialized cells functioning together.

**cell respiration.** The chemical process that occurs in all living cells whereby organic compounds are broken down to release energy required for life processes. For respiration, higher plants and animals require oxygen and release carbon dioxide and water as waste products, but certain microorganisms do not require oxygen. (See **anaerobic respiration.**)

**cellulose.** The organic macromolecule that is the prime constituent of plant cell walls and hence the major molecule in wood, wood products, and cotton. Cellulose is composed of glucose molecules, but because it cannot be digested by humans, its dietary value is only as fiber, bulk, or roughage.

**center pivot irrigation.** An irrigation system consisting of a spray arm several hundred meters long, supported by wheels pivoting around a central well from which water is pumped.

**centrally planned economy.** An economic system in which a ruling class makes most of the basic decisions about how the economy will be structured; typical of communist countries.

**CFCs.** See **chlorofluorocarbons.**

**chain reaction.** Nuclear reaction wherein each atom that fissions (splits) causes one or more additional atoms to fission.

**channelization/channelized.** The straightening and deepening of stream or river channels to speed water flow and reduce flooding. A waterway so treated is said to be channelized.

**chemical barrier.** A chemical aspect of a plant that makes it resist attacks from certain pests. Used in genetic pest control.

**chemical energy.** The potential energy that is contained in certain chemicals; most importantly, the energy contained in organic compounds such as food and fuels and that may be released through respiration or burning.

**chemical technology.** The use of pesticides and herbicides to control or eradicate agricultural pests.

**chemosynthesis.** Process whereby some microorganisms utilize the chemical energy contained in certain reduced inorganic chemicals (e.g., hydrogen sulfide) to produce organic material.

**chlorinated hydrocarbons.** Synthetic organic molecules in which one or more hydrogen atoms have been replaced by chlorine atoms. Chlorinated hydrocarbons are hazardous compounds because they tend to be nonbiodegradable and therefore to bioaccumulate; many have been shown to be carcinogenic. Also called organochlorides.

**chlorination.** The disinfection process of adding chlorine to drinking water or sewage water in order to kill microorganisms that may cause disease.

**chlorine cycle.** In the stratosphere, a cyclical chemical process in which chlorine monoxide breaks down ozone.

**chlorofluorocarbons (CFCs).** Synthetic organic molecules that contain one or more of both chlorine and fluorine atoms and that are known to cause ozone destruction.

**chlorophyll.** The green pigment in plants responsible for absorbing the light energy required for photosynthesis.

**chromosome.** In a living organism, one of a number of structures on which genes are arranged in a linear fashion.

**CITES (Convention on Trade in Endangered Species).** An international treaty conveying some protection to endangered and threatened species by restricting trade in those species or their products.

**Clean Air Act of 1970.** Amended in 1977 and 1990, the act is the foundation of U.S. air pollution control efforts.

**Clean Water Act of 1972.** The cornerstone federal legislation addressing water pollution.

**clear-cutting.** In forestry, cutting every tree, leaving the area completely clear.

**climate.** A general description of the average temperature and rainfall conditions of a region over the course of a year.

**climax ecosystem.** The last stage in ecological succession. An ecosystem in which populations of all organisms are in balance with each other and with existing abiotic factors.

**cogeneration.** The joint production of useful heat and electricity. For example, furnaces may be replaced with gas turbogenerators that produce electricity while the hot exhaust still serves as a heat source. An important avenue of conservation, cogeneration effectively avoids the waste of heat that normally occurs at centralized power plants.

**cold condensation.** The process whereby volatile organic pollutants condense on the snowpack in cold climates.

**combined-cycle natural-gas unit.** A recent technology for generating electricity, employing both a natural-gas turbine and a steam turbine that uses excess heat coming from the gas turbine.

**combined heat and power (CHP).** Sometimes called cogeneration, this is a process where a facility installs a small power plant that generates electricity, and the waste heat from the power plant is used to heat the facility. Close to 80% efficiency can be achieved with such a system.

**combustion.** The practice of disposing of wastes by incineration in a special facility designed to handle large amounts of waste; modern combustion facilities also capture some of the energy by generating electricity on the site.

**command-and-control strategy.** The basic strategy behind most public policy having to do with air and water pollution. The strategy involves setting limits on pollutant levels and specifying control technologies that must be used to achieve those limits.

**commons, common pool resources.** Resources (usually natural ones) owned by many people in common or, as in the case of the air or the open oceans, owned by no one, but open to exploitation.

**compaction.** Packing down. *Soil:* Packing and pressing out air spaces present in the soil. Compaction reduces soil aeration and infiltration and thus diminishes the capacity of the soil to support plants. *Trash:* Packing down trash to reduce the space that it requires.

**competitive exclusion principle.** The concept that when two species compete very directly for resources, one eventually excludes the other from the area.

**composting/compost.** The process of letting organic wastes decompose in the presence of air. A nutrient-rich humus, or compost, is the resulting product.

**composting toilet.** A toilet that does not flush wastes away with water, but deposits them in a chamber where they will compost. (See **composting.**)

**compound.** Any substance (gas, liquid, or solid) that is made up of two or more different kinds of atoms bonded together. (Contrast **element.**)

**Comprehensive Environmental Response, Compensation, and Liability Act of 1980.** See **Superfund.**

**condensation.** The collecting of molecules from the vapor state to form the liquid state, as, for example, when water vapor condenses on a cold surface and forms droplets. Opposite: **evaporation.**

**conditions.** Abiotic factors (e.g., temperature) that vary in time and space, but are not used up by organisms.

**confusion technique.** Pest control method in which a quantity of sex attractant is applied to an area so that males become confused and are unable to locate females. The actual quantities of pheromones applied are very small because of their extreme potency.

**conservation.** The management of a resource in such a way as to assure that it will continue to provide maximum benefit to humans over the long run. Conservation may include various degrees of use or protection, depending on what is necessary to maintain the resource over the long run. *Energy*: Saving energy. Energy conservation entails not only cutting back on the use of heating, air-conditioning, lighting, transportation, and so on, but also increasing the efficiency of energy use—that is, developing and instigating the means of doing the same jobs (e.g., transporting people) with less energy.

**Conservation Reserve Program (CRP).** A federal program whereby farmers are paid to place highly erodible cropland in a reserve.

**consumers.** In an ecosystem, those organisms that derive their energy from feeding on other organisms or their products.

**consumptive use.** The harvesting of natural resources in order to provide for people's immediate needs for food, shelter, fuel, and clothing.

**consumptive water use.** Use of water for such things as irrigation, wherein the water does not remain available for potential purification and reuse.

**containment building.** Reinforced concrete building housing a nuclear reactor. Designed to contain an explosion, should one occur.

**contour farming.** The practice of cultivating land along the contours across, rather than up and down, slopes. In combination with strip cropping, contour farming reduces water erosion.

**contraceptive.** A device or method employed by couples during sexual intercourse to prevent the conception of a child.

**control rods.** Part of the core of a nuclear reactor; the rods of neutron-absorbing material that are inserted or removed as necessary to control the rate of nuclear fissioning.

**convection.** The vertical movement of air due to atmospheric heating and cooling.

**Convention on Biological Diversity.** The Biodiversity Treaty signed by 158 nations at the Earth Summit in Rio de Janeiro in 1992 calling for various actions and cooperative steps between nations to protect the world's biodiversity.

**conveyor system.** The giant pattern of oceanic currents that moves water masses from the surface to the depths and back again, producing major effects on the climate.

**cooling tower.** A massive tower designed to dissipate waste heat from a power plant (or other industrial facility) into the atmosphere.

**corrosion.** In a nuclear power plant, the deterioration of pipes receiving hot, pressurized water in a circulation system that conveys heat from one part of the unit to another.

**cosmetic spraying.** Spraying of pesticides to control pests that damage only the surface appearance of fruits and vegetables.

**cost–benefit analysis.** See **benefit–cost analysis.**

**cost–benefit ratio/benefit–cost ratio.** The value of the benefits to be gained from a project, divided by the costs of the project. If the ratio is greater than unity, the project is economically justified; if the ratio is less than unity, the project is not economically justified.

**cost-effective.** Pertaining to a project or procedure that produces economic returns or benefits that are significantly greater than the costs incurred.

**coupled general circulation model.** A computer-based model for simulating long-term climatic conditions that combines global atmospheric circulation patterns with ocean circulation and cloud-radiation feedback.

**covalent bond.** A chemical bond between two atoms, formed by sharing a pair of electrons between the atoms. Atoms of all organic compounds are joined by covalent bonds.

**credit associations.** Groups of poor people who individually lack the collateral to assure loans, but collectively have the wherewithal to assure each other's loans. Associated with microlending.

**Criteria Maximum Concentration (CMC).** A water quality standard; the highest single concentration of a water pollutant beyond which environmental impacts may be expected.

**criteria pollutants.** Certain pollutants whose levels are used as a gauge for the determination of air (or water) quality.

**Criterion Continuous Concentration (CCC).** A water quality standard; the highest sustained concentration of a water pollutant beyond which undesirable impacts may be expected.

**critical level.** The level of one or more pollutants above which severe damage begins to occur and below which few, if any, ill effects are noted.

**critical number.** The minimum number of individuals of a given species that is required to maintain a healthy, viable population of the species. If a population falls below its critical number, it will almost certainly become extinct.

**crop rotation.** The practice of alternating the crops grown on a piece of land—for example, corn one year, hay for two years, and then back to corn. (Contrast **monocropping.**)

**crude birthrate.** Number of births per 1,000 individuals per year.

**crude death rate.** Number of deaths per 1,000 individuals per year.

**cultivar.** A cultivated variety of a plant species. Genetically, all individuals of the cultivar are highly uniform.

**cultural control.** A change in the practice of growing, harvesting, storing, handling, or disposing of wastes that reduces their susceptibility or exposure to pests. For example, spraying the house with insecticides to kill flies is a chemical control; putting screens on the windows to keep flies out is a cultural control.

**cultural eutrophication.** The process of natural eutrophication accelerated by human activities. (See **eutrophication.**)

**DDT (dichlorodiphenyltrichloroethane).** The first and most widely used of the synthetic organic pesticides belonging to the chlorinated hydrocarbon class.

**debt crisis.** Exigency brought about by the great debt incurred by many less developed nations. These nations are so heavily in debt that they may not be able to meet their financial obligations (e.g., interest payments).

**debt relief.** Steps, including cancellation of debts and other means of reducing poverty, taken by the World Bank and donor nations to alleviate the heavy debt of many poor nations.

**declining tax base.** The loss of tax revenues that occurs when affluent taxpayers and businesses leave an area and property values subsequently decline. Also referred to as an eroding tax base.

**decommissioning.** The taking of nuclear power plants out of service after 25–35 years because the effects of radiation will gradually make them inoperable.

**decomposers.** Organisms whose feeding action results in decay or rotting of organic material. The primary decomposers are fungi and bacteria.

**deep-well injection.** A technique used for the disposal of liquid chemical wastes that involves putting them into deep dry wells where they permeate dry strata.

**deforestation.** The process of removing trees and other vegetation covering the soil, leading to erosion and loss of soil fertility.

**Delaney clause.** A provision of the Federal Food, Drug and Cosmetic Act of 1958 (now superceded) that prohibited the use of any food additive that induces cancer in people or animals.

**demographic transition.** The transition of a human population from a condition of a high birthrate and a high death rate to a condition of a low birthrate and a low death rate. A demographic transition may result from economic or social development.

**demography/demographer.** The study of population trends (growth, movement, development, and so on). People who perform such studies and make projections from them.

**denitrification.** The process of reducing oxidized nitrogen compounds present in soil or water back to nitrogen gas in the atmosphere. Denitrification is a natural process conducted by certain bacteria (see discussion of the nitrogen cycle in the text), and is now utilized in the treatment of sewage effluents.

**density-dependent.** Attribute of population-balancing factors, such as predation, that increase and decrease in intensity in proportion to population density.

**Department of Transportation regulations (DOT regs).** Regulations intended to reduce the risk of spills, fires, and poisonous fumes by specifying the kinds of containers and methods of packing to be used in transporting hazardous materials.

**dependency ratio.** In a human population, the ratio of the non-working-age population (under 15 and over 65) to the working-age population.

**deregulation.** In the electric power industry, a policy that requires utility companies to divest themselves of power-generating facilities and allows the market to determine electrical costs.

**desalinization.** Process that purifies seawater into high-quality drinking water via distillation or microfiltration.

**desertification.** The formation and expansion of degraded areas of soil and vegetation cover in arid, semiarid, and seasonally dry areas, caused by climatic variations and human activities.

**desertified.** Said of land whose productivity has been significantly reduced (25% or more) because of human mismanagement. Erosion is the most common cause of desertified land.

**desert pavement.** A covering of stones and coarse sand protecting desert soils from further wind erosion. The covering results from the differential erosion of finer material.

**detritus.** The dead organic matter, such as fallen leaves, twigs, and other plant and animal wastes, that exists in any ecosystem.

**detritus feeders.** Organisms such as termites, fungi, and bacteria that obtain their nutrients and energy mainly by feeding on dead organic matter.

**deuterium** ($^2H$). A stable, naturally occurring isotope of hydrogen that contains one neutron in addition to the single proton normally in the nucleus.

**developed countries.** Industrialized countries—the United States, Canada, Western European nations, Japan, Australia, and New Zealand—in which the gross domestic product exceeds $9,206 per capita.

**developing countries.** All countries in which the gross domestic product is less than $9,206 per capita. The category includes nations of Latin America, Africa, and Asia, except Japan.

**differential reproduction.** Within a population, the more successful reproduction of certain individuals compared with that of others.

**dioxin.** A synthetic organic chemical of the chlorinated hydrocarbon class. Dioxin is one of the most toxic compounds known to humans, having many harmful effects, including carcinogenic and teratogenic effects, even in extremely minute concentrations. Because of the use of certain herbicides that contain dioxin as a contaminant, the chemical has become a widespread environmental pollutant.

**disability-adjusted life year (DALY).** 1 DALY equals the loss of one healthy year of a person's life (whether through disability or loss of life).

**discharge permit.** (Technically called an NPDES permit.) A permit that allows a company to legally discharge certain amounts or levels of pollutants into air or water.

**discount rate.** In economics, a rate applied to some future benefit or cost in order to calculate its present real value.

**disinfection.** The killing (as opposed to removal) of microorganisms in water or other media where they might otherwise pose a health threat. For example, chlorine is commonly used to disinfect water supplies.

**dissolved oxygen (DO).** Oxygen gas molecules ($O_2$) dissolved in water. Fish and other aquatic organisms depend on dissolved oxygen for respiration. Therefore, the concentration of dissolved oxygen is a measure of water quality.

**distillation.** A process of purifying water or some other liquid by boiling the liquid and recondensing the vapor. Contaminants remain behind in the boiler.

**distributive justice.** The ethical process of making certain that all parties receive equal rights regardless of their economic status.

**disturbance.** A natural or human-induced event or process (e.g., a forest fire) that interrupts ecological succession and creates new conditions on a site.

**DNA (deoxyribonucleic acid).** The natural organic macromolecule that carries the genetic or hereditary information for virtually all organisms.

**donor fatigue.** The tendency of donor organizations or countries to withhold their aid because of long-term misuse of aid funds in the recipient countries.

**dose.** The mathematical product of the concentration of a hazardous material and the length of exposure to it. The effects of any given material or radiation correspond to the dose received.

**doubling time.** The time it will take a population to double in size, assuming that the current rate of growth continues.

**drip irrigation.** Method of supplying irrigation water through tubes that literally drip water onto the soil at the base of each plant.

**drought.** A local or regional lack of precipitation such that the ability to raise crops and water animals is seriously impaired.

**drylands.** Ecosystems characterized by low precipitation (25 to 75 cm per year) and often subject to droughts; some 5.2 billion hectares are classified as drylands.

**earth-sheltered housing.** Housing that makes use of the insulating properties of earth materials and that is also oriented for passive solar heating; often involves earth berms built up against the building walls.

**easement.** In reference to land protection, an arrangement whereby a landowner gives up development rights into the future, but retains ownership of the land.

**ecdysone.** The hormone that promotes molting in insects.

**ecological economist.** An economist who thoroughly integrates ecological and economic concerns; part of a new breed of economist who disagrees with classical economic theory.

**ecological footprint.** A concept for measuring the demand placed on Earth's resources by individuals from different parts of the world, involving calculations of the natural area required to satisfy human needs.

**ecological pest management.** Natural control of pest populations through understanding various ecological factors and, so far as possible, utilizing those factors, as opposed to using synthetic chemicals.

**ecological restoration.** See **restoration ecology.**

**ecological succession.** Process of gradual and orderly progression from one ecological community to another.

**ecologists.** Scientists who study ecology.

**ecology.** The study of any and all aspects of how organisms interact with each other and with their environment.

**economic exclusion.** The cutting of access of certain ethnic or economic groups to jobs, a good education, and other opportunities and thus preventing them from entering the economic mainstream of society—a condition that prevails in poor areas of cities.

**economic threshold.** The level of pest damage that, to be reduced further, would require an application of pesticides that is more costly than the economic damage caused by the pests.

**ecosystem.** A grouping of plants, animals, and other organisms interacting with each other and with their environment in such a way as to perpetuate the grouping more or less indefinitely. Ecosystems have characteristic forms, such as deserts, grasslands, tundra, deciduous forests, and tropical rain forests.

**ecosystem capital.** One of the major integrative themes of the book, this is the sum of goods and services provided by natural and managed ecosystems, provided free of charge and essential to human life and well-being.

**ecosystem management.** The management paradigm, adopted by all federal agencies managing public lands, that involves a long-term stewardship approach to maintaining the lands in their natural state.

**ecotone.** A transitional region between two adjacent ecosystems that contains some of the species and characteristics of each one and also certain species of its own.

**ecotourism.** The enterprises involved in promoting tourism of unusual or interesting ecological sites.

**El Niño.** A major climatic phenomenon characterized by the movement of unusually warm surface water into the eastern equatorial Pacific Ocean. El Niño results in extensive disruption of weather around the world.

**electrolysis.** The use of electrical energy to split water molecules into their constituent hydrogen and oxygen atoms. Hydrogen gas and oxygen gas result.

**electrons.** Fundamental atomic particles that have a negative electrical charge and a very small mass, approximately $9.1 \times 10^{-31}$ kg. Electrons surround the nuclei of atoms and thus balance the positive charge of protons in the nucleus. A flow of electrons in a wire is identical to an electrical current.

**element.** A substance that is made up of one and only one distinct kind of atom. (Contrast **compound.**)

**embrittlement.** Becoming brittle. Pertains especially to the reactor vessel of nuclear power plants gradually becoming prone to breakage or snapping as a result of continuous bombardment by radiation. Embrittlement is the prime factor forcing the decommissioning of nuclear power plants.

**emergent vegetation.** Aquatic plants whose lower parts are under water, but whose upper parts emerge from the water.

**emission allowance/standards.** See **discharge permit.**

**endangered species.** A species whose total population is declining to relatively low levels such that if the trend continues, the species will likely become extinct.

**Endangered Species Act.** The federal legislation that mandates the protection of species and their habitats which are determined to be in danger of extinction.

**endocrine disrupters.** Any of a class of organic compounds, often pesticides, that are suspected of having the capacity of interfering with hormonal activities in animals.

**energy.** The capacity to do work. Common forms of energy are light, heat, electricity, motion, and the chemical bond energy inherent in compounds such as sugar, gasoline, and other fuels.

**energy flow.** The movement of energy through ecosystems, starting from the capture of solar energy by primary producers and ending with the loss of heat energy.

**enrichment.** With reference to nuclear power, the separation and concentration of uranium-235 so that, in suitable quantities, it will sustain a chain reaction.

**entomologist.** A scientist who studies insects, their life cycles, physiology, and behavior, and so on.

**entropy.** Degree of disorder; increasing entropy means increasing disorder.

**environment.** The combination of all things and factors external to the individual or population of organisms in question.

**environmental accounting.** A process of keeping national accounts of economic activity that includes gains and losses of environmental assets.

**environmental impact.** Effect on the natural environment caused by human actions. Includes indirect effects, for example, through pollution, as well as direct effects such as cutting down trees.

**environmental impact statement.** A study of the probable environmental impacts of a development project. The National Environmental Policy Act of 1968 (NEPA) requires such studies prior to proceeding with any project receiving federal funding.

**environmental justice.** The fair treatment and meaningful involvement of all people, regardless of race, color, national origin, or income, with respect to the development, implementation, and enforcement of environmental laws and regulations.

**environmental movement.** The upwelling of public awareness and citizen action regarding environmental issues that began during the 1960s.

**environmental resistance.** The totality of factors such as adverse weather conditions, shortages of food or water, predators, and diseases that tend to cut back populations and keep them from growing or spreading. (Contrast **biotic potential.**)

**Environmental Revolution.** In the view of some, a coming change in the adaptation of humans to the rising deterioration of the environment. The Environmental Revolution will purportedly bring about sustainable interactions with the environment.

**environmental science.** The branch of science concerned with environmental issues.

**environmentalism.** An attitude or movement involving concern about pollution, resource depletion, population pressures, loss of biodiversity, and other environmental issues. Usually also implies action to address those concerns.

**environmentalist.** Any person who is concerned about the degradation of the natural world and is willing to act on that concern.

**EPA (U.S. Environmental Protection Agency).** The federal agency responsible for the control of all forms of pollution and other kinds of environmental degradation.

**epidemiology, epidemiological study.** The study of the causes of disease (e.g., lung cancer) through an examination and comparison of large populations of people living in different locations or following different lifestyles or habits (e.g., smoking versus nonsmoking).

**epidemiologic transition.** In human populations, the pattern of change in mortality from high death rates to low death rates; contributes to the demographic transition.

**epiphytes.** Air plants that are not parasitic, but that "perch" on the branches of trees, where they can get adequate light.

**equilibrium theory.** The theory that ecosystems are maintained over time by natural checks and balances; equilibrium theory is challenged by many ecologists.

**erosion.** The process of soil particles' being carried away by wind or water. Erosion moves the smaller particles first and hence degrades the soil to a coarser, sandier, stonier texture.

**estimated reserves.** See **reserves.**

**estuary.** A bay or river system open to the ocean at one end and receiving fresh water at the other. In the estuary, fresh and salt water mix, producing brackish water.

**ETS (environmental tobacco smoke).** "Secondhand" tobacco smoke to which nonsmokers are exposed in the presence of smokers.

**euphotic zone.** In aquatic systems, the layer or depth of water through which an adequate amount of light penetrates to support photosynthesis.

**eutrophic.** Characterized by nutrient-rich water supporting an abundant growth of algae or other aquatic plants at the surface. Deep eutrophic water has little or no dissolved oxygen.

**eutrophication.** The process of becoming eutrophic.

**evaporation.** Process whereby molecules leave the liquid state and enter the vapor or gaseous state as, for example, when water evaporates to form water vapor. Opposite: **condensation.**

**evapotranspiration.** The combination of evaporation and transpiration that restores water to the atmosphere.

**evolution.** The theory that all species now on Earth descended from ancestral species through a process of gradual change brought about by natural selection.

**experimental group.** In an experiment, the group that receives the experimental treatment, in contrast to the control group, used for comparison, which does not receive the treatment. Synonym: **test group.**

**exotic species.** A species introduced into a geographical area to which it is not native.

**exponential increase.** The growth produced when a base population increases by a given percentage (as opposed to a given amount) each year. An exponential increase is characterized by doubling again and again, each doubling occurring in the same period of time. It produces a J-shaped curve.

**extended product responsibility.** Voluntary or government programs that assign some responsibility for reducing the environmental impact of a product at various stages of its life cycle—especially at disposal.

**externality/external cost.** Any effect of a business process not included in the usual calculations of profit and loss. Pollution of air or water is an example of a *negative* externality—one that imposes a cost on society that is not paid for by the business itself.

**extinction.** The death of all individuals of a particular species. When this occurs, all the genes of that particular line are lost forever.

**extractive reserves.** As now established in Brazil, forest lands that are protected for native peoples and rubber tappers who harvest natural products of the forests, such as latex and Brazil nuts.

**exurban migration.** The pronounced trend since World War II of relocating homes and businesses from the central city and older suburbs to more outlying suburbs.

**exurbs.** New developments beyond the traditional suburbs, but from which most residents still commute to the associated city for work.

**facilitation.** A mechanism of succession where earlier species create conditions that are more favorable to newer occupants, and are replaced by them.

**family planning.** Making contraceptives and other health and reproductive services available to couples in order to enable them to achieve a desired family size.

**famine.** A severe shortage of food accompanied by a significant increase in the local or regional death rate.

**FAO.** Food and Agriculture Organization of the United Nations.

**farm cooperatives.** An association of consumers who jointly own and manage a farm for the production of produce specifically for their own consumption.

**Farmer-centered Agricultural Resource Management (FARM) Program.** A U.N. FAO-administered program in Asian countries fostering sustainable agriculture.

**FDA.** Federal Food and Drug Administration, with jurisdiction over foods, drugs, and all products that come in contact with the skin or are ingested.

**fecal coliform test.** A test for the presence of *Escherichia coli,* the bacterium that normally inhabits the gut of humans and other mammals. A positive test result indicates sewage contamination and the potential presence of disease-causing microorganisms.

**Federal Agricultural Improvement and Reform (FAIR) Act of 1996.** Major legislation removing many subsidies and controls from farming.

**fermentation.** A form of respiration carried on by yeast cells in the absence of oxygen. Fermentation involves a partial breakdown of glucose (sugar) that yields energy for the yeast and the release of alcohol as a by-product.

**fertility rate.** See **total fertility rate.**

**fertility transition.** The pattern of change in birthrates in a human society from high rates to low; a major component of the demographic transition.

**fertilizer.** Material applied to plants or soil to supply plant nutrients, most commonly nitrogen, phosphorus, and potassium, but possibly others. *Organic* fertilizer is natural organic material, such as manure, that releases nutrients as it breaks down. *Inorganic* fertilizer, also called chemical fertilizer, is a mixture of one or more necessary nutrients in inorganic chemical form.

**field scouts.** Persons trained to survey crop fields and determine whether applications of pesticides or other pest-management procedures are actually necessary to avert significant economic loss.

**FIFRA (Federal Insecticide, Fungicide, and Rodenticide Act).** The key U.S. legislation passed to control pesticides.

**fire climax ecosystems.** Ecosystems that depend on the recurrence of fire to maintain the existing balance.

**first-generation pesticides.** Toxic inorganic chemicals that were the first used to control insects, plant diseases, and other pests. These pesticides included mostly compounds of arsenic and cyanide and various heavy metals, such as mercury and copper.

**first law of thermodynamics.** The empirical observation, confirmed innumerable times, that energy is never created or destroyed but may be converted from one form to another (e.g., electricity to light). Also called the **law of conservation of energy.** (See also **second law of thermodynamics.**)

**fishery.** Fish species being exploited, or a limited marine area containing commercially valuable fish.

**fish ladder.** A stepwise series of pools on the side of a dam where the water flows in small falls that fish can negotiate.

**fission.** The splitting of a large atom into two atoms of lighter elements. When large atoms such as uranium or plutonium fission, tremendous amounts of energy are released.

**fission products.** Any and all atoms and subatomic particles resulting from splitting atoms in nuclear reactors or nuclear explosions. Practically all such products are highly radioactive.

**fitness.** State of an organism whereby it is adapted to survive and reproduce in the environment it inhabits.

**flat-plate collector.** A solar collector that consists of a stationary, flat, black surface oriented perpendicular to the average angle of the sun. Heat absorbed by the surface is removed and transported by air or water (or some other liquid) flowing over or through the surface.

**flood irrigation.** Technique of irrigation in which water is diverted from rivers by means of canals and is flooded through furrows in fields.

**food aid.** Food of various forms that is donated or sold below cost to needy people for humanitarian reasons.

**food chain.** The transfer of energy and material through a series of organisms as each one is fed upon by the next.

**food guide pyramid.** A graphic presentation of six basic food needs arranged in a pyramid to indicate the relative proportions of each type of food needed for good nutrition.

**food security.** For families, the ability to meet the food needs of everyone in the family, providing freedom from hunger and malnutrition.

**food web.** The combination of all the feeding relationships that exist in an ecosystem.

**fossil fuels.** Energy sources—mainly crude oil, coal, and natural gas—that are derived from the prehistoric photosynthetic production of organic matter on Earth.

**FQPA (Food Quality Protection Act of 1996).** Legislation that removed the Delaney clause and replaced many provisions of FIFRA.

**fragmentation.** The division of a landscape into patches of habitat by road construction, agricultural lands, or residential areas.

**Framework Convention on Climate Change.** A result of the 1992 Earth Summit, this international treaty was a start in negotiating agreements on steps to prevent future catastrophic climate change.

**free-market economy.** In its purest form, an economy in which the market itself determines what and how goods will be exchanged. The system is wholly in private hands.

**freshwater.** Water that has a salt content of less than 0.1% (1,000 parts per million).

**front.** The boundary where different air masses meet.

**fuel assembly.** In a nuclear reactor, the assembly of many rods containing the nuclear fuel, usually uranium, positioned close together. The chain reaction generated in the fuel assembly is controlled by rods of neutron-absorbing material between the fuel rods.

**fuel elements.** The pellets of uranium or other fissionable material that are placed in tubes, which, together with the control rods, form the core of the nuclear reactor.

**fuel rods.** See **fuel elements.**

**fuelwood.** The use of wood for cooking and heating; the most common energy source for people in developing countries.

**fungus, pl. fungi.** Any of numerous species of molds, mushrooms, brackens, and other forms of nonphotosynthetic plants. Fungi derive energy and nutrients by consuming other organic material. Along with bacteria, they form the decomposer component of ecosystems.

**fusion.** The joining together of two atoms to form a single atom of a heavier element. When light atoms such as hydrogen are fused, tremendous amounts of energy are released.

**gasohol.** A blend of 90% gasoline and 10% alcohol that can be substituted for straight gasoline. Gasohol serves to stretch gasoline supplies.

**gene.** Segment of DNA that codes for one protein, which in turn determines a particular physical, physiological, or behavioral trait.

**gene pool.** The sum total of all the genes that exist among all the individuals of a species.

**genetic bank.** The concept that natural ecosystems with all their species serve as a repository of genes that may be drawn upon to improve domestic plants and animals and to develop new medicines, among other uses.

**genetic control.** Selective breeding of a desired plant or animal to make it resistant to attack by pests. Also, attempting to introduce harmful genes—for example, those that cause sterility—into the pest populations.

**genetic engineering.** The artificial transfer of specific genes from one organism to another.

**genetically modified organism (GMO).** Any organism that has received genes from a different species through the techniques of genetic engineering; a **transgenic organism.**

**genetics.** The study of heredity and the processes by which inherited characteristics are passed from one generation to the next.

**genetic variation.** An expression of the range of genetic (DNA) differences that occur among individuals of the same species.

**geothermal.** Refers to the naturally hot interior of Earth, where heat is maintained by naturally occurring nuclear reactions.

**geothermal energy.** Useful energy derived from water heated by the naturally hot interior of Earth.

**geothermal heat pump.** An energy system whereby heat can be obtained or deposited via pipes buried in the ground and an exchange process that allows heated or cooled air to be circulated via ductwork.

**global climate change.** The cumulative effects of various impacts of rising levels of greenhouse gases on Earth's climate. These effects include global warming, weather changes, and a rising sea level.

**globalization.** A major integrative theme of this book, this refers to the accelerating interconnectedness of human activities, ideas, and cultures, especially evident in economic and information exchange.

**glucose.** A simple sugar, the major product of photosynthesis. Glucose serves as the basic building block for cellulose and starches and as the major "fuel" for the release of energy through cell respiration in both plants and animals.

**goods.** Products, such as wood and food, that are extracted from natural ecosystems to satisfy human needs.

**grassroots movement.** An environmental movement that begins at the level of the populace, often in response to a perceived problem not being addressed by public policy.

**gravitational water.** Water that is not held by capillary action in soil, but that percolates downward by the force of gravity.

**graying.** The increasing average age in populations in developed countries and in many developing countries that is occurring because of decreasing birthrates and increasing longevity.

**gray water.** Wastewater, as from sinks and tubs, that does not contain human excrement. Such water can be reused without purification for some purposes.

**greenhouse effect.** An increase in the atmospheric temperature caused by increasing amounts of carbon dioxide and certain other gases that absorb and trap heat, which normally radiates away from Earth.

**greenhouse gases.** Gases in the atmosphere that absorb infrared energy and contribute to the air temperature. These gases are like a heat blanket and are important in insulating Earth's surface. Among the greenhouse gases are carbon dioxide, water vapor, methane, nitrous oxide, chlorofluorocarbons, and other halocarbons.

**green fee.** An added cost for a particular service or product that has the effect of internalizing the external costs and that reflects the real environmental cost of the service or product.

**green manure.** A legume crop such as clover that is specifically grown to enrich the nitrogen and organic content of soil.

**green revolution.** The development and introduction of new varieties of (mainly) wheat and rice that has increased yields per acre dramatically in many countries since the 1960s.

**green water.** In the hydrologic cycle, water that is evaporated or transpired and returned as water vapor to the atmosphere.

**green products.** Products that are more environmentally benign than their traditional counterparts.

**grit chamber.** Part of preliminary treatment in wastewater-treatment plants; a swimming pool–like tank in which the velocity of the water is slowed enough to let sand and other gritty material settle.

**gross domestic (national) product per capita.** The total value of all goods and services exchanged in a year in a country, divided by its population. A common indicator of the average level of development and standard of living for a country.

**gross primary production.** The rate at which primary producers in ecosystems are fixing organic matter.

**groundwater.** Water that has accumulated in the ground, completely filling and saturating all pores and spaces in rock or soil. Groundwater is free to move more or less readily, it is the reservoir for springs and wells, and is replenished by infiltration of surface water.

**groundwater remediation.** The repurification of contaminated groundwater by any of a number of techniques.

**gully erosion.** Soil erosion produced by running water and resulting in the formation of gullies.

**habitat.** The specific environment (woods, desert, swamp) in which an organism lives.

**habitat conservation plan.** Under the Endangered Species Act, a plan that may be drafted by the U.S. Fish and Wildlife Service in working with landowners to help mitigate conflicts resulting from an application of the act.

**Hadley cell.** A system of vertical and horizontal air circulation predominating in tropical and subtropical regions and creating major weather patterns.

**half-life.** The length of time it takes for half of an unstable isotope to decay. The length of time is the same regardless of the starting amount. Also refers to the amount of time it takes compounds to break down in the environment.

**halogenated hydrocarbon.** Synthetic organic compound containing one or more atoms of the halogen group, which includes chlorine, fluorine, and bromine.

**hard water.** Water that contains relatively large amounts of calcium or certain other minerals that cause soap to precipitate. (Contrast **soft water.**)

**hazard.** Anything that can cause (1) injury, disease, or death to humans, (2) damage to property, or (3) degradation of the environment. *Cultural* hazards include factors that are often a matter of choice, such as smoking or sunbathing. *Biological* hazards are pathogens and parasites that infect humans. *Physical* hazards are natural disasters like earthquakes and tornadoes. *Chemical* hazards refer to the chemicals in use in different technologies and household products.

**hazard assessment.** The process of examining evidence linking a particular hazard to its harmful effects.

**hazardous material (HAZMAT).** Any material having one or more of the following attributes: ignitability, corrosivity, reactivity, and toxicity.

**heavy metal.** Any of the high-atomic-weight metals, such as lead, mercury, cadmium, and zinc. All may be serious pollutants in water or soil because they are toxic in relatively low concentrations and they tend to bioaccumulate.

**herbicide.** A chemical used to kill or inhibit the growth of undesired plants.

**herbivore.** An organism such as a rabbit or deer that feeds primarily on green plants or plant products such as seeds or nuts. Such an organism is said to be herbivorous. (Synonym: **primary consumer.**)

**herbivory.** The feeding on plants that occurs in an ecosystem. The total feeding of all plant-eating organisms.

**heterotroph.** Any organism that consumes organic matter as a source of energy. Such an organism is said to be heterotrophic.

**household hazardous wastes.** An important component of the hazardous-waste problem in the United States.

**Highway Trust Fund.** The monies collected from the gasoline tax and designated for the construction of new highways.

**hormones.** Natural chemical substances that control the development, physiology, and behavior of an organism. Hormones are produced internally and affect only the individual organism. Hormones are coming into use in pest control. (See also **pheromones.**)

**host.** In feeding relationships, particularly parasitism, refers to the organism that is being fed upon (i.e., the organism that is supporting the feeder).

**host–parasite relationship.** The relationship between a parasite and the organism upon which it feeds.

**host-specific.** Attribute whereby insects, fungal diseases, and other parasites are unable to attack species other than their particular host.

**Human Poverty Index (HPI).** Based on life expectancy, literacy, and living standards, this index is used by the UNDP to measure progress in alleviating poverty.

**human resources.** One component of the wealth of nations. Comprises the *human capital*, or the population and its attributes, and the *social capital*, or the social and political environment that people have created in a society.

**human system.** The entire system that humans have created for their own support, consisting of agriculture, industry, transportation, communications networks, etc.

**humidity.** The amount of water vapor in the air. (See also **relative humidity.**)

**humus.** A dark brown or black, soft, spongy residue of organic matter that remains after the bulk of dead leaves, wood, or other organic matter has decomposed. Humus does oxidize, but relatively slowly. It is extremely valuable in enhancing the physical and chemical properties of soil.

**hunger.** Condition wherein the basic food required for meeting nutritional and energy needs is lacking and the individual is unable to lead a normal, healthy life.

**hunter–gatherers.** Humans surviving by hunting wild game and gathering seeds, nuts, berries, and other edible things from the natural environment.

**hybrid.** A plant or animal resulting from a cross between two closely related species that do not normally cross.

**hybrid electric vehicle (HEV).** An automobile with a small gasoline motor and batteries that is capable of getting over 50 mpg and that produces only one-tenth the pollution of a comparable gasoline-run car.

**hybrid lighting.** Experimental solar energy technology wherein photovoltaic cells are coupled with optical fibers in rooftop units.

**hybridization.** Cross-mating between two more or less closely related species.

**hydrocarbon emissions.** Exhaust of various hydrogen–carbon compounds due to incomplete combustion of fuel. Hydrocarbon emissions are a major contribution to photochemical smog.

**hydrocarbons.** *Chemistry:* Natural or synthetic organic substances that are composed mainly of carbon and hydrogen. Crude oil, fuels from crude oil, coal, animal fats, and vegetable oils are examples. *Pollution:* A wide variety of relatively small carbon–hydrogen molecules resulting from incomplete burning of fuel and emitted into the atmosphere. (See **volatile organic compounds.**)

**hydroelectric dam.** A dam and an associated reservoir used to produce electrical power by letting the high-pressure water behind the dam flow through and drive a turbogenerator.

**hydroelectric power.** Electric power that is produced from hydroelectric dams or, in some cases, natural waterfalls.

**hydrogen bonding.** A weak attractive force that occurs between a hydrogen atom of one molecule and, usually, an oxygen atom of another molecule. Hydrogen bonding is responsible for holding water molecules together to produce the liquid and solid states.

**hydrogen ions.** Hydrogen atoms that have lost their electrons (chemical symbol, $H^+$).

**hydrologic cycle.** The movement of water from points of evaporation, through the atmosphere, through precipitation, and through or over the ground, returning to points of evaporation.

**hydroponics.** The culture of plants without soil. The method uses water with the required nutrients in solution.

**hydroxyl radical.** The hydroxyl group (OH), missing the electron. The hydroxyl radical is a natural cleansing agent of the atmosphere. It is highly reactive, readily oxidizes many pollutants upon contact, and thus contributes to their removal from the air.

**hypothesis.** A tentative guess concerning the cause of an observed phenomenon that is then subjected to experiment to test its logical or empirical consequences.

**hypoxia.** The absence of oxygen in sediments and deeper levels of the water column brought about by microbial breakdown of organic matter; hypoxia has created the "dead zone" in the Gulf of Mexico.

**ICPD (International Conference on Population and Development)** A U.N.-sponsored conference held in 1994 in Cairo, Egypt, that produced a program of action which was drafted and later adopted by the United Nations.

**incremental value.** The value of some finite change in a natural service; used in calculating the value of natural services performed by ecosystems.

**indicator organism.** An organism, the presence or absence of which indicates certain conditions. For example, the presence of *Escherichia coli* indicates that water is contaminated with fecal wastes and that pathogens may be present; the absence of *E. coli* indicates that the water is free of pathogens.

**individual quota system.** A system of fishery management wherein a quota is set and individual fishers are given or sold the right to harvest some proportion of the quota.

**industrialized agriculture.** The use of fertilizer, irrigation, pesticides, and energy from fossil fuels to produce large quantities of crops and large numbers of livestock with minimal labor for domestic and foreign sale.

**industrialized countries.** See **developed countries.**

**Industrial Revolution.** During the 19th century, the development of manufacturing processes using fossil fuels and based on applications of scientific knowledge.

**industrial smog.** The grayish mixture of moisture, soot, and sulfurous compounds that occurs in local areas in which industries are concentrated and coal is the primary energy source.

**infant mortality.** The number of babies that die before one year, per 1,000 babies born.

**infiltration.** The process in which water soaks into soil as opposed to running off the surface of the soil.

**infiltration–runoff ratio.** The ratio of the amount of water soaking into the soil to that running off the surface. The ratio is obtained by dividing the first amount by the second.

**infrared radiation.** Radiation of somewhat longer wavelengths than red light, which comprises the longest wavelengths of the visible spectrum. Such radiation manifests itself as heat.

**infrastructure.** The sewer and water systems, roadways, bridges, and other facilities that underlie the functioning of a city and that are owned, operated, and maintained by the city.

**inherently safe reactor.** In theory, a nuclear reactor that is designed in such a way that any accident would be automatically corrected with no radioactivity released.

**inorganic compounds/molecules.** *Classical definition:* All things such as air, water, minerals, and metals, that are neither living organisms nor products uniquely produced by living things. *Chemical definition:* All chemical compounds or molecules that do not contain carbon atoms as an integral part of their molecular structure. (Contrast **organic compounds.**)

**insecticide.** Any chemical used to kill insects.

**instrumental value.** The value that living organisms or species have in virtue of their benefit to people; the degree to which they benefit humans. (Contrast **intrinsic value.**)

**insurance spraying.** Spraying of pesticides when it is not really needed, in the belief that it will ensure against loss due to pests.

**integrated pest management (IPM).** A program consisting of two or more methods of pest control carefully integrated together and designed to avoid economic loss from pests. The objective of an IPM is to minimize the use of environmentally hazardous synthetic chemicals. Such chemicals may be used in IPM, but only as a last resort to prevent significant economic losses.

**integrated waste management.** The approach to municipal solid waste that provides for several options for dealing with wastes, including recycling, composting, waste reduction, and landfilling and incineration where unavoidable.

**International Whaling Commission (IWC).** The international body that regulates the harvesting of whales; the IWC placed a ban on all whaling in 1986.

**intrinsic value.** The value that living organisms or species have in their own right; in other words, organisms and species do not have to be useful to have value. (Contrast **instrumental value.**)

**invasive species.** An introduced species that spreads out and often causes harmful ecological effects on other species or ecosystems

**inversion.** See **temperature inversion.**

**ion.** An atom (or a group of atoms) that has lost or gained one or more electrons and, consequently, has acquired a positive or negative charge. Ions are designated by "+" or "−" superscripts following the chemical symbol for the element(s) involved.

**ion-exchange capacity.** See **nutrient-holding capacity.**

**ionic bond.** The bond formed by the attraction between a positive and a negative ion.

**IPCC (Intergovernmental Panel on Climate Change).** The U.N.-sponsored organization charged with continually assessing the science of global climate change, its potential impacts, and the means of responding to the threat.

**irrigation.** Any method of artificially adding water to crops.

**isotope.** A form of an element in which the atoms have more (or less) than the usual number of neutrons. Isotopes of a given element have identical chemical properties, but differ in mass (weight) as a result of the superfluity (or deficiency) of neutrons. Many isotopes are unstable and radioactive. (See **radioactive decay, radioactive emissions,** and **radioactive materials.**)

**ISTEA (Intermodal Surface Transportation Efficiency Act).** Legislation that provides for funding of alternative transportation (e.g., mass transit or bicycle paths) using money from the Highway Trust Fund.

**Jubilee 2000.** A worldwide coalition of religious groups and individuals dedicated to promoting debt relief efforts on the part of creditor nations.

**junk science.** Information presented as valid science, but unsupported by peer-reviewed research. Often, politically motivated and biased results are selected to promote a particular point of view.

**juvenile hormone.** The insect hormone that, at sufficient levels, preserves the larval state. Pupation requires diminished levels of juvenile hormone; hence, artificial applications of the hormone may block pupal development.

**keystone species.** A species whose role is essential for the survival of many other species in an ecosystem.

**kinetic energy.** The energy inherent in motion or movement, including molecular movement (heat) and the movement of waves (hence, radiation and therefore light).

**knowledge assets.** A component of human capital, the codified or written fund of knowledge that can readily be transferred to others across space or time.

**Kyoto Protocol.** An international agreement among the developed nations to curb greenhouse gas emissions. The Kyoto Protocol was forged in December 1997.

**La Niña.** A complete reversal of El Niño conditions in the tropical Pacific Ocean. Easterly trade winds are intense, and global weather patterns are often the reverse of those brought on by an El Niño event.

**Lacey Act.** Passed in 1900, the first national act that gave protection to wildlife by forbidding interstate commerce in illegally killed animals.

**landfill.** A site where municipal, industrial, or chemical wastes are disposed of by burying them in the ground or placing them on the ground and covering them with earth.

**landscape.** A group of interacting ecosystems occupying adjacent geographical areas.

**land subsidence.** The gradual sinking of land. The condition may result from the removal of groundwater or oil, which is frequently instrumental in supporting the overlying rock and soil.

**land trust.** Land that is purchased and held by various private organizations specifically for the purpose of protecting the region's natural environment and the biota that inhabit it.

**larva, pl. larvae.** A free-living immature form that occurs in the life cycle of many organisms and that is structurally distinct from the adult. For example, caterpillars are the larval stage of moths and butterflies.

**law of conservation of energy.** See **first law of thermodynamics.**

**law of conservation of matter.** Law stating that, in chemical reactions, atoms are neither created, nor changed, nor destroyed; they are only rearranged.

**law of limiting factors.** Law stating that a system may be limited by the absence or minimum amount (in terms of that needed) of any required factor. Also known as Liebig's law of minimums. (See **limiting factor.**)

**leachate.** The mixture of water and materials that are leaching.

**leaching.** The process in which materials in or on the soil gradually dissolve and are carried by water seeping through the soil. Leaching may eventually remove valuable nutrients from the soil, or it may carry buried wastes into groundwater, thereby contaminating it.

**legumes.** The group of pod-bearing land plants that is virtually alone in its ability to fix nitrogen; legumes include such common plants as peas, beans, clovers, alfalfa, and locust trees, but no major cereal grains. (See **nitrogen fixation.**)

**Liebig's law of minimums.** See **law of limiting factors.**

**lifeboat ethic.** An argument to the effect that we should limit the amount of food aid to high-population countries in order to prevent further population growth.

**limiting factor.** A factor primarily responsible for determining the growth or reproduction of an organism or a population. The limiting factor in a given environment may be a physical factor such as temperature or light, a chemical factor such as a particular nutrient, or a biological factor such as a competing species. The limiting factor may differ at different times and places.

**limits of tolerance.** The extremes of any factor (e.g., temperature) that an organism or a population can tolerate and still survive and reproduce.

**lipids.** A class of natural organic molecules that includes animal fats, vegetable oils, and phospholipids, the last being an integral part of cellular membranes.

**litter.** In an ecosystem, the natural cover of dead leaves, twigs, and other dead plant material. This natural litter is subject to rapid decomposition and recycling in the ecosystem, whereas human litter, such as bottles, cans, and plastics, is not.

**livability.** A subjective index of how enjoyable a city is to live in.

**loam.** A soil texture consisting of a mixture of about 40% sand, 40% silt, and 20% clay; loam is usually good soil.

**longevity.** The average life span of individuals of a given population.

**low-till farming.** Similar to no-till farming, except that instead of using herbicides, farmers plow just once over residues of a previous crop and plant their new crop.

**LULU ("locally unwanted land use").** Acronym expressing the difficulty in siting a facility that is necessary, but that no one wants in his or her immediate locality.

**macromolecules.** Very large organic molecules, such as proteins and nucleic acids, that constitute the structural and functional parts of cells.

**Maximum achievable control technology** (MACT). The best technologies available for reducing the output of especially toxic industrial pollutants.

**Magnuson Act.** An act passed in 1976 that extended the limits of jurisdiction over coastal waters and fisheries of the United States to 200 miles offshore.

**malnutrition.** The lack of essential nutrients such as vitamins, minerals, and amino acids. Malnutrition ranges from mild to severe and life threatening.

**mass number.** The number that accompanies the chemical name or symbol of an element or isotope. The mass number represents the number of neutrons and protons in the nucleus of the atom.

**material safety data sheets (MSDS).** Documents containing information on the reactivity and toxicity of chemicals and that must accompany the shipping, storage, and handling of over 600 chemicals.

**materials recycling facility (MRF).** A processing plant in which regionalized recycling is carried out. Recyclable municipal solid waste, usually presorted, is prepared in bulk for the recycling market.

**matter.** Any gas, liquid, or solid that occupies space and has mass. (Contrast **energy.**)

**Maximum Contaminant Level (MCL).** A drinking water standard; the highest allowable concentration of a pollutant in a drinking water source.

**maximum sustainable yield.** The maximum amount of a renewable resource that can be taken year after year without depleting the resource. The maximum sustainable yield is the maximum rate of use or harvest that will be balanced by the regenerative capacity of the system—for example, the maximum rate of tree cutting that can be balanced by regrowth.

**meltdown.** The event of a nuclear reactor's getting out of control or losing its cooling water so that it melts from its own production of heat. The melted reactor would continue to produce heat and could melt its way out of its containment vessel and eventually down into groundwater, where it would cause a violent eruption of steam that could spread radioactive materials over a wide area.

**metabolism.** The sum of all the chemical reactions that occur in an organism.

**methane.** A gas, $CH_4$. Methane is the primary constituent of natural gas and is also a product of fermentation by microbes. Methane is one of the greenhouse gases.

**microbe.** Any microscopic organism, although primarily a bacterium, a virus, or a protozoan.

**microclimate.** The actual conditions experienced by an organism in its particular location. Owing to numerous factors, such as shading, drainage, and sheltering, the microclimate may be quite distinct from the overall climate.

**microfiltration.** A process for purifying water in which the water is forced under very high pressure through a membrane that is fine enough to filter out ions and molecules in solution; microfiltration used by small desalination plants to filter salt from seawater. Also called reverse osmosis.

**microlending.** The process of providing very small loans (usually \$50–\$500) to poor people to facilitate their starting a small enterprise and becoming economically self-sufficient.

**microorganism.** See **microbe.**

**midnight dumping.** The illicit dumping of materials, particularly hazardous wastes, frequently under the cover of darkness.

**Milankovitch cycle.** A cycle of major oscillations in the Earth's orbit, taking place over frequencies of thousands of years and known to influence the distribution of solar radiation and therefore global weather patterns.

**Millennium Development Goals.** A comprehensive set of goals aimed at addressing the most important needs of people in the developing countries to improve their well-being, adopted by the United Nations at the UN Millennium Summit in 2000.

**Minamata disease.** A disease named for a fishing village in Japan where an "epidemic" was first observed. Symptoms, which included spastic movements, mental retardation, coma, death, and crippling birth defects in the next generation, were found to be the result of mercury poisoning.

**mineral.** Any hard, brittle, stonelike material that occurs naturally in Earth's crust. All minerals consist of various combinations of positive and negative ions held together by ionic bonds. Pure minerals, or crystals, are one specific combination of elements. Common rocks are composed of mixtures of two or more minerals.

**mineralization.** The process of gradual oxidation of the organic matter (humus) present in soil that leaves just the gritty mineral component of the soil.

**mixture.** A combination of elements in which there is no chemical bonding between the molecules. For example, air contains (is a mixture of) oxygen, nitrogen, and carbon dioxide.

**moderator.** In a nuclear reactor, any material that slows down neutrons from fission reactions so that they are traveling at the right speed to trigger another fission. Water and graphite are two types of moderators.

**molecule.** The smallest unit of two or more atoms forming a compound. A molecule has all the characteristics of the compound of which it is a unit.

**monoculture/monocropping.** The practice of growing the same crop year after year on the same land. (Contrast **crop rotation** and **polyculture.**)

**monsoon.** The seasonal airflow created by major differences in cooling and heating between oceans and continents, usually bringing extensive rain.

**Montreal Protocol.** An agreement made in 1987 by a large group of nations to cut back the production of chlorofluorocarbons by 50% by the year 2000 in order to protect the ozone shield. A 1990 amendment called for the complete phaseout of these chemicals by 2000 in developed nations and by 2010 in less developed nations.

**morbidity.** The incidence of disease in a population.

**mortality.** The incidence of death in a population.

**municipal solid waste (MSW).** The entirety of refuse or trash generated by a residential and business community. Distinct from agricultural and industrial wastes, municipal solid waste is the refuse that a municipality is responsible for collecting and disposing of.

**mutagenic.** Causing mutations.

**mutation.** A random change in one or more genes of an organism. Mutations may occur spontaneously in nature, but their number and degree are vastly increased by exposure to radiation or certain chemicals. Mutations generally result in a physical deformity or metabolic malfunction.

**mutualism.** A close relationship between two organisms from which both derive a benefit.

**mycelia.** The threadlike feeding filaments of fungi.

**mycorrhiza, pl. mycorrhizae.** The mycelia of certain fungi that grow symbiotically with the roots of some plants and provide for additional nutrient uptake.

**NASA.** National Aeronautics and Space Administration.

**National Ambient Air Quality Standards (NAAQS).** The allowable levels of ambient air pollutants set by EPA regulation.

**National Emission Standards for Hazardous Air Pollutants (NESHAPS).** The standards for allowable emissions of certain toxic substances.

**national forests.** Public forests and woodlands administered by the National Forest Service for multiple uses, such as logging, mineral exploitation, livestock grazing, and recreation.

**national parks.** Lands and coastal areas of great scenic, ecological, or historical importance administered by the National Park Service with the dual goals of protection and providing public access.

**National Pollution Discharge Elimination System (NPDES).** An EPA-administered program that addresses point-source water pollution through the issuance of permits that regulate pollution discharge.

**national priorities list (NPL).** A list of the chemical waste sites presenting the most immediate and severe threats. Such sites are scheduled for cleanup ahead of other sites.

**natural.** Produced as a normal part of nature, apart from any activity or intervention of humans. Opposite of artificial, synthetic, human made, or caused by humans.

**natural capital.** The natural assets and the services they perform. One form of the wealth of a nation is its complement of natural capital.

**natural chemical control.** The use of one or more natural chemicals, such as hormones or pheromones, to control a pest.

**natural control methods.** Any of many techniques of controlling a pest population without resorting to the use of synthetic organic or inorganic chemicals. (See **biological control, cultural control, genetic control, hormones**, and **pheromones.**)

**natural enemies.** All the predators and parasites that may feed on a given organism. Organisms used to control a specific pest through predation or parasitism.

**natural increase.** The number of births minus the number of deaths in a given population. The natural increase does not take into account immigration and emigration and is the percent of growth (or decline) of a given population during a year. It is found by subtracting the crude death rate from the crude birthrate and changing the result to a percent.

**natural laws.** Generalizations derived from our observations of matter, energy, and other phenomena. Though not absolute, natural laws have been empirically confirmed to a high degree and are often derivable from higher level theory.

**natural resources.** Features of natural ecosystems and species that are of economic value and that may be exploited. Also, features of particular segments of ecosystems, such as air, water, soil, and minerals.

**natural selection.** The process whereby the natural factors of environmental resistance tend to eliminate those members of a population which are least well adapted to cope with their environment and thus, in effect, tend to select those best adapted for survival and reproduction.

**natural services.** Functions performed free of charge by natural ecosystems, such as control of runoff and erosion, absorption of nutrients, and assimilation of air pollutants.

**Neolithic Revolution.** The development of agriculture begun by human societies around 12,000 years ago, leading to more permanent settlement and population increases.

**net metering.** The practice of allowing individuals on the power grid to subtract the cost of the energy they get from photovoltaics or wind turbines from their electric bills.

**net national product (NNP).** The production of goods and services (GDP or GNP) minus the depreciation of produced assets, human capital and natural capital.

**net primary production.** The rate at which new organic matter is made available to consumers by primary producers (equals gross primary production minus plant respiration).

**neutron.** A fundamental atomic particle found in the nuclei of atoms (except hydrogen) and having one unit of atomic mass, but no electrical charge.

**new forestry.** Now part of the Forest Service's management practice, a forestry management strategy that places priority on protecting the ecological health and diversity of forests rather than maximizing the harvest of logs.

**New Source Review.** The EPA program that reviews and issues permits for new utilities and industrial facilities expected to release significant air pollutants; it also applies to existing facilities that wish to upgrade their equipment.

**niche (ecological).** The total of all the relationships that bear on how an organism copes with the biotic and abiotic factors it faces.

**NIMBY ("Not in my backyard").** A common attitude regarding undesirable facilities such as incinerators, nuclear facilities, and hazardous waste treatment plants whereby people do everything possible to prevent such facilities from being located near their residences.

**NIMTOO ("Not in my term of office").** Attitude expressing the reluctance of officeholders to make unpopular decisions on matters such as siting waste facilities.

**nitrogen fixation.** The process of chemically converting nitrogen gas ($N_2$) from the air into compounds such as nitrates ($NO_3^-$) or ammonia ($NH_3$) that can be used by plants in building amino acids and other nitrogen-containing organic molecules.

**nitrogen oxides ($NO_x$).** A group of nitrogen–oxygen compounds formed when some of the nitrogen gas in air combines with oxygen during high-temperature combustion. Nitrogen oxides are a major category of air pollutants and, along with hydrocarbons, are a primary factor in the production of ozone and other photochemical oxidants that are the most harmful components of photochemical smog. Nitrogen oxides also contribute to acid precipitation. The major nitrogen oxides are nitric oxide (NO), nitrogen dioxide ($NO_2$), and nitrogen tetroxide ($N_2O_2$).

**nitrous oxide.** A gas, $N_2O$. Nitrous oxide, which comes from biomass burning, fossil fuel burning, and the use of chemical fertilizers, is of concern because it is a greenhouse gas in the troposphere and it contributes to ozone destruction in the stratosphere.

**NOAA.** National Oceanic and Atmospheric Administration.

**nonbiodegradable.** Not able to be consumed or broken down by biological organisms. Nonbiodegradable substances include plastics, aluminum, and many chemicals used in industry and agriculture. Particularly dangerous are human-made nonbiodegradable chemicals that are also toxic and tend to accumulate in organisms (i.e., nonbiodegradable synthetic organic compounds). (See **biodegradable** and **bioaccumulation.**)

**nonconsumptive water use.** Use of water for such purposes as washing and rinsing, wherein the water, albeit polluted, remains available for further uses. With suitable purification, such water may be recycled indefinitely.

**nongovernmental organization (NGO).** Any of a number of private organizations involved in studying or advocating action on different environmental issues.

**nonpersistent.** Said of chemicals that break down readily to harmless compounds, as, for example, natural organic compounds break down to carbon dioxide and water.

**nonpoint sources.** Sources of pollution such as the general runoff of sediments, fertilizer, pesticides, and other materials from farms and urban areas, as opposed to specific points of discharge such as factories. Also called diffuse sources. (Contrast **point sources.**)

**nonrenewable resources.** Resources, such as ores of various metals, oil, and coal, that exist as finite deposits in Earth's crust and that are not replenished by natural processes as they are mined. (Contrast **renewable resources.**)

**no-till agriculture.** The farming practice in which weeds are killed with chemicals (or other means) and seeds are planted and grown without resorting to plowing or cultivation. The practice is highly effective in reducing soil erosion.

**NRCS.** Natural Resources Conservation Service, formerly the SCS (U.S. Soil Conservation Service).

**nuclear power.** Electrical power generated by using a nuclear reactor to boil water and produce steam that in turn, drives a turbogenerator.

**Nuclear Regulatory Commission (NRC).** The agency within the Department of Energy that sets and enforces safety standards for the operation and maintenance of nuclear power plants.

**nucleic acids.** The class of natural organic macromolecules that function in the storage and transfer of genetic information.

**nucleus.** *Biology*: The large body residing in most living cells that contains the genes or hereditary material (DNA). *Physics:* The central core of atoms, which is made up of neutrons and protons. Electrons surround the nucleus.

**nutrient.** *Animal*: Material such as protein, vitamins, and minerals required for growth, maintenance, and repair of the body and material such as carbohydrates required for energy. *Plant:* An essential element in a particular ion or molecule that can be absorbed and used by the plant. For example, carbon, hydrogen, nitrogen, and phosphorus are essential elements; carbon dioxide, water, nitrate ($NO_3^-$), and phosphate ($PO_4^{3-}$) are the respective ions or molecules containing the nutrient element.

**nutrient cycle.** The repeated pathway of particular nutrients or elements from the environment through one or more organisms and back to the environment. Nutrient cycles include the carbon cycle, the nitrogen cycle, the phosphorus cycle, and so on.

**nutrient-holding capacity.** The capacity of a soil to bind and hold nutrients (fertilizer) against their tendency to be leached from the soil.

**observations.** Things or phenomena that are perceived through one or more of the basic five senses in their normal state. In addition, to be accepted as factual, observations must be verifiable by others.

**ocean thermal-energy conversion (OTEC).** The concept of harnessing the difference in temperature between surface water heated by the sun and colder deep water to produce power.

**oil field.** The underground area in which exploitable oil is found.

**oil sand.** Sedimentary material containing bitumen, a tarlike hydrocarbon that is subject to exploitation under favorable economic conditions.

**oil shale.** A natural sedimentary rock that contains kerogen, a material that can be extracted and refined into oil and oil products.

**oligotrophic.** Nutrient poor and hence unable to support much phytoplankton (said of a body of water).
**omnivore.** An animal that feeds on both plant material and other animals.
**OPEC.** Organization of Petroleum Exporting Countries, a cartel of oil-producing nations that attempts to control the market price of oil.
**optimal population.** The population of a harvested biological resource that yields the greatest harvest for exploitation; according to maximum-sustained-yield equations, the optimal population is half the carrying capacity.
**optimal range.** With respect to any particular factor or combination of factors, the maximum variation that still supports optimal or near-optimal growth of the species in question.
**optimum.** The condition or amount of any factor or combination of factors that will produce the best result. For example, the amount of heat, light, moisture, nutrients, and so on that will produce the best plant growth.
**organically grown.** Grown without the use of hard chemical pesticides or inorganic fertilizer. Any produce grown according to USDA standards for organic foods is organically grown.
**organic compounds/molecules.** *Classical definition*: All living things and products that are uniquely produced by living things, such as wood, leather, and sugar. *Chemical definition:* All chemical compounds or molecules, natural or synthetic, that contain carbon atoms as an integral part of their molecular structure. The structure of organic compounds is based on bonded carbon atoms with hydrogen atoms attached. Organic compounds can be biodegradable or nonbiodegradable. (Contrast **inorganic compounds.**)
**organic fertilizer.** See **fertilizer.**
**organic gardening/farming/food.** Gardening or farming without the use of inorganic fertilizers, synthetic pesticides, or other human-made materials; food raised with organic farming techniques.
**organic phosphate.** Phosphate ($PO_4^{-3}$) bonded to an organic molecule.
**organism.** Any living thing—plant, animal, or microbe.
**orphan site.** The location of a hazardous waste site abandoned by former owners.
**OSHA (Occupational Safety and Health Administration).** Federal agency that promulgates regulations to protect workers.
**osmosis.** The phenomenon whereby water diffuses through a semipermeable membrane toward an area where there is more material in solution (i.e., where there is a relatively lower concentration of water). Osmosis has particular application in the salinization of those soils in which plants are unable to grow because of osmotic water loss.
**outbreak.** A population explosion of a particular pest. Often caused by an application of pesticides that destroys the pest's natural enemies.
**overcultivation.** The practice of repeated cultivation and growing of crops more rapidly than the soil can regenerate, leading to a decline in soil quality and productivity.
**overgrazing.** The phenomenon of animals' grazing in greater numbers than the land can support in the long term. There may be a temporary economic gain in the short term, but the grassland (or other ecosystem) is destroyed, and its ability to support life in the long term is vastly diminished.
**oxidation.** Chemical reaction that generally involves a breakdown of some substance through its combining with oxygen. Burning and cellular respiration are examples of oxidation. In both cases, organic matter is combined with oxygen and broken down to carbon dioxide and water.
**ozone.** A gas, $O_3$, that is a pollutant in the lower atmosphere, but that is necessary to screen out ultraviolet radiation in the upper atmosphere. May also be used for disinfecting water.
**ozone hole.** First discovered over the Antarctic, a region of stratospheric air that is severely depleted of its normal levels of ozone during the Antarctic spring because of CFCs from anthropogenic (human-made) sources.
**ozone shield.** The layer of ozone gas ($O_3$) in the upper atmosphere that screens out harmful ultraviolet radiation from the Sun.
**PANs (peroxyacetylnitrates).** A group of compounds present in photochemical smog that are extremely toxic to plants and irritating to the eyes, nose, and throat membranes of humans.
**parasites.** Organisms (plant, animal, or microbial) that attach themselves to another organism, the host, and feed on it over a period of time without killing it immediately, but usually doing harm to it. Parasites are commonly divided into ectoparasites—those that attach to the outside of the host—and endoparasites—those that live inside their hosts.
**parent material.** Rock material whose weathering and gradual breakdown is the source of the mineral portion of soil.
**particulates.** (See **$PM_{2.5}$ and $PM_{10}$** and **suspended particulate matter.**)
**parts per million (ppm).** A frequently used expression of concentration. The number of units of one substance present in a million units of another. Equivalent to milligrams per liter.
**passive safety features.** Those safety features of nuclear facilities which involve processes that are not vulnerable to operator intrusion or electrical power failures. Passive safety features enhance the degree of safety of nuclear reactors. (See **active safety features.**)
**passive solar-heating system.** A solar-heating system that does not use pumps or blowers to transfer heated air or water. Instead, natural convection currents are used, or the interior of the building itself acts as the solar collector.
**pasteurization.** The process of applying enough heat to milk or some other substance to kill pathogens and sufficient other bacteria to extend the shelf life of the product.
**pastoralist.** One involved in animal husbandry, usually in subsistence agriculture.
**pathogen.** An organism, usually a microbe, that is capable of causing disease. Such an organism is said to be pathogenic.
**percolation.** The process of water seeping downward through cracks and pores in soil or rock.
**permafrost.** The ground of arctic regions that remains permanently frozen. Defines tundra, since only small herbaceous plants can be sustained on the thin layer of soil that thaws each summer.
**persistent.** Attribute of pesticides or other chemicals that are nonbiodegradable and highly resistant to breakdown by other means. Such chemicals therefore remain present in the environment for periods of years or longer.
**persistent organic pollutants (POPs).** Any members of a class of organic pollutants that are resistant to biodegradation and that are often toxic; for example, DDT, PCBs, and dioxin are persistent organic pollutants.
**pest.** Any organism that is noxious, destructive, or troublesome; usually an agricultural pest.
**pesticide.** A chemical used to kill pests. Pesticides are further categorized according to the pests they are designed to kill—for example, herbicides kill plants, insecticides kill insects, fungicides kill fungi, and so on.
**pesticide treadmill.** The idea that use of chemical pesticides simply creates a vicious cycle of "needing more pesticides"

to overcome developing resistance and secondary outbreaks caused by the pesticide applications.

**pest-loss insurance.** Insurance that a grower can buy that will pay in the event of loss of a crop due to pests.

**petrochemical.** A chemical made from petroleum (crude oil) as a basic raw material. Petrochemicals include plastics, synthetic fibers, synthetic rubber, and most other synthetic organic chemicals.

**pH.** Scale used to designate the acidity or basicity (alkalinity) of solutions or soil, expressed as the logarithm of the concentration of hydrogen ions ($H^+$). pH 7 is neutral; values decreasing from 7 indicate increasing acidity, values increasing from 7 increasing basicity. Each unit from 7 indicates a tenfold increase over the preceding unit.

**pheromones.** Chemical substances secreted externally by certain members of a species that affect the behavior of other members of the same species. The most common examples are sex attractants, which female insects secrete to attract males. Pheromones are coming into use in pest control. (See also **hormones.**)

**phosphate.** An ion composed of a phosphorus atom with four oxygen atoms attached. Denoted $PO_4^{3-}$, phosphate is an important plant nutrient. In natural waters, it is frequently the limiting factor; therefore, additions of phosphate to natural water are often responsible for algal blooms.

**photochemical oxidants.** A major category of secondary air pollutants, including ozone, that are highly toxic and damaging, especially to plants and forests. Formed as a result of interactions between nitrogen oxides and hydrocarbons driven by sunlight.

**photochemical smog.** The brownish haze that frequently forms on otherwise clear, sunny days over large cities with significant amounts of automobile traffic. Photochemical smog results largely from sunlight-driven chemical reactions among nitrogen oxides and hydrocarbons, both of which come primarily from auto exhausts.

**photosynthesis.** The chemical process carried on by green plants through which light energy is used to produce glucose from carbon dioxide and water. Oxygen is released as a by-product.

**photovoltaic cells.** Devices that convert light energy into an electrical current.

**physical barrier.** A genetic feature on a plant, such as sticky hairs, that physically blocks attack by pests.

**phytoplankton.** Any of the many species of photosynthetic microorganisms that consist of single cells or small groups of cells that live and grow freely suspended near the surface in bodies of water.

**phytoremediation.** The use of plants to accomplish the cleanup of some hazardous chemical wastes.

**planetary albedo.** The reflection of solar radiation back into space due to cloud cover, contributing to cooling of the atmosphere.

**plankton.** Any and all living things that are found freely suspended in the water and that are carried by currents, as opposed to being able to swim against currents. Plankton includes both plant (phytoplankton) and animal (zooplankton) forms.

**plant community.** The array of plant species, including their numbers, ages, and distribution, that occupies a given area.

**$PM_{2.5}$ and $PM_{10}$.** Standard criterion pollutants for suspended particulate matter. $PM_{10}$ refers to particles smaller than 10 micrometers in diameter, $PM_{2.5}$ for particles smaller than 2.5 micrometers in diameter. The smaller particles are readily inhaled directly into the lungs.

**point sources.** Specific points of origin of pollutants, such as factory drains or outlets from sewage-treatment plants. (Contrast **nonpoint sources.**)

**policy life cycle.** The typical course of events occurring over time in the recognition, formulation, implementation, and control of an environmental problem requiring public policy action.

**pollutant.** A substance that contaminates air, water, or soil.

**pollution.** Contamination of air, water, or soil with undesirable amounts of material or heat. The material may be a natural substance, such as phosphate, in excessive quantities, or it may be very small quantities of a synthetic compound such as dioxin, which is exceedingly toxic.

**polychlorinated biphenyls (PCBs).** A group of widely used industrial chemicals of the chlorinated hydrocarbon class. Polychlorinated biphenyls have become serious and widespread pollutants because they are extremely resistant to breakdown and are subject to bioaccumulation. They also are known to be carcinogenic.

**polyculture.** The growing of two or more species together. (Contrast **monoculture.**)

**poor.** Economically unable to afford adequate food or housing.

**population.** A group within a single species whose individuals can and do freely interbreed.

**population density.** The number of individuals per unit of area.

**population equilibrium.** A state of balance between births and deaths in a population.

**population explosion.** The exponential increase observed to occur in a population when conditions are such that a large percentage of the offspring are able to survive and reproduce in turn. A population explosion frequently leads to overexploitation and eventual collapse of the ecosystem.

**population momentum.** Property whereby a rapidly growing human population may be expected to grow for 50–60 years after replacement fertility (2.1 live births per female) is reached. Momentum is sustained because of increasing numbers entering reproductive age.

**population profile.** A bar graph that shows the number of individuals at each age or in each five-year age group, starting with the youngest ages at the bottom of the profile.

**population structure.** See **age structure.**

**potential energy.** The ability to do work that is stored in some chemical or physical state. For example, gasoline is a form of potential energy because the ability to do work is stored in the chemical state and is released as the fuel is burned in an engine.

**poverty trap.** In a developing country, the situation where people are so poor that they are unable to earn or work their way out of their poverty and their condition deteriorates.

**power grid.** The combination of central power plants, high-voltage transmission lines, poles, wires, and transformers that makes up an electrical power system.

**precautionary principle.** The principle that says that where there are threats of serious or irreversible damage, the absence of scientific certainty shall not be used as a reason for postponing cost-effective measures to prevent environmental degradation.

**precipitation.** Any form of moisture condensing in the air and depositing on the ground.

**predator.** An animal that feeds on another living organism, either plant or animal.

**predator–prey relationship.** A feeding relationship existing between two kinds of organisms. The predator is the animal feeding on the prey. Such relationships are frequently instrumental in controlling populations of herbivores.

**preliminary treatment.** The removal of debris and grit from wastewater by passing the water through a coarse screen and a grit-settling chamber.
**primary consumer.** An organism, such as a rabbit or deer, that feeds more or less exclusively on green plants or their products, such as seeds and nuts. (Synonym: **herbivore.**)
**primary energy sources.** Fossil fuels, radioactive material, and solar, wind, water, and other energy sources that exist as natural resources subject to exploitation.
**primary pollutants.** Pollutants released directly into the atmosphere mainly as a result of burning fuels and wastes, as opposed to secondary pollutants.
**primary producers/primary production.** Photosynthetic organisms. The activities of these organisms in creating new organic matter in ecosystems.
**primary recovery.** In an oil well or oil field, the oil that can be removed by conventional pumping. (See **secondary recovery.**)
**primary standard.** The maximum tolerable level of a pollutant. The standard is intended to protect human health.
**primary succession.** See **succession.**
**primary treatment.** The process that follows preliminary sewage treatment and that consists of passing the water very slowly through a large tank so that the particulate organic material can settle out. The settled material is **raw sludge.**
**prior informed consent (PIC).** A U.N.-approved procedure whereby countries exporting pesticides must inform importing countries of actions the exporting countries have taken to restrict the use of the pesticides.
**produced assets.** The stock of buildings, machinery, vehicles, and other elements of a country's infrastructure that are essential to the production of economic goods and services. One component of the wealth of nations.
**producers.** In an ecosystem, those organisms (mostly green plants) which use light energy to construct their organic constituents from inorganic compounds.
**productive use.** The exploitation of ecosystem resources for economic gain.
**property taxes.** Taxes that the local government levies on privately owned properties, proportional to the value of the property. Property taxes are the major source of revenue for local governments.
**protein.** Organic macromolecules made up of amino acids; the major structural component of all animal tissues, as well as enzymes in both plants and animals.
**proton.** Fundamental atomic particle with a positive charge, found in the nuclei of atoms. The number of protons present equals the atomic number and is distinct for each element.
**protozoon, pl. protozoa.** Any of a large group of eukaryotic microscopic organisms that consist of a single, relatively large complex cell or, in some cases, small groups of cells. All have some means of movement. Amoebae and paramecia are examples.
**proven reserves.** See **reserves.**
**proxies.** These are measurable records, like tree rings, ice cores, marine sediments, etc, that can provide extensions back in time of climatic factors like temperatures and greenhouse gases.
**RACT (reasonably available control technology).** Applied to the goals of the Clean Air Act, EPA-approved forms of technology that will reduce the output of industrial air pollutants. (See **MACT.**)
**radiative forcing.** Refers to the influence of internal or external factors of the climate system that can influence the climate, usually to promote warming or cooling of the atmosphere as they affect the Earth's energy balance.
**radioactive decay.** The reduction in radioactivity that occurs as an unstable isotope (radioactive substance) gives off radiation, ultimately to become stable.
**radioactive emissions.** Any of various forms of radiation or particles that may be given off by unstable isotopes. Many such emissions have very high energy and can destroy biological tissues or cause mutations leading to cancer or birth defects.
**radioactive materials.** Substances that are or that contain unstable isotopes and that consequently give off radioactive emissions. (See **isotope** and **radioactive emissions.**)
**radioactive wastes.** Waste materials that contain, are contaminated with, or are radioactive substances. Many materials used in the nuclear industry become radioactive wastes because of such contamination.
**radioisotope.** An isotope of an element that is unstable and that gives off radioactive emissions. (See **isotope** and **radioactive decay.**)
**radon.** A radioactive gas produced by natural processes in Earth that is known to seep into buildings. Radon can be a major hazard within homes and is a known carcinogen.
**rain shadow.** The low-rainfall region that exists on the leeward (downwind) side of a mountain range. This rain shadow is the result of the mountain range's causing precipitation on the windward side.
**range of tolerance.** The range of conditions within which an organism or population can survive and reproduce—for example, the range from the highest to the lowest temperature that can be tolerated. Within the range of tolerance is the optimum, or best, condition.
**raw sludge.** The untreated organic matter that is removed from sewage water by letting it settle. Raw sludge consists of organic particles from feces, garbage, paper, and bacteria.
**raw wastewater.** (See **raw sludge.**)
**reactor vessel.** Steel-walled vessel that contains a nuclear reactor.
**recharge area.** The area over which groundwater will infiltrate and resupply an aquifer.
**recruitment.** The maturation and successful entry of young into an adult breeding population.
**recycling.** Recovery of materials that would otherwise be buried in landfills or be combusted. Primary recycling involves remaking the same material from the waste; in secondary recycling, waste materials are made into different products.
**reformulated gasoline.** Gasoline with oxygen-containing additives (MTBE, ethanol) to enhance clean burning.
**relative humidity.** The percentage of moisture in the air, compared with how much the air can hold at the given temperature.
**rem.** An older unit of measurement of the ability of radioactive emissions to penetrate biological tissue. (See **sievert.**)
**remediation.** A return to an original uncontaminated state. (See **bioremediation** and **groundwater remediation.**)
**renewable energy.** Energy sources—namely, solar, wind, and geothermal—that will not be depleted by use.
**renewable portfolio standard.** A policy that requires a percentage of power from renewable energy sources
**renewable resources.** Biological resources, such as trees, that may be renewed by reproduction and regrowth. Conservation to prevent overcutting and protection of the environment are still required, however. (Contrast **nonrenewable resources.**)
**replacement fertility/level.** The fertility rate (level) that will just sustain a stable population.
**replacement migration.** The migration of people from developing countries to developed countries that effectively supplies

workers from overpopulated countries to countries where workers are becoming scarce due to ageing.

**reproductive health.** Health care directed primarily towards the needs of women and infants, strongly emphasized in the 1994 Cairo population conference.

**reproductive isolation.** One of the processes of speciation, involving anything that keeps individuals or subpopulations from interbreeding.

**reserves.** The amount of a mineral resource (including oil, coal, and natural gas) remaining in Earth that can be exploited using current technologies and at current prices. Usually given as *proven* reserves—those which have been positively identified—and *estimated* reserves—those which have not yet been discovered, but are presumed to exist.

**resilience.** The tendency of ecosystems to recover from disturbances through a number of processes known as resilience mechanisms (e.g., succession after a fire).

**resistance.** The development of more hardy pests through the effects of the repeated use of pesticides in selecting against sensitive individuals in a pest population.

**resource partitioning.** The outcome of competition by a group of species where natural selection favors the division of a resource in time or space by specialization of the different species.

**resources.** Biotic and abiotic factors that are consumed by organisms.

**Resources Conservation and Recovery Act of 1976 (RCRA).** The cornerstone legislation to control indiscriminate land disposal of hazardous wastes.

**respiration.** See **cell respiration.**

**restoration ecology.** The branch of ecology devoted to restoring degraded and altered ecosystems to their natural state.

**resurgence.** The rapid comeback of a population (especially of pests after a severe dieoff, usually caused by pesticides) and the return to even higher levels than before the treatment.

**reuse.** The practice of continuing to use items, as opposed to throwing them away and producing new items (e.g., bottles collected and refilled in recycling efforts).

**riparian woodlands.** The strip of woody plants that grow along natural watercourses.

**risk.** The probability of suffering injury, disease, death, or some other loss as a result of exposure to a hazard.

**risk analysis.** The process of evaluating the risks associated with a particular hazard before taking some action. Often called risk assessment.

**risk characterization.** The process of determining the level of a risk and its accompanying uncertainties after hazard assessment, dose-response assessment, and exposure assessment have been accomplished.

**risk management.** The task of regulators whose job is to review risk data and make regulatory decisions based on the data. The process often is influenced by considerations of costs and benefits, as well as by public perception.

**risk perception.** Nonexperts' intuitive judgments about risks, which often are not in agreement with the level of risk as judged by experts.

**rooftop garden.** A form of above-ground garden on the top of a flat residential roof, using shallow soil beds and containers such as old automobile tires or plastic wading pools.

**runoff.** The portion of precipitation that runs off the surface, as opposed to soaking into the ground.

**Safe Drinking Water Act of 1974.** Legislation to protect the public from the risk that toxic chemicals will contaminate drinking-water supplies. The act mandates regular testing of municipal water supplies.

**safety net.** In a society, the availability of food and other necessities extended to people who are unable to meet their own needs for a variety of reasons.

**salinization.** The process whereby soil becomes saltier and saltier until, finally, the salt prevents the growth of plants. Salinization is caused by irrigation, because salts brought in with the water remain in the soil as the water evaporates.

**saltwater intrusion, saltwater encroachment.** The phenomenon of seawater's moving back into aquifers or estuaries. Such intrusion occurs when the normal outflow of freshwater is diverted or removed for use.

**sand.** Mineral particles 0.2 to 2.0 mm in diameter.

**sanitary sewer.** Separate drainage system used to receive all the wastewater from sinks, tubs, and toilets.

**SARA (Title III).** A section of the Superfund Amendments and Reauthorization Act that promulgates community Right-to-Know requirements. Also known as Emergency Planning and Community Right-to-Know Act of 1986. (See **toxics release inventory.**)

**savanna.** A type of grassland typical of subtropical regions, particularly in Africa, usually dotted with trees and supported by wet and dry seasons and frequent natural fires.

**secondary air pollutants.** Air pollutants resulting from reactions of primary air pollutants resident in the atmosphere. Secondary air pollutants include ozone, other reactive organic compounds, and sulfuric and nitric acids. (See **ozone, PANs**, and **photochemical oxidants.**)

**secondary consumer.** An organism such as a fox or coyote that feeds more or less exclusively on other animals that feed on plants.

**secondary energy source.** A form of energy such as electricity that must be produced from a primary energy source such as coal or radioactive material.

**secondary pest outbreak.** The phenomenon of a small and therefore harmless population of a plant-eating insect suddenly exploding to become a serious pest problem. Often caused by the elimination of competitors through the use of pesticides.

**secondary recovery.** In an oil well or oil field, the oil that can be removed by manipulating pressure in the oil reservoir by injecting brine or other substances; more costly than **primary recovery.**

**secondary succession.** See **succession.**

**secondary treatment.** Also called biological treatment. A sewage-treatment process that follows primary treatment. Any of a variety of systems that remove most of the remaining organic matter by enabling organisms to feed on it and oxidize it through their respiration. Trickling filters and activated-sludge systems are the most commonly used secondary treatment methods.

**second-generation pesticides.** Synthetic organic compounds used to kill insects and other pests. Started with the use of DDT in the 1940s.

**second law of thermodynamics.** The empirical observation, confirmed innumerable times, that in every energy conversion (e.g., from electricity to light), some of the energy is converted to heat and some heat always escapes from the system because it always moves toward a cooler place. Therefore, in every energy conversion, a portion of energy is lost, and since, by the first law of thermodynamics, energy cannot be created, the functioning of any system requires an energy input.

**secondary energy source.** A source of energy—especially electricity—that depends on a primary energy source for its origin.

**secure landfill.** A landfill with suitable barriers, leachate drainage, and monitoring systems such that it is deemed secure against contaminating groundwater with hazardous wastes.

**sediment.** Soil particles—namely, sand, silt, and clay—carried by flowing water. The same material after it has been deposited. Because of different rates of settling, deposits generally are pure sand, pure silt, or pure clay.

**sedimentation.** The filling in of lakes, reservoirs, stream channels, and so on with soil particles, mainly sand and silt. The soil particles come from erosion, which in turn generally results from poor or inadequate soil conservation practices in connection with agriculture, mining, or development. Also called **siltation.**

**sediment trap.** A device for trapping sediment and holding it on a development or mining site.

**seep.** An area where groundwater seeps from the ground. Contrast with a spring, which is a single point from which groundwater exits.

**selective pressure.** An environmental factor which causes individuals with certain traits that are not the norm for the population to survive and reproduce more than the rest of the population. The result is a shift in the genetic makeup of the population. For example, the presence of insecticides provides a selective pressure to increase pesticide resistance in the pest population. Selective pressure is a fundamental mechanism of evolution.

**self-thinning.** A form of intraspecific competition where crowding occurs in young plants or sessile animals, and subsequent growth of some leads to death of others.

**septic system.** An onsite method of treating sewage that is widely used in suburban and rural areas. Requires a suitable amount of land and porous soil.

**services.** Ecosystem functions that are essential to human life and economic well-being, such as waste breakdown, climate regulation, erosion control, etc. These can be further categorized as regulating, supporting, and provisioning services.

**sex attractant.** A natural chemical substance (see **pheromones**) secreted by the female of many insect species that serves to attract males for the function of mating. Sex attractants may be used by humans in traps or to otherwise confuse insect pests in order to control them.

**shadow pricing.** In cost–benefit analysis, a technique used to estimate benefits when normal economic analysis is ineffective. For example, people could be asked how much they might be willing to pay monthly to achieve some improvement in their environment.

**shaping principles.** These are the conscious and unconscious values and assumptions scientists bring to their work, which can often strongly influence the outcome of scientific investigations.

**sheet erosion.** The loss of a more or less even layer of soil from the land surface due to the impact of rain and runoff from a rainstorm.

**shelterbelts.** Rows of trees around cultivated fields for the purpose of reducing wind erosion.

**sievert.** A unit of measurement of the ability of radioactive emissions to penetrate biological tissues. 1 sievert = 100 rem.

**silt.** Soil particles between the size of sand particles and clay particles, namely, particles 0.002 to 0.2 mm in diameter.

**siltation.** See **sedimentation.**

**sinkhole.** A large hole resulting from the collapse of an underground cavern.

**slash-and-burn agriculture.** The practice, commonly found in tropical regions, of cutting and burning vegetation to make room for agriculture. The process destroys soil humus and may lead to rapid degradation of the soil.

**sludge cake.** Treated sewage sludge that has been dewatered to make a moist solid.

**sludge digesters.** Large tanks in which raw sludge (removed from sewage) is treated through anaerobic digestion by bacteria.

**Smart Growth.** A movement that addresses urban sprawl by protecting sensitive lands and directing growth to limited areas.

**smog.** See **industrial smog** and **photochemical smog.**

**social modernization.** A process of sustainable development that leads developing countries through the demographic transition by promoting education, family planning, and health improvements rather than simply economic growth.

**soft water.** Water with little or no calcium, magnesium, or other ions in solution that will cause soap to precipitate. (The soap would otherwise form a curd that makes a "ring" around the bathtub.) (Contrast **hard water.**)

**soil.** A dynamic terrestrial system involving three components: mineral particles, detritus, and soil organisms feeding on the detritus.

**soil classes.** Taxonomically arranged major groupings of soil that define the properties of the soils of different regions.

**soil erosion.** The loss of soil caused by particles' being carried away by wind or water.

**soil fertility.** Soil's ability to support plant growth; often refers specifically to the presence of proper amounts of nutrients. The soil's ability to fulfill all the other needs of plants is also involved.

**soil horizons.** Distinct layers within a soil that convey different properties to the soil and that derive from natural processes of soil formation.

**soil profile.** A description of the different naturally formed layers, called horizons, within a soil.

**soil structure.** The composition of soil in terms of particles (sand, silt, and clay) stuck together to form clumps and aggregates, generally with considerable air spaces in between. Soil structure affects infiltration and aeration and develops as organisms feed on organic matter in and on the soil.

**soil texture.** The relative size of the mineral particles that make up the soil. Generally defined in terms of the soil's sand, silt, and clay content.

**solar cells.** See **photovoltaic cells.**

**solar energy.** Energy derived from the Sun; includes direct solar energy (the use of sunlight directly for heating or the production of electricity) and indirect solar energy (the use of wind, which results from the solar heating of the atmosphere, and biological materials such as wood, which result from photosynthesis).

**solar-trough collectors.** Reflectors, in the shape of a parabolic trough, that reflect sunlight onto a tube of oil at the focal point of the trough. Thus heated, the oil is used to boil water to drive a steam turbine.

**solid waste.** The total of materials discarded as "trash" and handled as solids, as opposed to those that are flushed down sewers and handled as liquids.

**solubility.** The degree to which a substance will dissolve and enter into solution.

**solution.** A mixture of molecules (or ions) of one material in another, most commonly, molecules of air or ions of various minerals in water. For example, seawater contains salt in solution.

**sound science.** The results of scientific work based on peer-reviewed research. As one of the unifying themes of this text, sound science is the basis for our understanding of how the world works and how human systems interact with it.

**special-interest group.** In politics, people organized around a particular issue who lobby for their cause and exert their influence on the political process to bring about changes they favor.

**specialization.** In evolution, the phenomenon whereby species become increasingly adapted to exploit one particular niche, but thereby are less able to exploit other niches.

**speciation.** The evolutionary process whereby populations of a single species separate and, through being exposed to different forces of natural selection, gradually develop into distinct species.

**species.** All the organisms (plant, animal, or microbe) of a single kind. The "single kind" is determined by similarity of appearance or by the fact that members do or can mate and produce fertile offspring. Physical, chemical, or behavioral differences block breeding between species.

**splash erosion.** The compaction of soil that results when rainfall hits bare soil.

**springs.** Natural exits of groundwater to the surface.

**standing crop biomass.** The biomass of primary producers in an ecosystem at any given time.

**starvation.** The failure to get enough calories to meet energy needs over a prolonged period of time. Starvation results in a wasting away of body tissues until death occurs.

**Statement on Forest Principles.** One of the treaties signed at the 1992 Earth Summit whereby the signatory countries agreed to a number of nonbinding principles stressing sustainable management.

**Sterile-male technique.** Saturating an infested area with males of a pest species that have been artificially reared and sterilized by radiation. Matings between normal females and sterile males render the eggs infertile.

**steward/stewardship.** A steward is one to whom a trust has been given. Stewardship is an attitude of active care and concern for natural lands. As one of the unifying themes of this text, stewardship is the ethical and moral framework that informs our public and private actions.

**stoma, pl. stomata.** A microscopic pore in a leaf, mostly on the undersurface, that allows the passage of carbon dioxide and oxygen into and out of the leaf and that also permits the loss of water vapor from the leaf.

**storm drains.** Separate drainage systems used for collecting and draining runoff from precipitation in urban areas.

**stormwater.** In cities, the water that results directly from rainfall, as opposed to municipal water and sewage water piped to and from homes, offices, and so on. The extensive hard surfacing in cities creates a vast amount of stormwater runoff, which presents a significant management problem.

**stormwater management.** Policies and procedures for handling stormwater in acceptable ways to mitigate the problems of flooding and erosion of streambanks.

**stormwater retention reservoirs.** Reservoirs designed to hold stormwater temporarily and let it drain away slowly in order to mitigate the problems of flooding and streambank erosion.

**stratosphere.** The layer of Earth's atmosphere between 10 and 40 miles above the surface that contains the ozone shield. This layer mixes only slowly; pollutants that enter it may remain for long periods of time. (See **troposphere.**)

**strip cropping.** The practice of growing crops in strips alternating with grass (hay) at right angles to prevailing winds or slopes in order to reduce erosion.

**strip mining.** A mining procedure in which all the earth covering a desired material, such as coal, is stripped away with huge power shovels in order to facilitate removal of the material.

**submerged aquatic vegetation (SAV).** Aquatic plants rooted in bottom sediments and growing under water. Submerged aquatic vegetation depends on the penetration of light through the water for photosynthesis.

**subsistence farming.** Farming that meets the food needs of farmers and their families, but little more. Subsistence farming involves hand labor and is practiced extensively in the developing world.

**subsoil.** In a natural situation, the soil beneath topsoil. In contrast to topsoil, subsoil is compacted and has little or no humus or other organic material, living or dead. In many areas, topsoil has been lost or destroyed as a result of erosion or development, and subsoil is at the surface.

**succession.** The gradual or sometimes rapid change in the species that occupy a given area, with some species invading and becoming more numerous while others decline in population and disappear. Succession is caused by a change in one or more abiotic or biotic factors that benefits some species at the expense of others. *Primary succession:* The gradual establishment, through a series of stages, of a climax ecosystem in an area that has not been occupied before (e.g., a rock face). *Secondary succession:* The reestablishment, through a series of stages, of a climax ecosystem in an area from which it was previously cleared.

**sulfur dioxide** ($SO_2$). A major air pollutant and toxic gas formed as a result of burning sulfur. The major sources are burning coal (in coal-burning power plants) that contains some sulfur and refining metal ores (in smelters) that contain sulfur.

**sulfuric acid** ($H_2SO_4$). The major constituent of acid precipitation. Formed when sulfur dioxide emissions react with water vapor in the atmosphere. (See **sulfur dioxide.**)

**Superfund.** The popular name for the Comprehensive Environmental Response, Compensation, and Liability Act of 1980. This act is the cornerstone legislation that provides the mechanism and funding for the cleanup of potentially dangerous hazardous waste sites and the protection of groundwater.

**surface impoundments.** Closed ponds used to collect and hold liquid chemical wastes.

**surface water.** All bodies of water, lakes, rivers, ponds, and so on that are on Earth's surface. Contrast with **groundwater,** which lies below the surface.

**surge flow irrigation.** Irrigation under the control of microprocessors that release water periodically, thus cutting its use substantially.

**suspended particulate matter (SPM).** A category of major air pollutants consisting of solid and liquid particles suspended in the air. (See **$PM_{2.5}$** and **$PM_{10}$**.)

**suspension.** A system in which materials are contained in or carried by water and are kept "afloat" only by the water's agitation. The materials settle as the water becomes quiet.

**sustainability.** Property whereby a process can be continued indefinitely without depleting the energy or material resources on which it depends. As one of the unifying themes of the text, sustainability is the practical goal toward which our interactions with the natural world should be working.

**sustainable agriculture.** Agriculture that maintains the integrity of soil and water resources such that it can be continued indefinitely.

**sustainable development.** Development that provides people with a better life without sacrificing or depleting resources or causing environmental impacts that will undercut the ability of future generations to meet their needs.

**sustainable forest management.** Management of forests as ecosystems wherein the primary objective is to maintain the biodiversity and function of the ecosystem.

**sustainable society.** A society that functions in a way so as not to deplete the energy or material resources on which it depends. Such a society interacts with the natural world in ways that sustain existing species and ecosystems.

**sustainable yield.** The taking of a biological resource (e.g., fish or forests) that does not exceed the capacity of the resource to reproduce and replace itself.

**symbiosis.** The intimate living together or association of two kinds of organisms.

**synergism.** The phenomenon whereby two factors acting together have a greater effect than would be indicated by the sum of their effects separately—as, for example, the sometimes fatal mixture of modest doses of certain drugs in combination with modest doses of alcohol.

**synfuels, synthetic fuels.** Fuels similar or identical to those that come from crude oil or natural gas. Synfuels are produced from coal, oil shale, or tar sands.

**tar sands.** Sedimentary bitumen-containing material that can be "melted out" using heat and then refined in the same way as crude oil.

**taxonomy.** The science of identifying and classifying organisms according to their presumed natural relationships.

**temperature inversion.** The weather phenomenon in which a layer of warm air overlies cooler air near the ground and prevents the rising and dispersion of air pollutants.

**teratogenic.** Causing birth defects.

**terminator technology.** A transgenic technique that renders seeds incapable of germinating, in order to force farmers to purchase new seeds every year.

**terracing.** The practice of grading sloping farmland into a series of steps and cultivating only the level portions in order to reduce erosion.

**territoriality.** The behavioral characteristic exhibited by many animal species, especially birds and mammalian carnivores, to mark and defend a given territory against other members of the same species.

**theory.** A conceptual formulation that provides a rational explanation or framework for numerous related observations.

**thermal pollution.** The addition of abnormal and undesirable amounts of heat to air or water. Thermal pollution is most significant with respect to discharging waste heat from electric generating plants—especially nuclear power plants—into bodies of water.

**third world.** See **developing countries.**

**threatened species.** A species whose population is declining precipitously because of direct or indirect human impacts.

**threshold level.** The maximum degree of exposure to a pollutant, drug, or some other factor that can be tolerated with no ill effect. The threshold level varies, depending on the species, the sensitivity of the individual, the length of exposure, and the presence of other factors that may produce synergistic effects.

**tidal wetlands.** Areas of marsh grasses and reeds along coasts and estuaries where the ground is covered by high tides, but drained at low tide.

**tilth.** In farming, the ability of a soil to support crop growth.

**top-down regulation.** Basic control of a population (or species) occurs as a result of predation.

**topsoil.** The surface layer of soil, which is rich in humus and other organic material, both living and dead. As a result of the activity of organisms living in the topsoil, it generally has a loose, crumbly structure, as opposed to being a compact mass. In many cases, because of erosion, development, or mining activity, the topsoil layer may be absent.

**total allowable catch (TAC).** In fisheries management, a yearly quota set for the harvest of a species by managers of fisheries.

**total fertility rate.** The average number of children that would be born alive to each woman during her total reproductive years if her fertility was average at each age.

**Total Maximum Daily Load (TMDL) program.** An EPA-administered program to address non-point-source water pollution that sets pollution limits according to the ability of a body of water to assimilate different pollutants.

**total product life cycle.** The sum of all steps in the manufacture of a product, from the obtaining of raw materials through the production, use, and, finally, disposal of the product. By-products and the pollution resulting from each step are taken into account in a consideration of the total product life cycle.

**Toxic Substances Control Act of 1976 (TSCA).** A law requiring the assessment of the potential hazards of a chemical before the chemical is put on the market.

**toxicology.** The study of the impacts of toxic substances on human health and the pathways by which such substances reach humans.

**toxics release inventory.** An annual record of releases of toxic chemicals to the environment and the locations and quantities of toxic chemicals stored at all U.S. sites. Required by the Emergency Planning and Community Right-to-Know Act (EPCRA) of 1986.

**trace elements.** Those essential elements, like copper or iron, that are needed in only very small amounts.

**trade winds.** The more or less constant winds blowing in horizontal directions over the surface as part of Hadley cells.

**traditional agriculture.** Farming methods as they were practiced before the advent of modern industrialized agriculture.

**tragedy of the commons.** The overuse or overharvesting and consequent depletion or destruction of a renewable resource that tends to occur when the resource is treated as a commons—that is, when it is open to be used or harvested by any and all with the means to do so.

**trait.** Any physical or behavioral characteristic or talent that an individual is born with.

**transgenic organism.** Any organism with genes introduced from other species via biotechnology to convey new characteristics.

**transpiration.** The loss of water vapor from plants. Water evaporates from cells within the leaves and exits through the stomata.

**trapping technique.** The use of sex attractants to lure male insects into traps.

**treated sludge.** Solid organic material that has been removed from sewage and treated so that it is nonhazardous.

**trickling filter system.** System in which wastewater trickles over rocks or a framework coated with actively feeding microorganisms. The feeding action of the organisms in a well-aerated environment results in the decomposition of organic matter. Used in secondary or biological treatment of sewage.

**tritium ($^3H$).** An unstable isotope of hydrogen that contains two neutrons in addition to the usual single proton in the nucleus. Tritium does not occur in significant amounts in nature.

**trophic level.** Feeding level with respect to the primary source of energy. Green plants are at the first trophic level, primary consumers at the second, secondary consumers at the third, and so on.

**trophic structure.** The major feeding relationships between organisms within ecosystems, organized into trophic levels.

**troposphere.** The layer of Earth's atmosphere from the surface to about 10 miles in altitude. The tropopause is the boundary between the troposphere and the stratosphere above. The troposphere is well mixed and is the site and source of our weather, as well as the primary recipient of air pollutants. (See also **stratosphere.**)

**turbid.** Cloudy due to particles present. (Said of water and water purity.)

**turbine.** A rotary engine driven at a very high speed by steam, water, or exhaust gases from combustion.

**turbogenerator.** A turbine coupled to and driving an electric generator. Virtually all commercial electricity is produced by such devices.

**ultraviolet radiation.** Radiation similar to light, but with wavelengths slightly shorter and with more energy than violet light. Its greater energy causes ultraviolet light to severely burn and otherwise damage biological tissues.

**underground storage tanks.** Tanks used to store petroleum products at service stations, now subject to mandated protection against leakage. (See **UST legislation.**)

**undernutrition.** A form of hunger in which the individual lacks adequate food energy, as measured in calories. Starvation is the most severe form of undernutrition.

**underweight.** The world's number one health risk factor, this refers to the effects of undernutrition on children that prevents their normal growth.

**upwelling.** In oceanic systems, the upward vertical movement of water masses, caused by diverging currents and offshore winds, bringing nutrient-rich water to the surface.

**urban blight/decay.** The general deterioration of structures and facilities such as buildings and roadways, in addition to the decline in quality of services, such as education, that has occurred in inner-city areas.

**urban sprawl.** The rapid expansion of metropolitan areas through building houses and shopping centers farther and farther from urban centers and lacing them together with more and more major highways. Widespread development that has occurred without any overall land-use plan.

**UST legislation.** Amendments to the Resources Conservation and Recovery Act of 1976 passed in 1984 to address the mounting problem of leaking underground storage tanks (USTs).

**value.** In ethical considerations, a property attributed to objects, species, or individuals that implies a moral duty to the one assigning the value; intrinsic and instrumental value are two types of value.

**vector.** An agent, such as an insect or tick, that carries a parasite from one host to another.

**vitamin.** A specific organic molecule that is required by the body in small amounts, but that cannot be made by the body and therefore must be present in the diet.

**volatile organic compounds (VOCs).** A category of major air pollutants present in the air in the vapor state. The category includes fragments of hydrocarbon fuels from incomplete combustion and evaporated organic compounds such as paint solvents, gasoline, and cleaning solutions. Volatile organic compounds are major factors in the formation of photochemical smog.

**Waste Isolation Pilot Plant (WIPP).** A facility built by the Department of Energy in New Mexico to receive defense-related nuclear wastes.

**waste-to-energy.** A process of combustion of solid wastes that also generates electrical energy.

**water cycle.** See **hydrologic cycle.**

**water-holding capacity.** The ability of a soil to hold water so that it will be available to plants.

**waterlogging.** The total saturation of soil with water. Waterlogging results in plant roots' not being able to get air and dying as a result.

**watershed.** The total land area that drains directly or indirectly into a particular stream or river. The watershed is generally named from the stream or river into which it drains.

**water table.** The upper surface of groundwater, rising and falling with the amount of groundwater.

**water vapor.** Water molecules in the gaseous state.

**weather.** The day-to-day variations in temperature, air pressure, wind, humidity, and precipitation mediated by the atmosphere in a given region.

**weathering.** The gradual breakdown of rock into smaller and smaller particles, caused by natural chemical, physical, and biological factors.

**wet cleaning.** A water-based alternative to dry cleaning that avoids the use of hazardous chemicals.

**wetlands.** Areas that are constantly wet and are flooded at more or less regular intervals. Especially, marshy areas along coasts that are regularly flooded by tide.

**wetland system.** A biological aquatic system (usually a restored wetland) for removing nutrients from treated sewage wastewater and returning it virtually pure to a river or stream. Wetland systems are sometimes used when the application of treated wastewater for irrigation is not feasible.

**Wilderness Act of 1964.** Federal legislation that provides for the permanent protection of undeveloped and unexploited areas so that natural ecological processes can operate freely in them. Most human intrusions are excluded from such areas, which now total 104 million acres in the United States.

**wind farms.** Arrays of numerous, modestly sized wind turbines for the purpose of producing electrical power.

**windrows.** Piles of organic material extended into long rows to facilitate turning and aeration in order to enhance composting.

**wind turbines.** "Windmills" designed for the purpose of producing electrical power.

**wise use.** A form of environmental backlash reacting against regulations and restrictions on the use of public lands for recreation and extractive activities such as mining, forestry, and grazing.

**work.** Change associated with the motion or state of matter. Any such change requires the expenditure of energy.

**workability.** The relative ease with which a soil can be cultivated.

**World Bank.** A branch of the United Nations that acts as a conduit for handling loans to developing countries.

**worldview.** A set of assumptions that a person holds regarding the world and how it works.

**xeriscaping.** Landscaping with drought-resistant plants that need no watering.

**yard wastes.** Grass clippings and other organic wastes from lawn and garden maintenance.

**Younger Dryas event.** A rapid change in global climate between 10,000 and 12,000 years ago that brought on 1,500 years of colder weather.

**zones of stress.** Regions where a species finds conditions tolerable, but suboptimal. The species survives, but under stress.

**zooplankton.** Any of a number of animal groups, including protozoa, crustaceans, worms, molluscs, and cnidarians, that live suspended in the water column and that feed on phytoplankton and other zooplankton.

# index

Note: Pages in **boldface** locate the definition of the term.